Handbook of Optoelectronics
Second Edition

Series in Optics and Optoelectronics

Series Editors:
E. Roy Pike, Kings College, London, UK
Robert G. W. Brown, University of California, Irvine, USA

Handbook of Optoelectronics
Second Edition
Concepts, Devices, and Techniques
Volume 1

Edited by
John P. Dakin
Robert G. W. Brown

CRC Press
Taylor & Francis Group
Boca Raton London New York

CRC Press is an imprint of the
Taylor & Francis Group, an **informa** business

CRC Press
Taylor & Francis Group
6000 Broken Sound Parkway NW, Suite 300
Boca Raton, FL 33487-2742

First issued in paperback 2020

ISBN-13: 978-1-4822-4178-5 (hbk)
ISBN-13: 978-0-367-73567-8 (pbk)

Library of Congress Cataloging-in-Publication Data

Names: Dakin, John, 1947- editor. | Brown, Robert G. W., editor.
Title: Handbook of optoelectronics / edited by John P. Dakin, Robert G. W. Brown.
Description: Second edition. | Boca Raton : Taylor & Francis, CRC Press, 2017. | Series: Series in optics and optoelectronics ; volumes 30-32 | Includes bibliographical references and index. Contents: volume 1. Concepts, devices, and techniques -- volume 2. Enabling technologies -- volume 3. Applied optical electronics.
Identifiers: LCCN 2017014570 | ISBN 9781482241785 (hardback : alk. paper)
Subjects: LCSH: Optoelectronic devices--Handbooks, manuals, etc.
Classification: LCC TK8320 .H36 2017 | DDC 621.381/045--dc23
LC record available at https://lccn.loc.gov/2017014570

Visit the Taylor & Francis Web site at
http://www.taylorandfrancis.com

and the CRC Press Web site at
http://www.crcpress.com

Dedications

Special Tribute to Alan Rogers

The editors would like to extend a special tribute to Professor Alan Rogers, who made a notable impact on the first edition with his excellent introductory chapters on optics, and has since passed away. Alan was well known to many of us as a gentleman of science, a kind and well-loved colleague who made many contributions to the field, particularly in the area of optical fiber sensors, including the important areas of current sensing and fully distributed sensors. In this second edition, Dr. Vincent Handerek, Alan's former research student (then a close friend and work colleague for many years) has, with the full blessing of Alan's widow, Wendy, kindly assisted with providing updates to Alan's chapters. Alan is missed by all of us.

Special Tribute to John Love

The editors would also like to extend a similar special tribute to Professor John Love, who wrote a noteworthy chapter on the theory of optical fiber waveguides, an area where he has made an outstanding contribution, both alone and with his prior well-known colleague, Prof Alan Snyder. Together, in 1983, they wrote one of the seminal textbooks in this area, "Optical Waveguide Theory" (ISBN 978-1-4613-2813-1).

Since their time together, and his untimely passing away, John had continued to make major advances in the field. Because of the completeness of his chapter in our first edition, and the relatively steady state of waveguide theory since his work, this chapter will be reprinted in its original form.

Contents

Series Preface

This international series covers all aspects of theoretical and applied optics and optoelectronics. Active since 1986, eminent authors have long been choosing to publish with this series, and it is now established as a premier forum for high-impact monographs and textbooks. The editors are proud of the breadth and depth showcased by published works, with levels ranging from advanced undergraduate and graduate student texts to professional references. Topics addressed are both cutting edge and fundamental, basic science and applications-oriented, on subject matter that includes: lasers, photonic devices, nonlinear optics, interferometry, waves, crystals, optical materials, biomedical optics, optical tweezers, optical metrology, solid-state lighting, nanophotonics, and silicon photonics. Readers of the series are students, scientists, and engineers working in optics, optoelectronics, and related fields in the industry.

Proposals for new volumes in the series may be directed to Lu Han, executive editor at CRC Press, Taylor & Francis Group (lu.han@taylorandfrancis. com).

Introduction to the Second Edition

There have been many detailed technological changes since the first edition of the *Handbook* in 2006, with the most dramatic changes can be seen from the far more widespread applications of the technology. To reflect this, our new revision has a completely new Volume III focused on applications and covering many case studies from an ever-increasing range of possible topics. Even as recently as 2006, the high cost or poorer performance of many optoelectronics components was still holding back many developments, but now the cost of many high-spec components, particularly ones such as LEDs, lasers, solar cells and other optical detectors, optoelectronic displays, optical fibers and components, including optical amplifiers, has reduced to such an extent that they are now finding a place in all aspects of our lives. Solid-state optoelectronics now dominates lighting technology and is starting to dominate many other key areas like power generation. It is revolutionizing our transport by helping to guide fully autonomous vehicles, and CCTV cameras and optoelectronic displays are seen everywhere we go.

In addition to the widespread applications now routinely using optoelectronic components, since 2006 we have witnessed growth of various fundamentally new directions of optoelectronics research—and likely new component technologies for the near future. One of the most significant new areas of activity has been in nano-optoelectronics; the use of nanotechnology science, procedures, and processes is to create ultraminiature devices across the entire optoelectronics domain: laser and LED sources, optical modulators, photon detectors, and solar cell technology. Two new chapters on silicon photonics and nanophotonics, and graphene optoelectronics attempt to cover the wide range of nanotechnology developments in optoelectronics this past decade. It will, however, be a few years before the scale-up to volume-manufacturing of nano-based devices becomes an economically feasible reality, but there is much promise for new generations of optoelectronic technologies to come soon.

Original chapters of the first edition have been revised and brought up-to-date for the second edition mostly by the original authors, but in some cases by new authors, to whom we are especially grateful.

Introduction to the Second Edition

Introduction to the First Edition

Optoelectronics is a remarkably broad scientific and technological field that supports a multibillion US$ per annum global industry, employing tens of thousands of scientists and engineers. The optoelectronics industry is one of the great global businesses of our time.

In this *Handbook*, we have aimed to produce a book that is not just a text containing theoretically sound physics and electronics coverage, nor just a practical engineering handbook, but a text designed to be strong in both these areas. We believe that, with the combined assistance of many world experts, we have succeeded in achieving this very difficult aim. The structure and contents of this *Handbook* have proved fascinating to assemble, using input from many leading practitioners of the science, technology, and art of optoelectronics.

Today's optical telecommunications, display, and illumination technologies rely heavily on optoelectronic components: laser diodes, light emitting diodes, liquid crystal and plasma screen displays, etc. In today's world, it is virtually impossible to find a piece of electrical equipment that does not employ optoelectronic devices as a basic necessity—from CD and DVD players to televisions, from automobiles and aircraft to medical diagnostic facilities in hospitals and telephones, from satellites and space-borne missions to underwater exploration systems—the list is almost endless. Optoelectronics is in virtually every home and business office in the developed modern world, in telephones, fax machines, photocopiers, computers, and lighting.

"Optoelectronics" is not precisely defined in the literature. In this *Handbook*, we have covered not only optoelectronics as a subject concerning devices and systems that are essentially electronic in nature, yet involve light (such as the laser diode), but we have also covered closely related areas of electro-optics, involving devices that are essentially optical in nature but involve electronics (such as crystal light modulators).

To provide firm foundations, this *Handbook* opens with a section covering "Basic Concepts." The "Introduction" is followed immediately by a chapter concerning "Materials," for it is through the development and application of new materials and their special properties that the whole business of optoelectronic science and technology now advances. Many optoelectronic systems still rely on conventional light sources rather than semiconductor sources, so we cover these in the third chapter, leaving semiconductor matters to a later section. The detection of light is fundamental to many optoelectronic systems, as are optical waveguides, amplifiers, and lasers; so, we cover these in the remaining chapters of the "Basic Concepts" section.

The "Advanced Concepts" section focuses on three areas that will be useful to some of our intended audience, both now, in advanced optics and photometry—and now and increasingly in the future concerning nonlinear and short-pulse effects.

"Optoelectronics Devices and Techniques" is a core foundation section for this *Handbook*, as today's optoelectronics business relies heavily on such knowledge. We have attempted to cover all the main areas of semiconductor optoelectronics devices and materials in the 11 chapters in this section, from light emitting diodes and lasers of great variety to fibers, modulators, and amplifiers. Ultrafast and integrated devices are increasingly important, as are organic electroluminescent devices and photonic bandgap and crystal fibers. Artificially engineered materials provide a rich source of possibilities for next generation optoelectronic devices.

At this point the *Handbook* "changes gear"—and we move from the wealth of devices now available to us—to how they are used in some of the most important optoelectronic systems available today. We start with a section covering "Communication," for this is how the developed world talks and communicates by internet and email today—we are all now heavily dependent on optoelectronics. Central to such optoelectronic systems are transmission, network architecture, switching, and multiplex architectures—the focus of our chapters here. In "Communication," we already have a multi-tens-of-billions-of-dollars-per-annum industry today.

'Imaging and displays' is the other industry measured in the tens of billions of dollars per annum range at the present time. We deal here with most if not all of the range of optoelectronic techniques used today from cameras, vacuum and plasma displays to liquid crystal displays and light modulators, from electroluminescent displays and exciting new 3-D display technologies just entering the market place in mobile telephone and laptop computer displays—to the very different application area of scanning and printing.

"Sensing and Data Processing" is a growing area of optoelectronics that is becoming increasingly important—from noninvasive patient measurements in hospitals to remote sensing in nuclear power stations and aircraft. At the heart of many of today's sensing capabilities is the business of optical fiber sensing, so we begin this section of the *Handbook* there, before delving into remote optical sensing and military systems (at an unclassified level—for herein lies a problem for this Handbook—much of the current development and capability in military optoelectronics is classified and unpublishable because of its strategic and operational importance). Optical information storage and recovery is already a huge global industry supporting the computer and media industries in particular; optical information processing shows promise but has yet to break into major global utilization. We cover all of these aspects in our chapters here.

"Industrial, medical, and commercial applications" of optoelectronics abound, and we cannot possibly do justice to all the myriad inventive schemes and capabilities that have been developed to date. However, we have tried hard to give a broad overview within major classification areas, to give you a flavor of the sheer potential of optoelectronics for application to almost everything that can be measured. We start with the foundation areas of spectroscopy—and increasingly important surveillance, safety, and security possibilities. Actuation and control—the link from optoelectronics to mechanical systems is now pervading nearly all modern machines: cars, aircraft, ships, industrial production, etc.—a very long list is possible here. Solar power is and will continue to be of increasing importance—with potential for urgently needed breakthroughs in photon to electron conversion efficiency. Medical applications of optoelectronics are increasing all the time, with new learned journals and magazines regularly being started in this field.

Finally, we come to the art of practical optoelectronic systems—how do you put optoelectronic devices together into reliable and useful systems, and what are the "black art" experiences learned through painful experience and failure? This is what other optoelectronic books never tell you—and we are fortunate to have a chapter that addresses many of the questions we should be thinking about as we design and build systems—but often forget or neglect at our peril.

In years to come, optoelectronics will develop in many new directions. Some of the more likely directions to emerge by 2010 will include optical packet switching, quantum cryptographic communications, 3-D and large-area thin-film displays, high-efficiency solar-power generation, widespread biomedical and biophotonic disease analyses and treatments, and optoelectronic purification processes. Many new devices will be based on quantum dots, photonic crystals, and nano-optoelectronic components. A future edition of this *Handbook* is likely to report on these rapidly changing fields currently pursued in basic research laboratories.

We are confident you will enjoy using this *Handbook of Optoelectronics*, derive fascination and pleasure in this richly rewarding scientific and technological field, and apply your knowledge in either your research or your business.

Editors

John P. Dakin, PhD, is professor (Emeritus) at the Optoelectronics Research Centre, University of Southampton, UK. He earned a BSc and a PhD at the University of Southampton and remained there as a Research Fellow until 1973, where he supervised research and development of optical fiber sensors and other optical measurement instruments. He then spent 2 years in Germany at AEG Telefunken; 12 years at Plessey, research in Havant and then Romsey, UK; and 2 years with York Limited/York Biodynamics in Chandler's Ford, UK before returning to the University of Southampton.

He has authored more than 150 technical and scientific papers, and more than 120 patent applications. He was previously a visiting professor at the University of Strathclyde, Glasgow.

Dr. Dakin has won a number of awards, including "Inventor of the Year" for Plessey Electronic Systems Limited and the Electronics Divisional Board Premium of the Institute of Electrical and Electronics Engineers, UK. Earlier, he won open scholarships to both Southampton and Manchester Universities.

He has also been responsible for a number of key electro-optic developments. These include the sphere lens optical fiber connector, the first wavelength division multiplexing optical shaft encoder, the Raman optical fiber distributed temperature sensor, the first realization of a fiber optic passive hydrophone array sensor, and the Sagnac location method described here, plus a number of novel optical gas sensing methods. More recently, he was responsible for developing a new distributed acoustic and seismic optical fiber sensing system, which is finding major applications in oil and gas exploration, transport and security systems.

Robert G. W. Brown, PhD, is at the Beckman Laser Institute and Medical Clinic at the University of California, Irvine. He earned a PhD in engineering at the University of Surrey, Surrey, and a BS in physics at Royal Holloway College at the University of London, London. He was previously an applied physicist at Rockwell Collins, Cedar Rapids, IA, where he carried out research in photonic ultrafast computing, optical detectors, and optical materials. Previously, he was an advisor to the UK government, and international and editorial director of the Institute of Physics. He is an elected member of the European Academy of the Sciences and Arts (Academia Europaea) and special professor at the University of Nottingham, Nottingham. He also retains a position as adjunct full professor at the University of California, Irvine, in the Beckman Laser Institute and Medical Clinic, Irvine, California, and as visiting professor in the department of computer science. He has authored more than 120 articles in peer-reviewed journals and holds 34 patents, several of which have been successfully commercialized.

Dr. Brown has been recognized for his entrepreneurship with the UK Ministry of Defence Prize for Outstanding Technology Transfer, a prize from Sharp Corporation (Japan) for his novel laser-diode invention, and, together with his team at the UK Institute of Physics, a Queen's Award for Enterprise, the highest honor bestowed on a UK company. He has guest edited several special issues of *Applied Physics* and was consultant to many companies and government research centers in the United States and the United Kingdom. He is a series editor of the CRC Press "Series in Optics and Optoelectronics."

Contributors

Zbigniew Bielecki
Institute of Optoelectronics
Military University of Technology
Warsaw, Poland

Anders Bjarklev
Department of Photonics Engineering
Technical University of Denmark
Lyngby, Denmark

Nikolaus Boos
EADS Eurocopter SAS
Marignane, France

Robert G. W. Brown
Beckman Laser Institute and Medical Clinic
University of California, Irvine
Irvine, California

Chien-Jen Chen
Onetta Inc.
San Jose, California

Krzysztof Chrzanowski
Institute of Optoelectronics
Military University of Technology
Warsaw, Poland

Nadir Dagli
Electrical and Computer Engineering
    Department
University of California, Santa Barbara
Santa Barbara, California

Xavier Daxhelet
Ecole Polytechnique de Montreal
Montreal, Quebec, Canada

Sasan Fathpour
CREOL, The College of Optics and Photonics
University of Central Florida
Orlando, Florida

Corin B. E. Gawith
Optoelectronics Research Centre
University of Southampton
Southampton, United Kingdom

Martin Grell
Department of Physics and Astronomy
University of Sheffield
Sheffield, United Kingdom

Vincent Handerek
Office of the CTO
Fotech Solutions Ltd.
Hampshire, United Kingdom

Hidehiro Kume
Opto-Mechatronix, Inc.
Hamamatsu, Shizuoka, Japan

Suzanne Lacroix
Department of Engineering Physics
École Polytechnique de Montréal
Montreal, Quebec, Canada

Jesper Lægsgaard
Department of Photonics Engineering
Technical University of Denmark
Lyngby, Denmark

Christian Lerminiaux
Corning SA–CERF
Avon, France

John Love
Australian National University
Canberra, Australia

Janusz Mikolajczyk
Institute of Optoelectronics
Military University of Technology
Warsaw, Poland

Jayanta Mukherjee
University of Surrey
Kaiam Corporation
Newton Aycliffe, United Kingdom

Tanya Munro
University of South Australia
Adelaide, Australia

Johan Nilsson
Optoelectronics Research Centre
University of Southampton
Southampton, United Kingdom

Yoshi Ohno
National Institute of Standards and Technology
Sensor Science Division
Gaithersburg, Maryland

Antoni Rogalski
Institute of Applied Physics
Military University of Technology
Warsaw, Poland

Alan Rogers
University of Surrey
Guildford, United Kingdom

Neil J. Ross
Harrogate, United Kingdom

Peter G. R. Smith
Optoelectronics Research Centre
University of Southampton
Southampton, United Kingdom

Günter Steinmeyer
Max Born Institute
Berlin, Germany

Klaus Streubel
OSRAM Opto Semiconductors GmbH
Regensberg, Germany

Yan Sun
Onetta Inc.
San Jose, California

Stephen Sweeney
Department of Physics
University of Surrey
Surrey, United Kingdom

David O. Wharmby
Ilkley, United Kingdom

William S. Wong
Onetta Inc.
San Jose, California

PART I

Basic concepts

1

An introduction to optoelectronics

ALAN ROGERS
University of Surrey

VINCENT HANDEREK
Fotech Solutions Ltd.

1.1 OBJECTIVE

In this chapter, we take a general look at the nature of photons and electrons (and of their interactions) in order to gain familiarity with their overall properties, in so far as they bear upon our subject. Clearly, it is useful to acquire this "feel" in general terms before getting immersed in some of the finer details, which, whilst very necessary, do not allow the interrelationships between the various aspects to remain sharply visible. The intention is that the familiarity acquired by reading this chapter will facilitate an understanding of the other chapters in the book.

Our privileged vantage point for the modern views of light has resulted from a laborious effort of many scientists over many centuries, and a valuable appreciation of some of the subtleties of the subject can be obtained from a study of that effort. A brief summary of the historical development is thus our starting point.

1.2 HISTORICAL SKETCH

The ancient Greeks speculated on the nature of light from about 500 BC. The practical interest at that time centered, inevitably, on using the sunlight for military purposes; and the speculations, which were of an abstruse philosophical nature, were too far removed from the practicalities for either to have much effect on the other.

The modern scientific method effectively began with Galileo (1564–1642), who raised experimentation to a properly valued position. Prior to his time, experimentation was regarded as a distinctly inferior, rather messy activity, definitely not for true gentlemen. (Some reverberations from this period persist, even today!) Newton was born in the year in which Galileo died, and these two men laid the basis for the scientific method that was to serve us well for the following three centuries.

Newton believed that light was corpuscular in nature. He reasoned that only a stream of projectiles, of some kind, could explain satisfactorily the fact that light appeared to travel in straight lines. However, Newton recognized the difficulties in reconciling some experimental data with this view and attempted to resolve them by ascribing some rather unlikely properties to his corpuscles; he retained this basic corpuscular tenet, however.

Such was Newton's authority, resting as it did on an impressive range of discoveries in other branches of physics and mathematics, that it was not until his death (in 1727) that the views of other men such as Euler, Young, and Fresnel began to gain their due prominence. These men believed that light was a wave motion in a "luminiferous aether," and between them, they developed an impressive theory, which well explained all the known phenomena of optical interference and diffraction. The wave theory rapidly gained ground during the late eighteenth and early nineteenth centuries.

The final blow in favor of the wave theory is usually considered to have been struck by Foucault (1819–1868), who, in 1850, performed an experiment that proved that light travels more slowly in water than in air. This result agreed with the wave theory and contradicted the corpuscular theory.

For the next 50 years, the wave theory held sway until, in 1900, Planck (1858–1947) found it mathematically convenient to invoke the idea that light was emitted from a radiating body in discrete packets, or "quanta," rather than continuously as a wave. Although Planck was at first of the opinion that this was no more than a mathematical trick to explain the experimental relation between emitted intensity and wavelength, Einstein (1879–1955) immediately grasped the fundamental importance of the discovery and used it to explain the photoelectric effect, in which light acts to emit electrons from matter: the explanation was beautifully simple and convincing. It appeared, then, that light really did have some corpuscular properties.

In parallel with these developments, there were other worrying concerns for the wave theory. From early in the nineteenth century, its protagonists had recognized that "polarization" phenomena, such as those observed in crystals of Iceland spar, could be explained if the light vibrations were transverse to the direction of propagation. Maxwell (1831–1879) had demonstrated brilliantly (in 1864), by means of his famous field equations, that the oscillating quantities were electric and magnetic fields.

However, there arose persistently the problem of the nature of the "aether" in which these oscillations occurred and, in particular, how astronomical bodies could move through it, apparently without resistance. A famous experiment in 1887, by Michelson and Morley, attempted to measure the velocity of the earth with respect to this aether, and consistently obtained the result that the velocity was zero. This was very puzzling in view of the earth's known revolution around the sun. It thus appeared that the medium in which light waves propagate did not actually exist!

The null result of the aether experiment was incorporated by Einstein into an entirely new view of space and time in his two theories of relativity: the special theory (1905) and the general theory (1915). Light, which propagates in space and oscillates in time, plays a crucial role in these theories.

Thus, physics arrived (ca. 1920) at the position where light appeared to exhibit both particle (quantum) and wave aspects, depending on the physical situation. To compound this duality, it was found (by Davisson and Germer in 1927, after a suggestion by de Broglie in 1924) that electrons, previously thought quite unambiguously to be particles, sometimes exhibited a wave character, producing interference and diffraction patterns in a wave-like way.

The apparent contradiction between the pervasive wave–particle dualities in nature is now recognized to be the result of trying to picture all physical phenomena as occurring within the context of the human scale of things. Photons and electrons appear to behave either as particles or as waves to us only because of the limitations of our modes of thought. We have been conditioned to think in terms of the behavior of objects such as sticks, stones, and waves on water, the understanding of which has been necessary for us to survive, as a species, at our particular level of things.

In fact, the fundamental atomic processes of nature are not describable in these terms, and it is only when we try to force them into our more familiar framework that apparent contradictions, such as the wave–particle duality of electrons and photons, arise. Electrons and photons are neither waves nor particles but entities whose true nature is somewhat beyond our conceptual powers. We are very limited by our preference (necessity, almost) to have a mental picture of what is going on.

Present-day physics with its gauge symmetries and field quantizations rarely draws any pictures at all, but that is another story…

1.3 THE WAVE NATURE OF LIGHT

In 1864, Clerk Maxwell was able to express the laws of electromagnetism known at that time in a way that demonstrated the symmetrical interdependence of electric and magnetic fields. To complete the symmetry, he had to add a new idea: that a changing electric field (even in free space) gives rise to a magnetic field. The fact that a changing magnetic field gives rise to an electric field was already well known, as Faraday's law of induction.

Since each of the fields could now give rise to the other, it was clearly conceptually possible for the two fields mutually to sustain each other, and thus, to propagate as a wave. Maxwell's equations formalized these ideas and allowed the derivation of a wave equation.

This wave equation permitted free-space solutions that corresponded to electromagnetic waves with a defined velocity; the velocity depended on the known electric and magnetic properties of free space, and thus, could be calculated. The result of the calculation was a value so close to the known velocity of light as to make it clear that light could be identified with these waves, and thus, light was established as an electromagnetic phenomenon.

All the important features of light's behavior as a wave motion can be deduced from a detailed study of Maxwell's equations. We shall limit ourselves here to a few of the basic properties.

If we take Cartesian axes Ox, Oy, Oz (Figure 1.1), we can write a simple sinusoidal solution of the free-space equations in the form:

$$E_x = E_0 \exp\left[i(\omega t - kz) \right]$$
$$H_y = H_0 \exp\left[i(\omega t - kz) \right].$$

(1.1)

These two equations describe a wave propagating in the Oz direction with electric field (E_x) oscillating sinusoidally (with time t and distance z) in the xz plane and the magnetic field (H_y) oscillating in the yz plane. The two fields are orthogonal in direction and have the same phase, as required by the form of Maxwell's equations: only if these conditions are maintained can the two fields mutually sustain each other. Note also that the two fields must oscillate at right angles to the direction of propagation, Oz. Electromagnetic waves are transverse waves.

The frequency of the wave described by Equation 1.1 is given by

$$f = \frac{\omega}{2\pi}$$

and its wavelength by

$$\lambda = \frac{2\pi}{k},$$

where ω and k are known as the angular frequency and propagation constant, respectively. Since f intervals of the wave distance λ pass each point on

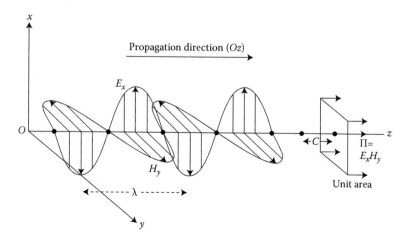

Figure 1.1 Sinusoidal electromagnetic wave.

the Oz axis per second, it is clear that the velocity of the wave is given by

$$c = f\lambda = \frac{\omega}{k}.$$

The free-space wave equation shows that this velocity should be identified as follows:

$$c_0 = \frac{1}{(\varepsilon_0\mu_0)^{1/2}}, \qquad (1.2)$$

where ε_0 is a parameter known as the electric permittivity, and μ_0, the magnetic permeability, of free space. These two quantities are coupled, independently of Equation 1.2, by the fact that both electric and magnetic fields exert mechanical forces, a fact that allows them to be related to a common force parameter, and thus to each other. This "force–coupling" permits a calculation of the product $\varepsilon_0\mu_0$, which, in turn, provides a value for c_0, using Equation 1.2. (Thus, Maxwell was able to establish that light in free space consists of electromagnetic waves.)

We can go further, however. The free-space symmetry of Maxwell's equations is retained for media that are electrically neutral and that do not conduct electric current. These conditions are realized for a general class of materials known as dielectrics; this class contains the vast majority of optical media. In these media, the velocity of the waves is given by

$$c = \left(\varepsilon\varepsilon_0\mu\mu_0\right)^{-1/2}, \qquad (1.3)$$

where ε is known as the relative permittivity (or dielectric constant) and μ the relative permeability of the medium. ε and μ are measures of the enhancement of electric and magnetic effects, respectively, which are generated by the presence of the medium. It is, indeed, convenient to deal with new parameters for the force fields, defined by

$$D = \varepsilon\varepsilon_0 E$$
$$B = \mu\mu_0 H,$$

where D is known as the electric displacement and B the magnetic induction of the medium. More recently, they have come to be called the electric and magnetic flux densities, respectively.

The velocity of light in the medium can (from Equation 1.3) also be written as

$$c = \frac{c_0}{(\varepsilon\mu)^{1/2}}, \qquad (1.4)$$

where c_0 is the velocity of light in free space, with an experimentally determined value of 2.997925×10^8 ms^{-1}. For most optical media of any importance, we find that $\mu \approx 1$, $\varepsilon > 1$ (hence, the name "dielectrics"). We have already noted that they are also electrical insulators. For these, then, we may write Equation 1.4 in the form:

$$c \approx \frac{c_0}{\varepsilon^{1/2}} \qquad (1.5)$$

and note that, with $\varepsilon > 1$, c is smaller than c_0. Now, the refractive index, n, of an optical medium is a measure of how much more slowly light travels in the medium compared with free space, and is defined by

$$n = \frac{c_0}{c}$$

and thus

$$n = \varepsilon^{1/2}$$

from Equation 1.5.

This is an important relationship because it connects the optical behavior of the optical medium with its atomic structure. The medium provides an enhancement of the effect of an electric field because that field displaces the atomic electrons from their equilibrium position with respect to the nuclei; this produces an additional field and thus, an effective magnification of the original field. The detailed effect on the propagation of the optical wave (which, of course, possesses an electric component) will be considered in Chapter 2 but we can draw two important conclusions immediately. First, the value of the refractive index possessed by the material is clearly dependent upon the way in which the electromagnetic field of the propagating wave interacts with the atoms and molecules of the medium. Second, since there are known to be resonant frequencies associated with the binding of electrons in atoms, it follows that we expect ε to be frequency dependent. Hence, via Equation 1.5, we expect n also to be frequency dependent. The variation of n (and thus of optical wave velocity) with frequency is a phenomenon known as optical dispersion and is very important in optoelectronic systems, not least because all practical optical sources emit a range of different optical frequencies, each with its own value of refractive index.

We turn now to the matters of momentum, energy, and power in the light wave. The fact that a light wave carries momentum and energy is evident from a number of its mechanical effects, such as the forced rotation of a conducting vane in a vacuum when one side is exposed to light (Figure 1.2). A simple wave picture of this effect can be obtained from a consideration of the actions of the electric and magnetic fields of the wave when it strikes the conductor. The electric field will cause a real current to flow in the conductor (it acts on the "free" electric charges in the conductor) in the direction of the field. This current then comes under the influence of the orthogonal magnetic field of the wave. A current-carrying conductor in a magnetic field that lies at right angles to the current flow experiences a force at right angles to both the field and the current (motor principle) in a direction that is given by Fleming's left-hand rule (this direction turns out to be, fortunately, the direction in which the light is traveling!). Hence, the effect on the conductor is equivalent to that of energetic particles striking it in the direction of travel of the wave; in other words, it is equivalent to the transport of momentum and energy in that direction.

We can take this description one stage further. The current is proportional to the electric field

and the force is proportional to the product of the current and the magnetic field; hence, the force is proportional to the product of electric and magnetic field strengths. The flow of energy, that is, the rate at which energy is transported across unit area normal to the direction of propagation, is just equal to the vector product of the two quantities;

$$\Pi = E \times H$$

(the vector product of two vectors gives another vector whose amplitude is the product of the amplitudes of the two vectors multiplied by the sine of the angle between their directions (in this case $\sin 90° = 1$) and is in a direction orthogonal to both vectors, and along a line followed by a right-handed screw rotating from the first to the second vector. Vectors often combine in this way; so, it is convenient to define such a product).

Clearly, if E and H are in phase, as for an electromagnetic wave traveling in free space, then the vector product will always be positive. Π is known as the Poynting vector. We also find that, in the case of a propagating wave, E is proportional to H, so that the power across unit area normal to the direction of propagation is proportional to the square of the magnitude of either E or H. The full quantitative relationships will be developed in later chapters, but we may note here that this means that a measurement of the power across unit area, a quantity known as the intensity of the wave (sometimes, the "irradiance") provides a direct measure of either E or H (Figure 1.1). This is a valuable inferential exercise since it enables us, via a simple piece of experimentation (i.e., measurement of optical power), to get a handle on the way in which the light will interact with atomic electrons, for example. This is because, within the atom, we are dealing with electric and magnetic fields acting on moving electric charges.

The units of optical intensity, clearly, will be Wm^{-2}.

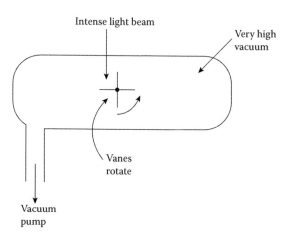

Figure 1.2 Force exerted by light falling on a conducting vane.

1.4 POLARIZATION

The simple sinusoidal solution of Maxwell's wave equation for E and H given by Equation 1.1 is only one of an infinite number of such solutions, with E and H lying in any direction in the xy plane, and with ω taking any value greater than zero.

It is customary to fix attention on the electric field to investigate general electromagnetic wave behavior, primarily because the effect of the electric field on the electrical charges within atoms tends to be more direct than that of the magnetic field. But the symmetry which exists between the E and H fields of the electromagnetic wave means that conclusions arrived at for the electric field have close equivalence for the magnetic field. It is simply convenient only to deal with one of them rather than two.

Suppose that we consider two orthogonal electric field components of a propagating wave, with the same frequency but differing phases (Figure 1.3a):

$$E_x = e_x \cos(\omega t - kz + \delta_x)$$

$$E_y = e_y \cos(\omega t - kz + \delta_y).$$

From Figure 1.3, we can see that the resulting electric field will rotate as the wave progresses, with the tip of the resulting vector circumscribing (in general) an ellipse. The same behavior will be apparent if attention is fixed on one particular value of z and the tip of the vector is now observed as it progresses in time. Such a wave is said to be elliptically polarized. (The word "polarized," being associated, as it is, with the separation of two dissimilar poles, is not especially appropriate. It derives from the attempt to explain crystal–optical effects within the early corpuscular theory by regarding the light corpuscles as rods with dissimilar ends, and it has persisted.) Of notable interest are the special cases where the ellipse degenerates into a straight line or a circle (Figure 1.3b and c). These are known as linear and circular polarization states, respectively, and their importance lies not least in the fact that any given elliptical state can be resolved into circular and linear components, which can then be dealt with separately. Light will be linearly polarized, for example, when either e_x or $e_y = 0$, or when $\delta_y - \delta_x = m\pi$. It will be circularly polarized only when $e_x = e_y$ and $\delta_y - \delta_x = (2m+1)\pi/2$, where m is a positive or negative integer: circular polarization requires the component waves to have equal amplitude and to be in phase quadrature. A sensible, identifiable polarization state depends crucially on the two components maintaining a constant phase

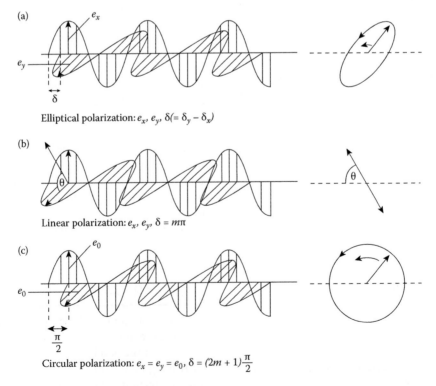

(a) Elliptical polarization: e_x, e_y, $\delta (= \delta_y - \delta_x)$

(b) Linear polarization: e_x, e_y, $\delta = m\pi$

(c) Circular polarization: $e_x = e_y = e_0$, $\delta = (2m + 1)\dfrac{\pi}{2}$

Figure 1.3 Linear and circular polarization as special cases of elliptical polarization.

and amplitude relationship. All of these ideas are further developed in Chapter 7.

The polarization properties of light waves are important for a number of reasons. For example, in crystalline media, which possess directional properties, the propagation of the light will depend upon its polarization state in relation to the crystal axes. This fact can be used either to probe the crystal structure or to control the state of the light via the crystal. Furthermore, the polarization state of the light can provide valuable insights into the restrictions imposed on the electrons that gave rise to it. Noncrystalline optical materials such as glasses might be expected to be free from polarization-dependent effects due to their amorphous structures being free from directional effects. Materials free from directional properties are called "isotropic" materials. However, in practice, glasses typically suffer from non-uniform internal strains that introduce small constraints on the motion of electrons in the material, and this leads to polarization-dependent propagation effects. Also, when light strikes a boundary between optical materials of different refractive index, polarization-dependent reflections will occur if the light is traveling in any direction other than perpendicular to the boundary, even if the materials themselves are perfectly isotropic.

Wherever there is directionality (i.e., the properties of the medium vary with spatial direction) in the medium in which the light is traveling, the polarization state of the light will interact with it; this is an extremely useful attribute with a number of important applications. For example, the propagation path taken by light can be controlled via polarization-dependent reflection in passive components such as optical isolators and circulators. These can be used for preventing unwanted instabilities and noise in active optical devices such as optical amplifiers and lasers, and can also be used to optimize efficiency in optical reflectometers by separating reflected from transmitted light without introducing any significant optical power loss. Also, when the directionality of the medium is controlled by electric or magnetic fields, the resulting effects can be used for polarization control and modulation of light. Examples of polarization-based modulation of light are given in Sections 7.7.7 and 7.7.8. Today, probably the most common use of polarization modulation is in liquid crystal displays, as discussed in Volume 2, Chapter 6. Optical communication links can benefit from a combined use of polarization control and optical path separation to double the communication capacity of a link in a technique called polarization division multiplexing.

1.5 THE ELECTROMAGNETIC SPECTRUM

Hitherto, in this chapter, we have dealt with optical phenomena in fairly general terms and with symbols rather than numbers. It may help to fix ideas somewhat if some numbers are quoted.

The wave equation allows single-frequency sinusoidal solutions and imposes no limit on the frequency. Furthermore, the equation is still satisfied when many frequency components are present simultaneously. If they are phase-related, then the superposition of the many waveforms provides a determinable time function via the well-known process of Fourier synthesis. If the relative phases of the components vary with time, then we have "incoherent" light; if the spread of frequencies in this latter case exceeds the bandwidth of the optical detector (e.g., the human eye), we sometimes call it "white" light.

The electromagnetic spectrum is shown in Figure 1.4. In principle, it ranges from (almost) zero frequency to infinite frequency. In practice, since electromagnetic wave sources cannot be markedly smaller than the wavelength of the radiation that they emit, the range is from the very low frequency ($\sim10^3$ Hz) radio waves ($\lambda\sim300$ km) to the very high frequency ($\sim10^{20}$ Hz) gamma radiation, where the limit is that of the very high energy needed for their production.

The most energetic processes in the universe are those associated with the collapse of stars and galaxies (supernovae, black holes), and it is these that provide the radiation of the highest observable frequencies.

Visible radiation lies in the range of 400–700 nm (1 nm $= 10^{-9}$ m), corresponding to a frequency range of 7.5×10^{14} to 4.3×10^{14} Hz. The eye has evolved a sensitivity to this region as a result of the fact that it corresponds to a broad maximum in the spectral intensity distribution of sunlight at the earth's surface: survival of the species is more likely if the sensitivity of the eye lies where there is most light!

The infrared region of the spectrum lies just beyond 700 nm and is usually taken to extend to

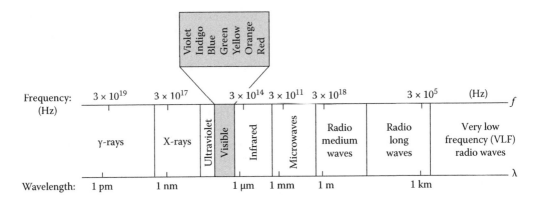

Figure 1.4 The electromagnetic spectrum.

about 300,000 nm (≡300 μm; we usually switch to micrometers for the infrared wavelengths, in order to keep the number of noughts down).

The ultraviolet region lies below 400 nm and begins at about 3 nm. Clearly, all of these divisions are arbitrary, since the spectrum is continuous.

It is worth noting that the refractive index of silica (an important optical material) in the visible range is ~1.47, so the velocity of light at these wavelengths in this medium is close to $2 \times 10^8 \, ms^{-1}$. Correspondingly, at the given optical frequencies, the wavelengths in the medium will be ~30% less than those in air, in accordance with the relation: $\lambda = c/f$. (The frequency will remain constant.)

It is important to be aware of this wavelength change in a material medium, since it has a number of noteworthy consequences that will be explored in Chapter 2.

1.6 EMISSION AND ABSORPTION PROCESSES

So far in our discussions, the wave nature of light has dominated. However, when we come to consider the relationships between light and matter, the corpuscular or (to use the modern word) "particulate" nature of light begins to dominate. In classical (i.e., pre-quantum theory) physics, atoms were understood to possess natural resonant frequencies resulting from a conjectured internal elastic structure. These natural resonances were believed to be responsible for the characteristic frequencies emitted by atoms when they were excited to oscillate by external agencies. Conversely, when the atoms were irradiated with electromagnetic waves at these same frequencies, they were able to absorb energy from the waves, as with all naturally resonant systems interacting with sympathetic driving forces. This approach seemed to provide a natural and reasonable explanation for both the emission and absorption spectral characteristics of particular atomic systems.

However, it was soon recognized that there were some difficulties with these ideas. They could not explain why, for example, in a gas discharge, some frequencies were emitted by the gas and yet were not also absorbed by it in its quiescent state; neither could they explain why the energy with which electrons were emitted from a solid by ultraviolet light (in the photoelectric effect) depends not on the quantity of absorbed light energy but only on the light's frequency.

We now understand the reasons for these observations. We know that atoms and molecules can exist only in discrete energy levels. These energy levels can be arranged in order of ascending value: $E_1, E_2, ..., E_m$ (where m is an integer) and each such sequence is characteristic of a particular atom or molecule. The highest energy level corresponds to the last level below the one at which the atom becomes ionized (i.e., loses an electron).

Fundamental thermodynamics (classical!) requires that under conditions of thermal equilibrium, the number, N_i, of atoms having energy E_i is related to the number N_j having energy E_j by the Boltzmann relation:

$$\frac{N_i}{N_j} = \exp\left[-\frac{(E_i - E_j)}{kT}\right]. \tag{1.6}$$

Here k is Boltzmann's constant ($1.38 \times 10^{-23} \, JK^{-1}$) and T is the absolute temperature.

The known physics now states that light of frequency v_{ij} can be either emitted or absorbed by the system only if they correspond to a difference between two of the discrete energy levels, in accordance with the relation

$$hv_{ij} = E_i - E_j,$$

where h is Planck's quantum constant (6.626×10^{-34} Js). The more detailed interpretation is that when, for example, an atom falls from an energy state E_j to E_i, a "particle" of light with energy hv_{ij} is emitted. This "quantum" of light is called the photon; we use the symbol v to denote frequency rather than f (or $\omega/2\pi$) to emphasize that light is now exhibiting its particulate, rather than its wave, character.

Thus, the relationship between light and matter consists of the interaction between atoms (or molecules) and photons. An atom either absorbs/emits a single photon, or it does not. There is no intermediate state.

The classical difficulties to which reference was made earlier are now resolved. First, some lines are emitted from a gas discharge that are not present in the absorption spectrum of the quiescent gas because the energetic conditions in the discharge are able to excite atoms to high energy states from which they can descend to some lower states; if these states are not populated (to any measurable extent) in the cold gas, however, there is no possibility of a corresponding incoming frequency effecting these same transitions and hence, being absorbed. Second, for an incoming stream of photons, each one either interacts or does not interact with a single atom. If the photon energy is higher than the ionization energy of the atom, then the electron will be ejected. The energy at which it is ejected will be the difference between the photon energy and the ionization energy. Thus, for a given atom, the ejection energy will depend only on the frequency of the photon.

It should be mentioned that modern optoelectronic devices make much use of electronic transitions between different energy levels in order both to generate and to detect light. Concentrating for the moment on light generation, cathode ray tubes were dominant for image displays until the middle 2000s. These devices rely on high energy electron beams exciting the electrons in phosphor materials by direct collision. Light is subsequently emitted when the electrons fall back to lower energy

levels in the phosphor. Since about 2010, cathode ray tubes have been almost completely superseded for image display tasks by pixelated liquid crystal display panels. However, liquid crystals are not light emitters themselves. Instead, the liquid crystal elements modulate filtered white light from flat panel "backlight" sources. Very often, these sources incorporate lamps where the electrons in a mixture of colored phosphors are excited by absorption of high energy photons emitted from a gas discharge; the electrons in the atoms of the gas are themselves excited to high energy by passing an electric current though the gas. The phosphor mixture is adjusted to produce white light. At the time of writing, there is competition between the use of gas discharges and semiconductor light-emitting diode (LED) sources to generate the primary high energy photons. LEDs will be discussed later, but the use of a mixture of phosphors to create an impression of white light remains an essential part of the technology!

Clearly, in light–matter interactions, it is convenient to think of light as a stream of photons. If a flux of p photons of frequency v crosses unit area in unit time, then the intensity of the light (defined by the Poynting vector) can be written as

$$I = phv. \tag{1.7}$$

It is not difficult to construct any given quantity in the photon approach that corresponds to one within the wave approach. However, there does still remain the more philosophical question of reconciling the two approaches from the point of view of intellectual comfort. The best that can be done at present is to regard the wave as a "probability" function, where the wave intensity determines the probability of "finding" a photon in a given volume of space. This is a rather artificial stratagem that does, however, work very well in practice. It does not really provide the intellectual comfort that we seek, but that, as has been mentioned earlier, is a fault of our intellect, not of the light!

Finally, it may be observed that, since both the characteristic set of energy levels and the return pathways from an excited state are peculiar to a particular atom or molecule, it follows that the emission and/or absorption spectra can be used to identify and quantify the presence of species within samples, even at very small partial concentrations. The pathway probabilities can be calculated from quantum principles, and this whole subject is a sophisticated, powerful, and sensitive tool for quantitative materials analysis. It is not, however, within the scope of this chapter.

1.7 PHOTON STATISTICS

The particulate view of light necessitates the representation of a light flux as a stream of photons "guided" by an electromagnetic wave. This immediately raises the question of the arrival statistics of the stream.

To fix ideas, let us consider the rate at which photons arrive at the sensitive surface of a photodetector.

We begin by noting that the emission processes that gave rise to the light in the first place are governed by probabilities, and thus, the photons are emitted, and therefore also arrive, randomly. The light intensity is a measurable, constant (for constant conditions) quantity that, as we have noted, is to be associated with the arrival rate p according to Equation 1.7, that is, $I = phv$. It is clear that p refers to the mean arrival rate averaged for the time over which the measurement of I is made. The random arrival times of the individual particles in the stream imply that there will be statistical deviations from this mean, and we must attempt to quantify these if we are to judge the accuracy with which I may be measured.

To do this, we begin with the assumption that atoms in excited states emit photons at random when falling spontaneously to lower states. It is not possible to predict with certainty whether any given excited atom will or will not emit a photon in a given, finite time interval. Added to this, there is the knowledge that for light of normal, handleable intensities, only a very small fraction of the atoms in the source material will emit photons in sensible detection times. For example, for a He–Ne laser with an output power of 5 mW, only 0.05% of the atoms will emit photons in 1 s.

Thus, we have the situation where an atom may randomly either emit or not emit a photon in a given time, and the probability that it will emit is very small: this is the prescription for Poisson statistics, that is, the binomial distribution for very small event probability (see, for example, Kaplan, 1981).

Poisson statistics is a well-developed topic, and we can use its results to solve our photon arrival problem.

Suppose that we have an assemblage of N atoms and that the probability of any one of them emitting a photon of frequency v in time τ is q, with $q = 1$.

Clearly, the most probable number of photons arriving at the detector in time τ will be Nq, and this will thus also be the average (or mean) number detected, the average being taken over various intervals of duration τ. But the actual number detected in any given time τ will vary according to Poisson statistics, which states that the probability of detecting r photons in time τ is given by (Kaplan, 1981)

$$P_r = \frac{(Nq)^r}{r!} \exp(-Nq).$$

Hence, the probability of receiving no photons in τ is $\exp(-Nq)$, and that of receiving two photons is $[(Nq)^2/2!]\exp(-Nq)$ and so on.

Now the mean optical power received by the detector clearly is given by

$$P_m = \frac{Nqhv}{\tau} \tag{1.8}$$

and P_m is the normally measured quantity. Hence, Equation 1.8 allows us to relate the mean of the distribution to a measurable quantity, that is,

$$Nq = \frac{P_m \tau}{hv} = \frac{P_m}{hvB},$$

where B is the detector bandwidth ($B = 1/\tau$). Now we need to quantify the spread of the distribution in order to measure the deviation from the mean, and this is given by the standard deviation which, for the Poisson distribution, is the square root of the mean. Thus, the deviation of the arrival rate is

$$D = (Nq)^{1/2} = \left(\frac{P_m}{hvB}\right)^{1/2}.$$

This deviation will comprise a "noise" on the measured power level and will thus give rise to a noise power.

$$P_{noise} = \left(\frac{P_m}{hvB}\right)^{1/2} \frac{hv}{\tau} = (P_m hvB)^{1/2}.$$

Thus the signal-to-noise ratio will be given by

$$\text{SNR} = \frac{P_m}{P_{noise}} = \left(\frac{P_m}{h\nu B}\right)^{1/2}.$$

This is an important result. It tells us the fundamental limit on the accuracy with which a given light power can be measured. We note that the accuracy increases with $(P_m/h\nu)^{1/2}$, and is thus going to be poor for low rates of photon arrival. This we would expect intuitively, since the "granular" nature of the process will inevitably be more noticeable when there are fewer photons arriving at any given time. It will also be poor for large optical frequencies, since this means more energy per photon, and thus fewer photons for a given total light energy. Again, the "granular" nature will be more evident. For good SNR, therefore, we need large powers and low frequencies. Radio wave fluxes from nearby transmitters are easy to measure accurately; gamma rays from a distant galaxy are not.

Finally, it should be remembered that the aforementioned conclusions only apply strictly when the probability q is very small. For the very intense emissions from powerful lasers ($\sim 10^6\,\text{Wm}^{-2}$, say), a substantial proportion of the atoms will emit photons in a typical detection time. Such light is sometimes classed as non-Poissonian (or sub-Poissonian) for reasons that will now be clear.

1.8 THE BEHAVIOR OF ELECTRONS

Our subject is optoelectronics, and so far we have been concerned almost exclusively with just one half of it: optics. The importance of our subject stems from the powerful interaction between optics and electronics, so we should now evidently gain the necessary equivalent familiarity with electronics, to balance our view. We shall, therefore, now look at the general behavior of electrons.

A free electron is a fundamental particle with negative electrical charge (e) equal to $1.602\times10^{-19}\,\text{C}$ and mass (m) equal to $9.11\times10^{-31}\,\text{kg}$.

All electrical charges exert forces on all other charges and, for any given charge, q, it is convenient to summarize the effect of all other charges by defining the electric field, E, via the value of the force F_E that the field exerts on q:

$$F_E = qE.$$

A magnetic field exerts no force on a stationary charge. When the charge moves with velocity v with respect to a magnetic field of induction B, however, the force on the charge is given by

$$F_B = q(v \times B)$$

where $v\times B$ denotes the vector product of v and B, so that the force is orthogonal to both the vectors v and B. Of course, a uniformly moving charge comprises an electrical current, so that $v\times B$ also describes the force exerted by a magnetic field on a current-carrying conductor. The two forces are combined in the Lorentz equation:

$$F = q(E + v \times B), \tag{1.9}$$

which also is a full classical description of the behavior of the electron in free space, and is adequate for the design of many electron beam devices (such as the cathode ray tube of television sets) where the electron can be regarded as a particle of point mass subject to known electromagnetic forces.

If an electron (or other electrical charge) is accelerating, then it comprises an electric current which varies with time. Since a constant current is known to give rise to a constant magnetic field, a varying current will give rise to a varying magnetic field, and this, as we have seen, will give rise in turn to an electric field. Thus, an accelerating electron can be expected to radiate electromagnetic waves. For example, in a dipole antenna (Figure 1.5), the electrons are caused to oscillate sinusoidally along a conducting rod. The sinusoidal oscillation comprises accelerated motion, and the antenna radiates radio waves.

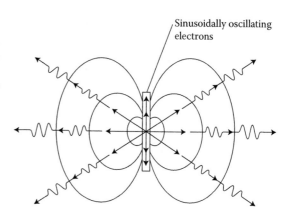

Sinusoidally oscillating electrons

Figure 1.5 The radiating dipole.

However, the electron itself also exhibits wave properties. For an electron with momentum p, there is an associated wavelength λ given by

$$\lambda = \frac{h}{p},$$

which is known as the de Broglie wavelength, after the Frenchman who, in 1924, suggested that material particles might exhibit wave properties. (The suggestion was confirmed by experiment in 1927.) Here, h is, again, the quantum constant.

The significance assigned to the wave associated with the electron is just the same as that associated with the photon: the intensity of the wave (proportional to the square of the amplitude) is a measure of the probability of finding an electron in unit volume of space. The wave is a "probability" wave. The particle/wave duality thus has perfect symmetry for electrons and photons. One of the direct consequences of this duality, for both entities, is the uncertainty principle, which states that it is fundamentally impossible to have exact knowledge of both momentum and position simultaneously, for either the photon or the electron. The uncertainty in knowledge of momentum, Δp, is related to the uncertainty in position, Δx, by the following expression:

$$\Delta p \Delta x \approx \frac{h}{2\pi}.$$

There is a corresponding relation between the uncertainty in the energy (ΔE) of a system and the length of time (Δt) over which the energy is measured:

$$\Delta E \Delta t \approx \frac{h}{2\pi}.$$

The interpretation in the wave picture is that the uncertainty in momentum can be related to the uncertainty in wavelength, that is,

$$p = \frac{h}{\lambda}$$

so that

$$\Delta p = \frac{-h\Delta\lambda}{\lambda^2}$$

and hence

$$\Delta x = \frac{h}{2\pi\Delta p} = \frac{\lambda^2}{2\pi\Delta\lambda}.$$

Hence, the smaller the range of wavelengths associated with a particle, the greater the uncertainty in its position (Δx). In other words, the closer is the particle's associated wave function to a pure sine wave, having constant amplitude and phase over all space, the better is its momentum known: if the momentum is known exactly, the particle might equally well be anywhere in the universe!

The wave properties of the electron have many important consequences in atomic physics. The atomic electrons in their orbits around the nucleus, for example, can only occupy those orbits that allow an exact number of wavelengths to fit into a circumference: again, the escape of electrons from the atomic nucleus in the phenomenon of β-radioactivity is readily explicable in terms of the "tunneling" of waves through a potential barrier. But probably the most important consequence of these wave properties, from the point of view of our present discussions, is the effect they have on electron behavior in solids, for the vast majority of optoelectronics is concerned with the interaction between photons and electrons in solid materials. We shall, therefore, need to look at this a little more closely.

The primary feature that solids possess compared with other states of matter (gas, liquid, and plasma) is that the atoms or molecules of which they are composed are sufficiently close together for their electron probability waves to overlap. Indeed, it is just this overlap that provides the interatomic bonding strength necessary to constitute a solid material, with its resistance to deformation.

When two identical atoms, with their characteristic set of energy levels, come close enough for their electronic wave functions (i.e., their waves of probability) to overlap, the result is a new set of energy levels, some lower, some higher than the original values (Figure 1.6). The reason for this is analogous to what happens in the case of two identical, coupled, mechanical resonant systems, say two identical pendulums that are allowed to interact by swinging them from a common support rod (Figure 1.7). If one pendulum is set swinging, it will set the other one in motion, and eventually, the second will be swinging with maximum amplitude while the first has become stationary. The process then reverses back to the original condition and this complete cycle recurs with frequency f_B. The system, in fact, possesses two time-independent normal modes: one is

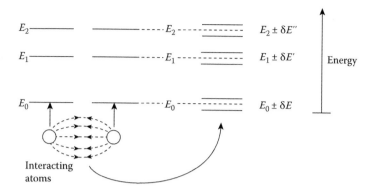

Figure 1.6 Splitting of energy levels for two interacting atoms.

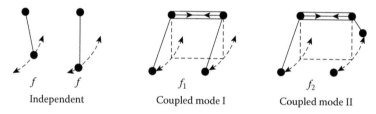

Figure 1.7 Interacting pendulums.

where both pendulums are swinging with equal amplitude and are in phase; the other with equal amplitudes in antiphase. If these two frequencies are f_1 and f_2, we find

$$f_1 - f_2 = f_B$$

and the frequency of each pendulum when independent, f, is related to these by

$$f_1 = f + \frac{1}{2} f_B$$

$$f_2 = f - \frac{1}{2} f_B,$$

that is, the original natural frequency of the system, f, has been replaced under interactive conditions by two frequencies, one higher (f_1) and one lower (f_2) than f.

It is not difficult to extend these ideas to atoms and to understand that when a large number of identical atoms are involved, a particular energy level becomes a band of closely spaced levels. Hence, in a solid, we may expect to find bands separated by energy gaps, rather than discrete levels separated by gaps; and that, indeed, is what is found.

The band structure of solids allows us to understand quite readily the qualitative differences between the different types of solid known as insulators, conductors, and semiconductors, and it will be useful to summarize these ideas.

We know from basic atomic physics that electrons in atoms will fill the available energy states in ascending order, since no two electrons may occupy the same state: electrons obey the Pauli exclusion principle. This means that at absolute zero temperature, for N electrons, the lowest N energy states will be filled (Figure 1.8a). At a temperature above absolute zero, the atoms are in thermal motion and some electrons may be excited to higher states, from which they subsequently decay, setting up a dynamic equilibrium in which states above the lowest N have a mean level of electron occupation. The really important point here is that it is only those electrons in the uppermost states that can be excited to higher levels, since it is only for those states that there are empty states within reach (Figure 1.8b). This fact has crucial importance in the understanding of solid state behavior. The electrons are said to have a Fermi–Dirac distribution among the energy levels at any given temperature, rather than the Maxwell–Boltzmann distribution they would have if they were not constrained

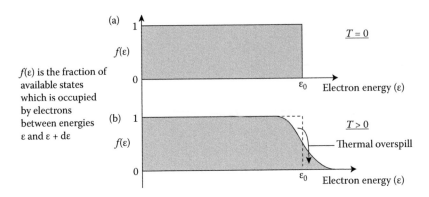

Figure 1.8 The Fermi–Dirac distribution for electrons in solids.

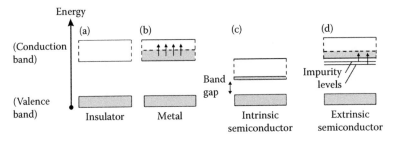

Figure 1.9 Energy-level schematic for the three main classes of solid ($T > 0$).

within the solid, and which is possessed by freely moving gas molecules, for example.

Consider now the energy band structure shown in Figure 1.9a. Here the lower band is filled with electrons and there is a large energy gap before the next allowable band, which is empty. The available electrons thus have great difficulty in gaining any energy. If an electric field is applied to this solid it would have very little effect on the electrons, since in order to move in response to the force exerted by the field, they would need to gain energy from it, and this they cannot do, since they cannot jump the gap. Hence the electrons do not move; no current flows in response to an applied voltage; the material is an insulator.

Consider now the situation in Figure 1.9b. Here the upper band is only half full of electrons. (The electrons in this band will be those in the outer reaches of the atom, and hence will be those responsible for the chemical forces between atoms, that is, they are valence electrons. Consequently, the highest band to contain any electrons is usually called the valence band.) The situation now is quite different from the previous one. The electrons near the top of the filled levels now have an abundance

of unfilled states within easy reach and can readily gain energy from external agencies, such as an applied electric field. Electric currents thus flow easily in response to applied voltages; the material is a metallic conductor.

The third case, Figure 1.9c, looks similar to the first; the only difference being that the gap between the filled valence band and the next higher unoccupied band is now much smaller. As a result, a relatively small number of electrons can be excited into the higher band (known as the conduction band) by thermal collisions and, once there, they can then move freely in response to an applied electric field. Hence, there is a low level of conductivity and the material is a semiconductor; more specifically, it is an intrinsic semiconductor. It is clear that the conductivity will rise with temperature since more energetic thermal collisions will excite more electrons into the conduction band. This is in contrast to metallic conductors in which the conductivity falls with temperature (owing to greater interference from the more strongly vibrating fixed atoms). There is a further important feature in the behavior of intrinsic semiconductors. When an electron is excited from the valence band into the conduction band, it leaves behind an unfilled state in the valence band. This creates mobility in the valence band, for electrons there that previously had no chance of gaining energy can now do so by moving into the empty state, or hole created by the promotion of the first electron. Further, the valence electron that climbs into the hole, itself leaves behind another hole that can be filled in turn. The consequence of all this activity is that the holes appear to drift in the opposite direction to the electrons when an electric field is applied, and thus, they behave like positive charges. (This is hardly surprising because they are created by the absence of negative charge.) Hence, we can view the excitation of the electron to the conduction band as a process whereby an electron/hole pair is created, with each particle contributing to the current that flows in response to an applied voltage.

Finally, we come to another very important kind of semiconductor. It is shown in Figure 1.9d. Here we note that there are discrete energy levels within the region of energy "forbidden" to states, the gap between bands. These are due to intruders in the solid, to "impurities."

To understand what is going on, consider solid silicon. Silicon atoms are tetravalent (i.e., they have a valency of four), and in the solid state, they sit comfortably in relation to each other in a symmetrical three-dimensional lattice (Figure 1.10). Silicon is an intrinsic semiconductor with an energy gap between the filled valence band and the empty (at absolute zero) conduction band of 1.14 eV. (An electron volt is the kinetic energy acquired by an electron in falling through a potential of 1 V, and is equal to 1.6×10^{-9} J.) The Boltzmann factor (Equation 1.6) now allows us to calculate that only about one in 10^{20} electrons can reach the conduction band at room temperature; but since there are of order 10^{24} electrons per cm^3 in the material as a whole, there are enough in the conduction band to allow it to semiconduct.

Suppose now that some phosphorus atoms are injected into the silicon lattice. Phosphorus is a pentavalent (valency of five) atom, so it does not sit comfortably within the tetravalent (valency of four) silicon structure. Indeed, it finds itself with a spare valence electron (it has five as opposed to silicon's four) after having satisfied the lattice requirements. This electron is loosely bound to the phosphorus atom and thus is easily detached from it into one of

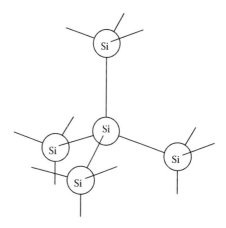

Figure 1.10 Structure of silicon lattice.

the conduction band states, requiring little energy for the excitation. Effectively, then, the electron sits in a state close to the conduction band (as shown in Figure 1.9d) and, depending on the density of phosphorus atoms (i.e., the "doping" level), can provide significantly greater conductivity than is the case for pure silicon. Such impurity-doped materials are called extrinsic semiconductors.

As the impurity we chose donated an electron to the conduction band (as a result of having one spare), it is called an n-type semiconductor, since it donates negative charge carriers. Conversely, we could have doped the silicon with a tervalent (valency of three) element, such as boron, in which case, it would sit in the lattice in need of an extra electron, since it has only three of its own. The consequence of this will be that a neighboring silicon valence electron can easily be excited into that vacant state, leaving a positive hole in the valence band as a consequence. This hole now enhances the electrical conductivity, leading to p-type ("positive carrier") semiconductivity. It is now easy to understand why "pentavalent" elements are said to give rise to "donor" energy levels and "tervalent" elements to "acceptor" levels (in silicon).

There are several reasons why extrinsic semiconductors are so important. The first is that the level of conductivity is under control, via the control of the dopant level. The second is that p-type and n-type materials can be combined with great versatility in a variety of devices having very valuable properties, the most notable of which is the transistor: many thousands of these can now be integrated on to electronic chips.

We are now in a position to understand, in general terms, the ways in which photons can interact with electrons in solids.

Consider again the case of an intrinsic semiconductor, such as silicon, with a band-gap energy E_g. Suppose that a slab of the semiconductor is irradiated with light of frequency v such that

$$hv > E_g.$$

It is clear that the individual photons of the incident light possess sufficient energy to promote electrons from the valence band to the conduction band, leaving behind positive "holes" in the valence band. If a voltage is now applied to the slab, a current, comprising moving electrons and holes, will flow in response to the light: we have a *photoconductor*. Moreover, the current will continue to flow for as long as the electron can remain in the conduction band, and that includes each electron that will enter the slab from the cathode whenever one is taken up by the anode. Hence, the number of electrons and holes collected by the electrodes per second can far exceed the number of photons entering the slab per second, provided that the lifetime of the carriers is large. In silicon, the lifetime is of the order of a few milliseconds (depending on the carrier density) and the electron/photon gain can be as large as 10^4. However, this also means that the response time is poor, and thus, photoconductors cannot measure rapid changes in light level (i.e., significant changes in less than a few milliseconds).

Small band-gap materials such as indium antimonide must be used to detect infrared radiation since the corresponding photon energy is relatively small. An obvious difficulty with a narrow band gap is that there will be a greater number of thermally excited carriers, and these will constitute a noise level; hence, these infrared detectors usually must be cooled for satisfactory performance, at least down to liquid nitrogen temperatures (i.e., <77 K).

In order to increase the speed with which the photoconduction phenomenon can be used to make measurements of light level, we use a device consisting of a combination of n- and p-type semiconductor materials. The two types of material are joined in a "pn junction" that forms a "photodiode" (Figure 1.11). Near the junction, electrons can diffuse from the electron-rich n-type material into

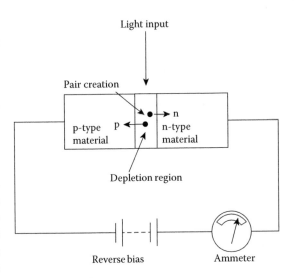

Figure 1.11 Schematic view of a p–n junction photodiode.

the p-type material, where they become trapped at electron-hungry dopant sites. Likewise, holes are annihilated if they diffuse into the n-doped region. Thus, the junction region becomes depleted of free charge carriers, and accordingly, this region is called a "depletion" region. The local migration of charges sets up an electric field across this region. Typically, the depletion region may be only a few microns wide. In this case, electron/hole pairs created in the depletion region by incident photons are immediately separated by the electric field and drift in opposite directions across the junction, thus giving rise to a measurable current as before. The electrons and holes each drift across the depletion layer toward the material where they form the majority of free charge carriers. Since the depletion region is so narrow, the time taken for the photo-generated charges to register in an external electrical circuit is very much reduced compared to the case of a slab of photoconductor. These pn photodiodes, in addition to being fast, are compact, rugged, cheap, and operate at low voltage. They are not generally as sensitive as photoconductive devices, however, since they do not allow "gain" in the way described for these latter devices (unless used in an "avalanche" mode, of which more in later chapters). However, one common method of improving the sensitivity of photodiodes, particularly for longer wavelengths with photon energies near to the band-gap energy, is to modify the device structure by using a wide region of intrinsic or very lightly doped material between the p and n doped materials. This results in a widened depletion region where the probability of absorption of incoming photons is maximized. These structures are known as p–i–n photodiodes and offer a further response speed advantage over p–n photodiodes because the wider depletion region confers reduced junction capacitance on these devices. With careful design, responses in times of order tens of picoseconds may be achieved. An important measure of the effectiveness of a photodetector is the "quantum efficiency" of the device. This is the ratio of the number of electrons produced at the output of a detector, divided by the number of photons arriving at the detector. A perfect device without gain should deliver 100% quantum efficiency. Modern p–i–n detectors used in optical communication systems deliver greater than 90% quantum efficiency across most of their operating wavelength range.

The pn detection process can also be used in reverse, in which case the device becomes a light emitter. For this action, electrons are injected into the pn junction by passing a current through it using, now, a "forward" bias. The electrons combine with holes in the region of transition between p and n materials, and, in doing so, release energy. If conditions are arranged appropriately, this energy is in the form of photons and the device becomes an emitter of light—a LED. Again, this has the advantages of ruggedness, compactness, cheapness, and low voltage operation. LEDs are in widespread use.

1.9 LASERS

Finally, in our general view of optoelectronics, we must have a quick glance at the laser, for that is from where it really all derives.

Our subject began (effectively) with the invention of the laser (in 1960) because laser light is superior in so many ways to nonlaser light. In "ordinary," so-called incoherent sources of light, each photon is emitted from its atom or molecule largely independently of one another, and thus the parameters that characterize the overall emission suffer large statistical variations and are ill-defined: in the case of the laser, this is not so. The reason is that the individual emission processes are correlated via a phenomenon known as stimulated emission, where a photon that interacts with an excited atom can cause it to emit another similar photon that then goes on to do the same again, etc. This "coupling" of the emission processes leads to emitted light that has sharply defined properties such as frequency, phase, polarization state and direction, since these are all correlated by the coupling processes. The sharpness of definition allows us to use the light very much more effectively. We can now control it, impress information upon it and detect it with much greater facility than is the case for its more random counterpart. Add to this facility the intrinsic controllability of electrons via static electric and magnetic fields and we have optoelectronics.

1.10 SUMMARY

Our broad look in this chapter at the subject of optoelectronics has pointed to the most important physical phenomena for a study of the subject and has attempted to indicate the nature of the relationships between them in this context.

Of course, in order to practice the art, we need much more than this. We need to become familiar with quantitative relationships and with a much finer detail of behavior. These the succeeding chapters will provide.

ACKNOWLEDGMENTS

Large parts of this chapter were first published (by the same author) in *Essentials of Optoelectronics*, Chapman & Hall (1997), and these are included here with permission.

FURTHER READING

Bleaney, B.I. and Bleaney, B. 1985. *Electricity and Magnetism*, 3rd ed. (Oxford: Oxford University Press) (for a readily digestible account of classical electricity and magnetism, including wave properties).

Cajori, F. 1989. *A History of Physics* (New York: Macmillan) (for those interested in the historical developments).

Chen, C.-L. 1996. *Elements of Optoelectronics and Fiber Optics* (New York: McGraw-Hill) (a treatment of the subject at a more detailed analytical level).

Ghatak, A.K. and Thyagarajaran, K. 1989. *Optical Electronics* (Cambridge: Cambridge University Press) (for general optoelectronics at a more advanced level than this chapter).

Goldin, E. 1982. *Waves and Photons, an Introduction to Quantum Theory* (New York: Wiley) (for the basics of photon theory).

Kaplan, W. 1981. *Advanced Mathematics for Engineers* (Reading, MA: Addison Wesley) p. 857 (for a good treatment of Poisson statistics).

Pollock, C.R. 1995. *Fundamentals of Optoelectronics* (New York: McGraw-Hill) (a more mathematical approach to the subject).

Richtmeyer, F.K., Kennard, E.H. and Lauritsen, T. 1955. *Introduction to Modern Physics* (New York: McGraw-Hill) (for the physical ideas concerning photons and electrons).

Smith, F.G. and King, A. 2001. *Optics and Photonics* (New York: Wiley) (a good treatment of the optics/photonics interface).

Solymar, L. and Walsh, D. 1993. *Lectures on the Electrical Properties of Materials*, 5th ed. (Oxford: Oxford University Press) (for a clear treatment of general properties of electrical materials).

Introduction to optical materials

NEIL J. ROSS
University of Southampton

2.1 INTRODUCTION

Optoelectronics is, in essence, concerned with the interactions between light and the electrons within materials through which the light is propagating. This paper reviews the basic solid-state physics that is necessary to understand the behavior of many optoelectronic devices. The emphasis is on the physical models that are used to understand and predict the behavior of materials. It is assumed that the reader has some knowledge of the basic principles of quantum mechanics, but no attempt will be made to formulate the models in a rigorous mathematical form. Only inorganic materials will be considered in this paper, as polymers and organic materials are considered elsewhere. As the interaction between light and a material is primarily through the electrons, it will be necessary to review the behavior of electrons in a solid material in some detail, with particular emphasis on semiconductors, because of their technological importance.

2.2 OPTICAL PROPERTIES OF SOME COMMON MATERIALS

Before considering the underlying physics of materials for use, it is appropriate to consider the optical properties of some commonly used materials. The two most fundamental optical properties of a material are its *transmission window*, i.e., the range of wavelengths over which it is able to transmit light, and its refractive index.

Refractive index is defined as the ratio of the speed of light in vacuum to its speed in the material. Strictly, this is the ratio of the *phase velocities,* rather than the *group velocities,* of the electromagnetic wave, but the difference is rarely significant. For most optical materials, which are transparent in the visible region of the spectrum, this ratio has a value that is within the range of about 1.3–2.0. The refractive index of some different types of optical glass for use in the ultraviolet (UV), visible, and near-infrared regions of the spectrum is shown in Figure 2.1. Much higher values are often found for materials transmitting in the infrared. For example, zinc selenide has a refractive index of about 2.4, silicon about 3.4, and germanium about 4.0. Refractive index is important, because it determines the reflection and refraction at the boundaries between materials. This makes it possible to produce familiar components, e.g., lenses, and also guided wave devices, e.g., optical fibers.

For some materials (usually crystalline), the refractive index depends on the polarization of the light and on the direction of propagation. Such materials are said to show *birefringence.* Examples of birefringent materials are quartz (SiO_2) and calcite ($CaCO_3$). The maximum difference in refractive index between the two orthogonal linear polarizations is quite small in quartz (0.009) but quite large in calcite (0.17). Isotropic materials, such as most glasses and some crystals, do not normally show birefringence. However, when an isotropic material is subjected to mechanical strain or an electric field, which introduces some anisotropy, birefringence may be induced.

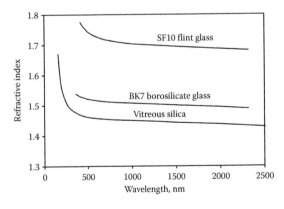

Figure 2.1 The refractive index of some optical glasses.

The attenuation of light as it travels through a medium is due to *scattering* and *absorption.* Scattering arises from inhomogeneities, possibly on an atomic or molecular scale, and microscopic voids or inclusions in the material. Light is scattered in all directions, rather like the beam of a vehicle headlight in fog. This process leads to an attenuation of the light beam, as the light lost from the beam diffuses out in all directions. Scattering is not usually a limiting factor for many applications of optical materials, because the materials have been chosen for their clarity and optical homogeneity. The limiting factor on transmission is usually absorption. However, for optical fibers, where very high purity, ultra-low-absorption glasses are used, it is the residual scattering from the microscopic fluctuations in the refractive index that ultimately limits the transmission. These fluctuations in refractive index arise from the thermal density fluctuations that occured in the molten glass. As the glass solidifies, these fluctuations become frozen.

The other loss mechanism is that the light may be absorbed by the material. Generally, the region of good optical transparency is bounded at the short wavelength end by strong electronic absorption where electrons within the material are excited to higher energy states. At the long wavelength end, it is generally the excitation of molecular vibrations or *phonons* (vibrational or elastic waves) that provides the limit to the region of optical transparency.

Figure 2.2 shows the transmission of various optical materials in the visible and infrared parts of the spectrum. The transmission curves show values of the external transmission. That is, the measured transmission includes not only the losses within the material but also the combined reflection losses at the two faces of the sample. Often, some of the absorption may be due to impurities in the material. This is the case for the absorption spectrum of UV grade vitreous silica shown in Figure 2.2. The strong absorption lines at around 1400, 2300, and 2800 nm are due to hydroxyl (OH) radicals (water). This is a consequence of the method of fabrication of the glass, and alternative methods of production lead to much lower absorption in this region, at the cost of increased absorption in the UV. The OH radicals conveniently act to reduce the absorption at short wavelengths by interacting with absorbing *color centers,* which can otherwise occur in the silica glass. The color centers are sites where there are nonbridging (i.e., missing) Si–O bonds and they introduce

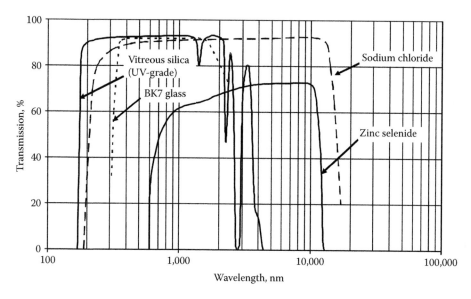

Figure 2.2 The external transmission of various optical materials for a 10 mm path length.

extra electronic states, which lead to additional absorption. Hydroxyl (OH) groups may become attached at these sites, reducing the UV absorption.

2.3 CRYSTALLINE AND AMORPHOUS MATERIALS

Although optical glasses are widely used for passive optical components (e.g., lenses, windows, and optical fibers), crystalline materials are generally used for many active electro-optic components (light sources, detectors, modulators, etc.). Crystalline materials are also the simplest and the best-understood materials. The essential feature of a crystal is that the atoms and molecules are arranged in a periodic pattern. An abstract concept of a periodic structure, the *lattice*, is used to describe the location of the molecules within the crystal [1, Chapter 1]. At each so-called *lattice point,* the atoms surrounding the point form the same pattern with the same orientation. Associated with each lattice point is a fixed volume, the *primitive cell* or *unit cell.* There are many possible choices for the shape of the primitive cell, but in each case, the volume is the same. The group of atoms associated with each lattice point is the *basis* of the cell. Together, the lattice and the basis define the structure of the crystal. Of particular significance is that they determine its symmetry properties with respect to refraction and reflection of light. These properties are of interest when considering the polarization and nonlinear properties of optical materials.

The lattice of an ideal single crystal extends throughout the crystal. In practice, defects, which disrupt the regular pattern, are often found in the structure of real-life crystals. These defects will, in general, change the physical properties of the material. For electronic or electro-optic devices, it is usually necessary to minimize or eliminate such defects. While some materials, e.g., common salt, are easily identified as crystals, because of the regular shape in which they form, some, e.g., most metals, are not. This is generally because, in the latter, the bulk material is polycrystalline and made up of many small crystallites with varying orientations. These crystallites are not readily visible but can be seen by suitable treatment of the surface, e.g., chemical etching.

Amorphous materials differ from crystals in that there is no long-range order [1, Chapter 17]. Around each atom, there may be some semblance of order, in that its nearest neighbors are approximately in the same pattern for all atoms; but this order rapidly decreases as the distance increases. Examples of amorphous materials would be soot, or, of more relevance to electro-optics, amorphous silicon. Some of these are soft materials with little mechanical strength. Glasses are also amorphous materials, which may be considered to be composed of one large macromolecule, with strong bonding between atoms, but no long or medium range order or periodicity. They are of considerable importance in optics. Glasses have no well-defined melting point but they progressively soften and become less viscous over a range of temperatures. Although there is no long-range order, there is some degree of structure at short distances.

2.4 ATOMIC BONDING

In a crystalline material, strong forces bond the atoms together. This force is primarily due to the interaction of the outer electrons of the atoms (valence electrons). The number of valence electrons determines the chemical properties and the position in the periodic table. Several types of bond are commonly identified.

2.4.1 Ionic bonds

In ionic crystals, two dissimilar atoms are bonded by charge transferred from one atom to the other, leaving a positive ion and a negative ion, with a consequent electrical force bonding the atoms together. A typical example would be common salt, NaCl.

2.4.2 Covalent bonds

In this case, the bonding occurs with electrons being shared among adjacent atoms. A typical example would be silicon. Such bonds are strong and generally have a well-defined direction, which will determine the structure of the unit cell.

2.4.3 Metallic bonds

In the case of metals, electrons are also shared among atoms, but unlike covalent bonds, the electrons involved are free to move through the whole crystal, giving high electrical and thermal conductivity and, generally, very low optical transmission. Because of the high conductivity, the reflectivity for light and longer wavelength electromagnetic waves is usually high.

2.4.4 Van der Waal bonds

These are much weaker bonds arising from a dipole interaction among the atoms. This type of bonding may be important in organic materials.

The attractive forces pull the atoms together, but as they approach more closely, a repulsive force arises due to the interaction of the inner core electrons. Thus, as two isolated atoms are brought

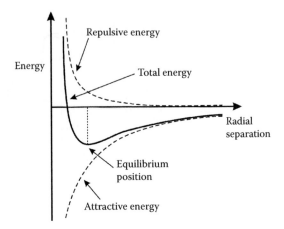

Figure 2.3 The binding energy between two atoms.

together, the potential energy will fall to a minimum and then rise again (Figure 2.3). The equilibrium separation will be at the location of the minimum. For small displacements from the minimum, the potential energy curve will be approximately quadratic with displacement. Thus, if the atoms are displaced, they will oscillate with simple harmonic motion. This vibrational mode of oscillation generally occurs at frequencies corresponding to the infrared region of the spectrum and may determine the infrared properties of a material.

2.5 THE FREE ELECTRON MODEL FOR METALS

A good place to start when looking at the electrons in solid-state materials is with metals. Although metals are not generally thought of as optical materials, starting the discussion here enables some of the key theoretical concepts to be introduced in a relatively simple way.

In a metal, the valence electrons are free to move through the body of the material. This permits the use of a simple model, which is to treat the electrons as a "gas" of free particles. This provides a good starting point for considering the electronic properties of solid materials and introduces some of the essential concepts.

The properties of this electron gas are analyzed by assuming that the electrons are completely free (no potential energy) up to the boundaries of the material, when the potential becomes infinite [1–5]. Using this "electron in a box" model, the energy states are found by solving the Schrödinger wave equation. Of course, there is no net electric charge. It is assumed that the positive charge of the ionized atoms is uniformly distributed through the whole "box." The model also ignores the interaction among the electrons and the consequent collective effects. This is a rather odd assumption given the long-range Coulomb forces. Hook and Hall [3, Chapter 13] discuss the reasons why this simplified model works.

Inside the box, the electrons can be analyzed as simple plane waves, with the walls of the box imposing boundary conditions that require the wave function to be zero outside the box. This requires standing wave solutions. The permitted solutions are characterized by their *wave vector*, *k*, which has a magnitude, *k*, of $2\pi/\lambda$, where λ is the *de Broglie wavelength* of the electron. Usually, *k* is called the *wave number*, although this term is also frequently used for the inverse of the wavelength. The direction of *k* is the direction in which the electron is moving. Details of the analysis can be found in any good text on solid-state physics [1–4].

From this model, it is possible to calculate the *density of states*, *N(E)*:

$$N(E) = \frac{V}{\pi^2 \hbar^3}(2m^2 E)^{1/2}, \qquad (2.1)$$

where *N(E)*d*E* is the number of states with energy between *E* and *E+*d*E* for a volume *V* of the metal, *m* is the mass of an electron, and $\hbar$ is the Planck's constant multiplied by 2π.

For a metal at very low temperature, the states will fill up from the bottom. The Pauli exclusion principle requires that each state can be occupied by only one electron. The maximum energy is, therefore, found by integrating *N(E)* from zero to E_F and equating this number of states to the number of electrons. The *Fermi energy*, E_F, is given by

$$E_F = \frac{\hbar^2}{2m}(3\pi^2 n_e)^{2/3}, \qquad (2.2)$$

where n_e is the number of electrons per unit volume. The value of *k* corresponding to the Fermi energy forms a spherical surface in three-dimensional *k*-space. This is the *Fermi surface*.

To give some idea of the magnitude of the Fermi energy, consider potassium, a simple metal for which the free electron theory works well. A mole

of potassium has a volume of $4.54 \times 10^{-5}\,\mathrm{m}^2$ and contains 6.0×10^{23} atoms, and each atom contributes one electron, so $n_e = 1.32 \times 10^{28}$ electrons m^{-3}. Substituting this into Equation 2.2 gives the Fermi energy as 2.04 eV.

At finite temperatures, electrons will be excited to energies somewhat higher than the Fermi energy and unfilled states will be left at lower energies. The Fermi–Dirac distribution function, *f(E)*, governs the probability that any particular state is occupied by an electron:

$$f(E) = \frac{1}{e^{(E-\mu)/k_B T} + 1}, \qquad (2.3)$$

where k_B is the Boltzmann's constant, *T* the absolute temperature, and μ the *chemical potential*. At absolute zero temperature, μ is equal to E_F. At finite temperatures, the value will vary but only slowly. Provided $k_B T$ is much less than E_F, μ and E_F may be assumed equal. In Figure 2.4, *f(E)* is plotted for a chemical potential of 2 eV, at temperatures of 0, 300, and 600 K. In practice, the transition from 1 to 0 is usually quite sharp because the thermal energy *kT* is usually much less than μ. The chemical potential is also commonly referred to as the *Fermi level*.

The number of electrons with energies lying between *E* and *E+*d*E* is given by

$$n(E) = N(E)f(E)\,dE. \qquad (2.4)$$

The energy of the electrons may also be related to their momentum or wave number. Classically, the energy of an electron is given by

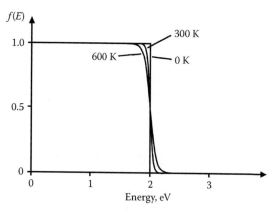

Figure 2.4 The Fermi distribution function at three temperatures.

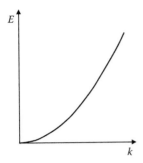

Figure 2.5 The parabolic variation of electron energy with k for the free electron model.

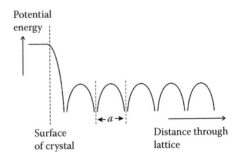

Figure 2.6 A representation of the periodic potential experienced by an electron due to the attraction of atomic nuclei.

$$E = \frac{p^2}{2m},$$

where E is the electron energy, p the electron momentum, and m the electron mass. The quantum mechanical equivalent is obtained by replacing p with M. This gives the relationship between energy, E, and the electron wave number, k, for free electrons:

$$E = \frac{\hbar^2 k^2}{2m}. \qquad (2.5)$$

Diagrams plotting E against k are useful for understanding band structure, particularly in semiconductors. The E–k diagram for free electrons (Figure 2.5) is the simplest example.

At zero temperature, with no electric field, the electrons will lie within the Fermi surface in k-space. At finite temperatures, the edge of this sphere will not be distinct but the distribution will still be spherically symmetric. When an electric field is applied to the metal, the electrons will acquire a drift velocity in a direction determined by the field. The net effect is that the whole distribution of electrons in k-space will be shifted slightly, while still maintaining its spherical symmetry. This shift corresponds to a small increase in momentum in the direction of motion and a small increase in total energy.

2.6 ELECTRONS IN A PERIODIC LATTICE

The free electron model has neglected the presence of the atoms associated with the crystal lattice. These atoms will provide a potential that will attract electrons at a large distance from the core of the atom and repel electrons that move close enough to interact with the tightly bound electrons of the atom's core. A one-dimensional representation of the potential along a line through a series of atoms is illustrated in Figure 2.6. The most significant feature of the potential is the periodic attractive potential due to the charged atomic cores, separated by a distance a. If the electrons have a de Broglie wavelength equal to the distance between atoms, then strong reflections may be expected. The solution of the Schrödinger wave equation is clearly more complex with this modulated potential.

The *nearly free electron model* provides insight into the behavior of electrons in a periodic potential [1–5]. Assuming that the interaction of the electrons with the periodic potential is weak, the electrons behave essentially as free electrons, unless their wavelength is close to the separation of the atoms, when traveling electron waves will interact coherently with the periodic potential and will be reflected. For a one-dimensional model of a crystal, with a line of atoms separated by a distance a, this reflection will occur when the electron wave number is given by

$$k = n\pi/a, \qquad (2.6)$$

where n is an integer. The variation of energy with k would be expected to follow that for free electrons (Equation 2.5), except in the region of k given by Equation 2.6. This simple model is able to predict that the electrons may only have energy within certain bands, separated by energy gaps.

In order to investigate the band structure, it is necessary to postulate a form for the potential energy and to solve the Schrödinger equation. One such model is the *Kronig–Penney model* [5]. This assumes a simple rectangular model for the potential along the one-dimensional line of atoms (Figure 2.7). The Schrödinger equation is solved, assuming that the rectangular potential reduces to a series of delta functions (b tends to zero, keeping bV_0 constant). Solutions of the wave equation are not possible for all values of the energy E. Discontinuities in the E–k diagram occur at $k=n\pi/a$, where n is an integer. Away from these values of k, the solution is approximately parabolic, as for the free electron model. This is illustrated in Figure 2.8a. As k approaches $k=n\pi/a$, the gradient of the E–k curve approaches zero.

Usually, the E–k diagram is modified by mapping all the bands to lie within the range of k between $-\pi/a$ and $+\pi/a$ (referred to as the *first Brillouin zone*), and usually only the positive values of k are included because the negative values give a mirror image. This mapping is referred to as the *reduced zone scheme* [1] and is illustrated in Figure 2.8b. The mapping is

mathematically valid because of the periodicity in k-space of the electron wave functions.

The one-dimensional, nearly free electron model demonstrates that the periodicity of a crystal results in electrons being confined to bands separated by energy gaps. It also enables the dynamics of the electron to be analyzed. The free electron equations may be used provided that the electron mass is replaced by an effective mass. This effective mass takes account of the interaction between the electrons and the lattice (see, for example, [3, Chapter 4]). The effective mass of an electron is given by

$$m^* = \hbar^2 \left(\frac{\mathrm{d}^2 E}{\mathrm{d}k^2} \right)^{-1}. \qquad (2.7)$$

Thus, the effective mass of an electron depends on the curvature of the band and hence will take positive values in the lower part of a band, negative values in the upper part of the band, and will become infinite at some intermediate point.

Another important result from the one-dimensional model is the total number of states in a band. It may be shown [4] that for a line of atoms of density N_a atoms per unit length and length L, the total number of states is $2N_a L$. Hence, at low temperature and filling up the bands from the lowest energy, all the bands will be full if the atoms have an even number of electrons, and if the number of electrons is odd, the final band will be half full.

Extending the model to three dimensions clearly increases its complexity but the same general features exist. Obviously, the periodicity will vary with direction through the crystal. The wave number must be replaced by the wave vector and the value of k at which the discontinuities occur will now be surfaces in three dimensions. The E–k diagram is still useful, but of course it must now be a plot of E against the wave vector in a particular direction. Frequently, E will be plotted against two different wave vector directions on one diagram, one direction for the positive axis and the other for the negative axis. For real materials, the band structure is much more complex than for the simple one-dimensional model. One particular feature is that, unlike the bands predicted by the simple model, the bands do not necessarily align. The maximum of one band does not necessarily occur at the same k value as the minimum of the next. The significance of this will become apparent shortly.

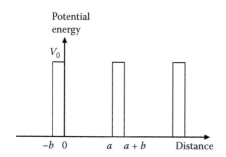

Figure 2.7 The periodic potential used in the Kronig–Penney model.

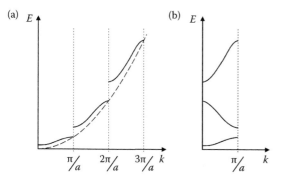

Figure 2.8 (a) Sketch of the E–k diagram obtained using the Kronig–Penney model and (b) the same data mapped using the reduced zone scheme.

2.7 METALS, INSULATORS, AND SEMICONDUCTORS

Conduction of electricity is only possible if there are electrons in a band having vacant states. A simple way to see this is to recall that, for free electrons, the effect of applying an electric field was to shift the whole electron distribution slightly to reflect the change in electron momentum when an electric current flows. For a full band, this process is not possible, and the distribution of electrons within the band is fixed. A full band cannot contribute to conduction.

A metal must have free electrons; it must, therefore, have an unfilled band (or bands). For the one-dimensional model discussed earlier, this implies that the number of electrons must be odd, in which case, the highest energy band will be only half full. Such a metal will be well described by the free electron model. This simple model works well for the alkali metals, sodium and potassium, which have an odd number of electrons. Many metals, however, e.g., magnesium or lead, have an even number of electrons and the simple one-dimensional theory does not work. This is because there is an overlap between bands, hence filling up the available energy states from the lowest available states gives two partially filled bands. These two scenarios are illustrated schematically in Figure 2.9a and b.

There is a significant difference between the two cases. For the simple metal, the electrons in the conduction band behave very much as predicted by the free electron model; however, in the case of overlapping bands, one band is nearly full and the other nearly empty. Near the Fermi surface, the curvature of the E–k curve for the almost full band

will be negative. Hence, the effective mass given by Equation 2.7 will also be negative. Rather than considering the effective mass to be negative, it is conventional to introduce the concept of *holes*. In essence, in an almost full band, an electron with an effective negative mass behaves like a positively charged particle with positive effective mass [1, Chapter 8, or 3, Chapter 5]. A more physical way to view a hole is that it is the absence of an electron from an atom, leaving a positively charged core, hence the name. The hole is able to move through the lattice as electrons move in the opposite direction, conserving the charge. The hole effectively carries the positive charge and behaves like a positively charged particle. Although these two models for a hole are very different, they are equivalent. In materials where there are two bands contributing to conduction, one almost full and one with only a few of its available states occupied, conduction will effectively be bipolar, with both electrons and holes contributing to the conduction.

If, at low temperature, the electrons exactly fill a number of bands, and there is an energy gap before the next band, then conduction will not be possible and the material is an insulator. At nonzero temperature, the Fermi distribution function (Equation 2.3) will still apply and the probability of electrons in the next band will be nonzero. If the energy gap is large, the number in this higher band will be very small and the material is an insulator. If, however, the energy gap is not so large (~1 eV), then there will be significant excitation at room temperature, and the material will be a semiconductor. The distinction between an insulator and a semiconductor is, therefore, not clearly defined in this model. However, diamond with a band gap of 5.4 eV would usually be considered an insulator, whereas silicon with a band gap of 1.17 eV would be a semiconductor. The full band(s) (at low temperature) below the energy gap is referred to as the *valence* band(s), whereas the band above the energy gap is the *conduction* band.

2.8 CARRIERS, CONDUCTION, AND DOPING IN SEMICONDUCTORS

In a semiconductor, the Fermi level or chemical potential lies within the energy gap. For a pure (intrinsic) semiconductor, the location of the Fermi level is determined by the need to balance the population of electrons in the conduction band with the holes in the valence band. Taking the zero

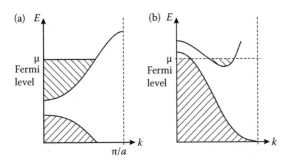

Figure 2.9 Schematic showing partially filled bands in a metal: (a) a simple metal with partially filled conduction band and (b) metal with overlapping bands.

of energy as the top of the valence band, μ is given by [3, Chapter 5]

$$\mu = \frac{1}{2}E_G + \frac{4}{3}k_B T \ln\left(\frac{m_h}{m_e}\right), \qquad (2.8)$$

where E_G is the energy gap and m_e and m_h the effective electron and hole masses at the bottom of the conduction band and the top of the valence band, respectively. This will generally be close to the center of the gap as the second term is small. The density of electrons in the conduction band, n, and holes in the valence band, p, is related by the equation

$$np = n_i^2 = = 4\left(\frac{k_B T}{2\pi\hbar^2}\right)^3 (m_h m_e)^{3/2} \exp\left(\frac{E_G}{k_B T}\right). \qquad (2.9)$$

In an intrinsic semiconductor, $n=p=n_i$.

When an electric field is applied, both the electrons and holes will drift under the influence of the applied field. The drift velocity depends on the strength of the applied field, the carrier density, and the rate at which the carriers lose energy to the atoms in the lattice. Generally, it is assumed that the current density j is proportional to the carrier density and to the applied field E (Ohm's law is obeyed). Then, the total current density is given by

$$j = (ne\mu_e + pe\mu_h)E, \qquad (2.10)$$

where μ_e and μ_h are the electron and hole mobilities, respectively, and e the electronic charge. Generally, the electron mobility will be greater than the hole mobility. For silicon, at room temperature, the electron and hole mobilities are about $\mu_e=1500\,\text{cm}^2\,\text{V}^{-1}\,\text{s}^{-1}$ and $\mu_h=450\,\text{cm}^2\,\text{V}^{-1}\,\text{s}^{-1}$. These values may be significantly reduced by impurities in the silicon.

In semiconductors, as in some metals, conduction is due to both the motion of electrons and holes. However, in semiconductors, adding low concentrations of certain impurity atoms (doping) may be used to control both the conductivity and the dominant carriers. In an elemental semiconductor with valency 4, such as silicon or germanium, adding pentavalent impurities such as phosphorous or arsenic will greatly increase the concentration of free electrons giving "n type" material. The impurity atoms can fit into the lattice reasonably well, but the extra electron is not required for the covalent bonding and is only weakly bonded to the donor atom. At room temperature, many of these donor atoms are thermally ionized, releasing their electrons into the conduction band where they are free to move. In a similar way, adding trivalent atoms such as boron or aluminum will provide acceptor sites, where an electron may be removed from the valence band and trapped. The hole created in the valence band is free to move and will contribute to electrical conduction. Material with this type of doping is "p type." If the electrons or holes from the ionized impurities dominate conduction, the semiconductor is said to be extrinsic. If the dominant mechanism by which the electrons and holes are produced is a direct thermal excitation of carriers from the valence to conduction band, then the semiconductor is said to be intrinsic. The latter usually exhibits a low conductivity for most semiconductors at room temperature. However, the conductivity of an intrinsic semiconductor increases rapidly with rising temperature.

From the point of view of the energy, the donors or acceptors introduce extra energy levels within the energy gap (Figure 2.10). The donor or acceptor levels at E_D or E_A lie close to the conduction or valence bands, respectively, and, when ionized, will either donate an electron to the conduction band or accept an electron from the valence band (leaving a hole). Taking the energy of the top of the valence band as zero and the energy gap as E_g, then the energies E_A and $E_g - E_D$ are typically of order 0.05 eV for the dopants used in silicon.

Equation 2.9 will still hold for an extrinsic semiconductor. Thus, if we consider an n type material, increasing the concentration of donors will increase the number of majority carriers (electrons) and decrease the concentration of minority carriers (holes). For p type materials, the holes will be the majority carriers.

As the temperature is increased from absolute zero, a number of changes occur in a doped

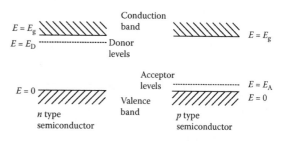

Figure 2.10 The location of the donor and acceptor levels within the band gap of a semiconductor.

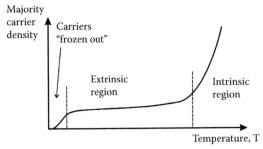

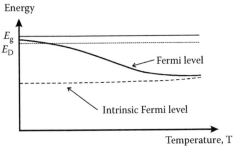

Figure 2.11 A schematic representation of the variation of majority carrier density and Fermi Level in an n type semiconductor.

semiconductor. At very low temperature, the impurity atoms are not ionized, the carrier concentrations are very small and the Fermi level lies midway between E_D and E_G for an n doped material. For a p doped material, the Fermi level will lie at an energy level of $E_A/2$. At very low temperature, the material behaves as an insulator. As the temperature rises, the donors or acceptors are ionized and the majority carrier density increases until all the impurities are ionized (extrinsic region). At the same time, the Fermi level decreases for n doped material and increases for p doped material. Increasing the temperature still further, direct thermal ionization across the energy gap becomes significant and the semiconductor moves into intrinsic conduction, with the Fermi level moving toward its intrinsic value (Equation 2.8). These changes are illustrated schematically in Figure 2.11 for an n doped material.

2.9 THE INTERACTION BETWEEN LIGHT AND MATERIALS

2.9.1 Refraction

Classical electromagnetic theory gives the interaction between a light wave and a medium in terms of the dielectric constant, ε_r, and the conductivity, σ. The solution of Maxwell's equations for a plane wave of angular frequency ω traveling in the +z direction is of the form [6, Chapter 8]:

$$E_x = E_{x0} \exp\left[j\omega(t - nz/c) \right], \quad (2.11)$$

where E_x is the electric field, n is the (complex) refractive index that is given by

$$n = n' - jn'' = \sqrt{\mu_r \varepsilon_r - j\left(\frac{\sigma\mu_r}{\omega\varepsilon_0}\right)}. \quad (2.12)$$

In Equation 2.12, μ_r is the relative permeability (generally very close to 1 for nonferromagnetic materials), ε_r the dielectric constant, σ the conductivity, and ε_0 the permittivity of free space.

Substituting a complex index of refraction into Equation 2.11 gives

$$E_x(\omega) = E_{x0} \exp\left(-\omega n'' z/c \right) \exp\left(j\omega(t - n'z) \right)$$

from which it can be seen that the absorption coefficient, α is given by

$$\alpha = \omega n''/c. \quad (2.13)$$

In a highly conducting material such as a metal, where there are plenty of free electrons, the second term in Equation 2.12 will be dominant and n will be of the form

$$n = n' - jn'' = \sqrt{\left(\frac{\sigma\mu_r}{\omega\varepsilon_0}\right)}(1 - j). \quad (2.14)$$

The absorption coefficient is, therefore, given by

$$\alpha = \frac{\omega n''}{c} = \left(\frac{\sigma\mu_r\mu_0\omega}{2}\right)^{1/2} = \delta^{-1}, \quad (2.15)$$

where c has been replaced by $(\varepsilon_0\mu_0)^{-1/2}$ and δ is the *skin depth*. The skin depth is a measure of the penetration of the field into a conductor. For aluminum, with a resistivity of 6.65×10^{-8} Ω m^{-1}, the skin depth at a wavelength of 500 nm (green light) is about 3.3 nm, corresponding to an attenuation coefficient 3×10^8 m^{-1}.

For a metal, the conductivity term generally dominates the absorption and the term in ε_r can be neglected. For an insulating material, the conductivity term may be neglected but not ε_r. The dielectric constant, ε_r, is a measure of the induced polarization of the material by the applied field. The polarization may be due to either the physical alignment of polar molecules, which generally only occurs at frequencies much below optical frequencies, or it may be due to dipole moments in the atoms or molecules induced by the applied field. In dense materials, the local field at a particular molecule or atom is distorted by the adjacent, induced dipoles. Taking account of this distortion (the *Lorentz field*), the Clausius–Mossotti equation may be deduced:

$$\frac{\varepsilon_r - 1}{\varepsilon_r + 2} = \frac{1}{3\varepsilon_0} \sum \alpha_i, \tag{2.16}$$

where α_i is the polarizability of the ith atom or molecule [3, Chapter 9]. The α_i here must not be confused with the absorption coefficient. This relation applies to isotropic materials, and needs to be treated with caution. The formula works well for gases but less well for solid-state materials, where the interaction between the molecular dipoles is stronger.

The simplest model for the polarizability of the atoms is to treat them as classical harmonic oscillators, comprising an electron bound to a fixed atomic core. Using this model, α_i will in general be complex. Close to resonance, the imaginary part of α_i, and hence ε_r, will be significant, corresponding to stronger absorption. Well away from the resonance, the polarization will be real and given by

$$\alpha_i = \frac{e^2}{m(\omega_0^2 - \omega^2)}. \tag{2.17}$$

The dc polarizability of each atom is given by $e^2/(m\omega_0^2)$. Combining Equations 2.16 and 2.17 and replacing ε_r by the square of the refractive index, the refractive index may be calculated as a function of normalized angular frequency ω/ω_0. The results are plotted in Figure 2.12, where it has been assumed that the low frequency refractive index is 1.5.

The simple model using classical harmonic oscillators clearly has weaknesses. A damping or loss term should be included to avoid the infinity at $\omega = \omega_0$ that can never occur in real materials. Treating the atoms as quantum mechanical oscillators modifies Equation 2.17, by introducing a constant, the oscillator strength, f_i. Also, it is not appropriate to treat all the oscillators as having a single resonant frequency. In solid materials, there are usually at least two regions of the frequency spectrum in which such resonances occur. In the infrared, vibrational modes of the lattice (phonons) lead to absorption and dispersion, while at shorter wavelengths, electronic or interband transitions

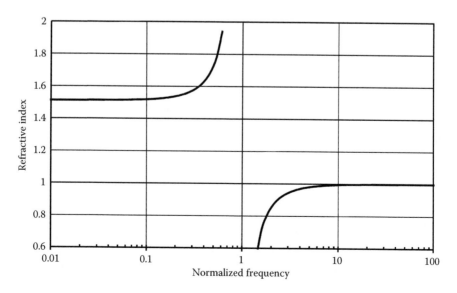

Figure 2.12 The refractive index close to resonance as predicted by Equations 2.16 and 2.17, with the dc refractive index adjusted to give a refractive index of 1.5 at low frequency.

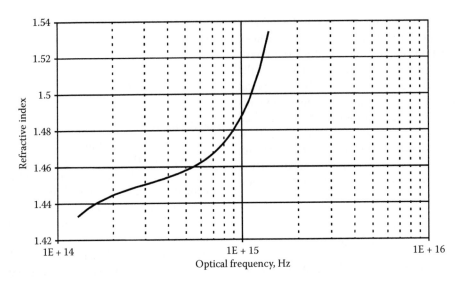

Figure 2.13 The refractive index of vitreous silica as a function of optical frequency. (Data from Kaye, G.W.C. and Laby, T.H. *Tables of Physical Constants*, Longman, London, 1978.)

lead to a further region of strong absorption with associated dispersion.

For visually transparent optical materials, the strong electronic absorption lines occur in the UV part of the spectrum and, using the simple model, a slowly rising refractive index would be expected over the range of frequencies at which the material is transparent. The refractive index of vitreous silica is shown in Figure 2.13 as a function of optical frequency. The data shown cover a range of frequencies that correspond to a wavelength ranging from 2325 to 213.9 nm. As expected from the basic theory, the refractive index rises as the frequency approaches the range in which the silica starts to absorb strongly by electronic transitions. The more rapid fall at the low frequency end of the spectrum occurs as the frequency approaches the near-infrared absorption arising from the excitation of vibrational states. In this case, approaching the absorption from the high frequency side, a dip in refractive index would be expected from Figure 2.12, as is observed in practice.

2.9.2 Absorption and emission

For isolated atoms, the light is absorbed when an electron in an atom is excited from one energy state to a higher state (Figure 2.14a). Not all transitions are possible as there are selection rules, arising from the need to conserve angular momentum and spin. The interactions only occur if the photon energy matches the difference between the two well-defined states and the line width is generally narrow. When atoms bond to form larger molecules, many additional states are introduced, due to both vibration of the bonds between atoms and to rotation of the molecules. In the condensed state (liquid or solid state), the atoms interact strongly with both near and more distant neighbors and the electronic energy levels of the atoms are spread out into bands and hence the absorption spectrum of liquid- and solid-state materials generally consists of broad bands, although they may have well-defined edges as discussed later. Absorption also occurs when the optical field excites quantized vibrational waves (phonons), which leads to absorption in the infrared part of the spectrum.

An atom in an excited state may emit a photon as it relaxes to a lower energy state. This emission may occur spontaneously (Figure 2.14b) or may be stimulated by another photon of the same energy (Figure 2.14c). This process of stimulated emission creates a photon that has the same direction and effective phase as the wave function of the incident photon. Stimulated emission and absorption oppose each other, and the net result is that light may be absorbed or amplified, depending on the relative populations of the atoms in the upper and lower states. For a material close to thermodynamic equilibrium, the population of the lower

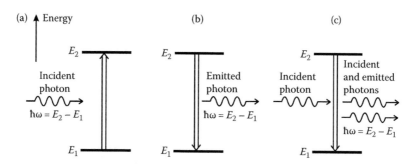

Figure 2.14 Energy diagrams illustrating (a) absorption, (b) spontaneous emission, and (c) stimulated emission.

state will exceed that of the upper state and the material will absorb light. In an optical amplifier, or in a laser that includes an optically amplifying region, the upper state is selectively excited and a *population inversion* occurs, with the population of the upper state exceeding that of a lower state. This provides optical gain.

2.9.3 Fluorescence

Fluorescence is the process by which radiation is absorbed at one wavelength and then reemitted at another, longer wavelength. Frequently, the excited state generated by the absorption relaxes rapidly by nonradiative processes leaving the system in a somewhat lower energy state with a longer lifetime, before radiating and relaxing to a lower energy state (possibly the ground state). If the lifetime of the excited state of the radiative transition is long, then the fluorescence may continue for a significant period (microseconds to milliseconds) after the exciting radiation has been cut off. Many materials show fluorescence, some with a high efficiency.

2.9.4 Scattering

The final interaction to be considered here is scattering. In this case, the material does not absorb the light but the incident radiation is scattered into all directions. The simplest form of scattering does not involve any interchange of energy and hence is called *elastic scattering*. Elastic scattering will occur from any small variations in the refractive index of the medium. For example, the scattering of light from a clear transparent glass is usually from tiny regions of the glass, where the refractive index differs from its mean value. These refractive index fluctuations arise from statistical density fluctuations, which occur in the liquid phase and then are frozen in as the glass solidifies. These regions of higher or lower refractive index are much smaller than the wavelength of light. Scattering from particles or regions of varying optical density that have dimensions much less than the wavelength of light ($<\lambda/10$) is called *Rayleigh* scattering. Rayleigh scattering occurs in all materials and in all phases, gas, liquid, or solid. The strength of the scattering depends inversely on the fourth power of the wavelength of the light (λ^{-4}). Thus, visible or UV light is scattered much more strongly than infrared.

Raman scattering is, by contrast, an inelastic process in which the energy is exchanged with the scattering material. This energy is in the form of molecular vibrations or optical phonons in the quantum explanation of the effect. The wavelength shift depends on the vibrational energy of the molecule or the phonon energy. The photon may lose energy (creating a phonon) as it is scattered, decreasing its frequency and increasing its wavelength (Figure 2.15a). The decrease in the frequency, which is usually expressed in wave numbers, is called the Stokes shift. Alternatively, a photon may gain energy by absorbing energy from a phonon or vibrating molecule (Figure 2.15b), giving a shift to a higher frequency (anti-Stokes shift). Note that the wave numbers used here and generally in spectroscopy are $1/\lambda$ and do not include the 2π term included in quantum mechanics. Raman scattering is normally a weak process, generally several orders of magnitude weaker than Rayleigh scattering, but it is useful as a chemical diagnostic technique. The effect is used in some optoelectronic devices but in most cases here it is stimulated Raman scattering that is used. Stimulated Raman scattering is a nonlinear process that occurs at high optical intensities, such as may

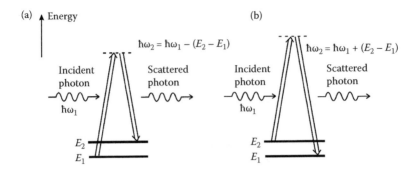

Figure 2.15 Energy diagrams illustrating Raman scattering: (a) Stokes scattering and (b) anti-Stokes scattering.

occur in optical fibers, even at moderate power levels. The spontaneous Raman effect may be enhanced if there is an excited state close to resonance with the exciting photon.

The third scattering mechanism is *Brillouin* scattering. This is an inelastic process similar to Raman scattering, except the energy states with which the exchange occurs are lower energy acoustic mode phonons (sound waves). In a simplified classical description, it can be thought of as light scattered from moving regions of acoustic wave-induced compression and rarefaction, which behave like a diffraction grating. The movement of this "grating" at the acoustic velocity induces a Doppler shift in the scattered light. The frequency shift is much smaller than for Raman scattering and again the effect is generally weak. It should, however, be noted that, as with Raman scattering, in a high optical field, with a long interaction length, the mechanism may lead to a nonlinear effect that results in strong stimulated scattering. This can occur at quite modest power levels (a few milliwatts) if light from a high coherence laser is launched into a long length of low-loss optical fiber.

The typical spectrum of scattered light from a glass, at moderate frequency resolution, is shown in Figure 2.16. The central peak at the excitation wavelength is strong and primarily due to the Rayleigh scattering but also includes unresolved Brillouin scattering. The frequency shift for the Brillouin scattering is only about $10\,\text{GHz}$ and hence can only be resolved by high-resolution spectroscopic techniques. The relative magnitude of the Stokes and anti-Stokes bands depends on the number of thermally excited phonons and hence the temperature.

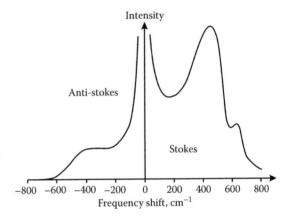

Figure 2.16 Raman spectrum at room temperature for a typical silica-based glass.

2.10 THE ABSORPTION AND EMISSION OF LIGHT BY SEMICONDUCTORS

2.10.1 Absorption and photoconductivity

In the previous discussion on semiconductors, only thermal excitation of electrons between the valence and conduction bands has been considered. Electrons may also be excited by the absorption of electromagnetic radiation. For materials with an energy gap, photons may only excite electrons across the gap if the photon energy exceeds the energy band gap.

$$\hbar\omega \gg E_g,$$

where ω is the optical angular frequency. There is a very small possibility of multiphoton absorption, but this will be neglected.

When a photon is absorbed, the usual conservation laws (energy, momentum, and spin) must all be satisfied. Thus, for a simple transition in which a photon is absorbed and an electron excited, the change in the energy of the electron will be $\Delta E = \hbar\omega$ and the change in the wave vector (momentum) of the electron will be equal to the wave vector of the photon. The wave vector of the photon will, in general, be very small compared with the wave vectors of the electrons. Hence, a simple absorption results in a very small change in k. To illustrate this, consider typical values. The lattice spacing for a typical crystal, a, is of the order 10^{-10} m, so the wave vector at the limit of the first Brillouin zone has a magnitude (π/a) of 3.1×10^{10}. The wave vector for a photon of wavelength $1\,\mu$m (i.e., an energy close to the band gap of silicon) has a magnitude of only 6.3×10^6. It follows that the absorption or emission of a photon will transfer an insignificant amount of momentum to or from the electrons. To preserve momentum, therefore, the transition is, therefore, essentially vertical on the E–k diagram.

For a direct gap semiconductor (such as GaAs), the maximum of the valence band lies directly below the minimum of the conduction band (Figure 2.17a). In this case, once the photon energy exceeds the band gap, the absorption in the semiconductor will rise rapidly. For an indirect gap semiconductor (such as Si or Ge), there is a significant change of k between the maximum of the valence band and the minimum of the conduction band (Figure 2.17b). Transitions between these two points can only occur if a third "particle" is available to enable momentum (k) to be conserved. This particle is a phonon, a quantized, vibrational excitation of the crystal lattice [1, Chapter 8; 4, Chapter 8]. The phonon may be created in the interaction and carry off the momentum and some of the energy or the interaction may be with an existing phonon. However, the energy associated with the phonon will be much less than the energy of the photon for the same k value, so most of the energy is transferred to the electron. This *three-body* interaction is much less probable than the simple interaction, where there is no change of electron momentum. Direct band gap semiconductors have an absorption that rises very rapidly near the band edge (photon energy corresponding to the band gap). By contrast, indirect gap semiconductors show a slow rise in absorption until the energy reaches a value that permits direct transitions (Figure 2.18).

The balance of the excitation rate and the decay rate will determine the steady-state carrier density produced by the radiation and may be written as

$$n_e = n_h = R\tau,$$

where τ is the carrier lifetime and R the excitation rate, which will depend on the intensity of the light

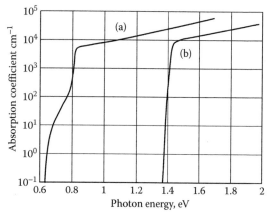

Figure 2.18 The absorption of (a) germanium and (b) gallium arsenide close to the band edge. (Data taken from Cassey, H.C., Sell, D.D. and Wecht, K.W. Concentration dependence of the absorption coefficient for *n*- and *p*-type GaAs between 1.3 and 1.6 eV. *J. Appl. Phys.* 46, 250–257, 1975. Dash, W.C. and Newman, R. Intrinsic absorption in single-crystal germanium and silicon at 77 K and 300 K. *Phys. Rev.* 99, 1151–1155, 1955.)

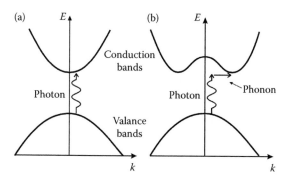

Figure 2.17 Diagram illustrating absorption in (a) a direct band gap semiconductor and (b) an indirect band gap semiconductor.

and the absorption coefficient of the material. The induced carriers will, in turn, lead to increased conductivity (photoconductivity), which will build up or decay in a time determined by the carrier lifetime.

2.10.2 Emission

If a semiconductor is able to absorb light, exciting electrons from valence band to conduction band, then it will also be possible for light to be emitted by the radiative decay from the conduction band to the valence band. Quantum mechanics shows that the two processes are closely linked. The excited electrons will generally relax very quickly to the bottom of the conduction band, and in the same way, the available holes will float to the top of the valence band. As with absorption, emission of radiation requires that k is conserved, and a high probability of radiative decay requires a simple two-body interaction, hence the electron cannot change its k value. The electrons rapidly decay to the bottom of the conduction band by nonradiative processes, and similarly, the holes rise to the top of the valence band. Hence, in an indirect band gap semiconductor, light emission will require the interaction with a phonon of suitable k value. This three-body interaction is, again, of low probability and recombination is more likely to occur by nonradiative processes. This implies that only direct gap semiconductors can be efficient emitters of light. Obviously, electrons must be excited to the conduction band in order that the semiconductor may emit light. This excitation may be optical, in which case the emission is fluorescence, or it may be due to the injection of minority carriers across a p–n junction.

2.11 POLYCRYSTALLINE AND AMORPHOUS SEMICONDUCTORS

In the earlier discussion, it has been assumed that the material is in the form of a perfect crystal. However, as mentioned earlier, crystals usually contain *point defects* and *dislocations*. These are disruptions to the regular lattice and may take a variety of forms [1–3]. Point defects occur when an atom is missing or displaced from its position within the lattice, they may also arise from impurity

atoms (as in the deliberate doping of semiconductors). Dislocations are due to imperfections in the lattice, which are not localized at a point but extend through the crystal. A simple example is the edge dislocation, when part of a row or plane of atoms is missing, creating stress around the dislocation and leaving *dangling bonds* (Figure 2.19a). Dislocations and defects may diffuse through the crystal, especially at elevated temperature. They will modify the band structure introducing extra energy states, which may trap carriers. Semiconductor devices are usually made from materials, which are as free as possible from dislocations and defects (other than deliberate doping).

In a polycrystalline material, the lattice does not extend unbroken throughout the whole sample, but there are many crystallites or crystal grains with their lattices orientated at different angles. Bonds exist across the grain boundaries but obviously there is a complex loss of order. For small angles between the lattices at a grain boundary, the boundary may be made up of a row or plane of edge dislocations [1, Chapter 20]. This is illustrated schematically for two dimensions in Figure 2.19b. Grain boundaries will introduce many surface states, which will generally seriously impair the performance of electronic devices. There are, however, cases where the impaired performance may be balanced by a greater need for reduced cost. An example of this is the polycrystalline solar cell, where the lower cost of producing a large area polycrystalline device, as opposed to a large area single crystal cell, makes the reduced performance economically acceptable.

Amorphous semiconductors may be formed as thin films by evaporation or sputtering. Such films have some short-range order due to the

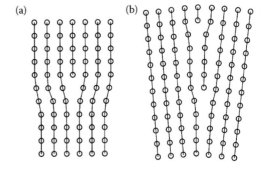

Figure 2.19 (a) Schematic diagram illustrating an edge dislocation (in cross section) and (b) a small angle grain boundary.

directionality of the bonds between atoms but possess no long-range order. Amorphous semiconductor films differ from polycrystalline films, in that order is maintained over a much shorter distance. Because of the disorder, there will be many dangling bonds, which result in many extra states within the energy gap. For amorphous silicon, it has proved possible to neutralize the effect of the unpaired electrons by depositing the amorphous silicon in an atmosphere of hydrogen. The hydrogen attaches to the dangling bonds and results in a material, which may be doped *p* or *n* type and which can be used to fabricate *p–n* junctions for solar cells. Although amorphous silicon is relatively cheap to produce, it performs much less well than crystalline silicon.

2.12 GLASSES

Glasses are a particular type of amorphous material and are particularly important for optics. They are generally produced from the liquid state by cooling the material. Unlike crystalline materials that have a well-defined melting point, the transition between solid and liquid occurs smoothly over a range of temperatures (the glass transition). While most commercially important glasses are very stable over long periods of time, some less stable glasses may revert to a crystalline state more quickly, particularly if held at high temperature or if they are cooled slowly from the molten state. Glasses, unlike crystals, have no long-range order, but because of the preferred directions of the bonds between the constituent atoms, there is an element of short-range order over a few atomic separations. A consequence of the lack of long-range structure is that there will be density fluctuations that are frozen in as the glass solidifies.

The most common glasses are based on silica (SiO_2) with additions of sodium and calcium oxide (and other metallic oxides) to reduce the temperature at which a glass is formed. Glasses of this type have been produced since prehistoric times and the Romans had a well-developed glass technology. Glasses with special properties are produced by varying the composition. Adding lead oxide increases the refractive index to produce *flint* glass (or lead crystal glass), whereas the replacement of some of the sodium oxide with boric oxide produces a glass with low thermal expansion (borosilicate glass, e.g., Pyrex). The addition of transition metal oxides, semiconductors, or other materials leads to colored glasses, which are frequently used as optical filters. Also, by adding materials such as neodymium or erbium, glasses may be produced that are able to operate as lasers or optical amplifiers.

This ability to control the optical and mechanical properties of glasses accounts for their importance. The development of optical fibers for communications in the 1980s led to the development of new glasses with very low optical attenuation. Such glasses are generally based predominantly on silica, with germania (GeO_2) and B_2O_3 dopants to control the refractive index. In order to achieve the very low attenuation required, it is necessary to ensure very high purity, particularly avoiding transition metals. For this reason, the glass is generally fabricated by a vapor phase reaction to generate the basic material, which is usually formed in a molten or finely divided state. The molten glass is then cooled to form the fiber preform rod, which is drawn into a fiber. Using these methods, the attenuation is reduced to a level limited by the scattering from the residual density fluctuations in the glass and the absorption due to the silica. The loss in an optical fiber decreases rapidly with increasing wavelength as the scattering becomes less, until it rises again due to the long wavelength absorption of the glass. Figure 2.20 shows an attenuation spectrum for a typical silica-based fiber. The absorption peak around 1400 nm arises from residual hydroxyl radicals in the glass. The success in producing very-low-loss optical fibers is largely a result of improved fabrication methods, reducing the concentration of these radicals.

While the common forms of glass are all based on oxides, there has recently been much interest in infrared transmitting glasses, for example, chalcogenide glasses that are oxygen free. These glasses are based on the chalcogen group of elements: sulfur, selenium, and tellurium, combined with arsenic, antimony, or germanium and/or halide elements. Chalcogenide glasses are of interest because of their transparency at longer wavelengths than oxide-based glasses, their semiconducting properties, and their optical nonlinearity. In addition, their low phonon energy helps prevent nonradiative decay in infrared optical fiber amplifiers. They are, however, much less robust than conventional glasses softening at much lower temperature (less than 200°C) and showing poor chemical durability.

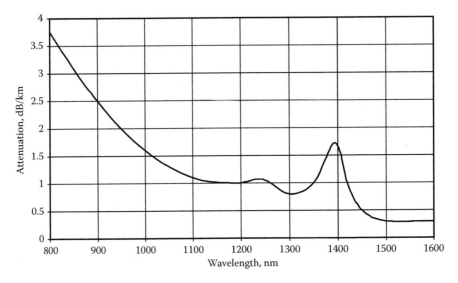

Figure 2.20 The attenuation spectrum of a typical silica-based optical fiber.

2.13 ANISOTROPY AND NONLINEAR OPTICAL PROPERTIES OF CRYSTALS

2.13.1 Anisotropy

In discussing the refractive index of materials, it was tacitly assumed that the polarizability of a material is a scalar quantity. This must be true for unstrained isotropic materials, such as glasses, but is not generally true for crystals. Simple cubic crystals, e.g., sodium chloride, are isotropic but crystals with more complex crystal structures, e.g., quartz, are not. The relationship between the electric displacement vector, D, and the electric field, E, must be expressed in the more general form:

$$D = \varepsilon_0 \varepsilon_r \cdot E = \varepsilon_0 \begin{pmatrix} \varepsilon_{xx} & \varepsilon_{xy} & \varepsilon_{xz} \\ \varepsilon_{yx} & \varepsilon_{yy} & \varepsilon_{yz} \\ \varepsilon_{zx} & \varepsilon_{zy} & \varepsilon_{zz} \end{pmatrix} \quad (2.18)$$

It may be shown that the dielectric constant matrix, ε_r, is symmetric and that by suitable choice of major axes, the *principle dielectric axes*, ε_r becomes diagonal. Note that unless the direction of the electric field, E, lies along one of these principle axes, the direction of the displacement vector, D, and the electric field is not in general parallel. The solution of Maxwell's equations for electromagnetic waves in an anisotrophic medium shows that the propagation depends on the direction and on the polarization state of the wave. For a given direction of propagation, there are generally two normal modes with orthogonal linear polarization, which propagate with different phase velocities. If the wave launched into the crystal is a mixture of the two normal modes, the polarization state will constantly change through a range of elliptical states as the wave propagates.

2.13.2 Electro-optic and nonlinear processes

In discussing the effect of refraction, it was also tacitly assumed that the induced polarization is proportional to the instantaneous electric field. However, for some materials, the interaction between the material and the electric field may be more complex. In the *linear electro-optic* or *Pockels effect*, the interaction is manifest as a change of refractive index with applied electric field. Whether a material shows the linear electro-optic effect or not is determined by the symmetric properties of the crystal. An isotropic material or a crystal with inversion symmetry cannot show the linear effect [10, Chapter 6], although it may show the, quadratic, *Kerr effect*. Thus, electro-optic crystals are those having lower symmetry, e.g., quartz or potassium dihydrogen phosphate. This electro-optic effect finds a use in optical modulators, where electric fields are applied to such a crystal to modulate the polarization or optical phase delay. (A subsequent polarizer or interferometric mixer can convert polarization or phase change to intensity changes.)

If the optical intensity is large, the nonlinearity between the polarization and the applied field may also manifest itself in terms of harmonic generation or optical mixing. The polarization is generally thought of in terms of a power series expansion

$$P = \varepsilon_0 \chi_1 E + \varepsilon_0 \chi_2 E^2 + \varepsilon_0 \chi_3 E^3 + \ldots, \quad (2.19)$$

where χ_1 is the linear susceptibility and χ_2 and χ_3 the second- and third-order nonlinear susceptibilities, respectively. The wave equation for propagation of light in a nonlinear medium may be written in the form [10]:

$$\nabla^2 E - \mu_r \varepsilon_0 \varepsilon_r \frac{\partial^2 E}{\partial t} = \mu_0 \mu_r \frac{\partial^2 P_{NL}}{\partial t^2}, \quad (2.20)$$

where P_{NL} contains all the nonlinear polarization terms. The second-order term $\chi_2 E^2$ acts as a driving term, which will contain components at twice the frequency of the electric field of the original wave, leading to the generation of light at a frequency twice that of the initial wave. The second-order component also leads to optical mixing. The third-order term leads to other nonlinear phenomena such as intensity-dependent refractive effects. Nonlinear effects are generally weak and require high optical intensity and/or a long interaction length.

Many optical devices take advantage of these electro-optic and nonlinear optical effects. For example, the Pockels effect may be used to create an optical phase modulator, or in combination with polarizers, to produce an optical amplitude modulator. Nonlinear optical effects, although usually weak, have long been used to change the wavelength of lasers by frequency doubling or optical mixing. The high optical intensity and long interaction length that is possible in optical fibers have extended the range of nonlinear effects that may be used without the need for very large optical power. They are considered in more detail in Vol 1, Part II, Chapters 7 and 9.

2.14 SUMMARY

This Chapter has attempted to provide a brief introduction to the electronic and optical behavior of some of the materials used. Many of the subjects raised will be taken up again in later chapters and treated in greater depth, notably Vol 1, Parts II and III. Clearly, this is a very extensive field and due to lack of space in this introduction, many issues have been omitted or only treated superficially. One significant area of omission is that of organic and polymeric materials. These materials are becoming increasingly important in a number of areas of optoelectronics, particularly in display technology, which is also the basis of a later chapter.

REFERENCES

1. Kittel, C. 1996. *Introduction to Solid State Physics*, 7th ed. (New York: Wiley).
2. Myers, H.P. 1990. *Introductory Solid State Physics* (London: Taylor & Francis).
3. Hook, J.R. and Hall, H.E. 1991. *Solid State Physics*, 2nd ed. (Chichester: Wiley).
4. Solymar, L. and Walsh, D. 1998. *Electrical Properties of Materials*, 6th ed. (Oxford: Oxford University Press).
5. Tanner, B.K. 1995. *Introduction to the Physics of Electrons in Solids* (Cambridge: Cambridge University Press).
6. Bleaney, B.I. and Bleaney, B. 1976. *Electricity and Magnetism*, 3rd ed. (Oxford: Oxford University Press).
7. Kaye, G.W.C. and Laby, T.H. 1978. *Tables of Physical Constants*, 14th ed. (London: Longman).
8. Cassey, H.C., Sell, D.D. and Wecht, K.W. 1975. Concentration dependence of the absorption coefficient for n- and p-type GaAs between 1.3 and 1.6 eV. *J. Appl. Phys.* 46, 250–257.
9. Dash, W.C. and Newman, R. 1955. Intrinsic absorption in single-crystal germanium and silicon at 77 K and 300 K. *Phys. Rev.* 99, 1151–1155.
10. Smith, S.D. 1995. *Optoelectronic Devices* (London: Prentice-Hall).

3

Incandescent, discharge, and arc lamp sources

DAVID O. WHARMBY

3.1 OVERVIEW OF SOURCES

There is a very wide range of incandescent and discharge lamps. The majority of these are sold as general lighting sources, but many are suited to optoelectronic applications. The major lamp companies, and numerous specialty lamp manufacturers also make lamps for applications other than general illumination. Examples of these applications are: projection, video, film, photographic, architectural, entertainment and other special effects, fiber optic illumination including numerous medical and industrial applications, photobiological processes, photochemical processing, microlithography, solar simulation, suntanning, disinfection, ozone generation, office automation, scientific applications, heating etc.

LED sources are covered in detail in Chapter 10. Section 3.10 of this chapter makes some brief comments on the applications in which LEDs are competing with conventional lamps.

This chapter will concentrate on principles and will be illustrated by a number of examples. These principles should make it possible to understand the wealth of information in manufacturer's websites and catalogues. A selected list of manufacturers is given in Appendix.

There are a number of useful books about light sources. The book by Elenbaas [1] is an excellent overview of the science of light sources, whilst for discharge lamps the book by Waymouth [2] contains clear and detailed explanations of many discharge phenomena. Coaton and Marsden [3] give a comprehensive introduction recent enough to cover many modern developments; their Appendix gives a useful generic table of lamp data for nearly every commercial source used for illumination. Zukauskas et al. [4] give an up to date review of the use of LEDs in lighting.

3.2 LIGHT PRODUCTION

Most optical radiation is the result of accelerating electrons and causing them to make *inelastic* collisions with atoms, ions, molecules or the lattice structure of solids. In the UV, visible and near IR, the photons are the result of *electronic* transitions between energy levels of these materials.

There are exceptions; in synchrotron radiation and related processes emission is from accelerated electrons.

As particle densities increase in the source, the spectral features broaden out until, in incandescent sources the spectrum is continuous. Discharge sources generally emit spectral lines of atoms and molecules that are broadened to an extent depending on the pressure. Lamps of various types therefore emit a wide range of spectral features ranging from narrow atomic lines to a full continuum. The types of spectra are often critical for optical applications [5] (see Appendix—Oriel Instruments for a selection of spectra).

In incandescent lamps, the radiation is from the surface of a hot material. In discharge lamps, conduction is the result of ionization of the gas; any light emission is a volume process. The task of the lamp designer is to ensure that this ionization is also accompanied by copious radiation of the correct quality for the application.

3.3 RADIATION FUNDAMENTALS

3.3.1 Full radiator radiation and limits on emission

Both in incandescent and discharge lamps, electron motion is randomized. In all cases of practical interest, the drift velocity of the electrons in the applied electric field is much less than the mean velocity. An electron energy distribution function is established that can usually be characterized by an electron temperature T_e. The distribution function may be far from Maxwellian when particle densities are low, or under transient conditions. It is the electrons in the high-energy tail of the distribution that excite the atoms, with subsequent emission of radiation.

The spectral radiance L_e (λ, T) of the full radiator or black body is given by Planck's equation (Chapter 8, where radiometric and photometric quantities are also defined). The spectral radiance is plotted in Figure 3.1 for temperatures typical of those found in incandescent and discharge lamps. Convenient units for spectral radiance are Wm^{-2} sr^{-1} nm^{-1}, obtained by multiplying the value of c_1 in Chapter 8 by 10^{-9}.

For incandescent or high-pressure (HP) discharge sources the electron temperature T_e is close in value to the temperature T of the solid or vapor, but for low-pressure (LP) discharges in which collisions between electrons and heavy particles are comparatively rare, T_e may be very much higher than the gas temperature. The Planck equation therefore forms a fundamental limit to the radiance that may be obtained from any source in which the electron motion is randomized. This sets a fundamental limit on the spectral distribution, the energy efficiency and the radiance of the source.

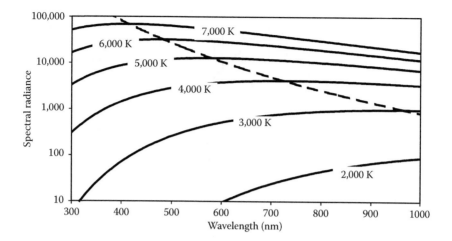

Figure 3.1 Spectral radiance of a full radiator (Wm⁻² sr⁻¹ nm⁻¹). The broken line is Wien's displacement law showing the shift in peak radiance to shorter wavelengths as the temperature increases.

3.3.2 Absorption and emittance

For radiation falling on a surface

$$\alpha(\lambda, T, \theta) + t(\lambda, T, \theta) + r(\lambda, T, \theta) = 1 \quad (3.1)$$

where the fractions $\alpha(\lambda, T, \theta)$, $t(\lambda, T, \theta)$ and $r(\lambda, T, \theta)$ are known as absorbance, transmittance and reflectance, respectively. In general, they depend on the wavelength, temperature and angle θ between a ray and the normal to the surface.

The spectral emittance $\varepsilon(\lambda, T, \theta)$ is the ratio of the thermal emission from the surface to that of a full radiator (black body) at the same temperature, wavelength and angle. This quantity is also known as spectral emittance. Derived from very general thermodynamic arguments, Kirchhoff's law [6] states that

$$\varepsilon(\lambda, T, \theta) = a(\lambda, T, \theta). \quad (3.2)$$

For a perfect absorber, $\alpha(\lambda, T, \theta) = 1$. Therefore, the spectral emittance of a full radiator is unity; a good approximation can be made by forming a cavity from an absorbing material.

All real materials have $\varepsilon(\lambda, T, \theta) < 1$. The best characterized material is tungsten (Figure 3.2) [1]. *Selective emittance* is characteristic of most materials; in metals the emittance tails off at long wavelengths, whereas refractory oxides usually have a region of high emittance in the IR.

3.3.3 Étendue

For all optical systems geometry determines how much of the radiation generated by the source can be used by the optical system. This behavior depends on a very general concept called étendue ε, also known as geometric extent [6–8].

A definition of étendue is

$$\varepsilon = \iint \cos\theta \, dA \, d\Omega \, (\mathrm{m^2 sr}) \quad (3.3)$$

where $\cos\theta \, dA$ is the projected area of the source under consideration, and $d\Omega$ is the solid angle into which it is radiating. Notice that the units are geometric, with no mention of amounts of radiation. A more general form is used when refractive indices are >1 [6]. Energy conservation requires that étendue is conserved in a lossless optical system; if there are losses caused by aberrations, scattering, or diffraction, étendue increases through the system. The étendue of a bundle of rays passing through an optical system either stays the same (ideal) or increases, but never decreases.

A simple example demonstrates some of the issues. Imagine projecting an image of the sun onto a surface. The diameter of the sun is about 1.4×10^9 m with an area $A_S \approx 1.5 \times 10^{18}$ m². Our distance from the sun is about 1.5×10^{11} m. Suppose the lens has a focal length of $f = 100$ mm and a diameter of 10 mm so that its area $A_L \approx 8 \times 10^{-5}$ m². The solid angle Ω_0 subtended by the lens at the sun is therefore about 3.5×10^{-27} sr. In this simple geometry the étendue

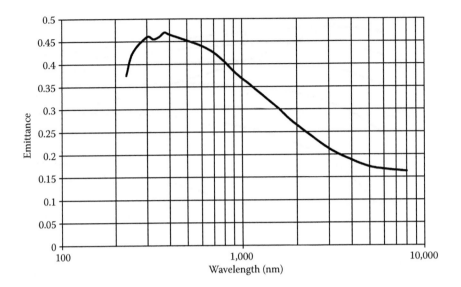

Figure 3.2 Spectral emittance of tungsten at 2800 K at normal incidence. (From Elenbaas, W., *Light Sources*, Macmillan, London, 1972.)

$\varepsilon = A_S\Omega_0 \approx 5.4 \times 10^{-9}\,\text{m}^2$ sr. The image is brought to a focus at a distance f in a converging beam of solid angle $\Omega_L = A_L/f^2 = 8 \times 10^{-3}$ sr. Assuming a perfect optical system so that étendue is conserved, the image area is therefore $A_I = \varepsilon/\Omega_L \approx 7 \times 10^{-7}\,\text{m}^2$, giving an image diameter of about 0.5 mm.

If we want to focus the sun onto a smaller spot, a lens of the same area needs to have a shorter focal length. Aberrations in a nonideal lens then cause some of the light to fall outside the area predicted above, increasing étendue. Scattering and diffraction are also losses that increase étendue. In general, the integration in Equation 3.3 has to be done numerically, e.g., by using an optical design code.

Étendue is also the quantity that determines how the power Φ (W) in the beam is related to the radiance L (Wm^{-2} sr^{-1}), as inspection of the units will confirm:

$$\Phi = L\varepsilon t \, (\text{W}) \qquad (3.4)$$

t is the transmittance of the lens (and related optics). Conservation of étendue and of energy means that radiance can never be increased by an optical system.

In a projector, there is always some component that has the smallest (limiting) étendue. Often this will be the film or light gate with its associated projection lens. If the étendue of the source is greater than this, some light will miss the light gate and be wasted. On the other hand, if the étendue of the light gate is much larger than that of the source then the gate will not be fully illuminated. The aim must therefore be to reduce the étendue of the source as far as possible, since it is usually much greater than the limiting étendue. This will minimize the amount of light that misses the light gate. Suppose that a projector lamp has a source of area A_S that radiates in all directions so that the solid angle is 4π and the source étendue is $\varepsilon_S = 4\pi A_S$ The limiting étendue ε_L of the system will be usually be that of the light gate. In order that £ s does not greatly exceed ε_L, with consequent wastage of light, the area of the source must be very small because the source solid angle is so large. Major advances in projector lamps have been to use HP arcs with an arc gap as small as 1 mm (see Section 3.7.4) and an effective area in the region of 0.1 mm^2.

The étendue concept is very general. It applies to any illumination system from fiber optics to street lanterns. For example, one of the benefits of LEDs is that their low étendue allows efficient use of the relatively low radiated fluxes; this is a reason why LED headlights for cars are a possibility.

3.3.4 Use of light in systems

The luminous flux in lumens (lm) [3, Chapter 1]

$$\Phi_v = 683 \int_{380}^{780} \Phi_{e\lambda} V(\lambda)\,\mathrm{d}\lambda \qquad (3.5)$$

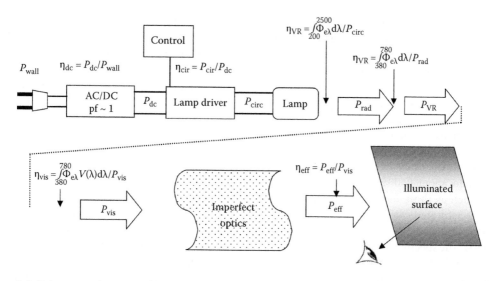

Figure 3.3 Schematic diagram of lighting system. The efficiency at each conversion is shown. Power from the wall P_{wall} is converted to dc power P_{dc}, which is used by the lamp circuit to input P_{cir} to the lamp. The broad arrows represent various aspects of the radiation from the lamp. The lamp converts P_{cir} to radiation P_{rad}; a convenient measurement and integration range is 200–2500 nm, which encompasses most of the radiation emitted. A fraction of this P_{VR} is in the visible region. This is then converted into visible power by weighting with the eye sensitivity curve to give power P_{vis}; the luminous flux is $683 \times P_{vis}$. That power is transmitted/reflected by an imperfect optical system onto a surface that reflects light into the eye. All powers P here are in watts and the spectral powers $P(\lambda)$ are in W nm^{-1}.

where $\Phi_{e\lambda}$ is the spectral radiant flux in W nm^{-1} and $V(\lambda)$ is the spectral luminous efficiency for photopic vision (Chapter 8). The factor 683 (lm W^{-1}) converts power to luminous flux. It is also useful to define the *luminous efficiency of radiation*

$$K = \Phi_v/\Phi_e. \qquad (3.6)$$

The (luminous) efficacy of a source is

$$\eta_v = \Phi_v/P_{in} \left(\text{lm W}^{-1} \right). \qquad (3.7)$$

For many commercial lamps, the input power P_{in} is defined as the power into terminals of the lamp, whereas self-contained sources (such as compact fluorescent lamps), or lamps sold as a system (such as some electrodeless lamps) P_{in} is taken to be the power coming from the electricity supply P_{wall}. The latter power is greater because it contains the losses in the lamp circuit; users should be aware of this possibility for confusion.

Many lighting systems are driven and controlled by electronics and this trend will be maintained in the future. Figure 3.3 shows a schematic view of a complete lighting system. To generate light that eventually reaches the eye, every system includes most

or all the steps shown. In order to work in terms of power so that we can calculate efficiencies, the quantity in Figure 3.3 $P_{vis} = \Phi_v/683$. For each stage in there is a loss and the system efficiency is then

$$\eta_{sys} = \eta_{dc} \times \eta_{cir} \times \eta_{rad} \times \eta_{VR} \times \eta_{vis} \times \eta_{eff} \quad (3.8)$$

The various terms are defined in Figure 3.3. Each stage in this chain of light production needs to be examined to discover how system efficiency can be improved. Notice that Equation 3.8 applies equally well to a street lamp, a projector, a self-ballasted lamp, a fiber-optic illuminator and, if the conversion from mains to ac is omitted, to battery operated lighting system.

3.3.5 Color properties and color temperature of sources

Definitions of quantities mentioned below related to color are given in Chapter 8. A comprehensive discussion of color in lighting is also given by Coaton and Marsden [3, Chapter 3].

An important color property of any source is *color appearance* or *chromaticity* (specified by the chromaticity coordinates). The color appearance of any source can be matched by mixture of three sources of different color appearance (for example, by red, green and blue sources, or by three spectral sources). The space of all possible colors is bounded by the spectral colors. For general illumination and for some optoelectronic applications such as projection, the preferred color of sources is "white"; the chromaticity of these sources is then close to that of a black body having a color temperature (see below) in the range from about 2800 (yellowish white) to about 8500 K (bluish white). Other sources, such as those used for signaling, usually have more saturated colors (that is, colors such as red, green, amber, etc.) that are close to the spectral colors). The specifications for these sources are closely controlled [9].

Color temperature is defined only for those sources having a color appearance close to that of a black body. The quantity most often used is the *correlated color temperature* (CCT) [10,11], defined in Chapter 8. A few examples help to set a scale. The glowing embers of a fire have a CCT in the region of 1000 K whilst a candle flame has a CCT of about 2000 K. Incandescent lamps, depending on type, have CCTs between 2400 and 3400 K. The CCT of the sun is about 6000 K. Discharge lamps for general illumination mostly have CCTs between 3000 and 6500 K. Xenon arcs and flash lamps have CCTs in excess of 6000 K.

Sources of a given chromaticity (that is, having the same color appearance) may have very different spectral distributions. A commonly observable example (at least in Europe) is that the color of an amber traffic signal and of the commonly used orange low-pressure sodium street lights are almost identical; the sodium lamp emits only at about 589 nm whereas the traffic signal is a filtered tungsten lamp that emits over a broad spectral range from the yellow through to the red. The value of K (Equation 3.6) for light from the LP sodium lamp also greatly exceeds that for the traffic signal.

Not surprisingly a surface illuminated by these two sources appears to have very different colors. The *color rendering* capability of a light source is an important measure. For general task illumination colors need to appear "natural"; this means that surfaces such as skin, fabric, building materials, etc. should not appear distorted when compared with their appearance under natural light or incandescent light, which both have continuous spectra. Along with high efficacy or high luminance, this is a major requirement for commercial light sources. The measure of color rendering used is the CIE General Color Rendering Index (CRI) or R_a (see Chapter 8) [12,23]. R_a is computed from the color shifts shown by a series of colored surfaces when illuminated by the test illuminant as compared to their color when illuminated by natural and Planckian reference illuminants.

The better the color rendering, the lower the efficacy of the lamp. One might think that since the sources that give "perfect" color rendering have continuous spectra, then high quality lamps should too, and this is usually the case. However, simultaneous optimization of K and R_a at constant color temperature has shown a surprising result; both quantities are maximized if the light is emitted in narrow bands at 450, 540 and 610 nm. This feature of human vision, confirmed by experiment, has been exploited in the *triphosphor* fluorescent lamps that are now standard in all new installations. The lamps use narrow band phosphors that emit close to the critical wavelength. Similar techniques are now being used to optimize white light LEDs (Zukauskas et al. [4] give a useful review of optimization.)

The CIE CRI is defined so that tungsten and daylight sources have $R_a = 100$. For general lighting in commercial premises requiring high-quality illumination, restaurants and homes, should be 80 or higher. Good quality sources for interior lighting such as triphosphor fluorescent lamps and HP ceramic metal halide (CMH) lamps have $R_a \geq 80$. Lower cost halophosphate fluorescent lamps have R_a around 50–60, as do HP mercury lamps with phosphor coatings. High-pressure sodium (HPS) lamps used for street lighting have R_a around 25.

Human vision is extremely sensitive to small differences in color [13] particularly in peripheral vision. This has proved to be a major challenge for lamp manufacturers, especially where lamps are used in large installations such as offices and stores. Not only should the initial spread in color be very small, but also the color shift during life must be very small otherwise when lamps are replaced it will be very obvious. Amongst the lamps for high-quality illumination triphosphor fluorescent and CMH are preeminent in this respect; such color differences as they have, are barely noticeable.

3.3.6 Radiation from atoms and molecules in extended sources

In a discharge lamp, each elementary volume of plasma emits optical radiation. In a volume source an atom or molecule with an upper state of energy E_u (J) can make a transition to a lower state E_l (J) with a transition probability of A_{ul} (s^{-1}). The emitted wavelength λ (m) is then given by

$$\frac{hc}{\lambda} = E_u - E_l(J) \qquad (3.9)$$

where h is Planck's constant in Js^{-1} and c is the velocity of light in ms^{-1}.

Since the emission is isotropic, the *emission coefficient* $\varepsilon\lambda(x)$ from a volume element at position x containing N_u atoms or molecules in the excited state is

$$\varepsilon_\lambda(x) = \frac{10^{-9}}{4\pi} N_u(x) A_{ul}$$
$$\times \frac{hc}{\lambda} P(\lambda) \,(\mathrm{Wm^{-3}\,sr^{-1}\,nm^{-1}}). \qquad (3.10)$$

$P(\lambda)$ is the line shape function having an area normalized to unity. Do not confuse the emission coefficient $\varepsilon\lambda$ [14] with the spectral emittance $\varepsilon(\lambda, T, \theta)$ of a surface, which is a dimensionless quantity.

Suppose, we view a nonuniform extended source of depth D. The spectral radiance along a line of sight for a spectral line at wavelength λ is

$$L(\lambda) = 10^{-9}$$
$$\times \int_0^D \varepsilon_\lambda(x)\,dx \,(\mathrm{Wm^{-2}sr^{-1}nm^{-1}}). \qquad (3.11)$$

This is only an approximation. When absorption is present the radiance does not depend linearly on atom density and is given by the radiation transport equation [14]. Examples of this important phenomenon are described in Sections 3.6.1 and 3.7.1.

3.4 INCANDESCENT LAMPS

3.4.1 Emission

Tungsten is the preeminent material for the manufacture of incandescent lamps. It has a melting point of 3680 K and it can be drawn into the fine wire necessary for making lamps. In normal

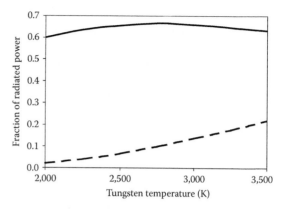

Figure 3.4 Calculation of fractions of power emitted from a typical coiled tungsten filament as a function of temperature. The dashed line is the fraction of power radiated at <750 nm (mostly visible but with a very small fraction of UV radiation). The full line is the IR fraction between 750 and 2000 nm. The remaining fraction is at wavelengths >2000 nm. (From Bergman, R.S. and Parham, T.G. *IEE Proc.*, 140A, 418–428, 1993.)

household bulbs the filament is operated at a temperature in the region of 2800 K, depending on the type. The comparatively low temperature is chosen to limit evaporation and give an acceptable life. This section will concentrate on the higher temperature tungsten–halogen lamps that have many optical applications.

A substantial fraction of radiation from a tungsten filament is emitted between 750 nm and the glass or silica cut-off in the IR. Figure 3.4 shows that the fraction of power radiated in the region 750–2000 nm is approximately independent of the tungsten temperature, whilst the visible fraction (<750 nm) doubles for an increase of 500 K in temperature.

3.4.2 Tungsten–halogen lamps

Use of a halogen chemical transport cycle [15] allows tungsten filaments to be operated at higher temperatures than in the standard household bulb. For lamps of similar wattage and life the filament can be operated 100 K higher in a halogen lamp compared with a conventional lamp [16].

The halogen—usually a fraction of a μmol cm^{-3} of iodine or bromine—is added to the lamp before it is sealed. During operation of the lamp the halogen reacts with evaporated tungsten in the cooler regions. The tungsten halide thus produced is a vapor that is transported by diffusion and convection to hotter regions, where it dissociates depositing tungsten and releasing halogen for further cleanup. The dissociation mainly takes place at a region of the filament lead. The net effect of the cycle is therefore to transport the tungsten from the wall to regions of the lamp that do not affect light output.

Because the lamp walls remain clean, the bulb can be made very small and strong. High pressures of inert gas of high molecular weight suppress evaporation. With smaller, stronger bulbs containing high pressures of Kr, or even Xe, the tungsten may be operated at temperatures of up to about 3500 K.

The higher the filament temperature the greater the rate of evaporation and the shorter the life of the lamp. Filaments operating at a color temperature of 3400 K (filament temperature ≈ 3330 K) will have a life of a few tens of hours. Life is also strongly dependent on operating voltage; manufacturers' data should be consulted for information.

Tungsten–halogen lamps have the advantage over all other sources of having excellent stability. For best stability, lamps should be operated from a dc constant current supply with current controlled to 1 part in 10^4—this is the technique used for operating calibration lamps. The current should be set to ensure that the voltage rating of the lamp is not exceeded. When lamp stability is at premium (as for example in standards of spectral irradiance) optical equipment suppliers select particularly stable lamps (Appendix—Oriel Instruments, Ealing).

3.4.3 Varieties of tungsten–halogen lamps

The development of tungsten–halogen lamps has resulted in thousands of new products being introduced. For optical applications the most important consideration is often the ability to focus the light into a tight beam. This is affected by the size of the filament, the tightness and evenness of winding of the coil, whether the coil is a flat or cylindrical, whether the coil is concentric with the bulb axis or normal to it, the quality and thickness of the bulb wall, and the type of glass used (hard glass can have better optical quality than fused silica). Examples are shown in Figure 3.5. High color temperature versions with powers in the range 25–1000 W or even greater are available. In some cases, these are

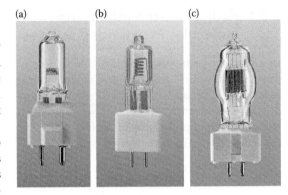

Figure 3.5 Examples of tungsten–halogen lamps (not to scale). All are mounted in a ceramic base that is pre-focused to allow accurate replacement. (a) and (b) operate from low voltage at a CCT of 3000 K or more and in some cases as high as 3500 K. Lamp (a) has a flat filament especially suitable for projection. Lamp (b) has an axial filament suitable for use in reflectors for video applications. Lamp (c) is a mains voltage lamp for use in overhead projectors; available in ratings up to 900 W and CCT is 3200 K. (Philips photographs.)

made from silica that is doped to prevent emission of short wave UV. Consult manufacturers' websites for "special" lamps designed for particular optical applications.

3.4.4 Lamps with integral reflectors

There is a wide range of tungsten–halogen lamps built into small reflectors. The reflectors may be aluminized, or have a dichroic (interference filter) coating allowing some IR radiation to escape from the rear of the reflector; this means that the beam is comparatively cool. They may also be fitted with cover glasses that reduce the already small amount of short wave UV that is emitted by fused silica tungsten–halogen lamps. Reflector diameters vary from 50 down to 35 mm. Versions are made with beam divergences from a few degrees up to 40°. Typically wattages vary from 12 to 75 W with a color temperature of about 3000 K.

These reflector lamps are used in large numbers for all sorts of commercial displays and accent lighting and therefore they are relatively inexpensive. In addition, all the major lamp manufacturers make special versions that are used in a number of optical applications such as overhead projection, microfilm and fiber optic illuminators. Figure 3.6 shows examples.

(a) (b)

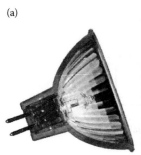

Figure 3.6 Examples of low voltage tungsten–halogen lamps in integral prefocused reflectors. In (a) the reflectance of the coating has been reduced to show the positioning of the axial filament in the reflector (Osram). Assembly (b) has been specially designed for fiber optic illumination (Philips). Lamps for specific optical purposes are also available from most manufacturers.

3.4.5 Lamps with IR reflectors

Over 90% of the radiation from tungsten–halogen lamps is in the IR region and so is wasted. Many attempts have been made to return some of this radiation to the filament where it can be absorbed. Commercial success was eventually achieved by using multilayer interference filters deposited by LP CVD [16]. Less input power is needed to maintain the tungsten coil at the design temperature. The main benefit therefore is a saving in power for a given light output. The beam is cooler since there is less IR radiation emitted although optical quality is degraded slightly by the coating. The main benefit is an improvement of up to 40% in the efficiency of generation of visible light.

3.4.6 IR sources

Incandescent lamps using either tungsten or carbon emitters make use of the IR radiation in industrial heating processes (Appendix—Heraeus). The main benefit is a heat source that can be controlled precisely and has a much shorter response time than a conventional oven.

The Nernst source is an example of a ceramic emitter electrically heated to 2000 K, used as an IR illuminator in spectrophotometers. This makes use of selective emittance in the IR. More recent versions of similar devices are given in manufacturers' data (see Appendix—Oriel Instruments). There are also low heat capacity carbon emitters

that can be modulated at low frequencies (Appendix—Hereaus).

3.5 DISCHARGE LAMPS WITH ELECTRODES

One way is to group discharge lamps into LP (low-pressure) and HP (high-pressure) types. In LP discharges, the electrons make relatively few collisions per second with the gas atoms and so electron temperature ≫ gas temperature. In HP discharges, relatively frequent collisions between electrons and gas atoms ensure that both temperatures are approximately equal. The same physical processes occur in LP and HP discharges. Section 3.5.1 is concerned with the common features of both types. The electrode regions are described in Section 3.5.2. Later sections describe their unique features.

Another way to group discharges is by the manner of coupling to the power supply. Most discharges have electrodes in which the cathode is hot; electrons are released into the plasma by thermionic emission. The term *arc* is not uniquely defined, but it is often taken to mean a discharge in which the cathode emits thermionically—examples are all HP discharge lamps and hot cathode fluorescent lamps. In *cold cathode* lamps, the electrodes emit as a result of ion bombardment of the cathode surface. Other discharges (Section 3.9) are operated at high frequency using induction or microwave sources. Dielectric barrier discharges (DBDs) are transient and self-limiting with little or no emission of electrons from the cathode (Section 3.9.2).

3.5.1 Stable discharge operation of discharges with electrodes

To start a discharge, a high voltage must be applied to make the gas conducting, and (an electron) current from an external circuit must be passed from cathode to anode through the conducting gas. A by-product of causing the gas to conduct is the production of radiation. To demonstrate the main effects we will consider dc discharges although the majority of commercial lamps operate on ac (Section 3.8.3).

We will illustrate the main features of a dc discharge using the LP mercury-rare-gas discharge of the type used in fluorescent lamps as an example (Figure 3.7). Other lamps including HP lamps have

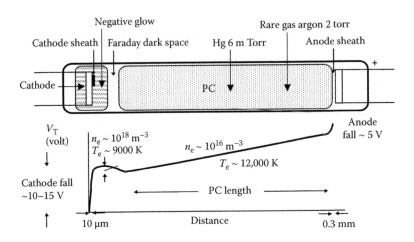

Figure 3.7 Structure of a dc discharge. This schematic diagram shows features visible in a typical fluorescent lamp discharge, but they are also present in other discharges. The upper picture shows the positive column (PC), which may be any length, together with the anode and cathode regions in which dimensions are dependent on vapor and pressure. The lower diagram shows the voltage drop V_T along the lamp. The cathode fall field adjusts so that sufficient electrons are extracted to maintain a stable current. Typical electron densities n_e and electron temperatures T_e are shown for the fluorescent lamp case.

similar features, but the regions around the electrodes have dimensions that are usually too small to see. The bulk of discharge in Figure 3.7—the positive column (PC)—is a plasma, so there are equal number of electrons and ions per unit volume. Some discharges such as neon indicators or deuterium lamps used for producing UV are so small that the PC does not exist.

In the PC, electrons form a near Maxwellian distribution of energies. Once the discharge has been established, the applied electric field causes the electrons to drift towards the anode and the ions to drift towards the cathode; because their mobility is much greater than that of the ions and the current is carried mainly by the electrons. Therefore, current density is approximately

$$j = n_e |e| \mu_e E \left(A \ m^{-2} \right) \quad (3.12)$$

where E is the electric field, μ_e the electron mobility, $|e|$ the electron charge and n_e the electron density.

The PC can be any length as long as sufficient open-circuit voltage is available from the supply (think of commercial display signs). A condition for stable operation is that the rate of loss of electrons by recombination with ions must be equal to the rate of gain caused by ionization. In LP discharges, most of the recombination occurs

after the carriers have diffused to the wall; in HP discharges, particle densities are high enough for volume recombination to dominate.

The electric field in the column adjusts itself so that electrons are accelerated to a mean energy in the region of 0.5–1.5 eV corresponding to an electron temperature of about T_e of 6000–18000 K. The electron energy distribution then contains enough high-energy electrons to ionize atoms, replacing the electrons lost by recombination. Figure 3.7 shows that the electric field in the PC is constant, so in a given gas, the longer the lamp the higher the voltage.

3.5.2 Electrode regions

Adjacent to the anode the voltage usually increases (Figure 3.7). This is a result of a space charge sheath. If there was no sheath then the anode would only collect the random current. Normally the anode area is too small; to collect the current required it charges positively to attract electrons.

The cathode is more complex [2, Chapter 4]. The conditions at the cathode surface have to adjust themselves so that each electron that leaves the cathode initiates events that cause the emission of at least one more electron from the cathode, otherwise the discharge will not be self-sustaining. Electrons emitted thermionically (hot cathode case) or by ion bombardment (cold cathode case) are accelerated in the high field of the cathode fall (CF) region. A beam of electrons from the CF region penetrates the cathode edge of the negative glow (NG) causing the production of positive ions that are accelerated through the cathode sheath. A fraction of these (~0.1) knocks further electrons out of the cathode. The process is entirely self-regulating; if the work function increases, the CF increases and the resulting extra ion bombardment heats the cathode surface to higher temperatures, producing more thermionic emission.

The velocities of electrons leaving the CF are strongly directed toward the anode. This beam is gradually randomized in the direction in the NG region. By the end of the NG they have lost enough energy for the excitation of atomic levels to decrease. This region of comparatively little light is known as Faraday dark space (FDS). At this point, electron motion has been randomized giving a near Maxwellian distribution. Finally as the electrons start to gain energy from the field again, excitation increases and this marks the start of the PC. The NG and FDS therefore serve to change the highly anisotropic electron distribution function coming from the CF into the random distribution in the PC.

In hot cathode lamps, the CF is usually a little greater than the ionization potential of the most easily ionized species (see Figure 3.7). In cold cathode lamps, the CF is much larger because electrons must be extracted by secondary processes such as ion bombardment. Cold CF voltages are typically in the region of 100–200 V. The CF in cold cathode lamp can be reduced by using hollow cathodes [17].

3.6 TYPES OF LP DISCHARGES

By far the most important type is the LP mercury rare-gas discharge used in fluorescent lamps and in UV sources for photochemical and photobiological

purposes (Section 3.6.1). Other LP discharges not described here are LP sodium, used for street lighting of very high luminous efficiency, deuterium lamps used as UV illuminators and LP hollow cathode spectral sources for chemical analysis. There are also a wide variety of LP laser discharges.

3.6.1 Low-pressure mercury rare-gas discharges

LP mercury lamps contain a rare gas, usually argon, krypton or neon or mixtures of these, at a pressure of a few hundred pascal (a few torr). Mercury is added as a small drop of liquid weighing a few milligrams, which collects at the coolest place in the lamp. At typical wall temperatures, the mercury evaporates from the liquid drop at the pressure of about 0.8 Pa (0.6 m Torr). Despite the relatively low number density of the mercury atoms they dominate the properties of the discharge. The fluorescent lamp discharge is a highly efficient emitter of UV in the mercury resonance lines at 254 and 185 nm (>70%).

Phosphors are used to convert UV to visible radiation [3, Chapter 7]. Lamp phosphors are ionic materials doped with activators that absorb at short wavelengths and then reemit at longer wavelengths. The energy deficit in this *Stokes'* shift is converted into lattice vibrations. In fluorescent lamps used in lighting the conversion loss is typically 50%. There is a very large range of phosphors [3, Chapter 7], and fluorescent lamps giving white light of many different CCTs and other color properties are available. Particularly important are the ionic rare-earth based phosphors as these emit at the wavelengths that combine high efficacy and color rendering index (R_a—Section 3.3.5). The principle ones are noted in Table 3.1; notice how close the peaks are to the 450, 540 and 610 nm wavelengths that optimize color rendering and efficacy.

Table 3.1 Phosphors commonly used in fluorescent lamps

Name	Formula	Wavelength of peak output (nm)
YEO	$Y_2O_3:Eu^{3+}$	611
CAT	$Ce_{0.65}Tb_{0.35}MgAl_{11}O_{19}$	543
LAP	$LaPO_4:Ce^{3+}, Tb^{3+}$	544
CBT	$GdMgB_5O_{10}:Eu^{2+}$	545
BAM	$BaMg_2Al_{16}O_{27}:Eu^{2+}$	450
Halophosphate	$Ca_5(PO_4)_3(F, Cl):Sb^{3+}, Mn^{2+}$	Broad bands

The notation in Table 3.1 is the chemical composition of host lattice:activator. The activator is an ion added deliberately at relatively small concentrations to absorb UV and emit visible light. In some cases, the host lattice has this same function. Quantum efficiencies are close to unity.

One very important benefit of rare-earth phosphors is their resistance to degradation by mercury discharges at high power leadings. It is this property that made possible the development of compact fluorescent lamps. The disadvantage is the high cost of rare-earth phosphors compared with the halophosphates that they have largely replaced. The complexity of the materials in Table 3.1 is such that phosphor research is still largely empirical, so the existence of each of these phosphors represent many man-years of painstaking research.

The mercury vapor pressure is a dominant factor in controlling the amount of radiation emitted and the efficiency with which it is generated. When a mercury atom is excited near the center of lamp, the emitted photon is at exactly the correct energy to be absorbed by a ground state atom nearby. The photon is absorbed and reabsorbed many times before it finally reaches the wall in a random walk. When the mercury pressure is high, there are so many steps in the random walk that the chance of losing the excitation energy nonradiatively in a collision increases. When the mercury vapor pressure is low the initial excitation energy can escape in a small number of steps, but then the fraction of collisions that lead to excited mercury atoms is low. (The related process in HP lamps is described in Section 3.7.1.)

This means there is a mercury vapor pressure at which the efficiency of generation of UV radiation is at a maximum. This optimum pressure is achieved by having a small amount of liquid mercury present at about 42°C. When using fluorescent lamps it is important to arrange for the fixture or unit holding the lamps to operate so that the mercury pressure is close to optimum. Lamps are designed to run close to optimum in commercial lighting fixtures. For other uses, such as backlighting some cooling may be necessary. Some types of multilimb compact fluorescent lamps are designed for operation in hot fixtures. In these, the mercury is dosed as a solid amalgam containing, for example, bismuth and indium. The vapor pressure of mercury above the amalgam is less than that

above free mercury, but the use of an amalgam also substantially increases the ambient temperature range over which the mercury pressure is close to optimum [18].

3.6.2 Applications of LP mercury discharges

The fluorescent lamp discharge lends itself to many different formats [3, Chapter 7]. The most familiar are the long thin lamps used in ceiling lighting in nearly all commercial and industrial premises. There are also a wide variety of compact fluorescent (CFL) designed as a high efficiency replacement for incandescent lighting.

Other than illumination, important applications for fluorescent lamps are in office equipment (copiers, fax machines, etc.) and in the backlighting of displays. Cold cathode fluorescent lamps have a number of benefits: they can be small in diameter allowing screens to be very thin; at the low powers needed they are efficient enough for the purpose; lives are long; they can be switched frequently; and low cost, efficient power supplies are readily incorporated in the end product. Hot cathode fluorescent lamps produce more light and can be used to backlight displays that are used in high ambient light levels such as ATM machines. Short wave radiation from hot cathode mercury rare-gas discharges is used in photochemical or photobiological processes; or it can be converted using a phosphor to UVA (as in "black light" sources) that show up fluorescence in materials.

3.7 HP DISCHARGES

There are many variants of HP discharges. Most of them are used for street lighting and interior illumination of stores and offices and other commercial premises, in which high luminous flux, high efficacy, good color quality and long life are at a premium. Lamps exist in *single-ended* (both connections at one end) and *double-ended* (one connection at each end) configurations to suit different applications. Many other types of HP discharges are used in which light must be projected and high brightness is needed. Some of the properties of HP discharges are described below. The two main classes of lamp are those that use volatile or gaseous elements, and those that use metal halides to introduce radiating species into the vapor.

3.7.1 General features of HP discharge lamps

We will illustrate the operation of HP discharges by using the HPS (high-pressure sodium) lamp as an example. An HPS lamp has electrodes inserted into a narrow arc tube made from translucent alumina, resistant to attack from sodium. As with many HP lamps the arc tube is contained within a glass outer bulb. These lamps are used as highly efficient (120 lm/lamp watt) long-lived (>20,000 h) street lights that give a pleasant golden light with CCT=2000 K, albeit with rather poor color rendering properties (R_a=25).

The dimensions of the tube are typically 7 mm internal diameter with 70 mm length between the electrode tips for 400 W rating, with dimensions decreasing for lower wattage lamps. They contain a small pressure of rare gas and a few milligram of sodium metal. On applying a voltage the rare gas breaks down. The resulting discharge heats and evaporates sodium until its pressure is about 1.4×10^4 Pa (100 Torr). A radial temperature profile develops in which the center temperature is about 4000 K and the wall temperature is about 1500 K. Most of the length of the discharge is a positive column uniform along the axial direction. Sodium lamps usually also contain about 10^5 Pa (760 Torr) of mercury vapor. This reduces thermal conduction and increases axis temperature, thus increasing spectral radiance.

The positive column is approximately in local thermodynamic equilibrium (LTE) [14]. This means that the properties are dependent on the local temperature in the plasma. The electron density is given by a version of the law of mass action called the Saha equation [19]

$$\frac{n_e n_i}{n_a} = S(T) \, (\text{m}^{-3}) \tag{3.13}$$

where n_e and n_i are the electron and ion densities, n_a is the density of atoms (number per m³) and

$$S(T) = 4.83 \times 10^{21} (U_i/U_a) T^{3/2}$$
$$\times \exp(-E_i/kT)(\text{m}^{-3}) \tag{3.14}$$

where the U factors are partition functions for the ion and atom. $S(T)$ depends strongly on temperature through the exponential factor, where E_i (J) is the ionization potential (including corrections for high electron density) and k is Boltzmann's constant. Since the hot gas is a plasma $n_e = n_i$. The atom density n_a in an elementary volume at temperature T is given by the gas law so $n_a = P/kT$ where P is the gas pressure. Table 3.2 shows values of ne in sodium vapor at various temperatures. Since the current density is proportional to n_e it is clear that the current flow is mainly in the high temperature region.

The population n_u of an energy level of an atom (labeled u) is given by another LTE formula:

$$n_u = \frac{g_u}{g} n_0 \exp\left(\frac{-E_u}{kT}\right) (\text{m}^{-3}) \tag{3.15}$$

where n_0 is the density of atoms in the ground state, E_u is the energy (J) of the upper state of the atom, whilst g_0 and g_u are the statistical weights of ground and upper states, respectively. The number of atoms excited to the upper state depends

Table 3.2 Shows how the plasma temperature affects the number density (m⁻³) of excited states and ions

	Plasma temperature (K)			
	2000	3000	4000	5000
Number density of sodium atoms	4.8×10^{23}	3.2×10^{23}	2.4×10^{23}	1.9×10^{23}
Fraction of sodium atoms excited to the states radiating at 589	1.5×10^{-5}	8.8×10^{-4}	6.4×10^{-3}	2.3×10^{-2}
Fraction of sodium atoms that are ionized	7.3×10^{-6}	1.9×10^{-3}	3.5×10^{-2}	2.2×10^{-1}

Note: Electron density is equal to ion density. There are two excited states at about 2.1 eV giving rise to the characteristic orange sodium D radiation. The ionization potential is 5.14 eV before correction is made for lowering of the value at high electron densities. The arc operates so that the electron density is sufficient to carry the current and the plasma temperature adjusts to make this so. For steady state sodium arcs this sets the maximum plasma temperature to about 4000 K. Calculated using Equations 3.13 and 3.15.

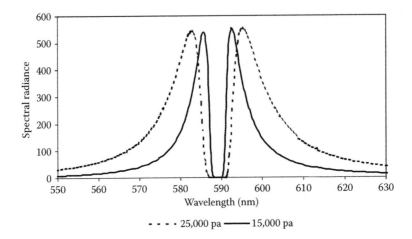

Figure 3.8 The formation of self-reversed lines in high-pressure sodium lamps at two sodium pressures. The calculation has been done for a parabolic radial temperature profile for a center temperature of 4000 K and a wall temperature of 1500 K. Comparison with Figure 3.1 shows that the peak radiance is substantially lower than that for a black body at the maximum temperature.

exponentially on temperature. Because of the exponential *Boltzmann factor* in Equation 3.15, the fraction of atoms in the excited state u is very small even at the highest temperatures. Only in the hottest parts of the discharge are significant numbers of atoms excited; the resulting "corded" appearance is a characteristic feature of an LTE arc. When a HP discharge operates horizontally convection bows the bright part upwards—the origin of the term arc. The importance of Equations 3.13 and 3.15 is shown in Table 3.2.

Self-absorption dominates the spectrum of many HP discharge lamps and is especially dominant in HP sodium discharges. As Figure 3.8 shows there is no significant radiation at 589 nm, the wavelength at which sodium radiates at low pressures. In an HPS lamp, the sodium pressure is so high that photons from excited sodium atoms can only travel about 10^{-7} m at the line center before being absorbed by a ground state atom. However, there is a chance that very close collisions with other sodium atoms can perturb the radiating atom sufficiently so that it radiates at wavelengths far from the line center at 589 nm. The hot plasma can therefore be considered as storing excitation energy until the energy can escape from an atom having strongly perturbed energy levels. The higher the pressure, the further from the line center the wavelength has to be, before the light can escape (Figure 3.8). This behavior is called self-reversal and it has a dominating effect on the operation of

many HP discharges [3, Section 5.6.3]. The cover of the book by de Groot and van Vliet [19] shows beautiful color photographs of the self-reversal of the sodium D lines at different pressures.

3.7.2 HP metal halide lamps

There are very few elements that have well-placed spectral lines and sufficiently high vapor pressures to be operated as HP discharges, the most important being mercury, sodium, sulfur, and the permanent gases (of which Xe is by far the most important).

There are perhaps 50 elements that have metal halides that are sufficiently volatile to be used in HP lamps. The principal ones are as follows:

- Na, I, In, Tl, Ga halides in which the metals have relatively few strong atomic lines.
- Sc, Fe, Dy (and other rare earth) halides in which the metals have many relatively weak visible lines so close together that the spectrum appears continuous at moderate resolution.
- Sn, Pb, and similar halides that form relatively stable monohalide molecules that emit a spectrum that appears continuous at low resolution.

The halide is usually the iodide, which has the least reactive chemistry.

Metal halide lamps are extensively used as efficient white light sources of good color quality for general illumination; all major lamp manufacturers make them (Appendix). Because the spectrum can be tailored to use, metal halide lamps are used extensively for production of UV for photopolymerization processes, such as ink drying or glue curing (Appendix—Heraeus). These lamps are installed as part of large production processes in the printing and packaging industries. The speed of curing is often the bottleneck in the processes, so with suitable UV sources productivity can be increased. HP mercury lamps are also used for similar processes. Other uses of metal halide lamps include special versions for medical conditions such as psoriasis (Appendix—Osram).

3.7.3 Operating principles of metal halide lamps

Many metal halide lamps contain thallium iodide (TlI). TlI is considered as a simple case to illustrate how the light is produced in metal halide discharges. Figure 3.9 shows a schematic diagram of a HP TlI discharge. When the lamp is made, a few milligram of solid TlI and a rare gas for starting are added. Usually enough mercury is added to give a partial pressure of about 10^6 Pa (10 bar) to reduce thermal conduction and to adjust operating voltage (Section 3.7.1). When the lamp is operated, the rare gas discharge heats up the Hg and TlI causing them to evaporate. In higher temperature regions, the TlI dissociates into Tl and I atoms. At higher temperatures still the Tl is excited and emits intense green light of high efficiency that can be useful for underwater illumination. Finally, near the axis the Tl is ionized producing the electrons needed to carry the current. This progressive evaporation, dissociation, excitation and ionization occurs in all metal halide discharges.

With mixtures of halides the ratio of salts has to be chosen with due consideration to the chemistry of the liquids and vapors. For example, one of the first types of metal halide lamp used mixtures of indium, thallium and sodium iodides that emit blue, green and orange self-reversed spectral lines. Altering the proportions of these can provide white light discharges of different color temperatures and quite good efficacy (luminous efficiency). However, their color rendition is rather poor.

Metal halide arc tubes are generally shorter than HPS lamps (for which the length to diameter ratio is more constrained by requirements of optimization)

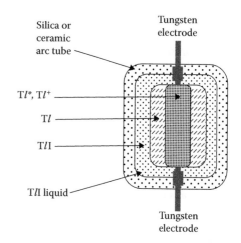

Figure 3.9 The principle of operation of a metal halide discharge. In this example solid thallium iodide (TlI) is dosed into the lamp along with a rare gas. The discharge starts in the rare gas, melting and then vaporizing the TlI. In the steady state the current is provided by ionizing Tl atoms. For this to happen the temperature on axis needs to be above 5000 K. At this temperature Tl atoms radiate strongly with their characteristic green line at 535 nm. The boundaries between the various regions are not sharp as shown schematically here, but blend into each other as the temperature increases from the wall to the axis. In a practical lamp, mercury vapor is also introduced at several bars to reduce thermal conduction losses.

and may even be close to spherical in shape. This has an effect on étendue and may make fixtures using these lamps more efficient at using the light.

There has been extensive research and development over the last 40 years that has produced mixed metal halide lamps with much improved color performance and efficacy. The halides used, their vapor pressures and their relative proportions all have a strong influence on the initial color properties and efficacy.

It is important that these properties stay constant through lives of 10,000 h or more. Reactions between the various components and the tube walls occur at different rates; all metal halide lamps show some color shift during life. Detailed R&D has improved the color stability of metal halide lamps in silica arc tubes so that it is acceptable in critical applications such as the lighting of stores and offices. A recent major improvement has been in the use of translucent alumina ceramic arc tubes for containing Na, Dy, Tl, HgI metal halide arcs.

Metal halide reactions with the envelope are much slower than with silica and this has provided a further major improvement in initial color uniformity and color stability through life.

Most metal halide lamps are used for illumination where the transparency of the arc tube is not an issue, but the scattering by the arc tube is a major disadvantage for projection. For projector and automotive head lamps, silica arc tubes are universally used. Because the axis temperature of metal halides is usually around 5500 K, the radiance of the gas close to the axis can be very high (see Figure 3.1).

3.7.4 Applications of HP discharge lamps

Table 3.3 gives information about HP lamps used for general illumination. It is intended to show the range of types available with some idea of the best characteristics to be expected. All types exist in more than one power rating, but the rating given is a fairly typical one for the indicated application and type. Generally efficacy increases as power rating increases [3, Appendix 1].

Most of the lamps that are in the table are arcs with positive columns of length of several centimeters that are stabilized by the tube wall. The CMH lamp is a short arc lamp in which the arc is mainly stabilized by the electrodes.

Manufacturers' websites give many examples of applications other than for illumination. All the major lamps manufacturers make a variety of metal-halide and xenon short arc lamps for projection and related uses (Appendix). In short arc lamps (length < few millimeters) there are usually regions close to the electrodes that have particularly high arc temperature. This region of high arc temperature forms because the electrodes cool the arc, and the field close to the electrodes has to increase to maintain conduction; moreover the current density normally increases as the arc contracts toward the cathode hot spot. The combined increase in current density and field means that the power per unit volume of arc is greatest just adjacent to the electrode. Although this generally leads to a reduction of efficacy there may be an increase in luminance. Figure 3.10 shows examples of lamps that are used for a variety of projection and entertainment applications and other more specialized applications such as solar simulators.

Short arc metal-halide lamps can readily be integrated into prefocused parabolic or elliptical reflectors (Appendix—Welch Allyn, Ushio and others). Various beam divergences are available to suit different applications. A relatively recent development is shown in Figure 3.11 (Appendix—Philips, Osram and GE Lighting). This very high pressure (1.5×10^7 Pa or 150 bar) has extremely high luminance because of its extremely high arc temperature. The reason for the high arc temperature is that 130 W are dissipated in an arc of length hardly more than a millimeter. Spectral lines show extreme broadening and there is an intense continuum giving good color rendition. With an arc gap of only 1.2 mm the étendue is very small. The lamps are designed to be operated in a prefocused reflector and the whole assembly used in data or video projectors.

Table 3.3 Indicative characteristics of HP discharge lamps used for general lighting

Lamp type	Application examples	Power (W)	Initial (lm W^{-1})	Life (10^{3}h)	CCT (K)	R_a
HPS	Road lighting	400	125	30	2000	25
High CRI	Prestige town lighting	400	100	24	2200	60
HP mercury vapor+phosphor	Road lighting	400	60	24	3500	55
Metal halide	Prestige outdoor, stores	400	90	24	4000	70
CMH	Commercial interiors	100	90	12.5	3000	85

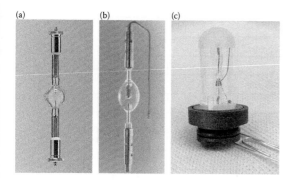

(a) (b) (c)

Figure 3.10 The large range of powers possible with HP discharges is shown. Not to scale. (a) An example of a metal halide lamp for entertainment applications with a CCT of about 6000 K, available in ratings between 575 and 12000 W (Philips). (b) An example of a high pressure xenon lamp operated from dc with a CCT of 6000 K, available in ratings between 450 and 12000 W (Osram). (c) A 10W metal halide lamp operating with CCT around 6000 K and ratings between 10 W (Welch-Allyn).

3.8 ELECTRICAL CHARACTERISTICS OF DISCHARGES

3.8.1 Breakdown and starting in discharge lamps

The gas in the lamp must be converted from an excellent insulator into a good conductor with a resistance that can be as low as a few ohms.

Figure 3.12 shows the voltage across the lamp as a function of current over a very wide range of currents. After breakdown, the current increases rapidly until finally it stabilizes at the value needed to satisfy the circuit equations. In order to start the lamp, the circuit must be able to provide a voltage in excess of the highest lamp voltage in this diagram.

In order to achieve breakdown some source of electrons is necessary. If not provided by other means, they result from ionization by cosmic rays or natural radioactivity in the materials of the lamp. In other cases reliable breakdown is aided by the addition of small amounts of radioactive materials such as Kr^{85}, or by photoemission from surfaces caused by a small external source of UV. In hot cathode fluorescent lamps the electrodes can be heated before the voltage is applied: at low temperatures the field-enhanced thermionic emission provides enough electrons [2]. In HP lamps, a third trigger electrode is often included adjacent to the main electrode. When the voltage is applied across this small gap, breakdown is assured; this gap then provides initial electrons for the main gap. A general rule is that the fewer the initial electrons the higher the starting voltage needs to be, and the longer the time lag before breakdown.

The majority of lamps operate with hot cathodes that emit thermionically. Once breakdown has been achieved the transition from abnormal glow to arc in Figure 3.12 must be achieved quickly and cleanly. Staying too long in the region of cold cathode operation, in which the electrons released

Figure 3.11 (a) A prefocused projection unit designed for LCD projectors. The reflector is carefully designed to keep étendue of the prefocused unit as low as possible and has a dichroic filter to reduce the amount of IR in the beam. (b) The arc tube used in the reflector. It operates at 130 W and a CCT of about 6200 K with a mercury pressure of more than 150 bar. The arc gap of about 1.2 mm with consequently low étendue. (Philips, Osram, GE Lighting.)

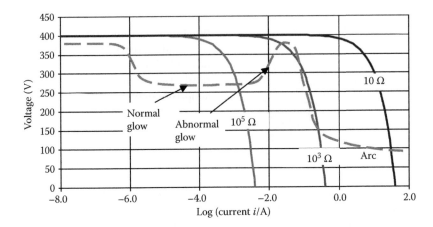

Figure 3.12 Discharge voltage as a function of current (dashed line) over a wide range of currents. The well-known discharge regions are shown. The voltage in the arc region is small because of the low cathode fall resulting from hot cathode operation. The load lines for various values of series resistance are shown with intersections in the normal glow, abnormal glow and arc regions. The intersection in the abnormal glow region may not be stable.

by ion bombardment can be very damaging and shortens lamp life, sometimes dramatically.

There are various schemes for starting fluorescent lamps [2] in which the electrodes are tungsten coils coated with electron emission mix. One of the most common is to use a starter switch in parallel with the lamp. The starter switch is wired so that when closed, current limited by the ballast is passed through both electrodes. Initially the starter switch is closed. This preheating process raises the temperature of the electrode to about 1000 K before the starter switch opens. On opening the switch the open circuit voltage plus the self-induced voltage across the inductance is applied across the lamp, causing breakdown. The main purpose of preheat ensures that thermionic emission occurs very soon after breakdown. The switch is usually a relatively inexpensive bimetallic type. The tolerances on starter switch operation are closely constrained according to the lamp type. Increasingly fluorescent lamps are operated from electronic ballasts. It is a relatively simple matter to include a precision electronic preheat circuit to enhance the lamp life. At the low cost end, preheat is not included; after breakdown the electrode is heated rapidly by ion bombardment until it reaches thermionic emitting temperatures. Up to this time secondary emission dominates but the ion bombardment heats the cathode; these so-called instant start lamps generally have shorter life than lamps that are preheated.

For HP lamps, preheating is not an option. Starting from cold when the pressure is around 10^4 Pa (0.1 bar), breakdown voltages in the region of some kilovolts are needed. After an HP lamp has stabilized the pressure may be many atmospheres; on turning off it will require some tens of kV to restart immediately. Various types of pulse ignitor are used. The speed of transition from the glow to the arc is sensitive to many factors related to electrode design, lamp fill, processing quality and the open circuit voltage available. Ballast and lamp designers work together to ensure that the glow to arc transition in Figure 3.12 occurs rapidly and cleanly to ensure long life.

3.8.2 Steady state electrical characteristics

For an *ohmic* conductor the number of carriers is independent of current, so changing the voltage simply changes the mean drift velocity of the carriers. As long as the temperature remains constant the current is proportional to voltage—Ohm's law. If the temperature increases the carrier mobility decreases and the resistance increases. This is what happens in tungsten lamps in which the hot resistance can be 15 or more times higher than the cold resistance.

Discharges show strongly *nonohmic* behavior (Figure 3.12). A discharge is a current-controlled device; and the voltage between the terminals sets itself to maintain this current. An additional impedance called a *ballast* is necessary to control the current. In response to increasing current hot cathode discharges respond by decreasing the voltage across their terminals. This is the so-called negative, or falling V–I characteristic. The rate of decrease of voltage with current is usually quite small. This corresponds to the arc region on the right hand end of Figure 3.12 where the lamp voltage is comparatively low.

Figure 3.12 shows what happens to voltage as the current is increased over many orders of magnitude. Increasing from low values of current the voltage decreases to a plateau region in which a glow is visible on the cathode. On increasing the current, the glow increases in area whilst the voltage remains constant, implying that the current density at the surface of the cathode is constant. This is called the *normal glow* regime. As the current is increased further, the glow finally covers all the cathode area and often the leads as well. At this point, the current density has to increase and the voltage across the terminals increases. This is called the *abnormal glow* region. In both the normal and abnormal regions, the major part of the lamp voltage is dropped across the CF. The resulting ion bombardment increases the cathode temperature and the cathode begins to emit thermionically and makes a transition to the arc regime, which has a low CF. The abnormal glow region has a positive resistance characteristic, but this is not stable unless there is a ballast in the circuit.

For a dc discharge a series resistance R is needed to stabilize the current I. If the supply voltage is V_S and the lamp voltage V_T then

$$V_T = V_S - IR(V) \qquad (3.16)$$

The right hand side of this equation is called the load line. Load lines for three resistances are shown in Figure 3.12. For the highest resistance the intersection point is in the normal glow region (typical of a neon indicator lamp). With the lowest resistance the intersection is in the arc region. The intermediate resistance has two intersections, the one at the lowest current is in the abnormal glow region. If the heating of the cathode is insufficient to cause a transition to an arc, then the lamp remains in the abnormal glow condition. In some cases when starting an arc the discharge sticks in the abnormal glow with a high cathode fall; the sputtering can then cause very rapid blackening of the walls and premature failure.

Despite what the manufacturers' data sheets may say, Figure 3.12 suggests that there is no specific power at which a discharge must operate; adjusting lamp current by using ballast impedance and supply voltage means lamps may be operated at a wide range of powers—at least for a time. The consequences of operating at powers different from the rated power are usually a reduction in life; properties such as color temperature and color rendering and efficacy will also change. Nevertheless,

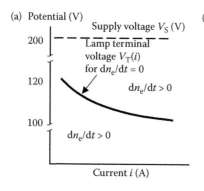

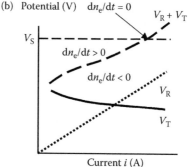

Figure 3.13 (a) A section of the arc region of Figure 3.9. The solid curve is the locus of points for which $dn_e/dt = 0$. If a discharge without a ballast could be prepared at a point on this line small fluctuations would cause the current to increase without limit or decrease to zero. (b) Shows the effect of a stabilizing ballast resistor. The voltage across the circuit is now the sum of the lamp voltage V_T and the resistor voltage V_R. If the current fluctuation increases the current, the sum of resistor and lamp voltage increases into a region where $dn_e/dt < 0$ and the current immediately decreases again until $V_R + V_T = V_S$. This operating point is therefore stable against lamp current fluctuations.

for specific applications this is an option that the user can consider.

The reason for needing a ballast is best explained by using an argument given by Waymouth [2, Chapter 2]. Figure 3.13a shows the falling arc characteristic (part of the right hand end of Figure 3.12). This characteristic is the locus of points for which $dn_e/dt=0$. The further above this line the more the rate of ionization exceeds the loss, so the current increases, and this increases dn_e/dt, with the result that the current continues to increase. If the applied voltage is below the line the loss exceeds production, the current decreases and the discharge extinguishes. Figure 3.13b shows the effect of a series ballast resistor. The total circuit voltage $V_T + IR$ now intersects the supply voltage at a certain current. If a fluctuation causes the current to increase, then the total circuit voltage moves into a region where $dn_e/dt < 0$, thus immediately decreasing the current. If the current decreases, then the total circuit voltage decreases so that $dn_e/dt > 0$, thus increasing the current again.

3.8.3 AC operation

Resistive ballasts work satisfactorily, but are lossy. Commercial lamps operate from the ac mains supply using magnetic inductances as ballasts [3, Chapter 17]. Figure 3.14 shows the lamp voltage and current waveforms for a fluorescent lamp on a resistive ballast. At 50 Hz there is an appreciable restriking voltage after current zero. This extra voltage is needed to restore the electron density after it has decayed during the latter part of the previous cycle. If this restrike voltage exceeds the supply voltage the lamp will extinguish. The phase relationships in an inductive circuit mean that a large voltage is available at the time that the current reverses, so extinction is less likely. For stability on ac mains supplies with a series inductance, the rms lamp voltage should not exceed about half of the rms mains voltage.

Most lamps are now developed to operate from electronic power supplies. Although more expensive than magnetic ballasts, there are a number of benefits: in fluorescent lamps there is an improvement in efficiency of UV production because of reduction in electrode loss and an increase in PC efficiency; electronic circuits can also provide programmed start and run-up

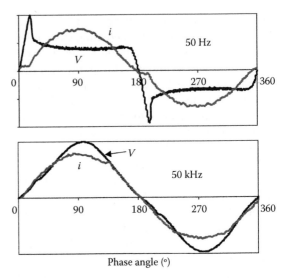

Figure 3.14 Measured voltage and current waveforms for a fluorescent lamp operated at frequencies of 50 Hz and 50 kHz. Noise on the waveforms is caused by oscilloscope digitization.

sequences that prolong lamp life; there is no perceptible 50 or 100 Hz flicker from lamps run from electronic circuits at high frequency; and there is no restrike peak. Figure 3.14 shows typical waveforms at 50 kHz. In the case of HP discharges, operation at high frequencies can cause acoustic resonances that result in gross movements or distortions of the arc [19]. The electronic option is then to operate the lamp from a commutated dc—a square wave with fast transition times at frequency in the region of 90–500 Hz. For HP discharges, the lack of flicker and the ability to control lamp power (and thus color) over life are important benefits.

The optical radiation from discharge sources fluctuates by a percent or two. Part of this is caused by small changes in the cathode termination resulting in arc movement. It has recently been found that modified square wave supply waveforms can reduce the movement of the arc termination on current reversal [20]. A great improvement in stability can also be achieved by measuring the light output and using it to adjust the power into the lamp (Appendix—Oriel Instruments, Light Intensity Control System). A similar device can also be used to control the already excellent stability of tungsten-halogen lamps.

3.9 OTHER METHODS OF EXCITATION OF DISCHARGES

3.9.1 Pulsed light sources

A number of lamps are designed for pulsed operation [21]. The obvious example is the xenon flash tube used for photography, laser pumping, as warning beacons and as a transient source for scientific studies. The duration of the flash is of the order of microseconds with repetition frequencies up to hundreds of hertz. Operation is by discharging a capacitor through the lamp. Peak currents may reach thousands of amperes and electrodes must be constructed accordingly. The effects of using pulsed or transient output are that electron temperature can reach substantially higher values than in steady state. The result is usually due to an enhancement of the short wavelength radiation and an increase in peak radiance.

3.9.2 Dielectric barrier discharges

A form of transient discharge, DBDs have been used for large-scale industrial processes such as ozone generation for water purification and for generating far UV radiation for photochemical processes. DBDs may be operated in the pressure range from about 10^2 to 10^5 Pa [22].

Recently a DBD light source, the Osram Planon lamp has been developed. This provides a very uniformly lit tile-shaped area of reasonable luminous efficiency. At present, lamps are made in square format with diagonals up to 540 mm having a uniform luminance of >6000 cd m^{-2} (Appendix— Osram) [23].

The operating principle of a DBD is as follows. High voltage pulses (of some kilovolts) are applied between two electrodes, at least one of which is covered with an insulator of high breakdown strength such as glass. On applying a high voltage pulse, electrons are accelerated towards the anode and form an avalanche that breaches the gap. Electrons arriving at the anode charge up the surface, thus reducing and finally reversing the electric field. Electron current flows first from cathode to anode and then, when the anode charges up, from anode to cathode. The discharge lasts for a time ~μs. During the off period the ionization decays, providing the starting conditions for the next pulse. The discharge therefore comprises a series of microdischarges with lateral extent approximately equal to the electrode spacing. Microdischarges occur every time the pulse is turned on. DBDs have extremely non-Maxwellian energy distributions in which there are many high energy electrons. Because of this the excitation of resonance states of rare-gas atoms and molecules is favored, leading to high efficiency of UV production.

The Planon lamp is formed from two glass plates. On the lower plate, a metal cathode interlaced with a metal anode structure is deposited. Both electrodes are coated with glass to form the barrier layers. This form of electrode structure results in very uniform illumination. The lamp is operated from an electronic power supply designed to produce the optimized pulse sequence that is necessary for high efficacy. The two plates are held apart by spacers and the whole structure is sealed and filled with Xe at about 1.4×10^4 (100 Torr). Xe forms an excimer Xe_2^* (excited dimer) that radiates efficiently in the vacuum UV at about 172 nm. Phosphor on the inner walls converts the UV to visible radiation. The use of Xe means that the output is almost independent of the lamp temperature so the lamp works just as well outside in cold weather as it does in the confines of office equipment.

The main applications are in displays and office equipment applications where a uniform and high luminance is a requirement. Cylindrical lamps based on the same technology are used in multifunction copiers.

3.9.3 Excitation by induction and by microwaves

In the last decade, a number of inductively coupled lamps have become available commercially from the major lamp manufacturers [3, Chapter 11]. All are variants on the fluorescent lamp discharge. Figure 3.15 shows a particularly compact example. The coil in the center is driven at a frequency of about 2.6 MHz. The rate of change of magnetic flux induces a voltage in the azimuthal direction. This causes a current to flow in a torus surrounding the coil. The ballasting is the result of the internal impedance of the supply. Benefits are long life and compactness. Other versions are Philips QL which, with a life of 100,000 h, is designed for use in inaccessible fixtures. Typically these will be high-bay fixtures with lumen packages between 2800 and 9600 lm. The Osram Endura or Icetron lamp which

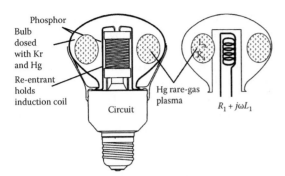

Phosphor
Bulb
dosed
with Kr
and Hg
Re-entrant
holds
induction coil
Circuit
Hg rare-gas
plasma
$R_1 + j\omega L_1$

Figure 3.15 Inductively coupled discharge lamp (GE Genura). The schematic diagram on the left shows that the plasma is a toroid with inductance L_a and resistance R_a that acts as a secondary to the excitation coil. The primary of the circuit is an impedance $R_1 + j\omega L_1$ that includes the effects of the plasma impedance, which depends on the power dissipated in the plasma.

has a stretched torus configuration has higher efficacy and packages of 8000–12,000 lm and a rated life of 80,000 h.

Microwaves can also be used to excite discharges. Fusion Lighting has pioneered a HP sulfur discharge in which the radiation is emitted by S_2 molecules. The light is white with a CCT in the region of 6000 K and the efficiency of generation can be up to 170 lm/microwave watt—higher than any other white light source. The overall efficiency is reduced because of the relatively poor efficiency of generation of microwave power. Light output levels are very high so the source is used in lighting large buildings. The very high radiance of such sources means that optical means can be used to distribute the light efficiently around buildings.

3.10 LEDs FROM THE PERSPECTIVE OF CONVENTIONAL LIGHTING

Chapter 10 gives detailed information about LEDs. The generation of light by conventional lamps is limited by the black body radiance at the electron temperature. The reason is that the electron motion is randomized. In an LED, the motion of the carriers into the recombination region is far from random; the maximum radiance is therefore not limited by the Planck distribution. Already trichromatic LED [24] assemblies reaching 100 lm W^{-1} have been made. (It is not clear in [4], and in many other LED publications, if this is the flux

per lamp watt or flux per wall plug watt.) Note that these lamps are colorimetrically similar to the triphosphor lamps mentioned in Section 3.3.5. A further advantage of LEDs for many applications is their low étendue (Section 3.3.3) which means that the light can be directed more efficiently to where it is needed.

LEDs have already made a substantial impact on the conventional lamp manufacturing businesses. There are a number of applications where LEDs are far superior to conventional lamps. The obvious example is traffic signals. The LEDs generate the colored light only at the wavelengths needed, as compared with the filtered tungsten lamps used until recently. The required signal luminance values can therefore be met at much lower power consumption. It has been estimated that if all traffic lights in the USA were converted to LEDs, electricity consumption would be reduced by 0.4 GW [4]. Moreover, the LEDs are particularly well suited to withstand the vibration they experience as a result of heavy traffic and wind, and they may be frequently switched without damage. But the main advantage is their long life, which dramatically reduces maintenance costs compared with conventional traffic lights. A further advantage is that catastrophic failure of an LED does not have to cause the complete failure of the traffic signal, so safety is improved. All these advantages pay for the extra cost of the LED systems, so we can expect to see a complete takeover of the signal lamp market in land, sea and air transport signs.

White light LEDs are now finding many applications in decorative, aesthetic and artistic lighting where their properties are stimulating designers to produce interesting new ways of using light. There are also other niche markets such as highly localized task lighting where LEDs will make inroads, such as car interiors, desk, stairwell, path lighting, etc. In any application requiring relatively low flux levels, conventional lamps can now probably be replaced by LEDs as long as the substitution is not too expensive.

The ultimate target for the LED industry must be the replacement of the huge numbers of fluorescent lamps for lighting offices and other commercial premises. Much is made of the possibility of exceeding the efficiency of present day fluorescent lamps. The future is far from clear however: the issue is the low luminous flux from LEDs. No single LED approaches the level of flux required, so any competitive installation will require many LEDs to produce the required flux levels, although the étendue advantage may reduce the flux levels required for LED installations. But the hard fact is that the cost of making an LED light source increases approximately linearly with flux, whereas the cost of making conventional lamps is very weakly dependent on flux. It is expected that manufacturing costs in both industries will be driven down by competition to comparable levels. If so, replacement of fluorescent lamps by LED equivalents on a global scale may come down to the cost of the materials used to make the lamps. In the case of fluorescent lamps, most of the material cost is in the rare-earth phosphors. Can the semi-conductors used in making LEDs ever approach those costs per lumen? Will the potentially low cost organic light emitters ever be able to meet the flux requirements of commercial lighting? We shall see.

APPENDIX 3A

Selected manufacturers and suppliers of lamps

The manufacturers on this list give particularly helpful data in catalogues and/or websites for lamps with actual or potential electro-optical applications.

Cathodeon	Lamps for scientific instruments
Ealing Electro-Optics	Lamp units for integration into optical systems
Fusion Lighting	Microwave discharge lamps
GE Lighting	Full range of lamps for illumination and special purposes
Harrison Electrical	Cold cathode fluorescent
Heraeus Noblelight	Special lamps mainly for industrial and scientific processes
Iwasaki	Full range of lamps for illumination and special purposes
Osram	Full range of lamps for illumination and special purposes
Oriel Instruments	Lamp units for integration into optical systems, spectra of lamps
Philips Lighting	Full range of lamps for illumination and special purposes
Stanley	Cold cathode fluorescent
Toshiba Lighting	Full range of lamps for illumination and special purposes
Ushio	Wide range lamps for audio-visual, entertainment, photographic, scientific/media and industrial processes
Welch Allyn	Lamps for special applications

REFERENCES

1. Elenbaas, W. 1972. *Light Sources* (London: Macmillan).
2. Waymouth, J.F. 1971. *Electric Discharge Lamps* (Cambridge: MIT Press).
3. Coaton, J.R. and Marsden, A.M. 1997. *Lamps and Lighting*, 4th ed. (the 3rd edition edited by Cayless and Marsden includes more and wider range spectra of lamps) (London: Arnold).
4. Zukauskas, A., Shur. M.S. and Caska, R. 2002. *Introduction to Solid-State Lighting* (New York: Wiley).
5. Cayless, M.A. and Marsden, A.M. 1983. *Lamps and Lighting*, 3rd ed. (London: Arnold).
6. Grum, F. and Becherer, R.J. 1979. *Optical Radiation Measurements: Vol. 1 Radiometry* (New York: Academic Press).
7. Boyd, R.W. 1983. *Radiometry and the Detection of Optical Radiation* (New York: Wiley).
8. Stupp, E.H. and Brennesholtz, M.S. 1999. *Projection Displays* (New York: John Wiley).
9. Commision Internationale de l'Éclairage (CIE) 2001 Colours of light signals. *CIE* S004/E:2001.
10. Kelly, K.L. 1963. Line of constant cor-related color temperature based on MacAdam's (u,v) uniform transformation of the CIE diagram. *J. Opt. Soc. Am.*, 53, 999–1002.

11. Rutgers, G.A.W. and de Vos, J.C. 1954. Relationship between brightness temperature, true temperature and colour temperature of tungsten. *Physica*, 20, 715–720.

12. Commision Internationale de l'Éclairage (CIE), 1995. Method of measuring and specifying colour rendering properties of light sources. *CIE* 13.3.

13. Wyszecki, G. and Stiles, W.S. 2000. *Color Science: Concepts and Methods, Quantitative Data and Formulae* (New York: Wiley).

14. Richter, J. 1968. Radiation from hot gases. In *Plasma Diagnostics*, ed. W. Lochte-Holtgreven (Amsterdam: North-Holland), pp. 1–65.

15. Coaton, J.R. and Fitzpatrick, J.R. 1980. Tungsten–halogen lamps and regenerative mechanisms. *IEE Proc.*, 127A, 142–148.

16. Bergman, R.S. and Parham, T.G. 1993. Applications of thin film reflecting coating technology to tungsten filament lamps. *IEE Proc.*, 140A, 418–428.

17. Weston, G.F. 1968. *Cold Cathode Discharge Tubes* (London: Iliffe).

18. Bloem, J., Bouwknegt, A. and Wesselink, G.A. 1977. Some new mercury alloys for use in fluorescent lamps. *J. Illum. Eng. Soc.*, 6, 141–147.

19. de Groot, J.J. and van Vliet, J.A.J.M. 1986. *The High-Pressure Sodium Lamp* (Deventer: Kluwer)

20. Derra, G., Fischer, H.E. and Mönch, H. 1997. High pressure lamp operating circuit with suppression of flicker. *US Patent Specification* 5,608,295, 4 March 1997.

21. Rehmet, M. 1980. Xenon lamps. *IEE Proc.*, 127A, 142–148.

22. Kogelschatz, U., Eliasson, B. and Egli, W. 1999. From ozone generators to flat television screens and future potential of dielectric barrier layer discharges. *Pure Appl. Chem.*, 71, 1819–1828.

23. Hitzschke, L. and Vollkommer, F. 2001. Product families based on dielectric barrier discharge. In *Proceedings 9th International Symposium on Science and Technology Light Sources*, ed. R.S. Bergman (Ithaca, NY: Cornell University Press), pp. 411–421.

24. Cayless, M.A. 1980. Future developments in lamps. *IEE Proc.*, 127A, 211–218.

4

Detection of optical radiation

ANTONI ROGALSKI, ZBIGNIEW BIELECKI, AND JANUSZ MIKOLAJCZYK
Military University of Technology

The birth of photodetectors can be dated back to 1873 when Smith discovered photoconductivity in selenium. Progress was slow until 1905, when Einstein explained the newly observed photoelectric effect in metals and Planck solved the blackbody emission puzzle by introducing the quanta hypothesis. Applications and new devices soon flourished, pushed by the dawning technology of vacuum tube sensors developed in the 1920s and 1930s culminating in the advent of television. Zworykin and Morton, the celebrated fathers of videonics, on the last page of their legendary book *Television* (1939) concluded that "when rockets will fly to the moon and to other celestial bodies, the first images we will see of them will be those taken by camera tubes, which will open to mankind new

horizons." Their foresight became a reality with the Apollo and Explorer missions. Photolithography enabled the fabrication of silicon monolithic imaging focal planes for the visible spectrum beginning in the early 1960s. Some of these early developments were intended for a picture phone, other efforts were for television cameras, satellite surveillance, and digital imaging. Infrared (IR) imaging has been vigorously pursed in parallel with visible imaging because of its utility in military applications. More recently (1997), the charge-coupled device (CCD) camera aboard the Hubble space telescope delivered a deep-space picture, a result of 10 days integration, featuring galaxies of the 30th magnitude—an unimaginable figure even for astronomers of our generation. Probably, the next effort will be in the big-band age. Thus, photodetectors continue to open to mankind the most amazing new horizons.

Before proceeding to detailed description of detection of optical radiation, it is now appropriate to digress on system considerations concerning photodetection. We would like to determine how good performance is in view of fundamental limits of sensitivity and speed of response, irrespective of the actual type of detector used. Next, different application circuits used in direct detection systems together with elucidation of the design of front-end circuits and discussion of their performance are presented. Third part of the chapter is devoted to advanced techniques in photodetection covering topics not usually found in textbooks and demonstrating how photodetection is far from being a completely explored field. Next, the classification of two types of detectors (photon and thermal detectors) is done on the basis of their principle of operation. In the last part, the updated information devoted to readout of signals from detector arrays and focal plane arrays (FPAs) is included. It is shown that detector focal plane technology has revolutionized many kinds of imaging in the past 25 years.

4.1 DETECTION REGIMES AND FIGURES OF MERIT

A common problem of any type of photon detector (yielding emitted electrons or internal electron–hole pairs as a response to incoming photons) is how to terminate the photodetector with a suitable load resistor and to trade off the performance between bandwidth and signal-to-noise ratio (SNR). This is necessary for a wide family of detectors, including phototubes, photoconductors, photodiodes, CCDs, vidicon targets, etc., all of which are described by a current generator I_{ph} with a stray capacitance C across it.

Let us consider the equivalent circuit of a photodetector ending on a load resistor R_L, as shown in Figure 4.1. We indicate current noise generator with diamond shape and asterisk. This is the basic circuit for detection, but, as will be shown later, it is not the best one to give a good compromise between bandwidth and noise. The output signal in voltage $V = IR_L$ or in current I has a bandwidth or 3-dB high-frequency cutoff given by

$$\Delta f = \frac{1}{2\pi R_L C}. \tag{4.1}$$

Two noise contributions are added to the signal. One is the Johnson (or thermal) noise of the resistance R_L, with a quadratic mean value

$$I_{nR}^2 = \frac{4kT\Delta f}{R_L}, \tag{4.2}$$

where k is the Boltzmann constant and T is the absolute temperature.

The total current $I = I_{ph} + I_d$ is the sum of the signal current and the dark current. With this current is associated the quantum (or shot) noise arising from the discrete nature of electrons and photoelectrons. Its quadratic mean value is given by

$$I_n^2 = 2q\left(I_{ph} + I_d\right)\Delta f, \tag{4.3}$$

where q is the electron charge and Δf is the observation bandwidth, as in Equation 4.1.

A general noise equivalent circuit for a photodetector is shown in Figure 4.1b. The above two

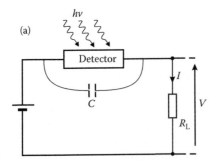

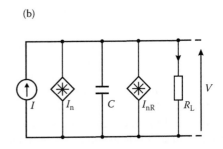

Figure 4.1 General circuit of (a) a photodetector and its equivalent circuit (b) including noise generators (noise generators shown with an asterisk).

fluctuations are added to the useful signal and the corresponding noise generators are placed across the device terminals. Because the two noises are statistically independent, it is necessary to combine their quadratic mean values to give the total fluctuation as

$$I_n^2 = 2q(I_{ph} + I_d) + \frac{4kT\Delta f}{R_L}. \quad (4.4)$$

From Equations 4.1 through 4.4, it can be seen that bandwidth and noise optimization impose opposite requirements on the value of R_L. To maximize Δf, one should use the smallest possible R_L, whereas to minimize I_n^2 the largest possible R_L is required. A photodetector can have a good sensitivity using very high load resistances (up to GΩs), but only modest bandwidths (≈kHz or less), or it can be made fast by using low load resistances (e.g., $R_L = 50\,\Omega$(but at the expense of sensitivity.

Using Equation 4.4, we can evaluate the relative weight of the two terms in total noise current. In general, the best possible sensitivity performance is achieved when the shot noise is dominant compared to the Johnson noise, i.e.,

$$2q(I_{ph} + I_d)\Delta f \geq \frac{4kT\Delta f}{R_L}$$

which implies the condition

$$R_{Lmin} \geq \frac{2kT/q}{I_{ph} + I_d} \quad (4.5)$$

From this equation at room temperature, then

$$R_{Lmin} \geq \frac{50\,mV}{I_{ph} + I_d}.$$

So, at low signal levels (for $I_{ph} < I_d$), very high values of resistance are required; for example, if $I_d = 5\,pA$, not an unusually low dark current, then $R_{Lmin} = 10\,G\Omega$. However, if we are using a resistor termination value $R_L < R_{Lmin}$ then, the total noise can be written as

$$I_n^2 = 2q(I_{ph} + I_d)\Delta f\left(1 + \frac{R_{Lmin}}{R_L}\right). \quad (4.6)$$

This means that the noise performance is degraded by factor R_{Lmin}/R_L compared to the intrinsic limit allowed by the dark current level. Thus, using

$R_L < R_{Lmin}$ means that the shot noise performance is reached at a level of current not less than

$$I_{ph} + I_d \geq \frac{2kT/q}{R_L}. \quad (4.7)$$

From this equation, we see that for a fast photodiode with a 50 Ω load at $T = 300$ K, this current has a very large value of 1 mA.

The above considerations are valid for photodetectors without any internal gain, G. Let us now extend the calculation of the SNR (S/N) to photodetectors having internal gain. In this case, the shot noise can be expressed in the form

$$I_n^2 = 2q(I_{ph} + I_d)\Delta f G^2 F, \quad (4.8)$$

where F is the excess noise factor to account for the extra noise introduced by the amplification process. Of course, in a nonamplified detector, $F = 1$ and $G = 1$. The total noise is a sum of shot noise and thermal noise,

$$N = \left[2q(I_{ph} + I_d)\Delta f G^2 F + \frac{4kT\Delta f}{R_L}\right]^{1/2} \quad (4.9)$$

and then we obtain a S/N ratio

$$\frac{S}{N} = \frac{I_{ph}}{\left[2q(I_{ph} + I_d)\Delta f F + (4kT\Delta f/R_L G^2)\right]^{1/2}}. \quad (4.10)$$

If we now introduce a critical value I_{ph0} called the *threshold of quantum regime*,

$$I_{ph0} = I_d + \frac{2kT/q}{R_L F G^2}$$

then, Equation 4.10 becomes

$$\frac{S}{N} = \frac{I_{ph}}{\left[2q(I_{ph} + I_{ph0})\Delta f F\right]^{1/2}}. \quad (4.11)$$

Analyzing Equation 4.11, two detection regimes can be found, according to whether the signal I_{ph} is larger or smaller than I_{ph0}. For the signals, $I_{ph} > I_{ph0}$ and $F = 1$

$$\frac{S}{N} = \left(\frac{I_{ph}}{2q\Delta f}\right)^{1/2}. \quad (4.12)$$

This S/N is called the *quantum noise limit* of detection. This limitation cannot be overcome by any detection system, whether operating on coherent or incoherent radiation. In fact, Equation 4.12 is a direct consequence of the quantitative nature of light and the Poisson photon arrival statistics.

In the small signal regime, $I_{ph} < I_{pho}$, we have

$$\frac{S}{N} = \frac{I_{ph}}{\left(2qI_{pho}\Delta f\right)^{1/2}}. \qquad (4.13)$$

That is, the S/N ratio is proportional to the signal, and the noise has a constant value, primarily given by the load resistance. This is the thermal regime of detection.

Figure 4.2 shows the trend of the S/N ratio (standardized to $(2q\Delta f)^{1/2}$) as a function of the signal amplitude I_{ph}/I_{pho}. We can notice that in the thermal regime, the slope is 20 dB per decade up to the threshold $I_{ph}/I_{pho} = 1$, and from here onward the slope becomes 10 dB per decade in the quantum regime. The effect of an excess noise factor F is also shown in Figure 4.2.

The threshold of the quantum regime $I_{ph} = I_{pho} = I_d + (2kT/q)/R_L FG^2$ is the signal level that corresponds to the break point between thermal and quantum regimes. When the dark current is very small, the second term is the dominant one; any eventual internal gain greatly helps because it scales the load resistance as G^2.

In all cases, I_{pho} can be interpreted as the equivalent dark current level of the photodetector and has a value of

$$I_{pho} = \frac{2kT/q}{R_{eq}G^2} \qquad (4.14)$$

with

$$\frac{1}{R_{eq}} = \frac{1}{R_F} + \frac{q}{2kT}G^2 I_d.$$

We define R_{eq} as the noise equivalent load resistance of the photodetector.

To provide ease of comparison between detectors, certain figures of merit, computed from the measured data, have been defined.

The voltage (or analogous current) responsivity is given by

$$R = \frac{Q_u}{P}, \qquad (4.15)$$

where Q_u is the output quantity supplied by the detector (e.g., a current I_u, a voltage V_u, or any other physical quantity) and P is the incident radiant power.

At equal responsivity, the detector with the smallest output noise g_n on the useful signal is the most sensitive. Therefore, the first figure of merit for a detector is the NEP—noise equivalent

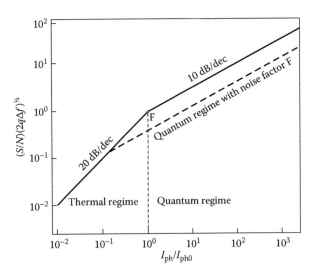

Figure 4.2 The S/N ratio of a photodetector, as a function of the input signal, in the thermal and quantum regimes of detection. (Reproduced from Donati, S., *Photodetectors. Devices, Circuits, and Applications*, Prentice Hall Inc., Upper Saddle River, 1999.)

power—defined as the ratio of output noise to responsivity:

$$NEP = \frac{g_n}{R}. \quad (4.16)$$

So, the NEP represents the input power that gives a unity SNR ($S/N = 1$) at the output, that is, a marginal condition of detection.

The better the detector performance, the smaller the NEP. Therefore, it is more convenient to define its inverse as a merit figure. In addition, it should be taken into consideration that whatever is the noise source, it can be expected that the noise quadratic total value is proportional to observation bandwidth Δf and detector area A. Thus, it is even better to take, as the intrinsic noise parameter of a detector, the ratio $NEP/(A\Delta f)^{1/2}$ normalized to unit area and bandwidth. In order to simplify the comparison of different detectors and to have a parameter that increases as the performance improves, the detectivity D^* (called D-star) is defined as

$$D^* = \frac{(A\Delta f)^{1/2}}{NEP}. \quad (4.17)$$

This is the fundamental figure of merit used for detectors. It can be transformed to the following equation:

$$D^* = \frac{(A\Delta f)^{1/2}}{P} \frac{S}{N}. \quad (4.18)$$

D^* is defined as the rms SNR (S/N) in a 1 Hz bandwidth, per unit rms incident radiation power, per square root of detector area. D^* is expressed in cmHz$^{1/2}$ W^{-1}, a unit that has recently been called a "Jones."

When detectors are operated in conditions where the background flux is less than the signal flux, the ultimate performance of detectors is determined by the signal fluctuation limit (SFL). It is achieved in practice with photomultipliers operating in the visible and ultraviolet region, but it is rarely achieved with solid-state devices, which are normally detector noise or electronic noise limited. This limit is also applicable to longer wavelength detectors when the background temperature is very low.

Figure 4.3 illustrates the spectral detectivities over the wavelength range from 0.1 to 4 μm assuming a background temperature of 290 K and a 2π steradian field of view (FOV) (applicable only

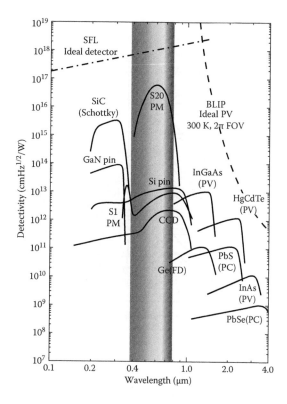

Figure 4.3 Detectivity vs. wavelength values of 0.1–4 μm photodetectors. *PC*, photoconductive detector; *PD*, photodiode; *PM*, photomultiplier.

to the background fluctuation limit). Note that the intersections of curves for signal fluctuation and background fluctuation limits lie at about 1.2 μm. At wavelengths below 1.2 μm, the SFL dominates; the converse is true above 1.2 μm. Below 1.2 μm, the wavelength dependence is small. Above 1.2 μm, it is very large, due to steep dependence of detectivity upon wavelength of the short wavelength end of the 290 K background spectral distribution.

As mentioned already, the ultimate performance of IR detectors is reached when the detector and amplifier noise are low compared to the photon noise. The photon noise is fundamental, in the sense that it arises not from any imperfection in the detector or its associated electronics but rather from the detection process itself, as a result of the discrete nature of the radiation field. The radiation falling on the detector is a combination of that from the target (signal) and that from the background. The practical operating limit for most IR detectors is the background fluctuation limit also known as the background limited infrared photodetector (BLIP) limit.

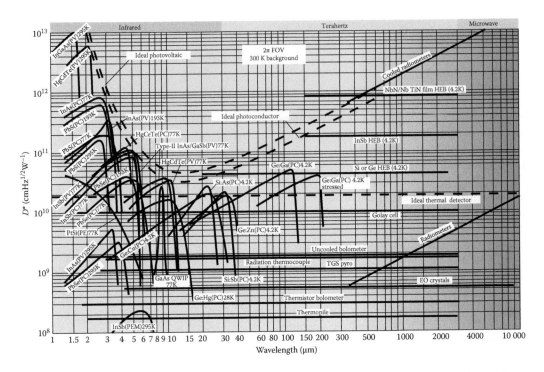

Figure 4.4 Comparison of the D^* of various available detectors when operated at the indicated temperature. Chopping frequency is 1000 Hz for all detectors except the thermopile (10 Hz), thermocouple (10 Hz), thermistor bolometer (10 Hz), Golay cell (10 Hz), and pyroelectric detector (10 Hz). Each detector is assumed to view a hemispherical surrounding at a temperature of 300 K. Theoretical curves for the background-limited D^* (dashed lines) for ideal photovoltaic and photoconductive detectors and thermal detectors are also shown. *PC*, photoconductive detector; *PV*, photovoltaic detector; *PEM*, photoelectromagnetic detector; *HEB*, hot electron bolometer.

Figure 4.4 compares typical D^*s of different detectors as a function of wavelength. Also the BLIP and the dark current limits are indicated.

4.2 DIRECT DETECTION SYSTEMS

4.2.1 Introduction

A receiver of optical radiation consists of a photodetector, preamplifier, and signal processing circuit (Figure 4.5). In a photodetector, the optical signal is converted to an electrical one, which is amplified before further processing. The sensitivity of an optical detection system depends primarily on the first stage of a photoreceiver, i.e., the photodetector and preamplifier.

A preamplifier should have low noise and a sufficiently wide bandwidth to ensure faithful reproduction of the temporal shape of an input signal. Figure 4.5 shows, the so-called, direct

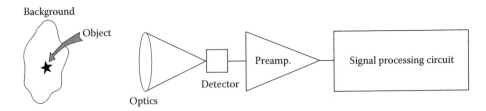

Figure 4.5 Block diagram of an optical radiation receiver.

detection system. It is necessary to minimize the noises from various sources, i.e., background noise, photodetector noise, biasing resistors noise, and any additional noises of signal processing. If further minimization of noise of the first photoreceiver stages is not possible, advanced methods of optical signal detection can sometimes be used to still recover information carried by optical radiation signals of extremely low power. Heterodyne and homodyne detection can be used to reduce the effects of amplifier noise. Postdetection methods are phase-sensitive detection and synchronous integration of a signal.

4.2.2 Selection of active amplifying elements

Several types of discrete devices or integrated circuits (ICs) are suitable for the active element in preamplifiers: bipolar junction transistor (BJT) or field-effect transistor (FET) or an IC with an input bipolar, FET, or metal-oxide semiconductor field-effect transistor (MOSFET) transistor can be used.

The most important parameter of each receiving device is its SNR (S/N). Because low-level signals reach the photoreceiver, the noise optimization of a system, i.e., to obtain maximum S/N is a very important problem [2]. Optimum design of a preamplifier can be obtained by analysis of particular noise sources in a detector–preamplifier circuit. The equivalent input noise V_{ni} will be used to represent all noise sources. A scheme of detector–preamplifier noise circuit is shown in Figure 4.6. A level of equivalent noise at the input of this circuit is determined by the detector noise V_{nd}, the background noise V_{nb}, and the preamplifier noise. The preamplifier noise is represented completely by the zero impedance voltage generator V_n in series with the input port and the infinite impedance current generator I_n in parallel with the input. Typically, each of these terms is frequency dependent. For noncorrelated noise components, the equivalent noise at the input of a photodetector–preamplifier circuit is described by the formula

$$V_{ni}^2 = V_{nd}^2 + V_{nb}^2 + V_n^2 + I_n^2 R_d^2, \qquad (4.19)$$

where R_d is the detector resistance.

This single noise source, located at V_s, can be substituted for all sources of the system noise. Note that V_{ni}^2 is independent of the amplifier gain and its input impedance. Thus, V_{ni}^2 is the most useful index against which the noise characteristics of various amplifiers and devices can be compared.

The first transistor of a preamplifier is a dominant noise contributor of the input noise of signal processing circuits (Figure 4.6). The effective contribution of this transistor noise also depends on the detector impedance. If the transistor of the input stage has a high noise current, it will be a bad choice to use a high-resistance (current source) detector. A low-resistance detector operates best with a preamplifier of low noise voltage.

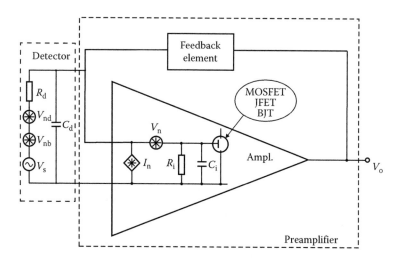

Figure 4.6 Noise equivalent diagram of a photodetector–preamplifier circuit; R_i and C_i are the input resistance and the capacity of a preamplifier, respectively, A is the voltage gain of the preamplifier.

Figure 4.7 shows the dependence of the ratio of preamplifier (or input transistor) noise to detector thermal noise as a function of detector resistance, for bipolar, junction field-effect transistor (JFET), and MOSFET amplifiers, in a common emitter or common source configuration [3]. It can be noticed that there are some ranges of detector resistance for which the preamplifier noise is lower than the detector thermal noise. For any given detector, the values of the thermal noise voltage V_t and the resistance R_d are determined by the detector type (they cannot be changed). However, it is possible to change the parameters V_n and I_n of the designed preamplifier. Thus, minimization of the input noise of the circuit is possible. Changes in the values V_n and I_n are made by choosing adequate elements in the preamplifier circuit. For low-resistance detectors (from tens of Ω to 1 kΩ), circuits with bipolar transistors at the input are usually used as they have low values of V_n. Sometimes, in order to decrease the value of optimal resistance R_o, a parallel connection of several active elements or even parallel connection whole amplifiers is advantageous [4]. Within the range of average detector resistances from 1 kΩ to 1 MΩ, the preamplifiers FET input stages can be used. However, for connection with high-resistance detectors (above 1 MΩ), especially recommended are the transistors of low I_n values. Such requirements fulfill JFET and MOSFET transistors.

4.2.3 First stages of photoreceivers

There are two general types of photon detectors without internal gain: photoconductive and junction devices (photovoltaic ones). These photodetectors are used with many types of preamplifiers. The choice of circuit configuration for the preamplifier is largely dependent upon the system application. The two basic preamplifier structures—voltage amplifiers type and transimpedance (current in–voltage out type) will be discussed. The voltage preamplifiers can be either low impedance or high impedance ones. A simplified circuit diagram is shown in Figure 4.8. Bias voltage is supplied by V_b. The detector signal is developed as a voltage drop across R_b. The variable resistance component of the detector R_d is represented by the incremental resistance ΔR_d. The load resistance R_L provides a bias path for the amplifier input.

The simplest preamplifier structure is the low-input impedance voltage preamplifier. This design is usually implemented using a bipolar transistor. Either common emitter or grounded emitter input stages may be designed with a reasonably low input impedance. In the low-input impedance preamplifiers, the signal source is loaded with a low-impedance (e.g., 50 Ω) input stage. The time constant of the temporal response is determined by the combined load resistance and input capacitance

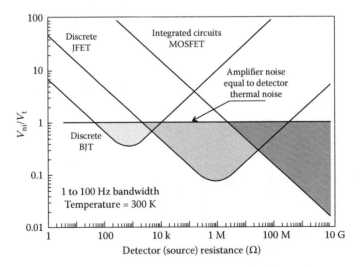

Figure 4.7 Dependence of the ratio of preamplifier input voltage noise to detector thermal noise, as a function of detector resistance. (Reproduced from Vampola, J.L., Readout electronics for infrared sensors. In *The Infrared and Electro-Optical Systems Handbook*, vol. 3, ed. W.D. Rogatto, SPIE Press, Bellingham, 1999.)

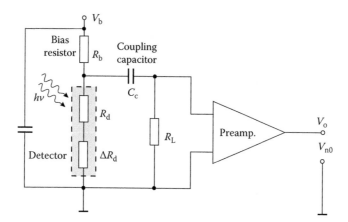

Figure 4.8 A simple circuit of voltage mode preamplifier.

of the detector and preamplifier and this determines the detection bandwidth. Preamplifiers of low input resistance can provide high bandwidth but not very sensitive photoreceiver. Photoreceivers with preamplifiers of low input resistance are, therefore, usually used when wide transmission band is required.

The high impedance preamplifier gives a significant improvement in sensitivity over the low impedance preamplifier, but it requires considerable electronic frequency equalization to compensate for its high-frequency roll-off. The preamplifier has problems of limited dynamic range, being easily saturated at higher input power levels. When a highly sensitive photoreceiver with a low dynamic range is needed, preamplifier of high input impedance is recommended. However, for this, there is a better configuration that we shall now discuss. This is called the transimpedance receiver.

The transimpedance preamplifier finds many applications in optical signals detection. A schematic of this is shown in Figure 4.9. In this circuit, R_d is the detector resistance, and it can be a photodiode or a photoresistor. Depending on detector type and required application, the detector can be biased from V_b or connected directly across the input without bias. The current produced by the detector flows through the resistor R_f located in a feedback loop. The optional potentiometer R_b is

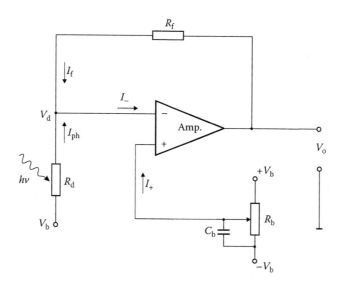

Figure 4.9 Schematic diagram of transimpedance amplifier.

used for setting the zero value of the output voltage without detector illumination [5]. It can be used for compensation of a dc level at the output of the circuit resulting from background radiation. Usually, the input bias current is negligible, so $I_{ph} = -I_f$.

The voltage at the preamplifier output is given by the formula

$$V_o = -\frac{R_f}{R_d} V_b = -I_{ph} R_f \qquad (4.20)$$

proper for low frequencies. The great advantage of this configuration is that the preamplifier gives low-noise performance without the severe limitation on bandwidth imposed by high-input impedance preamplifier. The bandwidth is much higher because the effective load resistance is very low and has a value of $R_f /(A+1)$, where A is the gain of the amplifier at the appropriate frequency. It also provides greater dynamic range than the high-input impedance structure. It is possible to linearly detect and process optical signals many orders of powers magnitude.

At low frequencies, using a photodiode in unbiased configuration, across the input of transimpedance preamplifier, avoids the noise and dc offset problems that dark current might otherwise cause (noninverting input is on the ground). In this configuration, there is the small price of increased photodiode capacitance, which would be far more serious at high frequencies. For InGaAs, photodiodes can have substantially lower quantum efficiencies if unbiased (due to much thinner depletion region), although the benefit to avoid dark current can be more useful here.

The effective input impedance R_i at the amplifier input as a result of the effects of the feedback circuit is given simply by

$$R_i = \frac{R_f}{A(f)+1} \approx \frac{R_f}{A(f)}, \qquad (4.21)$$

where $A(f)$ is the frequency-dependent gain of the preamplifier. As this input impedance is effectively in parallel with the capacitance $(C_d + C_j)$, the 3 dB breakpoint f_{-3dB} in the frequency response is given by

$$f_{-3dB} = \frac{1}{2\pi R_i (C_d + C_i)} = \frac{A(f)}{2\pi R_f (C_d + C_i)}. \qquad (4.22)$$

If $A(f)$ is described in terms of gain–bandwidth product $A(f)\Delta f$ (or GBP) such that GBP $= \Delta f A(f)$, then

$$f_{-3dB} = \frac{\text{GBP}/f_{-3dB}}{2\pi R_f (C_d + C_i)} \qquad (4.23)$$

therefore,

$$f_{-3dB} = \left[\text{GBP}/2\pi R_f (C_d + C_i) \right]^{\frac{1}{2}}. \qquad (4.24)$$

A practical problem with many transimpedance designs is to achieve a sufficiently high open-loop gain product, to keep the f_{-3dB} value high, yet still maintain a sufficiently low open-loop phase shift to maintain stability. It often proves necessary, in practice, to use a value of R_L, than desired on low-noise grounds, in order to achieve the necessary f_{-3dB}. This gives rise to the widely held belief that transimpedance designs are inherently noisier than high-impedance input design. However, this belief is not valid with good high-gain–bandwidth designs. Moreover, even with suboptimum amplifier design, achieving poorer gain–bandwidth, it is always possible to perform an engineering compromise, by choosing a high value of R_L, to maintain low noise and use small amount of equalization after the transimpedance amplifier to compensate for the resulting roll-off.

The preamplifiers described earlier are used with semiconductor detectors. Now, we would like to present exemplary preamplifiers to the photomultiplier tubes.

The operating principles, construction, and characteristics of a photomultiplier are given in Section 12 (Part III). Now, we would like to analyze matching the preamplifier to the photomultiplier. Different pulse processing techniques are typically employed, depending on whether the arrival time or the amplitude (energy) of the detected event must be measured. Three basic types of preamplifiers are available: the current-sensitive preamplifier, the parasitic-capacitance preamplifier, and the charge-sensitive preamplifier. The simplified schematic of the current-sensitive preamplifier is shown in Figure 4.10. The $R = 50\ \Omega$ input impedance of the current-sensitive preamplifier provides proper termination of the 50 Ω coaxial cable and converts the current pulse from the detector to a voltage pulse. The amplitude of the voltage pulse at the preamplifier output will be

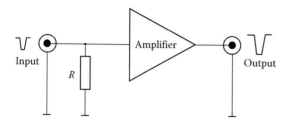

Figure 4.10 A simplified schematic of the current-sensitive preamplifier.

$$V_o = RI_iA, \qquad (4.25)$$

where I_i is the amplitude of the current pulse from the detector.

For counting applications, this signal can be fed to a fast discriminator, the output of which is recorded by a counter-timer. For timing application, the dominant limitation on timing resolution with photomultiplier tubes is fluctuation in the transit times of the electrons. This causes a jitter in the arrival time of the pulse at the detector output. If the detector signals are small enough to require a current-sensitive preamplifier, the effect of a preamplifier input noise on time resolution must also be considered. It is important to choose a current-sensitive preamplifier, the rise time of which is much faster than detector rise time. Unfortunately, the faster preamplifier does contribute extra noise because of the unnecessarily wide bandwidth. The excess noise will increase the timing jitter. Choosing a preamplifier rise time that is much slower than the detector rise time reduces the preamplifier noise contribution but causes degradation in pulse rise time and its amplitude. Consequently, the timing jitter becomes worse. The optimum chosen depends on the rise time and amplitude of the detector signal.

Most current-sensitive preamplifiers designed for timing applications have ac coupled.

The most effective preamplifier for these detectors is the parasitic-capacitance preamplifier shown in Figure 4.11. It has a high input impedance (above 1 MΩ).

The parasitic capacitance is presented by the detector and the preamplifier input (typically 10–50 pF). The resulting signal is a voltage pulse having an amplitude proportional to the total charge in the detector pulse. This type of preamplifier is sensitive to small changes in the parasitic capacitance. The rise time is equal to the duration of the detector current pulse. A resistor connected in parallel with the input capacitance causes an exponential decay of the pulse. An amplifier follower is included as a buffer to drive the low impedance of a coaxial cable at the output. These preamplifiers are highly recommended for photomultiplier tubes, microchannel plate MPTs, and scintillation detectors.

For most of the energy spectroscopy applications, charge-sensitive preamplifier is preferred (Figure 4.12). This preamplifier integrates the charge on the feedback capacitor. Its gain is not sensitive to a change in detector capacitance. The output voltage from the preamplifier has the amplitude V_o and the decay time constant τ_f, given, respectively, by

$$V_o = \frac{Q_d}{Q_f} = \frac{Eq}{\varepsilon C_f}10^6 (\text{mV}) \text{ and } \tau_f = R_fC_f, \qquad (4.26)$$

where Q_d is the charge of the detector, C_f is the feedback capacitor, E is the energy in MeV of the incident radiation, q is the charge of an electron, ε is the amount of energy (eV) required to produce an electron–hole pair in the detector, and 10^6 converts MeV to eV. The resistor R_f should be made as large as possible consistent with the signal

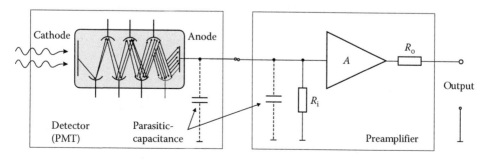

Figure 4.11 A simplified diagram of the parasitic-capacitance preamplifier.

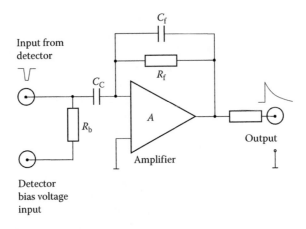

Figure 4.12 A simplified diagram of the charge-sensitive preamplifier.

energy-rate product and the detector leakage current. The input capacitance must be much greater than the other sources of capacitance connected to the preamplifier input in order for the preamplifier sensitivity to be unaffected by external capacitance changes. The stability of the preamplifier sensitivity is dependent on the stability of the feedback capacitor (C is selected for good temperature stability) and the preamplifier open-loop gain. The open-loop gain will be very large so that small changes in the C_f can be neglected.

The rise time of the output pulse of the preamplifier, in the ideal case, is equal to the charge collection time of the detector. When detectors with very fast collection times or large capacitances are used, the preamplifier itself may limit the rise time of V_o.

4.2.4 Photon counting techniques

Photomultiplier tubes are also used as a photon counting. Photon counting is one effective way to use a photomultiplier tube for measuring very low light (e.g., astronomical photometry and fluorescence spectroscopy). A number of photons enter the photomultiplier tube creating an output pulse signal (Figure 4.13a). The actual output pulse obtained by the measurement circuit is a dc with fluctuation (Figure 4.13b).

When the light intensity becomes so low that the incident photons are separated as shown in Figure 4.13c, this condition is called a single-photon event. The number of output pulses is in direct proportion to the amount of incident light and this pulse counting method has advantageous in SNR and stability over the dc method averaging all the pulses. This counting technique is called the photon counting method.

Because the photomultiplier tube output contains a variety of noise pulses in addition to signal pulses representing photoelectrons, simply counting of the pulses, without some form of noise elimination, will not result in an accurate measurement. The most effective approach to noise elimination is to investigate the height of the output pulses. Figure 4.14 shows the output pulse and discriminator level.

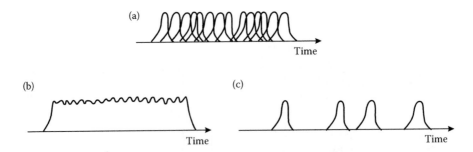

Figure 4.13 Photon counting technique: a number of photons enter photomultiplier tube (a), an actual output pulse obtained by measurement circuit (b), an incident photons for low light intensity (c).

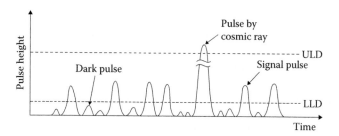

Figure 4.14 Output pulse and discriminator level.

A typical pulse height distribution (PHD) of the output of the photomultiplier tubes is shown in Figure 4.15. In this PHD method, the low-level discrimination (LLD) is set at the valley and the upper-level discrimination (ULD) at the foot. Most pulses smaller than the LLD are noises and pulses larger then ULD result from interference (e.g., cosmic rays). Therefore, by counting the pulses between the LLD and UPD, accurate light measurements are made possible. In the PHD, H_m is the mean height of pulses. It is recommended that the LLD be set at 1/3 of H_m and the ULD a triple H_m.

In addition, the avalanche photodiodes (APDs) can get an output pulse for each detected photon and thus potentially very high sensitivities comparable to that of photomultipliers. They are selected based on having extremely low-noise and low-bulk dark current. They are intended for ultra-low light level application (optical power less than 1 pW) and can be used in either their normal linear mode (polarization voltage V_R < breakdown voltage, V_{BR}) or as photon counters in the "Geiger" mode ($V_R > V_{BR}$), where a single photoelectron may trigger an avalanche pulse of about 10^8 carriers.

In the linear mode operation, the APD is well suited for application that requires high sensitivity and fast response time; for example, laser range-finders, fast receiver modules, light detection and ranging (LIDAR), and ultrasensitive spectroscopy.

When biased above the breakdown voltage, an APD will normally conduct a large current. However, if this current is limited to less than the APDs "latching" current, there is a strong statistical probability that the current will fluctuate to zero in the multiplication region, and the APD will then remain in the "off" state until an avalanche pulse is triggered by either a bulk or photogenerated carrier. If the number of bulk carrier generated pulses is low, the APD can, therefore, be used to count individual current pulses from incident photons. The value of the bulk dark current is, therefore, a significant parameter in selecting an APD for photocounting and can be reduced exponentially by cooling.

The APDs can be used in the Geiger mode using either "passive" or "active" pulse quenching circuits.

A passive quench takes the current from the diode and passes it through a load resistor and a

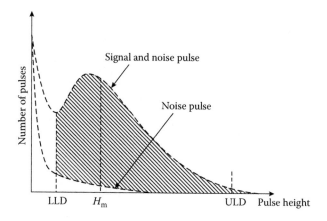

Figure 4.15 Typical PHD.

series resistor, causing the bias voltage to drop. An active quench uses a transistor for lower and raises the bias voltage. Passive quench circuits, although simple, limit the rate at which photons can be counted because time has to be allowed for the quench and for the return of the voltage to break down before the next photon arrives. This characteristic time is basically given by RC, where R is the series resistance and C is device capacitance. Because R must be sufficiently large to trigger the quench, the time is limited by device capacitance to typically hundreds of nanoseconds, and time resolution is at best about 400 ps.

The simplest, and in many cases a perfectly adequate method of quenching a breakdown pulse, is through the use of a current-limiting load resistor. An example of such a "passive" quenching circuit is shown in Figure 4.16a. The load line of the circuit is shown in Figure 4.16b.

To be in the conducting state at V_{BR}, two conditions must be met:

- The avalanche must have been triggered by either a photoelectron or a bulk-generated electron entering the avalanche region of the APD, to continue to be in the conducting state.
- A sufficiently large current, called latching current, must be passing through the device so that there is always an electron or hole in the avalanche region.

For the currents $(V_R - V_{BR})/R_L$ much greater than I_{latch}, the diode remains conducting. If the current $(V_R - V_{BR})/R_L$ is much less than I_{latch}, the diode switches almost immediately to the nonconducting state. When R_L is large, the photodiode is nonconducting and the operating point is at $V_R - I_d R_L$ in the nonconducting state. Following an avalanche breakdown, the device recharges to the voltage $V_R - I_d R_L$ with the time constant CR_L, where C is the total device capacitance including stray capacitance.

The rise time is fast (e.g., 5–50 ns), decreases as $V_R - V_{BR}$ increases, and is very dependent on the capacitances of the load resistors, leads, etc. The jitter is typically the same order of magnitude as the rise time.

The advantages are simplicity, low cost, and small area occupation allowing the fabrication of large arrays with small loss in fill factor (FF). Some modifications of the passive quenching circuit have also been designed. For example, counting limit rate is increased using an Active Reset configuration—see Figure 4.17.

To avoid an excessive dead time when operating at large voltage above V_{BR}, an "actively quenched" circuit can be used. Active quenching can greatly increase this performance and has become commercially available over the last few years. Now, recharging can be very rapid through a small load resistor. Alternatively, the bias voltage can be maintained but the load resistor is replaced by a transistor, which is kept off for a short time after an avalanche and turned on for a period sufficient to recharge the photodiode—see Figure 4.18.

The response time is limited by the transistor switching rather than an RC circuit and has been reduced to as low as 50 ns, and the timing of photons can be made with resolution as high as 20 ps. In this mode, no amplifiers are necessary

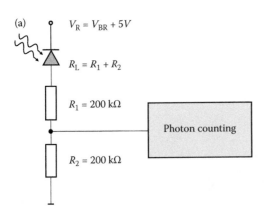
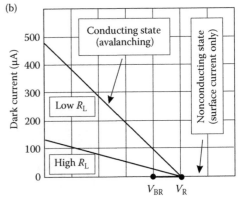

Figure 4.16 Passive quenched circuit (a) and the load line of the circuit (b).

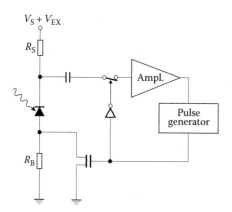

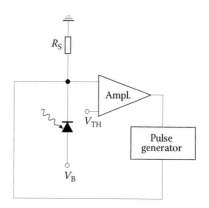

Figure 4.17 Passive quenched circuit modified to introduce an Active Reset. (Reproduced from Gallivanoni, A., Rech, I., Ghioni, M., Progress in quenching circuits for single photon avalanche diodes. *IEEE Trans. Nucl. Sci.*, 57, 3815–3826, 2010.)

Figure 4.19 Active quenched circuit with active components. (Reproduced from Gallivanoni, A., Rech, I., Ghioni, M., Progress in quenching circuits for single photon avalanche diodes. *IEEE Trans. Nucl. Sci.*, 57, 3815–3826, 2010.)

reduces the FF of detector arrays. This is the reason why fully integrated purely active quenching circuits are rarely used. Instead, mixed active–passive quenching circuits are generally preferred.

The most employed one is the mixed active–passive quenching with an active reset. In this setup, the APD is connected both to a high impedance and to the active quenching and reset circuitry. A basic diagram using switches as the active circuitry is shown in Figure 4.20. The main disadvantage of a mixed quenching circuit, compared to a passive one, is the much larger area

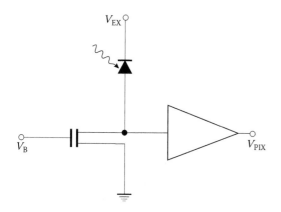

Figure 4.18 Active quenched circuit.

and single-photon detection probabilities of up to approximately 50% are possible.

Another active quenching technique is based on detection of the avalanche and reaction back on the device by controlling its bias voltage. Both quenching and reset are carried out using active components (pulse generators or fast active switches)—see Figure 4.19.

In this setup, quenching transition is not affected by avalanche multiplication fluctuations and is faster than in passive one. But the actual quenching time might not be so small because at first the active circuit has to detect the avalanche and intervene. Furthermore, reduction of quenching time is also obtained using direct integration of the active circuit with the detector but such design

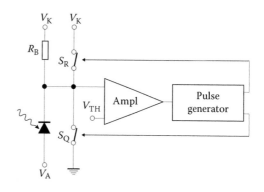

Figure 4.20 Mixed active–passive quenching circuit. (Reproduced from Mosconi, D., Stoppa, D., Pancheri, L., Gonzo, L. and Simoni, A. CMOS single-photon avalanche diode array for time-resolved fluorescence detection. *Solid-State Circuits Conference, ESSCIRC 2006 Proceedings of the 32nd European*, September 2006, pp. 564–567, 2006.)

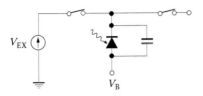

Figure 4.21 Digital APD circuit. (Reproduced from Gallivanoni, A., Rech, I., Ghioni, M., Progress in quenching circuits for single photon avalanche diodes. *IEEE Trans. Nucl. Sci.*, 57, 3815–3826, 2010.)

occupation that would limit their usage in large and closely packed arrays [7].

Apart from traditional passive, active, and mixed active–passive quenching circuits, other circuit topologies have been also developed. All this circuits are intended to be integrated with the detector in order to take advantage of the smaller both parasitic capacitance and propagation delay. To avoid active quenching circuits in order to obtain a densely packed array and passive quenching circuits to provide high count rates, a novel quenching mechanism has been designed. It was called Digital APD—see Figure 4.21. The digital APD takes advantage of the parasitic capacitance of the detectors to operate the devices in a manner similar to a dynamic RAM. But it is still based on passive quenching and gating.

4.2.5 High-speed photoreceivers

A variety of different applications and measurement techniques have been developed for high-speed detection. Most of them are described in Section 12 (Part III). Here, we focus mainly on the speed limitation of the response of p–i–n junction and Schottky barrier—two types of photodiodes that have the highest speed response.

In general, in comparison with Schottky barriers, the p–n junction photodiodes indicate some important advantages. The thermionic emission process in Schottky barrier is much more efficient than the diffusion process and therefore for a given built-in voltage, the saturation current in a Schottky diode is usually several orders of magnitude higher than in the p–n junction. In addition, the built-in voltage of a Schottky diode is smaller than that of a p–n junction with the same semiconductor. However, high-frequency operation of p–n junction photodiodes is limited by the

minority-carrier storage problem. In other words, the minimum time required to dissipate the carriers injected by the forward bias is dictated by the recombination lifetime. In a Schottky barrier, electrons are injected from the semiconductor into the metal under forward bias if the semiconductor is n-type. Next, they thermalize very rapidly ($\approx 10^{-14}$ s) by carrier–carrier collisions, and this time is negligible compared to the minority-carrier recombination lifetime.

In p–i–n photodiode, an undoped i-region (p^- or n^-, depending on the method of junction formation) is sandwiched between p^+ and n^+ regions. Because of the very low density of free carriers in the i-region and its high resistivity, any applied bias drops entirely across the i-region, which is fully depleted at zero bias or very low value of reverse bias.

The response speed of p–i–n photodiode is ultimately limited

- By transit time across i-region
- By circuit parameters (RC time constant)

Influence of diffusion time (τ_d) of carriers to the depletion region, which is inherently a relatively slow process, can be neglected because the generation of carriers occurs mainly in the high-field i-region. If the photodetector is not fully depleted, this time is taken for collection of carriers generated outside the depletion layer. Carrier diffusion may result in a tail (a deviation between input and output signal) in the response time characteristics.

The transit time of the p–i–n photodiode is shorter than that obtained in a p–n photodiode even though the depletion region is longer than in the p–n photodiode case because carriers travel at near their saturation velocity virtually the entire time they are in the depletion region (in p–n junction the electric field peaked at the p–n interface and then rapidly diminished).

The transit time of carriers across i-layer depends on its width and the carrier velocity. Usually, even for moderate reverse biases that carriers drift across the i-layer with saturation velocity. The transit time can be reduced by reducing the i-layer thickness. The fast photodiode must be thin to minimize the time that it takes for the photogenerated electrons and holes to traverse the depletion region. A thin photodiode has high capacitance per unit area and therefore must also be small in diameter to reduce the RC time constant.

The second component of the response time (t_r) is dependent mainly on the photodetector capacitance and input resistance of the preamplifier. The detector capacitance is a function of the area (A), the zero bias junction potential V_o, dielectric constant $(\varepsilon_o \varepsilon_s)$, impurity concentrations of acceptors (N_a) and donors (N_d), and the bias voltage (V_b). If we assume that the external bias V_b is large compared to the V_o, and we have p^+-n abrupt junction, then the diode has a capacitance given by

$$C_d = \frac{A}{2}\left(2q\varepsilon_s\varepsilon_o N_d\right)^{1/2} V_b^{-1/2}. \qquad (4.27)$$

In this case, lower RC time constant and, therefore, improved bandwidth can be achieved with the use of smaller detectors area and higher bias voltage. In practice, junctions are rarely abrupt; however, it still remains true that the capacitance decreases with increasing reverse bias.

The time constant is given by

$$\tau_{RC} = \frac{\left(R_L + R_s\right)R_{sh}}{R_L + R_s + R_{sh}}C = R_{eq}C, \qquad (4.28)$$

where R_s is the series resistance of the photodiode, C is the sum of the photodiode and input preamplifier capacitances, and R_{eq} is the equivalent resistance of the photodiode and load resistance.

The rise time depended on time constant is described by the formula

$$t_r = 2R_{eq}C. \qquad (4.29)$$

The p–i–n photodiode has a "controlled" depletion layer width, which can be tailored to meet the requirements of photoresponse and bandwidth. A tradeoff is necessary between response speed and quantum efficiency. For high response speed, the depletion layer width should be small but for high quantum efficiency (or responsivity), the width should be large. Increases in bias voltage will usually increase carrier velocities and, therefore, reduce transit times but result in higher dark current and noise.

The detector response time is of the form

$$t_T = \left(t_r^2 + t_t^2 + t_d^2\right)^{1/2}. \qquad (4.30)$$

If we take Equation 4.29, the 3 dB cutoff frequency is given by

$$f_{-3dB} = \left(2\pi R_{eq}C\right)^{-1}. \qquad (4.31)$$

Usually, a designer can increase the bandwidth only by using smaller detectors and/or by reducing the amplifier's input resistance.

A good figure of merit for comparing sensitivity to high-speed signals of similar photodiodes is the *Response factor × Bandwidth product*. Response factor is the photodiode responsivity multiplied by the impedance seen by the photodiode. Assuming that two photodiodes have single-pole high-frequency roll-off, the one with the highest *Response factor × Bandwidth product* will provide the greatest response to an ultrafast pulse.

4.2.6 Noise models of first stages of photoreceivers

A noise equivalent circuit of the first stage of a photoreceiver with a voltage-type preamplifier is shown in Figure 4.22. The signal current generator I_{ph} represents the detected signal. Noises in a

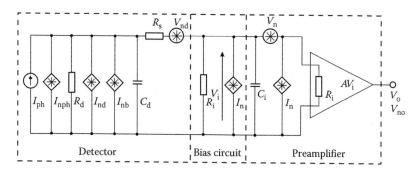

Figure 4.22 Equivalent scheme of photodetector–voltage-type preamplifier circuit (C_d is the detector capacity, and R_i and C_i are the input resistance and capacitance of the preamplifier, respectively).

detector (photodiode) are represented by three noise generators: I_{nph} is the shot noise originating from a photocurrent, I_{nd} is the shot noise of a dark current, and I_{nb} is the shot noise from a background current. If the input resistance of a preamplifier is high, the value of the load resistance depends on the chosen bias resistor. The load (bias) resistor R_L affects both the level of the detector signal and its noise. The noise current generator I_{nR} is the thermal noise current and excess noise of the load resistance R_L. Because the thermal noise of I_{nR} is inversely related to the square root of the resistance, R_L must be large. For the lowest-noise system, at very low frequency, the detector will be the dominant noise source, but at higher frequencies amplifier noise becomes increasingly important.

Expression for SNR at the first stage of a photoreceiver with voltage-type preamplifier results from Scheme 4.22:

$$\frac{S}{N} = \frac{I_{ph}}{\left[I_{nph}^2 + I_{nd}^2 + I_{nb}^2 + I_n^2 + \frac{4kT\Delta f}{R_L} + \left(\frac{V_n}{R_L}\right)^2 \right]^{1/2}}.$$

(4.32)

The numerator represents the photocurrent and the denominator represents the equivalent input noise of photodetector–preamplifier circuit. The first three components determine the noise originating from the photocurrent, the dark current, and the background; the fourth component is the current noise of the preamplifier; the fifth is the thermal noise of the resistance R_L; and the last one is the voltage noise contribution of the preamplifier.

The noise in electrical circuits is often a function of frequency. For high frequencies, the noise equivalent signal current, $I_{n\,total}$, is given by

$$I_{n\,total}^2 = I_{nph}^2 + I_{nd}^2 + I_{nb}^2 + V_{nd}^2\omega^2 C_d^2$$

$$+ I_n^2 + \frac{4kT\Delta f}{R_L} + V_n^2\omega^2\left(C_d^2 + C_i^2\right)^2, \quad (4.33)$$

where V_{nd} is the voltage noise of the serial resistance, R_s.

The input capacitance of the preamplifier, C_i, may be considered to lie across the amplifier input on the photodiode side of the noise generator, V_n. Thus, with this assumption, the noise generator,

V_n, is not a true input noise generator as generally understood, as it should normally lie on the input side of all components (except, possibly, the noise current generator, which would have the same effect whichever side V_n it were to be connected to). The justification for this approach is first that the amplifier capacitance is conveniently grouped with C_d, but second, it enables V_n to be a "white" noise generator.

Rewriting Equation 4.33 we obtain,

$$I_{n\,total}^2 = \left[I_{nph}^2 + I_{nd}^2 + I_{nb}^2 + I_n^2 + \frac{4kT\Delta f}{R_L} \right]$$

$$+ \omega^2\left[V_d^2 C_d^2 + V_n^2\left(C_d + C_i\right)^2 \right]. \quad (4.34)$$

There are thus two terms, a "white" noise term in the first setoff square brackets and a second term, which gives a noise current increasing in proportional to frequency. Although a capacitor does not add noise, the detector noise voltage (V_{nd}) and preamplifier noise voltage (V_n) are increased by C_d and $C_d + C_i$, respectively, as is evident from the coefficient of that term in Equation 4.34. Analyzing Equation 4.34, we see that for matching an amplifier to a detector, it is important to minimize the sum of $I_n + V_n^2\omega^2\left(C_d + C_i\right)^2$.

The sensitivity of an optical receiver is most conveniently expressed in terms of its NEP. This is defined as the optical power necessary to make the signal current, I_{ph} equal to the noise current, $I_{n\,total}$, i.e.,

$$NEP = \left(I_{n\,total}^2\right)^{1/2} \frac{h\nu}{\eta q}. \quad (4.35)$$

Many data sheets show the NEP figure for a detector, unfortunately. If we connected a preamplifier, the performance will often be dependent on the amplifier noise source, critically combined with parasitic features of the photodiode, particularly its capacitance and parallel resistance (see Equation 4.33).

It is possible to achieve a high value of the SNR at the first stage of a photoreceiver when using a voltage-type preamplifier by using a high resistance value for R_L and ensuring low current noise I_n and low voltage noise V_n of the preamplifier. Of course, high resistance value of a R_L causes narrowing of a photoreceiver bandwidth.

Figure 4.23 presents the noise equivalent scheme of the first stage of a photoreceiver using a transimpedance preamplifier. In this circuit, a noise source of a detector is identical as for the case shown in Figure 4.16, where R_{sh} is the shunt resistance of a detector. Preamplifier noise is represented by the voltage source V_n and the current source I_n. Thermal noises from a feedback resistor are represented by the current source I_{nf}.

From the arrangement in Figure 4.22, it can be shown that the equivalent input noise is the square root of the sum of squares of noise components from the photocurrent I_{ph}, the dark current of a detector I_d, the background current I_b, thermal noise of the resistor R_f, the current I_n, and the voltage V_n noise from a preamplifier. Thus, the SNR is of the form

$$\frac{S}{N} = \frac{I_{ph}}{\left[I_{nph}^2 + I_{nd}^2 + I_{nb}^2 + I_n^2 + \dfrac{4kT\Delta f}{R_f} + \left(\dfrac{V_n}{R_f} \right)^2 \right]^{1/2}}.$$

(4.36)

For high frequencies, the last term of the denominator should contain parallel combination of all impedances across the input of the preamplifier, e.g., R_f, R_d, $(\omega C_d)^{-1}$, and $(\omega C_i)^{-1}$.

In a photoreceiver using transimpedance preamplifier, an identical bandwidth can be obtained by choosing a feedback resistance R_f much higher than the resistance $R = R_d \| R_L \| R_i$, which would be possible in a simpler photoreceiver with a voltage-type preamplifier. Thus, comparing formulae 4.32 and

4.36, it can be noticed that for the same bandwidth, the SNR is higher for transimpedance preamplifier than for a classic one. In practice, it means that transimpedance amplifiers can have wider bandwidths and yet retain the low-noise characteristics of high impedance preamplifiers.

In p–n and p–i–n photodiodes, the basic source of noise is shot noise originating from the photocurrent I_{ph}, the dark current I_d, and the background radiation current I_b. In these photodetectors, the thermal noises of detector resistance and noises of active elements of a preamplifier can also play a significant role.

All the preamplifier noises V_n and I_n can be substituted for one equivalent current noise:

$$I_a^2 = \frac{1}{\Delta f} \int_0^{\Delta f} \left(I_n^2 + V_n^2 |Y|^2 \right) df,$$

(4.37)

where Y is the input admittance of an amplifier.

A stated earlier, photodiodes can operate with either voltage or transimpedance preamplifiers. For the equivalent schemes shown in Figures 4.22 and 4.23, the SNR is of the form:

$$\frac{S}{N} = \frac{I_{ph}}{\left[2q \left(I_{ph} + I_d + I_b \right) \Delta f + \dfrac{4kT\Delta f}{R_L} + I_a^2 \right]^{1/2}}.$$

(4.38)

The first term represents the shot noise component of the photocurrent, the dark current, and the background, whereas the second term is the

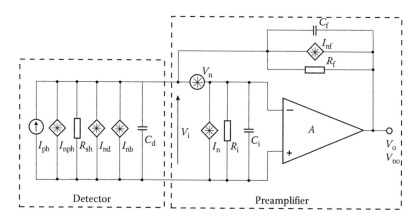

Figure 4.23 Equivalent scheme of the first stage of a photoreceiver with transimpedance preamplifier. (Reproduced from Bielecki, Z., Maximisation of signal to noise ratio in infrared radiation receivers. *Opto-Electron. Rev.*, 10, 209–216, 2002.)

thermal noise of load resistance of a photodetector, and the third term is the preamplifier noise. For the circuit shown in Figure 4.23, the resistance R_L is the feedback resistor, R_f.

Let us consider a few special cases. Assuming that the signal current is higher than the dark current, I_d can be omitted in Equation 4.38. This is valid when the dark current is insignificant or if the received optical power is high. The assumption can also be made that the shot noise significantly exceeds thermal noise when the optical power has a high level. For low-frequency FET transimpedance receivers, the shot noise limit is usually obtained when the output signal exceeds 50 mV. It means that the term $4kT\Delta f/R_L$ can be omitted. If additionally $I_a^2 \ll 2q\Delta f(I_{ph}+I_d)$, the expression for SNR can be simplified to the form:

$$\frac{S}{N}=\left(\frac{I_{ph}}{2q\Delta f}\right)^{1/2}=\left(\frac{\eta\Phi_e\lambda}{2hc\Delta f}\right)^{1/2}=\left(\frac{\eta AE_e\lambda}{2hc\Delta f}\right)^{1/2},$$

$$(4.39)$$

where λ is the wavelength of incident radiation, Φ_e is the incident radiation flux (in W), and E_e is the detector's irradiance. A photoreceiver, the SNR of which is described by formula 4.39, is limited only by a shot noise. This noise is also called quantum limited one—see Equation 4.12. Unfortunately, it is not always the case that the high optical power reaches a photoreceiver. If the power of an optical signal is low, the shot noise is negligible in relation to the thermal noise, then

$$\frac{S}{N}=\frac{\eta q\Phi_e}{h\nu}\left(\frac{R_L}{4kT\Delta f}\right)^{1/2}.$$

$$(4.40)$$

It is evident that when a photoreceiver is limited by the thermal noise, it is thermally dependent (see Equation 4.13).

When thermal noise is limited, it can be seen by analyzing Equation 4.40 that the SNR increases directly in proportion to the received optical power. Thus, in the range of a thermally dependent photoreceiver, small changes, e.g., of path transmission efficiency will cause significant differences in the SNR of the received signal. In the quantum limited systems, an increase in the optical power by $\Delta\Phi_e$ [dB] gives an improvement in the SNR of only half the change $\Delta\Phi_e$ when expressed in dB.

It was mentioned that in energy spectroscopy, the integration of detector current pulses is performed using preamplifier based on charge-sensitive amplifier (CSA). Figure 4.24 presents noise equivalent circuit of CSA where the total noise is dominated by its input transistor (thermal and $1/f$ noises). Additionally, the circuit includes two noise sources derived from leakage currents of gate (I_g) and detector (I_d) and from the feedback resistance.

The noise of the transistor is described by

$$V_{ni}=\frac{8}{3}kT\frac{1}{g_m}+\frac{K_f}{C_{OX}^2WLf},$$

$$(4.41)$$

where g_m is transistor transconductance, K_f is the $1/f$ noise coefficient, C_{OX} is gate oxide capacitance, W and L are the dimensions of the transistor.

Based on the CSA model, the total noise output voltage is given as

$$V_{no}^2=V_{ni}^2\left(1+\frac{C_{in}}{C_f}\right)^2+\left[I_{in}^2+\left(\frac{V_{nf}}{R_f}\right)^2\right]\frac{1}{(sC_f)^2},$$

$$(4.42)$$

where $I_{in}^2=2(I_g+I_d)$ is the total leakage current, $V_{nf}^2=4kTR_f$ is the thermal noise of the feedback resistor, and $C_{in}=C_{det}+C_f+C_g$ is the input capacitance of the CSA.

The first term (thermal noise) is constant over the entire range of frequency. It is amplified by the noise gain, which is given by

$$\text{Noise gain}=\left(1+\frac{C_{in}}{C_f}\right)^2.$$

$$(4.43)$$

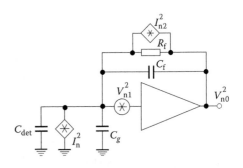

Figure 4.24 Noise equivalent circuit of the CSA. (Reproduced from Amin, F.U., *On the Design of an Analog Front-End for an X-Ray Detector*, Lamberd Academic Publishing, Linköping, 2009.)

The second noise component decreases with increasing frequency and is not dependent on the input capacitance. In the detector readout electronics, a shaper stage is also integrated. This stage is inserted to filter the CSA output signal to improve SNR value, to add more gain, to shorten the pulse duration, and to reduce the possibility of pileup pulses.

One way of overcoming the preamplifier noise is to use a detector having a high degree of internal gain, prior to connection to this device. There are two main types of detector that can be used: the photomultiplier, which is rather large and inconvenient, and the APD. The main advantages are apparent at high frequency, when amplifier noise contribution is more troublesome, because of stray capacitances lowering the impedance across the amplifier input.

Let us consider the SNR in the first stage of photoreceiver with an APD. The high sensitivities of APDs can be obtained due to a phenomenon of avalanche multiplication, which significantly increases the current signal generated at the output of the detector and improves the SNR. It does not of course influence the noises from the load resistance and the amplifier noises but increases the signal as that usually only noise originating from the signal current and dark current (quantum noise) is dominant. Unfortunately, at high gain levels, the random mechanism of carrier multiplication (M) introduces excess noise, in the form of higher shot noise, which eventually exceeds the noise level resulting only from the primary generation of unequilibrium carriers.

At moderate gain levels, the dominant source of the noise in an APD is the signal and dark current shot noises, which are multiplied. However, if the signal power increases by a factor M^2, the noise power increases by a factor M^{2+x}. The factor x is typically between 0.3 and 0.5 for silicon APDs and between 0.7 and 1.0 for germanium and III–V alloy APDs.

Knowing the total noise, a SNR can be determined as

$$\frac{S}{N} = \frac{MI_{ph}}{\left\{ 2q\Delta f \left[I_{ph}M^{2+x} + I_s + (I_b + I_{db})M^{2+x} \right] + \frac{4kT\Delta f}{R_L}F_n \right\}^{1/2}}.$$

(4.44)

The numerator of this equation determines the photocurrent and denominator noises. The first term of the denominator is a shot noise term and the second one represents the thermal noise of load resistance with a preamplifier noise (F_n is the noise factor of preamplifier, I_{db} is the bulk leakage current component of primary dark current, and I_s is the surface leakage current of a dark current). Shot noise components except the surface leakage component of the dark current are multiplied, so the photocurrent, current from a background, and bulk leakage current components of the primary dark current are multiplied. When M is large, the thermal and amplifier noise term becomes insignificant and the S/R ratio decreases with increasing M. Therefore, an optimum value of the multiplication factor M_{op} exists, which maximizes the S/N ratio.

Figure 4.25 presents the typical dependence of signal current, thermal noise, shot noise, and total noise on the avalanche multiplication factor [8]. For low values of multiplication factor, the signal amplitude is usually lower than the total noise amplitude. In this range, thermal noises are dominant at the low optical input levels usually used with APDs. For high multiplication values, thermal noises become less important but shot noises are then significant. There is a supply voltage for which a distance between line representing the signal and the curve representing the total noise is the largest, corresponding to when the S/N ratio is maximum. Of course, these parameters will change with temperature because the dark current of a photodiode and the avalanche multiplication factor are both strongly thermally dependent.

4.3 ADVANCED METHOD OF SIGNAL DETECTION

4.3.1 Signal averaging

When measuring a small repetitive or steady-state signal in the presence of noise, we can often improve the results by making a number of measurements and taking their average. Figure 4.26 presents a scheme of signal detection system for steady-state signals with an analog integrator.

If radiation from both a signal source and the background is incident on a detector surface, the desired signal will appear with noise at the detector output. It is amplified in a preamplifier and next reaches the integrator input. Let us assume that before beginning the measurement cycle, the

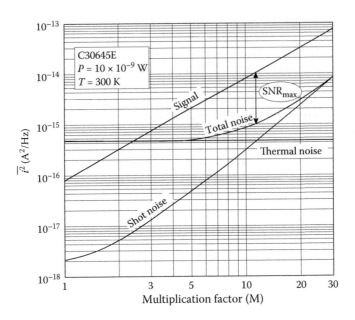

Figure 4.25 Dependence of signal, thermal noise, shot noise, and total noise on avalanche multiplication factor.

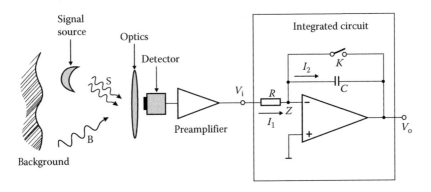

Figure 4.26 Analog integrator used to collect the detected signal level.

voltages at inputs of both the differential amplifier and the integrator and the integrator output are all equal to zero. If a voltage V_i appears at the integrator input, a current I_1 will flow through the resistor R. This current causes changes in the capacitor C to give a voltage $V_o(T)$ at the system output:

$$V_o(T) = -\frac{1}{RC}\int_0^T V_i \, dt + V_o(0),$$ (4.45)

where $V_o(0)$ is the initial voltage at the capacitor at $t=0$ and $RC=\tau$ is the time constant of the integrator.

The voltage at the system output is inversely proportional to the time constant. In practice,

the capacitor C is short-circuited at the beginning of a measuring cycle by means of the switch K. So, the second constant term of Equation 4.45 can be omitted. Opening the switch K initiates the measuring cycle, during which the signal integration occurs. If the assumption is taken that there is a constant voltage of the value v at the system input, then after a time t, the output voltage will be directly proportional to the product vt. Thus, the SNR at the integrator output takes a form

$$\frac{S}{N} = v\left(\frac{2t}{S_n}\right)^{1/2},$$ (4.46)

where S_n is the input white noise power density. This equation tells that the SNR can increase linearly with the square root of the integration time t (for $t \times RC$).

The integrating process in mathematical form is equal taking a series of measurements and summing each of the received values. If we make N measurements, each integrated over the period t, then the received results are added and the SNR is described as

$$\frac{S}{N} = v \left(\frac{2Nt}{S_n} \right)^{1/2}. \qquad (4.47)$$

In this case, SNR increases proportionally to the square root of a number of the performed measurements (or the total measurement time Nt). Similarly, as in Equation 4.46, the time constant of an integrator does not affect the value of a SNR. In real conditions, the value of the time constant should be chosen to ensure the desired level of the output signal after an appropriate integration time t [10].

One disadvantage of the simple integrator is that $1/f$ noise, electronic offsets, and background light level charges can all degrade the performance. We shall now describe another signal recovery method that avoids these problems.

4.3.2 Lock-in amplifier

A lock-in amplifier uses phase-sensitive detection to improve the SNR in continuous wave (cw) experiments. For phase-sensitive detection, the analog signal should be modulated at some reference frequency. This enables the noise component to be filtered out, even when it may initially be many times stronger than the signal itself. If the detected signal of interest is modulated by a carrier signal of defined phase and frequency, it is possible to separate this signal from the noise component. This requires the availability of a low-noise, stable reference frequency ω_{ref} of the same frequency as the carrier signal. In optical measurements, it is usually achieved by initially modulating a stable light source with this reference frequency ω_{ref}. The light transmitted, reflected, or scattered from the test sample, now heavily attenuated but still modulated with ω_{ref}, is then measured. Measurement information is provided by the amplitude of the signal, which falls on the detector (see Figure 4.27). The reference signal is connected to the lock-in amplifier via the reference input (Figure 4.28). The first stage of the lock-in amplifier is usually a programmable ac amplifier, which matches the signal amplitude to a suitable level, ensuring that it does not overload (saturate or "clip") the following electronic components.

In the diagram below, the reference signal is a square wave of frequency ω_{ref}. This might be the clock output from a function generator. If the sine output from the function generator is used to excite the experiment, the response might be the signal waveform shown below. In order to simplify the analysis, we shall assume that the system is noise free.

The signal at the optical detector is

$$V_d = V_S \sin(\omega t + \varphi) + V_n, \qquad (4.48)$$

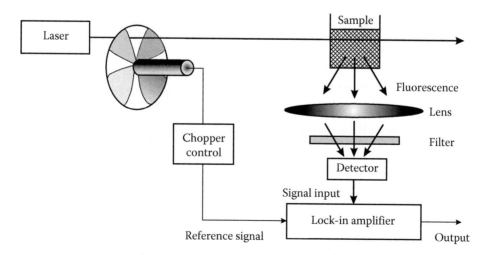

Figure 4.27 Use of a lock-in amplifier in fluorescence detection.

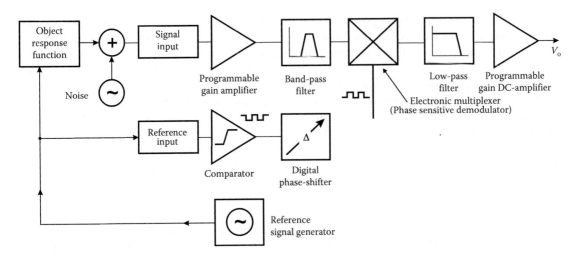

Figure 4.28 Schematic diagram of a lock-in amplifier. (Reproduced from http://www.lasercomponents.de.)

where V_S is the signal amplitude and ϕ is the signal phase.

The band-pass filter, centered at ω_{ref}, removes all noise frequencies, which are outside a defined bandwidth around ω_{ref}. This is to give some prefiltering to stop the following circuits being so easily overloaded by high levels of noise. The next stage is the phase-sensitive demodulator (PSD). Here, the filtered signal is multiplied by a square wave signal of the same frequency ω_{ref} and, ideally, the same phase as the signal at this point.

The internal reference is

$$V_H = V_L \sin(\omega_L t + \varphi_{ref}). \quad (4.49)$$

The output of the PSD is simply the product of two sine waves:

$$V_{psd} = V_S V_L \sin(\omega t + \varphi) \sin(\omega_L t + \varphi_{ref})$$

$$= \frac{1}{2} V_S V_L \cos\left[(\omega - \omega_L)t + \varphi - \varphi_{ref}\right]$$

$$- \frac{1}{2} V_S V_L \cos\left[(\omega + \omega_L)t + \varphi + \varphi_{ref}\right]. \quad (4.50)$$

The PSD output is two ac signals, one at the difference frequency $(\omega - \omega_L)$ and other at the sum frequency $(\omega + \omega_L)$. The phase difference ϕ between the square wave and the signal is important here. If both signals are exactly in phase, the resulting output signal reaches a maximum (Figure 4.29). If this is not the case, the desired ac component will

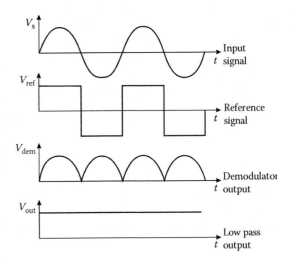

Figure 4.29 Signal processing where input signal and reference signal are in phase.

not be so well detected in the subsequent processes and will actually give zero output if the phase difference is 90°. For this reason, lock-in amplifiers are equipped with a phase shifter, which is used to bring the two signals into phase either manually or automatically. Sometimes these lock-ins are used with the references set 90° apart or "in-phase" and "quadrature" signals are obtained. Then the desired signal is the square root of sum of the squares of in-phase and quadrature components, and budding is not possible. The resulting signal includes contribution from the sum and difference frequencies of the two components.

A low-pass filter is then used to remove any ac components from the dc signal. The lower the cut-off frequency of the filter, the better suppression of the unwanted noise components. However, this also leads to a longer filter time constant and therefore a longer measurement time. Any changes in optical signal level that occur faster than the filter time constant can, of course, no longer be observed. The low-pass filter effectively removes any high-frequency noise remaining following the demodulator process. However, if ω_{ref} equals ω_L, the difference frequency component will be a dc signal. In this case, the filtered PSD output will be

$$V_{out} = \frac{1}{2}V_S V_L \cos(\varphi - \varphi_{ref}). \qquad (4.51)$$

The dc signal is proportional to the signal amplitude.

The final stage of the lock-in consists of a programmable dc amplifier, which further amplifies the smoothed signal. At the output, the result is ideally a noise-free dc signal whose amplitude is directly proportional to the strength of ac signal at the lock-in input. A further advantage of the lock-in amplifier is that it removes $1/f$ noise components, as only noise at frequencies close to the modulation frequency is effectively observed after the electronic processing.

In the above considerations, we have assumed that the chopped signal and the reference output share the same phase. This may not always be the case. In practice, the lock-in amplifier works not only in the so-called "$1f$" operation mode but also has application in which the second and the third harmonics are analyzed. In these measurements, advanced modulation techniques are usually used. Their main task is to determine the relationship of the studied phenomenon with a specific parameter that can be modulated. Functional diagram of this technique is shown in Figure 4.30.

In this diagram, the phenomenon is described by the expression of

$$A(\lambda) = A[\lambda(t)], \qquad (4.52)$$

and is related to the parameter λ.

If it is possible to perform the modulation of λ by time-dependent function

$$V(t) = \langle V \rangle + V_0 \sin(2\pi ft), \qquad (4.53)$$

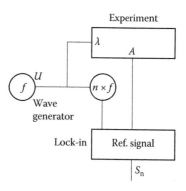

Figure 4.30 Functional diagram of demodulation technique with lock-in amplifier.

it will be possible to process the signal described by

$$S_n = \frac{1}{T} \int_0^T \sin(2\pi nft + \Delta\phi) A[\lambda(t)]dt. \qquad (4.54)$$

By analyzing the different harmonics of the signal, one can specify the virtues of the function describing the investigated phenomenon A, e.g., slope, optimum points, and inflection. Thus, the measurement of the selected harmonics can clearly define both possibility and range of the phenomenon analysis. This method is applied in wavelength modulation spectroscopy, which is able to detect very weak absorption line in the presence of noises as well and of a high baseline. The "$1f$" signal analysis is not effective because it is sensitive to the signal level and assumes the minimum value corresponding to the maximum value of absorption lines. In the case of low-level signal, the influence of any noises on measurements is more significant. Better results are obtained when analyzing "$2f$," which increases the SNR (e.g., from value of 23.5 dB to 35 dB) [12]. In this setup, $3f$ reference channel signal is also employed for locking the laser frequency to the peak of absorption line of the target analysis.

4.3.3 Boxcar detection systems

Phase-sensitive detection systems are used for measurements of steady periodic signals or ones that have a relatively slowly varying level. Frequently, it is desired to measure the amplitude envelope of periodical signals. For this purpose, various measuring methods can be used. One of these methods is the so-called boxcar detection system, which performs synchronous integration (Figure 4.31).

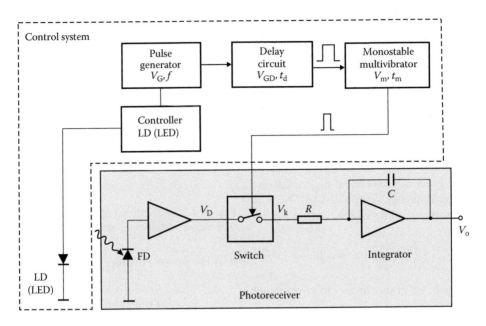

Figure 4.31 Analog synchronous integration system.

This method allows the measurement of periodical signals of complex shapes, even when the signal amplitude is lower than the level of the first stage noise of a photoreceiver. To apply this method, two basic conditions must be fulfilled:

- The measured signal should be periodic and repeatable.
- A trigger signal should be available, which can be used to tell the measurement system when each signal cycle begins (this latter signal could be derived using a phase-lock-loop if the periodic signal is available for a long enough time).

The detection system consists of a signal source, detector–preamplifier system, switch, integrator, and a control system. The control system includes a clock, delay circuit, and a multivibrator. Pulses from the clock generator are used to drive simultaneously an optical signal source and a delay circuit in the signal processor. The output signal from the delay circuit triggers the multivibrator, which in turn closes a switch. The switch on time of the key switch depends on the pulse duration of the multivibrator. A semiconductor laser (LD) or electroluminescent diode (LED) is usually used as the optical source.

We shall assume that the control generator operates at a frequency $f = 1/T$. The delay time of the clock pulse controlling initiation of the multivibrator is t_d and the key switch on time is t_m as determined by

the monostable multivibrator. The leading edge of each clock pulse starts the light pulse generation. Figure 4.32 shows the temporal variation courses of voltages at critical points of the boxcar detection system. A signal from the detector output is sent after amplification to an electronic switch and then to an integrator to average the signal.

The signal $V_k(t)$ reaching the integrator is described by the formula

$$V_k(t) \equiv V_D(t) \quad \text{for} \quad t_d \leq t \leq t_d + t_m$$
$$V_k(t) \equiv 0 \quad \text{for} \quad \text{other } t. \quad (4.55)$$

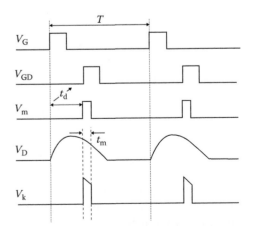

Figure 4.32 Control and data waveforms in phase-sensitive detection systems.

We shall now calculate the voltage at the output of the integrator. We assume that the voltage at the integrator output is initially equal to zero and that we make measurements for m periods of an input waveform. In order to simplify the initial analysis, we shall assume that the system is noise free. For this assumption, the voltage at the system output is

$$V_o(t_d) = -\frac{1}{RC} m \int_0^T V_k(t)dt = -\frac{1}{RC} m \int_{t_d}^{t_d+t_m} V_D(t)dt$$

(4.56)

where R and C determine the resistance and the capacitance values of the integrator, respectively. The minus sign in this equation is present because the voltage signal is applied to an inverting input. If the time interval is sufficiently small, the signal level will not change significantly between the times, t_d and t_d+t_m. Thus, formula 4.56 takes the form

$$V_o(t_d) = -\frac{m}{RC} V_D(t)t_m$$

(4.57)

It results from this formula that the voltage value at the output of the integrator $V_o(t)$ is directly proportional to the instantaneous value of the voltage at the preamplifier output $V_D(t)$, signal cycles m, and pulse duration t_m, but is inversely proportional to the time constant of the integrator. The boxcar detection system exhibits an increase in the amplitude of the output signal in proportion to the number of measuring cycles m.

The SNR of the boxcar detection system if a white noise input voltage source is now considered to be present is given by

$$\frac{S}{N} = \frac{V_D(t)}{V_{ni}} m(2\Delta ft_m)^{1/2}.$$

(4.58)

An increase in the SNR is obtained with a larger number of measuring cycles and longer time of key switch on, the improvement being in proportion to m.

The above measuring system is called a boxcar detection system because the samples are added synchronously with the measuring cycles of a signal like loading wagons (or boxcars) on a train. Boxcar detection systems are very effective at recovering information from repetitive signals when the noise level is quite high. Improvement in

the SNR obtained due to signal measurement for m cycles is, of course, at the expense of greater measurement time being given by mT.

A drawback of the method is that the integrator is disconnected from the input for most of the time. A voltage measurement at the photodetector–preamplifier system is performed only during the time interval t_m/T (i.e., when the key is switched on). So, to measure a pulse shape, we have to repeat the measurement process up to T/t_m times for each t_d value. The time required to measure a pulse shape will be mT^2/t_m. The total measurement time can be reduced by increasing the time of a single sampling t_m. However, it constraints observation of the pulse details.

The boxcar detection system is not efficient because most of the signal power is ignored during the key switch off. This drawback can be avoided by using a parallel processing scheme, involving a multiplexed array of boxcar detection systems. To achieve this, an analog multiplexer can be used to separate into parallel channels or, more frequently with modern technology, a system with the digital path of signal processing is applied. Many digital oscilloscopes are now equipped with such parallel averaging schemes.

4.3.4 Coherent detection

So far, we have considered systems that were based on the modulation of the light intensity in a transmitter and direct detection of this in a photoreceiver. It was not essential for the modulated light wave to be a coherent wave and its spectrum could be wide. These systems are simple and cheap but they have constrains on their transmission possibilities. The photoreceivers decoded only the information connected with the intensity or with the square of the electromagnetic field amplitude, whereas information can be carried also by its phase and frequency. The possible photoreceiver sensitivity results from the basic noise limits as the noise of the photodetector, preamplifier, and background.

To improve the SNR, it would be an advantage to increase the photocurrent at the detector output. We have already noted that using photoreceivers with APDs had constraints resulting from additional multiplication noises. To avoid problems with direct detection, coherent detection, a method of receiving based on interference of two beams of coherent laser radiation, can be used [13–15]. Figure 4.33 illustrates the differences between coherent detection and direct detection.

Figure 4.33a presents a direct detection system. To narrow down the received optical bandwidth, we have applied a filter to limit the spectral range of radiation reaching the detector—Figure 4.33b. For example, optical filters based on the Bragg effect can have 5 nm bandwidth for 1.56 μm wavelengths, corresponding to a detection band, Δf, equal to 600 GHz [16]. This wide bandwidth, of what is a reasonably narrow optical filter, illustrates that application of an optical filter cannot easily ensure narrow wavelength selection of a photoreceiver. A widely spaced Fabry–Perot filter can achieve bandwidth of several hundred MHz but is costly and difficult to stabilize and is so impractical for most systems. However, the simple addition of a beam splitter and an additional coherent light source, a so-called, local oscillator, provides a coherent detection scheme that can use heterodyne and homodyne detection systems (see Figure 4.33c).

Heterodyne detection technique has been commonly used for many years in commercial and domestic radioreceivers and also for the microwave range of an electromagnetic spectrum. The main virtues of this method detection are higher sensitivity, higher and more easily obtained selectivity, plus the possibility of detection of all types of modulation and easier tuning over wide range [17]. Coherent optical detection has been developed since 1962, but compact and stable production of this system is more difficult, and the system is more expensive and troublesome than its radio-technique equivalent. The basic block diagram of heterodyne optical receiver is shown in Figure 4.34.

Laser radiation containing information, is after being passed through an input optical filter and beam splitter, arranged to coherently combine or "mix" with a light beam of a local oscillator at the detector surface. A beam splitter can be made in many ways, the simplest being a glass plate with adequate refraction coefficient. In a general case, a device fulfilling such a role is called a direction coupler, as an analogy to microwave or radio devices. A detector used for signal mixing has to have a square-law characteristic to detect the electronic field of the light, but this is conveniently typical of most optical detectors (photodiode,

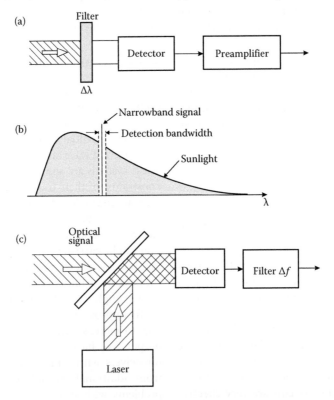

Figure 4.33 Comparison of coherent versus incoherent optical detection: (a) direct detection system, (b) radiation bandwidth limitation with an optical filter, (c) heterodyne and homodyne detection systems.

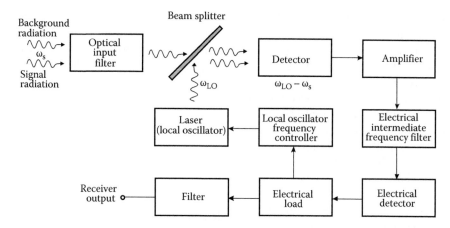

Figure 4.34 Block diagram of heterodyne detection optical receiver.

photoconductor, photomultiplier, APD, etc.). This signal is next amplified. An electrical filter of intermediate frequency (IF) extracts the desired difference component of the signal, which next undergoes a demodulation process. The design and operation principle of the subsequent electrical detector depends on the nature of the modulation of a signal. The signal from a load resistance passes through an output filter to a receiver output and by means of a local oscillator frequency controller it controls a laser. A frequency control loop is used for the local oscillator laser to maintain a constant frequency difference $\omega_L - \omega_s = \omega_p$ with the input signal. An indispensable condition for efficient coherent detection is to match the polarization and to the shape of both waveforms of both beams to match the profile of the detector surface.

Expression for the SNR at the heterodyne detection system with APD for $P_L < P_S$ is given by

$$\frac{S}{N} = \frac{2R_i M \sqrt{P_S P_L}}{\left(2q\Delta f M^{2+x} R_i P_L + \frac{4kT\Delta F}{R_L}\right)^{1/2}} \approx \left(\frac{R_i P_S}{q\Delta f M}\right)^{1/2}.$$

(4.59)

Assuming a photodiode responsivity $R_i = \eta q/h\nu$, usually in A/W, we have

$$\frac{S}{N} = \frac{\eta P_S}{h\nu\Delta f M^x} = 2\left(\frac{S}{N}\right)_{quanta}.$$ (4.60)

Figure 4.35 shows a comparison of coherent detection sensitivity (solid lines) with the sensitivity of direct $p–i–n$ photodiode detection ($M=1$) for the

same values of SNR. Significant improvement in sensitivity can be observed for weak signals. Higher sensitivity of a detector ensures qualitatively better detection as increased information bit rate can permit longer communications links to be used between each regenerator circuit. In long-distance fiber telecommunications, however, the use of optical fiber amplifier has taken much of the impetus from development of the coherent receiver,

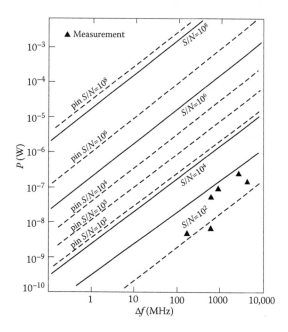

Figure 4.35 Sensitivity of the coherent photoreceiver (dashed line) and $p–i–n$ photodiode (solid line). (Reproduced from Bielecki, Z. and Rogalski, A., *Detection of Optical Radiation*, WNT: Warsaw (in Polish), 2001.)

although the latter still have the unique advantage of highly selective narrowband detection.

In heterodyne detection, the spectrum of laser modulation was shifted into an IF range, so selectivity of a photoreceiver depends on the bandwidth of the IF amplifier. This is arranged electronically, so it can easily be sufficiently narrow. Having narrow IF circuit bandwidth is especially important for the detection of multichannel signals.

In practice, the technique of heterodyne detection is used for construction of Doppler velocimeters and laser rangefinders as well as in spectroscopy (particular LIDAR systems). It may yet find application in more telecommunications systems.

If a signal frequency is equal to the frequency of a local oscillator, the IF frequency equals zero. It is a special case of coherent detection, the so-called homodyne detection.

In a homodyne detection optical receiver (Figure 4.36), the incoming laser carrier is again combined with a reference wave from a local laser on a photodiode surface, but in this case both frequencies are the same. It does not contain two blocks, filter of IF frequency and demodulator, which were in heterodyne receiver.

The photodetector current in a homodyne receiver is given by

$$I_{\text{hom}} = R_i M \left(P_S + P_L \right) + 2 R_i M \left(P_S P_L \right)^{1/2} \cos \varphi_p(t).$$
(4.61)

The first component is a direct-current component but the second one contains the useful information regarding the optical signal. The current at the detector output increases with increase in local oscillator power and with the optical receiver responsivity.

If the local oscillator power is high, the shot noise originating from a signal current, thermal noise, and dark current noise can be omitted. For amplitude modulation, we have

$$\frac{S}{N} = \frac{\left(2 R_i M (P_S P_L)^{1/2} \right)^2 R_L}{\left(2q \Delta f M^{2+x} R_i P_L + 4kT \Delta f F / R_L \right) R_L}$$

$$\approx \frac{2 R_i P_S}{q \Delta f M^x} = \frac{2 \eta P_S}{q \Delta f M^x}.$$
(4.62)

As can be seen, the SNR for homodyne detection is twice as high as heterodyne detection. This is basically because the homodyne detector allows direct addition or subtraction of the electrical fields, depending on whether the signal and local oscillators are in phase or 180° out of phase. With heterodyne detection, the relative phases change linearity with time and mixing of signals is not effective when signals have 90° or 270° phase difference.

As it results from Equation 4.61, homodyne detection derives the baseband modulation signal carrying the information directly. Thus, further electronic demodulation is not required.

Homodyne receivers are used in the most sensitive coherent systems. In practice, construction of such receivers is difficult because

- The local oscillator must be locked to keep a constant zero phase difference to the incoming optical signal, and this requires excellent spectral purity.
- Power fluctuation of the local laser must be eliminated.

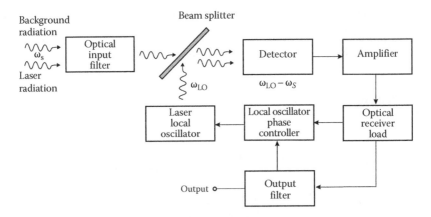

Figure 4.36 Block diagram of homodyne detection optical receiver.

Constant difference of both laser phases can be achieved using an optical phase-locked loop.

The requirements for spectrum purity are less critical in the diversity systems [14,16] in which the cosine component current expressed by Equation 4.61 and also the quadrature current component proportional to $\sin \phi_p t$ is produced. A sum of vectors of these currents makes it possible to avoid influence of ϕ_p phase change, but as with the heterodyne system, the SNR suffers when phases are not identical.

It has been assumed that the local laser has no amplitude noise. In practice, this type of noise is often the limiting factor because the local laser power is strong compared with signal component $2(P_S P_L)^{1/2}$. If detecting these output waves by means of photodiodes give the corresponding currents,

$$I_1 \propto a^2 P_S + b^2 P_L + 2ab \left(P_S P_L\right)^{1/2}$$

$$\times \cos\left[\omega_S t - \omega_L t + \varphi_S(t) - \varphi_L(t) + \pi/2\right], \quad (4.63)$$

$$I_2 \propto b^2 P_S + a^2 P_L - 2ab \left(P_S P_L\right)^{1/2}$$

$$\times \cos\left[\omega_S t - \omega_L t + \varphi_S(t) - \varphi_L(t) + \pi/2\right]. \quad (4.64)$$

This noise is contained in the terms $b^2 P_L$ and $a^2 P_L$, respectively. If we assume a symmetrical beam combiner ($a = b$), and the currents I_1 and I_2 are subtracted, then the terms containing P_S and P_L cancel out, and so do their amplitude fluctuations. Due to the opposite sign of the third term in Equations 4.63 and 4.64, the output signal from the subtractor is doubled. Using the two outputs in this way produces a balanced mixing receiver, using a beam splitter or fiber coupler (Figure 4.37).

In order to make the interference of the signal wave and the local oscillator wave that is received more efficient, their polarization states must coincide. Due to random vibration in the fiber and temperature changes, mechanical strain in the fiber introduces birefringence, which changes with time. As a consequence, the polarization state of the signal received changes randomly.

The problems caused by a polarization mismatch can be overcome in the following ways by

- Using a polarization state controller
- Polarization scrambling (the polarization state is deliberately changed at the transmitting end)
- Use of polarization-maintaining fibers (this solution is more expensive)
- Using polarization diversity (both the local optical wave and signal wave received are split into two orthogonal polarization states)

4.4 PHOTON AND THERMAL DETECTORS

In this chapter, optical radiation is considered as a radiation over the range from vacuum ultraviolet to the submillimeter wavelength (25 nm to 3000 μm). The terahertz (THz) region of electromagnetic spectrum (see Figure 4.38) is often described as the final unexplored area of spectrum and still presents a challenge for both electronic and photonic technologies. It is frequently treated as the spectral region within frequency range $\nu \approx 0.1$–10 THz ($\lambda \approx 3$ mm–30 μm) and is partly overlapping with loosely treated submillimeter (sub-mm) wavelength band $\nu \approx 0.1$–3 THz ($\lambda \approx 3$ mm–100 μm).

The majority of optical detectors can be classified in two broad categories: photon detectors (also called quantum detectors) and thermal detectors.

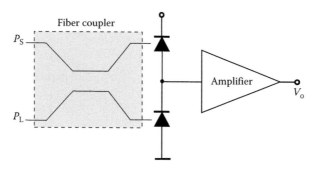

Figure 4.37 Balanced mixing receiver with fiber coupler and series connection of the photodiodes.

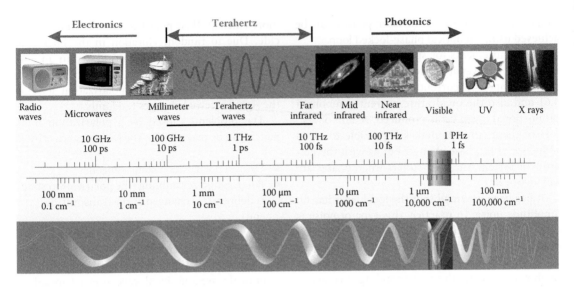

Figure 4.38 The electromagnetic spectrum.

4.4.1 Photon detectors

In photon detectors, the radiation is absorbed within the material by interaction with electrons either bound to lattice atoms or to impurity atoms or with free electrons. The observed electrical output signal results from the changed electronic energy distribution. The fundamental optical excitation processes in semiconductors are illustrated in Figure 4.39. The photon detectors show a selective wavelength dependence of response per unit incident radiation power (see Figure 4.40). They exhibit both good SNR performance and a very fast response. But to achieve this, the photon IR detectors require cryogenic cooling. This is necessary to prevent the thermal generation of charge carriers. The thermal transitions compete with the optical ones, making noncooled devices very noisy.

Depending on the nature of the interaction, the class of photon detectors is further subdivided into different types. The most important are intrinsic detectors, extrinsic detectors, and photoemissive detectors (Schottky barriers) [19–21]. Different types of detectors are briefly characterized in Table 4.1. Figures 4.3 and 4.4 show spectral detectivity curves for a number of commercially available IR detectors.

The most widely used photovoltaic detector is the p–n junction, where a strong internal electric field exists across the junction even in the absence of radiation. Photons incident on the junction

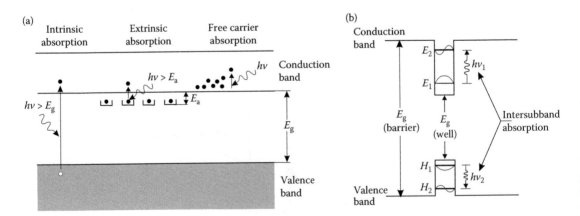

Figure 4.39 Optical excitation processes in bulk semiconductors (a) and in quantum wells (b).

Table 4.1 Photon detectors

Mode of operation	Schematic of detector	Operation and properties
Photoconductor	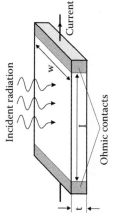	It is essentially a radiation-sensitive resistor, generally a semiconductor either in thin-film or bulk form. A photon may release an electron–hole pair or an impurity-bound charge carrier, thereby increasing the electrical conductivity. In almost all cases, the change in conductivity is measured by means of electrodes attached to the sample. For low resistance material, the photoconductor is usually operated in a constant current circuit. For high resistance photoconductors, a constant voltage circuit is preferred and the signal is detected as a change in current in the bias circuit.
BIB detector	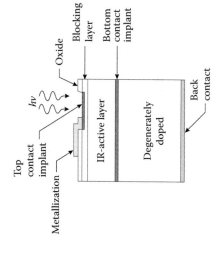	The active region of BIB detector structure, usually based on epitaxially grown n-type material, is sandwiched between a higher doped degenerate substrate electrode and an undoped blocking layer. Doping of active layer is high enough for the onset of an impurity band in order to display a high quantum efficiency for impurity ionization (in the case of Si:As BIB, the active layer is doped to $\approx 5 \times 10^{17}$ cm^{-3}). The device exhibits a diode-like characteristic, except that photoexcitation of electrons takes place between the donor impurity and the conduction band. The heavily doped n-type IR-active layer has a small concentration of negatively charged compensating acceptor impurities. In the absence of an applied bias, charge neutrality requires an equal concentration of ionized donors. Whereas the negative charges are fixed at acceptor sites, the positive charges associated with ionized donor sites (D$^+$ charges) are mobile and can propagate through the IR-active layer via the mechanism of hopping between occupied (D^0) and vacant (D$^+$) neighboring sites. A positive bias to the transparent contact creates a field that drives the preexisting D$^+$ charges toward the substrate, while the undoped blocking layer prevents the injection of new D$^+$ charges. A region depleted of D$^+$ charges is therefore created with a width depending on the applied bias and on the compensating acceptor concentration.

(Continued)

Table 4.1 (*Continued*) Photon detectors

Mode of operation	Schematic of detector	Operation and properties
p–n junction photodiode		It is the most widely used photovoltaic detector, but rather rarely used as THz detector. Photons with energy greater than the energy gap create electron–hole pairs in the material on both sides of the junction. By diffusion, the electrons and holes generated within a diffusion length from the junction reach the space-charge region where they are separated by the strong electric field; minority carriers become majority carriers on the other side. This way a photocurrent is generated causing a change in voltage across the open-circuit cell or a current to flow in the short-circuited case. The limiting noise level of photodiodes can ideally be $\sqrt{2}$ times lower than that of the photoconductor, due to the absence of recombination noise. Response times are generally limited by device capacitance and detector–circuit resistance.
MIS photodiode		The MIS device consists of a metal gate separated from a semiconductor surface by an insulator of thickness t_i and dielectric constant ε_i. By applying a negative voltage V_G to the metal electrode, electrons are repelled from the *I–S* interface, creating a depletion region. When incident photons create hole–electron pairs, the minority carriers drift away to the depletion region and the volume of the depletion region shrinks. The total amount of charge that a photogate can collect is defined as its well capacity. The total well capacity is decided by the gate bias, the insulator thickness, the area of the electrodes, and the background doping of the semiconductor. Numerous such photogates with proper clocking sequence form a CCD imaging array.

(*Continued*)

Table 4.1 (*Continued*) Photon detectors

Mode of operation	Schematic of detector	Operation and properties
Schottky barrier photodiode		Schottky barrier photodiode reveal some advantages over $p-n$ junction photodiode: fabrication simplicity (deposition of metal barrier on $n(p)$ semiconductor), absence of high-temperature diffusion processes, and high speed of response. Since it is a majority carrier device, minority-carrier storage and removal problems do not exist and therefore higher bandwidths can be expected. The thermionic emission process in Schottky barrier is much more efficient than the diffusion process and therefore for a given built-in voltage, the saturation current in a Schottky diode is several orders of magnitude higher than in the $p-n$ junction.

Schematic labels: Incident radiation; Semitransparent Schottky contact; Depletion region; $p(n)$-type; Ohmic contact; t

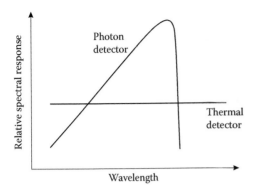

Figure 4.40 Relative spectral response for a photon and thermal detector.

produce free hole–electron pairs that are separated by the internal electric field across the junction, causing a change in voltage across the open-circuit cell or a current to flow in the short-circuited case. Due to the absence of recombination noise, the limiting p–n junction's noise level can ideally be $\sqrt{2}$ times lower than that of the photoconductor.

Photoconductors that utilize excitation of an electron from the valence to conduction band are called intrinsic detectors. Instead, those that operate by exciting electrons into the conduction band or holes into the valence band from impurity states within the band (impurity-bound states in energy gap, quantum wells, or quantum dots) are called extrinsic detectors. Intrinsic detectors are most common at the short wavelength, below 20 μm. In the more long wavelength region, the photoconductors are operated in an extrinsic mode. One advantage of photoconductors is their current gain, which is equal to the recombination time divided by the majority-carrier transit time. This current gain leads to higher responsivity than is possible with nonavalanching photovoltaic detectors. However, series problem of photoconductors operated at low temperature is the nonuniformity of detector element due to recombination mechanisms at the electrical contacts and its dependence on electrical bias.

Recently, interfacial work function internal photoemission detectors, quantum well, and quantum dot detectors, which can be included to extrinsic photoconductors, have been proposed especially for IR and THz spectral bands [21]. The very fast time response of quantum well and quantum dot semiconductor detectors makes them attractive for heterodyne detection.

Figure 4.41 shows the quantum efficiency of some of the detector materials used to fabricate arrays of ultraviolet (UV), visible, and IR detectors. Photocathodes and AlGaN detectors are being developed in the UV region. Silicon p–i–n diodes are shown with and without antireflection coating. Lead salts (PbS and PbSe) have intermediate quantum efficiencies, whereas PtSi Schottky barrier types and quantum well infrared photodetectors (QWIPs) have low values. InSb can respond from the near UV out to 5.5 μm at 80 K. A suitable detector material for near-IR (1.0–1.7 μm) spectral range is InGaAs lattice matched to the InP. Various HgCdTe alloys, in both photovoltaic and photoconductive configurations, cover from 0.7 μm to over 20 μm. InAs/GaSb strained layer superlattices (SLs) have emerged as an alternative to the HgCdTe. Impurity-doped (Sb, As, and Ga) silicon blocked impurity band (BIB) detectors operating at 10 K have a spectral response cutoff in the range of 16–30 μm. Impurity-doped Ge detectors can extend the response out to 100–200 μm

A key difference between intrinsic and extrinsic detectors is that extrinsic detectors require much cooling to achieve high sensitivity at a given spectral response cutoff in comparison with intrinsic detectors. Low-temperature operation is associated with longer-wavelength sensitivity in order to suppress noise due to thermally induced transitions between close-lying energy levels. The long wavelength cutoff can be approximated as

$$T_{max} = \frac{300\,K}{\lambda_c(\mu m)}.$$

The general trend is illustrated in Figure 4.42 for five high-performance detector materials suitable for low-background applications: Si, InGaAs, InSb, HgCdTe photodiodes, and Si:As BIB detectors and extrinsic Ge:Ga unstressed and stressed detectors.

4.4.2 Thermal detectors

The second class of detectors is composed of thermal detectors. In a thermal detector shown schematically in Figure 4.43, the incident radiation is absorbed to change the material temperature and the resultant change in some physical property is used to generate an electrical output. The detector is suspended on lags, which are connected to the heat sink. The signal does not depend upon the photonic

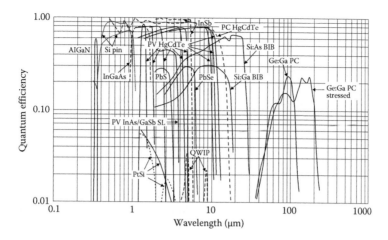

Figure 4.41 Quantum efficiency of different detectors. (Reproduced from Norton, P. Detector FPA technology. In *Encyclopedia of Optical Engineering*, ed. R. Driggers, Marcel Dekker Inc., New York, pp. 320–348, 2003.)

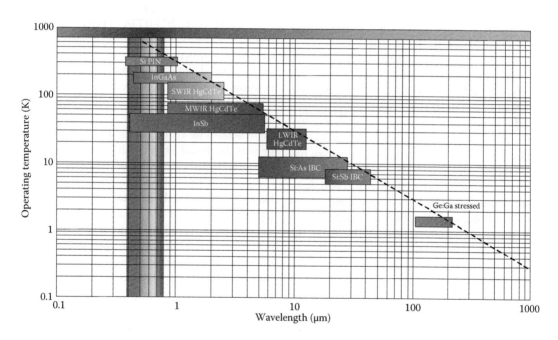

Figure 4.42 Operating temperatures for low-background material systems with their spectral band of greatest sensitivity. The dashed line indicates the trend toward lower operating temperature for longer wavelength detection.

nature of the incident radiation. Thus, thermal effects are generally wavelength independent (see Figure 4.40); the signal depends upon the radiant power (or its rate of change) but not upon its spectral content. Because the radiation can be absorbed in a black surface coating, the spectral response can be very broad. Attention is directed toward three

approaches that have found the greatest utility in IR technology, namely, bolometers, pyroelectric, and thermoelectric effects. The thermopile is one of the oldest IR detectors and is a collection of thermocouples connected in series in order to achieve better temperature sensitivity. In pyroelectric detectors, a change in the internal electrical polarization

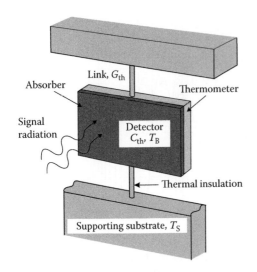

Figure 4.43 Schematic diagram of thermal detector.

is measured, whereas in the case of thermistor bolometers a change in the electrical resistance is measured. For a long time, thermopiles were slow, insensitive, bulky, and costly devices. But with developments in semiconductor technology, thermopiles can be optimized for specific applications. Recently, thanks to conventional complementary metal-oxide-semiconductor (CMOS) processes, the thermopile's on-chip circuitry technology has opened the door to mass production.

Usually bolometer is a thin, blackened flake or slab, whose impedance is highly temperature dependent. Bolometers may be divided into several types. The most commonly used are the metal, the thermistor, and the semiconductor bolometers. A fourth type is the superconducting bolometer.

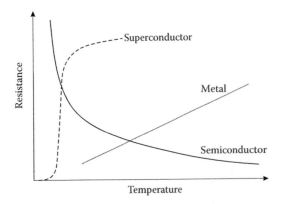

Figure 4.44 Temperature dependence of resistance of three bolometer material types.

This bolometer operates on a conductivity transition in which the resistance changes dramatically over the transition temperature range. Figure 4.44 shows schematically the temperature dependence of resistance of different types of bolometers.

Many types of thermal detectors are operated in a wide spectral range of electromagnetic radiation. The operation principles of thermal detectors are briefly described in Table 4.2.

The microbolometer detectors are now produced in larger volumes than all other IR array technologies together. At present, VO_x microbolometer arrays are clearly the most used technology for uncooled detectors. VO_x is winner of the battle between the amorphous silicon bolometers and hybrid barium strontium titanate (BST) ferroelectric detectors.

4.5 OVERVIEW OF FOCAL PLANE ARRAY ARCHITECTURE

In the last four decades, different types of detectors are combined with electronic readouts to make detector arrays [23]. The progress in IC design and fabrication techniques has resulted in continued rapid growth in the size and performance of these solid-state arrays. In the IR technique, these devices are based on a combination of a readout array connected to an array of detectors.

The term "focal plane array" refers to an assemblage of individual detector picture elements ("pixels") located at the focal plane of an imaging system. Although the definition could include one-dimensional ("linear") arrays as well as two-dimensional (2-D) arrays, it is frequently applied to the latter. Usually, the optics part of an optoelectronic images device is limited only to focusing of the image onto the detectors array. These so-called staring arrays are scanned electronically using circuits integrated with the arrays. The architecture of detector–readout assemblies has assumed a number of forms, which are discussed below. The types of readout integrated circuits (ROICs) include the function of pixel deselecting, antiblooming on each pixel, subframe imaging, output preamplifiers, and yet other functions. IR imaging systems, which use 2-D arrays, belong to the so-called "second-generation" systems.

Development in FPA technology has revolutionized many kinds of imaging. From γ rays to the IR and even radio waves, the rate at which images can be acquired has increased by more than a

Table 4.2 Thermal detectors

Mode of operation	Schematic of detector	Operation and properties
Thermopile	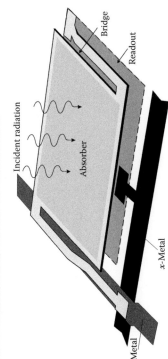	The thermocouple is usually a thin, blackened flake connected thermally to the junction of two dissimilar metals or semiconductors. Heat absorbed by the flake causes a temperature rise of the junction, and hence a thermoelectric electromotive force is developed which can be measured. Although thermopiles are not as sensitive as bolometers and pyroelectric detectors, they will replace these in many applications due to their reliable characteristics and good cost/performance ratio. Thermocouples are widely used in spectroscopy.
Bolometer Metal Semiconductor Superconductor Hot electron		The bolometer is a resistive element constructed from a material with a very small thermal capacity and large temperature coefficient so that the absorbed radiation produces a large change in resistance. The change in resistance is like to the photoconductor, however, the basic detection mechanisms are different. In the case of a bolometer, radiant power produces heat within the material, which in turn produces the resistance change. There is no direct photon–electron interaction. Most bolometers in use today are of the thermistor type made from oxides of manganese, cobalt, or nickel. Their construction is very rugged for system applications. Some extremely sensitive low-temperature semiconductor and superconductor bolometers are used in THz region.

(Continued)

Table 4.2 (*Continued*) Thermal detectors

Mode of operation	Schematic of detector	Operation and properties
Pyroelectric detector	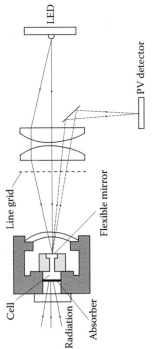	The pyroelectric detector can be considered as a small capacitor with two conducting electrodes mounted perpendicularly to the direction of spontaneous polarization. During incident of radiation, the change in polarization appears as a charge on the capacitor and a current is generated, the magnitude of which depends on the temperature rise and the pyroelectrical coefficient of the material. The signal, however, must be chopped or modulated. The detector sensitivity is limited either by amplifier noise or by loss-tangent noise. Response speed can be engineered making pyroelectric detectors useful for fast laser pulse detection, however with proportional decrease in sensitivity.
Golay cell		The Golay cell consists of a hermetically sealed container filled with gas (usually xenon for its low thermal conductivity) and arranged so that expansion of the gas under heating by a photon signal distorts a flexible membrane on which a mirror is mounted. The movement of the mirror is used to deflect a beam of light shining on a photocell and so producing a change in the photocell current as the output. In modern Golay cells, the photocell is replaced by a solid-state photodiode and light emitting diode is used for illumination. The performance of the Golay cell is only limited by the temperature noise associated with the thermal exchange between the absorbing film and the detector gas, consequently the detector can be extremely sensitive with $D^* \approx 3 \times 10^9$ cmHz$^{1/2}$ W^{-1}, and responsivities of 10^5 to 10^6 V/W. The response time is quite long, typically 15 ms.

Labels in pyroelectric detector schematic: Top electrodes, Plug metal, Equipotential plane, Mirror

Labels in Golay cell schematic: LED, Line grid, Cell, Radiation, Absorber, Flexible mirror, PV detector factor of a million in many cases. Figure 4.45 illustrates the trend in array size over the past 40 years. Imaging FPAs have developed in proportion to the ability of silicon ICs technology to read and process the array signals and with an ability to display the resulting image. The progress in arrays has been steady and has paralleled the development of dense electronic structures such as dynamic random access memories (DRAMs). FPAs have nominally the same growth rate as DRAM ICs, which

have had a doubling-rate period of approximately 18 months; it is a consequence of Moore's law, but lag behind in size by about 5–10 years. The graph in the insert of Figure 4.45 shows the log of the number of pixels per a sensor chip assembly (SCA) as a function of the year first used on astronomy for Mid-Wave Infrared (MWIR) SCAs. CCDs with close to 2 gigapixels offer the largest formats.

A number of architectures are used in the development of FPAs. In general, they may be classified

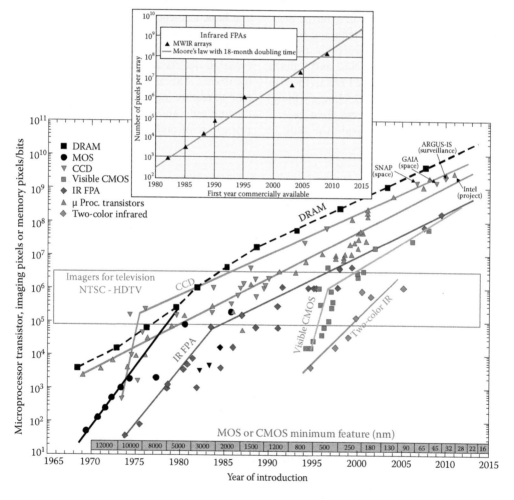

Figure 4.45 Imaging array formats compared with the complexity of silicon microprocessor technology and dynamic access memory (DRAM) as indicated by transistor count and memory bit capacity. The timeline design rule of MOS/CMOS features is shown at the bottom. CCDs with close to 2 gigapixels offer the largest formats. Note the rapid rise of CMOS imagers, which are challenging CCDs in the visible spectrum. The number of pixels on an infrared array has been growing exponentially, in accordance with Moore's law for 30 years with a doubling time of approximately 18 months. Imaging formats of many detector types have gone beyond that required for high definition TV. (Adapted from Norton, P., Detector FPA technology. In *Encyclopedia of Optical Engineering*, ed. R. Driggers, Marcel Dekker Inc., New York, pp. 320–348, 2003 with completions.)

as hybrid and monolithic, but these distinctions are often not as important as proponents and critics state them to be. The central design questions involve performance advantages versus ultimate producibility. Each application may favor a different approach depending on the technical requirements, projected costs, and schedule.

4.5.1 Monolithic arrays

In the monolithic approach, both detection of light and signal readout (multiplexing) is done in the detector material rather than in an external readout circuit. The integration of detector and readout onto a single monolithic piece reduces the number of processing steps, increases yields, and reduces costs. Common examples of these FPAs in the visible and near infrared (NIR) (0.7–1.0 μm) are found in camcorders and digital cameras. Two generic types of silicon technology provide the bulk of

devices in these markets: CCDs and CMOS imagers. CCD technology has achieved the highest pixel counts or largest formats with numbers above 10^9 (see Figure 4.45). This approach to image acquisition was first proposed in 1970 in a paper written by Bell Lab researchers W.S. Boyle and G.E. Smith. CMOS imagers are also rapidly moving to large formats and at present are competed with CCDs for the large format applications. Figure 4.46 shows different architectures of monolithic FPAs.

4.5.1.1 CCD DEVICES

The basic element of a monolithic CCD array is a metal–insulator–semiconductor (MIS) structure. Used as part of a charge transfer device, a MIS capacitor detects and integrates the generated photocurrent. Although most imaging applications tend to require high charge handling capabilities in the unit cells, an MIS capacitor fabricated in a narrow-gap semiconductor material (e.g., HgCdTe and

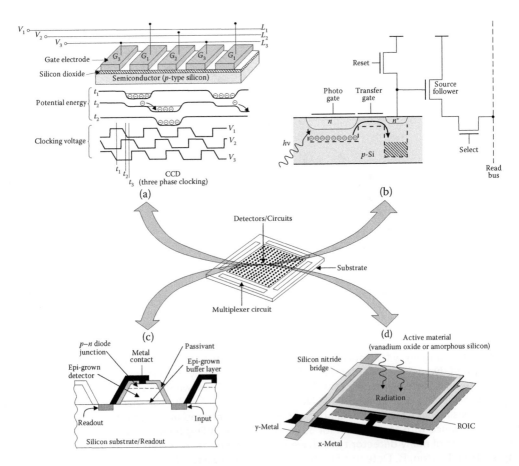

Figure 4.46 Monolithic FPAs: (a) CCD, (b) CMOS, (c) heteroepitaxy-on-silicon, and (d) microbolometer.

InSb) has a limited charge capacity because of its low background potential as well as more severe problems involving noise, tunneling effects, and charge trapping when shifting charge through the narrow bandgap CCD to accomplish the readout function. Because of the nonequilibrium operation of the MIS detector, much larger electric fields are set up in the depletion region than in the $p–n$ junction, resulting in defect-related tunneling current that is orders of magnitude larger than the fundamental dark current. The MIS detector required much higher material quality than $p–n$ junction detectors, which still has not been achieved. So, although efforts have been made to develop monolithic FPAs using narrow-gap semiconductors, silicon-based FPA technology is the only mature technology with respect to fabrication yield and attainment of near-theoretical sensitivity.

A metal-oxide semiconductor (MOS) capacitor typically consists of an extrinsic silicon substrate on which an insulating layer of silicon dioxide (SiO_2) is grown. When a bias voltage is applied across p-type

MOS structure, majority charge carriers (holes) are pushed away from the Si–SiO_2 interface directly below the gate, leaving a region depleted of positive charge and available as a potential energy well for any mobile minority charge carriers (electrons) (see Figure 4.46a). Electrons generated in the silicon through absorption (charge generation) will collect in the potential energy well under the gate (charge collection). Linear or 2-D arrays of these MOS capacitors can, therefore, store images in the form of trapped charge carriers beneath the gates. The accumulated charges are transferred from potential well to the next well by using sequentially shifted voltage on each gate (charge transfer). One of the most successful voltage-shifting schemes is called three-phase clocking. Column gates are connected to the separate voltage lines (L_1, L_2, L_3) in contiguous groups of three ($G_1, G_2,$ and G_3). The setup enables each gate voltage to be separately controlled.

Figure 4.47a shows the schematic circuit for a typical CCD imager. The photogenerated carriers

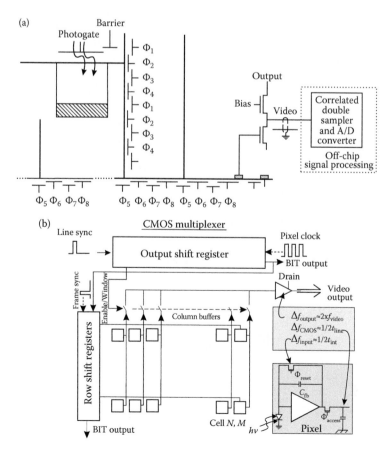

Figure 4.47 Typical readout architecture of CCD (a) and CMOS (b) images.

are first integrated in an electronic well at the pixel and subsequently transferred to slow and fast CCD shift registers. At the end of the CCD register, a charge carrying information on the received signal can be readout and converted into a useful signal (charge measurement).

The process of readout from the CCD consists of two parts:

- Moving charge packets (representing pixel values) around the sensor
- Converting the charge packet values into output voltages

The charge-to-voltage converter at the CCD output is basically a capacitor with single- or multistage voltage follower and a switch to preset the capacitor voltage to a "known" level. In simplest video systems, the switch is closed in the beginning of each pixel readout—that presets the capacitor voltage as well as the output level. After the pixel charge packet is transferred to the capacitor its voltage changes and the output signal represents the pixel value. Due to switch's finite residual conductivity, the capacitor is precharged to an unknown value and it adds the output signal. A way to compensate for this precharge uncertainty is readout technique method—correlated double sampling (CDS). In this method, the output signal is sampled twice for each pixel—just after precharging capacitor and after the pixel charge packet is added.

Figure 4.48 shows a preamplifier, in this example—the source follower per detector (SFD— see also Figure 4.54), the output of which is connected to a clamp circuit. The output signal is initially sampled across the clamp capacitor during the onset of photon integration (after the detector is reset). The action of the clamp switch and capacitor subtracts any initial offset voltage from the output waveform. Because the initial sample is made before significant photon charge has been integrated, by charging the capacitor, the final integrated photon signal swing is unaltered. However, any offset voltage or drift present at the beginning of integration is, by the action of the circuit, subtracted from the final value. This process of sampling each pixel twice, once at the beginning of the frame and again at the end, and providing the difference is called CDS. More information about readout techniques used in CCD devices (CDS, floating diffusion amplifier in each pixel, and floating gate amplifier) is described in detail, e.g., in References. [21,24,25].

The first CCD imager sensors were developed about 40 years ago primarily for television analog image acquisition, transmission, and display. With increasing demand for digital image data, the traditional analog raster scan output of image sensors is of limited use, and there is a strong motivation to fully integrate the control, digital interface, and image sensor on a single chip.

The most popular CCD consists of silicon sensor operating in visible and NIR wavelength ranges. These spectra could be extended into the UV using delta doping and antireflection coating. In this way, device stability and external quantum efficiency to 50%–90% at the wavelengths of 200–300 nm are obtained. CCD for scientific applications are routinely made with pixel counts exceeding 20 megapixels and visible 50 megapixel arrays are now available with digital output

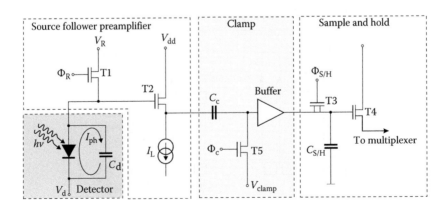

Figure 4.48 CDS circuit.

shaving ROIC noise levels of less than 10 electrons and offering a sensitivity advantage over consumer products [28].

4.5.1.2 CMOS DEVICES

An attractive alternative to the CCD readout is coordinative addressing with CMOS switches. In particular, silicon fabrication advances now permit the implementation of CMOS transistor structures that are considerably smaller than the wavelength of visible light and have enabled the practical integration of multiple transistors within a single picture. The configuration of CCD devices requires specialized processing, unlike CMOS imagers, which can be built on fabrication lines designed for commercial microprocessors. CMOS have the advantage that existing foundries, intended for application-specific ICs, can be readily used by adapting their design rules. Design rules of 14 nm are currently in production, with preproduction runs of 10 nm design rules. As a result of such fine design rules, more functionality has been designed into the unit cells of multiplexers with smaller unit cells, leading to large array sizes. Figure 4.45 shows the timelines for minimum circuit features and the resulting CCD, IR FPA, and CMOS visible imager sizes with respect to number of imaging pixels. Along the horizontal axis is also a scale depicting the general availability of various MOS and CMOS processes. The ongoing migration to even finer lithography will thus enable the rapid development of CMOS-based imagers having even higher resolution, better image quality, higher levels of integration, and lower overall imaging system cost than CCD-based solutions. The pixel's architecture is changed to improve the resolution by shrinking pixel size. Figure 4.49 is a roadmap of where CMOS pixel pitch became smaller than CCD due to the described technological development in 2010 [29]. CMOS imagers are also rapidly moving to large formats and at present are competing with CCDs for the large format applications. The silicon wafer production infrastructure, which has put high-performance personal computers into many homes, makes CMOS-based imaging in consumer products such as video and digital still cameras widely available.

A typical CMOS multiplexer architecture (see Figure 4.47b) consists of fast (column) and slow (row) shift registers at the edges of the active area, and pixels are addressed one by one through the selection of a slow register, while the fast register scans through a column, and so on. Each image sensor is connected in parallel to a storage capacitor located in the unit cell. A column of diodes and storage capacitors is selected one at a time by a digital horizontal scan register and a row bus is selected by the vertical scan register. Therefore, each pixel can be individually addressed.

CMOS-based imagers use active and passive pixels [26,27] as shown, in a simplified form, in Figure 4.46b. In comparison with passive pixel sensors (PPSs), active pixel sensors (APSs) apart from read functions exploit some form of amplification at each pixel. The PPS consists of three transistors: a reset FET, a selective switch, and a source follower (SF) for driving the signal onto the column bus. As a result, circuit overhead is low and the optical collection efficiency (FF) is high even for monolithic devices. Microlenses, typically used in CCD and CMOS APS imagers for visible application, concentrate the incoming light into the photosensitive region when they are accurately deposited over each pixel. Unfortunately, microlenses are less effective when used in low F/# imaging systems and may not be appropriate for all applications.

Figure 4.50 compares the principle of CCDs and CMOS sensors. Both detector technologies use a photosensor to generate and separate the charges in the pixel. Beyond that, however, the two

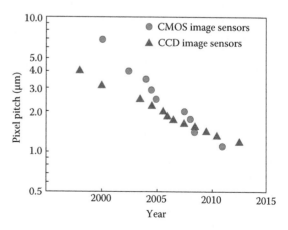

Figure 4.49 A roadmap of CMOS pixel pitch development. (Reproduced after Hirayama, T., The evolution of CMOS image sensors. *IEEE Asian Solid-State Circuits Conference*, 5–8, 2013.)

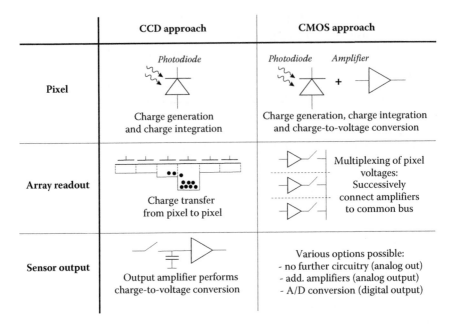

	CCD approach	CMOS approach
Pixel	*Photodiode* Charge generation and charge integration	*Photodiode* *Amplifier* Charge generation, charge integration and charge-to-voltage conversion
Array readout	Charge transfer from pixel to pixel	Multiplexing of pixel voltages: Successively connect amplifiers to common bus
Sensor output	Output amplifier performs charge-to-voltage conversion	Various options possible: - no further circuitry (analog out) - add. amplifiers (analog output) - A/D conversion (digital output)

Figure 4.50 Comparison between the CCD-based and CMOS-based image sensor approach. (From Hoffman, A., Loose, M. and Suntharalingam, V., CMOS detector technology. *Exp. Astron.*, 19, 111–134, 2005.)

sensor schemes differ significantly. During CCD readout, the collected charge is shifted from pixel to pixel all the way to the perimeter. Finally, all charges are sequentially pushed to one common location (floating diffusion), and a single amplifier generates the corresponding output voltages. On the other hand, CMOS detectors have an independent amplifier in each pixel (APS). The amplifier converts the integrated charge into a voltage and thus eliminates the need to transfer charge from pixel to pixel. The voltages are multiplexed onto a common bus line using integrated CMOS switches. Analog and digital sensor outputs are possible by implementing either a video output amplifier or an analog-to-digital (A/D) converter on the chip.

The processing technology for CMOS is typically two to three times less complex than standard CCD technology. In comparison with CCDs, the CMOS multiplexers exhibit important advantages due to high circuit density, fewer drive voltages, fewer clocks, much lower voltages (low power consumption), and packing density compatible with many more special functions, lower cost for both digital video, and still camera applications. The minimum theoretical read noise of a CCD is limited in large imagers by the output

amplifier's thermal noise after CDS is applied in off-chip support circuits. The alternative CMOS paradigm offers lower temporal noise because the relevant noise bandwidth is fundamentally several orders of magnitude smaller and better matches the signal bandwidth. While CCD sensitivity is constrained by the limited design space involving the sense node and the output buffer, CMOS sensitivity is limited only by the desired dynamic range and operating voltage. CMOS-based imagers also offer practical advantages with respect to on-chip integration of camera functions including command and control electronics, digitization, and image processing. CMOS is now suitable also for time-delay and integration type (TDI-type) multiplexers because of the availability from foundries of design rules lower than 1.0 μm, more uniform electrical characteristics, and lower noise figures.

4.5.2 Hybrid arrays

In the case of hybrid technology (see Figure 4.51), we can optimize the detector material and multiplexer independently. Other advantages of hybrid-packaged FPAs are near-100% FFs and increased signal-processing area on the multiplexer chip.

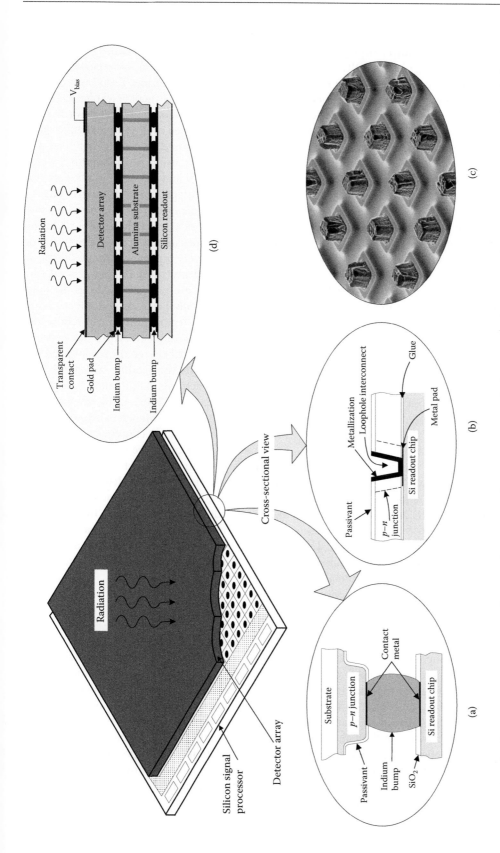

Figure 4.51 Hybrid IR FPA interconnect techniques between a detector array and silicon multiplexer: (a) indium bump technique, (b) loophole technique, (c) SEM photo shows mesa photodiode array with indium bumps, and (d) layered-hybrid design suitable for large format far IR and sub-mm arrays.

Photodiodes with their very low power dissipation, inherently high impedance, negligible $1/f$ noise, and easy multiplexing via the ROIC can be assembled in 2-D arrays containing a very large number of pixels, limited only by existing technologies. Photodiodes can be reverse biased for even higher impedance and can, therefore, be better matched electrically with compact low-noise silicon readout preamplifier circuits. The photoresponse of photodiodes remains linear for significantly higher photon flux levels than that of photoconductors, primarily because of higher doping levels in the photodiode absorber layer and because the photogenerated carriers are collected rapidly by the junction. Development of hybrid packaging technology began in the late 1970s and took the next decade to reach volume production. In the early 1990s, fully 2-D imaging arrays provided a means for staring sensor systems to enter the production stage. In the hybrid architecture, indium bump bonding with readout electronics provides for multiplexing the signals from thousands or millions of pixels onto a few output lines, greatly simplifying the interface between the vacuum-enclosed cryogenic sensor and the system electronics.

Different hybridization approaches are in use today. The most popular is flip-chip interconnect using bump bond (see Figure 4.51a and c). In this approach, indium bumps are formed on both the detector array and the ROIC chip. The array and the ROIC are aligned and force is applied to cause the indium bumps to cold-weld together. In the other approach, indium bumps are formed only on the ROIC; the detector array is brought into alignment and proximity with the ROIC, the temperature is raised to cause the indium to melt, and the contact is made by reflow.

IR hybrid FPA detectors and multiplexers are also fabricated using loophole interconnection—see Figure 4.51b. In this case, the detector and the multiplexer chips are glued together to form a single chip before detector fabrication. The photovoltaic detector is formed by ion implantation and loopholes are drilled by ion milling and electrical interconnection between each detector and its corresponding input circuit is made through a small hole formed in each detector. The junctions are connected down to the silicon circuit by cutting the fine, few micrometers in diameter holes through the junctions by ion milling, and then backfilling the holes with metallization. A similar type of hybrid technology called VIP™ (vertically integrated photodiode) was reported by DRS Infrared Technologies (former Texas Instruments).

It is difficult to make small pixel pitches (below 10 μm) using bump-bonding interconnect technique, especially when high yield and 100% pixel operability are required. A new facility gives 3-D integration process using wafer bonding, where such materials as Si and InP have been monolithically integrated with pixels size down to 6 μm.

The detector array can be illuminated from either the front side (with the photons passing through the transparent silicon multiplexer) or back side (with photons passing through the transparent detector array substrate). In general, the latter approach is most advantageous as the multiplexer will typically have areas of metallization and other opaque regions, which can reduce the effective optical area of the structure. The epoxy is flowed into the space between the readout and the detectors to increase the bonding strength. In the case of backside detector illumination, transparent substrates are required. When using opaque materials, substrates must be thinned to below 10 μm to obtain sufficient quantum efficiencies and reduce cross talk. In some cases, the substrates are completely removed. In the "direct" back side illuminated configuration, both the detector array and the silicon ROIC chip are bump mounted side-by-side onto a common circuit board. The "indirect" configuration allows the unit cell area in the silicon ROIC to be larger than the detector area and is usually used for small scanning FPAs, where stray capacitance is not an issue.

4.5.3 Readout ICs

The development of FPAs using IC techniques together with development of new material growth techniques and microelectronic innovations began about 40 years ago. The combination of the last two techniques gives many new possibilities for imaging systems with increased sensitivity and spatial resolution. Key to the development of ROICs has been the evolution in input preamplifier technology. This evolution has been driven by increased performance requirements and silicon processing technology improvements.

Readout circuit wafers are processed in standard commercial foundries and can be constrained in size by the die size limits of the photolithography step and repeat printers. Because of field size limitations in those photography systems, CMOS imager chip sizes must currently be limited to standard lithographic field sizes of less than $32 \times 26\,mm$ for submicron lithography. To build larger sensor arrays, a new photolithographic technique called *stitching* can be used to fabricate detector arrays larger than the reticle field of photolithographic steppers. The large array is divided into smaller sub-blocks.

The direct injection (DI) circuit was one of the first integrated readout preamplifiers and has been used as an input to CCDs and visible imagers for many years. The DI is also a commonly used input circuit for IR tactical applications, where the backgrounds are high and detector resistances are moderate. The goal is to fit as large a capacitor as possible into the unit cell, where SNRs can be obtained through longer integration times. Photon current in DI circuits is injected, via the source of the input transistor, onto an integration capacitor (Figure 4.52a). As the photon current charges the capacitor throughout the

frame, a simple charge integration takes place (Figure 4.52b). Next, a multiplexer reads out the final value and the capacitor voltage is reset prior to the beginning of the frame. To reduce the detector noise, it is important that a uniform, near-zero voltage bias be maintained across all the detectors.

The DI circuit is widely used for simplicity; however, it requires a high-impedance detector interface and is not generally used for low IR backgrounds due to injection efficiency issues. Many times, the strategic applications have low backgrounds and require low-noise multiplexers interfaced to high-resistance detectors. A commonly used input circuit for strategic applications is the capacitive transimpedance amplification (CTIA) input circuit.

The CTIA amplifier is reset integrator and addresses broad range of detector interface and performance requirements across many applications. It consists of an inverting amplifier with a gain of A, the integration capacitance C_f placed in a feedback loop, and the reset switch K (Figure 4.53). The photoelectron charge causes a slight change in

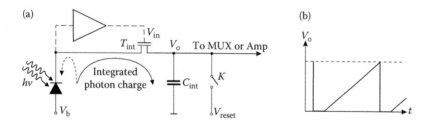

Figure 4.52 DI readout circuit: (a) schematic and (b) charge integration of photocurrent. (Reproduced from Hewitt, M.J., Vampola, J.L., Black, S.H. and Nielsen, C.J., Infrared readout electronics: A historical perspective. *Proc. SPIE*, 2226, 108–119, 1994.)

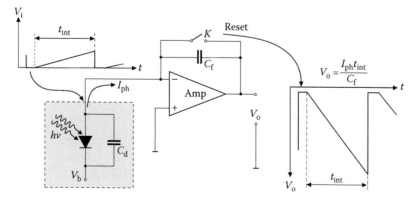

Figure 4.53 Schematic of a capacitive transimpedance amplifier unit cell.

a voltage at the inverting input node of an amplifier. The amplifier responds with a sharp reduction in output voltage. As the detector current accumulates over the "frame time," uniform illumination results in a linear ramp at the output. At the end of integration, the output voltage is sampled and multiplexed to the output bus. Because the input impedance of the amplifier is low, the integration capacitance can be made extremely small, yielding low-noise performance. The feedback, or integration, capacitor sets the gain. The switch K is cyclically closed to achieve reset. The CTIA provides low input impedance, stable detector bias, high gain, high-frequency response, and a high-photon current injection efficiency. It has very low noise from low to high backgrounds.

Besides the DI and CTIA inputs mentioned above, we can distinguish other multiplexers; the most important are SFD, buffered direct injection, and gate modulation input circuits.

The combined SFD unit cell is shown in Figure 4.54. The unit cell consists of an integration capacitance, a reset transistor ($T1$) operated as a switch, the SF transistor ($T2$), and selection transistor ($T3$). The integration capacitance may just be the detector capacitance and transistor $T2$ input capacitance. The integration capacitance is reset to a reference voltage (V_R) by pulsing the reset transistor. The photocurrent is then integrated on the capacitance during the integration period. The ramping input voltage of the SFD is buffered by the SF and then multiplexed, via the $T3$ switch, to a common bus prior to the video output buffer. After the multiplexer read cycle, the input node is reset and the integration cycle begins again. The switch

must have very low current leakage characteristics when in the open state or this will add to the photocurrent signal. The dynamic range of the SFD is limited by the current voltage characteristics of the detector. As the signal is integrated, the detector bias changes with time and incident light level. The SFD has low noise for low bandwidth applications such as astronomy and still has acceptable SNR at very low backgrounds (e.g., a few photons per pixel per 100 ms). It is nonlinear at medium and high backgrounds, resulting in a limited dynamic range. The gain is set by the detector responsivity and the combined detector plus source-follower-input capacitance. The major noise sources are the kTC noise (resulting from resetting the detector), MOSFET channel thermal, and MOSFET $1/f$ noise.

Table 4.3 provides a description of the advantages and disadvantages of DI, CTIA, and SFD circuit. As is mentioned above, the DI circuit is used in higher-flux situations. The CTIA is more complex and higher power but is extremely linear. The SFD is most commonly used in large format hybrid astronomy arrays as well as commercial monolithic CMOS cameras.

In Table 4.4 are gathered specifications of a family of large-format ROICs designed by FLIR and Raytheon Vision Systems (RVS). FLIR products provide an off-the-shelf solution for the most demanding applications. The large arrays include a variety of pixel ranging from 25 to 15 μm for customers with a wide range of optical design, dewar/cooler configurations, and resolution requirements. On the contrary, RVS has a rich heritage of developing astronomy FPAs. As the demand for both finer resolution and large FOV for astronomy

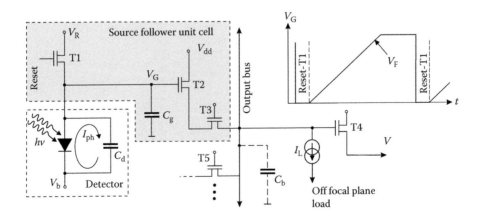

Figure 4.54 Schematic of SFD unit cell.

Table 4.3 Comparison of attributes of the three most common input circuits

Circuit	Advantages	Disadvantages	Comments
DI	Large well capacity Gain determined by ROIC design (C_{int}) Detector bias remains constant Low FET glow Low power	Poor performance at low flux	Standard circuit for high flux
CTIA	Very linear Gain determined by ROIC design (C_f) Detector bias remains constant	More complex circuit FET glow Higher power	Very high gains demonstrated
SFD	Simple Low noise Low FET glow Lowe power	Gain fixed by detector and ROIC input capacitance Detector bias changes during integration Some nonlinearity	Most common circuit in IR astronomy

Source: Hoffman, A., Loose, M. and Suntharalingam, V. CMOS detector technology. *Exp. Astron.*, 19, 111–134, 2005.

imagery has increased, the RVS sensor chip utilizes various detector materials and silicon readouts with an array size up to 4096×4096 pixels.

4.6 FRONTIERS IN FOCAL PLANE ARRAY DEVELOPMENT

As is marked in the previous session, imaging FPAs have developed in proportion to the ability of silicon ICs technology. In this section, we shortly describe the most spectacular achievements in fabrication FPAs in different spectral regions.

At present, the largest single-chip CCD arrays exceed 100 megapixels. DALSA announces that it has successfully produced a 252 megapixel CCD (see Figure 4.55). The active are measures approximately 4×4 in. and 17,216×14,656 pixels with 5.6 μm size [28].

The development of mosaics of area arrays to produce large format (above one gigapixel) frame image is an intriguing idea. One of them, the 1.4 gigapixel CCD imager used in PanSTARRS is comprised of 60 chips, each of 22 megapixel. An another example is a wide area persistent surveillance program such as the Autonomous Real-time Ground Ubiquitous Surveillance—Imaging System (ARGUS-IS), where extremely large mosaics of visible FPAs is used (see Figure 4.56). The 1.8 gigapixel video sensor produces more than 27 gigapixels per second running at a frame rate of

15 Hz. The airborne processing subsystem is modular and scalable providing more than 10 teraops of processing.

Another milestone step in the development of large CCD arrays is the Gaia camera for space mission. Gaia, funded by European Space Agency (ESA) with EADS Astrium manufacturer subsidiary of the European Aeronautic Defence and Space Company (EADS) as the prime contractor, is an ambitious space observatory designed to measure the positions of around 1 billion stars with unprecedented accuracy. The Gaia's FPA is populated with 106 back-illuminated devices, each with an active area of 45×59 mm corresponding to 4500×1966 pixels, each 10×30 μm in size. All of the Gaia CCDs are large area, back-illuminated, and full frame devices. The Gaia telescope and camera were launched on December 19, 2013. The Gaia spacecraft is parked at the Earth–Sun L2 Lagrange point, which is a spot 1.5 million km (932,057 miles) behind the earth, when viewed from the sun. Gaia is operating at a temperature of ~115°C. This low temperature will be maintained by passive thermal control, including the cold radiator on the focal plane assembly and a giant sunshade attached to the top of the spacecraft.

Visible hybrids have also been built for special applications to take advantage of the materials flexibility and larger surface area with nearly 100%

Table 4.4 Large-format readout ICs

	FLIR						RVS				
	ISC9803	ISC002	ISC9901	ISC0402	ISC0403	ISC0404	Aladdin	Orion	Virgo	Phoenix	Aquarius
Format	640×512	640×512	640×512	640×512	640×512	1024×1024	1024×1024	1024×1024, 2048×2048	1024×1024, 2048×2048	1024×1024, 2048×2048	1024×1024
Pixel size (μm)	25	25	20	20	15	18	27	25	20	25	30
ROIC type	DI	CTIA	DI	DI	DI	DI	SFD	SFD	SFD	SFD	SFD
Operating temperature (K)	80–310	80–310	80–310	80	80	80	10–30	30	77	10–30	4–10
Integrated capacity (e⁻)	1.1×10^7	2.5×10^6	7×10^6	1.1×10^7	6.5×10^6	1.2×10^7	2.0×10^5	3.0×10^5	$>3.5\times10^5$	3×10^5	1 or 15×10^6
ROIC noise (e⁻)	≤550	≤360	≤350	≤1279	≤760	≤1026	10–50	<20	<20	6–20	<1000
Full frame rates (Hz)	30	30	30	>30	>30	>30					
Number of outputs	1, 2 or 4	1, 2 or 4	1, 2 or 4	1, 2 or 4	1, 2 or 4	4, 8 or 16	32	64	4 or 16	4	16 or 64
Packaging	LCC	LCC	LCC	LCC	LCC	LCC	LCC	Module – 2 side buttable	Module – 3 side buttable	LCC	Module – 2 side buttable
Detector	p-on-n	p-on-n	p-on-n	p-on-n	p-on-n	p-on-n					
Compatible detectors	InSb or QWIP	InGaAs or HgCdTe	InSb or QWIP	InSb, InGaAs, HgCdTe or QWIP	InSb	InSb	InSb, HgCdTe, or impurity-blocked-conduction (IBC)	InSb	HgCdTe	InSb or IBC	IBC

LCC, leadless chip carrier.

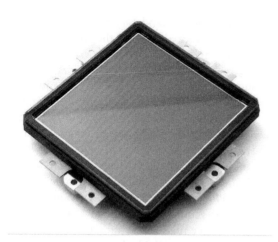

Figure 4.55 Teledyne Dalsa 252 mega-pixel CCD-array with 17,216×14,656 pixels (96.4×82.1 mm) with 5.6-μm size. (Reproduced from http://www.ipi.uni-hannover.de/uploads/tx_tkpublikationen/2011_GISOSTRAVA_KJ.pdf.)

FF. Si $p-i-n$ detector arrays for the astronomy and civil space communities in hybrid configuration with size as large as 4096×4096 have been demonstrated (see Figure 4.57). This design is scalable to an array format up to 16K×16K.

Although efforts have been made to develop monolithic structures using a variety of IR detector materials (including narrow-gap semiconductors) over the past 40 years, only a few have matured to a level of practical use. These included PtSi, and more recently PbS, PbTe, and uncooled silicon microbolometers. Other IR material systems (InGaAs, InSb, HgCdTe, InAs/GaSb SL, GaAs/AlGaAs QWIP, and extrinsic silicon) are used in hybrid configurations.

Pixel sizes as small as 10 μm have been demonstrated in hybrid systems. A general trend has been to reduce pixel sizes, and this trend is expected to continue. Systems operating at sorter wavelengths are more likely to benefit from small pixel sizes because of the smaller diffraction-limited spot size. Diffraction-limited optics with low f-numbers (e.g., $f/1$) could benefit from pixels on the order of one wavelength across about 10 μm in the LWIR. Over sampling the diffractive spot may provide some additional resolution for smaller pixels, but this saturates quickly as the pixel size is decreased.

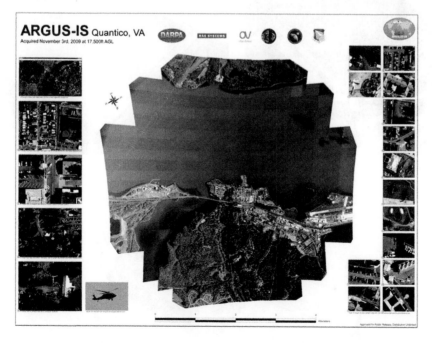

Figure 4.56 Sample of ARGUS-IS imagery. Mounted under a YEH-60B helicopter at 17,500 ft. over Quantico, VA., ARGUS-IS images an area more than 4 km wide and provides multiple 640×480-pixel real-time video windows. (Reproduced from *Seeing Photons: Progress and Limits of Visible and Infared Sensor Arrays*, Committee on Developments in Detector Technologies; National Research Council, 2010, http://www.nap.edu/catalog/12896.html.)

Figure 4.57 Teledyne's hybrid silicon *p–i–n* CMOS sensor (4K×4K H1RG-10 HyViSi) with 10 μm pixel pitch. (Reproduced after Bai, Y., Bajaj, J., Beletic, J.W. and Farris, M.C., Teledyne imaging sensors: Silicon CMOS imaging technologies for X-ray, UV, visible and NIR. *Proc. SPIE*, 7021, 702102, 2008.)

Pixel reduction is mandatory also to cost reduction of a system (reduction of the optics diameter, dewar size and weight, together with the power and increase the reliability).

Short Wave Infrared (SWIR), MWIR, and Long Wave Infrared (LWIR) electronically scanned HgCdTe arrays with CMOS multiplexer are commercially available from several manufactures. Most manufacturers produce their own multiplexer designs because these often have to be tailored to the applications. Figure 4.58 shows an example of large HgCdTe FPAs. Because the size of individual arrays continues to grow, the very large FPAs are required for many space missions by mosaic assembly of a large number of individual arrays. An example of a large mosaic developed by Teledyne Imaging Sensors is a 147 megapixel FPA that is comprised of 35 arrays, each with 2048×2048 pixels.

State-of-the-art QWIP and HgCdTe FPAs provides similar performance figure of merit

Figure 4.58 Large HgCdTe FPAs: (a) a mosaic of four Hawaii-2RG-18s (4096 × 4096 pixels, 18 μm pitch) and (b) Hawaii-4RG-10 (4096 × 4096 pixels, 10 μm pitch) as is being used for astronomy observations and (c) 16 2048×2048 HgCdTe arrays assembled for the VISTA telescope.

because they are predominantly limited by the readout circuits. The very short integration time of LWIR HgCdTe devices of typically below 300 μs is very useful to freeze a scene with rapidly moving objects. The integration time of QWIP devices must be 10–100 times longer for that, and typically it is 5–20 ms.

The BIB devices, in large staring array formats, are also now becoming commercially available. Impressive progress has been achieved especially in Si:As BIB array technology with formats as large as 2048×2048 and pixels as small as 18 μm, operated in spectral band up to 30 μm at about 10 K. The pixel size of 18 μm is smaller than the wavelength at Q band (17–24 μm), however, this does not pose a problem because an imager operating at these wavelengths will typically spread the beam out over many pixels to be fully sampled.

Multicolor detector capabilities are highly desirable for advanced IR imaging systems because they provide enhanced target discrimination and identification, combined with lower false-alarm rates. Systems that collect data in separate IR spectral bands can discriminate both absolute temperature and unique signatures of objects in the scene. These multicolor systems, now being developed, belong to the so-called third-generation systems.

The unit cell of integrated multicolor FPAs consists of several colocated detectors, each sensitive to a different spectral band (see Figure 4.59). Radiation is incident on the shorter band detector, with the longer wave radiation passing through to the next detector. Each layer absorbs radiation up to its cutoff, and hence transparent to the longer wavelengths, which are then collected in subsequent layers. In the case of HgCdTe, this device

architecture is realized by placing a longer wavelength HgCdTe photodiode optically behind a shorter wavelength photodiode.

In this class of systems, three detector technologies are developed: HgCdTe, QWIPs, and antimonide-based type-II SLs. Two-color FPAs are fabricated from multilayer materials using both sequential mode or simultaneous mode operations. The simplest two-color HgCdTe detector, and the first to be demonstrated, was the bias selectable $n–P–N$ triple-layer heterojunction, back-to-back photodiode. Many applications require true simultaneous detection in the two spectral bands. This has been achieved in a number of ingenious architectures considered in Reference. [33].

Large two-color FPAs are fabricated by Raytheon, Sofradir, and Selex. RVS has developed two-color, 1280×720 large format MWIR/LWIR FPAs with 20×20 μm unit cells (see Figure 4.60). The ROICs share a common chip architecture and incorporate identical unit cell circuit designs and layouts; both FPAs can operate in either dual-band or single-band modes. Excellent high-resolution IR camera imaging with $f/2.8$ FOV broadband refractive optics at 60 Hz frame rate has been achieved.

Recently, type-II InAs/GaInSb SLs have emerged as a candidate for third-generation IR detectors. Over the past few years, type-II SL-based detectors have been also made rapid progress in fabrication of dual-band FPAs. As an example, the excellent imagery delivered by the 288×384 InAs/GaSb simultaneously operated dual-color camera is presented in Figure 4.61. The image is a superposition of the images of the two channels coded in the complimentary colors cyan and red for the detection ranges of 3–4 μm and 4–5 μm,

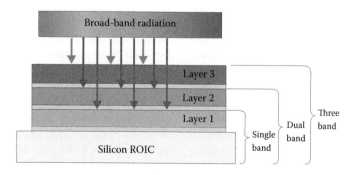

Figure 4.59 Structure of a three-color detector pixel. Infrared flux from the first band is absorbed in Layer 3, whereas longer wavelength flux is transmitted through the next layers. The thin barriers separate the absorbing bands.

Figure 4.60 Dual-band megapixel MW/LW FPAs: (a) RVS 1280 × 720 format HgCdTe FPAs mounted on dewar platforms and (b) JPL 1024 × 1024 format QWIP FPA mounted on a 124-pin lead less chip carrier. (Reproduced after King, D.F., Graham, J.S., Kennedy, A.M., Mullins, R.N., McQuitty, J.C., Radford, W.A., Kostrzewa, T.J., Patten, E.A., Mc Ewan, T.F., Vodicka, J.G. and Wootana, J.J., 3rd-Generation MW/LWIR sensor engine for advanced tactical systems. *Proc. SPIE*, 6940, 69402R, 2008 and Gunapala, S., Bandara, S.V., Liu, J.K., Mumolo, J.M., Ting, D.Z., Hill, C.J., Nguyen, J., Simolon, B., Woolaway, J., Wang, S.C., Li, W., LeVan, P.D. and Tidrow, M.Z., Demonstration of megapixel dual-band QWIP focal plane arrays. *IEEE J. Quantum. Electron.*, 46, 285–293, 2010.)

Figure 4.61 Bispectral infrared image of an industrial site taken with a 384×288 dual-color InAs/GaSb SL camera. The two-color channels 3–4 and 4–5 μm are represented by the complementary colors cyan and red, respectively. (Reproduced from Rehm, R., Walther, M., Schmitz, J., Rutz, F., Wörl, A., Scheibner, R. and Ziegler, J., Type-II superlattices: The Fraunhofer perspective. *Proc. SPIE*, 7660, 76601G, 2010.)

respectively. The red signatures reveal hot CO_2 emissions in the scene, whereas water vapor, e.g., from steam exhausts or in clouds, appears cyan due to the frequency dependency of the Rayleigh scattering coefficient.

Initially developed for the military market by US defense companies, IR uncooled cameras are now widely used in many commercial applications. Currently, the microbolometer detectors are produced in larger volumes than all other IR array technologies together. Their cost will be drastically dropped (about 15% per year). It is expected that

commercial applications in surveillance, automotive, and thermography will reach total volumes more than 1.1 million units in 2016 ($3.4 B in value).

Development of 17 μm pixel pitch FPAs is being extended to both smaller arrays (320×240) and arrays larger than 3 megapixel. Thermal image obtained with 1024×768 a-Si microbolometer detector shows both high sensitivity and resolution as shown in Figure 4.62. This device can detect temperature variations smaller than 50 mK.

Currently, the largest microbolometer array fabricated by Raytheon is shown on a wafer in

(a)

(b)

Figure 4.62 Ulis 17-µm pitch 1024×768 FPA: (a) packaging and (b) thermal image. (Reproduced from http://www.sofradir-ec.com/ wp-uncooled-detectors-achieve.asp.)

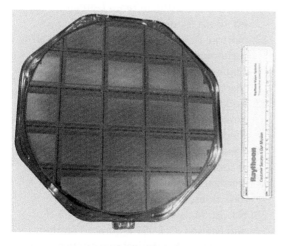

Figure 4.63 2048×1536 uncooled VO$_x$ microbolometers with 17 µm pixel pitch on a 200 mm wafer. (Reproduced after Black, S.H., Sessler, T., Gordon, E., Kraft, R., Kocian, T., Lamb, M., Williams, R. and Yang, T., Uncooled detector development at Raytheon. *Proc. SPIE*, 8012, 80121A, 2011.)

Figure 4.63. In the fabrication of 2048×1536 staring arrays and associated ROIC circuits, a stitching technique has been used. Each 200 mm wafer contains nine-2048×1536 uncooled detector die, which represents an 80% increase in yield over an equivalent 150 mm wafer.

The detection of far-IR and THz radiation is resistant to the commonly employed techniques in the neighboring microwave and IR frequency bands. In THz detection, the use of solid-state detectors has been hampered for the reasons of transit time of charge carriers being larger than the time of one oscillation period of THz radiation. Also the energy of radiation quanta is substantially smaller than the thermal energy at room temperature and even liquid nitrogen temperature.

Particular attention in development of THz imaging systems is devoted to the realization of sensors with a large potential for real-time imaging while maintaining a high dynamic range and room temperature operation. CMOS process technology is especially attractive due to their low price tag for industrial, surveillance, scientific, and medical applications. However, CMOS THz imagers developed thus far have mainly operated single detectors based on lock-in technique to acquire raster-scanned imagers with frame rates on the order of minutes. With this mind, much of recent developments are directed toward three types of focal plane sensors [39]:

- Schottky barrier diodes compatible with CMOS process
- FETs relay on plasmonic rectification phenomena
- Adaptation of IR bolometers to the THz frequency range

4.7 CONCLUSIONS

This chapter provides an overview of the important techniques for detection of optical radiation from the ultraviolet through visible to IR and far-IR spectral regions. In the beginning, single-point devices are considered; next, direct detector systems and advanced techniques, including coherent detection, and finally, image counterparts containing FPAs are considered. The reader should be able to gain a good understanding of the similarities and contrasts, the strengths and weaknesses of the great number of approaches that have been developed over a century of effort to improve our ability to sense photons. The emphasis is always upon the methods of operation and limitations of different techniques. In addition, currently achieved performance levels are also briefly described.

This chapter offers a rather wide coverage of detection techniques. However, for a full understanding of the technical content, basic courses in electronic devices and circuits and the very fundamentals of semiconductors and noise are a prerequisite.

REFERENCES

1. Donati, S. 1999. *Photodetectors. Devices, Circuits, and Applications* (Upper Saddle River: Prentice Hall Inc.)
2. Motchenbacher, C.D. and Connelly, J.A. 1995. *Low-Noise Electronic System Design* (New York: John Wiley & Sons, Inc.)
3. Vampola, J.L. 1999. Readout electronics for infrared sensors. In *The Infrared and Electro-Optical Systems Handbook*, vol. 3, ed. W.D. Rogatto (Bellingham: SPIE Press), pp. 285–342.
4. Bielecki, Z., Kołosowski, W., Dufrene, R. and Borejko, M. 2003. Proceedings of 11th European Gallium Arsenide and other Compound Semiconductors Application Symposium, Munich 137–140.
5. Rieke, G.H. 1994. *Detection of Light: From Ultraviolet to the Submillimeter* (Cambridge: Cambridge University Press).
6. Gallivanoni, A., Rech, I. and Ghioni, M. 2010. Progress in quenching circuits for single photon avalanche diodes. *IEEE Trans. Nucl. Sci.*, 57, 3815–3826.
7. Mosconi, D., Stoppa, D., Pancheri, L., Gonzo, L. and Simoni, A. 2006. CMOS single-photon avalanche diode array for time-resolved fluorescence detection. *Solid-State Circuits Conference, ESSCIRC 2006 Proceedings of the 32nd European*, September 2006, pp. 564–567.
8. Bielecki, Z. 2002. Maximisation of signal to noise ratio in infrared radiation receivers. *Opto-Electron. Rev.*, 10, 209–216.
9. Amin, F.U. 2009. *On the Design of an Analog Front-End for an X-Ray Detector* (Linköping: Lamberd Academic Publishing).
10. http://opticb.uoregon.edu.
11. http://www.lasercomponents.de.
12. Lins, B., Zinn, P., Engelbrecht, R. and Schmauss, B. 2010. Simulation-based comparison of noise effects in wavelength modulation spectroscopy and direct absorption TDLAS, *Appl. Phys. B*, 100, 367–376.
13. Skolnik, M.I. 1990. *Radar Handbook* (New York: Mc. Graw-Hill).
14. Jakobs, S.F. 1988. Optical heterodyne (coherent) detection. *Am. J. Phys.*, 56, 235–245.
15. Keiser, G. 1979. *Optical Fiber Communications* (New York: McGraw-Hill).
16. Barry, J.R. and Lee, E.A. 1990. Performance of coherent optical receivers. *Proc. IEEE*, 78, 1369–1990.
17. Brun, R. 1990. Gallium arsenide eyesafe laser rangefinder. *Proc. SPIE*, 1207, 172–181.
18. Bielecki, Z. and Rogalski, A. 2001. *Detection of Optical Radiation* (Warsaw: WNT) (in Polish).
19. Rogalski, A. 2003. Infrared detectors: Status and trends. *Prog. Quantum Electron.*, 27, 59–210.
20. Rogalski, A. 2003. Photon detectors. In *Encyclopedia of Optical Engineering*, ed. R. Driggers (New York: Marcel Dekker Inc.), pp. 1985–2036.
21. Rogalski, A. 2010. *Infrared Detectors*, 2nd ed. (Boca Raton: CRC Press).
22. Norton, P. 2003. Detector focal plane array technology. In *Encyclopedia of Optical Engineering*, ed. R. Driggers (New York: Marcel Dekker Inc.). pp. 320–348.
23. Rogalski, A. 2012. Progress in focal plane arrays technologies. *Prog. Quantum Electron.*, 36, 342–473.
24. Fossum, E.R. and Pain, B. 1993. Infrared readout electronics for space science sensors: State of the art and future directions. *Proc SPIE*, 2020, 262–285.
25. Hewitt, M.J., Vampola, J.L., Black, S.H. and Nielsen, C.J. 1994. Infrared readout electronics: A historical perspective. *Proc. SPIE*, 2226, 108–119.
26. Fossum, E.R. 1993. Active pixel sensors: Are CCD's dinosaurs? *Proc. SPIE*, 1900, 2–14.
27. Hoffman, A., Loose, M. and Suntharalingam, V. 2005. CMOS detector technology. *Exp. Astron.*, 19, 111–134.

28. http://www.ipi.uni-hannover.de/uploads/tx_tkpublikationen/2011_GISOSTRAVA_KJ.pdf.

29. Hirayama, T. 2013. The evolution of CMOS image sensors. *IEEE Asian Solid-State Circuits Conference*, Singapore, pp. 5–8.

30. *Seeing Photons: Progress and Limits of Visible and Infared Sensor Arrays*, Committee on Developments in Detector Technologies; National Research Council, 2010, http://www.nap.edu/catalog/12896.html.

31. Bai, Y., Bajaj, J., Beletic, J.W. and Farris, M.C. (2008). Teledyne imaging sensors: Silicon CMOS imaging technologies for X-ray, UV, visible and near infrared. *Proc. SPIE*, 7021, 702102.

32. Beletic, J.W., Blank, R., Gulbransen, D., Lee, D., Loose, M., Piquette, E.C., Sprafke, T., Tennant, W.E., Zandian, M. and Zino, J. (2008). Teledyne imaging sensors: Infrared imaging technologies for astronomy and civil space. *Proc. SPIE*, 7021, 70210H.

33. Rogalski, A., Antoszewski, J. and Faraone, L. (2009). Third-generation infrared photodetector arrays. *J. Appl. Phys.*, 105, 09110.

34. King, D.F., Graham, J.S., Kennedy, A.M., Mullins, R.N., McQuitty, J.C., Radford, W.A., Kostrzewa, T.J., Patten, E.A., Mc Ewan, T.F., Vodicka, J.G. and Wootana, J.J. (2008). 3rd-Generation MW/LWIR sensor engine for advanced tactical systems. *Proc. SPIE*, 6940, 69402R.

35. Gunapala, S., Bandara, S.V., Liu, J.K., Mumolo, J.M., Ting, D.Z., Hill, C.J., Nguyen, J., Simolon, B., Woolaway, J., Wang, S.C., Li, W., LeVan, P.D. and Tidrow, M.Z. (2010). Demonstration of megapixel dual-band QWIP focal plane arrays. *IEEE J. Quantum. Electron.*, 46, 285–293.

36. Rehm, R., Walther, M., Schmitz, J., Rutz, F., Wörl, A., Scheibner, R. and Ziegler, J. (2010). Type-II superlattices: The Fraunhofer perspective. *Proc. SPIE*, 7660, 76601G.

37. http://www.sofradir-ec.com/wp-uncooled-detectors-achieve.asp.

38. Black, S.H., Sessler, T., Gordon, E., Kraft, R., Kocian, T., Lamb, M., Williams, R. and Yang, T. (2011). Uncooled detector development at Raytheon. *Proc. SPIE*, 8012, 80121A.

39. Rogalski, A. (2013). Far-infrared semiconductor detectors and focal plane arrays. *Opto-Elecron. Rev.*, 21, 406–426.

5

Propagation along optical fibers and waveguides

JOHN LOVE
Australian National University

In this part of the *Handbook*, a description of the transmission of light along dielectric optical fibers and waveguides is presented in terms of a ray analysis for multimode propagation and in terms of a modal analysis for single- and few-mode propagation. While the emphasis will be on fiber propagation, the methods presented here are applicable to any type of waveguide, irrespective of material composition. To cater to a wider range of backgrounds in optics and electromagnetism, there is a basic introduction leading into more advanced topics. Accordingly, some readers may wish to skip the earlier Sections 5.1 and 5.2. This part is relatively self-contained and explanatory, and referencing to other material has been kept to a minimum.

5.1 HISTORICAL PERSPECTIVE

5.1.1 Light propagation

Light, as we know, is what we see and comes in a range of colors or, equivalently, wavelengths or frequencies that our eyes detect and our brain interpolates. Whatever we are looking at, whether it is this page or the distant stars, light travels in straight lines from the object to our eyes, regardless of the wavelength. This phenomenon occurs because light is a form of electromagnetic radiation and its propagation when considering waves in a uniform dielectric medium, such as air or vacuum,

is described by Maxwell's equations. Sometimes though, light can play tricks on us, such as the apparently sloping water level in the swimming pool shown in Figure 5.1.

While light travels in straight lines, a finite beam of light, whether it comes from a large-aperture torch or the narrow output face of a micron-size coherent semiconductor laser, tends to spread out with this effect being the more noticeable over short distances with the torch beam and over longer distances with the laser. If we combine this spread with the fundamental need for straight-line propagation, it is easy to see why it is not very convenient to use light in air to transmit information over long distances—it will not go around corners nor will it even follow the curvature of the earth. On top of this limitation, rain, fog, smoke, buildings, topographical features, and other obstacles

Figure 5.1 Swimming pool with an apparently sloping water surface.

ensure that the beam will be strongly attenuated because of absorption, reflection, and scattering.

These impediments notwithstanding, chains of semaphore repeater stations with large movable wooden arms on the tops of towers were built by a number of European countries in the eighteenth century on the peaks of hills within direct line of sight of one another, that is, about 20–25 km apart. These links provided the first purely optical, low-bandwidth, long-distance transmission systems between a country's capital and its ports, primarily for military and administrative applications.

5.1.2 Light pipes

However, to provide an optical transmission system with higher bandwidth, a more effective medium was needed that would be able to steer light flexibly to its destination and to insulate it from deleterious environmental effects. Following the invention of the laser in the 1950s, early attempts in the 1960s to solve this problem were based on the use of a long, evacuated light pipe, wherein a series of lenses periodically refocused the slowly diverging output beam from a laser located at the beginning of the pipe, as shown schematically in Figure 5.2. The lenses could also be used to steer the beam along new directions to follow the local topography.

While this technology offered high bandwidth and very low attenuation because light propagation would be predominantly in vacuum, the scheme relied on the maintenance of a good alignment over long distances, regardless of whether the pipe was supported above ground or buried beneath it. This presented a significant challenge because of ground movement due to natural and artificial phenomena. Thus, it was clear that a more flexible and less environmentally sensitive optical guide was needed.

5.1.3 Optical fibers

Optical fibers have been known since glass was first discovered several thousands of years ago, simply as a consequence of pulling on a piece of heat-softened glass, something that most of us have probably tried in a laboratory at school or university. However, any thoughts on the use of glass fibers for transmitting light for information purposes had to wait until the 20th century and the development of suitable light sources.

The potential for guiding light along a purely dielectric optical waveguide was first demonstrated in public in 1841 by Professor Daniel Colladon at the University of Geneva in Switzerland [1]. Colladon used a jet of water emerging from the side of a barrel and illuminated the jet by focusing sunlight onto it. Water has a higher refractive index of 1.33 compared to the surrounding air with an index of 1.0. Because Snell's laws predict total internal reflection of light from the water–air boundary, the relative indices ensure that sunlight is guided along the water jet. The jet of water curves downward because of gravity and the strong confining property of the jet determines that much of the light follows its curved path. Using more modern light sources, such as a colored diodes, the appearance of the effect can be enhanced. We now know that Colladon's experiment is equivalent to light propagation along a highly multimode, bent optical waveguide with water as the core and air as the cladding.

Glass fibers were developed intermittently over the next 100 years more as a curiosity than a technology solution. The first serious application of glass fibers to light transmission was motivated through medicine leading to the development of the endoscope in the 1950s with achievable transmission over one or two meters through a dense bundle of very thin fibers. However, there were two existing limitations that inhibited the use of the endoscope fibers for long-distance communications. First, early fibers

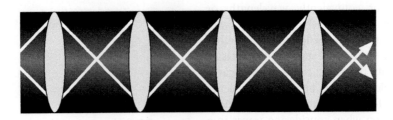

Figure 5.2 Schematic of a longitudinal section of a light pipe.

Figure 5.4 Terrestrial (left) and deep-ocean submarine (right) single-mode fiber cables used in long-distance optical communications systems.

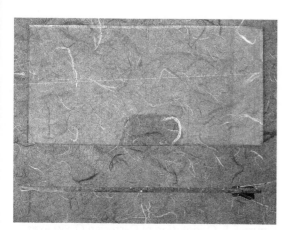

Figure 5.3 Top and side views of a sheet of ordinary window glass.

the goal of low-loss, single-mode fibers and cables with much higher bandwidths than multimode fibers was realized. Contemporary fiber cables for terrestrial and submarine applications are shown in Figure 5.4.

5.1.4 Scope

In this part of the *Handbook*, we develop the analysis of propagation along optical fibers and other waveguiding structures to produce a basic description of light transmission. The practical development of fibers was paralleled by major theoretical developments. In the 1960s and early 1970s, much effort was extended to produce increasingly sophisticated ray tracing and local plane-wave techniques that could adequately describe and quantify propagation and loss mechanisms in multimode fibers. Ray tracing is an accurate technique for multimode fibers because the relatively large core ensures that the diffraction effects associated with the relatively short but finite wavelengths used in communications are extremely small compared to the light-guiding effect based on the variation in the transverse refractive index profile. Conversely, the small core size associated with single-mode fibers requires a full electromagnetic analysis of propagation because diffraction effects become comparable with the confining effect of the profile (see Section 10.2 of Reference [3]).

The early ray tracing and electromagnetic analyses of fibers concentrated predominantly on analytical solutions of the governing ray tracing and Maxwell equations because of the limited capability of early computers. Initially, these solutions necessarily relied on refractive index profiles, together with fiber and waveguide cross-sectional geometries, for which analytical solutions of these equations were available in terms of simple or special mathematical functions. Commercial software routines were available for the quantification of

were of a single material and because light is distributed over the entire cross section of the material, it was susceptible to transmission loss wherever the single material touched external supports. Second, the glass was very impure and therefore, strongly absorbed and scattered light over longer distances. One can get an idea of the degree of light absorption by looking through a sheet of window glass sideways, as illustrated in Figure 5.3. Even a very low level of impurities in the glass, of the order of parts per million, results in a very high extinction level for the propagating light after only a few meters.

The transition from Colladon's water jet to the modern long-distance, low-loss optical communications system relied on (a) the development of two-layer, high–low-index glass optical fibers that not only transmit light but also confine it to within the central glass core well away from the outer surface and, more importantly, (b) the development of techniques for fabricating fibers with extremely low light loss from very pure materials.

The first limitation was overcome by using the now standard core-plus-cladding fiber to locate the propagating light around the fiber axis and isolate it from the fiber coating and other external influences. The second problem required the development of new fabrication techniques to produce the very pure silica-based glasses required for low-loss fibers. These goals were enshrined by the classic paper written by Kao and Hockham [2], published in 1966, and generally regarded as the catalyst for the ensuing optical telecommunications revolution.

The first such fibers that appeared in the 1960s were multimode and it was not until the 1970s that some of these special functions; otherwise, home-based software had to be developed.

Over the last 20 years, this situation has been almost completely reversed with the ready availability of commercially developed software, particularly that for determining ray tracing and electromagnetic propagation for a wide range of fiber and waveguide profiles and geometries. It is fair to say that this evolution has considerably simplified the development and understanding of specialized fiber and waveguide designs, such as holey fibers and photonic bandgap waveguides, by moving the emphasis to the exploitation of physical phenomena for light guidance rather than being limited by the shortcomings of analytical and numerical techniques.

5.2 LIGHT PROPAGATION, PLANE WAVES, AND RAYS

Before launching into the different descriptions of light propagation and guidance along fibers and waveguides, it is helpful and insightful to examine the propagation of electromagnetic plane waves in free space as a precursor to their development into rays and modes. Light rays, whether propagating in straight lines or along curved trajectories, are formally the solution of Maxwell's equations in the limit of zero wavelength; but for practical purposes, they can be thought of as local plane waves propagating with a small but finite wavelength. This section provides an elementary background to the more advanced material presented in Sections 5.3 and 5.5.

5.2.1 Plane waves

Consider an infinite, unbounded medium of uniform, real refractive index n that is lossless, that is, nonabsorbing and nonscattering. Introduce the triad of Cartesian axes O-xyz, shown in Figure 5.5, such that the z-axis defines the direction of propagation.

Assume an artificial, infinitely extended monochromatic (single-frequency) uniform light source in the xy plane with angular frequency ω or, equivalently, wavelength $\lambda = 2\pi c/\omega$, where c is the speed of light in vacuum. An electromagnetic plane wave can propagate parallel to the z-axis with vector electric E and magnetic H fields that are everywhere uniform and have only a sinusoidal dependence on distance z and time t. Accordingly, a simple solution of Maxwell's equations predict that these fields have the forms:

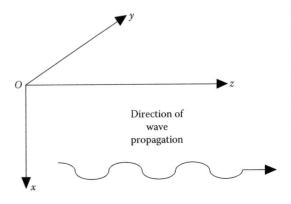

Figure 5.5 Orientation of Cartesian O-xyz axes.

$$E(x,y,z,t) = e \exp\{i(knz - \omega t)\};$$
$$H(x,y,z,t) = h \exp\{i(knz - \omega t)\}, \tag{5.1}$$

where $k = 2\pi/\lambda$ is the wave number, kn is the propagation constant, and e and h are *constant* orthogonal vectors. Note that the propagation constant is a continuous function of the source wavelength and decreases as the wavelength increases.

These fields propagate parallel to the z-direction with a *phase velocity* $v_{ph} = \omega/kn = c/n$, that is, the speed of light in a medium of refractive index n, which is a constant *independent* of the source wavelength or frequency, provided the medium itself is not dispersive. The forward direction of propagation corresponds to the negative sign in the exponent of Equation 5.1. Furthermore, the power in the plane wave propagates at the *group velocity*, $v_g = d\omega/dk = \omega/kn = c/n$, that is, identical to the group velocity for a plane wave.

A plane wave is an unphysical entity because it has constant amplitude everywhere and therefore, the total power propagating in the mode is infinite. Nevertheless, it provides the simplest example of light propagation in a uniform medium. It will be seen in Section 5.3 that the introduction of a core into the uniform medium model modifies the nature of plane-wave propagation in a constructive manner.

5.2.2 Polarization

The two constant vectors e and h each have only one *non-zero* Cartesian field component transverse to the z-direction of propagation. These components lie in the xy plane.

(a)

(b)

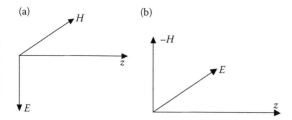

Figure 5.6 Polarized plane wave electromagnetic fields in the (a) x- and (b) y-direction.

Relative to the Cartesian axes, these vectors can be chosen to have either of the following two orthogonal forms:

$$e = (e,0,0) \, h = (0,h,0) \text{ or}$$
$$e = (0,e,0) \, h = (-h,0,0), \tag{5.2}$$

where e and h are scalar constants. Note that the second choice of components is simply the first choice rotated 90° about the z-axis. Either of these two solutions consists of electric and magnetic fields that are orthogonal to the z-direction of propagation and are also orthogonal to one another. Accordingly, they form *orthogonal triads* as shown in Figure 5.6.

The direction of the electric field vector E in each case defines the *polarization* of the plane wave, that is, it is either x-polarized or y-polarized, respectively. The description of the plane wave for either polarization can be thought of as a triad of vectors (E, H, and the z-axis or direction of propagation) propagating parallel to the z-axis at a speed equal to the phase velocity v_{ph}.

5.2.3 Local plane waves, rays, and waves

Plane waves are associated with an infinite medium of uniform index n; they travel in straight lines and have the same electromagnetic description everywhere. However, if the refractive index of the medium varies with position so that it is *graded*, this description is modified. The plane wave can be replaced by the concept of the *local plane wave*, which describes propagation in a small region of space in terms of a wave whose fields, phase and group velocities, and polarization are determined by the local value and variation of the index.

When propagation is described in terms of ray tracing for the multimode waveguides and fibers in Section 5.5, there is a simple relationship between rays and waves. A ray represents the local direction of propagation of a plane wave in a uniform medium or a local plane wave in a graded medium. Further, the ray direction also determines the direction of local power flow. Although this propagation description is based on the zero-wavelength limit of Maxwell's equations, it provides an accurate description for the relatively short wavelengths encountered in optical communications.

5.3 ELECTROMAGNETIC PROPAGATION IN NONUNIFORM DIELECTRIC MEDIA

5.3.1 Maxwell's equations and monochromatic sources

Propagation in dielectric media is governed by Maxwell's equations (a portrait of Maxwell is given in Figure 5.7) so that in the absence of currents and charges, the equations take the four-dimensional spatial–temporal forms:

Figure 5.7 James Clerk Maxwell, 1831–1879.

$$\nabla \times E = \frac{\partial B}{\partial t}; \quad \nabla \times H = -\frac{\partial D}{\partial t}, \qquad (5.3)$$

where $E = E(x, y, z, t)$ is the vector electric field and $H = H(x, y, z, t)$ is the vector magnetic field. The temporal dependence is denoted by t and the spatial dependence can be expressed in terms of any convenient orthogonal coordinate system, for example, Cartesian coordinates (x, y, z) for slab waveguides, cylindrical polar coordinates (r, ϕ, z) for circular fibers, and so on. The magnetic induction vector B and the displacement vector D are related to the magnetic and electric fields, respectively, by

$$B = \mu H; \quad D = \varepsilon E = \varepsilon_0 n^2 E, \qquad (5.4)$$

where μ is the magnetic permeability, $\varepsilon = \varepsilon(x, y, z)$ is the dielectric constant, and $n = n(x, y, z)$ is the refractive index distribution. For optical materials, μ normally takes its free-space value μ_0, and ε_0 is the free-space dielectric constant. For fibers and waveguides, the refractive index is normally assumed to be uniform (z-independent) along the fiber or waveguide and is also taken to be independent of the field amplitude, that is, ignoring nonlinear material effects as well as the source wavelength when ignoring material dispersion. Material dispersion is addressed in Section 5.6.4 and nonuniform, z-dependent propagation is discussed in Section 5.9.3.

5.3.1.1 MONOCHROMATIC SOURCES

In modeling the excitation from lasers or diodes when used as sources for fibers and waveguides, it is initially assumed that their output is exactly sinusoidal and monochromatic with a fixed wavelength. Practical sources, however, have an output with a finite but small *spectral width*. For lasers, this width is typically of the order of a nanometers or less. The effect of this width is of paramount importance when considering pulse dispersion (see Section 5.6).

Accordingly, we ascribe an angular frequency ω to the monochromatic source such that each component of the electromagnetic field is assumed to contain the implicit sinusoidal dependence $\exp(-i\omega t)$. The choice of sign is arbitrary, but here

is taken to be negative for convenience. If λ denotes the corresponding free-space wavelength and k is the free-space wave number, then these quantities are related by the expressions

$$k = \frac{2\pi}{\lambda}; \quad \lambda = \frac{2\pi c}{\omega}; \quad c = \frac{\omega}{k}, \qquad (5.5)$$

where c is the free-space speed of light (3×10^8 ms^{-1}). Using Equations 5.3 through 5.5, the monochromatic time dependence enables us to recast Maxwell's equations so that the spatial dependence is governed by

$$\nabla \times E = i\omega\mu_0 H = i\left(\frac{\mu_0}{\varepsilon_0}\right) kH;$$
$$\qquad (5.6)$$
$$\nabla \times H = i\omega\varepsilon_0 n^2 E = -i\left(\frac{\varepsilon_0}{\mu_0}\right)^{1/2} kn^2 E,$$

where $E = E(x, y, z)$ and $H = H(x, y, z)$ and implicitly contain the time dependence $\exp(-i\omega t)$.

5.3.2 Translational invariance, longitudinal, and transverse fields

For modeling fibers and waveguides, it is usual to assume that (1) the refractive index profile and the cross-sectional geometry do not vary with longitudinal distance z and (2) the fiber or waveguide is straight and essentially infinitely long. In this situation, the fibers and waveguides have *translational invariance* and hence, the z-dependence in Maxwell's equations becomes separable from the transverse dependence. Accordingly, we may set

$$E(x, y, z) = e(x, y)e^{i\beta z};$$
$$H(x, y, z) = h(x, y)e^{i\beta z}, \qquad (5.7)$$

where $e(x, y)$ and $h(x, y)$ are vector expressions that denote the transverse field dependence and the parameter β in the exponential or phase term is called the *propagation constant*. The choice of Cartesian (x, y) coordinates is appropriate for analyzing slab waveguides, while polar coordinates (r, ϕ) are more appropriate for fibers.

It is sometimes convenient for analytical and other purposes to split the fields into longitudinal and transverse components. This is equivalent to decomposing the vector dependences of e and h into transverse (subscript "t") and longitudinal (subscript "z") components, respectively, perpendicular and parallel to the z-axis, according to

$$
\begin{aligned}
e(x,y) &= e_t(x,y) + e_z(x,y)\hat{z}; h(x,y) \\
&= h_t(x,y) + h_z(x,y)\hat{z},
\end{aligned}
\tag{5.8}
$$

where e_t and h_t are vector quantities, e_z and h_z are scalar quantities, and z is the unit vector parallel to the fiber axis.

5.3.3 Power density and flow

For a monochromatic source with a sinusoidal time variation, the power flow density and direction at any position in space is determined by the time-averaged Poynting vector S

$$
S = \frac{1}{2}\mathrm{Re}(E \times H^*),
\tag{5.9}
$$

where $*$ denotes the complex conjugate, x, the vector cross product, and Re is the real part. Lossless propagation of light in fibers and waveguides is equivalent to the direction of S anywhere in the cross section being parallel to the z-axis, that is, S had only a z-component S_z. This property can be readily verified for the specific examples considered in Section 5.4. Finally, the total guided power flow is determined by integrating S_z over the infinite cross section of the fiber or waveguide.

5.3.4 Boundary conditions

For dielectric fibers and waveguides in the absence of currents, the boundary conditions for any bound solution of Maxwell's equations between regions of different index can be stated as follows:

- Continuity of all three components of the magnetic field at any interface
- Continuity of the two *tangential* components of the electric field at any interface

- Continuity of the *normal* component of the displacement vector at any interface
- The electric and magnetic fields decrease exponentially to zero at infinite distance from the fiber or waveguide axis

In the case of an interface between one region of uniform index and a second region of varying index, where the index values are equal on the interface, all six components of the electric and magnetic fields will be continuous across the interface.

5.3.5 Electromagnetic normal modes

Starting with an unbounded uniform index medium, consider an infinitely long uniform slab or cylinder of material of higher but uniform *core* refractive index n_{co} that is introduced into the infinite medium parallel to the z-axis. If the surrounding infinite medium is now referred to as the *cladding* with index $n = n_{cl} < n_{co}$, then the core and cladding constitute a dielectric slab waveguide or fiber with the step refractive index profile cross section shown in Figure 5.8. Although practical fibers or waveguides normally have more complex geometrical and profile structures in the cross section, the step profile is the simplest profile to analyze that is of some practical interest and is used in single-mode telecommunications fibers such as SMF28.

Propagation along a fiber or waveguide is governed formally by Maxwell's equations, but some of the salient physical features can be elucidated by considering propagation of electromagnetic waves along this structure parallel to the z-axis.

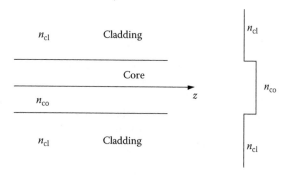

Figure 5.8 Step refractive index profile for a fiber or waveguide.

The simple plane-wave description of propagation in the uniform medium discussed in Section 5.2 provides a conceptual base, but is now modified in a number of ways because of the presence of the waveguide structure.

1. The uniformity of the structure in the z-direction or *translational invariance* ensures that waves propagate with a periodic longitudinal and temporal dependence and therefore, a well-defined phase velocity v_{ph}.
2. The value of the propagation constant is dependent on both the source wavelength λ and the length parameters of the core geometry and the refractive index values.
3. The propagation constant is no longer uniquely specified and can take *one or more discrete values* for a given set of values of all the earlier mentioned parameters.
4. The transverse dependence of the electromagnetic field vectors e and h for each discrete value of the propagation constant is now *spatially dependent* and varies with position (x, y) in the cross section.
5. The transverse electric and magnetic fields are no longer uniform, but are concentrated within and close to the core and decrease exponentially to zero at infinite distance from the axis.
6. The *infinite* power traveling in the infinite cross section of the plane wave is replaced by the *finite* power propagating along the waveguide in each propagation state.
7. Each discrete propagation state is known as a *bound mode* or *normal mode* and is a solution (or eigenfunction) of the *electromagnetic boundary value* problem for Maxwell's equations for the waveguide or fiber.

5.3.5.1 MECHANICAL ANALOG

Normal modes are two-dimensional electromagnetic vibrations in the cross section of the waveguide or fiber and are analogous to the mechanical vibrations of a flexible membrane fixed along its periphery, for example, a drum. Given the source frequency, each electromagnetic vibration state corresponds to a discrete value of the propagation constant β. Each value of β corresponds to a normal or bound mode where its total longitudinal power flow (parallel to the z-axis) is constant and

where its electromagnetic fields decrease exponentially with increasing distance from the z-axis and vanish at infinity.

The values of β are the discrete solutions of an *eigenvalue equation*, derived from solutions of Maxwell's equations together with appropriate boundary conditions. If the fiber core radius ρ or the waveguide cross-section 2ρ is small enough, typically a few microns for a silica-based fiber or waveguide, then as will be shown in Sections 5.4.4 and 5.4.5, there is only one solution for β, that is, a *single-mode* waveguide or fiber.

5.3.6 Mode orthogonality, normalization, and orthonormal modes

Each bound mode is orthogonal to all other bound modes on a uniform, straight waveguide. This means that if only one mode is excited, it cannot excite any other modes as it propagates. Mathematically, *orthogonality* between the jth and kth bound modes is expressed by the vanishing of the integral of a triple scalar product of their vector fields over the infinite cross section of the fiber or waveguide

$$\int_{A_\infty} e_j \times h_k^* \cdot \hat{z} \, dA = 0, \qquad (5.10)$$

where A_∞ is the infinite cross section, ž is the unit vector parallel to the axis, and * denotes the complex conjugate.

If we assume that the refractive index profile $n(x, y)$ is independent of field intensity, as is the case for *linear materials*, then Maxwell's equations constitute, in general, a set of coupled linear equations. The solutions of these equations together with the boundary conditions can only determine the fields for each mode to within an *arbitrary amplitude constant*. The value of this constant for each mode is normally determined from the source of excitation as discussed in Section 5.8.3.

For some special applications, such as in tapers, it is useful to be able to specify the value of this constant in an unambiguous manner independent of the source of excitation. This can be achieved by using the *normalization N* of a mode. This scalar quantity

$$N = \int_{A_\infty} e \times h^* \cdot \hat{z} \, dA \qquad (5.11)$$

is defined in terms of the integral over the infinite cross section of the triple scalar product of the electric field e, magnetic field h, and the unit vector parallel to the z-axis.

An *orthonormal mode* is then defined to be a normal mode with electric and magnetic field amplitudes such that it has unit normalization, that is, $N=1$. If the fields e and h of the bound mode are replaced by the following quantities:

$$\hat{e} = \frac{e}{N^{1/2}}; \quad \hat{h} = \frac{h}{N^{1/2}}, \qquad (5.12)$$

then substitution into Equation 5.11 gives $N=1$ and the fields are orthonormal.

5.3.7 Modal phase and group velocities

The longitudinal and temporal dependence of the fields of a mode is contained in the common phase factor $\exp[i(\omega t - \beta z)]$. A constant value of this expression defines a *phase front*, which is a plane of constant phase orthogonal to the z-axis. For a constant phase value, z varies linearly with t, the constant of proportionality being the ratio $\omega/\beta = v_{ph}$ or the *phase velocity* of the mode. This is the speed at which the phase front propagates.

For pulse propagation, the speed with which energy is transmitted by a mode is given by the *group velocity* with the standard definition $v_{gr} = d\omega/d\beta$. Since different modes have different values of the propagation constant β for the same source frequency, it follows that the phase and group velocities will also differ between modes. The modal group velocity determines pulse dispersion, as discussed in Section 5.6.

5.3.8 Propagation constant, effective index, and cut-off

For a given refractive index profile, the propagation constant β for every bound mode occupies a defined range of values in terms of the maximum n_{co} and minimum n_{cl} refractive index values and the source wavelength. If we assume that the modal

phase velocity v_{ph} must lie between the maximum and minimum values of the speed of light in the cladding and core, that is, $c/n_{co} < v_{ph} < c/n_{cl}$, then on setting $c = \omega/k$, it follows that

$$kn_{cl} < \beta < kn_{co}, \qquad (5.13)$$

where $k = 2\pi/\lambda$ is the free-space wave number. This result applies to any waveguide independent of the number of bound modes that can propagate.

Sometimes, it is more convenient physically to discuss propagation constant values in terms of an *effective index* value. If the effective index n_{eff} is defined through the relationship $\beta = kn_{eff}$, then Equation 5.13 is replaced by

$$n_{cl} < n_{eff} < n_{co}. \qquad (5.14)$$

In other words, each bound mode has an effective index value that must lie between the minimum and maximum refractive index values.

When $\beta = kn_{cl}$, or equivalently, $n_{eff} = n_{cl}$, a mode becomes *cut-off* and for $\beta < kn_{cl}$ or $n_{eff} < n_{cl}$, a mode is said to be *below cut-off*.

5.3.9 Waveguide, fiber, and modal parameters

Waveguide and fibers are commonly characterized in terms of a number of standard dimensionless parameters that combine various basic parameters. In the following definitions, it is assumed that n_{co} denotes the uniform core index in the case of a step profile or the maximum core index in the case of a graded profile, and n_{cl} denotes the uniform cladding index. The source wavelength is λ, $k = 2\pi/\lambda$ is the free-space wave number and ρ is the core radius in the case of a circular fiber or the core half-width of a slab or square-core waveguide. Note that some definitions assume that ρ is the full width of a slab or square-core waveguide.

5.3.9.1 RELATIVE INDEX DIFFERENCE

The uniform cladding index n_{cl} and the uniform or maximum core index n_{co} are often combined to define the *relative index difference* Δ

$$\Delta = \frac{n_{co}^2 - n_{cl}^2}{2n_{co}^2} \cong \frac{n_{co} - n_{cl}}{n_{co}}, \qquad (5.15)$$

where the second expression is obtained by factorizing the numerator in the first expression and assuming that $n_{co} \sim n_{cl}$. This representation is appropriate within the weak-guidance approximation discussed in Section 5.4.

5.3.9.2 NUMERICAL APERTURE

A measure of the light-capturing capacity of a waveguide or fiber is provided by the *numerical aperture* (NA), which is defined as

$$NA = (n_{co}^2 - n_{cl}^2)^{1/2} = n_{co} 2\Delta^{1/2}. \qquad (5.16)$$

A physical interpretation of NA is presented in Section 5.5.6.

5.3.9.3 WAVEGUIDE AND FIBER PARAMETER OR FREQUENCY

A second parameter V combines all the key parameters of a waveguide or fiber into a single normalized quantity known as the *waveguide* or *fiber parameter* or *frequency* that can be expressed in a number of equivalent ways

$$V = \frac{2\pi}{\lambda}(n_{co}^2 - n_{cl}^2)^{1/2} = k\rho(n_{co}^2 - n_{cl}^2)^{1/2}$$
$$= k\rho n_{co}(2\Delta)^{1/2}. \qquad (5.17)$$

This is an important parameter for its value determines, in particular, whether a fiber or waveguide is single mode, as will be discussed in Sections 5.4.4 and 5.4.5.

5.3.9.4 MODAL PARAMETERS

In the examples of modal analyses of waveguides and fibers presented in Sections 5.4.4 and 5.4.5, respectively, it is convenient to introduce normalized *modal parameters* U and W for the core and cladding of a fiber or waveguide, respectively, which incorporate the propagation constant or equivalently the effective index according to

$$U = k\rho(n_{co}^2 - \beta^2)^{1/2} = k\rho(n_{co}^2 - n_{eff}^2)^{1/2};$$
$$W = k\rho(\beta^2 - n_{cl}^2)^{1/2} = k\rho(n_{eff}^2 - n_{cl}^2)^{1/2}. \qquad (5.18)$$

It then follows from the definition of V in Equation 5.17 that

$$U^2 + W^2 = V^2. \qquad (5.19)$$

The values of U and W are necessarily discrete for bound modes.

5.3.10 Radiation modes, leaky modes, and super-modes

The range of propagation constant values for bound modes satisfies Equation 5.14. If $\beta > kn_{co}$, propagation is not possible, but if $\beta < kn_{cl}$, it is possible to analyze the propagation of the unguided field in the fiber or waveguide using one of three different descriptions. These descriptions depend on the particular physical model employed and whether the cladding is unbounded or finite, but large, compared to the core size. In each case, the particular method is focused on determining the propagation characteristics of light within the waveguide that is not guided by the bound modes.

5.3.10.1 RADIATION MODES

In the case of a fiber or waveguide with an *unbounded cladding*, there are no bound modes with propagation constants in the range $0 < \beta < kn_{cl}$; instead, a continuum of *radiation modes* can be derived, each mode having a continuously varying propagation constant value in this range. These modes can be used to analyze nonguided propagation in terms of an integration over each radiation mode and a sum over the integrals for different radiation modes (see Chapter 25 of Reference [4]). This approach is analytically complex and in view of the ready availability of vector-based beam propagation methods and other techniques, a numerical analysis may be easier to implement.

5.3.10.2 LEAKY MODES

The continuum of radiation mode solutions can be approximated by a discrete summation of so-called *leaky modes*. A leaky mode is defined by the analytic continuation of a bound mode to propagation constant values beyond the mode's cut-off, that is, for $\beta < kn_{cl}$. However, unlike a bound mode, where the power flow is parallel to the z-axis everywhere, the local power flow is at an angle to the axis so that power flows away from the core as the mode propagates. This divergence corresponds to a complex value of the leaky mode propagation constant. Furthermore, the power of a leaky mode is unbounded, so that a proper quantitative analysis of propagation cannot be undertaken; nevertheless, leaky modes can provide useful physical insight into propagation characteristics when radiation is present (see Chapter 24 of Reference [4]).

5.3.10.3 SUPER-MODES

The cladding of any practical waveguide is necessarily finite in cross section and may be surrounded by air or by a protective coating in the case of a fiber. In either case, it is normally possible to describe the transient field of a waveguide excited by a source using a superposition of the complete set of bound modes of the complete core–cladding–air structure. To distinguish these modes from the modes guided by the core–cladding refractive index profile, the former are often referred to as *super-modes*. The dimension of the complete cross section is relatively large compared to that of the core ensuring that the number of super-modes is very large. In the limit of an infinitely thick cladding, the super-mode and radiation mode solutions approach one another [5].

5.3.10.4 HOLEY FIBERS

Currently, there is significant interest in the light-guiding properties of *holey fibers*, which have certain attributes that are quite different from those of solid material fibers. Holey fibers differ from conventional fibers in that they are normally fabricated from a single material, such as silica or a polymer with a uniform refractive index, compared with conventional fiber that has an index contrast between the core and the cladding materials.

Light guidance along holey fibers depends on the presence of concentric circular arrays of small longitudinal holes about the fiber axis (Figure 5.9). Each ring of holes can be regarded as defining an annular region in which the average index is smaller than the material index and hence provides an effective index contrast with the material region around the fiber core and outside the ring.

However, the core and ring of holes do not support bound modes. The concentric region outside the ring of holes is of the same index as the core and this enables any modal field within the holes to gradually leak from the center across the ring of holes to the outer region. In other words, holey fibers only support *leaky modes*. Nevertheless, by judicious choice of the size, number, and distribution of air holes, it is possible to design a fiber

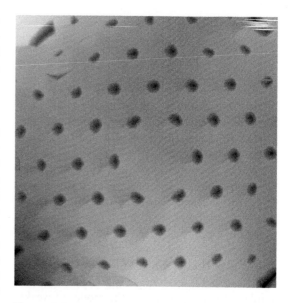

Figure 5.9 Cross section of a holey fiber showing the solid core and surrounding rings of air holes.

whereby the fundamental mode has virtually zero attenuation leakage but all higher order modes have relatively large leakage rates so that their fields rapidly disappear from the central region as they propagate along the fiber. For practical purposes, the holey fiber then behaves like a single-mode fiber, even over very long distances.

Photonic bandgap or *crystal* fibers also guide light, but the guidance mechanism is different from that of holey fibers. The physical basis for guidance along such fibers relies on a special arrangement of many rings of holes about the z-axis such that the rings provide a bandgap in the radial direction, that is, a barrier to the propagation of light away from the axis. One way to think of the bandgap effect qualitatively is to smear out the index contrast between the holes and the fiber material in each ring into an average reduced index. Then the quasi-periodic radial variation between the rings and the fiber material constitutes an effective radial Bragg reflection grating. At the effective Bragg wavelength in the fiber cross section, the grating inhibits propagation and sets up a radial evanescent field that decreases radially outward. In other words, for suitably chosen fiber parameters, the layers of rings can support the evanescent field of the fundamental mode propagating along the fiber with the majority of its field close to the fiber axis. Chapter B10 provides a detailed description and analysis of both kinds of fiber.

5.3.11 Polarization, mode nomenclature, and birefringence

When a mode propagates along a fiber or waveguide, the magnitude and direction of the transverse components of its electric and magnetic fields remain fixed and thereby define the orientation of the modal field. The *polarization* of a mode is defined by the direction of the transverse electric field vector at each position in the waveguide or fiber cross section. Generally, this direction will vary with position so that the contours of the field direction are normally curved, but in the case of two-dimensional slab waveguides, *all modes* have a transverse electric field that is parallel to a fixed direction, that is, each mode is *plane polarized*. A similar situation pertains to the modes of weakly guiding fibers where transverse electric fields are also plane polarized.

5.3.11.1 MODE NOMENCLATURE

For two-dimensional symmetric slab waveguides, there are two distinct classes of modes, depending on their polarization. Transverse electric or TE_j modes have only a single transverse electric field component and transverse magnetic or TM_j modes have only a single transverse magnetic field component, where $j = 0, 1, 2, 3 \ldots$ denotes the mode order. The even and odd values of j denote modes with even and odd field symmetry, respectively. As the value of j increases, the number of extrema in the field patterns also increases.

In the case of circular fibers, the situation is more complex because of the radial and azimuthal directions. There is a class of cylindrically symmetric TE_j and TM_j modes that have electric fields with a single radial or azimuthal transverse component, respectively, where $j = 1, 2, 3$. In addition, there is the class of hybrid HE_{ij} and EH_{ij} modes that can be regarded as linear combinations of both TE and TM field components and therefore have a more complex electric field structure in the fiber cross section.

Here, the subscript "i" relates to the order of the azimuthal angular dependence and "j" refers to the radial order, the value of which increases with the number of extrema in the field. The azimuthal dependence of each mode has a $\cos(m\varphi)$ or $\sin(m\varphi)$ variation where $m = 0, 1, 2, \ldots$

In the case of weakly guiding circular fibers, a second complimentary mode nomenclature is commonly used whereby each mode is labeled as LP_{ij}, the "LP" denoting *linearly polarized*. The subscripts "i" and "j" denote azimuthal and radial orders, respectively, and in terms of HE modes in the weak-guidance approximation, the relationship can be expressed as $LP_{ij} \leftrightarrow HE_{i+1, j}$.

Within the two nomenclatures, the fundamental mode is equivalent to the HE_{11} mode for arbitrary index difference circular fibers and to either the HE_{11} or LP_{01} mode for weakly guiding fibers. For other waveguide and fiber geometries, there is no generally accepted nomenclature for categorizing modes.

5.3.11.2 BIREFRINGENCE

On a circular weakly guiding fiber, the fields of the fundamental mode are rotationally invariant about the fiber axis, so that any pair of orthogonal directions can be chosen for its two polarization states that have identical propagation constants. In this situation, the fundamental mode is said to be *degenerate*.

However, in the case of noncircular fibers, such as the elliptical core fiber shown in Figure 5.10, this is no longer the situation. Working within the weak-guidance approximation, the two polarization directions of the planar transverse electric field are parallel to one of the *optical axes* of the fiber. For the elliptical cross section, the optical axes coincide with the major x-axis and minor y-axis and the fundamental mode has respective propagation constants β_x and β_y. Since $\beta_x > \beta_y$, the fundamental mode is *nondegenerate* and the fiber is said to be *birefringent* because of the difference in propagation constant for the two polarization states introduced by the noncircular core geometry.

Birefringence is a measure of the difference in the two propagation constants and is usually expressed in terms of the normalized parameter B that is defined by

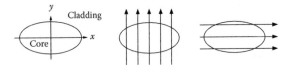

Figure 5.10 Cross section and polarization states of the fundamental mode of an elliptical core fiber.

$$B = \frac{\beta_x - \beta_y}{k} n_{ex} - n_{ey}, \qquad (5.20)$$

where $\beta_x = kn_{ex}$ and $\beta_y = kn_{ey}$ in terms of the equivalent effective indices, and k is the free-space wave number. If both fundamental mode polarizations are launched simultaneously in the fiber, beating will occur between them because of the difference in propagation constants. The *beat length* or distance over which the superposition of the two modal fields repeats periodically along the length of the fiber is denoted by z_b

$$z_b = \frac{2\pi}{\beta_x - \beta_y}; \quad B = \frac{2\pi}{z_b k} = \frac{\lambda}{z_b} \qquad (5.21)$$

and is therefore inversely proportional to the birefringence. Hence, a measurement of the beat length together with the source wavelength will determine the birefringence.

5.3.12 Attenuation due to absorption and scattering loss

When light propagates in a bound mode along a straight fiber or waveguide, there is no loss of modal power provided that the materials are perfectly lossless. However, in practical fibers and waveguides, a propagating mode steadily loses power because of two basic physical effects: (1) *bulk absorption* of optical power by the materials constituting the fiber core and cladding and (2) *scattering* of light by material and surface inhomogeneities, due principally to the distribution of dopants in the core of fibers and surface roughness between the core and cladding in waveguides.

In long-distance telecommunications fibers, Rayleigh scattering is the major cause of loss, absorption loss having been reduced to almost zero at the operating wavelength. The net effect of these two processes is to reduce the total light in the mode as it propagates along the length of the fiber, i.e., modal power is *attenuated*. The loss of power can be accounted for by adding an imaginary part to the refractive index value, that is, n becomes the complex index $n^{(r)} + in^{(i)}$, where superscripts r and i denote real and imaginary parts, respectively. Although the real part of the index must necessarily vary over the core cross section to provide guidance, the imaginary part is normally assumed to take a constant value.

5.3.12.1 MODAL ATTENUATION

In the determination of the propagation constant β of each bound mode, a real refractive index distribution $n(x, y)$ leads to a real value of β from the eigenvalue equations as is evident in the modal analyses of the step-profile slab waveguide and fiber in Sections 5.4.4 and 5.4.5, respectively. If n now becomes complex, it is evident from these eigenvalue equations that the modal parameters U and W and, therefore, the propagation constant must also become complex. Accordingly, we set

$$\begin{aligned} U &= U^{(r)} + iU^{(i)}; \\ W &= W^{(r)} + iW^{(i)}; \\ \beta &= \beta^{(r)} + i\beta^{(i)} \end{aligned} \qquad (5.22)$$

where $\beta^{(i)} > 0$. If we recall that the electric and magnetic field components all have the common longitudinal dependence

$$\begin{aligned} \exp(i\beta z) &= \exp\left(i\{\beta^{(r)} + i\beta^{(i)}\}\right) \\ &= \exp\left(i\beta^{(r)}z\right)\exp\left(-\beta^{(i)}z\right), \end{aligned}$$

it follows that the fields are attenuated as $\exp(-\beta^{(i)}z)$ and the local power density is attenuated as $\exp - (2\beta^{(i)}z)$. Further, since the total power density flow in a mode is the integral of the local power density over the cross section, it follows that the power in all modes is attenuated according to

$$P(z) = P(0)\exp\left(-2\beta_z^{(i)}\right), \qquad (5.23)$$

where $P(z)$ is the mode power distance z along the fiber or waveguide. The expression $2\beta^{(i)}$ is the *power attenuation coefficient*. The measure of attenuation is commonly expressed in decibels, or dB, where the dB value is calculated as

$$\begin{aligned} \mathrm{dB} &= -10\log_{10}\left(\frac{P(z)}{P(0)}\right) = 20\beta^{(i)}z\log_{10}e \\ &= 8.686\beta^{(i)}z, \end{aligned} \qquad (5.24)$$

The determination of the exact values of the complex propagation constant from the eigenvalue equation for a particular fiber or waveguide with a complex refractive index profile can generally only be undertaken numerically, and formally poses a two-dimensional root-finding problem.

However, for the majority of problems, $\beta^{(i)} \ll \beta^{(r)}$, so that $\beta^{(r)}$ can be determined straightforwardly by assuming that to the lowest order, n is pure real. For example, if the imaginary part of the index $n^{(i)}$ is a constant everywhere throughout the fiber or waveguide, then $\beta^{(i)} \sim kn^{(i)}$, which means that the dB loss is given, approximately, by $55n^{(i)}z/\lambda$, where λ is the source wavelength. This expression is identical to the power attenuation of a plane wave propagating in a uniform medium with imaginary index $n^{(i)}$. For more general situations, a perturbation approach can be adopted to determine an accurate approximation for $\beta^{(i)}$, as quantified by the example in Section 5.9.1.1.

For standard silica-based single-mode fibers used for telecommunications, there is a minimum loss of about $0.2\,\mathrm{dB\,km^{-1}}$ occurring at a wavelength of 1550 nm. Using the plane wave expression, this leads to an imaginary index value of approximately 5.6×10^{-12}.

5.3.13 Analytical and numerical solutions

There are only a small number of waveguide geometries and refractive index profiles for which there are exact analytical solutions of Maxwell's equations for the bound mode fields and analytical expressions for the eigenvalue equation that determines the values of the propagation constant. The latter are generally transcendental and can only be solved numerically. Of these solutions, the most practical and simplest example is the step-profile slab waveguide and the step-profile circular fiber. The former has modal fields and eigenvalue equations expressed in terms of trigonometric and exponential functions while the latter has modal fields and eigenvalue equations expressed in terms of Bessel functions and modified Bessel functions (see Chapter 12 of Reference [4]). These solutions are valid for arbitrary relative index difference. In the weak-guidance approximation, the corresponding analytical solutions of the scalar wave equation are less complex and are derived in detail in Sections 5.4.4 and 5.4.5.

Although there are analytical solutions of Maxwell's equations or the scalar wave equation that can be derived for certain other profiles and geometries, they involve special mathematical functions the properties of which are less familiar compared to the situation just 25 years ago. A major factor that has reduced their familiarity is the development and ready availability of reliable commercial and in-house software routines that can solve Maxwell's equations for the fields and propagation constants of bound modes given almost arbitrary refractive index profiles and waveguide geometries.

5.3.13.1 PLANAR AND RIB WAVEGUIDES

There is a class of waveguides referred to generically as *planar waveguides*. This name actually refers to the planar substrate on which waveguides with a variety of core cross sections are fabricated. One class relates to *buried channel waveguides* that generally have a nominally square or rectangular core surrounded by an effectively unbounded cladding. Another class includes *rib waveguides* that have a core where the cross section has the shape of an inverted "T" and is commonly surrounded by air above and a lower refractive index layer below. Provided the core index is greater than the surrounding cladding index, these waveguides support one or more bound modes, depending on the waveguide parameters and source wavelength.

For both scalar modes and vector modes, the waveguide geometry requires a numerical solution for the modal fields and propagation constants, although some analytical approximation methods, such as the *effective index method* can be applied [6].

5.4 WEAK-GUIDANCE APPROXIMATION

The solution of Maxwell's equations for the bound modes of a dielectric fiber or waveguide with arbitrary cross-sectional geometry and refractive index profile, whether investigated by analytical or numerical means, generally involves the simultaneous determination of all six scalar components of the electric and magnetic field vectors. Such solutions are therefore necessarily complex. However, there is a set of practical fiber and waveguide problems for which it is possible to make a significant simplification and replace the set of six coupled Maxwell equations with a single scalar equation for just one component of the fields. This is the basis of the *weak-guidance approximation* [7,8].

5.4.1 Scalar electromagnetic fields and power flow

Many waveguides, such as solid silica-based optical fibers and planar waveguides, have a refractive index profile $n(x, y)$ across the core and cladding where the maximum variation in index is relatively small, typically below 1%. Now consider the vector wave equation satisfied by the electric field. This equation is generated by eliminating the magnetic field H between the two Maxwell Equations 5.3 and leads to

$$(\nabla_t^2 + k^2 n^2 - \beta^2)E = -\nabla_t(E_t \cdot \nabla_t \ln n^2), \quad (5.25)$$

where ∇_t^2 is the transverse *vector* Laplace operator, $k = 2\pi/\lambda$ is the free-space wave number, and β is the propagation constant. If the overall variation in index is small, then the term on the right-hand side can be neglected.

In a general orthogonal coordinate system, the vector operator ∇_t^2 couples the scalar components of E_t. However, if the components of E_t are chosen to be Cartesian, that is, $E_t = (e_x, e_y) = \exp(i\beta z)$, then the component equations decouple. The respective scalar components e_x or e_y for these states independently satisfy the same scalar wave equation

$$(\nabla_t^2 + k^2 n^2 - \beta^2)e_x = 0;$$
$$(\nabla_t^2 + k^2 n^2 - \beta^2)e_y = 0, \quad (5.26)$$

where ∇_t^2 is now the two-dimensional *scalar* Laplace operator the analytical form of which depends on the waveguide or fiber geometry.

Once the electric field component of a mode has been determined from the scalar wave Equation 5.26, together with the appropriate boundary conditions, the corresponding component of the magnetic field for the respective polarization states is given by a simple algebraic relationship

$$h_y = n_{co}\left(\frac{\varepsilon_0}{\mu_0}\right)^{1/2} e_x; \quad h_x = -n_{co}\left(\frac{\varepsilon_0}{\mu_0}\right)^{1/2} e_y, \quad (5.27)$$

where ε_0 and μ_0 are, respectively, the free-space dielectric constant and permeability, and n_{co} is the maximum core index. Note that in the weak-guidance approximation, all other field components are very small compared to the dominant transverse electric and magnetic field components and can normally be ignored.

It follows from Equations 5.9, 5.26, and 5.27 that, in weak guidance, the z-directed power flow density for the x- and y-polarized fields is given, respectively, by

$$S_z = \frac{1}{2} n_{co}\left(\frac{\varepsilon_0}{\mu_0}\right)^{1/2} e_x^2; \quad S_z = \frac{1}{2} n_{co}\left(\frac{\varepsilon_0}{\mu_0}\right)^{1/2} e_y^2. \quad (5.28)$$

There is a strong analogy between the modes and plane waves of Section 5.2 in terms of the nonzero field components and the orthogonal electric field polarization directions. For the two mode polarizations, the field components are

$$E = (e_x, 0, 0)e^{i\beta z}, H = (0, h_y, 0)e^{i\beta z};$$
$$E = (0, e_y, 0)e^{i\beta z}, H = (h_x, 0, 0)e^{i\beta z} \quad (5.29)$$

corresponding to x- and y-polarizations, respectively.

5.4.2 Transverse nature of the electromagnetic field

The modal electromagnetic field in the core can be considered at each position in the core cross section as a plane wave. The propagation constant β is the z-component of the local planar wave vector, that is,

$$\beta = kn_{co} \cos(\theta_z), \quad (5.30)$$

where θ_z is the direction that the local plane wave vector makes with the z-axis in the core of the waveguide or fiber, and k is the free-space wave number. The propagation constant β is bounded by

$$kns_{cl} < \beta < kn_{co} \quad (5.31)$$

and the core and cladding indices are similar, that is, $\beta \sim kn_{co} \sim kn_{cl}$ leading to $\theta_z \sim 0$.

The wave vector is approximately parallel to the z-direction and hence, both the vector electric and

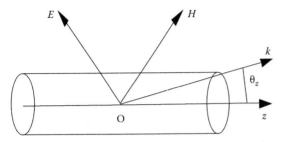

Figure 5.11 Triad of electric and magnetic field vectors and the local wave vector.

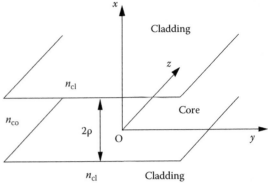

Figure 5.12 Geometry and parameters for the symmetric step-profile slab waveguide.

magnetic fields are approximately transverse to the z-axis. Both the longitudinal electric and magnetic field components are negligible in magnitude compared to the transverse fields. The Cartesian components of the electric and magnetic fields for either mode polarization, together with the z-axis comprise an orthogonal triad (Figure 5.11), that is, the same relationship satisfied by the corresponding components of a plane wave.

5.4.3 Slab and circular geometries

In this section, we will delineate the geometry, profiles, field representation, governing equations, coordinates, and parameters to analyze the modes on symmetric slab waveguides and circular fibers in the weak-guidance approximation.

5.4.3.1 SYMMETRIC SLAB WAVEGUIDE

For the one-dimensional slab waveguide of Figure 5.12, the core–cladding interfaces are parallel to the yz plane, so that there is only the x-variation in the transverse direction, together with the usual z-dependent phase term in the z-direction of propagation. Thus, for the x- and y-polarizations of the TE and TM modes, respectively, the scalar electric field components can be expressed as

$$E_x(x,z) = e_x(x)e^{i\beta z}; E_y(x,z) = e_y(x)e^{i\beta z}, \quad (5.32)$$

where $e_x(x)$ and $e_y(x)$ are both solutions of the one-dimensional scalar wave equation

$$\left(\frac{d^2}{dx^2} + k^2 n^2(x) - \beta^2\right) e(x) = 0, \quad (5.33)$$

obtained from Equation 5.26, where $n(x)$ denotes the profile.

5.4.3.2 CIRCULAR FIBER

Note that while either e_x and e_y is the single Cartesian *component* of the vector electric field e, their spatial variation does not necessarily have to be described in *Cartesian coordinates*. For the circular fiber in Figure 5.13, the spatial dependence of the modal fields in the cross section is best described by polar coordinates (r, ϕ) based on the z-axis of the fiber. Thus, for the x- or y-polarized modes, we may set

$$e_x = e_x(r,\phi); e_y = e_y(r,\phi). \quad (5.34)$$

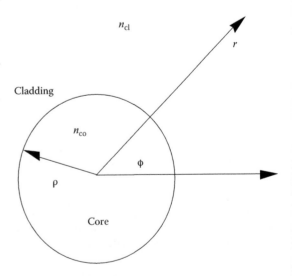

Figure 5.13 Geometry and parameters for circular fibers.

Here, $e_x(r, \phi)$ and $e_y(r, \phi)$ are both solutions of the two-dimensional scalar wave equation

$$\left(\frac{\partial^2}{\partial r^2} + \frac{1}{r}\frac{\partial}{\partial r} + \frac{1}{r^2}\frac{\partial^2}{\partial\phi^2} + k^2 n^2(r) - \beta^2 \right) e(r,\phi) = 0, \quad (5.35)$$

where $n(r)$ denotes an axisymmetric profile for circularly symmetric fibers.

5.4.3.3 BOUNDARY CONDITIONS

If ψ denotes either e_x or e_y in the scalar wave equation, then the boundary conditions satisfied by the bound mode solutions of the scalar wave equation in the core and cladding are the weak-guidance limit of the boundary conditions for Maxwell's equations. In this limit, all six components of the electromagnetic field are continuous everywhere. Accordingly:

- $\Psi \to 0$ exponentially far from the fiber or waveguide core
- Ψ is continuous across the core–cladding interface
- All first derivatives of ψ are continuous across the core–cladding interface

For the one-dimensional slab waveguide, both Ψ and $\mathrm{d}\Psi/\mathrm{d}x$ are continuous across both core–cladding interfaces, while for the two-dimensional fiber, Ψ $\mathrm{d}\Psi/\mathrm{d}r$ and $\mathrm{d}\Psi/\mathrm{d}\Psi$ are continuous across the core–cladding interface. The solution of the scalar wave equation together with the boundary conditions constitutes an *eigenvalue problem* for the modal propagation constants, and the *eigenfunctions* determine the spatial dependence of a scalar transverse electric field. The propagation constants are determined by the *eigenvalue equation*.

5.4.4 Step-profile slab waveguide

The spatial dependence of the modal fields in Equation 5.32 and the eigenvalue equation for the modal propagation constants are obtained by solving Equation 5.33 in the core and the cladding, and then matching the two solutions using the boundary conditions on the core–cladding interfaces. Because the slab waveguide profile is symmetric about the yz plane, the solutions of the scalar wave equation for $\psi(x)$ will be accordingly either symmetric or antisymmetric in x.

5.4.4.1 EVEN AND ODD MODES

In the core, $n(x) = n_{\mathrm{co}}$ and it is convenient to introduce the core modal parameter $U = \rho(k^2 n_{\mathrm{co}}^2 - \beta^2)^{1/2}$, since $\beta < k n_{\mathrm{co}}$ for bound modes, and the normalized coordinate $X = x/\rho$. With these substitutions, the scalar wave equation can be written as

$$\left(\frac{\mathrm{d}^2\Psi}{\mathrm{d}X^2} + U^2 \right)\Psi = 0. \quad (5.36)$$

This is a harmonic equation and its solutions are proportional to $\cos(UX)$ or $\sin(UX)$. Since the former is symmetric in x, we set for the even modes:

$$\Psi(X) = A\cos(UX), \quad (5.37)$$

for $0 < |X| < 1$, where A is a constant. In the cladding, $n(x) = n_{\mathrm{cl}}$ and it is convenient to work with the cladding modal parameter $W = \rho(\beta^2 - k^2 n_{\mathrm{cl}}^2)^{1/2}$ since $\beta > k n_{\mathrm{cl}}$ for bound modes and the normalized $X = x/\rho$. With these substitutions, the scalar wave equation becomes

$$\left(\frac{\mathrm{d}^2\Psi}{\mathrm{d}X^2} + W^2 \right)\Psi = 0. \quad (5.38)$$

The solution of this equation is exponential and proportional to either $\exp(WX)$ or $\exp(-WX)$. Since we require a symmetric solution that decreases to zero as $X \to +\infty$ and $X \to -\infty$, we set

$$\Psi(X) = B\mathrm{e}^{-W|X|}, \quad (5.39)$$

for $|X| > 1$, where B is a constant.

5.4.4.2 EIGENVALUE EQUATIONS

The boundary conditions on $(X=1)(x=\rho)$ and $(X=-1)(x=-\rho)$, which link the core and cladding solutions, are the continuity of Ψ and $\mathrm{d}\Psi/\mathrm{d}X$, which lead, respectively, to

$$A\cos U = B\mathrm{e}^{-W}; -AU\sin U = -BW\mathrm{e}^{-W}, \quad (5.40)$$

so that on dividing the two equations, the constants A and B are eliminated and we obtain one equation linking the core and cladding modal parameters U and W. A second equation follows from the definitions of U, V, and W. Hence, the eigenvalue equations for the even modes are

$$W = U\tan U; W^2 + U^2 = V^2. \qquad (5.41)$$

The derivation of the corresponding equations for the odd modes is identical to that for the even modes provided the core modal field has the anti-symmetric form

$$\Psi(X) = A\sin(UX). \qquad (5.42)$$

This leads to the eigenvalue equations for the odd modes

$$W = -U\cot U; W^2 + U^2 = V^2. \qquad (5.43)$$

Both pairs of eigenvalue equations can be further simplified by eliminating U between each pair and for the even and odd modes this leads, respectively, to

$$V = \pm\frac{U}{\cos U}; \quad V = \pm\frac{U}{\sin U}, \qquad (5.44)$$

but care needs to be exercised with the signs to identify correct solutions.

5.4.4.3 SOLUTION OF THE EIGENVALUE EQUATIONS

The eigenvalue equations for both even and odd modes are transcendental and do not possess closed-form analytical solutions for U (or W) in terms of a given value of V. Accordingly, solutions must be determined numerically and each value of U corresponds to one bound mode. This can be undertaken using a hand calculator incorporating trigonometric functions by using trial and error, that is, given V, we guess U, or use a more sophisticated computer program.

However, it may be simpler to calculate the value of V, given the value of U in Equation 5.44 since these are single-valued and explicit. This approach generates the V–U curve for each and every mode of the waveguide as shown in Figure 5.14. In these plots, an even or odd subscript corresponds to a mode with an even or odd field, respectively, and also indicates the mode order.

5.4.4.4 SINGLE-MODE WAVEGUIDE

In Figure 5.14, every mode, except the fundamental mode, has a finite cut-off value of U when $U = V$ below which it cannot propagate as a bound mode

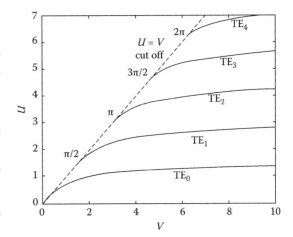

Figure 5.14 Plots of U against V for the modes of the symmetric step-profile slab waveguide.

and becomes a leaky mode. For the mode with subscript "m," the cut-off value is $V = V_{co} = m\pi/2$.

The waveguide is *single mode* if $V < \pi/2$, and only the fundamental mode can propagate in either of its two polarization states (TE_0 or TM_0) with the same propagation constant β. The fundamental mode has the largest value of propagation constant or equivalently the smallest value of U for a given value of V. The fundamental mode corresponds to the TE_0 or TM_0 mode.

5.4.4.5 MODAL FIELDS

For the TE (transverse electric) modes, the electric field direction is parallel to the y-axis and transverse to the z-axis while for the TM (transverse magnetic) modes, the magnetic field is parallel to the y-axis and transverse to the z-axis. In the weak-guidance approximation, the pair of TE_j and TM_j modes has an identical propagation constant while their transverse electric and magnetic field components have identical spatial distribution, and are y- and x-polarized, respectively.

The spatial distribution of the transverse electric or magnetic fields for the first four TE or TM modes are plotted in Figure 5.15. Note that the TE_m and TM_m mode fields have $m+1$ extrema, and the even mode has a maximum at the center of the waveguide where the odd mode fields vanish. The total number of modes that can propagate for a given value of V is given by the nearest integer value below $4V/\pi$ allowing for the two possible polarization states.

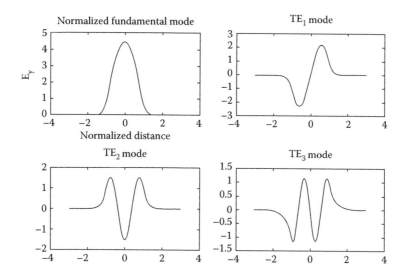

Figure 5.15 Plots of the transverse electric field for the first four modes of the symmetric step-profile slab waveguide.

5.4.5 Step-profile fiber

The spatial dependence of the modal fields in Equation 5.34 and the eigenvalue equation for the modal propagation constants are obtained by solving Equation 5.35 in the core and the cladding, and then matching the two solutions using the boundary conditions on the core–cladding interface. If $\Psi(r, \phi)$ denotes either the x-polarized electric field $e_x(r, \phi)$ or the y-polarized electric field $e_y(r, \phi)$ in cylindrical polar coordinates, then, within weak guidance, $\Psi(r, \phi)$ satisfies the scalar wave equation (Equation 5.34), which is more conveniently written in the normalized form

$$\left\{\frac{\partial^2}{\partial R^2}+\frac{1}{R}\frac{\partial}{\partial R}+\frac{1}{R^2}\frac{\partial^2}{\partial \phi^2}+\rho^2(k^2n(R)^2-\beta^2)\right\}\Psi=0, \quad (5.45)$$

where $R=r/\rho$ is the normalized radial coordinate, $n(R)$ is the refractive index profile, and ρ is the core radius. This equation is a second-order partial differential equation and as $n(R)$ is independent of ϕ, it can be solved using *separation of variables*.

5.4.5.1 SEPARATION OF VARIABLES

Using this technique, the two-dimensional spatial dependence of ψ is expressed as the product of two single-variable functions

$$\Psi(R,f)=F(R)G(f), \quad (5.46)$$

where $F(R)$ is a function of R only and $G(\phi)$ is a function of ϕ only. On substituting into the scalar wave equation (Equation 5.38), dividing by Ψ, and multiplying by R^2, it follows

$$\frac{R^2}{F}\left\{\frac{d^2F}{dR^2}+\frac{1}{R}\frac{dF}{dR}+\rho^2\left[k^2n^2(R)-\beta^2\right]F\right\}+\frac{1}{G}\frac{d^2G}{d\phi^2}=0. \quad (5.47)$$

The right-hand expression is a function of ϕ only and can be held fixed while R varies arbitrarily. Because periodicity of the solution is required in the azimuthal direction as a result of the circular geometry, we set

$$\frac{d^2G}{d\phi^2}=-v^2G; \quad G=\left\{\begin{array}{l}\sin(v\phi)\\\cos(v\phi)\end{array}\right., \quad (5.48)$$

where v is a constant. This is a harmonic equation with sinusoidal solutions that are single-valued, provided $v=0, 1, 2, 3$. Substituting this back into the scalar wave equation (Equation 5.38) leads to the second-order ordinary differential equation

$$\left\{\frac{d^2}{dR^2}+\frac{1}{R}\frac{d}{dR}-\frac{v^2}{R^2}+\rho^2[k^2n^2(R)-\beta^2]\right\}F=0 \quad (5.49)$$

that is solved for the core and cladding regions separately.

5.4.5.2 CORE SOLUTION

In the core, $n=n_{co}$, so that Equation 5.49 for F can be written as

$$\left\{\frac{d^2}{dR^2}+\frac{1}{R}\frac{d}{dR}-\frac{v^2}{R^2}+U^2\right\}F=0, \quad (5.50)$$

where the core modal parameter is defined by $U=\rho(k^2 n_{co}^2-\beta^2)^{1/2}$. We then multiply this equation by R^2 and set $s=UR$ to obtain

$$\left\{s^2\frac{d^2}{ds^2}+s\frac{d}{ds}+s^2-v^2\right\}F=0. \quad (5.51)$$

This is *Bessel's equation* with the general solution

$$F(R)=AJ_v(s)+BY_v(s), \quad (5.52)$$

where J_v and Y_v are *Bessel functions* of the first and second kinds, respectively, and A and B are constants. The solution is bounded throughout the core and as Y_v is singular on the fiber axis $R=0$, then $B=0$. Hence, $F(R)=AJ_v(UR)$ and together with Equation 5.48, the complete core solution dependence $F(R)G(\phi)$ is given by

$$\Psi=AJ_v(UR)\begin{cases} \sin(v\phi) \\ \cos(v\phi) \end{cases}, \quad (5.53)$$

where $v=0, 1, 2, \dots$.

5.4.5.3 CLADDING SOLUTION

In the cladding, $n=n_{cl}$, so that the corresponding equation for F becomes

$$\left\{s^2\frac{d^2}{ds^2}+s\frac{d}{ds}-(v^2+s^2)\right\}F=0, \quad (5.54)$$

where $s=WR$ and the cladding modal parameter is defined by $W=\rho(\beta^2-k^2 n_{cl}^2)^{1/2}$. This is the *modified Bessel equation* with the general solution

$$F(R)=CI_v(s)+DK_v(s), \quad (5.55)$$

where I_v and K_v are *modified Bessel functions* of the first and second kinds, respectively, and C and D are constants. This solution is bounded throughout

the cladding and since I_v is singular as $R\to\infty$, then $C=0$. Hence, $F(R)=DK_v(WR)$ and together with Equation 5.48, the complete cladding solution dependence, $F(R)G(\phi)$ is

$$\Psi=DK_v(WR)\begin{cases} \sin(v\phi) \\ \cos(v\phi) \end{cases}, \quad (5.56)$$

where $v=0, 1, 2, \dots$.

5.4.5.4 BOUNDARY CONDITIONS

Boundary conditions at the core–cladding interface require continuity of the solution of the scalar wave equation and all first derivatives, equivalent to continuity of $\Psi(R, \phi)$ and $d\Psi(R, \phi)/dR$ on $R=1$. The latter, when applied to Equations 5.53 and 5.56, gives

$$AJ_v(U)=DK_v(W); \quad AUJ_v'(U)=DWK_v'(W), \quad (5.57)$$

where the prime denotes differentiation with respect to the argument. On dividing these two equations, we obtain one eigenvalue equation for U and W:

$$\frac{UJ_v'(U)}{J_v(U)}=W\frac{K_v'(W)}{K_v(W)}. \quad (5.58)$$

It is more convenient for computational purposes to recast this equation into an alternative form avoiding the use of derivatives by using the following recurrence relations satisfied by Bessel functions of different orders:

$$J_v'=\frac{v}{U}J_v-J_{v+1}; \quad K_v'=\frac{v}{W}K_v-K_{v+1}, \quad (5.59)$$

where the prime denotes differentiation with respect to the argument. Thus, the final form of the eigenvalue equations for U and W are

$$U\frac{J_{v+1}(U)}{J_v(U)}=W\frac{K_{v+1}(W)}{K_v(W)}; \quad V^2=U^2+W^2, \quad (5.60)$$

where $v=0, 1, 2, \dots$.

5.4.5.5 SOLUTIONS OF THE EIGENVALUE EQUATIONS

The left-hand term in Equation 5.60 is transcendental and can be solved only numerically. There

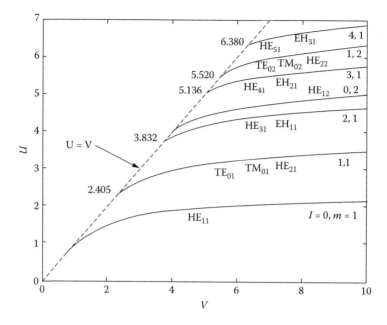

Figure 5.16 Plots of U against V for modes of the step-profile circular fiber.

are standard library routines for evaluating the two types of Bessel functions required. Given the value of V, the eigenvalue equations can be solved by expressing W in terms of U and V using the second term in Equation 5.60 and then solving the resulting equation numerically for the discrete values of U. Each value of U corresponds to a particular bound mode. For each value of v, there are a finite number of bound solutions of the eigenvalue equation, the number increasing with the value of V according to the closest integer below $V^2/2$ for the step profile.

Figure 5.16 plots the values of U as a function of V for each bound mode solution. Given the value of U, the value of the propagation constant can be determined from the definition of U given earlier. Every mode, except the fundamental mode, has a finite cut-off value of U when $U=V$. Below this value, the mode cannot propagate as a bound mode and becomes a leaky mode. The fiber is multimode when $V \gg 1$, and many modes can propagate. For a given, large value of V, the number of bound modes that can propagate is approximately $V^2/2$.

5.4.5.6 MODE NOMENCLATURE

Because of the cylindrical geometry of the fiber, there are four possible types of bound mode solution. The TE_{01} and TM_{01} solutions, like the

corresponding solutions for the slab waveguide in Section 5.4.4, indicate modal electric or magnetic fields that have only a radial electric or magnetic field component, respectively, and are axisymmetric. The EH_{pq} and HE_{pq} modes can be regarded as hybrid modes, their fields containing both electric and magnetic radial components. The first and second subscripts p and q on each mode designator indicate, respectively, the azimuthal symmetry and the radial order. The former has the dependence $\cos([p-1]\phi)$ or $\sin([p-1]\phi)$ for $p=1, 2, 3, \ldots$, while the latter has the number of extrema in the field profile.

There is a second mode nomenclature in use that was devised specifically for weakly guiding fibers and is based on the linear polarization (LP) property of these modes. In this system, the fundamental HE_{11} mode is known as the LP_{01} mode with axisymmetric field components, and the TE_{01}, TM_{01} and HE_{21} set of second modes with antisymmetric $\sin \phi$ or $\cos \phi$ azimuthal dependence are known as LP_{11} modes.

5.4.5.7 SINGLE-MODE FIBER

The fiber is single mode for $V < 2.405$ when only one mode, the fundamental mode, can propagate in either of its two orthogonal but otherwise arbitrary polarization states with the same propagation

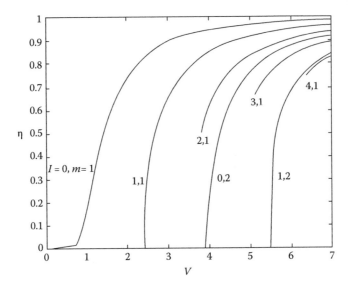

Figure 5.17 Fraction of modal power propagating in the fiber core.

constant β. The fundamental mode has the largest value of propagation constant or equivalently, the smallest value of U for a given value of V. The fundamental mode corresponds to the HE_{11} or equivalent LP_{01} mode.

5.4.5.8 FRACTIONAL OF MODAL POWER IN THE CORE

For a particular fiber and source wavelength, the power of each propagating mode is distributed between the core and the cladding in a ratio that depends on the mode order. It is useful to define the parameter η as the fraction of total mode power that propagates in the core, so that $0 < \eta < 1$. Figure 5.17 plots this fraction as a function of fiber parameter V for the first six low-order modes of the weakly guiding step profile. As V increases, the value of η for each mode approaches 1 asymptotically, consistent with the zero-wavelength geometric optics limit. Note that as V approaches the mode cut-off, the fraction of core power does not always approach zero.

5.4.6 Fundamental mode

The mode of most interest for single-mode fibers is the fundamental HE_{11} or LP_{01} mode, that is, the only mode that can propagate on a fiber for V values below the cut-off of the second mode, namely, for $V < 2.405$ in the case of the step-profile fiber.

If λ_{co} denotes the cut-off wavelength of the second mode, the fundamental mode alone can propagate for wavelengths such that $\lambda > \lambda_{co}$.

5.4.6.1 INTENSITY DISTRIBUTION

The normalized fundamental mode electric field intensity $|\Psi|^2$ for typical graded- or step-index profiles has a characteristically axially symmetric bell-shaped distribution about the fiber axis. This characteristic shape is exemplified by the plots of intensity against normalized radius, $R=r/\rho$, shown in Figure 5.18 calculated from Equations 5.28, 5.53, and 5.56 for the step-profile fiber for various values of V. In the geometric optics limit, when V becomes unbounded, that is, $\lambda \to 0$, all the modal power is concentrated in the core. The normalization is such that the geometric optics limit corresponds to unit intensity on the fiber axis.

5.4.6.2 FIBER CLADDING THICKNESS

As V decreases, an increasing fraction of fundamental mode power spreads into the cladding, as is clear from Figures 5.17 and 5.18. For a practical single-mode step-profile fiber operating around $V=2.3$, the fraction of modal power in the cladding is about 25% and the cladding field, which decreases approximately exponentially with increasing radius, needs to be sufficiently small at the outer edge of the cladding so that the absorption due to the fiber coating is negligibly small. It is

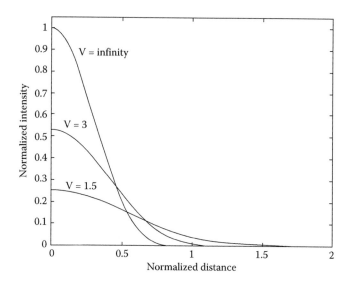

Figure 5.18 Plots of the fundamental mode intensity distribution on a step-profile fiber for various values of V.

5.4.7 Spot size and Gaussian approximation

Single-mode fibers and waveguides are often characterized by the *spot size* of the fundamental mode intensity distribution, which is an experimental measurement of the diameter of the brightest part of this distribution as it appears in either the near field or far field at the end of a long length of single-mode fiber. The plots of the local power density of the fundamental mode transverse intensity distribution, as shown in Figure 5.18, have the common characteristic of a peak on the vertical axis and a monotonic decrease with increasing radius. Accordingly, for a given wavelength, the *spot size* can be defined as the half-width of this distribution at an appropriate position on the vertical axis while the *mode field diameter* is the full width and therefore equal to twice the spot size.

As there is no immediately obvious position to choose, a number of different definitions of spot size have been proposed over the years, each being tailored to provide, indirectly, an accurate value of a particular fundamental mode property from the spot size measurement. In this chapter,

we introduce a definition known as the *Gaussian approximation* based solely on the shape of the plots in Figure 5.18 (see Chapters 15 and 16 of Reference [4]).

5.4.7.1 GAUSSIAN APPROXIMATION

Generally, there is no analytical solution of the scalar wave equation for the fundamental mode field of a practical profile, but this equation can of course always be solved numerically. However, a simple but approximate analytical representation for this field can be derived by examining the intensity distribution of the fundamental mode in Figure 5.18. At a sufficiently large normalized radial distance R from the fiber axis, the asymptotic form of the square of the modified Bessel function $K_0(WR)$ intensity decreases approximately exponentially as $\exp(-2WR)$, where W is the cladding mode parameter. Furthermore, the intensity also exhibits an approximately parabolic variation with R around its maximum on the fiber axis. This is evident for a general index profile by examining a power series solution of Equation 5.51, and in the case of the step-profile fiber, the analytical dependence of the square of the core field on $J_0(UR)$ in Equation 5.53 exhibits this property for small values of the argument of the Bessel function.

A standard mathematical function that reflects these attributes and is algebraically tractable for integration in particular is the *Gaussian function*.

It is used as an approximate representation of the fundamental mode field according to

$$\Psi(R) = A\exp\left(\frac{1}{2}\frac{r^2}{s^2}\right) = A\exp\left(\frac{1}{2}\frac{R^2}{2S^2}\right), \quad (5.61)$$

where A is an arbitrary constant, s is the spot size, and $S = s/\rho$ is the normalized spot size. The factor of 1/2 is purely for convenience in subsequent algebraic manipulations. This function has an approximately parabolic variation in the neighborhood of the fiber axis. In the far field at $R \gg 1$, the function has an exponential decrease that varies as R^2 rather than linearly but since the fraction of the modal power in this region is small, the additional error will be small.

The normalized spot size S is determined from the fiber profile, using standard variational methods to derive the following implicit equation (see Section 15.1 of Reference [4]):

$$\frac{1}{V^2} = \int_0^\infty R^2 \frac{df(R)}{dR} \exp\left(-\frac{R^2}{S^2}\right) dR; \quad (5.62)$$

$$n^2(R) = n_{co}^2\{1 - 2\Delta f(r)\}.$$

where $f(R)$ describes the profile variation, V is the fiber parameter, and Δ the relative index difference. For the step profile and the Gaussian profil with $f(R) = 1 - \exp(-R^2)$ over $0 < R < \infty$, this leads, respectively, to

$$S = \frac{1}{2\ln V^{1/2}}; \quad S = \frac{1}{(V-1)^{1/2}}, \quad (5.63)$$

with the limitation that $V > 0$ for the former and $V > 1$ for the latter. These expressions can then be used to generate analytical expressions for other modal properties of interest.

5.4.8 Depressed-cladding and W-fibers, fundamental mode cut-off

The fibers and waveguides studied so far have a uniform cladding index and a core refractive index profile that is either uniform or graded but has a minimum index equal to or greater than the cladding. This type of profile is known as a *matched-cladding* profile.

5.4.8.1 DEPRESSED-CLADDING FIBERS

In contrast, a *depressed-cladding* fiber has a depressed ring-shaped region between the core and cladding that is relatively wide compared to the core radius, as shown in Figure 5.19a. The term arises from the index mismatch that occurs between the deposited silica and the silica of the starting preform tube used in the fiber fabrication process. The depressed ring has a slightly lower index n_{dep} compared with that of the cladding.

5.4.8.2 W-FIBERS

A second type of fiber with a depressed region is known as a W-fiber. It is characterized by a depression that is relatively narrow compared to the core radius, but that has a much lower index n_{dep} compared with that of the depressed-cladding fiber, as indicated in Figure 5.19b. W-fibers were originally developed as they offer better dispersion characteristics than a matched-cladding fiber, in keeping with the discussion of pulse dispersion and multilayered fibers for dispersion flattening in Section 5.6.4.

5.4.8.3 FUNDAMENTAL MODE CUT-OFF

W-fibers and depressed-cladding fibers are normally designed to be single-mode relative to the core–cladding index difference for a particular

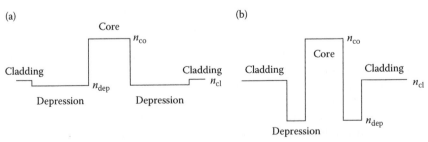

Figure 5.19 Schematics for depressed-cladding and W-profiles.

wavelength. Thus, they can support only the fundamental bound mode with an effective index value $n_{\rm eff}$ such that $n_{\rm cl} < n_{\rm eff} < n_{\rm co}$. On matched-cladding fibers, the fundamental mode does not have a finite cut-off wavelength below which it cannot propagate as a bound mode, whereas on depressed-cladding and W-fibers, the fundamental mode may have a finite cut-off wavelength depending on the width and depth of the depressed region.

There is a simple algebraic condition that determines whether or not the fundamental mode has a finite cut-off wavelength on a depressed-cladding or W-fiber. This condition may be expressed in terms of the following integrals for circular fibers and slab waveguides, respectively:

$$I = \int_0^\infty \left\{ n^2(r) - n_{\rm cl}^2 \right\} r \, dr; \quad I = \int_{-\infty}^\infty \left\{ n^2(x) - n_{\rm cl}^2 \right\} dx, \quad (5.64)$$

where n is the complete refractive index profile and $n_{\rm cl}$ is the uniform cladding index. If $I > 0$, the fundamental mode does not have a finite cut-off wavelength; but if $I < 0$, there is a finite cut-off wavelength. Note that this condition does not determine the value of the cut-off wavelength; this needs to be evaluated from the appropriate eigenvalue equation by setting $\beta = kn_{\rm cl}$ and determining the corresponding wavelength. This result has obvious applications in the design of fibers.

5.5 RAY TRACING

The propagation of guided light along any fiber or waveguide is governed by Maxwell's equations and can be formally analyzed exactly in terms of a superposition of bound modes as described in Sections 5.3 and 5.4. However, when there are a large number of modes present, as occurs in multimode waveguides and fibers, there is an alternative and accurate approach to the analysis of propagation using *ray tracing*.

Ray tracing is based on the zero-wavelength limit of Maxwell's equations and assumes that: (1) propagation is governed only by the local value of the refractive index and its variation within the core and (2) it provides a continuum solution as opposed to a summed discretized modal solution.

For real sources, the wavelengths used in optoelectronics are relatively small but nonzero, and

where there are many modes present, the modal superposition and ray solutions are almost equivalent, the relative error decreasing with increasing values of V. For these situations, rays can be regarded as local plane waves. One advantage of ray tracing is that in some situations, it provides a more physical description and interpretation of propagation compared to a summed modal analysis.

5.5.1 Snell's laws

For refractive index distributions that are piecewise continuous, propagation in the uniform media away from interfaces is described by straight-line ray trajectories equivalent to plane-wave propagation. At interfaces between regions of uniform but different indices, the transmission, reflection, and refraction of incident rays is described by *Snell's laws*.

Consider the planar interface shown in Figure 5.20 between uniform media of refractive indices $n_{\rm co}$ and $n_{\rm cl}$, where $n_{\rm co} > n_{\rm cl}$. A ray is incident in the medium of index $n_{\rm co}$ at angle θ_z relative to the interface. If $\theta_z > \theta_c$, where θ_c is the *complementary critical angle* relative to the interface (rather than the normal) and is defined by

$$\sin(\theta_c) = \left(1 - \frac{n_{\rm cl}^2}{n_{\rm co}^2} \right)^{1/2}; \quad \cos(\theta_c) = \frac{n_{\rm cl}}{n_{\rm co}}, \quad (5.65)$$

then the incident ray is known as a *refracting ray* and is both reflected and refracted as shown in Figure 5.20a. The angle of reflection is equal to the angle of incidence and the angle of transmission θ_t satisfies $n_{\rm co} \cos \theta_z = n_{\rm cl} \cos \theta_t$. Since $n_{\rm co} > n_{\rm cl}$, the angle of transmission is smaller than the angle of incidence. An important consequence of refraction is that only a fraction of the incident ray power is reflected and the balance is transmitted.

If $\theta_z < \theta_c$, then the incident ray is a *reflecting ray* and is only reflected as shown in Figure 5.20b, where the angle of reflection is equal to the angle of incidence. In this situation, all the power in the incident ray enters the reflected ray.

In the weak-guidance approximation, $n_{\rm co}$ is only slightly larger than $n_{\rm cl}$, so the complementary critical angle is relatively small and only rays incident at small angles relative to the interface are totally reflected.

5.5.2 Eikonal equation

In graded media, where the refractive index varies continuously with position, the ray paths are no longer straight lines and a generalized form of Snell's laws is required to determine the curved ray trajectories. Inasmuch that Snell's laws can be derived using plane-wave incidence, reflection, and transmission across the interface in Figure 5.20, the governing *eikonal equation* for curved ray paths can be derived from Maxwell's equations for a graded-index distribution in the limit of zero wavelength [9]. In Equation 5.66, the first expression gives this equation in general three-dimensional form where $r(x, y, z)$ is the position vector, $n(r)$ is the refractive index distribution, and s is the distance along the curved path of the ray. The second and third expressions apply to a slab waveguide with a transverse variation of index $n(x)$:

$$\frac{d}{ds}\{n(r)\}\frac{dr}{ds} = \nabla n(r); \quad \frac{d}{ds}\{n(x)\}\frac{dx}{ds} = \frac{dn(x)}{ds};$$

$$\frac{d}{ds}\{n(x)\}\frac{dz}{ds} = 0.$$

(5.66)

Similar forms apply to circular fibers.

5.5.3 Step-profile multimode slab waveguide

Assuming a symmetric, step-profile slab waveguide with a core of uniform refractive index n_{co} surrounded on either side by a cladding of uniform but lower index n_{cl}, there are two types of ray paths that can propagate along the waveguide.

5.5.3.1 BOUND RAY PATHS

These are zig-zag, straight-line paths that Snell's laws predict are totally reflected from alternate core–cladding interfaces (Figure 5.21) at a constant angle θ_z relative to the z-direction provided that this angle is smaller than the complementary critical angle, that is, $\theta_z < \theta_c$. If the waveguide materials are lossless, bound rays can propagate indefinitely along the straight waveguide and hence, the total power carried by all bound rays must remain constant everywhere along the length of the waveguide.

5.5.3.2 REFRACTING RAY PATHS

These are zig-zag, straight-line paths that are both reflected and refracted at successive core–cladding interfaces (Figure 5.22) and make a constant angle θ_z relative to the z-direction, such that $\theta_z > \theta_c$. However, the effect of refraction is to remove a constant fraction of power from the ray

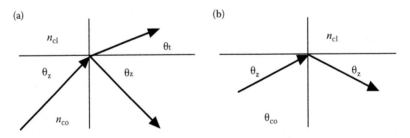

Figure 5.20 Snell's laws at the core–cladding interface.

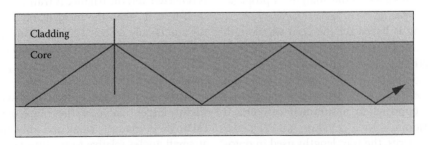

Figure 5.21 Zig-zag bound ray paths on a step-profile slab waveguide.

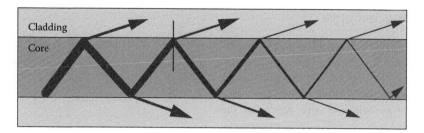

Figure 5.22 Zig-zag refracting ray paths on a step-profile slab waveguide.

propagating within the waveguide at each reflection, which is equivalent to an approximately exponential decrease in power with distance z along the waveguide. Accordingly, refracted rays can be referred to as *leaky rays* as their power decreases along the length of the waveguide, that is, they are attenuated.

Refracting rays with a propagation angle θ_z that is not too close to the complementary critical angle generally attenuate very quickly. This means that after a short distance along the waveguide, virtually all their power is lost to the cladding. Rays with θ_z close to the complementary critical angle attenuate more slowly and persist farther along the waveguide. After a sufficient distance, essentially only the power in bound rays remains.

5.5.4 Graded-profile multimode slab waveguide

Given a particular graded-index profile, the ray paths along a slab waveguide are determined from the solution of the two-dimensional form of the eikonal equations (Equation 5.66). In regions where the index is decreasing away from the index, ray paths propagating away from the waveguide axis curve back toward the waveguide axis and conversely, ray paths propagating toward the axis curve away from the axis. Put together, this means that bound rays follow a quasi-sinusoidal path that is periodic along the waveguide.

The distance over which the path repeats itself is known as the *ray period*, and the extremity of each path touches a line on either side of the axis known as the *ray caustic*. This caustic and its counterpart on the opposite side of the waveguide z-axis constitute the bounding domain for all ray paths with the same angle relative to the z-axis regardless of the longitudinal starting position. In general, the ray period will vary with the angle that the ray path

makes with the waveguide axis. However, there is a profile for which all ray directions have exactly the same ray period.

5.5.4.1 HYPERBOLIC SECANT PROFILE

The hyperbolic secant profile is symmetric about the waveguide axis and is defined by

$$n^2(x) = n_{co}^2 \sec^2(\kappa x), \qquad (5.67)$$

where x is the transverse coordinate, n_{co} is the maximum index on the waveguide axis, and κ is a scaling constant. The two-dimensional eikonal equations (Equation 5.66) can be solved analytically for this profile and confirm that every ray path has the *same ray period* z_p regardless of the angle it makes with the z-axis, where $z_p = \pi/\kappa$ (Figure 5.23). Such a waveguide is sometimes referred to as a GRIN (for gradient index) waveguide.

These waveguides also have the property that the ray half-period $z_p = \pi/\kappa$ is independent of wavelength. Furthermore, the hyperbolic secant profile produces exactly zero intermodal dispersion in the slab waveguide as discussed in Section 5.6.5, so that waveguide and material dispersion may need to be taken into account.

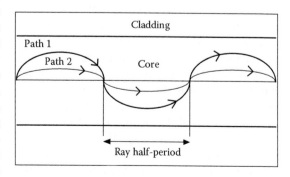

Figure 5.23 Bound ray paths with a common ray period on the hyperbolic-profile slab waveguide.

5.5.5 Multimode fibers

Ray tracing in step- and graded-profile multimode fibers comprises a straightforward generalization of the ray path tracing techniques for slab waveguides from two to three dimensions, but the reflection of straight-line core rays from the curved core–cladding interface of a step-profile fiber depends on wavelength. In the limit of zero wavelength, rays are reflected/refracted depending on the angle the ray path makes with the tangential plane on the interface, independent of the curvature of the interface. However, if the wavelength is nonzero, a third class of ray paths, known as *tunneling rays*, appears.

5.5.5.1 TUNNELING RAYS

If a ray, or local plane wave with a small but finite wavelength, is incident on the curved interface between the core and cladding of a step-profile fiber, it will be refracted if the angle of incidence α_i that it makes relative to the normal at the reflection point is less than the classical critical angle, that is, $\alpha_i < \sin^{-1}(n_{cl}/n_{co})$. For $\alpha_i < \sin^{-1}(n_{cl}/n_{co})$, there are two possible classifications of rays, depending on the angle the ray direction makes with the z-axis of the fiber, that is, θ_z. If $\theta_z < \theta_c$, where θ_c is the complementary critical angle, then the ray is totally internally reflected, but if $\theta_z < \theta_c$ such that $\alpha_i < \sin^{-1}(n_{cl}/n_{co})$, then the ray is partially reflected and partially transmitted and is known as a *tunneling ray*. Accordingly, refracting and tunneling rays are commonly referred to as *leaky rays*, since they lose power at each reflection when propagating along the core of the fiber.

The delineation between the three classes of rays is depicted in Figure 5.24, where the dark half-cone denotes refracting rays, the light half-cone denotes reflecting rays, and the two regions between them denote tunneling rays. The left-hand curved surface denotes the core–cladding interface.

Tunneling rays are partially reflected from the interface such that the angles of incidence and reflection are equal and the transmitted part reappears at a finite distance into the cladding (see Chapter 7 of Reference [4]). These rays are the analog of leaky modes that have an evanescent field between the core–cladding interface and the position at which the field-radiating field becomes oscillatory (see Section 36.11 of Reference [4]).

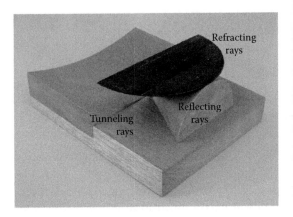

Figure 5.24 Representation of the three types of ray incident at a point on the curved core–cladding interface of a step-profile fiber.

5.5.5.2 STEP-PROFILE FIBER

The zig-zag ray paths of the slab waveguide become *meridional rays* in the circular fiber and lie in planes that intersect the axis of the fiber. There is a second class of rays that follows helical-like zig-zag paths around the fiber axis without intersecting it known as *skew rays*. A characteristic of skew rays on the step-profile fiber is that there is an inner cylindrical surface known as the *inner caustic* that every straight-line segment of the ray touches periodically along its path. The radius of this caustic varies with the skewness of the ray path. The characteristic dimensions of all these ray paths can be readily quantified using simple geometry (see Chapter 2 of Reference [4]).

5.5.5.3 GRADED-PROFILE FIBER

The sinusoidal-like ray paths of the graded-profile slab waveguide become *meridional rays* in the graded-index fiber, lie in planes that intersect the axis of the fiber, and are bounded by the *ray caustic*. All other rays follow continuous skew paths that have a helical-like shape, each ray path being bounded between the *outer* and *inner ray caustics* (see Chapter 2 of Reference [4]). The characteristics of each path are determined from the three-dimensional eikonal equation in cylindrical polar coordinates.

Unlike the hyperbolic-profile slab waveguide, there is no known refractive index profile that produces exactly zero dispersion in a multimode fiber for all bound ray directions. An approximately parabolic index profile fiber comes close to satisfying this goal and is known as a graded index or GRIN fiber.

5.5.6 Numerical aperture

Ray tracing provides a physical explanation for the NA of a fiber as defined by Equation 5.16. Using ray tracing, consider a ray from a source incident in air on the end face of a step-profile fiber at angle θ_α, as shown in Figure 5.25. Snell's laws give

$$\sin(\theta_\alpha) = n_{co} \sin(\theta_z), \qquad (5.68)$$

for the angle θ_z, the transmitted ray makes with the fiber z-axis. Recalling that the largest angle a bound ray can subtend relative to the axis is the complementary critical angle, the maximum angle of incidence in air for confinement within the core, θ_{max}, is given by Equation 5.68 with $\theta_z = \theta_c$. Again recalling the definition of NA from Equation 5.16, gives

$$\sin(\theta_{max}) = n_{co} \sin(\theta_c) = NA. \qquad (5.69)$$

Assuming that NA is small, $\theta_{max} \sim NA$ radians.

When applied to multimode fibers, this result shows that the maximum power input into bound rays (or modes) occurs when the NA is filled in a cone of angles in the range $0 < \theta_\alpha < \theta_{max}$ at every position on the end face of the core. For single-mode fibers, a ray analysis is not accurate but, nevertheless, this result suggests that filling the NA of the fiber over the core cross section should help maximize the bound mode power.

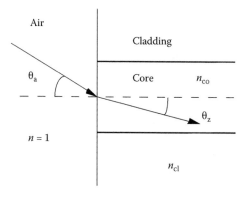

Figure 5.25 Ray paths refracting from air into the core of a fiber.

5.6 PULSE DISPERSION

5.6.1 Pulse propagation

Encoded binary information is transmitted along fibers as a sequence of discrete light pulses normally generated by a semiconductor laser and received by a semiconductor detector at the far end of the system. Each pulse is generated by modulating the light output from the laser source, or by modulating the input into the laser itself. The bandwidth of the fiber corresponds to the number of pulses, or bits, per second, for example, a 10 Gigabit s^{-1} bandwidth corresponds to 10^9 bits of information or pulses per second.

Dispersion is the phenomenon whereby each pulse changes its shape and effective length due to the physical properties of both the source and the fiber. As each pulse propagates along the fiber, the pulse power spreads out due to the dispersion of the fiber and its constituent materials; it is also attenuated because of the losses from the fiber materials due to absorption and scattering. At the low light powers considered here (of the order of milliwatts), nonlinear material effects can be neglected, that is, dispersion caused by the effect of intense light on the fiber material, which changes the refractive index value.

The maximum distance a train of pulses can propagate data is limited because (1) the pulse spread may result in an overlap with the preceding or following pulses, so that a pulse is not unambiguously detected at the end of the fiber, leading to an error and (2) attenuation due to fiber loss is so large that a pulse has insufficient power to be detected at the end of the fiber.

The material presented here provides a background to Part C1 of this book.

5.6.1.1 SOURCE WAVELENGTH VARIATION

Dispersion arises because the laser source is not purely monochromatic or of a single wavelength; it has a finite, but small, spectral width (the variation of output power intensity with wavelength) as measured by the full width at half maximum (FWHM; Figure 5.26). A typical laser has a spectral width that is less than one nanometer (10^{-9} m).

To provide a qualitative description of the effect of dispersion, consider a square pulse excited by the source of excitation that has just two monochromatic wavelengths λ_1 and λ_2 (Figure 5.27). At the

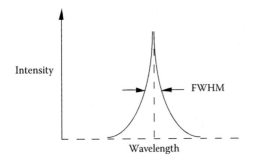

Figure 5.26 Schematic of laser output showing intensity variation with wavelength.

beginning of the fiber, the pulses coincide. Light at the two wavelengths travels at slightly different speeds (group velocity) along the fiber, such that the pulse components at wavelengths λ_1 and λ_2 also have different speeds and thus, the pulse spreads out with distance (or time). The total power in the two pulses is conserved in the absence of scattering or absorption by the fiber and the initial pulse shape becomes elongated and flattened.

To quantify the effects of dispersion on a pulse, the modal phase velocity, group velocity, and group velocity dispersion need to be defined and quantified.

5.6.2 Modal phase and group velocities

5.6.2.1 MODAL PHASE VELOCITY

Recall that the complete spatial and temporal dependence of the scalar fundamental mode field in the weak-guidance approximation for a monochromatic source has the form

$$E(x,y,z,t) = \psi(x,y)e^{i(\beta z - \omega t)}, \quad (5.70)$$

where ω is the source frequency related to the source (free-space) wavelength λ by $\omega = 2\pi c/\lambda$ and c is the speed of light in free space (in vacuo). The

modal *phase velocity*, v_{ph} is the speed with which the planar phase front of a mode in the fiber cross section propagates along the waveguide and is expressed as

$$v_{ph} = \frac{\omega}{\beta} = \frac{2\pi c}{\lambda \beta} = \frac{c}{n_{eff}}, \quad (5.71)$$

where n_{eff} is the effective refractive index. The modal phase velocity is equal to the phase velocity of an infinite plane wave propagating in a uniform medium of index n_{eff}. Since the modal propagation constant β, or, equivalently, the effective index n_{eff}, depends on the source wavelength λ (or color) through the eigenvalue equation for the fiber, the phase velocity is also wavelength dependent, that is, $v_{ph} = v_{ph}(\lambda)$.

5.6.2.2 GROUP VELOCITY

The speed with which the power in a pulse propagates along the fiber is given by the *group velocity*, v_g. and *not* by the phase velocity. Modal group velocity is defined by the differential

$$v_g = \frac{d\omega}{d\beta}. \quad (5.72)$$

In terms of the source wavelength λ, this definition is equivalent to

$$v_g = \frac{d\omega}{d\beta} = \frac{d\lambda}{d\beta}\frac{d\omega}{d\lambda} = \frac{d\lambda}{d\beta}\frac{d}{d\lambda}\left(\frac{2\pi c}{\lambda}\right) = -\frac{2\pi c}{\lambda^2}\frac{d\lambda}{d\beta}. \quad (5.73)$$

Setting $\beta = k n_{eff} = 2\pi n_{eff}/\lambda$ and differentiating this expression with respect to wavelength

$$\frac{d\beta}{d\lambda} = \frac{2\pi}{\lambda}\frac{dn_{eff}}{d\lambda} - \frac{2\pi n_{eff}}{\lambda^2}. \quad (5.74)$$

Substitution into Equation 5.73 leads to the expression for the modal group velocity in terms of effective index

Figure 5.27 Pulse spread due to source spectral width.

$$v_g = \frac{c}{n_{eff} - \lambda \dfrac{dn_{eff}}{d\lambda}}. \quad (5.75)$$

The second term in the denominator accounts for the waveguide dispersion due to the finite spectral width of the source. If this term were not present, the group and phase velocities would be equal, which is the case for an infinite plane wave propagating in a uniform medium of index n_{eff}. Since β, or n_{eff}, depends on the source wavelength λ through the waveguide eigenvalue equation, the group velocity is wavelength dependent, that is, $v_g = v_g(\lambda)$. As the group velocity depends on the solution of the eigenvalue equation, its value is determined numerically for a general fiber profile.

5.6.3 Transit time and pulse spread

If the group velocity at the central wavelength λ_c of the source of excitation is denoted by $v_g(\lambda_c)$, then the *transit time* t of the center of the pulse over length L of fiber is given by

$$t = \frac{L}{v_g(\lambda_c)}. \quad (5.76)$$

In terms of the effective index n_{eff}, it is straightforward to show from Equation 5.75 that this expression is equivalent to

$$t = \frac{Ln_{eff}}{c} - \frac{L}{c}\lambda \frac{dn_{eff}}{d\lambda}, \quad (5.77)$$

where the first term on the right is the transit time for a pulse propagating the distance L in a medium of uniform index n_{eff} and the second term accounts for the dispersion caused by the waveguide.

5.6.3.1 GROUP VELOCITY DISPERSION

Now consider a light pulse with an initial time width τ that is excited at the beginning of the fiber. The temporal spread $\delta\tau$ in t after propagating distance L along the waveguide is due to the source spectral width $\delta\omega_s$ (in frequency units) and the corresponding spread δv_g in the range of values about the mean value of the group velocity. This spread is sometimes referred to as *group velocity dispersion*. The expression for $\delta\tau$ is obtained by differentiating

the transit time expression for t given in Equation 5.77. Thus

$$\delta\tau = \frac{dt}{d\omega}\delta\omega = -\frac{L}{v_g^2}\frac{dv_g}{d\omega}\delta\omega_s = L\frac{d^2\beta}{d\omega^2}\delta\omega_s, \quad (5.78)$$

where the final expression on the right follows from the definition of the modal group velocity given in Equation 5.72.

In terms of the equivalent spectral width $\delta\lambda_s$ (in wavelength units) of the source, this expression can be recast as

$$\delta\tau = \frac{dt}{d\lambda}\delta\lambda = -\frac{L}{v_g^2}\frac{dv_g}{d\lambda}\delta\lambda_s. \quad (5.79)$$

If v_g is expressed in terms of the effective index n_{eff} of the mode using Equation 5.75, it is straightforward to show that

$$\delta\tau = -\frac{L\lambda}{c}\frac{d^2n_{eff}}{d\lambda^2}\delta\lambda. \quad (5.80)$$

If there is no net dispersion, that is, n_{eff} is either independent of λ or varies linearly with λ, then there is no pulse dispersion.

5.6.4 Sources of dispersion

When an optical pulse propagates along a practical fiber with an arbitrary index profile and cross-sectional geometry, the *total pulse dispersion* can be a combination of several different contributing effects (Figure 5.28).

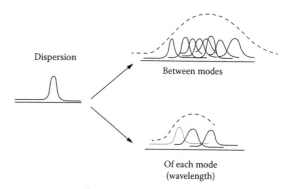

Figure 5.28 Dispersion includes intermodal dispersion between different modes and the dispersion of individual modes.

5.6.4.1 INTERMODAL DISPERSION

This form of dispersion occurs when two or more different modes propagate simultaneously and arises because each mode has a different group velocity for the same wavelength. This phenomenon occurs regardless of the spectral width of the source. Intermodal dispersion contributes the most to dispersion in multimode fibers, but can be minimized using multimode fibers with a graded-index fiber as discussed in Section 5.5.5.

5.6.4.2 MATERIAL DISPERSION

In any optical material, the refractive index of the bulk material varies slightly with wavelength, that is, $n=n(\lambda)$, in addition to any spatial variation. For example, the refractive index of the pure silica cladding of a fiber decreases from 1.458 at 633 nm (wavelength of a HeNe laser) to 1.447 at 1300 nm, and to 1.444 at 1550 nm. The same relative decrease in index is normally assumed for the lightly doped core material of the fiber.

5.6.4.3 WAVEGUIDE OR FIBER DISPERSION

This form of dispersion occurs in each and every mode of a fiber because of the wavelength dependence of the modal group velocity and the slight spectral width of the source. The combined effects of material and waveguide dispersion are sometimes termed *chromatic dispersion.*

5.6.4.4 POLARIZATION-MODE DISPERSION

In a noncircular single-mode fiber, such as a weakly guiding elliptical fiber, the two polarizations of the fundamental mode are parallel to the major and minor axes and generally have distinctly different propagation constants. Dispersion is then dominated by the intermodal dispersion between these two polarizations. In a perfectly circular single-mode fiber, the fundamental mode propagation constant is independent of its polarization state. Further, the polarization of the mode does not change as it propagates along the fiber as long as it remains translationally invariant along its entire length.

Practical telecommunications grade single-mode fibers are neither perfectly circular nor perfectly concentric, that is, there are very slight nonuniformities that occur along their entire length. These nonuniformities have the effect of introducing random coupling between the two polarization states and, as a result, the accumulated group delay in each state can differ and therefore, cause an overall dispersion known as *polarization-mode dispersion.* The delay difference cannot be predicted theoretically and has to be measured. In practice, it is offset actively. This type of dispersion is only significant in very long single-mode fiber transmission systems over thousands of kilometers, that is, to submarine systems, as terrestrial systems are generally, at most, a few hundreds of kilometers long.

5.6.5 Zero-dispersion, dispersion-shifted, flattened, and compensating fibers

5.6.5.1 ZERO-SHIFTED FIBERS

In principle, intermodal dispersion can be avoided altogether if only single-mode fibers are employed. This arrangement only became a practical proposition with the development of techniques for producing low-loss single-mode fibers in the 1970s. It was then possible to design a single-mode fiber so that the combination of the fiber dispersion together with the material dispersion is zero at just one wavelength. This involves a complex design procedure, as fiber and material dispersion interact with one another, that is, they cannot be calculated separately and then simply added together. The first such designs provided a *zero-dispersion* fiber at the first fiber transmission loss minimum of 1300 nm, as shown in Figure 5.29a.

5.6.5.2 DISPERSION-SHIFTED FIBERS

With the introduction of optical transmission systems operating close to the absolute minimum of silica fiber transmission loss around 1550 nm, a single-mode fiber with zero dispersion at this wavelength was required. As material dispersion is essentially fixed, an approximately triangular-shaped profile was developed to meet this need, as shown in Figure 5.29b. Such fibers are known as *dispersion-shifted* fibers. By judicious profile design, it is also possible to produce fibers that are wavelength flattened, that is, have close to zero dispersion over the entire wavelength range of 1300–1550 nm.

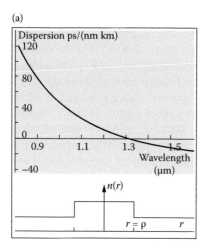
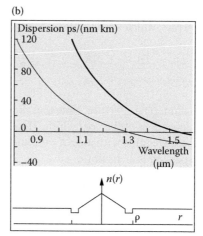

Figure 5.29 Plots of total dispersion against wavelength for (a) standard fiber and (b) dispersion-shifted fiber (right).

5.6.5.3 DISPERSION-COMPENSATING FIBERS

At the opposite extreme, by combining the effects of fiber and material dispersion in such a way to enhance dispersion, single-mode fibers can be produced that have a very high and negative dispersion, of the order of -200 ps nm^{-1} km^{-1}. Such fibers have application, for example, in 1300 nm transmission systems that are operated at 1550 nm to take advantage of erbium-doped optical amplifiers. The fibers in these systems have zero dispersion at 1300 nm but a dispersion of about 20 ps nm^{-1} km^{-1} when operated at 1550 nm. By inserting lengths of the large negative dispersion fiber equal to 10% of the length of the 1300 nm fiber between repeaters, the introduced dispersion can be completely offset.

5.6.6 Dispersion in multimode waveguides and fibers

By definition, multimode waveguides and fibers support a large number of bound modes with propagation constants that take on almost a continuum of values in the range $kn_{cl} < \beta < kn_{co}$, where n_{cl} is the uniform cladding index and n_{co} is the maximum core index. Accordingly, pulse dispersion is strongly influenced by intermodal dispersion but insensitive to source wavelength. There was a flurry of research into minimizing dispersion in multimode fibers in the early 1970s prior to

the development of techniques for fabricating low-loss single-mode fibers.

5.6.6.1 STEP-PROFILE MULTIMODE SLAB WAVEGUIDES

A simple analytical example of intermodal dispersion is provided by the step-profile slab waveguide. Consider excitation of the waveguide shown in Figure 5.30. Assuming an incoherent source, such as a light emitting diode (LED), all bound modes are excited approximately equally, which is well modeled by assuming that all bound ray directions are equally excited within the core. As shown in Section 5.5.3, the range of bound ray directions for the step profile satisfies $0 < \beta_z < \beta_c$, where β_z is the angle with the z-axis.

As light propagates as a local plane wave, its phase and group velocities in the core are equal, that is, $v_{ph} = v_g = c/n_{co}$. Accordingly, the accumulated dispersion, δt, over length L of the waveguide is given by

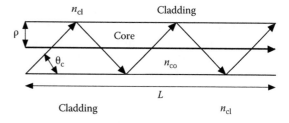

Figure 5.30 Bound ray path on a step-index slab waveguide.

the difference in transit times of the slowest ($\theta_z = \theta_c$) and fastest ($\theta_z = 0$) rays, which leads to

$$\delta t = \frac{Ln_{co}^2}{cn_{cl}} - \frac{Ln_{co}}{c} = \frac{L}{c}\frac{NA^2}{n_{cl}}, \tag{5.81}$$

where NA is the numerical aperture. Note that this expression is independent of the core width. Thus, for example, for a waveguide with an NA = 0.1 and $n_{cl} = 1.45$, the dispersion is 23 ns km⁻¹.

5.6.6.2 ZERO INTERMODAL DISPERSION MULTIMODE SLAB WAVEGUIDE

Intermodal dispersion in slab waveguides can be totally suppressed through a judicious choice of the refractive index profile, as discussed in Section 5.5.4. There is a unique profile, the *hyperbolic secant* profile of Equation 5.67 for which the transit time for all off-axis ray paths is equal to the on-axis transit time. Note that if intermodal dispersion is totally suppressed, both waveguide and material dispersion are still present.

5.6.6.3 MINIMIZING INTERMODAL DISPERSION IN MULTIMODE FIBERS

Using ray tracing, intermodal dispersion in circular fibers can be readily determined, by analogy, with the aforementioned analysis for the slab waveguide. Like the step-profile slab waveguide, intermodal dispersion in a multimode step-profile fiber is relatively large. However, unlike the hyperbolic secant slab waveguide, there is no known profile for the fiber that leads to exactly zero intermodal dispersion when all ray directions, both meridional and skew, are excited.

Nevertheless, it is possible to minimize intermodal dispersion using a core profile that is approximately parabolic, as might be anticipated from the approximately parabolic variation of the hyperbolic secant profile. The term *graded-index fiber* is often used to denote a multimode fiber with this core profile surrounded by the normal uniform cladding. However, the bandwidth of a single-mode fiber is still significantly larger than that of a graded-index fiber.

5.7 BEND LOSS

A mode propagating on a straight fiber or waveguide fabricated from nonabsorbing, nonscattering materials will in principle propagate indefinitely

without any loss of power. However, if a bend is introduced, the translational invariance is broken and power is lost from the mode as it propagates into, along and out of the bend. This applies to the fundamental mode in the case of single-mode fibers and waveguides and to all bound modes in the case of bent multimode fibers or waveguides.

It is conventional to distinguish between the two types of losses that can occur on a bend. *Transition loss* is associated with the abrupt or rapid change in curvature at the beginning and the end of a bend, and *pure bend loss* is associated with the loss from the bend of constant curvature in between.

5.7.1 Transition loss

Consider an abrupt change in the curvature κ from the straight waveguide ($\kappa = 0$) to that of the bent waveguide of constant radius R_b ($\kappa = 1/R_b$). The fundamental mode field is shifted slightly outward in the plane of the bend, thereby causing a mismatch with the field of the straight waveguide. The fractional loss in fundamental mode power, $\delta P/P$, can be calculated from the overlap integral between the fields. Within the Gaussian approximation to the fundamental mode field and assuming that the spot size s and core radius or half-width ρ are approximately equal, this gives

$$\frac{\delta P}{P} \approx \frac{1}{16}\frac{V^4}{\Delta^2}\frac{\rho^2}{R_b^2}. \tag{5.82}$$

where V is the fiber or waveguide parameter and Δ is the relative index difference. For typical values of $V = 2.5$, $\Delta = 0.005$, $\rho = 5$ mm and a bend radius of 5 mm, the transition loss is about 9.8% of the fundamental mode power.

5.7.1.1 MINIMIZING TRANSITION LOSS

There are a couple of strategies for significantly reducing transition loss. In the case of planar waveguides, it is often possible to fabricate the bend so that there is an abrupt offset between the cores of the straight and bent waveguides in the plane of the bend. In Figure 5.31, this can be seen as being equivalent to displacing the bent core downward so that the two fundamental mode fields overlap. Alternatively, if a gradual increase in curvature is introduced between the straight and uniformly bent sections, the fundamental field of the straight

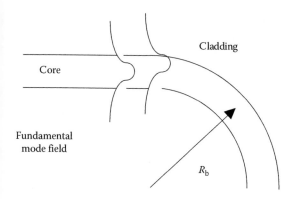

Figure 5.31 Outward shift in the fundamental mode electric field on entering a bend.

waveguide will evolve approximately adiabatically into the offset field of the uniformly bent section.

5.7.2 Pure bend loss

In following the curved path of the core of constant radius R_b, the fundamental mode continuously loses power to radiation, known as *pure bend loss*. In assuming that the cladding is essentially unbounded, the physical effect is analogous to radiation from a bent antenna in free space. The radiation increases rapidly with decreasing bend radius and occurs predominantly in the plane of the bend; in any other plane, the effective bend radius is larger and hence the loss is very much reduced (Figure 5.32).

The physical basis of bend loss can be interpreted in terms of the motion of the planar phase front of the fundamental mode as it rotates about an axis through the center of the bend at C with

angular velocity Ω. On the straight waveguide, the phase front travels with phase velocity $v_\phi = \omega/\beta$, where ω is the source frequency and β the propagation constant. By matching the phase velocities of the straight and bent waveguides on the fiber axis at the beginning of the bend

$$R_b\Omega = \frac{\omega}{\beta} \Rightarrow \Omega = \frac{\omega}{\beta R_b}$$
$$= \frac{2\pi c}{\lambda \beta R_b} = \frac{c}{n_{eff} R_b}, \qquad (5.83)$$

in terms of the source wavelength λ and the fundamental mode effective index n_{eff}. Along the planar phase front through C, the phase velocity increases monotonically with the distance from C, until at a radius R_{rad}, the phase velocity is equal to the speed of light in the cladding, that is, when

$$\Omega R_{rad} = \frac{c}{n_{cl}} \Rightarrow R_{rad} = \frac{c}{\Omega n_{cl}} = \frac{n_{eff}}{n_{cl}} R_b. \qquad (5.84)$$

Since by definition $n_{eff} > n_{cl}$, it follows that $R_{rad} > R_b$.

5.7.2.1 RADIATION CAUSTIC

The phase velocity anywhere on the modal phase front rotating around the bend cannot exceed the speed of light in the cladding. Hence, beyond radius R_{rad}, the modal field must necessarily radiate into the cladding, the radiation being emitted tangentially. The interface between the guided portion of the modal field around the bend and the radiated portion at R_{rad} is known as the *radiation caustic*, and is the apparent origin of radiation.

Between the core and the radiation caustic, the modal field is evanescent and decreases approximately exponentially with increasing radial distance from C. As the bend radius increases, the radiation caustic moves farther into the cladding, and the level of radiated power decreases. Conversely, as the bend radius decreases, the radiation caustic moves closer to the core and the radiation loss increases. In practical fibers, the radiated power from the bend is either absorbed by the acrylic coating surrounding the outside of the cladding or propagates through the coating into free space.

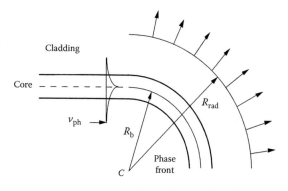

Figure 5.32 Rotation of the mode phase front along a bend.

5.7.2.2 MODE ATTENUATION

If z is the distance from the beginning of the bend relative to the fiber axis, then the total fundamental mode power $P(z)$ attenuates according to

$$P(z) = P(0)e^{-\gamma z}, \qquad (5.85)$$

where $P(0)$ is the total fundamental mode power entering the bend and γ is the power attenuation coefficient. In decibels, this relationship is equivalent to

$$dB = 10\log_{10}\left(\frac{P(z)}{P(0)}\right) = 10\log_{10}(e^{-\gamma z}) \qquad (5.86)$$

$$= (20\log_{10}e)\gamma z = 8.686\gamma z,$$

indicating that the loss of power per unit length of bent fiber is 8.686γ dB.

5.7.2.3 STEP-PROFILE FIBER

In terms of the core and cladding modal parameters U and W, respectively, relative index difference A, core radius ρ, fiber parameter V, and the bend radius R_b, an approximate expression for γ for the fundamental mode of a step-profile fiber has the form (see Chapter 23 of Reference [4])

$$\gamma = \left(\frac{\pi\rho}{R_b}\right)^{1/2}\frac{V^2 W^{1/2}}{2\rho U^2}$$

$$\exp\left\{-\frac{4}{3}\Delta\frac{R_b}{\rho}\frac{W^3}{V^2}\right\}, \qquad (5.87)$$

where R_b is necessarily large compared to ρ because it is not possible to bend a fiber into a radius much below 1 cm without breakage. The pure bend loss coefficient is most sensitive to the expression inside the exponent because $R_b \gg \rho$. Loss decreases very rapidly with increasing values of R_b or Δ or V (since W also increases with V), and becomes arbitrarily small as $R_b \to \infty$.

Consider, for example, a standard single-mode fiber such as the Corning SMF28, for which $\Delta = 0.3\%$, $\rho = 4\,\mu m$, and NA $= 0.12$. At a wavelength of 1.3 µm, the fiber parameter is $V = 2.32$. The eigenvalue equation for the step profile gives $U = 1.66$, $W = 1.62$, and the power attenuation coefficient of Equation 5.87 reduces to

$$\gamma = \frac{3.48 \times 10^4}{\sqrt{R_b}}\exp(-0.78 R_b), \qquad (5.88)$$

where R_b is in millimeters and γ has units of m^{-1}. Accordingly, the fractional power loss from the fundamental mode due to pure bend loss in a single loop of fiber is expressible as

$$\frac{P(0) - P(2\pi R_b)}{P(0)} = 1 - \exp\left(-\frac{\pi\gamma R_b}{500}\right), \qquad (5.89)$$

where R_b is in millimeters. Thus, if $R_b = 150$ mm (the radius of the standard drum on which fiber is normally spooled), then $\gamma \sim 10^{47}$ m^{-1} and loss is totally negligible, but if $R_b = 10$ mm, then $\gamma = 4.5$ m^{-1} and the loss is almost 25% in one loop of fiber.

5.7.3 Bend loss in multimode fibers and waveguides

Uniformly bent multimode waveguides lose power because each mode is attenuated as it propagates around the bend. Like the fundamental mode, the power attenuation coefficient for each mode is dominated by the common exponential dependence

$$\exp\left\{-\frac{4}{3}\Delta\frac{R_b}{\rho}\frac{W^3}{V^2}\right\}$$

$$= \exp\left\{-\frac{4}{3}kR_b\Delta\frac{(n_{eff}^2 - n_{cl}^2)^{3/2}}{(n_{co}^2 - n_{cl}^2)}\right\}, \qquad (5.90)$$

where the right-hand expression follows by setting $\beta = kn_{eff}$ and $V = k\rho(n_{co}^2 - n_{cl}^2)^{1/2}$.

Recall that the effective index of every bound mode of the straight fiber mode lies in the range $n_{cl} < n_{eff} < n_{co}$, the lower order modes having n_{eff} values closer to n_{co} and the higher order modes having n_{eff} values closer to n_{cl}. It follows from Equation 5.90 that the higher order modes are attenuated much more quickly than the lower order modes. However, the overall attenuation along the bent multimode waveguide is determined by the attenuation of each mode. Since the equilibrium distribution of power between modes also depends on the slight variations along the waveguide, the overall attenuation around a bend must take account of this distribution. As a result, there is no simple analytical expression for bend loss in multimode waveguides.

Numerical quantification of bend loss for multimode slab waveguides and fibers with step and parabolic profiles have been undertaken using a

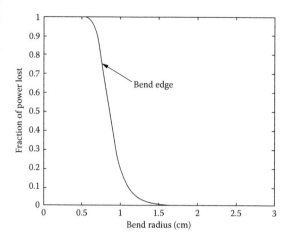

Figure 5.33 Bend loss as a function of bend radius.

combination of ray tracing and loss coefficients (Figure 5.33). The loss coefficients account for bend loss at reflections from the core–cladding interface, in the case of the step profile, and from the outer caustic in the case of the parabolic profile. For the slab waveguide, this involves a two-dimensional integration over the core width and the range of ray directions, while for the fiber, it involves a four-dimensional integration of the core cross-sectional area and all meridional and skew ray paths (see Chapter 9 of Reference [4]).

5.8 EXCITATION, REFLECTION, MISMATCH, OFFSET, AND TILT

Waveguides and fibers are illuminated by a variety of sources, including lasers, LEDs, or light propagating out of another fiber or waveguide. Because of the relatively small core size of fibers and waveguides, it is necessary to use an intense source to ensure that sufficient light enters the fiber to be detectable at the far end. The goal is normally to maximize the fraction of incident light that can excite the fundamental mode in the case of a single-mode fiber or all the bound modes in the case of a multimode fiber.

However, an analysis of excitation is a complex issue because of (1) reflection from the fiber end face, (2) mismatch between the source and fiber modes, (3) offset between the source and the fiber, and (4) tilt between the source and the fiber. Further, an exact mathematical formulation of the problem, even within the weak-guidance

approximation, would require an appropriate representation of both the reflected and transmitted bound mode and radiation fields, and the solution of an extremely complex and large matrix of equations. Fortunately for most practical applications, it is not necessary to undertake such an analysis. Here, we provide some insight and methodology that provides an accurate quantification of all these effects for most practical applications.

5.8.1 End-face reflection

If a fiber is excited by light propagating in air, the light will be partially reflected because of the difference in indices between air and the fiber material. Consider the simple situation shown in Figure 5.34 where a plane wave, representing the incident light, impinges normally on the end face of the fiber with a scalar transverse electric field E_i. A fraction of the field is reflected into the backward-propagating field E_r and the balance is transmitted into the fiber or waveguide.

A simple way to estimate the fraction of reflected light is to approximate the fiber or waveguide core–cladding cross section by a uniform, infinite medium of refractive index approximately equal to the core index n_{co} since the core and cladding indices are similar in the case of weak guidance. For the incident plane wave, Fresnel's laws determine the fraction of reflected light power according to

$$\left(\frac{n_{co} - 1}{n_{co} + 1} \right)^2. \tag{5.91}$$

For a typical fiber core with index $n_{co} = 1.46$, this formula gives around 3.5% for the fraction of incident power reflected. It should also be noted that the same fraction of light will be reflected from the

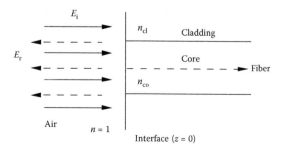

Figure 5.34 Reflection of light from the end of a fiber or waveguide.

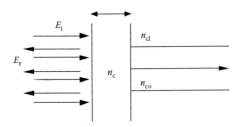

Figure 5.35 Schematic of an anti-reflecting coating on the end of a fiber or waveguide.

propagating light impinging on the end face at the far end of the fiber, assuming it is also located in air.

5.8.2 Anti-reflecting coating

The reflection of light from the end face of a fiber can be suppressed by depositing a thin antireflection coating with index different from that of the fiber, as shown schematically in Figure 5.35. The coating has uniform thickness δ and refractive index n_c. A simple plane-wave analysis of propagation across this layer at normal incidence from air into the fiber shows that reflection can be totally suppressed from the end face between the coating and air provided that the following two conditions are satisfied:

$$d = \frac{m\lambda}{4n_c}, m = 1,3,5,\ldots; \quad n_c = \sqrt{n}, \quad (5.92)$$

where $m = 1, 3, 5, \ldots$, n is the mean fiber index and λ is the source wavelength.

5.8.3 Mode excitation

Assume that the transmitted scalar electric field of the source on the fiber or waveguide side of the interface at $z=0$ is given by the scalar expression $\psi_s(x, y)$, where x, y are Cartesian coordinates in the cross section (Figure 5.36). This function is assumed to take into account any reflection loss from the interface and also contains the implicit monochromatic time dependence $\exp(-i\omega t)$.

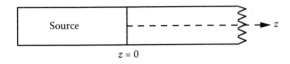

Figure 5.36 Schematic of a source abutting the end of a fiber or waveguide at $z=0$.

The complete field everywhere within the core and cladding for $z > 0$ consists, in general, of the sum of the discrete set of bound modes and the radiation field. Since radiated power is lost from the fiber through the cladding, the total bound mode field, $\psi_s(x, y)$, is expressible as a sum over all bound modes

$$\psi_s(x,y) = \sum_{m=1}^{M} a_m \psi_m(x,y)e^{i\beta_m z}, \quad (5.93)$$

where the a_m are amplitude constants, $\psi_m(x, y)$ is the solution of the scalar wave equation for the field of the mth mode, β_m is the propagation constant for the mth mode, and the summation is over all M bound modes. For a single-mode fiber, $M=m=1$, and the polarization of the exciting field is assumed to be parallel to the mode polarization.

5.8.3.1 MODAL AMPLITUDES

The expansion over bound modes and the radiation field ψ_{rad} must match the transmitted field on the end face at $z=0$. Hence

$$\psi_s(x,y) = \sum_{m=1}^{M} a_m \psi_m(x,y) + \psi_{rad}. \quad (5.94)$$

The amplitude of the kth bound mode is then obtained by multiplying this equation on both sides by $\psi_k(x, y)$, and integrating over the infinite cross-section A_∞. Since, by definition, any bound mode field is necessarily orthogonal to the radiation field

$$\int_{A_\infty} \Psi_s(x,y)\psi_k(x,y)dxdy$$

$$= \sum_{m=1}^{M} a_m \int_{A_\infty} \psi_m(x,y)\psi_k(x,y)dxdy. \quad (5.95)$$

The amplitude coefficient a_m of each mode is then determined by using the orthogonality property of the scalar modal fields

$$\int_{A_\infty} \psi_m(x,y)\psi_k(x,y)\,dxdy = 0 \quad \text{if } m \neq k. \quad (5.96)$$

Accordingly, the right-hand side of Equation 5.95 reduces to a single term when $k=m$ and the amplitude of the kth excited mode is given by

$$a_k = \int_{A_\infty} \Psi_s(x,y)\psi_k(x,y)\,dxdy \Bigg/ \int_{A_\infty} \psi_k(x,y)^2\,dxdy. \quad (5.97)$$

Note that for a circular fiber, the Cartesian element of area $dxdy \to rdrd\phi$ in polar coordinates with $0 < r < \infty$ and $0 < \phi < 2\pi$.

5.8.3.2 MODAL POWER

Equation 5.28 determines that the total power P_k propagating in each excited mode is proportional to

$$(a_k)^2 \int_{A_\infty} (\psi k)^2 \, dxdy, \qquad (5.98)$$

where A_∞ is the infinite cross section of the fiber or waveguide and $k = 1, 2, \ldots, M$. On substituting a_k from Equation 5.97, the fraction of transmitted power in each excited mode relative to the total power exciting the fiber is given by

$$\frac{\left(\int_{A_\infty} \psi_k \Psi_s \, dxdy \right)^2}{\int_{A_\infty} (\psi_k)^2 \, dxdy \int_{A_\infty} (\Psi_s)^2 \, dxdy}. \qquad (5.99)$$

The total fraction of incident power that is transmitted into bound modes is given by summing Equation 5.99 over all values of k. Thus, the fraction of total exciting power lost as radiation is given by

$$1 - \sum_{k=1}^{M} \frac{\left(\int_{A_\infty} \psi_k \Psi_s \, dxdy \right)^2}{\int_{A_\infty} (\psi_k)^2 \, dxdy \int_{A_\infty} (\Psi_s)^2 \, dxdy}. \qquad (5.100)$$

5.8.4 Fiber splicing, mismatch, offset, and tilt

Optical fibers are normally spliced together by removing the coating (mechanically or chemically),

cleaving the fiber ends so that their end faces are flat and perpendicular to the fiber axis and butting them together in a V-groove. The fibers are then heated by an electric arc to soften the glass and fuse them together to form a smooth continuous joint. Splicing fibers using this technique essentially requires fibers with the same cladding diameter, commonly referred to as the *outer diameter* or *o/d*, although their core refractive index profiles and radii can differ quite significantly.

5.8.4.1 SPLICE LOSS

The difference between the cross-sectional profile and geometry of the cores of the two fibers is known as the *mismatch*. For single-mode fibers, the mismatch accounts for the loss of power at the splice as the fundamental mode fields of the two fibers will not be identical. Additional losses will occur if the fibers are *offset* transversely or if their axes are no longer parallel and have a relative *tilt*.

Splice loss, due to any of the three loss mechanisms shown schematically in Figure 5.37, can formally be calculated and quantified using the analysis of mode excitation in the previous section but with the input source being the field of the fundamental mode of the input fiber. However, such an analysis would require detailed knowledge of the refractive index profiles of both fibers and a formal solution of the scalar wave equation (numerically) to determine the fields of the fundamental mode of each fiber. A simpler approach, which requires knowledge of only each fiber's spot size (taken from fiber measurement) and gives an approximate but accurate measure of splice loss, is to use the Gaussian approximation of Section 5.4.7 to represent the field of the fundamental mode of each fiber.

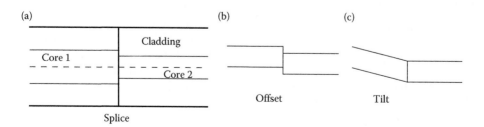

Figure 5.37 (a) Mismatch, (b) offset, and (c) tilt between fiber cores.

5.8.4.2 MISMATCH LOSS

Accordingly, if ψ_1 and ψ_2 denote the fundamental fields of fibers 1 and 2, respectively, in the weak-guidance approximation, then the representation of the Gaussian approximation is

$$\psi_1(r) = a_1 \exp\left\{-\frac{r^2}{2s_1^2}\right\};$$

$$\psi_2(r) = a_2 \exp\left\{-\frac{r^2}{2s_2^2}\right\}, \tag{5.101}$$

where s_1 and s_2, a_1 and a_2 are, respectively, the spot sizes and amplitudes for fiber 1 and fiber 2, and r is the cylindrical radial coordinate measured from the fiber axis.

The power P_1 propagating in the fundamental mode in fiber 1 is obtained by integrating the local power flow density parallel to the fiber axis over the infinite cross section relative to polar coordinates (r, ϕ) in the fiber cross section. In weak guidance, the power flow density in the z-direction is given by Equation 5.28. Since ψ_1 is axisymmetric, the ϕ integration is readily performed and leads to

$$P_1 = \frac{n_{co}}{2}\left(\frac{\varepsilon_0}{\mu_0}\right)^{1/2}\int_2^{2\pi} r\psi_1^2 dr d\phi$$

$$= \pi n_{co}\left(\frac{\varepsilon_0}{\mu_0}\right)^{1/2}\int_0^{\infty} r\psi_1^2 dr. \tag{5.102}$$

Substituting for ψ_1 leads to a standard integral

$$P_1 = \pi(a_1)^2 n_{co}\left(\frac{\varepsilon_0}{\mu_0}\right)^{1/2}\int_2^{\infty} r\exp\left\{-\frac{r^2}{s^2}\right\}dr$$

$$= \frac{\pi n_{co}}{2}\left(\frac{\varepsilon_0}{\mu_0}\right)^{1/2} s_1^2 a_1^2. \tag{5.103}$$

Similarly, the power P_2 propagating in the fundamental mode in fiber 2 is given by

$$P_2 = \frac{\pi n_{co}}{2}\left(\frac{\varepsilon_0}{\mu_0}\right)^{1/2} s_2^2 a_2^2. \tag{5.104}$$

There is negligible reflected power at the splice because both fibers are weakly guiding and their refractive index profile values are very similar. In the pure silica cladding, the indices are of course identical.

The amplitude of the transmitted mode, a_2, is determined by using the fundamental mode field Ψ_1 in fiber 1 as the source of excitation (ψ_s) of the fundamental mode ψ_2 (ψ_k) for fiber 2 in Equation 5.97, leading to

$$a_2 = \frac{\displaystyle\int_{A_\infty} \Psi_1 \psi_2 \, dx \, dy}{\displaystyle\int_{A_\infty} \psi_2^2 \, dx \, dy}$$

$$= a_1 \frac{\displaystyle\int_0^{\infty} r\exp\left(-\frac{r^2}{2}\left[\frac{1}{s_1^2}+\frac{1}{s_2^2}\right]\right)dr}{\displaystyle\int_0^{\infty} r\exp\left(-\frac{r^2}{s_2^2}\right)dr}. \tag{5.105}$$

By analogy with the derivation of the aforementioned power integrals, these integrals are readily evaluated and lead to

$$a_2 = a_1 \frac{2s_1^2}{s_1^2 + s_2^2}. \tag{5.106}$$

Using Equations 5.105 and 5.106, the fractions of transmitted power and the fraction of power lost are given, respectively, by

$$\frac{P_2}{P_1} = \frac{s_2^2 a_2^2}{s_1^2 a_1^2} = \frac{4s_1^4}{(s_1^2+s_2^2)^2}\frac{s_2^2}{s_1^2}; \quad 1-\frac{P_2}{P_1}$$

$$= 1 - \frac{4s_1^4}{(s_1^2+s_2^2)^2}\frac{s_2^2}{s_1^2} = \left(\frac{s_1^2-s_2^2}{s_1^2+s_2^2}\right)^2. \tag{5.107}$$

Thus, for example, a 1% relative difference in the two spot sizes gives rise to a power loss of only 0.01%, whereas a 10% relative difference gives a 1% loss.

5.8.4.3 OFFSET AND TILT LOSSES

If two identical single-mode fibers with the same spot size 5 are offset parallel to one another by distance d, a similar analysis to the Gaussian approximation for the mismatch loss gives the first expression in Equation 5.108 for the fraction of power loss. Similarly, if the same two fibers have a relative tilt angle θ_s, the second expression in Equation 5.108 determines the fraction of power loss:

$$1-\exp\left(-\frac{d^2}{2s^2}\right); \quad 1-\exp\left(-\frac{(kn_{co}\theta_s)^2}{2}\right), \tag{5.108}$$

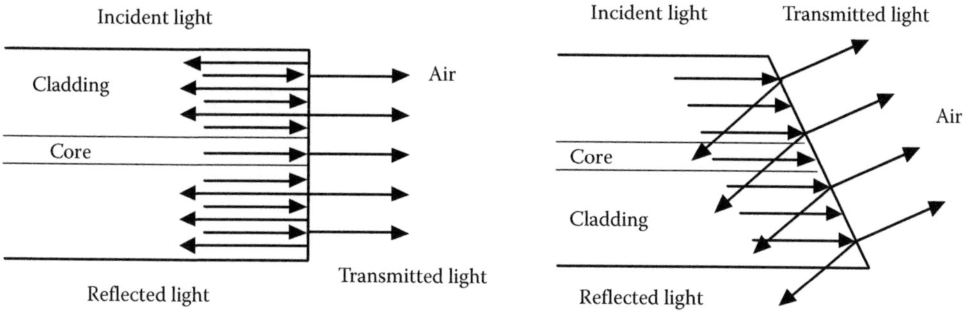

Figure 5.38 Reflection from the far end face of a fiber without and with an angle face.

where k is the free-space wave number and n_{co} is the core index value. If both offset and tilt losses are present, then the two expressions in Equation 5.108 are multiplied together to determine the overall loss. Similarly, if mismatch loss is also present, then a third multiplicative factor given by the second expression in Equation 5.107 should be included.

5.8.5 Near and far fields

5.8.5.1 NEAR FIELD

The *near field* refers to the field of a fiber or waveguide on the far end face in air. In the case of a sufficiently long length of single-mode fiber, the near field will be virtually identical in shape to the fundamental mode field within the core and cladding because the power in all higher order cladding modes will have been absorbed and scattered by the coating material. However, the amplitude of the near field will be reduced by about 1.75% to account for the 3.5% of reflected power from the end face back along the fiber. For multimode fibers, ray tracing and Snell's laws can be used to determine the near-field pattern.

5.8.5.2 SUPPRESSION OF END-FACE REFLECTION

When light propagating in the excited bound mode(s) of a fiber reaches the far end of the fiber, it is partly reflected back along the fiber because of the change in index from glass to air at the end face. 3.5% of reflected power propagates back to the front face and all but 3.5% of this light is transmitted back into the source laser where it can interfere with the operation of the laser.

This reflection problem can be avoided in practice by cleaving the end face at a sufficiently large angle, typically $10°$, to deflect the reflected light into the cladding and out of the fiber core as illustrated in Figure 5.38.

5.8.5.3 FAR FIELD

The *far field* of a fiber or waveguide is the field emerging from its far end into air as measured at a sufficiently large distance from the end face. In the case of multimode fibers, the shape of the far field can be predicted using ray tracing by tracking the rays exiting the end face and taking into account their refraction at the end face.

For single-mode fibers and waveguides with relatively small core sizes of the order of a few wavelengths of the source light, the far field is *diffracted* at the end face. Because of the confining effect of the cladding on light in the core, this effect is analogous to the diffraction of a beam of light passing through a small orifice and then spreading laterally to form a series of lobes when viewed sufficiently far from the end face.

If the first lobe makes a far-field diffraction angle θ_d with the fiber axis, as shown in Figure 5.39, the standard diffraction theory can be used to relate this angle to the fiber parameters.

5.9 PERTURBED AND NONUNIFORM FIBERS AND WAVEGUIDES

There are two sets of problems in which (1) the cross-sectional geometry and/or the refractive index profile of a fiber or waveguide are subject to uniform longitudinal perturbations and (2) the translational invariance of a waveguide or fiber no longer holds, but a quasi-modal description of propagation is still applicable. In the first set, translational invariance

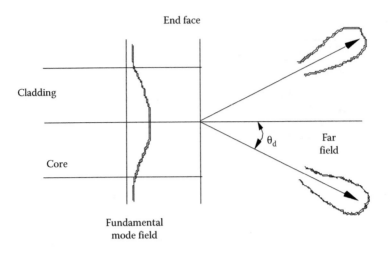

Figure 5.39 Fundamental mode evolving into lobes in the far field.

still holds so that the structure supports bound modes, whereas in the second set, which includes longitudinal tapering, mode coupling necessarily occurs. In keeping with the thrust of earlier sections, an outline of the methodology employed in each of these techniques will be presented within the weak-guidance approximation.

5.9.1 Uniform perturbations: Single core

When a fiber or waveguide has a single core for which the field and propagation constant are known for any bound mode, a simple perturbation technique enables the propagation constant to be determined when a uniform perturbation in the cross-sectional geometry and/or refractive index profile is applied to the length of the fiber or waveguide.

Consider a waveguide or fiber with known profile $n(x, y)$, scalar modal field $\psi(x, y)$, and propagation constant β, respectively. If the waveguide or fiber is slightly but uniformly distorted along its length so that it now has profile $\bar{n}(x,y)$, field $\bar{\psi}(x,y)$, and propagation constant $\bar{\beta}$, respectively, then the perturbed propagation constant can be expressed approximately in terms of the unperturbed quantities according to

$$\bar{\beta}^2 = \beta^2 + k^2 \frac{\int_{A_\infty} (\bar{n}^2 - n^2)\Psi^2 \, dA}{\int_{A_{co}} \psi^2 \, dA}, \quad (5.109)$$

where k is the free-space wave number and the integrals are over the infinite cross section of the waveguide or fiber (see Chapter 18 of Reference [4]).

5.9.1.1 SLIGHT ABSORPTION

To illustrate the usefulness of the aforementioned result, consider a step-profile fiber with a lossless cladding and core that is slightly absorbing, that is, n_{co} is replaced by the complex expression $n_{co} + i\delta$, where $\delta \ll 1$ is a constant. Substituting into Equation 5.109 to the lowest order in δ gives

$$\bar{\beta}^2 = \beta^2 + 2i\delta k^2 n_{co} \frac{\int_{A_{co}} \psi^2 \, dA}{\int_{A_{co}} \psi^2 \, dA} \quad (5.110)$$

$$= \beta^2 + 2i\delta\eta k^2 n_{co},$$

where A_{co} is the core cross-sectional area and η is the fraction of mode power propagating in the core. Upon taking the square root, setting $\beta \propto k n_{co}$, and ordering δ

$$\bar{\beta} = \beta \left(1 + \frac{2i\delta k^2 n_{co}}{\beta^2}\right)^{1/2}$$

$$\cong \beta + \frac{i\delta\eta k^2 n_{co}}{\beta} \cong \beta + i\delta\eta k. \quad (5.111)$$

Hence, the mode power $P(z)$ at a distance z along the fiber is attenuated according to

$$P(z) = P(0)\exp(-2\delta\eta kz); \quad dB = 8.686\delta\eta kz. \quad (5.112)$$

Thus, the attenuation increases with increasing absorption but decreases with increasing mode order because a larger fraction of power propagates in the cladding. In other words, the effect of absorption is most pronounced for the fundamental mode.

5.9.2 Uniform perturbations: Multiple cores

When more than one core is present in a fiber or waveguide, there are two essentially equivalent techniques for analyzing propagation: *super-modes* and *coupled-mode theory*. This is a brief background for the more detailed applications to light processing devices discussed in Chapter B3 of this book.

5.9.2.1 SUPER-MODE ANALYSIS

By analogy with the discussion in Section 5.3.10, super-modes are the bound modes associated with multiple cores such that the propagation of guided optical power can be expressed in terms of a summation over the super-modes. For example, in the case of two identical cores that in isolation from one another are single mode, there are two super-modes with even and odd symmetry, respectively, relative to the midplane between the two cores. If both modes are excited simultaneously, there will be interference or beating between them because of the difference in their propagation constants. Physically, this leads to a coupling of optical power between the two cores.

5.9.2.2 COUPLED-MODE ANALYSIS

An alternative approach is to employ coupled-mode theory, the details of which are described in more detail in Section 5.9.4. Physically, this approach can be thought of as follows. When there is only one core present, the fundamental mode propagates unperturbed. If a second parallel core is now introduced supporting a second fundamental mode with a propagation constant identical to that in the first core, this situation gives rise to a resonance phenomenon and power is transferred from the first core to the second core and back again while the modes propagate. This periodic transfer of power between the two cores can be described quantitatively using coupled-mode equations and the results are formally identical to those derived using super-modes.

5.9.3 Tapered fibers and waveguides

A waveguide or fiber taper corresponds to a variation of the core cross section and/or refractive index profile with distance z along the fiber, that is, $\rho = \rho(z)$ and/or $n = n(z)$, where ρ is either the half-side of a square-core waveguide or the core radius of a step-profile fiber and n is the profile. If the taper is single mode, then at each position z along the taper, only the fundamental mode is bound, but because of the variation in $\rho(z)$, the fundamental mode must lose power by coupling to radiation. However, it is intuitive that if the core radius $\rho(z)$ varies sufficiently slowly along the taper, than the radiation loss should be minimal. If there were no power loss, then the taper would be described as *adiabatic*; with a small but finite loss, the taper can be described as *approximately adiabatic*.

5.9.4 Local modes

If the taper is approximately adiabatic, and the fiber or the waveguide is single mode along its entire length, it is intuitive that the fundamental mode defined by the core cross section at the beginning of the taper will have virtually the same power as the fundamental mode at the end of the taper. However, its scalar field $\psi(x, y)$ and propagation constant β at the ends of the taper can be quite different; so, how can we accommodate this change and at the same time retain the fundamental mode description?

The answer is the concept of a *local mode*. Such a mode has the property that, at each position along the taper, its modal field $\psi(x, y, z)$ and propagation constant $\beta(z)$ adapt so that they are determined by the local cross-sectional geometry and index profile. Hence, at cross-section AA′ in Figure 5.40, the core radius is, say, $\rho(z_A)$, while at BB′ it will be $\rho(z_B)$. The solutions of the scalar wave equation for the field and propagation constant are then determined for the step profile with radius $\rho(z_A)$ at AA′ and radius $\rho(z_B)$ at BB′. To connect these solutions at each end, consider the local mode phase and amplitude variation, separately.

5.9.4.1 LOCAL MODE PHASE AND POWER DEPENDENCE

In a uniform fiber, phase increases linearly with distance as βz. If β now varies continuously with z, that is, $\beta \to \beta(z)$, it is intuitive that the linear

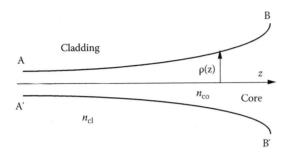

Figure 5.40 Schematic of a tapered core.

accumulation of phase is replaced by a continuous summation that can be represented by the integral

$$\int_z \beta(z) \, dz \qquad (5.113)$$

that reduces to βz when $\beta(z)$ is a constant.

In weak guidance, the total power P propagating in the fundamental mode along a uniform waveguide or fiber is given in terms of its amplitude a_0 and field ψ_0 by

$$P_0 = \frac{n_{co}}{2}\left(\frac{\varepsilon_0}{\mu_0}\right)^{1/2} |a_0|^2 \int_{A\infty} |\psi_0|^2 dA, \quad (5.114)$$

where the integration is over the infinite cross section. Suppose we assume that this expression is invariant at every position along the taper, that is, the total power remains fixed for all values of z. The field ψ_0 will then vary with the change in cross section, that is, $\psi_0 = \psi_0(z)$, and Equation 5.114 determines that the modal amplitude a_0 must also vary as

$$a_0(z) = \left(\frac{P_0}{\dfrac{n_{co}}{2}\left(\dfrac{\varepsilon_0}{\mu_0}\right)^{1/2} \displaystyle\int_{A\infty} |\psi_0(z)|^2 \, dA}\right)^{1/2}$$

$$= \left(\frac{P}{N_0}\right)^{1/2}. \qquad (5.115)$$

The denominator in the first expression is identical to the definition of the normalization N_0 in Equation (5.11), so on recalling the definition of the field of an *orthonomal mode* in Equation 5.12, it follows that

$$\hat{\psi}_m = \frac{\hat{\psi}_m}{N_m^{1/2}}; \quad a_m \psi_m = P\frac{\hat{\psi}_m}{(N_m)^{1/2}}$$

$$= P\hat{\psi}_m. \qquad (5.116)$$

In other words, the orthonormal fundamental mode describes the modal field everywhere along the taper if it is adiabatic and is a good approximation if the taper is approximately adiabatic.

5.9.5 Mode coupling

In situations where there is a loss of power from a mode of a waveguide or fiber due to either an external influence, as is the case with multiple cores in Section 5.9.2, or to the longitudinal change in the profile or cross-sectional geometry due to tapering in Section 5.9.3, the variation in the power of the mode can be quantified using a set of *coupled-mode equations* [10]. An individual mode of any order propagates without change in its total power, field, and propagation constant along a translationally invariant, lossless waveguide or fiber.

If the waveguide or fiber is now subject to a length-dependent perturbation, then the mode generally will couple its power to all other bound modes and to the radiation field, in a multimode guidance situation, but only to the radiation field in a single-mode guidance situation, assuming an infinite cladding. For a finite-cladding cross section, the radiation field comprises guided cladding modes together with a radiation field propagating into the medium beyond the cladding (normally air). Accordingly, only coupling between bound (core and cladding) modes will be included from here on.

5.9.5.1 FORWARD- AND BACKWARD-PROPAGATING MODES

The coupling of a particular mode, say the jth forward-propagating mode, is generally to all other forward-propagating modes and to all backward-propagating modes, including the jth backward-propagating mode. The jth forward-propagating mode in the weak-guidance approximation has a scalar transverse electric field E_j with the complete temporal–spatial dependence

$$E_j(x,y,z,t) = a_j(z)\psi_j(x,y)$$
$$\exp[i(-\beta_k z - \omega t)], \qquad (5.117)$$

where $a_j(z)$ is the z-dependent modal amplitude, which accounts for the coupling of power to/or from the jth mode. The kth backward-propagating mode has the same field as the kth forward-propagating mode, and its propagation constant has the same absolute value, but is opposite in sign to that of the forward-propagating mode to account for the negative phase velocity, that is,

$$E_k(x,y,z,t) = a_k(z)\psi_k(x,y)$$
$$\exp\left[i(-\beta_k z - \omega t)\right]. \qquad (5.118)$$

If $k = j$, this gives the form of the jth backward-propagating mode.

5.9.5.2 MODE NORMALIZATION

If $\psi_j(x, y)$ is a solution of the scalar wave equation satisfying the usual boundary conditions, then any multiple $p\psi_j$ is also a solution, where p is an arbitrary constant. Thus, ψ_j in the aforementioned expressions are not uniquely specified, which in turn affects the value of the amplitude coefficients $a_j(z)$. To remove this potential ambiguity, it is convenient to replace ψ_j, and $\psi-k$ by their orthonormal forms, that is, by

$$\psi_j(x,y) \rightarrow \frac{\psi_j(x,y)}{\sqrt{N_j}};$$
$$\psi - k(x,y) \rightarrow \frac{\psi - k(x,y)}{\sqrt{N_k}}, \qquad (5.119)$$

where N_j and N_k are the respective mode normalization defined by Equation 5.11. Note that normalization is independent of the direction of propagation.

5.9.5.3 COUPLED-MODE EQUATIONS

The coupled-mode equations comprise a set of linear, first-order differential equations that determine the z-dependence of the modal amplitude of each forward- or backward-propagating mode. It is convenient to combine the z-dependent amplitude and phase of each mode by defining new coefficients $b_j(z)$ and $b_{-k}(z)$ such that (see Chapter 27 of Reference [4])

$$b_j(z) = a_j(z)\exp(i\beta_j z);$$
$$b - k = a - k(z)\exp(i\beta_k z). \qquad (5.120)$$

The coupled equations can then be expressed as

$$\frac{db_j}{dz} = i\beta_j b_j = i\sum_k C_{jk} b_k;$$
$$\frac{db_j}{dz} + i\beta_j b_{-j} = -ii\sum_k C_{-jk} b_k \qquad (5.121)$$

for the jth forward- and backward-propagating modes, respectively, where the summation in k is over all forward- and backward-propagating modes. The $C_{jk} = C_{jk}(x, y)$ coupling coefficients, are defined by

$$C_{jk} = \frac{k}{2n_{co}} \int_{A_\infty} [\bar{n}^2(x,y,z) - n^2(x,y)] \qquad (5.122)$$
$$\psi_j \psi_k dA,$$

where A_∞ is the infinite cross section, $n^2(x, y)$ is the refractive index profile of the unperturbed waveguide or fiber, $n^2(x, y, z)$ is the refractive index profile of the perturbed waveguide or fiber, and ψ values are orthonormal.

5.9.5.4 APPLICATIONS OF COUPLED-MODE EQUATIONS

The set of coupled-mode equations is useful for solving a range of practical propagation problems where there is a significant exchange of power between modes. Such an analysis requires a set of propagation conditions along the fiber that does not change or only changes very slightly so that the field and propagation constant of each mode remains constant. Examples where coupled-mode equations offer a method of solution include multiple-core fibers and waveguides, as discussed in Section 5.9.2, Bragg reflection gratings, and long-period gratings.

5.9.5.5 COUPLED LOCAL-MODE EQUATIONS

Problems in which the translational invariance of the modes no longer holds, such as well-tapered fibers, waveguides, or devices, are not appropriate for a coupled-mode equation analysis. In these cases, an alternative analysis based on an analogous set of coupled local-mode equations can be used (see Chapter 28 of Reference [4]).

In practice, however, problems involving tapered fibers, waveguides, and devices based on single- or two-mode propagation generally have a sufficiently slow rate of taper such that coupling between local modes may be negligible so that each local mode propagates approximately adiabatically and almost conserves its power.

The earliest optical data links developed in the 1970s were based on multimode fibers, as good quality, low-loss, single-mode fibers were yet to be realized. However, because of the high level of dispersion in multimode fibers with all modes simultaneously excited by a single LED on the end of the fiber, these links were limited to relatively short distances in order to be able to realize a useful bandwidth, for example, between adjacent urban telephone exchanges. Low-loss, single-mode fibers were developed in the 1980s and paved the way for long-distance land- and submarine-based fiber systems. Today, single-mode fibers carry more than 95% of the world's national and international communications traffic, predominantly Internet traffic.

The single-mode fiber's profile design, material properties, and manufacture have been continually refined over the intervening years in order to maximize data transmission, including the multiplexing of several different wavelength channels carried by the fundamental mode. Other important parameters, such as cost, transmission loss, cabling loss, bend loss, nonlinear effects, have all been minimized. However, the world's rapidly increasing demand for bandwidth has now reached a critical stage where the total capacity of a wavelength-multiplexed, single-mode fiber operating within the 1550 nm wavelength window has almost reached its practical limit. To try and avoid the relatively high cost of simply installing more and more of the same single-mode fiber transmission systems to meet this demand, fibers containing more than one core or transmitting more than one mode are currently being investigated as alternative and potentially more economic solutions.

5.9.6 Multicore fibers

There is significant interest in optical fibers whose cross sections contain more than one single-mode core. A popular design is a seven-core fiber where six cores form a symmetric hexagon around the central seventh core, with all cores embedded in a common cylindrical cladding. There has been a lot of research undertaken to determine optimal practical designs for such a fiber that minimizes propagation loss, bend loss, and crosstalk between the seven cores, and is also straightforward and economical to fabricate. Each single-mode fiber in this array supports wavelength multiplexing, so that the data capacity of the seven-core fiber would be seven times that of the single-mode fiber. The excitation and detection of light in each core relies on splaying out the seven cores at the beginning and end of the multicore fiber length for splicing to sources and detectors, respectively. Furthermore, the splicing of successive lengths of multicore fiber in a system requires careful alignment so that the cores in one fiber can align accurately with the cores of the second fiber [11].

5.9.7 Few-mode fibers

An alternative and possibly more straightforward strategy to increase fiber data capacity employs fiber that is few mode by using each mode as a separate data channel. Since fiber modes are formally orthogonal to one another, in principle, there should be no crosstalk at all between them. For example, a two-mode, step-profile fiber with the same o/d as the single-mode fiber is a straightforward modification of the standard single-mode fiber design if the core radius is increased by a few percent, and would double the fiber data capacity.

Multimode fibers are inappropriate for this purpose as it is not possible to excite or detect light in individual modes when there is very close spacing between the propagation constants of successive higher order modes. Over the lengths of fiber required for systems, adjacent pairs of modes will couple significantly with one another because of the slight nonuniformities that are inevitably introduced into the fibers along their length during fabrication. In a few-mode fiber, the propagation constants are sufficiently well spaced to essentially avoid this problem.

So how can individual higher order modes in a few-mode fiber be addressed or detected? There have been a number of strategies proposed to deal with this problem based on bulk optics, filters, etc., but a recent waveguide approach relies on the use of mode-selective fiber couplers to excite and detect individual modes [12]. A traditional single-mode coupler is composed of two parallel, identical single-mode fibers in close proximity within a common cladding. If light is introduced into the fundamental mode of one core, it will necessarily couple 100% with the fundamental mode in the second core as they propagate since they have identical propagation constants. If the coupler is long enough, light will then couple back from the second core into the first core and so on.

Now consider the same situation but with a *second-order or higher order mode in the second core that has the same propagation constant as the fundamental mode in the first core.* The two fibers have a common cladding index. These two modes will also couple 100% when the fundamental mode is excited in the first core, regardless of its polarization. When the situation is reversed, a third single-mode core is needed in a plane at right angles to the plane of the first two cores to ensure that all the power in the second mode is detected in the two single-mode cores, regardless of the orientation of the antisymmetric field of the second mode field (see Figure 5.41.).

Such couplers have been successfully fabricated using a pulsed laser, direct-write system into special glass. By appropriately tapering the cores in these couplers, the coupling action becomes relatively wavelength and position insensitive [12].

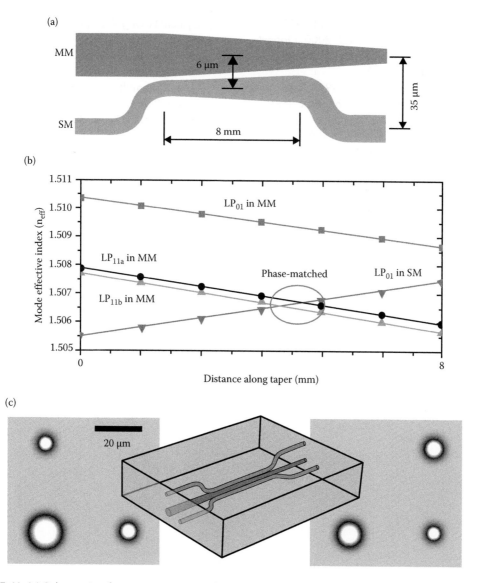

Figure 5.41 (a) Schematic of a two-core tapered mode-selective coupler comprising a tapered multimode (MM) and a counter-tapered single-mode waveguide (SM). (b) Evolution of the guided-mode propagation constants along the taper showing the crossover between the LP_{01} mode of the single-mode waveguide and the LP_{11} mode of the multimode waveguide. (c) Bright field microscope images of the end faces (writing laser incident from the top) and 3D sketch of the fabricated coupler.

REFERENCES

1. Hecht, J. 1999. *City of Light—The Story of Fiber Optics* (Oxford: Oxford University Press).
2. Kao, K.C. and Hockham, G.A. 1966. Dielectric-fiber surface waveguides for optical frequencies. *Proc. IEEE*, 113, 1151–1158.
3. Snyder, A.W. and Love, J.D. 1975. Reflection at a curved dielectric interface—Electromagnetic tunnelling. *IEEE Trans. Microwave Theory Tech.*, 23, 134–141.
4. Snyder, A.W. and Love, J.D. 2000. *Optical Waveguide Theory* (Dordrecht: Kluwer Academic Publishers).
5. Besley, J.A. and Love, J.D. 1997. Supermode analysis of fiber transmission. *Proc. IEEE*, 144, 411–419.
6. Ladouceur, F. and Love, J.D. 1996. *Silica-Based Buried Channel Waveguides and Devices* (London: Chapman & Hall).
7. Snyder, A.W. 1969. Asymptotic expressions for eigenfunctions and eigenvalues of dielectric or optical waveguide. *IEEE Trans. Microwave Theory Tech.*, 17, 1130–1138.
8. Gloge, D. 1971. Weakly guiding fibers. *Appl. Optics*, 10, 2252–2258.
9. Born, M. and Wolf, E. 1980. *Principles of Optics* (London: Pergamon Press), Section 3.2.1.
10. Marcuse, D. 1974. *Theory of Dielectric Optical Waveguides* (San Diego: Academic Press).
11. van Uden, R.G.H. et al. 2014. Ultra-high-density spatial division multiplexing with a few-mode multi-core fiber. *Nat. Photonics*, 8, 865–870.
12. Gross, S. et al. 2014. Three-dimensional ultra-broadband integrated tapered mode multiplexers. *Laser and Photonics Rev.*, 8, L81–L85.

6

Introduction to lasers and optical amplifiers

WILLIAM S. WONG, CHIEN-JEN CHEN, AND YAN SUN
Onetta Inc.

6.1 INTRODUCTION

Although Albert Einstein did not invent the laser, his work laid the foundation for its development. In 1917, Einstein was the first to explain how radiation could induce, or stimulate, more radiation when it interacts with an atom or a molecule [1]. A few years later, Richard Tolman discussed stimulated emission and absorption in his paper [2], realizing the important fact that stimulated emission is coherent with the incoming radiation. In other words, the electric dipoles in the atoms oscillate with the incoming photons, which in turn, reradiate photons that have a fixed phase relationship with incoming photons. If the reradiated photons are in-phase with the incoming ones, they add constructively to amplify the incoming photons. Thus, the general idea of coherent amplification via stimulated emission was understood since the 1920s. However, it was not until the 1950s when the concept of the "maser," which is an acronym for microwave amplification by simulated emission of radiation, was developed and demonstrated by Charles Townes and his coworkers at Columbia University [3, 4]. They directed excited ammonia molecules into a cavity whose resonance frequency is tuned to the 24 GHz transition frequency of ammonia [5]. A sufficient number of these excited molecules will initiate an oscillating microwave field in the cavity, part of which will be coupled out of the cavity (see Figure 6.1). It is interesting to note that maser operation was first demonstrated in the microwave region. Since the spontaneous radiative lifetime is inversely proportional to the third power of the transition frequency, at microwave transition frequencies, the radiative lifetime of the ammonia molecules is about 1×10^{12} longer than it would be at optical frequencies, which allows the system to achieve population inversion easily with a reasonable amount of pump power.

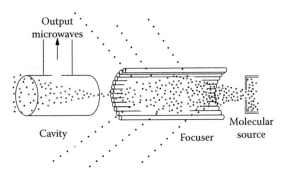

Figure 6.1 The ammonia maser. Molecules diffuse from the source into a focuser where the excited molecules (shown as open circles) are focused into a cavity and molecules in the ground state (shown as solid circles) are rejected. A sufficient number of excited molecules will initiate an oscillating electromagnetic field in the cavity, which is emitted as the output microwaves. (From Townes, C.H., *Nobel Lectures Physics 1963–1970*, Elsevier, Amsterdam, pp. 58–85, 1964.)

Following the invention of the maser, researchers attempted to create a laser by extending the maser action to optical frequencies, where "laser" is an acronym for light amplification by simulated emission of radiation. Schawlow and Townes published a paper to explore the possibility of laser action in the infrared and visible spectrum [6]. Although many researchers at the time speculated that a system containing alkali vapors would be the most likely candidate that could be made to oscillate at optical frequencies, they were surprised to learn that a system involving optical pumping of chromium ions in ruby, invented by Maiman [7], was the first to produce coherent optical radiation. His laser consisted of a ruby crystal surrounded by a helicoidal flash tube enclosed within a polished aluminum cylindrical cavity (see Figure 6.2). When pumped by a very intense pulse of light from a flash lamp, the ruby laser operated in pulsed mode at 694 nm.

Great progress has been made since the 1960s. Today, lasers have become ubiquitous in our lives. They are used in a wide range of applications in communications, medicine, consumer electronics, and military systems. Although it has been fewer than 50 years since the laser was first invented, we have made, and will continue to make, great strides in advancing the state of the art in lasers.

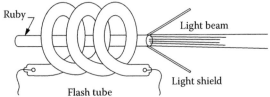

Figure 6.2 Schematic diagram of a ruby laser. When the gas flash tube is activated, electromagnetic oscillations occur within the ruby rod. Some of the visible radiation is emitted in a beam through one partially reflecting end of the rod. (From Townes, C.H., *Nobel Lectures Physics 1963–1970*, Elsevier, Amsterdam, pp. 58–85, 1964.)

6.1.1 A two-level atomic system

Before we explain the operation of a laser, we need to understand how radiation interacts with matter. Atoms, molecules, and solids have specific energy levels that are determined by the laws of quantum mechanics. A photon may interact with an atom if its energy matches the difference between two of the energy levels.

Let us focus our attention on two energy levels, E_1 and E_2 of an atom. If we assume that $E_2 > E_1$, we can refer to the levels 1 and 2 as lower and upper levels, respectively. When the energy of the photon $h\nu$ matches the atomic energy-level difference, i.e., $h\nu = E_2 - E_1$, the photon interacts with the atom where processes such as spontaneous emission, absorption, and stimulated emission take place.

Spontaneous emission of a photon takes place when the atom transitions from level 2 to level 1 without any external excitation (Figure 6.3). The energy difference is released in the form of a photon with energy $h\nu = E_2 - E_1$. The rate of this spontaneous transition, which is independent of the number of photons that is present, is the reciprocal

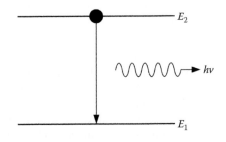

Figure 6.3 The process of spontaneous emission.

of the spontaneous radiative lifetime of the $2 \to 1$ transition. Typical values of radiative lifetimes of specific transitions range from picoseconds to several minutes.

In the presence of pumping, the atom will transition from the lower level to the upper level. For example, in the He–Ne laser, the He atoms, initially excited by collisions with electrons, transfer the excitation to the Ne atoms. In addition, in the presence of a photon with energy $h\nu = E_2 - E_1$, the atom in the lower level will also transition to the upper level, while absorbing the photon at the same time (Figure 6.4). It is an induced transition since the rate of absorption is *proportional* to the intensity of the radiation. As an example, cesium atomic clocks, which utilize the principle of absorption, have the ability to measure time with an accuracy as good as 1 s in over 20 million years. Cesium is the best choice of atom for such a measurement, because all of its 55 electrons, except the outermost one, are confined to orbits in stable shells of electromagnetic force. The energy difference $E_2 - E_1$ is attributed to the cesium atom's outermost electron transitioning from a lower to a higher orbit. After tuning a microwave beam to the resonant absorption frequency (9, 192, 631, and 770 Hz) of a collection of cesium atoms, one measures the resulting cycles of oscillations in the microwave signal to establish a time/frequency standard.

If the atom is in the upper level in the presence of radiation, it will make a downward transition to the lower level while emitting a photon of energy $h\nu$ (Figure 6.5). The emission of the new photon is induced, or stimulated, by the incident photon. The new photon also has the same frequency, polarization, and phase as the incident photon. In an optical amplifier, this process of stimulated emission takes place continuously along the length of the gain medium, resulting in a large number of output photons that are replicas of the input photons.

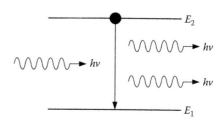

Figure 6.5 The process of stimulated emission.

In general, in a spontaneous process, such as spontaneous emission, each atom acts independently to produce a noise-like output. On the other hand, in a stimulated process, the atoms in the medium act collectively and oscillate in phase, which makes coherent amplification possible.

Although no photons are involved, nonradiative decays are also an important process for understanding the operation of a laser. If an atom is initially in the upper energy level, it can decay to the lower energy level without emitting a photon. Instead, the energy-level difference $E_2 - E_1$ appears in a phonon, where either the rotational energy or the translational energy of the material system increases, resulting in heating effects.

At thermal equilibrium, the ratio of the atoms in levels 1 and 2 is given by the Boltzmann factor:

$$N_2/N_1 = \exp[-(E_2 - E_1)/kT], \qquad (6.1)$$

where N_1 and N_2 are the number of atoms in the lower and upper levels, respectively, $k_B \approx 1.38054 \times 10^{-23}$ JK^{-1} is the Boltzmann constant, and T is the temperature in kelvin. Since $E_2 > E_1$ by assumption, the above equation tells us that at thermal equilibrium, $N_2 > N_1$; in other words, there are more absorbers than emitters. Therefore, at a normal temperature where $T > 0$, the two-level system always absorbs more incoming radiation with energy $h\nu$ than it will emit. In order to have a net amplification of radiation, it is necessary to achieve a population inversion, where $N_2 > N_1$. Interestingly, the corresponding system is said to have a "negative temperature" since T will have to be a negative quantity in kelvin in order to satisfy Equation 6.1.

6.1.2 Multilevel laser systems

Although, "laser" is an acronym for "light amplification by stimulated emission of radiation,"

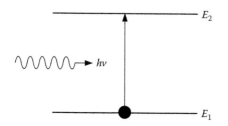

Figure 6.4 The process of stimulated absorption.

lasers are actually optical oscillators (generators or sources of light) and not optical amplifiers (devices for increasing the strength of an input lightwave signal). Despite the variety in the types of lasers, nearly all lasers possess the following three essential elements:

A lasing medium, which can be a solid, liquid, gas, semiconductor, or others

A pumping process that excites the atoms in the medium into a higher energy level

An optical resonator to enable the light to bounce back and forth

If the amount of coherent amplification inside the resonator exceeds the internal loss, coherent laser oscillation will take place.

It is impossible to achieve continuous laser oscillation in a two-level atomic system. Imagine that we are trying to excite as many atoms into the upper level as possible. While doing so, we are also increasing the rate of stimulated absorption. At best, both upper and lower levels in the two-level system will be populated equally, failing the requirement that we need to have a population inversion ($N_2 > N_1$). To overcome this problem, most laser systems utilize three or more energy levels. A three-level system is shown in Figure 6.6. Atoms are excited from level 1, the ground state, to level 3, which decay quickly to level 2, also known as a metastable state. Provided that there is a population inversion, stimulated emission, which takes place from level 2 to level 1, will dominate. This condition is met if the nonradiative lifetime from level 3 to level 2 is short, so that it will not give the pumping process a chance to repopulate the ground state. The ruby laser invented by Maiman is an example of a three-level laser system.

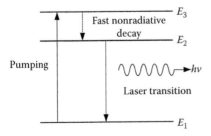

Figure 6.6 A typical three-level laser.

6.1.3 Laser cavity

A major element of a laser, in addition to the gain medium, is the laser cavity, which provides optical feedback so that laser oscillation can be achieved. It works like an electronic oscillator where stored energy, amplification, feedback, and output coupling are all engineered for oscillations to take place. It is straightforward to form a laser cavity. In fact, in the simplest form, two parallel mirrors facing each other form the most basic laser cavity—the Fabry–Perot cavity [8–11]. Lightwave bounces back and forth upon both mirrors, gaining energy each time it passes through the gain medium. Typically, one mirror is highly reflective and the other is partially transmitting to provide output coupling. The gain, loss, and coupling ratios of the mirrors determine the output power of a laser.

Since light is circulating inside the cavity, the phase of the reflected light, after a complete round trip, needs to match that of the original light so that a constructive interference or a standing wave can form. This condition is satisfied only at certain frequencies and these resonant frequencies are referred to as the longitudinal (or axial) modes of the laser cavity. For a cavity with a fixed length, the longitudinal modes are separated by a constant frequency that is equal to the inverse of the cavity round trip time. Within the gain bandwidth of the gain medium of a laser, there may exist multiple longitudinal modes. The lasing threshold is reached when the gain medium is pumped, in which the total round-trip gain of the mode (closest to the gain peak) equals the total round-trip loss. When the threshold is exceeded, the laser oscillation starts in this axial mode. Depending on the property of the gain medium, when the pump power increases, a homogeneously broadened gain medium, which maintains the same gain shape when saturated, can sustain one stable lasing axial mode only, while an inhomogeneous broadened medium can have multiple modes coexisting in a laser.

Similar to a RLC electronic resonator, the quality factor Q of a laser cavity is defined as 2π times the ratio of the stored energy to the energy loss per cycle. Since an optical cycle is very short (a few femtoseconds), energy loss per cycle is relatively small and the quality factor of an optical cavity is normally much greater than that of an electronic resonator.

An important method to generate short laser pulses is to "switch" the cavity quality factor artificially from very low to normal so that the stored energy in the gain medium can be released in a short time period, otherwise known as the Q-switching method. Lowering the cavity Q can be achieved by increasing the cavity loss or decreasing the feedback from the mirrors. The stored energy, in the form of large inversion in the gain medium, grows as pump power is applied, but the laser oscillation does not happen because of the high cavity loss or the lack of optical feedback. When the cavity Q is restored, the laser threshold is exceeded and the lightwave inside the cavity experiences a large optical gain due to the large inversion and builds up rapidly. The huge optical energy then quickly consumes the large population inversion causing the optical power starts to decrease. In addition, when the cavity Q is lowered at the same time, the optical power diminishes further as the laser is operating below its threshold. The cycle is then repeated for subsequent pulse generation. The Q switching technique is widely applied on all types of lasers, from solid-state lasers, gas lasers, to semiconductor lasers, to produce short pulses.

In addition to the Fabry–Perot cavity, there are different ways to construct a laser cavity. One can replace the planar mirrors with concave or convex spherical mirrors to achieve better confinement of the optical energy inside the cavity [12]. An important consideration of a laser cavity is whether a stable mode exists that corresponds to the mirror geometry. This problem can be approached in a setting of the paraxial wave equation, which describes wave propagation in the paraxial limit, and its eigen solutions, i.e., the Hermit–Gaussian modes. Another approach is ray tracing, which is described by the ABCD matrices. Both methods lead to the same stability criterion written explicitly as $0 \leq g_1 g_2 \leq 1$ [8–11], where the g parameter of a mirror is defined as $g = 1 - L/R$, where L is the distance between the two mirrors and R is the radius of curvature of the mirror.

As described by the same analysis, higher order Hermit–Gaussian modes, or the transverse modes, can exist in a laser cavity to compete for gain, degrade mode profile, and cause output power fluctuation. Since the higher order transverse modes have larger spatial profiles, they can be eliminated by restricting the active region of the gain media to fit only the fundamental mode. In some high gain lasers, unstable cavities, in which off-axis light

diverges, can be used as they have advantages of a larger spatial profile, more efficient gain, ease of alignment, and single modalness.

There are other types of laser cavities, such as a ring cavity consisting of a length of optical fiber. In semiconductor lasers, cleaved facets can be used as mirrors; in addition, linear cavities formed using distributed Bragg reflectors are also common. Depending on the available technology and requirements, laser cavities are chosen to suit their purposes.

6.2 SPECIFIC TYPES OF LASERS

6.2.1 Solid-state lasers

Since the invention of the very first ruby laser in 1960s [7,13], solid-state lasers, which are composed of gain media in their solid phase by definition, remained as one of the most important and versatile lasers for academia and industry. When pumped to threshold population inversion levels, the three-level or four-level solid-state gain media provide sufficient gain to compensate for losses in the laser cavity. The laser cavity forms a resonator, analogous to the electronic version of an oscillator, which determines the output power, operation modes, and the beam profile of the solid-state laser.

Generally, a solid-state gain medium consists of a crystalline (e.g., YAG) or an amorphous (e.g., glass) host doped with a certain ion (e.g., several percent of Cr^{3+}, Nd^{3+}, Er^{3+}, etc.) that acts as the active lasing material. Most commercially available solid-state lasers are pumped by gas discharge flash lamps (e.g., krypton or xenon) or by semiconductor lasers for greater reliability and efficiency. The choice of the gain media depends on lasing wavelength, optical properties (birefringence, dispersion, optical nonlinearity), and mechanical properties (material strength, thermal conductivity). Important solid-state gain media include Nd^{3+}-doped YAG (Nd^{3+}:YAG), Nd^{3+}:glass, Ti^{3+}:sapphire, and color-center lasers, etc.

The Nd^{3+}:YAG laser is a four-level laser and is typically operating at the $1.064\,\mu m$ wavelength range with a narrow linewidth of a few nanometers [14,15]. Its good thermal conductivity and temperature-insensitive threshold make it suitable for both pulsed and continuous-wave operations. Typically, it is pumped optically with flash lamps or laser diodes, and water cooled in high output power applications. It is popular for various laboratory applications such as pumping another laser or being frequency doubled to provide green light. The energy-level diagram of the Nd^{3+}:YAG laser and its emission spectrum are shown in Figure 6.7.

Compared with Nd^{3+}:YAG, the Nd^{3+}:glass is easier to manufacture and process. It can be made into a larger size gain medium to achieve higher output powers. Depending on the host glass types, it typically operates at around 1.06 μm with a wider linewidth of tens of nanometers [16,17]. Using a large size Nd^{3+}:glass gain medium and the Q-switch technique, one can generate pulses with terawatt peak power. However, due to its low thermal conductivity, the Nd^{3+}:glass laser operates at a low repetition rate (10 pulses per second). For comparison, we show the fluorescent emission spectra of Nd^{3+} in various glass hosts in Figure 6.8.

Among the solid-state gain media, Ti^{3+}:sapphire is characterized by its broad gain bandwidth from 0.66 μm to greater than 1.0 μm (more than 100 THz of bandwidth when modelocked) [18]. It is used in tunable continuous wave (CW) lasers for broad tuning ranges or in mode-locked configuration to produce ultrashort pulses (down to several femtoseconds). The Ti^{3+}:sapphire lasers are often pumped by other lasers such as laser diodes or argon ion gas lasers due to its short excited state lifetime (3.2 μs).

Two common operation modes of solid-state lasers are the CW mode and the pulsed mode. In the CW operation, solid-state lasers generate monochromatic, highly coherent, and high-intensity light. A widely used mechanism to produce pulses in solid-state and other lasers is the Q-switching method [19,20]. By changing the cavity Q mechanically, electrically, or optically, the stored energy in the laser gain medium can be released with a short time period to generate a pulse (in the order of nanoseconds) with a high peak intensity [10, 11,21] (see also Section 6.1.3 for a description of Q switching). Short pulses provide advantages in manufacturing and in medical applications such as micromachining/microfabrication and laser ablation since the size of the heat-affected zone can be reduced. As the demand on shorter pulse duration and higher peak intensity increases, different techniques such as mode-locking techniques [22,23]

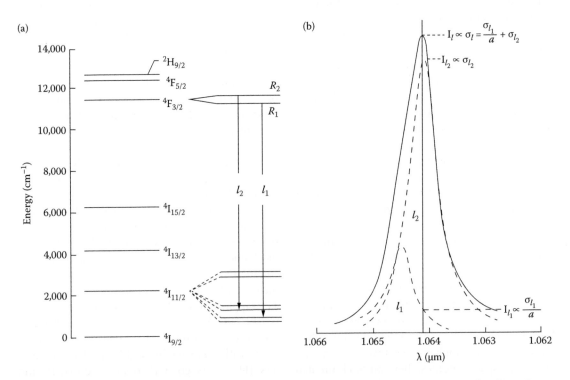

Figure 6.7 (a) Energy-level diagram of Nd^{3+}:YAG. (b) Spontaneous emission spectrum of Nd^{3+}:YAG near 1.064 μm at room temperature. The two Lorentzian lines contributing to the laser transition are shown by dashed lines. (From Kushida, T. et al., *Phys. Rev.*, 167, 289, 1968.)

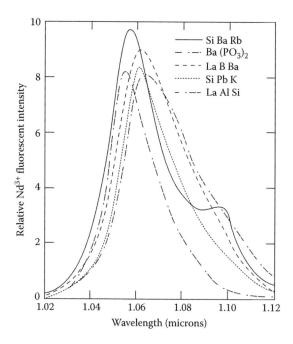

Figure 6.8 Spontaneous emission spectrum of Nd^{3+} in various glass hosts near 1.06 mm. (From Snitzer, E. and Young C.G., *Lasers*, Dekker, New York, p. 191, 1968.)

and chirped pulse amplification (CPA) techniques [24] were applied on solid-state lasers to generate ultrashort pulses (down to sub-picosecond ranges). In fact, solid-state lasers, due to their wide bandwidth and excellent optical properties, generated many of the record-breaking short pulses. These ultrashort pulses are suitable for applications in the study of ultrafast phenomena, spectroscopy, and telecommunication.

6.2.2 Gas lasers

As another important family of lasers, gas lasers were fast growing in the laser industry due to their simple pumping schemes and the wide availability of gain media. In gas lasers, the population inversion is typically created by electric discharge, which is relatively simple to construct and operate. Although the gain media are in the gaseous phase, they can be made up of neutral atoms, ions, or gas molecules. As examples, we will briefly introduce the three most popular gas lasers: the He–Ne laser, the Argon ion laser (Ar$^+$), and the CO_2 laser.

The He–Ne laser, one of the best known lasers, operates at the famous 632.8 nm wavelength [25, 26]. The gain medium consists of a mixture of two noble gases, He and Ne, with a population ratio of about 10–1. The He atoms are first excited to higher energy levels by the electrical discharge and then pass energy to the Ne atoms through inelastic collisions. This is possible due to similar energy levels between the He and the Ne atoms (see Figure 6.9). The excited Ne atoms provide gain for the laser operation in the He–Ne lasers, which also radiate at wavelengths of 1.15 µm and of 3.39 µm, in addition to the 632.8 nm. In fact, optical filtering in the laser cavity is often applied in the He–Ne lasers to reduce the internal gain in the infrared since the optical gain at 632.8 nm is relatively low. The He–Ne lasers were widely used in optical alignment, survey, and bar-code scanning, etc., before the advent of inexpensive laser diodes.

The Ar$^+$ lasers are capable of producing high power in the visible wavelength range. Hundreds of wavelengths could exist in an Ar$^+$ laser cavity; however, the 488.0 and 514.5 nm are two of the most prominent wavelengths [27,28]. The pumping mechanism in Ar$^+$ lasers is complicated, including multiple collisions between electrons, between Ar atoms, and between Ar$^+$ ions. Since the active laser medium consists of Ar$^+$ ions, whose ground state level is 16 eV higher than that of the Ar atoms, much of the pump power is wasted in providing ionization of the Ar atoms. Because of the ground

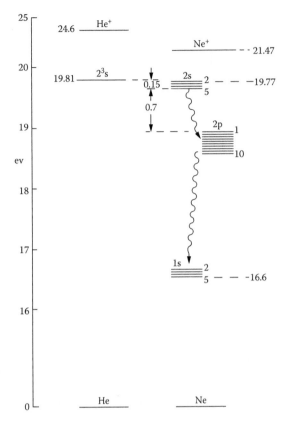

Figure 6.9 Energy level diagram of the He and Ne atoms. (From Javan, A. et al., *Phys. Rev. Lett.*, 6, 106, 1961.)

level difference, which does not contribute to the laser oscillation, the Ar^+ laser threshold current density is typically high and the efficiency of the laser is low. Due to the low efficiency, bulky water-cooling subsystems are used to cool high-power Ar^+ lasers.

The CO_2 lasers are gas lasers and are well known for their high efficiency and high output powers [29,30]. In contrast with the He–Ne laser and the Ar^+ laser, in which transitions in electron energy levels provide laser gain in the cavity, CO_2 lasers operate on molecular vibration modes. The vibration modes are represented by three quantum numbers $(n_1 - n_3)$ where the ground state is at (0, 0, 0). Typically, N_2 molecules and He atoms are also present in the CO_2 laser cavity. The N_2 molecules store the energy from the electrical discharge in the fundamental vibration mode and then transfer the energy to the asymmetric vibration (0, 0, 1) mode through collision since both vibration modes have similar energy levels. The CO_2 laser gain is then

produced from the transition of the asymmetric vibration mode to lower level vibration modes, including wavelengths of 10.6 and 9.6 μm. The He atoms, moving at higher speed, effectively reduce the lifetime of the vibration modes at lower levels of the laser and remove heat. This process makes population inversion possible and increases the overall efficiency. Due to its high output power, the CO_2 laser is used in industries such as metal processing and machining.

6.2.3 Semiconductor lasers

A semiconductor (e.g., silicon, germanium, and gallium arsenide) is a material whose electrical properties are between those of a conductor and an insulator. The electrons in a semiconductor are found in bands that are separated by a band gap—the lower band is the valence band while the upper band is called the conduction band. For example, GaAs has a bandgap of 1.424 eV. Furthermore, by doping intrinsic semiconductor materials with impurities, one can make n-type or p-type materials that have more or fewer negative current carriers.

Shortly after the invention of the first laser, the first semiconductor laser was demonstrated independently in 1962 by four research groups in the United States [31–34]. These early devices are homojunctions operated at liquid nitrogen temperatures. When a current (defined as the flow of *positive* charges) is injected so that it flows through the junction from a *p*-type material into an *n*-type material, electrons from the *n*-type material will recombine with the holes from the *p*-type material, releasing a form of energy known as recombination energy. In an indirect-bandgap material such as silicon, this energy is released as vibrational energy and heat; on the other hand, in a direct-bandgap material such as gallium arsenide, radiation is emitted whose frequency is a function of the bandgap energy. In the absence of optical feedback, this device functions as a light-emitting diode (LED) where its output consists of incoherent spontaneous emission. However, feedback from reflective surfaces is made possible using the cleaved facets. Since the refractive index of a semiconductor is usually greater than 3.0, the reflectivity of each cleaved facet can be as high as 25% without any coatings.

By the late 1970s, semiconductor lasers are able to operate in CW mode at room temperature. These

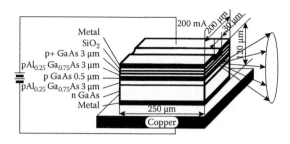

Figure 6.10 Schematic view of the structure of the first injection double-heterostructure laser operating in the CW regime at room temperature. (From Alferov et al., *Sov. Phys. Semicond.*, 4, 1573–1575, 1971.)

lasers utilize a double heterostructure (Figure 6.10) to improve the confinement of light in the active region, resulting in lower threshold currents [35, 36]. Later, the incorporation of thin quantum wells into the heterostructure offers additional advantages such as low threshold current density, high efficiency, and high differential gain. Figure 6.11 shows that the threshold current density has been dramatically reduced by a factor of 10,000 since the early 1960s.

The technology of semiconductor lasers has been advanced to a point where, without realizing it, an average person owns at least a few of them. Examples are lasers used in CD or DVD players, in laser printers, and in laser pointers. In addition, in fiber optics communications, semiconductor lasers

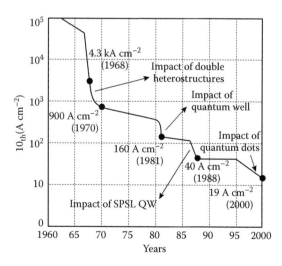

Figure 6.11 Improvements of the threshold current of semiconductor lasers over the last 40 years. (From Alferov, Z., *IEEE J. Sel. Top. Quant. Electron.*, 6, 832–840, 2000.)

are used as transmitters and as EDFA and Raman pump lasers.

6.2.4 Short-pulse lasers

When the gain medium of a Fabry–Perot laser is homogenously broadened, only one single longitudinal mode oscillates. On the other hand, if the gain medium is inhomogeneously broadened, for example, via spatial hole burning or spectral hole burning, a few longitudinal modes will oscillate, provided that their round-trip gain is greater than unity. These longitudinal modes are spaced evenly in the frequency domain with frequency spacing $\Delta \upsilon = c/2nL$, where n is the average refractive index inside the Fabry–Perot cavity, and c is the speed of light in vacuum. If we are able to lock these modes together such that their phases are fixed relative to one another, we can create a train of laser pulses at the output of a laser. This technique, known as modelocking, can be initiated actively using a time-varying amplitude modulation, or it can be triggered using a saturable absorber in the cavity whose transmissivity decreases with increasing optical intensity. Very short optical pulses can be generated from modelocked lasers because the electric field from the various longitudinal modes are coherent—they interfere constructively at the peak of the pulse while cancelling one another at the wings of the pulse.

To capture fast-moving images on film, a photographer uses a fast shutter setting. Since the speed of mechanical shutters is limited to milliseconds, improvement is made to capture faster events by illuminating the object with a stroboscope. As one of the pioneers in strobe-light photography, Edgerton captures the dramatic image of a rifle bullet piercing an apple in one of his high-speed photographs (Figure 6.12). The photos are taken in a room with the lights turned off. When the bullet is fired, its shock wave is detected by a crystal microphone, which then triggers the strobe light. Edgerton's technique enables him to freeze a microsecond event onto a photographic negative. With the advancement of ultrafast laser research, nanosecond, picosecond, and subsequently, femtosecond laser pulses are available for scientists to "freeze" and study physical and chemical processes in a very short time scale. In fact, some used to believe that the various chemical and biological processes, such as the breaking of chemical bonds and vision, were slow processes and did not occur

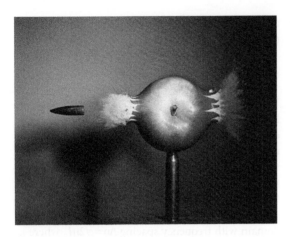

Figure 6.12 A microsecond exposure of a bullet traveling 2800 ft s⁻¹ while piercing an apple. (From http://www.geh.org/taschen/htmlsrc4/m199603470007_ful.html.)

in the femtosecond time scale, until photochemistry experiments using short laser pulses proved otherwise. These results can be explained by the fact that, since molecular motions occur over very short distances, they can be very fast.

Given the existence of the various techniques to produce femtosecond pulses, how does one measure the duration of these pulses? In general, to measure a fast event, one needs an even faster event in order to capture it, which is not often possible since the laser pulse in interest might be the shortest event available. Likewise, researchers working in scanning–tunneling microscopy face a similar problem when they want to establish the spatial resolution of their system—the fine tip they use to probe the sample is often the finest man-made object that is available to them. In 1967, Weber found a partial solution; he suggested measuring the laser pulse width with the pulse itself by performing an intensity autocorrelation [38]. The optical pulse is split into two identical pulses using a beam splitter. The two pulses are then focused onto a nonlinear crystal that is capable of generating a second harmonic. The second harmonic generated will be collected in a photomultiplier tube while the temporal delay between the two pulses is varied. For example, the largest amount of signal is obtained when the two pulses overlap temporally. This operation is identical to performing an autocorrelation function mathematically, from which the approximate pulse duration can be deduced. Although it is a clever technique, the

autocorrelation method does not yield complete details about the intensity profile of the pulse. Nor does it provide any information regarding the phase or the chirp of the optical pulse.

Ideally, one would like to get pulse information in both time and frequency domains, which is usually referred to as a spectrogram. It is analogous to the musical score for a symphony, informing musicians which notes to play at a given time. It was not until 1993 that researchers developed a novel method to retrieve to obtain the spectrogram of an optical pulse. Using a technique called frequency-resolved optical gating [39], or FROG, they measured the optical spectrum of the autocorrelation and then applied a phase retrieval algorithm to obtain the final spectrogram. Their algorithm works because knowledge of only the magnitude of a two-dimensional Fourier transform of a function of two variables uniquely determines the function (both phase and magnitude), provided that the function is well behaved. Figure 6.13 shows that

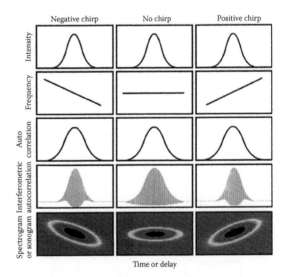

Figure 6.13 The intensity versus time, the frequency versus time, the intensity autocorrelation versus delay, the interferometric autocorrelation versus delay, and spectrograms (or sonograms) of negatively chirped, unchirped and positively chirped Gaussian-intensity pulses. In the spectrograms, the vertical axis is frequency and the intensity is color coded. Note that the autocorrelation and interferometric autocorrelation cannot distinguish positive from negative chirp, while the spectrogram and sonogram can. (From Walmsley, I. and Trebino, R., Measuring fast pulses with slow detectors. *Opt. Photon. News*, 7, 23, 1996.)

both the intensity autocorrelation and the interferometric autocorrelation cannot distinguish positively chirped pulses from negatively chirped ones. On the other hand, the spectrogram extracted using FROG resolves this ambiguity as it contains complete amplitude and phase information about the optical pulse.

6.3 OPTICAL AMPLIFIERS

6.3.1 Basics of optical amplifiers

Optical amplifiers are used to boost the optical power of the input optical wave. Optical amplifiers are essentially "single-pass" lasers, or "lasers" without the two (reflective) mirrors. Just as in the case of lasers, the gain medium is "inverted" and provides stimulated emission when there is an input optical wave. In principle, all types of lasers can be converted into optical amplifiers. For simplicity, we will use optically pumped amplifiers in the following discussions. In such amplifiers, three optical waves are present, the pump wave that provides the inversion in the gain medium, the input optical wave that is the input signal to be amplified, and the output wave that has been amplified in the amplifier.

There are different applications for various types of optical amplifiers. Accordingly, the parameters of interest also vary. For example, the gain is an important parameter for laser fusion, while noise is essential for telecom applications. In general, the three commonly used parameters are gain, output power, and noise figure.

Gain is defined to be the ratio between the output power and input power, where it is usually measured in decibels. Unity gain, or 0 dB, means no gain or loss. On the other hand, most practical optical amplifiers operate in gain regions much higher than 1. A typical gain curve is shown in Figure 6.14 as a function of output power for a fixed pump power. The gain is also commonly plotted as a function of input power. There are four regions in this saturation curve: the small signal region, the transition region, the saturated region, and the transparent region.

In the small signal region, the input power is low, and the gain remains constant when input power changes. The inversion in the gain medium is determined only by the amount of pump power and is independent of the weak input optical power.

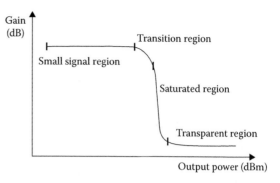

Figure 6.14 Gain versus output power for a fixed pump power.

An advantage for this operation region is that the amplifier's gain value remains constant. However, the photon conversion efficiency from pumps to signals is low. Most of the pump photons are not converted into signal photons, but into spontaneous emission or heat or other forms of energy.

As input power increases into the transition region, the interaction between the signal and the gain medium gets stronger. In this region, the signal starts to deplete the gain medium and the inversion level starts to drop. This causes the gain to decrease when input power increases.

When the input power is high, the amplifier works in the saturated region where the gain drops sharply with increased input power. The inversion level is low and the photon conversion efficiency from the pump wave to the signal wave is high. This is a common operation region for amplifier design in order to make efficient use of the pump power. In this highly saturated region, the output power changes very little as input power varies, as shown in Figure 6.14.

In the limit of high input power, the amplifier moves into the transparent region where the pump is relatively weak and the input optical wave essentially bleaches through the medium. There are approximately equal number of atoms in the upper level and in the lower level. No practical amplifiers are designed to work in this region.

During the amplification process, spontaneous emission from the gain medium may add to the optical signal wave. Noise figure is the parameter to measure how much noise is added in the amplifier. A low noise figure is important for telecom applications, as we desire to minimize the degradation of the signal-to-noise ratio when a signal is amplified. The quantum limit for a practical optical amplifier is 3 dB [41]. Under the assumption of high-gain operation for the amplifier, high inversion level allows for low noise figure while low inversion level yields high noise figure.

There are other parameters that need to be considered when designing an amplifier, such as reliability, size, cost, etc. We will not discuss the details here as they are beyond the scope of this chapter. In the following sections, we will cover four types of amplifiers—erbium-doped fiber amplifiers (EDFAs), Raman optical amplifiers (ROAs), semiconductor optical amplifiers (SOAs), and amplifiers that are built to amplifier short pulses.

6.3.2 Erbium-doped fiber amplifiers (EDFAs)

Perhaps the most well-known optical amplifier is the EDFA that was first reported in 1987 [42, 43]. Traditionally, the so-called O–E–O signal regeneration was used in optical communication systems, where optoelectronic regenerators are installed between terminals to convert signals from the optical domain to the electrical domain, and then back to the optical domain. Since its invention, EDFA has revolutionized optical communications. Unlike optoelectronic regenerators, this optical amplifier does not require high-speed electronic circuitry and is transparent to data rate and format, which dramatically reduces system cost. EDFAs also provide high gain, high output power, and low noise figure. We will introduce several new important EDFA features and parameters below.

Energy levels. The energy levels of the erbium ion and the associated spontaneous lifetime in the fiber glass host are shown in Figure 6.15. The energy difference between the upper level and the lower level is such that the photons generated are in the 1.5 μm transmission window of an optical fiber. The gain spectra at different inversion levels are shown in Figure 6.16. Erbium-doped fiber can be pumped by semiconductor lasers at either 980 or 1480 nm. The rapid improvements in semiconductor pump lasers have made the EDFA possible for practical applications. A three-level model can be used for 980 nm pumps while a two-level model usually suffices for 1480 nm pumps [44,45]. Complete inversion can be achieved with 980 nm pumping but not with 1480 nm pumping [45]. Because of the photon energy difference, the quantum efficiency is higher with 1480 pumping.

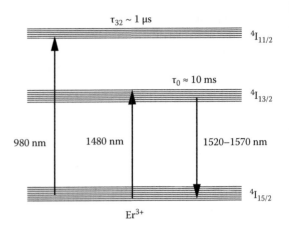

Figure 6.15 The energy levels of erbium ion in optical fiber. (From Sun, Y. et al., *IEEE J. Sel. Top. Quant. Electron.*, 3, 991–1007, 1997.)

Dynamics. The spontaneous lifetime of the excited energy level is about 10 ms, which is much slower than the signal bit rates of practical interest. Because of the slow dynamics, an EDFA only experiences the averaged optical power for the practical date rates. As a result of the slow dynamics, inter-symbol distortion and inter-channel crosstalk are negligible even when the EDFAs are working in the saturated region. Besides, since all of the optical signal channels can be amplified simultaneously in one erbium-doped fiber, the EDFA has become a key enabler for the widely used wavelength-division multiplexing (WDM) technology. This is a significant advantage for EDFAs.

Bandwidth. For WDM applications, since uniform gain is desired for all the signal channels, bandwidth is another important parameter for EDFAs. From Figure 6.16, we can see that the gain is flat somewhere between 1540 and 1560 nm for an inversion level of approximately 50%. Actually, it is this generic flat gain band that was used in initial WDM systems.

Significant progress has been made in amplifier design to achieve excellent performance. Several key techniques are discussed below:

Multistage design. In order to achieve both low noise figure and high output power, two or more gain stages are usually used where the input stage is kept at a high inversion level and the output stage is kept at a low inversion level [46,47]. For optical amplifiers with two or more gain stages, the overall noise figure is

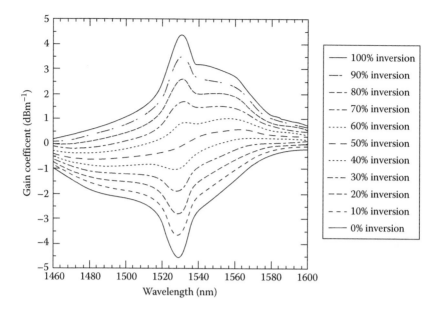

Figure 6.16 Gain spectra of erbium-doped fiber. (From Sun, Y. et al., *IEEE J. Sel. Top. Quant. Electron.*, 3, 991–1007, 1997.)

mainly decided by the high gain input stage and the output power is basically determined by the strongly saturated output stage. The passive components have minimal impact on the noise figure and the output power when they are in the middle stage.

Gain equalization filter. A wide bandwidth is desired to accommodate a large number of optical channels. To fully utilize the gain band between 1530 and 1565 nm, gain equalization filters (GEFs) can be used to flatten the gain spectrum. Several technologies have been investigated to fabricate GEFs, including thin film filters, long period gratings, short period gratings, silica waveguide structure, fused fibers, and acoustic filters. Depending on the design, a bandwidth of 35–40 nm can be obtained in the C-band [48]. This kind of amplifier with 35 nm of flat bandwidth was used in the long distance transmission of 32 and 64 channels at 10 Gb s⁻¹ [49], and has since become the commonly used bandwidth for C-band communication systems.

L-band operation. Because the gain drops sharply on both sides of C-band at the inversion level discussed above, it is not practical to further increase the bandwidth with a GEF. On the other hand, a flat gain region between 1565 and 1615 (L-band) can be obtained at a much lower inversion level [50]. To achieve comparable gain as the C-band, a longer piece of erbium-doped fiber or higher erbium doping would be needed. By combining C-band and L-band, a much wider bandwidth can be realized [51]. Such ultrawide-band optical amplifiers have made possible high-capacity communication systems at terabits/s or higher [52].

Large dynamic range. For commercial systems deployed in the field, the fiber span length and loss varies from location to location. A large dynamic range in the amplifier gain is required for in-line optical amplifiers. Optical attenuators can be used in the middle stage to increase the dynamic range of EDFAs. Such amplifiers can provide flat spectrum and good noise figure when span loss varies.

Dispersion compensation. Dispersion compensation is needed for high-speed optical channels and can be done with dispersion compensating fiber. Such compensation is best done on a span-by-span basis. The dispersion compensation modules (DCM) are usually inserted in a middle stage in the EDFAs to optimize the overall optical performance. Typically a middle-stage loss of about 10 dB is reserved for the plug-in of DCM.

In the above discussion, we mainly dealt with the static features of EDFA. In the event of either a network reconfiguration or a failure, the number of WDM signals traversing the amplifiers would change. As a result, the power of the surviving channels would increase or decrease due to the cross-saturation effect in amplifiers. Dropping channels can give rise to surviving channel errors since the power of these channels may surpass the threshold for nonlinear effects such as Brillouin scattering. The addition of channels can cause bit errors by depressing the power of the surviving channels to below the receiver sensitivity. To overcome such error bursts in surviving channels in the network, the signal power transients have to be controlled. Because of the saturation effect, the speed of the gain dynamics in a single EDFA is in general much faster than the spontaneous lifetime of about 10 ms [45]. The time constant of gain dynamics is a function of the saturation caused by the pump power and the signal power. The time constant of gain recovery on single-stage amplifiers was reported to be between 110 and 340 μs [53].

In a recent work, the phenomena of fast power transients in an EDFA chain was reported [54]. Even though the gain dynamics of an individual EDFA is unchanged, the rate of change of the channel power at the end of the system becomes faster for longer amplifier chains. This fast gain dynamics results from the effects of the collective behavior in a chain of amplifiers. The output of the first EDFA attenuated by the fiber span loss acts as the input to the second EDFA. Since both the output of the first EDFA and the gain of the second EDFA increase with time, the output power of the second amplifier increases at a faster rate. This cascading effect results in faster and faster transients as the number of amplifiers increase in the chain. To prevent performance penalties in a large scale WDM optical network, surviving channel power excursions must be limited to certain values depending on the system margin. Several control schemes have been studied, including pump control [55], link control [56], and laser control [57].

Considerable progress has been made in optical amplifiers in recent years. The bandwidth of amplifiers has increased nearly seven times and flat gain amplifiers with 84 nm bandwidth have been demonstrated. This has been made possible by the addition of the L-band branch. With the advent of these amplifiers, commercial terabits/s lightwave systems will be realizable. Research is underway to develop amplifiers outside the erbium fiber band. Raman amplifiers and semiconductor amplifiers are also potential candidates for amplification across the entire silica fiber transmission band. Progress has also been made in the understanding of gain dynamics of amplifiers. Several control schemes have been successfully demonstrated to mitigate the signal impairments due to fast power transients in a chain of amplifiers and will be implemented in lightwave networks. The terrestrial lightwave systems have been increasing in transmission capacity. To meet the enormous capacity demand the currently available 400 Gb s^{-1} capacity system with 40 channels will soon be followed by systems having terabits/s and higher capacity on a single fiber.

6.3.3 Raman optical amplifiers

Distributed Raman amplification using optical fibers is an old and yet emerging technology that can supplement the functionality of EDFAs in high-speed ($\geq$10 Gbs^{-1}) long-haul transmission systems. Before the advent of the EDFAs, ROAs were used to reamplify solitons in a recirculating loop (Figure 6.17) in order to demonstrate the feasibility of long-haul soliton transmission [58]. Major

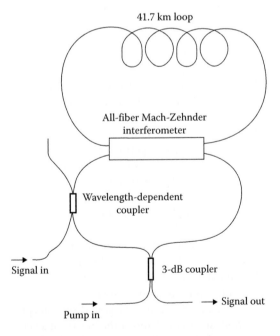

Figure 6.17 The first optical transmission system using Raman optical amplifiers. (From Boyd, G.D. and Kogelnik, H., *Bell Syst. Tech. J.*, 41, 1347, 1962.)

advantages of using ROAs include low-noise characteristics in distributed mode operation, flexible gain allocation, wide gain bandwidth, and simple construction.

The amount of Raman scattering and its spectrum depend on the material. For silica, which is an amorphous material, the peak Raman gain is at 13.2 THz lower than the pump frequency, which corresponds to a wavelength of about 100 nm longer than the pump at 1450 nm [59,60]. One can modify the operating gain shape and flatness of the Raman optical amplifier by choosing pump wavelengths and pump powers. For example, one can readily make Raman amplifiers to operate in the S-band (1485–1525 nm) or in the L+ (1604–1640 nm) band. The pump scheme is relatively simple because one only needs to launch the pump light into the fiber via a wavelength-dependent coupler (WDC) or an optical circulator.

The process of Raman scattering can be viewed as the modulation of light (the pump) by molecular vibrations in a material, which are referred to as optical phonons [61]. Upper and lower sidebands appear in the scattered light spectrum because the frequency of the pump is both upshifted and downshifted by the optical phonon. At high temperatures, the two sidebands are of equal intensity; however, at room temperature, the lower frequency sideband is favored. So far we have described the process of spontaneous Raman scattering, which was discovered by Sir Raman in 1928 [62]. One can stimulate the process by injecting both the pump and the downshifted signal into the medium. In this case, the two optical waves will beat together to stimulate the optical phonon, which in turn, will cause more pump to convert into the signal. This is known as stimulated Raman scattering, which was observed 34 years later by Woodbury and Ng [63]. One direct consequence is that the input is amplified coherently and the medium acts like a distributed optical amplifier.

The distributed nature of the Raman amplification provides an advantage to the overall noise performance. As much as 5 dB of improvement in the optical signal-to-noise ratio (OSNR) can be achieved with 10–15 dB of distributed Raman gain. Signal is amplified along the fiber length such that the amount of noise generated in the amplification process is less than that of a lumped amplifier placed at the end of the same fiber span, where signal strength is the weakest. The Raman on–off

gain (in dB), defined as the signal gain with, and without pumps, is proportional to the Raman gain coefficient, the effective length, and the pump intensity. It is also inversely proportional to the fiber effective area (see Figure 6.18).

Although ROAs hold promise for telecommunication applications, they also have some drawbacks in certain applications compared with those of EDFAs. First, the low efficiency in converting pump power to signal power implies that there is a need for high pump powers. Note that high-power semiconductor pump lasers, which were only available recently, was the main driver for the reemerging of ROAs. The high-pump-power requirement is generally not desirable for several reasons, such as personnel safety concerns and components reliability concerns. Secondly, the relatively long length of fiber in a ROA allows Rayleigh-related reflection to degrade the performance of a telecommunication system. Typically, the fiber lengths of ROAs are of the order of several kilometers. These Rayleigh-related effects include multiple-path interference (MPI), double-Rayleigh backscattering noise (DRS), and ROA instability, all of which become more severe when pump power is increased [64]. In addition, the fast Raman response time causes time-dependent deleterious effects such as inter-channel cross-talk and the transfer of pump noise to the signal [65]. This is an important constraint for ROAs where the pumps are co-travelling in the same direction as that of the signals. Lastly, similar to EDFAs, ROAs also exhibit transient effects when the input signal power fluctuates [66].

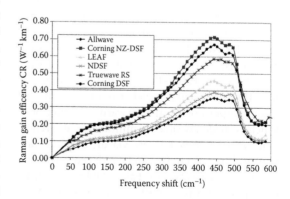

Figure 6.18 Raman gain as a function of fiber type. (From Fludger, C. et al., *Proceedings of Optical Fiber Communications Conference* (Baltimore, MD, 2000) paper FF2, 2000.)

Through careful amplifier and system design, we can overcome some of the problems in ROAs. Nowadays, it is common, especially in laboratories, to combine ROAs and EDFAs to achieve the optimal performance in optical transmission. Several recent 40 Gb s^{-1} transmission experiments rely on distributed Raman gain in order to achieve record-breaking performance. For example, multi-terabits/second transmission experiments over several thousand kilometers were carried out [67–70].

The quest for higher capacity is, and will still be the focus of transmission research since higher capacity transmission may provide an economic advantage. Increasing the signal-to-noise-ratio using Raman amplifiers is one effective approach. Together with other technologies such as sophisticated coding schemes (forward-error correction, etc.), new modulation format, new multiplexing schemes, and higher-quality fibers, the transmission capacity is constantly increasing. Recently, research teams demonstrated 40 Gb s^{-1} transmission using a 100 GHz channel spacing for ultra-long-haul (>3000 km) systems, yielding a high spectral efficiency of 40 Gb s^{-1}/100 GHz = 0.4 b s^{-1} Hz^{-1}. In the years to come, even more efficient use of available fiber spectrum and higher spectral efficiency are likely to take place. Transmission of multi-terabits/second over long distances will be achieved.

6.3.4 Semiconductor optical amplifiers

SOAs, because of their small physical size, low cost, and electrical pumping, are ideal amplifiers to be used in low-cost telecommunication systems. To achieve gain in the medium, the semiconductor material is excited to achieve population inversion. There exists two types of SOAs—travelling-wave amplifiers, and Fabry–Perot amplifiers. In the former, the input signal travels from the input port to the output port in the forward direction only, which is made possible by minimizing the reflectivity at the cleaved facets via an antireflection coating. The reflectivity can be reduced further by using angled facets. For Fabry–Perot amplifiers, the input signal is partially reflected at each facet as naturally cleaved facets are used. Because of its resonant nature, the output power of Fabry–Perot SOAs is limited to −10 dBm, while the output power of travelling-wave SOAs can exceed +13 dBm.

The gain spectrum of SOAs can be engineered to be as broad as 100 nm. Moreover, through the choice of the material composition (GaAlAs, InGaAlP, InGaAsP, etc.), they can operate in the visible or in the near-infrared (1.3–1.6 μm). Since the typical noise figures of SOAs are in the range of 7–9 dB, the use of SOAs has been limited to short-haul (100–200 km) transmission experiments [71,72].

In a quest to boost output power of a laser, a master-oscillator power-amplifier (MOPA) design is used, where the output of a laser, acting as an oscillator, is amplified subsequently in a booster section. Greater than 1 W of output power is achievable in monolithically integrated MOPAs [73].

SOAs possess undesirable characteristics that must be overcome with clever engineering solutions. For example, the polarization-dependent gain (PDG) of an SOA can be as high as 6 dB. One method to mitigate the polarization effect is to use polarization diversity, where the input optical signal traverses an SOA twice, one in each orthogonal polarization. Another method of growing strained multiple quantum wells in the active region reduces the amount of PDG to below 1 dB [74].

Because the carrier lifetime in SOAs is as short as 500 ps, its fast gain dynamics, together with its relatively low saturation power, induce cross-gain saturation, when the input signal consists of multiple wavelength channels. In other words, if the SOA is used in a WDM system, the instantaneous gain of a given wavelength channel is saturated not only by its channel power, but also by the combined instantaneous power (bit pattern) of the remaining channels. This deleterious effect can be counteracted by (1) operating the SOA in the quasi-linear regime, hence reducing the output power per channel [71], (2) adding a strong saturating tone to the input signal to act as a reservoir [72], (3) using polarization multiplexing [75], and (4) not using intensity modulation on the optical carrier, but using frequency or phase modulation instead [76].

6.3.5 Short-pulse amplification

Although the intensity of a moderate laser beam can be high when it is focused to a spot, on-going work has never stopped in building laser amplifiers to generate higher power laser beams. One limiting factor in amplifying short pulses is the excessive nonlinearity generated when the short pulse passes through the gain medium. Nonlinear effects encountered include (1) self-focusing, which causes the laser beam to collapse to a focal point with catastrophic results; (2) spatial filamentation of the input beam; (3) excessive self-phase modulation, where the intensity-dependent phase shift degrades the temporal quality of the output pulse. In other words, in typical high-power short-pulse laser systems, it is the peak intensity, rather than the energy or the fluence, that causes pulse distortion or laser damage. Previously, large laboratories, such as the Lawrence Livermore National Laboratory, construct lasers and amplifiers with large beam diameters as an expensive way to increase the laser power while keeping the intensity below nonlinear effects. CPA overcomes the problem by prestretching the input pulse using a dispersive diffraction grating pair and then compressing the output pulse using another pair of diffraction gratings (Figure 6.19) [24]. Since the optical pulse remains dispersed as it traverses the amplifying medium, the amount of nonlinearity experienced by the optical pulse is minimized. By 1990, this technique has been used to boost the peak power of an incoming laser pulse to over 10 TW (10×10^{12} W) levels (Figure 6.20). Since the maximum intensity is limited to 10^{12} W m^{-2} without using CPA, the pulse energy is therefore boosted 10 billion-fold. Although this optical pulse lasts a few tens of femtoseconds only, its instantaneous peak power is equal to the power output from the entire electrical grid in the world. When focused onto a small spot, one can

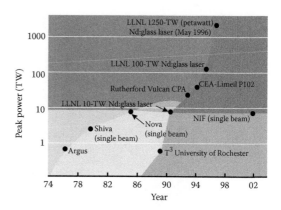

Figure 6.20 Milestones in the advancement of laser peak power. The darkest shaded region at upper right indicates that the laser pulses are amplified using chirped pulse amplification. (From http://www.llnl.gov/str/Petawatt.html.)

create high-energy pulses with intensities as high as 10^{22} W m^{-2}, thus opening up new and exciting areas in nonlinear optics. For example, electrons are accelerated to relativistic velocities in the so called "table-top accelerators." To avoid excessive gain saturation in the amplifier, the repetition rate of the incoming pulse train is often reduced using a pulse-selecting Pockels cell. Incidentally, the concept of CPA is also used in chirped radar systems—in order to avoid saturating the power amplifier in the transmitter, the millimeter radar pulse is stretched in time (linearly chirped) by a surface acoustic wave (SAW) acting as a dispersive delay line, before it is amplified and emitted. A pulse-compression filter is then used in the receiver to demodulate and compress the return signal into a shorter pulse, which results in a higher range resolution than radar systems not using chirped pulses.

Using the latest technological advances in pump laser, rare-earth-doped fiber, control electronics, and packaging, vendors are now offering turn-key table-top systems utilizing CPA. For example, IMRA's oscillator-amplifier laser system can produce high-energy femtosecond pulses with peak powers as much as 10 MW (Figure 6.21). The diode-pumped system uses CPA in a large-core Yb-doped fiber amplifier. Since optical fiber is an excellent medium to be used as a temperature sensor, the entire system is packaged athermally to achieve good frequency stability and low pulse-timing jitter.

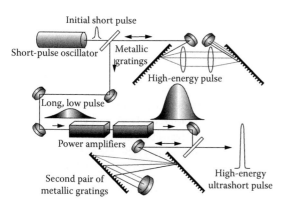

Figure 6.19 The concept of chirped pulse amplification. (From http://www.llnl.gov/str/Petawatt. html.)

Figure 6.21 IMRA America, Inc.'s turn-key oscillator-amplifier laser system. (From www.imra.com.)

REFERENCES

1. Einstein, A. 1917. On the quantum theory of radiation. *Physikalische Zeitschrift*, 18, 121.
2. Tolman, R.C. 1924. Duration of molecules in upper quantum states. *Rev. Mod. Phys.*, 23, 693–709.
3. Gordon, J.P., Zeiger, H.J. and Townes, C.H. 1954. Molecular microwave oscillator and new hyperfine structure in the microwave spectrum of NH_3. *Phy. Rev.*, 94, 282–284.
4. Gordon, J.P., Zeiger, H.J. and Townes, C.H. 1955. The Maser—New type of microwave amplifier, frequency standard, and spectrometer. *Phy. Rev.*, 95, 1264–1274.
5. Townes, C.H. 1964. Production of coherent radiation by atoms and molecules. In *Nobel Lectures Physics 1963–1970* (Amsterdam: Elsevier), pp. 58–85.
6. Schawlow, A.L. and Townes, C.H. 1958. Infrared and optical masers. *Phys. Rev.*, 112, 1940–1949.
7. Maiman, T.H. 1960. Stimulated optical radiation in ruby. *Nature*, 187, 493.
8. Haus, H.A. 1984. *Waves and Fields in Optoelectronics* (Englewood Cliffs: Prentice-Hall).
9. Saleh, B.E.A. and Teich, M.C. 1991. *Fundamental of Photonics* (New York: Wiley).
10. Siegman, A.E. 1986. *Lasers* (Mill Valley: University Science Books).
11. Yariv, A. 1985. *Optical Electronics*, 3rd edn (New York: Holt, Rinehart, and Winston).
12. Desurvire, E., Simpson, J.R. and Becker, P.C. 1987. High-gain erbium-doped traveling-wave fiber amplifier. *Opt. Lett.*, 12, 888–890.
13. Maiman, T.H. 1960. Optical and microwave-optical experiments in ruby. *Phys. Rev. Lett.*, 4, 564.
14. Geusic, J.E., Marcos, H.M. and Van Uitert, L.G. 1964. Laser oscillations in Nd-doped yttrium aluminum, yttrium gallium and gadolinium garnets. *Appl. Phys. Lett.*, 4, 182.
15. Kushida, T., Marcos, H.M. and Geusic, J.E. 1968. Laser transition cross section and fluorescence branching ratio for Nd^{3+} in yttrium aluminum garnet. *Phys. Rev.*, 167, 289.
16. Koechner, W. 1976. *Solid State Engineering* (New York: Springer).
17. Snitzer, E. and Young, C.G. 1968. Glass lasers. In *Lasers*, vol. 2, ed. A.K. Levine (New York: Dekker), p. 191.

18. Moulton, P.F. 1986. Spectroscopic and laser characteristics of Ti:Al$_2$O$_3$. *J. Opt. Soc. Amer. B*, 3, 125–133.

19. Vuylsteke, A.A. 1963. Theory of laser regeneration switching. *J. Appl. Phys.*, 34, 1615.

20. Wagner, W.G. and Lengyel, B.A. 1963. Evolution of the giant pulse in a laser. *J. Appl. Phys.*, 34, 2040.

21. Verdeyen, J.T. 1989. *Laser Electronics*, 2nd edn (Englewood Cliffs: Prentice-Hall).

22. Haus, H.A. 1975. A theory of forced modelocking. *IEEE J. Quant. Electron.*, 11, 323–330.

23. Smith, P.W. 1970. Mode-locking of lasers. *Proc. IEEE*, 58, 1342–1359.

24. Strickland, D. and Mourou, G. 1985. Compression of amplified chirped optical pulses. *Opt. Commun.*, 55, 447.

25. Hall, R.N., Fenner, G.E., Kingsley, J.D., Soltys, T.J. and Carlson, R.O. 1962. *Phys. Rev. Lett.*, 9, 366.

26. Javan, A., Bennett, W.R.Jr. and Herriott, D.R. 1961. Population inversion and continuous optical maser oscillation in a gas discharge containing a He–Ne mixture. *Phys. Rev. Lett.*, 6, 106.

27. Bridges, W.B. 1964. Laser oscillation in singly ionized argon in the visible spectrum. *Appl. Phys. Lett.*, 4, 128.

28. Hayashi, I., Panish, M.B., Foy, W. and Sumski, S. 1970. *Appl. Phys. Lett.*, 17, 109.

29. Patel, C.K.N. 1964. Introduction of CO2 optical maser experiments. *Phys. Rev.*, 136, A1187.

30. Patel, C.K.N. 1968. High power carbon dioxide lasers. *Sci. Am.*, 219, 22–23.

31. Bennett, W.R. 1962. Gaseous optical masers. *Appl. Opt.*, Suppl. 1 Optical Masers, 24.

32. Kane, D.J. and Trebino, R. 1993. Characterization of arbitrary femtosecond pulses using frequency-resolved optical gating. *IEEE J. Quant. Electron.*, 29, 571–579.

33. Nathan, M.I., Dumke, W.P., Burns, G., Dill, F.H. Jr. and Lasher, G. 1962. *Appl. Phys. Lett.*, 1, 62.

34. Quist, T.M., Rediker, R.H., Keyes, R.J., Krag, W.E., Lax, B., McWhorter, A.L. and Zeiger, H.J. 1962. *Appl. Phys. Lett.*, 1, 91.

35. Alferov, Z.I., Andreev, V.M., Garbuzov, D.Z., Zhilyaev, Y.V., Morozov, E.P., Portnoi, E.L. and Trofim, V.G. 1971. Investigation of the influence of the AlAs–GaAs heterostructure parameters on the laser threshold current and the realization of continuous emission at room temperature. *Sov. Phys. Semicond.*, 4, 1573–1575.

36. Gordon, E.I., Labuda, E.F. and Bridges, W.B. 1964. Continuous visible laser action in singly ionized Argon, Krypton and Xenon. *Appl. Phys. Lett.*, 4, 178.

37. Alferov, Z. 2000. Double heterostructure lasers: Early days and future perspectives. *IEEE J. Sel. Top. Quant. Electron.*, 6, 832–840.

38. Weber, H.P. 1967. Method for pulsewidth measurement of ultrashort light pulses generated by phase-locked lasers using nonlinear optics. *J. Appl. Phys.*, 38, 2231–2234.

39. Massicott, J.F., Armitage, J.R., Wyatt, R., Ainslie, B.J. and Craig-Ryan, S.P. 1990. High gain, broadband, 1.6 µm Er^{3+} doped silica fiber amplifier. *Electron. Lett.*, 20, 1645–1646.

40. Walmsley, I. and Trebino, R. 1996. Measuring fast pulses with slow detectors. *Opt. Photon. News*, 7, 23.

41. Agrawal, G.P. 2002. *Fibre-Optic Communication Systems*, 3rd edn (New York: Wiley).

42. Mollenauer, L.F. and Smith, K. 1988. Demonstration of soliton transmission over more than 4000 km in fiber with loss periodically compensated by Raman gain. *Opt. Lett.*, 13, 675–677.

43. Mears, R.J., Reekie, L., Jauncey, I.M. and Payne, D.N. 1987. Low-noise erbium-doped fiber amplifier operating at 1.54 µm. *Electron. Lett.*, 23, 1026–1028.

44. Giles, C.R. and Desurvire, E. 1991. Modeling erbium-doped fiber amplifiers. *J. Lightwave Technol.*, 9, 271–283.

45. Sun, Y., Zyskind, J.L. and Srivastava, A.K. 1997. Average saturation level, modeling, and physics of erbium-doped fiber amplifiers. *IEEE J. Sel. Top. Quant. Electron.*, 3, 991–1007.

46. Delavaux, J.-M.P. and Nagel, J.A. 1995. Multi-stage erbium-doped fiber amplifier designs. *J. Lightwave Technol.*, 13, 703–720.

47. Smart, R.G., Zyskind, J.L. and DiGiovanni, D.J. 1994. Two-stage erbium-doped fiber amplifiers suitable for use in long-haul soliton systems. *Electron. Lett.*, 30, 50–52.

48. Wysocki, P.F., Judkins, J.B., Espindola, R.P., Andrejco, M. and Vengsarkar, A.M. 1997. Broad-band erbium-doped fiber amplifier flattened beyond 40 nm using long-period grating filter. *IEEE Photon. Technol. Lett.*, 9, 1343–1345.

49. Sun, Y. et al. 1997. Transmission of 32-WDM 10-Gb/s channels over 640 km using broadband, gain-flattened erbium-doped silica fiber amplifiers. *IEEE Photon Technol. Lett.*, 9, 1652–1654.

50. Holonyak, N. Jr. and Bevacqua, S.F. 1962. *Appl. Phys. Lett.*, 1, 82.

51. Sun, Y. et al. 1997. Ultra-wide-band erbium-doped silica fiber amplifier with 80 nm of bandwidth. In *Optical Amplifiers and Their Applications* (Victoria, BC, 1997) postdeadline paper PD-2.

52. Srivastava, A.K. et al. 1998. 1 Tb/s transmission of 100 WDM 10 Gb/s channels over 400 km of TrueWave™ fiber. In *OFC '98 Technical Digest* (San Jose, CA, 1998) postdeadline paper PD-10.

53. Giles, C.R., Desurvire, E. and Simpson, J.R. 1989. Transient gain and cross-talk in erbium-doped fiber amplifier. *Opt. Lett.*, 14, 880–882.

54. Zyskind, J.L., Sun, Y., Srivastava, A.K., Sulhoff, J.W., Lucero, A.L., Wolf, C. and Tkach, R.W. 1996. Fast power transients in optically amplified multiwavelength optical networks. In *Proceedings of the Optical Fiber Communications Conference* (San Jose, CA, 1996) postdeadline paper PD-31.

55. Desurvire, E., Zirngibl, M., Presby, H.M. and DiGiovanni, D. 1991. Dynamic gain compensation in saturated erbium-doped fiber amplifiers. *IEEE Photon. Technol. Lett.*, 3, 453–455.

56. Srivastava, A.K. et al. 1997. Fast-link control protection of surviving channels in multiwavelength optical networks. *IEEE Photon. Technol. Lett.*, 9, 1667–1669.

57. Zirngibl, M. 1991. Gain control in erbium-doped fiber amplifiers by an all-optical feedback loop. *Electron. Lett.*, 27, 560–561.

58. Boyd, G.D. and Kogelnik, H. 1962. Generalized confocal resonator theory. *Bell Syst. Tech. J.*, 41, 1347.

59. Fludger, C., Maroney, A., Jolley, N. and Mears, R. 2000. An analysis of improvements in OSNR from distributed Raman amplifiers using modern transmission fibers. In *Proceedings of the Optical Fibre Communications Conference* (Baltimore, MA, 2000) paper FF2.

60. Stolen, R.H. 1980. *Proc. IEEE*, 68, 1232.

61. Hellwarth, R.W. 1963. *Phys. Rev.*, 130, 1850.

62. Raman, C.V. 1928. *Indian J. Phys.*, 2, 387.

63. Woodbury, E.J. and Ng, W.K. 1962. *Proc. IRE*, 50, 2347.

64. Lewis, S.A.E., Chernikov, S.V. and Taylor, J.R. 2000. Characterization of double Rayleigh scatter noise in Raman amplifiers. *IEEE Photon. Technol. Lett.*, 12, 528.

65. Chraplyvy, A.R. and Henry, P.S. 1983. Performance degradation due to stimulated Raman scattering in wavelength-division-multiplexed optical fiber systems. *Electron. Lett.*, 19, 641–643.

66. Chen, C.-J. and Wong, W.S. 2001. Transient effects in saturated Raman amplifiers. *Electron. Lett.*, 37, 371.

67. Charlet, G. et al. 2002. 6.4 Tb/s (159×42.7 Gb/s) capacity over 21×100 km using bandwidth-limited phase-shaped binary transmission. In *Proceedings of the European Conference on Optical Communication* (Copenhagen, Denmark, 2002) paper PD4.1.

68. Foursa, D.G. et al. 2002. 2.56 Tb/s (256×10 Gb/s) transmission over 11000 km using hybrid Raman/EDFAs with 80 nm of continuous bandwidth. In *Proceedings of the Optical Fiber Communications Conference* (Anaheim, CA, 2002) postdeadline paper FC3.

69. Gnauck, A.H. et al. 2002. 2.5 Tb/s (64×42.7 Gb/s) transmission over 40×100 km NZDSF using RZ-DPSK format and all-Raman-amplified spans. In *Proceedings of the Optical Fiber Communications Conference* (Anaheim, CA, 2002) postdeadline paper FC2.

70. Grosz, D.F. et al. 2002. 5.12 Tb/s (128×42.7 Gb/s) transmission with 0.8 bit/s/ Hz spectral efficiency over 1280 km of standard single-mode fiber using all-Raman amplification and strong signal filtering. In *Proceedings of the European Conference in Optical Communication* (Copenhagen, Denmark, 2002) paper PD4.3.

71. Spiekman, L.H., Wiesenfeld, J.M., Gnauck, A.H., Garret, L.D., van den Hoven, G.N., van Dongen, T., Sander-Jochem, M.J.H. and

Binsma, J.J.M. 2000. 8×10Gb/s DWDM transmission over 240km of standard fiber using a cascade of semiconductor optical amplifiers. *IEEE Photon Technol. Lett.*, 12, 1082–1084.

72. Sun, Y., Srivastava, A.K., Banerjee, S., Sulhoff, J.W., Pan, R., Kantor, K., Jopson, R.M. and Chraplyvy, A.R. 1999. Error-free transmission of 32×2.5Gb/s DWDM channels over 125km using cascaded in-line semiconductor optical amplifiers. *Electron. Lett.*, 35, 1863–1865.

73. Mehuys, D., Parke, R., Waarts, R.G., Welch, D.F., Hardy, A. and Streifer, W. 1991. Characteristics of multistage monolithically integrated master oscillator power amplifiers. *IEEE J. Quant. Electron.*, 27, 1574–1581.

74. Cole, S., Cooper, D.M., Devlin, W.J., Ellis, A.D., Elton, D.J., Isaak, J.J., Sherlock, G., Spurdens, P.C. and Stallard, W.A. 1989. *Electron. Lett.*, 25, 314–315.

75. Srivastava, A.K., Banerjee, S., Eichenbaum, B.R., Wolf, C., Sun, Y., Sulhoff, J.W. and Chraplyvy, A.R. 2000. A polarization multiplexing technique to mitigate WDM crosstalk in SOAs. *IEEE Photon. Technol. Lett.*, 12, 1415–1416.

76. Kim, H.K., Chandrasekhar, S., Srivastava, A., Burrus, C.A. and Buhl, L. 2001. 10Gbit/s based WDM signal transmission over 500km of NZDSF using semiconductor optical amplifier as the in-line amplifier. *Electron. Lett.*, 37, 185–187.

PART II

Advanced concepts

7

Advanced optics

ALAN ROGERS
University of Surrey

VINCENT HANDEREK
Fotech Solutions Ltd.

7.1 INTRODUCTION

The practice of optoelectronics at the level of device and system design necessitates a familiarity with some quite advanced, and sometimes quite subtle, optical physics. This chapter will deal, at a fairly fundamental level, with those topics which the author considers most important for an appreciation of present-day optoelectronics.

The topics chosen naturally are concerned with the properties of light and its control by waveguides and with those properties of solid materials relevant to their interactions with light radiation.

Many of the basic ideas that are needed to understand these processes have already been introduced in earlier chapters; in this chapter, however, we shall need to extend and refine the coverage to the point where we can understand the structure and operation of specific optoelectronic devices. These devices depend, in most cases, on the physics of optical radiation and of the solid state, in a fairly detailed way.

It is thus necessary to gain a more detailed familiarity with this physics because it lies at the heart of the subject, and without this knowledge it would be impossible either to understand properly existing optoelectronics or to progress to new devices and new systems beyond our present-day thinking.

We shall begin by looking at the physics of radiation.

7.2 THE PHYSICS OF RADIATION

7.2.1 Black-body radiation

All matter, provided that it has a temperature other than absolute zero, emits radiation. This is a consequence of the fact that a temperature above absolute zero implies that the atoms or molecules are in motion and are thus colliding with each other constantly. These collisions not only transfer kinetic energy of motion but also sometimes excite the atomic system to a higher state from which it may relax by emitting a photon. This is a consequence of a very basic principle in physics, the "law of equipartition of energy," which states that the energy of a system, in equilibrium, will be distributed equally among all possible degrees of freedom: the kinetic energy of a material is one such degree, excited states represent another. By assigning a temperature to a body, we require it

to be in thermal equilibrium with its surroundings (i.e., there is no net heat gain from, or loss to, the surroundings over time), so equipartition must apply.

The first question that naturally arises now is, how much radiation is emitted by a body at a given temperature? And the second (perhaps not quite as naturally!) is, what is the distribution of this emitted radiation over the wavelength spectrum?

In answering these questions, we shall explore ideas that are valuable for a whole range of topics in optoelectronics, and more general physics, so it is worthwhile taking some time over them.

Classical thermodynamics assumed that atoms emitted light as a result of radiation by electrons that were oscillating at natural resonant frequencies within the atoms. It further assumed (it had no reason to assume otherwise) that these oscillations could occur with any amount of energy, depending on the strength of the stimulus.

The other piece of information that the classical thermodynamicists needed before they could proceed was the Boltzmann factor. This tells us the ratio of the number of atoms that have energy E_1 compared to those with energy E_2 in a system in equilibrium at absolute temperature T, and takes the form [1]

$$N_2/N_1 = \exp[-(E_2 - E_1)/kT],$$

where k is Boltzmann's constant, with value 1.38×10^{-23} JK^{-1}. This factor had already been derived, using classical statistical thermodynamics, by means of an exquisite argument (which we do not have space to develop, but see any text on statistical mechanics).

Let us now derive the classical result for the radiation emitted by a "perfect" body, i.e., a body capable of emitting or absorbing radiation of any wavelength and thus containing oscillators (atoms or molecules) capable of oscillating at any frequency. Such a body is called a "black" body because it absorbs, rather than reflects, all light falling upon it, and therefore looks "black" (until it emits!). Such a body is a valuable idealization, because we can categorize "real" bodies according to how closely they approximate to it.

Let us suppose that within this black body, the oscillators can have any energy and (for reasons which will become clear later) these energies will be described as a set distributed as

$$0, dE, 2dE, \ldots, ndE, \ldots$$

where dE is infinitesimally small so that the distribution is continuous. The ratios of numbers of oscillators within each of these energy bands will comprise the set (according to the Boltzmann factor):

$$1 : \exp\left(-\frac{dE}{kT}\right) : \exp\left(-\frac{2dE}{kT}\right) : \ldots : \exp\left(-\frac{ndE}{kT}\right) : \ldots$$

Thus, if there are N oscillators with zero energy, the total number of oscillators will be

$$N_T = N\left[1 + \exp\left(-\frac{dE}{kT}\right) + \exp\left(-\frac{2dE}{kT}\right)\right.$$

$$\left. + \cdots + \exp\left(-\frac{ndE}{kT}\right) + \cdots\right].$$

This is a geometrical progression, which is easily summed to give

$$N_T = \frac{N}{1 - \exp\left(-\frac{dE}{kT}\right)} \quad (7.1)$$

Also, the total energy of all the oscillators is given by multiplying each term by its energy allocation:

$$E_T = N\left[dE\exp\left(-\frac{dE}{kT}\right) + 2dE\exp\left(-\frac{2dE}{kT}\right)\right.$$

$$\left. + \cdots + ndE\exp\left(-\frac{ndE}{kT}\right) + \cdots\right]$$

giving, on summation

$$E_T = \frac{NdE}{\exp\left(\frac{dE}{kT}\right)\left[1 - \exp\left(-\frac{dE}{kT}\right)\right]^2}. \quad (7.2)$$

On dividing Equation 7.2 by Equation 7.1, we obtain the mean energy per oscillator as

$$\bar{E} = \frac{dE}{\exp\left(\frac{dE}{kT}\right) - 1}. \quad (7.3)$$

We may now let $dE \to 0$ to discover the physical value for this mean energy, whereupon we find (expanding the exponential in the denominator)

$$\bar{E} = \lim_{dE \to 0} \frac{dE}{1 + \frac{dE}{kT} + \frac{1}{2}\left(\frac{dE}{kT}\right)^2 + \cdots - 1}$$

or

$$\bar{E} = kT. \quad (7.4)$$

(This is entirely in accordance with other "equipartitional" approaches, which allow $kT/2$ of energy per degree of freedom. In our case, there are two degrees of freedom per oscillator, one for kinetic energy, the other for potential energy, giving kT in all.)

The final piece of information we need is the number of independent oscillations, which can occur within a given volume of material. Clearly, the fact that the volume is finite means that there are boundaries and these impose boundary conditions on the oscillations, just as a string stretched between two fixed points is bounded by the fact that any oscillation of the string must have zero amplitude at the points of fixation.

Let us simplify things by taking the volume to be a cube of side l (Figure 7.1a): suppose now that oscillations occur within the cube and that the velocity with which these propagate is c. The walls of the cube impose a zero-amplitude boundary condition for these oscillations, so the resonant oscillations can only occur parallel with the sides of the cube with frequencies $nc/2l$, n being a positive integer and $c/2l$ the fundamental.

Now waves can, of course, travel in many directions and we can best represent any given wave by its wave vector k, which has the same direction of the wave and an amplitude $2\pi/\lambda$, where λ is the wavelength. We may now write the frequencies of the waves traveling parallel to the sides of the cube as

$$f_n = \frac{nc}{2l}$$

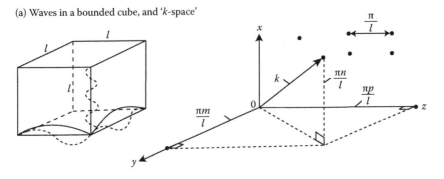

(a) Waves in a bounded cube, and 'k-space'

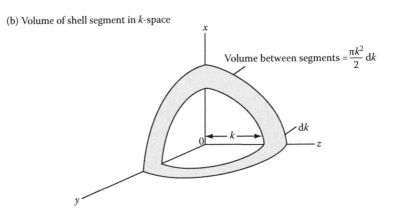

(b) Volume of shell segment in k-space

Volume between segments $= \dfrac{\pi k^2}{2} dk$.

Figure 7.1 k-space diagram.

and the wave numbers as

$$k_n = \frac{2\pi}{c} f_n = \frac{\pi n}{l}. \tag{7.5}$$

Let us now take axes Ox, Oy, and Oz parallel with the sides of the cube and plot in three dimensions a lattice of points corresponding to all the k_n in Equation 7.5. It is easily seen that *any* oscillation for the cube can be represented by its wave vector from the origin of axes to one of the points we have plotted. The plot is often called "k-space," for obvious reasons.

We now ask the question: how many oscillations can the cube support with wavelengths between k and $k+dk$? This, we can see, will correspond to the number of points on our plot, which lie in the volume between spheres of radius k and $k+dk$, respectively. This volume in k-space is, $4\pi k^2 dk$ (Figure 7.1b). However, because only positive values of n are valid, we only need that octant of the spherical shell where all the axes are positive, which is one-eighth of the total, i.e., $\pi k^2 dk/2$. To find the number of points in this volume, we divide

by one elementary volume in our lattice, defined by the interval between points, i.e., a cube of side π/l, volume π^3/l^3.

Hence, the number of oscillations between k and $k+dk$ is

$$N_0' = \frac{\frac{1}{2} \pi k^2 dk}{\pi^3/l^3} = \frac{k^2 l^3 dk}{2\pi^2}$$

and if we allow two orthogonal linear polarizations per oscillation (any electromagnetic wave can always be resolved into two such components), this becomes

$$N_0 = \frac{k^2 l^3 dk}{\pi^2}.$$

Because the volume of our original cube is l^3, we can express this in the form of a number of oscillations per unit volume:

$$N_v = \frac{k^2 dk}{\pi^2}. \tag{7.6a}$$

It is now more convenient to write N_v in terms of frequency (since frequency is more directly related with energy). We have

$$k = \frac{2\pi}{\lambda} = \frac{2\pi f}{c}.$$

Hence

$$N_v = \frac{8\pi f^2}{c^3}\mathrm{d}f. \tag{7.6b}$$

This is an important result in itself and appears in many aspects of laser theory; it should be noted carefully.

We may use it immediately for our present purposes to derive the classical result for the energy spectrum of a black-body radiator. From Equation 7.4, we observed that each oscillation has mean energy kT. Hence, the energy density (i.e., energy per unit volume) lying between frequencies f and $f+\mathrm{d}f$ is given by

$$\rho_f\mathrm{d}f = \frac{8\pi f^2}{c^3}kT\mathrm{d}f. \tag{7.7}$$

This is the classical result, the so-called Rayleigh–Jeans equation. *It is wrong!* It has to be. We can see this immediately by calculating the total energy density emitted over all wavelengths of the spectrum:

$$\rho_T = \int_0^\infty \frac{8\pi f^2}{c^3}kT\,\mathrm{d}f$$

$$= \frac{8\pi f^2}{c^3}\left[\frac{f^3}{3}\right]_0^\infty \to \infty!$$

The answer is thus that an infinite amount of total energy is emitted by any black body; this is, of course, quite impossible. This result caused much head scratching among classical physicists around the beginning of the 20th century. When the spectrum of a black body (or as close to one as could be realized in practice) was measured, the shape was as shown in Figure 7.2. The agreement with the Rayleigh–Jeans expression (Equation 7.7) was good at low frequencies, but the two diverged wildly at the higher frequencies: the problem was thus dubbed the "ultraviolet (i.e., high frequency) catastrophe."

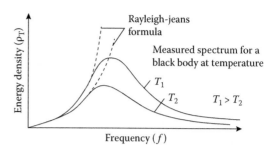

Figure 7.2 The black-body spectrum.

7.2.2 The quantum result

Max Planck, in 1900, saw a very simple way to avoid the above problem. He suggested that an oscillator could not possess *any* value of energy but only a value which was an integer times a certain minimum value. If this latter was ε, then the only possible values for the energy of the oscillation were ε, 2ε, 3ε,..., $n\varepsilon$. This changed things completely, as we shall now show.

We can now, very conveniently, return to Equation 7.3 for the mean energy of an oscillator:

$$\bar{E} = \frac{\mathrm{d}E}{\exp\left(\dfrac{\mathrm{d}E}{kT}\right)-1}$$

$\mathrm{d}E$ can now be identified with ε, and now it does *not* tend to zero, but always remains nonzero. Now, ε is the minimum energy of the oscillator, which emits radiation and thus can be identified, in turn, with the quantity $h\nu$, where ν is the lowest frequency of radiation it emits, remembering that the oscillator can have any of the energies $nh\nu$, where n is any positive integer. (We now use ν for frequency, rather than f, to remind ourselves that we are dealing with quantum phenomena rather than continuous events, h is the quantum constant (Planck's constant) with value 6.626×10^{-34} Js.)

Thus, we have, in the quantum case

$$\bar{E} = \frac{h\nu}{\exp\left(\dfrac{h\nu}{kT}\right)-1}$$

and the energy density lying between ν and $\nu+\mathrm{d}\nu$ now will be, using Equation 7.7:

$$\rho_v dv = \frac{8\pi v^2}{c^3} \frac{hv}{\exp\left(\dfrac{hv}{kT}\right) - 1} dv. \qquad (7.8)$$

This is the celebrated Planck's radiation formula, and it solves all our problems, for it agrees with the experimental spectrum (Figure 7.2).

If integrated over all frequencies it remains finite and gives the result:

$$E_T = \frac{2\pi^5 k^4}{15c^2 h^3} T^4. \qquad (7.9)$$

Equation 7.9 represents the Stefan–Boltzmann law for the total energy emitted by a black body; classical thermodynamics was able to show that this quantity should be proportional to the fourth power of the absolute temperature, but was unable to predict the value of the constant of proportionality; quantum physics has provided the answer to this.

Similarly, classical thermodynamics was able to prove Wien's displacement law, which states that the value of the wavelength associated with the energy maximum in the spectrum (Figure 7.2) is inversely proportional to the absolute temperature, i.e.,

$$\lambda_m = \frac{\Omega}{T}$$

but was unable to determine the value of the constant Ω. By differentiating Equation 7.8, we easily find that

$$\Omega = \frac{ch}{4.9651 k}.$$

The above results had a profound effect. Although Planck at first felt that his quantum hypothesis was no more than a mathematical trick to avoid the ultraviolet catastrophe, it soon became clear that it was fundamentally how the universe did, in fact, behave: quantum theory was born.

7.2.3 "Black-body" sources

The concept of a black body is that of a body that emits and absorbs all frequencies of radiation. We know now that the quantum theory requires us to limit the frequency to multiples of a certain fundamental frequency, but, in practice, owing to the particular molecular structure of any given body, the quantum (and classical) "black body" remains an idealization, and real bodies, when hot, will not yield a spectrum in strict accordance with Planck's radiation law but only an approximation to it (sometimes a very close approximation, however).

Nevertheless, we can very conveniently measure the temperature of a radiating body by measuring the wavelength at which the spectrum peaks, using Wien's law, or, if the peak is not at a convenient (for our detector) position in the spectrum, by measuring the total energy emitted (using a bolometer) and applying the Stefan–Boltzmann law. Very often we require a source that emits over a broad range of frequencies, and a convenient way to obtain this is to create a discharge in a gas. An electrical discharge creates a large number of free, energetic electrons, which cause a large range of atomic excitations, thus giving rise to radiation over a broad frequency range. Intensities can be quite high, so that the experimenter or designer can then pick out those frequencies that are needed, with frequency-selective optical components such as prisms or diffraction gratings.

However, the importance of the idealization known as a black body lies primarily in the fact that it allows an insight into the fundamental nature of electromagnetic radiation and the quantum laws that it obeys. This is crucial to our understanding of its role in optoelectronics, and especially to our understanding of laser radiation, which is the next topic for consideration.

7.2.4 The theory of laser action

7.2.4.1 THE RATE EQUATIONS AND THE GAIN MECHANISM

The elements of laser action were introduced in Chapter 1, Section 1.9. Lasers are extremely important in optoelectronics, as has been stressed, and it is necessary now to deal with laser action in more quantitative detail.

Let us consider two energy levels of an atomic system E_1 and E_2, with $E_2 > E_1$ (Figure 7.3).

We know that the system can be raised from E_1 to E_2 by absorption of a photon with frequency ν_{12}, where

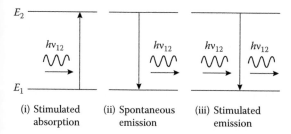

Figure 7.3 Two-level photon transitions.

$$hv_{12} = E_2 - E_1$$

and we also know that the system, after having been excited, will eventually, spontaneously, revert to its ground state E_1 by emitting a photon of energy hv_{12}.

However, the excited state E_2 can also be stimulated to decay to the state E_1 by the action of another photon of energy hv_{12}. This process is called stimulated emission. Thus, now we are considering three distinct processes:

1. Stimulated absorption ($E_1 \rightarrow E_2$)
2. Spontaneous emission ($E_2 \rightarrow E_1$)
3. Stimulated emission ($E_2 \rightarrow E_1$)

There can be no spontaneous "absorption" because this would violate the law of conservation of energy.

In order to calculate the relationships between atoms and radiation in equilibrium (i.e., blackbody radiation), Einstein used a very simple argument: consider the atoms to be in equilibrium with each other and with the radiation in a closed system. The rate (per unit volume) at which atoms are raised to the upper state is proportional to the density of photons, ρ_v, at energy hv_{12} and to the density of atoms N_1 (number per unit volume) in state E_1, that is,

$$R_{12} = N_1 \rho_v B_{12} \text{ (stimulated absorption)}$$

where B_{12} is a constant.

Similarly, the rate at which atoms in state E_2 are stimulated to return to state E_1 is given by

$$R_{21} = N_2 \rho_v B_{21} \text{ (stimulated emission)}$$

where N_2 is the density of atoms in state E_2. Now spontaneous emission from state E_2 to E_1 occurs

after a characteristic delay determined by the detailed atomic characteristics and is governed by quantum rules. Its rate, therefore, is proportional to N_2, the constant of proportionality, comprising, essentially, the reciprocal of the decay time. Thus, we have

$$S_{21} = N_2 A_{21} \text{ (spontaneous emission)}.$$

The constants A_{21}, B_{12}, and B_{21} are called the Einstein coefficients.

Clearly, in equilibrium, we must have

$$N_1 \rho_v B_{12} = N_2 \rho_v B_{21} + N_2 A_{21} \quad (7.10)$$

because the total upward and downward transition rates must be equal.

Hence, from Equation 7.10

$$\rho_v = \frac{(A_{21}/B_{21})}{(B_{12}N_1/B_{21}N_2) - 1}.$$

However, we know from the Boltzmann relation that

$$\frac{N_1}{N_2} = \exp\left(-\frac{E_1 - E_2}{kT}\right)$$

and also that $E_2 - E_1 = hv_{12}$.

Hence, generalizing from v_{12} to v

$$\rho_v = \frac{(A_{21}/B_{21})}{(B_{12}/B_{21})\exp\left(\dfrac{hv}{kT}\right) - 1}. \quad (7.11)$$

Now it was shown in Section 7.2.2 that for equilibrium (black body) radiation (Equation 7.8)

$$\rho_v = \frac{8\pi hv^3}{c^3} \frac{1}{\exp\left(\dfrac{hv}{kT}\right) - 1}.$$

Hence, it follows, by comparing this with Equation 7.11

$$B_{12} = B_{21} \quad (7.12a)$$

$$A_{21} = B_{21} \frac{8\pi hv^3}{c^3}. \quad (7.2b)$$

Relations 7.12a and 7.12b are known as the Einstein relations and are very important determinants in the relationships between atoms and radiation. For example, it is clear that, under these conditions, the ratio of stimulated to spontaneous emission from E_2 to E_1 is given by

$$S = \frac{R_{21}}{S_{21}} = \frac{\rho_v N_2 B_{21}}{N_2 A_{21}} = \frac{\rho_v c^3}{8\pi h v^3}$$

and using the expression for ρ_v from Equation 7.8

$$S = \frac{1}{\exp\left(\dfrac{hv}{kT}\right) - 1}.$$

If, for example, we consider the specific case of the He–Ne discharge at a temperature of 370 K with $\lambda = 632.8$ nm ($v = 4.74 \times 10^{14}$ Hz) then we find

$$S \approx 2 \times 10^{-27}.$$

Stimulated emission is thus very unlikely for equilibrium systems.

Another point worthy of note is that, for given values of N_2 (density of atoms in upper state E_2) and ρ_v (density of photons), the rate of stimulated emission (B_{21}) is proportional to $1/v^3$. This follows from Equation 7.12b because

$$B_{21} = \frac{A_{21} c^3}{8\pi h v^3}$$

and A_{21} is an atomic constant, representing the reciprocal of the spontaneous decay time.

This means that the higher the frequency the more difficult is the laser action, for this depends upon stimulated emission. Ultraviolet, X-ray, and γ-ray lasers present very special problems that, hopefully, will preclude the possibility of "death-ray" weapons (X-rays and γ-rays are very damaging to living tissues).

However, we do wish to use lasers at lower frequencies, visible and infrared, for example, for purposes of communication, display, and measurement, and the equation for R_{21} tells us that the way to increase the stimulated emission is to increase the values of N_2 and ρ_v.

We know that, in equilibrium, $N_2 < N_1$, from the form of the Boltzmann factor, and ρ_v is given by

Equation 7.8. Hence, we shall have to disturb the equilibrium to achieve significant levels of stimulated emission.

One way in which this can be done is to inject radiation at frequency v, so that ρ_v is increased above its equilibrium value. Suppose that this is done until the stimulated emission greatly exceeds the spontaneous emission (which does not, of course, depend upon ρ_v), i.e., until

$$N_2 \rho_v B_{21} \gg N_2 A_{21}.$$

The condition for this, clearly, is that

$$\rho_v \gg \frac{A_{21}}{B_{21}}$$

which, from Equation 7.12b, means that

$$\rho_v \gg \frac{8\pi h v^3}{c^3}.$$

However, increasing ρ_v does also increase the stimulated absorption. In fact, Equation 7.10 becomes, when ρ_v is large

$$N_1 \rho_v B_{12} = N_2 \rho_v B_{21}.$$

But we also know from Equations 7.12a and b that $B_{12} = B_{21}$; hence, $N_1 = N_2$ under these conditions. In other words, an incoming photon at frequency v is just as likely to cause a downward transition (stimulated emission) as it is an upward one (stimulated absorption). Hence, we cannot increase the population N_2 above that of N_1 simply by pumping more radiation, at frequency v, into the system. Clearly, we must change tack if we are to enhance the stimulated emission and produce a laser.

Consider a three-level rather than a two-level system (Figure 7.4a). Suppose that light at frequency v_{13} is injected into this system, so that there is a large amount of stimulated absorption from E_1 to E_3. Spontaneous decays will occur from E_3 to E_2 and then $E_2 \rightarrow E_1$ with also $E_3 \rightarrow E_1$; but if the levels are chosen appropriately according to the quantum rules, the $E_3 \rightarrow E_2$ decay can be fast and the $E_2 \rightarrow E_1$ relatively much slower. Clearly, the result of this will be that atoms will accumulate in level E_2. Now the really important point is that, unlike the previous two-level case, atoms in level E_2 are

(a) Three-level laser

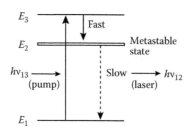

(b) Four-level laser

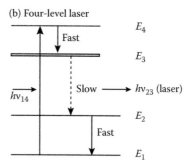

Figure 7.4 Energy-level diagrams for laser action.

immune from stimulated emission by photons at frequency ν_{13}. Hence, we can now increase the number of atoms in level E_2, at the expense of those in E_1, by increasing the intensity of the radiation at frequency ν_{13}.

We can thus soon ensure that

$$N_2 > N_1$$

and we have an "inverted population" (i.e., more atoms in a higher energy state than a lower one) as a result of the "pump" at frequency ν_{13}. This inverted population can now be exploited to give optical amplification at frequency ν_{12}.

Let us quantify this amplification via the rate equations we have developed. Suppose that photons at frequency ν_{12} are injected into the medium in a certain direction. These will meet the inverted population in energy state E_2 and will stimulate the downward transition $E_2 \rightarrow E_1$, producing more photons at frequency ν_{12} in so doing (this is, of course, the origin of the amplification). We assume, quite confidently, that the medium is being sufficiently strongly pumped for the stimulated photons to be well in excess of any spontaneous emission from E_2 to E_1. Now suppose that, under these conditions, the number of photons per unit volume when the injected radiation enters the system is p_{12}. Then, the rate at which p_{12} increases will depend upon the difference between upward and downward transition rates between levels 1 and 2, and hence we write

$$\frac{dp_{12}}{dt} = N_2 \rho_{\nu12} B_{21} - N_1 \rho_{\nu12} B_{12}.$$

Now $\rho_{\nu12}$ is the energy density of photons, hence

$$\rho_{\nu12} = p_{12} h\nu_{12}.$$

Also we know that $B_{12} = B_{21}$ (from Equations 7.12a and 7.13) and thus

$$\frac{1}{h\nu_{12}} \frac{d\rho_{\nu12}}{dt} = B_{12} \rho_{\nu12} (N_2 - N_1). \quad (7.13)$$

We shall now write $\rho_{\nu12}$ as ρ_ν to avoid cluttered equations and, integrating Equation 7.13

$$\rho_\nu = \rho_{\nu,0} \exp\left[(N_2 - N_1)t\right]$$

where $\rho_{\nu,0} = \rho_\nu$ at $t = 0$.

If the injected wave is traveling at velocity c in the medium, we can transfer to a distance parameter via $s = ct$ and obtain

$$\rho_\nu = \rho_{\nu,0} \exp\left[\frac{h\nu}{c} B_{12}(N_2 - N_1)s\right].$$

This is to be compared with the standard loss/gain relation for propagation in an interactive medium, i.e.,

$$I = I_0 \exp(gx)$$

and it is clear that the gain coefficient g can be identified as

$$g = \frac{h\nu}{c} B_{12}(N_2 - N_1) \quad (7.14a)$$

which is the gain coefficient for the medium (fractional increase in intensity level per unit length) and will be positive (i.e., gain rather than loss) provided that $N_2 > N_1$, as will be the case for an inverted population. Hence, this medium is an optical amplifier. The injected radiation at frequency ν_{12} receives gain from the optical pump of the amount:

$$G = \frac{I}{I_0} = \exp(gs)$$

so that it increases exponentially with distance into the medium. Clearly, g in Equation 7.14a is proportional to (N_2-N_1). In a three-level system such as we are considering the lower level of the amplifying transition is the ground state, which is initially heavily populated. It follows that more than half the atoms must be excited by the pump before population inversion can be achieved $(N_2 > N_1)$. It is quite hard work for the pump to excite all these atoms. Consider, however, the *four-level* system shown in Figure 7.4b. Here, the pump is at ν_{14}; there is a quick decay to level 3 and a slow one to levels 2 and 1. The decay from 2 to 1 is again fast. Clearly, the consequence of this is that it is relatively easy to provide level 3 with an inverted population over level 2 because level 2 was not well populated in the first place (being above the ground state), and atoms do not accumulate there because it decays quickly to ground. Hence, we can ensure that

$$N_3 > N_2$$

with much less pump power than for $N_2 > N_1$ in the three-level case. The amplification at ν_{32} is thus much more efficient and the four-level system makes for a more efficient amplifier.

7.2.4.2 THE LASER STRUCTURE

Having arranged for efficient amplification to take place in a medium, it is a relatively straightforward matter to turn it into an oscillator, i.e., a laser source. To do this for any amplifier, it is necessary to provide positive feedback, i.e., to feed some of the amplified output back into the amplifier in reinforcing phase. This is done by placing parallel mirrors at each end of a box containing the medium to form a Fabry–Perot cavity. The essential physics of this process is that any given photon at v12 will be bounced back and forth between the mirrors, stimulating the emission of other such photons as it does so, whereas without the mirrors it would make only one such pass.

An important condition for any system to oscillate under these circumstances is that the gain should be in excess of the loss for each cycle of oscillation. The total loss for a photon executing a double passage of the cavity (Figure 7.5) will depend not only on the loss per unit length in the medium (due to scattering, excitations to other states, wall losses, etc.) but also on the losses at the mirrors and it must be remembered that one of the mirrors has to be a partial mirror in order to let some of the light out otherwise we could not use the laser oscillator as a source! Hence, the condition for oscillation is

$$\frac{I_f}{I_i} = R_1 R_2 \exp[(g-\alpha)2l] > 1 \qquad (7.14b)$$

where I_f and I_i refer to the final and initial intensities for the double passage of the cavity, R_1 and R_2 are the reflectivities for the two mirrors, respectively, α is the loss per unit length in the medium,

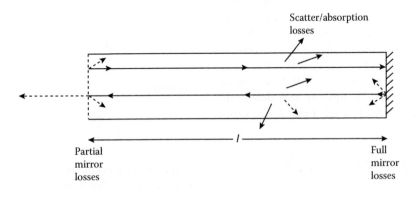

Figure 7.5 Loss mechanisms in laser cavity.

and *l* is the cavity length. (The factor 2 in the exponential refers to the double passage of the photon.)

One further word of warning: the value of *g* must correspond to the population inversion while oscillation is taking place, not the value before the feedback is applied. Clearly, the value of N_2/N_1 will be very different, once stimulated emission starts to occur, from its value when the system is simply being pumped into its inverted state. This has implications for pumping rates and the balancing of rate equations, which we shall not pursue: the principles, hopefully, are clear, however.

The simple arrangement of a pumped medium lying between two parallel mirrors (one partial) will, under the correct pump conditions, therefore, lead to radiation emerging with the following properties:

1. Narrow linewidth, because only one energy of transition is involved in the laser action; and the mirrors, if wavelength selective, will block any spontaneous light, which is emitted in addition.
2. The output direction of the light will be exactly normal to the (accurately parallel) planes of the mirrors and thus will be highly collimated in one direction.
3. When a photon is emitted via stimulated emission by another photon, it is emitted with the same phase as the original photon (remember the driving force/resonating system analogy), thus all the laser photons are locked in phase: we have coherent light (within the limitations only of the linewidth of the transition).
4. The light can be very intense because all the "light amplification by stimulated emission of radiation" from a long length of medium with small cross-sectional area can be collimated into one direction.

The above important features summarize the basic properties of laser light: it is pure (in wavelength and phase), intense, well-collimated light. It is thus easy to control and modulate; it is a powerful tool.

In order to enhance its usefulness as a tool there are two quite simple additions, which can be made to the basic design: The Fabry–Perot cavity formed by the two parallel mirrors will possess defined longitudinal "modes." Waves propagating in opposite directions within the cavity, normal to the mirrors, will interfere and reinforce to give rise to an allowable stable mode only when

$$2L = m\lambda$$

where *L* is the length of the cavity and *m* is an integer.

From this, we can also write

$$\lambda = \frac{2L}{m}; \quad f = \frac{cm}{2L}.$$

At all other wavelengths, there is destructive interference. Now, the stimulated emission occurs over a small range of wavelengths. This range is determined by the spectral width of the downward transition. The width depends upon a number of factors but primarily (unless cooled to very low temperatures) on the Doppler shift caused by the thermal motion of the molecules. Clearly, at any given time, some molecules will be moving toward the stimulating photon and others away, leading to a spread of Doppler shifts around the central line for the stationary molecule (at absolute zero of temperature!).

The output spectrum of the laser light is thus the result of combining these two features, as shown in Figure 7.6. Here, we can see the Fabry–Perot mode structure enveloped by the natural linewidth of the transition. In order to fix ideas somewhat, let us insert some real numbers into this. Suppose we have a He–Ne gas laser with length 0.5 m. Because in a gas at less than atmospheric pressure, we have $c \sim 3 \times 10^8 \ \mathrm{ms^{-1}}$, we see that

$$\frac{c}{2L} = 300 \ \mathrm{MHz}$$

which is the separation of the modes along the frequency axis. Now the Doppler linewidth of the 632.8 nm transition at 300 K is ~1.5 GHz; hence, the number of modes within this width is

$$\frac{1.5 \times 10^9}{3 \times 10^8} \sim 5$$

so that we have just five modes in the output spectrum.

So far we have dealt only with longitudinal modes; but off-axis rays also may interfere (Figure 7.7). The reinforcement condition now depends also on the angle that the ray makes with the long axis, and the result is a variation in intensity over the cross

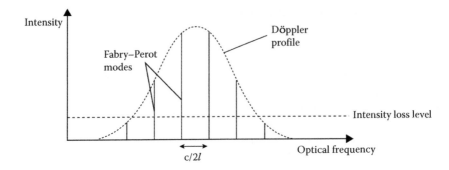

Figure 7.6 Laser-cavity spectrum.

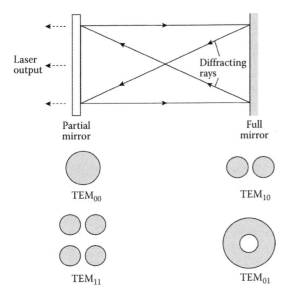

Figure 7.7 Transverse cavity modes.

section of the cavity, and thus over the cross section of the output laser beam (Figure 7.7). (The notation used to classify these variations will be described in more detail when we deal with wave guiding (Section 7.5) but TEM stands for "transverse electromagnetic" and the two suffixes refer to the number of minima in the pattern in the horizontal and vertical directions, respectively.)

7.2.4.3 MODE-LOCKING

Let us return now to the longitudinal mode structure of the laser cavity. Normally, these longitudinal modes are entirely independent because they result from wholly independently acting interference conditions. Suppose, however, that they were to be locked into a constant phase relationship. In that case, we would have a definite relationship,

in the frequency domain, between the phases and the amplitudes of the various components of the frequency spectrum. If we were to translate those relationships into the time domain, by means of a Fourier transform, the result would be a series of pulses spaced by the reciprocal of the mode frequency interval, with each pulse shape the Fourier transform of the mode envelope (Figure 7.8). All we are really saying here, in physical terms, is that if each frequency component bears a constant phase relationship to all the others, when all frequency components are superimposed, then there will be certain points in time where maxima occur (the peaks of the pulses) and others where minima occur (the troughs between pulses). If there is no fixed phase relationship between components both maxima and minima are "washed out" into a uniform-level, randomized continuum.

Now a series of evenly spaced pulses is often a very useful form of laser output, so how can it be achieved?

We must lock the phases of the longitudinal modes. One way of doing this is to include, within the cavity, an amplitude modulator, and then modulate (not necessarily sinusoidally) the amplitudes of the modes at just the mode frequency interval, $c/2L$. Then, each mode generates a series of sidebands at frequencies $mc/2L$, which corresponds to the frequencies of the other modes. The result of this is that all the modes are "pulled" mutually into phase by the driving forces at the other frequencies, and complete phase locking occurs. The inserted modulator thus has the effect of producing, from the laser output, a pulse stream with pulse repetition rate $2L/c$. For example, with the He–Ne laser quoted in section 7.2.4.2, the repetition rate is 300 MHz, and each pulse has a width (see Figure 7.8) given by the following expression.

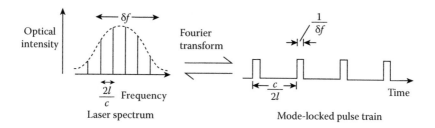

Figure 7.8 Mode-locking Fourier transform: spectrum/pulse train.

$$\sim \frac{1}{1.5\times10^9} \sim 0.67 \text{ ns.}$$

The laser is now said to be "mode-locked" and the pulse stream is a set of "mode-locked pulses." Sometimes when a laser is being pumped quite hard and the output levels are high, the laser will "self-mode lock." This is because the medium has been driven into the nonlinear regime (see Section 7.7) and the modes generate their own harmonics as a result of the induced optical nonlinearities. Clearly, this will depend upon the medium as well as the driving level because it will depend on which particular nonlinear threshold is exceeded by the pumping action.

7.2.4.4 Q-SWITCHING

The "Q" or "quality factor" of an oscillator refers to its purity, or "sharpness of resonance." The lower the loss in an oscillator the narrower is its resonance peak and the longer it will oscillate on its own after a single driving impulse. The equivalent quantity in a Fabry–Perot cavity (an optical oscillator) is the "finesse," and the two quantities are directly related. From these ideas, we can readily understand that if the loss in a resonator is varied then so is its "Q."

Suppose we have a laser medium sitting in its usual Fabry–Perot cavity but with a high loss; this means that a large fraction of the light power oscillating between the mirrors is lost per pass: we might, for example, have one of the mirrors with very low reflectivity.

Now the oscillator can only oscillate if the gain, which the light receives per double pass between the mirrors, exceeds the loss per double pass (Section 7.2.4.2), and we shall suppose that the loss is very high, so that as we pump more and more molecules of the medium up into the excited state of the inverted population, the loss still exceeds the gain

for as hard as our pump source can work. The result is that the inversion of the population becomes very large indeed, for there are very few photons to cause stimulated emission down to the lower state—they are all being lost by other means (e.g., a poor mirror at one end). Having achieved this very highly inverted population suppose that the loss is now suddenly reduced by means of an intracavity switch ("Q" switch) by, for example, speedily rotating to a high-reflectivity mirror (Figure 7.9). The result is that there is suddenly an enormous number of photons to depopulate the inverted population, which then rapidly de-excites to emit all its accumulated energy in one giant laser pulse—the Q-switched pulse. Thus, we have the means by which very large energy, very high intensity pulses can be obtained, albeit relatively infrequently (~25 pps).

Three further points should be noted concerning Q-switching:

1. At the end of the pulse, the lasing action ceases completely, because the large number of photons suddenly available completely depopulates the upper laser state.
2. The switching to the low loss condition must take place in a time that is small compared with the stimulated depopulation time of the upper state, so as to allow the pulse to build up very quickly.
3. The pumping rate must be large compared with the spontaneous decay rate of the upper state so as to allow a large population inversion to occur.

Q-switching can produce pulses with several millijoules of energy with only a few nanoseconds duration. Thus, peak powers of several megawatts can result. Such powers take most media into their nonlinear regimes (many will be evaporated!), so Q-switching is very useful for studying the nonlinear optical effects, which will be considered in Section 7.7.

(a) Inversion of the atomic population before lasing

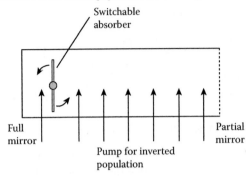

(b) Rapid depopulation of inverted states on removing absorption

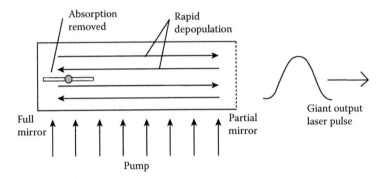

(c) A practical Q-switched laser cavity

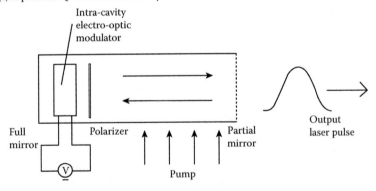

Figure 7.9 Q-switching.

Both mode-locking and Q-switching require intracavity modulation devices. These can take a variety of forms.

7.2.5 Summary

In this section, we have examined those aspects of the physics of radiation relevant to optoelectronics. It is necessary to understand the essential quantum nature of radiation in order to appreciate fully the way in which light interacts with matter and, in particular, the action of the laser in providing a source of pure light. It was the invention of the laser (in 1960), which launched the subject of optoelectronics.

7.3 OPTICAL POLARIZATION

7.3.1 Introduction

The essential idea of optical polarization was introduced in Chapter 1 but we must now consider this important topic in more detail. We know that the electric and magnetic fields, for a freely propagating light wave, lie transversely to the propagation direction and orthogonally to each other.

Normally, when discussing polarization phenomena, we fix our attention on the electric field, because it is this that has the most direct effect when the wave interacts with matter.

In saying that an optical wave is polarized we are implying that the direction of the optical field is either constant or is changing in an ordered, prescribable manner. In general, the tip of the electric vector circumscribes an ellipse, performing a complete circuit in a time equal to the period of the wave, or in a distance of one wavelength. Clearly, the two parameters are equivalent in this respect.

As is well known, linearly polarized (LP) light can conveniently be produced by passing any light beam through a sheet of polarizing film. This is a material which absorbs light of one linear polarization (the "acceptance" direction) to a much smaller extent (~1000 times) than the orthogonal polarization, thus, effectively, allowing just one linear polarization state to pass. The material's properties result from the fact that it consists of long-chain polymeric molecules aligned in one direction by stretching a plastic and then stabilizing it. Electrons can move more easily along the chains than transversely to them, and thus the optical wave transmits easily only when its electric field lies along this acceptance direction. The material is cheap and allows the use of large optical apertures. It thus provides a convenient means whereby, for example, a specific linear polarization state can be defined; this state then provides a ready polarization reference that can be used as a starting point for other manipulations. In order to study these manipulations and other aspects of polarization optics, we shall begin by looking more closely at the polarization ellipse.

7.3.2 The polarization ellipse

In Chapter 1 the most general form of polarized light wave propagating in the Oz direction was derived from the two LP components in the Ox and Oy directions (Figure 7.10):

$$E_x = e_x \cos(\omega t - kz + \delta_x)$$
$$E_y = e_y \cos(\omega t - kz + \delta_y) \qquad (7.15a)$$

Polarization ellipse

Figure 7.10 Components for an elliptically polarized wave.

If we eliminate $(\omega t - kz)$ from these equations, we obtain the following expression:

$$\frac{E_x^2}{e_x^2} + \frac{E_y^2}{e_y^2} + \frac{2E_x E_y}{e_x e_y}\cos(\delta_y - \delta_x)$$
$$= \sin^2(\delta_y - \delta_x) \qquad (7.15b)$$

which is the ellipse (in the variables E_x, E_y) circumscribed by the tip of the resultant electric vector at any one point in space over one period of the combined wave. This can only be true, however, if the phase difference $(\delta_y - \delta_x)$ is constant in time, or, at least, changes only slowly when compared with the speed of response of the detector. In other words, we say that the two waves must have a large mutual "coherence." If this was not so then relative phases and hence resultant field vectors would vary randomly within the detector response time, giving no ordered pattern to the behavior of the resultant field and thus presenting to the detector what would be, essentially, unpolarized light.

Assuming that the mutual coherence is good, we may investigate further the properties of the polarization ellipse.

Note, first, that the ellipse always lies in the rectangle shown in Figure 7.11 but that the axes of the ellipse are not parallel with the original x, y directions.

The ellipse is specified as follows: with e_x, e_y, $\delta(=\delta_y - \delta_x)$ known, then we define $\tan\beta = e_y/e_x$. The orientation of the ellipse, α, is given by

$$\tan 2a = \tan 2\beta \cos\delta.$$

Semimajor and semiminor axes a, b are given by

$$e_x^2 + e_y^2 = a^2 + b^2 \sim I.$$

The ellipticity of the ellipse (e) is given by $e = \tan \chi = \pm b/a$ (the sign determines the sense of the rotation) where

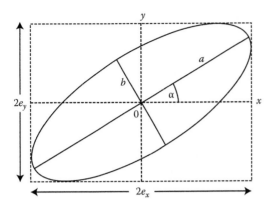

Figure 7.11 The polarization ellipse.

$$\sin 2\chi = -\sin 2\beta \sin \delta.$$

We should note also that the electric field components along the major and minor axes are always in quadrature (i.e., $\pi/2$ phase difference, the sign of the difference depending on the sense of the rotation).

Linear and circular states of polarization may be regarded as special cases where the polarization ellipse degenerates into a straight line or a circle, respectively.

A linear state is obtained with the components in Equation 7.15a when either

$$\left.\begin{array}{l} e_x = 0 \\ e_y \neq 0 \end{array}\right\} \text{linearly polarized in } Oy \text{ direction}$$

$$\left.\begin{array}{l} e_x \neq 0 \\ e_y = 0 \end{array}\right\} \text{linearly polarized in } Ox \text{ direction}$$

or

$$\delta_y - \delta_x = m\pi$$

where m is an integer. In this latter case, the direction of polarization will be at an angle

$$+\tan^{-1}\left(\frac{e_y}{e_x}\right) \quad m \text{ even}$$

$$-\tan^{-1}\left(\frac{e_y}{e_x}\right) \quad m \text{ odd}$$

with respect to the Ox axis.

A circular state is obtained when

$$e_x = e_y$$

and

$$(\delta_y - \delta_x) = (2m+1)\pi/2$$

i.e., in this case, the two waves have equal amplitudes and are in phase quadrature. The waves will be right-hand circularly polarized when m is even and left-hand circularly polarized when m is odd.

Light can become polarized as a result of the intrinsic directional properties of matter: either the matter that is the original source of the light or the matter through which the light passes. These intrinsic material directional properties are the result of directionality in the bonding that holds together the atoms of which the material is made. This directionality leads to variations in the response of the material according to the direction of an imposed force, be it electric, magnetic, or mechanical. The best known manifestation of directionality in solid materials is the crystal, with the large variety of crystallographic forms, some symmetrical, some asymmetrical. The characteristic shapes that we associate with certain crystals result from the fact that they tend to break preferentially along certain planes known as cleavage planes, which are those planes between which atomic forces are weakest.

It is not surprising, then, to find that directionality in a crystalline material is also evident in the light which it produces, or is impressed upon the light which passes through it.

In order to understand the ways in which we may produce polarized light, control it, and use it, we must make a gentle incursion into the subject of crystal optics.

7.3.3 Crystal optics

Light propagates through a material by stimulating the elementary atomic dipoles to oscillate and thus to radiate. In our previous discussions, the forced oscillation was assumed to take place in the direction of the driving electric field, but in the case of a medium whose physical properties vary with direction, an anisotropic medium, this is not necessarily the case. If an electron in an atom or molecule can move more easily in one direction than another, then an electric field at some arbitrary angle to the preferred direction will move the electron in a direction that is not parallel with the field direction (Figure 7.12). As a result, the direction in which the oscillating dipole's radiation is maximized (i.e., normal to its oscillation direction) is not the same as that of the driving wave.

The consequences, for the optics of anisotropic media, of this simple piece of physics are complex.

Immediately we can see the already-discussed (see Chapter 1) relationship between the electric displacement D and the electric field E, for an *isotropic* (i.e., no directionality) medium

$$D = \varepsilon_R \varepsilon_0 E$$

must be more complex for an anisotropic medium; in fact the relation must now be written in the form (for any, arbitrary three orthogonal directions Ox, Oy, and Oz)

$$D_x = \varepsilon_0 \left(\varepsilon_{xx} E_x + \varepsilon_{xy} E_y + \varepsilon_{xz} E_z \right)$$
$$D_y = \varepsilon_0 \left(\varepsilon_{yx} E_x + \varepsilon_{yy} E_y + \varepsilon_{yz} E_z \right)$$
$$D_z = \varepsilon_0 \left(\varepsilon_{zx} E_x + \varepsilon_{zy} E_y + \varepsilon_{zz} E_z \right).$$

Clearly, what is depicted here is an array, which describes the various electric field susceptibilities in the various directions within the crystal: ε_{ij} (a scalar quantity) is a measure of the effect which an electric field in direction j has in direction i within the crystal, i.e., the ease with which it can move electrons in that direction and thus create a dipole moment.

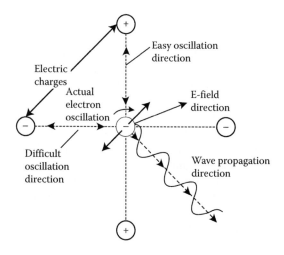

Figure 7.12 Electron response to an electric field in a dielectric medium.

The array can be written in the abbreviated form

$$D_i = \varepsilon_0 \varepsilon_{ij} E_j \, (i, j = x, y, z)$$

and ε_{ij} is now a *tensor* known, in this case, as the permittivity tensor. A tensor is a physical quantity that characterizes a particular physical property of an anisotropic medium, and takes the form of a matrix. Clearly D is not now (in general) parallel with E, and the angle between the two also will depend upon the direction of E in the material.

Now it can be shown from energy considerations that the permittivity tensor is symmetrical, i.e., $\varepsilon_{ij} = \varepsilon_{ji}$. Also, symmetrical tensors can be cast into their diagonal form by referring them to a special set of axes (the principal axes), which are determined by the crystal structure [2]. When this is done, we have

$$\begin{pmatrix} D_x \\ D_y \\ D_z \end{pmatrix} = \varepsilon_0 \begin{pmatrix} \varepsilon_{xx} & 0 & 0 \\ 0 & \varepsilon_{yy} & 0 \\ 0 & 0 & \varepsilon_{zz} \end{pmatrix} \begin{pmatrix} E_x \\ E_y \\ E_z \end{pmatrix}.$$

The new set of axes, Ox, Oy, and Oz, is now this special set.

Suppose now that $E = E_x \mathbf{i}$, i.e., we have, entering the crystal, an optical wave whose E field lies in one of these special crystal directions.

In this case, we simply have

$$D_x = \varepsilon_0 \varepsilon_{xx} E_x$$

as our tensor relation and ε_{xx} is, of course, a scalar quantity. In other words, we have D parallel with E, just as for an isotropic material, and the light will propagate, with refractive index $\varepsilon_{xx}^{1/2}$, perfectly normally. Furthermore, the same will be true for

$$E = E_y \mathbf{j}, \text{(refractive index } \varepsilon_{yy}^{1/2}).$$

$$E = E_z \mathbf{k}, \text{(refractive index } \varepsilon_{zz}^{1/2}).$$

Before going further we should note an important consequence of all this: the refractive index varies with the direction of E. If we have a wave traveling in direction Oz, its velocity now will depend upon its polarization state: if the wave is LP in the Ox direction it will travel with velocity $c_0/\varepsilon_{xx}^{1/2}$, whereas if it is LP in the Oy direction its velocity will be $c_0/\varepsilon_{xx}^{1/2}$. Hence, the medium is offering two refractive indices to the wave traveling in this direction: we have the phenomenon known as double refraction or "birefringence." A wave that is LP in a direction at 45° to Ox will split into two components, LP in directions Ox and Oy, the two components traveling at different velocities. Hence, the phase difference between the two components will steadily increase and the composite polarization state of the wave will vary progressively from linear to circular and back to linear again.

This behavior is a direct consequence of the basic physics that was discussed earlier: it is easier, in the anisotropic crystal, for the electric field to move the atomic electrons in one direction than in another. Hence, for the direction of easy movement, the light polarized in this direction can travel faster than when it is polarized in the direction for which the movement is more sluggish. Birefringence is a long word, but the physical principles that underlie it really are very simple. It follows from these discussions that an anisotropic medium may be characterized by means of three refractive indices, corresponding to polarization directions along Ox, Oy, Oz, and that these will have values $\varepsilon_{xx}^{1/2}, \varepsilon_{yy}^{1/2}, \varepsilon_{zz}^{1/2}$, respectively. We can use this information to determine the refractive index (and thus the velocity) for a wave in any direction with any given linear polarization state.

To do this we construct an "index ellipsoid" or "indicatrix," as it is sometimes called (see Figure 7.13), from the form of the permittivity tensor for any given crystal. This ellipsoid has the following important properties.

Suppose that we wish to investigate the propagation of light, at an arbitrary angle to the crystal axes (polarization as yet unspecified). We draw a line, OP, corresponding to this direction within the index ellipsoid, passing through its center O (Figure 7.13). Now we construct the plane, also passing through O, which lies at right angles to the line. This plane will cut the ellipsoid in an ellipse. This ellipse has the property that the directions of its major and minor axes define the directions of linear polarization for which D and E are parallel for this propagation direction, and the lengths of these axes OA and OB are equal to the refractive

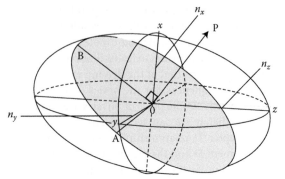

OA and OB represent the linearly polarized eigenstates for the direction OP

Figure 7.13 The index ellipsoid.

indices for these polarizations. Because these two linear polarization states are the only ones that propagate without change of polarization form for this crystal direction, they are sometimes referred to as the "eigenstates" or "polarization eigenmodes" for this direction, conforming to the matrix terminology of eigenvectors and eigenvalues.

The propagation direction we first considered, along Oz, corresponds, to one of the axes of the ellipsoid, and the two refractive indices $\varepsilon_{xx}^{1/2}$ and $\varepsilon_{yy}^{1/2}$ are the lengths of the other two axes in the central plane normal to Oz.

The refractive indices $\varepsilon_{xx}^{1/2}, \varepsilon_{yy}^{1/2}, \varepsilon_{zz}^{1/2}$ are referred to as the *principal* refractive indices and we shall henceforth denote them n_x, n_y, n_z. $OxOy$, $OyOz$, and $OzOx$ are the principal planes.

Several other points are very well worth noting. Suppose, first, that

$$n_x > n_y > n_z.$$

It follows that there will be a plane that contains Oz for which the two axes of interception with the ellipsoid are equal (Figure 7.14). This plane will be at some angle to the yz plane and will thus intersect the ellipsoid in a circle. This means, of course, that, for the light propagation direction corresponding to the normal to this plane, all polarization directions have the same velocity; there is no double refraction for this direction. This direction is an *optic axis* of the crystal and there will, in general, be two such axes, because there must also be such a plane at an equal angle to the yz plane on the other side (see Figure 7.14). Such a crystal with two optic axes is said to be biaxial.

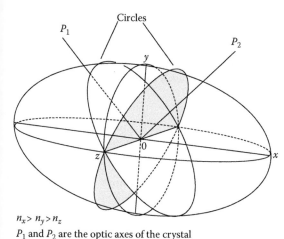

$n_x > n_y > n_z$

P_1 and P_2 are the optic axes of the crystal

Figure 7.14 The ellipsoid for a biaxial crystal.

Suppose now that

$$n_x = n_y = n_o \text{ (say)}, \text{ the "ordinary" index}$$

and

$$n_z = n_e \text{ (say)}, \text{ the "extraordinary" index.}$$

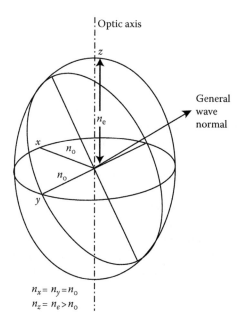

$n_x = n_y = n_o$
$n_z = n_e > n_o$

Figure 7.15 The ellipsoid for a positive uniaxial crystal.

In this case, one of the principal planes is a circle and it is the only circular section (containing the origin) that exists. Hence, in this case, there is only one optic axis, along the Oz direction. Such crystals are said to be uniaxial (Figure 7.15). The crystal is said to be positive when $n_e > n_o$ and negative when $n_e < n_o$. For example, quartz is a positive uniaxial crystal and calcite a negative uniaxial crystal. These features are, of course, determined by the crystal class to which these materials belong.

It is clear that the index ellipsoid is a very useful device for determining the polarization behavior of anisotropic media.

7.3.4 Circular birefringence

So far we have considered only linear birefringence, where two orthogonal linear polarization eigenstates propagate, each remaining linear, but with different velocities. Some crystals also exhibit circular birefringence. Quartz (again) is one such crystal and its circular birefringence derives from the fact that the crystal structure spirals around the optic axis in a right-handed (dextrorotatory) or left-handed (levorotatory) sense depending on the crystal specimen: both forms exist in nature.

It is not surprising to find, in view of this knowledge, and our understanding of the easy motions of electrons, that light that is right-hand circularly polarized (clockwise rotation of the tip of the electric vector as viewed by a receiver of the light) will travel faster down the axis of a matching right-hand spiraled crystal structure than left-hand circularly polarized light. We now have circular birefringence: the two circular polarization components propagate without change of form (i.e., they remain circularly polarized) but at different velocities. They are the circular polarization eigenstates for this case.

The term "optical activity" has been traditionally applied to this phenomenon, and it is usually described in terms of the rotation of the polarization direction of a LP wave as it passes down the optic axis of an "optically active" crystal. This fact is exactly equivalent to the interpretation in terms of circular birefringence because a linear polarization state can be resolved into two oppositely rotating circular components (Figure 7.16). If these travel at different velocities, a phase difference is inserted between them. As a result of this, when recombined, they again form a resultant that is LP but rotated

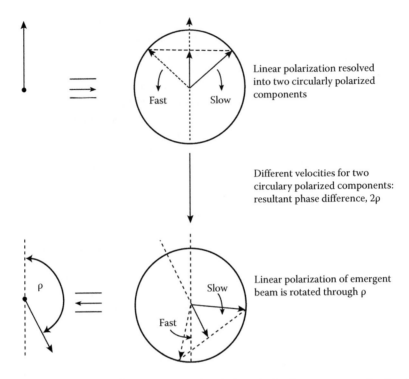

Linear polarization resolved into two circularly polarized components

Different velocities for two circulary polarized components: resultant phase difference, 2ρ

Linear polarization of emergent beam is rotated through ρ

Figure 7.16 Resolution of linear polarization into circularly polarized components in circular birefringence (2ρ).

with respect to the original direction (Figure 7.16). Hence, "optical activity" is equivalent to circular birefringence. In general, both linear and circular birefringence might be present simultaneously in a material (such as quartz). In this case, the polarization eigenstates, which propagate without change of form (and at different velocities) will be elliptical states, the ellipticity and orientation depending upon the ratio of the magnitudes of the linear and circular birefringences, and on the direction of the linear birefringence eigenaxes within the crystal.

It should, again, be emphasized that only the polarization eigenstates propagate without change of form. All other polarization states will be changed into different polarization states by the action of the polarization element (e.g., a crystal component). These changes of polarization state are very useful in optoelectronics. They allow us to control, analyze, modulate, and demodulate polarization information impressed upon a light beam and to measure important directional properties relating to the medium through which the light has passed. We must now develop a rigorous formalism to handle these more general polarization processes.

7.3.5 Polarization analysis

As has been stated, with both linear and circular birefringence present, the polarization eigenstates (i.e., the states which propagate without change of form) for a given optical element are elliptical states, and the element is said to exhibit elliptical birefringence because these eigenstates propagate with different velocities.

In general, if we have, as an input to a polarization-optical element, light of one elliptical polarization state, it will be converted, on emergence, into a different elliptical polarization state (the only exceptions being, of course, when the input state is itself an eigenstate). We know that any elliptical polarization state can always be expressed in terms of two orthogonal electric field components defined with respect to chosen axes Ox, Oy, that is,

$$E_x = e_x \cos(\omega t - kz + \delta_x)$$

$$E_y = e_y \cos(\omega t - kz + \delta_y)$$

or, in complex exponential notation:

$$E_x = |E_x| \exp(i\varphi_x); \varphi_x = \omega t - kz + \delta_x.$$

$$E_y = |E_y| \exp(i\varphi_y); \varphi_y = \omega t - kz + \delta_y.$$

When this ellipse is converted into another by the action of a lossless polarization element, the new ellipse will be formed from components that are linear combinations of the old because it results from directional resolutions and rotations of the original fields. Thus, these new components can be written

$$E'_x = m_1 E_x + m_4 E_y$$

$$E'_y = m_3 E_x + m_2 E_y$$

or, in matrix notation

$$E' = ME$$

where

$$M = \begin{pmatrix} m_1 & m_4 \\ m_3 & m_2 \end{pmatrix} \quad (7.16)$$

and m_n are, in general, complex numbers. M is known as a "Jones" matrix after the mathematician who developed an extremely useful "Jones calculus" for manipulations in polarization optics [3]. Now in order to make measurements of the input and output states in practice, we need a quick and convenient experimental method.

A convenient method for this practical determination is to use a linear polarizer and a quarter-wave plate, and to measure the light intensities for a series of fixed orientations of these elements.

Suppose that $I(\vartheta, \varepsilon)$ denotes the intensity of the incident light passed by the linear polarizer set at an angle ϑ to Ox, after the Oy component has been retarded by an angle ε as a result of the insertion of the quarter-wave plate with its axes parallel with O_x, O_y. We measure what are called the four Stokes parameters, as follows:

$$S_0 = I(0°, 0) + I(90°, 0) = e_x^2 + e_y^2$$

$$S_1 = I(0°, 0) - I(90°, 0) = e_x^2 - e_y^2$$

$$S_2 = I(45°, 0) - I(135°, 0) = 2e_x e_y \cos\delta$$

$$S_3 = I\left(45°, \frac{\pi}{2}\right) - I\left(135°, \frac{\pi}{2}\right) = 2e_x e_y \sin\delta$$

$$\delta = \delta_y - \delta_x.$$

If the light is 100% polarized, only three of these parameters are independent, because

$$S_0^2 = S_1^2 + S_2^2 + S_3^2.$$

S_0 being the total light intensity.

If the light is only partially polarized, the fraction

$$\eta = \frac{S_1^2 + S_2^2 + S_3^2}{S_0^2}$$

defines the degree of polarization. In what follows we shall assume that the light is fully polarized ($\eta = 1$). It is easy to show [4a] that measurement of the S_n provides the ellipticity, e, and the orientation, α, of the polarization ellipse according to the relations:

$$e = \tan\chi$$

$$\sin 2\chi = \frac{S_3}{S_0}$$

$$\tan 2\alpha = \frac{S_2}{S_1}.$$

Now, the above relations suggest a geometrical construction that provides a powerful and elegant means for description and analysis of polarization optical phenomena. The Stokes parameters S_1, S_2, S_3 may be regarded as the Cartesian coordinates of a point referred to axes Ox_1, Ox_2, Ox_3. Thus, every elliptical polarization state corresponds to a unique point in three-dimensional space. For a constant intensity (lossless medium) it follows that all such points lie on a sphere of radius S_0: the Poincaré sphere (Figure 7.17). The properties of the sphere are quite well known [5]. We can see that the equator will comprise the continuum of LP states, whereas the two poles will correspond to the two oppositely handed states of circular polarization.

It is clear that any change, resulting from the passage of light through a lossless element, from one polarization state to another, corresponds to a rotation of the sphere about a diameter. Now any such rotation of the sphere may be expressed as a unitary 2×2 matrix M. Thus, the conversion from one polarization state E to another E' may also be expressed in the form:

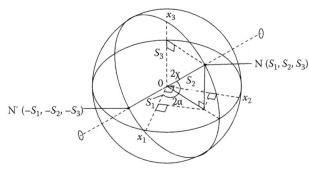

Figure 7.17 The Poincaré sphere: the eigenmode diameter (NN).

$$E' = ME$$

or

$$\begin{pmatrix} E'_x \\ E'_y \end{pmatrix} = \begin{pmatrix} m_1 & m_4 \\ m_3 & m_2 \end{pmatrix} \begin{pmatrix} E_x \\ E_y \end{pmatrix}$$

i.e.,

$$E'_x = m_1 E_x + m_4 E_y$$

$$E'_y = m_3 E_x + m_2 E_y$$

where

$$M = \begin{pmatrix} m_1 & m_4 \\ m_3 & m_2 \end{pmatrix}$$

and M may be immediately identified with our previous M (Equation 7.16).

M is a Jones matrix [3] which completely characterizes the polarization action of the element and is also equivalent to a rotation of the Poincaré sphere.

The two eigenvectors of the matrix correspond to the eigenmodes (or eigenstates) of the element (i.e., those polarization states which can propagate through the element without change of form). These two polarization eigenstates lie at opposite ends of a diameter (NN') of the Poincaré sphere, and the polarization effect of the element is to rotate the sphere about this diameter (Figure 7.18) through an angle Δ that is equal to the phase that the polarization element inserts between its eigenstates.

The polarization action of the element may thus be regarded as that of resolving the input

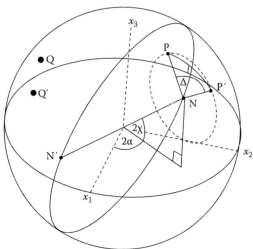

Figure 7.18 Rotation of the Poincaré sphere about an eigenmode diameter, NN'.

polarization state into the two eigenstates with appropriate amplitudes, and then inserting a phase difference between them before recombining to obtain the emergent state. Thus, a pure rotator (e.g., optically active crystal) is equivalent to a rotation about the polar axis, with the two oppositely handed circular polarizations as eigenstates. The phase velocity difference between these two eigenstates is a measure of the circular birefringence. Analogously, a pure linear retarder (such as a wave plate) inserts a phase difference between orthogonal linear polarizations, which measures the linear birefringence. The linear retarder's eigenstates lie at opposite ends of an equatorial diameter.

It is useful for many purposes to resolve the polarization action of any given element into its linear and circular birefringence components. The Poincaré sphere makes it clear that this may always be done because any rotation of the sphere can always be resolved into two subrotations, one about the polar diameter and the other about an equatorial diameter.

From this brief discussion, we can begin to understand the importance of the Poincaré sphere. It is a construction that converts all polarization actions into visualizable relationships in three-dimensional space.

To illustrate this point graphically, let us consider a particular problem. Suppose that we ask what is the smallest number of measurements necessary to define completely the polarization properties of a given lossless polarization element, about which we know nothing in advance. Clearly, we must provide known polarization input states and measure their corresponding output states, but how many input/output pairs are necessary: one, two, more?

The Poincaré sphere answers this question easily. The element in question will possess two polarization eigenmodes and these will be at opposite ends of a diameter. We need to identify this diameter. We know that the action of the element is equivalent to a rotation of the sphere about this diameter, and through an angle equal to the phase difference which the element inserts between its eigenmodes. Hence, if we know one input/output pair of polarization states (NN'), we know that the rotation from the input to the output state must have taken place about a diameter that lies in the plane which perpendicularly bisects the line joining the two states (see Figure 7.18). Two other input/output states (QQ') will similarly define another such plane and thus the required diameter is clearly seen as the common line of intersection of these planes.

Further, the phase difference Δ inserted between the eigenstates (i.e., the sphere's rotation angle) is easily calculated from either pair of states, once the diameter is known.

Hence, the answer is that *two* pairs of input/output states will define completely the polarization properties of the element. Simple geometry has provided the answer. A good general approach is to use the Poincaré sphere to determine (visualize?) the nature of the solution to a problem, and then to revert to the Jones matrices to perform the precise calculations. Alternatively, some simple results in spherical trigonometry will usually suffice.

7.3.6 Summary

In this section, we have looked closely at the directionality possessed by the optical transverse electric (TE) field, i.e., we have looked at optical polarization. We have seen how to describe it, to characterize it, to control it, to analyze it, and how, in some ways, to use it.

We have also looked at the ways in which the TE and magnetic fields interact with directionalities (anisotropies) in material media through which the light propagates. In particular, we first looked at ways in which the interactions allow us to probe the nature and extent of the material directionalities, and thus to understand better the materials themselves.

Second, we looked briefly at the ways in which these material interactions allow us to control light: to modulate it and perhaps to analyze it.

We shall find later that the knowledge we have gained bears upon more advanced phenomena, such as those that allow light to switch light and to process light, opening up a new range of possibilities in the world of very fast (femtosecond: 10^{-15} s) phenomena.

7.4 OPTICAL COHERENCE

7.4.1 Introduction

In dealing with interference and diffraction, the assumption usually is made that each of the interfering waves bears a constant phase relationship to the others in both time and space. Such an assumption cannot be valid for all time and space intervals because the atomic emission processes that give rise to light are largely uncorrelated, except for the special case of laser emission. In this section, the topic of "coherence" will be dealt with. Clearly, it will have a bearing on interference phenomena.

The coherence of a wave describes the extent to which it can be represented by a pure sine wave. A pure sine wave has infinite extension in space and time, and hence cannot exist in reality. Perfect coherence is thus unachievable, but it is nevertheless a valuable concept.

Coherence, in general, is a valuable concept because it is a measure of the constancy of the relationships between one part of a wave (in time and/or space) and another; and between one wave and another. This is why it is so important from the point of view of interference: a wave can only interfere with itself or with another wave (of the same polarization) to produce a sensible interference pattern if the phases and amplitudes remain in constant relationship while the pattern is being sensed. Additionally, it is clear that if we wish to impose information on an optical wave by modulating one of its defining parameters (i.e., amplitude, phase, polarization or frequency), then the extent to which that information remains intact is mirrored by the extent to which the modulated parameter remains intact on the wave itself; thus, coherence is an important parameter in respect of the optical wave's information-carrying capacity, and our ability generally to control and manipulate it.

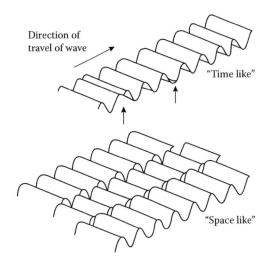

Figure 7.19 Illustrations for partial coherence.

Waves that quite accurately can be represented by sine waves for a limited period of time or in a limited region of space are called partially coherent waves. Figure 7.19 shows the examples of time-like and space-like partial coherence.

A normal light source emits quanta of energy at random. Each quantum conveniently can be regarded as a finite wave train having angular frequency ω_0, say, and duration $2\tau_c$ (Figure 7.20).

Fourier theory tells us that this wave train can be described in the frequency domain as a set of waves lying in the frequency range $\omega_0/2\pi \pm 1/\tau_c$. For a large number of randomly emitted wave packets, all the components at any given frequency will possess random relative phases. Spatial and temporal (time-like) coherence will thus only exist for, respectively, a distance of the order of the length of one packet ($2c\tau_c$) and a time of the order of its duration ($2\tau_c$). If the wave packets were of infinitely short duration (δ-function pulses), then Fourier theory tells us that all frequencies would be present in equal amounts and they would be completely uncorrelated in relative phase. This is the condition we call white light. Its spatial and temporal coherence are zero (i.e., there is perfect *incoherence*) and, again, it is an unachievable fiction. Between the two fictions of perfect coherence and perfect incoherence there lies the real world.

In this real world, we have to deal with real sources of light. Real sources always have a non-zero spectral width, i.e., the light power is spread over a range of frequencies, and the result of superimposing all the frequency components, as again we know from Fourier theory, is to produce a disturbance that is not a pure sine wave: an example is the wave packet we have just considered. Another is the infinitesimally narrow δ-function.

Hence, there is seen to be a clear connection between spectral width and coherence. The two are inversely proportional. The narrower is a pulse of light (or any other waveform) the less it is like a pure sine wave and, by Fourier theory, the greater is its spectral width.

In the practical case of a two-slit interference pattern, the pattern will be sharp and clear if the light used is of narrow linewidth (a laser perhaps); but if a broad-linewidth source, such as a tungsten filament lamp, is used, the pattern is multicolored, messy, and confused.

Of course, we need to quantify these ideas properly if we are to use them effectively.

7.4.2 Measure of coherence

If we are to determine the measurable effect which the degree of coherence is to have on optical systems, especially those which involve interference and diffraction phenomena, we need a quantitative measure of coherence. This must measure the extent to which we can, knowing the (complex) amplitude of a periodic disturbance at one place and/or time, predict its magnitude at another place

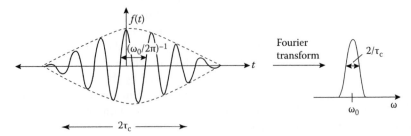

Figure 7.20 The optical wave packet.

and/or time. We know that this measure can be expected to have its maximum value for a pure sine wave, and its minimum value for white light. A convenient definition will render these values 1 and 0, respectively.

To fix ideas, and to simplify matters, let us consider first just temporal coherence.

We may sensibly postulate that for a time function $f(t)$, a knowledge of its value at t will provide us with some knowledge of its value up to a later time, say t', when it becomes completely independent of its value at t. If two time functions are completely independent, then we expect the average value of their product over a time that is long compared with the characteristic time constant for their variations (i.e., the reciprocal of bandwidth) to be equal to the product of their individual average values, i.e.,

$$\langle f(t)f(t')\rangle = \langle f(t)\rangle\langle f(t')\rangle \qquad (7.17)$$

and if, as is the case for the vast majority of optical disturbances, the functions oscillate about a zero mean, i.e.,

$$\langle f(t)\rangle = \langle f(t')\rangle = 0$$

then it follows from Equation 7.17 that

$$\langle f(t)f(t')\rangle = 0.$$

In words: the product of two independent functions, when averaged over a term that is long compared with the time over which each changes significantly, is zero.

On the other hand, if we set $t=t'$, our "delay average" above must have its maximum possible value, because a knowledge of the value of $f(t)$ at t enables us to predict its value at that time t with absolute certainty! Hence, we have

$$\langle f(t)f(t')\rangle = \langle f^2(t)\rangle$$

and this must be the maximum value of the "delay-average" function.

Clearly, then the value of this product (for all real-world functions) will fall off from a value of unity when $t'=t$ to a value of zero when the two variations are completely independent, at some other value of t' and it is also clear that the larger the value of t' for which this occurs, the stronger is

the dependence and thus the greater is the coherence. It might well be, then, that the quantity that we seek in order to measure the coherence is a quantity that characterizes the speed at which this product function decays to zero.

Suppose, for example, we consider a pure, temporal sine wave in this context. We know in advance that this is a perfectly coherent disturbance, and conveniently we would thus require our coherence measure to be unity.

Let us write the wave as

$$f(t) = a\sin\omega t.$$

To obtain the "delay-average" function, we first multiply this by a replica of itself, displaced by time τ (i.e., $t'=t+\tau$; Figure 7.21). We have

$$f(t)f(t+\tau) = a\sin\omega t\, a\sin\omega(t+\tau)$$

$$= \frac{1}{2}a^2[\cos\omega\tau - \cos(2\omega t + \omega\tau)].$$

We now average this over all time (zero bandwidth gives a characteristic time constant of infinity!) and, because $\langle\cos(2\omega t+\omega\tau)\rangle=0$, we have

$$\langle f(t)f(t+\tau)\rangle = \frac{1}{2}a^2\cos\omega\tau.$$

This quantity we call the self-correlation function (sometimes the autocorrelation function), $c(\tau)$, of the disturbance. In more formal mathematical terms, we would calculate it according to

$$c(\tau) = \lim_{T\to\infty}\frac{1}{T}\int_0^T f(t)f(t+\tau)dt \qquad (7.18)$$

i.e.,

$$c(\tau) = \lim_{T\to\infty}\frac{1}{T}\int_0^T \frac{1}{2}a^2[\cos\omega\tau - \cos(2\omega t + \omega\tau)]dt$$

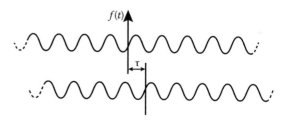

Figure 7.21 Self-correlation delay.

and thus, again

$$c(\tau) = \frac{1}{2}a^2 \cos\omega\tau.$$

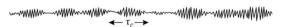

Figure 7.23 A random stream of wave packets.

This function does not decay but oscillates with frequency ω and constant *amplitude a²/2*. It is this latter amplitude that we take as our measure of coherence of the light wave because the sinusoidal term in ωτ will always be present for an oscillatory field such as an optical wave and provides no useful information on the coherence. The mathematical form of $c(\tau)$ is known as a "convolution" integral.

Because we require, for convenience, that this measure be unity for the sine wave, we choose to normalize it to its value at τ=0 (i.e., $c(0)=a^2/2$, in this case).

This normalized function we call the coherence function γ(τ), and hence for the case we have considered, γ(τ)=1 (perfect coherence).

Consider now the white light. We have noted that this is equivalent to a series of randomly spaced δ-functions, which are also randomly positive or negative (Figure 7.22). Clearly, this series must have a mean amplitude of zero if it is to represent a spread of optical sinusoids over an infinite frequency range, each with a mean amplitude of zero. If now to obtain the "delay-average" function, we multiply this set of δ-functions by a displaced replica of itself, then only a fraction of the total will overlap, and an overlap between two δ-functions of the same sign will have equal probability with that between two of the opposite signs. Consequently, the mean value of the overlap function will also be zero, always, regardless of the time delay. Hence,

for this case, $c(\tau)=0$ for all τ and γ (τ)=0 (i.e., perfect incoherence).

Consider, finally, a random stream of quanta, or wave packets (Figure 7.23).

The packets run into each other but each packet is largely coherent within itself. If this stream waveform is multiplied by a displaced replica of itself, the result will be of the form shown in Figure 7.24a. Only when the displacement exceeds the duration of one packet does the correlation fall essentially to zero. Thus, in this case, we have a decaying sine wave, and the quantity that characterizes the decay rate of its amplitude will be our measure of coherence (e.g., time to 1/e point for an exponential decay).

All of the above requirements are taken care of in the general mathematical expression for the coherence function:

$$\gamma(\tau) = \frac{\left| \int_0^\infty f(t)f^*(t+\tau)\,\mathrm{d}t \right|}{\int_0^\infty f(t)f^*(t)\,\mathrm{d}t} = \frac{|c(\tau)|}{|c(0)|}.$$

The integration performs the time averaging, the use of the complex form allows the complex conjugate in one of the functions to remove the oscillatory term in the complex exponential representation, the use of the modulus operation returns the complex value to a real value and the division effects the required normalization. The function $c(\tau)$, as defined in Equation 7.18, is called the correlation coefficient and is sometimes separately useful. Note that it is, in general, a complex quantity.

To cement ideas let us just see how these functions work, again for the pure sine wave.

We must express the sine wave as a complex exponential, so that we write

$$f(t) = a\exp(i\omega t).$$

Then, we have

$$c(\tau) = \int_0^\infty a\exp(i\omega t)a\exp[-i\omega(t+\tau)]\,\mathrm{d}t$$

$$= a^2 \exp(-i\omega t)[T]_0^\infty \to \infty$$

and

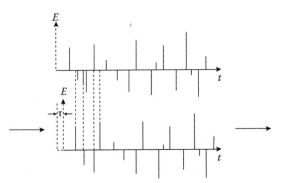

Figure 7.22 Autocorrelation for randomly spaced δ-functions.

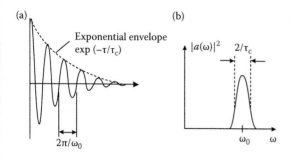

Figure 7.24 Stream correlation.

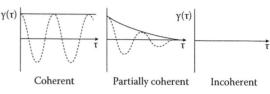

Figure 7.25 Coherence functions.

$$c(0) = \int_0^\infty a\exp(i\omega t)a\exp[-i\omega t)]\mathrm{d}t = a^2[T]_0^\infty \to \infty.$$

Hence

$$\gamma(\tau) = \frac{|c(\tau)|}{|c(0)|} = |\exp(i\omega \tau)| = 1$$

as required.

The coherence functions for the three cases we have been considering are shown in Figure 7.25. The sine wave's function does not decay and therefore its coherence time is infinite; the white light's function decays in zero time and thus its coherence time is zero; the "stream of wave packets" function decays in time τ_c, and thus it is partially coherent with coherence time τ_c. All other temporal functions may have their coherence time quantified in this way. It is clear, also, that the quantity $c\tau_c$ (where c is the velocity of light) will specify a coherence length.

The same ideas, fairly obviously, can also be used for spatial coherence, with τ replaced by σ, the spatial delay. Specifically, in this case

$$\gamma(\sigma) = \frac{\left| \int_0^\infty f(s)f^*(s+\sigma)\mathrm{d}s \right|}{\int_0^\infty f(s)f^*(s)\mathrm{d}s} = \frac{|c(\sigma)|}{|c(0)|}$$

and a "decay" parameter σ_c will define the coherence length.

Finally, the mutual coherence of two separate functions $f_1(t)$ and $f_2(t)$ may be characterized by a closely similar *mutual coherence* function:

$$\gamma_{12}(\tau) = \frac{\left| \int_0^\infty f_1(t)f_2^*(t+\tau)\mathrm{d}t \right|}{\int_0^\infty f_1(t)f_2^*(t)\mathrm{d}t} = \frac{|c_{12}(\tau)|}{|c_{12}(0)|}$$

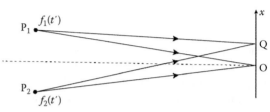

Figure 7.26 Two-source interference.

with t, τ again replaceable by s, σ, respectively, for the mutual spatial coherence case. γ_{12} is sometimes called the "degree of coherence" between the two functions.

7.4.3 Dual-beam interference

We shall first consider in more detail the conditions for interference between two light beams (Figure 7.26). It is clear from our previous look at this topic that interference fringes will be formed if the two waves bear a constant phase relationship to each other but we must now consider the form of the interference pattern for varying degrees of mutual coherence. In particular, we must consider the "visibility" of the pattern; in other words, the extent to which it contains measurable structure and contrast.

At the point O (in Figure 7.26), the (complex) amplitude resulting from the two sources P_1 and P_2 is given by

$$A = f_1(t'') + f_2(t'')$$

where $t'' = t' + \tau_0$, τ_0 being the time taken for light to travel from P_1 or P_2 to O. If f_1, f_2 represent the electric field amplitudes of the waves, the observed intensity at O will be given by the square of the modulus of this complex number. Remember that the modulus of the complex number: $A = a + ib$ is written

$$|a + ib| = \left(a^2 + b^2\right)^{1/2}$$

and a convenient stratagem for obtaining the square of the modulus of a complex number is to multiply it by its complex conjugate, i.e.,

$$AA^* = (a + ib)(a - ib) = (a^2 + b^2).$$

Hence, in this case, the optical intensity is given by

$$I_Q = \langle AA^* \rangle = \langle [f_1(t'') + f_2(t'')][f_1^*(t'') + f_2^*(t'')] \rangle$$

where the triangular brackets indicate an average taken over the response time of the detector (e.g., the human eye) and we assume that f_1 and f_2 contain the required constant of proportionality ($K^{1/2}$) to relate optical intensity with electric field strength, i.e., $I = KE^2$.

At point Q, the amplitudes will be

$$f_1\left(t'' - \frac{1}{2}\tau\right), \quad f_2\left(t'' + \frac{1}{2}\tau\right)$$

τ being the time difference between paths P_2Q and P_1Q.

Writing $t = t' - \frac{1}{2}\tau$, we have the intensity at Q

$$I_Q = \langle [f_1(t) + f_2(t + \tau)][f_1^*(t) + f_2^*(t + \tau)] \rangle$$

i.e.,

$$I_Q = \langle f_1(t)f_1^*(t) \rangle + \langle f_2(t)f_2^*(t) \rangle + \langle f_2(t + \tau)f_1^*(t) \rangle$$
$$+ \langle f_1(t)f_2^*(t + \tau) \rangle.$$

The first two terms are clearly the independent intensities of the two sources at Q. The second two terms have the form of our previously defined mutual correlation function, in fact

$$\langle f_1(t)f_2^*(t + \tau) \rangle = c_{12}(\tau) \langle f_1^*(t)f_2(t + \tau) \rangle = c_{12}^*(\tau).$$

We may note, in passing, that each of these terms will be zero if f_1 and f_2 have orthogonal polarizations because in that case neither field amplitude has a component in the direction of the other, there can be no superposition, and the two cannot interfere. Hence, the average value of their product is again just the product of their averages, each of which is zero, being a sinusoid.

If $c_{12}(f)$ is now written in the form

$$c_{12}(\tau) = |c_{12}(\tau)| \exp(i\omega\tau)$$

(which is valid provided that f_1 and f_2 are sinusoids in ωt) we have

$$c_{12}(\tau) + c_{12}^*(\tau) = 2|c_{12}(\tau)| \cos\omega\tau.$$

Hence, provided that we observe the light intensity at Q with a detector that has a response time very much greater than the coherence times (self and mutual) of the sources (so that the time averages are valid), then we may write the intensity at Q as

$$I_Q = I_1 + I_2 + 2|c_{12}(\tau)| \cos\omega t. \tag{7.19}$$

As we move along x we shall effectively increase τ, so that we shall see a variation in intensity whose amplitude will be $2|c_{12}(\tau)|$ (i.e., twice the modulus of the mutual coherence function) and which varies about a mean value equal to the sum of the two intensities (Figure 7.27). Thus, we have an experimental method by which the mutual coherence of the sources, $c_{12}(t)$, can be measured.

If we now define a fringe *visibility* for this interference pattern by

$$V = \frac{I_{max} - I_{min}}{I_{max} + I_{min}}$$

which quantifies the contrast in the pattern, i.e., the difference between maxima and minima as a fraction of the mean level, then, from Equation 7.19

$$V(\tau) = \frac{2|c_{12}(\tau)|}{(I_1 + I_2)}$$

and with, as previously defined

$$\gamma(\tau) = \frac{|c_{12}(\tau)|}{|c_{12}(0)|}$$

we note that

$$|c_{12}(0)| = \left| \int_0^\infty f_1(t)f_2^*(t)\,dt \right| = K\langle E_1 \rangle\langle E_2 \rangle = (I_1 I_2)^{1/2}$$

and thus

$$\gamma(\tau) = \frac{|c_{12}(\tau)|}{(I_1 I_2)^{1/2}}.$$

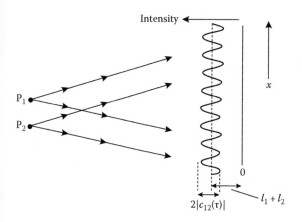

Figure 7.27 Mutual coherence function ($|c12(\tau)|$) from the two-source interference pattern.

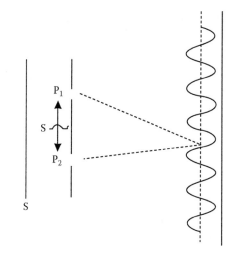

Figure 7.28 Extended-source interference.

Hence, the visibility function $V(\tau)$ is related to the coherence function $\gamma(\tau)$ by

$$V(\tau)=\frac{2(I_1 I_2)^{1/2}}{(I_1+I_2)}\gamma(\tau)$$

and if the two intensities are equal, we have

$$V(\tau)=\gamma(\tau)$$

i.e., the visibility and coherence functions are identical.

From this we may conclude that, for equal-intensity coherent sources, the visibility is 100% ($\gamma = 1$); for incoherent sources it is zero; and for partially coherent sources the visibility gives a direct measure of the actual coherence.

If we arrange that the points P_1 and P_2 are pinholes equidistant from and illuminated by a single source S, then the visibility function clearly measures the self-coherence of S. Moreover, the Fourier transform of the function will yield the source's power spectrum.

Suppose now that the two holes are placed in front of an extended source, S, as shown in Figure 7.28, and that their separation is variable. The interference pattern produced by these sources of light now measures the correlation between the two corresponding points on the extended source. If the separation is initially zero and is increased until the visibility first falls to zero, the value of the separation at which this occurs defines a spatial coherence dimension for the extended source. And

if the source is isotropic, a coherence area is correspondingly defined. In other words, in this case, any given source point has no phase correlation with any point that lies outside the circular area of which it is the center point.

7.4.4 Summary

In this section, we have looked at the conditions necessary for optical waves to interfere in a consistent and measurable way, with themselves and with other waves. We have seen that the conditions relate to the extent to which properties such as amplitude, phase, frequency, and polarization remain constant in time and space or, to put it another way, the extent to which knowledge of the properties at one point in time or space tells us about these properties at other points.

Any interference pattern will only remain detectable as long as coherence persists and, by studying the rise and fall of interference patterns, much can be learned about the sources themselves and about the processes that act upon the light from them.

Coherence also relates critically to the information-carrying capacity of light and to our ability to control and manipulate it sensibly. The design and performance of any device or system that relies on interference or diffraction phenomena must take into account the coherence properties of the sources that are to be used; some of these work to the designer's disadvantage, but others do not.

7.5 OPTICAL WAVEGUIDING

7.5.1 Introduction

The basic principles of optical waveguiding are quite straightforward. Waves are guided when they are constrained to lie within a channel between two other media, the refractive index of the channel material being slightly higher than those of the other media, so that the light can "bounce" along the channel by means of a series of total internal reflections (TIRs) at the boundaries between media. The case shown in Figure 7.29 is that where a channel of refractive index n_1 lies between two slabs, each with refractive index $n_2 (n_1 > n_2)$; this is the easiest arrangement to analyze mathematically, yet it illustrates all the important principles.

The other important point is that in order to progress down the guide indefinitely, the waves from the successive boundaries must interfere constructively, forming what is essentially a continuous, stable, interference pattern down the guiding channel. If the interference is not fully constructive, the wave will eventually "self-destruct," owing to the out-of-phase cancellations (although, clearly, if the phasings are almost correct, the wave might persist for a considerable distance, attenuating only slowly). The condition that must be imposed for constructive interference defines for us the guided wave parameters; in particular, those angles of bounce that can give rise to the "modes" of the waveguide, i.e., the various patterns of constructive interference that are allowed by the restrictions (boundary conditions) of the guide geometry.

The ideas involved in waveguiding are thus quite simple. In order to make use of them we need, as always, a proper mathematical description, so we shall in this section develop this description.

7.5.2 The planar waveguide

We begin by considering the symmetrical slab waveguide shown in Figure 7.29. The guiding channel consists here of a slab of material of refractive index n_1 surrounded by two outer slabs, each of refractive index n_2. The resultant electric field for light which is LP in a direction perpendicular to the plane of incidence (the so-called TE mode) is given by the sum of the upward and downward propagating rays:

$$E_T = E_i + E_r$$

where

$$E_i = E_0 \exp(i\omega t - ikn_1 x \cos\vartheta - ikn_1 z \sin\vartheta)$$

(i.e., a wave traveling in the xz plane at an angle ϑ to the slab boundaries which lie parallel to the yz plane) and

$$E_r = E_0 \exp(i\omega t + ikn_1 x \cos\vartheta - ikn_1 z \sin\vartheta + i\delta_s)$$

which is the wave resulting from the reflection at the boundary, and differs from E_i two aspects: it is now traveling in the negative direction of Ox. Hence, the change of sign in the x term, and there has been a change of phase at the reflection, hence, the $i\delta_s$ term. We must also remember that δ_s depends on the angle, ϑ, the polarization of the wave and, of course, n_1 and n_2 according to the Fresnel equation [4b]. Hence

$$E_T = E_i + E_r = 2E_0 \cos\left(kn_1 x \cos\vartheta + \frac{1}{2}\delta_s\right)$$
$$\times \exp\left(i\omega t - ikn_1 z \sin\vartheta + i\frac{1}{2}\delta_s\right)$$

(7.20a)

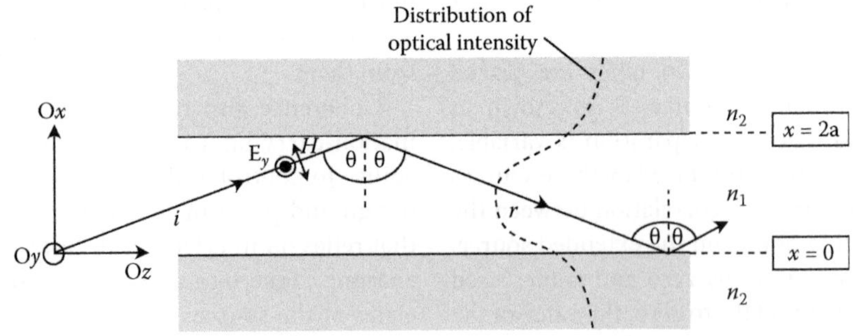

Figure 7.29 Optical slab waveguide.

which is a wave propagating in the Oz direction, but with amplitude varying in the Ox direction according to $2E_0\cos\left(kn_1 x\cos\vartheta + \frac{1}{2}\delta_s\right)$ (see Figure 7.29).

The symmetry of the arrangement tells us that the intensity (~square of the electric field) of the wave must be the same at the two boundaries and thus that it is the same at $x=0$ as at $x=2a$. Hence,

$$\cos^2\left(\frac{1}{2}\delta_s\right) = \cos^2\left(kn_1 2a\cos\vartheta + \frac{1}{2}\delta_s\right)$$

which implies that

$$2akn_1\cos\vartheta + \delta_s = m\pi \qquad (7.20b)$$

where m is an integer. This is our "transverse resonance condition" and it is a condition on ϑ (remember that δ_s also depends on ϑ), which defines a number of allowed values for ϑ (corresponding to the various integer values of m), which in turn define our discrete, allowable modes (or interference patterns) of propagation.

Now the wave number, $k=2\pi/\lambda$, for the free space propagation of the wave has suffered a number of modifications. First, the wavelength of the light is smaller in the medium than in free space (the frequency remains the same, but the velocity is reduced by a factor $n_{1,2}$), so we can conveniently define

$$\beta_1 = n_1 k \quad \beta_2 = n_2 k$$

as the wave numbers in the guiding and outer slabs, respectively. Second, however, if we choose to interpret Equation 7.20a as one describing a wave propagating in the Oz direction with amplitude modulated in the Ox direction, it is convenient to resolve the wave number in the guiding medium into components along Oz and Ox, i.e.,

Along Oz

$$\beta = n_1 k \sin\vartheta \qquad (7.21a)$$

Along Ox

$$q = n_1 k \cos\vartheta \qquad (7.21b)$$

Of these two components, β is clearly the more important because it is the effective wave number for the propagation down the guide. In fact, Equation 7.20a can now be written

$$E_T = 2E_0\cos\left(qx + \frac{1}{2}\delta_s\right)\exp i\left(\omega t - \beta z + i\frac{1}{2}\delta_s\right)$$

What can be said about the velocity of the wave down the guide? Clearly, the phase velocity is given by

$$c_p = \frac{\omega}{\beta}.$$

However, we know that this is not the end of the story for the velocity with which optical energy propagates down the guide is given by the group velocity [5], which, in this case is given by

$$c_g = \frac{d\omega}{d\beta}.$$

What, then, is the dependence of ω upon β? To answer this, let us start with Equation 7.21a, i.e.,

$$\beta = n_1 k \sin\vartheta.$$

The first thing to note is that, for all real ϑ, this requires

$$\beta \leq n_1 k.$$

Also, because the TIR condition requires that

$$\sin\vartheta \geq \frac{n_2}{n_1}$$

if follows that

$$\beta = n_1 k \sin\vartheta \geq n_2 k.$$

Hence, we have

$$n_1 k \geq \beta \geq n_2 k$$

or

$$\beta_1 \geq \beta \geq \beta_2.$$

In other words, the wave number describing the propagation along the guide axis always lies between the wave numbers for the guiding medium (β_1) and the outer medium (β_2). This we might have expected from the physics because the propagation lies partly in the guide and partly in the outer medium (evanescent wave). We shall be returning to this point later.

Remember that our present concern is about how β varies with ω between these two limits, so how else does Equation 7.21a help?

Clearly, the relation

$$k = \frac{\omega}{c_0}$$

where c_0 is the free space velocity, gives one dependence of β on ω, but what about $\sin \vartheta$? For a given value of m (i.e., a given mode), the transverse resonant condition (Equation 7.20b) provides the dependence of ϑ on k. However, this is quite complex because as we know, δ_s is a quite complex function of ϑ. Hence, in order to proceed further, this dependence must be considered.

The expressions for the phase changes that occur under TIR at a given angle are well known [4]

$$\tan \frac{1}{2}\delta_s = \frac{(n_1^2 \sin^2 \vartheta - n_2^2)^{1/2}}{n_1 \cos \vartheta}$$

for the case where the electric field is perpendicular to the plane of incidence and

$$\tan \frac{1}{2}\delta_p = \frac{n_1(n_1^2 \sin^2 \vartheta - n_2^2)^{1/2}}{n_2^2 \cos \vartheta}$$

for the case where it lies in the plane of incidence.

Note also that

$$\tan \frac{1}{2}\delta_p = \frac{n_1^2}{n_2^2} \tan \frac{1}{2}\delta_s.$$

Finally, let us define, for convenience, a parameter, p, where

$$p^2 = \beta^2 - n_2^2 k^2 = k^2(n_1^2 \sin^2 \vartheta - n_2^2). \quad (7.22)$$

The physical significance of p will soon become clear.

We now discover that we can cast our "transverse resonance" condition (Equation 7.20b) into the form

$$\tan\left(aq - \frac{1}{2}m\pi\right) = \frac{p}{q} \ (E_\perp) \quad (7.23a)$$

for the perpendicular polarization and

$$\tan\left(aq - \frac{1}{2}m\pi\right) = \frac{n_1^2}{n_2^2}\frac{p}{q} \ (E_\parallel) \quad (7.23b)$$

for the parallel polarization.

The conventional waveguide notation designates these two cases as "TE" for $E_\perp$ and "transverse magnetic (TM)" for $E_\parallel$. The terms refer, of course, to the direction of the stated fields with respect to the plane of incidence of the ray.

We can use Equations 7.23a and b to characterize the modes for any given slab geometry. The solutions of the equations can be separated into odd and even types according to whether m is odd or even. For odd m, we have

$$\tan\left(aq - \frac{1}{2}m_{odd}\pi\right) = \cot aq \quad (7.24a)$$

and for even m

$$\tan\left(aq - \frac{1}{2}m_{even}\pi\right) = \tan aq. \quad (7.24b)$$

Taking m to be even we may then write Equation 7.23a, for example, in the form

$$aq \tan aq = ap \ (E_\perp). \quad (7.25)$$

Now from the definitions of p and q, it is clear that

$$a^2p^2 + a^2q^2 = a^2k^2(n_1^2 - n_2^2). \quad (7.26)$$

Taking rectangular axes ap, aq this latter relation between p and q translates into a circle of radius $ak(n_1^2 - n_2^2)^{1/2}$ (Figure 7.30). If, on the same axes, we also plot the function $aq \tan aq$, then Equation 7.25 is satisfied at all points of intersection between the two functions (Figure 7.30). (A similar set of solutions clearly can be found for odd m.) These points, therefore, provide the values of ϑ which

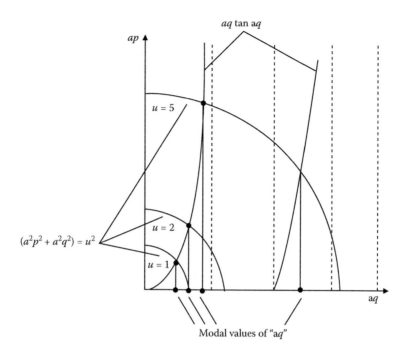

Figure 7.30 Graphical solution of the modal equation for the slab waveguide.

correspond to the allowed modes of the guide. Having determined a value for ϑ for a given k, β can be determined from

$$\beta = n_1 k \sin \vartheta$$

and hence β can be determined as a function of k (for a given m) for the TE modes. Now, finally, with

$$k = \frac{\omega}{c}$$

we have the relationship between β and ω that we have been seeking. For obvious reasons, these are called dispersion curves and are important determinants of waveguide behavior. They are drawn either as β versus k or as ω versus β. The three lowest order modes for a typical slab waveguide are shown in Figure 7.31a using the latter representation. Clearly, this is the more convenient form for determining the group velocity $d\omega/d\beta$ by simple differentiation (Figure 7.31b).

A final point of great importance should be made. As k decreases, so the quantity

(a) Dispersion diagram for slab waveguide

(b) Variation of group velocity with wavenumber

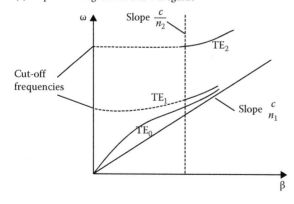

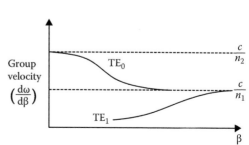

Figure 7.31 Dispersion and group velocity for a slab waveguide.

$$a^2p^2 + a^2q^2 = a^2k^2(n_1^2 - n_2^2)$$

decreases and the various modes are sequentially cut-off as the circle (Figure 7.30) reduces in radius. This is also apparent in Figure 7.31a because a reduction in k corresponds to a reduction in ω. Clearly, the number of possible modes depends upon the waveguide parameters a, n_1, and n_2. However, it is also clear that there will always be at least one mode because the circle will always intersect the tan curve at one point, even for a vanishingly small circle radius. If there is only one solution, then Figure 7.30 shows that the radius of the circle must be less than $\pi/2$, i.e.,

$$ak(n_1^2 - n_2^2)^{1/2} < \frac{1}{2}\pi$$

or

$$\frac{2\pi a}{\lambda}(n_1^2 - n_2^2)^{1/2} < 1.57. \qquad (7.27)$$

This quantity is another important waveguide parameter, for this and many other reasons. It is given the symbol V and is called the "normalized frequency," or, quite often, simply the "V number." Thus

$$V = \frac{2\pi a}{\lambda}(n_1^2 - n_2^2)^{1/2}.$$

Equation 7.27 is thus the single-mode condition for this symmetrical slab waveguide. It represents an important case because the existence of just one mode in a waveguide simplifies considerably the behavior of radiation within it, and thus facilitates its use in, for example, the transmission of information along it. Physically, Equation 7.27 is stating the condition under which it is impossible for constructive interference to occur for any ray other than that which (almost) travels along the guide axis.

Clearly, a very similar analysis can be performed for the TM modes, using Equation 7.23b.

Look again now at Figure 7.29. It is clear that there are waves traveling in the outer media with amplitudes falling off the farther we go from the central channel. This is a direct result of the necessity for fields (and their derivatives) to be continuous across the media boundaries. We know from Equation 7.20a that the field amplitude in the central channel varies as

$$E_x = 2E_0 \cos\left(kn_1 x \cos\vartheta + \frac{1}{2}\delta_s\right).$$

How does the field in the outer slabs vary? The evanescent field in the second medium, when TIR occurs, falls off in amplitude according to

$$E_x = E_a \exp\left(-\frac{2\pi x}{\lambda_2}\sinh\gamma\right); \; x > a$$

where
1. E_a is the value of the field at the boundary, i.e.,

$$E_a = 2E_0 \cos\left(kn_1 a\cos\vartheta + \frac{1}{2}\delta_s\right)$$

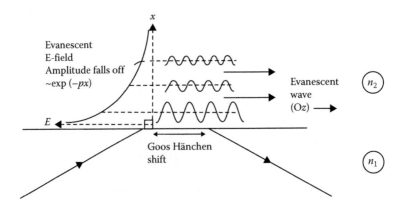

Figure 7.32 Evanescent wave decay.

2. λ_2 is the wavelength in the second medium, and is equal to λ/n_2

3. $2\pi \sinh\gamma/\lambda_2 = k(n_1^2 \sin^2\vartheta - n_2^2)^{1/2}$ and this can now be identified with p from Equation 7.22. Hence

$$E_x = E_a \exp(-px); \quad x > a$$

and we see that p is just the exponential decay constant for the amplitude of the evanescent wave (Figure 7.32) and, from Equation 7.22, we note that $p \sim 0.1\,k$. (It is a fact of any physical analysis that all parameters of mathematical importance will always have a simple physical meaning.)

So the evanescent waves are waves that propagate in the outer media parallel with the boundary but with amplitude falling off exponentially with distance from the boundary.

These evanescent waves are very important. First, if the total propagation is not to be disturbed the thickness of each outer slab must be great enough for the evanescent wave to have negligible amplitude at its outer boundary: the wave falls off as $\sim\exp(-x/\lambda)$, so at $x \sim 20\lambda$ it normally will be quite negligible ($\sim 10^{-9}$). At optical wavelengths, then the slabs should have a thickness $\geq 20\,\mu m$.

Second, because energy is traveling (in the Oz direction!) in the outer media, the evanescent wave properties will influence the core propagation, in respect, for example, of loss and dispersion.

7.5.3 Integrated optics

Planar waveguides find interesting application in integrated optics. In this, waves are guided by planar channels and are processed in a variety of ways. An example is shown in Figure 7.33. This is an electro-optic modulator and it utilizes electro-optic effect (see Section 7.7.7), whereby the application of an electric field to a medium alters its refractive index. However, the electric field is acting on a waveguide that, in this case, is a channel (such as we have just been considering) surrounded by "outer slabs" called here a "substrate." The electric field is imposed by means of the two substrate electrodes, and the interaction path is under close control, as a result of the waveguiding. The material of which both the substrate and the waveguide are made should, in this case, clearly be an electro-optic material, such as lithium tantalate (LiTaO$_3$) whose refractive index depends upon the applied electric field. The central waveguiding channel may be constructed by diffusing ions into it (under careful control); an example of a suitable ion is niobium (Nb), which will thus increase the refractive index of the "diffused" region and allow TIR to occur at its boundaries with the "raw" LiTaO$_3$. Many other functions are possible using suitable materials, geometries, and field influences. It is possible to fabricate couplers, amplifiers, polarizers, filters, etc., all within a planar "integrated" geometry.

One of the advantages of this integrated optics technology is that the structures can be produced to high manufactured tolerance by "mass production" methods, allowing them to be produced cheaply if required numbers are large as is likely to be the case in optical telecommunications, for example.

A fairly recent, but potentially very powerful, development is that of the "optoelectronic integrated circuit," which combines optical waveguide functions with electronic ones such as optical source control, photodetection, and signal processing, again on a single, planar, readily manufacturable "chip."

Note, finally, that in Figure 7.33, the "upper" slab (air) has a different refractive index from the lower one (substrate). This is thus an example of an asymmetrical planar waveguide, the analysis of which is more complex than the symmetrical one that we have considered. However, the basic principles are the same; the mathematics is just more cumbersome and is covered in many other texts [6].

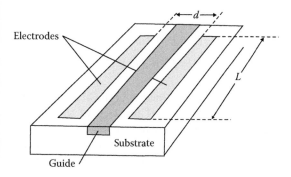

Figure 7.33 An integrated-optical electro-optic phase modulator.

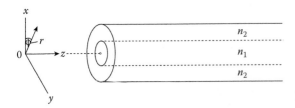

Figure 7.34 Cylindrical waveguide geometry.

7.5.4 Cylindrical waveguides

Let us now consider the cylindrical dielectric structure shown in Figure 7.34. This is the geometry of the optical fiber, the central region being known as the "core" and the outer region as the "cladding." In this case, the same basic principles apply as for the dielectric slab, but the circular, rather than planar, symmetry complicates the mathematics. We use, for convenience, cylindrical coordinates (r, φ, z) as defined in Figure 7.34. This allows us to cast Maxwell's wave equation for the dielectric structure into the form

$$\nabla^2 E = \frac{1}{r}\frac{\partial}{\partial r}\left(r\frac{\partial E}{\partial r}\right) = \frac{1}{r^2}\frac{\partial^2 E}{\partial \varphi^2} + \frac{\partial^2 E}{\partial z^2}$$

$$= \mu\varepsilon\frac{\partial^2 E}{\partial t^2}. \tag{7.28}$$

If we try a solution for E in which all variables are separable, we write

$$E = E_r(r)E_\varphi(\varphi)E_z(z)E_t(t)$$

and can immediately, from the known physics, take it that

$$E_z(z)E_t(t) = \exp\left[i(\beta z - \omega t)\right].$$

In other words, the wave is progressing along the axis of the cylinder with wave number β and with angular frequency ω. It follows, of course, that its (phase) velocity of progression along the axis is given by

$$c_p = \frac{\omega}{\beta}.$$

By substitution of these expressions into the wave Equation 7.28, we may rewrite it in the form

$$\frac{\partial}{\partial r}\left(r\frac{\partial(E_r E_\varphi)}{\partial r}\right) + \frac{1}{r^2}\frac{\partial^2(E_r E_\varphi)}{\partial \varphi^2} - \beta^2 E_r E_\varphi$$

$$+ \mu\varepsilon\omega^2 E_r E_\varphi = 0.$$

Now if we suggest a periodic function for $E\varphi$ of the form

$$E_\varphi = \exp\left(\pm il\varphi\right)$$

where l is an integer, we can further reduce the equation to

$$\frac{\partial^2 E_r}{\partial r^2} + \frac{1}{r}\frac{\partial E_r}{\partial r} + \left(n^2 k^2 - \beta^2 - \frac{l^2}{r^2}\right)E_r = 0.$$

This is a form of Bessel's equation, and its solutions are Bessel functions (see any advanced mathematical text, e.g., Reference [7]). If we use the same substitutions as for the previous planar case, i.e.,

$$n_1^2 k^2 = \beta^2 = q^2$$

$$\beta^2 - n_2^2 k^2 = p^2$$

we find for $r \le a$ (core)

$$\frac{\partial^2 E_r}{\partial r^2} + \frac{1}{r}\frac{\partial E_r}{\partial r} + \left(q^2 - \frac{l^2}{r^2}\right)E_r = 0$$

and for $r > a$ (cladding)

$$\frac{\partial^2 E_r}{\partial r^2} + \frac{1}{r}\frac{\partial E_r}{\partial r} + \left(p^2 + \frac{l^2}{r^2}\right)E_r = 0.$$

Solutions of these equations are (see Figure 7.35a)

$$E_c = Ec\, J_l(qr);\; r \le a$$

$$E_r = E_{cl}\, K_l(pr);\; r > a$$

where J_l is a "Bessel function of the first kind" and K_l is a "modified Bessel function of the second kind" (sometimes known as a "modified Hankel function"). The two functions must clearly be continuous at $r = a$, and we have for our full "trial" solution in the core

$$E = E_c J_l(qr)\exp\left(\pm il\varphi\right)\exp i(\beta z - \omega t)$$

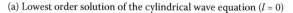

(a) Lowest order solution of the cylindrical wave equation ($l = 0$)

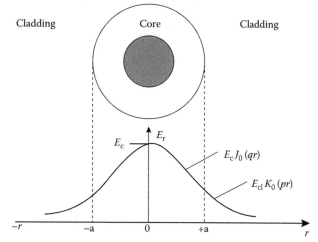

(b) The geometry of the weakly-guiding appoximation

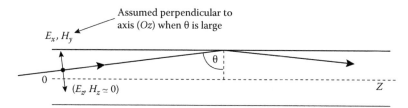

Figure 7.35 Solution for the cylindrical waveguide equation, and the weakly guiding approximation.

and a similar one for the cladding

$$E = E_{cl}J_l\left(pr\right)\exp\left(\pm il\varphi\right)\exp i\left(\beta z - \omega t\right)$$

Again, we can determine the allowable values for p, q, and β by imposing the boundary conditions at $r=a$ [8]. The result is a relationship that provides the β versus k, or "dispersion" curves, shown in Figure 7.36. The mathematical manipulations are tedious but are somewhat eased by using the so-called weakly guiding approximation. This makes use of the fact that if $n_1 \sim n_2$, then the ray's angle of incidence on the boundary must be very large, if TIR is to occur. The ray must bounce down the core almost at grazing incidence. This means that the wave is very nearly a transverse wave, with very small z components. By neglecting the longitudinal components H_z, E_z, a considerable simplification of the mathematics results (Figure 7.35b). Because the wave is, to a first approximation, transverse, it can be resolved conveniently into two LP components, just as for free space propagation. The modes are thus dubbed "LP" modes, and the notation that describes the profile's intensity distribution is the "LP" notation.

7.5.5 Optical fibers

The cylindrical geometry relates directly to the optical fiber. The latter has just the geometry we have been considering and for a typical fiber:

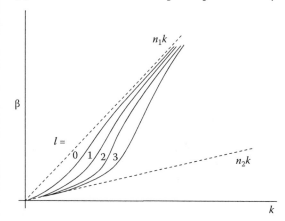

Figure 7.36 Dispersion curves for the cylindrical waveguide.

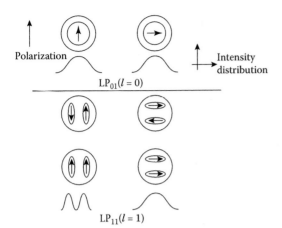

Polarization — Intensity distribution

$LP_{01}(l = 0)$

$LP_{11}(l = 1)$

Figure 7.37 Some low-order modes of the cylindrical waveguide (with weakly guiding mode labels).

$$\frac{n_1 - n_2}{n_1} \sim 0.01$$

so that the weakly guiding approximation is valid. Some of the low-order "LP modes" of intensity distribution are shown in Figure 7.37, together with their polarizations, and values for the azimuthal integer, 1. There are, then, two possible LP optical fiber modes. For the cylindrical geometry, the "single-mode condition" is (analogously to Equation 7.24 for the planar case)

$$V = \frac{2\pi a}{\lambda}(n_1^2 - n_2^2)^{1/2} < 2.405.$$

The value 2.405 derives from the value of the argument for which the lowest order Bessel function, J_0, has its first zero. Some important practical features of optical-fiber design can be appreciated by reversion to geometrical (ray) optics.

Let us consider, first, the problem of launching light into the fiber. Referring to Figure 7.38a, we have for a ray incident on the front face of the fiber at angle ϑ_0, and with refracted angle ϑ_1:

$$n_0 \sin \vartheta_0 = n_1 \sin \vartheta_1$$

where n_0 and n_1 are the refractive indices of air and the fiber core material, respectively. If the angle at which the ray then strikes the core/cladding boundary is ϑ_T, then, for TIR, we must have sin ϑ > n_2/n_1 where n_2 is the cladding index.

(a) Acceptance angle for an optical fiber

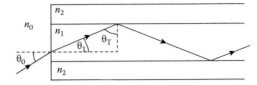

(b) Ray representations of fiber modes

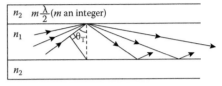

(c) Graded index ray paths

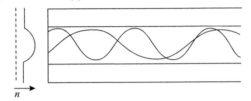

Figure 7.38 Ray propagation in optical fibers.

Because $\vartheta_T = (1/2)\pi - \vartheta_1$, the inequality is equivalent to

$$\cos \vartheta_1 > \frac{n_2}{n_1}$$

so from Snell's law expression above

$$\cos \vartheta_1 = \left(1 - \frac{n_0^2 \sin^2 \vartheta_0}{n_1^2}\right)^{1/2}$$

or

$$n_0 \sin \vartheta_0 < (n_1^2 - n_2^2)^{1/2}$$

The quantity on the right hand side (RHS) of this inequality is known as the numerical aperture (NA) of the fiber. It is a specification of the "acceptance" cone of light, this being a cone of apex half-angle ϑ_0. Clearly, a large refractive index difference between core and cladding is necessary for a large acceptance angle; for a typical fiber, $\vartheta_0 \sim 10°$.

The discrete values of reflection angle that are allowed by the transverse resonance condition (within the TIR condition) can be represented by the ray propagations shown in Figure 7.38b. This makes clear that for a large number of allowable rays (i.e., modes) the TIR angle should be large, implying a large NA. However, it is also clear, geometrically, that the rays will progress down the guide at velocities that depend on their angles of reflection: the smaller the angle, the smaller the velocity. This leads to large intermodal dispersion at large NA because if the launched light energy is distributed among many modes, the differing velocities will lead to varying times of arrival of the energy components at the far end of the fiber. This is undesirable in, for example, communications applications, because it will lead to a limitation on the communications bandwidth. In a digital system, a pulse cannot be allowed to spread into the pulses before or after it. For greatest bandwidth only one mode should be allowed, and this requires a small NA. Thus, a balance must be struck between good signal level (large NA) and large signal bandwidth (small NA).

A fiber design that attempts to attain a better-balanced position between these is shown in Figure 7.38c. This fiber is known as graded-index (GI) fiber and it possesses a core refractive index profile which falls off parabolically (approximately) from its peak value on the axis. This profile constitutes, effectively, a continuous convex lens, which allows large acceptance angle while limiting the number of allowable modes to a relatively small value. GI fiber is used widely in short and medium distance communications systems. For trunk systems single-mode fiber is invariably used, however. This ensures that the intermodal dispersion is entirely absent, thus removing this limitation on bandwidth. Single-mode fiber possesses a communications bandwidth, which is orders of magnitude greater than that of multimode fiber, though even this nominally ideal case is marred by the fact that in reality, a single mode is subject to the influence of fiber birefringence on propagation velocity. Intrinsic birefringence in optical fibers is caused by internal stress and noncircularity of the fiber's core. Thus, the fundamental mode of an optical fiber is split into two orthogonally polarized eigenmodes with different propagation velocities. This effect is similar to intermodal dispersion and is called polarization mode dispersion. This type of dispersion can present a significant limitation on the design of communication links using single mode optical fiber.

7.5.6 Summary

Optical waveguiding is of primary importance to the optoelectronic designer. With its aid, it is possible to confine light and direct it to where it is needed, over short, medium, and long distances.

Furthermore, with the advantage of confinement, it is possible to control the interaction of light with other influences, such as electric, magnetic, or acoustic fields, which may be needed to impress information upon it. Control also can be exerted over its intensity distribution, its polarization state, and its nonlinear behavior; this last topic is covered in Section 7.7.

In short, optical waveguiding is crucial to the control of light. For the designers of devices and systems (especially telecommunications systems, using optical fibers) this control is essential.

7.6 ELECTRONS IN SOLIDS

7.6.1 Introduction

In order to understand the mechanisms involved in the operation of important solid-state devices such as semiconductor lasers, light-emitting diodes, various types of photodetectors, light modulators, etc., it is necessary to look into some of the rather special features of the behavior of electrons in solid materials, and this is the subject of the present section.

A solid is a state of matter where the constituent atoms or molecules are held in a rigid structure as a result of the fact that the intermolecular forces are large compared with the forces of thermal motion of the molecules. This can only be true if the molecules are close together, for the molecules are electronically neutral overall, and forces can only exist between them if there is significant overlap among the wave functions of the outer electrons. This overlap leads to another important consequence: the energy levels in which the electrons lie are shared levels; they are a property of the material as a whole rather than of the individual molecules, as is the case for a gas, for example.

In order to gain a physical "feel" for the effect of the strong interaction on the energy level structure in a solid, consider what happens when two simple oscillators, such as two pendulums, interact. If two pendulums each of same length, and thus with the same independent frequency of oscillation, f, are strung from the same support bar, they will interact with each other via the stresses transmitted through the bar, as they swing. For the combined system there are two "eigenmodes" that is to say two states of oscillation that are stable in time. These are the state where the two pendulums swing together, in phase, and that where they swing in opposition, in antiphase. For all the other states, the amplitudes and relative phases of the two pendulums vary with time. The two eigenstates have difference frequencies f_p (in phase) and f_a (in antiphase) and we find that

$$f_p > f > f_a.$$

The original frequency f is not now a characterizing parameter of the system, having been replaced by two other frequencies, one higher and one lower. If just one of the two pendulums is set swinging it will do so at a frequency in the range f_p to f_a and will set the other pendulum swinging. The second pendulum will acquire maximum amplitude when the first has come to a stop and then the process will reverse. The energy will continuously transfer between the pendulums at a frequency $(f_p - f_a)$. Consider now *three* identical pendulums strung from the same bar. Now there are three eigenstates: (1) all in phase; (2) outer two in phase, central one in antiphase; (3) left-or right-hand two in phase, right-or left-hand one in antiphase. Each of these states has its own frequency of oscillation, so we now have three frequencies. It is an easy conceptual extrapolation to n pendulums, where there will be n frequencies centered on the original f, i.e., the original single frequency has become a band of n frequencies. If n is very large, as it is with the number of molecules in a solid, the frequencies are so close together as to comprise essentially a continuous *band* of frequencies, and thus also of electron energy levels. Thus, we can expect each discrete energy level of the isolated atom or molecule to form a separate band of allowable energies, and the bands will be separated by gaps which represent energies forbidden to electrons (Figure 7.39). This feature is crucial to the behavior of electrons in solids and accounts for most of the properties, which are important in optoelectronics. It is, therefore, necessary to study it in more detail before looking at why, exactly, it is so important to us.

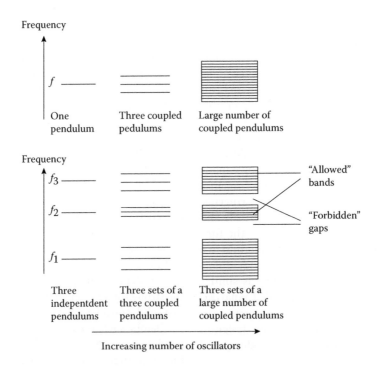

Figure 7.39 Band structure resulting from the coupling of oscillators.

7.6.2 Elements of the band theory of solids

Having understood why energy bands form in solids, it is necessary now to understand the ways in which electrons occupy them. This is as crucial to the understanding of the optoelectronic properties of solids as the formation of the bands themselves.

First, it is necessary to remember that electrons obey quantum rules. Associated with each electron is a wave whose wave number (~reciprocal of wavelength) is related to the momentum (p) of the electron.

If the electron is propagating freely, the relationship is

$$p = \frac{h}{\lambda} = \frac{hk}{2\pi}$$

and because the kinetic energy of a particle of mass m is related to its momentum p by

$$E_k = \frac{p^2}{2m}.$$

We have, in the case of the free electron

$$E_k = \frac{h^2 k^2}{8\pi^2 m}. \tag{7.29}$$

We shall need this shortly.

Another consequence of the quantum behavior of electrons is that they distribute themselves among available energy levels in a rather special way. We say that their distribution obeys "Fermi–Dirac" statistics, and although it is not necessary to go very deeply into this, it is necessary to understand the basic ideas.

All fundamental particles, such as electrons, protons, neutrons, mesons, quarks, etc., are indistinguishable particles, i.e., there is no way in which an electron, say, can be "labeled" at one time or place, in such a way that it is possible to recognize it as the same particle at another time or place; this is not just a "labeling" problem: it is quite fundamental—a consequence of quantum physics. Hence, if two identical (indistinguishable) particles are interchanged in any energy distribution within a system, there can be no change in any of the observable macroscopic properties of the system. Now these observable properties depend only on the square of

the modulus of the system's overall wave function (Section 1.8), which is formed from all of the individual electron wave functions, that is,

$$\psi = \psi(1)\psi(2)\cdots\psi(n).$$

If electrons 1 and 2 are interchanged, then $|\psi|^2$ must remain the same, that is,

$$\left|\psi_{12}\right|^2 = \left|\psi_{21}\right|^2$$

hence

$$\psi_{12} = \pm\psi_{21}.$$

This presents two possibilities: either the interchange leaves the sign of the wave function the same or it reverses it. Particles that leave the sign the same are called symmetrical particles; particles that reverse it are called antisymmetrical particles.

Now comes the vital point: two antisymmetrical particles cannot occupy the same quantum state because the interchange of two identical particles occupying the same quantum state cannot alter the wave function in any way at all, not even its sign. Hence, no two antisymmetrical particles can have the same set of "quantum numbers," numbers that define the quantum state uniquely. This is known as the Pauli exclusion principle. Electrons are antisymmetrical particles and thus obey the Pauli exclusion principle. In fact, all particles with "half-integral spin," $\left(n + \frac{1}{2}\right)h/2\pi$, obey the principle, for example, electrons, protons, neutrons, pu-mesons; these are called fermions (note the small f now!). Particles with integral spin, $nh/2\pi$, are symmetrical particles and obey "Bose–Einstein" statistics: e.g., photons, α-particles, and π-mesons; these are called bosons.

The fact that no two electrons can occupy the same quantum state is profound and is the single most important feature of the behavior of electrons, in regard to the optoelectronic properties of solids. It means that the available electrons will fill the available quantum states progressively and systematically from bottom to top, like balls in a vertical tube whose diameter is just sufficient to take one ball at a time.

Let us examine this "filling" process in more detail.

Each allowed energy level in any system contains (in general) more than one quantum state. The number of states that it contains is called the "degeneracy" of the energy level. (Remember also that each of the bands in the solid-state energy structure results from a large number of closely spaced energy levels, so there is also a kind of multiple degeneracy within a band.)

Now suppose, first, that the electrons within a given energy band are completely free to move around as if they were an electron "gas" in the solid. This is approximately true for electrons in a metal and the only restriction really is that the electrons are not free to leave the solid. How should we calculate the energy states available to the electrons in this case?

Well, fortunately, most of the work necessary for calculating the number of electron states that lie between energies E and $E+dE$ for this case has already been done, in Section 7.2.1, for atomic oscillators that give rise to electromagnetic waves; the analogy between electromagnetic waves in a box and electrons in a box is very close. The electron waves are restricted to the same set of discrete values by the box boundaries as were the electromagnetic waves. The only difference is that whereas we had to allow for two polarization states in the electromagnetic case, we now have to allow for two spin directions (e.g., up and down) in the electron case. In both cases, we must multiply by a factor of 2, so that Equation 7.6a remains valid; i.e., the number of electron states with k values between k and $k+dk$ is $g(k)$ where

$$g(k)dk = \frac{k^2 dk}{\pi^2}.$$

$g(k)$ is known as the degeneracy function. All that is necessary now is to express this in terms of the energy by substituting for k and dk from Equation 7.29, that is,

$$g(E_k)dE_k = \frac{1}{\pi^2} \frac{8\pi^2 m}{h^2} E_k \left(\frac{8\pi^2 m}{h^2} \right)^{1/2} \frac{1}{2} E_k^{-1/2} dE_k$$

$$= \frac{4\pi}{h^3} (2m)^{3/2} E_k^{1/2} dE_k \tag{7.30a}$$

This function is shown in Figure 7.40 and in solid-state parlance, it is usually called the "density of states" function. It represents the number of states between energies E_k and E_k+dE_k.

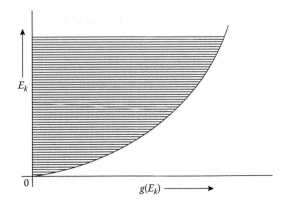

Figure 7.40 "Density of states" function for a metal.

Hence, these are the states that are going to be filled from the bottom up. Each range of E_k to E_k+dE_k will be filled sequentially like the balls in the tube. What, then, is the occupancy of these states? How do the electrons actually distribute themselves among them?

If each energy range is filled in turn, the best way to express this is to plot the fraction of the $g(E_k)$ levels that is filled by a total of N_T electrons. At the absolute zero of temperature, this occupancy function will look like variation A in Figure 7.41. All the states will be filled up to the level at which electrons are exhausted. Hence, up to that level the fractional occupancy is 1; above that level, it is 0. This level is known as the Fermi level, E_F, and is easily calculated if the total number of electrons, N_T, is known, for it is necessary only to integrate Equation 7.30a between 0 and E_F:

$$N_T = \frac{4\pi}{h^3} (2m)^{3/2} \int_0^{E_F} E_k^{1/2} dE_k.$$

Hence

$$N_F = \left(\frac{3N_T}{8\pi} \right)^{2/3} \frac{h^2}{2m}. \tag{7.31}$$

Using Equation 7.31, the density of states function 7.30a may now conveniently be expressed in the form

$$g(E)dE = \frac{3}{2} N_T \frac{E^{1/2}}{E_F^{3/2}} dE. \tag{7.30b}$$

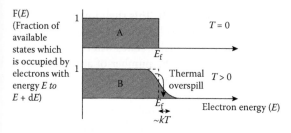

Figure 7.41 Fermi–Dirac "occupancy" distributions.

Suppose now that the temperature rises above absolute zero to a small value $T(>0)$. This makes available to each electron an extra energy $\sim kT(\ll E_F)$. However, most of the electrons cannot take advantage of this because the next available empty level for them is more than kT away. The only electrons that can gain energy are those at the top of the distribution, for they have empty states above them. The distribution thus changes to the one shown as B in Figure 7.41 for temperature T. Hence, the electrons behave very differently from a gas, say, where the average energy of *all* the molecules would increase by kT. (It is for this reason that the specific heat of metals is much smaller ($\sim$1%) than was predicted on a free electron theory of metallic conduction; this discrepancy was a great puzzle to physicists in the early years of the 20th century.)

The function that describes the occupancy of the levels at a given temperature is known as the Fermi–Dirac function. It is given by

$$F(E) = \frac{1}{\exp\left(\dfrac{E - E_F}{kT}\right) + 1}. \qquad (7.32a)$$

Note that it has the behavior, which has already been described

For $T=0$ and $E > E_F$:

$$\exp\left(\frac{E - E_F}{kT}\right) \to 0, \quad F(E) \to 1.$$

For $T=0$ and $E > E_F$:

$$\exp\left(\frac{E - E_F}{kT}\right) \to \infty, \quad F(E) \to 0.$$

This clearly corresponds to variation A in Figure 7.41.

As the temperature rises, the topmost electrons move to higher states and the function develops a "tail," whose width is $\sim kT$ (curve B in Figure 7.41). The energy E_F in this case corresponds to the energy for which $F(E) = 0.5$.

Now that we are in a position to make the final step: the density of electrons within a given small energy range will be the product of the density of quantum states and the actual occupancy of these states. It will be the product of the density of states function (7.30b) and the Fermi–Dirac function 7.32, i.e., $n(E)dE = g(E)F(E)dE$ or

$$n(E)dE = \frac{3}{2} N_T \frac{E^{1/2}}{E_F^{3/2}} \frac{dE}{\exp\left(\dfrac{E - E_F}{kT}\right) + 1} \qquad (7.33a)$$

where $n(E)dE$ is the number per unit volume of electrons with energies between E and $E+dE$. This function is shown in Figure 7.42 for $T=0$ and for $T \neq 0$.

It is interesting to note, before leaving this, that the Fermi–Dirac distribution is a prevalent feature primarily because, in a solid, the number of electrons is comparable with the number of quantum states, and therefore the electrons must be carefully packed according to the quantum rules. If the number of quantum states far exceeds the number of identical particles, as it does in a gas for example, the quantum rules are scarcely noticeable. To see this suppose that, in Equation 7.30b, $g(E) \gg N_T$, then $E^{1/2} \gg E_F^{3/2}$ and hence $E \gg E_F$. Equation 7.33a becomes

$$n(E)dE = \frac{3}{2} N_T \frac{E^{1/2}}{E_F^{3/2}} \exp\left(-\frac{E}{kT}\right) dE. \qquad (7.33b)$$

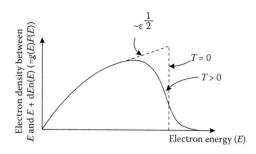

Figure 7.42 Electron density distributions for a metal.

Expressed in terms of molecular velocity, v, and remembering that the molecular energy in this case is purely kinetic energy of motion, i.e.,

$$E = \frac{1}{2}mv^2$$

we have

$$n(v)dv = Av^2 \exp\left(-\frac{mv^2}{2kT}\right)dv.$$

which is the Maxwell–Boltzmann gas velocity distribution as deduced from classical (i.e., nonquantum) statistical thermodynamics [1].

It turns out that Equation 7.33b often also represents a useful approximation in solid-state physics. In all cases where the electron distribution is being considered well above the Fermi level (i.e., $E \gg E_F$) the Fermi–Dirac distribution function of Equation 7.32a approximates to

$$F(E) = \exp\left(-\frac{E}{kT}\right) \qquad (7.32b)$$

which is, of course, the Boltzmann factor. We shall have several occasions to use this later.

7.6.3 Metals, insulators, and semiconductors

We are now in a position to understand, qualitatively at first, what it is that distinguishes metals, insulators, and semiconductors. It all depends upon the position of the Fermi level.

Consider the solid-state band structure in Figure 7.43. Suppose, first, that a solid consists of atoms or molecules with just one electron in the outermost energy shell. This shell forms a band of energy levels in the solid, as we have seen, and the total number of available states will be $2N$ per unit volume, where N is the number of atoms per unit volume (i.e., two electron spin states per quantum state). But there will be only N electrons per unit volume because there is only one electron per atom. Hence, the band is half-filled, and the Fermi level lies halfway up the second band, as in Figure 7.43a. The electrons at the top of the Fermi–Dirac distribution have easy access to the quantum states above them and can thus move freely in response to, for example, an applied electric field, by gaining small amounts of kinetic energy from it; they can also move to conduct heat energy quickly and easily: we have a metal.

Suppose, second, that there are two electrons in the outer shell of the atoms or molecules comprising the solid. The band formed from the shell is now just full and the Fermi level is above the top of the band, as in Figure 7.43b. The electrons at the top of the band now can only increase their energies by jumping up to the next band. If the energy gap is quite large neither moderate temperatures nor moderate electric fields can provide sufficient energy for this to happen. Hence, the material does not conduct electricity at all easily: we have an insulator.

Finally, consider the case shown in Figure 7.43c, again a case where the uppermost level is just full (which will, clearly, be the case for any even number of electrons in the outer shell). Here, the Fermi level lies about halfway up the gap between the valence and conduction bands and the gap is now relatively small, say less than $100kT$ for room temperature. (For example, the element silicon has a bandgap of 1.1 eV, compared with a value for kT, at room temperature, of ~2.5×10^{-2} eV. An electron

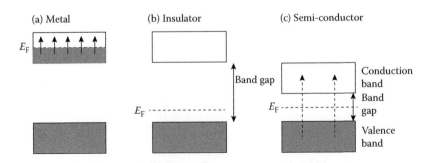

Figure 7.43 The Fermi level and the classification of solids.

would gain an energy of 1 eV in falling through a potential difference of 1 V.) In this case, although at low temperature the Fermi–Dirac "tail" does not extend into the upper, conduction band, at higher temperatures it does, giving a small number of electrons in the conduction band. These electrons can then move easily into the abundance of empty states now available to them in this band. Thus, the room temperature electrical conductivity is low but measurable; furthermore, it is clear that it will increase quite rapidly with temperature, as the "tail" extends. We have here a "semiconductor," more precisely an *intrinsic* semiconductor (it will become clear later why this adjective is necessary). For obvious reasons the upper band is called the conduction band and the lower one the valence band (because it is the stability of electrons in the lower band which provides the atomic forces holding the solid together). There is another important point to be made for the intrinsic semiconductor. When thermal agitation raises an electron from the valence band to the conduction band, it leaves behind an empty state. This state can be filled by another electron, in the valence band, which can then gain energy and contribute to the electrical conduction. These empty states, comprising, as they do, the absence of negative electric charge, are equivalent to positive "holes" in the valence band, and they effectively move like positive charges as the electrons in the valence band move in the opposite direction to fill them. Positive holes in the valence band comprise an important feature of semiconductor behavior, and we shall be returning to them shortly.

Before moving on it should be emphasized that the description above is a greatly simplified one in order to establish the ideas. Solids are complicated states of matter and are three-dimensional, so in general we must not deal just with a single Fermi level but with a three-dimensional Fermi surface, which will have a shape dependent upon the variation of the material's properties with direction. Many important properties of solids depend upon the particular shape that this surface assumes. Especially important is the fact that two energy bands can sometimes overlap, so that it is possible for some elements to behave as metals even though each of their atoms possesses an even number of electrons (the lower band feeds electrons into the middle of the upper band); examples are beryllium, magnesium, and zinc. However, this is the stuff of pure solid-state physics and, for more, interested readers must refer to one of the many specialist texts on solid-state physics [9].

It has become clear then that the position of the Fermi level in relation to the band structure for a particular solid material is vitally important. It is important not only for distinguishing between metals, insulators, and semiconductors but also for understanding the detailed behavior of any particular material.

We have seen how to calculate the Fermi level for the case of electrons moving freely within a solid. However, electrons are not entirely free to move within a crystal, or even a quasi-crystal structure, and this has important consequences that lie within a more detailed consideration of this topic. Space limitations do not allow this treatment here and the interested reader is referred to the literature [9].

7.6.4 Extrinsic semiconductors

Finally, we must consider another very important type of semiconductor material. This is the doped semiconductor, otherwise known as the "extrinsic" semiconductor. In these materials, the semiconducting properties can be both enhanced and controlled by adding specific impurities in carefully judged quantities. The effect of this is to alter the electron energy distribution in a controlled way.

We begin by considering a particular intrinsic semiconductor, silicon, because, with germanium, it is one of the two most commonly used materials for doping in this way. Both materials have a diamond-like structure, with each atom surrounded symmetrically by four others. Silicon is tetravalent having an even number of electrons in its valence shell. There will thus be $4N$ available electrons per unit volume. The first valence energy band will be filled with $2N$ electrons and the second band also with $2N$ electrons; thus, both lower bands are full and the next higher one is empty, at absolute zero (Figure 7.44). The gap between the upper valence band and the conduction band is quite small, only 1.1 eV, compared with a room temperature value of kT of $\sim 2.5 \times 10^{-2}$ eV so, although silicon is an intrinsic semiconductor, its semiconductivity is moderate, and it increases exponentially with temperature.

Suppose now that the silicon is doped with a small fraction (between 1 atom in 10^6 and 1 in

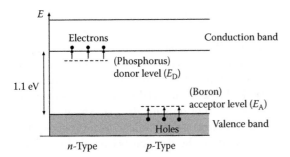

Figure 7.44 Energy level diagram for doped silicon.

10⁹) of a pentavalent (valency of 5) impurity atom such as phosphorus. This atom sits quite comfortably in the silicon lattice but has one extra electron compared with the silicon atoms by which it is surrounded and this electron is easily detached to roam through the lattice and add to the conductivity. In energy level parlance, we say that it needs little energy to raise it into the conduction band; in fact for this particular case it needs only 4.5×10^{-2} eV, equivalent to only about $2kT$ at room temperature. So the energy level structure looks like Figure 7.44 with the "donor" level, E_D, just below the conduction band. Note that the level remains sharp because the dopant atoms are scarce and spaced well apart in the lattice, and thus their wave functions cannot overlap to form a band structure, as do the atoms of the host lattice. Because, in this case, the effect of the phosphorus is to donate electrons (negative charges) to the conductivity, this is called an n-type semiconductor. The really important point is that the conductivity now is entirely under control via control of the doping level. The greater the concentration of the dopant atoms, the greater will be the concentration of electrons in the conduction band.

Consider, on the other hand, what happens if we dope the silicon with a tervalent (valency of 3) impurity such as boron. In this case, the impurity atom has one electron fewer than the surrounding silicon atoms, and electrons from the valence band in the silicon can easily move into the space so created. These "absent" (from the valence band) electrons create positive holes, as we have seen, and these also are effective in increasing the conductivity. For obvious reasons this is now called a p-type semiconductor (Figure 7.44) and the corresponding energy level is an "acceptor" level. The

"majority carriers" are holes in this case; in an n-type material the majority carriers are electrons.

Usually, the donor or acceptor dopings dominate the semiconductor behavior. In other words, it is normally the case that the dopant concentrations exceed the intrinsic carrier concentration n_i. If the dopant concentrations are N_d for donor and N_a for acceptor, it must be that for charge neutrality of the material:

$$n + N_a = p + N_d. \tag{7.33}$$

However, it is also true that, for all circumstances:

$$pn = n_i^2.$$

Hence, for an acceptor doping (p-type material), we have: $N_a \gg N_d, n_i$ and thus from Equation 7.33

$$p = N_a$$

$$n = \frac{n_i^2}{N_a}.$$

For a donor material (n-type): $N_a \gg N_d, n_i$ and thus

$$n = N_d$$

$$p = \frac{n_i^2}{N_d}.$$

It is clear, then, that a knowledge of n_i and the dopant level fixes the carrier concentrations and thus allows the main features of behavior to be determined. Where are the Fermi levels in these extrinsic semiconductors? We know that in the case of intrinsic semiconductors, the Fermi level lies about halfway between the valence and conduction bands. In an n-type semiconductor, the valence band is almost full and most of the conduction is due to electrons donated from the donor levels. Hence, it follows that the "50% electron occupancy" level, i.e., the Fermi level, will now lie about half way between the donor level and the bottom of the conduction band, because the top of the "valence" band can now be identified with the donor level (Figure 7.45).

Similarly, for p-type semiconductors, it will lie midway between the top of the valence band and the acceptor energy level. However, this can only be the case as long as the donor or acceptor mechanisms dominate. At higher temperatures, most

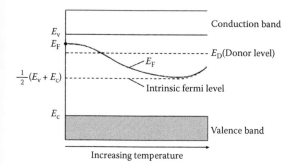

Figure 7.45 Fermi level in an n-type intrinsic semiconductor, and its variation with temperature.

of the donor and acceptor sites would have been exhausted and the true valence band then starts to dominate the conduction mechanism. Hence, the Fermi level will vary with temperature as shown in Figure 7.45, until at high enough temperatures, it reverts to the intrinsic value in both p and n cases.

7.6.5 Binary and ternary semiconductors

We cannot leave semiconductors without a mention of some important, relatively new, materials. These are alloys, made from two or more elements in roughly equal proportions and from different groups in the periodic table, and thus with differing numbers of electrons in the outermost shell.

The best known of these is gallium arsenide (GaAs), which, because Ga has a valency of III and As of V, is an example of what is called a III–V compound.

An important aspect of these compounds is that we can "tailor" the bandgap by varying the mix. The eight electrons in the two outer shells are shared to some extent and create some ionic bonding, through the absence from the parent atom (i.e., it creates a positive ion). The bandgap of GaAs when the two elements are present in equal proportions (i.e., same number of molecules per unit volume) is 1.4 eV but this can be varied by replacing the As by P (GaP, 2.25 eV) or Sb (GaSb, 0.7 eV), for example. Furthermore, the materials can be made p- or n-type by increasing the V(As) over the III(Ga) component, or vice versa.

Another very important aspect of GaAs is that it is a direct bandgap material: the minimum energy in the conduction-band Brillouin zone

occurs at the same k value as the maximum energy of the valence band (Figure 7.46a). This means that electrons can make the transition between the two bands without having to lose or gain momentum in the process. Any necessary loss or gain of momentum must always involve a third entity, a "phonon" (quantum of vibration), for example, and this renders the transition much less probable. Hence, a direct bandgap material is much more efficient than an indirect bandgap material (Figure 7.46b) and the processes are much faster, leading to higher device bandwidth.

Quite frequently even finer control is required over the value of the bandgap and, for this, "bandgap engineers" turn to ternary alloys, i.e., those involving three elements, where the ratio of III–V composition is still approximately 1:1. An example is the range of alloys, which is described by the

(a)

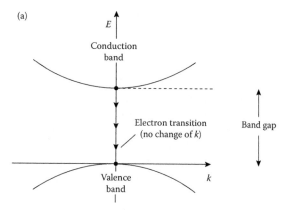

(b)

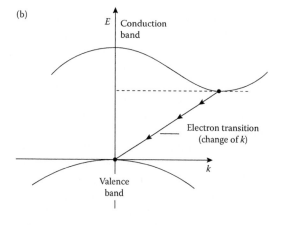

Figure 7.46 Energy level diagrams for direct and indirect bandgap semiconductors. (a) Direct bandgap semiconductor (e.g. GaAs) and (b) indirect bandgap semiconductor (e.g. Si).

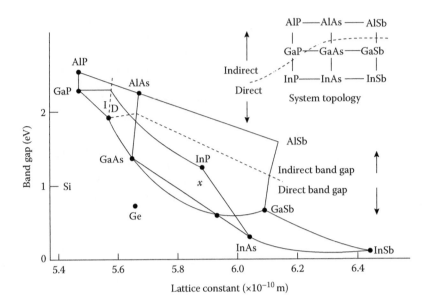

Figure 7.47 Alloy bandgap diagram.

formula $Al_xGa_{1-x}As$. By varying x, one can move along the line between GaAs and Aluminum arsenides (AlAs) on Figure 7.47 and thus vary the bandgap appropriately. Figure 7.47 shows other materials, which can be tailored in this way.

One final difficulty is that in order to grow the required material one needs a substrate on which to grow it, from either the gas phase (gas-phase epitaxy) or the liquid phase (liquid-phase epitaxy), and this requires that the desired material has approximately the same lattice spacing as the substrate. For the example of Al_xGa_{1-x} As, there is little difficulty because the GaAs/AlAs line is almost vertical (Figure 7.47), and thus the lattice spacing always is close to that of GaAs, which can thus be used as the substrate. For other materials, for example, $InAs_ySb_{1-y}$, this is clearly not the case because the InAs/InSb line is almost horizontal. This problem can be solved by going one stage yet further, to quaternary alloys that lie in the regions bounded by the lines in Figure 7.47. An example of a quaternary alloy is $In_xGa_{1-x} As_yP_{1-y}$ and one of these alloys is shown marked x on Figure 7.47. This has a lattice spacing similar to that of InP, which can thus be used as a satisfactory substrate.

Thus, a bandgap engineer generally will choose a substrate, then a suitable quaternary alloy that has the required bandgap, probably making sure it is a direct bandgap material, and then grow the semiconductor.

Bandgap engineering is now a sophisticated, and extremely valuable, technology for the provision of materials for optoelectronic devices; their behavior and performance depend critically on the material from which they are made.

7.6.6 Summary

Solids are complex states of matter, with their range of overlapping atomic wave functions. Many of the ideas are broadly unfamiliar to a nonspecialist physicist. However, many optoelectronic devices and systems rely on the behavior of electrons in solids, and we need to have a good understanding of this. In studying optoelectronics, we certainly need to draw repeatedly upon the ideas outlined in this section.

7.7 NONLINEAR OPTICS

7.7.1 Introduction

In all of the various discussions concerning the propagation of light in material media so far, we have been dealing with linear processes. By this we mean that a light beam of a certain optical frequency that enters a given medium will leave the medium with the same frequency, although the amplitude and phase of the wave will, in general, be altered.

The fundamental physical reason for this linearity lies in the way in which the wave propagates through a material medium. The effect of the electric field of the optical wave on the medium is to set the electrons of the atoms (of which the medium is composed) into forced oscillation; these oscillating electrons then radiate secondary wavelets (because all accelerating electrons radiate) and the secondary wavelets combine with each other and with the original (primary) wave, to form a resultant wave. Now the important point here is that all the forced electrons oscillate at the same frequency (but differing phase, in general) as the primary, driving wave, and thus we have the sum of the waves all of same frequency, but with different amplitudes and phases.

If two such sinusoids are added together

$$A_T = a_1 \sin(\omega t + \varphi_1) + a_2 \sin(\omega t + \varphi_2)$$

and we have, from simple trigonometry

$$A_T = a_T \sin(\omega t + \varphi_T)$$

where

$$a_T^2 = a_1^2 + a_2^2 + 2a_1 a_2 \cos(\varphi_1 - \varphi_2)$$

and

$$\tan \varphi_T = \frac{a_1 \sin \varphi_1 + a_2 \sin \varphi_2}{a_1 \cos \varphi_1 + a_2 \cos \varphi_2}$$

In other words, the resultant is a sinusoid of the same frequency but of different amplitude and phase. It follows, then, that no matter how many more such waves are added, the resultant will always be a wave of the same frequency, that is,

$$A_T = \sum_{n=0}^{N} a_n \sin(\omega t + \varphi_n) = \alpha \sin(\omega t + \beta)$$

where α and β are expressible in terms of a_n and φ_n.

It follows, further, that if there are two primary input waves, each will have the effect described above independently of the other, for each of the driving forces will act independently and the two will add to produce a vector resultant. We call this the "principle of superposition" for linear systems because the resultant effect of the two (or more) actions is just the sum of the effects of each one acting on its own. This has to be the case although the displacements of the electrons from their equilibrium positions in the atoms vary linearly with the force of the optical electric fields. Thus, if we pass into a medium, along the same path, two light waves, of angular frequencies ω_1 and ω_2, emerging from the medium will be two light waves (and only two) with those same frequencies, but with different amplitudes and phases from the input waves.

Suppose now, however, that the displacement of the electrons is *not* linear with the driving force. Suppose, for example, that the displacement is so large that the electron is coming close to the point of breaking free from the atom altogether. We are now in a nonlinear regime. Strange things happen here. For example, a given optical frequency input into the medium may give rise to waves of several different frequencies at the output. Two frequencies ω_1 and ω_2 passing in may lead to, among others, sum and difference frequencies $\omega_1 \pm \omega_2$ coming out.

The fundamental reason for this is that the driving sinusoid has caused the atomic electrons to oscillate nonsinusoidally (Figure 7.48). Our knowledge of Fourier analysis tells us that any periodic nonsinusoidal function contains, in addition to the fundamental component, components at harmonic frequencies, i.e., integral multiples of the fundamental frequency.

This is a fascinating regime. All kinds of interesting new optical phenomena occur here. As might be expected, some are desirable, some are not. Some are valuable in new applications; some just comprise sources of noise. But to use them to advantage, and to minimize their effects when they are a nuisance, we must understand them better. This we shall now try to do.

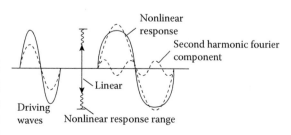

Figure 7.48 Nonlinear response to a sinusoidal drive.

(a) The Rayleigh distance for free-space focusing

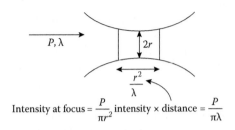

$$\text{Intensity at focus} = \frac{P}{\pi r^2}, \quad \text{intensity} \times \text{distance} = \frac{P}{\pi \lambda}$$

(b) Nonlinear facility in optical fibers

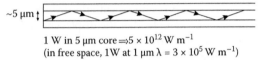

1 W in 5 μm core $\Rightarrow 5 \times 10^{12}$ W m^{-1}
(in free space, 1W at 1 μm $\lambda = 3 \times 10^5$ W m^{-1})

Figure 7.49 The intensity–distance product for nonlinearity. (a) The Rayleigh distance for free-space focusing and (b) nonlinear facility in optical fibers.

7.7.2 Nonlinear optics and optical fibers

Let us begin by summarizing the conditions, which give rise to optical nonlinearity.

In a semiclassical description of light propagation in dielectric media, the optical electric field drives the atomic/molecular oscillators of which the material is composed, and these oscillators become secondary radiators of the field; the primary and secondary fields then combine vectorially to form the resultant wave. The phase of this wave (being different from that of its primary) determines a velocity of light different from that of free space, and its amplitude determines a scattering/absorption coefficient for the material.

Nonlinear behavior occurs when the secondary oscillators are driven beyond the linear response; as a result, the oscillations become nonsinusoidal. Fourier theory dictates that, under these conditions, frequencies other than that of the primary wave will be generated (Figure 7.48).

The fields necessary to do this depend upon the structure of the material because it is this that dictates the allowable range of sinusoidal oscillation at given frequencies. Clearly, it is easier to generate large amplitudes of oscillation when the optical frequencies are close to natural resonances,

and one expects (and obtains) enhanced nonlinearity there. The electric field required to produce optical nonlinearity in materials therefore varies widely, from ~10^6 up to ~10^{11}Vm^{-1}, the latter being comparable with the atomic electric field. Even the lower of these figures, however, corresponds to an optical intensity of ~10^9 W m^{-2}, which is only achievable practically with laser sources. It is for this reason that the study of nonlinear optics only really began with the invention of the laser, in 1960.

The magnitude of any given nonlinear effect will depend upon the optical intensity, the optical path over which the intensity can be maintained, and the size of the coefficient that characterizes the effect.

In bulk media, the magnitude of any nonlinearity is limited by diffraction effects. For a beam of power P watts and wavelength λ focused to a spot of radius r, the intensity, $P/\pi r^2$, can be maintained (to within a factor of ~2) over a distance ~r^2/λ (Rayleigh distance), beyond which diffraction will rapidly reduce it. Hence, the product of intensity and distance is ~$P/\pi\lambda$, independent of r and of propagation length (Figure 7.49a).

However, in an optical fiber the waveguiding properties, in a small diameter core, serve to maintain a high intensity over lengths of up to several kilometers (Figure 7.49b). This simple fact allows magnitudes of nonlinearities, in fibers, which are many orders greater than in bulk materials. Further, for maximum overall effect, the various components' effects per elemental propagation distance must add coherently over the total path. This implies a requirement for phase coherence throughout the path which, in turn, implies a single propagation mode: monomode rather than multimode fibers must, in general, be used.

7.7.3 The formalism of nonlinear optics

Under "normal" propagation conditions we assume a linear relationship between the electric polarization (P) of a medium and the electric field (E) of an optical wave propagating in it, by taking

$$\chi = \frac{P}{E}$$

where χ is the volume susceptibility of the medium, and is assumed constant. The underlying assumption for this is that the separation of atomic positive and negative charges is proportional to the imposed field, leading to a dipole moment per unit volume (P), which is proportional to the field.

However, it is clear that the linearity of this relationship cannot persist for ever-increasing strengths of field. Any resonant physical system must eventually be torn apart by a sufficiently strong perturbing force and, well before such a catastrophe occurs, we expect the separation of oscillating components to vary nonlinearly with the force. In the case of an atomic system under the influence of the electric field of an optical wave, we can allow for this nonlinear behavior by writing the electric polarization of the medium in the more general form:

$$P(E)=\chi_1 E+\chi_2 E^2+\chi_3 E^3+\chi_j E^j+\dots \quad (7.34)$$

The value of χ_j (often written $\chi^{(j)}$) decreases rapidly with increasing j for most materials. Also the importance of the jth term, compared with the first, varies as $(\chi_j/\chi_1)E^{(j-1)}$ and so depends strongly on E. In practice, only the first three terms are of any great importance, and then only for laser-like intensities, with their large electric fields. It is not until one is dealing with power densities of ~10^9 W m^{-2}, and fields ~10^6 V m^{-1}, that $\chi_2 E^2$ becomes comparable with $\chi_1 E$.

Let us now consider the refractive index of the medium. From elementary electromagnetism we know that

$$\varepsilon=1+\chi, n^2=\varepsilon$$

where χ is the electric permittivity of the medium and n is its refractive index.

Hence

$$n=(1+\chi)^{1/2}=\left(1+\frac{P}{E}\right)^{1/2}$$

i.e.,

$$n=\left(1+\chi_1+\chi_2 E+\dots\chi_j E^{j-1}+\dots\right)^{1/2}. \quad (7.35)$$

Hence, we note that the refractive index has become dependent on E. The optical wave, in this nonlinear regime, is altering its own propagation

conditions as it travels. This is a central feature of nonlinear optics.

7.7.4 Second harmonic generation and phase matching

Probably, the most straightforward consequence of nonlinear optical behavior in a medium is that of the generation of the second harmonic of a fundamental optical frequency. To appreciate this mathematically, let us assume that the electric polarization of an optical medium is quite satisfactorily described by the first two terms of Equation 7.34, that is,

$$P(E)=\chi_1 E+\chi_2 E^2. \quad (7.36)$$

Before proceeding, there is an important point to make about Equation 7.36.

Let us consider the effect of a change in sign of E. The two values of the field, $\pm E$, will correspond to two values of P:

$$P(+E)=\chi_1 E+\chi_2 E^2$$

$$P(-E)=-\chi_1 E+\chi_2 E^2$$

These two values clearly have different absolute magnitudes. Now if a medium is isotropic (as is the amorphous silica of which optical fiber is made), there can be no directionality in the medium and thus the matter of the sign of E, i.e., whether the electric field points up or down, cannot be of any physical relevance and cannot possibly have any measurable physical effect. In particular, it cannot possibly affect the value of the electric polarization (which is, of course, readily measurable). We should expect that changing the sign of E will merely change the sign of P, but that the magnitude of P will be exactly the same: the electrons will be displaced by the same amount in the opposite direction, all directions being equivalent. Clearly, this can only be so if $\chi_2=0$. The same argument extended to higher order terms evidently leads us to the conclusion that all even-order terms *must* be zero for amorphous (isotropic) materials, i.e., $\chi_{2m}=0$. This is a point to remember. The corollary of this argument is that in order to retain any even order terms the medium must exhibit some anisotropy. It must, for example, have a crystalline structure without a center of symmetry. It follows that Equation 7.36 refers to such a medium.

Suppose now that we represent the electric field of an optical wave entering such a crystalline medium by

$$E = E_0 \cos \omega t.$$

Then substituting into Equation 7.36, we find

$$P(E) = \chi_1 E_0 \cos \omega t + \frac{1}{2}\chi_2 E_0^2 + \frac{1}{2}\chi_2 E_0^2 \cos 2\omega t.$$

The last term, the second harmonic term at twice the original frequency, is clearly in evidence. Fundamentally, it is due to the fact that it is easier to polarize the medium in one direction than in the opposite direction, as a result of the crystal asymmetry. A kind of "rectification" occurs.

Now the propagation of the wave through the crystal is the result of adding the original wave to the secondary wavelets from the oscillating dipoles which it induces. These oscillating dipoles are represented by P. Thus, $\partial^2 P/\partial t^2$ leads to e/m waves because radiated power is proportional to the acceleration of charges, and waves at all of P's frequencies will propagate through the crystal.

Suppose now that an attempt is made to generate a second harmonic over a length L of crystal. At each point along the path of the input wave a second harmonic component will be generated. But because the crystal medium will almost certainly be dispersive, the fundamental and second harmonic components will travel at different velocities. Hence, the successive portions of second harmonic component generated by the fundamental will not, in general, be in phase with each other, and thus will not interfere constructively. Hence, the efficiency of the generation will depend upon the velocity difference between the waves.

A rigorous treatment of this process requires a manipulation involving Maxwell's equations but a semianalytical treatment which retains a firm grasp of the physics will be given here.

Suppose that the amplitude of the fundamental (driving) wave between distances z and $z+dz$ along the optical path in the crystal is $e\cos(\omega t - kz)$. Then, from Equation 7.34, there will be a component of electric polarization (dipole moment per unit volume) of the form: $\chi_2 e^2 \cos^2(\omega t - kz)$ giving a time-varying second harmonic term $(1/2)\chi_2 e^2 \cos^2(\omega t - kz)$, as before. Consider, then, a slab, in the medium, of unit cross section, and thickness dz (Figure 7.50).

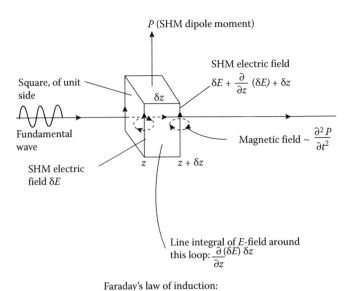

Figure 7.50 Infinitesimals for second-harmonic generation along the path of the fundamental.

For this slab, the dipole moment will be

$$P = \frac{1}{2}\chi_2 e^2 \cos 2(\omega t - kz)dz.$$

Now a time-varying dipole moment represents a movement of charge, and therefore an electric current. This current will create a magnetic field. A time-varying magnetic field (second derivative $\partial^2 P/\partial t^2$ of the dipole moment) will generate a voltage around any loop through which it threads (Faraday's law of electromagnetic induction). From Figure 7.50, it can be seen that if δE is the elemental component of the second harmonic electric field generated by the changing dipole moment in the thin slab, then this voltage is proportional to $\partial(\delta E)/\partial z$. Hence, we have

$$\frac{\partial(\delta E)}{\partial z} = A\frac{\partial^2 P}{\partial t^2} \qquad (7.37)$$

where A is a constant.

Hence, from Equation 7.37

$$\frac{\partial(\delta E)}{\partial z} = -A \cdot 2\omega^2 \chi_2 e^2 \cos 2(\omega t - kz)dz.$$

Integrating this with respect to z gives

$$\delta E = A\frac{\omega^2}{k}\chi_2 e^2 \sin 2(\omega t - kz)dz$$

and, with $\omega/k = c$, we have

$$\delta E = Ac\omega\chi_2 e^2 \sin 2(\omega t - kz)dz$$

as the element of the second harmonic electric field generated by the slab between z and $z+dz$. But the second harmonic component now propagates with wave number k_s say (because the crystal will have a different refractive index at frequency 2ω, compared with that at ω), so when this component emerges from the crystal after a further distance $L - z$, it will become:

$$\delta E_L = Ac\omega\chi_2 e^2 \sin\left[2\omega t - 2kz - k_s(L - z)\right]dz.$$

Hence, the total electric field amplitude generated over the length L of crystal will be, on emergence

$$E_L(2\omega) = \int_0^L Ac\omega\chi_2 e^2 \sin[2\omega t - 2kz - k_s(L-z)]dz.$$

Performing this integration gives

$$E_L(2\omega) = Ac\chi_2 e^2 L\omega \frac{\sin\left(k - \frac{1}{2}k_s\right)L}{\left(k - \frac{1}{2}k_s\right)L}$$

$$\sin[2\omega t - (2k + k_s)L].$$

The intensity of the emerging second harmonic will be proportional to the square of amplitude of this

$$I_L(2\omega) = B\chi_2^2 e^4 L^2 \omega^2 \left[\frac{\sin\left(k - \frac{1}{2}k_s\right)L}{\left(k - \frac{1}{2}k_s\right)L}\right]^2$$

where B is another constant. Now the intensity of the fundamental wave is proportional to e^2, so the intensity of the second harmonic is proportional to the square of the intensity of the fundamental, i.e.,

$$I_L(2\omega) = B'\chi_2^2 I_L^2(\omega) L^2 \omega^2 \left[\frac{\sin\left(k - \frac{1}{2}k_s\right)L}{\left(k - \frac{1}{2}k_s\right)L}\right]^2 \qquad (7.38)$$

where B' is yet another constant. From this, we can define an efficiency η_{SHG} for the second harmonic generation process as

$$\eta_{SHG} = \frac{I_L(2\omega)}{I_L(\omega)}.$$

Note that η_{SHG} varies as the square of the fundamental frequency and of the length of the crystal; note also that it increases linearly with the power of the fundamental.

From Equation 7.38 it is clear that, for maximum intensity, we require that the sinc² function has its maximum value, that is,

$$k_s = 2k_f.$$

This is the *phase-matching condition* for second harmonic generation. Now the velocities of the fundamental and the second harmonic are given by

$$c_f = \frac{\omega}{k_f}, \quad c_s = \frac{2\omega}{k_s}$$

These are equal when $k_s = 2k_f$, so the phase-matching condition is equivalent to a requirement that the two velocities are equal. This is to be expected because it means that the fundamental generates, at each point in the material, second harmonic components that will interfere constructively. The phase match condition usually can be satisfied by choosing the optical path to lie in a particular direction within the crystal. It has already been noted that the material must be anisotropic for second harmonic generation to occur; it will also, therefore, exhibit birefringence (Section 7.3.3). One way of solving the phase matching problem, therefore, is to arrange that the velocity difference resulting from birefringence is canceled by that resulting from material dispersion. In a crystal with normal dispersion, the refractive index of both the eigenmodes (i.e., both the ordinary and extraordinary rays) increases with frequency. Suppose we consider the specific example of quartz, which is a positive uniaxial crystal (see Section 7.3.3). This means that the principal refractive index for the extraordinary ray is greater than that for the ordinary ray, that is,

$$n_e > n_o$$

Because quartz is also normally dispersive, it follows that

$$n_e^{(2\omega)} > n_o^{(\omega)}$$

$$n_o^{(2\omega)} > n_o^{(\omega)}.$$

Hence, the index ellipsoids for the two frequencies are as shown in Figure 7.51a. Now it will be remembered from Section 7.3.3 that the refractive indices for the "o" and "e" rays for any given direction in the crystals are given by the major and minor axes of the ellipse in which the plane normal to the direction, and passing through the center of the index ellipsoid, intersects the surface of the ellipsoid. The geometry (Figure 7.51a) thus makes it clear that a direction can be found [10] for which

$$n_o^{(2\omega)}(\vartheta_m) = n_e^{(\omega)}(\vartheta_m)$$

so SHG phase matching occurs provided that

$$n_o^{(2\omega)} < n_e^{(\omega)}.$$

The above is indeed true for quartz over the optical range. Simple trigonometry allows ϑ_m to be determined in terms of the principal refractive indices as

$$\sin^2 \vartheta_m = \frac{(n_e^{(\omega)})^{-2} - (n_e^{(2\omega)})^{-2}}{(n_o^{(2\omega)})^{-2} - (n_e^{(2\omega)})^{-2}}.$$

Hence, ϑ_m is the angle at which phase matching occurs. It also follows from this that for second harmonic generation in this case, the wave at the fundamental frequency must be launched at an angle ϑ_m with respect to the crystal axis and *must have the "extraordinary" polarization;* and that the second harmonic component *will appear in the same direction and will have the "ordinary" polarization,* i.e., the two waves are collinear and have orthogonal linear polarizations! Clearly, other crystal direction and polarization arrangements also are possible in other crystals.

The required conditions can be satisfied in many crystals but quartz is an especially good one owing to its physical robustness, its ready obtainability with good optical quality and its high optical power-handling capacity.

Provided that the input light propagates along the chosen axis, the conversion efficiency ($\omega \to 2\omega$) is a maximum compared with any other path (per unit length) through the crystal. Care must be taken, however, to minimize the divergence of the beam (so that most of the energy travels in the chosen direction) and to ensure that the temperature remains constant (since the birefringence of the crystal will be temperature dependent).

The particle picture of the second harmonic generation process is viewed as an annihilation of two photons at the fundamental frequency, and the creation of one photon at the second harmonic frequency. This pair of processes is necessary in order to conserve energy, that is,

$$2h\nu_f = h(2\nu_f) = h\nu_s.$$

The phase-matching condition is then equivalent to conservation of momentum. The momentum of a photon wave number k is given by

$$p = \frac{h}{2\pi} k$$

(a)

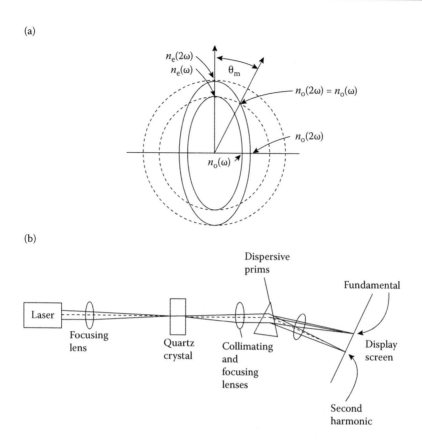

(b)

Figure 7.51 Conditions for second-harmonic generation in quartz. (a) Phase matching with the bire-fringence index ellipsoides and (b) schematic experimental arrangement for SHM generation.

and thus conservation requires that

$$k_s = 2k_f$$

as in the wave treatment.

Quantum processes that have no need to dispose of excess momentum are again the most probable, and thus this represents the condition for maximum conversion efficiency in the particle picture.

The primary practical importance of second harmonic generation is that it allows laser light to be produced at the higher frequencies, into the blue and ultraviolet, where conditions are not intrinsically favorable for laser action, as was noted earlier (Section 7.2.4.1). In this context we note again, from Equation 7.38, that the efficiency of the generation increases as the square of the fundamental frequency, which is of assistance in producing these higher frequencies.

7.7.5 Optical mixing

Optical mixing is a process closely related to second harmonic generation. If, instead of propagating just one laser wave through the same nonlinear crystal, we superimpose two (at different optical frequencies) simultaneously along the same direction, then we shall generate sum and difference frequencies, that is,

$$E = E_1 \cos\omega_1 t + E_2 \cos\omega_2 t$$

and thus again using Equation 7.36

$$P(E) = \chi_1 (E_1 \cos\omega_1 t + E_2 \cos\omega_2 t)$$
$$+ \chi_2 (E_1 \cos\omega_1 t + E_2 \cos\omega_2 t)^2$$

This expression for $P(E)$ is seen to contain the term

$$2\chi_2 E_1 E_2 \cos\omega_1 t \cos\omega_2 t = \chi_2 E_1 E_2 \cos(\omega_1 + \omega_2)t$$
$$+ \chi_2 E_1 E_2 \cos(\omega_1 - \omega_2)t$$

giving the required sum and difference frequency terms. Again, for efficient generation of these components, we must ensure that they are phase matched. For example, to generate the sum frequency efficiently, we require that

$$k_1 + k_2 = k_{(1+2)}$$

which is equivalent to

$$\omega_1 n_1 + \omega_2 n_2 = (\omega_1 + \omega_2)n_{(1+2)}$$

where n represents the refractive indices at the suffix frequencies. The condition again is satisfied by choosing an appropriate direction relative to the crystal axes.

This mixing process is particularly useful in the reverse sense. If a suitable crystal is placed in a Fabry–Perot cavity, which possesses a resonance at ω_1, say, and is "pumped" by laser radiation at $\omega_{(1+2)}$, then the latter generates both ω_1 and ω_2. This process is called parametric oscillation: ω_1 is called the signal frequency and ω_2 is the idler frequency. It is a useful method for "down conversion" of an optical frequency, i.e., conversion from a higher to a lower value.

The importance of phase matching in nonlinear optics cannot be overstressed. If waves at frequencies different from the fundamental are to be generated efficiently they must be produced with the correct relative phase to allow constructive interference, and this, as we have seen, means that velocities must be equal to allow phase matching to occur. This feature dominates the practical application of nonlinear optics.

7.7.6 Intensity-dependent refractive index

It was noted in Section 7.7.4 that all the even-order terms in expression 7.34 for the nonlinear susceptibility (χ) are zero for an amorphous (i.e., isotropic) medium. This means that in an optical fiber made from amorphous silica, we can expect that $\chi_{(2m)} = 0$, so it will not be possible to generate a second harmonic according to the principles outlined in Section 7.7.4. (However, second harmonic

generation has been observed in fibers [11] for reasons which took some time to understand!) It is possible to generate a third harmonic, however, because to a good approximation the electric polarization in the fiber can be expressed by

$$P(E) = \chi_1 E + \chi_3 E^3. \tag{7.39}$$

Clearly, though, if we wish to generate the third harmonic efficiently we must again phase match it with the fundamental, and this means that somehow we must arrange for the two relevant velocities to be equal, i.e., $c_\omega = c_{3\omega}$. This is very difficult to achieve in practice, although it has been done.

There is, however, a more important application of Equation 7.39 in amorphous media. It is clear that the effective refractive index in this case can be written

$$n_e = \left(1 + \chi_1 + \chi_3 E^2\right)^{1/2}$$

and, if $\chi_1, \chi_3 E^2 \ll 1$,

$$n_e \approx 1 + \frac{1}{2}\chi_1 + \frac{1}{2}\chi_3 E^2.$$

Hence

$$n_e = n_o + \frac{1}{2}\chi_3 E^2 \tag{7.40a}$$

where n_o is the "normal," linear refractive index of the medium. But we know that the intensity (power/unit area) of the light is proportional to E^2, so that we can write

$$n_e = n_o + n_2 I \tag{7.40b}$$

where n_2 is a constant for the medium. Equation 7.40b is very important and has a number of practical consequences. We can see immediately that it means that the refractive index of the medium depends upon the intensity of the propagating light: the light is influencing its own velocity as it travels.

In order to fix ideas to some extent, let us consider some numbers for silica. For amorphous silica $n_2 \sim 3.2 \times 10^{-20}$ m^{2}W^{-1}, which means that a 1% change in refractive index (readily observable) will occur for an intensity of $\sim 5 \times 10^{17}$ W m^{-2}. For a fiber with a core diameter of ~5 μm, this requires an optical power level of 10 MW. Peak power levels of this magnitude are readily obtainable, for short durations, with modern lasers.

It is interesting to note that this phenomenon is another aspect of the electro-optic effect. Clearly, the refractive index of the medium is being altered by an electric field. This will now be considered in more detail.

7.7.7 The electro-optic effect

When an electric field is applied to an optical medium the electrons suffer restricted motion in the direction of the field, when compared with that orthogonal to it. Thus, the material becomes linearly birefringent in response to the field. This is known as the electro-optic effect.

Consider the arrangement of Figure 7.52. Here, we have incident light that is LP at 45° to an electric field and the field acts on a medium transversely to the propagation direction of the light. The field-induced linear birefringence will cause a phase displacement between components of the incident light which lie, respectively, parallel and orthogonal to the field; hence, the light will emerge elliptically polarized.

A (perfect) polarizer placed with its acceptance direction parallel with the input polarization direction will of course, pass all the light in the absence of a field. When the field is applied, the fraction of light power that is passed will depend upon the form of the ellipse, which in turn depends upon the phase delay introduced by the field. Consequently, the field can be used to modulate the intensity of the light, and the electro-optic effect is, indeed, very useful for the modulation of light. The phase delay introduced

may be proportional either to the field (Pockels effect) or to the square of the field (Kerr effect). All materials manifest a transverse Kerr effect. Only crystalline materials can manifest any kind of Pockels effect, or longitudinal (E field parallel with propagation direction) Kerr effect. The reason for this is physically quite clear. If a material is to respond linearly to an electric field, the effect of the field must change sign when the field changes sign. This means that the medium must be able to distinguish (e.g.,) between "up" (positive field) and "down" (negative field). But it can only do this if it possesses some kind of directionality in itself, otherwise all field directions must be equivalent in their physical effects. Hence, in order to make the necessary distinction between up and down, the material must possess an intrinsic asymmetry, and hence must be crystalline. By a similar argument a longitudinal E field can only produce a directional effect orthogonally to itself (i.e., in the direction of the optical electric field) if the medium is anisotropic (i.e., crystalline) for otherwise all transverse directions will be equivalent. In addition to the modulation of light (phase or intensity/power) it is clear that the electro-optic effect could be used to measure an electric field and/or the voltage that gives rise to it.

7.7.8 Optical Kerr effect

The normal electro-optic Kerr effect is an effect whereby an electric field imposed on a medium induces a linear birefringence with slow axis parallel with the field (Figure 7.53a). The value of the induced birefringence is proportional to the square of the electric field. In the optical Kerr effect, the electric field involved is that of an optical wave, and

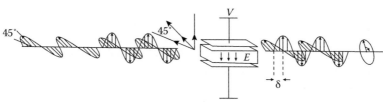

Linear polarization becomes elliptical by passing through an electro-optic medium with applied field E

Figure 7.52 The electro-optic effect.

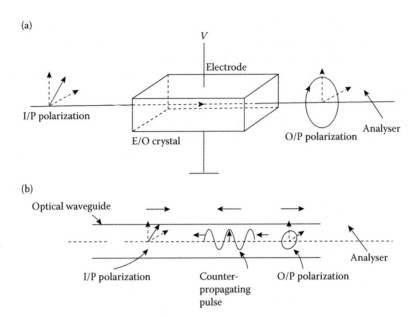

Figure 7.53 "Normal" and "optical" Kerr effects. (a) "Normal" electro-optic Kerr effect and (b) "Optical" Kerr effect: light acting on light.

thus the birefringence probed by one wave may be that produced by another (Figure 7.53b).

The phase difference introduced by an electric field E over an optical path L is given by

$$\Delta\varphi = \frac{2\pi}{\lambda}\Delta nL$$

where $\Delta n = KE^2$, K being the Kerr constant.

Now from Equations 7.40a and b, we have

$$\Delta n = n_2 I = \frac{1}{2}\chi_3 E^2 = KE^2 \qquad (7.41)$$

From elementary electromagnetism, we know that

$$I = c\varepsilon E^2.$$

Hence, we have, from Equation 7.41

$$K = n_2 c\varepsilon = \frac{1}{2}\chi_3$$

showing that the electro-optic effect, whether the result of an optical or an external electric field, is a nonlinear phenomenon, depending on χ_3. Using similar arguments it can easily be shown that the electro-optic Pockels effect also is a nonlinear effect, depending on χ_2.

The optical Kerr effect has several other interesting consequences. One of these is self-phase modulation (SPM), which is the next topic for consideration.

7.7.9 Self-phase modulation

The fact that refractive index can be dependent on optical intensity clearly has implications for the phase of the wave propagating in a nonlinear medium. We have

$$\varphi = \frac{2\pi}{\lambda}nL.$$

Hence, for $n = n_0 + n_2 I$

$$\varphi = \frac{2\pi L}{\lambda}(n_0 + n_2 I)$$

Suppose now that the intensity is a time-dependent function $I(t)$. It follows that φ also will be time dependent, and because

$$\omega = \frac{d\varphi}{dt}$$

the frequency spectrum will be changed by this effect, which is known as SPM.

In a dispersive medium, a change in the spectrum of a temporally varying function (e.g., a pulse) will change the shape of the function. For example, pulse broadening or *pulse compression* can be obtained under appropriate circumstances. To see this, consider a Gaussian pulse (Figure 7.54a). The Gaussian shape modulates an optical carrier of frequency ω_0, say, and the new instantaneous frequency becomes

$$\omega' = \omega_0 + \frac{d\varphi}{dt}$$

If the pulse is propagating in the Oz direction

$$\varphi = -\frac{2\pi z}{\lambda}(n_0 + n_2 I) \qquad (7.42a)$$

and we have

$$\omega' = \omega_0 - \frac{2\pi z}{\lambda}n_2\frac{dI}{dt}. \qquad (7.42b)$$

At the leading edge of the pulse $dI/dt > 0$, hence

$$\omega' = \omega - \omega_I(t).$$

At the trailing edge

$$\frac{dI}{dt} < 0$$

and

$$\omega' = \omega + \omega_I(t).$$

Hence, the pulse is now "chirped," i.e., the frequency varies across the pulse. Figure 7.54b shows an example of this effect.

Suppose, for example, a pulse from a mode-locked Argon laser, initial width 180 ps, is passed down 100 m of optical fiber. As a result of SPM the frequency spectrum is changed by the propagation. Figure 7.54c shows how the spectrum varies as the initial peak power of the pulse is varied. The peak power will lead to a peak phase change, according to Equation 7.42a and this phase change is shown for each of the spectra. It can be seen that the initial spectrum ($\Delta\varphi=0$) is just due to the modulation of the optical sinusoid (Fourier spectrum of a Gaussian pulse) and, as the value of $\Delta\varphi$ increases, the first effect is a broadening. At $\Delta\varphi=1.5\pi$ the spectrum has split into two clear peaks, corresponding to the frequency shifts at the back and front edges of the pulse. The spectra then develop multiple peaks.

(a)

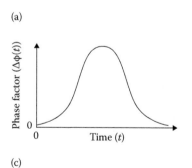

(b)

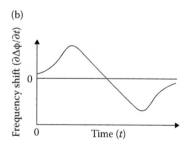

(c)

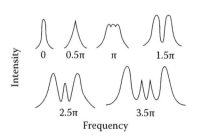

Figure 7.54 SPM for a Gaussian pulse. (a) Intensity-dependent phase factor for a Gaussian pulse, (b) the instantaneous frequency shift for (a), and (c) frequency spectra for (a) designated by maximum phase shift, at peak.

It is important to realize that this does not necessarily change the shape of the pulse envelope, just the optical frequency within it. However, if the medium through which the pulse is passing is dispersive, the pulse shape will change. This is an interesting possibility and it leads to the phenomenon of soliton propagation [12].

7.7.10 Inelastic scattering effects

The effects that we have considered so far in this survey of nonlinear optics are all observable at arbitrary wavelengths and do not involve any intrinsic net loss of energy to the medium itself. By contrast, in our discussion of laser action in Section 7.2.4 we saw that very strong nonlinear behavior can be generated at particular wavelengths in materials where we can exploit energy exchange with the medium through movements of electrons between different energy levels. There are also other effects that involve exchanges of energy with the optical medium and wherever the scattering mechanism involves some loss or gain of energy, respectively, to or from the medium, we refer to this as inelastic scattering. Some inelastic scattering mechanisms are not selective in terms of the wavelength of the incident light, and generate new wavelengths directly related to the incident wavelength. These effects result from the promotion of electrons to energy levels that exist due to the molecular structure of the medium. In optical fibers, the two effects of this type that are most important are Raman and Brillouin scattering.

We will first consider Raman scattering. We saw in Section 7.6 that energy levels in solid materials are widened into bands by interaction between the molecules that form the solid. Part of the width of the band represents vibrational states of the molecules. In a dielectric material such as the glass used to make an optical fiber, there is a large energy gap above the top of the valence band, so that electrons absorbing the energy of incident photons must remain in the valence band after the photon has passed by. However, it is possible for electrons to be very briefly promoted to a "virtual" energy level within the energy gap. Such "virtual" levels can only be occupied for a very short time $\Delta\tau$ given by the uncertainty principle $\Delta\tau\Delta\varepsilon \sim h$, the quantum constant. Once excited, the electron must immediately decay to the valence band, but it may decay to a level representing a higher vibrational energy than its initial level. In this case, part of the energy of the incident photon is transferred to vibration (heat) in the medium, and by conservation of energy, the resulting emitted photon must have lower energy than the incident photon. The scattered light then has a longer wavelength than the incident light and is referred to as "Stokes" radiation. The opposite energy exchange can also happen if the excited electron decays to a lower vibrational energy level compared to the initial state. Here, the medium loses vibrational energy and the emitted photon has a higher energy than the incident photon, thus the scattered light has a shorter wavelength than the incident light. The scattered light in this case is called "anti-Stokes" radiation. The energy levels and photons involved in Raman scattering are illustrated in Figure 7.55. Note that, unlike many of the scattering processes discussed earlier in this section, Raman scattering does not require phase matching between the incident and scattered waves.

The values and distribution of the molecular vibrational frequencies generating Raman scattering vary depending on the medium. Isolated

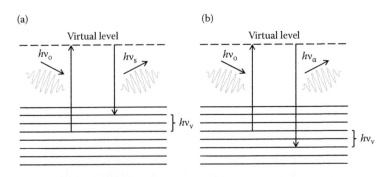

Figure 7.55 Raman energy level transitions: (a) Stokes and (b) anti-Stokes.

molecules such as gases can have very well-defined Raman spectra. On the other hand, disordered solids such as glasses contain molecules with a wide range of vibrational energies and chemical bonds of different strengths and orientations. This situation leads to very broad Raman spectra. Spontaneous Raman scattering is a very weak effect. Typically, only one incident photon in a million will be converted to Stokes or anti-Stokes radiation. However, the Stokes and anti-Stokes processes are not equally likely because of the thermal distribution of vibrational energies. The relative likelihood of production of a Stokes or anti-Stokes photon is proportional to the ratio of the numbers of molecules in the lower and upper energy states, respectively. In the typical case of a medium in thermal equilibrium, the ratio of the number of molecules N_1 in an excited state with energy E_1 will be lower than the number N_2 in a lower energy state E_2. The ratio of the numbers of molecules in these states is given by the Boltzmann distribution

$$\frac{N_1}{N_2} = \exp\left(-\frac{E_1 - E_2}{kT}\right) = \exp\left(-\frac{h\omega_v}{2\pi kT}\right),$$

where ω_v is the difference between the molecular vibration frequencies of the two states. For light incident at a frequency ω_0, conservation of energy requires that the frequency ω_s of the Stokes light is given by $\omega_s = \omega_0 - \omega_v$ and that of the anti-Stokes light by $\omega_a = \omega_0 + \omega_v$. Individual molecules generating Raman scattered light act as isolated dipoles, so their radiation efficiency is proportional to the fourth power of the frequency. Taking this into account together with the relative probability of the generation of Stokes and anti-Stokes photons, we can predict the relative intensity of the two emitted frequencies for the case of spontaneous Raman scattering to be

$$\frac{I_a}{I_s} = \left(\frac{\omega_a}{\omega_s}\right)^4 \exp\left(-\frac{h\omega_v}{2\pi kT}\right)$$

Brillouin scattering is similar to Raman scattering to the extent that this effect also involves energy exchanges with molecular vibrations, but in this case the vibrations concerned are the shared vibrational modes of large groups of molecules associated with bulk acoustic waves propagating in the medium. Such waves will always be present in a medium because the molecules are always in thermal motion. Some of this thermal energy naturally couples into the acoustic vibrational modes of the medium. Whenever an acoustic wave propagates in an optical medium, the variations in the acoustic pressure and the associated periodic strain produce corresponding local variations in refractive index. Light propagating in the medium will then be scattered from these refractive index variations but because the acoustic waves are themselves propagating, the light is being scattered from moving scatterers and an optical frequency shift must occur. The magnitude of this frequency shift is governed by the acoustic velocity of the medium. For a practical solid material such as fused silica, the optical frequency shift is of the order of 11 GHz, some two orders of magnitude smaller than the frequency shift for Raman scattering. Just as for Raman scattering, the incident optical wave can either gain energy or lose it to the medium, so both Brillouin Stokes and anti-Stokes processes can occur, leading, respectively, to a down-shift and an up-shift in the optical frequency of the scattered waves. Spontaneous Brillouin scattering is a weak effect because the thermally generated acoustic waves have low amplitude.

We have so far considered only spontaneous Raman and Brillouin scattering, where the populations of electrons in the different energy levels of a material are close to their equilibrium values. However, when the incident wave becomes sufficiently intense, it is possible to drive the energy level population distribution away from the normal thermal equilibrium condition, increasing the frequency of stimulated emission events for either type of scattering mechanism such that a new, stimulated scattering regime is produced. Stimulated Raman scattering provides an opportunity for creating broadband optical amplifiers in condensed matter, as we would expect from the energy band structure referred to earlier and the lack of any phase matching restrictions in the wave interactions. However, stimulated Brillouin scattering (SBS) does involve phase matching restrictions and is intrinsically narrow band.

It is worth examining the phenomenon of SBS in optical fiber in somewhat greater detail as this behavior leads to a potential for some advanced applications in optical waveguides. The principles of the analysis apply not only to optical fibers but also to integrated optics. In a single mode optical fiber, propagating modes can travel only forward or backward, parallel to the fiber's axis. Thus, spontaneous Brillouin scattering of an intense incident wave, known as a pump wave, traveling in a positive direction along the core of a fiber can only lead to guided scattered radiation traveling either forward or backward in the core. The backward propagating wave will interfere with the incident wave to produce a standing wave whose nodes and antinodes move at the acoustic velocity of the medium. The direction of the movement is positive for Stokes shifted backscatter and negative for anti-Stokes shifted backscatter, as depicted in Figure 7.56. In this figure, the pump wave has a radian frequency ω_p and the frequencies of the Stokes and anti-Stokes waves are represented as ω_s and ω_a, respectively. The radian frequency of the acoustic wave is denoted by Ω. The nodes and antinodes of the standing waves will produce modulation of the refractive index of the core of the fiber via electrostriction, thus each standing wave creates a moving Bragg grating in the core, which will tend to reinforce the magnitudes of the backscattered waves. A positive feedback effect is therefore established, which can lead to a large increase in the Brillouin backscatter. The Stokes frequency is given by $\omega_s = \omega_p - \Omega$ and the anti-Stokes frequency by $\omega_a = \omega_p + \Omega$. The frequency shift in Hertz, δf, is related to acoustic velocity v_a by the relation $\delta f = 2nv_a/\lambda_0$.

One key factor that limits the bandwidth of the SBS effect is the decay time of acoustic waves at the Brillouin frequency. In fused silica, this decay time is of the order of 5 ns. For optimum coherent scattering from the acoustic wave, the optical wave must have a coherence time similar to or larger than the acoustic decay time, thus the optical bandwidth should be less than ~100 MHz. Within this bandwidth, the strength of the effect can be sufficient to transfer a large proportion of the pump wave's energy into the stimulated Brillouin backscattered waves within only a few meters of fiber. This principle can be employed to produce Brillouin effect-based amplifiers and filters in optical fibers. Another potential application is to produce tunable delays by different methods. One of these involves the production of "slow light" by the stimulated Brillouin effect, and this will now be introduced.

Slow and fast light are terms that are used to describe the observation that the group velocity of a light wave in a medium can be substantially different from the velocity expected from the average bulk refractive index of that medium. The group velocity is the velocity with which energy and information are carried by a wave, so the possibility of creating media in which their arrival time can be delayed or advanced could have great practical value. This could be particularly useful if the group velocity could be controlled at will. One of the more attractive ways in which this might be

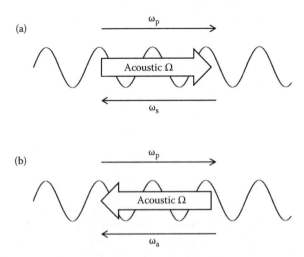

Figure 7.56 Wave interactions for Brillouin scattering in an optical fiber (a) Acoustic wave co-propagating with pump wave (b) Acoustic wave counter-propagating against pump wave.

done employs SBS in an optical waveguide, but the principles on which the phenomena depend are common to many different mechanisms, including the behavior of atomic resonances and optical microresonators, to name but two. The basic requirement is that the medium should possess a sharp, strong resonance at an optical frequency. The typical behavior of such a resonance is shown in Figure 7.57. The amplitude response of the medium is represented by the solid curve in the figure. The resonance is shown by the peaking behavior as the optical frequency changes. In the case of SBS, this peak represents the frequency variation of the backscattered intensity. However, whenever such a resonance exists, there must also be a characteristic variation in the refractive index of the medium, as shown by the dashed curve in the figure. The group velocity V_g of a modulated light wave is given by

$$V_g(\omega) = \frac{c}{n+\omega\dfrac{dn}{d\omega}}$$

where c is the vacuum velocity of light, n is the refractive index, and ω is the optical frequency. Clearly, as $dn/d\omega$ is changing across the range of frequencies shown, the group velocity will also be varying. The curves shown in Figure 7.57 indicate that the magnitude of the slope of $dn/d\omega$ is largest in the vicinity of the resonance peak, exerting the largest effect on the denominator in the expression for the group velocity. Thus, the modulated envelope of the light will be either most advanced

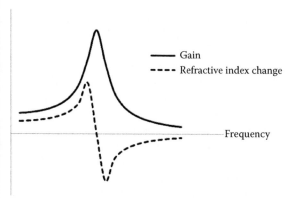

Figure 7.57 Gain and refractive index variation with frequency near to an optical resonance.

or retarded near to this frequency, depending on whether the peak represents gain or loss. In the case of stimulated Brillouin gain in an optical waveguide, the Stokes frequency experiences a large positive slope in the refractive index curve and so slow light is produced.

7.7.11 Summary

We have seen in this chapter that nonlinear optics has its advantages and disadvantages. When it is properly under control it can be enormously useful; but on other occasions it can intrude, disturb, and degrade.

The processes by which light waves produce light waves of other frequencies need very high optical electric fields and thus high peak intensities. It was for this reason that nonlinear optics only became a serious subject with the advent of the laser. Optical fibers provide a convenient means by which peak intensities can be maintained over relatively long distances and are thus very useful media for the study and control of nonlinear optical effects.

We must also remember that in order to cause one optical frequency component to generate another, the second must be generated in phase with itself along the generation path: phase matching is an important feature of such processes.

We have seen how the effects occur when electrons are stretched beyond the comfortable sinusoidal oscillations in their atoms or molecules can yield useful extra optical waves and can influence their own propagation conditions (optical Kerr effect, SPM).

Finally, we can also use light to alter, permanently or semipermanently, the optical properties of a medium, and thus provide the means whereby a new class of optical components, especially fiber components, can be fabricated.

There is a wealth of potential here. The exploration of possibilities for nonlinear optics, especially in regard to new, natural, or synthetic optical materials (e.g., organics, high T_c superconductors), etc., has not even really begun. The prospects, for example, for new storage media, fast switching of light by light, and three-dimensional television, which will be opened up in the future by such materials, are intriguing, and it could well be that nonlinear optical technology soon will become a powerful subject in its own right.

7.8 CONCLUSIONS

In this chapter, the aim has been to emphasize the principles that comprise the physical basis of opto-electronics. A thorough grasp of these principles should facilitate a better appreciation of the more advanced treatments of components and systems in later chapters of this handbook.

ACKNOWLEDGMENTS

The majority of the material presented in this chapter first appeared in "Essentials of Optoelectronics" (Rogers), published by Chapman & Hall, 1997, and is included here with permission.

REFERENCES

1. Zemansky, M.W. 1968. *Heat and Thermodynamics*, 5th ed. (New York: McGraw Hill), Chapter 6.
2. Nye, J.F. 1976. *Physical Properties of Crystals* (Oxford: Clarendon), Chapter 2.
3. Jones, R.C. 1941–1956. A new calculus for the treatment of optical systems. *J. Opt. Soc. Am.*, 31(46), 234–241.
4a. Born, M. and Wolf, E. 1975. *Principles of Optics*, 5th ed. (Oxford: Pergamon), p. 30.
4b. Born, M. and Wolf, E. 1975. *Principles of Optics*, 5th ed. (Oxford: Pergamon), p. 40.
5. Jerrard, H.G. 1954. Transmission of light through optically active media. *J. Opt. Soc. Am.*, 44, 634–664.
6. Syms, R. and Cozens, J. 1992. *Optical Guided Waves and Devices* (New York: McGraw-Hill), Chapter 6.
7. Kaplan, W. 1981. *Advanced Mathematics for Engineers* (Reading, MA: Addison-Wesley), Chapter 12.
8. Adams, M.J. 1981. *An Introduction to Optical Waveguides* (New York: Wiley), Chapter 7.
9. Kittel, C. 1968. *Introduction to Solid State Physics*, 3rd ed. (New York: Wiley).
10. Nye, J.F. 1976. *Physical Properties of Crystals* (Oxford: Clarendon), Chapter 13.
11. Fujii, Y., Kawasaki, B.S., Hill, K.O. and Johnson, D.C. 1980. Sum-frequency light generation in optical fibres. *Opt. Lett.* 5, 48.
12. Mollenauer, L.F., Stolen, R.H. and Gordon, J.P. 1980. Experimental observation of pico-second narrowing and solitons in optical fibres. *Phys. Rev. Lett.*, 45, 1095.

FURTHER READING

Agrawal, G.P. 1989. *Nonlinear Fiber Optics* (New York: Academic).
Andonovic, I. and Uttamchandani, D. 1989. Optical information processing, storage media integrated optics polarimeters. *Principles of Modern Optical Systems* (Boston, MA: Artech House), 608 pp.
Bjarklev, A. 1993.*Optical Fiber Amplifiers* (Boston, MA: Artech House).
Blaker, J.W. and Rosenblum, W.B. 1993. *Optics: An Introduction for Students of Engineering* (New York: Macmillan) (Instruments interferometry and holography).
Boyd, R.W. 1992. *Nonlinear Optics* (New York: Academic).
Collett, E. 1993.*Polarised Light: Fundamentals and Applications* (New York: Dekker).
Dakin, J.P. 1990. *The Distributed Fibre-Optic Sensing Handbook* (Berlin: Springer).
Grattan, K.T.V. and Meggitt, B.T. 1995. Interferometric sensors, optical-fibre current measurement, distributed optical-fibre sensors. *Optical-Fibre Sensor Technology* (London: Chapman and Hall).
Guenther, R.D. 1990. *Modern Optics* (New York: Wiley), chapter 15.
Heavens, O.S. and Ditchburn, R.W. 1991. *Insight into Optics* (New York: Wiley).
Hecht, E. 1987.*Optics*, 2nd ed. (Reading, MA: Addison-Wesley), Chapter 12.
Keiser, G. 1991. *Optical Fiber Communications* (New York: McGraw-Hill).
Kliger, D.S., Lewis, J.W. and Randall, C.E. 1990. *Polarised Light in Optics and Spectroscopy* (New York: Academic).
Lefevre, H. 1993. *The Fiber-Optic Gyroscope* (Boston, MA: Artech House).
Marz, R. 1995. *Integrated Optics: Design and Modelling* (Boston, MA: Artech House).
Midwinter, J.E. 1978. *Optical Fibres for Transmission* (New York: Wiley).
Najafi, S.I. 1992. *Introduction to Glass Integrated Optics* (Boston, MA: Artech House).

Ryan, S. 1995. *Coherent Lightwave Communications Systems* (Boston MA: Artech House).

Shurchiff, W.A. 1962. *Polarised Light: Production and Use* (Cambridge, MA: Harvard University Press) (an excellent introduction).

Siegman, A.E. 1986. *Lasers* (Mills Valley, CA: University Science Books).

Solymar, L. and Walsh, D. 1993. *Lectures on the Electrical Properties of Materials*, 5th ed. (Oxford Science Publications).

Sze, S.M. 1981. *Physics of Semiconductor Devices*, 2nd ed. (New York: Wiley).

Basic concepts in photometry, radiometry, and colorimetry

YOSHI OHNO
National Institute of Standards and Technology

8.1 INTRODUCTION

The term *photometry* refers to measurement of quantities for optical radiation as evaluated according to a standardized human eye response, and therefore, is limited to the visible spectral region (360–830 nm) [1]. Photometry uses either optical radiation detectors constructed to mimic the spectral response of the eye or spectroradiometry coupled with appropriate calculations for weighting by the spectral response of the eye. Typical photometric units include the lumen (luminous flux), the candela (luminous intensity), the lux (illuminance) and the candela per square meter (luminance). On the other hand, measurement of optical radiation at all wavelengths (approximately in the range from 10 nm to 1000 μm including ultraviolet, visible and infrared) is referred to as *radiometry*. The official definition of radiometry [1] is measurement of the quantities associated with radiant energy. Typical radiometric units include the watt (radiant flux), watt per steradian (radiant intensity), watt per square meter (irradiance) and watt per square meter per steradian (radiance). Radiometry often involves spectrally resolved measurements of these quantities as well as spectrally integrated measurements. Similar to photometry, measurement of color of light sources and objects also deals with broadband measurement of the visible radiation and is referred to as *colorimetry*. Colorimetry is ascribed to measurement of light spectra weighted by three standardized spectral weighting functions, one of which is identical to the standardized human eye response used in photometry.

Photometry and colorimetry are essential for evaluation of light sources used for lighting, signaling, displays and other applications where light is seen by the human eye. Light-emitting diodes, for example, are now produced in all color ranges and expected to gain wide acceptance in many applications. This chapter focuses on fundamentals of photometry, but introduces radiometry and colorimetry also because radiometry is closely related to photometry, and colorimetry is also important for evaluation of optoelectronic light sources. Some of the presented materials in this chapter are from Reference [2] by the same author. The terminology used in this chapter follows international standards and recommendations [1,3,4].

8.2 BASIS OF PHYSICAL PHOTOMETRY

8.2.1 Visual response

The primary aim of photometry is to measure light (visible optical radiation) in such a way that the results correlate with what the visual sensation is to a normal human observer exposed to that radiation. Until about 1940, visual comparison techniques of measurements were predominant in photometry. The intensity of one light source is matched to the intensity of another light source using human eyes. In modern photometric practice, measurements are made with photodetectors. This is referred to as physical photometry. In order to achieve the aim of photometry, one must take into account the characteristics of human vision. The relative spectral responsivity of the human eye was first defined by the Commission Internationale de l'Éclairage (CIE) in 1924 [5]. It is called *the spectral luminous efficiency for photopic vision*, with a symbol $V(\lambda)$, defined in the domain from 360 to 830 nm, and is normalized to one at its peak, 555 nm (Figure 8.1). This model has gained wide acceptance. The values were republished by CIE in 1983 [6], and adopted by Comité International des Poids et Mesures (CIPM) in 1983 [7] to supplement the 1979 definition of the candela. The tabulated values of the function at 1 nm increments are available in References [6–8]. In most cases, the region from 380 to 780 nm suffices for calculation with negligible errors because the value of the $V(\lambda)$ function falls below 10^{-4} outside this region.

As specified in the definition of the candela by Conference Générale des Poids et Mesures (CGPM) in 1979 [9] and a supplemental document from CIPM in 1982 [10], a photometric quantity X_v is now defined in relation to the corresponding radiometric quantity $X_{e,\lambda}$ by the equation:

$$X_v = K_m \int_{360\,nm}^{830\,nm} X_{e,\lambda}\, V(\lambda)\, d\lambda. \qquad (8.1)$$

The constant, K_m, relates the photometric quantities and radiometric quantities, and is called the

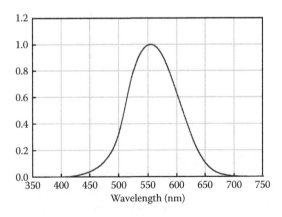

Figure 8.1 CIE $V(\lambda)$ function.

maximum spectral luminous efficacy (of radiation) *for photopic vision.* The value of K_m is given in the 1979 definition of candela, which defines the spectral luminous efficacy of radiation at the frequency 540×10^{12} Hz (at the wavelength 555.016 nm in standard air) to be 683 lm W^{-1}. Note that this is not exactly at the peak of the $V(\lambda)$, 555 nm. The value of K_m is calculated as $683 \times V(555.000$ nm$)/V(555.016$ nm$) = 683.002$ lm W^{-1} [6]. K_m is normally rounded to 683 lm W^{-1} with negligible errors.

It should be noted that the $V(\lambda)$ function is defined for the *CIE standard photometric observer for photopic vision*, which assumes additivity of sensation and a 2° field of view at relatively high luminance levels (higher than approximately 1 cd m^{-2}). The human vision in this level is called *photopic* vision. The spectral responsivity of human eyes deviates significantly at very low levels of luminance (at luminance levels less than approximately 10^{-3} cd m^{-2}) when the rods in the eyes are the dominant receptors. This type of vision is called *scotopic* vision. Its spectral responsivity, peaking at 507 nm, is designated as $V'(\lambda)$, and was defined by CIE in 1951 [11], recognized by CIPM in 1976 [12] and republished by CIPM in 1983 [7]. The human vision in the region between photopic vision and scotopic vision is called *mesopic* vision. While there have been active researches in this area [13], there is no internationally accepted spectral luminous efficiency function for the mesopic region yet. In current practice, almost all photometric quantities are given in terms of photopic vision, even at such low light levels. Quantities in scotopic vision are seldom used except for special calculations for research purposes. Further details of the contents in this section are available in Reference [6].

8.2.2 Photometric base unit, the candela

The history of photometric standards dates back to the early nineteenth century, when the intensity of light sources was measured in comparison with a standard candle using visual bar photometers [14]. At that time, the flame of a candle was used as a unit of luminous intensity that was called *the candle*. The old name for luminous intensity "candle power" came from this origin. Standard candles were gradually superseded by flame standards of oil lamps, and in 1920, the unit of luminous intensity, recognized as *the international candle*, was

adopted by the CIE. In 1948, it was adopted by the CGPM with a new Latin name "candela" defined as the luminous intensity of a platinum blackbody at its freezing temperature under specified geometry. Although the 1948 definition served to establish the uniformity of photometric measurements in the world, difficulties in fabricating the blackbodies and in improving accuracy were addressed. In 1979, the candela was redefined in relation to the optical power, watt, so that complicated source standards would not be necessary. The current definition of the candela adopted in 1979 by the CGPM [9] is

> The candela is the luminous intensity, in a given direction, of a source that emits monochromatic radiation of frequency 540×10^{12} Hz and that has a radiant intensity in that direction of (1/683) watt per steradian.

The value of K_m (683 lm W^{-1}) was determined in such a way that the consistency from the prior unit was maintained, and was determined based on the measurements by several national laboratories. Technical details on this redefinition of the candela are reported in References [15,16]. This 1979 redefinition of the candela has enabled the derivation of the photometric units from the radiometric units using various techniques.

8.3 QUANTITIES AND UNITS IN PHOTOMETRY AND RADIOMETRY

In 1960, the Système International (SI) was established, and the candela became one of the seven SI base units [17]. For further details on the SI, References [17–20] can be consulted. Several quantities and units, defined in different geometries, are used in photometry and radiometry. Table 8.1 lists the photometric quantities and units, along with corresponding radiometric quantities and units.

While the candela is the SI base unit, the luminous flux (lumen) is perhaps the most fundamental photometric quantity, as the other photometric quantities are defined in terms of lumen with an appropriate geometric unit. The definitions of these photometric quantities are described later. The descriptions given here are occasionally simplified from the definitions given in official reference [1]

Table 8.1 Quantities and units used in photometry and radiometry

Photometric quantity	Unit	Relationship with lumen	Radiometric quantity	Unit
Luminous flux	lm (lumen)		Radiant flux	W (watt)
Luminous intensity	cd (candela)	lm sr^{-1}	Radiant intensity	Wsr
Illuminance	lx (lux)	lm m^{-2}	Irradiance	Wm^{-2}
Luminance	cd m^{-2}	lm sr^{-1} m^{-2}	Radiance	Wsr^{-1} m^{-2}
Luminous exitance	lm m^{-2}		Radiant exitance	Wm^{-2}
Luminous exposure	lx s		Radiant exposure	Wm^{-2}s
Luminous energy	lm s		Radiant energy	J (joule)
Total luminous flux	lm (lumen)		Total radiant flux	W (watt)
Color temperature	K (kelvin)		Radiance temperature	K (kelvin)

for easier understanding. Refer to this reference for official, rigorous definitions.

8.3.1 Radiant flux and luminous flux

Radiant flux (also called *optical power* or *radiant power*) is the energy Q (in joules) radiated by a source per unit of time, expressed as

$$\Phi = \frac{dQ}{dt}. \tag{8.2}$$

The unit of radiant flux is the watt (W = Js^{-1}).

Luminous flux (Φ_v) is the time rate of flow of light as weighted by $V(\lambda)$. The unit of luminous flux is the lumen (lm). It is defined as

$$\Phi_v = K_m \int_\lambda \Phi_{e\lambda} V(\lambda) d\lambda \tag{8.3}$$

where $\Phi_{e\lambda}$ is the spectral concentration of radiant flux as a function of wavelength λ. The term, luminous flux, is often used in the meaning of total luminous flux (see Section 8.3.8) in photometry.

8.3.2 Radiant intensity and luminous intensity

Radiant intensity (I_e) or *luminous intensity* (I_v) is the radiant flux (luminous flux) from a point source emitted per unit solid angle in a given direction, as defined by

$$I = \frac{d\Phi}{d\Omega} \tag{8.4}$$

where $d\Phi$ is the radiant flux or luminous flux leaving the source and propagating in an element of solid angle $d\Omega$ containing the given direction. The unit of radiant intensity is W sr^{-1}, and that of luminous intensity is the candela (cd = lm sr^{-1}) (Figure 8.2).

The radiant intensity or luminous intensity of a real light source varies with the direction of emission, and is specified or measured for given direction(s) from the light source. In real measurements, the solid angle $d\Omega$ will be a finite solid angle defined by the area A of the detector surface and distance r from the source (see Figure 8.3), with $d\Phi$ being the flux incident on the detector surface. The distance r should be large enough so that the source can be assumed as a point source (far field condition) and the detector area A should be small enough so that $d\Omega$ is considered negligible. Measurement results tend to vary depending on geometrical conditions if these conditions are not met. See also Section 8.4.1 for practical aspects of the geometry and Reference [30] for the specific

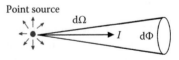

Figure 8.2 Radiant intensity and luminous intensity.

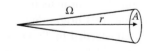

Figure 8.3 Solid angle.

measurement geometries recommended for measurement of luminous intensity of LEDs.

8.3.2.1 SOLID ANGLE

The solid angle (Ω) of a cone is defined as the ratio of the area (A) cut out on a spherical surface (with its center at the apex of that cone) to the square of the radius (r) of the sphere, as given by

$$\Omega = \frac{A}{r^2}. \tag{8.5}$$

The unit of solid angle is steradian (sr), which is a dimensionless unit.

8.3.3 Irradiance and illuminance

Irradiance (E_e) or *illuminance* (E_v) is the density of incident radiant flux or luminous flux at a point on a surface, and is defined as radiant flux or luminous flux per unit area, as given by

$$E = \frac{d\Phi}{dA} \tag{8.6}$$

where $d\Phi$ is the radiant flux or luminous flux incident on an element dA of the surface containing the point. The unit of irradiance is Wm^{-2}, and that of illuminance is lux (lx=lm m^{-2}) (Figure 8.4).

By definition, $d\Phi$ incident on dA from any angles within the 2π solid angle above the surface contributes to E. Thus, illuminance meters are normally designed to receive light from 2π solid angle ($\pm 90°$) with its angular responsivity tailored to follow the cosine function response.

8.3.4 Radiance and luminance

Radiance (L_e) or *luminance* (L_v) is the radiant flux or luminous flux per unit solid angle emitted from a surface element in a given direction, per unit projected area of the surface element perpendicular to the direction (Figure 8.5). The unit of radiance is

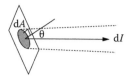

Figure 8.5 Radiance and luminance.

W sr^{-1} 1m^{-2}, and that of luminance is cd m^{-2}. These quantities are defined by

$$L = \frac{d^2\Phi}{d\Omega dA \cos\theta} \tag{8.7}$$

where $d\Phi$ is the radiant flux (luminous flux) emitted (reflected or transmitted) from the surface element and propagating in the solid angle $d\Omega$ containing the given direction. dA is the area of the surface element and θ the angle between the normal to the surface element and the direction of the beam. The term $dA \cos\theta$ gives the projected area of the surface element perpendicular to the direction of measurement.

In photometry, luminance is an important parameter in a sense that it represents how bright objects look to the human eyes. Luminance is important, e.g., to specify the brightness of visual displays. Luminance meters are normally constructed using an imaging optics (like a camera lens) to focus on an object surface and accept light only from a given cone angle (e.g., 0.3°, 1°, 3°, etc.) from the meter.

8.3.5 Radiant exitance and luminous exitance

Radiant exitance (M_e) or *luminous exitance* (M_v) is defined to be the density of radiant flux or luminous flux leaving a surface at a point. The unit of radiant exitance is W m^{-2} and that of luminous exitance is lm m^{-2} (but it is not lux) (Figure 8.6). These quantities are defined by

$$M = \frac{d\Phi}{dA} \tag{8.8}$$

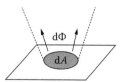

Figure 8.6 Radiant exitance and luminous exitance.

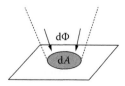

Figure 8.4 Irradiance and illuminance.

where $d\Phi$ is the radiant flux or luminous flux leaving the surface element. Luminous exitance is rarely used in the general practice of photometry.

8.3.6 Radiant exposure and luminous exposure

Radiant exposure (H_e) or *luminous exposure* (H_v) is the time integral of irradiance $E_e(t)$ or illuminance $E_v(t)$ over a given duration Δt, as defined by

$$H = \int_{\Delta t} E(t)\,dt. \qquad (8.9)$$

The unit of radiant exposure is J m^{-2}, and that of luminous exposure is lux second (lxs). These quantities are often used for pulsed radiation.

8.3.7 Radiant energy and luminous energy

Radiant energy (Q_e) or *luminous energy* (Q_v) is the time integral of the radiant flux $\Phi_e(t)$ or luminous flux $\Phi_v(f)$ over a given duration Δt, as defined by

$$Q = \int_{\Delta t} \Phi(t)\,dt. \qquad (8.10)$$

The unit of radiant energy is joule (J), and that of luminous energy is lumen second (lm s). These quantities are often used for pulsed radiation. Luminous energy is also called *quantity of light*, which is listed in Reference [1].

8.3.8 Total radiant flux and total luminous flux

Total radiant flux or *total luminous flux* (Φ_v) is the geometrically total radiant flux or luminous flux of a light source. It is defined as

$$\Phi = \int_{\Omega} I\,d\Omega. \qquad (8.11)$$

or

$$\Phi = \int_{A} E\,dA. \qquad (8.12)$$

where I is the radiant or luminous intensity distribution of the light source and E the irradiance or

illuminance distribution over a given closed surface surrounding the light source. If the radiant or luminous intensity distribution, or the irradiance or illuminance distribution, is given in polar coordinates (θ, ϕ), the total radiant flux or luminous flux Φ of the light source is given by

$$\Phi = \int_{\phi=0}^{2\pi} \int_{\theta=0}^{\pi} I(\theta,\phi)\sin\theta\,d\theta\,d\phi \qquad (8.13)$$

or

$$\Phi = r^2 \int_{\phi=0}^{2\pi} \int_{\theta=0}^{\pi} E(\theta,\phi)\sin\theta\,d\theta\,d\phi. \qquad (8.14)$$

For example, the total luminous flux of an isotropic point source having luminous intensity of 1 cd is 4π lumens.

Total luminous flux (lumen) is a very important quantity to specify lamp products. It represents how much visible light a lamp can produce (for the given wattage of the lamp), no matter what the intensity distributions are. The ratio of the total luminous flux to the input electrical power (lumens per watt) for a light source—called *luminous efficacy of a light source*—is an important parameter concerned with energy saving, and thus, the importance of accurate measurement of total luminous flux. The total luminous flux of light sources is normally measured either with an integrating sphere photometer (see Section 8.4.5) or a goniophotometer (see Reference [25]).

8.3.9 Radiance temperature and color temperature

Radiance temperature (unit: kelvin) is the temperature of the Planckian radiator for which the radiance at the specified wavelength has the same spectral concentration as for the thermal radiator considered. It is commonly used for blackbodies and spectral radiance standard lamps.

Color temperature (unit: kelvin) is the temperature of a Planckian radiator emitting radiation of the same chromaticity (see Section 8.6.2) as that of the light source in question. However, actual light sources other than blackbodies rarely have exactly the same chromaticity as a Planckian radiator. Therefore, for various lamps used in general lighting (such as fluorescent lamps and other discharge lamps), another term "correlated color temperature" is used. See Section 8.6.3 for further details.

Distribution temperature (unit: kelvin) is the temperature of a blackbody with a spectral power distribution closest to that of the light source in question, and is used for quasi-Planckian sources such as incandescent lamps. Refer to Reference [21] for details.

8.3.10 Relationship between SI units and inch–pound system units

The SI units as described earlier should be used in all radiometric and photometric measurements according to international standards and recommendations on SI units. However, some inch–pound system units are still being used in many application areas. The use of these non-SI units is discouraged. The definitions of such inch–pound system units used in photometry are given in Table 8.2 for conversion purposes only.

The definition of foot–lambert is such that the luminance of a perfect diffuser is 1 fL when

Table 8.2 Inch–pound system units and their definitions

| Foot–candle (fc) | Illuminance | Lumen per square foot (lm ft^{-2}) |
| Foot–lambert (fL) | Luminance | 1/π candela per square foot (π^{-1} cd ft^{-2}) |

Note: The use of these non-SI units is discouraged.

Table 8.3 Conversion between inch–pound system units and SI units

To obtain the value in	Multiply the value in	By
lx from fc	fc	10.764
fc from lx	lx	0.09290
cd m^{-2} from fL	fL	3.4263
fL from cd m^{-2}	cd m^{-2}	0.29186
m (meter) from feet	Feet	0.30480
mm (millimeter) from inch	Inch	25.400

Note: The use of these non-SI units is discouraged.

illuminated at 1 fc. In the SI unit, the luminance of a perfect diffuser would be 1/π (cd m^{-2}) when illuminated at 1 lx. For convenience of changing from these inch–pound system units to SI units, the conversion factors are listed in Table 8.3. For example, 1000 lx is the same illuminance as 92.9 fc, and 1000 cd m^{-2} is the same luminance as 291.9 fL. Conversion factors to and from many other units are available in Reference [22].

8.4 PRINCIPLES IN PHOTOMETRY AND RADIOMETRY

Several important theories in practical photometry and radiometry are introduced in this section.

8.4.1 Inverse square law

Illuminance E (lx) at a distance d (m) from a point source having luminous intensity I (cd) is given by

$$E = \frac{1}{d^2}. \qquad (8.15)$$

For example, if the luminous intensity of a lamp in a given direction is 1000 cd, the illuminance at 2 m from the lamp in this direction is 250 lx. Or, the luminous intensity I of a lamp is obtained by measurement of illuminance E at distance d from the light source. Note that the inverse square law is valid only when the light source is regarded as a point source. Sufficient distances relative to the size of the source are needed to assume this relationship.

8.4.2 Lambert's cosine law

The luminous intensity of a Lambertian surface element is given by

$$I(\theta) = I_n \cos\theta. \qquad (8.16)$$

Lambertian surface: A surface whose luminance is the same in all directions of the hemisphere above the surface. The total luminous flux Φ of a Lambertian surface shown in Figure 8.7 is given by

$$\Phi = \pi I_n = \pi La \qquad (8.17)$$

where L is the luminance of the surface and a the area of the surface.

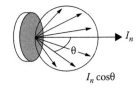

Figure 8.7 Lambert's cosine law.

Perfect (reflecting/transmitting) diffuser: A Lambertian diffuser with a reflectance (transmittance) equal to 1.

8.4.3 Relationship between illuminance and luminance

The luminance L(cd m^{-2}) of a perfect diffuser illuminated by E (lx) is given by (Figure 8.8)

$$L = \frac{E}{\pi} \qquad (8.18)$$

and, for a Lambertian surface of reflectance ρ,

$$L = \frac{\rho E}{\pi}. \qquad (8.19)$$

8.4.3.1 REFLECTANCE (ρ)

The ratio of the reflected flux to the incident flux in a given condition. The value of ρ can be between 0 and 1.

In the real world, there is neither existing perfect diffuser nor perfectly Lambertian surfaces, and Equation 8.19 does not apply for real surfaces. For real object surfaces, the following terms apply.

8.4.3.2 LUMINANCE FACTOR (β)

Ratio of the luminance of a surface element in a given direction to that of a perfect reflecting or

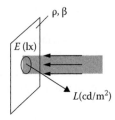

Figure 8.8 Relationship between illuminance and luminance.

transmitting diffuser under specified conditions of illumination. The value of β can be larger than 1. For a Lambertian surface, reflectance is equal to the luminance factor. Equation 8.19 for a real object is restated using β as

$$L = \frac{\beta E}{\pi}. \qquad (8.20)$$

The radiometric term for this quantity is *radiance factor*. Also *reflectance factor* is often used for the same meaning. However, reflectance factor is officially defined for reflected flux in a given cone angle, and it can mean radiance factor (solid angle=0) or reflectance (solid angle=2π). See Reference [1] for the details.

Note that radiance factor and luminance factor of a real surface vary depending on the illumination and viewing geometry, and thus the geometry must be specified. For example, $\beta_{0/45}$ means radiance (luminance) factor at a $0°$ incident angle and a $45°$ viewing angle. See Reference [40] for such geometrical notations. Luminance factor also depends on the spectral distribution of the illumination, and thus it must be specified. CIE standard illuminant A or standard illuminant D65 [41] is normally used. Furthermore, radiance factor is given as a function of wavelength.

8.4.3.3 LUMINANCE COEFFICIENT (q)

Quotient of the luminance of a surface element in a given direction by the illuminance on the surface element under specified conditions of illumination.

$$q = \frac{L}{E}. \qquad (8.21)$$

Using q, the relationship between luminance and illuminance is thus given by

$$L = qE. \qquad (8.22)$$

The radiometric term for this quantity is *radiance coefficient*, used either for light-reflecting or transmitting diffuser materials. The bidirectional reflectance distribution function (BRDF) is also used for the same definition as the radiance coefficient (but expressed as a function of angle of incidence and angle of reflection).

8.4.4 Planck's law

The spectral radiance of a blackbody at a temperature $T(K)$ is given by

$$L_e(\lambda, T) = c_1 n^{-2} \pi^{-1} \lambda^{-5} \left[\exp\left(\frac{c_2}{n\lambda T} \right) - 1 \right]^{-1} \quad (8.23)$$

where $c_1 = 2\pi hc^2 = 3.74177107 \times 10^{-16}$ W m^2, $c_2 = hc/k = 1.4387752 \times 10^{-2}$ m K (1998 CODATA from Reference [23]), h is Planck's constant, c the speed of light in vacuum, k Boltzmann's constant, n the refractive index of medium and A the wavelength in the medium, $n = 1.00028$ in standard air [6,24].

8.4.5 Principles of integrating sphere

An integrating sphere is a device to achieve spatial integration of luminous flux (or radiant flux) generated (or introduced) in the sphere. In the case of measurement of light sources, the spatial integration is made over the entire solid angle (4π). In Figure 8.9, assuming that the integrating sphere wall surfaces are perfectly Lambertian, luminance L of a surface element Δa (generated by uniform light incident on this element) creates the equal illuminance E on any part of the sphere surfaces, as given by

$$E = \frac{L\Delta a}{4R^2}. \quad (8.24)$$

In other words, the same amount of flux incident anywhere on the sphere wall will create an equal illuminance on the detector port. In the case of actual integrating spheres, the surface is not perfectly Lambertian, but due to interreflections of light in the sphere, the distribution of reflected light will be sufficiently uniform to approximate the condition assumed in Equation 8.24.

8.4.5.1 INTEGRATING SPHERE PHOTOMETER

By operating a light source in an integrating sphere as shown in Figure 8.10, the total luminous flux of the light source is measured using one detector on the sphere wall. Such a device is called an integrating sphere photometer (or Ulbricht sphere). The direct light from an actual light source is normally not uniform, and thus must be shielded from the detector using a baffle. When a light source with luminous flux Φ is operated in a sphere having reflectance ρ, the flux created by interreflections is given by

$$\Phi(\rho + \rho^2 + \rho^3 + \ldots) = \Phi \frac{\rho}{1-\rho}. \quad (8.25)$$

Then, the illuminance E_d created by all the interreflections is given by

$$E_d = \frac{\Phi\rho}{1-\rho} \frac{1}{4\pi R^2}. \quad (8.26)$$

The sphere efficiency (E_d/Φ) is strongly dependent on reflectance ρ due to the term $1-\rho$ in the denominator, where a high reflectance coating such as $\rho = 0.98$ is often used. For this reason, E_d cannot be predicted accurately enough to determine Φ. Real integrating sphere photometers are used as a relative device to measure test lamps against standard lamps whose luminous flux is known. For further details of the integrating sphere photometer, refer to References [25,26].

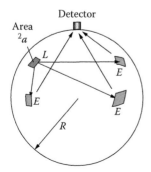

Figure 8.9 Flux transfer in an integrating sphere.

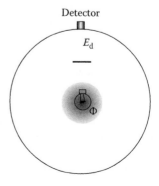

Figure 8.10 Integrating sphere photometer.

8.5 PRACTICE IN PHOTOMETRY AND RADIOMETRY

Photometry and radiometry are practiced in many different areas and applications, dealing with various light sources and detectors, and cannot be covered in this chapter. Various references are available on practical measurements in photometry and radiometry. References [26,27] provide the latest information on standards and practical aspects in photometry. There are a number of publications from CIE on many subjects in photometry including characterization of illuminance meters and luminance meters [28], luminous flux measurement [25], spectroradiometry [29], measurements of LEDs [30], etc. The latest list of CIE publications is available on-line [31]. A series of measurement guide documents are published from the Illuminating Engineering Society of North America (IESNA) for operation and measurement of specific types of lamps [32–34] and luminaires. The American Society for Testing and Materials (ASTM) provides many useful standards and recommendations on optical properties of materials and color measurements [35]. There are also a number of publications from the National Institute of Standards and Technology (NIST) on photometry [27], spectral irradiance [36], spectral reflectance [37], spectral responsivity [38], etc.

8.6 FUNDAMENTALS OF COLORIMETRY

8.6.1 Color matching functions and tristimulus values

The perception of color is a psychophysical phenomenon, and the measurement of color must be defined in such a way that the results correlate accurately with what the visual sensation of color is to a normal human observer. *Colorimetry* is the measurement science used to quantify and describe physically the human color perception. The basis of colorimetry was established by CIE in 1931 based on a number of visual experiments that defined a set of three spectral weighting functions [39]. These functions, shown in Figure 8.11, are called the *CIE 1931 XYZ color matching functions* denoted as $\bar{x}(\lambda), \bar{y}(\lambda), \bar{z}(\lambda)$. These functions were derived from a linear transformation of the

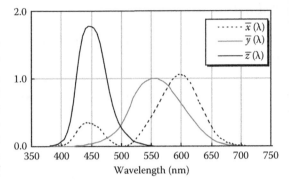

Figure 8.11 CIE 1931 XYZ color matching functions.

original set of color matching functions in such a way that $\bar{y}(\lambda)$ is equal to $V(\lambda)$.

By using the color matching functions, light stimuli having any spectral power distribution $\phi_\lambda(\lambda)$ can be specified for color by three values:

$$X = k \int_\lambda \phi_\lambda(\lambda)\bar{x}(\lambda)d\lambda, \quad Y = k \int_\lambda \phi_\lambda(\lambda)\bar{y}(\lambda)d\lambda,$$

$$Z = k \int_\lambda \phi_\lambda(\lambda)\bar{z}(\lambda)d\lambda \qquad (8.27)$$

where $\phi_\lambda(\lambda)$ is the spectral distribution of light stimulus and k a normalizing constant. These integrated values are called *tristimulus values*. Two light stimuli having the same tristimulus values have the same color even if the spectral distributions are different. For light sources and displays, $\phi_\lambda(\lambda)$ is given in quantities such as spectral irradiance and spectral radiance. If $\phi_\lambda(\lambda)$ is given in an absolute unit (such as $Wm^{-2}nm^{-1}$ Wm^{-2} $sr^{-1}nm^{-1}$) and $k=683$ lm W^{-1} is chosen, Y yields an absolute photometric quantity such as illuminance (in lux) or luminance (in cd m^{-2}). For object colors, $\phi_\lambda(\lambda)$ is given as

$$\phi_\lambda(\lambda) = R(\lambda)S(\lambda) \qquad (8.28)$$

where $R(\lambda)$ is the spectral reflectance factor of the object, $S(\lambda)$ the relative spectral distribution of the illumination, and

$$k = \frac{100}{\int_\lambda S(\lambda)\bar{y}(\lambda)d\lambda} \qquad (8.29)$$

so that $Y=100$ for a perfect diffuser and Y indicates the luminance factor (in %) of the object surface.

To calculate color of objects from spectral reflectance factor $R(\lambda)$, one of the standard illuminants (see Section 8.6.3) is used.

Tristimulus values can be obtained either by numerical summation of Equation 8.27 from the spectral data $\phi_\lambda(\lambda)$ obtained by a spectroradiometer or spectrophotometer, or by broadband measurements using detectors having relative spectral responsivity matched to the color matching functions. Such a device using three (or four) detector channels is called *tristimulus colorimeter*.

When applying colorimetric data for real visual color matching, it should be noted that the $\bar{x}(\lambda), \bar{y}(\lambda), \bar{z}(\lambda)$. color matching functions are based on experiments using $2°$ field of view and applicable only to narrow fields of view (up to $4°$). Such an ideal observer is called the *CIE 1931 standard colorimetric observer*. In 1964, the CIE defined a second set of standard color matching functions for a $10°$ field of view, denoted as $\bar{x}_{10}(\lambda), \bar{y}_{10}(\lambda), \bar{z}_{10}(\lambda)$, to supplement those of the 1931 standard observer. This is called the *CIE 1964 supplementary standard colorimetric observer*, and can be used for a field of view greater than $4°$. The $2°$ observer is used in most applications for colorimetry of light sources. The $10°$ observer is often used in object color measurements. For further details of colorimetry and color science, refer to official CIE publications [40–42] and many other general Reference [43].

8.6.2 Chromaticity diagrams

While the tristimulus values can specify color, it is difficult to associate what color it is from the three numbers. By projecting the tristimulus values onto a unit plane $(X+Y+Z=1)$, color of light can be expressed on a two-dimensional plane. Such a unit plane is known as the *chromaticity diagram*. The color can be specified by the *chromaticity coordinates* (x, y) defined by

$$x = \frac{X}{X+Y+Z}, \quad y = \frac{Y}{X+Y+Z}. \quad (8.30)$$

The diagram using the chromaticity coordinates (x, y), as shown in Figure 8.12a, is referred to as the *CIE 1931 chromaticity diagram*, or the *CIE(x, y) chromaticity diagram*. The boundaries of this horseshoe-shaped diagram are the plots of monochromatic radiation (called the *spectrum locus*).

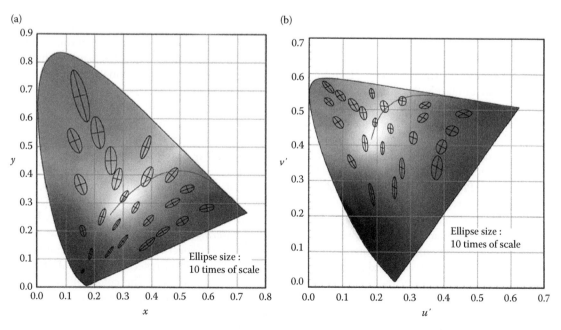

Figure 8.12 (a) MacAdam ellipses on CIE 1931(x, y) chromaticity diagram and (b) MacAdam ellipses on the CIE 1976 (u′v′) chromaticity diagram. The ellipses are plotted 10 times their actual size. The curve near the center region is the Planckian locus.

The (x, y) chromaticity diagram is significantly nonuniform in terms of color difference. The minimum perceivable color differences in the CIE (x, y) diagram, known as the *MacAdam ellipses*, are shown in Figure 8.12a. To improve this, in 1960, CIE defined an improved diagram—*CIE 1960 (u, v) chromaticity diagram* (now obsolete), and in 1976, a further improved diagram—*CIE 1976 uniform chromaticity scale (UCS) diagram*, as shown in Figure 8.12b, with its chromaticity coordinate (u', v') given by

$$u' = \frac{4X}{X+15Y+3Z}, \quad v' = \frac{9Y}{X+15Y+3Z}. \quad (8.31)$$

While the (u', v') chromaticity diagram is a significant improvement from the (x, y) diagram, it is still not satisfactorily uniform. Both of these diagrams are widely used. Note that these chromaticity diagrams are intended to present color of light sources (emitted light) and not color of objects (reflected light). Presentation of object colors requires a three-dimensional color space that incorporates another dimension—lightness (black to white). Refer to other publications [40, 43] for the details of object color specification.

8.6.3 Color temperature and correlated color temperature

Figure 8.13 shows the trace of the (x, y) chromaticity coordinate of blackbody radiation (see Section 8.4.4) at its temperature from 1600 to 20,000 K. This trace is called the *Planckian locus*. The colors on the Planckian locus can be specified by the blackbody temperature in kelvin and is called color *temperature* (see also Section 8.3.9). The colors around the Planckian locus from about 2500 to 20,000 K can be regarded as *white*, 2500 K being reddish white and 20,000 K being bluish white. The point labeled "Illuminant A" is the typical color of an incandescent lamp, and "Illuminant D65" the typical color of day light, as standardized by the CIE [41]. The colors of most traditional lamps for general lighting fall in the region between these two points (2800–6500 K). Strictly speaking, color temperature cannot be used for colors away from the Planckian locus, in which case *correlated color temperature* (CCT) is used. CCT is the temperature of the blackbody whose perceived color

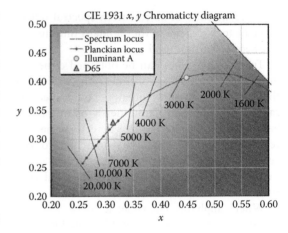

Figure 8.13 Planckian locus on (x, y) chromaticity diagram and iso-CCT lines.

most closely resembles that of the light source in question [2]. Due to the nonuniformity of the x, y diagram, the iso-CCT lines are not perpendicular to the Planckian locus on the x, y diagram (see Figure 8.13). To calculate CCT, therefore, one of the improved uniform chromaticity diagrams is used. Due to the long tradition, CIE specifies that the 1960 (u, v) diagram (now obsolete for other purposes) be used, where the iso-CCT lines are perpendicular to the Planckian locus by definition. From (u', v') coordinates, (u, v) can be obtained by $u=u'$, $v=2v'/3$. On the (u, v) diagram, find the point on the Planckian locus that is at the shortest distance from the given chromaticity point. CCT is the temperature of the Planck's radiation at that point. A practical way of computing CCT is available [44].

8.6.4 Color rendering index

For light sources for lighting applications, it is important to evaluate how well their illumination can render colors of objects. The CIE defines the *color rendering index* (CRI) [45]. The CRI is calculated from the spectral distribution of light under test and the spectral reflectance factor data of 14 Munsell color samples. The color difference ΔE_i (on the 1964 $W*U*V*$ uniform color space—now obsolete) of each sample illuminated by the light under test and by a reference source (Planckian radiation for CCT < 5000 K or a daylight illuminant for CCT ≥ 5000 K) is calculated taking into account chromatic adaptation. The *special color* *rendering index* R_i for each color sample is calculated by

$$R_i = 100 - 4.6\Delta E_i \qquad (8.32)$$

This gives an indication of color rendering for each particular color. The *general color rendering index*, R_a, is given as the average of the first eight color samples (medium saturation). With the maximum value being 100, R_a gives a scale that expresses well the visual impression of color rendering. For example, lamps having R_a values greater than 80 may be considered suitable for interior lighting, and R_a greater than 95 for visual inspection purposes, etc. See Reference [45] for further details.

8.6.5 Color quantities for LEDs

In addition to chromaticity coordinates x, y and u', v', the following quantities are used to specify the color and spectrum of LEDs. The definitions in this section follow Reference [30].

Peak wavelength λ_p: The wavelength at the maximum of the spectral distribution.

Spectral bandwidth (at half intensity level) $\Delta\lambda_{0.5}$: Calculated as the width between the wavelengths at half of the peak of spectral distribution, as shown in Figure 8.14. It is also denoted as $\Delta\lambda$(FWHM).

Centroid wavelength λ_c: Calculated as the "center of gravity wavelength," according to the equation

$$\lambda_c = \frac{\int_\lambda \lambda S(\lambda) d\lambda}{\int_\lambda S(\lambda) d\lambda} \qquad (8.33)$$

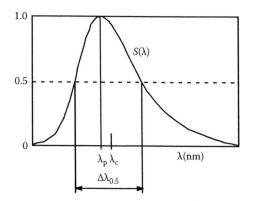

Figure 8.14 Typical relative spectral distribution of an LED.

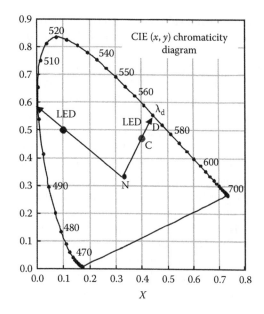

Figure 8.15 (x, y) chromaticity diagram showing the dominant wavelength and excitation purity.

Dominant wavelength λ_d: Wavelength of the monochromatic stimulus that, when additively mixed in suitable proportions with the specified achromatic stimulus, matches the color stimulus considered. Equal energy spectrum with $(x, y) = (0.3333, 0.3333)$ is used as the achromatic stimulus. See Figure 8.15, where N denotes the achromatic stimulus.

Excitation purity p_e: Defined as the ratio NC/ND in Figure 8.15. The value of excitation purity is unity if the chromaticity of the LED is on the spectrum locus.

8.6.6 Spectroradiometry for LED color measurement

Spectroradiometers are commonly used for color measurement of LEDs. Tristimulus colorimeters are rarely used for LED color measurements because errors tend to be too large for measurement of quasi-monochromatic sources such as LEDs. Major sources of error in spectroradiometric measurements of LEDs are bandwidth, wavelength error and stray light. To obtain the color quantities of LEDs accurately, a monochromator bandwidth (FWHM) of 5 nm or less is recommended. The scanning interval should be matched with the bandwidth, or much smaller than the bandwidth. If the bandwidth of the instrument is larger than

5 nm, the bandwidth error can be very large. The errors with a 10 nm bandwidth can be up to 0.005 in x, y depending on the LED peak wavelength [46]. A method for correcting the bandwidth error is available in the case where the bandwidth is a triangular shape and is matched with the scanning interval [47]. The wavelength accuracy is also critical for LED color measurements. An error for 1 nm in wavelength scale would lead to a maximum error of 0.02 in x, y, depending on the LED peak wavelength. The stray light of a monochromator is not very critical for color measurement of broadband white light sources such as fluorescent lamps, but it is critical for LEDs having narrowband emissions. The errors should be examined for single monochromators including diode-array spectroradiometers. For the details of spectroradiometry in general, refer to other References [29,48].

REFERENCES

1. *International Lighting Vocabulary.* 1987. (CIE Publication No.17.4 /IEC Publication 50 (845)).
2. Ohno, Y. 2001. Chapter 14, Photometry and radiometry—review for vision optics, Part 2 vision optics. In *OSA Handbook of Optics*, vol. 3 (New York: McGraw-Hill).
3. *International Vocabulary of Basic and General Terms in Metrology.* 1994. (BIsssPM, IEC, IFCC, ISO, IUPAC, IUPAP, and OIML).
4. *Quantities and Units: ISO Standards Handbook.* 1993. 3rd ed.
5. CIE Compte Rendu. 1924. p. 67.
6. *The Basis of Physical Photometry.* 1983. (CIE Publication No.18.2).
7. *Principles Governing Photometry.* 1983. (Sèvres, France: Bureau International Des Poids et Mesures (BIPM) Monograph, BIPM).
8. *CIE Disk D001 Photometric and Colorimetric Tables.* 1988.
9. *CGPM, Comptes Rendus des Séances de la 16e Conférence Générale des Poids et Mesures.* 1979. (Sèvres, France: Paris 1979, BIPM).
10. *CIPM, Comité Consultatif de Photométrie et Radiométrie 10e Session–1982.* 1982. (Sèvres, France: BIPM, Pavillon de Breteuil).
11. *CIE Compté Rendu.* 1951. vol. 3, Table II, pp. 37–39.
12. *CIPM Procés-Verbaux 44.* 1976. p. 4.
13. *Mesopic Photometry: History, Special Problems and Practical Solutions.* 1989. (CIE Publication No. 81).
14. Walsh, J. W. T. 1953. *Photometry* (London: Constable).
15. Blevin, W. R. and Steiner, B. 1975. *Metrologia*, 11, 97.
16. Blevin, W. R. 1979. The candela and the watt. *CIE Proc.* P-79-02.
17. *Le Système International d'Unité (SI), The International System of Units (SI).* 1991. 6th edn (Sèvres, France: Bur. Intl. Poids et Mesures).
18. Taylor, B. N. 1995. *Guide for the Use of the International System of Units (SI).* (National Institute of Standards and Technology Special Publication 811).
19. Taylor, B. N., ed. 1991. *Interpretation of the SI for the United States and Metric Conversion Policy for Federal Agencies.* (National Institute of Standards and Technology Special Publication 814).
20. *SI Units and Recommendations for the Use of Their Multiples and of Certain Other Units.* 1992. (ISO 1000:1992, International Organization for Standardization, Geneva, Switzerland).
21. *CIE Collection in Photometry and Radiometry.* 1994. (Publication No.114/4).
22. *Appendix, the Lighting Handbook 9th Edition, Illuminating Engineering Society of North America.* 2000.
23. *J. Phys. and Chem. Ref. Data.* 1999. 28, 1713–1852.
24. Blevin, W. R. 1972. Corrections in optical pyrometry and photometry for the refractive index of air. *Metrologia*, 8, 146.
25. *CIE Publication No. 84.* 1989. Measurements of Luminous Flux.
26. C. DeCusatis, ed. 1997. *OSA/AIP Handbook of Applied Photometry* (Woodbury, NY: AIP Press).
27. Ohno, Y. 1997. *Photometric Calibrations*, NIST Special Publication 250-37. This document is available at http://physics.nist.gov/photometry.
28. *Methods of Characterizing Illuminance Meters and Luminance Meters.* 1987. (CIE Publication 69).
29. *The Spectroradiometric Measurement of Light Sources.* 1984. (CIE Publication 63).

30. *Measurement of LEDs.* 1997. (CIE Publication 127).
31. CIE Central Bureau website: http://www.cie.co.at/cie.
32. *IES Approved Method for the Electric and Photometric Measurement of Fluorescent Lamps.* 1999. (IESNA LM-9).
33. *Electrical and Photometric Measurements of General Service Incandescent Filament Lamps.* 2000. (IESNA LM-45).
34. *Electrical and Photometric Measurements of Compact Fluorescent Lamps.* 2000. (IESNA LM-66).
35. *ASTM Standards on Colour and Appearance Measurement.* 1996. 5th ed.
36. Walker, J. H., Saunders, R. D., Jackson, J. K. and McSparron, D. A. 1987. *Spectral Irradiance Calibrations* (NBS Special Publication 250-20).
37. Barnes, P. Y., Early, E. A. and Parr, A. C. 1998. *Spectral Reflectance* (NIST Special Publication 250-41).
38. Larason, T. C., Bruce, S. S. and Parr, A. C. 1998. *Spectroradiometric Detector Measurements* (NIST Special Publication 250-48).
39. *CIE Compte Rendu.* 1931. Table II, pp. 25–26.
40. CIE Publ. No.15.2. 1986. *Colorimetry.* 2nd ed.
41. ISO10526/CIE5005: *CIE standard illuminants for colorimetry.* 1999.
42. ISO/CIE10527–1991, *CIE standard colorimetric observers.* 1991.
43. Wyszecki, G. and Stiles, W. S. 1982. *Colour Science: Concepts and Methods, Quantitative Data and Formulae* (New York: Wiley).
44. Robertson, R. 1968. Computation of correlated colour temperature and distribution temperature. *J. Opt. Soc. Am.* 58, 1528–1535.
45. *Method of Measuring and Specifying Colour Rendering Properties of Light Sources.* 1995. (CIE Publ. No. 13.3).
46. Jones, C. F. and Ohno, Y. 1999. Colorimetric Accuracies and Concerns in Spectroradiometry of LEDs. *Proceedings of CIE Symposium'99—75 Years of CIE Photometry, Budapest*, pp. 173–177.
47. Stearns, E. and Stearns, R. 1988. An example of a method for correcting radiance data for bandpass error. *Color Res. Application*, 13–4, 257–259.
48. Kostkowski, H. J. 1997. *Reliable Spectroradiometry* (La Plata, MD: Spectroradiometry Consulting).

Nonlinear and short pulse effects

GÜNTER STEINMEYER
Max Born Institute

PROLOGUE

The science and technology of ultrafast pulses has, in recent years, come of age, becoming ever more important in the real world. In the early days of lasers, efforts concentrated on getting lasers to operate as continuous light sources. Compared to conventional light sources, continuous-wave (cw) lasers perfectly concentrate light in the plane transverse to the propagation direction, which gives rise to the extreme high brightness of laser beams. Nevertheless, a continuous laser ignores the third dimension along the propagation direction for further possible concentration of energy. Using mode-locking, the light field circulating

through a mode-locked laser can be temporally focused within one millionth of the total cavity length, which translates to a pulse duration of a few femtoseconds compared to a total cavity roundtrip time of several nanoseconds. The peak power, even in an oscillator pulse, readily reaches hundreds of kilowatts. Employing the method of chirped-pulse amplification, one can easily generate pulses with multiple gigawatt peak power, which exceeds the continuous capabilities of several nuclear power plants for the duration of a few femtoseconds. The advantages of having lasers operating with very short pulses are currently being realized more and more in both industry and science. An ultrashort pulse can process material surfaces by action of its extreme electric field strength without causing internal heating or melting. This nonthermal interaction may be applied to ablate material for analysis or to create gaseous phases as desired. In spectroscopy, very short pulses can be used to investigate very fast chemical reactions or trigger fast photochemical reactions and observe features that would be impossible to observe using slower processes, leading to far greater understanding of the science and technology involved. This enables the interim states of such reactions to be observed, whereas previously only the starting materials and end products would be apparent. To make use of this, it is necessary to be able to generate ultrashort pulses and also to be able to measure the pulses that have been created. This chapter describes the science and technology in this rapidly developing field.

9.1 INTRODUCTION

In most parts of this handbook, we deal with *linear optical effects*. Linear optics means that the optical power at the outputs of an optical device always scales linearly with input power. The device may spectrally or spatially filter the input beam; it may split the input beam into a multitude of output beams; regardless of what the device does, the output power always relates linearly to the input power. Looking through textbooks on classical optics from the prelaser era, the impression may arise that the linearity of optical phenomena is a given thing as there is no mention of any *nonlinear effects*. This is in strong contrast, for example, to acoustics, where nonlinearities are so widespread that the art lies more in their avoidance than in the observation of nonlinearities. One may think of a cheap set of speakers just as one simple example. Increasing the volume, these speakers will start to sound increasingly annoying with more and more audible distortion. This distortion is not related to the frequency dependence of the speaker's transmission characteristics because this would be independent of volume. The distortion effect is related to unwanted *harmonics* of the input. These harmonics arise due to nonlinearities between the emitted acoustic wave and the input current to the speaker's solenoid. Beyond a certain drive amplitude, the speaker's membrane position does not linearly follow the current anymore. These harmonics are not necessarily bad. All musical instruments

also rely on acoustic nonlinearities, giving rise to a characteristic spectrum of overtones of the excited fundamental vibration of a chord. This characteristic spectrum allows us to distinguish different musical instruments. The omnipresence of nonlinearities in acoustics is in strong contrast to optics, where similar effects could not be observed until the advent of the laser.

In a nonlinear system, the output power has to scale nonlinearly with input power. One such effect in optics, which is similar to the harmonics in acoustics, is called *second harmonic generation (SHG)*. In this nonlinear optical effect, the *overtone* of an optical wave is generated inside a crystal, for example, converting the infrared light of a laser into bright green light with half the wavelength of the input light. The output intensity of the green light scales quadratically with the input power, which is a clear indication of a nonlinear optical process of second order. SHG is an example of a *degenerate* nonlinear optical process, i.e., all interacting waves carry the same wavelength. In a crystal with such a second-order nonlinearity, we may also observe *nondegenerate* processes. Nondegeneracy means *mixing* of two input waves with different wavelength, generating the *sum frequency* or the *difference frequency* of the two input waves. Now the intensity of the generated sum-frequency light scales with the product of the two input intensities. Note that all nonlinearities discussed so far only involve optical fields. That means that they are *all-optical nonlinear effects*. This has to be seen in contrast to nonlinearities involving one low-frequency electric field (as accessible by conventional electronics) and one optical field. This interaction may be thought of as a limiting case of sum-frequency generation (SFG) with one of the fields close to zero frequency. This mixing effect can be used for building *electro-optical modulators*. However, the *optical* output power of the device scales *linearly* with the *optical* input power. Therefore, electro-optic mixing is not considered a nonlinear optical effect. The same argument holds for *acousto-optics*, where there is in fact mixing between acoustic waves and the optical field. Clearly, acousto-optics is also not considered a nonlinear optical effect as power scaling is entirely linear for the optical field involved. In the case of the electro-optic effect, we note that this distinction may appear somewhat arbitrary looking at the underlying physics of the process.

This distinction will become much clearer from the different regime of applications of all-optical nonlinearities compared to optical modulators.

With the high intensities of laser sources, an all-optical nonlinear regime can easily be reached. The higher the peak power, the more pronounced the nonlinear behavior of certain optical materials. This makes it clear that a pulsed laser source generates much more abundant nonlinearities compared to a cw source, which is why we refer to nonlinear processes also as *short pulse effects*. In the high-peak-power regime, the response of optical materials starts to react nonlinearly, both in amplitude and phase, with incident optical power. A further simple example for nonlinear behavior may be the bleaching of an optical neutral density filter in a pulsed high-power laser beam. As soon as a certain intensity is surpassed, the attenuation of the filter starts to decrease because a substantial part of the carriers within the beam cross section is transferred into the conduction band. In turn, the density of carriers in the valence band is decreased. These *carrier-related* effects, therefore, cause a *saturation* of the absorption in this material.

In optoelectronics, such nonlinear optical behavior is particularly interesting for the construction of an *all-optical switch,* a device that allows controlling light with light. We will use the term *optical switch* as a general term for very different applications of optical nonlinearities, including optical gates, mode lockers, and optical limiters. For some applications, one may think of an optical switch acting like an *optical transistor* or *light valve,* employing the nonlinearity for modulation of one light beam by another one. Other applications rely on the *self-action* of intense short pulses in a nonlinear medium. Here, the pulse modulates its own amplitude or phase profile via a nonlinear optical effect. Consequently, one talks of *self-amplitude modulation (SAM)* or *self-phase modulation (SPM),* respectively. Quite generally all methods of all-optical switching outperform acousto-optic or electro-optic concepts in terms of speed. On the other hand, it is very challenging for an all-optical switch to reach anything close to the contrast and efficiency possible with electro-optic switches. This trade-off between speed and switching contrast appears to be the fundamental dilemma of all types of optical switches. In this chapter, we will first give a general overview of nonlinear optical processes.

Because of their importance in optoelectronics, we will then concentrate on applications of nonlinear optical processes as all-optical switches.

9.2 QUASI-INSTANTANEOUS NONLINEAR OPTICAL PROCESSES

In the introduction, we have already given examples for the two main types of nonlinear processes. Saturable absorption due to *carrier transfer* into the excited state is a process with *limited response times,* which typically lie in the *picosecond range. Nonresonant nonlinearities* such as SHG occur in totally *transparent dielectric* materials and can be significantly faster. As no direct transfer of carriers into any kind of excited state takes place, there is also no limitation by relaxation processes. In the nonresonant case, *bound electrons* contribute to the nonlinearity instead. One can coarsely estimate the order of magnitude of the response time τ of nonlinearities in dielectric materials as $\tau = 1/\Delta v$, where Δv is the transparency frequency range of the material. This immediately brings us to single femtosecond response times for materials that are transparent in the visible [1]. We will, therefore, refer to these as materials with *quasi-instantaneous* response. For simplicity, we will often relate to these nonlinear processes as *instantaneous* effects. Instantaneous effects comprise sum and difference-frequency generation (DFG), parametric interaction, and also phase effects such as SPM, see Figure 9.1.

Given the instantaneous nature of the electronic nonlinearities, one can directly write the polarization P in a dielectric material as a function of the electric field $\vec{E}$ [1,2]

$$P_i \sum_{j=1}^{3} \chi_{ij}^{(1)} E_j + \sum_{j,k=1}^{3} \chi_{ijk}^{(2)} E_j E_k$$

$$+ \sum_{j,k,l=1}^{3} \chi_{ijkl}^{(3)} E_j E_k E_l + \cdots. \tag{9.1}$$

The susceptibility coefficients $\chi^{(i)}$ are tensors of corresponding rank $i+1$. The linear optical properties of a material are all included in the first-order

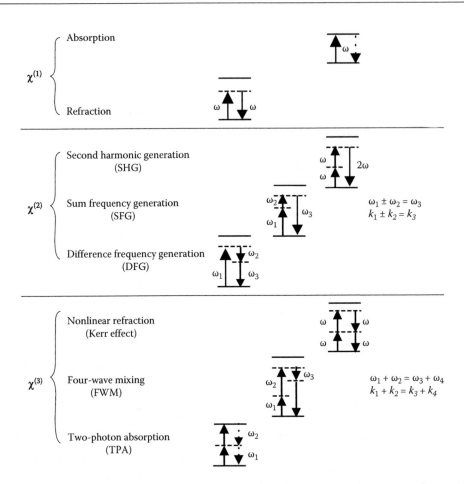

Figure 9.1 Linear and nonlinear optical processes in the photon picture. Real states are designated by solid horizontal lines; virtual states by dashed lines. Linear absorption is caused by a transition between two real states with the potential relaxation indicated by a dotted arrow, refraction is caused by absorption and delayed re-emission of a photon by a virtual state. In a virtual process, the energy of all photons is conserved. In the virtual $\chi^{(2)}$ processes, three waves interact in different sum and DFG schemes. In this picture, the $\chi^{(3)}$ process of nonlinear refraction (all-optical Kerr effect) can also be considered as the degenerate case of the more general four-wave mixing process. $\chi^{(2)}$ processes can only occur in anisotropic materials. $\chi^{(3)}$ processes occur in any kind of medium and are of particular interest in fiber optics. TPA ends in an excited state. The potential return path is again indicated by a dotted arrow.

coefficient $\chi^{(1)}$. In many materials, all tensor elements $\chi_{ij}^{(1)}$ will be identical and we can use simple scalar susceptibility coefficients $\chi^{(1)}$ instead. Only materials exhibiting birefringence or dichroism require a full-blown tensorial treatment of the susceptibility.

Otherwise, we can use a simplified version of Equation 9.1

$$\vec{P} = \chi^{(1)}\vec{E} + \left(\chi^{(2)}\left|\vec{E}\right|\vec{E} + \left(\chi^{(3)}\left|\vec{E}\right|^2\right)\vec{E}\right)EL. \qquad (9.2)$$

The complex index of refraction is related to the susceptibility via $n + ik = \sqrt{1 + X^{(1)}}$, i.e., $\chi^{(1)}$

contains information about both the refractive index n and the absorption coefficient k. Nonlinear optical processes are included in the higher-order coefficients and again may affect both the amplitude and the phase of the light. In isotropic materials, such as glasses, all tensor elements are degenerate and the susceptibility coefficients of even order vanish, i.e., $\chi^{(2)} = \chi^{(4)} = \ldots = 0$. Anisotropy can be found in many crystals and is often, but not necessarily, accompanied by birefringence. Tabulated second-order coefficients $\chi^{(2)}$ for nearly all relevant second-order materials can be found in Reference [2].

In Figure 9.1, some of the most commonly encountered optical nonlinearities up to order 3 are summarized. The linear processes of absorption and refraction are also included for comparison. For most of these nonlinear optical processes, $\sum_i \pm \hbar\omega_i = 0$ has to hold as a consequence of energy conservation (see Figure 9.1). This is not true for absorption and two-photon absorption (TPA), where energy is stored in an excited state. Here, $\sum_i \pm \hbar\omega_i = E_{abs}$ has to be fulfilled instead. This means that the sum of all photon energies equals the energy of the excited state E_{abs} in the case of one or multiphoton absorption. If only virtual states are involved, no energy is absorbed in the process and the sum has to vanish. The latter is the case for *SHG*, *DFG*, and *SFG*. Some of the $\chi^{(3)}$ processes shown in Figure 9.1 also only involve virtual excited states. In all these nonabsorptive cases, momentum conservation $\sum \pm k_i = 0$ also has to hold, where the wavenumber k_i is defined as $\omega_i n/c$. One may think of the real processes absorption and TPA as frustrated refraction or four-wave mixing processes, respectively. These frustrated processes end in a real rather than virtual state and can, therefore, not relax back to the ground state.

It is useful to relate Equation 9.2 to our initial acoustics example, where the electric field $\vec{E}$ takes the role of the solenoid drive current and the polarization $\vec{P}$ is related to the membrane position. As soon as a significant contribution from the nonlinear terms in Equation 9.2 appears, we would expect to see optical harmonics in analogy to acoustics. Let us come back to our original example for nonlinear optical behavior, i.e., SHG. This effect manifests itself as a parabolic susceptibility term $\chi^{(2)}$ in Equation 9.2. In such a medium, the polarization P does not proportionally follow the electric field $E \propto \sin(\omega t)$ of the input field. As illustrated in Figure 9.2, the resulting polarization

$$P \propto \chi^{(1)} \sin(\omega t) + \chi^{(2)} \sin^2(\omega t)$$
$$= \chi^{(1)} \sin(\omega t) + \frac{\chi^{(2)}}{2}[1 - \cos(2\omega t)] \quad (9.3)$$

clearly exhibits distortions compared to the sinusoidal input field, which manifest themselves as *frequency-doubled components* in the polarization

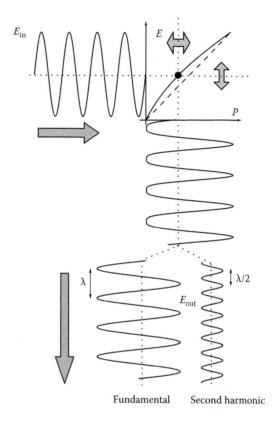

Figure 9.2 Generation of the second harmonic. If the polarization P inside a medium does not simply proportionally follow the input field E_{in} but $P(E)$ includes a parabolic term, the resulting polarization contains nonsinusoidal components, which reveal themselves as a harmonic contribution at double the frequency of the input field. Generally, a $\chi^{(i)}$ susceptibility term gives rise to generation of the ith harmonic in the output field.

[1,3]. The output field, therefore, consists of the fundamental and its second harmonic (and a constant term, which does not propagate). If we had investigated the $\chi^{(3)}$ term instead, we would find the third harmonic of the input field. Similarly, one can easily show that higher order terms $\chi^{(i)}$ lead to the generation of the ith harmonic in the polarization. The coefficients $\chi^{(i)}$, therefore, govern the characteristic spectrum of harmonics of a nonlinear material, similar to the relative strength of overtones in acoustics discussed in the introduction. In the nondegenerate case of mixing of two electric fields of different frequency, one observes both the sum and difference frequency mixing terms in the output spectrum.

9.3 SCATTERING PROCESSES— COUPLING TO THE LATTICE

The effects described so far only encompassed interaction of photons and electrons. A third category of nonlinear optical processes comprises scattering processes, in particular those involving phonons in solid-state materials. Coupling of lattice vibrations and the electric field give rise to *Brillouin scattering* (acoustic phonons) and *Raman scattering* (optical phonons) [3,4]. These processes can be understood similar to the DFG and SFG processes but with one photon replaced by a quantized lattice vibration, i.e., a phonon. Again, these processes involve virtual intermediate states (see Figure 9.1). The decay of one input photon into a photon of smaller energy and one phonon is referred to as the Stokes process. The wavelength shift accompanying this scattering process is called the *Stokes shift*. Therefore, looking at scattering processes from a photon point of view, energy conservation does not appear to hold as the excess energy is transferred into the phonon. The sum of all particle energies, however, is conserved. At high light intensities, the opposite process similar to the sum-frequency process may also show up. In this case, a previously generated phonon may add to the energy of one photon, with a resulting shorter-wavelength photon as the product of this so-called anti-Stokes process.

The scattering processes are difficult to categorize as they rely on the same kind of mechanism as SFG but are linear in terms of optical power. Therefore, they do not qualify as nonlinear optical processes similar to the electro-optic processes discussed in the introduction. Nevertheless, the appearance of Raman and Brillouin scattering is closely related to nonlinear processes as are the electro-optic processes.

At high pump intensities, a new phenomenon called stimulated scattering can be observed. Here, an exponential increase of the Stokes wave with input power is observed. This process is nonlinear, whereas the spontaneous scattering described above has to be considered a linear optical process. These processes have found many applications in fiber optics, for example, for Raman lasers and frequency conversion. For a detailed description of stimulated scattering processes and their applications, we refer the reader to specialized literature on nonlinear fiber optics [4].

9.4 SATURABLE ABSORBERS

9.4.1 Types of saturable absorbers

So far, we have mainly addressed processes that follow the polarization in the medium instantaneously. Many nonlinear processes, in particular, those involving *generation of free carriers,* exhibit a noninstantaneous response. For continuous light exposure or at long pulse durations, the nonlinearity of these processes can also be expressed as a perturbative series like Equation 9.2. On shorter time scales, however, these processes may not relax quickly enough, and the number of free carriers grows linearly with the accumulated exposure. As the number of carriers is responsible for the nonlinearity, there is also a cumulative effect for the nonlinearity of the process. For short pulses (short meaning faster than the relaxation time constant), the nonlinearity, therefore, rather scales with pulse energy than with peak power.

Saturable absorption is the prime example of such an effect. It is also a very useful effect for constructing nonlinear optoelectronic devices. A *saturable absorber* acts like a switch that allows, for example, discrimination of low-energy pulses inside a laser. As short pulses with sufficiently high-energy experience a decreased optical loss by the saturable absorption, a saturable absorber effectively favors pulsed operation of the laser over continuous operation. Such devices typically contain one or several saturable absorbers and a reflector. These devices have been named *saturable Bragg reflector* (SBR, [5,6]) or *semiconductor saturable absorber mirror* (SESAM, [7]) or sometimes simply *saturable absorber mirror*. As these optoelectronic devices have become an important element in the construction of mode-locked lasers, we will outline some fundamental physical processes contributing to the nonlinearity of such devices. We will also give an overview on the construction of these optoelectronic building blocks.

In principle, any absorbing material could be used to build a saturable absorber. Dyes have been very popular for that purpose in the 1970s and 1980s [8,9]. Dyes offer some adaptability by changing concentration and the center wavelength of the absorber. Several solid-state materials based on doped glasses have been proposed as an alternative [10,11]. As a crystalline material, Cr:YAG has found widespread applications as an intracavity switch [12–15]. The potential of semiconductor materials for electro-optical switches, in particular, that of quantum well absorbers, has already been recognized in the 1970s and 1980s [16–20]. As the optical properties of quantum wells can be tailored in the most flexible way of all materials discussed so far [21,22], this type of saturable absorber has become the method of choice for all-optical switches and has found widespread use.

9.4.2 The physics of saturable absorption

The fundamental processes involved in saturable absorption are schematically illustrated in Figure 9.3. Initially, a short laser pulse transfers electrons from the valence band of the absorber into the conduction band of the material. Again, short means that there is only insignificant relaxation back to the valence band within the pulse duration and sometime after. By depleting the valence band this process, therefore, reduces the density of carriers that are capable of absorbing the laser light, which effectively decreases absorption [21]. A short powerful laser pulse can generate a *nonthermal carrier distribution* high up in the valence band [22,23] as schematically shown in Figure 9.3b. These hot electrons are not in thermal equilibrium with the lattice and tend to rapidly cool via phonon scattering processes. Typically, this *intraband relaxation process* occurs on a time scale of $\tau_1 \approx 100\,\text{fs}$ (Figure 9.3c). The cooled carriers will fill the bottom of the conduction band, relaxing to the valence band on much longer time scales τ_2, which are typically of the order of nanoseconds. This is also called *interband relaxation*. The two different timescales of carrier cooling and recombination generate a marked *bitemporal behavior* [7] of the absorption versus time, as shown in Figure 9.3e.

$$L(t) = \begin{cases} L_{\text{NS}} - L_1\left[1 - \exp(-t/\tau_1)\right] \\ -L_2\left[1 - \exp(-t/\tau_2)\right] & t \geq 0 \\ L_{\text{NS}} & t < 0 \end{cases} \quad (9.4)$$

The time-dependent loss $L(t)$ includes two saturable contributions with strength L_1 and L_2, and a nonsaturable term L_{NS}. The nonlinear response of Equation 9.4 can be measured using *pump-probe spectroscopy*. Pump-probe spectroscopy employs a second weaker pulse to monitor the depletion at variable delays after the strong pump pulse at $t=0$. From a logarithmic plot of the measured response, one can easily retrieve the time constants τ_1 and τ_2. Figure 9.3e qualitatively shows the bitemporal behavior. The induced transmission is strongest when all excited carriers are still in resonance with the pump that created them. Cooling processes rapidly broaden the initial hot distribution of electrons, with more and more electrons being transferred to the bottom of the conduction band. This is the first effect reducing the induced transmission signal. On longer time scales, the signal is further weakened by recombination back into the valence band until it will eventually vanish, which typically happens on the nanosecond time scale.

For applications of saturable absorbers in lasers, the time constants τ_1 and τ_2 are important device parameters. These time constants are to be compared with the pulse duration of the laser. Despite the fact that it is possible to generate pulses that are more than one order of magnitude shorter than the dominant response time of the absorber [24,25], the usefulness of an absorber is greatly enhanced the faster its response. Clearly, a recombination time constant τ_2 of a few nanoseconds, i.e., approaching typical cavity roundtrip times, is problematic. Significant efforts went into *acceleration of the recombination process*. One method of doing so is the introduction of *additional mid-gap* states into the material as schematically shown in Figure 9.3d. These additional states can be generated by a variety of processes. The first observation that *defect-induced* additional states contribute to an accelerated relaxation of a saturable absorber was reported by Ippen et al. [26]. Later several methods for introducing defects in a controlled way were explored, including low-temperature growth [27,28], ion implantation [29], proton bombardment [30,31], or impurity implantation [32]. Lifetimes have been pushed down to one to several picoseconds, which constitutes an improvement of about three orders of magnitude compared to low-defect materials.

9.4.3 Design of saturable absorber mirrors

Apart from the time constants, the relative strength of the nonlinear absorption and its wavelength characteristics are further important design

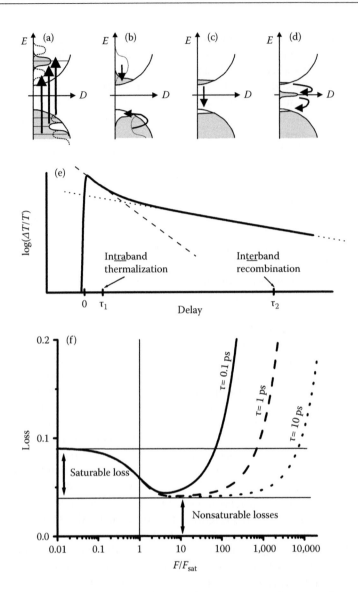

Figure 9.3 A detailed look at saturable absorption. After excitation of carriers from the valence band of a semiconductor into its conductance band (a), the initial nonthermal distribution tends to rapidly cool (b) with a time constant τ_1. On this time scale of intraband relaxation, the majority of carriers is still in the excited state and only relaxes into the valence band on longer time scale, which is called interband relaxation (c). Introducing additional mid-gap states as shown in (d) may greatly accelerate the interband recombination. Monitoring the induced transmission change $\Delta T/T$ as shown in (e) identifies intraband and interband relaxation by two distinct time constants, τ_1 and τ_2. (f) Shows the typical dependence of the induced increase of transmission as a function of input fluence. For input fluence well below the saturation fluence F_{sat}, the overall loss of the device amounts to the sum of saturable and nonsaturable losses. At higher fluences, the loss starts to roll off in the vicinity of F_{sat} and reaches a minimum value dictated by nonsaturable losses L_{NS}. At still higher fluence, one may observe an increased loss due to two-photon processes.

parameters of the SESAM. Both bulk absorption and *quantum wells* are used for generating saturable absorption in SESAMs. Quantum wells exhibit a relatively strong nonlinear response because of *quantum confinement* effects [21]. The number and position of intrinsic and excitonic states of a quantum well can be influenced both by the material and by the thickness of the quantum well.

Note that saturable absorption extends to below the nominal bandgap of a quantum well, as there are also strong *excitonic effects* in semiconductor heterostructures. Excitonic contributions tend to be faster but also weaker than contributions from band filling. The design of the quantum wells allows for some *tailoring of the wavelength response* and also of the strength of the saturable absorption. The latter can be further influenced by a multitude of design parameters, namely the number of quantum wells, their position relative to the nodes of the electric field, and also by the optical design of the SESAM [7]. If the quantum well is positioned at a maximum of the standing wave inside the laser cavity, its response is coupled much tighter to the electric field than if it were positioned at a node of the field (see Figure 9.4).

Figure 9.4 shows two different aspects of the design of a SESAM device, its *bandgap structure* and its *refractive index structure*. The

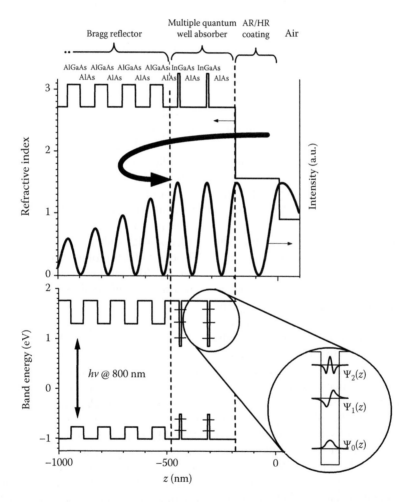

Figure 9.4 Schematic refractive index and bandgap structure of a saturable absorber mirror. A Bragg reflector at the bottom of the device provides a high reflectivity and is typically implemented as a quarter wave layer stack of two different semiconductor materials. The index of refraction is alternating in this stack. Both materials should have a bandgap larger than the photon energy in the device. Saturable absorption is produced by one or several quantum wells. In these quantum well regions, discrete states exist. Additionally excitonic effects may contribute to the absorption behavior of the quantum wells. The quantum well position relative to the nodes of the field allows us to tailor the strength of the nonlinear response. The field strength at the quantum wells can be additionally manipulated by an external Fabry–Perot structure. For this purpose an additional mirror is coated on top of the device. This mirror can either be HR (high-finesse case) or be an antireflection (AR) coating.

index structure is responsible for positioning of the nodes and the reflection properties [7]. The bandgap structure is engineered to provide suitable saturation properties, as was explained above. As the SESAM has to also act as a mirror, the absorber structure is deposited on a highly reflecting (HR) mirror reflector. In the simplest case, a *Bragg mirror*, i.e., a combination of quarter-wave layers of alternating index of refraction, is used as outlined in Figure 9.4. The bandgap of the materials used in the reflector is significantly higher than the photon energy; therefore, negligible absorption takes place in the mirror structure. One of the disadvantages of the Bragg mirror approach is its limited bandwidth, which is mainly caused by the weak index contrast of the available transparent semiconductor materials. Typically, a Bragg mirror consisting of AlGaAs and AlAs, as suggested in Figure 9.4, only covers a bandwidth of about 100 nm at an 800-nm center wavelength. Alternatively, metal mirrors have also been employed for providing more broadband reflectivity [33,34]. This approach, however, is much more demanding in terms of manufacturing.

Apart from the Bragg reflector and the absorber, a second mirror structure can be added on top of the absorber toward the interface to air. The reflectivity of the top mirror can reach from more than 50% down to an AR coating. The case of high reflectivity results in a high finesse Fabry–Perot-like structure. Typically, high-finesse SESAMs are used in antiresonance (*antiresonant Fabry–Perot saturable absorber*, A-FPSA [35]). The antiresonance decreases the intensity of the light inside the absorber relative to the main cavity. For example, a 95% reflectivity top mirror increases the saturation fluence by about a factor 100 [7]. In more recent SESAM configurations, lower reflectivities of the top reflector were used and scaling of the nonlinearity was obtained by other means already discussed above [36]. The top mirror or AR structure can be manufactured either by semiconductor technologies or using dielectric materials such as TiO_2 or SiO_2. Very simple but efficient AR coatings can be manufactured by depositing a single half-wave layer of a material with index of refraction $n_{AR} = \sqrt{n_{top}}$, where n_{top} is the top layer refractive index.

Depending on the pulse duration being larger or smaller than the dominant time constant of the SESAM, one either observes a functional dependence of the reflectivity with pulse peak power or pulse energy, respectively. On a femtosecond time scale, the response of the SESAM generally scales with energy, because significant interband relaxation has not set in yet. In this case, one can write the power-dependent reflectivity of an SESAM as

$$R(F) = R_0 + \Delta R\left(1 - \exp\left(-F/F_{sat}\right)\right) \quad (9.5)$$

where F is the fluence in Joule per unit area, F_{sat} is the saturation fluence, R_0 is the small-signal reflectivity, and ΔR is the modulation coefficient [7], which is connected to the coefficients L_1 and L_2 in Equation 9.4. The functional dependence of the reflectivity on the flux is shown in Figure 9.3f. At very high fluence $F > 10F_{sat}$, additional effects may contribute, which are treated below. For lower fluences, the saturable ΔR may range from L_2 to $L_1 + L_2$, depending on pulse duration. In the absence of nonsaturable losses, $R_0 + \Delta R$ should add up to unity reflectivity. In reality, *nonsaturable losses* $L_{NS} = 1 - R_0 - \Delta R$ are typically on the same order of magnitude as ΔR itself, i.e., a few percent. The onset of saturation is governed by the parameter F_{sat}. The saturation fluence is related to the *cross-section* σ of the absorber by the relation

$$F_{sat} = \frac{h\nu}{N\sigma} \quad (9.6)$$

i.e., the *saturation fluence* is reached when on average one photon with energy $h\nu$ impinges on the cross-sectional area σ. Inside a linear laser cavity, this may happen twice ($N=2$). In a ring cavity or otherwise $N=1$. The saturation fluence is a decisive engineering parameter of the SESAM and should be characterized carefully prior to use of an SESAM inside a laser. The saturation fluence of the absorber has to be carefully balanced with the saturation fluence of the laser gain material to avoid spiking of the laser. The gain saturation fluence is related to the gain cross section as defined in Equation 9.6 [37]. For clean mode-locked pulse trains, it is of paramount importance to scale spot sizes in the gain medium and on the absorber accordingly. Guidelines for a suitable layout of a laser cavity are given in Reference [38].

9.4.4 TPA in saturable absorber mirrors

For extremely high-peak intensities, the nonlinear response of a saturable absorber may be affected by TPA giving rise to an additional term in Equation 9.4 [39–41]. As TPA is a quasi-instantaneous non-linear process (see Figure 9.1), this mechanism always scales with peak intensity $I(t)$, regardless of the pulse duration regime:

$$\Delta R_{\mathrm{TRA}} = -b_2 I \qquad (9.7)$$

As TPA creates additional absorption for high enough peak powers, it defeats the effect of saturable absorption at extremely high-power levels (as shown in Figure 9.3f). At first sight, TPA may appear as a limitation of SESAMs. However, this effect can be beneficial causing a stabilization of the pulse duration and preventing pulse break up, stabilizing the laser pulse duration and the average power of the laser. This effect can also reduce the spiking tendency of the laser [41] and help to provide clean mode-locked pulse trains with stable pulse energy.

9.4.5 Devices closely related to saturable absorber mirrors

The SESAM has found widespread use as intracavity saturable absorbers for generating short optical pulse [7,23]. Several similar devices have been proposed and demonstrated. The SESAM's main application is SAM. In a slightly modified geometry, saturable absorption can also be used to modulate one laser beam with the aid of another one [42]. Here, the control light is fed into a waveguide and bleaches out an arrangement of several quantum wells. This saturable absorption modulates a light beam that crosses the quantum wells at normal incidence to the waveguide direction.

Other nonlinear optical switches have been proposed, which do not directly rely on the absorptive properties of quantum wells or semiconductor bulk materials. These devices embed the saturable absorber in between a high-finesse Fabry–Perot structure. Other than with the A-FPSA approach previously mentioned, the Fabry–Perot is used at resonance and relies on refractive index changes accompanying the creation of free carriers in the conduction band of the semiconductor. Although SESAMs can be understood as amplitude modulators, this type of *Fabry–Perot modulator* (FPM) relies on *phase modulation*. For inducing mode-locking inside a laser resonator, the SPM has to be converted back into an amplitude modulation, compare "nonlinear interferometers as switches" section. This is accomplished by the Fabry–Perot structure. The index change causes a tuning of the Fabry–Perot resonances. If the length is chosen such that the additional index change increases the overall reflectivity of the device, such a device can also be used as an optical switch for mode-locking [43]. Moreover, one can further exploit the phase modulation aspect of such Fabry–Perot structures for synchronizing lasers as shown in [44]. Using a laser with a photon energy above the bandgap of the spacer material in the FPM, one can modulate index of refraction for light below gap without inducing any losses. This index modulation allows control of the cavity roundtrip time and passive locking of cavity repetition rates.

9.4.6 Inverse saturable absorption— optical limiting

For many applications, an effect with exactly the reverse behavior of saturable absorption is desirable. One example could be protection of a sensitive photodetector from intense light pulses that might otherwise destroy the detector. Such a device requires a transmission that decreases with increasing light intensities, a characteristic that is referred to as *optical limiting*. The previous discussion already addressed TPA in semiconductor materials as one possible mechanism showing exactly the right sign of saturable absorption needed for an optical limiter (Reference [39–41], see Equation 9.7 and Figure 9.3f). In fact, semiconductor materials such as GaAs have been suggested early on for this purpose [45,46]. Additionally, *self-defocusing* may contribute to optical limiting in these devices, an effect that will be explained in the following section. Rather than using semiconductors, one can again also use dyes for the same purpose [44–46]. One of the major mechanisms for reverse saturable absorption in complex organic molecules is the generation of excited electronic states with an absorption cross section larger than the ground state absorption. *Excited state absorption* has been reported for several classes of organic chromophores [47]. Apart from sequential single-photon absorption, TPA may also contribute to optical limiting in dye molecules [48,49]. Different chemical compounds, such as fullerenes, organometallics, or carbon black suspensions have been discussed for optical limiting devices. Optically induced scattering, for example, by localized melting of the limiter material, has also been suggested as a mechanism for optical limiting [50]. However, despite the variety of nonlinearities, materials, and device configurations that have been used to implement passive optical limiters, no single device or combination of devices has yet been identified that will protect any given sensor from all potential optical threats.

9.5 THE ALL-OPTICAL KERR EFFECT AND NONLINEAR REFRACTION

9.5.1 Kerr-based switches and Kerr gates

From the point of view of optoelectronics, saturable absorption is probably the most interesting process among the variety of nonlinear optical processes, as it allows us to build an *all-optical switch*. Depleting carriers by means of a pump pulse, one can influence the transmission properties of a saturable absorber and thereby control the energy of a second probe pulse. In some way, this device may be thought as an all-optical transistor or *light valve*, despite the fact that one needs rather large light intensities to control relatively small ones. Apart from acting as a light valve, a technologically important aspect of an all-optical switch is that it forms the product of optical waveforms. Multiplying an unknown waveform with a known one is at the heart of optical sampling techniques and allows characterization of unknown optical pulse shapes with the aid of known ones. Unfortunately, with SESAMs one can do neither of these arbitrarily fast, as saturable absorption is a rather slow process with picosecond relaxation times even when midstate traps are used to accelerate this process. This calls for methods that benefit from *quasi-instantaneous nonlinearities* (cf. "Quasi-instantaneous nonlinear optical processes" section).

It is rather clear that the temporal response of carrier-related effects cannot be much further accelerated without sacrificing modulation depth. Instead, many approaches have been pursued to employ bound-electronic optical processes to build an all-optical switch. These approaches rely nearly exclusively on *nonlinear refraction*, i.e., the change of the index of refraction with instantaneous intensity. First and foremost, nonlinear refraction is a phase effect that requires *translation into an amplitude effect* to be effective as a nonlinear optical switch. In the following, we will first review the physics of nonlinear refraction and point out that it is inseparably connected to TPA and other nonlinear optical effects. In fact, this connection between nonlinear refraction and TPA poses a stringent limitation of all-optical valves. We will then give an overview of several architectures of all-optical switches based on nonlinear refraction. These switches have found widespread use in rather distinct applications of optics ranging from telecommunications to measurement techniques for short pulses.

9.5.2 The physics of nonlinear refraction

The propagation (phase) velocity of light v_φ in a dielectric medium is governed by the refractive index n according to $v_\varphi = c/n$. The refractive index n is a function of wavelength and can be computed from the complete knowledge of absorbing resonances in the ultraviolet and infrared spectral region using *Kramers–Kronig relationship* [51]. The slightest changes, for example, of the lattice constant or temperature of a solid-state material, will automatically affect the refractive index when they modify the spectral position or relative strength of the resonances. In dielectric media, this will cause the index of refraction change according to

$$n(I) = n_0 + n_2 I. \qquad (9.8)$$

This effect of an intensity-dependent refractive index is called the *all-optical Kerr effect* [1,3,4]. The coefficient n_2 is called the *nonlinear index of refraction* and is related to the real part of the susceptibility tensor $\chi^{(3)}$ via

$$n_2 = \frac{3}{8n} \mathrm{Re}\, X^{(3)} \qquad (9.9)$$

where it was assumed that the light is linearly polarized such that only one component of the third-rank tensor $\chi^{(3)}$ contributes. Often, $\chi^{(3)}$ is quoted using electrostatic units (esu). One can convert to the more useful W cm^{-2} employing the relation

$$n_2\left(cm^2W^{-1}\right)=\frac{12\pi^2}{n_0^2c}\,\mathrm{Re}\,X^{(3)}\left(esu\right). \qquad (9.10)$$

Typical values for n_2 of optical glasses lie in the range of a few 10^{-16} cm^2 W^{-1} [4,52,53]. The main effect in a dielectric medium far away from the bandgap is electronic polarization, which has a response time of 1 fs or faster. In other materials, additional effects such as molecular orientation or atomic resonances can create much stronger nonlinear refraction, at the expense of a greatly slowed response and a much narrower bandwidth. In the following, we will restrict ourselves on the Kerr effect in dielectrics and semiconductors and focus on nonlinear refraction induced by electronic polarization.

A power-dependent index of refraction means that the phase velocity of the light has become a function of intensity, with high peak power pulses traveling slower than a low peak power beam with identical cross-section would do, see Equation 9.8. This self-action of the light is called SPM:

$$\Delta_\phi(I)=\frac{2\pi}{\lambda}\,n_2IL, \qquad (9.11)$$

where λ is the length of the medium and L is the wavelength of the light. The importance of nonlinear refraction can be estimated from inverting Equation 9.11 to find the critical intensity that is necessary to create a macroscopic 2π phase shift

$$I_{crit}=\frac{\lambda}{n_2L}. \qquad (9.12)$$

Plugging in numbers for a 1 cm long piece of glass, one finds a critical intensity level on the order of 10^{11} W cm^{-2}. This makes it immediately clear that SPM only becomes important when laser pulses are either extremely tightly focused or when a very long interaction length can be used as in optical fibers. In fact, SPM is extensively used in *nonlinear fiber optics* [4].

9.5.3 Connection of nonlinear refraction and TPA

As nonlinear refraction is the most important effect for building quasi-instantaneous optical switches, a somewhat deeper insight into the underlying physics is important to gain an understanding on some of the limitations. The Kerr effect can be generated by a variety of different mechanisms, including molecular orientation, saturated absorption, and electrostriction [1]. In a wider sense, also thermal effects create refractive index changes. With response times of picosecond to milliseconds, application of these effects is much less interesting than use of the Kerr effect induced by electronic polarization changes. For all practical applications, this response can be considered instantaneous. Quite generally, there appears to be wide-ranging inverse connection between the response time of the effect and its relative strength. Investigations of the electronic Kerr effect, therefore, were motivated by finding a material that combines quasi-instantaneous response with maximum nonlinear susceptibility.

In semiconductors, an *inverse Kerr effect* has been observed, i.e., high intensities travel at higher velocity than low intensities, other than in dielectric materials. This sign change can normally be explained by the *generation of free carriers*, which is a strong indication for the onset of multiphoton absorption in the material. The magnitude of the nonlinear response shows a dramatic increase close to the bandgap of an optical material [54]. This renders semiconductors very interesting for building all-optical switches. Figure 9.5 shows the fundamental behavior of nonlinear refraction and TPA versus photon energy. This curve is based on the theoretical model of Sheik-Bahae et al. for nonlinear refraction. The computation is based on the *Kramers–Kronig relationship*, which allows us to calculate refraction changes directly from absorption changes [51,55]. The main effect contributing to the nonlinear absorption is TPA, which sets in above half the bandgap of the material. Other effects, such as the *Stark effect* and *Raman* contributions, also play a role and have been accounted for in Figure 9.5. It is important to understand that TPA poses a severe limitation on using the Kerr effect for a switching application. Excessive TPA will destroy the switch. From the behavior of

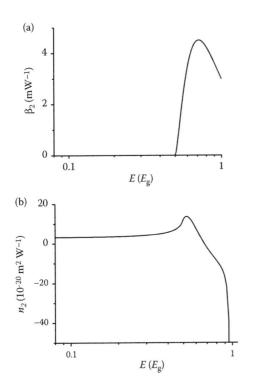

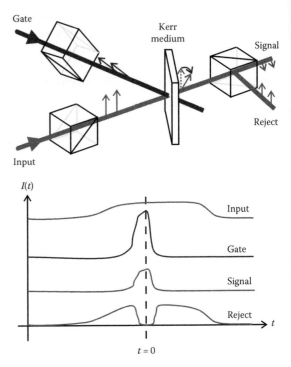

Figure 9.5 Behavior of TPA (a) and nonlinear refraction (b) as a function of photon energy. These calculations are based on material parameters of silica. The photon energy is scaled in units of the bandgap of the material. Nonlinear refraction is rather constant at low energies, exhibits a peak above half the bandgap and then turns negative close to the bandgap.

nonlinear refraction alone, one would tend to go as close to the bandgap as possible to create the maximum nonlinear response. Unfortunately, such a device would also suffer from strong possible TPA. For switching applications, it has been found that an optimum between both effects is reached when the photon energy is about 70% of the bandgap energy [55].

9.5.4 The Kerr lens as a nonlinear optical switch

The earliest switching method that was based on the Kerr effect is the so-called Kerr gate [56], which is illustrated in Figure 9.6. In a Kerr gate, a probe pulse is sent through a Kerr medium, in which its path crosses the path of a strong pump pulse. Crossed polarizers prevent detection of the probe light in the absence of an overlap with the pump.

Figure 9.6 Kerr gate based on polarization rotation. The gating pulse and the input pulse are crossed inside a Kerr medium. Both pulses are linearly polarized, with the polarization of the gating pulse rotated 45° relative to the input pulse. The Kerr effect causes a polarization rotation of the input pulse. A second crossed polarizer in the input beam line allows separation of the gated pulse from the rejected part of the input pulse. An example for the temporal behavior of the gating pulse, the input pulse, and the two output pulses are shown below for comparison.

Only if the pump induces a *polarization rotation* inside the Kerr medium, is some portion of the probe transmitted at the polarizer. This Kerr gate is instantaneous provided that the Kerr nonlinearity is sufficiently fast. In the more recent literature, interferometer-based switches are also referred to as optical gates (see, e.g., Reference [57]). We will treat these methods of translating an instantaneous phase nonlinearity into an amplitude nonlinearity below. Instantaneous optical gates can be built from Kerr induced polarization changes, Kerr lenses, or nonlinear interferometers.

A further Kerr-related effect that can be used for building a switch is *self-focusing*. This method is illustrated in Figure 9.7 [3,58,59]. When the

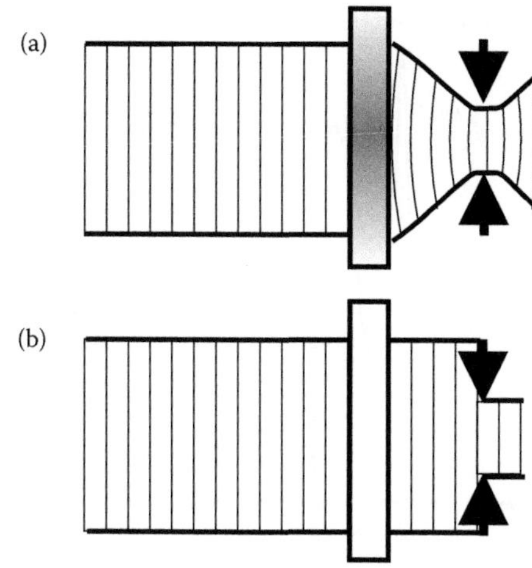

Figure 9.7 Kerr gate based on Kerr lensing. Schematically shown are propagating wave fronts of a high-intensity and a low-intensity beam, (a) and (b), respectively. For sufficiently high intensities, a transverse index profile is induced by the pulse as indicated by gray shades. The stronger the induced change of the refractive index, the stronger the corresponding part of the beam front is retarded. This causes an additional beam front curvature, i.e., the Kerr effect transversally acts like a lens. A subsequent aperture can be used to convert the lensing effect into an effective amplitude modulation. The combination of the Kerr lens and an aperture acts similar to the gate of Figure 9.6, but is not background free.

central most intense part of a beam experiences a phase delay according to Equation 9.8 this is equivalent to the focusing action of a convex lens. In a convex lens, the phase shift is generated by the larger amount of glass in the beam path of the central rays. Other than in an ideal conventional lens, however, the wave front change induced by the nonlinear Kerr lens is not exactly spherical but proportional to the spatial beam profile. For example, if the beam profile is Gaussian, only the central part of the Kerr lens can be approximated by a conventional lens. A nonlinear *gradient-index duct* [37,59,60] is normally a better approximation for the Kerr lens.

As the Kerr lens only focuses the most intense parts of a short pulse, one can use it to build an ultrafast optical switch, which is also illustrated in Figure 9.7. If an aperture is suitably placed in the focus of the Kerr lens, it will introduce losses for cw laser light of low intensity, whereas high

intensity light can pass through as it sees the focusing action of the Kerr lens. This kind of self-switching is used in lasers to discriminate short pulses and create losses for any kind of low intensity background. Placed in a cavity, the Kerr lens switch strips off a pulse pedestal, recleaning the pulse on every passage. This is used for generating some of the shortest pulses ever generated from a laser with a method called *Kerr-lens mode-locking* [59–62].

9.5.5 Nonlinear interferometers as switches

Another method to build a switch is based on SPM (compare Equation 9.11). Here, the Kerr effect is translated into an effective SAM in a *nonlinear interferometer*. Basic idea is illustrated in Figure 9.8 with a Michelson interferometer. If one introduces a Kerr medium in only one arm of the interferometer, high light intensities will experience a phase shift compared to low intensities. Provided the bias phase and the nonlinear phase are properly set, low intensities will be back reflected into the input port, whereas high intensities go into the opposite port. Sending pulses in such a nonlinear Michelson interferometer, only the central and most intense part of the pulse will go into the output port. The weaker parts of the pulse will be redirected into the input port, effectively leading to a compression of the input pulse. A Michelson interferometer is not the ideal configuration for such a nonlinear switch because it will require active means to stabilize the bias phase of the interferometer. Mach–Zehnder interferometers can be built in a microoptical configuration, dramatically reducing problems of phase drift [57]. A *Sagnac* configuration, as has first been suggested by Blow and Doran [63], is typically the method of choice. In the *nonlinear optical loop mirror*, the two interferometer arms consist of different direction of propagation in a fiber loop (see Figure 9.8). An imbalance of the Kerr effect between the two directions of propagation is simply induced by asymmetric beam splitting. Otherwise, the function is identical to the previously discussed case. The high-intensity portion of the pulse is switched into one port of the interferometer, the pedestal goes into the other port, effectively leading to pulse compression. Such Sagnac interferometers

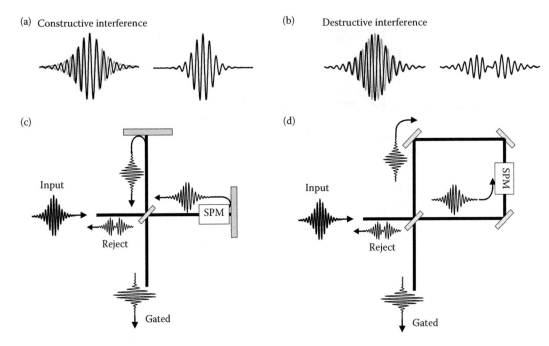

Figure 9.8 Kerr gate based on nonlinear interferometers. (a and b) Interference of an unmodulated pulse (gray line) with a self-phase modulated pulse (black line). If the bias phase at pulse center is chosen for constructive interference as in (a), destructive interference in the wings compresses the resulting pulse. The opposite case is shown in (b), resulting in a lengthened pulse. (c) Shows a Michelson configuration, where cases (a) and (b) appear in the two different output ports of the interferometer. A Sagnac configuration (d) works in a similar way at the added advantage that no stabilization of the bias phase is required.

have been used for pulse shape cleaning and also to induce mode-locking.

As the impression may arise that the exact type of interferometer does not matter for the principle of operation, it needs to be pointed out that the interferometer types discussed so far only exploit dual-beam interference. Using a Fabry–Perot interferometer instead, interference of many pulses may create complex behavior. A nonlinear Fabry–Perot switch has been extensively discussed as an all-optical switch as it exhibits bistability and hysteresis, a behavior analogous to what a *Schmitt trigger* does in electronics. Using nonlinear systems with large feedback is interesting for applications as it allows us to switch large light powers with relatively small ones. However, this method has not found very widespread use because of several disadvantages. The nonlinear Fabry–Perot has extremely rich nonlinear dynamics, including *period doubling* and *chaotic behavior* [64,65]. Generally, this type of behavior becomes more

pronounced the stronger the feedback inside the system. Another disadvantage of all nonlinear interferometers is the delay of the interferometer arms. In the case of two-pulse interference, the actual switching action happens after single propagation through the interferometer. In the case of the nonlinear Fabry–Perot, the switching time amounts to many roundtrips through the interferometer. However, despite these problems, nonlinear Fabry–Perot interferometers have been successfully incorporated into lasers to initiate and sustain mode-locking [66–69]. Compared to saturable-absorber mode-locking or Kerr-lens mode-locking, these lasers require a stabilization of the interferometer phase.

9.5.6 Applications of nonlinear optical switches

One important application of a nonlinear optical switch has already been addressed in the previous section—mode-locking. For generation of short pulses, the nonlinear optical switch is inserted into the laser cavity. Its main purpose is the *suppression of low light levels* and a preference for high light levels. Such a preferred transmission of high peak powers leads to a temporal focusing of the entire energy contents of the cavity into one short pulse. Continuous operation of the cavity is discouraged by the higher losses induced by the nonlinear switch. As the switch reduces the power in the pulse's pedestals compared to its peak power, one may also think of the switch *recompressing the pulse* every roundtrip [68]. The nonlinear switch initializes the mode-locking process from noise and stabilizes it in the steady-state regime. Additional processes such as dispersion and spectral broadening contribute during the steady-state operation and determine the pulse duration of the mode-locked laser.

A second application of a nonlinear switch is *pulse characterization*. One can use the geometry of the Kerr gate, for example, to sample one unknown pulse with the aid of another one, simply by varying the delay between pump and probe in the Kerr gate. This is known as *cross-correlation*. Moreover, one can also try to sample one pulse by itself, i.e., perform an *autocorrelation*. Pulse retrieval from autocorrelation is ambiguous and impossible without further knowledge on the pulse shape. Advanced methods to measure and characterize pulse shapes are discussed in Chapter 17.

Telecommunication also strongly relies on optical switches. For many applications, electro-optic switches are still fast enough for data processing. However, there is a strong tendency to substitute electro-optical methods by all-optical variants, in particular when approaching data rates in the tens of Gbits^{-1}. One potential application lies in the area of *time-domain multiplexing and demultiplexing* of optical data streams. One can use fiber-based switches similar to those previously discussed to selectively extract, for example, every fourth or sixteenth pulse from a stream of pulses. The extracted pulse train then has a much smaller data rate than the original one and can be processed with slower electro-optic components. An optical switch based on a polarization discriminating Mach–Zehnder interferometer has been demonstrated with up to 1500 GBs^{-1} data rates and 200-fs pulse duration

[70]. Many different switch architectures have been proposed for demultiplexing applications [42, 71–75]. Another application of all-optical switches is optical memories. It has been demonstrated that pulse sequences can be *stored* inside fiber ring resonators. A pulse sequence is launched into the ring and later extracted by an optical switch [76,77]. For launching or extracting an entire sequence at once, no particular fast switch is needed. With an all-optical switch, however, individual pulses can be launched or extracted from the optical ring buffer, which greatly enhances the utility of such optical memories.

9.6 SUMMARY

Optical nonlinearities are important building blocks in optoelectronic devices, which have a variety of different applications as optical switches. Two fundamentally different basic concepts have evolved. One of them is the *saturable absorber,* which is based on saturation of an electronic transition. Given the very strong interaction of quantum wells and light, extremely compact all-optical switches can be built using modern semiconductor manufacturing technology. These devices have found widespread application for mode-locking of lasers. Their only limitation is the limited switching speed, which cannot easily be pushed below 1 ps. A second concept employs *quasi-instantaneous nonlinearities,* namely the *Kerr effect,* to overcome speed limitations of the device. As this nonlinearity is a phase nonlinearity, some means of phase to amplitude conversion has to be incorporated into the device to make it an effective saturable absorber with practically unlimited switching speed. This translation can be accomplished either by an interferometer, by polarization switching, or by using the Kerr lens mechanism. As the Kerr effect is a third-order nonlinearity, it is typically much weaker than saturable absorption of a quantum well. Often, sufficient nonlinearity can only be accumulated by propagation through a long piece of fiber. Therefore, many concepts for using the Kerr nonlinearity as a switch involve relatively long delays compared to the more directly switching semiconductor-based saturable absorbers. It strongly depends on the application, which of the two fundamental concepts is to be preferred.

REFERENCES

1. Boyd, R.W. 1992. *Nonlinear Optics* (San Diego, CA: Academic Press).
2. Dmitriev, V.G., Gurzadyan, G.G. and Nikogosyan, D.N. 1999. *Handbook of Nonlinear Optical Crystals*, 3rd ed. Springer Series in Optical Sciences 64 (Berlin: Springer).
3. Yariv, A. 1989. *Quantum Electronics*, 3rd ed. (New York: Wiley).
4. Agrawal, G.P. 2001. *Nonlinear Fiber Optics*, 3rd ed. (San Diego, CA: Academic Press).
5. Keller, U., 'tHooft, G.W., Knox, W.H. and Cunningham, J.E. 1991. Femtosecond pulses from a continuously self-starting mode-locked Ti:sapphire laser. *Opt. Lett.*, 16, 1022–1024.
6. Tsuda, S., Knox, W.H., De Souza, E.A., Jan, W.Y. and Cunningham, J.E. 1995. Low-loss intracavity AlAs/AlGaAs saturable Bragg reflector for femtosecond mode-locking in solid-state lasers. *Opt. Lett.*, 20, 1406–1408.
7. Keller, U., Weingarten, K.J., Kärtner, F.X., Kopf, D., Braun, B., Jung, I.D., Fluck, R., Hönninger, C., Matuschek, N. and Aus der Au, J. 1996. Semiconductor saturable absorber mirrors (SESAM's) for femtosecond to nanosecond pulse generation in solid-state lasers. *IEEE J. Sel. Top. Quantum Electron.*, 2, 435–453.
8. Hillman, L.W. and Duarte, F.J. (eds.) 1990. *Dye Laser Principles: With Applications* (Boston, MA: Academic Press).
9. Duarte, F.J. (ed.) 1992. *Selected Papers on Dye Lasers*. SPIE Milestone Series 45 (Boca Raton, FL: SPIE Press).
10. Bilinsky, I.P., Fujimoto, J.G., Walpole, J.N. and Missaggia, L.J. 1999. InAs-doped silica films for saturable absorber applications. *Appl. Phys. Lett.*, 74, 2411–2413.
11. Wu, E., Chen, H., Sun, Z. and Zeng, H. 2003. Broadband saturable absorber with cobalt-doped tellurite glasses. *Opt. Lett.*, 28, 1692–1694.
12. Okhrimchuk, A.G. and Shestakov, A.V. 1994. Performance of YAG: Cr^{4+} laser crystal. *Opt. Mater.*, 3, 1–13.
13. Shimony, Y., Burshtein, Z. and Kalisky, Y. 1995. Cr^{4+}:YAG as passive Q-switch and Brewster plate in a pulsed Nd:YAG laser. *IEEE J. Quantum Electron.*, 31, 1738–1741.
14. Eilers, H., Hoffman, K.R., Dennis, W.M., Jacobsen, S.M. and Yen, W.M. 1992. Saturation of 1.064 μm absorption in Cr, $Ca:Y_3Al_5O_{12}$ crystals. *Appl. Phys. Lett.*, 61, 2958–2960.
15. Griebner, U. and Koch, R. 1995. Passively Q-switched Nd:glass fibre-bundle laser. *Electron. Lett.*, 21, 205–206.
16. Gibson, A.F., Kimmit, M.F. and Norris, B. 1974. Generation of bandwidth-limited pulses from a TEA CO_2 laser using p-type germanium. *Appl. Phys. Lett.*, 24, 306–307.
17. Boyd, G.D., Bowers, J.E., Soccolich, C.E., Miller, D.A.B., Chemla, D.S., Chirovsky, L.M.F., Gossard, A.C. and English, J.H. 1989. 5.5 GHz multiple quantum well reflection modulator. *Electron. Lett.*, 25, 558–560.
18. Chmielowski, M. and Langer, D.W. 1989. Quantum well vertical light modulator. *IEEE Photon. Technol. Lett.*, 1, 77–79.
19. Whitehead, M. and Parry, G. 1989. High-contrast reflection modulation at normal incidence in asymmetric multiple quantum well Fabry–Perot structure. *Electron. Lett.*, 25, 566–568.
20. Klingshirn, C.F. 1995. *Semiconductor Optics* (Berlin: Springer).
21. Weisbuch, C. and Vinter, B. 1991. *Quantum Semiconductor Structures—Fundamentals and Applications* (San Diego, CA: Academic Press).
22. Knox, W.H., Hirlimann, C., Miller, D.A.B., Shah, J., Chemla, D.S. and Shank, C.V. 1986. Femtosecond excitation of nonthermal carrier populations in GaAs quatum wells. *Phys. Rev. Lett.*, 56, 1191–1193.
23. Keller, U. 2003. Recent developments in compact ultrafast lasers. *Nature*, 424, 831–838.
24. Haus, H.A. 1975. Theory of mode-locking with a slow saturable absorber. *IEEE J. Quantum Electron.*, 11, 736–746.
25. Kärtner, F.X., Jung, I.D. and Keller, U. 1996. Soliton mode-locking with saturable absorbers. *IEEE J. Sel. Top. Quantum*, 2, 540–556.
26. Ippen, E.P., Eichenberger, D.J. and Dixon, R.W. 1980. Picosecond pulse generation by passive modelocking of diode lasers. *Appl. Phys. Lett.*, 37, 267–269.

27. Gupta, S., Whitaker, J.F. and Mourou, G.A. 1992. Ultrafast carrier dynamics in II–V-semiconductors grown by molecular beam epitaxy at very low substrate temperatures. *IEEE J. Quantum Electron.*, 28, 2464–2472.

28. Siegner, U., Fluck, R., Zhang, G. and Keller, U. 1996. Ultrafast high-intensity nonlinear absorption dynamics in low-temperature grown gallium arsenide. *Appl. Phys. Lett.*, 69, 2566–2568.

29. Tan, H.H., Jagadish, C., Lederer, M.J., Luther-Davies, B., Zou, J., Cockayne, D.J.H., Haiml, M., Siegner, U. and Keller, U. 1999. Role of implantation-induced defects on the response time of semiconductor saturable absorbers. *Appl. Phys. Lett.*, 75, 1437–1439.

30. van der Ziel, J.P., Tsang, W.T., Logan, R.A., Mikulyak, R.M. and Augustyniak, W.M. 1981. Subpicosecond pulses from passively modelocked GaAs buried optical guide semiconductor lasers. *Appl. Phys. Lett.*, 39, 525–527.

31. Gopinath, J.T., Thoen, E.R., Koontz, E.M., Grein, M.E., Kolodziejski, L.A., Ippen, E.P. and Donnelly, J.P. 2001. Recovery dynamics in proton-bombarded semiconductor saturable absorber mirrors. *Appl. Phys. Lett.*, 78, 3409–3411.

32. Delpon, E.L., Oudar, J.L., Bouché, N., Raj, R., Shen, A., Stelmakh, N. and Lourtioz, J.M. 1998. Ultrafast excitonic saturable absorption in ion-implanted InGaAs/InAlAs multiple quantum wells. *Appl. Phys. Lett.*, 72, 759.

33. Fluck, R., Jung, I.D., Zhang, G., Kärtner, F.X. and Keller, U. 1996. Broadband saturable absorber for 10-fs pulse generation. *Opt. Lett.*, 21, 743–745.

34. Zhang, Z., Nakagawa, T., Torizuka, K., Sugaya, T. and Kobayashi, K. 2000. Gold-reflector-based semiconductor saturable absorber mirror for femtosecond mode-locked Cr^{4+}: YAG lasers. *Appl. Phys. B*, 70, S59–S62.

35. Keller, U., Miller, D.A.B., Boyd, G.D., Chiu, T.H., Ferguson, J.F. and Asom, M.T. 1992. Solid-state low-loss intracavity saturable absorber for Nd:YLF lasers: an antiresonant semiconductor Fabry–Perot saturable absorber. *Opt. Lett.*, 17, 505–507.

36. Brovelli, L.R., Keller, U. and Chiu, T.H. 1995. Design and operation of antiresonant Fabry–Perot saturable semiconductor absorbers for mode-locked solid-state lasers. *J. Opt. Soc. Am. B*, 12, 311–322.

37. Siegman, A.E. 1986. *Lasers* (New York: University Science Books).

38. Hönninger, C., Paschotta, R., Morier-Genoud, F., Moser, M. and Keller, U. 1999. Q-switching stability limits of cw passive modelocking. *J. Opt. Soc. Am. B*, 16, 46–56.

39. Thoen, E.R., Koontz, E.M., Joschko, M., Langlois, P., Schibli, T.R., Kärtner, F.X., Ippen, E.P. and Kolodziejski, L.A. 1999. Two-photon absorption in semiconductor saturable absorber mirrors. *Appl. Phys. Lett.*, 74, 3927–3929.

40. Langlois, P., Joschko, M., Thoen, E.R., Koontz, E.M., Kärtner, F.X., Ippen, E.P. and Kolodziejski, L.A. 1999. High fluence ultrafast dynamics of semiconductor saturable absorber mirrors. *Appl. Phys. Lett.*, 75, 3841–3843.

41. Schibli, T.R., Thoen, E.R. and Kärtner, F.X. 2000. Suppression of Q-switched mode locking and break-up into multiple pulses by inverse saturable absorption. *Appl. Phys. B*, 70, S41–S49.

42. Guina, M.D., Vainionpaa, A., Orsila, L., Harkonen, A., Lyytikainen, J., Gomes, L.A. and Okhotnikov, O.G. 2003. Saturable absorber intensity modulator. *IEEE J. Quantum Electron.*, 39, 1143–1149.

43. Seitz, W., Ell, R., Morgner, U., Schibli, T.R., Kärtner, F.X., Lederer, M.J. and Braun, B. 2002. All-optical active mode locking with a nonlinear semiconductor modulator. *Opt. Lett.*, 27, 2209–2211.

44. Seitz, W., Schibli, T.R., Morgner, U., Kärtner, F.X., Lange, C.H., Richter, W. and Braun, B. 2002. Passive synchronization of two independent laser oscillators with a Fabry–Perot modulator. *Opt. Lett.*, 27, 454–456.

45. Boggess, T.F., Moss, S.C., Boyd, I.W. and Van Stryland, E.W. 1985. Optical limiting in GaAs. *IEEE J. Quantum Electron.*, 21, 488–494.

46. van Stryland, E.W., van her Zeele, H., Woodall, M.A., Soileau, M.J., Smirl, A.L., Guha, S. and Boggess, T.F. 1985.

2-Photon-absorption, nonlinear refraction, and optical limiting in semiconductors. *Opt. Eng.*, 24, 613–623.

47. Perry, J.W., Mansour, K., Marder, S.R., Perry, K.J., Alvarez, D. and Choong, I. 1994. Enhanced reverse saturable absorption and optical limiting in heavy-atom-substituted phtalocyanines. *Opt. Lett.*, 19, 625–627.

48. He, G.S., Xu, G.C., Prasad, P.N., Reinhardt, B.A., Bhatt, J.C. and Dillard, A.G. 1995. 2-Photon absorption and optical-limiting properties of novel organic-compounds. *Opt. Lett.*, 20, 435–437.

49. Ehrlich, J.E., Wu, X.L., Lee, I.Y.S., Hu, Z.Y., Rockel, H., Marder, S.R. and Perry, J.W. 1997. Two-photon absorption and broadband optical limiting with bis-donor stilbenes. *Opt. Lett.*, 22, 1843–1845.

50. Tutt, L.W. and Boggess, T.F. 1993. A review of optical limiting mechanisms and devices using organics, fullerenes, semiconductors and other materials. *Prog. Quantum Electron.*, 17, 299–338.

51. Nussenzweig, H.M. 1972. *Causality and Dispersion Relations* (New York: Academic Press).

52. Stolen, R.H. and Ashkin, A. 1973. Optical Kerr effect in glass waveguide. *Appl. Phys. Lett.*, 22. 294–296.

53. Sheik-Bahae, M., Said, A.A., Wei, T.H., Hagan, D.J. and Van Stryland, E.W. 1990. Sensitive measurement of optical nonlinearities using a single beam. *IEEE J. Quantum Electron.*, 26, 760–769.

54. De Salvo, R., Said, A.A., Hagan, D.J., Van Stryland, E.W. and Sheik-Bahae, M. 1996. Infrared to ultraviolet measurements of two-photon absorption and n_2 in wide bandgap solids. *IEEE J. Quantum Electron.*, 32, 1324–1333.

55. Sheik-Bahae, M., Hutchings, D.C., Hagan, D.J. and Van Stryland, E.W. 1991. Dispersion of bound electronic nonlinear refraction in solids. *IEEE J. Quantum Electron.*, 27, 1296–1309.

56. Duguay, M.A. and Hansen, J.W. 1969. An ultrafast light gate. *Appl. Phys. Lett.*, 15, 192–194.

57. Lattes, A., Haus, H.A., Leonberger, F.J. and Ippen, E.P. 1983. An ultrafast all-optical gate. *IEEE J. Quantum Electron.*, 19, 1718–1723.

58. Kelley, PL. 1965. Self-focusing of optical beams. *Phys. Rev. Lett.*, 15, 1005.

59. Salin, F., Squier, J. and Piché, M. 1991. Mode-locking of $Ti:Al_2O_3$ lasers and self-focusing—A Gaussian approximation. *Opt. Lett.*, 16, 1674–1676.

60. Magni, V., Cerullo, G., De Silvestri, S. and Monguzzi, A. 1995. Astigmatism in Gaussian-beam self-focusing and in resonators for Kerr-lens mode-locking. *J. Opt. Soc. Am. B*, 12, 476–485.

61. Spence, D.E., Kean, P.N. and Sibbett, W. 1991. 60-fsec pulse generation from a self-mode-locked Ti:sapphire laser. *Opt. Lett.*, 16, 42–44.

62. Haus, H.A., Fujimoto, J.G. and Ippen, E.P. 1992. Analytic theory of additive pulse and Kerr lens mode-locking. *IEEE J. Quantum Electron.*, 28, 2086–2096.

63. Blow, K.J., Doran, N.J. and Nayar, B.K. 1989. Experimental demonstration of optical soliton switching an all-fiber nonlinear Sagnac interferometer. *Opt. Lett.*, 14, 754–756.

64. Ikeda, K. 1979. Multiple-valued stationary state and its instability of the transmitted light by a ring cavity system. *Opt. Commun.*, 30, 257–261.

65. Steinmeyer, G., Jaspert, D. and Mitschke, F. 1994. Observation of a period-doubling sequence in a nonlinear-optical fiber ring cavity near zero dispersion. *Opt. Commun.*, 104, 379–384.

66. Haus, H.A. and Islam, M.N. 1985. *IEEE J. Quantum Electron.*, 21, 1172–1188.

67. Mollenauer, L.F. and Stolen, R.H. 1984. The soliton laser. *Opt. Lett.*, 9, 13–15.

68. Haus, H.A., Fujimoto, J.G. and Ippen, E.P. 1991. Structures for additive pulse mode-locking. *J. Opt. Soc. Am. B*, 8, 2068–2076.

69. Ippen, E.P., Haus, H.A. and Liu, L.Y. 1989. Additive pulse mode-locking. *J. Opt. Soc. Am. B*, 6, 1736–1745.

70. Nakamura, S., Ueno, Y. and Tajima, K. 1998. Ultrafast (200-fs switching, 1.5-Tb/s demultiplexing) and high-repetition (10 GHz) operations of a polarization-discriminating symmetric Mach–Zehnder all-optical switch. *IEEE Photon. Technol. Lett.*, 10, 1575–1577.

71. Hirano, A., Kobayashi, H., Tsuda, H., Takahashi, R., Asobe, M., Sato, K. and Hagimoto, K. 1998. 10 Gbit/s RZ all-optical

discrimination using refined saturable absorber optical gate. *Electron. Lett.*, 34, 198–199.

72. Whitaker, N.A., Gabriel, M.C., Avramopoulos, H. and Huang, A. 1991. All-optical, all-fiber circulating shift register with an inverter. *Opt. Lett.*, 16, 1999–2001.

73. Schiek, R. 1994. All-optical switching in the directional coupler caused by nonlinear refraction due to cascaded 2nd-order nonlinearity. *Opt. Quantum Electron.*, 26, 415–431.

74. Naoum, R. and Salah-Belkhodja, F. 1997. Opto-optical switch in nonlinear integrated optics. *Pure Appl. Opt.*, 6, L33–L36.

75. Ramamurthy, P. 2002. Ultrafast all-optical switch using LT-GaAs based on DFB. *Proc. SPIE*, 4643, 266–273.

76. Moores, J.D., Hall, K.L., Lepage, S.M., Rauschenbach, K.A., Wong, W.S., Haus, H.A. and Ippen, E.P. 1995. 20-GHz optical storage loop laser using amplitude-modulation, filtering, and artificial fast saturable absorption. *IEEE Photon. Technol. Lett.*, 7, 1096–1098.

77. Moores, J.D., Wong, W.S. and Hall, K.L. 1995. 50-Gbit/s optical pulse storage ring using novel rational-harmonic modulation. *Opt. Lett.*, 20, 2547–2549.

PART III

Optoelectronic devices and techniques

10

Light-emitting diodes (LEDs)

KLAUS STREUBEL
OSRAM Opto Semiconductors GmbH

10.1 INTRODUCTION

Light-emitting semiconductor diodes (LEDs) are light sources that were developed in the last few decades. For most of this time, they have been used as small indicator lights in a wide range of consumer applications. Some 10 years ago, two new material systems, AlGaInP and InGaN, entered the LED arena and gave birth to a new generation of light-emitting diodes: the high-brightness LEDs. This was an important breakthrough for the entire LED business, which enhanced the prospects of

LED use in a much wider range of applications. Now, with InGaN covering the emission range from blue to green and AlGaInP from yellow to red, the entire visible spectrum has become accessible to LED light (Figure 10.1). Furthermore, the continuously improving material quality, together with better chip and package designs, have led to much enhanced performances in terms of efficiency and total output power. Today, at certain wavelengths, LEDs achieve more than 50% energy efficiency in the laboratory and ones with more than 20% efficiency are commercially available. The internal conversion

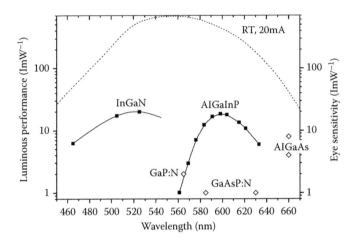

Figure 10.1 Luminous performance of commercial high-brightness InGaN and AlGaInP LEDs, compared with a few conventional GaP and GaAsP-based devices. The dotted line shows the human eye sensitivity.

of electrical power into light is sometimes close to 100% and the only task left is to extract as much light as possible out of the semiconductor material without it being lost internally. It can be projected that the good performance at certain colors will eventually be extended to the entire spectrum and the efficiency of commercial LEDs is expected to exceed 50%. Another attractive feature of LEDs is their very long lifetime, of at least some 10,000 h or several years of continuous operation. Finally, the availability of highly efficient LEDs covering the range from violet to red has now allowed the generation of white light and enabled the LED to enter the wide field of illumination and lighting.

This chapter will give a brief overview of the field of visible light-emitting diodes. The history of the development of visible diodes will be summarized, starting with the early GaAsP and GaP-based devices and continuing up to the development of high-brightness AlGaInP and InGaN LEDs. Next, some basic aspects of the physics of LEDs will be described, with emphasis on their optical and electrical properties. The major semiconductor material systems used for visible LEDs and their fundamental properties are then introduced. Technologies for enhanced light extraction, such as wafer bonding on transparent substrates (TSs), wafer-soldering for substrate-less devices or resonant-cavity and photonic-bandgap designs are discussed next, followed by a description of the most common ways of generating white LED light. Finally, the standard packages for LEDs and

some of the most important applications of high-brightness LEDs will be presented. The appendix A includes an introduction into the standard Commission Internationale de L'Eclairage (CIE) color system, a description of the human eye sensitivity curves and an overview of the most important radiometric and photometric units that are frequently used to describe the properties of LEDs.

At this point, it might be helpful to define some conventions for the following chapter. The most important parameters to characterize an LED are the efficiency and the output power. If not specified otherwise, efficiency measures the conversion of electrical to optical power and is usually expressed in per cent. For many applications, however, it is more relevant to measure the power of light as perceived by the human eye. In this case, it is convenient to use photometric units such as lumen and candela. The lumen is a measure of the total light power, weighed according to the human eye sensitivity. Efficiency is then expressed in lumens per (electrical) watt. Candelas are a measure of brightness emitted into a certain direction per unit of solid angle and is expressed in lumens per steradian. In the following, the performance of LEDs will be expressed in lumens and lumens per watt (lm W^{-1}) with a few exceptions where mW and per cents are used. Another parameter is the color of the LED emission. If nothing else is specified, the emission wavelength normally means the peak wavelength of the optical spectrum, which has to be distinguished from the dominant apparent wavelength when the human eye perception is taken into account (see Appendix A).

10.1.1 Historical review

In the very beginning of the last century in 1907, H.J. Round discovered a "curious phenomenon" when he applied a large voltage, of more than 100 V, to a SiC crystal: electroluminescence—the conversion from electrical current to light [1]. Many years later, in the early sixties, electroluminescence was studied extensively on III–V semiconductor alloys such as GaAs or GaAsP. In 1962, the first visible red emitting LEDs were made of GaAsP with a P-fraction of 40%. The devices were fabricated by vapor phase epitaxy on GaAs-substrates, with a pn-junction formed by selective Zn-diffusion into a thick n-type layer. For low P-fractions, GaAsP has a direct energy gap (see Section 10.3.1) and is almost lattice matched to GaAs. As the P-content is increased, in order to achieve shorter wavelengths, the lattice mismatch and thus the defect density in the material increases appreciably, leading to rather low efficiencies. In 1968, the first commercial red GaAsP LEDs, with luminous efficiencies around 0.2 lm W^{-1}, were introduced to the marketplace by Hewlett Packard. Shortly thereafter, more efficient devices based on GaP doped with Zn and O, were developed using liquid phase epitaxy (LPE) on GaP wafers. GaP:Zn–O LEDs emit at 700 nm, where the human eye sensitivity is down to 0.4% of its maximum value, resulting in luminous efficiencies around 0.4 lm W^{-1}. A breakthrough for the early LEDs was the discovery that isoelectronic impurities, such as nitrogen, can act as efficient radiative recombination centers in GaP and GaAsP. GaAsP:N and GaP:N based LEDs already showed efficiencies around 1 lm W^{-1}, about one order of magnitude more than the first GaAsP diodes.

In the early 1970s, AlGaAs/GaAs single heterojunction diodes were developed, also using LPE as the method of crystal growth. Starting with emission in the infrared spectral range, the emission wavelength was continuously reduced by increasing the Al-fraction, until red 660 nm emitting diodes were achieved in 1980. The next advance was the introduction of double heterostructures (DHs) for more efficient confinement of the electrical carriers in the active region of the device. The first commercial AlGaAs DH LED with an efficiency around 4 lm W^{-1} was introduced in 1985.

Although more efficient as a result of a better carrier confinement and direct band transitions, AlGaAs/GaAs LEDs had one obvious disadvantage compared to the early GaP-based diodes: unlike GaP, GaAs is not transparent to visible light. The introduction of LPE grown, transparent AlGaAs substrates solved this problem and enhanced the efficiency by almost a factor of two. The first TS LEDs were introduced in 1987 as "super-high-brightness devices" with efficiencies around 8 lm W^{-1}.

Because the AlGaAs bandstructure becomes indirect at high Al-fractions, it was not possible to fabricate AlGaAs LEDs emitting at wavelengths shorter than 660 nm. Therefore, a new material system, AlGaInP lattice matched to GaAs, was investigated intensively. Epitaxial growth was only possible using the new method of metal-organic vapor phase epitaxy (MOVPE). The quaternary composition of AlGaInP offered an additional advantage: the bandgap could be tuned from yellow-green to red while maintaining the same lattice constant thus allowing lattice matching on GaAs. With the new material, efficiencies in the order of 10 lm W^{-1} were achieved. As previously with the AlGaAs/GaAs LEDs, the efficiency of AlGaInP/GaAs devices was limited by absorption in the GaAs substrate. A TS AlGaInP LED was therefore developed by Hewlett Packard, using the novel technology of direct wafer bonding. After the growth of the AlGaInP layers on GaAs, the original GaAs substrate was removed and the thin AlGaInP film was transferred ("wafer-bonded") to a transparent GaP wafer. As before in the case of TS-AlGaAs LEDs, the efficiency of TS-AlGaInP LEDs was doubled compared to absorbing substrate (AS) LEDs and efficiencies of more than 20 lm W^{-1} were demonstrated on red devices. Another successful method for high efficiency developed at that time was the introduction of thick, transparent window layers giving better current spreading and improved light extraction. In commercial AlGaInP-LEDs, both AlGaAs and GaP windows were employed.

The performance of TS-AlGaInP LEDs was further improved by optimizing the size and optical properties of the chips, yielding efficiencies of more than 70 lm W^{-1} for red emission. Recently, this already impressive performance was further enhanced by shaping the TS-die into truncated inverted pyramids. Record high efficiencies of >100 lm W^{-1} (610 nm) or >50% (630 nm) were achieved with this technology. Alternatively, Osram-OS started to solder the AlGaInP epitaxial structure onto a new carrier and to remove the original GaAs substrate. The advantage of the wafer-soldering process is its suitability for mass production and its potential to achieve high yields on large diameter wafers. The best "substrate-less" AlGaInP LEDs achieve >60 lm W^{-1} (615 nm) or up to 40% efficiency (630 nm) with the potential for further improvements.

The early blue emitting LEDs were made of SiC with efficiencies of only 0.1–0.2 lm W^{-1}. In parallel, researchers started to investigate the electroluminescence of GaN-based devices in 1970 [2]. Green and "violet" emitting devices were fabricated but the devices were very inefficient, mainly due to the almost unsolvable problem of p-doping in GaN. The major breakthrough came in 1989, when a Japanese group achieved real p-conductivity by activating Mg dopants with low-energy electron-beam irradiation. Nichia Chemicals finally developed a new technique for efficient Mg-activation by high-temperature annealing and brought the first commercial blue, GaN-based LEDs to the market in 1994. The GaN was deposited by MOVPE on sapphire substrates. The bandgap of nitride-based LEDs can be tailored over a wide range, from 362 nm (3.4 eV) to 615 nm (2 eV), by adding additional In to the InGaN alloy. This allowed the fabrication of green LEDs and the first blue–green (500 nm) and green (520 nm) devices were also produced by Nichia, for application in traffic lights. As an alternative to the growth on sapphire, which is electrically insulating, InGaN can also be grown epitaxially on SiC wafers. This technology was developed by Cree Research and resulted in the "superbright" GaN/SiC LED chip introduced in 1995.

With efficient blue and green LEDs at hand to complement the high-brightness red emitters, it became possible to fabricate white LEDs. Several companies developed white LEDs, using a combination of blue, green and red emitting dies in a single package. An alternative approach to generate white light was developed independently in Japan and Germany: the conversion of blue LED light into white using a phosphor wavelength converter. Osram-OS (formerly part of Siemens)

developed a phosphor based on $Y_3Al_5O_{12}$ garnet, doped with cerium ions, that absorbed part of the blue emission of a GaN-LED and produced yellow luminescence. The combination of blue and yellow emission produces white light. The phosphor is either suspended in the epoxy resin used for encapsulation or is directly coated on the chip surface. Commercial white LEDs were introduced to the marketplace in 1998 by Osram-OS and Nichia.

10.2 PHYSICS OF LEDs

The basic function of an LED is to generate light following the injection of an electrical current into the semiconductor material. The conversion of electrical carriers (electrons) into light (photons) is called electroluminescence and involves two important processes: the excitation of electrons into higher energy states and the relaxation of excited electrons back to empty lower states. If the relaxing electrons release most or all of the energy difference in form of electromagnetic radiation, the process is called a radiative transition. In the case of non-radiative transitions, the energy difference in such a relaxation process is released in form of heat (phonons). The band structure of the semiconductor material plays an important role for the transition processes of excited electrons. In semiconductors with a direct band structure (Figure 10.2), the excited electrons in the conduction band can relax into states in the valence band under momentum conservation. In materials with an indirect band structure, the transition process requires the assistance of a phonon in order to conserve momentum. Indirect band transitions are therefore much less probable.

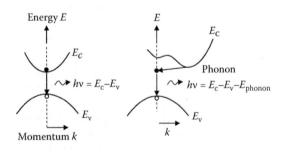

Figure 10.2 Left: radiative band–band transitions in a direct bandgap material; right: radiative transition in an indirect bandgap material.

10.2.1 Optical properties of LEDs

10.2.1.1 RADIATIVE RECOMBINATION

Figure 10.3 shows a number of possible radiative and non-radiative transitions in a simplified band diagram. The direct band–band transition is shown in Figure 10.3a. Because two particles, electrons n and holes p, are involved, this process is called bimolecular recombination. The recombination rate, R (recombinations, per unit volume, per second), is proportional to the density of carriers in the upper state (electrons) and the density of empty lower states (holes) and is given by

$$R = -\frac{dn}{dt} = -\frac{dp}{dt} = Bnp \qquad (10.1)$$

where n and p are the electron and hole concentrations. The proportionality factor B is a measure of the probability of radiative recombination and is called the bimolecular recombination coefficient. It can be calculated from a few basic material parameters such as the bandgap energy, the absorption coefficient and the refractive index [3] which are all experimentally accessible. For most III–V semiconductors, the bimolecular recombination coefficient is in the order of 1–10×10^{-10} cm^3 s^{-1}. At very high carrier concentrations, as for instance in laser diodes, B starts to decrease and Equation 10.1 has to be modified [4].

If the carrier concentrations deviate from the equilibrium, e.g., in the case of electrical or optical excitation, n and p can be written as $n = n_0 + \Delta n$ and $p = p_0 + \Delta p$, and the recombination rate can be split into two parts, an equilibrium recombination rate R_0 and an excess recombination rate ΔR:

$$R_0 + \Delta R = B(n_0 + \Delta n)(p_0 + \Delta p). \qquad (10.2)$$

The small term $\Delta n \Delta p$ in Equation 10.2 can be neglected. With $R_0 = B n_0 p_0$, the expression can be written as

$$\frac{\Delta R}{R_0} = \frac{\Delta n}{n_0} + \frac{\Delta p}{p_0} \qquad (10.3)$$

with the excess recombination rate given by

$$\Delta R = B(n_0 + p_0)\Delta n. \qquad (10.4)$$

In the case of low excitation, with an excess carrier concentration much smaller than $(n_0 + p_0)$, the excess carriers will decay exponentially with time

$$\Delta n(t) = \Delta n_0 \exp\left(-\frac{t}{\tau}\right) \qquad (10.5)$$

which gives the time-dependent recombination rate as

$$\Delta R(t) = -\frac{d\Delta n(t)}{dt} = \frac{\Delta n_0}{\tau} \exp\left(-\frac{t}{\tau}\right). \qquad (10.6)$$

Comparing Equations 10.4 and 10.6, the decay lifetime is obtained as

$$\tau = \frac{\Delta n}{\Delta R} = \frac{1}{B(n_0 + p_0)} = \frac{1}{R_0}\frac{n_0 p_0}{n_0 + p_0}. \qquad (10.7)$$

In the intrinsic case of an undoped semiconductor, with $n_0 = p_0 = n_i$, Equation 10.7 reduces to

$$\tau = \frac{1}{2 B n_i} = \frac{n_i}{2 R_0}. \qquad (10.8)$$

By inserting Equation 10.8 in Equation 10.1, the radiative recombination rate can be expressed as

$$R = Bnp = \frac{np}{2 n_i \tau}. \qquad (10.9)$$

For n-type materials, the electron is referred to as the majority carrier and the hole as the minority carrier,

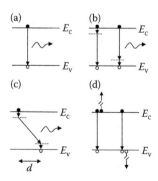

Figure 10.3 Recombination mechanisms in a simple band diagram: (a) radiative band–band transition, (b) band-impurity level transitions, (c) donor–acceptor transition, and (d) Auger recombination process.

since $n_0 \gg p_0$. For p-type semiconductors, the roles are reversed. For an n-type semiconductor, we have $n_0 \gg \Delta n \gg p_0$, meaning that the excess carrier concentration is much smaller than the majority carrier concentration, yet much larger than the minority carrier concentration. Because both types of carriers are required for the recombination process, the carrier lifetime that determines the radiative decay of the excess carriers, is the minority carrier lifetime. A very large fraction of majority carriers cannot find a minority carrier for recombination and thus the average majority carrier lifetime is much longer. In the case of doping, the carrier lifetimes become $\tau_n = 1/(Bp_0)$ and $\tau_p = 1/(Bn_0)$, for p-and n-doped semiconductors, respectively.

Intermediate states in the semiconductor bandgap may also contribute to the transition from the upper conduction band to the lower valence band, as shown in Figure 10.3b. Examples are transitions from the conduction band to acceptor states or transitions from donor states to the valence band. If the impurity level is deep inside the bandgap, such transitions can be highly probable. Because the impurity has a well-defined position in the crystal lattice, its wave function extends widely in the momentum space. This permits momentum-conserving transitions to occur with reasonable probability even in indirect bandgap materials such as GaP.

Apart from band-impurity level transitions, transitions between two impurity levels are also possible (Figure 10.3c). These donor–acceptor transitions can be rather effective and are very useful for light-emitting diodes. Photons generated in this process have an energy, $h\nu$, given by

$$h\nu = E_c - E_v - (E_D + E_A) \tag{10.10}$$

where E_D and E_A are the donor and acceptor binding energies. Coulomb interaction between the donor and acceptor atoms increases the energy of the excited states by an increment of $q^2/\varepsilon d$, where ε is the dielectric constant of the semiconductor host crystal and d the distance between donor and acceptor atoms. This additional energy increases inversely with the separation, d, between donors and acceptors. Because the spatial distance between donors and acceptors in the host crystal can vary largely, the Coulomb energy increment varies accordingly, causing a wide range of possible photon energies and thus a broad emission spectrum.

10.2.1.2 NON-RADIATIVE RECOMBINATION

Radiative transitions are electron–hole recombinations in which the excess energy is used to generate a photon. Non-radiative transitions are all other transitions having one feature in common, namely that no photon is generated in the transition process. In fact, in many semiconductors non-radiative carrier recombination is the dominant process. In semiconductors with indirect band structure, such as Si or Ge, the measured radiative lifetime is three orders of magnitude smaller than the calculated value, due to the much larger probability of non-radiative transitions. Experimentally, non-radiative transitions are much more difficult to characterize than radiative transitions, because no characteristic photons are generated. They can be measured only indirectly by analysing the radiative efficiency or the dynamics of the radiation process after external excitation.

The most relevant non-radiative process for optoelectronic devices is recombination via states related to crystal defects. Examples of such defects are dislocations, pores, grain boundaries, vacancies, inclusions or precipitates. Carriers within a diffusion length of the defect will usually be trapped by the defect states and recombine there. The recombination rate scales linearly with the carrier concentration:

$$R_{nr} = A * n \tag{10.11}$$

where the proportionality factor A^* increases with the density of non-radiative centers.

Usually a defect deforms the bandstructure, either by trapping charges or by deforming the lattice, thereby inducing local strain. In either case, the deformation produces a potential barrier of the height, E_{act}, which has to be overcome by the carriers. Therefore, the process recombination has a thermal dependence of the form:

$$R_{nr}(T) = R^* \exp\left(-\frac{E_{act}}{kT}\right) \tag{10.12}$$

where $R_{nr}(T)$ is the non-radiative recombination rate and R^* a temperature independent coefficient. The temperature dependence of the radiative transition can be neglected. Therefore, as the temperature is reduced, the non-radiative recombination rate decreases exponentially and the radiative processes become more dominant. The temperature dependence of the radiative recombinations over the total number of recombinations is then

$$\eta_{qi}(T) = \frac{R_r}{R_r + R_{nr}} = \frac{1}{1 + z\frac{R^*}{R_r}\exp\left(-\frac{E_{act}}{kT}\right)} \quad (10.13)$$

where the ratio R^*/R_r is constant. The quantity $\eta_{qi}(T)$ is the internal quantum efficiency.

So far, we have been looking at non-radiative processes inside the semiconductor material. Additionally, minority carriers reaching the surface of the LED are lost due to surface recombination [5]. At the semiconductor surface, the periodic arrangement of atoms, which is the basis for the band structure model, no longer exists. Thus, the surface is a perturbation of the crystal lattice with a strong impact on the band diagram. Due to the lack of neighbours, surface atoms have dangling bonds, partly filled electron orbitals, which can be described as deep or shallow energy levels in the bandgap which may act as recombination centers. The recombination of carriers via surface states is dependent on the semiconductor material and is phenomenologically described by a parameter called the surface recombination velocity, S. It is particularly high in GaAs ($S = 10^{6\,cm}$ s^{-1}) compared to InP ($S = 10^{3\,cm}$ s^{-1}) or Si ($S = 10^{1\,cm}$ s^{-1}). In Al-containing alloys such as AlGaInP, the surface recombination scales appreciably with the Al-fraction [6]. S increases from $10^{5\,cm}$ s^{-1} for $(Al_{0.1}Ga_{0.9})_{0.5}In_{0.5}P$ to $10^{6\,cm}$ s^{-1} for AlInP.

When designing LEDs, it is very important to consider surface recombination. Carrier injection into the active region of the device should take place several diffusion lengths away from any surface. Recombination at the top-surface of the LED is usually prevented by a high-bandgap confinement layer above the active region and a thick current-spreading or window layer. In some LEDs, light extraction shapes are etched deeply into these top layers. Theoretically, very high extraction efficiencies should be achievable if such shapes would penetrate the pn-junction, but then a substantial number of carriers could recombine at the surface states. However, as long as the active region remains planar, surface recombination can be neglected due to the lack of minority carriers above the confinement layers. In LEDs with a planar surface, carrier diffusion to the lateral edges is reduced by placing the top electrode in the center of the die.

Instead of generating a photon, the energy released during carrier recombination can also be transferred to other carriers (electrons or holes) and then be dissipated as phonons. This non-radiative process is called Auger recombination. Two examples for Auger processes are shown in Figure 10.3d, but many other processes are possible, depending on the nature and occupation of the involved electronic states. Since two carriers of one type and one carrier of the opposite type are required for the Auger process, the recombination rate is proportional to either np^2 or pn^2. The proportionality factor C is called the Auger coefficient and has typical values for III–V materials of 10^{-28}–10^{-29} cm^6 s^{-1}. It is described as

$$R_{Auger} = C_p np^2 + C_n pn^2 \quad (10.14)$$

or, in the intrinsic case:

$$R_{Auger} = Cn_i^3. \quad (10.15)$$

The Auger recombination thereby scales as the cube of the carrier concentration. For LEDs where typical carrier densities are low compared to lasers, the Auger recombination plays a minor role in the device efficiency. However, the Auger effect becomes more prominent in materials with small bandgap energies such as infrared emitting devices.

10.2.1.3 EMISSION WAVELENGTH

The center emission wavelength of an LED is mainly given by the bandgap energy $\lambda = hc/E_g$. The full width at half maximum (FWHM) of the emission is approximately $\Delta E \approx 2kT$ [7]. Thus, at a given temperature, $\Delta\lambda/\lambda$ scales with $2kT/E_g$ or $\Delta\lambda$ increases with λ^2. The linewidth can be increased by using high doping levels or having graded compositions in the active region of the LED. The emission linewidth is an important parameter for near-infrared LEDs for optical fiber transmission systems due to the dispersive nature of the fiber.

Light that is extracted through the top surface has to penetrate through only transparent material before reaching the semiconductor–air or –epoxy interface. A significant fraction of light is guided

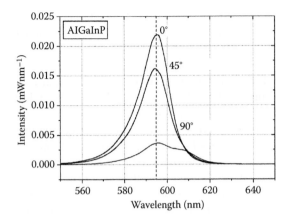

Figure 10.4 Spectrum of an AlGaInP LED for different directions of emission. 0°=normal surface emission, 90°=side emission.

along the active layers and extracted at the die sidewalls. Along this path, the emission wavelength is changed by absorption, which affects mostly the short-wavelength part of the spectrum. This effect is shown in Figure 10.4 for an AlGaInP LED. Here, the emission spectrum measured at the top-side (0°) differs substantially from the side emission (90°). While the effect is typical for all LEDs that employ band–band transitions, it is not present in LEDs which are based on impurity level transitions, such as used in GaP-based devices. Figure 10.4 shows the emission spectra of a GaP:Zn, O LED which is essentially independent of the emission direction.

10.2.1.4 LIGHT EXTRACTION

Although light can be efficiently generated inside an LED ($\eta_{qi} > 90\%$) only a few per cent of the generated photons can actually escape from the semiconductor material. Due to the large difference in the refractive index between the semiconductor ($n=3–3.5$) and the surrounding medium ($n=1–1.5$), the interface acts as a perfect mirror for most of the generated photons and prevents them from leaving the diode. The major contribution to the high reflectivity is total internal reflection which occurs at all angles larger than the critical angle ϑ_c. According to Snell's law, this angle is

$$\vartheta_c = \arcsin\left(\frac{n_m}{n_s}\right) \qquad (10.16)$$

where n_s and n_m are the refractive indices of the semiconductor and the surrounding medium. All angles are measured to the direction normal to the interface. Only rays of light with $\vartheta < \vartheta_c$ can escape from the diode. In 3D-space this range of angles forms a cone, which is frequently called the "escape cone." If the semiconductor chip (e.g., GaAs, $n=3.5$) is surrounded by air ($n=1$), the critical angle ϑ_c is 17°, in the case of epoxy encapsulation with a refractive index of $n=1.5$, ϑ_c becomes 26°. The fraction of escaping photons to the total number of generated photons is then given by

$$\eta_{ext} = \frac{2\pi(1-\cos\vartheta_c)}{4\pi} = \frac{1}{2}(1-\cos\vartheta_c) \qquad (10.17)$$

which is the ratio of the surface subtended by the spherical escape cone divided by the surface of the full sphere. Even for light that is incident perfectly normal to the chip surface some of the photons are reflected by Fresnel reflection:

$$RF_{resnel} = \frac{(n_m - n_s)^2}{(n_m + n_s)^2}. \qquad (10.18)$$

This relation describes the reflection at normal incidence. The correct angular dependence is described by the so-called Fresnel formulae which were first derived by Fresnel in 1823 [8]. For reasons of simplicity we approximate the Fresnel reflection within the light extraction cone by Equation 10.18. Then the total extraction efficiency is

$$\eta_{extr} = \frac{1}{2}(1-\cos\vartheta_c)\left(1 - \frac{(n_m - n_s)^2}{(n_m + n_s)^2}\right). \qquad (10.19)$$

For a GaAs-based LED the fraction of light that escapes the material is as low as ≈2% or ≈4% for air or epoxy as surrounding medium, respectively. Internal reflection is therefore a major obstacle for the fabrication of high efficiency AlGaInP/GaAs or AlGaAs/GaAs-based LEDs.

The typical destiny of a totally reflected photon is to be reflected a couple of times before finally being absorbed either in the active region or someplace else in the LED structure. Even without the presence of absorption, the angles of reflection in a typically cube-shaped LED die will continue to reproduce themselves, without ever falling inside an escape cone (Figure 10.5). A straightforward remedy for LEDs on TSs is to give the entire die a

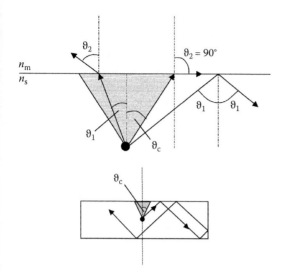

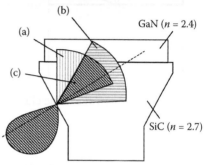

Figure 10.5 Top: total internal reflection and "escape cone" at the semiconductor–air interface. Bottom: reproduction of reflection angles in a cubically shaped LED die.

Figure 10.6 Above: InGaN "Aton" chip with inclined facets. Below: principle of light extraction in the "Aton"-die: (a) range of angles coupled from GaN into SiC, (b) range of angles that can transmit the SiC–air interface, (c) fraction of light in SiC that can be extracted.

geometrical shape which is more suitable for light extraction. The ideal solution would be a spherical or hemispherical chip with small active area in the center. Other geometries are LEDs with hemispherically shaped top side, inverted cones, truncated spheres or truncated cones. Already, in the early sixties, die shaping for high efficiency LEDs was extensively investigated [9,10]. At that time, the technology of more or less individually shaped dies turned out to be not practical, mostly because of the associated high costs. However, more than 30 years later, high brightness LEDs with geometrically shaped dies have now entered the market: the truncated inverted pyramid AlGaInP LED (see Section 10.4) and the InGaN/SiC "Aton" LED, offered by Lumileds and Osram-OS, respectively. A photograph of the Aton-chip is shown in Figure 10.6. Light is generated in the GaN-based structure on the SiC-substrate. Because the refractive index of SiC is larger than the index of GaN, light rays which enter the SiC material can only cover an angular range of 0°–68°, according to Snell's law. Since most of this light would be totally reflected at the SiC–air interface in a cube-shaped die, the side facets are inclined by 30°. This increases the angular range of extracted light rays appreciably and results in an almost doubled brightness compared to the standard rectangular die [11].

From a fabrication point of view, planar LED structures with the shape of a rectangular parallelepiped are favorable. One way to increase the efficiency of planar LEDs is to increase the thickness of the transparent layers above the active layer, until the escape cones parallel to the active layer are fully utilized. This is schematically shown in Figure 10.7. If the window layer thickness d is big enough, light generated in the center and emitted into the side escape cone can escape if

$$d_{\text{window}} = \frac{w}{2} \tan \vartheta_c \tag{10.20}$$

where w is the lateral size of the die. For an encapsulated GaP LED with $n_s = 3.4$ and $n_m = 1.5$, the critical angle ϑ_c is 26°. If the chip width is $w = 300\,\mu m$, the optimum window layer would be $d_{\text{window}} = 74\,\mu m$. If the substrate is transparent, it is theoretically possible to extract the escape cones from all six sides of the cube, which would give 24% extraction efficiency for an

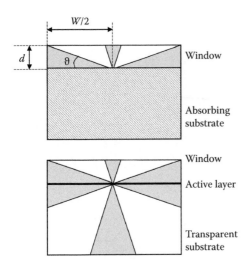

Figure 10.7 Escape cones in a cubic LED die. Top: LED with AS and thick window layer. Bottom: TS LED.

encapsulated LED. In case of an AS, the bottom cone and the bottom half of the four side cones are lost, leaving an extraction efficiency of 12%, if the window layer complies with Equation 10.20. This is of course only a very crude model to estimate the effect of transparent windows and substrates because important effects like multiple pass reflection, absorption and re-emission are neglected. Real LEDs with TS and thick window layer can achieve up to 32% external efficiency [12] exceeding even the theoretical value of 24% using the simplified model.

Bragg mirror—About half of the emission of an LED is directed towards the substrate. If the substrate is not transparent, this light is basically lost by absorption resulting in a substantial reduction in extraction efficiency. This loss can be partly reduced by the introduction of a Bragg mirror between the active layer and the AS. A Bragg mirror consists of a stack of layer pairs with different refractive indices, where each layer has the thickness of a quarter optical wavelength. In such a stratified layer structure, each interface provides a certain reflectivity, depending on the contrast in refractive index. The Fresnel reflectivity at each interface is given by

$$R_{\text{Fresnel}} = \frac{(n_{\text{low}} - n_{\text{high}})^2}{(n_{\text{low}} + n_{\text{high}})^2}. \tag{10.21}$$

Here, n_{low} and n_{high} are the refractive indices of the layer with higher and lower index, respectively. Because all layers have the thickness of a quarter wavelength, all reflections in that direction add in phase so that the total reflection for perpendicular incidence can be very high. If the number of layer pairs is m, the mirror reflectivity is

$$R_{\text{DBR}} = \left[\frac{1 - \left(\dfrac{n_{\text{low}}}{n_{\text{high}}}\right)^{2m}}{1 + \left(\dfrac{n_{\text{low}}}{n_{\text{high}}}\right)^{2m}} \right]^2 \tag{10.22}$$

This shows that a high mirror reflectivity can be achieved even for low index contrast materials, if the number of layer-pairs m is high enough. However, besides the practical drawback of growing large number of layer pairs, a low index contrast also results in a very narrow reflection bandwidth both in terms of wavelength and angular range. Therefore, Bragg mirrors are only of practical use if the material system offers a ratio $n_{\text{high}}/n_{\text{low}}$ sufficiently larger than 1.

In an LED, the Bragg mirror should be tuned to a maximum power reflectivity integrated over a range of angles within the escape cone. Note, that this situation is typically not achieved with a Bragg mirror optimized for perpendicular incidence. By shifting the center reflectivity of a DBR towards longer wavelength, the reflectivity for off-axis light rays is enhanced and, since the solid angle with a given angular increment in these directions is much larger, a significantly higher integral reflectivity is obtained. If we take a 630 nm AlGaInP-LED, with a 20 period $Al_{0.5}Ga_{0.5}As/AlAs$ DBR, the integral reflectivity for a center wavelength at 630 nm is in the order of 50%, whereas it is close to 75% if the mirror is tuned to 670 nm.

Surface structuring—Another approach to enhance the extraction efficiency of LEDs on ASs is to texture the surface [13]. Figure 10.8 shows a schematic drawing of a commercial surface-structured high-brightness AlGaInP-LED. The light extraction structure is etched into the window layers above the pn-junction. The basic idea is to inject the electrical current only along the edges of the die so that most of the light is generated close to the edge. The surface structure is then optimized for the extraction of light generated

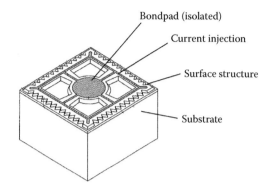

Figure 10.8 Sketch of a surface structured LED, with an etched light-extraction structure along the contacts at the chip edges.

below the electrical contacts. An additional benefit of this approach is that the light is generated far away from the bond pad, which would otherwise shadow a substantial fraction of the emission. Other approaches for high extraction efficiency are discussed in Section 10.4.

10.2.1.5 RADIATION PATTERNS

If external factors such as reflection, absorption or re-emission are excluded, the spontaneously generated photons in an electronic transition are emitted isotropically into all directions. The radiant intensity I_e of an LED is the divergence of optical power at a defined angular direction. In the case of ideal isotropic emission, the radiant intensity I_e would have the same value of $I_e = P_{out}/\Omega = P/4\pi$ in all directions. The solid angle Ω of a full sphere has the value $\Omega = 4\pi$. In the non-isotropic case, I_e is defined as the integral of $dP_{out}/d\Omega$ over all angles. Ideal isotropic radiation can only be achieved with point light sources and without the presence of internal reflection. In a real semiconductor diode, however, light is generated in an active region with the lateral extension of the die and is extracted through the semiconductor–air interface. In this case, the radiant intensity can be described by a cosine dependence:

$$I_e(\vartheta) \approx I_0 \cos(\vartheta) \qquad (10.23)$$

This is the Lambertian cosine law of radiation which applies to most planar LEDs. LEDs with TSs such as GaP-and AlGaAs/AlGaAs-LEDs or wafer-bonded AlGaInP/GaP devices, have enhanced side emission. Some novel types of LEDs show totally

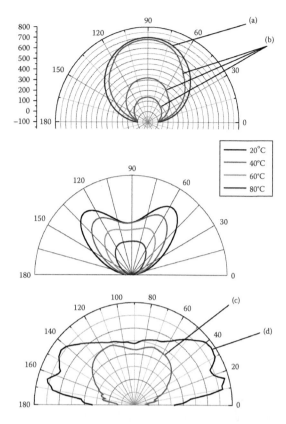

Figure 10.9 Emission pattern of different LED dies: top: ideal Lambertian radiator (a), AlGaInP LED at different current (b), middle: 650 nm resonant cavity LED, bottom: conventional SiC/InGaN LED (c) and InGaN "Aton" die (d).

different emission patterns. Examples are LEDs with shaped dies such as the truncated inverted pyramid LED [14] or the "Aton LED" [15]. Other examples are resonant-cavity LEDs, where the radiation pattern is given by the built-in optical resonator [16] (Figure 10.9). Besides the properties of the LED die, the radiation characteristics of encapsulated LEDs is strongly influenced by the used package.

10.2.2 Electrical properties electroluminescence

In a semiconductor diode, an electrical current composed of a flow of electrons and holes, is injected into the active region. A forward bias across the pn-junction reduces the potential barriers for electrons and enables them to spill over into the p-region (Figure 10.10). Similarly, the forward

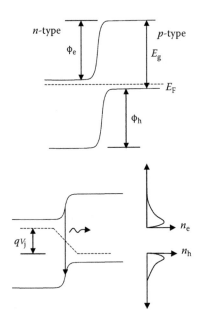

Figure 10.10 Band structure at the pn-junction of a simple homostructure LED. Above: with no bias and below: under forward bias. Lower right: electron and hole distributions.

bias reduces the potential barrier which blocked the flow of holes into the n-region. The overlapping concentration of electrons and holes allows spontaneous recombination.

The most simple diode structure is a pn-homojunction LED, where the p-and n-doped regions of a semiconductor material form a pn-junction. Examples of homojunction LEDs are blue GaN-LEDs where the pn-junction is created during the eptiaxial growth of n- and p-type GaN layers. Better carrier confinement is achieved, if the pn-junction is formed by two semiconductors having different bandgaps. Single heterojunction (SH) LEDs typically involve an n-type semiconductor with a narrow bandgap and a p-type semiconductor with a wider bandgap. SH active regions were first employed for LEDs in the AlGaAs system. Even better injection efficiency of electrons and holes and improved carrier confinement can be achieved if two heterojunctions are used to form the active region. Double heterostructure active regions of LEDs use a narrow bandgap semiconductor, sandwiched between two wide bandgap materials (Figure 10.11) [17,18]. In semiconductor lasers, the narrow bandgap layer has a thickness of a few ten nanometers, whereas in LEDs it can

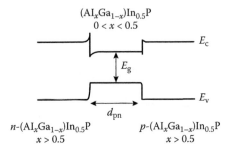

Figure 10.11 Schematic diagram of a double heterostructure LED active layer in the AlGaInP-system.

be up to some micrometers thick. High-brightness AlGaInP-and InGaN-LEDs employ double heterostructure active regions.

If A_{pn} and d_{pn} are the area and thickness of the active region, the generation density of carriers is

$$G = \frac{I_a/q}{A_{pn}d_{pn}} = \frac{J_a}{qd_{pn}} \quad (10.24)$$

with $J_a = I_a/A_{pn}$ as the current density. If the generation rate of carriers equals the recombination rate in the active layer of the LED, the system has reached an equilibrium:

$$G = \frac{J_a}{qd_{pn}} = An + Bn^2 + Cn^3. \quad (10.25)$$

The right-hand side of Equation 10.25 is the total recombination rate for a given density of injected electrons n. The coefficient A in the linear term accounts for surface recombination, as well as recombination via defect levels. For typical, undoped III–V materials, A is in the order of 10^8 s^{-1}. The quadratic term includes the bimolecular recombination coefficient B and accounts for radiative transitions. The cubic term is the Auger recombination.

According to Equation 10.25, the radiative recombination rate increases with the square of the carrier density, whereas the non-radiative recombination is basically a linear function of the carrier density (neglecting Auger). As the carrier concentration in the active region increases, the radiative process starts to dominate over the non-radiative process and the internal efficiency of the device will increase. High carrier density in the active layer is usually achieved with a double het-erostructure, where the active layer is sandwiched between large-bandgap barrier layers. The number of carriers that spill over the barriers is exponen-tially dependent on the barrier height, as well as on the temperature. Thus, to minimize this carrier leakage, the heterobarriers should be as high as possible, at least a few kT. While some alloy sys-tems, such as AlGaAs/GaAs or InGaN/GaN, offer the possibility to realize high potential barriers, some other systems, like AlGaInP, lack large direct bandgap compositions.

The typical ternary and quaternary III–V alloys used for LEDs consist of solid solutions of binaries with direct (GaAs, InP) and indirect (GaP, AlP) bandgaps. As the fraction of indirect materials is increased, the bandstructure can change from direct to indirect. The closer the composition of an active region is to the so-called cross-over point, where the indirect conduction band minimum (X) becomes lower than the direct minimum (Γ), the more carriers are leaking into the X-minimum. The Γ–X transfer of carriers is a non-radiative loss mechanism which substantially limits the internal efficiency of yellow and green emitting devices (see Section 10.3).

10.2.2.1 CURRENT–VOLTAGE CHARACTERISTICS

The current–voltage characteristic of pn-junctions is described by the Shockley equation [19,20], which is given by

$$I(V) = I_s \left(\exp \frac{qV}{kT} - 1 \right) \qquad (10.26)$$

where I_s is the current obtained with a reverse bias (V negative). I_s is called the saturation current and is controlled by the number of minority carriers which diffuse to the pn-junction. In this (ideal) case, the current density, I_s/A, can be calculated according to

$$\frac{I_s}{A} = \frac{qD_p p_{n0}}{(D_p \tau_p)^{1/2}} + \frac{qD_n n_{p0}}{(D_p \tau_n)^{1/2}} \qquad (10.27)$$

where A is the cross-sectional area of the diode junction, D_n and D_p are the electron and hole dif-fusion constants, n_{p0} and p_{n0} are the equilibrium electron and hole concentration on the p-and

n-side, respectively. Under forward bias for $V > 3kT/q$ the current rises exponentially with V.

To describe real diodes, with a series resistance R_s, and in the presence of recombination centers in the active material, the Shockley equation has to be modified:

$$I(V) = I_s \left(\exp \frac{qV - qR_s I}{\alpha kT} - 1 \right). \qquad (10.28)$$

The factor α is the so-called ideality factor. In the case of an ideal diode, α becomes 1, and, in the case of recombination only via defect states in the depletion region, α becomes 2. Since in real diodes both processes take place, the ideality factor α will take a value between 1 and 2. The series resistance R_s is typically of the order of a few ohms and can be neglected at small currents <1 mA. However, at higher currents R_s has an important influence on the $I(V)$ characteristics and on the energy efficiency.

A typical current–voltage characteristics of an AlGaInP-LED at room temperature is shown in Figure 10.12. A linear current scale is used for the upper half of Figure 10.12, whereas the lower part uses a logarithmic scale. At currents in the microampere region and below, non-radiative transitions dominate the recombination mecha-nisms and the ideality factor α is 2. For currents between approximately one microampere and one milliampere, more carriers are injected into the active region and the non-radiative transi-tions become saturated with carriers. At those currents, the desired radiative recombination can compete successfully with the transitions via non-radiative centers and the ideality factor α reduces to 1.3–1.5. At currents above some milli-amperes, the series resistance of the LED eventu-ally determines the slope of the current–voltage characteristics.

10.2.3 Efficiencies

In an ideal LED, every injected electron generates one photon with an energy of $h\nu = E_g$. In this case, the number of injected carriers I/q equals the num-ber of generated photons and the efficiency of this process is 1. However, in reality, not all injected electrons reach the active region, not all electrons in the active region can generate photons and not

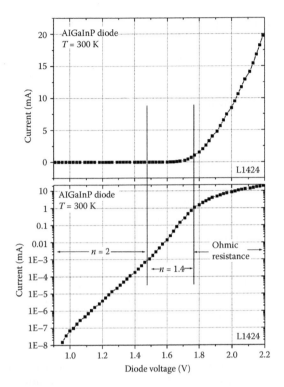

Figure 10.12 Current–voltage characteristics of an AlGaInP diode at room temperature on a linear (top curve) and logarithmic (bottom curve) current scale.

all photons can escape from the semiconductor. Therefore, all involved processes like carrier injection, carrier recombination or photon extraction are more or less efficient. Their efficiencies can be expressed either as quantum efficiency in terms of the number of photons and electrons involved or as power efficiency in terms of the optical or electrical power of all carriers involved.

In real LEDs, some of the electrical current is lost, e.g., due to conductive channels along the surfaces of the die or at crystal defects in the vicinity of the pn-junction. The fraction of current which is actually reaching the active region is

$$I_a = \eta_{inj} I_0 \qquad (10.29)$$

The ratio $\eta_{inj} = I_a/I_0$ is called the injection efficiency.

At the pn-junction, the carriers recombine, either via radiative or non-radiative transitions. The internal quantum efficiency η_{qi} is defined as the number of generated photons per unit time

over the number of electrons injected per unit time. In some cases, it is more convenient to express η_{qi} in terms of radiative and non-radiative recombination rates (Equation 10.13) or the corresponding carrier lifetimes:

$$\eta_{qi} = \frac{R_r}{R_r + R_{nr}} = \frac{\tau_{nr}}{\tau_r + \tau_{nr}} \qquad (10.30)$$

Only photons that leave the semiconductor chip are useful. Since only a small portion of photons generated in the die can be extracted, we have to define an extraction efficiency $\eta_{extraction}$ as the number of extracted photons over the total number of generated photons.

The external quantum efficiency of an LED is the product of the internal quantum efficiency and the extraction efficiency:

$$\eta_{ext} = \eta_{qit} \eta_{extraction} \qquad (10.31)$$

This is the ratio of the number of photons emitted from the LED per unit time to the number of injected electrons into the LED per unit time. The number of emitted photons from the LED is obtained by dividing the measured optical energy by the photon energy $h\nu$. The number of electrons per unit time is the injected electrical current divided by the elementary electron charge q. Thus, the external quantum efficiency, η_{ext}, can be calculated according to

$$\eta_{ext} = \frac{\left(\dfrac{P_{out}}{h\nu}\right)}{\left(\dfrac{I_0}{q}\right)} = \frac{P_{out}q}{h\nu I_0}. \qquad (10.32)$$

In an ideal diode, the forward voltage, V_f, of the device equals the bandgap energy divided by the elementary electron charge E_g/q. In real LEDs, however, the device structure contains additional series resistances and potential barriers which will increase the forward voltage. Therefore, we can define an external power efficiency or "wall-plug efficiency" as

$$\eta_{wp} = \frac{P_{out}}{I_0 V_f}. \qquad (10.33)$$

Table 10.1 Summary of different definitions of internal and external efficiencies

	Quantum efficiency	Power efficiency
Internal efficiency	$\eta_{qi} = \dfrac{\#\,of\,generated\,photons}{\#\,of\,injected\,electrons}$	$\eta_{pi} = \dfrac{power\,of\,one\,photon}{power\,of\,one\,electron}$
	$\eta_{qi} = \dfrac{R_r}{R_r + R_{nr}} = \dfrac{\tau_{nr}}{\tau_r + \tau_{nr}}$	$\eta_{pi} = \dfrac{h\nu}{qV}$
External efficiency	$\eta_{ext} = \dfrac{\#\,of\,extracted\,photons}{\#\,of\,injected\,electrons}$	$\eta_{wp} = \dfrac{extracted\,optical\,power}{injected\,electrical\,power}$
	$\eta_{ext} = \dfrac{P_{out}q}{h\nu I_0}$	$\eta_{wp} = \dfrac{P_{out}}{I_0 V_f} = \eta_{inj}\eta_{qi}\eta_{extraction}\dfrac{h\nu}{qV_f}$
Injection efficiency	$\eta_{inj} = \dfrac{\#\,of\,carriers\,at\,pn\,junction}{\#\,of\,injected\,carriers} = \dfrac{I_a}{I_0}$	
Extraction efficiency	$\eta_{extraction} = \dfrac{\#\,of\,extracted\,photons}{\#\,of\,generated\,photons}$	
	$\eta_{extraction} = \tfrac{1}{2}(1-\cos\vartheta_c)\left(1 - \dfrac{(n_m - n_s)^2}{(n_m + n_s)^2}\right)$	

Note: The last expression for the extraction efficiency is valid for LEDs with a planar structure.
I_0, total injected current; I_a, current flow into the pn-junction; R_{rad}, radiative recombination rate; R_{nr}, non-radiative recombination rate; τ_r, radiative carrier lifetime; τ_{nr}, non-radiative carrier lifetime; P_{out}, optical power emitted from the LED; V, applied voltage; ϑ_c, angle of total reflection; n_s, refractive index of the semiconductor material; n_m, refractive index of the surrounding medium; $\#$, number.

The "wall-plug" efficiency is the most important parameter for LED applications. It can be expressed in terms of several efficiencies defined above:

$$\eta_{wp} = \eta_{inj}\eta_{qi}\eta_{extraction}\frac{h\nu}{qV_f}. \qquad (10.34)$$

The different definitions are summarized in Table 10.1.

10.3 MATERIAL SYSTEMS FOR VISIBLE LEDs

10.3.1 GaP and GaAsP

The evolution of ternary III–V alloys started with the development of epitaxial growth methods like LPE or VPE. Binary alloys like GaAs or GaP were first available as bulk crystals and could be used as substrate materials. If P is added to an epitaxial layer of GaAs, the emission is shifted from the infrared to the visible spectrum (Figure 10.13)

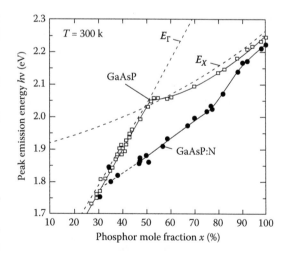

Figure 10.13 Emission energy versus phosphorous mole fraction of GaAsP and GaAsP:N LEDs. The solid lines are obtained from fits to experimental data at 5 A cm^{-2}. The dashed lines are calculated data for the direct and indirect bandgap [21].

and the lattice constant becomes larger than that of GaAs. For low P-fractions up to about 0.45 in GaAs$_{1-x}$P$_x$, the bandstructure remains direct and direct red emission is achieved [21]. However, the more P is added to the alloy, the larger the lattice mismatch to GaAs becomes and the more misfit dislocations are generated. The external efficiency of a GaAsP LED decreases by about a factor of 10 as the P-fraction is increased to the direct–indirect crossover point at $x=0.45$. Fortunately, the human eye response increases in this wavelength range by more than a factor of 50 so that the product of LED efficiency and eye sensitivity reaches a maximum at $x=0.4$ (≈ 650 nm, red). The fabrication process of these LEDs is rather simple. It includes a selective Zn-diffusion step into an n-type layer to form the pn-junction and subsequent contact formation.

The emission properties are significantly changed if the material is doped with an isoelectronic impurity such as nitrogen. As a group V-element, nitrogen does not produce additional carriers as standard dopants do, but introduces a carrier trapping level approximately 100 meV below the conduction band. Trapped electrons become localized in space which causes a large spread in the allowed momentum of the electron. Matching the momentum of holes in the valence band enhances dramatically the probability for radiative recombination and thus the generation of light. The problem of lattice-mismatch-induced dislocations is, of course, still present and limits the efficiency for higher P-fractions in GaAsP [22]. On the other hand, isoelectronically doped LEDs do not suffer from re-absorption to the same extent as intrinsic devices, which is a substantial advantage for higher efficiency. Consequently, the light extracted in different directions from the die does not show the variation in the wavelength spectrum that is observed in high-brightness LEDs (Figure 10.14).

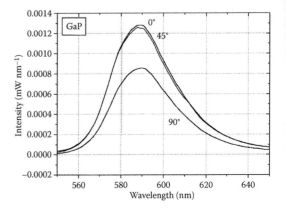

Figure 10.14 Emission spectrum from a red GaP:Zn, O LED showing a directionally independent emission spectrum.

Doping GaP simultaneously with the two group V-elements N and O also enhances the efficiency of GaP-LEDs. Introduced together in similar quantities, the two dopants tend to form complexes after proper annealing, and these function as isoelectronic traps. The Zn–O complex level lies deeper in the GaP band gap (approximately 0.3 eV below the conduction band) than the N-level and produces red photons at a wavelength of 700 nm. The low sensitivity of the human eye at 700 nm and the rapid saturation of the Zn–O centers with increasing current density limits the luminous efficiency to values below 0.5 lm W^{-1}.

Table 10.2 summarizes some characteristic features of the most common GaAsP and GaP-based LEDs. One property of transitions via impurity levels is, that the recombination rate saturates as more carriers are injected, which limits the maximum applicable current to the diode. The brightness of isoelectronically doped LEDs, such as green GaP:N devices, is still sufficient for many

Table 10.2 Properties of the most common types of GaAsP-and GaP-based LEDs

	Substrate	Color	Peak wavelength (nm)	Ext. efficiency (%)	Performance (lm W^{-1})
GaAs$_{06}$P	GaAs	Red	650	0.2	0.2
GaAs$_{0.35}$P$_{0.65}$:N	GaAs	Orange–red	630	0.7	1
GaAs$_{0.14}$P$_{0.86}$:N	GaAs	Yellow	585	0.2	1
GaP:N	GaP	Green	565	0.4	2.5
GaP:Zn–O	GaP	Red	700	2	0.4

applications such as indoor signs, indicators or illumination of small displays. However, although such LEDs are still sold in quantities of billions of devices per year, the performance is not sufficient for high brightness applications.

10.3.2 AlGaAs/GaAs

The first high-brightness LEDs on the market were red $Al_xGa_{1-x}As/GaAs$ devices. The ternary alloy $Al_xGa_{1-x}As$ is a solid solution of AlAs (E_g=2.168 eV) and GaAs (E_g=1.424 eV). Since AlAs (a=5.66 Å) and GaAs (a=5.65 Å) have basically the same lattice constant, AlGaAs is conveniently lattice matched to a GaAs substrate for the entire composition range. While GaAs is a direct bandgap material, AlAs has an indirect bandstructure. The bandgap of AlGaAs is direct for an Al-fraction up to 45% and indirect for higher values. This limits the efficient emission of visible light to the red spectral range above 620 nm. The composition dependence of the Γ-(direct gap) and the X-minima (indirect gap) of the Brillouin zone can be calculated according to Reference [23]:

$$E_{g\Gamma}(x) = 1.424 + 0.247x \text{ eV} \quad 0 \le x \le 0.45 \text{ (10.35)}$$

$$E_{g\Gamma}(x) = 1.424 + 0.247x$$
$$+1.147(x-0.45)^2 \text{ eV} \quad 0.45 \le x \le 1 \text{ (10.36)}$$

$$E_{gX}(x) = 1.9 + 0.125x$$
$$+0.143x^2 \text{ eV} \quad 0 \le x \le 1.0. \quad \text{(10.37)}$$

The energies of the AlGaAs conduction band minima of the Γ-, X- and L-bands, are shown in Figure 10.15 as a function of the Al-mole fraction. The band discontinuity at an AlGaAs/AlGaAs heterojunction splits in a ratio of roughly 60:40 between the conduction and valence bands [24]. Typical doping elements for p-type AlGaAs are zinc, carbon, beryllium and magnesium, while n-doping can be achieved with silicon, selenium or tellurium. The electrical conductivity of the material decreases rapidly with increasing Al-content.

Today, AlGaAs is a mature material system, readily fabricated using LPE or MOVPE growth [25]. High brightness visible LEDs emitting between 650 and 660 nm are routinely produced and are used worldwide. High performance AlGaAs-LEDs utilize an active layer with a double heterostructure

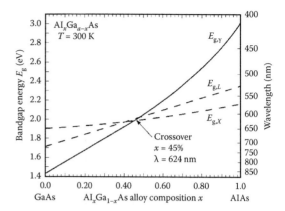

Figure 10.15 Composition dependence of the AlGaAs conduction band minima at the Γ-, X- and L-point.

confinement layer. Grown on absorbing GaAs substrates, these devices achieve luminous performances around 3 lm W^{-1}. The material system offers the possibility to grow thick (100–150 μm) epi-layers by LPE on GaAs substrates. With these very thick AlGaAs layers, it is possible to remove the original GaAs substrate by, e.g., selective etching, leaving a transparent "substrate" LED structure. Such TS AlGaAs LEDs are roughly twice as bright as devices on GaAs substrates with luminous performances of 8 lm W^{-1}. One major drawback of AlGaAs is its tendency to oxidize in the presence of oxygen or moisture. Because the standard epoxy materials cannot seal the LED die efficiently, the oxidation is also present after polymer encapsulation. The process is accelerated at elevated temperatures. Although visible AlGaAs LEDs are still used in large quantities, it is anticipated that they will be replaced in many outdoor applications by the more robust AlGaInP LEDs.

10.3.3 AlGaInP/GaAs

The alloy system $(Al_xGa_{1-x})_yIn_{1-y}P$ is a direct semiconductor material that covers the range of colors from red (650 nm) to yellow-green (560 nm). For higher brightness applications, it has widely replaced the indirect bandgap emitters GaP:N and GaAsP:N. The quaternary alloy AlGaInP can be regarded as a mixture of three binary materials, AlP, GaP and InP. The first two binaries have nearly the same lattice constant, which is appreciably different from the third involved material InP (Figure 10.16). In order to obtain lattice matching to GaAs, the In mole fraction is fixed at about 0.48, whereas the Al

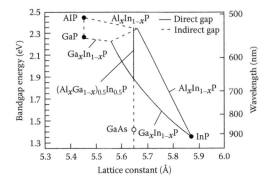

Figure 10.16 Bandgap versus lattice constant for the quaternary AlGaInP-system.

mole fraction x can be varied widely without affecting the lattice constant (Figure 10.17).

As usual for zinc blende type lattices, the band structure of the material has a single conduction band, with minima at the Γ-, X- and L-points, and three valence bands with degenerate heavy and light hole bands at the Γ-point. For low Al-content the bandgap is direct, but becomes indirect, with X being the lowest conduction band level for increasing Al-content. The bandgap energies have been determined by various methods [26], yielding slightly different results. A commonly accepted relation for the bandgap variation with composition at room temperature is

$$E_{g\Gamma}(X) = (1.900 + 0.61x)\ \text{eV} \qquad (10.38)$$

$$E_{gX}(X) = (2.204 + 0.085x)\ \text{eV}. \qquad (10.39)$$

These values indicate that the Γ–X crossover takes place for x=0.58, corresponding to an energy of 2.25 eV or a wavelength of 550 nm. As the crossover composition is approached from the low-energy side, the radiative efficiency of the material decreases drastically due to transfer of electrons from the Γ-to the X-valley, determining the maximum accessible range of wavelengths for LEDs. The conduction L-valley has not been seen directly in any experiment, but from measurements involving hydrostatic pressure [26], it has been estimated to be at least 125 meV above Γc. Although some theoretical work indicates a closer proximity to Γc [27], there are negligible effects on the performance of luminescent devices.

Various data indicate that the common 60:40 rule for the conduction and valence band offsets is also applicable in AlGaInP material [28]. The X-minimum of the conduction band changes accordingly, leading to a maximum band offset of about 200 meV in the conduction band. This conclusion has been nicely confirmed by transport experiments on n–i–n heterostructures [29].

For the active material, a double heterostructure design or a multiple quantum well (MQW) structure, embedded between larger bandgap layers, is used to optimize the carrier confinement. As the Al-concentration in the active region is increased to achieve shorter wavelength emission, electrons are thermally transferred from the Γ-minimum to the X-minimum, generating an additional non-radiative recombination channel. Beyond

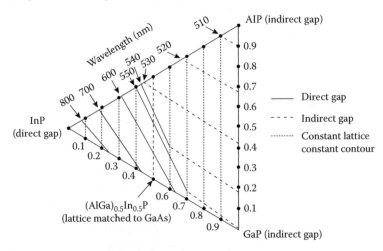

Figure 10.17 Two-dimensional representation of the AlGaInP composition ranges. Solid lines are lines of constant wavelength, dotted lines mark compositions of constant lattice parameter. Some indirect bandgap wavelengths are shown as dashed lines.

the Γ–X-crossover, the radiative efficiency drops to zero. Closed-form expressions for the non-radiative losses through Γ–X-transfer have been derived under certain approximations [28] but a fit to experimental data requires additional assumptions, specifically a reduction of carrier lifetimes with Al-concentration. While this assumption is reasonable (the residual oxygen incorporation, e.g., increases with the Al content), electron leakage and, to a minor extent, hole leakage across the barriers follow a similar relationship on current and temperature, making it difficult to distinguish between the mechanisms in simplified models.

Simulations of the device efficiency show that, in the red color range, the internal quantum efficiency is solely determined by non-radiative processes [30]. For emission wavelengths below 600 nm on the other hand, carrier leakage supersedes the non-radiative losses. Carrier leakage can be minimized by using larger active layers. Lower carrier densities in the active layer lead to higher effective barriers. At the same time, however, non-radiative processes become more favorable. Thus, as in many cases, the optimum configuration is determined by the material quality and different results have been obtained by different research groups [12,13]. Another way of suppressing the carrier leakage is to increase the barrier heights. The barriers are maximized by using doped AlInP for the confinement layers. Even larger barriers can be obtained by thin layers of AlInP, with tensile strain or multiple quantum barrier (MQB) structures [31]. In the latter case, a combination of barrier layer and short-period strained superlattice (SPSS) is used, such that the lowest SPSS miniband accessible for electrons tunneling out of the active layer is above the largest conduction band energy of any of the constitutant materials.

An important issue for the fabrication of optoelectronic devices is the tendency of the material to form ordered phases. Depending on the growth conditions, atomic ordering of In and Al/Ga along the (111) planes can occur on the group-III sublattice [32]. The modified crystal structure leads to a significant reduction of the bandgap energy, which has been measured to be up to 160 meV in partially ordered material [33] and is estimated to be as large as 471 meV for completely ordered material [34]. Moreover, in photoluminescence experiments, it has been found that peaks can be split and broadened [35] and luminescence intensities are decreased [36], possibly due to the formation of piezoelectric fields and non-radiative recombination centers at domain boundaries. Therefore, disordered material is more favorable for most LED devices, in particular when short wavelengths are needed. Generally, the use of misoriented substrates, e.g., (100) tilted towards the $\langle 110 \rangle$ or the $\langle 111 \rangle_A$ direction, and the proper choice of growth conditions and doping profiles allows the complete suppression of ordering in the material [37]. More details about the AlGaInP alloy system can be found in the literature [28,30,38].

10.3.4 InGaN

The group-III nitrides are the only III–V system that allows the generation of UV, blue and green photons. (Al)InGaN alloys can be described as a solution of the three binaries AlN, GaN and InN, with a very restricted solid miscibility due to thermodynamic constraints [40]. All three materials are direct semiconductors with band gaps of 6.2, 3.4 and 1.9 eV, respectively [41]. By adding In to GaN, the emission wavelength is shifted from the UV (365 nm) to the visible spectrum. Theoretically, the (Al)InGaN-system can cover a wide wavelength range up to 630 nm. The practical limit today, however, is somewhere in the green spectral range ($\approx$ 550 nm).

The alloys can crystallize in two different modifications, either in a cubic zinc blende structure or in a hexagonal wurzite structure with the latter being considered as thermodynamically more stable. Since neither GaN nor AlN are available as bulk crystals, the system lacks a suitable semiconductor substrate for epitaxial growth (Figure 10.18). The most common substrate today is sapphire, with a lattice mismatch of 16% to GaN. Although this large mismatch causes a very high dislocation density between 10^7 and 10^9 cm^{-2} in the epitaxial films [42], the system is still capable of producing highly efficient LEDs, for reasons that are not fully understood. Obviously, the dislocations that must occur do not act as efficient recombination centers, which, together with the low diffusion length of holes, do not appreciably affect the radiative recombination efficiency. An alternative substrate is SiC which has a lower lattice mismatch to GaN and offers a number of additional advantages such as good electrical and thermal conductivity or a high electrostatic

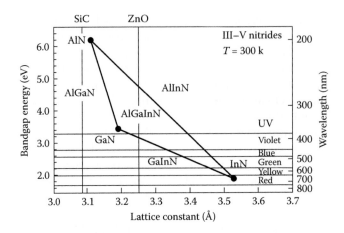

Figure 10.18 Bandgap energy versus lattice constant for III–V alloy systems of InGaN. The bowing parameters for the energy bands of AlGaN and InGaN are assumed to be 1.0 according to [39].

discharge hardness [43,44]. Compared to sapphire, SiC is about ten times more expensive and less transparent. Also, its refractive index is less suitable for light extraction.

It has been found that the growth of a GaN, AlN or AlGaN buffer layer is essential for the growth of high-quality layers on both mismatched substrates SiC and sapphire. This buffer is usually grown at low temperatures (500°C–600°C), before the temperature is raised to 1000°C–1100°C for the growth of the device structure. On sapphire substrates, the buffer layer is necessary to achieve a smooth surface morphology and good optical and electrical properties despite the large lattice mismatch [45–47]. On SiC-substrate, the buffer layer reduces the voltage drop across the SiC–GaN interface. Also, the buffer thickness directly impacts the quality of the subsequently grown layers. Maximum Hall mobility in n-type GaN is achieved with 20 nm thick GaN-buffer layer thickness [47]. From the different temperature dependence of the Hall mobility with GaN-and AlN-buffer layers, it might be concluded that the GaN-buffer reduces the effect of ionized impurity scattering [48].

Nominally undoped GaN is n-type, with a high residual electron concentration. P-doping is very difficult to achieve in III-N-alloys. The lack of p-type material, and hence of pn-junctions, has been a blocking point for the fabrication of light-emitting devices for a long time. A lot of the early work was done with Zn as acceptor [49,50], but today the preferred material for p-doping is

magnesium [51]. After the epitaxial growth, the acceptors are passivated by the formation of some sort of Mg–H complexes [52]. P-conductivity can be obtained after a post-growth treatment by thermal annealing or using low-energy electron irradiation (LEEBI), but also "in situ" by a suitable growth and cooling-down procedure. The activation energy of Mg acceptors in GaN, deduced from temperature-dependent Hall measurements, is assumed to be around 160 meV. In InGaN and AlGaN-alloys, the activation of Mg acceptors is dependent on the composition. Silicon or germanium is used for n-type doping. The donor activation energy of Si in GaN is around 33 meV.

As a consequence of the high ionicity of the nitrides, their refractive indices are relatively small. The values for the binaries are listed in Table 10.3. The index of (Al)InGaN can be changed appreciably by changing the composition and thus the bandgap. The refractive index is an important parameter for light extraction from the diode chip. Another favorable property of the III-nitrides is their high thermal conductivity. Some of the values are listed in Table 10.3. The common substrates such as sapphire and 6H-SiC have thermal conductivities of 0.439 and 4.9 W cm^{-1} K^{-1}, respectively.

The first blue emitting LEDs were homotype pn-junction GaN devices. The electroluminescence spectrum consists of a narrow acceptorband emission in the UV (around 370 nm) and a broad band-acceptor line at 420 nm. Better

Table 10.3 Material properties of binary III-nitrides and their potential substrates

Material	Lattice constant (Å)	Thermal expansion coefficient (10^{-6} K^{-1})	Thermal conductivity (W cm^{-1} K^{-1})	Refractive index
GaN	$a = 3.160\ldots3.190$[a] $c = 5.25\ldots5.190$	5.59 7.75	1.3	2.29 @ 1000 nm[a]
InN	$a = 3.5446$a $c = 5.7034$		–	2.56 @ 500 nm[a]
AlN	$a = 3.11$[a] $c = 4.98$	5.27[a] 4.15	2.0	
6H-SiC	$a = 3.0806$[a] $c = 15.1173$	4.9	4.89 (0.878)	$2.55378 + 3.417 \times 10^4 \, \lambda^{2a}$
Sapphire	$a = 4.758$[b]	7.70 (^c)	0.439	Ordinary ray: 1.78 @ 589 nm[c] (Extraordinary ray: −0.008)
		8.33 (‖c)		
GaAs	5.65325	5.8	0.46	3.5 @ 600 nm
Si	5.43102	2.29	1.48	3.94 @ 600 nm[d]

Note: GaAs and Si are listed as reference. Note that the thermal conductivity of SiC is anisotropic. The value in brackets gives the value in a different crystallographic direction.
[a]Landolt Börnstein data handbook, new series.
[b]S. Nakamura, Semiconductor and Semimetals, vol. 48.
[c]E.D. Palik, Handbook of Optical Constants of Solids II.
[d]E.D. Palik, Handbook of Optical Constants of Solids I.

carrier confinement and thus more efficient recombination are achieved in double hetero-structure devices. By adding Al to GaN, the bandgap is increased while the addition of In shifts the bandgap to lower energies. Thus, typical DH-devices involve AlGaN confinement layers around GaN or InGaN active layers. Due to the large difference of lattice constant between AlN and InN, the range of usable composition and layer thickness is limited. Similar to the impurity related transitions in GaP, simultaneous doping of InGaN with Zn and Si can generate relatively efficient donor–acceptor emission in the violet to green spectral range [53,54]. Using active material of quantum wells with a well thickness below the critical thickness the quality of the LED structure is improved significantly, which enables the use of band–band or near band–edge transitions as major recombination process [55]. As long as mismatch-induced strain and the quantum well thickness do not exceed the critical limits, the emission wavelength can be readily adjusted with the In-mole fraction. Nakamura et al. reported emission from violet to yellow for In-fractions between 0.15 and 0.7 [56].

10.4 HIGH EFFICIENCY LEDs AND NOVEL TECHNOLOGIES

The traditional design of an LED chip is a semiconductor die, with a planar metal contact at the bottom and a circular or square electrode at the top (Figure 10.19). Very simple devices use a homostructure pn-junction as active region, the more advanced ones include a single-or double-heterostructure. Several factors limit the external efficiency of such LEDs: total internal reflection, shadowing by the metal electrodes and absorption in the semiconductor material. With the exception of GaP-based LEDs and some AlGaAs-devices on TSs, a substantial amount of light is lost by absorption in the substrate material. Therefore, even with an efficient

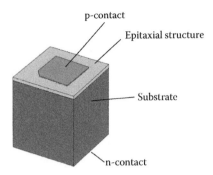

Figure 10.19 Conventional LED die on a conductive substrate. Typical dimensions are a chip area of 200×200 μm² and a height of 200 μm.

internal process of current-light conversion, the external efficiency of standard LEDs hardly exceeds a few per cent. This frustrating situation has stimulated a lot of work on ways to increase the efficiency of LEDs such as

- Fabrication of dices that are as transparent as possible, by the use of very thick transparent layers and a TS
- Die-shaping to enhance the extraction efficiency
- Fabrication of substrate-less LEDs, to avoid substrate absorption
- Modification of the spontaneous emission by using optical cavities

Each approach tackles the problem of light extraction in a different way. The first one applies rigorously the principle of extracting light from as many escape cones as possible. Die shaping is a step towards the ideal solution, that of a point light source in the center of a spherical semiconductor die. Most of such ideas are almost as old as the LED itself, but practical devices have not been generated until very recently. For obvious reasons, die shaping works best on LEDs with TSs. The (absorbing) substrate can be removed entirely, if the LED layers are transferred to a new carrier such as a semiconductor wafer or other materials such as metal or ceramic. If metal soldering is used to combine the LED structure with the carrier, the metal also acts as mirror. Because the reflected light has to pass through the entire layer structure, the efficiency of LEDs with such an integrated reflector is very sensitive to internal absorption (Table 10.4).

A completely different idea for high extraction efficiency is to change the fundamental process of spontaneous emission. Placing light generating material in an optical resonator, with spatial dimensions in the order of the emission wavelength, can change the spontaneous generation of light significantly [57,58]. The ultimate microcavity LED would allow only one single optical resonator mode which could be extracted as efficiently as the optical modes in a laser. While such ideal LEDs are still far from any practicably manufacturable device, some aspects of micro-cavity

Table 10.4 Summary of LEDs with the highest reported efficiencies

Device	Material system	Technology	Color range (nm)	Status
Transparent substrate LED	AlGaInP/GaAs	Direct wafer bonding	590–650	Prod.
Truncated inverted pyramid LED	AlGaInP/GaAs	Direct wafer bonding, die shaping	590–650	Prod.
Thin-film LED	AlGaInP/GaAs	Wafer soldering	590–650	Prod.
Surface textured thin-film LED	AlGaAs/GaAs	Epitaxial lift-off, surface roughness	850–980	Demo.
LED with tapered waveguide	AlGaAs/GaAs	Etched lateral taper	850–980	Demo.
	InGaAlP/GaAs		650	Demo.
Resonant cavity LED	AlGaAs	Standard technology	850–980	Demo.
	AlGaInP/GaAs		650	Prod.

Note: Listed are the material system, the technology used, the color range of emission and the status (productive or demonstrator). The different LEDs are described in more detail in the text.

LEDs have already found their way into commercial LEDs, e.g., red emitting (650 nm) RCLEDs.

10.4.1 Transparent substrate AlGaInP LEDs

The extraction efficiency of LEDs with TSs is substantially higher than that of LEDs on ASs. Examples of LEDs on TSs are the early GaP-devices or AlGaAs-LEDs fabricated on thick LPE grown transparent layers. Due to the large lattice mismatch to GaP, high-brightness AlGaInP must be grown on absorbing GaAs-wafers, which strongly reduces the extraction efficiency. In order to overcome this fundamental limitation, Kish et al. developed a technique to remove the absorbing GaAs substrate and to transfer the AlGaInP-LED layers to a transparent GaP wafer [59]. Direct wafer bonding allows the combination of various semiconductor materials more or less independently of their crystallographic lattice constant. In the case of AlGaInP–GaP, wafers with up to 3 in. diameter are bonded [60], the limit set by the diameter of the available GaP-wafers. The clean surfaces of both wafers are brought into contact under uniaxial pressure and heated up to temperatures of 750°C or more [61]. Under such conditions, and with proper crystallographic alignment, the two materials form a semiconductor heterointerface with covalent bonds between the two materials [60]. The interface is optically transparent and conducts both heat and electrical current. Excellent control over this complex process is required, in order to ensure low-resistance electrical contact over the entire bonded area. The feasibility of the AlGaInP–GaP wafer bonding process for the fabrication of an LED has been demonstrated by Hewlett-Packard and is now routinely used in high-volume production. The new class of AlGaInP LEDs is named TS LEDs to distinguish them from AS LED on GaAs-wafers.

With the very thick GaP current spreading layer on top of the AlGaInP active region and the transparent GaP-wafer below, the TS-LEDs achieve record high levels of light extraction. Very high efficiencies of up to 32% at 630 nm have been reported [12]. In terms of luminous efficiency, a maximum value of 74 lm W^{-1} has been achieved at shorter wavelengths (615 nm). The extraction efficiency of TS-LEDs can be further enhanced, if the die is shaped into cones, pyramids or spheres [62].

The highest reported efficiencies of AlGaInP LEDs are achieved with wafer-bonded TS-material which is cut into dices with the shape of a truncated inverted pyramid [14]. Light is generated at the base of the inverted pyramid and is extracted after a reduced number of reflections and with a low average photon path length within the semiconductor. The truncated inverted pyramid LED achieves a luminous efficiency of 102 and 68 lm W^{-1} at wavelengths of 610 and 598 nm, respectively. A peak external efficiency of 55% was measured at 650 nm.

10.4.2 Wafer-bonded thin-film LED

A viable alternative to direct wafer bonding is to solder wafers with an intermediate metal layer. This is well established in silicon technology and is commonly used for the fabrication of micro mechanical components. In the context of AlGaInP-LEDs, the idea is to transfer the epitaxial structure at wafer level to a new carrier by soldering followed by GaAs removal [63]. This process generates a new wafer, with a metal layer buried between the epitaxial LED layers and the carrier material. Both the high reflectivity of the metal-semiconductor interface and the possibility to form ohmic contacts are favorable for the functionality of the LED.

The process flow is shown schematically in Figure 10.20. After epitaxial growth, Au and AuSn are deposited on the AlGaInP-LED structure and the carrier wafer, respectively. Carrier and LED-structure are then brought into contact and soldered at 350°C. Then the original GaAs substrate is removed by selective wet chemical etching. Processing of the "new wafer" is then finished using standard LED processing technology.

Figure 10.21 shows the basic principle of a thin-film LED. The metal layer below the active layer serves both as a reflector and as the anode. Light which is not directly extracted is reflected either by total reflection at the semiconductor-air (or epoxy) interface or at the metal mirror. Extraction is significantly enhanced if the reproduction of reflection angles is suppressed by surface roughness. Other critical parameters for high efficiency are a sufficiently thin active layer for reduced self-absorption and a power reflectivity of more than 90% at the metal–semiconductor interface. Although metals like Au, Al or Ag offer a very high reflectivity on AlGaInP or

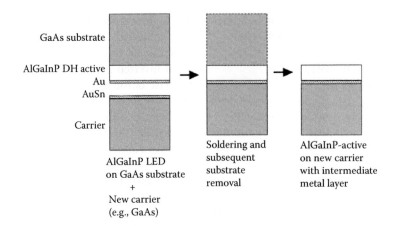

GaAs substrate

AlGaInP DH active
Au
AuSn

Carrier

AlGaInP LED
on GaAs substrate
+
New carrier
(e.g., GaAs)

Soldering and
subsequent
substrate
removal

AlGaInP-active
on new carrier
with intermediate
metal layer

Figure 10.20 Schematic layout of the thin-film process sequence. Left: metal layers of Au and AuSn are deposited on the epitaxial AlGaInP-structure and the carrier wafer, respectively. After bonding and GaAs removal, the LED structure together with the new carrier forms an artificial wafer (right) which can be processed using standard LED processing.

AlGaAs, the reflectivity is greatly reduced if the metals are alloyed to form good electrical contacts. Even on highly doped semiconductors contact alloying is necessary to reduce the contact resistance. For the thin-film LED, this problem is solved by locally separating areas with high reflective mirrors from areas with good electrical contacts to the semiconductor. A thin dielectric layer, such as Si_2N_3 or SiO_2, is deposited between the semiconductor and the metal layer(s). Openings in this dielectric film define the electrical contacts, whereas the rest of the area serves as a dielectric/metallic mirror. The layer system can be alloyed in order to lower the contact resistance, without significantly changing the optical properties of the mirror area.

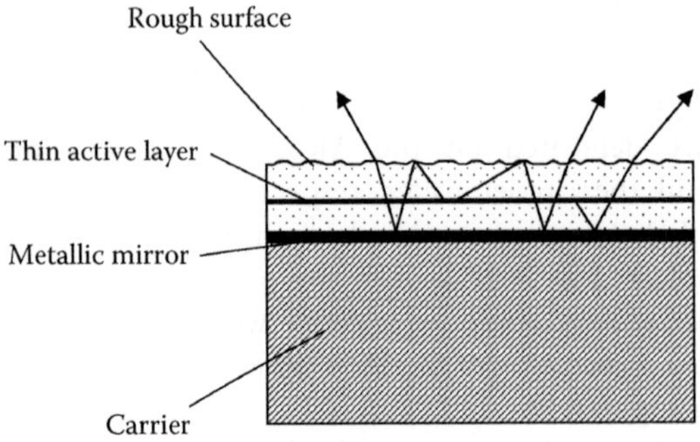

An attractive feature of wafer soldering for LEDs is the possibility to structure one of the surfaces before bonding. Geometrical shapes, such as cones, prisms or spheres, might be transferred into the AlGaInP-structure (or the carrier surface) by etching before they are covered with metal and embedded inside the LED structure. Generally, most of the geometries that have been used to shape dices, in order to enhance the extraction of light [62], can also be buried at the bonded interface of a thin-film LED. An example for an LED with an embedded array of cone-shaped micro reflectors is shown in Figure 10.22. Truncated cones are etched through the active layer and electrically contacted at openings in the dielectric layer at the soldered interface. The fact that the localized current injection leads to a preferred generation of light in the center of the etched cones is used to design the actual shape of the microstructure. Light rays that are totally reflected at the metal mirror along the cones are directed upwards towards the LED surface where they are extracted. The path length from the active layer to the top of the LED structure is only a few

Rough surface

Thin active layer

Metallic mirror

Carrier

Figure 10.21 Principle of operation of an LED with metallic reflector.

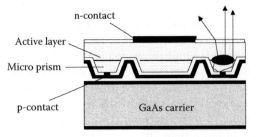

n-contact

Active layer

Micro prism

p-contact

GaAs carrier

Figure 10.22 Schematic cross-section of a thin-film LED with buried micro reflectors.

micrometer and does not include any absorbing material such as the active layer so that the effect of self-absorption inside the device is greatly reduced.

Figure 10.23 (left) shows the top view of a 615 nm device with a structured interface. Although the array of micro reflectors is located underneath the planar top-layers, it is clearly visible from the top. Current injection is facilitated from the isolated bond pad and the connected square frame along the edge via the top n-contact layer to the individual p-electrodes in the center of the individual reflectors. As shown in the illumination pattern (Figure 10.23, right) light is generated and extracted around the micro reflectors, confirming the principle of operation. Figure 10.24 depicts the optical characteristics of the device. Driven at a dc-current of 100 mA, the LED emits up to 9.7 lm, corresponding to an optical power around 32 mW. In the range of 10–20 mA of operation current, the luminous performance is above 50 lm W⁻¹ with a peak value of 53 lm W⁻¹ at 10 mA. One of the advantages of the thin-film technology is the low

ohmic resistance that can be achieved. Operated at a dc-current of 10 mA, the forward voltage of the 615 nm LEDs is still below 2.0 V.

Thin-film LEDs are also very attractive as large-area chips for high-current application. Contrary to most other high-brightness LEDs, they do not need a thick window layer for light extraction. Because the thickness of these GaP or AlGaAs window layers is directly related to the chip size (Equation 10.20), an increase of chip area usually results in a reduced extraction efficiency. The output characteristics of two large-area thin-film LEDs with 700 and 1000 μm chips are shown in Figure 10.25. Up to a drive current of several hundred mA, the efficiency of both devices is comparable. Only at very high currents, the output power of the smaller chip begins to roll over due to thermal problems. The 1 mm LED achieves 64 lm at 1 A.

The possibility of shaping the LED surface before bonding offers a whole range of new opportunities to optimize or tailor the device performance by combining novel micro-structures for light

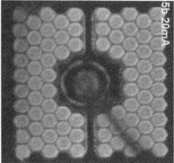

Figure 10.23 Top view (left) and illumination pattern (right) of a 615 nm thin-film LED.

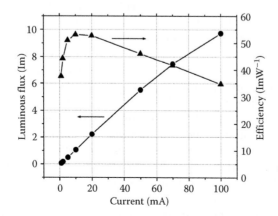

Figure 10.24 Optical performance of a 615 nm thin-film LED operated under dc-conditions.

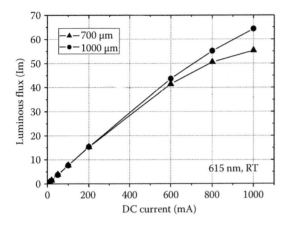

Figure 10.25 Light-output characteristics of a 1 mm thin-film chip emitting at 615 nm. The samples are mounted in a package for high-current operation.

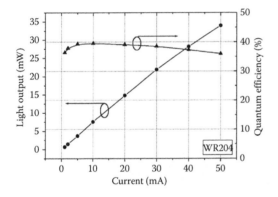

Figure 10.26 Characteristics of a 637 nm thin-film LED. The device gives 34 mW of output power at 50 mA dc-current. Maximum values of quantum- and wall-plug efficiency are 40% (10 mA) and 37.7% (2 mA), respectively.

extraction with new schemes for current injection, heat dissipation or emission profiles. Figure 10.26 shows the performance of a 637 nm thin-film LED with a modified reflector structure. At 50 mA, the device emits a power of 34 mW corresponding to a quantum efficiency of 36%. The maximum values of quantum and wall-plug efficiency are 39.5% and 38%, achieved at a drive current of 10 and 2 mA, respectively.

A different approach to fabricate thin-film AlGaInP-LEDs is to bond the epitaxial structure to an isolating wafer. In this case, two top electrodes are required. Horng et al. fabricated 600–620 nm AlGaInP LEDs on Si-wafers coated with

SiO_2 [64,65] using a metal combination of Au and AuBe for bonding. Despite the intermediate dielectric layer, the LEDs benefited from the good thermal properties of Si. A luminous intensity of 90 and 205 mcd (620 nm) was demonstrated at 20 and 50 mA of operation current, respectively.

Obviously, the thin-film approach is not restricted to a specific material system. Recently, a 460 nm InGaN thin-film LED was reported achieving an external quantum efficiency of 25% [66]. The device structure was grown on a sapphire and subsequently bonded to a GaAs wafer. A laser lift-off process was used to remove the original sapphire substrate.

10.4.3 Surface-textured thin-film LED

The idea of surface-textured thin-film LEDs dates back to the early 1990s, when Schnitzer et al. demonstrated an external efficiency as high as 72% with an optically pumped thin-film LED [67] and, shortly later, an electrically driven diode with an external efficiency of approximately 30% [68]. The devices are based on two new techniques for LEDs: "epitaxial lift-off" (ELO) [69] and surface roughening by "natural lithography" [70]. The idea of the ELO process is to insert a sacrificial layer, typically some 10 nm of AlAs, between the LED structure and the substrate. A selective wet chemical etching process removes the sacrificial layers and separates the epitaxial layers from the GaAs substrate. Subsequently, the thin epitaxial film is transferred to a new wafer by van der Waals bonding [71]. A metal film on the carrier serves as a highly reflective back mirror for the LED. Surface roughening is achieved by depositing a monolayer film of randomly ordered polystyrene spheres on the LED wafer. Then the random pattern is transferred into the semiconductor surface by dry etching.

The principle of operation is to randomize the angles of totally reflected light at the top surface with the high reflectivity of a metallic back mirror on the bottom side (Figure 10.21). Light that is not extracted through the surface is reflected back and forth between the back mirror and the rough surface. For high efficiency devices, the losses per round-trip through the LED structure have to be minimized by (i) reducing the thickness of all absorbing layers including the active layer and (ii) optimizing the reflectivity of the back mirror. Both can be achieved easily in the AlGaAs/InGaAs/GaAs material system. The band structure offers a sufficiently high carrier confinement to use only a few nanometers of active material for efficient light generation. Additionally the reflectivity of common metals like Au, Ag or Al is very high in the infrared regime.

The combination of epitaxial lift-off and surface roughening has been optimized by Windisch et al. on near infrared (850 nm) LEDs [72,73]. A schematic layout of the device is shown in Figure 10.27a. The LED employs a mesa structure, with a selectively oxidized current aperture to prevent the generation of light under the metal contact. Current injection is facilitated via an annular top-side p-contact and a lateral n-contact around the mesa. In order to extract some of the laterally guided light, not only the surface on top of the mesa but also the area between mesa and bottom contact was roughened. Very high quantum efficiencies of up to 43% and 54% were achieved before and after encapsulation, respectively [74]. Current densities up to 1000 A cm^{-2} were applied. Similar devices at 650 nm emission wavelength resulted in 24% external quantum efficiency (31% after encapsulation) [75]. Due to relatively high forward voltages, the wall-plug efficiency of the red LEDs ranges between 10% and 15% at 1 mA of operation current.

10.4.4 LED with tapered waveguide

A device that extracts solely the laterally guided modes inside the epitaxial structure is the LED with a radial tapered output coupler [76] shown in Figure 10.27b. The circular device structure comprises a central top contact, a circular symmetric out-coupling taper with the shape of a shallow truncated cone and a ring-contact along the taper perimeter. Light extraction occurs through the bottom side, where the GaAs substrate has been removed by wet chemical etching.

The principle of operation is to generate light in a small active area, defined by the geometry of the p-contact and to guide it radially towards the tapered area. The dimensions are chosen such that the light rays that hit the taper surface have a small azimuthal wave vector component which is an essential design parameter for efficient light extraction. Light that is reflected at the out-coupling surface is reflected again at the metal mirror on the taper and returns under a more favorable angle of incidence. The procedure is repeated until finally the angle of incidence falls within the escape cone. Absorption inside the waveguide layers which include unpumped active material and a reflectivity below 100% at the metal mirror on the taper limit the efficiency of this device.

With one and two InGaAs/GaAs compressively strained quantum wells as active material, Schmid et al. demonstrated 45% quantum efficiency at 980 nm on encapsulated devices [77]. Maximum efficiency was reached at 2 mA injection current and a forward voltage around 2 V. This resulted in a wall-plug efficiency of 30% and 44%, before and after encapsulation, respectively. The first attempt to apply the same technology to 650 nm GaInP/AlGaInP devices resulted in an external efficiency of 13%.

10.4.5 Resonant-cavity LEDs

The resonant-cavity LED or RCLED consists of a light-emitting active region between two Bragg mirrors which form an optical resonator. If the distance between the Bragg mirrors is set to a small multiple of half the optical wavelength, the cavity is in resonance with the emission and becomes transparent. In 1946, Purcell discovered that the spontaneous emission properties of an atom can be influenced by placing it inside a small (micro-)

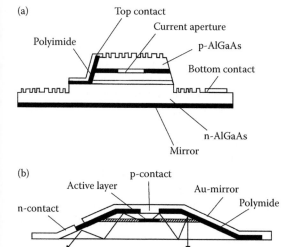

Figure 10.27 (a) Non-resonant LED with metallic back mirror and roughened surface. (b) LED with a radial tapered waveguide.

cavity [78]. The local strength of the electromagnetic modes and the density of states might then change the angular distribution as well as the emission rates. Schubert et al. [79] proposed in 1992 to use a one-dimensional resonant cavity with a thin active layer inside as LED and to apply the cavity to manipulate the spontaneous emission.

The cavity determines the optical properties of the RCLED. It can be designed to direct more spontaneous emission into the escape cone than in the case of isotropic emission. This results in a higher intensity emitted in the surface normal direction as shown in Figure 10.28. The different wavelengths within the emission linewidth fulfil the resonance conditions of the cavity at different angles of incidence at the top surface. Therefore, the RCLEDs spatial emission pattern contains every individual wavelength within the material linewidth in a specific angular direction. If the emission is spatially filtered, e.g., by coupling it into an optical fiber, also the range of emitted wavelengths is filtered, resulting in a narrow fiber coupled emission linewidth. However, when integrating over the entire halfsphere around the device, the measured emission linewidth is the same as that of the active material without the resonator.

A comprehensive introduction into the physics of RCLEDs can be found in the review of Benisty et al. [80,81]. Here, we will summarize only a few of the most important rules for the design of RCLEDs:

1. The cavity order should be as low as possible. Even if the spacing between the Bragg mirrors is one half or one wavelength, the penetration depth of light into the Bragg layers increases the effective cavity length. Hence, the index contrast should be as high as possible.
2. The quantum-well active layer should be placed in a maximum of the standing wave in the cavity.
3. The emission of the active material should be tuned to a shorter wavelength than the resonance wavelength of the cavity. This compensates for a possible wavelength shift as the temperature rises and enhances the fraction of extracted light. The latter is due to the fact that the shorter wavelengths are resonant in off-axis directions with a much larger solid-angle element than the exactly tuned wavelength.

Contrary to conventional LEDs, many properties of RCLEDs such as the temperature dependence of intensity, the radiation pattern or the coupling efficiency into some kind of optical system can be designed. The key parameter here is the wavelength tuning of the cavity resonance, given by the thickness between the two mirrors and the emission wavelength. According to point 3, it is usually favorable to have the active material emitting at a few nanometers shorter than the on-axis cavity resonance wavelength. As the temperature is increased, the emission wavelength is shifted towards the resonance and increases the amount of light extracted in the surface normal direction. This is schematically shown in Figure 10.29. Thus, for example the amount of fiber coupled light can increase with increasing temperature. At even higher temperatures, the emission will shift from the resonance to longer wavelengths and the surface normal intensity will decrease again. The wavelength tuning also affects the spatial radiation pattern of RCLEDs. If the peak intensity of the

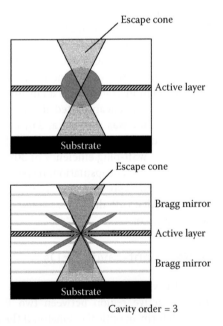

Figure 10.28 Spontaneous emission in a conventional LED (top) and in an RCLED (bottom). Because the spontaneous emission in the resonant cavity is no longer isotropic, the cavity can be designed to launch more emission into directions where the light is extracted (escape cone).

active layer emission is shifted to shorter wavelength than the cavity resonance, more light is extracted in off-normal directions. A two-dimensional-cross section of the resulting radiation pattern shows the typical double-lobe shape as shown in Figure 10.29.

RCLEDs have been fabricated mostly for the infrared spectral range, often with the intention to use them as efficient light source in fiber communication systems. For this application, the RCLED has to compete with vertical cavity lasers, which are even more suitable for high-speed communication. However, a few groups continued to optimize RCLEDs and achieved remarkable results using, e.g., metal-mirrors or fully oxidized Al_xO_y/GaAs DBRs [82]. An external quantum efficiency above 20% was achieved by De Neve et al. [83] with a bottom emitting 980 nm RCLED using an AlGaAs/GaAs-DBR and a metallic Ag-mirror. Also visible RCLEDs based on the AlGaInP/GaAs system have been demonstrated [16,84,85].

Only in the dark red around 650 nm, RCLEDs have made it to commercial products. At this wavelength there is a need for medium speed light sources for plastic fiber communication. RCLEDs are attractive because the directional output yields in a high fiber coupling efficiency and, due to the above-mentioned spatial filtering effect, in a narrow spectral linewidth in the fiber. Even with small devices and active diameters of less than 100 μm, levels of fiber coupled power well above 1 mW are achievable. If operated at some 10 mA of electrical current, the radiative carrier lifetime is reduced mainly by the high current density to rise-and fall-times of a few nanoseconds [86,87]. Figure 10.30 shows the structure of a commercial 650 nm RCLED. It consists of two $Al_{0.5}Ga_{0.5}As/Al_{0.95}Ga_{0.5}As$ Bragg mirrors and an active layer of compressively strained GaInP quantum wells inside the $1-\lambda$-cavity [88]. For high-modulation speed, the material is processed to devices

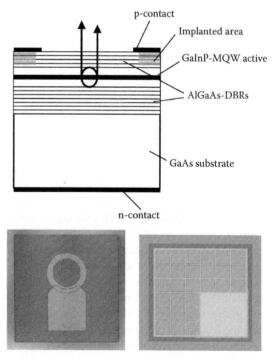

Figure 10.30 Top: layout of a 650 nm RCLED with annular top contact and proton implanted current confinement. The distance between the AlGaAs-Bragg mirrors corresponds to one optical wavelength ($1-\lambda$ cavity). Bottom: top view of two commercial RCLEDs. The chip dimensions are (200×250) μm² and (260×260) μm² for the left and right device. The smaller die has a light-emitting area with a diameter of 80 μm.

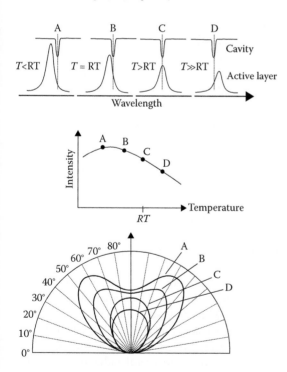

Figure 10.29 Effect of cavity tuning in RCLEDs. Top: emission wavelength of the active layer and cavity resonance at different temperatures. Middle: emitted intensity versus temperature. Bottom: radiation pattern for the four temperatures A–D.

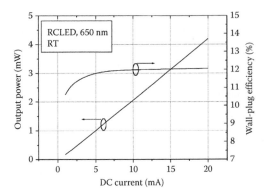

Figure 10.31 Wall-plug efficiency and optical output power of a 650 nm RCLEDs. The die size is 700×700 μm². At 20 mA, the device emits 4.2 mW.

with an 80 μm light opening within an annular p-electrode (bottom of Figure 10.29), but also conventional LED contact layouts are used. Large area devices achieve wall-plug efficiencies around 12% at 20 mA and 1.75 V forward voltage (Figure 10.31). The high-speed devices are less efficient (9.5%), mainly due to the lack of photon recycling [89].

10.4.6 Photonic bandgap approaches

Similar to the electronic bandgap, photonic bandgaps are defined as frequency bands over which all electromagnetic modes and spontaneous emission are suppressed [90–92]. They are a feature of periodically patterned materials with a strong dielectric constant contrast, so-called photonic crystals. The realization of photonic bandgap structures at optical frequencies requires the modulation of the dielectric constant on a nanometer scale, which is a challenging task for semiconductor processing technologies. In the context of optoelectronic device design, photonic bandgap structures are discussed as novel approaches for light confinement but their optical diffraction properties are also interesting.

Photonic bandgap structures can be employed for LEDs in different ways. One is to confine the photons in the device in two or three dimensions by using a periodic pattern in the semiconductor active material. This can be used to create waveguide structures with unique features on a very small scale but also to fabricate micro-cavities. The spontaneous emission in such a micro-cavity can be

controlled if at least one dimension of the cavity is in the order of the emission wavelength. Ultimately, the vision is to realize single-mode LED emission or threshold-less micro-cavity lasing operation [92,93]. In micro-cavities with metallic mirrors, two-dimensional photonic bandgap structures are used to create bandgaps for propagating surface waves or surface plasmon polaritons [94,95]. Unlike three-dimensional photonic crystals, two-dimensional structures are easier to fabricate and therefore may be more interesting for practical applications.

Similar to grating output couplers on top-emitting in-plane lasers [96,97], a two-dimensional photonic bandgap structure can be used to redirect laterally guided light inside an LED towards the surface normal direction. Photonic crystal light extractors have already been used in a number of different devices where the photonic bandgap structure was either etched into the top layer or all the way through the active LED structure. Erchak et al. demonstrated a sixfold enhancement of PL-intensity at 925 nm using a two-dimensional photonic bandgap in the upper cladding of an asymmetric LED structure [98]. Rattier [99] proposed a device where an unstructured LED-area for direct microcavity emission is surrounded by a deeply etched photonic bandgap for guided mode outcoupling. In both approaches, an oxidized Al_xO_y/GaAs Bragg reflector is used as bottom mirror, which offers a sufficiently high index contrast but is electrically isolating.

In order to combine an enhancement of the spontaneous emission rate with a mechanism for efficient light extraction, light can be generated in a thin film of active material and guided in a surface plasmon polariton mode [95]. These are guided optical modes that may exist at the interface between a dielectric and a metal and consist of an oscillating electromagnetic field coupled to an oscillating surface charge density. The rate of spontaneous emission can be increased by the coupling of the emitter to the enhanced electric fields associated with the guided modes (Purcell effect). A two-dimensional periodic photonic bandgap structure is used to create Bragg scattering and to couple the mode into useful radiation. A significant enhancement of PL extraction efficiency was demonstrated with optically pumped LEDs based on a thin-slab photonic crystal [100] and external quantum efficiencies of >70% were predicted with such a structure [94]. Most of the efficiency increase was associated with the improvement of

Bragg extraction and to a minor increase in the Purcell enhancement.

The potential performance of LEDs using surface plasmon polariton modes is difficult to estimate. Apart from all the technical problems of realizing such devices, there are still a number of open questions to answer. Dissipative losses related to the propagation of light along a dielectric/metal interface will affect the performance and must be circumvented. Also it is not obvious if an effective extraction of the guided mode can be achieved with a structure that efficiently couples the emitter to the surface plasmon polarity mode [101].

10.5 WHITE LED

With the availability of wide-bandgap InGaN-based semiconductors, it became possible to produce not only efficient ultraviolet, blue and green diodes but also white emitting LEDs. Before that, the lack of white light has been a major barrier for LEDs in a wide range of applications. With LEDs covering the full color spectrum including white, the ultimate goal of LED-based illumination and lighting became possible.

10.5.1 White light

It is important to define what is meant by "white" light, as it has to be compared with some kind of standard. The most fundamental model for white light is the solar spectrum shown in Figure 10.32,

but even then the sunlight changes with daytime and season of the year. It is usually therefore "idealized" by using an equivalent black body radiation spectrum. According to Wien's law, the peak maximum of the black-body spectrum is only a function of the temperature:

$$\lambda_{peak} = \frac{2880}{T} \mu m\, K. \tag{10.40}$$

As the temperature increases from room temperature to thousands of degrees kelvin, the peak wavelength of the black body radiation moves from infrared to visible wavelengths. In the chromaticity diagram, this is represented by a line from the red towards the center of the white colors (see Appendix A). Consequently, assuming true black-body radiation, it is sufficient to use only one parameter, the *color temperature*, to specify the spectral properties. The Commission Internationale de l'Eclairage (CIE) defined several standards for the so-called white, illuminants. The most commonly used standard for white light is the Illuminant C, describing overcast sunlight at a color temperature around 6770 K.

Besides the color temperature, the quality of white light can also be measured by the so-called *color-rendering index* R_a. R_a describes the color of objects illuminated with the light source as compared to the illumination with a reference light source. The CIE defined eight different sample objects, again using sunlight as reference. At first, eight special values R_a are deduced from these objects and then the average R_a is calculated as the arithmetic average [102]. By definition, sunlight has $R_a = 100$, incandescent bulbs achieve Ra up to 100, fluorescent lamps $R_a = 85$ and Na-vapor lamps $R_a = 20$. A selection of some application-specific requirements on R_a is given in Table 10.5. The color rendering index is a relative parameter and does not describe the real appearance of illuminated objects or the true color of the light source. Therefore, even bluish or reddish light sources can achieve the ideal value of 100. Table 10.6 summarizes the values of R_a for a selection of conventional and LED-based white light sources. Note that the color rendering index is very important for all illumination purposes, but is irrelevant for many other applications such as white signals or signs.

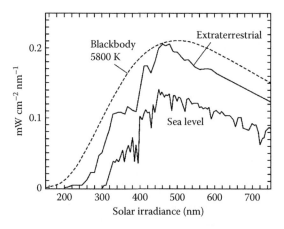

Figure 10.32 Spectrum of sunlight. The solid lines show the solar spectrum before and after transmitting the earth's atmosphere. The dashed line is the blackbody radiation at 5800 K.

Table 10.5 Color rendering index required for different applications

Application	R_a
Indoor/retail	90
Indoor office/home	80
Indoor work area	60
Outdoor pedestrian Area	≥60
Outdoor general Illumination	≤40

Source: ODIA workshop on LEDs for solid state lighting, October 26–27, 2000, Albuquerque, nm.

Table 10.6 Color rendering of different LED-based and conventional white light sources [111]

Technology	R_a
Blue LED+yellow phosphor	75–80
Blue LED+green+red phosphors	≥90
UV LED+blue+green+ red phosphor	≥90
Red+blue+green LED (610/470/560 nm)	60
Blue+yellow LED (500/595 nm)	40
Halogen W-filament incandescent lamp	100
Fluorescent lamp	85
Na vapor lamp	20–65
Hg vapor lamp	35–55

10.5.2 Phosphor-converted white LEDs

The first commercial white LEDs introduced by Nichia, used the same principle as white-emitting high pressure mercury lamps, that of adding a "color-correcting" phosphor to the bulb to convert part of the primary emission into yellow. These LEDs mixed the residual blue emission from the LED chip with the complementary yellow luminescence from phosphor. The first phosphor-converted white LEDs were developed by the Fraunhofer Institute for Solid State Physics (IAF) in Germany. Blue emitting GaN LEDs were used in conjunction with organic color convertors. The luminescent dyes, dissolved in the epoxy

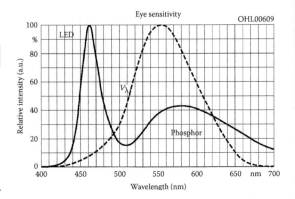

Figure 10.33 Spectral emission of a white LED based on phosphor conversion. Also shown is the human eye sensitivity with its maximum at 555 nm (green).

resin, absorbed the blue light and re-emitted luminescence at different colors [103]. Using the same principle of luminescence conversion, the blue GaN emission could also be used to fabricate green, yellow and red LEDs [104].

Today, white LED light is usually achieved by the use of an efficient inorganic phosphor and a single blue LED [105, 106]. The resulting emission spectrum, together with the human eye sensitivity is shown in Figure 10.33. The most common phosphors are based on yttrium aluminium garnet (YAG) doped with one or more rare earth elements or rare earth oxides. Optically active rare earth elements such as neodymium (Nd), erbium (Er), cerium (Ce) or thorium (Th) are widely used in light sources like lasers, light amplifiers or fluorescent tubes. A very common and extensively studied phosphor for blue–yellow conversion is YAG:Ce ($Y_3Al_5O_{12}$:Ce^{3+} (4f1)) [107]. Within the packaging procedure, the YAG phosphor powder is suspended in the epoxy and deposited directly on or around the LED die [108]. Color temperature and color rendering index of such a single-chip white LED can be adjusted by the amount of phosphor suspended in the epoxy and the emission wavelength of the LED die. From a production point of view, the packaging step is critical for maintaining consistency in the color characteristics, because subtle variations in concentration or spatial distribution of the phosphor coating changes the tint of white light between a more yellowish-or bluish-white.

Typically, the emission of such devices appears as a bluish or cold white due to the low level of absorption in the phosphor. Because the emission of the blue LED is directional, while the phosphor converted light radiates over a 2π solid angle, the appearance of white changes for an observer looking from the side. Compared to other approaches, phosphor converted LEDs have lower fabrication costs and offer a good control over quality and color. Color rendering with R_a around 75 is typically achieved at effective color temperatures around 5000 K. The energy conversion efficiency of a single phosphor is below 60%, which is still rather low. Another drawback is the lifetime of the encapsulating materials which can degrade in the presence of the near-UV radiation tail from the blue LED die.

White light can also be generated by the combination of a near-UV (380–400 nm) LED and a two-or three-color phosphor. The generation of white light by UV-pumped three-color phosphors is widely used in fluorescent tubes and phosphors for the emission of the gas-discharge process at 254 nm are readily available. The overall efficiency of such an LED is given by the efficiency of the UV-LED, the degree of absorption in the phosphors and the efficiency of the wavelength-conversion process. Despite the challenging material constraints in the AlGaN/InGaN system [109], UV LEDs with remarkably high quantum efficiencies of up to 32% at 390 nm and output power levels of 21 mW at 20 mA have been reported [110]. Although less efficient than these record values, UV InGaN-LEDs with a few mW of output power are already commercially available. The UV absorption in the phosphors is still an issue for optimization and much effort is dedicated to either developing LEDs that match the absorption of existing phosphors or developing appropriate phosphors absorbing at wavelengths where the LED is most efficient. However, even with 100% UV-absorption, a substantial amount of energy is lost in the actual wavelength conversion processes which markedly limits the overall luminous performance of such devices.

A potential advantage of the UV-pumped phosphor-based LED is that the visible emission is solely originated by phosphorescence with a broad spectral output. The superposition of three broad lines at red, green and blue (e.g., 610, 560 and 470 nm) results in white light with excellent CRI values (>90). Also, the white point in the CIE chromaticity diagram is independent of the LED characteristics. Thus, in terms of color rendering and color

stability the UV pumped phosphor-based LED is the most attractive source for white lighting.

10.5.3 Multi-chip white

By definition, the addition of two complementary colors produces white. Hence, white light can also be generated by mixing the emission of two different LEDs with complementary colors. In the chromaticity diagram, the positions of complementary colors lie on opposite sides of the achromatic point with $x=y=0.3333$. All white points along this line are accessible by adjusting the power ratio of the two monochromatic colors at the perimeter of the chromaticity diagram. An example for white light generated by a mixture of blue (490 nm) and amber (590 nm) is shown in Figure 10.34. LED-based white light sources using two discrete colors are very efficient in terms of luminous performance (lm W⁻¹) and costs/lumen. The main problem, however, is the very limited color rendering which excludes this solution for general lighting applications. Despite this drawback, two-chip white LEDs are commercially available and used, e.g., as map lights in the interior of automobiles. The luminous output of such lamps can be increased by using several LEDs of each color within the same package.

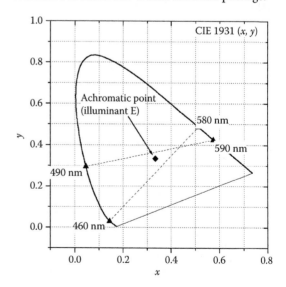

Figure 10.34 Examples for complementary colors in the CIE chromaticity diagram. The line between 490 and 590 nm is used for two-chip white LEDs. The 460–580 nm line is an example for white from phosphor-converted blue.

White with a better spectral distribution is generated by the mixture of three discrete colors of three different LED chips. The three-chip white LED has the advantage of providing no inherent loss mechanisms due to wavelength conversion to the emission from the primary sources. In the CIE chromaticity diagram, the three monochromatic colors span a triangle in which all mixed colors including white are accessible by the adjustment of the power ratios of the corner point colors. Figure 10.38 shows an arrangement of three LEDs (blue, green and red) in a single package. With individually addressable dices, such a multi-chip full color LED lamp is capable of producing various mixed colors, including a range of whites. As a white light source, the multi-chip LED achieves relatively good color rendering, color temperatures between 3000 and 7000 K, and high efficiencies in terms of luminous performance [112]. Unlike single-chip white LED, the color characteristics of multi-chip white LEDs can be altered after the packaging step. Used only as a source of white light, the multi-chip LED might be too expensive, but it is ideally suited for applications where variable-color pixels are required. Commercial multi-chip LEDs are offered with three or even five dices of discrete color in one lamp.

10.6 APPLICATIONS

10.6.1 Packaging

The packaging technology is becoming increasingly important for the performance of LEDs in many current and future applications. Some of the older, conventional packages today are inadequate for the rapidly improving high brightness AlGaInP or InGaN dices. Novel packages must consider better optical, electrical and thermal performance. The demand for high reliability puts stringent requirements on the chemical and thermal stability of the packaging, die attach and encapsulating materials as well as the selected processes. Devices with light converting phosphors such as white LEDs have to take additional care of the efficiency and stability of the phosphor materials. The best monochromatic LEDs today achieve >100 lm W^{-1} and the goal to achieve this and better efficiencies within the entire visible spectrum and, most importantly, in white LEDs will depend not only on the dices but also on the package used for assembly.

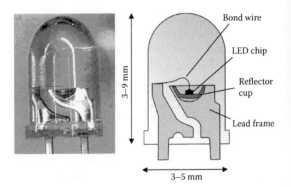

Figure 10.35 Radial LED lamp.

A simple radial LED package is shown in Figure 10.35. It was originally designed for low current indoor applications and has a thermal resistance of max. 280 K W^{-1} (R_{thjs}, thermal resistance at junction solder point), limiting the electrical input power to some hundred mW. The die has typically a lateral dimension of 200–300 μm. It is attached to the metal lead frame with epoxy-based conductive glue, so that the lead frame can act as electrical contact to the outside. This metal is shaped in a way that it can act as a mirror cup as well as heat sink for the die. A separate metal pin is connected to the bond-pad on the chip and acts as second electrode. Chip and lead-frame are encapsulated by epoxy which has the form of a dome in order to achieve a certain radiation characteristics. The epoxy almost doubles the extraction efficiency due to its favourable refractive index around 1.5 and the non-planar epoxy–air interface. Standard diameters of the epoxy dome are 3 and 5 mm and the package is also named simply "5 or 3 mm-LED package." About two-thirds of all high brightness LEDs are shipped in radial housings.

Packages for LEDs, like most electronic components, can be divided into two categories: through hole and surface mount. Through hole components like the radial package are loaded to a PC-board from one side and soldered from the other. Surface mount devices (SMDs) are loaded and soldered on the same side. This has several benefits for industrial production such as faster placing in automatic machines, smaller size, less parasitic effects and lower costs. In particular, in applications where space is limited such as in mobile phones, the surface mount technology (SMT) is superior. Figure 10.36 shows a standard SMT-package for LEDs with the lateral

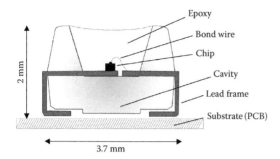

Figure 10.36 Schematic drawing of a surface mount LED package (TOPLED).

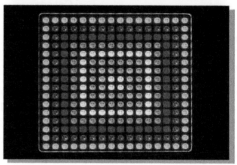

dimensions of 3.4×3.0 mm² and a height of 2.1 mm. The die is attached to the lead-frame with the pre-moulded housing of plastic (A-model). A bond wire connects the die's top electrode to the other part of the lead-frame. Finally, the housing is filled with encapsulating resin and, if necessary, one or more phosphor converters. The flat epoxy–air interface of SMT-packages results in a 10% lower extraction efficiency compared to radial packages with epoxy domes. However, for applications with the need for narrower emission profiles or more directionality, a transparent lens can be integrated (see Figure 10.37).

Depending on the application, the leads are folded either outwards or inwards under the housing. Some packages use specially bent leads so that

Figure 10.38 Three-chip LED in an SMT-package. Below: arrangement of MultiLEDs for a full-color display.

the LED can disappear in a hole in the PC-board while still being soldered from the top. SMT packages can be made very small. The smallest devices are 0.5–1 mm wide and high, which is just a little more than the dimensions of the die. The thermal resistance of SMT packages ranges from 300 to 500 K W⁻¹ This limits the maximum applicable current to 100–150 mA. SMT packages can also house several chips, e.g., for the generation of white light or as multiple-color LEDs. An example for a multi-chip SMT package is shown in Figure 10.38.

High-flux LEDs are designed for operation currents of 1 A or even more. For this current range, the package has to be capable of dissipating thermal powers of 1–2 W. High power packages, like the one shown in Figure 10.39, include a deep reflector and heat sinking metal base. The package can also include a lens-shaped epoxy dome to optimize the radiation pattern.

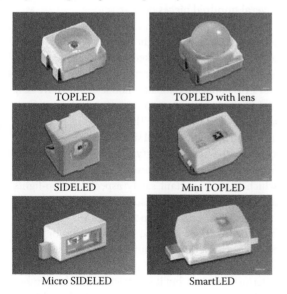

TOPLED

TOPLED with lens

SIDELED

Mini TOPLED

Micro SIDELED

SmartLED

Figure 10.37 Examples for SMT packages. Note the small size of, e.g., the SmartLED: 0.8×1.2 and 0.6 mm height.

10.6.2 Applications

For a long time, LEDs were low cost, low brightness devices suitable for single color, low power applications, e.g., to illuminate switches, indicators, small

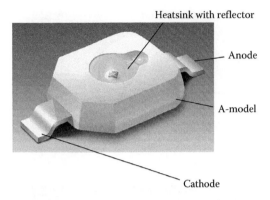

Heatsink with reflector

Anode

A-model

Cathode

Figure 10.39 High flux package. The package can take dices up to 1×1 mm² area and dissipate more than 1 W of excessive heat.

signs or to transmit information. With the availability of high brightness yellow and red AlGaInP LEDs and the development of blue and green InGaN-LEDs, the full color spectrum could be covered and a whole range of new applications opened up. A good example is signs and displays, where originally LEDs were only applied for monochromatic indoor numerical or text displays. Today, LED-based single or full color outdoor signs can be found as variable message signs on motorways or large area video panels for entertainment arenas and sports stadiums. One of the largest high resolution outdoor panels is the outer skin of the NASDAQ-building at New York's Time Square, which consumes 18 million LEDs. The LED sign market accounts for about one-third of the total market for high brightness LEDs (2000: $1.22 billion).

LCD backlighting—The rising market for cellular phones has also opened up new opportunities for LEDs. Here, LEDs are used to illuminate both the LCD display as well as the keypad. In 2000 about 400 million cellular phones were sold worldwide with an average number of approximately ten LEDs per unit. This application is still dominated by standard-type low brightness GaP yellow–green LEDs, but the use of high-brightness devices, in particular white, is growing rapidly. The arguments for LEDs here is mainly their small size but also the low power consumption, which helps to prolong the battery lifetime. White LEDs are also increasingly used to illuminate other types of LCD displays such as digital camera displays, handheld computers or, maybe in a longer perspective, even large area LCD screens.

Traffic lights—Traffic lights are a typical example for an application, where LEDs as efficient

monochromatic light sources can favourably compete with conventional solutions. A tungsten bulb with a red filter provides about $4 \, lm \, W^{-1}$ compared to more that $50 \, lm \, W^{-1}$ of a high-brightness AlGaInP-diode. The motivation to use LEDs is mainly driven by energy savings and reduced requirements for replacement. After the initial hurdles of standards and regulations were taken, high brightness AlGaInP-(red and yellow) and InGaN-LEDs (green) have an impressive degree of market penetration, especially in the US. In 2000 about 14% of all red and 6% of all green traffic signals in the US had been converted to high-brightness LEDs [113].

LEDs for automotive applications—Another application which increasingly consumes high numbers of LEDs is the interior and exterior illumination of automobiles. Inside a car, LEDs are used not onlly to backlight the dash panel, push buttons and indicator lamps but also LCD-displays, e.g., for the navigation system. Primarily, LEDs are used for dashboard illumination because their lifetime is significantly longer than that of conventional incandescent bulbs. Despite the higher price for LEDs, cost savings can be achieved, because no replacement has to be considered, but also because surface mount LED packages allow for cost efficient automated dashboard assembly. In total, the number of LEDs inside a car can easily sum up to several hundred pieces per car.

The first application in the exterior was the center high mounted stop light (CHMSL), which is still a steadily increasing market for AlGaInP red LEDs. While the CHMSL was for many years the only exterior LED application for automobiles, LEDs have now also started to penetrate other areas such as stop-or tail-lights and indicators. The benefits of LEDs here are their ruggedness, small size, low power consumption and the high reliability. They also offer a much faster turn-on time compared to incandescent bulbs, which is an important safety aspect for stop-lights and CHMSLs. Sometimes, LEDs are clustered in modules of 70 or 80 red and amber LEDs that combine the functions of stop-, tail-and indicator lights. So far only a few car manufacturers have started to equip some high-end car models with LED-based exterior functions, but it is expected that the number of LEDs in the exterior of automobiles and trucks will grow rapidly over the next years.

General lighting—The ultimate goal for the future of white LEDs is, of course, general lighting.

This market is valued to approximately $12 billion and represents a huge potential market for LEDs. In order to make this happen, LEDs do have to face a number of issues that are relevant to the lighting industry. The efficiency of LEDs as well as the light-output generated per device or module has to be raised to the level of conventional lighting systems. In terms of efficiency, the target for white LEDs is somewhere between 100 and 150 lm W^{-1}, the total generated flux should come up at least to levels of incandescent bulbs (1700 lm for a 100 W bulb). For indoor lighting, the color rendering index describing the appearance of colors when illuminated with LED light, should be higher than 80. The color temperature of LED-white should at least replicate the values of fluorescent or incandescent (2850 K) lamps. Color variations in different directions of radiation or variations from device to device have to stay within the acceptable ranges of conventional sources. The color stability of phosphor-converted white LEDs is mainly an issue for the packaging process, whereas multi-chip LEDs require very reproducible intensity ratios or an active adjustment of the color point. A long device lifetime is important to reduce the overall operating costs including replacement and maintenance. The crucial part for high reliability is the package including encapsulating and converter materials. Even if the diode chip alone has a lifetime of many 10,000 h, the encapsulating materials tend to degrade faster in the presence of UV-photons. Finally, it is of paramount importance to reduce the cost for white LED light to competitive levels.

ACKNOWLEDGMENT

I would like to thank all colleagues at Osram Opto Semiconductors who contributed to this chapter. In particular, I thank the AlGaInP R&D team: Norbert Linder, Ralph Wirth, Walter Wegleiter, Reiner Windisch, Peter Stauss, Christian Karnutsch, Wolfgang Schmid, Ines Pietzonka, Simone Thaler, Gertraud Huber, Monika Mändl, Kornelia Kruger and Monika Kuttenberger.

APPENDIX 10A

10A.1 The CIE color system

In 1931, the Commission Internationale de l'Eclairage (CIE) produced the "Color Standard Table" in order to create an objective method of determining colors [114]. As a basis for standardization, the CIE chose the response of the three sets of color receptors in the eye. Colors are measured by comparison with an additive mixture of three elementary colors and specified by the tristimulus values X, Y and Z. The next task was to find a way for a two-dimensional representation of colors in a color map, similar to geographical maps. This was achieved by calculating a new set of variables, the color-masses x, y and z from the measured tristimulus values by dividing each of them by their total sum: $x=X/(X+Y+Z)$, $y=Y/(X+Y+Z)$ and $z=Z/(X+Y+Z)$. With this conversion, only two values, e.g., x and y, remain independent and can be used as coordinates in a two-dimensional chart. A representation with the x-values on the horizontal axis and the y-values on the vertical axis is called the CIE chromaticity diagram. The saturated colors are located on the locus of the chromaticity diagram. A straight line, the "purple line" joins the red and violet ends of the spectral locus to form a closed diagram. The colors on the purple line are mixtures of the spectral colors 770 and 400 nm on the right-and left-hand side. All existing colors lie within the tongue shape delineated by the line of spectral colors and the purple line (Figure 10A.1).

The CIE defined the color coordinates of several illuminants to describe commonly used white light sources (Table 10A.1).

The special point E with the coordinates $x=y=1/3$ is denoted "equal energy" and locates

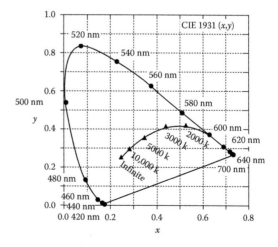

Figure 10A.1 1931 CIE diagram of colors. The solid line gives the black body radiation at temperatures from 1000 K to infinite.

Table 10A.1 Color coordinates and color temperature of CIE illuminants

	Illuminant	x	y	T (K)
Incandescent lamp	A	0.4476	0.4074	2856
Direct sunlight	B	0.3484	0.3516	4870
Overcast sunlight	C	0.3101	0.316	6770
Daylight	D_{65}	0.3128	0.32932	6504
Equal energy	E	0.3333	0.3333	

Note: Illuminant E is an imaginary color with equal values for x, y and z.

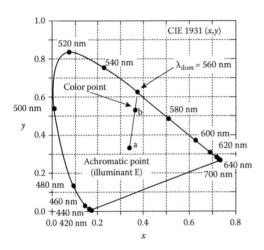

Figure 10A.2 Dominant wavelength and spectral purity of a light source. The color purity is defined as $a/(a+b)$.

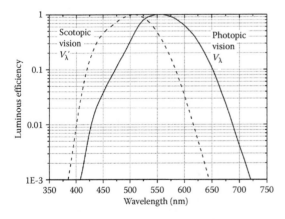

Figure 10A.3 Spectral eye sensitivity curves $V\lambda$ and V'_λ for light-adapted (photopic) and dark-adapted (scotopic) vision. The peak of V'_λ is shifted 43 nm towards shorter wavelength.

the achromatic point representing greys and white. Complementary colors lie on opposite sides of the achromatic point. An important definition for the color of LEDs is the *dominant wavelength*. It is defined as the spectral color that is perceived to be the same as the color of the LED. In the chromaticity diagram, the dominant wavelength can be determined by drawing a straight line from the achromatic point through the point representing LED color (Figure 10A.2). The intersection of the prolonged line with the perimeter of the diagram gives the spectral color that defines the dominant wavelength. This concept is also used to define the color purity of a light source. *Color purity* is a measure of how close the color point lies to the perimeter of the CIE diagram. It is defined as the ratio of the distance of the color point to the perimeter, weighted with the distance of the achromatic point to the same intersection point. Thus, the purity of a color ranges from 0 for the achromatic point to 1 for saturated colors at the perimeter.

The human eye sensitivity has been defined and standardized by the CIE in 1924. Usually two response curves $V\lambda$ and V'_λ for day and night vision (photopic and scotopic vision) are used (Figure 10A.3). In daylight vision, yellowish–green light at 555 nm stimulates the eye more than blue or red light. At this wavelength, the human eye can detect about 10 photons s^{-1} or a radiant power of 3.58×10^{-18} W. At blue (450 nm) and red (650 nm), the eye detection limits are 214 and 126 photons s^{-1} respectively [115].

10A.2 Photometric and radiometric units

Radiometry is the measurement of electromagnetic radiation at all frequencies, whereas photometry is restricted to the measurement of visible light. The

Table 10A.2 Radiometric and photometric Units

Radiometric units		Photometric units	
Radiant flux ϕ_e	W_{opt} (watt)	Luminous flux ϕ_v	lm (lumen)
Radiant efficiency	W_{opt}/W_{el}	Luminous efficiency	lm/W_{el}
		Luminous efficacy	lm/W_{opt}
Radiant intensity I_e	W_{el}/sr	Luminous intensity I_v	lm/sr = cd (candela)
Radiance L_e	W_{el}/sr m^2	Luminance L_v	cd/m^2

Note: Symbols for the radiometric quantities have the subscript e, and photometric quantities have the subscript v. The SI unit watt is suffixed with "opt" and "el" in order to denote the units of optical and electrical power. W_{opt}, unit of the radiant flux (watt); W_{el}, unit of electrical power (watt); sr, SI unit of the solid angle (steradian).

difference between the two methods is that in photometry everything is weighted by the human eye sensitivity.

In radiometry, the power flux of electromagnetic radiation is measured in watts (W). This flux is called the *radiant flux* ϕ_e of a light source. The photometric pendent to the radiant flux is the *luminous flux* ϕ_v and has the SI unit lumen (lm). The definition of lumen is related to the CIE eye response curve where 1 W of radiant flux at 555 nm wavelength is defined to have a luminous flux of 683 lm.

$$\phi_v = 683 V_\lambda \phi_e \qquad (A1)$$

Luminous efficiency is the ratio of luminous flux measured in lumen and the electrical power used to generate this flux measured in watt. One very confusing thing here is that watt as the general SI unit for power is used for optical and electrical measurements. Thus another term, the *luminous efficacy K* defined as the ratio of luminous flux and radiant flux has the same unit lm W^{-1}.

$$K = \frac{\phi_v}{\phi_e}. \qquad (A2)$$

Luminous efficiency is inevitably lower than the luminous efficacy, because some of the input power is lost in form of heat and does not appear as emitted radiant power. The efficiency of converting electrical power into radiant flux is measured by the *radiant efficiency* with the unit W W^{-1}.

The total radiant or luminous flux does not indicate how the LED output varies with direction. For many applications, it is important to know how much power in watts or lumen is concentrated within a narrow angular range in a particular direction. This is measured by the *radiant intensity* I_e or the luminous intensity I_v of the LED in that direction. The units for *luminous intensity* are W sr^{-1} (sr = steradian) and lm sr^{-1}, respectively. The latter unit, lm sr^{-1} is called candela (cd) and is equal to 1 lm sr^{-1}.

The concept of radiant and luminous intensity applies only to point sources. Sources with an extended light-emitting area are characterized by the luminance, defined as luminous intensity per unit light-emitting area. The unit of *luminance* is cd m^{-2}. Since light sources that emit the same luminous intensity from a smaller emitting area appear brighter, the term luminance is correlated with the qualitative term "brightness." The radiometric counterpart to luminance is *radiance*, which is the radiant flux per steradian per square meter of emitting surface area (Table 10A.2).

REFERENCES

1. Round, H. J. 1907. A note on carborundum. *Electrical World*. 49: 309.
2. Pankove, J. I., Maruska, H. P. and Berkeyheiser, J. E. 1970. Optical absorption of GaN. *Appl. Phys. Lett.* 17: 197–199.
3. Roosbroeck, W. V. and Shockley, W. 1954. Photon-radiative recombination of electrons and holes in Ge. *Phys. Rev. B.* 94: 1558–1560.
4. Olshanksy, R., Su, C., Manning, J. and Powazi, W. 1984. Measurement of radiative and nonradiative recombination rates in InGaAsP and AlGaAs light sources. *J. Quantum Electron.* 20: 838–854.

5. Henry, C. H., Logan, R. A. and Merrit, F. R. 1978. The effect of surface recombination on current in AlGaAs heterojunctions. *J. Appl. Phys.* 49: 3530–3542.

6. Boroditsky, M., Gontijo, I., Jackson, M., Vrijen, R. and Yablonovitch, E. 2000. Surface recombination measurements on III–V candidate materials for nanostructure light-emitting diodes. *J. Appl. Phys.* 87: 3497–3504.

7. Saul, R. H., Lee, T. P. and Burrus, C. A. 1985. *Light-Emitting-Diode Device Design in Semiconductors and Semimetals*, vol 22 (Bell Telephone Laboratories Inc.) (San Diego, CA: Academic).

8. Born, M. and Wolf, E. 1959. *Principles of Optics* (Cambridge: Cambridge University Press).

9. Carr, W. and Pittman, G. 1963. One-watt GaAs p–n junction infrared source. *Appl. Phys. Lett.* 3: 173–175.

10. Franklin, A. and Newman, R. 1964. Shaped electroluminescent GaAs diodes. *J. Appl. Phys.* 35: 1153–1155.

11. Osram enhances brightness of blue InGaN LEDs. *Compound Semicond.* 2001, 7: 7.

12. Gardner, N. F., Chui, H. C., Chen, E. I., Krames, M. R., Huang, J. W., Kish, F. A., Stockman, S. A., Kocot, C. P., Tan, T. S. and Moll, N. 1999. 1.4× efficiency improvement in transparent-substrate $(Al_xGa_{1-x})_{0.5}In_{0.5}P$ light-emitting diodes with thin (<2000 Å) active regions. *Appl. Phys. Lett.* 74: 2230–2232.

13. Linder, N., Kugler, S., Stauss, P., Streubel, K. P., Wirth, R. and Zull, H. 2001. High-brightness AlGaInP light-emitting diodes using surface texturing. *Proc. SPIE.* 4278: 19–25.

14. Krames, M. R. et al. 1999. High-power truncated-inverted-pyramid $(Al_xGa_{1-x})_{0.5}In_{0.5}P$/ GaP light-emitting diode exhibiting >50% external quantum efficiency. *Appl. Phys. Lett.* 75: 2365–2367.

15. Härle, V., Hahn, B., Lugauer, H., Brüderl, G., Eisert, D., Strauss, U., Lee, A. and Hiller, N. 2000. Developments in InGaN on SiC LEDs and lasers at Osram. *Compound Semicond.* 6: 81–85.

16. Streubel, K., Helin, U., Oskarsson, V., Bäcklin, E. and Johansson, A. 1998. High brightness visible (660 nm) resonant cavity light emitting diode. *Photon. Technol. Lett.* 10: 1685–1687.

17. Nishizawa, J., Koike, M. and Jin, C. C. 1983. Efficiency of GaAlAs heterostructure red light-emitting diode. *J. Appl. Phys.* 54: 2807–2812.

18. Ishiguro, H., Sawa, K., Nagao, S., Yamanaka, H. and Koike, S. 1983. High efficient GaAlAs light-emitting diodes of 660 nm with a double heterostructure on a EaAlAs substrate. *Appl. Phys. Lett.* 43: 1034–1036.

19. Shockley, W. 1950. *Electrons and Holes in Semiconductors* (Princeton, NJ: Van Nostrand-Reinhold).

20. Sze, S. M. 1981. *Physics of Semiconductor Devices* (New York: Wiley).

21. Craford, M. G., Shaw, R. W., Herzog, A. H. and Groves, W. O. 1972. Radiative recombination mechanisms in GaAsP diodes with and without nitrogen doping. *J. Appl. Phys.* 43: 4075–4083.

22. Campel, J. C., Holonyak, N., Craford, M. G. and Keune, L. D. 1974. Band structure enhancement and optimization of radiative recombination in $GaAs_{1-x}P_x$:N (and $In_{1-x}Ga_x$ P:N). *J. Appl. Phys.* 45: 4543–4553.

23. Casey, H. C. and Panish, M. B. 1978. *Heterostructure Lasers: Part A. Fundamental Principles* (New York: Academic).

24. Watanabe, M. O., Yoshida, J., Mashita, M., Nakashini, T. and Hojo, A. 1985. Band discontinuity for GaAs/AlGaAs heterojunction determined by C–V profiling technique. *J. Appl. Phys.* 57: 5340–5345.

25. Steranka, F. M. 1997. AlGaAs red light-emitting diodes. *Semiconductors and Semimetals, High Brightness Light Emitting Diodes*, vol 48, eds G B Stringfellow and M G Craford (San Diego, CA: Academic), pp. 65–95.

26. Meney, A., Prins, A., Philips, A., Sly, J., O'Reilly, E., Dunstan, D., Adams, A. and Valster, A. 1995. Determination of the band structure of disordered AlGaInP and its influence on visible laser characteristics. *J. Quantum Electron.* 1: 697–706.

27. Brennan, K. and Chiang, P. K. 1992. Calculated electron and hole steady-state drift velocities in lattice matched GaInP and AlGaInP. *J. Appl. Phys.* 71: 1055.

28. Kish, F. and Fletcher, R. 1997. AlGaInP light-emitting diodes. *Semiconductors and Semimetals, High Brightness Light Emitting Diodes* vol 48, eds G B Stringfellow and M G Craford(San Diego, CA: Academic), pp. 149–220.

29. Morrison, A. P., Lambkin, J. D., Poel, C. J. and Valster, A. 2000. Electron transport across bulk (AlGa)InP barriers determined from the *I–V* characteristics of n–i–n diodes measured between 60 and 310 K. *J. Quantum Electron.* 36: 1293–1298.

30. Streubel, K., Linder, N., Wirth, R. and Jaeger, A. 2002. High brightness AlGaInP LEDs. *J. Sel. Top. Quantum Electron.* 8: 321–332.

31. Chang, S. J., Chang, C. S., Su, Y. K., Chang, P. T., Wu, Y. R., Huang, K. H. and Chen, T. P. 1997. AlGaInP yellow-green light emitting diodes with a tensile strain barrier cladding layer. *Photon. Technol. Lett.* 9: 1199–1201.

32. Zunger, A. and Mahajan, S. 1995. *Handbook of Semiconductors* (Amsterdam: Elsevier).

33. Su, L. C., Ho, I. H., Kobayashi, N. and Stringfellow, G. B. 1994. Order/disorder heterostructure in $Ga_{0.5}In_{0.5}P$ with $\Delta E_g = 160$ meV. *J. Cryst. Growth* 145: 140–146.

34. Ernst, P., Geng, C., Scholz, F., Schweizer, H., Zhang, Y. and Mascarenhas, A. 1995. Band-gap reduction and valence-band splitting of ordered GaInP. *Appl. Phys. Lett.* 67: 2347–2349.

35. Ernst, P., Geng, C., Scholz, F. and Schweizer, H. 1996. Ordering in GaInP studied by optical spectrometry. *Phys. Status Solidi B* 193: 213.

36. Nasi, L., Salviati, G., Mazzer, M. and Zanotti-Fregonara, C. 1996. Influence of surface morphology on ordered GaInP structures. *Appl. Phys. Lett.* 68: 3263–3265.

37. Suzuki, T., Gomyo, A., Hino, I., Kobayashi, K., Kawata, S. and Iijima, S. 1988. P-type doping effects on band-gap energy for GaInP grown by metalorganic vapor phase epitaxy. *Jpn. J. Appl. Phys. Lett.* 27: L1549–L1552.

38. Vanderwater, D. A., Tan, I. H., Höfler, G. E., Defevere, D. C. and Kish, F. A. 1997. High-brightness AlGaInP light emitting diodes. *Proc. IEEE* 85: 1752–1764.

39. Koide, Y., Itoh, H., Khan, M. R. H., Hiramatsu, K., Sawaki, N. and Akasaki, I. 1987. Energy band-gap bowing parameter in an Al_xGa_{1-x} alloy. *J. Appl. Phys.* 61: 4540–4543.

40. Ho, I. H. and Stringfellow, G. B. 1996. Solid phase immiscibility in GaInN. *Appl. Phys. Lett.* 69: 2701–2703.

41. Wright, A. F. and Nelson, J. S. 1995. Bowing parameters for zinc-blende $Al_{1-x}Ga_xN$ and $Ga_{1-x}In_xN$. *Appl. Phys. Lett.* 66: 3051–3053.

42. Lester, S. D., Ponce, F. A., Craford, M. G. and Steigerwald, D. A. 1995. High dislocation densities in high efficiency GaN-based light-emitting diodes. *Appl. Phys. Lett.* 66: 1249–1251.

43. Härle, V., Hahn, B., Lugauer, H., Bader, S., Brüderl, G., Baur, J., Eisert, D., Strauss, U., Zehnder, U., Lell, A. and Hiller, N. 2000. GaN-based LEDs and lasers on SiC. *Phys. Status Solidi A* 180: 5–13.

44. Stath, N., Härle, V. and Wagner, J. 2001. The status and future development of innovative optoelectronic devices based on III-nitrides on SiC and on III-antimonides. *Mater. Sci. Eng. B* 80: 224–231.

45. Yoshida, S., Misawa, S. and Gonda, S. 1983. Improvements on the electrical and luminescent properties of reactive molecular beam epitaxially grown GaN films by using AlN-coated sapphire substrates. *Appl. Phys. Lett.* 42: 427–429.

46. Amano, H., Sawaki, N., Akasaki, I. and Toyoda, Y. 1986. Metalorganic vapor phase epitaxial growth of a high quality GaN film using an AlN buffer layer. *Appl. Phys. Lett.* 48: 353–355.

47. Nakamura, S. 1991. GaN growth using GaN buffer layer. *Jpn. J. Appl. Phys. Lett.* 30: 1705–1707.

48. Nakamura, S. 1997. Group III–V nitride-based ultraviolet LEDs and laser diodes. *Semiconductors and Semimetals, High Brightness Light Emitting Diodes*, eds G B Stringfellow and M G Craford (San Diego, CA: Academic), pp. 391–441.

49. Strite, S., Lin, M. E. and Morkoc, H. 1993. Progress and prospects for GaN and the III–V nitride semiconductors. *Thin Solid Films* 231: 197–210.

50. Maruska, H. P. and Tietjen, J. J. 1969. The preparation and properties of vapor-deposited single-crystal-line GaN. *Appl. Phys. Lett.* 15: 327–329.

51. Amano, H., Kitoh, M., Hiramatsu, K. and Akasaki, I. 1990. Growth and luminescence properties of Mg-doped GaN prepared by MOVPE. *J. Electrochem. Soc.* 137: 1639–1641.

52. Nakamura, S., Iwasa, N., Senoh, M. and Mukai, T. 1992. Hole compensation mechanism of P-type GaN films. *Jpn. J. Appl. Phys. Lett.* 31: 1258–1266.

53. Nakamura, S., Mukai, T. and Senoh, M. 1994. Candela-class high-brightness InGaN/AlGaN double-heterostructure blue-light-emitting diodes. *Appl. Phys. Lett.* 64: 1687–1689.

54. Nakamura, S., Mukai, T. and Senoh, M. 1994. High-brightness InGaN/AlGaN double-heterostructure blue-green-light-emitting diodes. *J. Appl. Phys.* 76: 8189–8191.

55. Akasaki, I., Amano, H., Koide, Y., Hiramatsu, K. and Sawaki, N. 1995. Crystal growth of column III nitrides and their applications to short wavelength light emitters. *J. Cryst. Growth* 146: 455–461.

56. Nakamura, S. 1995. High-brightness InGaN blue, green and yellow light-emitting diodes with quantum well structures. *Jpn. J. Appl. Phys. Lett.* 34: L797–L799.

57. Björk, G., Yamamoto, Y. and Heitman, H. 1995. *Confined Electrons and Photons* (New York: Plenum).

58. Joannoplolus, J. D., Meade, R. D. and Winn, J. N. 1995. *Photonic Crystals* (Princeton, NJ: Princeton University Press).

59. Kish, F. A. et al. 1994. Very high-efficiency semiconductor wafer-bonded transparent-substrate $(Al_xGai_{1-x})_{0.5}In_{0.5}P$/GaP light-emitting diodes. *Appl. Phys. Lett.* 64: 2839–2841.

60. Kish, F. A., Vanderwater, D. A., Peanasky, M. J., Ludowse, M. J., Hummel, S. G. and Rosner, S. J. 1995. Low-resistance ohmic conduction across compound semiconductor wafer-bonded interfaces. *Appl. Phys. Lett.* 67: 2060–2062.

61. Höfler, G. E., Vanderwater, D., DeFevere, D. C., Kish, F. A., Carnras, M., Steranka, F. and Tan, I. H. 1996. Wafer bonding of 50mm diameter GaP to AlGaInP-GaP light-emitting diode wafers. *Appl. Phys. Lett.* 69: 803–805.

62. Carr, W. N. 1966. Photometric figures of merit for semiconductor luminescent sources operating in spontaneous mode. *Infrared Phys.* 6: 1–19.

63. Illek, S., Jacob, U., Plößl, A., Stauss, P., Streubel, K., Wegleiter, W. and Wirth, R. 2002. Buried micro-reflectors boost performance of AlGaInP LEDs. *Compound Semicond.* 8: 39–42.

64. Horng, R. H., Wuu, D. S., Wei, S. C., Tseng, C. Y., Huang, M. F., Chang, K. H., Liu, P. H. and Lin, K. C. 1999. AlGaInP light-emitting diodes with mirror substrates fabricated by wafer bonding. *Appl. Phys. Lett.* 75: 3054–3056.

65. Horng, R. H., Wuu, D. S., Wei, S. C., Huang, M. F., Chang, K. H., Liu, P. H. and Lin, K. C. 1999. AlGAInP/AuBe/glass light-emitting diodes fabricated by wafer bonding technology. *Appl. Phys. Lett.* 75: 154–156.

66. Härle, V. H. 2003. Light extraction technologies for high efficiency GaInN-LED devices. *SPIE Photonics West'03* (San Jose, CA, January 2003).

67. Schnitzer, I., Yablonovitch, E., Caneau, C. and Gmitter, T. J. 1992. Ultra high spontaneous emission quantum efficiency, 99.7% internally and 72% externally, from AlGaAs/GaAs/AlGaAs double heterostructures. *Appl. Phys. Lett.* 62: 131–133.

68. Schnitzer, I. and Yablonovitch, E. 1993. 30% external quantum efficiency from surface textured, thin-film light-emitting diodes. *Appl. Phys. Lett.* 63: 2174–2176.

69. Yablonovitch, E., Gmitter, T., Harbison, J. P. and Bhat, R. 1987. Extreme selectivity in the lift-off of epitaxial GaAs films. *Appl. Phys. Lett.* 51: 2222–2224.

70. Deckman, H. W. and Dunsmuir, J. H. 1982. Natural lithography. *Appl. Phys. Lett.* 41: 377–379.

71. Yablonovitch, E., Hwang, D. M., Gmitter, T. J., Florenz, L. T. and Harbison, J. P. 1990. van der Waals bonding of GaAs epitaxial liftoff films onto arbitrary substrates. *Appl. Phys. Lett.* 56: 2419–2421.

72. Windisch, R., Kuijk, M., Dutta, B., Knobloch, A., Kiesel, P., Döhler, G. H., Borghs, G. and Heremans, P. 2000. Non-resonant cavity light-emitting diodes. *SPIE Proceedings of Photonics West*, 1–7.

73. Windisch, R., Dutta, B., Kuijk, M., Knobloch, A., Meinlschmidt, S., Schobert, S., Kiesel, P., Borghs, G., Döhler, G. H. and Heremans, P. 2000. 40% efficient thin-film surface-textured light-emitting diodes by optimization of natural lithography. *IEEE Trans. Electron. Devices* 47: 1492–1498.

74. Windisch, R., Rooman, C., Meinlschmidt, S., Kiesel, P., Zipperer, D., Döhler, G. H., Dutta, B., Kuijk, M., Borghs, G. and Heremans, P. 2001. Impact of texture-enhanced transmission on high-efficiency surface-textured light-emitting diodes. *Appl. Phys. Lett.* 79: 1–3.

75. Rooman, C., Windisch, R., D'Hondt, M., Dutta, B., Modak, P., Mijlemans, P., Borghs, G., Vounckx, R., Moerman, I., Kuijk, M. and Heremans, P. 2001. High-efficiency thin-film light emitting diodes at 650 nm. *Electron. Lett.* 37: 852–853.

76. Schmid, W. et al. 1999. Infrared light-emitting diodes with lateral outcoupling taper for high extraction efficiency. *SPIE Photonics West'99* (San Jose, CA, January 1999).

77. Schmid, W., Eberhard, F., Jäger, R., King, R., Miller, M., Joos, J. and Ebeling, K. J. 2000. 45% Quantum efficiency light-emitting diodes with radial outcoupling taper. *SPIE Proceedings of Photonics West* 3938: 90–97.

78. Purcell, E. M. 1946. Spontaneous emission probabilities at radio frequencies. *Phys. Rev. Lett.* 69: 681.

79. Schubert, E. F., Wang, Y. H., Cho, A. Y., Tu, L. W. and Zydzik, G. J. 1992. Resonant cavity light-emitting diode. *Appl. Phys. Lett.* 60: 921–923.

80. Benisty, H., Neve, H. D. and Weisbuch, C. 1998. Impact of planar microcavity effects on light extraction—Part I: Basic concepts and analytic trends. *J. Quantum Electron.* 34: 1612–1631.

81. Benisty, H., Neve, H. D. and Weisbuch, C. 1998. Impact of planar microcavity effects on light extraction—Part II: Selected exact simulations and role of photon recycling. *J. Quantum Electron.* 34: 1632–1643.

82. Huffaker, D. L., Lin, C. C., Shin, J. and Deppe, D. G. 1995. Resonant cavity light emitting diode with an Al_xO_y/GaAs reflector. *Appl. Phys. Lett.* 66: 3096–3098.

83. De Neve. H., Blondelle, J., Baets, R., Demeester, P., Daele, P. V. and Borghs, G. 1996. Resonant cavity LEDs. *Microcavities Photonic Bandgaps: Physics Appl.* 324: 333–342.

84. Jalonen, M., Toivonen, M., Köngäs, J., Savolainen, P., Salokatve, A. and Pessa, M. 1997. Oxide-confined AlGaInP/AlGaAs visible resonant cavity light-emitting diodes grown by solid source molecular beam epitaxy. *LEOS 10th Annual Meeting 97* (November 1997), pp. 239–240.

85. Dumitrescu, M., Toikkanen, L., Sipilä, P., Vilokkinen, V., Melanen, P., Saarinen, M., Orsila, S., Savolainen, P., Toivonen, M. and Pessa, M. 2000. Modeling and optimization of resonant cavity light emitting diodes grown by solid source molecular beam epitaxy. *Microelectron. Eng.* 51–52: 449–460.

86. Stevens, R., Risberg, A., Wurtemberg, M. V., Schatz, R., Ghisoni, M. and Streubel, K. 1999. High-speed visible VCSEL for POF Data Links. *POF Conference* (Japan, 1999).

87. Streubel, K. and Stevens, R. 1998. 250 Mbit/s plastic fibre transmission using 660 nm resonant cavity light emitting diodes *Electron. Lett.* 34: 1862–1863.

88. Wirth, R., Karnutsch, C., Kugler, S. and Streubel, K. 2001. High efficiency resonant-cavity leds emitting at 650 nm. *Photon. Technol. Lett.* 13: 421.

89. De Neve, H., Blondelle, J., Daele, P. V., Demester, P., Baets, R. and Borghs, G. 1997. Recycling of guided mode light emission in planar microcavity light emitting diodes. *Appl. Phys. Lett.* 70: 799–801.

90. Yablonovitch, E., Gmitter, T. J. and Bhat, R. 1988. Inhibited and enhanced spontaneous emission from optically thin AlGaAs/GaAs double heterostructures. *Phys. Rev. Lett.* 61: 2546–2549.

91. John, S. 1987. Strong localization of photons in certain disordered dielectric superlattices. *Phys. Rev. Lett.* 58: 2486–2489.

92. Baba, T. 1997. Photonic crystals and microdisk cavities based on GaInAsP-InP system. *J. Sel. Top. Quantum Electron.* 3: 808–830.

93. Yablonovitch, E. 1993. Photonic bandgap crystals. *J. Phys. Condens. Matter* 5: 2443–2460.

94. Boroditsky, M., Krauss, T. F., Coccioli, R., Vrijen, R., Bhat, R. and Yablonovitch, E. 1999. Light extraction from optically pumped light-emitting diode by thin-slab photonic crystals. *Appl. Phys. Lett.* 75: 1036–1038.

95. Barnes, W. L., Kitson, S. C., Preist, T. W. and Sambles, J. R. 1996. Photonic surfaces. *Microcavities Photonic Bandgaps: Physics Appl.* 324: 265–274.

96. Evans, G. A. et al. 1991. Characterization of coherent two-dimensional grating surface emitting diode laser arrays during cw operation. *J. Quantum Electron.* 27: 1594–1605.

97. Eriksson, N., Hagberg, M. and Larsson, A. 1995. Highly efficient grating-coupled surface-emitters with single outcoupling elements. *Photon. Technol. Lett.* 7: 1394–1396.

98. Erchak, A. A., Ripin, D. J., Fan, S., Rakich, P., Joannopoulos, J. D., Ippen, E. P., Petrich, G. S. and Kolodziejski, L. A. 2001. Enhanced coupling to vertical radiation using a two-dimensional photonic crystal in a semiconductor light-emitting diode. *Appl. Phys. Lett.* 78: 563–565.

99. Rattier, M. 2001. Diodes électro-luminescentes à cristaux photoniques: Extraction de la lumière guidée. *PhD Thesis* (Ecole Polytechnique de Paris).

100. Boroditsky, M., Vrijen, R., Krauss, T. F., Coccioli, R., Bhat, R. and Yablonovitch, E. 1999. Enhancement from thin-film 2-D photonic crystals. *J. Lightwave Technol.* 17: 2096–2112.

101. Barnes, W. L. 1999. Electromagnetic crystals for surface plasmon polaritons and the extraction of light from emissive devices. *J. Lightwave Technol.* 17: 2170–2182.

102. Wyszecki, G. and Stiles, W. S. 1982. *Colour Science Concepts and Methods, Quantitative Data and Formula*, 2nd ed. (New York: Wiley).

103. Schlotter, P., Schmidt, R. and Schneider, J. 1997. Luminescence conversion of blue light emitting diodes. *Appl. Phys. A* 64: 417–418.

104. Schlotter, P., Baur, J., Hielscher, C., Kunzer, M., Obloh, H., Schmidt, R. and Schneider, J. 1999. Fabrication and characterization of GaN/InGaN/AlGaN double heterostructure LEDs and their application in luminescence conversion (LUCOLEDs). *Mater. Sci. Eng. B* 59: 390–394.

105. Nakamura, S., Senoh, M., Iwasa, N., Nagahama, S., Yamada, T. and Mukai, T. 1995. Superbright green InGaN single-quantum-well-structure light-emitting diodes. *Jpn. J. Appl. Phys. Lett.* 34: L1332–L1335.

106. Nakamura, S., Pearton, S. and Fasol, G. 1997. *The Blue Laser Diode* (Berlin: Springer).

107. Baur, J., Schlotter, P. and Schneider, J. 1998. White light emitting diodes. *Adv. Solid State Phys.* 67: 67–78.

108. Bogner, G., Debray, A. and Höhn, K. 2000. High performance epoxy casting resins for SMD-LED packaging. *Light-Emitting Diodes: Research, Manufacturing, and Applications IV, Proc. SPIE* 3938: 249–261.

109. Crawford, M. H., Han, J., Chow, W. W., Banas, M. A., Figiel, J. J., Zhang, L. Z. and Shul, R. J. 2000. Design and performance of nitride-based UV LEDs. *Light-Emitting Diodes: Research, Manufacturing, and Applications IV, Proc. SPIE* 3938: 13–23.

110. Cree Lighting Inc. 2001. Press Release.

111. OSRAM. 1987. *Taschenbuch der Lampentechnik* (Berlin: Osram GmbH).

112. Mueller-Mach, R. and Mueller, G. O. 2000. White light emitting diodes for illumination. *Light-Emitting Diodes: Research, Manufacturing, and Applications, Proc. SPIE* 3938: 30–42.

113. *High-Brightness LED Market Review and Forecast* 2001 Strategies Unlimited.

114. *Commission Internationale de lEclairage Proceedings* 1931 (Cambridge: Cambridge University Press).

115. Ryer, A. 1998. *Light Measurement Handbook* (Newburyport, MA: International Light Inc.) http://www.intl-light.com.

11

Semiconductor lasers

JAYANTA MUKHERJEE
Kaiam Corporation

STEPHEN J. SWEENEY
University of Surrey

11.1 INTRODUCTION

The design and technology of semiconductor lasers has come a long way since their advent in the early 1960s, and as a result, these lasers have revolutionized communications, spectroscopy, data, and material processing [1–5]. The simple *pn* homojunction laser device [6] formed a key starting point but had characteristics, which meant they were not well suited to many applications. To realize the capability of semiconductor lasers, a convergence of many technologies had to be achieved. Advancements in crystal growth technologies [6,7], the development of the idea of the double heterojunction (DH) laser, [8] and subsequent laser designs incorporating quantum-confined active media such as quantum wells (QWs), wires and dots (QD), material passivation and efficient thermal management schemes, strained layer materials, breakthroughs in device designs including single-frequency, single-mode lasers and laser diode arrays, distributed feedback (DFB) lasers, and the simultaneous development of complementary technologies, the most significant of which is the rare earth doped fibers for fiber amplifiers (EDFAs) and fiber lasers, all contributed to one of the most enabling technological industries of today, that of semiconductor lasers [9–13]. In many ways, semiconductor lasers and amplifiers are second only to the transistor and integrated circuit due to their impact on today's high-technology market place. They form the backbone on which the Internet became economically feasible worldwide and are the key technological development of the 21st century on which the information age depends. Semiconductor lasers are now preferred over gas and solid-state lasers due to their wide tuning rage, high wall plug efficiency, compactness, low cost, and reliability. While low-power diode lasers are used in communication and data processing, their high-power counterparts are used in modern medical and surgical equipment, material processing, and optical pumping of solid-state lasers among other applications.

Within 10 years of their first proposal, semiconductor lasers had advanced to the point where DH lasers had been developed, resulting in reduced threshold continuous wave (CW) emission [14], together with advancements in single-frequency emission designs, although reproducibility still remained an issue. Initially, liquid phase epitaxy was used to fabricate these lasers [6], which resulted in poor material growth quality causing poor device performance and characteristics. However, with the development of two key crystal growth technologies, metal-organic chemical vapor deposition and molecular beam epitaxy, a new chapter was started in the field of semiconductor lasers [6]. Now it was possible to achieve crystal deposition control with reproducible atomic layer accuracy. This resulted in two important benefits: uniform material deposition and growth of quantum-confined active layers, both of which took semiconductor laser performance to new heights [7]. For modern semiconductor laser designs, improved crystal quality meant reduced propagation losses and decreased thermal resistance as the lasers could now be made thinner, broader, and longer without compromising on the crystal quality. Additionally, quantum-confined

active regions reduced the threshold current among other beneficial effects. Out of the various applications mentioned above, the use of semiconductor lasers and amplifiers in optical communications stands out. Extremely low bit-error rates are possible with these devices under high speed operation. Initial 10 Gbits/s systems have been now replaced by 40 and 100 Gbits/s systems and beyond will transform modern Internet technology and facilitate the so-called *Internet of Things* [9]. Recent advancement in high-power diode laser technology has replaced traditional solid state and gas laser in medical and material processing industries [4,5,10,15].

In the following sections, we will attempt to bring together the basic concepts behind the operation of semiconductor lasers while simultaneously following the history of their development. This will be followed by a review of different types of semiconductor lasers that have been developed for various applications. For catering to applications at different wavelengths, a variety of semiconductor alloys have been developed which we will discuss. Finally, we conclude by looking at the future prospects for semiconductor lasers.

11.2 BASIC THEORY OF SEMICONDUCTOR LASERS

The primary requirements for achieving laser action are (1) an *active medium*, (2) achieving *population inversion* in the active medium, and (3) an *optical cavity* to preferentially amplify generated radiation along its optical axis via positive feedback. We will start by understanding the basic functionality of an optical cavity before discussing the active medium and schemes of achieving population inversion in the cavity.

11.2.1 Optical cavity

The simplest type of optical cavity is a Fabry–Perot cavity [9–13,15], which is also the most common cavity configuration used in semiconductor diode lasers. A Fabry–Perot cavity consists of two reflecting planes (can be mirrors) at the end of its optical (z-) axis (Figure 11.1a and b). In the transverse (x, y) direction, the cavity walls may or may not be reflecting depending on the required cavity design (Figure 11.1a). Regardless of the generating source (the active medium), which we will discuss in the next section, for the generated optical radiation

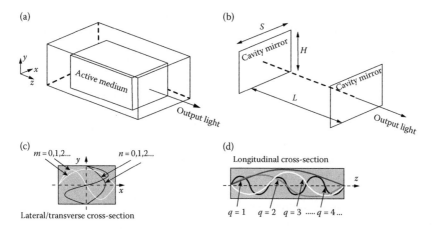

Figure 11.1 (a) Basic optical cavity with active medium inside showing a possible direction of coupling out the intracavity radiation. (b) An ideal Fabry–Perot cavity (without side walls), (c) and (d) cavity dimensions together with possible intersections of the intracavity field in the transverse (x, y) showing the (m, n) order spatial modes which can be sustained in the cavity. (e) Some of the possible longitudinal modes in the cavity of length L.

to remain confined inside the cavity, the tangential components of the longitudinally propagating intracavity electric field vector must be zero at the cavity mirrors, thus forming a standing wave [15]. This boundary condition determines the spatial profile and the possible frequencies of the optical field that can be sustained in the cavity. The spatial electric field profiles (*modes*) that can be sustained inside the cavity in the transverse (x, y) direction and along the optical (z-) axis are called the *spatial modes* and *longitudinal modes*, respectively. The spatial and longitudinal modes are sometimes also referred to in the literature as transverse and axial modes, respectively. These cavity modes are solutions of Maxwell's equations for a given optical cavity and are specific to the dimensions of the cavity. We will develop a simple intuitive picture below. If one of the cavity mirrors is made partly reflecting, then the cavity radiation can be coupled out of from this mirror (see Figure 11.1a and b).

Because the wavelength of optical radiation is generally much smaller than the dimensions of the cavity, several modes can be sustained by the cavity. Let us analyze the mode formation in the longitudinal direction first, where the formation for stationary modes requires that *standing waves* exist between the two cavity mirrors. It can be shown that an integral number q of half wavelengths can (simultaneously) satisfy the above-mentioned boundary condition along its optical axis (Figure 11.2a) [15]. In

the presence of an active medium having refractive index η (generally itself a function of wavelength), the wavelength of each mode is related to the cavity length L via the relation:

$$q \frac{\lambda_p}{2\eta_q} = L, \qquad (11.1)$$

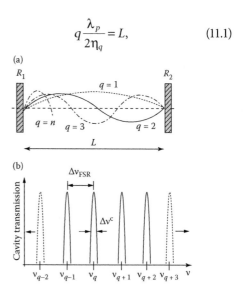

Figure 11.2 (a) Possible standing wave configurations, or modes, (three shown out of n) for intracavity fields which can be sustained simultaneously in the longitudinal direction of a Fabry–Perot cavity formed by two plane parallel mirrors of reflectivity R_1 and R_2. (b) Equidistant resonant frequency components as a function of cavity transmission (see text for further details).

where q is an integer (1, 2, 3, ...). The corresponding frequency is $v_q = c / \lambda_q$, and η_q is the refractive index corresponding to a discrete wavelength λ_q (since, in general, the refractive index is wavelength dependent). The electric field in the longitudinal direction labeled by q is called the qth longitudinal mode. Note that longitudinal modes differing in order are of different wavelengths. All these modes form *resonant modes* for the given cavity. For finite reflectivity of the mirrors, the cavity supports only specific frequency (wavelength) components to have high transmission for output coupling (OC), as shown in Figure 11.2b. The separation between the adjacent longitudinal modes, known as the free spectral range (FSR) of the cavity, is given by [15]

$$\Delta\lambda_{FSR} = \frac{\lambda_q^2}{2\eta_q L}, \quad (11.2)$$

which for typical semiconductor diode lasers is generally less than 0.5 nm. In the frequency domain, the FSR is given by

$$\Delta v_{FSR} = \frac{c}{2\eta_q L}. \quad (11.3)$$

In between the longitudinal modes, the cavity has high losses for intracavity radiation. The width (full width at half-maxima) of a single cavity resonance is related to the total losses α_{tot} in the cavity (see Section 11.6) [15]:

$$\Delta v_p^c = \frac{\alpha_{tot} c}{2\pi\eta_q} = \frac{1}{\pi\tau_{cav}}, \quad (11.4)$$

where τ_{cav} is the cavity decay time—the time it takes for the intracavity field (not intensity, see Equation 11.26) to decay to $1/e$ of its initial value. Δv_p^c is the line-width of the cavity resonance. Another useful parameter is the quality factor Q of the cavity, defined by [15, 21]

$$Q = \frac{v_p}{\Delta v_p^c}. \quad (11.5)$$

The higher the Q-factor, the better the cavity is in sustaining intracavity radiation of a particular frequency. We will discuss more about the cavity losses in a later section. Semiconductor lasers generally have lower Q-factors compared to the other solid-state and gas lasers owing to a poorer cavity resonance line-width Δv_q^c. In practice, the refractive index η_q should be substituted by the effective group index in the semiconductor, which takes into account the wavelength dependence of the refractive index influencing a propagating laser mode [9].

The transverse modes that can be sustained in a cavity with plane parallel mirrors symmetrical about the cavity optical axis can be obtained as solutions of Maxwell's equation as [6,15]

$$E(x,y,z) \propto H_m(ax)H_n(by)$$
$$\exp\left[-\left(\frac{x^2+y^2}{w^2}\right) - ikz\right], \quad (11.6)$$

where H_m and H_n are mth and nth order Hermite polynomials, respectively; a, b, and w are cavity-dependent parameters, k is the optical wave number, and w is called the spot size of the mode. The (m, n) order Hermite–Gaussian modes describe well the spatial modes in semiconductor lasers. Note that the n, m take values starting from zero (i.e., 0, 1, 2, ..., number of intersections of the intracavity field with the lateral and transverse cavity axes) unlike q for longitudinal modes, which starts from 1. The 0th order spatial mode ($m=0$, $n=0$) corresponds to the fundamental spatial mode. We will discuss spatial modes again in further detail later in this chapter.

11.2.2 Active medium

The source of optical radiation for directional amplification in the cavity described above is an active medium. Not all materials capable of generating luminescence can act as an active medium for laser action and special preparation and designs are generally required. Before we proceed to discuss more about the active medium, it is important to discuss the optical transitions in materials, which govern the properties of light emitters and absorber. The induced (stimulated) emission and absorption processes as well as spontaneous emission are fundamental to lasing action.

Note that the phenomenon of spontaneous emission can only be explained properly via pure quantum mechanical arguments, whereas the stimulated emission and absorption (both induced) processes can (also) be explained via semiclassical arguments. Einstein in his seminal 1917 paper *Zur Quantentheorie der Strahlung* (On the Quantum Theory of Radiation) [16], predicted, based on thermodynamic arguments, that together with induced absorption and spontaneous emission, another emission process (stimulated emission) ought to exist for a system in the presence of a radiation field to be in thermal equilibrium [16,17,89]. It is also interesting to note that a full quantum mechanical theory of spontaneous emission had not yet been developed in 1917 [90]. Einstein also asserted in the same paper, based on the requirement of momentum balance, that radiation as a result of stimulated emission should be identical in all respects to the stimulating radiation. We will build an intuitive picture of these three fundamental processes here without using a rigorous quantum mechanical treatment, which can be found elsewhere [16,17,90]. Consider a two-level system separated by energy $\Delta E = E_2 - E_1$ in the presence of a radiation field, as shown in Figure 11.3.

An incoming photon can *induce* a system to move to an excited state by *absorbing* a photon from the radiation field (Figure 11.3a). This induced process of *absorption* can occur, for example, in a semiconductor medium where a photon from incoming radiation is absorbed and an electron from the valence band (state E_1) is excited to the conduction band (state E_2). An excited state is generally an unstable state and the electron in this state can emit a photon spontaneously and return to the valence band, recombining with a hole. The spontaneously emitted photons are emitted in random directions with random phases and polarization [18,19]. There is no driving or inducing element for this spontaneous emission process and it is generally agreed that the omnipresent zero-point energy fluctuations (virtual photon interactions) of the system are responsible for the occurrence of this fundamental

emission process [19]. It is also possible that another photon in the radiation field *induces* the electron in state E_2 to return back to E_1 and in the process the system emits a photon in a state (quantum state) completely identical to the photon inducing (stimulating) this process. This process is called *stimulated emission* and is at the heart of the lasing phenomenon (Figure 11.3c). The reason why the photon resulting from stimulated emission is identical to the photon stimulating its emission can be intuitively understood by noting the fact that photons are bosons that have an affinity to occupy the same quantum states [18,20]. The fact that laser emission is coherent also stems from the aforementioned statement. Photons generated via stimulated emission tend to occupy the same quantum state of momentum and polarization, thereby creating a stream of spatially and temporally coherent radiation. The degree of coherence varies from one laser type to another and an appropriate understanding can only be developed via advanced treatments [21]. Semiconductor lasers are in general less coherent than, for example, gas lasers [15,21].

The concept of stimulated emission was put forward in 1917 by Einstein but it took almost half a century for the laser to be realized. The understanding that a resonant cavity (described above) in addition to a suitable active medium (radiation source) is required for directional amplification of stimulated emission took time to develop [6,15,21]. An interesting account on these developments can be found in the Nobel lecture by Prokhorov [89].

We now consider the type of semiconductors that are efficient sources of optical radiation. In semiconductors only *direct band-gap* materials (such as GaAs) have a high probability of radiative transitions (spontaneous and stimulated emission) resulting in *radiative recombination* (generation of optical radiation) between electrons in the conduction band and holes in the valence band, see Figure 11.4a. In indirect band semiconductors (such as Si, Ge), electron–hole recombination generally occurs via nonradiative routes (i.e., without the generation of optical radiation), such as via generation of phonons which heat-up the lattice (Figure 11.4b). The probability of radiative transitions is significantly lower in indirect band-gap semiconductors owing to the fact that the indirect semiconductor requires the presence of a phonon in order to conserve energy and momentum in the process. The amount of optical radiation that electron–hole pairs can generate is governed by the

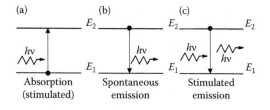

(a) Absorption (stimulated) (b) Spontaneous emission (c) Stimulated emission

Figure 11.3 Three fundamental processes for interaction of radiation with matter: (a) induced (stimulated) absorption, (b) spontaneous emission, and (c) induced (stimulated) emission process (see text for explanation of the three processes).

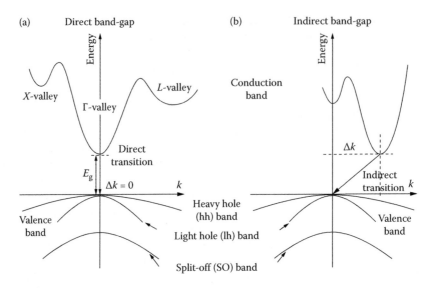

Figure 11.4 Schematic of $E(k)$ vs. k diagram for (a) direct and (b) indirect band semiconductors showing momentum balance in direct and indirect transitions, respectively. Γ (at $k=0$) and indirect (L, X) valleys are shown. The heavy-hole (hh) and light-hole (lh) bands are also shown. The energy difference between the conduction band minimum and valence band minimum is the band-gap, E_g.

competition between the radiative and nonradiative rates and the latter tends to dominate in indirect semiconductor materials. The underlying reason for this is that for any transition the momentum should be conserved in addition to the energy. The energy conservation requirement is met in both direct and indirect transitions in a straightforward manner. However, the momentum conservation requires the transition to be vertical, the so-called *k-conservation rule* ($\Delta \vec{k} = 0$) [9]. As the photon momentum is very small, the momentum conservation requirement is direct semiconductors is easily met. However, in indirect semiconductors a transition is possible only via the generation of another particle (a phonon) for the net change in momentum is zero (see Figure 11.4b); the absorption of a phonon for momentum balance is less probable owing to the many-body requirements of the process.

11.2.3 Population inversion

From the above discussion it is clear that in order to achieve a higher rate of stimulated emission in a semiconductor active medium (primary requirement for lasing action), one would need to constantly maintain a large number of electrons in an excited state (conduction band), i.e., the system needs to be in a state of *population inversion*.

In addition, the semiconductor should be a direct band-gap so that the radiative recombination is dominant. A state of inverted population can be made possible via optical pumping the system with radiation [at energy $> E_g$ ($= E_c - E_v$)], whereby the system constantly absorbs the radiation field and remains in an excited state. This scheme of optical pumping for achieving population inversion is used in a particular class of semiconductor lasers, the vertical external cavity semiconductor lasers (VECSEL), as discussed in Section 11.3. Most present day semiconductor laser designs however rely on the phenomenon of electrical injection to generate luminescence [electroluminescence (EL)], where injected electrical current is used to maintain the system in a state of population inversion (Section 11.4). Nevertheless, the notion of optical pumping allows us to intuitively understand the conditions required for achieving population inversion, gain, and amplification.

Let optical (pump-) radiation I_{pump} with variable energy $h\nu$ be incident on a slab of semiconductor material (assumed direct and intrinsic) and for simplicity let $T=0$ K (see Figure 11.5). If $h\nu < E_g$, no absorption takes place as electrons in the fully occupied valence band do not have enough energy to move to the conduction band. Absorption occurs (ignoring band tail effects [9–12]) if $h\nu \geq E_g$

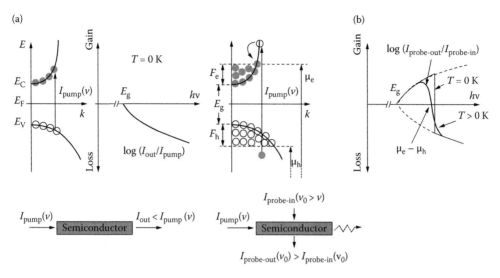

Figure 11.5 Schematic of the $E(k)$ vs. k diagram (for $k > 0$), optical absorption, inversion and gain in a semiconductor (considered direct and intrinsic. (a) Optical (pump) radiation with a variable energy $h\nu$ is assumed incident on a slab of semiconductor and the logarithm of the ratio of output vs. input intensity $\log(I_{out}(\nu)/I_{pump}(\nu))$ (absorption) is plotted as a function of energy $h\nu$. (b) Gain experienced under population inversion by a variable energy probing wave while a strong broad-band pump radiation is continuously incident on the semiconductor; $\log (I_{probe-out}(\nu)/I_{probe-in}(\nu))$ is plotted (see text for further explanation). Filled circles denote electrons, whereas open circles denote holes.

when electrons in the valence band obtain enough energy to move to the empty states in the conduction band. It is easy to see that the degree of absorption is proportional to the number of empty states (recall density of states [9–12], see Figure 11.12) available for the electrons in the valence band. As $h\nu$ increases further and beyond E_g, the absorption grows according to the increase in the density of states above E_g (see Figure 11.5a) and $\log(I_{out}/I_{pump})$ decreases according to the Beer–Lambert law. Now consider a slightly different situation where the semiconductor slab is bathed in a very stong I_{pump} and a large electron population is maintained in the conduction band. Such a nonequilibrium state of *population inversion* is depicted in Figure 11.5b, where electron and hole populations are now described by new individual chemical potentials μ_e and μ_h (generally synonymous to *quasi-Fermi levels* F_e and F_h provided they have the same zero reference energy [22,23], otherwise not [24]). Note that E_c, E_v, μ_e, and μ_h generally have a common zero reference energy, while F_e and F_h are measured from the respective band edges [22]. Also note that at $T=0$ K, the Fermi energy coincides with the chemical potential (maximum value of a single chemical potential) [23,25]. The concept of Fermi

energy is borrowed from the theory of metals [23]. Thus, usage of chemical potential is more appropriate and generally valid for understanding carrier population in semiconductors [22]. However, due to its widespread usage in the literature, we will use quasi-Fermi levels in the discussion that follows.

As interband relaxation (see Figure 11.5b curved arrow) is fast (~ps) compared to intraband transitions (~ns), the *quasi-Fermi levels* can be used to describe electron and hole populations *local* to the conduction and valence band, respectively [12]. Thus, at $T=0$ K all states in the conduction band up to F_e are filled and all states above F_e are empty. Similar arguments apply to the valence band. If a variable energy probing wave now passes through the semiconductor under population inversion, it experiences *gain* in intensity via the stimulated emission process rather than loss via absorption. The probing wave grows in intensity, again as per the Beer–Lambert law. As gain is just opposite to the absorption phenomenon, the gain profile is symmetric to the absorption profile (Figure 11.5b) about the energy axis.

The gain, however, is not experienced at all energies. If the probing wave energy $h\nu > \mu_e-\mu_h$ ($\mu_e-\mu_h=E_g+F_e+F_h=\mu_{e-h}$, the total chemical potential for the electron–hole plasma [22]), then we have the normal situation of filled states in the valence band and empty states in the conduction band and the probing wave suffers absoption. Similarly for $h\nu < E_g$ the semiconductor is completely transparent to the incident probing radiation. Therefore, the necessary condition for amplification of incident radiation at energy $h\nu$ in a semiconductor medium with inverted population is $E_g < h\nu < \mu_e - \mu_h$. This requirement was first put forward by Bernard and Duraffourg who derived the above relation in 1961 [26]. So far we have considered $T = 0$ K in the above analysis; for $T > 0$ K the sharp transition from gain to absorption (Figure 11.5b) is smoothed out due to the thermal distribution of electrons and holes. We will discuss more about the energy dependence of gain in Section 11.5.

11.2.4 Injection luminescence, carrier, optical, and current confinement

Consider a simple *pn* homojunction shown in Figure 11.6a and b. If electrons and holes are continuously injected from the *n*- and the *p*-contacts, respectively, a state of population inversion similar to the one described in the previous section can be created under high injection. The carriers (electrons, holes) can recombine in the vicinity of the depletion region resulting in radiation (injection luminescence or EL), which can experience optical gain.

We will discuss the wavelength of the generated radiation later but at this stage it is interesting to note that such an injected *pn*-junction, if placed

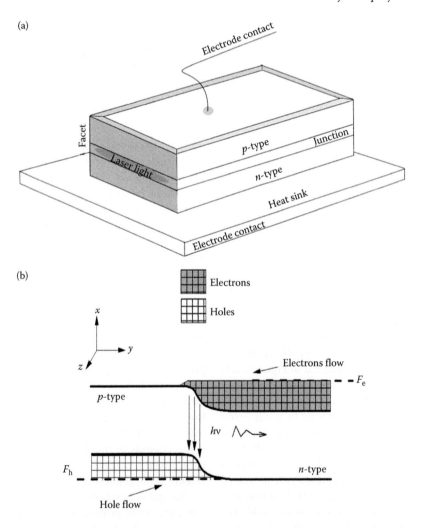

Figure 11.6 (a) A homojunction (injection) laser and (b) injected carriers recombine in the vicinity of depletion region.

inside a Fabry–Perot cavity, can result in laser emission. Indeed, the first *n*-GaAs-based (edge emitting) semiconductor laser with a Zn diffused p-region used the naturally cleaved facets of GaAs (33% reflective) as the Fabry–Perot cavity mirrors (Figure 11.6) [27]. However, this was a very inefficient device that generated laser emission only at cryogenic temperature, required very high electrical current for lasing emission to start and worked only under low-duty cycle (nonthermal) pulsed operation. Further research led to the realization that carrier leakage from the junction, inefficient confinement of the optical mode both vertically (perpendicular to the junction) and laterally (parallel to the junction) together with optical losses via absorption and due to inferior (epitaxial) material quality was behind the inefficiency and high current operation.

As the epitaxial growth technology improved and semiconductor materials with reduced defects and dislocations became available, the active region design of semiconductor lasers evolved from a homojunction to a single heterojunction and then to double heterojunction (DH) (see Figures 11.6 through 11.8) in order to get better

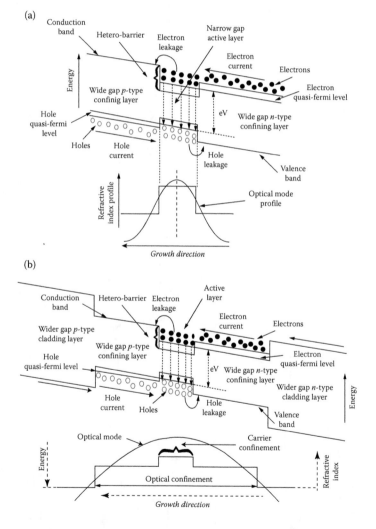

Figure 11.7 Schematic of active region designs for semiconductor lasers. (a) Double-heterojunction (DH) design showing simultaneous carrier confinement and optical mode confinement, e.g., narrow band-gap GaAs active region sandwiched between wide band-gap $Al_xGa_{1-x}As$. (b) Separate confinement heterostructure (SCH) with DH design used in modern semiconductor lasers: extra wide band-gap material for mode confinement. Tilted bands shown under high current injection (forward bias). Doping and space-charge interface effects are ignored for simplicity.

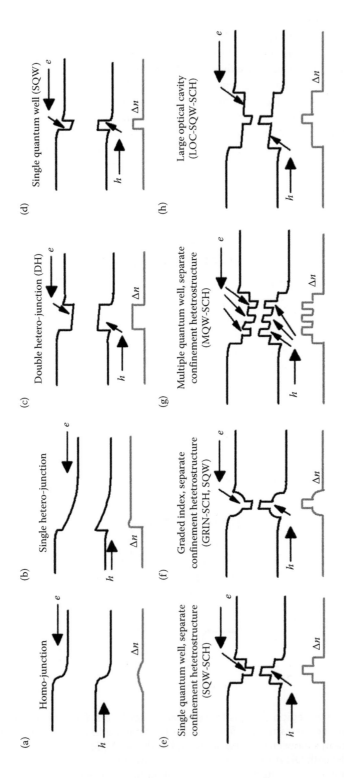

Figure 11.8 Evolution in the design of semiconductor laser active regions: (a) homojunction, (b) single heterojunction, (c) double heterojunction, (d) single-quantum well (SQW), (e) SQW with separate confinement heterostructure (SCH), (f) SQW with graded-index SCH, (g) multiquantum well with SCH (MQW-SCH), and (h) large optical cavity design with SQW-SCH (LOC-SQW-SCH) [11–13]. The corresponding refractive index profiles (perpendicular to the junction) are also shown. Direction of electron and hole injection into the active region are indicated by arrows. Diagrams not to scale.

confinement of the injected carriers and simultaneously an improved optical confinement of the generated radiation. It is a fortuitous fact of nature and a gift to the field of semiconductor lasers that, in general, as the band-gap of a semiconductor material decreases its refractive index increases (and vice versa) [11,12]. H. Kroemer in 1963 presented in his paper a DH design (similar to shown in Figure 11.7a), which automatically allowed for simultaneous optical and carrier confinement in semiconductor diode lasers [8]. In addition, it significantly reduced the optical losses via absorption in adjacent layers due to the wider band-gap, reducing the laser threshold. The active region design for the first room temperature CW semiconductor laser based on a lattice-matched $Al_xGa_{1-x}As/GaAs$ ($x \sim 0.3$) double heterostructure [14] is shown in Figure 11.7a. Such a lattice-matched material system (see Section 11.7) allowed for relatively defect-free material growth together with the flexibility of designing the required band offsets and refractive index profile (Figure 11.7a) for simultaneous carrier and optical mode confinement. For their contributions to the development of semiconductor lasers, Kroemer and Alferov were awarded the Nobel Prize in Physics in the year 2000.

With the proposal of the possibility of using a QW as the active region in a semiconductor laser in 1974 [28], researchers immediately realized that such a thin active region would benefit from a separate confinement heterostructure (SCH) that they had been working on [29,30], in addition to the potential benefits from the quantum-confined active region itself (Section 11.7). As QWs are too

thin to confine the optical mode, it was suggested that having a separate confinement for carriers (electron and holes) and that of the generated light via a SCH would result in lowering of the lasing threshold as well as increasing the efficiency of the lasing process (Figure 11.7b). In addition, the SCH design reduces carrier leakage from the active region and the optical power load at the facet preventing facet damage. Modern semiconductor lasers use variants of the SCH design and the active regions are generally undoped. Figure 11.8 shows the changes in semiconductor laser active region designs as they evolved toward achieving better simultaneous carrier and optical confinement.

Another crucial aspect of semiconductor laser design is the design for lateral (x-direction) current and spatial mode confinement. It was realized quite early that the injected carriers can easily undergo undesirable nonradiative recombination if allowed to reach the surface (side walls) of the device. Thus, a current injection stripe design was implemented [31,32] for laterally confining the injected carriers defining the emission width of the laser. Variants of this stripe geometry design are still commonly used in commercial edge-emitting semiconductor lasers.

As noted above, in the growth direction (y-axis) the optical (spatial) mode is index-guided (via total internal reflection) by the refractive index profile created by the epitaxial layers, whereas in the lateral direction (x-axis) the lateral current injection stripe design defines the lateral spatial mode guiding via either *gain-guiding* (Figure 11.9b) or *index-guiding* (Figure 11.9c). In a gain-guided

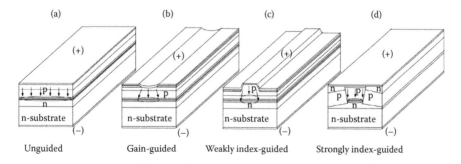

Figure 11.9 Different current injections schemes in semiconductor lasers: (a) unguided, (b) gain-guided, (c) weakly index guided, and (d) strongly index guided, a buried heterostructure design (see text for details). Top (anode) and bottom (cathode) metal contacts are also shown. Oxide insulation [gray strips on p-side in (b) and (c)] defines the lateral current path and spreading which in turn determines the lateral emitting aperture (in red) of the laser. On the other hand, the vertical emitting aperture is determined by the refractive index profile perpendicular to the junction, see Figure 11.8.

semiconductor laser, the light is laterally confined by optical gain created under the contact stripe, see Figure 11.9b, whereas in the strong index-guided design a lateral index step created generally by etching the p-side and regrowth (called buried heterostructure) in the transverse epitaxial structure, Figure 11.9c, which guides the mode. Weak index-guided designs involve just etching but no regrowth resulting in ridge-waveguide (also called mesastripe) structures [9–13].

11.2.5 Optical gain

We have discussed how population inversion in a semiconductor medium can result in optical gain for intracavity optical radiation. Before proceeding, it is important to understand the spectral, carrier, and temperature dependence of the optical gain in semiconductors. As we will see later, laser actions starts only when optical gain overcomes the losses in the system (cavity + active medium). Here, we present a simple but realistic approach for optical gain in bulk semiconductor material and qualitatively discuss the effects of quantum confinement. More sophisticated approaches to undertaking gain calculations can be found elsewhere [33,34]. Considering transitions only between conduction and heavy-hole valence band for simplicity, we start with the expression for the photon energy ($E_{ph} = h\nu$) and carrier density (N) dependent absorption α_{opt} [10,11,24], which is opposite to optical gain (also known as material gain) in an inverted semiconductor at $T > 0$ K:

$$g_{opt}\left(E_{ph}, N\right) = -\alpha_{opt}\left(E_{ph}, N\right) = \frac{8\pi\eta^2 E_{ph}^2}{h^3 c^2}$$

$$\left[1 - e^{\left(E_{ph} - E_g - F_e(N) - F_h(N)\right)/k_B T}\right] r_{spon}\left(E_{ph}, N\right). \quad (11.7)$$

Here, η is the refractive index of the gain medium (which itself is energy dependent, ignored here for brevity), E_g is the band-gap (neglecting carrier-induced band-gap shrinkage [9–12], the electron and hole quasi-Fermi levels are denoted by F_e and F_h, respectively). The energetic distribution of electrons and holes are given by the Fermi functions f_e and f_h, respectively [34].

In order to calculate the optical gain using the expression 11.7, we need to know the total

spontaneous emission $r_{spon}\left(E_{ph}, N\right)$. Electronic states in a semiconductor are broadened by inelastic scattering and the scattering rate (dephasing rate [9–12]) can be of the order of tens of ps^{-1} leading to several meV (of the order of $\mu_e - \mu_h$) of broadening, γ. This causes an intrinsic *homogeneous broadening* of the spontaneous emission spectrum. Homogeneous broadening can be modeled using a Lorentzian broadening function, which gives

$$r_{spon}\left(E_{ph}, N\right) = r \int_0^\infty f_e(N) f_h(N)$$

$$\frac{\gamma/2\pi}{\left(E_g + E - E_{ph}\right)^2 + \left(\gamma/2\right)^2} E^2 dE, \quad (11.8)$$

where we have ignored, as is done in practice, the crystal momentum ($E/\hbar$) dependence of the electron scattering rate [12]. r is a material-dependent constant (see Figure 11.11). f_e and f_h are carrier density-dependent Fermi-distribution functions for electrons and holes, respectively [34]. With Equations 11.7 and 11.8 one gets simple treatment for the behavior of optical gain in (direct bandgap) bulk semiconductors and a realistic behavior of optical gain variation with energy, carrier density, and temperature can be reproduced. Including homogeneous broadening directly in the spontaneous emission calculation ensures that the optical transparency occurs at $F_e + F_h + E_g$, which is thermodynamically correct, see Figure 11.10 (note x-axis is rescaled to $E_{ph} - E_g$, so optical transparency is at $F_e + F_h$). In contrast, incorporating homogeneous broadening as a multiplicative factor (line-shape function [15,21]) with the gain, a treatment found in many textbooks (e.g., [35,36]), which directly follows a generic laser physics-based formalism for bulk semiconductor gain calculations, results in optical transparency below $F_e + F_h + E_g$ (thermodynamically incorrect) in addition to substantial sub-band-gap absorption, which is not observed experimentally [24,34,37].

Most semiconductor gain media are homogeneously broadened at and above room temperature, where any change in carrier density equally (homogeneously) affects the gain over the whole spectral range. In contrast, inhomogeneous broadening causes each part of the gain medium to spectrally respond independently

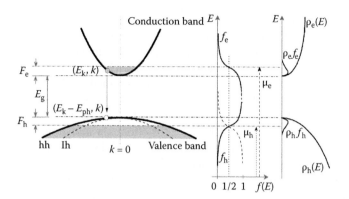

Figure 11.10 Schematic of the conduction band and valence band in an inverted semiconductor. Shaded area depicts occupied electronic states. Also shown are the electron occupation f_e and hole occupation f_h (f_e in valence band is the dotted line in valence band) energy distribution in each band a $T > 0$ K is also shown. The nonequilibrium chemical potentials μ_e and μ_h are defined. Also defined are the quasi-Fermi levels F_e and F_h. The density of states ρ and the carrier distribution $\rho_{e,h} f_{e,h}$ is also shown in the respective bands. Diagram not to scale. Also, see the discussion on the realistic position of F_e and F_h for a semiconductor active media in Section 11.7 [45].

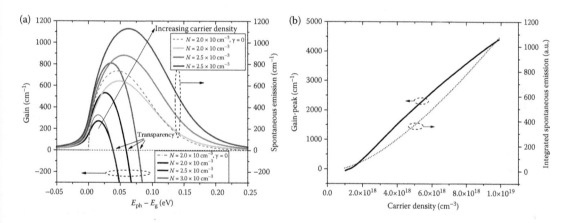

Figure 11.11 (a) Gain and spontaneous emission spectrum for bulk GaAs ($E_g = 1.4$ eV) at 300 K showing the dependence on carrier density. The Lorentzian broadening is taken as $\Upsilon = 15$ meV for all, except for two cases where it is considered zero (in red) for comparison. (b) Variation of gain peak and integrated spontaneous emission with increasing carrier density. Estimation of the optical transition matrix element gives $r = 300$ cm^{-1} [12].

to the intracavity radiation. Quantum dot and nitride-based gain media (Sections 11.7 and 11.4) show a varied degree of inhomogeneous broadening, but is significant only at low temperatures. The nature of gain broadening plays a crucial role in mode selection and formation in all lasers [15,21] and semiconductor lasers are no exception [34,38]. In addition, broadening affects the lasing threshold and the dynamic characteristics of semiconductor lasers [39]. Before proceeding further, it is useful to look at some typical

representative but useful results one can obtain from the simple gain model outlined above for the bulk gain medium.

A few things are worth noticing from the plots in Figure 11.11a. The gain spectrum is narrower than the spontaneous emission spectrum with a peak at lower energy than the latter. The peak for both shifts to higher energies (shorter wavelengths) with increasing carrier density (*band-filling*) at a constant temperature (300 K). Both have an asymmetric profile dictated by the density of states and carrier distribution functions. The effect of homogeneous broadening leads to a broadening of the spectra as expected, a reduction in peak value and some sub-band-gap gain and spontaneous emission. The optical transparency point also moves to higher energies with increasing carrier density. If we track the variation of the integrated (total) spontaneous emission and the peak gain with carrier density (Figure 11.11b), we find that that the total spontaneous emission has a stronger non-linear dependence on carrier density compared to peak gain (which is sublinear). Thus, the total spontaneous emission is generally modeled as BN^2, where B is taken as constant for a material (bimolecular recombination coefficient) while the gain is assumed to linearly depend on carrier density as

$$g_\lambda(N) = \tilde{g}_\lambda(N - N_{tr}) \qquad (11.9)$$

at a given energy (wavelength λ) and temperature. For QW lasers, however, a linear dependence of peak gain on carrier density is considered incorrect and a logarithmic dependence (see Section 11.10) is assumed instead [9]. In the above expression of peak gain, $\tilde{g}_\lambda$ is the *differential gain* $dg/dN_\lambda|$, and N_{tr} is the *transparency carrier density*, the carrier density at which optical gain balances the material loss, the stimulated emission and absorption rates become equal rendering the material transparent to radiation at wavelength λ. This brings us to the discussion on losses in a gain material and the concept of threshold in a laser cavity.

11.2.6 Carrier loss, photon loss, and laser threshold

Recombination of injected carriers resulting in spontaneous emission is a major route to carrier loss in semiconductor lasers. A laser operates by amplifying spontaneous emission leading to seeding and enhancement of stimulated emission. It is a fundamental process that cannot be suppressed completely and replaced by stimulated emission in a laser. Spontaneous emission plays a key role deciding the longitudinal mode structure, the line-width, and noise in semiconductor lasers. It is a bimolecular phenomenon where an electron and a hole is required for the process to occur and the coefficient B can be shown to have a (300 K/T)$^{3/2}$×($E_g/1.5$ eV)^2 dependence on temperature

and energy, respectively, in a bulk semiconductor [40]. Spontaneous emission clamps at a constant value at the onset of lasing due to *gain saturation* (see Sections 11.8 and 11.10) in homogeneously broadened gain media [15, 19]. Another loss mechanism for injected carriers is recombination via crystal growth induced defect states in the semiconductor. This is a monomolecular process (Shockley–Read–Hall [SRH] recombination [9–12]), which depends linearly on carrier density (as AN) with A as the corresponding recombination coefficient. The degree of carrier loss via defect recombination depends on the density of defects in the material and can be reduced by improving the crystal growth quality. A third process, Auger recombination is a dominant carrier loss mechanism in long wavelength semiconductor lasers [9–12]. is a three-carrier Columbic interaction whereby the energy of a recombining electron and hole is given to a third carrier (electron or hole), which is excited high in its respective band. The energy is then dissipated via phonons, heating the lattice. As a three-carrier process, it is assumed to have a cubic dependence on carrier density and is modeled as CN^3, with C as the Auger recombination coefficient. Finally, leakage of carriers (thermally activated or otherwise) from the active region causes carrier loss in the semiconductor active medium. Carrier leakage has an exponential dependence on the carrier density, but is frequently modeled as a higher order carrier density (N^x, $x \geq 3$) dependent process. Note that while spontaneous emission is a *radiative* loss mechanism, defect recombination, Auger recombination, and leakage are generally *nonradiative* loss mechanisms. All of these are *carrier recombination loss* processes. Photon (optical) loss processes such as *intervalence band absorption* (IVBA) [9–12,41] and *free-carrier absorption* [9–12] and absorption in the cladding layer (adjacent to the SCH layer) also contribute to the total optical loss in the system and are generally combined with losses due to scattering and diffraction in the cavity and collectively known as the internal loss α_{int} of the system. In addition, *optical losses* due to the (useful) light escaping from the finite reflectivity (R_1 and R_2) cavity mirrors (facets) can be written as $\alpha_m = -\ln(R_1 R_2)/2L$ [9–12], L being the cavity length. This term is known as the mirror loss and represents the useful light extracted from the laser.

At the transparency carrier density, optical gain balances the total material (carrier) loss.

Additional carrier injection is needed to increase the carrier density up to the threshold carrier density, N_{th} ($> N_{tr}$), when the *modal gain* equals the total photon loss, α_{tot}, which accounts for the light escaping from the mirrors, thus $\alpha_{tot} = \alpha_{int} + \alpha_m$ and laser threshold is attained. N_{tr} and N_{th} are in general a function of λ. The requirement of the threshold condition is written as [9–12]

$$\Gamma_y g_{th} = \alpha_{int} + \alpha_m = \alpha_{int} + \frac{1}{2L} \ln\left[\frac{1}{R_1 R_2}\right], \quad (11.10)$$

where Γ_y is the transverse *optical confinement factor* defined using the overlap integral as

$$\Gamma_y = \frac{\int_{-d/2}^{d/2} |\varphi(y)|^2 \, dy}{\int_{-\infty}^{\infty} |\varphi(y)|^2 \, dy}. \quad (11.11)$$

The product of Γ_y with the material gain is called the *modal gain*, which at threshold is $\Gamma_y g_{th}$. Γ_y is defined under the assumption that the lateral field amplitude $\psi(x)$ of a guided spatial mode $(\varepsilon(x,y) = \varphi(x)\psi(y))$ can be decoupled from its transverse counterpart (y) under the effective index approximation [42], which assumes that the lateral refractive index variation is much slower than its transverse variation. d is the thickness of the active region. Usually, the lateral confinement factor Γ_x is close to unity for most edge-emitting semiconductor laser designs and only Γ_y influences the modal gain. Semiconductor lasers have higher threshold modal gain typically 100 cm^{-1} or more compared to, e.g., He–Ne lasers for which g_{th} is typically $10^{-3} - 10^{-2}$ cm^{-1} [15]. The transverse electric (TE) spatial modes, i.e., spatial modes polarized in the plane of the active region, have a higher Γ_y compared to the orthogonal TM (transverse magnetic) modes [9–12]. In addition, the facet reflectivity for cleaved facets is also higher for TE modes than TM modes [43]. Hence, the TE modes satisfy the threshold condition Equation 11.10 at a lower gain. Thus, the laser emission primarily constitutes TE polarized spatial modes. However, with the advent of strained QW lasers it has been possible to increase the gain of TM polarized modes resulting in TM polarized emission from tensile trained QW gain media (more in Section 11.7) [9–12].

The injected current corresponding to the threshold carrier density is called the threshold current I_{th} and the threshold current divided by the active area ($I_{th}/L{\cdot}S$) is the threshold current density, J_{th}, where L and S are the cavity length and width of the current injection stripe, respectively. In steady state, the injected current through a semiconductor laser with an active region volume V ($=L{\cdot}d{\cdot}S$) can be defined in a straightforward manner as $I = q_e VN / \tau_c$ where N is the injected carrier density and τ_c the (total) carrier lifetime, that is the average time before injected carriers recombine (radiatively and nonradiatively) [9–12]. From the above discussion, we can define the carrier lifetime as

$$\frac{N}{\tau_c} = AN + BN^2 + CN^3 + \text{leakage rate} \quad (11.12)$$
$$+ \text{stim. emiss. rate.}$$

Stimulated emission rate depends on photon density, which is still weak below and near threshold. Hence, we can write the carrier lifetime as (for $N \leq N_{th}$)

$$\tau_c = \frac{1}{A + BN + CN^2} \equiv \tau_c(N), \quad (11.13)$$

where we have also assumed that carrier leakage is negligible. The A, B, C model used to define the carrier lifetime is only an empirical model [33], but works quite well in practice to quantify losses in semiconductor lasers. Multiplying Equation 11.9 with $q_e V/\tau_c$, setting $N = N_{th}$ and using Equation 11.10 for the laser threshold at a given wavelength, one can obtain the threshold current for the semiconductor laser using the above definition of injected current as

$$I_{th} = q_e V \frac{N_{tr}}{\tau_c} + q_e V \frac{\alpha_{int} + \alpha_m}{\tilde{g}\tau_c\Gamma_y} \quad (11.14)$$
$$= I_{tr} + I_{int+mirror}.$$

Note that the above expression is derived under the linear gain approximation Equation 11.9, but provides useful insights into the basic requirements of achieving low threshold in semiconductor lasers. As noted above, the threshold current constitutes of two parts, the transparency current I_{tr} required to render the active semiconductor transparent to radiation and an extra current $I_{int+mirror}$ required to overcome the internal and mirror losses. Obviously, I_{th} can be lowered by reducing the total active volume. $I_{int+mirror}$ is reduced mainly by increasing the differential gain $\tilde{g}$ and the optical confinement factor Γ_y while keeping the internal and cavity losses $(\alpha_{int} + \alpha_m)$ as low as possible. Note that a higher Γ_y can be achieved by increasing the active volume V, which in turn increases I_{th} (in practice, there is a trade-off). Both I_{tr} and $I_{int+mirror}$ are affected by the material (carrier) losses (via the carrier lifetime in Equation 11.13) and once the cavity and internal losses are optimized, the magnitude of I_{th} is essentially dominated by radiative and nonradiative recombination in the semiconductor below and up to threshold. Using the above discussion one can write

$$I_{th} \propto q_e V \left(AN + BN^2 + CN^3 \right) + I_{leakage}, \quad (11.15)$$

where we have added the leakage component back in Equation 11.14. It is also useful to compare the corresponding expression for QW lasers based on Equation 11.28 in Reference [9]. The above relationship between I_{th}, the injected carrier density N, and the radiative and nonradiative recombination provides a useful tool to understand the factors limiting the magnitude of laser threshold in semiconductor lasers (see Section 11.9).

11.2.7 Advantages of quantum confinement

In DH lasers, the typical active layer thickness is of the order of ~50–300 nm (with SCH designs) [9–12], resulting in a transverse confinement factor that is in the range 10%–70%. The density of electronic states $\rho(E)$ increases with the square root of energy at the band edge $\rho(E) \propto \sqrt{E - E_g}$ for gain media based on bulk semiconductors, Figure 11.10a (dashed lines). If the thickness of the active layer is shrunk to values of 20 nm or less (~de Broglie wavelength of an electron), the electronic wave functions in this QW show a quantization in a direction perpendicular to the junction plane resulting in discrete energy levels, see Figure 11.12a (solid lines). In this case, the density of electronic states $\rho(E)$ increases in steps located at the electronic energy levels of the QW [9–12]. $\rho(E) \propto m^*/L_z$ for QWs, where m^* is the effective mass of the particular band/sub-band, and L_z is the width of the QW. Thus, in principle

the density of states close to the lowest energy level of a QW is much higher than the density of states at the band edge of the bulk material. The density of electronic carriers at a given energy is the product of the density of states $\rho(E)$ and the probability of being occupied by electrons f_e or holes f_h (Figure 11.12b). Thus, the carrier distribution in a QW (active layer) laser structure has a higher maximum value and a smaller energy width. Thus, in addition to a smaller active volume, fewer states are required to be filled to reach inversion, hence the transparency carrier density is lower in a QW laser [9–12]. Thus, lower threshold currents can be obtained compared to DH lasers. Also, the material- and the differential-gain are much higher and the spectral shift of the gain curve is much smaller [9–12]. However, the thin QW results in a very low Γ_y and hence modal gain and in practice multiple QWs and a SCH layer is employed for an efficient carrier and optical confinement. Furthermore, the possibility of introduction of strained QWs extends the usable wavelength range of a particular material system. For these reasons, the use of (strained-) QWs is a rule rather than an exception in modern state-of-the-art semiconductor lasers.

Figure 11.12 shows a simplistic representation of a QW active medium, with the quasi-Fermi levels inside the bands under population inversion. In practice, the situation is different in an unstrained (lattice-matched) material as the effective mass for holes is in general larger than for electrons [9–12]. This affects the transport and density of states in such a way that as the injected carrier density approaches transparency, the quasi-Fermi level for electrons is already well into the conduction band before the quasi-Fermi level for holes has reached the valence band edge. This increases the density of electronic states in the valence band and a higher carrier injection is needed to achieve transparency and therefore lasing threshold. Nevertheless, due to the properties outlined above, unstrained QW lasers still have lower threshold than semiconductor lasers based on a bulk active medium. It was proposed independently by Adams [44] and by Yablonovitch and Kane [45] that lasing threshold in QW lasers can be further reduced by introducing biaxial (in-plane) strain into these thin QW layers, leading to additional beneficial characteristics in QW lasers such as an extended emission wavelength range, an increase in differential gain, increased differential quantum efficiency, reduced temperature sensitivity

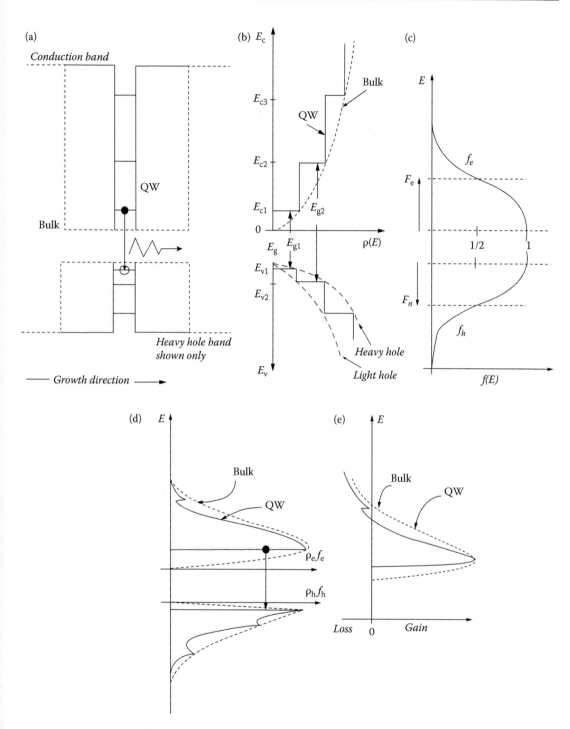

Figure 11.12 (a) Bulk (dashed lines) and QW (straight lines). (b) Electronic density of energy states, bulk (parabolic) and QW (step-like). (c) Electron and hole Fermi distributions f_e (E) and f_h (E). (d) Population density $\rho_e f_e$ and $\rho_h f_h$ of electrons and holes respectively. Also showing recombination between electrons and holes. (e) Normalized gain spectra for a bulk and QW laser (kink denotes excited state loss/gain) at a given carrier density and temperature ($T > 0$ K). Diagram not to scale. (Adapted from Wilcox, J.Z. et al., *Appl. Phys. Lett.*, 55, 825–827, 1989.)

of lasing threshold (Section 11.9), increased modulation bandwidth (see Section 11.11), and a decrease in phase–amplitude coupling (Section 11.10) [9–12]. Incorporation of biaxial strain (tensile or compressive) can be achieved by growing QWs with a slight lattice mismatch to the substrate material. Due to the thin layers, QWs can remain strained without getting plagued by dislocations and defects unlike thick bulk layers under strain. The critical thickness for growing a SQW under strain without forming dislocations is approximately $0.2(a/\Delta a)$ nm, where a is the lattice constant of the well material and Δa is the lattice-constant difference between the QW and the substrate material [9–12], implying that a 20 nm QW can be grown with 1% lattice mismatch, although we note that 20 nm would be relatively wide for a QW. Figure 11.13 shows how the introduction of biaxial strain lifts the degeneracy of the light- and the heavy-hole bands resulting in a reduced hole effective mass (for compressive strain), which in turn reduces the density of electronic states near the top of the valence band (at the Brillouin zone center [25]). This can be understood by recalling the effective mass for motion along the confinement direction in the QW determines the confinement energy, whereas the effective mass in the plane of the well determines the density of states. Now the hole quasi-Fermi level can move into the valence band fulfilling the Bernard and Duraffourg condition (Section 11.3) at a lower injected carrier density [9–12,44,45]. Thus, the transparency carrier density is reduced and hence the lasing threshold. A reduction in the effective hole mass also results in an increase in the differential gain [9–12]. This directly impacts the lasing threshold (see Equation 11.14) and has profound effects on the modulation bandwidth and phase–amplitude coupling as

indicated above. The lower transparency carrier density reduces the carrier-dependent losses such as Auger recombination, which is a major loss channel for injected carriers in long wavelength (infrared) lasers [9–12]. Additionally, carrier leakage to the valence (sub-) bands in the QW and IVBA is also reduced [9–12]. Tensile-strained QWs also offer reduced threshold currents and improved performance. However, the reason for this is different because the hole effective mass at the top of the valence band is not reduced. Instead, in the tensile strained case, the benefits arise due to the fact that a higher fraction of the valence band states can contribute to optical gain and uniquely allow lasing in the TM mode.

Application of both compressive and tensile strain has been shown to have beneficial effects in QW laser characteristics [9–12]. Compressively strained QWs show higher gain for the TE polarized intracavity radiation, whereas tensile-strained QWs have higher gain for TM polarized intracavity radiation [9–12]. Thus, one can design strained QWs for emission at a particular polarization. In general, both tensile and compressive strained QWs and barriers may be sandwiched appropriately in multiquantum well (MQW) lasers in order to get a net unstrained active medium, which is useful for long-term reliability of the devices.

A QW creates quantization effects only in one dimension (the crystal growth direction). It turns out that quantum confinement in two or all three dimensions (see Figure 11.14) results in very interesting and useful properties for semiconductor laser gain media. In the case of three-dimensional quantum confinement, such as with quantum dots [10,39], the density of states becomes a delta function, see Figure 11.14d, i.e., it becomes atom-like

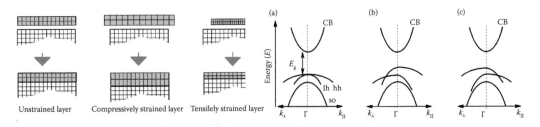

Figure 11.13 (Left) Illustrations of unstrained, compressively strained, and tensile-strained layers. For the strained layers the epilayer expands (contracts) in the plane while expanding (contracting) in the growth direction, corresponding to compressive (tensile) strain, respectively. (Right) Effects of strain for a bulk-like direct-band-gap semiconductor for (a) unstrained, (b) compressive strain, and (c) tensile strain.

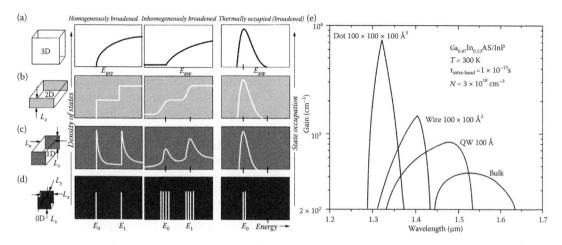

Figure 11.14 Schematic representation of the effect of reduction in dimensionality of confinement on the electronic density of states in a semiconductor medium. Inhomogeneous broadening leads to smearing out of the ideal density of states. A continuous density of states results in a broad temperature-dependent distribution for a given carrier density. For bulk (a), quantum well (b), and quantum wire (c), the density of states are continuous and carriers undergo thermal occupation. (d) Quantum dots (ideal) show a complete quantization resulting in discrete energy levels (E_0 and E_1). Self-assembled growth leads a dot size distribution which leads to inhomogeneous broadening. But at room temperatures only the deeper dots are occupied by carriers are making the medium homogeneously broadened [10,46]. (e) Schematic of the gain spectra for InGaAs/InP-based QWs, wires and dots. (Adapted from Asada, M. et al., *IEEE J. Quantum Electron.* 22, 1915–1933, 1986.)

resulting in many desirable properties such as decreased transparency carrier density (hence reduced laser threshold), increased characteristic temperature (infinite for ideal QDs), and a reduced (zero for ideal QDs) phase–amplitude coupling [39] (see Sections 11.9 and 11.10). A comparison of calculated gain spectra for ideal gain media with bulk (no quantum confinement), QW (1D confinement), wire (2D confinement), and QD (3D confinement) active regions is given in Figure 11.14e. In practice, growth of QD/wire type active regions leads to an inhomogeneous distribution of dot/wire sizes due to the self-organized (Stranski–Krastanov [10,39]) growth technique used. This leads to inhomogeneous broadening of the gain-spectrum and negates some of the ideal properties [91,92].

11.2.8 Modal structure in semiconductor lasers

Mode formation (both longitudinal and spatial) in a semiconductor laser is decided by a dynamical competition between gain and loss in the system. Material dispersion (refractive index) plays

an important part too as any change in gain at a given frequency affects the dispersion in the entire frequency range (Kramers–Kronig relation [23]). A complete understanding of mode selection and formation in semiconductor lasers can only be developed via sophisticated theoretical models, some of which are still being refined [38,47]. Most semiconductor lasers (e.g., for optical communications) are designed for single spatial (lateral and transverse) and single longitudinal mode operation. Single spatial mode emission ensures improved spatial coherence (e.g., improves light coupling to fibers), while single longitudinal mode results in an enhanced temporal coherence (improved laser line-width). However, for high-power generation these requirements are generally relaxed as comparatively wider stripe widths and cavity lengths are needed for achieving high gain and maximum heat dissipation (see Section 11.3). Wider stipe widths (wider gain regions) sustain many lateral modes, whereas longer cavities sustain several longitudinal modes. We will consider a single spatial mode design (e.g., multi-QW with ~2 μm wide current injection stripe, index-guided Fabry–Perot laser) to understand the longitudinal mode selection and formation first, before moving on to discuss spatial mode formation.

For a pumped active medium, useable gain is experienced at cavity (resonant) frequencies that fall within the gain bandwidth Δ_g, see Figure 11.15a. The number of longitudinal modes that can experience gain is given by the ratio between the gain and the $\Delta\nu_{FSR}$ (see Equation 11.3), FSR of the cavity (i.e., $\Delta_g/\Delta\nu_{FSR}$). Typically, the width of a semiconductor gain spectrum is an order of magnitude larger than the FSR of the cavity; hence, several longitudinal modes experience gain. However, the theory of gain saturation in a homogeneously broadened gain medium (such as a semiconductor QW laser) suggests that the gain should increase and eventually saturate homogeneously over the entire spectrum [15,21]. Thus, as pumping increases, photons at the cavity transmission frequency (cavity resonance) closest to the peak gain experience the maximum gain at the onset of laser emission, i.e., the laser threshold (see Equation 11.9 and Figure 11.15a). Photon density at adjacent cavity resonances can also grow but will be relatively weak compared to the photon density near the gain peak.

With further increase in pumping, photons at frequencies near the peak gain will extract most of the available gain depleting the gain at the adjacent cavity resonances (via gain narrowing—a characteristic of any optical amplifier [48]) as the gain clamps at threshold (homogenous gain saturation [15,21]). The laser should therefore emit a single longitudinal mode near the gain peak above threshold [15,21]. This however, does not happen in practice in any homogeneously (or inhomogeneously) broadened lasing medium, including semiconductor lasers, which emit more than one longitudinal mode unless specific design elements are in place (Section 11.3). Gain-guided semiconductor lasers tend to emit in more longitudinal modes in comparison to their index-guided counterparts (see below). Multilongitudinal mode emission in a homogenously broadened gain medium occurs primarily due to two effects: longitudinal *spatial hole-burning* [15] (see Figure 11.15b) and secondly due to spontaneous emission feeding the photons at cavity resonances adjacent to the gain peak [13]. An additional effect—*spectral hole burning*—affects mode selection in an inhomogeneously broadened gain medium [15,21,38], the details of which are beyond the scope of this introductory chapter.

Spatial hole burning in a laser is said to occur when the intracavity photon density depletes unused carriers (gain) in space (locally) thereby burning a local hole in the carrier pool [13,15]. Now consider the spatial distribution of the carriers along the optical axis of a cavity in the presence of intracavity radiation (Figure 11.15b). Adjacent longitudinal modes that form standing waves along the optical (z-) axis have their crests and toughs at slightly different spatial locations (separated by ~λ/4 for adjacent standing waves [15]) due to their different wavelengths (see Equation 11.1). Hence, adjacent longitudinal modes can extract gain from spatial locations where the gain is unused by the other

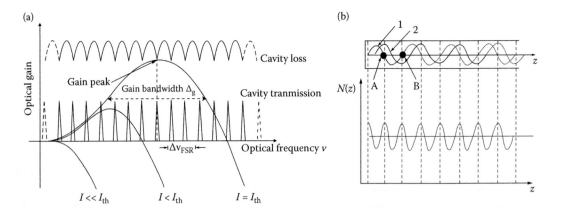

Figure 11.15 (a) Schematic of optical gain spectrum as a function of injected current in a semiconductor laser overlaid on the cavity resonances and loss. Laser emission occurs when gain compensates the optical losses in the cavity, and the photons in the longitudinal mode closest to the peak gain reaches threshold first. (b) Two adjacent longitudinal modes (1 and 2) along the optical axis of the laser cavity. The "unused" carriers (gain) from mode 1 can be used by mode 2 at points A, B, and so on.

longitudinal mode(s). In Figure 11.15b, if we tag the longitudinal mode at the gain peak by 1 (the central mode) then the adjacent longitudinal mode 2 can extract gain from the unused longitudinal carrier profile (Figure 11.15b, points A, B, and so on) and saturate the available gain to the value at threshold via longitudinal spatial hole burning. Hence, even though the longitudinal modes adjacent to the central mode do not experience gain spectrally, spatially they can experience gain and the spectrum of a Fabry–Perot laser shows multiple longitudinal modes. However, as the laser is pumped high above threshold, the central and adjacent longitudinal modes smear out the longitudinal carrier distribution and aided by gain narrowing, prevent the appearance of additional longitudinal (side-) modes.

This, however, does not explain the difference observed between the lasing spectra of gain- and index-guided semiconductor lasers, where the former emits in more longitudinal modes compared to the latter (Figure 11.16a and b). Clearly, all photons generated in the active medium via

spontaneous emission are not coupled to guided modes in the laser but a fraction of them are. It can be shown that the number of photons in a mode is directly proportional to the fraction of spontaneous emission feeding that mode and that this fraction is increased by a factor K (called Peterman's K-factor or astigmatism factor) [13], depending on the nature of the phase fronts of the spatial modes that the laser sustains. The K-factor is generally an order of magnitude larger for gain-guided lasers (curved phase fronts) compared to the index-guided counterparts (flat phase fronts, $K=1$ ideally). The underlying reason for this is the nonorthogonality of the spatial and longitudinal modes in a laser cavity (in any generic lasing system) [49]. A higher effective rate of spontaneous emission feeding the longitudinal modes in gain-guided semiconductor lasers is said to result in multilongitudinal mode emission compared to few longitudinal mode emission in index-guided lasers, as the pumping increases (Figure 11.16a and b) [13]. The two effects considered above explain the

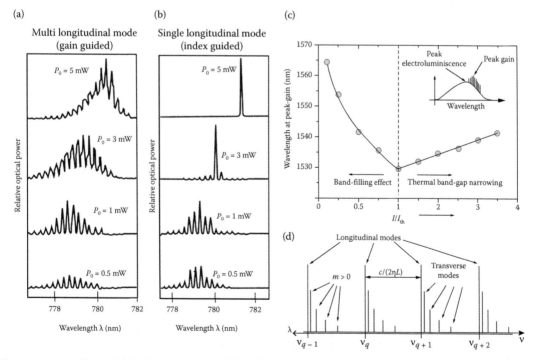

Figure 11.16 (a) Longitudinal mode spectrum in (a) gain- and (b) index-guided semiconductor lasers above threshold [50]. (c) Shift in emission wavelength peak tracked from below to above threshold (inset EL vs. lasing peak). (d) Fine structure of the longitudinal mode spectra when higher order spatial modes are allowed by laser design (e.g., wider stripe width). Spatial mode family (indexed by m) corresponding to each longitudinal mode (indexed by q) is shown. Each longitudinal mode is separated by a frequency separation of $c/2\eta L$.

basic mechanism for selection and oscillation of longitudinal modes in semiconductor lasers. More sophisticated theoretical models have been put forward in order to explain the complex longitudinal mode spectra observed in different semiconductor lasers and the reader is directed to specialized references for further reading [47].

As the gain peak is at a lower energy compared to the spontaneous emission (EL) peak (see Figure 11.11a), laser emission starts at the longer wavelength side of the EL spectrum at threshold, see Figure 11.16c inset. If one tracks the peak emission wavelength from below to above threshold (see Figure 11.16c), until threshold, the (EL peak-) wavelength shifts to shorter wavelengths (higher energies) due to band-filling (carrier injection filling states at higher energy) [9–12] and refractive index change via the free carrier plasma effect [10,13,34]. Beyond threshold, under CW operation, the peak lasing wavelength can shift to longer wavelengths due to temperature-induced band-gap shrinkage (shifting the gain peak to lower energies) and temperature-induced refractive index change. In order to understand longitudinal mode

structure in semiconductor lasers we considered a single spatial (lateral and transverse) mode design above, i.e., the laser sustaining only the fundamental spatial mode. If we relax the single spatial mode requirement in design, for example, by increasing the current injection stripe width, the laser can sustain higher order lateral modes. In this case, the longitudinal mode spectra develop a fine structure, where each longitudinal mode has an associated spatial mode family, see Figure 11.16d [15].

Similar to the discussion in Section 11.1, one can proceed and define the number of intersections, the transverse and the lateral modes made with the y- and x-axis as l and m, respectively. The transverse modes (along the y-axis) reflect the standing wave structure between the two heterojunctions confining the active layer and are generally guided via total internal reflection (SCH design) or refraction (GRINSCH (Graded Index Separate Confinement Heterostructure) design) [9,58] (see Figure 11.17). The active layer of modern semiconductor lasers with quantum-confined active regions are usually thin and allow only the fundamental transverse mode ($l=0$) to propagate. Thus, for semiconductor

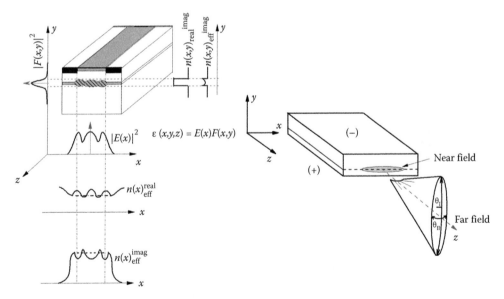

Figure 11.17 Left: typical wide-stripe gain-guided diode laser showing lateral $E(x)$ and transverse mode $F(x, y)$ profiles under pulsed (nonthermal) operation. Real and imaginary parts of the complex refractive are also shown. Notation used is based on the effective-index approximation [42]. Note that the transverse mode is a fundamental mode due to the transverse epitaxial design. In comparison the lateral direction sustains a higher order (second order) mode. The lateral effective index forms an antiguide for the sustained lateral mode due to carrier induced effective index change (Section 11.10). Right: diagram explaining near- and far-field of a typical diode laser. θ is the slow-axis and $\theta_\perp$ is the fast-axis far-field angle (see text for explanation).

lasers, lateral modes and spatial modes are generally regarded as synonymous. Near laser threshold the angular frequencies ω_{qm} of stationary and stable lateral modes sustained in a semiconductor laser having a stripe width S and cavity length L are given by [51]

$$\left[\frac{\eta_q \omega_{qm}}{c}\right]^2 = \left[\frac{\pi q}{L}\right]^2 + \left[\frac{\pi m}{S}\right]^2 \quad (11.16)$$

resulting in an optical frequency separation $\delta \upsilon_S$ between adjacent spatial modes as

$$\delta \upsilon_S = \frac{\omega_{qm} - \omega_{q1}}{2\pi} \approx \frac{\lambda L}{4\eta_q S^2}\left(m^2 - 1\right) \Delta \upsilon_{FSR}, \quad (11.17)$$

where $\Delta \upsilon_{FSR}$ is given by Equation 11.3 and $q > 1$ and $m > 0$ [$m = 0$ (fundamental lateral mode) corresponds to $q = 1$]. Note, a close inspection of Figure 11.16a shows higher-order spatial modes emerging in the longitudinal mode spectrum of the gain-guided laser at the shorter wavelength (higher frequency) side of each longitudinal mode as the output power (pumping) increases. As a rule of thumb, one can assume $\delta \upsilon_S : \Delta \upsilon_{FSR} : \Delta_g \equiv 1 : 10 : 100$ for QW lasers [15]. As noted in Section 11.4, the lateral mode structure gets defined by either gain guiding or via index guiding or a combination [10,13,31,38]. Pure gain guiding is usually a weak guiding mechanism and is not preferred. Instead quasi-index guided (weak lateral index guided) designs are generally used when wide stripe designs are required, e.g., for high-power diode lasers. Figure 11.17 describes the basics of lateral and transverse mode formation in gain-guided semiconductor lasers. In quasi-index guided and completely index-guided structures, lateral mode formation is generally tighter (reduced lateral spreading of the mode) [38]. Transverse mode formation essentially has the same mechanism in both gain- and index-guided designs.

The spatial distribution of light intensity at the emitting facet of the semiconductor laser is called the *near-field* intensity (often abbreviated to just *near-field*) (Figure 11.17e) [9–13,96]. Due to the asymmetric nature of wave guiding in a semiconductor laser, the emitted light diffracts asymmetrically as it propagates outwards. The diffraction is stronger perpendicular to the junction plane (hence called the fast axis) and weaker along the junction plane (called the slow axis) making the overall *far-field* intensity distribution elliptical, see Figure 11.17e). Mathematically, the far-field is the Fourier transform of the near-field [9–13,97]. The asymmetry in waveguiding (transverse vs. lateral) in a semiconductor laser also leads to astigmatism in the output field intensity [9–13]. Although the astigmatism can be corrected using cylindrical lenses, multilobed far fields occurring due to the presence of higher order spatial modes that cannot be easily controlled (see Section 11.3 for further discussion).

11.2.9 Light–current–voltage characteristics of semiconductor lasers

Under low injection, a semiconductor laser behaves similarly to a light emitting diode (LED) with the emission dominated by spontaneous emission. As the current injection is increased, the semiconductor laser emits amplified spontaneous emission (ASE [11]) in the longitudinal direction, i.e., along the cavity and exiting from the laser facets. Pure spontaneous emission continues to be emitted in other directions and is useful as a means of probing the fundamental behavior of the laser in experiments [93–95]. For conventional bulk and QW lasers (generally not for QD lasers [52]), spontaneous emission saturates as the injected current passes I_{th} (Section 11.10), beyond which almost all the additional injected carriers recombine via stimulated emission. The carrier density (hence gain) clamps at the threshold carrier density (threshold gain, Section 11.10) value and the output power continues to grow at a high rate giving rise to a light (power) current curve as shown in Figure 11.18a. The total output power (from both facets) P_{out} above I_{th} at a given current I can be written as [9,11,53,54]

$$P_{out} = \frac{hc}{\lambda q_e}\left[\frac{\alpha_m}{\alpha_m + \alpha_{int}}\right]\eta_{int}^d\left(I - I_{th}\right), \quad (11.18)$$

where q_e is the electronic charge and λ is generally taken as the peak emission wavelength. The internal differential quantum efficiency η_{int}^d, which quantifies the number of photons generated above threshold per injected carrier, is generally close to unity [53,54].

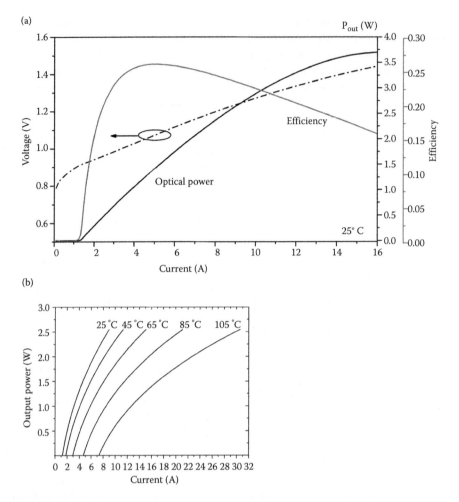

Figure 11.18 (a) Light power–current–voltage (*L–I–V*) plot for a typical high power broad area semiconductor laser (1.48 μm, based on InGaAsP/InP) showing the variation of optical power from and voltage drop across the device. Also shown is the variation of total conversion efficiency (P_{out}/P_{in}). (b) Dependence of the threshold current and external differential quantum efficiency for the device (junction) temperature rise (e.g., due to Joule heating). Note the power roll-over at high current that becomes stronger at higher junction temperatures.

The rate of change of output power with injected current per unit photon energy is defined as the external differential quantum efficiency as [53,54]

$$\eta_{ext}^d = \frac{dP_{out}/dI}{hc/\lambda q_e} = \left[\frac{\alpha_m}{\alpha_m + \alpha_{int}}\right]\eta_{int}^d, \quad (11.19)$$

where η_{ext}^d is also referred to as the slope efficiency. The above relation can be used to determine η_{int}^d and by plotting vs. *L* for identical lasers with different cavity lengths and known facet reflectivities [11]. Because the carrier density saturates at threshold, $\mu_e - \mu_h$ pins at its threshold value (commonly referred to as Fermi level pinning [11,52])

resulting in the saturation of the junction voltage V_j ($q_e V_j = \mu_e - \mu_h$). The junction voltage is related to the applied forward bias V_a via the series resistance R_s (internal to the laser plus external) in the circuit.

$$V_j = V_a - IR_s. \quad (11.20)$$

The slope of the *IV* curve above threshold in Figure 11.18a gives the series resistance in the circuit, which can be used to estimate the internal resistance of the laser provided the external resistance in the circuit is known. The total power balance above threshold in the laser can be written as [55]

$$IV_a = I^2 R_s + \frac{hc}{\lambda q_e} I_{th} + \frac{hc}{\lambda q_e}(I - I_{th})$$

$$(1 - \eta_{ext}^d) + \frac{hc}{\lambda q_e}(I - I_{th})\eta_{ext}^d, \tag{11.21}$$

where the first term in the right hand side is due to Ohmic losses (Joule heating), the second term is due to carrier recombination losses, the third term is due to internal optical losses, and the final one is due to the emitted power from the laser.

Under CW operation, Joule heating has a detrimental effect on almost all of the characteristics of semiconductor lasers. Mostly significantly, I_{th} increases, η_{ext}^d decreases, and the emission wavelength shifts, among other unwanted effects. Therefore, wherever possible, semiconductor lasers are operated under low duty (0.1%–1%), nonthermal (few hundred ns) pulsed conditions. Even then, in commercial applications the operating temperature conditions vary significantly, which influences the laser characteristics. Therefore, commercial devices employ thermo-electric coolers (TECs) in the semiconductor laser packages to control the device temperature. Unfortunately, in most cases, the TECs consume more electrical power than the semiconductor laser itself. Reducing the temperature sensitivity of semiconductor laser characteristics is still an open problem [9,11,12,93–95]. The temperature sensitivity of I_{th} is characterized by the characteristic temperature T_0, which are related via an empirical relationship

$$T_0 = \left(\frac{d\ln I_{th}}{dT}\right)^{-1}. \tag{11.22a}$$

Over a limited temperature range, if T_0 is constant, then the solution is a simple exponential dependence, thus

$$I_{th}(T) = I_0 \exp\left(\frac{T}{T_0}\right), \tag{11.22b}$$

where I_0 is a reference value. The characteristic temperature is generally measured in Kelvin and can be calculated using the following relation

$$T_0 = \frac{T_2 - T_1}{\ln\left[I_{th}(T_2)/I_{th}(T_1)\right]} \tag{11.23}$$

by measuring the laser threshold at two different temperatures, T_1 and T_2.

Strictly speaking, T_0 is a function of temperature itself over a wide temperature range, but is often taken as a constant in practice. Typical 1.3 and 1.5 μm emitting QW semiconductor lasers show a T_0 in the range of 50–60 K, whereas shorter wavelength semiconductor lasers emitting in visible and near infra-red show a higher T_0, in the 100–200 K range. The higher the value of T_0, the less temperature sensitive is the threshold current of the semiconductor laser. It can be shown that for an ideal QW laser a maximum $T_0 = 300$ K is possible at room temperature [10,56]. The temperature dependence of the threshold current originates from the temperature dependence of the various carrier recombination channels, leakage, and photon loss (via IVBA [41]) processes in the semiconductor material (see Equation 11.14).

One can separate the radiative (spontaneous emission) and nonradiative (SRH, Auger, and leakage) components of the threshold current and write $I_{th} = I_{th, rad} + I_{th, non-rad}$. As the rate of nonradiative recombination mechanisms depend on temperature, albeit differently, by monitoring I_{th} and $I_{th, rad}$ as a function of temperature, see Figure 11.19 [57], it can

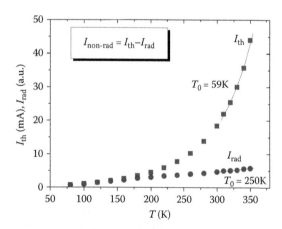

Figure 11.19 The laser threshold current and its radiative component current is measured as a function of device temperature. I_{th} equals I_{rad} at very low temperatures (70 K), from which one can extrapolate I_{rad} to room temperature and above by monitoring the power in the spontaneous emission spectrum as a function of temperature [57]. T_0 is shown for each current component. It can be seen that as the temperature of the device increases the threshold current is mainly dominated by the nonradiative current [57]. Data for a 1.55 μm emitting (InGaAsP/InP) laser diode.

be shown that the threshold current, for example, of 1.3 and 1.5 μm semiconductor lasers at and above room temperature is mainly dominated by nonradiative recombination [95].

11.2.10 Static and dynamic characteristics of semiconductor lasers

In this section we will outline the basic rate equations for semiconductor lasers and use them to study their static and dynamic characteristics. The rate equations described here are based on the photon density S and not the electric field (complex and a vector quantity); hence the phase of the optical field, which plays an important part in understanding many of the optical characteristics of the laser, cannot be treated self-consistently. Also they are valid only for time scales larger than the cavity decay time (see Equation 11.4), hence cannot be used in the ultrafast regime [58]. Nevertheless, the simple nonlinear rate equations defined below describe the basic characteristics of semiconductor lasers reasonably well. For a detailed understanding, however, one needs to rely on electric field-based equations together with an appropriate model for the gain and index variation in the semiconductor medium as a function of injection and temperature [33, 38]. The simple rate equations defined below assume that the laser emission occurs at a single spatial and longitudinal mode (at wavelength λ_L). Additionally, it is assumed that there is no variation of gain and loss along the laser cavity (z-axis) or laterally along the current injection stripe (x-axis). Thermal effects are ignored. With these assumptions and borrowing from the discussion in Section 11.6, we can write two rate equations describing the (coupled) carrier and photon density dynamics in a semiconductor laser:

$$\frac{dN}{dt} = \frac{I}{q_e V} - \frac{N}{\tau_c} - \frac{2c}{\eta} \Gamma_y g_{opt} S \quad (11.24)$$

and

$$\frac{dS}{dt} = \frac{2c}{\eta} \left(\Gamma_y g_{opt} - \frac{\alpha_{tot}}{2} \right) S + \beta R_{spon}, \quad (11.25)$$

where the factor $2c/\eta$ converts the difference between optical gain and total loss to a rate.

$c/\eta = v_g$ is the group velocity of the light in the laser. The factor 2 comes from going from electric field to photon density (intensity) while deriving Equations 11.24 and 11.25 [9,12,13,34] and is generally absorbed in the definition of gain but we have kept it for clarity. β is the spontaneous emission factor that supplies the laser mode [13]. The total spontaneous emission rate $R_{spon} = BN^2 = N\tau_{spon}^{-1}$, where τ_{spon} is the spontaneous emission lifetime. The photon losses can be related to a photon lifetime in the cavity as

$$\tau_{ph}^{-1} = v_g \alpha_{tot} = 2\tau_{cav}^{-1}. \quad (11.26)$$

This gives the time taken for the intracavity photon density to decay to $1/e$ of its initial value (compare with Equation 11.4). The remaining parameters have already been described in the sections above. For bulk semiconductor active media, one can define

$$g_{Bulk} = \tilde{g}_{Bulk}(N - N_{tr})(1 - \varepsilon_{Bulk}S) \quad (11.27)$$

and for QW gain media (see Section 11.5 and 11.7) [9]

$$g_{QW} = \tilde{g}_{QW} \ln(N / N_{tr})(1 - \varepsilon_{QW}S), \quad (11.28)$$

where $\tilde{g}$ is the differential gain corresponding to the respective g at λ_L. ε is a phenomenological parameter introduced here to account for intensity dependent gain suppression. Intensity-dependent gain saturation can only be explained adequately by applying sophisticated semiconductor laser models [58]. The right-hand side of Equation 11.24 has the current source term (pump), the carrier loss (via nonradiative and spontaneous emission) term followed by the carrier loss (via stimulated emission) term, which balances the total carrier generation and loss rate in the system.

The photon density rate balance equation (Equation 11.25) describes the photon density gain and loss with an additional source term resulting from spontaneous emission, which seeds the laser mode at λ_L. In steady state Equation 11.25 gives $(dS/dt = 0)$

$$S = \frac{\beta R_{spon}}{2v_g(\alpha_{tot} - \Gamma_y g_{opt})}, \quad (11.29)$$

which suggests that the optical power in the laser is seeded by a small fraction of the total spontaneous emission ($10^{-4} < \beta < 10^{-5}$ typically [12,13]).

As spontaneous emission is essentially white noise, it is easy to see that that laser amplification starts from noise. As current injection increases and the optical gain approaches the total loss, the denominator of Equation 11.29 approaches zero and the amplification of spontaneous emission increases. As the optical gain nears the loss, the photon density grows to such a high value that each additional injected carrier results in stimulated emission. It can be shown using Equations 11.24 and 11.25 that at threshold (when the denominator of Equation 11.29 equals zero), the carrier density pins at its threshold value and does not increase significantly upon further current injection, see Figure 11.20a. A close inspection of Figure 11.20a reveals that there is a very slow increase of the carrier density from its threshold value N_{th} as the current is increased beyond threshold. This happens because the small signal (unused) gain does not saturate at threshold and continues increasing even when the modal gain has saturated at threshold [59].

It is important to notice in Figure 11.20b that there is a delay in turn on of the laser emission even when the injected current is above threshold. This is due to the fact that the laser oscillations take time to build up in the cavity before the photon density reaches a steady-state value (typically more than 200 cavity round trips are required [60]). The turn on delay is of the order of τ_c and decreases with injected current above threshold according to [61]

$$\tau_{\text{turn-on}} = \tau_c \ln\left[I / \left(I - I_{th} \right) \right] \qquad (11.30)$$

It can be reduced by applying a small prebias to the semiconductor laser [62]. As the injection surpasses the threshold, there is a sharp increase in stimulated emission; hence, the emitted photon density depletes the carrier density, which had grown above its threshold value during the turn on delay. Note that the photon dynamics are faster (~ps) than the carrier dynamics (~ns); hence, the pump is unable to supply enough carriers in time and the carrier density drops below N_{th}. A lower carrier density reduces the photon density until the pump catches up and increases the carrier density above threshold again, which in turn increases the photon density back depleting the carriers again. This oscillation in carrier density continues and slowly damps to N_{th} together with a simultaneous damping of the counter correlated photon density to its steady state value S_{st}. The damping time τ_{damp} for this *relaxation oscillation* (RO) is

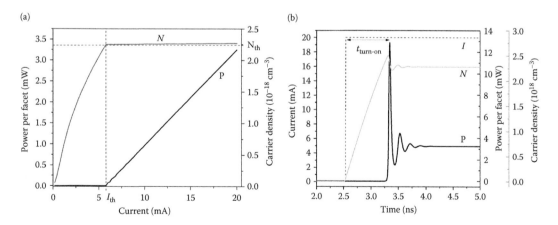

Figure 11.20 (a) Light power–current characteristics obtained by solving Equations 11.24 through 11.27 showing a steep increase in output power from one of the laser facets (equal output from the other) after threshold. The corresponding variation of carrier density pinning at its threshold value, N_{th}, is also shown. (b) Temporal large-signal response of the laser power and carrier density under step current (dotted line) injection at $t = 2.5$ ns (plot shown from 2 ns instead of zero ns for clarity) obtained by solving (Equations 11.24 through 11.27), using typical material and cavity parameters [11]. The turn-on delay, the ROs and its damping is clearly visible. Also the optical power P and steady-state photon density S_s are related as $P = hcVS_{st}v_g\alpha_m/\lambda$ (see text for details).

approximately double the turn on delay time and is determined by the differential gain and τ_c [9]. The frequency of the ROs v_{RO} before damping varies as the square root of the steady-state photon density and is generally between 1 and 10 GHz [9,12]. They are given as [9,12]

$$\tau_{damp}^{-1} = \gamma_{damp} = 2v_g \tilde{g} S_{st} + \tau_c^{-1} \quad (11.31)$$

and

$$v_{RO} = \frac{1}{2\pi} \sqrt{2v_g \tilde{g} S_{st} \tau_{ph}^{-1}}, \quad (11.32)$$

where γ_{damp} is the damping rate of the RO. In order to generate fast optical signals by modulating the injected current into the semiconductor laser, the optical pulse intensity should achieve a steady-state value S_{st} quickly (similar arguments apply after the current switch-off [9,12]). This would require a large damping rate γ_{damp} and the RO frequency v_{RO}. This can be achieved if the differential gain is large and both the carrier and photon lifetimes are short (see Equations 11.31 and 11.32). While a large differential gain lowers the laser threshold (see Equation 11.14), short carrier and photon lifetimes would lead to a higher threshold current. Thus, for simultaneously achieving low-threshold and high-speed modulation, one needs only high differential from the laser that can be readily achieved in strained QW lasers. It is worth noting that during the time τ_{damp}, oscillations in the carrier density (of the order of 10^{17} cm^{-3}) and hence in the gain and refractive index (via phase amplitude coupling, see below) lead to complex dynamics and shifting of the longitudinal modes (emission wavelength) selected before the intensity reaches S_{st}. Such *chirping* [9,12] of the laser emission leads to in an effective increase in the laser line-width. This is unwanted in long-haul large capacity fiber optic communication systems in which semiconductor lasers play a pivotal role. The line-width of the emission spectrum from a single (longitudinal and spatial) mode semiconductor laser emitting at a peak frequency v_L is inversely proportional to optical power output P_{out} and is given by [9,11,12]

$$\Delta v_L = \frac{h v_L (\alpha_{int} + \alpha_m) \alpha_m v_g^2}{8\pi P_{out}} n_{sp} (1+\alpha_H^2), \quad (11.33)$$

where $n_{sp} = \left[1 - e^{(E_{ph} - E_g - F_e(N) - F_h(N))/k_BT}\right]^{-1}$ is the spontaneous emission coefficient (see Equation 11.7) accounting for any incomplete inversion in the semiconductor laser and is typically between 1.5 and 2.5. Δv_L is the Schawlow–Townes semiconductor laser line-width Δv_L^{ST} [15,21] broadened by $(1+\alpha_H^2)$, where α_H is the line-width enhancement factor proposed and calculated by Henry (hence the subscript H) [63]. The line-width enhancement occurs due to the random fluctuating phases of the spontaneously emitted photons that feed a lasing mode in a semiconductor laser. Δv_L is the intrinsic laser line-width for a semiconductor laser (tens of MHz), which is further broadened via finite cavity losses, i.e., via the cavity resonance line-width Δv^c (see Equation 11.4) that is of the order of tens of GHz. The fact that Δv_L is narrower than Δv^c can be intuitively understood by noticing that the effective photon lifetime (Equation 11.26) in a cavity with gain is longer than a passive cavity photon lifetime [15]. Thus the effective Δv_L can be of the order of tens of MHz for single-mode Fabry–Perot lasers but can be reduced down to a few kHz for specifically designed semiconductor lasers such as DFB lasers (see Section 11.3). Still Δv_L in semiconductor lasers is significantly higher than what can be achieved using solid state and gas lasers (in 100s kHz or less) [9,12,13,15]. The line-width enhancement factor, also known as the antiguiding parameter typically, has an average value between 2 and 5 and plays an important role deciding the characteristics of modern semiconductor lasers and is briefly described next.

Unlike gas lasers there is a strong dependence of the local refractive index on the carrier density in semiconductor lasers, which affect mode discrimination and beam quality severely in a diode laser. α_H couples the phase (related to refractive index) and amplitude (related to gain and hence carrier density) of the intracavity electric field in a semiconductor laser. The local refractive index in the semiconductor active medium decreases as carrier density is increased locally (and vice versa) due to free-carrier absorption, band-filling, intraband transition and scattering, and band-gap shrinkage effects [22,23]. The carrier-induced refractive index changes are basically all a consequence of causality, governed by the Kramers–Kronig relations [23]. Generally, the intraband and free-carrier absorption components are the strongest among the factors contributing to phase amplitude coupling [34] and depending on the nature of quantum confinement and transverse waveguide design, the intraband or the free-carrier part can dominate over the other [34]. Instead of the Kramers–Kronig approach, the line-width enhancement factor can be obtained by direct evaluation of the real and imaginary parts of the carrier density-dependent complex dielectric susceptibility χ as [55]

$$\alpha_H = \alpha_H^{ib} + \alpha_H^{fc} = \frac{\mathrm{Re}[\chi]}{\mathrm{Im}[\chi]} = -\frac{4\pi}{\lambda_L} \frac{\partial\eta/\partial N}{\partial g/\partial N}, \quad (11.34)$$

where α_H^{ib} is the intraband transition component and α_H^{fc} is the free-carrier component [55]. $\partial\eta/\partial N$ is the differential index and $\partial g/\partial N = \tilde{g}$ the differential gain, both defined at the peak wavelength λ_L. It can be shown that the differential change in complex effective index is given as [55]

$$\delta n_{eff} = -\frac{\delta g_m [\alpha_H - \iota]}{2k_0}, \quad (11.35)$$

where δg_m is the modal differential gain and k_0 the vacuum wave number. The negative sign indicates that the local refractive index (real part of Equation 11.35) increases when the effective modal differential gain increases for a positive α_H. It can be shown that α_H is in general a function of wavelength, temperature, as well as carrier density [33,34], even though these dependencies are often ignored in the literature [9–12]. A nonzero and positive α_H not only causes chirping and an enhanced line-width of the emitted radiation in semiconductor lasers but also increases unwanted external optical feedback noise [64] and causes filamentation (or flame like beam flickering) of the emitted radiation from wide aperture semiconductor lasers used for high-power generation [38,96]. Compared to bulk devices, strained QW lasers have shown much smaller α_H due to the enhanced differential gain.

11.2.11 Modulation characteristics of semiconductor lasers

A major advantage of semiconductor lasers is that they can be directly modulated by simply varying the injected pump current. External modulation of the emitted optical beam, while keeping the laser output power constant, is also possible using

optical modulators [9,12]. This eliminates the issues associated with turn-on delay, ROs, and multimode operation, but increases the costs. Modern light-wave transmission systems use intensity modulation-direct detection schemes for short haul transmission links and coherent systems for long haul transmission. The former scheme uses an intensity modulated semiconductor laser coupled to an optical fiber where the output is detected via a photo-diode, which transforms the optical signal into an electrical signal. This is a cost-effective approach but induces noise in long distance communications, which is resolved using optical fiber amplifiers [9,10]. In contrast, coherent systems are based on amplitude shift keying, frequency shift keying, and phase shift keying or a combination where two semiconductor lasers are used in a master-slave configuration in its basic format and the generated optical beat signal is detected by the receiving photo diode [9,10]. This approach needs semiconductor lasers with closely matched emission wavelengths, narrow line-widths, and the same polarization and additionally polarization controllers and complicated receiver circuitry in the system, making it less cost effective than the direct detection scheme. We will now briefly discuss direct modulation of semiconductor lasers highlighting the basic characteristics of semiconductor lasers under modulation. Analogue modulation can be achieved by superimposing a small AC modulation over a DC (bias) offset. For digital modulation, the laser can be DC biased close to threshold and a digital modulation is superimposed, see Figure 11.21a. If the modulating bias is small compared to the DC offset, one can use small signal analysis to obtain the *modulation efficiency* $\delta(\omega)$ (with $\omega = 2\pi\nu$), which gives the number of photons generated per injected electron normalized to the value at $\omega = 0$ as [10,13]

$$\frac{\delta(\omega)}{\delta(0)} = \frac{\omega_{RO}^2}{\sqrt{\left(\omega^2 - \omega_{RO}^2\right)^2 + \omega^2\gamma_{damp}^2}}, \quad (11.36)$$

which clearly shows resonance characteristics with a resonance (angular) frequency at ω_{RO}. Figure 11.21b shows Equation 11.36 plotted as a function of $\nu = \omega/2\pi$ under different bias conditions. $R_{int}(\omega) = [\delta(\omega)/\delta(0)]^2$ gives the intrinsic frequency response of the laser. The maximum modulation bandwidth $\nu_{3\,dB} = \omega_{3\,dB}/2\pi$ of the laser

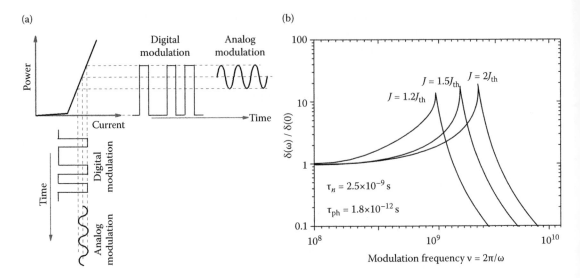

Figure 11.21 (a) Analogue and digital modulation. (b) Schematic of the resonance phenomena [13].

can be found by setting $R_{int}(\omega)=1/2$. It should be pointed out that the modulation bandwidth from the laser is also affected by the electrical resistance and capacitance of the device and the circuit. The maximum modulation frequency for analogue modulation is ν_{RO} but requires high linearity of the light-current characteristic of the laser. Analogue modulation can transmit more information than digital modulation but is plagued by optical feedback noise and other noises in the system, which can be tolerated only in short haul transmission links. Digital modulation uses on-off signal patterns on top of a fixed bias and hence reduces the noise in the system but is affected by carrier density-dependent pattern effects [10].

Noise in semiconductor lasers primarily originates from carrier- and photon-density fluctuations. The relative intensity noise (RIN) spectrum can be used to analyze the noise characteristics and the modulation bandwidth in high-speed semiconductor lasers. A simplified expression for RIN can be written as [10, 13]

$$\text{RIN}(\nu)=\frac{4}{\pi}\Delta\nu_L^{ST}\frac{\nu^2+\left(\tilde{\gamma}_{damp}^*/2\pi\right)^2}{\sqrt{\left(\nu^2-\nu_{RO}^2\right)^2+\left(\tilde{\gamma}_{damp}/2\pi\right)^2\nu^2}},$$

$$(11.37)$$

where $\tilde{\gamma}_{damp}^*$ and $\tilde{\gamma}_{damp}$ now additionally depend on carrier transport effects and the gain suppression parameter ε [10,13]. $\tilde{\gamma}_{damp}^*$ is the cavity loss independent component of $\tilde{\gamma}_{damp}$. RIN assessment is

generally believed to be free from parasitic resistance and capacitance noises in the system highlighting the actual modulation performance from the semiconductor lasers. RIN, however, is strongly affected by carrier transport effects [9,10,13]. A common way of measuring the RIN spectrum is to operate the laser in CW and detect the output via a wide-band photodetector in an electrical (RF) spectrum analyzer. The intensity noise in the laser results in an output power fluctuation, which in turn is translated to fluctuations in detected photocurrent measured as a function of frequency in the RF spectrum analyzer resulting in a RIN spectrum.

Because the dominant factor limiting the modulation response of high speed lasers are the parasitic losses, designs having low parasitic resistance and capacitance are of immense importance in practice. In order to gain a better understanding of these parasitic losses an electrical equivalent circuit of the laser is generally employed. The equivalent circuit includes the bond wire capacitance, p- and n-side series resistances, capacitances of the top and bottom contact area, and the intrinsic capacitance of the laser diode. For a typical laser diode (short cavity, low modulation frequency), the equivalent circuit can be represented as below in Figure 11.22. The roll off of the overall response due to parasitic effects is characterized by the total RC constant of the laser diode package. The length, diameter, and ground proximity of the bond wire

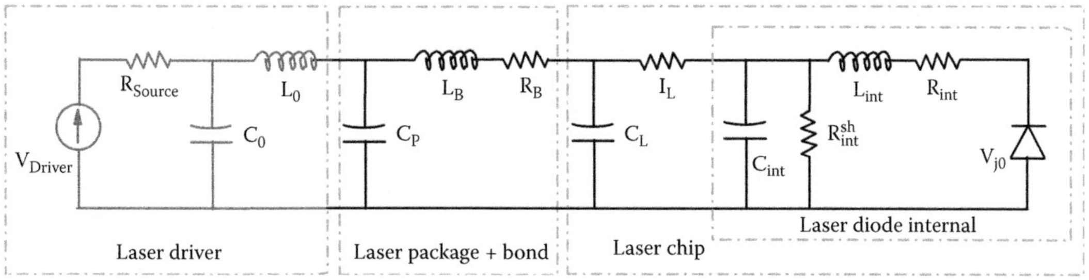

Figure 11.22 Equivalent circuit representation of a packaged laser diode together with it driver circuit [115]. vj_0 is the voltage drop across the laser diode junction, R_{int} is the internal dynamic resistance of the laser diode. Lj is the internal inductance. R_{int}^{Sh} and C_{int} are the shunt components which determine the damping and relaxation rate. R_L and C_L are determined by the laser structure and strongly depend on the current confinement design. C_P is the parasitic capacitance due to the packaging, whereas L_B and R_B correspond to the inductance and resistance of the bonding wires. Up to this point the circuit represents an RF equivalent circuit of a laser diode module. The driver circuit also presents itself as an inductance L_0 and a capacitance C_0 together with R_{Source} to the laser diode package.

are minimized. For modulation speeds above 20–30 GHz using QW lasers, more complicated equivalent circuit models are employed to analyze and optimize the parasitic effects [115].

For operating diode lasers in CW or under pulsed conditions, power supplies with appropriate voltage and current range are employed. As laser diodes typically have a low series resistance, the internal resistance of the power supply is designed to be low for maximum power transfer. Furthermore, as diode lasers are highly susceptible electrical surges, overvoltage, reverse bias, current upper limit, as well as electrostatic discharge protection form an essential element for diode laser driver designs. A simple working representation of the driver circuit is shown in Figure 11.22 is shown in Figure 11.23. For high-speed modulation of laser diodes or for high-power generation more complex driver circuits are used [115,116], the details of which are beyond the scope of this chapter.

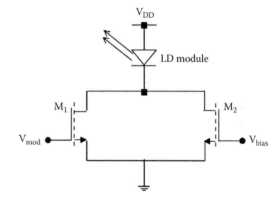

Figure 11.23 A typical simple laser diode (LD) module driver circuit based on NMOS (Negative-channel Metal Oxide Semiconductor) transistors (M1 and M2), which can be used for modulation of the laser diode by setting a modulation voltage V_{mod} at a given bias voltage V_{bias} [116]. V_{DD} is the drain voltage for the NMOS transistors.

occurs perpendicular to the active layer. A comprehensive discussion of each design is beyond the scope this chapter. Therefore, for a detailed understanding, the reader will be referred to more specialized references.

11.3 SPECIFIC SEMICONDUCTOR LASER DESIGNS FOR TARGETED APPLICATIONS

We will briefly discuss below specific designs that cater to particular applications of semiconductor lasers. They mainly fall under two categories: (1) edge emitting semiconductor lasers where the laser emission occurs in the plane of the active region, which follows on from the above discussion; (2) vertically emitting lasers, where the laser emission

11.3.1 Single-frequency and wavelength-tunable lasers

The Fabry–Perot edge-emitting lasers discussed above generally emit in multiple longitudinal modes unless special design features are in place. We have also noted above that multiple spatial mode operation can be controlled by shrinking the width of the emitting stripe to the order of the emission wavelength. For many applications, such as in atomic spectroscopy and long haul optical transmission links, etc., multilongitudinal mode operation is not desirable and special mode control features are incorporated into the laser structure to encourage single-mode operation. Longitudinal mode selection can be achieved by using a grating structure to selectively reflect and maximize emission at the desired wavelength (Figure 11.24a). This is done in a distributed Bragg reflected (DBR) laser design shown in Figure 11.24b, where the cavity formed by the end facets in a generic semiconductor laser is replaced by an effective cavity formed by repeating (corrugated) reflectors, which selectively reflects only the desired wavelength.

The corrugation does not necessary have to be near the facets; the grating structure can be defined throughout the active layer, e.g., in the waveguide or the cladding layer, forming a DFB laser design where the evanescent tail of the intracavity field interacts with the grating (Figure 11.24c). Both DBR and DFB lasers essentially have the same operating principle. The optical radiation generated in the active layer satisfying the Bragg condition will be reflected back efficiently. This condition is set by waves scattering off successive corrugations reflecting and constructively interfering. For a guiding layer with refractive index η, the Bragg condition is given by [9,55]

$$2\Lambda \sin\theta = m\lambda_0/\eta. \tag{11.38}$$

The minimum requirement for a positive optical feedback within the cavity is scattering at ~180° with respect to the direction of propagating mode resulting in $\theta \sim 90°$. One can include the angular dependence in the definition of the modal effective index η_{eff} for the guided mode giving

$$\Lambda = m\lambda_0/2\eta_{eff}, \tag{11.39}$$

i.e., the grating period required for the laser to operate at λ_0 depending on the order of diffraction m.

The grating structure can be produced via photolithography in which holographic techniques are used for projecting a periodic interference pattern on the surface to be etched [9,12,55]. For many near-infrared lasers, the requirement for the first-order grating period is in the 100–300 nm range, requiring resolutions available only with advanced lithographic equipment. To make fabrication easier, higher-order diffraction grating structures have been investigated although their efficiency tends to be lower due to lower-order diffraction loss. It can be shown that a DFB laser has two resonant frequencies around the desired cavity

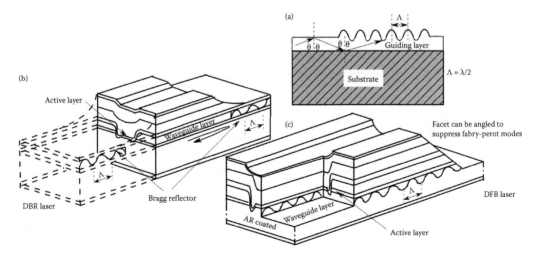

Figure 11.24 (a) Schematic of a distributed Bragg reflected (DBR)-based waveguide design on a substrate with grating period Λ. (b) A DBR laser design with a grating structure defined to selectively reflect the desired wavelength for oscillating in the active layer. (c) A distributed feedback laser design where the grating covers the entire cavity. Both lasers shown have a buried heterostructure lateral current confinement design [9,12].

resonance where the cavity oscillations can occur [65]. This is undesirable and is generally remedied by using a $\lambda_0/4$ phase-shifted design [66]. DBR lasers can also provide single-frequency operation with a high degree of stability and tunability. As the gain and Bragg reflection regions are separate, independent control of the carrier density in the grating region is possible which can be used to alter the effective index of the waveguide using the free carrier plasma effect [55], for fine frequency tuning. The above longitudinal mode control designs allow generation of very narrow laser line-widths (<0.1 nm), not possible in conventional Fabry–Perot designs. In addition, the emission wavelength in DBR and DFB lasers are less susceptible to temperature changes via the change in band-gap and depend more on the temperature dependence of the grating period and refractive index, which are comparatively less sensitive to temperature.

The DBR design can be adapted to design efficient and widely tunable lasers on a single semiconductor chip. Figure 11.25a shows a three-section laser design where each section can be pumped independently. This allows wavelength tuning via phase tuning or tuning of the effective grating period (both via refractive index tuning) by independent electrical pumping of the respective sections. A combination of both allows a continuous tuning of about 1% around the central wavelength [67]. This limitation of tuning range is imposed by the extent to which the effective refractive index can be altered via current injection, i.e., (mainly) via the free carrier plasma effect. In this effect, the free carriers under mutual Columbic interaction alter the dielectric polarization of the material, thereby altering its refractive index [67,68]. An order of magnitude increase in the tunability range can be achieved by using a superstructure grating (SSG) design (Figure 11.25b). A SSG can be approximated as an array of frequency independent scattering centers spaced at Λ. The SSG-DFB laser works using the Vernier effect between the reflection peaks of the two SSGs which are designed to have slight offset in their reflection peaks [67,68]. SSG-DFB lasers and related design variations are used as local oscillators in modern coherent detection systems forming an integral part of today's optical communication networks.

11.3.2 Vertical cavity surface emitting lasers (VCSELs)

We note from the discussion in Section 11.8 that if the frequency spacing between the adjacent longitudinal modes is wider that the gain bandwidth, a single longitudinal mode operation is possible. Equation 11.3 suggests that this can be achieved by having a very short cavity. In a vertical cavity surface emitting laser, the cavity length is made very short and the laser beam is emitted perpendicular to the wafer, i.e., in the same direction as that of the injected current. The short cavity and small active volume require a high-Q cavity to keep the lasing threshold current low [9,69]. VCSELs can be fabricated using conventional planar integrated circuit fabrication techniques and can provide emission over a wider

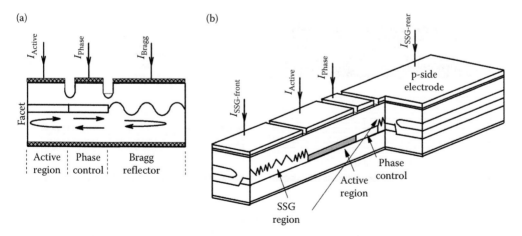

Figure 11.25 (a) A three section laser design for independent gain, phase and mode control. (b) A superstructure grating distributed Bragg reflected (SSG-DBR) tunable laser.

aperture if high-order spatial modes can be tolerated in the application. Optical interconnects based on narrow aperture VCSELs are used in many levels of modern light-wave communication systems.

The active layer in a VCSEL is sandwiched between a stack of Bragg mirrors, i.e., alternate layers of high and low refractive index materials each of which is designed to be a quarter wavelength thick at the desired laser wavelength (Figure 11.26a and b). The Bragg mirrors provide very high reflectivity of the light back into the active region, allowing the VCSEL to generate sufficient gain despite the small gain region and optical path length. Constructive and destructive interference within the Bragg stack results in a Bragg resonance that gives rise to a reflectivity, which increases with the number of layer pairs (one high index η_{high} and one low index η_{low}). The resultant reflectivity for the intracavity field can be approximated by

$$r_{Bragg} \approx \tanh\left[2m\left(\frac{\eta_{high} - \eta_{low}}{\eta_{high} + \eta_{low}} \right)\right] \quad (11.40)$$

$$= \tanh\left[2m\, \Delta\Lambda / \lambda_0 \right]$$

if the entry and exit material for the radiation is considered the same [70], where m is the number of layer pairs. For $\eta_{high} - \eta_{low} = 0.3$, greater than 98% intensity reflectance $\left(R = r_{Bragg}^2 \right)$ can be achieved

for 30 layers in a AlGaAs/GaAs-based Bragg stack for resonance wavelength around ~850 nm. For longer wavelengths, a higher number of (thicker) layer pairs is generally required [70]. A Bragg stack with a sufficient number of layer pairs results in a flat and broad reflectivity spectrum (called a stop-band) with a width $\Delta\Lambda$ around the cavity resonance wavelength λ_0 it is designed for, see Figure 11.27b. The top Bragg stack usually has a lower number of layer pairs for OC of the generated laser beam.

This emission wavelength from a VCSEL (or for that matter any single-frequency laser) is selected via a dynamic competition between the cavity resonance and material gain peak. For a high Q cavity, the emission wavelength is closer to the cavity resonance than the gain peak (mode/frequency pulling [15]). Thus, in general, the VCSEL cavity is designed so that there is a fixed detuning between the material and the cavity resonances at room temperature so that at the operating temperature the cavity resonance aligns with the peak gain and the desired wavelength is emitted with maximum gain available at the cavity resonant mode [101–103]. In addition to the advantage of being an inherently single-frequency laser, VCSELs offer several other advantages including ease of fabrication and on-wafer testing of single chips, ease of fabrication of planar arrays (hence lower costs), and their symmetric circular beam profile that makes coupling

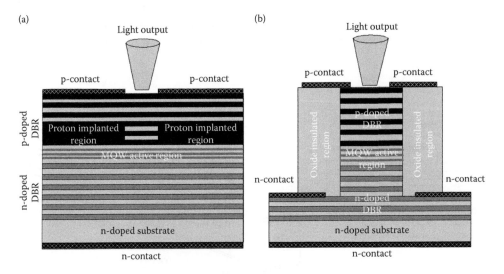

Figure 11.26 Cross-sections of two common cylindrical VCSEL designs for lateral current confinement: (a) a proton-bombarded top-bottom contact design, (b) an oxide confined side-contact design. The multiquantum well active region sandwiched between the distributed Bragg reflected stacks forming the cavity is also shown. The VCSEL being a cylindrical design emits a circular beam profile.

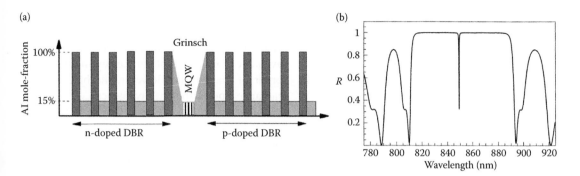

Figure 11.27 A multiquantum well active region in a AlGaAs/GaAs-based VCSEL sandwiched between two AlGaAs/GaAs Bragg stacks, which are differently doped for efficient carrier transport. (b) Typical stop-band characteristics in a VCSEL around the resonance wavelength.

of the generated light to an optical fiber easier and more efficient (lower coupling loss) compared to edge-emitting lasers. Note that the resonant cavity LED (RCLED) is a related device that provides surface emission in a VCSEL-like structure. However, such devices have a lower number of DBR layer pairs in the Bragg stack and do not achieve a lasing threshold and thus emit narrow band spontaneous emission. Such devices are, however, useful for applications such as plastic optical fiber-based visible light data links [118]. Other forms of surface emitting laser have also been designed where an angled facet is used to direct the laser beam perpendicular to the plane of the epistructure while using conventional current injection designs as in edge emitting semiconductor lasers [9,10,12]. Laser designs based on photonic band-gap materials (i.e., photonic crystal lasers [5]) also employ vertical emitting designs with carrier injection schemes similar to conventional VCSELs.

11.3.3 Designs for high-power generation

Direct and efficient conversion of electrical energy into coherent light provides a wealth of applications such as in solid state and fiber laser pumping, medical applications, optical communication via free space energy transfer and high quality material processing (cutting and welding) allowing the process to be controlled with high spatial and temporal resolution. Thus, interest in scaling the output power to the kilowatt range and beyond has been there from the very beginning of semiconductor lasers. There is a limit to the maximum optical power that can be extracted from an edge emitting stripe-geometry laser. This limit is set by the *thermal runaway* power load and secondly the optical power load at the facet which can be up to few $\sim$MW cm^{-2}, beyond which the facet succumbs to the generated heat and undergoes catastrophic optical damage (COD) [104,105]. With an in increase in carrier injection (pumping) the optical output power from the laser increases until the generated heat (via unused injected power) causes thermal escape of the carriers from the active region (thermal runaway [71]) resulting in a drop in the optical power upon further current injection. At sufficiently high injection the device no longer lases owing to the fact that the threshold current itself increases (due to joule heating) with injection until such point that the threshold current exceeds the injected current level. Thermal runaway is a reversible phenomenon, unlike COD, which irreparably damages the laser [71]. COD is sensitive to the quality of the facet and modern facet passivation approaches mean that COD is less problematic than it once was. Thermal runaway generally occurs at a much lower power load than facet damage in diode lasers and thus, under some circumstances, self-limits the device from undergoing COD. High-power diode lasers consequently require wide aperture designs; first, to achieve a higher gain volume for high-power generation and second, to withstand and efficiently dissipate the heavy thermal load generated by the device. Additionally, they are often operated under quasi-CW conditions (100 µs–1 ms pulse length) to limit thermal effects and to extract the maximum possible power.

The distinction between low-power and high-power semiconductor lasers is not very well defined. It may depend on the type of diode laser and the application for which it has been designed. Generally, 50 mW (or more) CW for single-mode, single-frequency lasers and 500 mW (or more) CW for broad-area, multimode lasers and laser arrays is generally considered to be high power [71]. For many applications, however, the power requirement from diode lasers is quite low. For example, compact disc players use a 3 mW range GaAs laser emitting at $\lambda = 780$ nm, supermarket scanners use visible ($\lambda = 670$ nm) 3 mW AlGaInP lasers and in optical fiber telecommunications 10–50 mW (in-fiber) InGaAsP lasers ($\lambda = 1.3$ μm or $\lambda = 1.55$ μm) are generally used, all emitting in single mode. However, an important class of applications requires much higher output powers in either pulsed or CW operation. Operation in a single spatial mode or at a single frequency is often desirable but raw power can also be very useful. One example where raw power output is most important is diode-pumped yttrium aluminum garnet (YAG)-based solid-state lasers, where the pumping laser array fabricated from AlGaAs material emits kilowatts of incoherent quasi-CW pulses. The wavelength of the AlGaAs laser array is well-tuned to the absorption spectrum of the YAG crystal (806–810 nm), providing much higher overall efficiency than flash-lamp pumping. Applications where both high-power and near-diffraction-limited beam quality (i.e., coherent emission) are required include many communication and most scientific applications, where transmission over long distances and/or the ability to focus the light to a

small spot of high-power density is desired, such as pumping EDFAs. In general, the power available from incoherent sources exceeds the power available from coherent sources by more than an order of magnitude. However, the brightness or radiance (power/beam divergence) of coherent, diffraction-limited sources is significantly higher. Tremendous advances have been made in scaling the output power of diode lasers and arrays, thanks to modern crystal growth, device cooling and facet passivation technologies.

Over the last two decades, the output power from diode lasers has increased by an order of magnitude [4,71]. The electro-optic (wall-plug) conversion efficiency measured under laboratory conditions is now >70% in GaAs-based lasers, which has reduced the thermal load immensely [72–75]. Under pulsed conditions more than 150 W of optical power output has already been demonstrated from a ~100 μm emitting stripe geometry laser [73]. Under CW operation, optical power in the range of tens of watts has been demonstrated before thermal runaway occurs [74]. Further improvements are being reported each year [74,75]. Multi-kW level raw power delivery is already possible from modern stacked laser diode arrays [76].

A diode laser array is essentially a multistripe semiconductor laser with stripes fabricated in close proximity to each other (Figure 11.28). Lateral current confinement can either be achieved via a mesa-stripe (gain guided) or via proton bombarded (index-guided) design [76–78]. It is common to have 1 cm wide arrays (called diode laser bars) in array stacks. The proximity of the

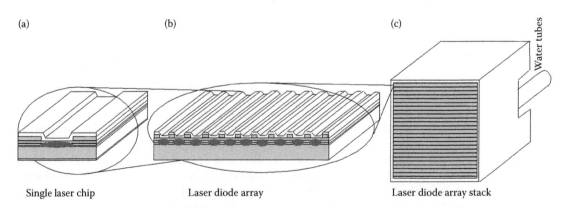

(a)　　　　　　(b)　　　　　　(c)

Single laser chip　　　Laser diode array　　　Laser diode array stack

Figure 11.28 Schematic for a high-power diode laser: (a) single chip, (b) array of diode lasers, and (c) a two-dimensional stack (requires active cooling).

individual stripes results in an overlap of oscillating intracavity field distributions from adjacent emitters (Figure 11.28b), and with appropriate design, these can phase-locked resulting in a combined, coherent wave front being emitted from the whole facet [76–78]. Vertical stacks of the 1 cm bars operate under active, e.g., liquid-based cooling (Figure 11.28c).

Optically pumped semiconductor slab/disc lasers (also known as VECSELs or optically pumped surface emitting lasers (OPSELs), see below) [79] are the only technology competing with stacked diode laser arrays. Vertical external cavity lasers or VECSELs are essentially optically pumped large aperture VCSELs which are designed with a pump absorbing layer in place of the top DBR stack (OC side), see Figure 11.29a. Hence, they are also called OPSELs or just disc or slab lasers. Diode laser array stacks are used for optical pumping, while an external mirror completes the cavity (Figure 11.28) [79].

The advantage of VECSELs over a stack of edge emitting laser bars is a symmetric beam profile, a narrow line-width and the possibility of intracavity frequency doubling [79]. The scalability of the aperture allows generation of high powers while maintaining the circular beam shape and narrow line-width. However, efficient thermal management is essential for VECSEL operation due to the generally poor thermal conductivity of the bottom DBR.

Despite the developments in high-power diode laser technology, achieving high-spatial coherence (beam quality) still remains an open challenge. Beam quality of high-power diode lasers (both edge emitting and large aperture VECSELs) is affected by nonlinear filamentation of the spatial modes and is far behind from what is achievable from modern solid state and gas lasers. Poor beam quality affects the focusability (brightness) of the beam and the coupling efficiency to optical fibers. A nonzero positive phase–amplitude coupling in the semiconductor gain medium is the primary reason behind the filamentation instability [71,78]. Thermal effects also play a key role [71,78,97–99]. Currently, a master oscillator power amplifier (MOPA) design or a derivative of it offers some promise in achieving high spatial and temporal coherence from a broad area diode laser [71,78]. However, the coherence properties start deteriorating at higher pumping due to thermal effects [38,96–100]. The working principle behind a MOPA design is as follows. The MOPA design consists of a narrow aperture ridge waveguide section with a inbuilt grating (a DBR laser type design) acting as the master oscillator, which injects a spatially and spectrally filtered beam into an adiabatically tapered section, which amplifies and scales the optical power (see Figure 11.29b). Some variants of the MOPA design do not use a grating when spectral purity is not important for the application. Beam spoiling grooves are sometimes used to attenuate any backward propagating wave in the tapered region for enhancing the beam stability [71,78].

High brightness, high-power diode laser sources are gaining in interest for applications previously dominated by solid-state lasers, because of their efficiency, compactness, and high reliability. Among

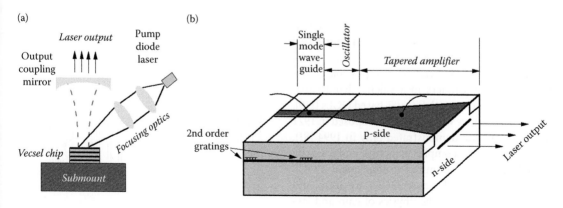

Figure 11.29 (a) Schematic of a vertical external cavity surface emitting laser (VECSEL) showing the VECSEL chip, the external cavity and the pumping scheme. (b) A master oscillator power amplifier (MOPA) laser design for high brightness (see text for details).

others, laser projection displays, free-space communication, frequency doubling, and direct material processing are rapidly emerging areas, where high output power together with diffraction-limited beam quality are either the key requirements or strongly enhance system performance. Production of high-power diode lasers with improved brightness will allow them to compete with the very best solid-state and gas lasers at much lower cost [4,71], and may enable future generations of compact, inexpensive, and portable laser systems.

11.3.4 Quantum cascade lasers

The semiconductor lasers that we have discussed so far emit wavelengths largely determined by the band-gap of the available (direct band-gap) semiconductor materials because the devices rely on so-called interband transitions. This restricts the use of semiconductors as a lasing medium in large parts of the mid- to far-infrared spectrum not covered by conventional direct semiconductor materials or for which processes, such as Auger recombination, which become increasingly important for narrow band-gaps, are problematic and limit device performance [106,107]. A quantum cascade laser (QCL) uses intraband transitions (instead of interband transitions used by conventional lasers) to generate photons in the mid- and the far infrared spectrum, including in the terahertz (THz) range. These intraband transitions are electronic transitions within the QWs and the emitted photon energy depends on the QW width and heterojunction band offsets. Therefore the desired emission wavelength can be varied just by varying the well-width. Thus, QCLs are unipolar lasers unlike conventional semiconductor lasers, which are bipolar devices. Almost all III–V semiconductor-based QCLs use electrons in the conduction band as the charge carriers; however, it is also possible to use holes as charge carriers in a valence band-based device as is the case with SiGe/Si-based QCLs [119]. The QCL structure was originally proposed in 1971 [80] and was first demonstrated only in 1994 [81], mainly due to the complex structure and accurate growth technology required to implement a QCL design in practice. Because QCLs are intraband and unipolar devises, the physics of gain and lasing is different from what we have discussed above for interband lasers and a detailed discussion is beyond the scope of this chapter. Nevertheless,

QCLs form an important class of modern semiconductor lasers so we will present a qualitative outline of their working principles.

The cavity and waveguide design used in both conventional semiconductor lasers and QCLs are essentially similar except for the doping (both cladding layers are n-doped in the majority of QCLs) [5,80,82]. The main difference in QCLs lies in the active layer which employs intraband transitions only in one band. Because electron charge carrier in conduction band-based devices form the vast majority of QCLs, we will focus on these. This can be seen in Figure 11.30, which shows a section of the active layer, which employs MQWs in two stages: an injector region and an active region all under an electric field. The closely spaced QWs and narrow barriers in the injector region result in the formation of mini-bands in a superlattice [5,80,82]. Quantum-mechanical tunneling assisted transport of the injected electrons through the minibands into the adjacent active region creates an effective population inversion in the active QW. This leads to an intraband transition of the electron from the upper level to lower level of the QW resulting in a photon being emitted. Subsequently, the electron thermalizes to a lower state and tunnels into the next injector region cascading through the structure (hence the name of the laser) producing many more photons. A typical QCL has 20–80 such stages each having several MQW structures. Typically, there are several hundred to over a thousand epilayers in QCLs. QCLs have been demonstrated from just below 3 μm up to 67 μm, opening-up the THz range and its associated applications [82,83]. High-power and DFB QCLs have also been demonstrated [5,82]. It

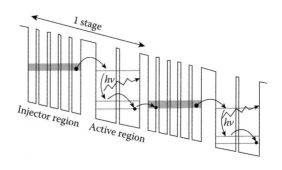

Figure 11.30 Schematic of the conduction band design for two successive stages inside the active region of a quantum cascade laser (QCL).

is interesting to note that the *quantum* efficiency of QCLs can be far above 100% owing to the fact that one electron produces many photons. However, this is not breaking any law of physics, as the power (wall-plug) efficiency is typically <50% owing to the large voltages that are required to produce the electric field across the structure, which allows the cascading process to occur.

11.4 SEMICONDUCTOR MATERIALS FOR MODERN SEMICONDUCTOR LASERS

The lasing characteristics and reliability of semiconductor lasers largely depend on materials used in their design. The band-gap and the lattice constant both play a vital part as well as the maturity of the growth technology of the chosen semiconductor. While the band-gap for most types of laser largely determines the emission wavelength, the lattice constant determines whether the semiconductor

material can be grown lattice matched on the substrate of choice, hence reducing the density of defects. The latter directly affects the performance and reliability of the device. As discussed earlier, a small amount of strain can be beneficial if kept within a thin QW layer [9,10]. However, in general, the overall semiconductor laser structure should be relatively strain free to minimize defect formation. A simple SQW-SCH active region design requires a minimum of five epilayers on the substrate to make a laser structure (typically there are more than ten layers). One can see that this immediately limits the number of semiconductor alloys with the desired band-gap and lattice constants that can be used in a typical laser design. Although many new semiconductor alloys have been proposed in recent years to fill the existing gaps, there are only a few with mature enough growth technology that can be used for commercial devices, which require high reliability. Figure 11.31 lists some semiconductor alloys with their band-gaps and lattice constants [84,85].

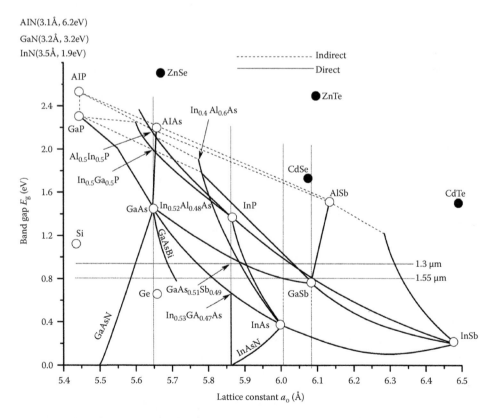

Figure 11.31 Band-gap energies and lattice constants of common semiconductor alloys used for making semiconductor lasers [84]. Open symbols refer to indirect band-gap materials, whereas closed circles are direct band-gap materials. AlN, GaN, and InN indicated separately. Alloys are indicated by the connecting curves (solid are direct band-gaps and dashed are indirect band-gaps).

Table 11.1 List of some example active and cladding layer compositions used in modern semiconductor lasers

Cladding layer/active layer	Substrate	Emission wavelength range (μm)
$GaN/In_xGa_{1-x}N$	Sapphire, GaN, SiC	~0.35–0.50
$Al_xGa_{1-x}N/In_xGa_{1-x}N$	Sapphire, GaN, SiC	~0.40–0.60
$Cd_xZn_{1-x}S/CdS$	CdS	~0.35–0.50
$(Al_xGa_{1-x})_{0.5}In_{0.5}P/Ga_yIn_{1-y}P$	GaAs	~0.60–0.72
$Al_xGa_{1-x}As/GaAs$	GaAs	~0.60–0.92
$Al_xGa_{1-x}As/In_yGa_{1-y}As$	GaAs	~0.90–1.10
$InP/Ga_xIn_{1-x}As_yP_{1-y}$	InP	~1.1–1.65
$Al_xGa_{1-x}As/In_yGa_{1-y}As_{1-x}N_x$	GaAs	~1.24–1.30
$Al_xGa_{1-x}Sb_{1-y}As_y/In_xGa_{1-x}Sb_{1-y}As_y$	GaSb or InAs	~1.7–4.0

Note: The substrate material and approximate emisson wavelength range is also indicated in References [6,7].

Materials falling on a vertical line are lattice matched to each other. There is a great impetus toward achieving efficient semiconductors active regions grown on silicon substrates which will allow on-chip integration of lasers with silicon-based optical wave-guide components for all-optical communications and data processing. Being an indirect band-gap semiconductor, silicon itself is a poor emitter of light; hence, the integration of high optical quality direct band-gap semiconductors on cheap silicon substrates is highly desirable [86]. The primary limiting issue to achieve this is the lattice mismatch between silicon and standard III–V materials such as InP- or GaAs-based alloys (see Figure 11.31), which makes it difficult to grow high-quality alloys directly on silicon.

There have been tremendous advancements in hybrid approaches to this where strain relaxed (metamorphic) growth techniques have been used to grow relatively defect-free working laser structures [87,108], through the use of quantum dot [109] and exotic QW [110] active regions and through wafer fusion approaches [108]. This remains an active area of ongoing research as noted in a number of reviews on this topic [120,121]. Entirely new materials for semiconductor lasers continue to being investigated to help push the wavelength limits of semiconductor lasers and to improve the efficiency and temperature sensitivity of devices. Recently, for example, GaAsBi-based semiconductor lasers have been demonstrated [87,111–114]. Such alloys have the potential overcome the ill effects of Auger recombination, which have limited the maximum achievable power and efficiency of near- and mid-infrared semiconductor lasers and have the potential to significantly reduce the energy demands of optical fiber communications [88,111–114].

Table 11.1 lists some of the common active and cladding layer compositions used in modern semiconductor lasers together with the substrate material and the emission wavelength range. The list is incomplete as many materials and compositions still under research are not included [88]. The values of mole-fractions indicated in Table 11.1 are either for direct band-gap or strain-balanced designs. Semiconductor lasers emitting in the range 0.35–0.92 μm find applications in holographic storage, image recorders, fax machines, printers, solid state, and fiber laser pumping, barcode readers, blu-ray discs and other disc read write devices, material processing, medical therapeutics, range finders, and aerospace. 0.9–1.10 μm semiconductor lasers are used as optical fiber amplifier pumps, in material processing and medicine. 1.1–1.65 μm lasers find used in optical fiber communication systems, transceivers, and Raman amplifiers. 1.24–1.30 μm are also used in aerospace and military range finders. 1.7–4.4 μm lasers are used in spectroscopic sensing of humidity, gas impurities, drugs and optical fiber, and free-space communications [5].

11.5 CONCLUDING REMARKS

In this chapter, we have presented a brief outline of the essential working principles of semiconductor lasers. We have also discussed the many and varied applications of semiconductor lasers, various semiconductor laser designs and materials used for making these fascinating devices. As it is not possible to cover all the relevant details of semiconductor lasers and their novel applications such as mode- and injection locking, etc. [5–13,34,39] in a short chapter, the reader is encouraged to refer to more specialized articles and monographs highlighted in the references. The field of semiconductor lasers is an amazing field of science where one is able to bring fundamental quantum physics directly in connection with everyday life. The Internet as we see it today would have been very different without the advent and development of semiconductor lasers. The field is fast advancing as new ideas and semiconductor materials emerge and can be expected to be a fruitful area of research in science, engineering, and technology development for many years to come.

ACKNOWLEDGMENTS

The authors gratefully acknowledge the support of various members of the photonics group at Surrey and our collaborators over the years who have contributed to the work on semiconductor lasers. We are particularly grateful to the Engineering and Physical Sciences Research Council (EPSRC), UK, Innovate UK, and the European Community for funding the majority of the work on semiconductor lasers at Surrey, and in particular through EPSRC grant EP/H005587/1.

REFERENCES

1. Murphy, E. 2010. The semiconductor laser: Enabling optical communication. *Nat. Photon.*, 4, 287.
2. Won, R. 2012. Celebrating semiconductor lasers. *Nat. Photon.*, 6, 508–510.
3. Overton, G., Nogee, A. and Holton, C. 2014. *Laser Marketplace 2014: Lasers forge 21st century innovations.* (Tulsa, OK: Laser Focus World).
4. Treusch, H.G., Ovtchinnikov, A., He, X., Kanskar, M., Mott, J. and Yang, S. 2000. High-brightness semiconductor laser sources for materials processing: Stacking, beam shaping and bars. *IEEE J. Sel. Top. Quantum Electron.*, 6, 601–614.
5. Coleman, J.J., Bryce, A.C. and Jagadish, C. 2012. *Advances in Semiconductor Lasers* (New York: Academic Press).
6. Casey, H.C. Jr. and Panish, M.B. 1978. *Heterostructure Lasers: Part A and B* (New York: Academic Press).
7. Razeghi, M. 2011. *The MOCVD Challenge: A Survey of GaInAsP-InP and GaInAsP-GaAs for Photonic and Electronic Device Applications* (New York: CRC Press).
8. Kroemer, H. 1963. A proposed class of hetero-junction injection lasers. *IEEE Proc.*, 51, 1782–1783.
9. Coldren, L.A., Corzine, S.W. and Mashanovitch, M.L. 2012. *Diode Lasers and Photonic Integrated Circuits* (New York: Wiley Wiley).
10. Kapon, E. (ed.) 1998. *Semiconductor Lasers I and II* (New York: Academic Press).
11. Chuang, S.L. 2009. *Physics of Photonic Devices* (New York: Wiley).
12. Ebeling, K.J. 1993. *Integrated Optoelectronics* (New York: Springer-Verlag).
13. Peterman, K. 1991. *Laser Diode Modulation and Noise* (Dordrecht: Kluwer Academic).
14. Hayashi, I., Panish, M.B., Foy, P.W. and Sumski, S. 1970. Junction lasers which operate continuously at room temperature. *Appl. Phys. Lett.*, 17, 109–111.
15. Svelto, O. 2009. *Principles of Lasers*, 5th ed. (Berlin: Springer).
16. Einstein, A. 1917. Zur Quantentheorie der Strahlung. *Physikalische Zeitschrift*, 18, 121–128. English translation: Quantum theory of radiation in *The Old Quantum Theory* (New York: Pergamon, 1967), 167–183.
17. Friedberg, R. 1994. Einstein and stimulated emission: A completely corpuscular treatment of momentum balance. *Am. J. Phys.*, 62, 26–32.
18. Sakurai, J.J. 1967. *Advanced Quantum Mechanics* (Reading, MA: Addison-Wesley).
19. Milonni, P.W. 1984. Why spontaneous emission. *Am. J. Phys.*, 52, 340–343.
20. Holland, M., Burnett, K., Gardiner, C., Cirac, J. and Zoller, P. 1996. Theory of an atom laser. *Phys. Rev. A*, 54, R1757–R1760.
21. Milonni, P.W. and Eberly. J.H. 2010. *Laser Physics* (Chapter 13) (New York: Wiley).

22. Klingshirn, C. 2007. *Semiconductor Optics*, 3rd ed. (Berlin: Springer).

23. Peyghambarian, N., Koch, S.W. and Mysyrowicz, A. 1993. *Introduction to Semiconductor Optics* (Englewood Cliffs, NJ: Prentice Hall), 340–342.

24. Chuang, S.L., Gorman, J.O. and Levi, A.F.J. 1993. Amplified spontaneous emission and carrier pinning in laser diodes. *IEEE J. Quantum Electron.*, 29, 1631–1639.

25. Levy, R.A. 1998. *Principles of Solid State Physics* (New York: Academic Press).

26. Bernard, M.G.A. and Duraffourg, G. 1961. Laser conditions in semiconductors. *Phys. Stat. Solidi*, 1, 699.

27. Hayashi, I., Panish, M.B. and Foy, P.W. 1969. A low-threshold room-temperature injection laser. *IEEE J. Quantum Electron.*, 5, 211–212.

28. Dingle, R., Wiegmann, W. and Henry, C.H. 1974. Quantum states of confined carriers in very thin $Al_x Ga_{1-x}As$–$GaAs$–$Al_xGa_{1-x}As$ heterostructures. *Phys. Rev. Lett.*, 33, 827–830.

29. Thompson, G.H.B. and Kirkby, P.A. 1973. (GaAl) As lasers with a heterostructure for optical confinement and additional hetero-junctions for extreme carrier confinement. *IEEE J. Quantum Electron.*, 9, 311–318.

30. Panish, M.B., Casey, H.C. Jr., Sumski, S. and Foy, P.W. 1973. Reduction of threshold current density in $GaAs$–$Al_xGa_{1-x}As$ heterostructure lasers by separate optical and carrier confinement. *Appl. Phys. Lett.*, 22, 590–591.

31. Dyment, J.C. 1967. Hermite–Gaussian mode patterns in GaAs junction lasers. *Appl. Phys. Lett.*, 10, 84–86.

32. Dyment, J.C., D'Asaro, L.A., North, J.C., Miller, B.I. and Ripper, J.E. 1972. Proton-bombardment formation of stripe-geometry heterostructure lasers for 300 K CW operation. *Proc. IEEE*, 60, 726.

33. Moloney, J.V., Hader, J. and Koch, S.W. 2007. Quantum design of semiconductor active materials: Laser and amplifier applications. *Laser Photon. Rev.*, 1, 24–43 (and references therein).

34. Chow, W.W. and Koch, S.W. 1999. *Semiconductor Laser Fundamentals* (Berlin: Springer).

35. Nagourney, W. 2010. *Quantum Electronics for Atomic Physics* (Oxford: Oxford University Press) p. 153.

36. Connelly, M.A. 2002. *Semiconductor Optical Amplifiers* (Dordrecht: Kluwer), p. 48.

37. Yamanishi, M. and Lee, Y. 1987. Phase dampings of optical dipole moments and gain spectra in semiconductor lasers. *IEEE J. Quantum Electron.*, 23, 367–370.

38. Mukherjee, J. and McInerney, J.G. 2009. Spatial mode dynamics in wide-aperture quantum-dot lasers. *Phys. Rev. A*, 79(5), 053813.

39. Ustinov, V.M., Zhukov, A.E., Egorov, A. Yu and Maleev, N.A. 2003. *Quantum Dot Lasers* (Oxford: Oxford University Press).

40. Garbuzov, D.Z. 1982. Radiation effects, lifetimes and probabilities of band-to-band transitions in direct A_3B_5 compounds of GaAs type. *J. Lumin.*, 27, 109–112.

41. Adams, A.R., Asada, M., Suematsu, Y. and Arai, S. 1980. The temperature dependence of the efficiency and threshold current of $In_{1-x}Ga_xAsyP_{1-y}$ lasers related to inter-valence band absorption. *Jpn J. Appl. Phys.*, 19, L621–L624.

42. Streifer, W. and Kapon, E. 1979. Application of the equivalent-index method to DH diode lasers. *Appl. Opt.*, 18, 3724–3725.

43. Ikegami, T. 1972. Reflectivity of mode at facet and oscillation mode in double-heterostructure injection lasers. *IEEE J. Quantum Electron.*, 8, 470–476.

44. Adams, A.R. 1986. Band-structure engineering for low-threshold high-efficiency semiconductor lasers. *Electron. Lett.*, 22, 249–250.

45. Yablonovitch, E. and Kane, E.O. 1986. Reduction of lasing threshold current density by the lowering of valence band effective mass. *J. Lightwave Technol.*, 4, 504–506.

46. Asada, M., Miyamoto, M. and Suematsu, Y. 1986. Gain and the threshold of three dimensional quantum dot lasers. *IEEE J. Quantum Electron.* 22, 1915–1933.

47. Gil, L. and Lippi, G.L. 2014. Beyond the standard approximations: An analysis leading to a correct description of phase instabilities in semiconductor lasers. *Proc SPIE*, 9134 (arXiv:1403.4835).

48. Smith, W.V. and Sorokin, P.P. 1966. *The Laser* (New York: McGraw Hill), p. 68.

49. Siegman, A.E. 2000. Excess quantum noise in non-normal oscillators. *Front. Laser Phys. Quantum Opt.*, 31–38.

50. *Laser Diode User's Manual* (Osaka: Sharp Corp). 1988.

51. Stelmakh, N. and Flowers, M. 2006. Measurement of spatial modes of broad-area diode lasers with 1-GHz resolution grating spectrometer. *IEEE Photon. Technol. Lett.*, 18(15), 1618–1620.

52. Marko, I.P., Adams, A.R., Massé, N.F., Sweeney, S.J. 2014. Effect of non-pinned carrier density above threshold in InAs quantum dot and quantum dash lasers. *IET Optoelectron.*, 8, 88–93.

53. Smowton, M.P. and Blood, P. 1997. The differential efficiency of quantum well lasers. *IEEE J. Sel. Top. Quantum Electron.*, 3, 491–498.

54. Tansu, N. and Mawst, L.J. 2005. Current injection efficiency of 1300-nm InGaAsN quantum-well lasers. *J. Appl. Phys.*, 97, 054502.

55. Amann, M.C. and Buus, J. 1998. *Tunable Laser Diodes* (Boston, MA: Artech).

56. Adams, A.R., Silver, M. and Allam, J. 1998. Semiconductor optoelectronic devices. In *High Pressure in Semiconductor Physics II, Semiconductors and Semimetals*, vol. 55 (London: Academic Press), p. 301.

57. Marko, I. and Sweeney, S.J. 2015. Optical and electronic processes in semiconductor materials for device applications. *Excitonic and Photonic Processes Mater., Springer Series in Mater. Sci.*, 203, 253–297.

58. Yao, J., Agrawal, G.P., Gallion, P. and Bowden, C. 1995. Semiconductor laser dynamics beyond the rate-equation approximation. *Opt. Commun.*, 119, 246–255.

59. Yang, W. 2007. Picosecond dynamics of semiconductor Fabry–Perot lasers: A simplified model. *IEEE J. Sel. Top. Quant. Electron.*, 13, 1235.

60. O'Gorman, J., Levi, A.F.J., Coblentz, D., Tanbun-Ek, T. and Logan, R.A. 1992. Cavity formation in semiconductor lasers. *Appl. Phys. Lett.*, 61, 889–891.

61. Konnerth, K. and Lanza, C. 1994. Delay between current pulse and light emission of a gallium arsenide injection laser. *Appl. Phys. Lett.*, 4, 120–121.

62. Dixon, R.W. and Joyce, W.B. 1979. Generalized expressions for the turn-on delay in semiconductor lasers. *J. Appl. Phys.*, 50, 4591–4595.

63. Henry, C.H. 1982. Theory of the linewidth enhancement factor of semiconductor lasers. *IEEE J. Quantum Electron.*, 18, 259–264.

64. Huyet, G., White, J.K., Kent, A.J., Hegarty, S.P., Moloney, J.V. and McInerney, J.G. 1999. Dynamics of a semiconductor laser with optical feedback. *Phys. Rev. A*, 60, 1534–1537.

65. Yamaguchi, M., Numai, T., Koizumi, Y., Mito, I. and Kobayashi, K. 1987. Stable single-longitudinal-mode operation in λ/4 shifted DFB-DC-PBH laser diodes. *Optical Fibre Communication Conference* January 19, 1987, *Reno, NV*, TUC4.

66. Kobayashi, K. and Mito, I. 1987. Progress in narrow-linewidth tunable laser sources. *Optical Fiber Communication Conference Record* January 19, 1987, *Reno, NV*, WC1.

67. Venghaus, H. and Grote, N. (ed.) 2012. *Fibre Optic Communication: Key Devices* (Berlin: Springer).

68. Sarlet, G., Morthier, G. and Baets, R. 2000. Control of widely tunable SSG-DBR lasers for dense wavelength division multiplexing. *J. Lightwave Tech.*, 18(8), 1128–1138.

69. Iga, K. 2008. Vertical cavity surface emitting laser: Its conception and evolution. *Jpn J. Appl. Phys.*, 47, 1–10.

70. Wilmsen, C.W., Temkin, H. and Coldren, L.A. (eds) 1999. *Vertical-Cavity Surface-Emitting Lasers* (Cambridge: Cambridge University Press).

71. Bachmann, F., Loosen, P. and Popware, R. 2007. *High Power Diode Lasers: Technology and Applications* (Berlin: Springer-Verlag).

72. Crump, P., Wenzel, H., Erbert, G., Ressel, P., Zorn, M., Bugge, F., Einfeldt, S., Staske, R., Zeimer, U., Pietrzak, A. and Tränkle, G. 2008. Passively cooled TM polarized 808-nm laser bars with 70% power conversion at 80-W and 55-W peak power per 100-μm stripe width. *IEEE Photon. Technol. Lett.*, 20, 1378–1380.

73. Vinokurov, D.A., Kapitonov, V.A., Lyutetskii, A.V., Nikolaev, D.N., Pikhtin, N.A., Rozhkov, A.V., Rudova, N.A., Slipchenko, S.O., Stankevich, A.L., Fetisova, N.V., Khomylev,

M.A., Shamakhov, V.V., Borshchev, K.A. and Tarasov, I.S. 2006. Studying the characteristics of pulse-pumped semiconductor 1060-nm lasers based on asymmetric heterostructures with ultra-thick waveguides. *Tech. Phys. Lett.*, 32(8), 712–715.

74. Crump, P.H., Wenzel, H., Erbert, G. and Trankle, G. 2002. Progress in increasing the maximum achievable output power of broad area diode lasers. *Proc. SPIE*, 8241, 82410U.

75. Crump, P., Frevert, C., Hösler, H., Bugge, F., Knigge, S., Pittroff, W., Erbert, G., Tränkle, G. 2013. Efficient high-power laser diodes. *IEEE J. Sel. Top. Quantum Electron.*, 19(4), 1501211.

76. Schreiber, P., Hoefer, B., Dannberg, P. and Zeitner, U.D. 2005. High-brightness fiber-coupling schemes for diode laser bars. *Proc. SPIE*, 5876, 1–10.

77. Botez Dand Scifres, D.R. 1994. *Diode Laser Arrays* (Cambridge: Cambridge University Press).

78. Carlson, N.W. 1994. *Monolithic Diode-Laser Arrays* (Berlin: Springer-Verlag).

79. Okhotnikov, O. 2010. *Semiconductor Disk Lasers-Physics and Technology* (Weinheim: Wiley VCH).

80. Kazarinov, R.E. and Suris, R.A. 1971. Possibility of the amplification of electromagnetic waves in a semiconductor with a superlattice. *Soviet Phys. Semiconductors*, 5, 707–709.

81. Faist, J., Capasso, F., Sivco, D.L., Sirtori, C., Hutchinson, A.L. and Cho, A.Y. 1994. Quantum cascade laser. *Science*, 264, 553–556.

82. Faist, J. 2013. *Quantum Cascade Lasers* (Oxford: Oxford University Press).

83. Kohler, R., Tredicucci, A., Beltram, E., Beere, H.E., Linfield, E.H., Davies, A.G., Ritchie, D.A., Lotti, R.C. and Rossi, E. 2002. Terahertz semiconductor heterostructure laser. *Nature*, 417, 156–159.

84. Adachi, S. 2005. *Properties of Group-IV, III-V and II-VI Semiconductors* (Chichester: John Wiley).

85. Vurgaftman, I., Meyer, J.R. and Ram-Mohan, L.R. 2001. Band parameters for III–V compound semiconductors and their alloys. *J. Appl. Phys.*, 89, 5815–5875.

86. Liebich, S., Zimprich, M., Beyer, A., Lange, C., Franzbach, D.J., Chatterjee, S., Hossain, N., Sweeney, S.J., Volz, K., Kunert, B., Stolz, W. 2011. Laser operation of Ga(NAsP) lattice-matched to (001) silicon substrate. *Appl. Phys. Lett.*, 99(7), 071109.

87. Sweeney, S.J., Marko, I.P., Jin, S.R., Hild, K., Batool, Z., Ludewig, P., Natterman, L., Bushell, Z., Stolz, W., Volz, K., Broderick, C.A., Usman, M., Harnedy, P.E., O'Reilly, E.P., Butkute, R., Pacebutas, V., Geiutis, A. and Krotkus, A. 2014. Electrically injected GaAsBi quantum well lasers. *Semiconductor Laser Conference (ISLC) Proceedings*, 80–81.

88. Sweeney, S.J. and Jin, S.R. 2013. Bismide-nitride alloys: Promising for efficient light emitting devices in the near- and mid-infrared. *J. Appl. Phys.*, 113, 043110.

89. Prokhorov, A.M. 1965. Quantum electronics: Noble lecture 1964. *Reproduced in Science*, 149, 828.

90. Dirac, P.A.M. 1947. The quantum theory of the emission and absorption of radiation. *Proc. Roy. Soc.*, A114, 243.

91. Crowley, M.T., Marko, I.P., Masse, N.F., Andreev, A.D., Sweeney, S.J., O'Reilly, E.P. and Adams, A.R. 2009. The importance of recombination via excited states in InAs/GaAs 1.3 μm quantum dot lasers. *IEEE J. Sel. Top. Quantum Electron.*, 15, 799–807.

92. Masse, N.F., Sweeney, S.J., Marko, I.P., Adams, A.R., Hatori, N. and Sugawara, M. 2006. Temperature dependence of the gain in p-doped and intrinsic 1.3 μm InAs/GaAs quantum dot lasers. *Appl. Phys. Lett.*, 89, 1191118–1191120.

93. Sweeney, S.J., Phillips, A.F., Adams, A.R., O'Reilly, E.P. and Thijs, P.J.A. 1998. The effect of temperature dependent processes on the performance of 1.5 μm compressively strained InGaAs(P) MQW semiconductor diode lasers. *IEEE Photonics Technol. Lett.*, 10(8), 1076–1078.

94. Fehse, R., Tomić, S., Adams, A.R., Sweeney, S.J., O'Reilly, E.P., Andreev, A. and Riechert, H. 2002. A quantitative study of radiative, Auger and defect related recombination processes in 1.3 μm GaInNAs-based quantum-well lasers. *IEEE Sel. Top. Quantum Electron.*, 8, 801–810.

95. Sweeney, J. 2004. Novel experimental techniques for semiconductor laser characterisation and optimisation. *Phys. Scr.*, T114, 152–158.

96. Hempel, M., Tomm, J.W., Baeumler, M., Konstanzer, H., Mukherjee, J. and Elsaesser, T. 2011. Near-field dynamics of broad area diode laser at very high pump levels. *AIP Adv.*, 1, 042148.

97. Pagano, R., Mukherjee, J., Sajewicz, P. and Corbett, B. 2011. Above threshold estimation of alpha (Henry) parameter in stripe lasers using near- and far-field intensity measurements. *IEEE J. Quantum Electron.*, 47(4), 439–446.

98. Mukherjee, J. and McInerney, J.G. 2007. Electro-thermal analysis of CW high power broad area laser diodes: A comparison between 2D and 3D modeling. *IEEE Sel. Top. Quantum Electron.*, 13(5), 1180–1187.

99. Mukherjee, J., Ziegler, M., LeClech, J., Tomm, J.W., Corbett, B., McInerney, J.G., Reithmaier, J.P., Deubert, S. and Forchel, A. 2009. Bulk temperature mapping of broad area quantum dot lasers: Modeling and micro-thermographic analysis. *Proc. SPIE*, 7230, 72300W.

100. LeClech, J., Ziegler, M., Mukherjee, J., Tomm, J.W., Elsaesser, T., Landesman, J.P., Corbett, B., McInerney, J.G., Reithmaier, J.P., Deubert, S., Forchel, A., Nakwaski, W. and Sarzala, R.P. 2009. Micro-thermography of diode lasers: The impact of light propagation on image formation. *J. Appl. Phys.*, 105, 014502.

101. Knowles, G., Sweeney, S.J., Sale, T.E. and Adams, A.R. 2001. Self-heating effects in red (665 nm) VCSELs. *IEE Proc. Optoelect.*, 148, 256–260.

102. Sale, T.E., Sweeney, S.J., Knowles, G. and Adams, A.R. 2001. Gain-cavity alignment in efficient visible (660 nm) VCSELs studied using high pressure techniques. *Phys. Status Solidi B*, 223, 587–591.

103. Ikyo, A.B., Marko, I.P., Adams, A.R., Sweeney, S.J., Bachmann, A., Kashani-Shirazi, K. and Amann, M.C. 2009. Gain peak-cavity mode alignment optimisation in buried tunnel junction mid-infrared GaSb VCSELs using hydrostatic pressure. *IET Optoelectron.* 3, 305–309.

104. Hempel, M., Tomm, J.W., Ziegler, M., Elsaesser, T., Michel, N. and Krakowski, M. 2010. Catastrophic optical damage at front and rear facets of diode lasers. *Appl. Phys. Lett.* 97, 231101 and references therein.

105. Sweeney, S.J., Lyons, L.J., Adams, A.R. and Lock, D.A. Direct measurement of facet temperature up to melting point and COD in high power 980 nm semiconductor diode lasers. *IEEE Sel. Top. Quantum Electron.*, 9, 1325–1332.

106. O'Brien, K., Sweeney, S.J., Adams, A.R., Murdin, B.N., Salhi, A., Rouillard, Y. and Joullie, A. 2006. Recombination processes in mid-infrared InGaAsSb diode lasers emitting at 2.37 μm. *Appl. Phys. Lett.*, 89, 051104–051106.

107. Ikyo, B.A., Marko, I.P., Adams, A.R., Sweeney, S.J., Canedy, C.L., Vurgaftman, I., Kim, C.S., Kim, M., Bewley, W.W. and Meyer, J.R. 2011. Temperature dependence of 4.1 μm mid-infrared type II "W" interband cascade lasers. *Appl. Phys. Lett.*, 99, 021102.

108. Volz, K., Beyer, A., Witte, W., Ohlmann, J., Németh, I., Kunert, B. and Stolz, W. 2011. GaP-nucleation on exact Si (001) substrates for III/V device integration. *J. Cryst. Growth*, 315(1), 37–47.

109. Liu, H., Wang, T., Jiang, Q., Hogg, R., Tutu, F., Pozzi, F. and Seeds, A. 2011. Long-wavelength InAs/GaAs quantum-dot laser diode monolithically grown on Ge substrate. *Nat. Photon.*, 5, 416–419.

110. Srinivasan, S., Tang, Y., Read, G., Hossain, N., Liang, D., Sweeney, S.J. and Bowers, J.E. 2013. Hybrid silicon devices for energy efficient transmitters. *IEEE Microelectron.*, 33, 22–31.

111. Sweeney, S.J. 2010. Light-emitting semiconductor device. Patent WO 2010/149978.

112. Jin, S. and Sweeney, S.J. 2013. InGaAsBi alloys for efficient mid-IR devices on InP. *J. Appl. Phys.*, 114, 21303–21307.

113. Ludewig, P., Knaub, N., Hossain, N., Reinhard, S., Nattermann, L., Marko, I. P., Jin, S.R., Hild, K., Chatterjee, S., Stolz, W., Sweeney S.J. and Volz, K. 2013. Electrical injection Ga(AsBi)/(AlGa)As single quantum well laser. *Appl. Phys. Lett.* 102, 242115–242117.

114. Marko, I.P., Ludewig, P., Bushell, Z.L., Jin, S.R., Hild, K., Batool, Z., Reinhard, S., Nattermann, L., Stolz, W., Volz, K. and Sweeney, S.J. 2014. Physical properties of GaBiAs/(Al)GaAs based near-infrared laser diodes grown by MOVPE with up to 4.4% Bi. *J. Phys. D*, 47, 345103.

115. Gao, J. 2011. *Optoelectronic Integrated Circuit Design* (Singapore: John Wiley).

116. Mathine, D.L., Droopad, R. and Maracas, G.N. 1997. A vertical-cavity surface-emitting laser appliqued to a 0.8-μm NMOS driver. *IEEE Photonics Technol. Lett.*, 9(7), 869.

117. Wilcox, J.Z., Ou, S., Yang, J.J., Jansen, M. and Peterson, G.L. 1989. Dependence of external differential efficiency on laser length and reflectivities in multiple quantum well lasers. *Appl. Phys. Lett.*, 55, 825–827.

118. Hild, K., Sale, T.E., Hosea, T.J.C., Hirotani, M., Mizuno, Y. and Kato, T. 2001. Spectral and thermal properties of red AlGaInP RCLEDs for polymer optical fibre applications. *IEE Proc. Optoelectron.*, 148, 220.

119. Paul, D.J. 2010. The progress towards terahertz quantum cascade lasers on silicon substrates. *Laser Photon. Rev.*, 4(5), 610–632.

120. Liang, D. and Bowers, J.E. 2010. Recent progress in lasers on silicon. *Nat. Photon.*, 4, 511–517.

121. Read, G., Marko, I.P., Hossain, N. and Sweeney, S.J. 2015. Physical properties and characteristics of III–V lasers on silicon. *J. Sel. Top. Quantum Electron.*, 21(6), 1502208. (in press).

FURTHER READING

Agrawal, G.P. and Dutta. N.K. 1993. *Semiconductor Lasers* (New York: Van Nostrand Reinhold).

Carroll, J.E., Whiteaway, J.E.A. and Plumb, R.G.S. 1998. *Distributed Feedback Semiconductor Lasers* (Stevenage: IEE).

Morthier, G. and Wankwikelberge, P. 1997. *Handbook of Distributed Feedback Laser Diodes* (Noorwood, MA: Artech House).

Petermann, K. 1988. *Laser Diode Modulation and Noise* (Dordrecht: Kluwer Academic).

Yariv, A. and Yeh, P. 2007. *Photonics: Optical Electronics in Modern Communications*, 6th ed. (New York: Oxford University Press).

Zory, P.S. ed. 1993. *Quantum Well Lasers* (Boston, MA: Academic Press).

Optical detectors and receivers

HIDEHIRO KUME
Opto-Mechatronix, Inc.

12.1 INTRODUCTION

This chapter describes the operating principles, construction, and characteristics of major optical detectors.

Optical detectors (often called photodetectors) can be classified by their principle of light detection as shown in Table 12.1. Typical optical detectors utilizing physical or chemical changes are photographic films, but these are now seldom used in photometric applications. Methods utilizing solid or gas ionization and scintillation are usually limited to detection in the X-ray and gamma-ray regions.

Optical detectors making use of photoelectric effects are widely used as UV (ultraviolet) to IR (infrared) sensors in various applications including measurement instruments, industrial production equipment, and communication devices. To make the contents of this chapter more practical, we will chiefly discuss optical detectors utilizing photoelectric effects.

Optical detectors utilizing photoelectric effects can be further divided by their detection principle into two groups: one using external photoelectric effects and the other using internal photoelectric effects. Table 12.2 shows typical optical detectors utilizing these effects and their features. As can be seen from Table 12.2, optical detectors also fall under two categories: point detectors that merely detect light intensity and two-dimensional detectors including position sensors and image sensors. Spectral response range (detectable wavelength band) also differs depending on the type of optical detector. In general, optical detectors utilizing the external photoelectric effect are represented by

photomultiplier tubes (PMTs) and exhibit fast time response and high sensitivity. On the other hand, optical detectors using the internal photoelectric effect, such as photodiodes (PDs) and photoconductive cells, offer a wide spectral response range, compact size, and easy operation. Along with recent trends toward image measurements, the product quality and quantity of two-dimensional detectors are improving and increasing.

12.2 PHOTOELECTRIC EFFECTS

As stated earlier, photoelectric conversion is roughly divided into external photoelectric effects [1] by which bound electrons inside a semiconductor thin film are released into a vacuum when light strikes the semiconductor and internal photoelectric effects [2], where photoelectrons are generated inside a semiconductor by light and excited into the conduction band. The photocathode used as the photoemissive surface of a PMT has the former function. The photoconductive effect and photovoltaic effect take place by the latter principle.

12.2.1 External photoelectric effect

Semiconductor thin films having a photoemissive surface are usually called "photocathodes" [1,3]. They have a band model structure like that shown in Figure 12.1. Inside a semiconductor, there exists a valence band occupied by electrons, a forbidden band that cannot be occupied by electrons and a conduction band where electrons are free to move. When photons strike a photocathode, electrons in the valence band absorb photon energy $h\nu$, become

Table 12.1 Classification of optical detectors

Principle	Method	Detectable electromagnetic radiation
Physical or chemical changes	Dosimeters	Charged particles
	Photographic films	UV, visible, IR radiation, charged particles
	Cloud chambers, bubble chambers	X-rays, gamma-rays, charged particles
Solid or gas ionization	Proportional counters, etc.	Charged particles, X-rays, gamma-rays
Scintillation	Solid-state or liquid scintillators+optical detectors	Charged particles, X-rays, gamma-rays
Photoelectric effect	Optical detectors utilizing external photoelectric effect (PMT, etc.)	UV, visible, near-IR radiation
	Optical detectors utilizing internal photoelectric effect (PDs, etc.)	UV, visible, near-IR radiation, X-rays, gamma-rays

Table 12.2 Types and classification of optical detectors utilizing photoelectric and thermal effects

Detecting principle		Detectors (point detectors)	Two-dimensional detectors	Spectral response range
External photoelectric effect		Phototubes	IIs	Vacuum UV to near-IR
		PMTs	Streak cameras	
Internal photoelectric effect	Photoconductive effect	Photoconductive cells	Photoconductive type camera tubes (vidicons, etc.)	Visible to IR
	Photovoltaic effect	PDs	Semiconductor image sensors (CCD, etc.)	UV to near-IR
		Phototransistors Avalanche PDs	Semiconductor position sensors (PSD, etc.)	
Thermal effect	Pyroelectric effect	Pyroelectric IR detectors	Pyroelectric image sensors	Near-IR to far-IR
	Photovoltaic type	Thermocouples		
	Conductivity type	Bolometers Thermistors		

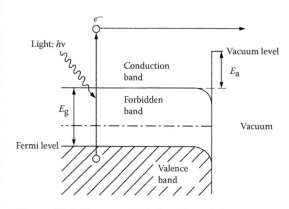

Figure 12.1 Photocathode band model.

excited into the conduction band and diffuse toward the photocathode surface. When the diffused electrons have enough energy to overcome the vacuum level barrier, they are emitted into the vacuum as photoelectrons. Photocathodes can be classified by the photoelectron emission process into reflection mode and transmission mode photocathodes. The reflection mode photocathode is usually formed on a metal plate and photoelectrons are emitted from the photocathode in the opposite direction to the incident light. The transmission mode photocathode is usually deposited as a thin film on an optically transparent flat plate and photoelectrons are emitted in the same direction as that of the incident light. Most

photocathodes are made of a compound semiconductor chiefly consisting of alkali metals with a low work function such as Cs_3Sb and Na_2KSb. Changing the photocathode materials, it is possible to achieve sensitivity in various bands of the spectrum from soft X-rays to near-IR radiation (described in Section 12.3 in detail).

PMTs [3] are typical optical detectors utilizing the external photoelectric effect. They are vacuum tubes with a glass envelope and consist of a photoemissive cathode (photocathode), an electron multiplier, and an electron collector (anode) in a vacuum tube. Figure 12.2 shows the schematic construction of a PMT. Light, which enters a PMT, is detected and produces an output signal through the following processes.

1. Light passes through the input window and enters the photocathode in a vacuum.
2. It excites the electrons in the photocathode so that photoelectrons are emitted into the vacuum (external photoelectric effect).
3. Photoelectrons are accelerated and focused by the focusing electrode onto the first electrode called a "dynode" in the electron multiplier section, where they are multiplied by means of secondary electron emission. This secondary emission is repeated at each of the successive dynodes. (A dynode is an electrode capable

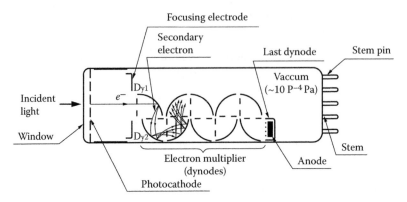

Figure 12.2 Construction of a PMT.

of emitting secondary electrons. An electron multiplier used in a PMT usually has about 10 stages of dynodes.)

4. A bunch of multiplied secondary electrons emitted from the last dynode are ultimately collected by the anode and output to an external circuit as an electrical signal. Since the PMT is a kind of electron tube, it is relatively large in size, but superior in sensitivity and response speed, making PMTs useful as optical detectors in a variety of applications such as analytical instruments, medical equipment, and industrial measurement systems.

12.2.2 Photovoltaic effect

Internal photoelectric effects will now be explained using a PD as an example.

A cross section of a typical PD is shown in Figure 12.3a. The p-layer at the light sensitive surface and the n-layer at the substrate form a p–n junction that serves as a photoelectric converter. The usual p-layer for a silicon PD is formed by selective diffusion of boron to a thickness of approximately 1 μm. The neutral region at the junction between the p-layer and the n-layer is known as the depletion layer. By varying and controlling the thickness of the outer p-layer, substrate n-layer and bottom N^+-layer as well as the doping concentration, the PD's spectral response and frequency response can be controlled.

A band model for photoelectric conversion that occurs at the p–n junction of a PD is shown in Figure 12.3b. When light strikes a PD and its energy is greater than the bandgap energy E_g, the electrons in the valence band are excited and pulled up into the conduction band, leaving holes

in their place in the valence band. These electron–hole pairs are generated throughout the p-layer, depletion layer, and n-layer. In the p-layer and depletion layer, the electric field accelerates the electrons toward the n-layer and the holes toward the p-layer. Of the electron–hole pairs generated in the n-layer, the electrons are left in the n-layer conduction band along with electrons that have arrived from the p-layer, while the holes diffuse through the n-layer up to the p–n junction and collect in the p-layer valence band while being accelerated. In this manner, electron–hole pairs generated in proportion to the amount of incident light are accumulated in the n-layer and p-layer, resulting in a positive charge in the p-layer and a negative charge in the n-layer. When an external circuit is connected between the p-layer and n-layer, electrons will flow from the n-layer and holes from the p-layer toward the opposite electrode, respectively.

12.2.3 Photoconductive effect

When light strikes some kinds of semiconductor, electron–hole pairs are generated and their internal electric conductivity increases. This phenomenon is called the "photoconductive effect." Photoconductive detectors are divided into intrinsic detectors and extrinsic detectors doped with impurities.

Figure 12.4 shows the operation models of photoconductive detectors. A phenomenon called "intrinsic photoconduction" is presented in Figure 12.4a. When photons $h\nu$ with energy greater than the energy bandgap E_g in the forbidden band enter an intrinsic detector, electron–hole pairs (carriers) are generated and the number of conductive charges in

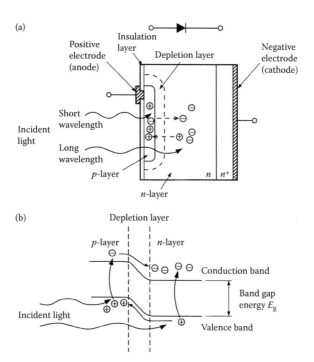

Figure 12.3 (a) PD cross section and (b) PD p–n junction state.

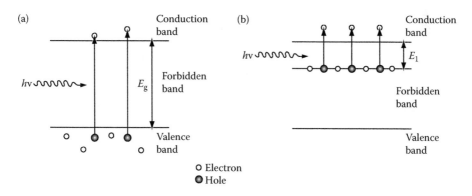

Figure 12.4 Operation models of photoconduction. (a) Intrinsic photoconduction and (b) impurity photoconduction.

the conduction band changes. Figure 12.4b shows the operation of an extrinsic detector doped with impurity atoms. In this extrinsic detector, an impurity level is formed at a relatively deep energy level in an n-type semiconductor. When the input photons have energy higher than the ionization energy E_1, they mainly act on the impurities and create free electrons and bound holes. The free electrons and free holes contribute to changes in the electric conductivity. This phenomenon is called "extrinsic photoconduction." Since $E_g \gg E_1$, extrinsic detectors

are used for IR detection at longer wavelengths when compared to intrinsic detectors.

Figure 12.5 shows how a photoconductive sensor is used to detect light. When light enters the photoconductive sensor, its internal resistance changes to produce an increase in electric current, ΔI, which is added to the constant bias current I_b. This change in the electric current is detected as a signal. Various types of material are used to fabricate photoconductive sensors, such as CdS for visible light detection and PbS, PbSe, and InSb for IR detection.

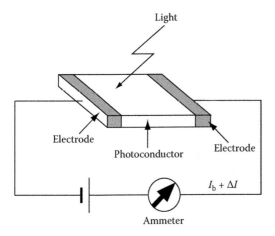

Figure 12.5 Photoconductive sensor operation.

Although photoconductive sensors have the disadvantages of slow response speeds, they are still widely used because of their useful features such as small size, light weight, and wide spectral response range (from the visible to the far IR region).

12.3 SPECTRAL RESPONSE CHARACTERISTICS OF PHOTODETECTORS

It is essential to select an appropriate type of photodetector that matches the wavelength region of the light to be measured. This section explains photodetectors used to detect wavelengths in the X-ray to IR region.

12.3.1 Definition of photodetector sensitivity

A wavelength spectrum of light from the X-ray to IR ray region is shown in Figure 12.6 along with the name of each spectral region and unit systems, which will be used throughout this chapter.

In this section, we will define sensitivity used to evaluate photodetectors. If light at a certain intensity level (W) enters a photodetector and a certain amount of photocurrent (A) flows after photoelectric conversion, then the following terms are usually used to define the photoelectric sensitivity of the photodetector.

12.3.1.1 RADIANT SENSITIVITY OR PHOTOSENSITIVITY S

Radiant sensitivity or photosensitivity S is the photocurrent A divided by the incident light level W. It is represented as

$$S = \frac{A}{W}(AW^{-1}). \qquad (12.1)$$

12.3.1.2 QUANTUM EFFICIENCY

Quantum efficiency (QE) is the number of electrons or holes extracted. It is photocurrent divided by the number of incident photons, and generally expressed in percent (%). QE and radiant sensitivity S (A W^{-1}) have the following relationship at a given wavelength λ (nm):

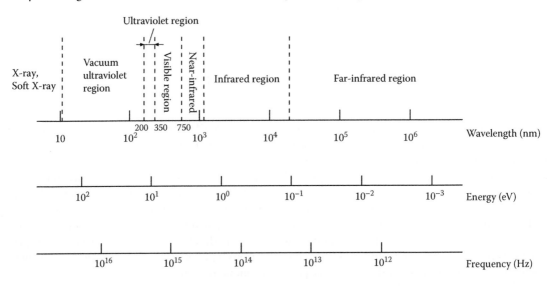

Figure 12.6 Wavelength spectrum of light and unit conversion scale.

$$QE = \frac{S \times 1240}{\lambda} \times 100 \ (\%). \qquad (12.2)$$

12.3.1.3 NOISE EQUIVALENT POWER

This is the amount of light equivalent to the noise level of the photodetector. In other words, it is the light level required to obtain a signal-to-noise (S/N) ratio of 1. Noise equivalent power (NEP) is usually defined as shown in Equation 12.3, using the wavelength λ_p of maximum sensitivity and a bandwidth of 1 Hz:

$$NEP = \frac{\text{Noise current(A Hz}^{-1/2})}{\text{Photosensitivity at } \lambda_p (\text{A W}^{-1})} (\text{W Hz}^{-1/2}).$$
$$(12.3)$$

12.3.1.4 D* VALUE

This indicates the value of the S/N ratio of a detector when radiant energy of 1W enters the detector. D^* is normalized by a sensitive area of 1 cm^2 and noise bandwidth of 1 Hz so that detector materials can be compared regardless of the size and shape of the detector element. An optical chopper is usually used to measure D^* for passing and interrupting a beam of incident light. D^* is normally expressed in the format $D^*(A, B, C)$, where A is the color temperature (K) of the light source, B is the chopping frequency (Hz), and C is the noise bandwidth (Hz). D^* is therefore represented in units of cm Hz$^{1/2}$. The higher the D^* value, the better the detector. D^* is given by

$$D^* = \frac{(S/N)\Delta f^{1/2}}{PA^{1/2}}, \qquad (12.4)$$

where S is the signal, N is the noise, P is the incident light energy (W cm^{-2}), A is the sensitive area (cm^2), and Δf is the noise bandwidth (Hz). The following relation is established by D^* and NEP:

$$D^* = \frac{A^{1/2}}{NEP}. \qquad (12.5)$$

Radiant sensitivity and QE are usually used to define photodetector sensitivity for UV to visible photodetectors, while NEP and D^* are frequently used for IR detectors.

12.3.2 X-ray to vacuum UV photodetectors

Light tends to behave as particles more actively in the X-ray to vacuum UV (VUV) region. Absorption of light by any substance also changes a great deal in some wavelength bands of this region. Because of this, in this region, it is important to consider the stopping power (efficiency) versus light and the performance characteristics of window materials as well as photoelectric conversion sensitivity. Typical transmittance characteristics of window materials [5] used in the VUV, soft X-ray, and X-ray regions are shown in Figure 12.7. As can be seen from this figure, organic films and light metals such as Be (beryllium) and Al (aluminum) can be used as a window in the X-ray region at several kiloelectron volts or less, although there are currently no window materials available for some bands of the X-ray region. Figure 12.8 shows the typical spectral transmittance of optical windows used in the UV to VUV region. Quartz glass and UV-transmitting glass are preferably used as

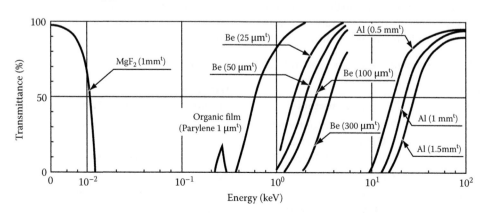

Figure 12.7 Transmittance of window materials used in VUV, soft X-ray, and X-ray regions.

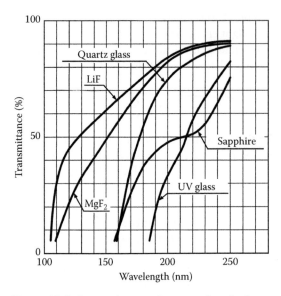

Figure 12.8 Spectral transmittance of optical windows used in the VUV to UV region.

transparent window materials in the UV region. MgF$_2$ (magnesium fluoride) and LiF (lithium fluoride) crystals are chiefly used as windows for detectors in the VUV region where significant absorption of light by oxygen and nitrogen occurs. Table 12.3 shows a list of typical photodetectors [6] used in the X-ray to VUV region.

12.3.2.1 PHOTODETECTORS UTILIZING GAS IONIZATION

Proportional counter tubes and Geiger–Mueller (GM) tubes using gas ionization have been employed as photodetectors for quite a long time. The operating principles of these tubes are slightly different from each other. Therefore, proportional counter tubes are capable of detecting both radiation energy and dose rate, while GM tubes are used only to measure dose rate. Due to the transmission bands of the window material (Be window) and fill-gas absorption characteristics, proportional counter tube applications are usually limited to a radiation energy region of 1–25 keV. Likewise, applications of GM tubes are determined by the transmission bands of the window material (Fe–Cr alloy) and by fill-gas absorption characteristics. GM tubes are therefore usable at a radiation energy of 50 keV–1 MeV. The multiwire proportional counter (MWPC) is a position-sensitive proportional counter tube and is used as a two-dimensional detector in X-ray scattering, diffraction, and synchrotron radiation experiments.

12.3.2.2 SCINTILLATION PHOTODETECTORS

In scintillation counting [7], a scintillator is used to convert radiation into visible light and this light is detected with a PMT or PD. Figure 12.9 shows a scintillation photodetector consisting of a scintillator coupled to a PMT. The amount of light emitted from a scintillator is proportional to the energy level of incident radiation, and this light emission is measured with a PMT or PD. Table 12.4 lists typical characteristics and applications of major scintillators that are made of inorganic or organic materials. Scintillation counting is excellent in energy resolution when measuring radiation so it is used in nuclear medical equipment, nuclear physics experiments, and oil well logging.

An X-ray image intensifier (II) [8] is a kind of scintillation photodetector. It is a two-dimensional X-ray detector having a phosphor-coated window for X-ray to visible light conversion and a photocathode sensitive to visible light. X-ray IIs can be fabricated with a large sensitive area, making them useful in X-ray medical equipment and industrial nondestructive inspection.

Table 12.3 Typical photodetectors used in X-ray to VUV regions

Detecting principle	Detectors	Detectable region	Energy (wavelength) region
Gas ionization	Proportional counter tube, GM tubes, MWPC	X-rays, γ-rays	A few keV to 1 MeV
Scintillation	Scintillator + PMT (or PD)	X-rays, γ-rays	A few keV to several dozen MeV
	X-ray IIs	X-rays	20–150 MeV
Photoelectric effect	Semiconductor detectors	X-rays, γ-rays	A few keV to several dozen MeV
	PMT without window	Soft X-rays	A few to 100 nm
	PMT with window	VUV	100 nm or more

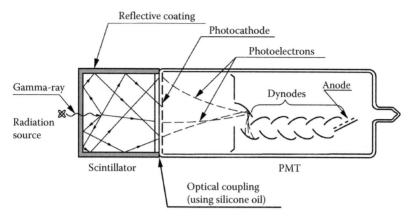

Figure 12.9 Scintillation photodetector for radiation measurement.

12.3.2.3 PHOTOCONDUCTIVE DETECTORS

Among the photoconductive detectors, semiconductor radiation detectors [9] are classified by the manufacturing method into the following groups:

1. p–n junction type
2. Surface barrier type
3. Lithium (Si(Li)) drift type
4. High-purity germanium type

Each type of these detectors makes use of the depletion layer that is formed when a reverse bias voltage is applied to the junction of p- and n-type semiconductors. Different photosensitive materials, size, and window materials are used depending on the wavelength band (energy) to be detected.

Si(Li) drift type and high-purity germanium type detectors are mainly used in the soft X-ray to X-ray region. Although the actual wavelengths that can be detected differ according to the

Table 12.4 Typical characteristics and applications of scintillators

	Density (g cm⁻³)	Relative emission intensity (NaI(Tl)=100)	Emission time (ns)	Emission peak wavelength (nm)	Applications
NaI(Tl)	3.67	100	230	410	Survey meter, area monitor, gamma camera, SPECT
BGO	7.13	15	300	480	PET, X-ray CT
CsI(Tl)	4.51	45–50	1000	530	Survey meter, area monitor, X-ray CT
Pure CsI	4.51	<10	10	310	High energy physics experiments
BaF₂	4.88	20	0.9/630	220/325	TOF, PET
GSO:Ce	6.71	20	30	310/430	Area monitor, oil well logging
Plastic	1.03	25	2	400	Area monitor, neutron detector
LSO	7.35	70	40	420	PET
PWO	8.28	0.7	15	470	High energy physics experiments
YAP	5.55	40	30	380	Survey meter, compact gamma camera

detector material and structure, these can usually detect X-ray energy ranging from 1 keV to 10 MeV. These detectors are not easy to handle because liquid nitrogen cooling is required, but they offer the advantage of high-energy resolution. Currently, these detectors are available with a sensitive area from 2 or 3 up to 10 cm in diameter, and their detection sensitivity is constantly being improved. Multielement detector arrays are also being developed.

Photocathodes (cathodes made of photoemissive materials) having external photoelectric effects are also used for detection in the UV to X-ray region. In a wavelength region where optically transparent windows are available, semi-transparent (transmission mode) photocathodes can be used, but at even shorter wavelength regions, where proper window materials are unavailable, reflection mode photocathodes are exclusively used without windows. Typical photoemissive materials used to detect wavelengths shorter than the UV region are alkali halide metals, pure metals and metal oxide. Figure 12.10 shows the typical spectral response characteristics of photoemissive materials that can be used in the soft X-ray region. These materials can be applied to an X-ray to electron conversion surface such as the first stage of electron multipliers or to a reflection mode photocathode by vacuum deposition onto the input edge surface of microchannel plates (MCPs) (described in Section 12.6.1).

12.3.3 UV detectors

Photodetectors used to measure radiant energy of UV rays can be divided into photovoltaic and photoemissive types.

PDs using SiC, GaN, AlGaN, and diamond thin films with a large bandgap are now being developed as photovoltaic type semiconductor UV sensors [10]. In actual applications, UV sensors using a general-purpose PD combined with a UV filter are frequently used. Figure 12.11 shows a typical spectral response curve of a UV sensor that can be obtained by the combination of a UV–visible range sensitive GaAsP PD and a UV-transmitting filter. Major applications of this type of UV sensor are UV measurement of sunlight, mercury lamp monitoring, etc.

Phototubes are also used in UV detection. They are made up of a photocathode using the external photoelectric effect and an anode for collecting photoelectrons emitted from the photocathode (see Figure 12.12). Phototubes can be fabricated with various types of spectral response characteristics by changing the combination of photoemissive materials [11]. Figure 12.13 shows the typical spectral response characteristics of photocathodes specifically selected and processed to have sensitivity only in the UV range. The spectral response characteristics on the short-wavelength side are determined by absorption characteristics of window materials. Metals and alkali halide compounds having a large bandgap

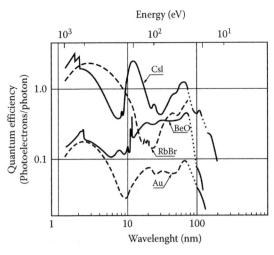

Figure 12.10 Typical spectral response characteristics of photoemissive materials in the soft X-ray region.

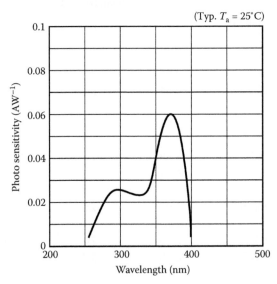

Figure 12.11 Typical spectral response curve of a semiconductor UV sensor.

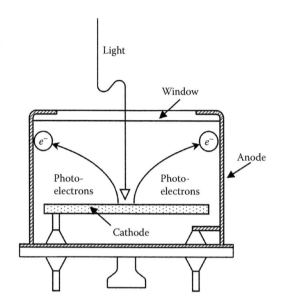

Figure 12.12 Construction of a phototube.

response (nanosecond rise time) and low noise due to extremely high output impedance. Since UV phototubes are sensitive only in the UV region (insensitive to visible and longer wavelengths and often called "solar blind"), they have found applications in colorimeters, pollution monitors, densitometers, and UV laser detection.

12.3.4 UV to visible region photodetectors

Light is most actively utilized in this wavelength region. Compared to other regions of the electromagnetic spectrum, there are numerous types of photodetectors available in the UV to visible region, in terms of both quantity and quality. Phototubes and PMTs utilizing the external photoelectric effect and PDs having the internal photovoltaic effect are most frequently used in this region.

12.3.4.1 PHOTODETECTORS UTILIZING EXTERNAL PHOTOELECTRIC EFFECT (PHOTOTUBES, PMTs)

Phototubes and PMTs are widely used in the UV to visible region. As stated earlier, these detectors use a semiconductor thin film that is a photoelectric

are used as UV sensitive photocathode materials. Sensitivity in the longer wavelength region is determined by the bandgap of each material. In addition to having various spectral response characteristics selectable by the combination of a window and photocathode materials, phototubes offer outstanding features such as high-speed

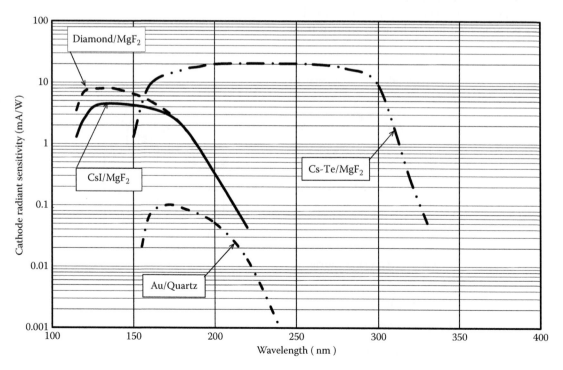

Figure 12.13 Typical spectral response characteristics of UV sensitive photocathodes.

converter surface called a "photocathode." In this section, we will briefly discuss the basic operating principle and operating method of a PMT. As shown in Figure 12.14, a high voltage is applied between the photocathode and the anode. Photoelectrons emitted from the photocathode are accelerated and focused onto the first dynode (Dy_1) to produce secondary electrons, which are then accelerated toward the subsequent dynodes (Dy_2 to Dy_n). When the accelerated electrons strike a dynode, secondary electrons are emitted with a secondary emission ratio (δ). Since this process is repeated up to the last dynode (Dy_n), the electrons are multiplied and increase. To make the operating circuit simpler, voltage dividing resistors are usually placed between the photocathode and the anode to distribute the supply voltage to each dynode, as shown in Figure 12.14.

The secondary electron emission ratio δ is a function of the interstage voltage E between each dynode and is given by

$$\delta_1 = \alpha E^k, \qquad (12.6)$$

where α is a constant, and k is a coefficient determined by the material and structure of the electrode (dynode). δ has a value 0.7–0.8.

The photoelectron current I_k (photocurrent that flows per unit light flux in lumens) emitted from the photocathode strikes the first dynode where a secondary electron current (I_{d1}) is emitted. At this point, the secondary electron emission ratio 8 of the first dynode is given by

$$\delta_1 = \frac{I_{d1}}{I_k}. \qquad (12.7)$$

These electrons are multiplied in a cascade process from the first dynode → second dynode →···→ the nth dynode. The secondary emission ratio δ_n of the nth stage can be calculated as

$$\delta_n = \frac{I_{dn}}{I_{d(n-1)}}. \qquad (12.8)$$

The anode current I_p is given by

$$I_p = I_k \delta_1 \delta_2 \delta_n, \qquad (12.9)$$

so that

$$\frac{I_p}{I_k} = \delta_1 \cdot \delta_2 \cdots \delta_n. \qquad (12.10)$$

The product of $\delta_1, \delta_2, ..., \delta_n$ is called the gain (μ) and is given by

$$\mu = \delta_1 \cdot \delta_2 \cdot \delta_n. \qquad (12.11)$$

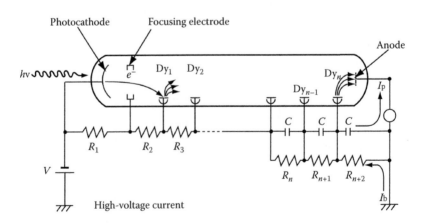

I_b: Bleeder current $= \dfrac{V}{R_1 + R_2 + ... + R_{n+2}}$

I_p: Anode current $=$
V: Overall supply voltage
E: Interstage voltage ($= V/n$)

Figure 12.14 PMT operation circuit.

If the number of dynodes in a PMT, which is operated using an equal-division voltage divider, is n, then changes in the gain μ versus the supply voltage V can be obtained by

$$\mu = (\alpha E^k)^n = \alpha^n \left(\frac{V}{n+1} \right)^{kn} = A V^{kn}, \quad (12.12)$$

where A should be equal to $\alpha^n/(n+1)^{kn}$. From this equation, it is clear that the gain μ is proportional to the kn exponential power of the supply voltage. Typical gain versus supply voltage is shown in Figure 12.15 along with luminous sensitivity. Since this figure is expressed in a logarithmic scale for both horizontal and vertical axes, the slope of the straight line becomes kn. The gain increases with an increase in the supply voltage, so a high gain of up to 10^7 or more can be obtained in most cases. Figure 12.16 shows various types of dynode structures. The circular-cage dynode is mainly used in side-on PMTs and has high gain and fast time response. The box-and-grid dynode has high gain, and the linear-focused dynode offers fast time response. The Venetian-blind dynode features high gain and compactness. Recently, the mesh-type dynode and MCP have been put to practical use as an electron multiplier of PMTs. These dynode types each have merits and demerits in terms of time response, dynamic range, and gain, so they should be carefully selected according to the application.

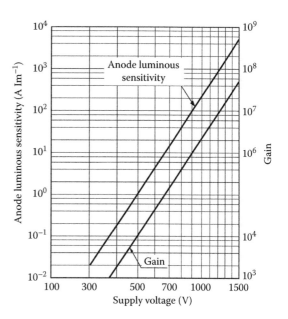

Figure 12.15 Luminous sensitivity and gain versus supply voltage.

A PMT is usually operated with a supply voltage of 1000–2000 V in order to obtain a high gain of 10^6–10^7, and can be used in photon counting mode (explained in Section 12.4.2). PMTs have excellent capability for low-light-level detection and also deliver extremely fast response with rise times of nanoseconds (PMTs using normal dynode types) to subnanoseconds (PMTs using MCP).

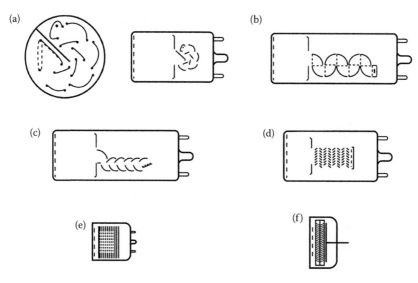

Figure 12.16 Types of dynodes. (a) Circular-cage type, (b) box-and-grid type, (c) linear-focus type, (d) Venetian-blind type, (e) fine-mesh type, and (f) MCP.

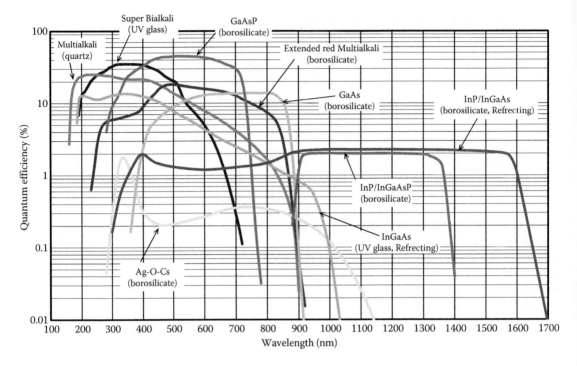

Figure 12.17 Typical photocathode spectral response characteristics.

PMT spectral response is determined by the photocathode material on the long-wavelength side and the transmittance of the window material on the short-wavelength side. Typical photocathode spectral response characteristics are shown in Figure 12.17. Up until now, photocathode sensitivity has been limited to the UV to near-IR region. However, recent developments in semiconductor crystal materials [12] have extended photocathode

sensitivity up to the IR region. Figure 12.18 shows a photograph of typical PMT products.

12.3.4.2 PHOTODETECTORS UTILIZING INTERNAL PHOTOVOLTAIC EFFECT (E.G., PDS)

Various types of semiconductor sensors have also been developed and used for light detection in the UV to visible region. Among these, PDs [4] are most

Figure 12.18 Typical PMT products.

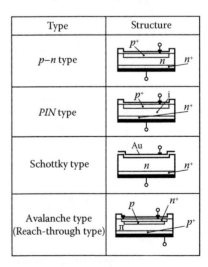

Figure 12.19 Types of PDs.

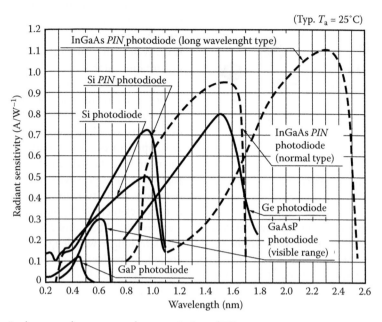

Figure 12.20 Typical spectral response characteristics of PDs.

extensively used. PDs can be classified according to their manufacturing method and structure as shown in Figure 12.19. In the p–n type, light falling on a PD is absorbed in the depletion layer near the internal p–n junction after passing through the p^+-layer on the PD surface and is converted into an electrical signal of electron–hole pairs. The p–n type PDs have a planar structure that allows a relatively large active area with better sensitivity not only in the visible to IR region, but also in the UV region. However, these have a larger junction capacitance so response speed is limited. The PIN type PDs have a high resistance intrinsic (i) layer formed between the p- and n-layers. This i-layer reduces the junction capacitance so response time is improved while maintaining high sensitivity. This type of PD exhibits an even faster response time when used with a reverse voltage and so is designed with low leak current.

Schottky-type PDs have a structure in which a thin gold (Au) coating is sputtered onto the surface of an n-type semiconductor to form a Schottky effect p–n junction. Since the distance from the outer surface to the junction is small, UV sensitivity is high. Avalanche photodiodes (APDs) are used with a reverse voltage applied to the p–n junction so as to form a high electric field within the depletion layer. When light enters in this state,

the generated electrons are accelerated by the electric field and collide with atoms to produce secondary electrons. The process occurs repeatedly so signals are amplified. This phenomenon is known as the avalanche effect and is ideal for detecting low-level light.

Typical spectral response characteristics of PDs are shown in Figure 12.20. Various types of PDs are currently available with sensitivity in the UV, visible, and also near-IR regions. Recently, InGaAs PDs have become widely used as promising receivers in IR optical communications. Semiconductor photodetectors are small and compact yet have high sensitivity, so they are now used in large numbers in general electronics products and also other diverse applications. Figure 12.21 shows a photograph of typical PD products.

12.3.5 IR detectors

IR radiation covers the electromagnetic spectrum at wavelengths from 0.8 μm to 1 mm. The wavelength region of 0.8–3 μm is called the near-IR region, the wavelength region of 3–15 μm, the mid-IR, and the wavelength region of 15 μm–1 mm, the far-IR.

Table 12.5 shows the types of IR detectors and their characteristics [14]. In the IR region,

Figure 12.21 Typical PD products.

thermal-type and quantum-type (photovoltaic and photoconductive) detectors are commonly used. Thermal-type detectors make use of IR energy as heat, their responsivity is independent of wavelength and cooling is not required. However, their response speed is slow and detection sensitivity

is not so high. Quantum-type detectors, on the other hand, have higher responsivity and faster time response, while their responsivity depends on wavelength. They often have to be cooled to ensure more stable operation.

Figure 12.22 shows the typical spectral response characteristics of IR detectors. As can be seen from this figure, IR detectors using various kinds of photoelectric material have been developed to cover a broad spectral range from the near-IR to far-IR radiation. It should be noted that spectral responsivity of InGaAs, Ge, and InSb detectors shifts to shorter wavelengths when the detector element is cooled, while spectral responsivity of PbS, PbSe, and MCT (HdCdTe) detectors shifts to longer wavelengths when the element is cooled. Pyroelectric detectors are thermal-type IR detectors made of pyroelectric materials such as $LiTaO_3$, TGS, and PZT. Unlike quantum-type detectors, pyroelectric detectors operate at

Table 12.5 Types of IR detectors and their characteristics

	Type	Detector	Spectral response (μm)	Operating temperature (K)
Thermal type	Thermocouple, Thermopile	Golay cell,	Depends on	300
	Bolometer	condenser-	window	300
	Pneumatic cell	microphone	material	300
	Pyroelectric detector	PZT, TGS, $LiTaO_3$		300
Quantum type	Intrinsic type Photoconductive type	PbS	1–3.6	300
		PbSe	1.5–5.8	300
		InSb	2–6	213
		HgCdTe	2–16	77
	Photovoltaic type	Ge	0.8–1.8	300
		InGaAs	0.7–1.7	300
		Ex InGaAs	1.2–2.55	253
		InAs	1–3.1	77
		InSb	1–5.5	77
		HgCdTe	2–16	77
		Ge:Au	1–10	77
		Ge:Hg	2–14	4.2
		Ge:Cu	2–30	4.2
		Ge:Zn	2–40	4.2
		Si:Ga	1–17	4.2
		Si:As	1–23	4.2

Source: Hamamatsu Photonics; Technical information SD-12/Characteristics and use of IR detectors, Cat. No. KIRD9001E04, 2011.

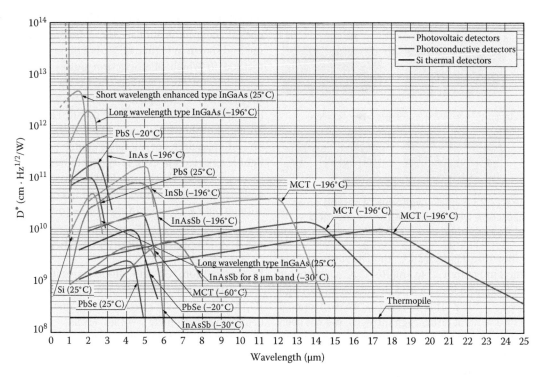

Figure 12.22 Typical spectral response characteristics of IR detectors. (From Hamamatsu Photonics; Technical information SD-12/Characteristics and use of IR detectors, Cat. No. KIRD0001E08, 2015.)

room temperatures and their responsivity is not dependent on wavelength. One of their applications, recently the focus of much attention, is the "human body sensor" that is finding wide applications in many fields such as home automation, security, and energy saving.

IR detectors are used in diverse application fields including general electronics, security, disaster prevention, industrial measurement, communications, remote sensing, medical equipment, and analytical instruments.

12.4 PHOTODETECTOR SIGNAL PROCESSING METHODS AND *S/N* RATIO

To operate photodetectors in actual applications, optimum detectors must first be selected by considering the light wavelength to be detected, the required time response, and *S/N* ratio. In addition, the signal processing circuit connected to photodetectors must also be optimized. Particularly, when the incident light intensity is very low, it is important to take countermeasures against

external noise and design measurement systems that will maintain a satisfactory *S/N* ratio.

12.4.1 Incident light intensity and signal processing method

Figure 12.23 shows output signal waveforms of a photodetector (a PMT is used here) observed on an oscilloscope while changing the intensity of light emitted from a pulse-driven light emitting diode. When the light intensity is high, the photoelectron pulses generated after photoelectric conversion overlap each other and create an analog waveform as shown in Figure 12.23a. When the light intensity is reduced slightly, the output waveform will contain more AC components than DC components like those shown in Figure 12.23a through c. If the light intensity is reduced further, the output signal will be discrete pulses as shown in Figure 12.23d. This is the so-called photon counting region (digital count mode).

In this way, the output signal waveform differs depending on the incident light intensity so the subsequent signal processing [15,16] may use various methods. Typical optical measurement methods

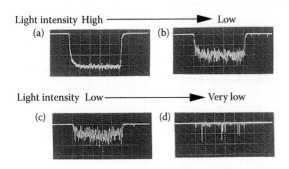

Light intensity High ⟶ Low

(a)　　　　　　(b)

Light intensity Low ⟶ Very low

(c)　　　　　　(d)

Figure 12.23 Signal waveforms observed on an oscilloscope when light intensity is changed. (a–c) Analog mode and (d) digital mode.

are shown in Figure 12.24. The DC measurement method in Figure 12.24a amplifies DC components from the photodetector and detects them through a low-pass filter. This method is used in optical measurements at rather high intensity and has been extensively used for many years. In the AC or pulse measurement method in Figure 12.24b, only the AC components in the output are extracted via a capacitor, amplified by a pulse amplifier, and converted into digital signals by a high-speed AD converter. This method is frequently used for demodulating

pulsed light signals of wide bandwidth such as in optical communications. The photon-counting method shown in Figure 12.24c is a pulse-counting method in which photoelectron pulses from the photodetector are amplified one by one, and only the pulses with an amplitude higher than the preset discrimination pulse height are counted as signals. Though not shown in Figure 12.24, there are other optical measurement methods suitable for low-light-level detection even in applications subject to excessive noise. These include the boxcar method and lock-in detection method used in conjunction with an optical chopper in spectrophotometry.

12.4.2 Photon counting method

In this section, we will discuss specific circuit configurations used to perform photon counting [17,18], which is an effective technique for light detection at extremely low-light levels.

Figure 12.25 shows a typical circuit configuration for photon counting and a pulse waveform obtained at each circuit. In a PMT, input photons are converted into photoelectrons, which are then multiplied by the dynodes up to an order of 10^6–10^7. The multiplied pulses output from the

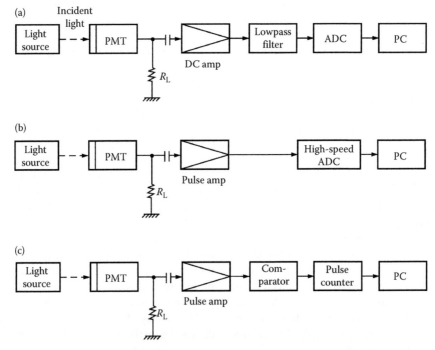

Figure 12.24 Typical light measurement methods. (a) DC measurement, (b) pulse measurement, and (c) photon counting.

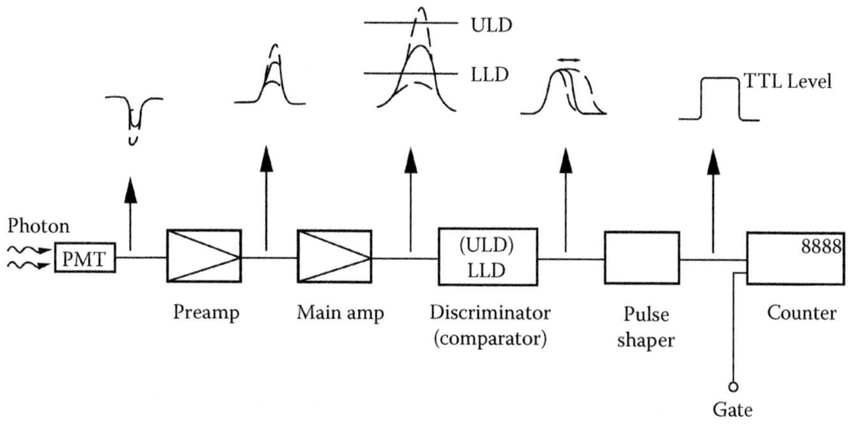

Figure 12.25 Circuit configuration for photon counting.

PMT are converted into voltage pulses by a wide-band amplifier and amplified. These voltage pulses are fed to the discriminator and counted by a counter via a pulse shaper. The discriminator usually employs a comparator IC that compares the input voltage pulses with the preset reference voltage (threshold level) and eliminates those pulses with amplitudes lower than this value. In general, the lower level discrimination is set at the lower pulse height side and the upper level discrimination is set at the higher pulse height side to eliminate noise pulses with higher amplitudes. The counter usually has a gate circuit to set the desired measurement timing and intervals. This photon-counting method is most effective in detecting extremely low-level light.

To enable photon counting, photodetectors must have the following performance characteristics.

1. Photodetectors must have adequate gain. Usually, a gain of 10^5 or more is required in consideration of the relation between the next circuit's amplifier noise level and noise index.
2. High QE at wavelengths to be measured, and few noise pulses in order to attain a high *S/N* ratio.
3. Narrow pulse height distribution.
4. Large photosensitive area (when measuring diffused light).

PMTs satisfy almost all of the aforementioned prerequisites, so they are widely used in low-light-level photometric equipment such as photon counters. In recent years, PMTs for photon counting have been improved significantly, leading to the development of PMTs [19] that deliver

an exceptionally low noise pulse count around 0.1 cps. This allows measurement at extremely low-light levels equivalent to nearly 1 photon per second.

In fluorescence lifetime measurement, even changes in the light emission that persist for a very short time must be measured along with the photon counting, so sophisticated techniques like time-resolved photon counting and TCPC are used.

12.4.3 Signal-to-noise ratio

In optical measurements, *S/N* ratio [20,21] is a critical factor in determining the lower detection limit of photodetectors.

Figure 12.26 shows a typical model of a photodetector connected to an external circuit. Light is converted into electrons in the photoelectric section of the photodetector and the electrons are amplified if the photodetector has an electron multiplier function. The output from the photodetector is then amplified by the externally connected electronic circuit and extracted as an output signal.

Figure 12.27 shows an analog output waveform generally obtained from a photodetector. This output contains both signal components produced by the incident light and the noise components. These components can be defined as follows:

Mean value of noise component: I_d
AC component of noise: i_d (rms)
Mean value of signal (including noise component): I_{p+d}
AC component of signal (including noise component): i_{p+d} (rms)

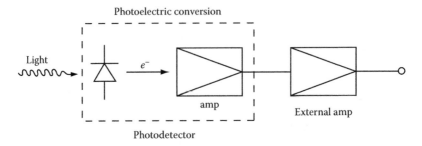

Figure 12.26 Typical model of a photodetector connected to an external circuit.

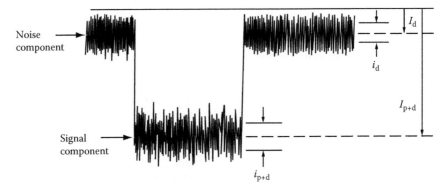

Figure 12.27 Analog output waveform obtained from a photodetector.

Using these factors, the S/N ratio is given by

$$S/N = \frac{I_p}{i_{p+d}}, \tag{12.13}$$

where I_p is the mean value of the signal component obtained by subtracting I_p from I_{p+d}.

The extent of the degradation in the S/N ratio during the multiplication process is commonly expressed in terms of the noise figure (NF). NF is defined by

$$\mathrm{NF} = \frac{(S/N)^2_{\mathrm{in}}}{(S/N)^2_{\mathrm{out}}}. \tag{12.14}$$

In general, the S/N ratio in analog measurement regions can also be calculated by

$$S/N = \frac{1}{\sqrt{2eF_1(I_k + I_d) + \dfrac{4kTF_2}{R_L M^2}}} \frac{1}{\sqrt{B}}, \tag{12.15}$$

where I_k is the mean value of photocurrent generated by signal, e is the electron charge, F_1 is the NF of the photodetector, I_d is the mean value of dark current, k is Boltzmann's constant, T is the absolute temperature, F_2 is the NF of the externally connected amplifier, R_L is the equivalent resistance for connection to the externally connected amplifier, M is the gain of the photodetector, and B is the bandwidth of the entire system. The numerator in Equation 12.15 is the signal component and the denominator is the sum of the AC components (including the signal and noise) and the noise generated in the externally connected electronic circuit. The NF indicates the increase in noise ratio caused during the multiplication process in the photodetector. Since PMTs have high gain, the noise in the externally connected electron circuit can be ignored in most cases. The NF is about 1.3 for PMTs and 2–3 for APDs.

The second term of the noise in the numerator can be mostly ignored in PMTs since they have a high gain that can be regarded as infinite ($M = \infty$). On the other hand, this noise cannot be ignored in PDs because they have no gain ($M = 1$). If the bandwidth is widened, the S/N ratio will degrade.

In photon counting, the average value of the counts of individual photoelectron pulses is treated as a signal and the fluctuations in the count values are treated as noise. The S/N ratio in photon counting is given by

$$S/N = \frac{N_s \sqrt{T}}{\sqrt{N_s + N_d}}, \qquad (12.16)$$

where N_s is the average value for the signal pulse count per second, N_d is the average value for the noise pulse count per second, and T is the counting time in seconds.

When comparing the S/N ratio of the photon counting and analog methods (Equations 12.15 and 12.16), the extent of the NF makes it clear that the photon-counting method is superior to the analog method.

12.5 HIGH-SPEED PHOTODETECTORS

In recent years, there has been a shift away from conventional methods and toward using light in many applications. For example, light is extensively used to analyze substances and properties, make measurements and control various devices. A variety of different observation and measurement techniques have been developed for high-speed light measurement [22,23] and put to practical use. To analyze even higher speed phenomenon and achieve higher speed information processing, there is a demand to create light sources producing

even shorter pulses and demodulators (photodetector and electrical circuit) with high-speed time response [24]. In this section, high-speed photodetectors will be described after touching briefly on general methods for high-speed light measurement.

12.5.1 High-speed light measurement techniques

Typical methods for measuring short light pulse signals from high-speed phenomena are shown in Table 12.6. The analog measurement sampling technique is a method for finding the waveform of a high-speed PD or PMT on an oscilloscope or other device. The response boundaries of this method are determined by the response of the high-speed photodetector, and the frequency bandwidth of the oscilloscope. Current technology has achieved a time resolution of around 10 ps. Among the real-time measurement methods, the streak method [25] has the best time resolution (picosecond to subnanosecond range).

The auto-correlation method attains a time resolution of 10 fs and excellent high-speed characteristics yet cannot find the optical waveform itself due to its measurement principle.

Table 12.6 Typical methods for measuring short light pulses

Measurement method	Detectors	Time resolution	Features	
Analog measurement sampling method	PDs	Nanoseconds	Merits:	Easy to use, relatively fast response
	Biplanar phototubes			
	PMTs MCP-PMT	Subnanoseconds		Narrow dynamic range
Streak method	Streak camera	Femtoseconds	Merit:	Ultra-fast response
Auto-correlation method	SHG correlator	Femtoseconds	Merit:	Measurement of ultra-fast phenomenon
			Demerit:	Optical waveform cannot be measured directly
Time-resolved photon-counting method	PMTs	Subnanoseconds	Merits:	Fast response, wide dynamic range
			Demerit:	Longer measurement time than analog method

The TCPC method [26], along with having a comparatively fast time resolution, also has high sensitivity to low-level light, and a wide dynamic range (10^5–10^6); the only drawback is that it has a long measurement time.

12.5.2 High-speed photodetectors

High-speed photodetector types and their time response characteristics are shown in Table 12.7. Besides showing whether or not the detector has an amplification mechanism, this table also reveals large differences in photodetector response speeds. The biplanar phototube has a rise time of 60 ps but no amplification mechanism, so detection is limited to strong pulsed light. Conventional PMTs have response in the nanosecond range but PMTs using MCPs (MCP-PMT) have a time resolution of 25 ps and a gain of 10^6.

Semiconductor photodetectors are widely used in measurement, control, and analysis equipment as well as optical communications. Time response characteristics of semiconductor photodetectors are determined by carrier transit time in the depletion layer, the delay time set by the CR time constant [CR time constant meant constant defined by C (capacitance) and R (resistance) of the circuit used]; in the case of APDs, the time response is additionally determined by the avalanche rise time. Figure 12.28 shows the effective area versus frequency characteristics of PIN-PDs. As shown in the figure, the photosensitive area of the PIN-PDs is nearly proportional to the frequency response. A response up to 300 MHz is shown for an area size of 1.5 mm in diameter but even smaller photosensitive areas have a better response of several gigahertz.

High-speed photodetectors are particularly indispensable in the optical communications field due to their large data communication capacity

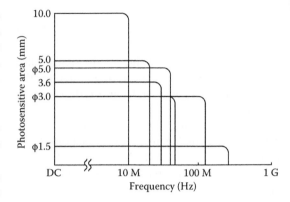

Figure 12.28 Effective area versus frequency characteristics of PIN-PDs.

and high speed. Current technology has achieved frequency characteristics up to 10 GHz using a long-wavelength band InGaAs PD, yet photodetectors with response characteristics of 50–90 GHz on up to the terahertz range will be in actual use.

High-speed devices like the metal–semiconductor–metal photodetectors (or MSM-PD) using GaAs as the substrate are well known among semiconductor photodetectors. The MSM has a response of 20 ps (rise time) in actual operation; it is reported that a high-speed response in the subpicosecond [27] level can be attained.

Streak tubes have the highest speed among currently used photosensors, delivering a time resolution of 2 ps, but in recent years, some have also attained the subpicosecond range. With development, femtosecond streak tubes have attained a time resolution of 360 fs and just recently, a time resolution of 180 fs has reportedly been reached [28]. Streak tubes are used for researching chemical reaction dynamics, measuring the semiconductor device relaxation process, and observation of super-high-speed phenomena such as implosions in laser nuclear fusion.

Table 12.7 High-speed photodetector types and their time response characteristics

Detectors	Amplification	Response speed
Biplanar phototubes	No	60 ps
PMTs	Yes	0.3–20 ns
MCP-PMT	Yes	160 ps
PDs	No	Subnanoseconds to microseconds
APDs	Yes	Subnanoseconds to microseconds
Streak tubes	Yes	0.2–20 ps

12.5.3 Photodetectors for optical communication

In this section, we will discuss photodetectors used in optical communication. Digital communication [29] using light has become a useful tool in a wide range of applications mainly due to technical advances such as long-wavelength optical fibers (1.3 and 1.5 μm bands) having low transmission loss as well as rapid progress in wavelength multiplexing technology. High-speed digital communication using optical fibers requires photodetectors [30] with fast response speeds and high sensitivity that can be easily coupled to fiber-optic cables.

12.5.3.1 OPTICAL FIBER COMMUNICATION

A typical block diagram for information transmission using an optical fiber is shown in Figure 12.29. An information source is converted from electrical signals to optical signals (E/O conversion) after being modulated by various means; it is then transmitted to the receiver side through the optical fiber. At the receiver end, optical signals are converted to electrical signals (O/E conversion) and then demodulated back into the original information. Optical fiber communication has the following advantages compared to conventional information transmission by electrical signals.

1. High resistance to noise since there is no electromagnetic induction.
2. Good insulation.
3. Lower transmission loss means longer distance communication.
4. Optical fibers are lightweight and take up less space.
5. Huge bandwidth.

12.5.3.2 PERFORMANCE REQUIRED OF PHOTODETECTORS IN OPTICAL FIBER COMMUNICATION

Figure 12.30 shows a typical configuration for a fiber-optic receiver. Light transmitted through an optical fiber is detected by a photodetector via an optical connector and converted to electrical signals. These signals are electrically processed by an analog/digital circuit and then output. The following characteristics are required of photodetectors for use in optical fiber communication.

1. High sensitivity at signal light wavelength.
2. Time response fast enough for system transmission speed.
3. Low noise level.
4. High stability and long service life under continuous operation.

Besides optical connectors such as that shown in Figure 12.30 for coupling between the optical fiber and the photodetector, fiber-to-fiber direct coupling using pigtail type photodetector modules (see Figure 12.31) is often used.

12.5.3.3 RECEIVER MODULES FOR OPTICAL FIBER COMMUNICATION

In optical fiber communication, photodetectors generally come attached to optical fibers (pigtail type) rather than as discrete components, or as the so-called optical receiver modules with preamplifiers built in for signal amplification. Various types of packaged receiver modules are shown in Figure 12.31 [31]. This photograph shows typical products including coaxial types with internal preamplifiers, coaxial types without internal preamplifiers, and a fiber jointed device (pigtail type). Long-distance communication uses laser diodes on the 1.3 and 1.5 μm bands, so InGaAs compound semiconductor photodetectors are used rather than silicon photodetectors. Due to the need for optical coupling with the fibers and high-speed response, APDs or PDs with small sensitive areas 20–80 μm in diameter are generally used.

Device characteristics of the most recent receiver modules of Si APDs, Si PIN-PDs, InGaAs PIN-PDs, and InGaAs APDs are shown in Table 12.8. The highest speed modules are used in 10 Gbps long-distance communication.

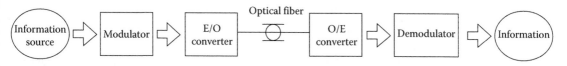

Figure 12.29 Typical block diagram for information transmission using an optical fiber.

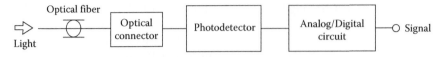

Figure 12.30 Typical configuration of a fiber-optic receiver.

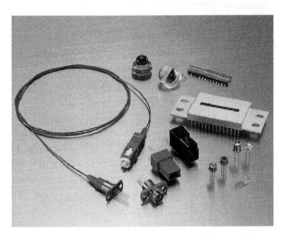

Figure 12.31 Various types of packaged receiver modules.

Even further advances in optical communication technology are predicted. Technical innovations will expand communication now attainable at 10–40 Gbps and further future advances will take optical communication to speeds in the terabit per second range.

12.6 IMAGE SENSORS

Up to here in this chapter, we have defined photodetectors as sensors that only detect the intensity of light. We now define image sensors as photodetectors that not only detect the intensity of light but also detect one-dimensional or two-dimensional information.

Table 12.9 shows categories of image sensors. Image sensors can be roughly classified into electron tube image sensors, solid-state image sensors, and hybrid image sensors combining both types. Electron tube image sensors are further divided into those using the external photoelectric effect or the internal photoelectric effect. IIs are typical electron tube image sensors using the external photoelectric effect. They are extremely high-sensitivity image sensors since they have an internal electron multiplier called MCP that delivers a multiplication factor (gain) in the order of 10^3–10^6. IIs were primarily developed for low-light imaging such as at night time so they are still referred to as night vision tubes. In addition to low-light imaging applications, IIs are applied to high-speed shutter

Table 12.8 Typical characteristics of receiver modules for optical communications

	Active area (mm)	Sensitivity (A W^{-1}) Typical	Wavelength (nm)	Bandwidth fc (GHz) Minimum	Preamplifier	Package	Data rate
InGaAs PIN-PD	40	1.00	1550	8.0	Built in	17-PIN mini butterfly	10 Gbps
	50	0.89 0.94	1310 1550	2.5	No	Coaxial	2.5 Gbps
	80	0.89 0.94	1310 1550	2.5	No	Coaxial	≤622 Mbps
InGaAs APD	20	0.7	1550	7.0	Built in	17-PIN mini butterfly	10 Gbps
	50	0.94 0.96	1310 1550	2.5	Built in	Coaxial	2.5 Gbps
	50	0.94 0.96	1310 1550	1.0	No	Coaxial	≤622 Mbps
	50	0.94 0.96	1310 1550	2.5	No	Coaxial	2.5 Gbps

Table 12.9 Categories of image sensors

Detecting principle	Readout method	Typical product	Sensitivity	Wavelength range
Image Sensors				
External photoelectric (electron tube sensor)	Photoemissive	II	Super-high sensitivity	VUV to near-IR
	Photoconductive	Vidicon	Sensitive to invisible light	Visible to IR
Internal photoelectric (solid-state sensor)	Address type	MOS CID	High sensitivity	UV to near-IR
Hybrid sensor	Charge transfer	CCD	High sensitivity	UV to near-IR
		ICCD	Super-high sensitivity	VUV to near-IR
		EBCCD		

cameras because an electronic shutter that works in subnanoseconds can be implemented.

Typical solid-state image sensors are charge transfer type CCD (charge-coupled device) image sensors and CID image sensors, and addressed type MOS (metal–oxide–semiconductor) image sensors. More recently, hybrid image sensors using an electron tube image sensor and a semiconductor image sensor have been developed and put to practical use.

Image sensors can also be classified by application into sensors used for quantitative measurement such as for spectrophotometry and pattern recognition; and sensors used for general imaging such as home video cameras. Since only limited space is available, this section will only describe image sensors designed for measurement applications.

12.6.1 Electron tube image sensors

Camera tubes, the best known electron tube image sensors, have been used in many imaging applications. In recent years, however, camera tubes have been almost completely replaced by semiconductor image sensors; they are now seldom used in the visible light region.

IIs [32,33] are electron tube image sensors with ultra-high sensitivity and a high-speed electronic shutter mechanism.

Figure 12.32 illustrates the structure of a proximity-focused II. In operation, an optical image from a low-light-level object is converted into

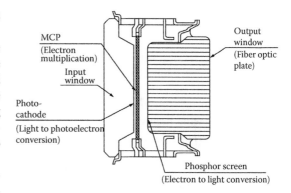

Figure 12.32 Structure of a proximity-focused II.

photoelectrons by the photocathode, multiplied by the MCP, and reconverted into an amplified optical image on the phosphor screen. An MCP consists of a multitude of glass capillaries (channels) 10–20 μm in inner diameter, fused together and formed into the shape of a thin disc 0.1–1.0 mm thick. The inner wall of each channel is coated with a secondary emissive material with a proper resistance value, so each channel serves as an independent, nondiscrete secondary electron multiplier. For example, in an MCP of 18 mm outside diameter, about 10^6 channels are arrayed in two dimensions so that each channel corresponds to a pixel when used as a two-dimensional sensor. This MCP multiplication function allows IIs to have a high gain of 10^3–10^6 (depending on the number of MCP stages), so that they are used in night vision devices and low-level-light image sensors. Figure 12.33 shows a circuit for driving a high-speed electronic shutter (gate) built into an II. By applying

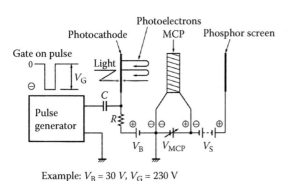

Example: $V_B = 30$ V, $V_G = 230$ V

Figure 12.33 Electronic shutter operation.

Figure 12.34 II products.

a negative pulse between the photocathode and the MCP, the photoelectrons emitted from the photocathode can be controlled so as to have an electronic shutter function [34] that turns the II operation on or off. In Figure 12.33, the shutter turns on when a negative voltage pulse is applied to the photocathode. This electronic shutter is used at speeds of nanoseconds to microseconds in most applications, although special shutters are designed to operate at ultra-high speeds of 50 ps [35]. When viewing highly repetitive phenomena, phosphor materials [36] having a short decay time τ designated as P47 ($\tau = 0.1$ μs), P46 ($\tau = 0.2$ μs), or P24 ($\tau = 8$ μs) are used. But, in general applications, P43 ($\tau = 1$ ms) is used because of its high emission efficiency.

By selecting the window and photocathode materials, IIs can be used for imaging in various wavelength regions of the spectrum from soft X-rays to IR radiation. II products are shown in Figure 12.34.

12.6.2 Semiconductor image sensors

CCD image sensors and MOS image sensors are typical semiconductor image sensors [37,38]. They are broadly divided by structure into one-dimensional arrays (linear image sensors) and two-dimensional array sensors (area image sensors). Semiconductor image sensors are generally mass-produced for image capturing applications but in this section, we will discuss their use in making measurements.

Semiconductor image sensors are also grouped into address types and charge transfer types according to their scan method.

The address type shown in Figure 12.35a is a two-dimensional array of pixels consisting of

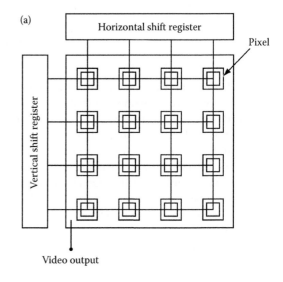

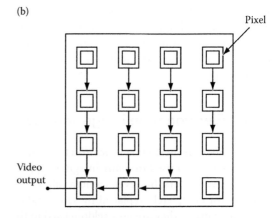

Figure 12.35 Operating principle of semiconductor image sensors. (a) Address type and (b) charge transfer type.

photoelectric elements such as PDs. Each pixel is, respectively, connected to a vertical and horizontal shift register. To read out signal charges from pixels, an external pulse is sent in sequence to the pixels, and the signal charge accumulated at the intersecting pixel is read out. In the charge transfer type shown in Figure 12.35b, signal charges of pixels comprising PDs are sequentially transferred in parallel vertically, and also transferred one line at a time horizontally. The signal charge is then read out. MOS image sensors are typical address type sensors, while CCD image sensors are typical charge transfer type sensors.

The CCD has high sensitivity, low noise and is capable of handling large numbers of pixels. It is the most dependable yet least expensive device among currently available image sensors and is used in large numbers. MOS image sensors offer a large photosensitive area capable of handling a large storage charge and have low power consumption. One disadvantage of MOS image sensors is their large capacitive noise compared to CCD; the S/N ratio gets worse under low-level light.

In this section, we cover image sensors intended mainly for optical measurement. For information on general-purpose image sensors, such as for video cameras, refer to other texts [39].

12.6.2.1 LINEAR IMAGE SENSORS [40]

Linear image sensors are image sensors segmented in only one dimension. A typical application of CCD linear image sensors is for facsimile readout. Image sensors are also used in multichannel wavelength detectors in polychromators for spectrophotometry.

Figure 12.36 shows the structure of an n-MOS linear image sensor. The photosensitive area is a $p-n$ junction PD consisting of a p-type silicon substrate on which an n-type diffusion layer is formed. The sensor has a photoelectric function to convert the optical signal into an electrical signal and a function to temporarily store the acquired signal charge.

Table 12.10 shows typical structure and device characteristics of linear image sensors having a photoelectric surface made of InGaAs or silicon material. The photosensitive layers of InGaAs or silicon, respectively, match the visible-to-near-IR and the IR regions. Pixel size is several millimeters in height and up to 100 μm in width, with some hundreds to thousands of pixels fabricated as channels. Figure 12.37 shows an equivalent circuit of an n-MOS linear image sensor. The

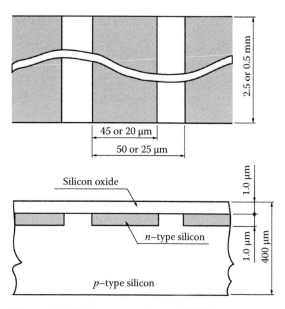

Figure 12.36 Structure of an n-MOS linear image sensor.

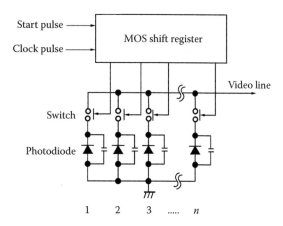

Figure 12.37 Equivalent circuit of an n-MOS linear image sensor.

photosensitive area of the PD, as shown in the figure, comprises a switching section to read out the PD signal, and a shift register to address that switch. The linear image sensor using silicon may be fabricated to have a maximum of 2048 channels. These image sensors are usually used singly but may often be used in combination with an II as described later.

The image sensor using InGaAs is utilized for spectrophotometry in the IR range on a wavelength of approximately 0.5–2.55 μm. As a photodetector for near-IR multichannel spectrophotometry, this

Table 12.10 Typical structure and device characteristics of linear image sensors

Detector material	Window material	Spectral response range	Cooling	Pixel size	Number of pixels
Si	Quartz	200–1000 nm	No	50 μm×0.5 mm	128, 256, 512
				50(25) μm×2.5 mm	128, 256, 512, (1024)
				14 μm×1.0 mm	2048×1
			Yes	50(25) μm×2.5 mm	256, 512, (1024)
InGaAs	Sapphire	0.5–1.7 μm	No	25 μm×500 μm	512
		0.9–1.67 μm	Yes	50 μm×500 μm	256, 512
		0.9–2.55 μm	Yes	25 μm×250 μm	512

has applications in the inspection of farm products by the IR absorption technique, radiation thermometers, and nondestructive inspections.

12.6.2.2 AREA IMAGE SENSORS

Two-dimensional image sensors (area image sensors) for measurement applications often use CCD image sensors [41]. The transfer method, structure, and characteristics of CCD image sensors are described as follows.

CCD image sensor transfer method. CCD image sensors can typically be grouped into one of the following three types of transfer methods.

1. *Frame transfer (FT) type*: The FT type CCD shown in Figure 12.38 comprises two vertical shift registers made up of a photosensitive area and charge storage area, one horizontal shift register, and an output section. A transparent electrode such as polycrystalline silicon is generally used as the metal electrode for the photosensitive area. The areas other than the photosensitive area are covered with a nontransparent electrode such as aluminum so that light will not enter those areas.
2. *Full frame transfer (FFT) type*: The FFT type CCD in Figure 12.39 is basically the FT type CCD without the charge storage area. Since there is no charge storage section, it must

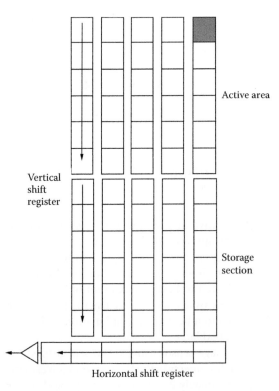

Figure 12.38 Structure of a FFT CCD.

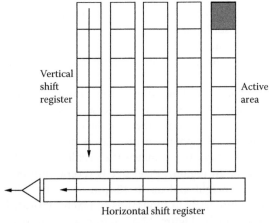

Figure 12.39 Structure of a full FFT CCD.

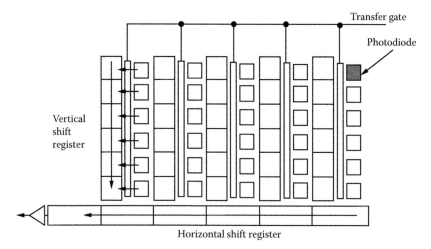

Transfer gate

Photodiode

Vertical
shift
register

Horizontal shift register

Figure 12.40 Structure of an IT type CCD.

normally be used with some kind of external shutter mechanism. This limitation makes it difficult to use the CCD as a video camera so its use is mainly limited to measurement applications.

3. *Interline transfer (IT) type*: In the IT type CCD shown in Figure 12.40, the photosensitive area is formed separately from the transfer section. Here, the vertical shift register is formed on the sides so as to enclose the PD that constitutes the photosensitive area. Because of this structure, the numerical aperture is low; so, this type is not suited for measurement applications.

To sum up the aforementioned described methods, the IT type is used in video cameras, while the FT and FFT types are mainly used in measurement applications.

Structure of area image sensors for measurement work. To function effectively in measurement applications, the area image sensor must have high sensitivity, a wide dynamic range and be multichannel. Table 12.11 shows pixel formats of typical area image sensors used in measurement applications. The structure of an FFT (or FT) type CCD image sensor is generally used. A rectangular format is generally used for spectrophotometry as a precondition for binning (described later); a square format is used for imaging. Actual CCD image sensor products are shown in Figure 12.41.

To improve measurement performance of the CCD image sensor, low noise is achieved by cooling (to −20°C), sensitivity is improved by back-illumination, and scanning is used in the binning operation.

Figure 12.42 shows the structure of a back-illumination CCD [42] (also called "back-thinned CCD" since the backside of the CCD chip is thinned).

Table 12.11 Pixel formats of typical area image sensors used in measurement applications

Applications	Number of pixels ($H \times V$)	Pixel size
Spectrophotometry	2048 (1024)×122, 128, 250, 506	12 μm×12 μm
	2048 (1024)×16, 64	14 μm×14 μm
	1024 (512)×4, 58, 60, 122, 124	24 μm×24 μm
	1536×128	48 μm×48 μm
Imaging	1280×1024	7.4 μm×7.4 μm
	659×494	9.9 μm×9.9 μm
	2048×2048	21 μm× 12 μm
	1024×1024	24 μm× 24 μm

Figure 12.41 CCD image sensor products.

As shown in the figure, light enters from the front surface on a normal CCD. However, light loss occurs when light is input from the front due to absorption of light by the oxide film and the electrodes. In contrast, in the back-illumination CCD, there is no such light loss, high sensitivity is obtained, and a QE as high as 90% is achieved. Even in the UV region, a high QE of 40% or more is achieved. Spectral response characteristics of the back-illumination CCD are shown in Figure 12.43, along with characteristics of the front-illumination CCD for comparison. A great increase in sensitivity can be seen in the UV region as well as in the visible region.

The binning operation unique to CCD image sensors will be described next. Figure 12.44 shows the principle of the binning operation in the CCD image sensor. In the FFT-CCD, the signal charges are stored in the potential well

during the integration time, and when the integration time ends, the signal charges are stored two-dimensionally. Here, signals can be summed in the line direction (line binning) as shown in the figure, by transferring charges separately to the vertical shift register and horizontal shift register. Besides vertical binning that adds signals perpendicularly in this way, binning can also be performed horizontally.

Binning allows the use of a two-dimensional image sensor as a one-dimensional image sensor (line binning); it allows changing the number of effective channels (horizontal binning).

Spectrophotometry applications of semiconductor image sensors are expanding to fluorescence spectrometry, induction coupled plasma, Raman spectrometry, and multichannel spectrometry. Imaging applications include fields such as bio-imaging, semiconductor device analysis, and astronomical observation.

12.6.3 Latest advanced high sensitivity/high-speed image sensors

Recently, in the field of biology and physics, the need for imaging of very weak luminescence and fluorescence is increasing. Therefore, scientific measurement cameras must feature a higher sensitivity, resolution, and frame rate than general industrial application cameras.

This chapter describes the high-sensitivity technology and fundamental knowledge of this

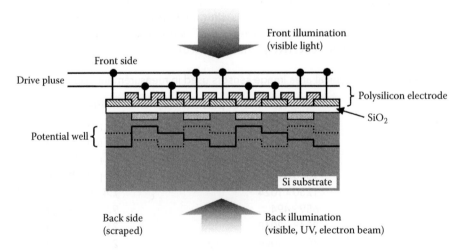

Figure 12.42 Structure of a back-illumination CCD.

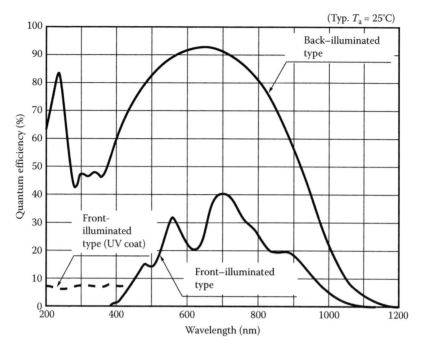

Figure 12.43 Spectral response characteristics of a back-illumination CCD (without window). (From Hamamatsu Photonics; Image Sensors, Cat No. KMPD0002E01, 2011.)

scientific measurement camera, and introduces the basic technology and camera characteristics of the EM-CCD (electron multiplying CCD) camera and scientific CMOS (sCMOS) camera, both of which have seen improved performance recently.

There are two methods for improving the sensitivity of CCD and CMOS cameras: amplifying the signal itself or reducing the noise completely. The amplifying method relates to the EM-CCD camera and reducing the noise relates to both the cooled CCD and sCMOS cameras. The amplifying camera obtains the signal while the light (photon)

entering into the sensor is converted into charge (electron) then amplified by the amplifier in the sensor. Therefore, super high-sensitivity detection is required. It features super high-sensitivity though it may not improve the S/N depending on the light level for the sample since it amplifies the noise factor as well. Conversely, the low noise camera does not have an amplifier but by reducing the noise inside the sensor or noise (electron) produced during the signal read-out, it can detect a very weak signal level. This can be explained by the following example: stars cannot be observed under daylight due to the strong scattering light from the sun, but can be observed in the nighttime without the noise factor of sunlight scattering. For a low noise camera, the exposure time can be extended as a method for obtaining a higher S/N image.

12.6.3.1 EM-CCD camera

The advances made in CCD technology are quite impressive, as a result, very sensitive and low-noise CCDs has become available. Although, the detection limit is defined by the read-out noise, in order to detect the signal below the noise level, the signal is made above the noise level by accumulating the input light on the CCD. To overcome this problem, amplified type sensors that amplify the detected

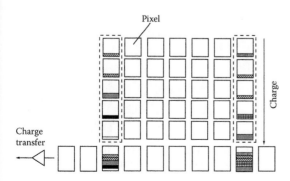

Figure 12.44 Principle of binning operation in the CCD image sensor.

input light inside the sensor and then produce output signals have been developed such as the II tubes and electron bombarded CCD (EB-CCD). To provide amplification within the sensor itself, there were many technical barriers to overcome. Recently, the EM-CCD has been introduced to provide a solution for the problem; it is widely used as an effective method in the field of low-light-level detection.

Figure 12.45a [43] illustrates the functionality of EB-CCD. The basic function is the same as the FT-CCD, and is backside illuminated. Also similar to FT-CCD, electrons created at the image area (photon detection area) are transferred to the storage area; then, the signal carriers are forwarded to a horizontal serial transfer resister and taken as a signal at each pixel. With EB-CCD, an amplifying resister (charge multiplying) is placed behind the horizontal serial transfer resister. This amplifying resister is applied a higher voltage than the horizontal transfer voltage, therefore the signal electrons are amplified. Figure 12.45b [43] illustrates the amplification theory of an amplifying resister. Each gate electrode is applied a higher voltage from 30 to 40 V (varies according to the CCD); this higher electric field accelerates the signal electron to generate an electron–hole pair. This effect is called impact ionization. The probability (g) is very low, 1.0%–1.6% as an average value. Inside the amplifying resister, this effect is repeated, thus a high EM gain is obtained. Normally, the amplification stage number (N) is 400–600; thus, the equation to obtain the EM gain becomes

$$M = \left(1 + g\right)^{N},$$

where

g probability of producing an electron–hole pair at each pixel, N amplification stage number.

The probability of producing an electron–hole pair at each pixel (g) is highly dependent on the supply voltage to amplify the resister and the temperature of the sensor. Therefore, a stable supply voltage and temperature control of the sensor become very important when a quantitative measurement is performed.

The key technical point of the EM-CCD is that the electric charge inside the CCD is amplified within the amplifying resister before the charge is converted into voltage in the photodiode array (PDA). As described earlier, the major defining factor of the detection limit of the CCD is the read-out noise induced in the PDA. This noise is frequency dependent so that the noise increases in proportion to the half power of the driving speed of CCD when the latter is increased in order to take high speed into account. Therefore, it is not practical to observe a low illuminated object in real time at high speed.

With the EM-CCD, the signal is amplified by the amplifying resister before it is read; therefore, the signal exceeds the noise level after the amplification, the noise induced within the PDA becomes less than 1 electron relatively, and it becomes possible to detect single photon level phenomenon in real time. This is

(a)

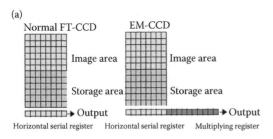

(b)

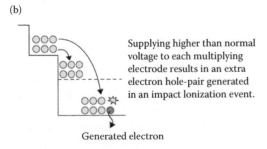

Supplying higher than normal voltage to each multiplying electrode results in an extra electron hole-pair generated in an impact Ionization event.

Generated electron

Figure 12.45 (a) Functionality of EB-CCD and (b) amplification theory of amplifying resistor.

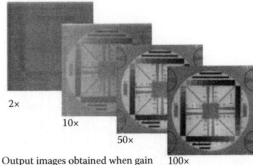

Output images obtained when gain was varied with light level kept constant.

Figure 12.46 Output images at different amplifications of EM-CCD.

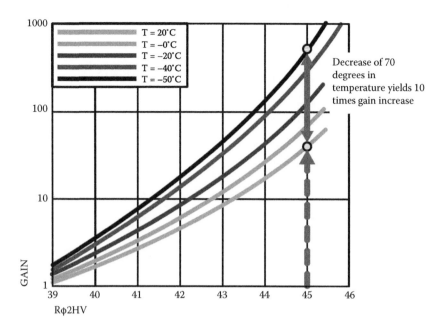

Figure 12.47 Temperature dependency of amplification of EM-CCD.

the main feature of EM-CCD that makes real-time high-sensitivity measurement possible by amplifying the signal in the amplifying resister under high-speed driving mode. An actual imaging sample is shown in Figure 12.46 [43]; the image are taken with no EM gain level by the signal being above noise level while amplification is gradually increased.

As noted earlier, for the EM-CCD, the EM gain in the amplifying resister has large temperature dependency so that the stability of the sensor temperature becomes very important. Figure 12.47 [43] shows the temperature dependency of the amplification. As shown, the gain varies approximately 10 times when the sensor temperature varies 70°C. Further, if the sensor temperature decreases, the gate voltage dependency of the gain curve becomes rather steep so that the temperature dependency becomes large even at the same gate voltage. Accordingly, not only does the temperature of the camera have to be cooled, it is important to maintain the temperature stability.

12.6.3.2 sCMOS CAMERA

In the past, CMOS sensors were known to have higher noise level than CCD. Thus, it was not suitable for use in scientific measurement application. But, recently, design and manufacturing technology of CMOS has advanced so much that its performance and image quality have overcome the historical common sense

of the CMOS sensor, and it has been used in various optical imaging and measurement applications from very weak light levels through fluorescence light.

The latest sCMOS sensor has an on-chip microlens on each pixel, which improves the light collection efficiency. Thus, the sensitivity has increased compared to former conventional CMOS image sensors. Figure 12.48 [44] shows the wavelength dependency of QE.

Input–output characteristics (linearity) is given proper linearity between the input incident light and output signal, as shown in Figure 12.49 [44], by the most suitable circuit designed for the applicable sensor. As for the read-out noise, which defines the detection limit, is decreased drastically by applying appropriate semiconductor process for best suited element amplifier, high gain and reduction of variation of element amplifiers. Further, a CDS circuit, which is an important factor for the reduction of noise in the CCD image sensor, is installed on the element for lowering the noise. In addition, two line parallel sequential readout, column amplifiers, and A/D converters, allow both low noise and high-speed readout. The latest sCMOS sensor realizes very low readout noise level of around 1.3 electrons, which is far below that of conventional cooled CCD image sensors. Nevertheless, despite the lower readout noise capability, it achieves high-speed readout of 100 frame/s at 400 megapixels,

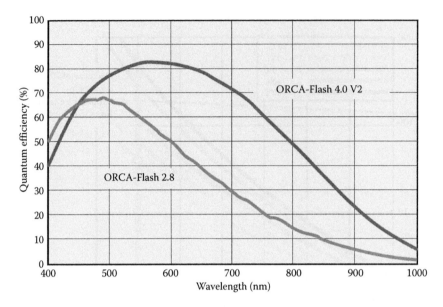

Figure 12.48 Spectral QE of EM-CCD.

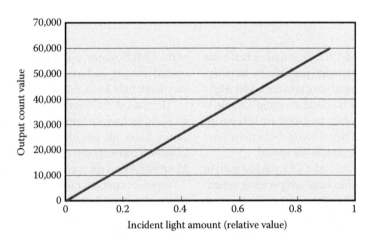

*The amount of light incident on the Y-axis is the brightness minus the dark.

Figure 12.49 Linearity characteristics on input light intensity of EM-CCD.

which is impossible for conventional CCD image sensors. As for the dynamic range, the readout noise is around 1.3 electrons against the saturation electric charge of 30,000 electrons. Therefore, the dynamic range is very large, that is, 23,000:1, which is almost the same level as a slow-scan cooled CCD camera (known to have a large dynamic range among conventional cameras).

The cameras introduced in this chapter each have specific features. It is important to select the appropriate kind depending on the application best suited for the feature noted. In general, it is recommended that for the measurement of very low light level detection in photon counting, the EM-CCD camera [45,49] should be chosen; and for the measurement of weak fluorescent light at high frame rate, one should choose the sCMOS camera [46–48]. In the future, improvements in the performance of CMOS sensors may be more drastic; it is expected to have much higher sensitivity, speed, and resolution allowing expansion of usage to other fields of application.

12.7 FUTURE PROSPECTS

This chapter has discussed categories, basic principles, and various characteristics of photodetectors or optical detectors. Future years will show accelerated progress in applying light to measurement, machining, control technology, communications, general electronics, and basic science fields. Along with these advances, extensive effort will be poured into developing practical photodetectors with higher sensitivity and time response, greater miniaturization or integration, and multichannel detection capabilities.

REFERENCES

1. Sommer, A.H. 1980. *Photoemissive Materials* (Malabar, FL: Krieger).
2. Bube, R.H. 1992. *Photoelectronic Properties of Semiconductors* (Cambridge: Cambridge University Press).
3. Hamamatsu Photonics K.K. Editorial Committee. 1999. *Photomultiplier Tubes— Basics and Applications*, 2nd ed. (Japan: Hamamatsu).
4. Driscoll, W.G. 1978. *Handbook of Optics* (New York: McGraw-Hill).
5. Samson, J.A.R. 1967. *Techniques of Vacuum Ultraviolet Spectroscopy* (New York: Wiley).
6. Knoll, G.F. 2000. *Radiation Detection and Measurement*, 3rd ed. (New York: Wiley).
7. Endo, T. et al. 1993. *SPIE*, 1982, 186.
8. Kosel, P.B. et al. 1994. *SPIE*, 2685, 140.
9. Razeghi, M. et al. 1996. *SPIE*, 2685, 114.
10. Breskin, A. 1996. *Nucl. Instrum. Meth. Phys. Res.*, A371, 116.
11. Niigaki, M. et al. 1997. *Appl. Phys. Lett.*, 71–17, 2493.
12. Dereniak, E.L. 1984. *Optical Radiation Detectors* (New York: Wiley).
13. Keyes, R.J. 1977. *Optical and Infrared Detectors* (Berlin: Springer).
14. Willardson, R.K. 1977. *Semiconductors and Semimetals*, vol. 12 (New York: Academic).
15. Carr, J.J. 1997. *Electronic Circuit Guidebook: Electro-Optics* (Indianapolis, IN: Prompt Publications Div. Sams).
16. Wobschall, D. 1979. *Circuit Design for Electronic Instrumentation* (New York: McGraw-Hill).
17. Hamamatsu Photonics K.K. 2000. *Photon Counting*, TPHO 9001E02.
18. Candy, B.H. 1985. *Rev. Sci. Instrum.*, 56, 194.
19. Theodórsson, P. 1996. *Appl. Radiat. Isot.*, 47, 827.
20. Vetokhin, S.S. et al. 1987. *Sov. J. Opt. Technol.*, 54, 754.
21. Pruett, H.D. 1972. *Appl. Opt.*, 11(11), 2529.
22. Shapiro, S.L. (ed.) 1977.*Ultrashort Light Pulses, Picosecond Techniques and Applications, Topics in Applied Physics*, vol. 18 (Berlin: Springer).
23. Harris, C.B., Ippen, E.P., Mourou, G.A. and Zewail, A.H. (ed.) 1990.*Ultrafast Phenomena VII. Springer Series in Chemical Physics*, vol. 53 (Berlin: Springer).
24. Alfano, R.R. (ed.) 1982. *Biological Events Probed by Ultrafast Laser Spectroscopy* (New York: Academic).
25. Tsuchiya, Y. 1991. *SPIE*, 1599, 244.
26. Yamazaki, I. et al. 1985. *Rev. Sci. Instrum.*, 56, 1187.
27. Chou, S.Y., Liu, Y. and Fischer, P.B. 1991. *IDEM*, 91, 745.
28. Takahashi, A. et al. 1994. *Proc. SPIE*, 2116, 275.
29. *ITU-T Recommendation*, G671 to be approved in 2002.
30. Tan, I.H. et al. 1995. *IEEE Photon Technol Lett.*, 7, 1477.
31. NEC Compound Semiconductor Devices, Ltd. 2002. *Optical Semiconductor Devices for Fiber Optic Communications Selection Guide*. Document No. PX10161EJOIVOPF Hamamatsu Photonics 2001 Optical communication device KOTH0005E01.
32. Csorba, I.P. 1985. *Image Tubes* (Indianapolis, IN: Sams).
33. Biberman, L.M. 2000. *Electro-Optical Imaging* (Bellingham, WA: SPIE).
34. Kume, H. et al. 1990. *SPIE*, 1358, 1444.
35. Thomas, S. 1990. *SPIE*, 1358, 91.
36. Shionoya, S. and Yen, W.M. 1999. *Phosphor Handbook* (Boca Raton, FL: CRC Press).
37. Janesick, J.R. 2001. *Scientific Charge-Coupled Devices* (Bellingham, WA: SPIE).
38. Séquin, C.H. 1975. *Charge Transfer Devices* (New York: Academic).
39. Schroder, O.K. 1987. *Advanced MOS Devices* (Reading, MA: Wesley)

40. Sweedler, J.V. et al. 1989. *Appl. Spectrosc.*, 43–46, 953.
41. Theuwissen, A.J.P. 1995. *Solid-State Imaging with Charged-Coupled Device* (Dordrecht: Kluwer).
42. Muramatsu, M. et al. 1997. *SPIE*, 3019, 2.
43. Tsuchiya, Y., Inuzuka, E., Kurono, T. and Hosoda, M. 1985. Photon-counting imaging and its application. *Adv. Electron. Electron Phys.*, 64A, 21.
44. Hirano, A. et al. 1993. *Jpn. J. Appl. Phys.*, 32, 3300.
45. Xuanze, C. et al. 2016. Superior performance with sCMOS over EMCCD in super resolution optical fluctuation imaging. *J. Biomed. Opt.*, 21(6), 10.
46. Rodrigues, R.M. et al. 2011. Autofluorescence microscopy: a non-destructive tool to monitor mitochondrial toxicity. *Toxicol Lett.*, 206(3), 281–288.
47. Fullerton, S. et al. 2012. Optimization of precision localization microscopy using CMOS camera technology. *Proc. SPIE*, 8228, 82280T–1.
48. Fullerton, S. et al. 2012 .Camera simulation engine enables efficient system optimization for super resolution imaging,*Proc. SPIE*, 8228, 8228–8229.
49. Long, F. et al. 2012. Localization based super resolution microscopy with an sCMOS camera. *Opt Express*, 19(20), 19156–19168.

Optical fiber devices

SUZANNE LACROIX AND XAVIER DAXHELET
Ecole Polytechnique de Montreal

13.1 INTRODUCTION

Maximizing the capacity of the fiber as a transmitting medium is the major challenge met by network operators. Solitonic propagation as well as wavelength division multiplexing (WDM) are among the solutions presently proposed to increase the flow of information propagating in these networks. This progress is made possible thanks to the development of such essential components as the erbium doped fiber amplifier (EDFA) and frequency stabilized laser sources. However, other components to perform all the functions (routing, filtering, dispersion compensation, etc.) are not less essential. Different approaches to design and realize the corresponding components are commonly used. The advantage of the all-fiber approach over its competitors (micro- and integrated optics) lies undoubtedly in the fact that all-fiber components are readily integrated to the network without significant splicing loss. In addition, their polarization sensitivity (which would induce loss and dispersion) is intrinsically much smaller than that of their integrated optic counterparts.

The present chapter content is restricted to all-fiber components, their micro-optics and integrated-optics equivalents being treated in Chapters A16 and B8, respectively. Unless expressly mentioned, these all-fiber components are made of single-mode fiber. The emphasis is put on the way they function and their intrinsic limitations, leaving out details about fabrication and packaging problems. Besides, only passive components are considered. Optical amplifiers and lasers can be found in Chapter 16. Some other more advanced components, such as wavelength converters, are based on non-linear effects (see Section 2.4). They usually take advantage of the inherent third-order non-linear effects, but second-order effects can also be induced. But these components are not treated in the present chapter.

In the following section, we present the basic technologies that are involved. The subsequent section gives concepts that determine the behavior of these all-fiber components. Example components are presented in the third section.

13.2 TECHNOLOGIES

Two main technologies are presently used to manufacture all-fiber components:

1. The fusion and tapering technology
2. The inscription of gratings including both short-period gratings (usually referred to as Bragg gratings) and long-period gratings (LPGs)

13.2.1 Fusion and tapering

The tapering technique consists of locally heating and stretching a fiber using a micro-torch or a CO_2 laser, thus creating a biconical structure such as that of Figure 13.1.

Before tapering, one may laterally fuse two (or more) fibers together so as to create a more complex transverse structure in order to transfer power from one guide to another. Such structures are referred to as couplers.

Irrespective of the transverse structure, the behavior of the component is largely determined by the slopes of the longitudinal structure. In the case of a tapered single fiber, however, the angle is made small enough everywhere so that only a negligible leakage of power from the fundamental mode as it propagates along the structure is ensured. In such cases, the propagation and, by extension, the taper itself is said to be *adiabatic*, where the fiber transmission is not affected by the tapering process [1]. In contrast, when the slopes are abrupt, such as those of the structure shown in Figure 13.1, one may observe large oscillations in the transmitted power, as the fiber is elongated. In addition, for a given elongation, similar oscillations are seen in the transmission as a function of wavelength (Figure 13.20) or of the refractive index surrounding the tapered part of the fiber [2].

While adiabaticity is usually required for couplers, non-adiabaticity of tapered single fibers can be used to design a variety of all-fiber spectral filter.

13.2.2 Gratings

All-fiber short- and long-period gratings are periodically perturbed fibers. The number of periods ranges from ten to several thousands depending on the intensity of the index perturbation (up to 10^{-2}). Both short- and long-period gratings operate_according to a resonant couphng effect resulting in a condition relating to the modal wave vectors $\vec{\beta}_1$ and $\vec{\beta}_2$ with the grating one $\vec{\beta}_B$

$$\vec{\beta}_1 = \vec{\beta}_2 + \vec{\beta}_B. \tag{13.1}$$

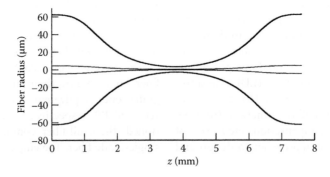

Figure 13.1 Profile of an abruptly tapered fiber.

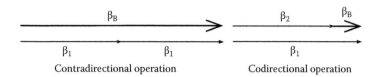

Contradirectional operation   Codirectional operation

Figure 13.2 Wave vector conservation in fiber gratings. Contradirectional coupling corresponds to large grating wave vectors β_B and thus small grating periods Λ (typically half a µm), while codirectional coupling corresponds to small grating wave vectors β_B and thus large grating periods Λ (several hundred µm).

Wave vector resonance conditions are shown in Figure 13.2 for contra-and codirectional coupling. Short periods $\Lambda = 2\pi/\beta_B$ are of the order of 0.5 µm and long ones 500 µm. As their periods are of the order of half the wavelength in silica, short-period gratings reflect the fundamental mode into itself (through contradirectional coupling) at a given wavelength ($\bar{\beta}_2 = -\bar{\beta}_1$). They also reflect higher-order cladding modes at shorter wavelengths with lesser efficiency, unless especially designed for this purpose. On the other hand, long-period gratings couple, at a given wavelength, the fundamental mode into copropagating higher-order modes (through codirectional coupling), which are eventually absorbed by the fiber jacket. Both types of grating are thus rejection filters in transmission.

13.2.2.1 SHORT-PERIOD GRATINGS

There are essentially two methods to manufacture this type of grating. Both are, however, based on photosensitivity—a stable refractive index change of the order of a few 10^{-4} up to 10^{-2} that the Gedoped core of the fiber experiences, when exposed to near UV-radiation.

1. The first method consists of exposing the fiber to an interference field created by two locally planar waves of a UV-laser such as doubled Ar+, dye lasers or copper vapor lasers. The period Λ is then given by

$$\Lambda = \frac{\lambda_{UV}}{2\sin(\theta/2)} \quad (13.2)$$

where λ_{UV} is the writing wavelength and θ the angle between the interfering waves. This method referred to as the holographic method is very flexible because any periodicity may be realized. A number of practical implementations have been designed [3,4]. However, as for

any holographic setup, stability is an issue. It is thus easier to implement it in a pulsed regime.

2. The alternative method uses a phase mask as schematized in Figure 13.3. It is a phase grating having a periodicity Λ_m, twice that of the grating to be written in the fiber ($\Lambda_m = 2\Lambda$). Its depth e and refractive index n (at λ_{UV}) are related by $2(n-1)e = \lambda_{UV}$ ensuring that the zero order is not transmitted. Waves from orders $+1$ and -1 thus interfere in a similar manner to that described above for the holographic method.

The main advantage of the phase mask method over the holographic one is that the UV-source may have a limited spatial and temporal coherence. This is the case of excimer lasers (KrF lasing at $\lambda_{UV} = 248$ nm and ArF at $\lambda_{UV} = 193$ nm), which are frequently used for fiber grating inscriptions. The fiber and the mask are then separated by only a few micrometers. The disadvantage of this method is that the Bragg period is determined by that of the phase mask, which is an expensive component.

This drawback may, however, be overcome by a combination of both methods in set-ups such as the Talbot interferometer in which the phase mask is used as a splitting plate and the fringe periodicity is varied through a mirror angle [3,4].

13.2.2.2 LONG-PERIOD GRATINGS

As their period is long ($\Lambda = 300$–800 µm), this type of grating usually does not require sophisticated set-ups such as those of their short-period counterparts. Most of the time they use step-by-step methods.

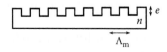

Figure 13.3 Phase mask.

1. The fiber may be irradiated by a UV-source just as in the short-period case. One may use the step-by-step method or an amplitude mask.
2. CO_2-laser irradiation ($\lambda = 10, 6\ \mu m$) also induces index changes of the same order of magnitude, but the mechanism responsible for this change, although not completely identified, is certainly different as the perturbation persists at higher temperatures than photosensitivity [5].
3. Larger index changes (up to a few 10^{-3}) may also be induced by an electrical discharge. The grating is also stable at high temperatures and its inscription is clearly related to the fiber inner stresses [5].
4. Femtosecond intense Ti:sapphire lasers are also known to induce large refractive index changes that may be used to manufacture long-period gratings.
5. Finally, one may manufacture LPGs by periodical tapering of the fiber.

All these fabrication techniques (fusion-tapering and grating inscription) may be viewed from a theoretical viewpoint as perturbations to the longitudinal invariance of the fiber structure. These perturbations in turn induce coupling between the modes and thus power the transfer from the fundamental mode to others. These effects are described by the coupled mode formalism, which is now presented.

13.3 COUPLED MODE THEORY

Except for the propagation factor $e^{i\beta_j z}$ along the guide, modes of an unperturbed guide, identified by their indices j, are z-invariant solutions of Maxwell equations. Any solution, i.e., any electromagnetic field propagating along the guide, is a superposition of these modes with amplitudes a_j.

When the guide is perturbed, these modes are no longer solutions and the optical power may transfer from one mode to another. This is described by permitting the modal amplitudes a_j to vary with z.

Whatever the perturbation, periodical or not, resonant or not, between co- or contradirectional modes, it always consists of a variation of the

refractive index, i.e., of the polarization vector. Coupled mode equations may be demonstrated from Maxwell equations by expanding the electromagnetic field on a modal basis [6]. These equations may be written as

$$\frac{da_j}{dz} - i\beta_j a_j = i\sum_\ell C_{j\ell}(z)a\ell \qquad (13.3)$$

where a_j and β_j are, respectively, the amplitude and the propagation constant of mode j and $C_{j\ell}(z)$ the coupling coefficient between mode j and mode ℓ. This coupling coefficient may be real or complex depending on the waveguide perturbation. As the set of all the modes (guided as well as radiation modes) of an unperturbed guide constitutes a basis for the expansion of any electromagnetic field, it is natural to choose a_j to be amplitudes of the normal modes of a guide. The choice of the guide used for the basis depends on the problem as will be shown in the following section.

13.3.1 Ideal mode coupling

As shown in Figure 13.4, whenever the perturbation is a slight departure from a z-invariant guide, referred to as the *ideal guide,* the natural basis is the set of modes of this ideal guide. One also refers to this expansion as the *ideal mode* expansion. The coupling coefficient in the scalar approximation [6] is shown to be

$$C_{j\ell} = \frac{k^2}{2\sqrt{|\beta_j \beta_\ell|}} \int_{A\infty} (n^2 - \bar{n}^2)\hat{\psi}_j^* \hat{\psi}_\ell \, dA \qquad (13.4)$$

where $n(x, y)$ is the perturbed guide index profile, $\bar{n}(x, y)$ the unperturbed one, A_∞ the guide cross-section area $\hat{\psi}_j$ and $\hat{\psi}_\ell$ the normalized *scalar* normal fields obeying

$$\int_{A\infty} \hat{\psi}_j^* \hat{\psi}_\ell \, dA = \delta_{j\ell} \qquad (13.5)$$

with $\delta_{j\ell}$ the Kroonecker symbol.

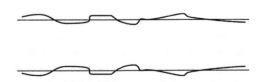

Figure 13.4 Perturbation of an ideal guide.

Figure 13.5 Perturbation of local guide.

13.3.2 Local mode coupling

Whenever, as shown in Figure 13.5, the guide varies slowly along z, one can define, in each z cross-section plane, a guide which locally coincides with the perturbed guide. Then, the so-called *local mode* expansion is performed. In this case, as opposed to the former one, the basis modes depend on z in particular through their propagation constants $\beta_j(z)$. The coupling coefficient, in the scalar approximation, is shown when one neglects longitudinal variations of the modal field

$$C_{j\ell} = -i \frac{k^2}{2\sqrt{\beta_j \beta_\ell}} \frac{1}{\beta_j - \beta_\ell} \int_{A\infty} \frac{\partial n^2}{\partial z} \hat{\psi}_j^* \hat{\psi}_\ell dA. \quad (13.6)$$

13.3.3 Individual guide coupling

In the case of couplers made of single-mode waveguides, as shown in Figure 13.6, the amplitudes a_j and a_ℓ of Equation 13.3 are the fundamental mode amplitudes of the unperturbed guides (i.e., without the other guides). Although obviously not constituting a basis, these modes serve as an expansion set, which permits us to describe the power exchange between the guides quite accurately. The coupling coefficient takes the same form as in Equation 13.4 although, here, $\hat{\psi}_j$ and $\hat{\psi}_\ell$ are the normalized modal fields of individual guides j and ℓ the unperturbed index profile $\bar{n}(x,y)$ is that of the individual guide supposed alone and $n(x, y)$ the index profile of the whole coupler.

Note, however, that in the case of a fused coupler, single-mode guides are not defined for any

cross-section so that one must abandon this concept and use supermodes instead, as described in more detail in Section 13.3.1.

13.3.4 Beating and coupling lengths

The beating and coupling lengths are the key parameters to predict the behavior of components.

Let $C_{j\ell}$ be the coupling coefficient between modes j and ℓ. Whenever the coupling coefficients do not depend on z, solutions of the coupled mode Equation 13.3 are sinusoidal functions. The coupling length L_C is defined as the length for which a complete *power* transfer cycle takes place (transfer from mode j to mode ℓ and back to mode j). One thus defines

$$L_C = \frac{2}{|C_{j\ell}|}. \quad (13.7)$$

The beating length z_b between these two modes is defined as the length along which the modes accumulate a 2π phase difference. One thus has

$$z_b = \frac{2}{|\beta_j - \beta_\ell|}. \quad (13.8)$$

These length scales permit us to define a criterion to discriminate between slowly varying adiabatic structures and abruptly varying ones for which modes are coupled. A rule of the thumb is given by the condition

$$L_c = z_b \quad (13.9)$$

1. If the coupling length of two given modes is larger than their beating length $L_C > z_b$, the modes accumulate a phase difference through the beating phenomenon without an exchange of power. The process may be considered as an adiabatic one.

2. If the coupling length of two given modes is smaller than their beating length $L_C < z_b$, the

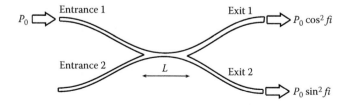

Figure 13.6 Diagram of a 2 × 2 coupler.

modes exchange power through the coupling process over such distances that their propagation phase difference is negligible. The process is then non-adiabatic.

13.4 ALL-FIBER PASSIVE COMPONENTS

Passive components are defined as components that do not necessitate an external source of energy, as opposed to lasers, amplifiers and other active components. Most of the time, all-fiber passive components are based on the technologies such as fusion-tapering and grating inscription described earlier. The fabrication parameters of an individual structure, tapered fiber coupler or grating, may be varied to obtain a prescribed behavior. In addition, several individual components may be concatenated, cascaded or mixed to accomplish more complex functionalities. The following sections give the essential basic concepts to understand the principles underlying the operation of all these components and describe some of the example components based on these principles.

These components are characterized by several parameters. Essential parameters are IL, CD, PDL and DGD.

1. The *insertion loss* (IL) is the transmittance in dB, defined by IL = 10 log T.
2. The *chromatic dispersion* (CD) is the pulse broadening per unit wavelength and length of propagation, resulting from both material and waveguide contributions. For fibers, it is expressed in ps nm^{-1} km^{-1} and, for step index profiles, the main contribution comes from the material CD, i.e., the dependence of the index of silica on wavelength. In components, it is expressed in ps nm^{-1} and is usually small except for gratings, which may be designed to compensate the dispersion accumulated over several kilometers of fiber.
3. The *polarization dependent loss* (PDL) is due to the component birefringence. This birefringence results an IL depending on polarization. The PDL parameter is defined by the difference of the ILs for the best and worst polarization states and is thus expressed in dB.
4. The *differential group delay* (DGD) characterizes the variation of the group delay with polarization. It is defined by the difference

between group delays for the fastest and slowest polarization states, and is usually expressed in ps. Note that the length of fiber is characterized by its *polarization mode dispersion* (PMD) expressed in ps km$^{-1/2}$, as the DGD accumulates statistically as the square root of distance of propagation.

13.4.1 Splitters and combiners

All-fiber splitters and combiners are based on couplers which are, most of the time, made by the fusion and tapering technique. Although a coupler may consist of an arbitrary number of fused possibly different fibers, most of them are 2 × 2 couplers made of identical fibers. Two geometric parameters determine their behavior:

- Their degree of fusion f that may theoretically vary from zero (tangent fibers) to one (completely fused fibers resulting in a circular cross-section) [7].
- Their longitudinal profile that is described by the variation of the inverse taper ratio (ITR-) parameter along z. In a first approximation, the tapering process is assumed to preserve the structure respective dimensions, so that one can characterize the transverse reduction by the ratio of a given length (e.g., the core radius or the coupler width) measured after and before the tapering process. This is the definition of the ITR-parameter.

13.4.1.1 PRINCIPLE OF OPERATION
13.4.1.1.1 Concept of supermodes

As a fused and tapered coupler is not longitudinally invariant, one must work with the *local modes* (or in other words, *local supermodes)* of the fused structure. As explained later, in identical fiber couplers, the power exchange from the main to the secondary branch, and vice versa, takes place via the *beating* phenomenon between supermodes. Supermode *coupling*, which would arise from very abrupt slopes, is undesirable as it would create loss. Symmetric fused couplers are thus, most of the time, adiabatic structures.

1. At the coupler entrance, the fibers are not tapered: modes of individual guides are confined in the cores, and one can consider that the supermodes are combinations of individual

guide modes. The two supermodes of a symmetrical 2×2 coupler, i.e., made of two identical fibers, are

$$\left|\psi_+\right\rangle = \frac{\left|\psi_1\right\rangle + \left|\psi_2\right\rangle}{\sqrt{2}} \text{ and } \left|\psi_-\right\rangle = \frac{\left|\psi_1\right\rangle - \left|\psi_2\right\rangle}{\sqrt{2}} \quad (13.10)$$

where the Dirac notation $\left|\psi_1\right\rangle$ and $\left|\psi_2\right\rangle$ is used to describe the fields of the fundamental (scalar) LP01 modes of guides 1 and 2. The supermodes corresponding to $\left|\psi_+\right\rangle$ and $\left|\psi_-\right\rangle$ are, respectively, called SLP01 and SLP11 after their circular two-layer fiber counterparts as can be seen in Figure 13.7 for degree of fusion $f=1$. Whenever one excites $\left|\psi_1\right\rangle$, the fundamental mode of fiber 1 at the entrance of the coupler, one actually excites a superposition of both supermodes, which can be written as

$$\left|\psi_1\right\rangle = \frac{\left|\psi_+\right\rangle + \left|\psi_-\right\rangle}{\sqrt{2}} \text{ and } \left|\psi_2\right\rangle = \frac{\left|\psi_+\right\rangle - \left|\psi_-\right\rangle}{\sqrt{2}} \quad (13.11)$$

All these modes are degenerate, i.e., the supermode propagation constants, β_+ and β_-, are identical to that of the each individual guide fundamental mode, β ($\beta = \beta_1 = \beta_2 = \beta_+ = \beta_-$)

Figure 13.8 shows, as an example, supermode fields for two different degrees of fusion ($f=0.1$ and 1) in the case of a core guiding structure. For a coupler made of standard

telecommunication fibers, supermodes may be approximated by superpositions of individual fiber modes as long as ITR > 0.4.

2. In the central region of the coupler, the individual guides lose their identity and the cores in their guiding roles. Only supermodes keep a physical meaning. Figure 13.7 shows the supermodes for the same degrees of fusion as in Figure 13.8, but for ITR=0.1. As shown in Figures 13.9 and 13.10 for two different degrees of fusion, the effective indices $n_{\text{eff}} = n_\pm$ (and thus propagation constants $\beta_\pm = 2\pi n_\pm/\lambda$) of the first two supermodes (SLP01 and SLP11) split apart in the central region of the coupler (for which ITR < 0.4). Supermodes accumulate a phase difference as they propagate along the coupler structure. In other words, it is the supermode *beating* phenomenon which governs the power exchange process.

3. At the coupler exit, the situation resembles that of the entrance, i.e., the guides are sufficiently separated so that the supermodes are again superpositions of individual guide modes and are consequently degenerated. It is thus the phase difference accumulated in the central region that determines the power splitting ratio at the exit of the coupler. Expressions of the individual modes (Equation 13.11) show that whenever the supermodes are in phase, the power is recovered in branch 1 and, whenever

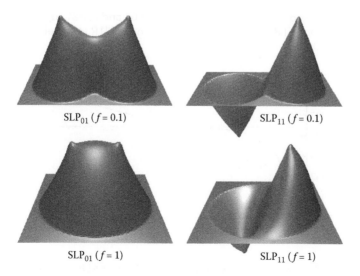

SLP$_{01}$ ($f = 0.1$)   SLP$_{11}$ ($f = 0.1$)

SLP$_{01}$ ($f = 1$)   SLP$_{11}$ ($f = 1$)

Figure 13.7 Fields of the first two supermodes of a 2×2 coupler for two different values of the degree of fusion and for an ITR=0.1.

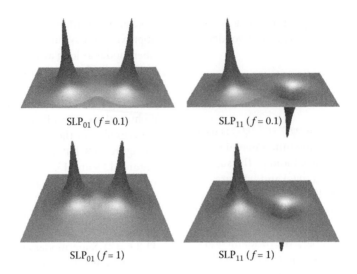

Figure 13.8 Fields of the first two supermodes of a 2 × 2 coupler for two different values of the degree of fusion and for an ITR=0.5.

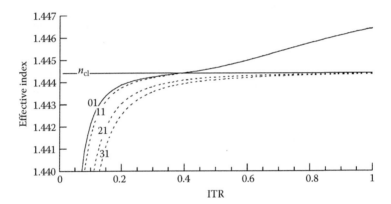

Figure 13.9 Effective indices of the first two symmetric SLP01 and SLP21 supermodes and of the first two antisymmetric SLP11 and SLP31 supermodes as functions of ITR at wavelength λ=1550 nm for a slightly fused coupler (f=0.1).

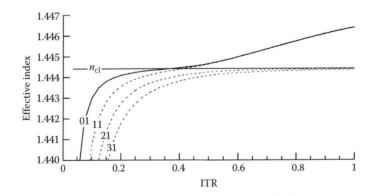

Figure 13.10 Effective indices of the first two symmetric SLP_{01} and SLP_{21} supermodes and of the first two antisymmetric SLP_{11} and SLP_{31} supermodes as functions of ITR at wavelength λ=1550 nm for a strongly fused coupler (f=1).

they are out of phase, it is recovered in branch 2. A 50/50 splitting ratio—which corresponds to the so-called 3 dB coupler—occurs whenever the supermodes exit the coupler with a $\pi/2$ phase difference.

13.4.1.2 BIREFRINGENCE

As detailed in the following section, a coupler is characterized by its transfer matrix, which relates individual guide amplitudes at the exit of the coupler to the entrance ones. This matrix depends on the accumulated supermode phase difference and thus on their local propagation constants β_+ (z) and β_- (z).

As the transverse cross-section of a coupler does not show the circular symmetry (even though $f=1$, the cores break off the symmetry), two propagation constants must be attributed to each supermode $SLP_{\ell m}$ corresponding to both polarizations along the symmetry axes. As a result, two transfer matrices must be defined, one for each polarization. Thus, from a general viewpoint, couplers are birefringent and their transmissions are polarization dependent. The less fused the coupler, the more obvious the departure from circular symmetry. One then understands that the birefringence is larger for slightly fused couplers than for those approaching $f=1$. If the birefringence is negligible, the scalar approximation is sufficient for the calculations. Otherwise, polarization corrections (see Chapter 13 of [6]) or exact (i.e., vectorial) calculations must be made.

13.4.1.3 TRANSFER MATRICES

The coupler transmissions as functions of elongation and wavelength greatly depend on the fabrication parameters, degree of fusion and longitudinal profile. These fabrication parameters determine the transfer matrix, which relates individual guide amplitudes at the exit of the coupler to the entrance ones. For a coupler of length L, one has [7]

$$M_{2\times2}(\alpha) = e^{i\bar{\alpha}} \begin{bmatrix} \cos\alpha & i\sin\alpha \\ i\sin\alpha & \cos\alpha \end{bmatrix} \quad (13.12)$$

with the average propagation phase

$$\bar{\alpha} = \int_0^L \bar{\beta}(v)dz, \quad \bar{\alpha}(z) = \frac{\beta_+(z)+\beta_-(z)}{2} \quad (13.13)$$

and the supermode accumulated phase difference

$$\alpha = \int_0^L \frac{\beta_+(z)+\beta_-(z)}{2}dz. \quad (13.14)$$

Here β_+ (z) and β_- (z) are the propagation constants of the local fundamental supermode SLP_{01} and of the first antisymmetric supermode SLP_{11}, respectively. The parameter 2α is the phase difference accumulated by these two supermodes along the coupler.

From the transfer matrix, one may, for a given entrance condition, calculate the transmitted power in both branches. For example, excitation in branch 1 corresponds to a column vector

$$\begin{bmatrix} 1 \\ 0 \end{bmatrix}$$

and, as a result, in a power transmission in the same branch

$$T_1 = \cos^2\alpha = \frac{1}{2}[1+\cos 2\alpha] \quad (13.15)$$

and, in the secondary branch

$$T_2 = \sin^2\alpha = \frac{1}{2}[1+\cos 2\alpha]. \quad (13.16)$$

Whenever the modes are in phase ($2\alpha=0+2p\pi$ with p an integer) at $z=L$, the transmission is maximum in the main branch ($T_1=1$). Whenever they are out of phase ($2\alpha=\pi+2p\pi$) at $z=L$, the transmission is minimum ($T_1=0$), corresponding to a complete power transfer in the secondary branch ($T_2=1$). This power exchange for a strongly fused 2×2 coupler is shown in Figure 13.11, which displays an experimental response as well as the theoretical one predicted from the fabrication parameters. The polarization effects are clearly visible in this example recording. From a more general viewpoint, the overall transmission is a superposition of both the polarization transmissions. For a polarization entering at $45°$ from the cross-section symmetry axes x and y, the transmission in the main branch may be written as

$$T_1 = \frac{T_{1x}+T_{1y}}{2} = \frac{\cos^2\alpha_x + \cos^2\alpha_y}{2}$$

$$= \frac{1}{2}\left[1+\cos(\alpha_x - \alpha_y)\cos[2(\alpha_x + \alpha_y)]\right] \quad (13.17)$$

with α_x and α_y the supermode accumulated phase difference for polarization x and y, respectively. In Figure 13.11, the birefringence appears as an

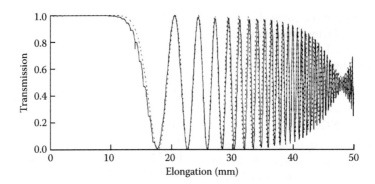

Figure 13.11 Recording (plain line) of the main branch transmission of a strongly fused 2 × 2 coupler as a function of elongation at a given wavelength (λ = 1550 nm). The transmission in the secondary branch, if the losses are negligible, is the complement. The dashed line is the corresponding numerical calculation. Loss of contrast is due to the coupler birefringence.

overmodulation of the power exchange with nodes, whenever both polarization transmissions are out of phase $(2\alpha_x - 2\alpha_y = \pi \pm 2m\pi)$. In the first power exchange cycles, the birefringence effect is negligible, which permits us to manufacture 3 dB polarization-independent couplers. In contrast, slightly fused couplers present strong birefringence [7].

Oscillations visible in Figure 13.11 as a function of elongation also occur as a function of wavelength, which confer to couplers spectral filtering or multiplexing applications as shown in the following sections.

13.4.1.4 COMPONENTS

13.4.1.4.1 Power splitters

The simplest component to be conceived is the power splitter. According to the Figure 13.11 recording,

any polarization-independent splitting ratio may be obtained for a relatively short elongation. They are, however, wavelength dependent, which may be undesirable for a number of applications.

Wavelength-independent splitters may be realized using 2 × 2 dissymmetric couplers. As opposed to the symmetric coupler, modes of individual guides of a dissymmetric coupler are not degenerated. When the individual guides are sufficiently separated, at the coupler entrance and exit, the supermodes are the individual guide modes. Quite paradoxically, an adiabatic dissymmetrical coupler would not experience any power exchange. It is thus a supermode *coupling* process (as opposed to a *beating* process) which governs the power transfer in a dissymmetric coupler. This power transfer may be complete or do not depend, for a given asymmetry, on the slopes of the longitudinal structure. The

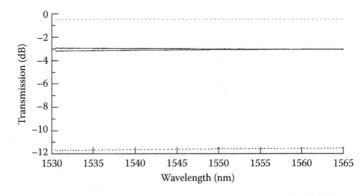

Figure 13.12 Experimental transmissions in dB (also referred to as IL) in both branches of wavelength-independent fused fiber 2 × 2 couplers as a function of wavelength. The plain lines are the transmission of a 3 dB wavelength-independent coupler while the dotted lines are those of a coupler with a large coupling ratio (around 12 dB). (By Courtesy of ITF Optical Technologies.)

calculation of the adiabaticity criterion as a function of the ITR-parameter is then critical to design a coupler with a prescribed response [8].

An alternative technique is by concatenating two 3dB couplers in a Mach–Zehnder (MZ) arrangement with a π phase difference between the two branches [9].

Both solutions are used in practice. Any coupling ratio is attainable with these techniques as exemplified in Figure 13.12.

13.4.1.4.2 Polarization splitters/ Combiners and depolarizers

Although considered a nuisance for power splitters, one may take advantage of the intrinsic polarization dependence of couplers, to realize polarization splitters—the all-fiber equivalent of polarizing cubes. These all-fiber polarizers consist of couplers elongated until the polarization node (obtained for an elongation of about 47 mm in Figure 13.11) is attained. Used in the reverse configuration, they serve as polarization combiners. These polarization splitters/combiners, however, suffer from being narrowband. A wideband alternative consists in an all-fiber MZ structure with a short length (half the beatlength) of a polarization maintaining fiber in one branch. An example response of a wideband polarization splitter/combiner is shown in Figure 13.13. These components are particularly useful as polarization pump combiners in Raman amplifiers. For a given power of the pump sources, they allow us to double the pump power at a given wavelength and reduce the

impact of the inherent polarization dependence of the Raman gain.

An alternative solution to obtain a polarization-independent Raman amplifier is to depolarize the pump source. This can be done with the help of a depolarizer that consists of an unbalanced interferometer. The arm length difference is longer than the polarization coherence length (which coincides with the usual light source coherence length) so as to uncorrelate the two polarizations. Different designs of depolarizers may be found in [10]. The highest-performance device takes advantage of the MZ polarization combiner (described in the previous section) combined with a fiber ring delay line working in a non-interferometric operation, i.e., with a loop length much longer than the coherence length of the light source. This type of device is characterized by its degree of polarization (DOP) defined as follows

$$DOP = \sqrt{1 - 4\left[\frac{I_x I_y}{(I_x + I_y)^2}\right]\left(1 - |g_{xy}|^2\right)} \quad (13.18)$$

where I_x and I_y are the autocorrelation functions for perpendicular polarizations and g_{xy} the cross-correlation normalized coefficient.

An example response of such a depolarizer is shown in Figure 13.14. It features a DOP of the order of 10% over a 120 nm spectral band for any temperature between 0°C and 70°C. This type of device is also of use for test and measurement applications.

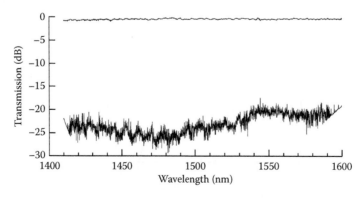

Figure 13.13 Experimental transmissions in one branch for both polarizations of a wideband polarization pump combiner as a function of wavelength. Transmissions in the second branch are similar with polarizations exchanged. (By Courtesy of ITF Optical Technologies.)

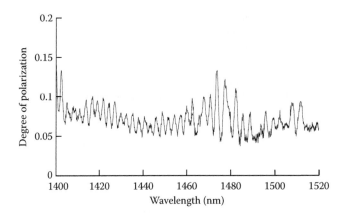

Figure 13.14 DOP as a function of wavelength of a wideband depolarizer. The ripples are due to imperfect alignment of the components in the MZ polarization combiner. (By Courtesy of ITF Optical Technologies.)

13.4.1.4.3 Wavelength splitters/ combiners

Fused couplers present an almost inherent sinusoidal wavelength dependence, with a period that decreases when the elongation increases. This wavelength dependence is the basis of several WDM applications. However, their intrinsic properties limit their application.

1. Even for a relatively large wavelength separation, the inherent sinusoidal spectral response might be a limitation. This is the case for a 1480 nm pump–1550 nm signal multiplexer, useful for Er-doped amplifiers/lasers.
2. The shortest wavelength period attainable is limited by the polarization dependence of

the couplers as mentioned earlier. They are consequently not suitable to realize dense wavelength division multiplexing (DWDM) interleavers able to de/multiplex channels as close as 1 nm or less.

In both the cases, alternative solutions may be found by concatenating couplers in MZ arrangements.

1. Figure 13.15 shows, as an example, the spectral response of an MZ 1480–1550 nm WDM allowing a large band around 1550 nm.
2. Dense interleavers are also usually made of an unbalanced MZ structure. Simple MZ consists of two concatenated 3 dB couplers, but more sophisticated non-sinusoidal responses

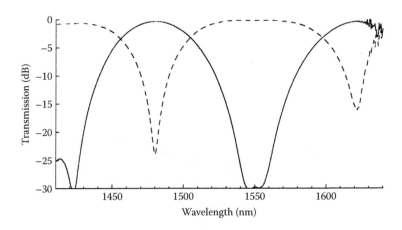

Figure 13.15 Spectral response of a pump–signal WDM combiner for Er-doped amplifiers/lasers. (By Courtesy of ITF Optical Technologies.)

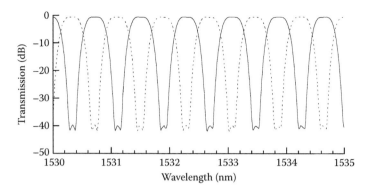

Figure 13.16 Spectral response of a DWDM interleaver with a wavelength spacing of 0.4 nm (50 GHz). (By Courtesy of ITF Optical Technologies).

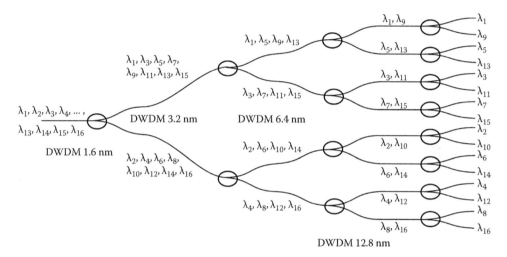

Figure 13.17 Cascade of 2^n-1 WDM splitters to de/multiplex 2^n channels (example wavelength spacing 1.6 nm).

may be obtained by concatenating several MZ structures. Figure 13.16 shows an example of spectral response of such a device. These interleavers are then cascaded in a tree arrangement to effectively separate (or combine) the channels as shown in Figure 13.17. Challenges are to make them temperature insensitive with a high isolation, and to respect, over a large spectral range, the ITU grid which is characterized by constant frequency intervals.

Single couplers are nevertheless useful, among other things, to realize de/multiplexers for well-separated signal channels (coarse WDM such as that shown in Figure 13.18), pump–signal multiplexers and pump combiners for Er-doped and Raman amplifiers.

13.4.1.4.4 Mode splitters/combiners

When made of two-mode fibers, fused couplers may be designed to separate the modes, LP_{01} staying in the main branch while LP_{11} is transferred to the secondary branch [11]. This modal splitter in combination with a reflecting modal converter Bragg grating opens up the possibility to get rid of inherently lossy micro-optic circulators.

13.4.2 Spectral filters and dispersion compensators

13.4.2.1 PRINCIPLES OF OPERATION

Most of these components, as opposed to couplers, have only one entrance and one exit branch. Some are based on the tapering technology, others on

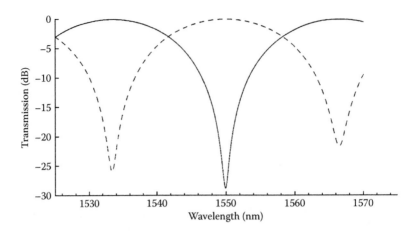

Figure 13.18 Spectral response of a coarse WDM. Degradation due to polarization dependence is visible. (By Courtesy of ITF Optical Technologies).

the grating inscription technology. Combination of both the technologies, as in the optical add and drop multiplexers (OADM) described later, may also be used. The tapering technology takes advantage of the sinusoidal wavelength responses of tapered fibers while gratings are resonant devices: they couple, at a given wavelength λ, two modes of propagation constants β_1 and β_2 through the matching condition of Equation 13.1

$$(\beta_1 - \beta_2)\Lambda = 2p \qquad (13.19)$$

where Λ is the grating period. Mode 1 is usually the fundamental core mode, while mode 2 is, in the case of a short-period grating, the counter-propagating fundamental mode ($\beta_2 = -\beta_1$) and, in the case of a long-period grating, a copropagating cladding mode.

13.4.2.2 TAPERED FIBERS

13.4.2.2.1 Oscillatory transmissions

When the slopes of a tapered single-mode fiber are abrupt, such as those of Figure 13.1, one observes oscillations in the transmitted power as a function of elongation (see Figure 13.19). For a given elongation, similar oscillations occur in the transmission as a function of wavelength (see Figure 13.20). This behavior may be explained in terms of coupling and beating of the local modes of the tapered structure [12].

When the downtaper is so abrupt that the adiabaticity criterion is not fulfilled, the fundamental mode LP_{01} is unable to adapt its field and propagation constant to the guide change and it is transformed through the coupling process into a superposition of modes of the same symmetry (LP_{01}, LP_{02}, LP_{03}, …), which propagate along the adiabatic central region, thus accumulating phase differences. When they

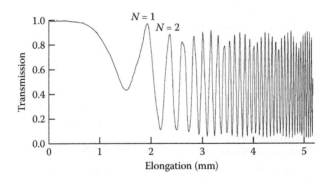

Figure 13.19 Fundamental mode transmission of a tapered single-mode fiber as a function of elongation.

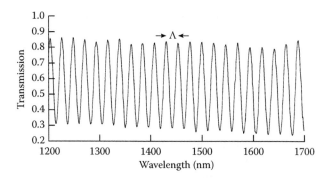

Figure 13.20 Spectral response of the tapered fiber of Figure 13.19.

enter the uptapered region, they again experience a coupling process. Power is then recovered in the core. The core transmission depends on the relative phase of the different modes. If they exit the component in phase, transmission is 1. Otherwise, power is dispatched between the fundamental core mode and cladding modes. Cladding mode power is eventually lost in the jacket.

This coupling-beatings-coupling process thus confers to a tapered fiber an oscillatory transmission. Large-amplitude oscillations seen in Figure 13.20 come from LP_{01} and LP_{02} modes, while small-amplitude and larger-frequency oscillations only occur if LP03 and possibly higher-order modes participate in the process [13]. Besides, the $LP_{01} - LP_{02}$ period A of the tapered fiber spectral response may be predicted from N, the number of oscillations observed during elongation performed at wavelength λ. For sufficiently large N, one has

$$\Lambda = \frac{\lambda}{N}. \tag{13.20}$$

This oscillatory predictable spectral response is the basis for spectral filtering devices made of tapered fibers.

13.4.2.2.2 Long-period gratings (LPGs)

Gratings are also useful in terms of their filtering applications, but as opposed to tapered fibers their behavior is based on a resonant coupling resulting in a wavelength peak in their spectral response.

- *Transfer matrix*: A sinusoidal perturbation of period $\Lambda = 2\pi/\beta_B$ induces a coupling coefficient

$$C_{12} = 2c \, \cos(\beta_B z + f) \tag{13.21}$$

which, around resonance condition (13.19), couples two codirectional modes of propagation

constants β_1 and β_2. One can show, by solving coupled mode equations, that the transfer matrix is (13.22)

$$M_{LPG} = \begin{bmatrix} \left[\cos(\gamma\ell) + i\frac{\Delta}{\gamma}\sin(\gamma\ell) \right] e^{i(\beta_1 - \Delta)\ell} & i\frac{c}{\gamma}\sin(\gamma\ell) e^{i[\beta_1 - \Delta]\ell + \beta_B z_0 + \phi} \\ i\frac{c}{\gamma}\sin(\gamma\ell) e^{i[\beta_2 + \Delta]\ell - \beta_B z_0 - \phi} & \left[\cos(\gamma\ell) - i\frac{\Delta}{\gamma}\sin(\gamma\ell) \right] e^{i(\beta_1 + \Delta)\ell} \end{bmatrix}. \tag{13.22}$$

Here ℓ is the length of the grating extending from z_0 to $z_0 + \ell$ and the other parameters are defined as follows

$$\Delta = \frac{\beta_1 - \beta_2 - \beta_B}{2} = \frac{\beta_1 - \beta_2}{2} - \frac{\pi}{\Lambda} \text{ and}$$

$$\gamma^2 = \Delta^2 + c^2. \tag{13.23}$$

As for the coupler, if one assumes injection in mode 1 ($a_2(z_0) = 0$), the power transfer to mode 2 is readily calculated to be

$$T_2(z_0 + \ell) = |M_{21}|^2 = \frac{c^2}{\gamma^2} \sin^2(\gamma\ell). \tag{13.24}$$

Remaining power in mode 1 can be calculated by $T_1 = |M_{11}|^2$ or using the energy conservation condition $T_1 + T_2 = 1$.

The power transfer to mode 2 is sinusoidal as a function of ℓ, but a sinc function of $\gamma = \pm\sqrt{\Delta^2 + c^2}$. The peak-to-peak amplitude of the transmission is

$$T_{2MAX} = \frac{c^2}{\gamma^2} = \frac{c^2}{\Delta^2 + c^2}. \tag{13.25}$$

It is always smaller than 1, being maximum for $\Delta=0$, i.e., at resonance. Figure 13.21 shows the oscillatory fundamental mode transmission along z.

- *Bandwidth*: Let the LPG length ℓ equal half the coupling length defined by Equation 13.7: $\ell=L_C/2=\pi/(2c)$ so that, at resonance, the power transfer is complete. (This would occur for any odd number of $L_C/2$.) Figure 13.22 shows, for a component of length $L_C/2$, the power transfer from mode 1 to mode 2 as a function of the detuning $|\Delta|$.
- The LPG used as a filter may be characterized by its bandwidth, i.e., by the FWHM of its transmission, or approximately of T_{2MAX}, which in terms of Δ is equal to $2c$. It is readily converted in terms of wavelength to give

$$\delta\lambda = \frac{2\lambda^2}{L_C|n_{g1}-n_{g2}|} \qquad (13.26)$$

where n_{g1} and n_{g2} are the group indices defined by

$$n_{gi} = n_i - \lambda\frac{dn_i}{d\lambda}. \qquad (13.27)$$

The following approximation

$$\delta\lambda \approx \frac{2\lambda\Lambda}{L_C} = \frac{\lambda}{N} \qquad (13.28)$$

is valid inasmuch as $n_{g1} - n_{g2}\approx n_1 - n_2$. Here, n_i are the modal effective indices ($n_i=2\pi/\beta_i/\lambda$), λ is the peak wavelength and N the total number of steps. As a result, the greater the number of steps, the more selective the grating will be. For a given grating step (determined by the pair of chosen modes at λ), the longer the grating (and as a consequence the smaller the coupling coefficient c, since $L=L_C/2$), the narrower its bandwidth.

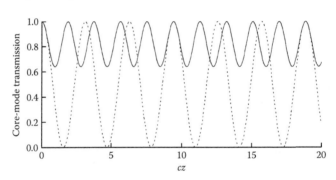

Figure 13.21 Fundamental mode transmission along the grating for different values of the detuning parameter Δ. The oscillation amplitude is maximum (equal to 1) for $\Delta=0$ and the frequency minimum (dotted line). As Δ increases, γ increases as well, making the amplitude decrease and the frequency increase: the transmission is shown for $\Delta/c=4/3$ (plain line).

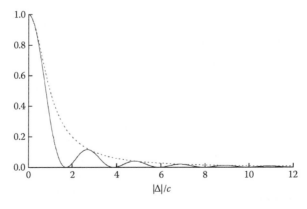

Figure 13.22 Power transfer T_2 as a function of the detuning $|\Delta|$ in a component of length $L_C/2$ (plain line). Its amplitude T_{2MAX} is the dotted line.

- *Fundamental mode transmission:* The cladding modes being absorbed by the jacket, most of the time, one only has access to T_1. LPGs are thus rejection band filters. Actually, for a given grating step Λ, several pairs of modes (i, j) may be resonant in a wavelength range inasmuch as the following condition is fulfilled.

$$\beta_i - \beta_j = \frac{2\pi}{z_{ij}} = \frac{2\pi}{\lambda}(n_i - n_j) = \frac{2\pi}{\Lambda}. \quad (13.29)$$

Here, z_{ij} is the beating length of modes i and j and n_i and n_j their effective indices, which depend on the wavelength λ. This equation generalizes the matching condition written in Equation 13.19. If, as in tapered fibers, the perturbation preserves the circular symmetry, only modes of the same symmetry are involved. Figure 13.23 shows the beating lengths z_{1j} for LP_{01}–LP_{0j} modal pairs as functions of λ. This graph permits us to find the resonance wavelengths for a given step Λ. Alternatively, it gives the periods that may be chosen to realize a filter at a given wavelength.

Figure 13.24 gives, as a function of wavelength, the dB transmission in the fundamental mode of a grating of step $\Lambda = 525\,\mu m$ in a standard telecommunication fiber.

Note finally that, with slanted gratings, it is also possible to transfer power from the fundamental core mode LP_{01} to cladding modes with different symmetry such as LP_{11} and LP_{21}.

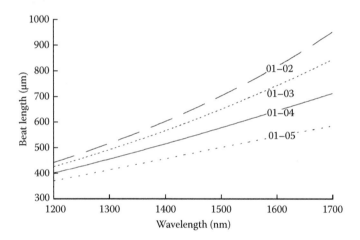

Figure 13.23 Beating lengths of LP_{01}–LP_{0j} modal pairs for a typical telecommunication fiber.

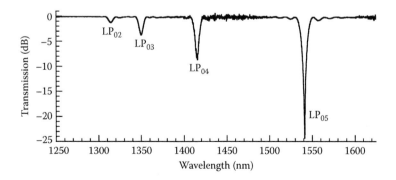

Figure 13.24 Fundamental mode transmission of a long-period grating. Experimental response (plain line) is compared to the calculated one (dotted line). The grating was made by irradiating a standard telecommunication fiber with a CO_2 laser. It consists of 118 steps of 525 μm. The peaks are identified by the 0m indices of the LP cladding mode, which is coupled with the fundamental core mode.

The main limitation of the LPGs used as filtering components is their temperature dependence of the order of 100–150 pm°C^{-1} (10–15 nm/100°C). This high figure, which largely depends on the fiber, comes from the fact that the principle of operation of an LPG is based on the coupling of a core mode with a cladding mode. Being made up of different materials, these two layers are subjected during hot drawing and subsequent cooling of fiber to different stresses, which result in a different thermal sensitivity. LPGs made of standard SMF28 *Corning* fibers exhibit a thermal dependence of 50 pm°C^{-1}.

13.4.2.2.3 Short-period gratings

Let the period $\Lambda = 2\pi/\beta_B$ of the sinusoidal perturbation couple two contradirectional modes of the same propagation constant $\beta_1 = -\beta_2 = \beta$. The corresponding coupling coefficient may be written as

$$C_{12} = 2ic \cos(\beta_B z + \phi). \tag{13.30}$$

With a notation similar to that used in the codirectional case, one has

$$\Delta = \frac{\beta_1 - \beta_2}{2} - \frac{\beta_B}{2} = \beta - \frac{\beta_B}{2} = \beta - \frac{\pi}{\Lambda}.$$

- *Transfer matrix:* By solving the coupled mode equations, one then finds the transfer matrix of a grating extending from z_0 to $z_0 + \ell$

$$M_{SPG} = \begin{bmatrix} \left[\cosh(\gamma\ell) + i\dfrac{\Delta}{\gamma}\sinh(\gamma\ell)\right]e^{i(\beta-\Delta)\ell} \\ -i\dfrac{c}{\gamma}\sinh(\gamma\ell)e^{-i[(\beta-\Delta)\ell+\beta_B z_0 - \phi]} \end{bmatrix}$$

$$\begin{bmatrix} i\dfrac{c}{\gamma}\sinh(\gamma\ell)e^{i[(\beta-\Delta)\ell+\beta_B z_0 + \phi]} \\ \left[\cosh(\gamma\ell) - i\dfrac{\Delta}{\gamma}\sinh(\gamma\ell)\right]e^{-i(\beta-\Delta)\ell} \end{bmatrix}. \tag{13.31}$$

where

$$\gamma^2 = c^2 - \Delta^2 \tag{13.32}$$

1. If $\Delta \leq c$,

$$\gamma = \sqrt{c^2 - \Delta^2} \tag{13.33}$$

and solutions are those written in Equation 13.31 with γ real.

2. If $\Delta \geq c$,

$$\gamma = i\sqrt{\Delta^2 - c^2} \tag{13.34}$$

so that solutions can be written as in Equation 13.31, but with $\gamma = i|\gamma|$ an imaginary number. Using then $\cosh(i|\gamma|z_0) = \cos(|\gamma|z_0)$ and $\sinh(i|\gamma|z_0) = i\sin(|\gamma|z_0)$, one can rewrite them as sinusoidal functions.

- *Fundamental mode reflection and transmission:* The reflection and transmission coefficients may then be calculated for given limit conditions. Inversion of the matrix M_{SPG} gives

$$\begin{bmatrix} a_1(z_0) \\ a_2(z_0) \end{bmatrix} = \begin{bmatrix} M_{22} & -M_{12} \\ -M_{21} & M_{11} \end{bmatrix} \begin{bmatrix} a_1(z_0+\ell) \\ a_2(z_0+\ell) \end{bmatrix} \tag{13.35}$$

Let $z_0 = 0$. Usually one has $a_2(\ell) = 0$, so that one finds the amplitude reflection and transmission coefficients

$$r = \frac{a_2(0)}{a_1(0)} = -\frac{M_{21}}{M_{22}} \text{ and } t = \frac{a_1(\ell)}{a_1(0)} = \frac{1}{M_{22}} \tag{13.36}$$

and the power coefficients

$$R = R(0) = \frac{|M_{21}|^2}{|M_{22}|^2} = \frac{|M_{21}|^2}{|M_{11}|^2} \text{ and } T$$
$$= T(\ell) = \frac{1}{|M_{22}|^2} = \frac{1}{|M_{11}|^2} \tag{13.37}$$

Explicitly, one has to distinguish the cases depending on the detuning with respect to the coupling coefficient.

1. If $\Delta \leq c$

$$R(z) = \frac{|a_2(z)|^2}{|a_1(0)|^2} = \frac{c^2}{\gamma^2} \frac{\sinh^2[\gamma(\ell-z)]}{\cosh^2(\gamma\ell) + \dfrac{\Delta^2}{\gamma^2}\sinh^2(\gamma\ell)} \tag{13.38}$$

$$T(z) = \frac{|a_1(z)|^2}{|a_1(0)|^2} = \frac{\cosh^2[\gamma(\ell-z)] + \dfrac{\Delta^2}{\gamma^2}\sinh^2[\gamma(\ell-z)]}{\cosh^2(\gamma\ell) + \dfrac{\Delta^2}{\gamma^2}\sinh^2(\gamma\ell)} \tag{13.39}$$

with $\gamma = \sqrt{c^2 - \Delta^2}$.

2. If $\Delta \geq c$

$$R(z) = \frac{|a_2(z)|^2}{|a_1(0)|^2} = \frac{c^2}{\gamma^2} \frac{\sin^2[\gamma(\ell-z)]}{\cos^2(\gamma\ell) + \dfrac{\Delta^2}{\gamma^2}\sin^2(\gamma\ell)} \tag{13.40}$$

$$T(z)=\frac{|a_1(z)|^2}{|a_1(0)|^2}=\frac{\cos^2[\gamma(\ell-z)]+\dfrac{\Delta^2}{\gamma^2}\sin^2[\gamma(\ell-z)]}{\cos^2(\gamma\ell)+\dfrac{\Delta^2}{\gamma^2}\sin^2(\gamma\ell)}$$

(13.41)

with $\gamma=\sqrt{\Delta^2-c^2}$.

The minimum transmission (corresponding to a maximum reflection) takes place at resonance: $\Delta=0$ or $\gamma=c$ (Figure 13.25). For a grating of length ℓ one calculates

$$R_{\text{Max}}=\tanh^2(c\ell)\text{ and }T_{\min}=\frac{1}{\cosh^2(c\ell)}$$

(13.42)

Figures 13.26 and 13.27 show the reflection and transmission behavior as a function of z depending on the value of the detuning parameter Δ. The transmission coefficient $T(z)$ may be locally > 1. This local energy accumulation, similar to that observed in a Fabry–Perot interferometer, is due to the resonance or stationary waves between the grating ends. This effect, which gives rise to side lobes in the spectral response, is undesirable for spectral filtering applications. It may be suppressed by apodizing techniques as explained in a subsequent section.

- *Spectral response*: As for LPGs, the wavelength response is determined by the transmission and reflection coefficients as functions of the Δ parameter. The magnitude of Δ compared to that of the coupling coefficient c determines γ as per Equation 13.32, which is real close to the resonance condition, but purely imaginary

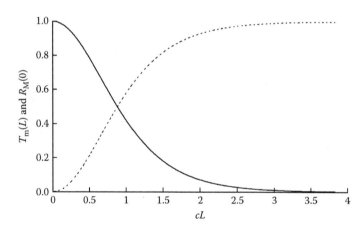

Figure 13.25 Reflection and transmission in the fundamental core mode at resonance ($\Delta=0$) as a function of the grating length.

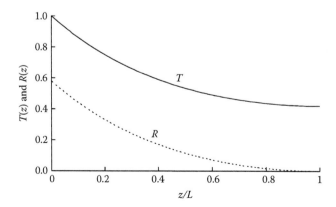

Figure 13.26 Reflection $R(z)$ and transmission $T(z)$ in the fundamental mode along the grating of length $\ell=1/c$ at resonance, i.e., for $\Delta=0$. Since $\Delta < c$, the amplitude varies exponentially.

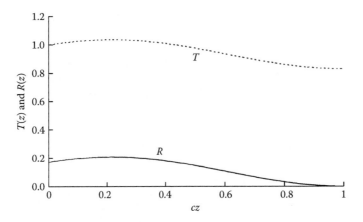

Figure 13.27 Reflection $R(z)$ and transmission $T(z)$ in the fundamental mode along the grating for a far from resonance case $\Delta > c$. In this case, the amplitude shows a sinusoidal variation. The calculation is made for $\Delta/c = \sqrt{17}$.

as soon as $\Delta > c$. Transmission then becomes oscillatory, resulting in the typical spectral side lobes. The $\Delta = c$ condition, as in the LPG case, may thus be used to define the peak width. An example spectral response in reflection is shown in Figure 13.28.

- *Apodization*: For most applications, side lobes seen in the spectral response are undesirable. They originate from the multiple reflections that take place between the grating ends as in a Fabry–Perot resonator. One can also

understand this effect by remembering that, in a first approximation (i.e., when the perturbation is weak), the spectral response of a grating is the Fourier transform of the amplitude c of the refractive index modulation. The Fourier transform of a rectangular function is a sinc function. The larger the rectangular function, the narrower the sinc function. Similar observations are valid for the grating: the longer the grating, the narrower the spectral width and its side lobes.

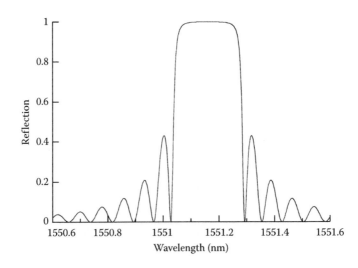

Figure 13.28 Power reflection coefficient of a short-period grating as a function of wavelength (simulation).

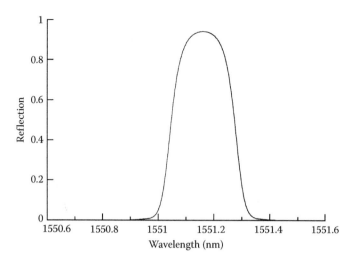

Figure 13.29 Power reflection coefficient of an apodized short-period grating as a function of wavelength (simulation).

One can reduce these side lobes by smoothing the coupling coefficient, giving it a gaussian envelope. This results in smoothing the wavelength response (in other words, apodizing). To be efficient both on the short- and long-wavelength sides, the average refraction index all over the grating length should be uniform. Figure 13.29 shows the effect of apodization on the grating of Figure 13.28.

The thermal dependence of SPGs is much weaker than that of LPGs. It is typically 10 pm°C (1 nm/100°C), i.e., approximately ten times less than that of LPGs, and does not depend on the fiber. Nevertheless, the thermal sensitivity of these gratings, irrespective of their application, must be compensated by an appropriate packaging.

13.4.2.3 SPECTRAL FILTERS

Spectral filtering is a key function of WDM networks. Some applications are described below.

13.4.2.3.1 Narrow-width spectral filters

The most immediate application of the Bragg SPGs is to use them as selective mirrors in all-fiber lasers in linear (as opposed to loop) configuration. They are currently found in doped fiber lasers as well as cascaded Raman lasers used to pump Raman gain amplifiers.

Bragg SPGs are also used to stabilize both the wavelength and power of semiconductor pump laser diodes. In this case, the grating, with a reflection coefficient of only a few per cent, is integrated in the pigtail at a distance exceeding the coherence length of the laser.

Finally, filters are needed to enhance isolation between the pump and the signal or between two adjacent signal channels. They are also used in combination with a circulator (which is a micro-optic device) to drop the signal of a particular wavelength channel. The ideal spectral amplitude shape to accomplish this type of function is then a rectangular response with a prescribed bandwidth. The Bragg reflection grating remains well suited for this type of filtering, especially if narrowness is an issue: although not perfectly rectangular, well-known standard apodization techniques are used to eliminate undesirable side lobes and give rise to excellent spectral characteristics. However, as can be seen in Figure 13.30 (top), such a standard grating filter has a parabolic group delay profile that adversely affects signals at the transmission rates of 10 and 40 Gb s^{-1}. More complex apodization profiles with phase shifts allow for correction and equalization of this parabolic group delay characteristic and result in an ultra-low dispersion over the filter passband. As shown in Figure 13.30 (bottom), group delay can be minimized to a ripple function of less than ±5 ps in amplitude [14].

13.4.2.3.2 Large-width spectral filters

Other filters with larger bandwidths are also needed. Among them, the most often used is the gain flattening filter (GFF) designed to equalize

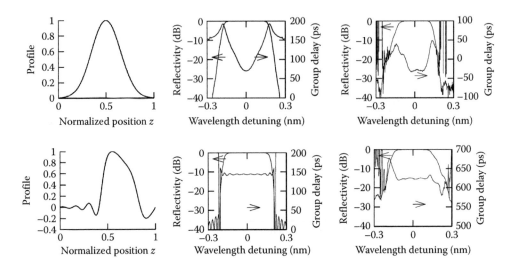

Figure 13.30 Comparison of standard (top) and low-dispersion (bottom) designs of Bragg SPG 50 GHz filters. The apodization profile (left column), theoretical reflectivity and group delay spectra (center column) and experimental measurements (right) are compared. (By Courtesy of TeraXion.)

the channels amplified by Er-doped fiber gain. The following all-fiber technologies have been explored and implemented sometimes in conjunction.

1. Fusion-tapering technology
2. Short-period gratings (SPGs), straight or slanted
3. Long-period gratings (LPGs), straight or slanted
4. Thin-film technology

However, the best fits to the inverse gain curve are obtained by using concatenated tapered structures or chirped short-period unslanted Bragg gratings in the transmission mode. Both types of GFF offer similar performance in terms of error function (< ±0.1 dB), PMD (< 0.05 ps). Thermal stability is better for gratings (< 0.5 pm°C^{-1}) than for tapered fiber filters (< 2 pm°C^{-1}). While the grating solution is more compact, it suffers back reflection, which needs an additional isolator. Group delay ripples, which result from undesired weak reflections occurring along the grating, are negligible with a typical amplitude of ±0.3 ps. They both are commercially available. Example responses are shown in Figures 13.31 and 13.32.

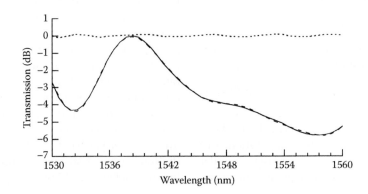

Figure 13.31 Transmission of a GFF made with the fusion-tapering technique. The plain line shows the target filter transmission, while the dashed one is that of the manufactured filter. The dotted line around 0 dB shows the error function. Its amplitude is less than ±0.1 dB. (By Courtesy of ITF Optical Technologies.)

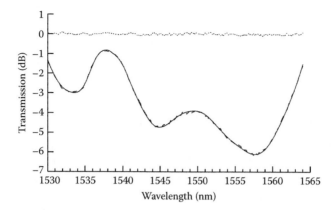

Figure 13.32 Transmission of a GFF made of a chirped Bragg grating. The plain line shows the target filter transmission, while the dashed one is that of the manufactured filter. The dotted line around 0 dB shows the error function. Its amplitude is less than ±0.1 dB. (By Courtesy of TeraXion.)

13.4.2.3.3 Mode converters

At the resonance wavelength, an LPG is a mode converter operating in the codirectional scheme. Similarly, an SPG can also be used as a mode converter, but operates in the contradirectional scheme, since not only the fundamental mode, but other contradirectional modes may also be reflected through a short-period Bragg grating [15]. By choosing the step, the length and the grating tilt with respect to the fiber axis, one can choose the wavelength and the mode(s) in which there is conversion. Figure 13.33 shows the spectral response of an LP_{01}–LP_{11} converter. To avoid undesirable conversion to cladding modes, a special fiber should be used. Used in conjunction with mode splitters, the mode converters open up the design of new all-fiber components, which would get rid of micro-optic devices such as circulators.

13.4.2.3.4 OADMs

A signal at the Bragg wavelength rather than reflected from a short-period grating back to the source can be extracted. It is then a selective wavelength demultiplexer, which has a significant role to play in future communication networks. Indeed, the strategy that is implemented to currently increase the data flow in a fiber link is the DWDM. Channel frequency spacing is typically 100 GHz (corresponding to 0.8 nm around $\lambda = 1550$ nm), which makes it possible to put 40 channels in the

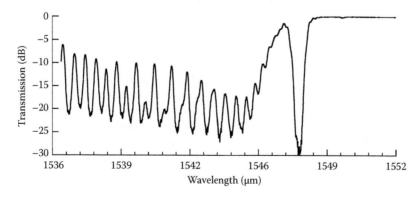

Figure 13.33 Experimental transmission of a mode converter as a function of wavelength. The component consists in a slanted short-period grating written in a bimodal fiber. The fundamental LP_{01} core mode is converted into the LP_{11} core mode at $A = 1547.5$ nm. At shorter wavelengths conversion to higher-order LP_1m, LP_2m, LP_3m cladding modes can be seen.

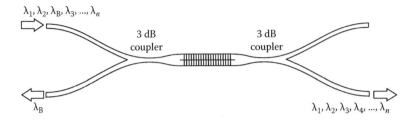

Figure 13.34 Principle of operation of an all-fiber OADM.

EDFA gain bandwidth (the C-band) and thus to multiply the current flow by 40, which is of 2.5 Gbit s^{-1} Narrower spacings such as 50 and 25 GHz (0.4 and 0.2 nm) are also used (see Figure 13.16 for an example 50 GHz spacing component).

To extract the reflected signal, one can use a circulator, which is made up of micro-optic components. An alternative all-fiber solution is shown in Figure 13.34, which makes use of identical gratings in an MZ arrangement. Cross-talk between channels must be avoided by apodizing the spectral response and by a proper design of the fiber to get rid of the cladding mode reflections. From the fabrication viewpoint, the symmetry of the component is critical for its correct operation. Alternative solutions making use of a single grating written in the central region of a single coupler were also studied [16].

13.4.2.4 DISPERSION COMPENSATORS

Dispersion in standard fibers mainly comes from material CD, which happens to be zero around

$\lambda = 1.3\,\mu m$. In the C-band (around $\lambda = 1.55\,\mu m$), dispersion causes a pulse to spread in time, the highest frequencies (shortest wavelengths) arriving before the lowest ones. CD is inherently troublesome and solutions to eliminate it completely by using dispersion-shifted fiber were rapidly abandoned because four-wave mixing (a non-linear effect, which induces cross-talk between channels) is automatically phase matched around the zero-dispersion wavelength. It is thus better to compensate for dispersion by using additional devices.

The simplest way is to use a dispersion compensation module (DCM), which consists of a length of a different fiber with a carefully designed index profile to tailor the modal propagation constant. A relatively long fiber length of compensating fiber must be used (typically 20 km of compensating fiber for 100 km of standard fiber), and it only partially compensates for higher-order dispersion effects (typically 60% of the slope), which is the channel-to-channel variation of the dispersion.

Because of their compactness, Bragg SPGs are alternative promising devices that may be used for dispersion compensation. Chirped SPGs (CSPGs) shown in Figure 13.35 reflect different frequency components of a pulse at different locations. Its first step being longer than the last, with a careful design, it can compensate for the delay of the low frequencies on the high ones. To recover the reflected signal, a circulator (a micro-optic component) is necessary. An SPG can have a dispersion of several orders of magnitude higher than a similar length of fiber, so that a few cm long dispersion compensator compensates for the dispersion of many kilometers of optical fiber. While relatively narrowband in nature, CSPGs can be made wideband by making them multi-channel. A multichannel CFBG-based dispersion compensator can be made by writing many CSPG components in separate sections of a fiber or by superimposing many CSPG components on the same section of

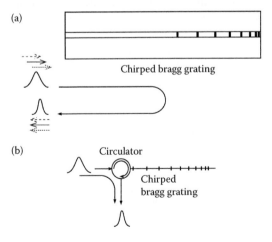

Figure 13.35 (a) Principle of operation of a chirped SPG to compensate pulse dispersion and (b) dispersion compensator made of a chirped SPG associated with a circulator.

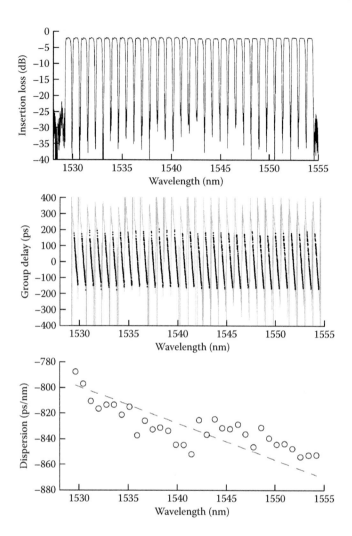

Figure 13.36 Reflectivity spectrum (top), group delay spectrum (center), and dispersion (bottom) of a 32-channel dispersion compensation grating. (By Courtesy of TeraXion).

fiber, the latter solution resulting in a very compact component. Another advantage of the SPGs over standard broadband compensation techniques such as DCF is that they are inherently independent of each other, which allows the design of complex structures including channel skipping and channel-to-channel variation of the dispersion.

Using the superimposed approach, compensation over up to 32 channels has been achieved with 100 GHz spacing and allowing for compensation of both the second- and third-order CD accumulated over 50 km of SMF-28 fiber [14]. Due to the possible adjustment of the dispersion on a per-channel basis, the wideband compensation based on a superimposed grating is a good candidate for compensating such a residual CD. As an example,

Figure 13.36 shows the reflectivity and group delay spectra of a 32-channel dispersion compensation grating with a 100 GHz spacing in which the dispersion varies from channel to channel: from −860 to −800 ps nm^{-1}.

CSPGs can also be used in conjunction with DCM to fully manage the dispersion over a large number of WDM channels.

13.5 CONCLUSION

This chapter has highlighted the importance of all-fiber components, which are an integral part of the current telecommunication networks. They already perform extremely diversified functions and their performance, thanks to the maturity of manufacturing technologies, is excellent. Depending on applications, they must sometimes compete with their integrated or micro-optics equivalents. The future communication networks will undoubtedly take advantage of all the available technologies, but the all-fiber will keep a privileged position thanks to the inherent low excess loss.

ACKNOWLEDGMENTS

The authors gratefully acknowledge the contribution of Dr Nawfel Azami and Dr Nicolas Godbout from ITF Optical Technologies Inc. and of Dr Yves Painchaud from TeraXion.

REFERENCES

1. Stewart, W. J., and Love, J. D. 1985. Wavelength-flattened fused couplers. *Electron. Lett.* 21, 742–743.
2. Lacroix, S., Gonthier, F., and Bures, J. 1988. Fibres unimodales effilées. *Ann. Télécomm.* 43, 43–47.
3. Kashyap, R. 1999. Fiber Bragg gratings. *Optics and Photonics*, ed. T. Tamir (New York: Academic).
4. Othonos, A., and Kalli, K. 1999. *Fiber Bragg Gratings, Fundamentals and Applications in Telecommunication and Sensing* (Boston, MA: Artech).
5. Perron, D., Orsini, P., Daxhelet, X., Lacroix, S., and Verhaegen, M. 2000. Long-period-grating fabrication techniques. *ICAPT 2000—Photonics North.*
6. Snyder, A. W., and Love, J. D. 1983. *Optical Waveguide Theory* (London: Chapman and Hall).
7. Lacroix, S., Gonthier, F., and Bures, J. 1994. Modeling of symmetric 2×2 fused-fiber couplers. *Appl. Opt.* 33, 8361–8369.
8. Birks, T. A., and Hussey, C. D. 1989. Wavelength-flattened couplers: Performance optimisation by twist-tuning. *Electron. Lett.* 25, 407–408.
9. Gonthier, F., Ricard, D., Lacroix, S., and Bures, J. 1991. Wavelength-flattened 2×2 splitters made of identical single-mode fibers. *Opt. Lett.* 16, 1201–1203.
10. Azami, N., Villeneuve, E., Villeneuve, A., and Gonthier, F. 2003. All-sop all-fiber depolarizer linear design. *Optical Fiber Conference TuK5*, pp. 230–231.
11. Shou, Y., Bures, J., Lacroix, S., and Daxhelet, X. 1999. Mode separation in fused fiber coupler made of two-mode fibers. *Opt. Fiber Technol.* 5, 92–104.
12. Love, J. D., Stewart, W. J., Henry, W. M., Black, R. J., Lacroix, S., and Gonthier, F. 1991. Tapered single-mode fibres and devices: Part 1. Adiabaticity criteria. *IEE Proc. J: Optoelectron.* 18, 343–354.
13. Lacroix, S., Bourbonnais, R., Gonthier, F., and Bures, J. 1986. Tapered monomode optical fibers: Understanding large power transfer. *Appl. Opt.* 25, 4424–4429.
14. Lachance, R. L., Painchaud, Y., and Doyle, A. 2002. Fiber Bragg gratings and chromatic dispersion. *ICAPT 2002—Photonics North*, pp. 1009–1016.
15. Vaillancourt, I., Daxhelet, X., Lacroix, S., and Godbout, N. 2001. A 99.9% efficient lp_{01}–lp_{11} mode converter without lp_{01} back reflection: a detailed analysis. *Bragg Gratings, Photosensitivity, and Poling in Glass Waveguides, OSA*, pp. PD4–1.
16. Bakhti, F., Sansonetti, P., Sinet, C., Gasca, L., Martineau, L., Lacroix, S., Daxhelet, X., and Gonthier, F. 1997. Novel optical add/drop multiplexer based on uv-written Bragg grating in a fused 100% coupler. *Electron. Lett.* 33, 803–804.

Optical modulators

NADIR DAGLI
University of California

14.1 INTRODUCTION

Optical modulators accept as input the continuous wave (CW) or pulsed output of the laser diode and generate as output, a modulated optical waveform. The modulating signal is an electrical input voltage either in digital or analog format. Using an external modulator allows the laser to operate independently. Hence, its output power and frequency should be controlled very accurately. There are many different applications that require modulators. One very important application is the digital fiber optic communication systems. The advent and widespread use of Internet has increased the data transmission rates drastically. There is an ongoing demand for more bandwidth. Fiber optic communication networks have the capability to deliver such demand. At present, 40 Gbit/s fiber optic transmission systems are being developed and installed all around the globe. This requires optical signals to be modulated to such rates. Since most data generated is in electrical form, some form of electrical to optical modulation is required. This can be achieved by directly modulating the output of a laser diode through its drive current. There are laser diodes with small signal modulation bandwidths approaching 40 GHz [1]. Although this approach is simple, it is not generally used over 2.5 Gbit/s mainly because of the chirp of the directly modulated laser output [5]. The undesired frequency modulation associated with amplitude modulation severely limits the transmission distance over the fiber at high bit rates. This difficulty necessitated the development of external optical modulators. Transmission of analog signals also requires external modulators. Analog transmission is typically used to carry and distribute analog cable TV signals. Such distribution eliminates frequent microwave amplification that is needed on a coaxial distribution system, thus, improving the reach and reliability of the transmission [2]. Subcarrier multiplexing is another scheme that allows transmission independent of data format. It can be combined with digital transmission to improve functionality, such as optical labeling and header recognition [3]. Modulators are also used in military applications, especially in phased array radar [4]. Control signals to different radiating elements can be carried over the optical fiber, allowing for large distances between the antenna and the control site of the radar. Furthermore, many other functions in the microwave domain, such as tunable filtering, tunable delaying, and high-speed analog-to-digital conversion, can be performed using photonics techniques. All these microwave photonics applications rely on optical modulators for electrical-to-optical conversion. However, the required properties of the modulator for each one of these applications are different. Digital applications typically require low drive voltages, wide bandwidth, and adjustable chirp. On the other hand, analog applications require high degree of linearity and very low insertion loss. These requirements are most often conflicting and impose significant challenges on the design and fabrication of the modulator.

In this chapter, the criteria used to characterize an optical modulator are first described. Next, the fundamentals of phase and amplitude modulators are given. Then, the traveling wave modulation technique, which is universally used in all high-speed modulators, is described. This is followed by the description of the physical effects used in optical modulation. Both electro-absorptive and electro-optic effects are described. Then, specific examples of modulators in different material systems are given. These are electro-absorption (EA), $LiNbO_3$, III–V compound semiconductor, and polymer modulators. This is followed by a description of silicon and graphene modulators.

14.2 MODULATOR SPECIFICATIONS

A modulator, which is described by the block diagram shown in Figure 14.1, is characterized using certain specifications. These are defined and described briefly in the following subsections.

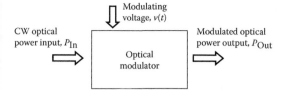

Figure 14.1 Block diagram of an optical modulator.

Their ranges and specifics for different technologies are described in detail in Section 14.7.

14.2.1 Insertion loss

Insertion loss is the optical power loss in the on state. It is typically given in units of decibels and is defined as

$$10 \log \left(\frac{(P_{Out})_{On}}{P_{In}} \right), \qquad (14.1)$$

where $(P_{Out})_{On}$ and P_{In} are the output power in the on state and input power, respectively.

14.2.2 Extinction ratio

This is the ratio of the off state output power, $(P_{Out})_{Off}$, to the on state output power and typically defined in decibel units as

$$10 \log \left(\frac{(P_{Out})_{Off}}{(P_{Out})_{On}} \right), \qquad (14.2)$$

14.2.3 Drive voltage

This is the voltage required to switch the modulator from on to off state. It is most often referred to as V_π. V_π is desired to be as low as possible, especially for high-speed operations, where the generation of voltages larger than a few volts may be very difficult. Most commercial modulators require a modulator driver, which amplifies the voltage available to a level sufficient to drive the modulator at the required impedance level.

14.2.4 Chirp

In a modulator, the amplitude and phase changes are coupled. This means that every time the amplitude changes, so does the phase and vice versa. This coupling could be due to material properties and/or modulator geometry. Since frequency is the time rate of phase change, phase modulation accompanying amplitude modulation changes the instantaneous frequency of the optical wave. This is known as chirping. Chirping combined with dispersion can severely limit the transmission distance of high-speed pulses in a communication system [5]. The chirp of a modulator is quantified using a chirp parameter, σ, which is the ratio of phase modulation to amplitude modulation:

$$\sigma = \frac{\dfrac{d\phi}{dt}}{\dfrac{1}{|E|} \dfrac{d|E|}{dt}}, \qquad (14.3)$$

where ϕ and $|E|$ are the phase and the amplitude of the optical electric field at the output of the modulator. The chirp parameter is expected to be zero for a pure amplitude modulator and infinity for a pure phase modulator.

14.2.5 Polarization dependence

Polarization dependence quantifies the performance of a modulator for different polarizations. Since any incoming polarization can be thought of as the superposition of two mutually orthogonal polarizations, the performance of the modulator is specified with respect to two mutually orthogonal polarizations. These polarizations are chosen as transverse electric (TE) and transverse magnetic (TM) in a guided wave modulator. The physical effect used in most modulators is observed only at certain polarizations of the optical mode. In other words, the operation of the modulator is polarization dependent. This is not a major concern in lasers where the modulator is used right after the laser diode. However, it could become an important issue if the modulator is used after some fiber transmission, which can generate random polarization at the input of the modulator. Such difficulty can be dealt with by using polarization diversity schemes that separate the random output polarization into two mutually orthogonal polarizations and deal with each polarization component separately.

14.2.6 Bandwidth

The bandwidth specifies the range of modulation frequencies over which the device can be operated.

It is usually taken as the difference between the upper and lower frequencies at which the modulation depth falls to 50% of its maximum value.

14.2.7 Bias stability and drift

For some modulators, the bias voltage required to keep the modulator operating properly may change over time. Therefore, constant monitoring of the bias point and its continuous adjustment using a feedback circuitry may be required.

14.3 PHASE MODULATORS

As described in Section 14.6, the index of refraction of a material can be changed using external perturbations. Changes in the index of refraction create changes in the phase velocity and the phase of the optical wave. Therefore, an optical wave propagating in a material whose index of refraction is modulated is phase modulated. Phase modulation can be achieved either in the bulk material or in an optical waveguide. In either case, any component of the electric field of the propagating optical wave can be expressed as

$$E(z,t) = E_0 \cos(\omega t - \beta z) = E_0 \cos\left(\omega t - \frac{2\pi}{\lambda_0} nz\right).$$

$$(14.4)$$

If the index of refraction n is perturbed using a time-varying external voltage $v(t)$,

$$n(t) = n_0 + \Delta n = n_0 + Kv(t), \qquad (14.5)$$

where K is a proportionality constant depending on the physical effect, geometry, and the material used. Specific K values depend on the technology and are described in detail in Section 14.7. With this perturbation, the electric field at the end of the propagation through the material of length L becomes

$$E(L,t) = E_0 \cos$$

$$\left(\omega t - \frac{2\pi}{\lambda_0} n_0 L - \frac{2\pi}{\lambda_0} KLv(t)\right)$$

$$= E_0 \cos(\omega t - mv(t)) \qquad (14.6)$$

Clearly the output wave is phase modulated with a modulation index $m = \left(\frac{2\pi}{\lambda_0}\right) KL$. $\left(\frac{2\pi}{\lambda_0}\right) n_0 L$ term

was dropped, because it is a fixed phase delay. In other words, by choosing the time origin appropriately, it can always be eliminated. The optical spectrum of such a phase-modulated wave can be quite complicated. For example, for a simple sinusoidal modulating voltage

$$E(L,t) = E_0 \cos(\omega t + m \sin \omega_m t) = E_0 [J_0(m) \cos \omega_0 t$$

$$+ J_1(m) \cos[(\omega_0 + \omega_m)t] - J_1(m) \cos[(\omega_0 - \omega_m)t]$$

$$+ J_2(m) \cos[(\omega_0 + 2\omega_m)t] + J_2(m) \cos[(\omega_0 - 2\omega_m)t]$$

$$+ J_3(m) \cos[(\omega_0 + 3\omega_m)t] - J_3(m) \cos[(\omega_0 - 3\omega_m)t]$$

$$+ J_4(m) \cos[(\omega_0 + 4\omega_m)t] - J_4(m) \cos[(\omega_0 - 4\omega_m)t]$$

$$+ \ldots],$$

$$(14.7)$$

where J_ms are the Bessel functions of the first kind of order m. As this expression shows, as soon as a single-frequency optical wave enters an index-modulated medium, its spectrum broadens and it is no longer a single frequency waveform due to phase modulation.

14.4 AMPLITUDE MODULATORS

Amplitude modulation can be achieved by absorption or index changes. Amplitude modulation based on index changes requires converting phase modulation into amplitude modulation. This is most commonly done using either Mach–Zehnder interferometers or directional couplers, which are described in the next section.

14.4.1 Directional coupler amplitude modulators

An optical directional coupler consists of two single-mode optical waveguides brought in close proximity over a length L as illustrated schematically in Figure 14.2. If the separation between the waveguides is sufficiently small, the evanescent tail of the optical waveguide mode of one waveguide does not completely decay to zero before reaching the other waveguide. As a result, coupling between the waveguides occurs. Hence, part of the wave in one waveguide can cross over or couple to the other waveguide. This situation is typically analyzed using the coupled mode theory [6,12]. The field over the coupled section can be approximated as

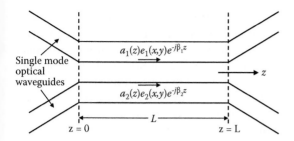

Figure 14.2 Schematic of an optical directional coupler.

$$E(x,y,z)=a_1(z)e_1(x,y)e^{-j\beta_1 z}+a_2(z)e_2(x,y)e^{-j\beta_2 z}.$$
$$(14.8)$$

This is the superposition of individual normalized waveguide modes, $e_1(x,y)$ and $e_2(x,y)$, with propagation constants β_1 and β_2. The amplitudes of the individual waveguide modes $a_1(z)$ and $a_1(z)$ are z dependent due to coupling between the waveguides. If only one of the waveguides is excited at the input, such that $a_1(0)=1$ and $a_2(0)=0$

$$E_1(x,y,L)=e_1(x,y)\left(\cos\left(\sqrt{\delta^2+\kappa^2}\,L\right)\right.$$
$$+j\frac{\delta}{\sqrt{\delta^2+\kappa^2}}\sin\left(\sqrt{\delta^2+\kappa^2}\,L\right)\right)e^{-j\left(\frac{\beta_1+\beta_2}{2}\right)L},\ (14.9)$$

$$E_2(x,y,L)=e_2(x,y)$$
$$\left(\frac{j\kappa}{\sqrt{\delta^2+\kappa^2}}\sin\left(\sqrt{\delta^2+\kappa^2}\,L\right)\right)e^{-j\left(\frac{\beta_1+\beta_2}{2}\right)L},$$
$$(14.10)$$

where $\delta=(\beta_1-\beta_2)/2$ is the detuning between the waveguides and κ is the coupling coefficient. κ depends on the degree of spatial overlap between the individual waveguide modes [6]. Hence, over a given length, a certain fraction of the input power will cross over to the other waveguide. In particular, if $\beta_1=\beta_2=\beta_0$

$$|a_1(z)|^2=\cos^2(\kappa z)\text{ and }|a_2(z)|=\sin^2(\kappa z).$$

Hence, for

$$\kappa L=\frac{\pi}{2}\text{ or }L=\frac{\pi}{2\kappa},\qquad(14.11)$$

complete cross over takes place. It is possible to eliminate complete cross over by modulating the index of refraction and hence the propagation constant of one or both waveguides. For example, if $\sqrt{\delta^2+\kappa^2}\,L=\pi$, the optical power will switch back to the input waveguide. The required detuning is found after eliminating L using Equation 14.11 as

$$\delta=\frac{\beta_1-\beta_2}{2}=\sqrt{3}\kappa.\qquad(14.12)$$

If we express β_1 and β_2 as

$$\beta_1=\beta_0+\Delta\beta_1=\frac{2\pi}{\lambda_0}(n_0+\Delta n_{1\text{eff}})=\frac{2\pi}{\lambda_0}(n_0+K_1v(t)),$$
$$(14.13)$$

$$\beta_2=\beta_0+\Delta\beta_2=\frac{2\pi}{\lambda_0}(n_0+\Delta n_{2\text{eff}})=\frac{2\pi}{\lambda_0}(n_0+K_2v(t)),$$
$$(14.14)$$

then

$$\delta=\frac{\pi}{\lambda_0}(K_1-K_2)v(t)\qquad(14.15)$$

and the required external voltage is found as

$$V_\pi=\frac{\sqrt{3}\lambda_0\kappa}{\pi(K_1-K_2)}=\frac{\sqrt{3}\lambda_0}{2L(K_1-K_2)}.\qquad(14.16)$$

The required voltage magnitude is minimized if $K_1=-K_2=K$. This requires an index change of equal magnitude and opposite sign in the coupled waveguides and is known as the push–pull drive. Depending on the applied voltage, different degrees of coupling will occur. The variation of the coupled power as a function of the applied voltage is known as the transfer function of the modulator. Mathematically, this would be expressed in the cross-over waveguide as

$$\frac{|a_2(L)|^2}{|a_1(0)|^2}=\frac{1}{1+\left(\sqrt{3}\dfrac{v(t)}{V_\pi}\right)^2}\sin^2\left(\frac{\pi}{2}\sqrt{1+\left(\sqrt{3}\dfrac{v(t)}{V_\pi}\right)^2}\right),$$
$$(14.17)$$

where V_π is the required voltage swing to turn the modulator off and is given by

$$V_\pi=\frac{\sqrt{3}\lambda_0}{4LK}.\qquad(14.18)$$

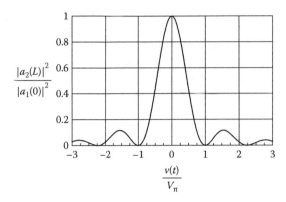

Figure 14.3 Transfer function of a directional coupler modulator.

The transfer function of a directional coupler modulator is shown in Figure 14.3.

One can also use the result of this analysis to determine the chirp of a directional coupler modulator. Examining Equation 14.10, it is seen that the phase of $E_2(x,y,L)$ can be kept constant if the drive is a push–pull, i.e., if β_1 increases a certain amount and β_2 decreases the same amount, which leaves $(\beta_1-\beta_2)/2$ unchanged. This results in a chirp-free operation. A chirp-free operation is not possible for $E_1(x,y,L)$ since its phase depends on the drive conditions. Therefore, if a chirp-free operation is desired when using a directional coupler modulator, modulated power should be taken out of the cross-over waveguide and push–pull drive should be used [7].

14.4.2 Mach–Zehnder amplitude modulators

Another commonly used amplitude modulator is the Mach–Zehnder interferometer shown in Figure 14.4. This is the integrated optics version of the Michelson interferometer. The optical wave in the incoming waveguide is split into two equal parts using a Y-branch and made to propagate along the arms of the interferometer. At the other end, the waves in the two arms are combined into the output waveguide using another Y-branch. It is also possible to use 3 dB couplers in place of the input and output Y-branches. For proper operation of the interferometer, all the optical waveguides should be of single mode. The mode amplitude in the output waveguide can be written as

$$a_{10} = \left[s^2 e^{-j(\beta_1 L + 2\varphi_1)} + (1-s^2)e^{-j(\beta_2 L + 2\varphi_2)} \right]a_1. \tag{14.19}$$

where s^2 is the power splitting ratio in the Y-branch, φ_1 and φ_2 are the phase shifts along the arms of the Y-branch, and β_1 and β_2 are the propagation constants of the arms of the interferometer. Any phase shift due to a length difference between the arms can also be included in φ_1 and φ_2, allowing us to assume that the arms have the same length L.

Ideally for equal power splitting, $s^2 = 1/2$. However, due to imperfections during fabrication, there could be slight deviations from the ideal value. In this case, we can express s^2 as $s^2 = 1/2 + \xi$. Using this definition, Equation 14.19 can be expressed as

$$a_{10} = e^{-j\left(\left(\frac{\beta_1+\beta_2}{2}\right)L + (\varphi_1+\varphi_2)\right)} \left[\cos\left(\left(\frac{\beta_1-\beta_2}{2}\right)L + (\varphi_1-\varphi_2)\right) -2j\xi\sin\left(\left(\frac{\beta_1-\beta_2}{2}\right)L + (\varphi_1-\varphi_2)\right) \right]a_1.$$

$$(14.20)$$

The ratio of the output power to the input power is known as the transfer function of the Mach–Zehnder interferometer. Assuming that β_1 and β_2

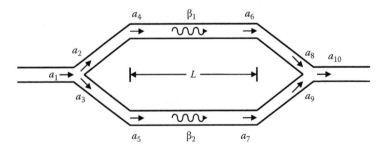

Figure 14.4 Schematic of a Mach–Zehnder interferometer.

can be perturbed with external voltages as shown in Equations 14.13 and 14.14, and $\xi = 0$ and $\varphi_1 = \varphi_2$, the transfer function becomes

$$\frac{|a_{10}|^2}{|a_1|^2} = \cos^2\left(\frac{\pi}{\lambda_0}(K_1 - K_2)Lv(t)\right). \quad (14.21)$$

This transfer function is periodic with respect to the applied voltage. The modulator is turned off when $(\pi/\lambda_0)((K_1 - K_2)Lv(t)) = (\text{Odd integer})\pi/2$. The lowest required voltage to turn the modulator off, known as V_π, is given by

$$V_\pi = \frac{\lambda_0}{2L(K_1 - K_2)} \quad (14.22)$$

Comparing this equation with Equation 14.16, we observe that the voltage swing required to turn a Mach–Zehnder modulator from on to off state is $\sqrt{3}$ times less than that required to turn a directional coupler modulator from on to off state for a given technology. Again, V_π is minimized if $K_1 = -K_2 = K$, i.e., for a push–pull drive. In terms of this definition of V_π, the transfer function becomes

$$\frac{|a_{10}|^2}{|a_1|^2} = \cos^2\left(\frac{\pi}{2}\frac{v(t)}{V_\pi}\right). \quad (14.23)$$

The transfer function of a Mach–Zehnder modulator under push–pull drive is plotted in Figure 14.5.

While considering the operation of a Mach–Zehnder modulator, it is important to have a physical explanation of what is happening in the off state. In the analysis, a lossless modulator is assumed. Equation 14.21 predicts that there is no power output

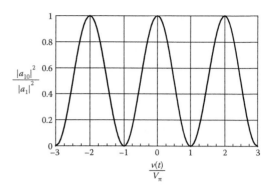

Figure 14.5 Transfer function of a Mach–Zehnder interferometer.

from the modulator in the off state even though there is input power. Since there are no reflections and optical absorption, one wonders what causes the power loss. This question can be answered with the help of Figure 14.6, which describes the operation of a Mach–Zehnder modulator. In the on state, the incoming mode of the single-mode waveguide splits with equal amplitude at the input Y-branch to both arms of the interferometer. If $\beta_1 L = \beta_2 L$, the waves in both arms arrive at the output Y-branch with equal phase and combine to form the mode of the single mode output waveguide. However, in the off state, $\beta_1 L - \beta_2 L = \pi$, and the waves in both arms arrive at the output Y-branch with a π phase shift as shown in Figure 14.6b. As a result, when they are gradually combined by the Y-branch in the output waveguide, a mode with a null in the middle is excited. This would be the higher order mode of the output waveguide. If the output waveguide is of single mode, the higher order mode would radiate

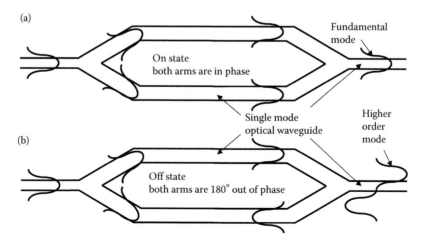

Figure 14.6 The physical description of a Mach–Zehnder modulator. (a) On state and (b) off state.

out of the waveguide. As a result, there will be no power left in the output waveguide after a certain length. This explanation shows that in the off state, power is not lost. It simply radiates out of the output waveguide. It also shows that for proper operation, the waveguides in the interferometer should be of single mode, and the output waveguide should be sufficiently long for the radiation to take place in the off state. If the waveguides are not of single mode in the off state, only mode conversion between the fundamental and the higher order mode takes place.

Examining Equation 14.20, it is observed that if $\xi = 0$ and $\varphi_1 = \varphi_2$

$$a_{10} = e^{-j\left(\frac{\beta_1+\beta_2}{2}\right)L}\left[\cos\left(\left(\frac{\beta_1-\beta_2}{2}\right)L\right)\right]a_1. \quad (14.24)$$

If the drive is a push–pull so that $(\beta_1+\beta_2)/2$ is constant during modulation, the phase of the output signal does not change and a chirp-free operation is achieved. Therefore, a push–pull driven Mach–Zehnder intensity modulator has no chirp.

14.4.3 EA amplitude modulators

In some materials, the absorption at a given wavelength can be controlled using external voltages. Hence, it becomes possible to change the transmission through a waveguide with external voltages. This results in a very simple modulator, which is typically a short waveguide with an EA layer in it.

However, critical control of the material composition and thickness is required. These devices are described in detail in Section 14.7.1.

14.5 TRAVELING WAVE MODULATORS

In optical modulators, the physical effects used to create index changes are very weak. As a result, the K coefficients used in Equations 14.18 and 14.22 are very small. This necessitates making the modulator electrode very long in order to have low drive voltages that can be supplied at high frequencies. Typical electrode lengths for electro-optic modulators are in the order of several centimeters. However, this increases the capacitance of the electrode drastically. The resulting RC product limits the bandwidth of modulation to less than a few GHz. These conflicting requirements on the electrode length can be eliminated by using the traveling wave electrode concept. In this approach, the electrode is designed as a transmission line. Therefore, electrode capacitance is distributed and does not create an RC limit on the modulator speed. A schematic of a traveling wave Mach–Zehnder modulator is shown in Figure 14.7.

In this figure, the electrode is designed as a coplanar waveguide. An electrical signal is applied from a voltage generator of internal impedance Z_s Ω. In this case, the signal is an electrical pulse. Both the electrical pulse and the CW optical signal travel in the same direction. On the part of the

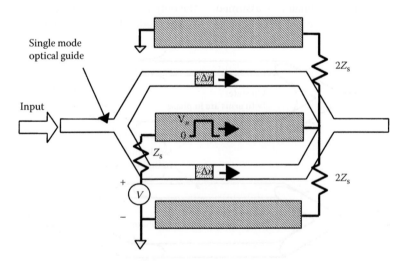

Figure 14.7 Schematic of a traveling wave Mach–Zehnder modulator.

electrode where the voltage is present, the index of the material is modified. In this case, a push–pull drive is considered; hence, the index increases a certain amount in one arm of the interferometer and decreases the same amount on the other arm. Clearly, if the optical signal travels at the same velocity as the electrical signal, it will experience the same index change all along the electrode. Hence, the phase shift it experiences will be integrated over the electrode length. Therefore, a very small index change may create a large phase shift. Any velocity mismatch between the electrical and the optical signals will reduce the net phase shift. The small signal modulation response of a traveling wave modulator whose electrode is terminated by its characteristic impedance is given by [8]

$$
M(f)=e^{-\left(\frac{\alpha_m L}{2}\right)}\left[\frac{\sinh^2\left(\frac{\alpha_m L}{2}\right)+\sin^2\left(\frac{\pi f\left(n_\mu-n_o\right)L}{c}\right)}{\left(\frac{\alpha_m L}{2}\right)^2+\left(\frac{\pi f\left(n_\mu-n_o\right)L}{c}\right)^2}\right]^{\frac{1}{2}},
$$

(14.25)

where α_m and L are the loss coefficient and the length of the electrode, respectively. f is the electrical frequency and c is the speed of light in vacuum. n_μ and n_o are the microwave and optical indices and are related to the microwave and optical velocities through well-known expressions. The electrical waveform applied to the electrode of the modulator has a certain shape and contains many frequency components. The CW light that enters the modulator also becomes phase modulated as soon as it starts to interact with the electrical signal and exhibits a spectral range. As a result, the center of gravity of both the electrical and optical signals should match [9]. This requires group velocity matching; hence, the microwave and optical indices defined earlier should be the group indices. Usually, the modulator electrode is a quasi-transverse electric magnetic (TEM) transmission line and has no or very little dispersion. That makes the group and the phase velocities on the electrode the same. If the electrode exhibits significant dispersion, complications may arise. For example, if the electrode is driven by a CW microwave signal like in a pulse generation application, microwave phase and optical group velocities should be matched. However, if a digital electrical signal is applied to the electrode,

optical and microwave group velocities should be matched. This can create velocity mismatch depending on the application. Furthermore, optical group velocities may change due to material dispersion. For example, a modulator velocity matched at 1.5 μm can be mismatched at 1.3 μm [10,11]. This is a problem especially for semiconductor modulators, where material dispersion is much more significant compared to LiNbO$_3$.

Even in the case of perfect velocity matching, the bandwidth is limited by the microwave loss of the electrode. At high microwave frequencies, the microwave loss increases; this reduces the voltage on the line. Hence, modulation is no longer as effective as it was at lower frequencies where microwave loss was lower. Based on Equation 14.25, if there is no velocity mismatch, the 3 dB bandwidth will be at a frequency where the total electrode loss becomes 6.34 dB. Therefore, a low-loss, velocity-matched electrode is essential for the realization of a very wide bandwidth traveling wave modulator. Another very important consideration is impedance matching the modulator electrode. If the electrode is terminated with impedance having a value different from that of the electrode impedance, there is an impedance mismatch at the end of the modulator. This mismatch creates a reflected voltage wave traveling in the opposite direction. The reflected wave interferes with the wave traveling in the same direction as the optical wave and a standing wave is formed on the electrode. As a result, voltage on some parts of the electrode is higher than the rest. These parts modulate effectively; whereas, modulation is less effective on the parts having less voltage amplitude. Variation of the standing wave voltage is frequency dependent and the modulation efficiency could significantly vary as a function of frequency. In modulators with long electrodes, counter propagating part of the standing wave is badly mismatched and contributes to modulation only at low frequencies. In these cases, modulation efficiency drops rapidly over a few GHz.

Based on this discussion, we note that the following requirements should be satisfied to take full advantage of the traveling wave idea:

1. The propagation loss of the optical guide should be low so that a long modulator can be realized. This helps to significantly reduce the drive voltage.

2. The microwave electrode mode should be quasi-TEM; hence, the phase and the group velocities are the same for the electrode.
3. Microwave and optical group velocities should be matched.
4. The electrode microwave loss coefficient should be low, so that a long modulator can be realized. As described before, the total electrode loss should be lower than 6.34 dB at the desired 3 dB bandwidth point.
5. The electrode should be terminated by its characteristic impedance so that there is no standing wave on the electrode. It is desirable to have an electrode characteristic impedance of 50 Ω, but this may not always be possible.

14.6 PHYSICAL EFFECTS USED IN OPTICAL MODULATORS

The physical mechanisms used in optical modulators are change in absorption and the index of refraction under external perturbations. Change in absorption and the index of refraction are fundamentally related. The most commonly used external perturbation is an externally applied electric field, which can be generated electronically by applying voltages to the material. Modulators utilizing index changes and absorption changes under an external electric field are known as electro-optic and EA modulators, respectively. Index and absorption changes can also be generated using pressure or strain and magnetic fields. The strain is typically generated using acoustic waves. Modulators using acoustic waves to induce index changes due to strain are known as acousto-optic modulators. Similarly, modulators using magnetic fields to induce index changes due to magneto-optic effect are known as magneto-optic modulators. Absorption is most important in III–V compound semiconductors, such as GaAs, AlAs, InP, and their alloys AlGaAs and InGaAsP.

14.6.1 Change of absorption in semiconductors under external electric fields

In semiconductors, the sources of optical absorption are band-to-band transitions, free carriers (FCs), intraband transitions, excitons, transitions between band tails, transitions between bands and impurities, and acceptor-to-donor transitions.

Of these, band-to-band absorption, FC absorption, intraband absorption, and excitonic absorption are the most dominant processes in good quality material. These processes can be perturbed by external electric fields and are briefly described in the following sections.

14.6.1.1 BAND-TO-BAND ABSORPTION AND FRANZ–KELDYSH EFFECT

In semiconductors, an incoming photon can excite an electron from the valence band to the conduction band under certain conditions as illustrated in Figure 14.8. Since the photon momentum h/λ, where h is Planck's constant and λ is the wavelength of the light, is much smaller than the crystal momentum h/b, where b is the lattice constant, momentum conservation requires that the momentum of the initial and final states remain unchanged. Hence, only vertical transitions are allowed. Furthermore, in order to absorb a photon, the initial state in the valence band should be filled and the final state in the conduction band should be empty. Energy of the photon should be larger than the energy difference between the filled initial state and the empty final state of the semiconductor. This energy difference is typically the bandgap of the material, but depending on the doping type and level, the required photon energy could be larger than the bandgap. The absorption coefficient α due to band-to-band absorption in a direct bandgap semiconductor can be expressed as

$$\alpha = A|R(h\upsilon)|^2 \rho_r(h\upsilon)(F_1 - F_2),$$

where R is the matrix element, $h\upsilon$ is the photon energy, F_1 and F_2 are the Fermi factors for the valence and conduction bands, respectively, ρ_r is the reduced density of states, and A is a coefficient that depends on fundamental constants [12]. The matrix element R is proportional to the overlap of the electronic wave functions in the conduction and valence bands. The value of R depends on the material type and its dimensionality and the polarization of the optical wave. Polarization dependence typically appears in lower dimensional systems such as quantum wells (QWs). ρ_r also depends on the material dimensionality. For bulk materials, absorption can be expressed as [12]

$$\alpha = B\left(h\upsilon - \varepsilon_g\right)^{\frac{1}{2}},$$

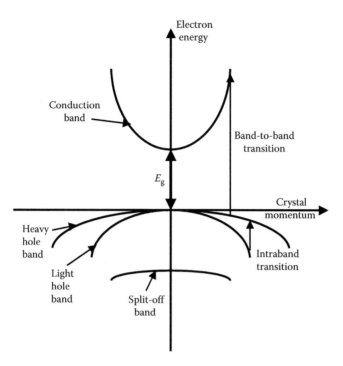

Figure 14.8 Band structure of a typical III–V compound semiconductor in the vicinity of the zone center. Various possible transitions responsible for dominant optical absorption are also illustrated.

assuming $F_1 = 1$, and $F_2 = 0$ as in undoped bulk material. In this expression, ε_g is the bandgap of the material.

In the presence of an electric field, energy bands are tilted. Figure 14.9 shows the variation of the conduction band minimum and valence band maximum in space for different values of applied electric field. If there is no external electric field and the material properties such as doping and composition are uniform, band extreme are flat. Increasing external electric field values tilt the energy bands, which increases the penetration of the electron and hole wave functions into the bandgap as shown in Figure 14.9. This in turn increases the overlap of the electrons and holes in the bandgap. In Figure 14.9, points A and B are the classical turning points for the electron and hole wave functions. In other words, at these points, the wave functions change from oscillatory to exponentially decaying behavior. Due to penetration of the wave functions into the bandgap, it is possible to excite an electron from the valence band to the conduction band with a photon of energy $h\upsilon$ as seen in Figure 14.9. This energy is clearly less than the bandgap of the material. Hence, it becomes possible to absorb at photon

energies less than the bandgap energy and this absorption increases with increasing external field. As a result, the absorption tail extends to shorter energies or longer wavelengths. This phenomenon is known as the Franz–Keldysh effect [13,14]. It is possible to modulate the absorption from low to high absorption in the long-wavelength side with external fields. This situation is similar to tunneling through a potential barrier under an applied electric field. The increasing external electric field effectively reduces the width of the barrier due to more and more penetration of the wave functions into the bandgap. The absorption coefficient under an external applied field, α_E, can be expressed as [15,16]

$$\alpha_E = B\sqrt{h\upsilon_E}\,\pi \int_{\frac{\varepsilon_g - h\upsilon}{h\upsilon_E}}^{\infty} Ai(x)^2 \,dx,$$

where

$$\upsilon_E = \left(\frac{e^2 E^2 \hbar^2}{2m_r}\right)^{\frac{1}{3}}.$$

In this equation, $Ai(x)$ is the Airy function and m_r is the reduced effective mass. The absorption is also

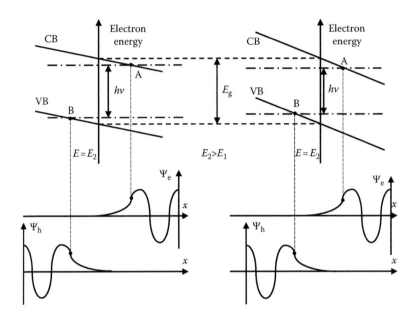

Figure 14.9 Variation of the conduction band minimum and valence band maximum under different applied external electric fields in a material with uniform doping and composition. Penetration of the electron and hole wave functions (Ψ_e and Ψ_h) into the bandgap is also illustrated.

modified for photon energies larger than the bandgap energy. In this case, absorption becomes oscillatory since electron and hole wave functions are oscillatory above the bandgap. Therefore, bringing them closer changes their overlap in an oscillatory manner. Below the bandgap, the wave functions are exponentially decaying; hence, bringing them closer increases their overlap monotonically. Absorption changes above the bandgap are not used for modulation due to large background absorption.

14.6.1.2 EXCITONIC ABSORPTION IN BULK AND QW MATERIAL AND THE QUANTUM-CONFINED STARK EFFECT

In a material, free electrons and holes attract one another and form a bond similar to a hydrogen atom. This is called an exciton. An exciton has a strong absorption somewhat similar to an atomic absorption. In an exciton, electron–hole interaction is treated as a Coulomb interaction between two point charges. Using the hydrogen atom model, the binding energy for an exciton can be expressed in eV as $\varepsilon_{xn} = (13.6/n_x^2)(m_r/\varepsilon_r^2)$, where n_x is an integer representing different energy levels, i.e., the quantum number, m_r is the reduced effective mass, and ε_r is the relative

dielectric constant. Another important parameter is the exciton dimension. Again using the hydrogen atom model, the exciton radius can be expressed in Å as $a_{xn} = (0.53\varepsilon_r/m_r)(n_x^2)$. In a bulk compound semiconductor, ε_{x1} is about 5 meV and a_{x1} is about 150 Å. As a result, a bulk exciton covers many lattice sites and is weakly bound. Therefore, it is observed only at low temperatures. However, in a QW, electrons and holes are confined in the same physical space. As a result, they overlap and interact strongly and form an exciton confined in the QW. This strong confinement increases the binding energy or the energy required to ionize an exciton into an electron–hole pair. For a purely two-dimensional exciton, the increase in the binding energy would be factor 4 [17]. However, in a QW, the exciton wave function penetrates into the barriers and the dimensionality of the exciton is somewhere in between two and three. For a very thin QW, this penetration could be excessive and the exciton behaves like a three-dimensional exciton. The increase in the binding energy due to quantum confinement makes the binding energy of the exciton larger than the broadening due to phonon scattering at room temperature. As a result, excitonic absorption is observed at room temperature. Spectra of such strong excitonic absorption are very sharp, and are localized in

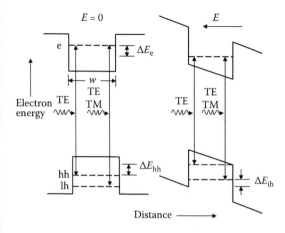

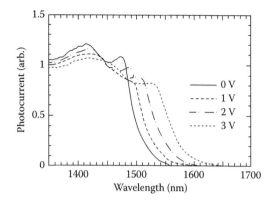

Figure 14.10 Schematic of an unstrained QW energy diagram with and without an applied electric field *E*. e, hh, and lh designate energy levels for electrons, heavy holes, and light holes, respectively. (Dagli, N., *IEEE Trans. Microwave Theory Tech.*, Figure 3. ©1999 IEEE.)

Figure 14.11 Photocurrent spectra of an unstrained MQW modulator as a function of bias voltage. The incident light is TE polarized. (Ido, T. et al., *IEEE Photon. Technol. Lett.*, Figure 2. ©1994 IEEE.)

the vicinity of wavelengths corresponding to the bandgap of the QW. This energy is larger than the bandgap of the QW material due to quantization of the electron and hole energy levels as shown in Figure 14.10. When an external electric field is applied, electrons and holes are forced to opposite ends of the QW and are physically separated as seen in Figure 14.10. The applied field shifts the energy levels in the QW. This situation is similar to the shift in the energy levels of an atom under an applied electric field, which is known as the Stark effect. Hence, the corresponding transition energy shift in a QW is known as the quantum-confined Stark effect (QCSE). In QCSE, the spatial overlap or interaction of the electron and the hole is reduced and excitonic absorption is decreased and broadened. This makes it possible to modulate the absorption around a narrow wavelength range very strongly with external fields.

Figure 14.11 shows the photocurrent spectra of an unstrained multi quantum well (MQW) material as a function of wavelength at different applied voltages [18], where device photocurrent is proportional to optical absorption. Two peaks are resolved in the absorption spectra. These are due to excitons formed between electrons and heavy holes (hhs) and electrons and light holes (lhs). Transition energies of hh and lh excitons are different due to different effective mass of lhs and hhs as seen in Figure 14.11. Furthermore, the excitons

interact with different optical polarizations. The hh excitons interact with TE polarized light and lh excitons interact with both TE and TM polarized light. In Figure 14.11, the first absorption peak has a lower transition energy or higher optical wavelength and corresponds to the lowest hh exciton indicated in Figure 14.10. The second absorption peak corresponds to the lh exciton. As bias voltage increases, absorption characteristics broaden and peak absorption decreases and moves toward longer wavelengths. For example, at 1.55 μm, absorption is modulated strongly when bias changes between 0 and 3 V. For low bias voltages, the shift in the ground state energy $\Delta\varepsilon_i$ of a particle in a QW can be approximated as [19]

$$\Delta\varepsilon_i = Cm_i w^4 E^2, \quad i = e, hh, lh \quad (14.26)$$

where m_i is the effective mass of the particle, w is the width of the QW, and C is a constant. Hence, wide QWs are desired for efficient operation.

14.6.1.3 FC ABSORPTION

FCs are electrons and holes free to move in the conduction and valence bands. They can interact with photons and make transitions to a higher energy in the same band by absorbing a photon. This is known as an intraband absorption. Since such transitions are not vertical, they require additional interactions to conserve momentum. The required momentum change can be provided by phonons or scattering from ionized impurities. There is no energy threshold for this transition; its spectrum is typically monotonic and covers a very wide range. Using a simple Drude model, which models an electron in an oscillatory field as a damped oscillator, absorption coefficient due to FCs, α_f, can be expressed as [20]

$$\alpha_f = \frac{Nq^2\lambda^2}{8\pi^2 m^* nc^3\tau},\tag{14.27}$$

where N is the electron concentration, τ is the carrier relaxation time, m^* is the effective mass, n is the index of refraction, c is the speed of light in vacuum, and λ is the free space wavelength. Other formulations in which the momentum is provided by acoustic phonons, optical phonons, or ionized impurities yield a wavelength dependence of λ^p, where p can range from 1.5 to 3.5 [20]. Since one of the elements that affect τ is ionized impurity concentration, its value depends on the doping level and type. At high doping concentrations, the carrier concentration dependence of α_f could be more like $N^{3/2}$. Around 1.55 μm, $\alpha_f = 1\times10^{-18}N$ (cm^{-3}) in GaAs [21]. The situation is similar for intraband absorption involving holes. However, there is another source of absorption due to the possibility of transition between light and heavy hole bands in p-type material. The most likely source for this absorption is vertical or near vertical transitions at longer wavelengths—intraband absorption—and is indicated schematically in Figure 14.8.

14.6.2 Electro-optic effects

The real and imaginary parts of a complex function are related to one another if the function has no poles in the lower or upper complex plane. If one describes the index of refraction of a material as a complex function, the real and imaginary parts will be related. In this representation, the real and imaginary parts of the complex index of refraction are the index of refraction and the absorption of the material. As a result, a change in absorption will generate a change in the refractive index and vice versa. This relationship is known as the Kramers–Kronig relation and can be expressed as

$$\Delta n(\varepsilon, E) = \frac{hc}{\pi} \int_0^\infty \frac{\Delta\alpha(\varepsilon', E)}{\varepsilon'^2 - \varepsilon^2} \, d\varepsilon'.\tag{14.28}$$

Hence, any effect that creates an absorption change $\Delta\alpha$ under an applied electric field E at photon energy ε will also create an index change Δn. These index changes are known as electro-optic effects and are described in the next section.

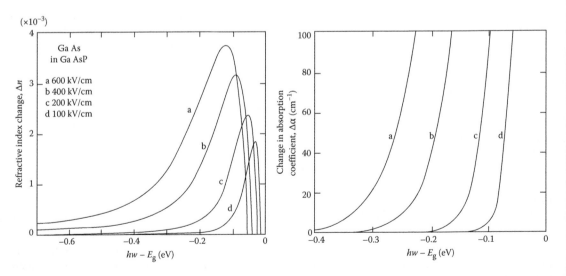

Figure 14.12 The calculated change in the index of refraction and absorption coefficient as a function of photon energy for GaAs and InGaAsP at different electric field strengths. (Reprinted with permission from Alping, A. and Coldren, L.A., J. Appl. Phys., 61, 2430–2433, 1987, Figures 1 and 5. Copyright 1987, American Institute of Physics.)

14.6.2.1 ELECTRO-REFRACTIVE EFFECT

Electro-refractive effect is the phenomena of index change accompanied with Franz–Keldysh and QCSE. Electro-refraction in bulk GaAs and InGaAsP were calculated in Reference [22]. The calculated change in the index of refraction and absorption coefficient as a function of photon energy for GaAs and InGaAsP at different electric field strengths are shown in Figure 14.12.

Index change increases rapidly toward the bandgap. The peak of the increase shifts away from the band edge with increasing electric field. For wavelengths very close to the bandgap index, change decreases and it should eventually be negative. Figure 14.13 shows the change in the refractive index in $In_{0.76}Ga_{0.24}As_{0.52}P_{0.48}$ as a function of the applied electric field at two different wavelengths. For photon energies far below the bandgap, the refractive index for a given electric field can be expressed as

$$\Delta n = G(\lambda)E^2. \qquad (14.29)$$

There is a quadratic dependence on the electric field and the wavelength-dependent coefficient $G(\lambda)$ can be expressed for GaAs as

$$G(\lambda) = 3.45 \times 10^{-16} \exp\left(\frac{3}{\lambda^3}\right) cm^2\, V^{-2}, \qquad (14.30)$$

where wavelength is expressed in micrometers [23]. For InGaAsP, $G(1.55\ \mu m) = 5.8 \times 10^{-15}\ cm^2\, V^{-2}$. At high electric fields, it is possible to raise the refractive index by about 3×10^{-3} at photon energies near the bandgap. However, this index change is accompanied with a large absorption change due to the Franz–Keldysh effect. In order to get a predominantly electro-optic effect, absorption change should be kept at minimum. The relative variation of the real and imaginary parts of the index of refraction with applied field is quantified using the figure of merit $\Delta n/\Delta k$, where Δk is the change in the imaginary part of the refractive index. Δk and $\Delta\alpha$ are related by the equation, $\Delta\alpha = 4\pi\Delta k/\lambda$. As described earlier, it is possible to modify absorption in a QW close to the band edge strongly by QCSE. This also creates an accompanying index change. Δn and $\Delta n/\Delta k$ as functions of the electric field for two MQW samples at two different wavelengths are shown in Figure 14.14 [24]. In both cases, Δn shows a quadratic dependence on the applied field. $G(1.537\ \mu m) = 6.73 \times 10^{-13}\ cm^2\, V^{-2}$ for an 85 A InGaAsP QW of bandgap energy 1.57 μm within 85 A InP barriers. $G(1.306\ \mu m) = 7.32 \times 10^{-13}\ cm^2\, V^{-2}$ for a 70 A InGaAsP QW of bandgap energy 1.33 μm within 250 A InP barriers [24]. These values are more than two orders of magnitude larger than the corresponding bulk material values. But this does not necessarily make a much better modulator since there is a few percent overlap of the optical mode with the QW material in a typical modulator. However, the spectral width of the absorption in a QW material is much narrower compared to bulk material. Furthermore, QCSE red shifts this relatively narrow resonance; hence, for a given wavelength, detuning $\Delta n/\Delta k$ value is larger in the QW material compared to the bulk material.

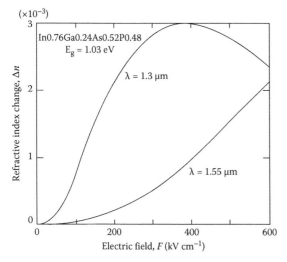

Figure 14.13 The change in the refractive index Δn in $In_{0.76}Ga_{0.24}As_{0.52}P_{0.48}$ as a function of the applied electric field at two different wavelengths. (Reprinted with permission from Alping, A. and Coldren, L.A., J. Appl. Phys., 61, 2430–2433, 1987, Figure 3. Copyright 1987, American Institute of Physics.)

14.6.2.2 PLASMA EFFECT

The plasma effect is the accompanying index change associated with the FC absorption. For an n-type GaAs of carrier concentration N, the index change Δn_N with respect to an undoped material is

$$\Delta n_N = -9.6 \times 10^{-21} \frac{N}{n\varepsilon^2},$$

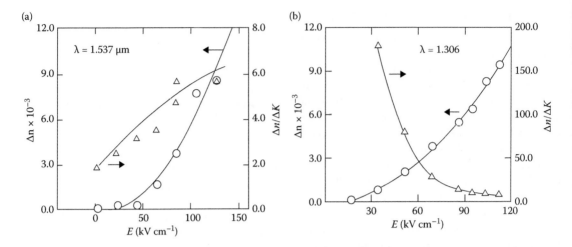

Figure 14.14 Change in the refractive index, Δn, and the ratio of the changes in the real and imaginary parts of the refractive index, $\Delta n/\Delta k$, as functions of the electric field for two MQW samples (a) ten periods of 85 A InP barriers and 85 A InGaAsP wells of bandgap energy 1.57 μm at 1.537 μm and (b) five periods of 250 A InP barriers and 70 A InGaAsP wells of bandgap energy 1.33 μm at 1.16 μm. (Reprinted with permission from Zucker, J.E. et al., *Appl. Phys. Lett.*, 54, 1, 10–12, 1989, Figure 3. Copyright 1989, American Institute of Physics.)

where n is the index of refraction and ε is the photon energy. For a p-type GaAs of carrier concentration P, the corresponding index changes for intraband and interband transitions are

$$\Delta n_{P\text{inter}} = -6.3 \times 10^{-22}\, \frac{P}{\varepsilon^2} \text{ and}$$

$$\Delta n_{P\text{intra}} = -1.8 \times 10^{-21}\, \frac{P}{n\varepsilon^2}.$$

As these expressions indicate, removal of electrons or holes in a doped semiconductor reduces FC absorption, which in turn increases the index of refraction.

14.6.2.3 BAND-FILLING EFFECT

In a doped material, the position of the Fermi level moves depending on the doping level. In very heavily doped material, the Fermi level can move closer or even into the conduction or valence bands. Therefore, the absorption threshold in doped material could be different from that in undoped material. Therefore, at photon energies near the bandgap, absorption could be stronger for intrinsic material compared to heavily doped material. This creates an index change due to the Kramers–Kronig relationship between doped and undoped materials. This index change can be expressed as [23,25]

$$\Delta n(\varepsilon) = H(\varepsilon) N.$$

The coefficient H strongly peaks around photon energy ε corresponding to the bandgap of the material [23]. Below the bandgap, its value is about 5×10^{-21} cm^{-3} for GaAs. This relationship holds as long as the semiconductor is not very heavily doped and the doping level is less than about 5×10^{17} cm^{-3}. For heavier doping, band tails that develop at the conduction band edge and bandgap shrinkage start to dominate the absorption. This increase in the absorption starts to give an index change that opposes the band-filling effect [26]. For a p-type material, the variation of the Fermi level with doping concentration is much less due to the heavier hole effective mass. Hence, the band-filling effect is much weaker in a p-type material.

14.6.2.4 LINEAR ELECTRO-OPTIC EFFECT

Linear electro-optic effect can be thought of as the accompanying index change due to absorption at ultraviolet wavelengths. However, the tail of absorption at such high photon energies is very weak around 1.55 μm. Hence, this effect is thought of as a pure index change and modeled in a different way. In a dielectric material, electric field $\vec{E}$ and electric flux density $\vec{D}$ are related through the dielectric tensor, which can be expressed as

$$
\begin{bmatrix} E_x \\ E_y \\ E_z \end{bmatrix} = \frac{1}{\varepsilon_0} \begin{bmatrix} \left(\frac{1}{n^2}\right)_{xx} & \left(\frac{1}{n^2}\right)_{xy} & \left(\frac{1}{n^2}\right)_{xz} \\ \left(\frac{1}{n^2}\right)_{xy} & \left(\frac{1}{n^2}\right)_{yy} & \left(\frac{1}{n^2}\right)_{yz} \\ \left(\frac{1}{n^2}\right)_{xz} & \left(\frac{1}{n^2}\right)_{yz} & \left(\frac{1}{n^2}\right)_{zz} \end{bmatrix} \begin{bmatrix} D_x \\ D_y \\ D_z \end{bmatrix}.
$$

$$(14.31)$$

Index of refraction of the crystal n is defined as $n = \sqrt{\varepsilon_r}$. This tensor has only six independent components [27]. The electric energy density W_e in a medium is given as $W_e = \frac{1}{2}\vec{E}\cdot\vec{D}$. Using Equation 14.31, this can be written as

$$
2\varepsilon_0 W_e = \left(\frac{1}{n^2}\right)_{xx} D_x^2 + \left(\frac{1}{n^2}\right)_{yy} D_y^2
$$

$$
+ \left(\frac{1}{n^2}\right)_{zz} D_z^2 + 2\left(\frac{1}{n^2}\right)_{yz} D_y D_z + 2\left(\frac{1}{n^2}\right)_{xz} D_x D_z + 2\left(\frac{1}{n^2}\right)_{xy} D_x D_y.
$$

$$(14.32)$$

It should be noted that in Equation 14.32, $\left(1/n^2\right)_{xx} \neq \left(1/n_{xx}^2\right)$ unless the principle axes are used as the coordinate system. Since the left-hand side of Equation 14.32 is a constant, the following association can be made:

$$
x = D_x / \sqrt{C}, \, y = D_y / \sqrt{C}, \text{ and } z = D_z / \sqrt{C},
$$

where

$$
C = \sqrt{2\varepsilon_0 W_e}.
$$

With this association and using the abbreviated notation in which $xx \equiv 1, yy \equiv 2, zz \equiv 1, yz \equiv 4, xz \equiv 5$, and $xy \equiv 6$, Equation 14.32 can be rewritten as

$$
\left(\frac{1}{n^2}\right)_1 x^2 + \left(\frac{1}{n^2}\right)_2 y^2 + \left(\frac{1}{n^2}\right)_3 z^2 + 2\left(\frac{1}{n^2}\right)_4 yz
$$

$$
+ 2\left(\frac{1}{n^2}\right)_5 xz + 2\left(\frac{1}{n^2}\right)_6 xy = 1.
$$

$$(14.33)$$

In this equation, x, y, and z relate to the polarization of the optical field. For example, $y = z = 0$, means an x polarized optical field. Equation 14.33 represents an ellipse and is known as the index

ellipsoid or optical indicatrix. Using this equation, one can find the phase velocity of an optical wave propagating in the crystal in any arbitrary direction [27].

If an external electric field is applied to the material, the coefficients of the index ellipsoid change. This external electric field is the modulating field and typically extends in frequency from DC to millimeter wave range. This change in the coefficients is equivalent to modulating the velocity of the optical wave and is used to modulate the optical wave. It is linearly proportional to the applied field; hence, the relationship between the six coefficients in Equation 14.33 and the three electric field components can be expressed as

$$
\begin{pmatrix} \Delta\left(\frac{1}{n^2}\right)_1 \\ \Delta\left(\frac{1}{n^2}\right)_2 \\ \Delta\left(\frac{1}{n^2}\right)_3 \\ \Delta\left(\frac{1}{n^2}\right)_4 \\ \Delta\left(\frac{1}{n^2}\right)_5 \\ \Delta\left(\frac{1}{n^2}\right)_6 \end{pmatrix} = \begin{pmatrix} r_{11} & r_{12} & r_{13} \\ r_{21} & r_{22} & r_{23} \\ r_{31} & r_{32} & r_{33} \\ r_{41} & r_{42} & r_{43} \\ r_{51} & r_{52} & r_{53} \\ r_{61} & r_{62} & r_{63} \end{pmatrix} \begin{pmatrix} E_x \\ E_y \\ E_z \end{pmatrix}.
$$

$$(14.34)$$

The 6×3 matrix multiplying the electric field vector is known as the electro-optic tensor and its elements r_{ij} are known as the electro-optic coefficients. The typical magnitude of these coefficients is in the order of 10^{-12} m V^{-1}. With these changes, the index ellipsoid is written as

$$
\begin{pmatrix} x & y & z \end{pmatrix} \begin{vmatrix} \left(\left(\frac{1}{n^2}\right)_1 + \Delta\left(\frac{1}{n^2}\right)_1\right) & \left(\left(\frac{1}{n^2}\right)_6 + \Delta\left(\frac{1}{n^2}\right)_6\right) & \left(\left(\frac{1}{n^2}\right)_5 + \Delta\left(\frac{1}{n^2}\right)_5\right) \\ \left(\left(\frac{1}{n^2}\right)_6 + \Delta\left(\frac{1}{n^2}\right)_6\right) & \left(\left(\frac{1}{n^2}\right)_2 + \Delta\left(\frac{1}{n^2}\right)_2\right) & \left(\left(\frac{1}{n^2}\right)_4 + \Delta\left(\frac{1}{n^2}\right)_4\right) \\ \left(\left(\frac{1}{n^2}\right)_5 + \Delta\left(\frac{1}{n^2}\right)_5\right) & \left(\left(\frac{1}{n^2}\right)_4 + \Delta\left(\frac{1}{n^2}\right)_4\right) & \left(\left(\frac{1}{n^2}\right)_3 + \Delta\left(\frac{1}{n^2}\right)_3\right) \end{vmatrix} \begin{pmatrix} x \\ y \\ z \end{pmatrix} = 1.
$$

$$(14.35)$$

Crystal symmetry imposes certain restrictions on the form of the index ellipsoid. Figure 14.15 shows the form of the electro-optic tensor for 3 m and

(a)

$$\begin{bmatrix} 0 & -r_{22} & r_{13} \\ 0 & r_{22} & r_{13} \\ 0 & 0 & r_{33} \\ 0 & r_{51} & 0 \\ r_{51} & 0 & 0 \\ -r_{22} & 0 & 0 \end{bmatrix}$$

(b)

$$\begin{bmatrix} 0 & 0 & 0 \\ 0 & 0 & 0 \\ 0 & 0 & 0 \\ r_{41} & 0 & 0 \\ 0 & r_{41} & 0 \\ 0 & 0 & r_{41} \end{bmatrix}$$

Figure 14.15 Electro-optic tensor (a) for 3m crystal symmetry, and (b) for cubic crystal symmetry.

cubic crystal symmetries. More complete representation of the electro-optic tensor for other crystal symmetries can be found in Reference [28].

14.7 SPECIFIC EXAMPLES OF DIFFERENT MODULATOR TECHNOLOGIES

14.7.1 EA modulators

14.7.1.1 LUMPED EA MODULATORS

Earlier it was shown that in a QW, it is possible to modulate the absorption very strongly with external fields around a narrow wavelength range due to QCSE. If such QWs are embedded in an optical waveguide, the insertion loss of this waveguide can be modulated by applying an electric field and changing the absorption of the QWs through the QCSE [29]. Figure 14.16 shows such a modulator [1,30]. Typically, MQWs are used to increase absorption and are embedded in the i region of a reverse biased p-i-n diode. The typical thickness of the i region, d_i, is in the 0.1–0.5 μm range. Therefore, it is possible to apply very strong electric fields with a few volts of reverse bias. In this case, the well and barrier material are InGaAs and InAlAs, respectively. The index of refraction of the top p and bottom n-InAlAs is lower than that of MQW. This forms a slab waveguide. By deeply etching this slab waveguide, a channel optical guide is formed. Waveguide widths are in the order of 1–3 μm and the etch depths are about 1–2 μm. Typical device lengths are in the 50–300 μm range. Ohmic contacts are formed at the top and bottom.

Etched areas underneath the bonding pads are filled with a low dielectric constant polyimide to reduce device capacitance. Since only the absorption modulation is utilized, optical waveguide need not be of single mode. Therefore, the profile of the optical waveguide can be optimized for improved coupling with the incoming fiber mode. However, critical control of the composition and thickness of the epitaxial layers are required.

The transmission through such a modulator as a function of applied voltage can be expressed as

$$T(V) = T_0 \exp[-\Gamma \alpha(V) L], \qquad (14.36)$$

where Γ is the overlap between the optical mode and the MQW, $\alpha(V)$ is the absorption coefficient as a function of applied voltage V, L is the length, and T_0 is the coupling coefficient between the fiber and the optical waveguide. The on/off ratio of an EA modulator in dB can be expressed as

$$10 \log \left[\frac{T(V)}{T(0)} \right] = \frac{\alpha(V) - \alpha(0)}{\alpha(0)} 10 \log \left[e^{-\alpha(0)L} \right]$$

$$= \frac{\Delta \alpha}{\alpha} \left[\text{Propagation loss (dB)} \right].$$

$$(14.37)$$

Optical propagation loss of EA modulators is large, typically in the 15–20 dB/mm range. The main components of this loss are the FC absorption, especially in the p layers, and band-to-band absorption, both described in earlier sections. The second loss component can be made smaller by increasing the separation between the wavelength of operation and the absorption peak, which is called detuning. Typical detuning values are about 20–50 nm. Typical $\Delta \alpha / \alpha$ values are in the 3–10 range. Large on/off ratio devices can be obtained using long devices, but these also increase the insertion loss. For the typical EA modulator lengths in the 50–300 μm range, propagation loss is 1–3 dB. Therefore, to get large extinction ratios with low device insertion loss, $\Delta \alpha / \alpha$ should be maximized. Furthermore, efficient modulation requires a large $\Delta \alpha / \Delta v$ or $(1/d_i)(\Delta \alpha / \Delta E)$, where E is the applied electric field. In other words, large Stark shifts are required. For low bias voltages, the shift in the ground state energy ΔE_i of a particle in a QW is proportional to the fourth power of the

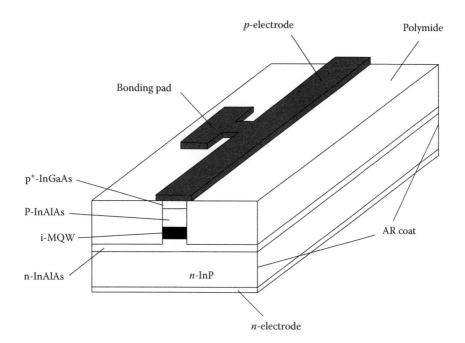

p-electrode

Polymide

Bonding pad

p⁺-InGaAs

P-InAlAs

i-MQW

AR coat

n-InAlAs

n-InP

n-electrode

Figure 14.16 Schematic of an MQW EA modulator. (Ido, T. et al., *IEEE Photon. Technol. Lett.*, Figure 1. ©1994 IEEE.)

well width as shown in Equation 14.26 [19]. Hence, wide QWs are desired for efficient operation.

For fiber optic communication applications, the desired wavelength is around 1.55 μm. This requires alloys of InP and GaAs as active material grown on InP substrates. Quaternary alloys such as InGaAsP and InGaAlAs can be grown lattice matched to InP; this makes it possible to change the bandgap energy and well thickness independently [31,32]. Thick QWs up to 12 nm have been reported in InGaAsP/ InGaAsP MQW resulting in drive voltages as low as 1.2 V [31]. Similarly, InGaAlAs/InAlAs MQWs with well widths as large as 19.6 nm yielded very low voltage EA modulators at 1.55 μm, requiring about 1 V for 10 dB on/off ratio [33]. Larger electron confinement due to increased conduction band discontinuity in this material system makes such wells comparable to narrower wells in the InGaAsP system. It is possible to obtain similar results using lattice matched InGaAs ternary QWs. However, as the wells get thicker, the absorption edge shifts to longer wavelengths and operation around 1.55 μm becomes difficult. This difficulty in the ternary material can be eliminated by using tensile strained QW [18]. A commonly used material design consists of a 0.6 μm n-InAlAs buffer, an

undoped strained MQW absorption layer, a 2 μm p-InAlAs cladding layer, and a p⁺-InGaAs contact layer [34]. The strained MQW layer contains ten 8.8 nm In$_{0.48}$Ga$_{0.52}$As wells and 5 nm In$_{0.53}$Al$_{0.47}$As barriers. The wells are under 0.35% tensile strain and the barriers are under 0.5% compressive strain. The strain in the barriers is used for strain compensation. Proper amounts of alternating compressive and tensile strain can make the total strain in the MQW near zero, allowing large periods of strained MQW without exceeding the critical layer thickness. Fiber-to-fiber insertion loss of a typical high-speed EA modulator as a function of external bias at different wavelengths is shown in Figure 14.17 [34]. QCSE is most pronounced for photon energies near the bandgap of the material and shows strong wavelength dependence as seen in Figure 14.17. At shorter wavelengths, modulation becomes more efficient, but insertion loss also increases. EA modulators are very short devices and, hence have small device capacitance. Therefore, when driven as a lumped circuit element, their speed of operation will be limited by the RC time constant of the circuit. Therefore, reducing the capacitance by shortening the device increases the speed of operation. Typically, a 2.5 μm wide and 150 μm long device has a capacitance

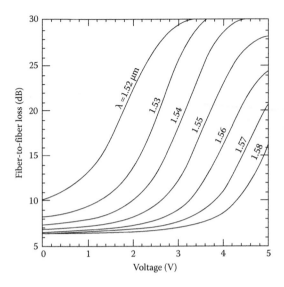

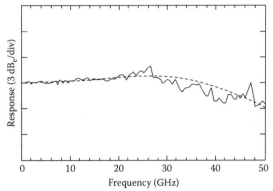

Figure 14.18 Frequency response of an EA modulator with integrated input and output optical waveguides under 1.5 V reverse bias and +5 dBm optical input power at 1.55 μm. The dashed line shows the calculated data. (Ido, T. et al., *J. Lightwave Technol.*, Figure 11. ©1996 IEEE.)

Figure 14.17 Fiber-to-fiber insertion loss of a 40 Gb/s EA modulator for various wavelengths. The input light was TE polarized. (Ido, T. et al., *J. Lightwave Technol.*, Figure 10. ©1996 IEEE.)

of about 0.33 pF, which is low enough for 20 GHz bandwidth when driven by a 50 Ω source with a 50 Ω terminating resistance across the device. Higher bandwidths are possible using a lower terminating resistance. This reduces the RC time constant, and hence increases the bandwidth at the expense of modulation efficiency. Electrical 3 dB bandwidths approaching 60 GHz have been reported for lumped devices [34–37]. High-speed operation requires not only a short device but also good impedance matching and proper microwave packaging. Although unpackaged devices demonstrated very high-speed operation, it was difficult to duplicate these results in packaged modules. The main source of the difficulty was cleaving and packaging devices as short as 50 μm. One solution to this difficulty is to integrate passive input and output waveguides to a very short EA section [34]. This requires etching of the MQW region outside the EA modulator section and regrowing low loss passive waveguide sections. This technique was used to obtain overall device lengths of 1 mm, with EA modulator sections as short as 50 μm. The measured frequency response of such a 1.55 μm modulator is illustrated in Figure 14.18, which shows a 3 dB electrical bandwidth of 50 GHz. At 1.3 μm, bandwidths as large as 38 GHz have been demonstrated [38].

In an unstrained material, QCSE is polarization dependent. However, using strain compensated QWs can significantly reduce the polarization dependence [18,39,40]. The most common approach is to use about 0.5% tensile strained InGaAs wells and about 0.5% compressive strained InAlAs barriers [34,39].

14.7.1.2 TRAVELING WAVE EA MODULATORS

One of the emerging directions in EA device research is to reduce the drive voltage and increase the bandwidth. Lumped device performance seems to be saturated around 2 V drive voltage and 50 GHz bandwidth. The main reasons for this are the conflicting requirements on the length of the device. Low drive voltages require longer devices yet wide bandwidths require shorter devices. One way to address this problem is to design the device as a traveling wave modulator. As described earlier for a traveling wave device, the RC time constant is not the bandwidth limit and the device can be made longer without sacrificing bandwidth. However, the special structure of an EA modulator presents interesting issues. First, the optical propagation loss of an EA device is rather high. Therefore, increasing the length over even a few hundred microns introduces excessive loss. Large device capacitance per unit length also makes it difficult to make a 50 Ω transmission line with matched velocity. Since the device is rather short, velocity matching is not an issue up to frequencies well into the sub-millimeter-wave range. The bandwidth is typically limited by the microwave loss [41]. For traveling wave EA (TW-EA) modulator electrodes, measured microwave loss coefficients of about 60–80 dB/cm at 40 GHz were reported [41]. This excessive loss is due to heavily doped layers inside the device and is another factor limiting the length of the device. Characteristic impedance values are about 25 Ω [41,42]. However, designing the device as part of a 25 Ω transmission line makes longer devices possible and packaging easier. Recently, a 200 μm long TW-EA device with a bandwidth over 54 GHz [43] and a 300 μm long device with a bandwidth of 25 GHz [42] were reported. Both devices were polarization insensitive and operated at 1.55 μm. For 20 dB on/off ratio, the drive voltage of the 200 and 300 μm devices were 3 and 1.9 V, respectively.

14.7.2 LiNbO₃ modulators

Among all external modulators, LiNbO₃ traveling wave modulators have the most mature technology. Such modulators are commercially available from several manufacturers and are used in many commercial applications. LiNbO₃ is a ferroelectric crystal that is readily available commercially.

A schematic of a typical LiNbO₃ phase modulator is shown in Figure 14.19. On the surface, there are two electrodes across which the modulating voltage is applied. An optical waveguide is also formed in the crystal. The most common way of fabricating optical waveguides in LiNbO₃ is Ti indiffusion. Typically, Ti stripes of about 3–8 μm wide and about 0.1 μm thick are patterned on the surface of LiNbO₃ using the lift-off process. These Ti stripes are subsequently driven into LiNbO₃ at around 1000°C for about 10 h in an oxygen atmosphere. The resulting Ti diffusion profile translates itself into a higher index of refraction and optical

guiding results. The index steps generated this way are in the order of 0.01, and typical optical propagation loss coefficients are less than 0.2 dB/cm. Furthermore, the optical modes in such waveguides match rather well with the mode of a single-mode optical fiber. As a result, 5 dB fiber-to-fiber insertion is quite common for LiNbO₃ traveling wave modulators with 5 cm long electrodes. Other techniques of fabricating optical waveguides in LiNbO₃ include ion exchange, proton exchange, and Ni diffusion. The effect used for modulation is the linear electro-optic effect and its specifics for LiNbO₃ are described next.

14.7.2.1 ELECTRO-OPTIC EFFECT IN LINBO₃

LiNbO₃ has 3 m crystal symmetry. When an external modulating field is applied to LiNbO₃, its index ellipsoid is perturbed and observing Equations 14.34, 14.35 and Figure 14.15, it can be expressed as

$$
\begin{pmatrix} x & y & z \end{pmatrix}
\begin{bmatrix}
\frac{1}{n_o^2} - r_{22}E_y + r_{13}E_z & -r_{22}E_x & r_{51}E_x \\
-r_{22}E_x & \frac{1}{n_o^2} + r_{22}E_y + r_{13}E_z & r_{51}E_y \\
r_{51}E_x & r_{51}E_y & \frac{1}{n_e^2} + r_{33}E_z
\end{bmatrix}
\begin{pmatrix} x \\ y \\ z \end{pmatrix} = 1.
$$

(14.38)

where n_o and n_e are known as the ordinary and extraordinary indices of refraction, respectively. Their values are $n_e = 2.21$ and $n_o = 2.3$. The electro-optic coefficients are given as $r_{22} = 3.4 \times 10^{-12}$ m/V, $r_{13} = 8.6 \times 10^{-12}$ m/V, $r_{33} = 30.8 \times 10^{-12}$ m/V, and $r_{51} = 28 \times 10^{-12}$ m/V. In this case, depending on the direction of the modulating electric field, different possibilities exist. For example, if the modulating field is z directed and $E_x = E_y = 0$, Equation 14.38 reduces to

$$
\begin{pmatrix} x & y & z \end{pmatrix}
\begin{vmatrix}
\frac{1}{n_o^2} + r_{13}E_z & 0 & 0 \\
0 & \frac{1}{n_o^2} + r_{13}E_z & 0 \\
0 & 0 & \frac{1}{n_e^2} + r_{33}E_z
\end{vmatrix}
\begin{pmatrix} x \\ y \\ z \end{pmatrix} =
$$

$$
\begin{pmatrix} x & y & z \end{pmatrix}
\begin{bmatrix}
\frac{1}{(n_o + \Delta n_x)^2} & 0 & 0 \\
0 & \frac{1}{(n_o + \Delta n_y)^2} & 0 \\
0 & 0 & \frac{1}{(n_e + \Delta n_z)^2}
\end{bmatrix}
\begin{pmatrix} x \\ y \\ z \end{pmatrix} = 1.
$$

(14.39)

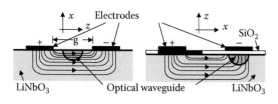

Figure 14.19 The most commonly used LiNbO₃ modulator configurations using (a) an x-cut and (b) a z-cut crystal.

In this case, the index ellipsoid remains diagonal. The index of refraction for an optical wave polarized in the z direction and propagating in either the x or y direction can be found using Equation 14.39 as

$$\frac{1}{\left(n_e+\Delta n_z\right)^2}=\frac{1}{n_e^2}+r_{33}E_z \text{ or } \left(n_e+\Delta n_z\right)=\frac{n_e}{\sqrt{1+r_{33}n_e^2E_z}}.$$

Since $r_{33}n_e^2E_z \ll 1$, $\left(n_e+\Delta n_z\right)\approx n_e\left(1-\frac{1}{2}r_{33}n_e^2E_z\right)$. Hence

$$\Delta n_z=-\frac{1}{2}r_{33}n_e^3E_z. \qquad (14.40)$$

Since r_{33} is the largest electro-optic coefficient, this arrangement generates the largest index change, while keeping the index ellipsoid diagonal. For this reason, a z-directed modulating field, modulating a z polarized optical wave is the most commonly used arrangement for $LiNbO_3$ electro-optic modulators.

The two most commonly used configurations taking advantage of this arrangement are shown in Figure 14.19. For x-cut crystals, the optical waveguide is in the y crystal orientation. A horizontal electric field parallel to the surface of the crystal, i.e., an electric field along the z-axis is utilized. This configuration modulates the TE polarized optical mode, in which the main electric field component of the optical waveguide mode is parallel to the surface of the crystal, i.e., it is along the z-axis of the crystal. For z-cut crystals, an electric field vertical to the surface of the crystal is used. Optical waveguides are along the y-axis of the crystal as shown in Figure 14.19. This field modulates the TM polarized optical mode most efficiently. For TM modes, the main electric field component of the optical waveguide mode is also perpendicular to the surface of the crystal, i.e., it is along the z-axis of the crystal. For both configurations, the modulating external field and the main electric field component of the optical mode are parallel to each other and to the z-axis of the crystal. In z-cut crystals, a low index buffer layer, such as SiO_2, is usually used under the electrode. This buffer layer is used to isolate the optical mode from the metal electrode in order to keep the optical propagation loss low.

The presence of the applied electric field modifies the index of the material as described earlier. This in turn modifies the propagation constant of the optical mode. Since the effect is very small, this perturbation can be found using a perturbation analysis. The result is [12]

$$\Delta\beta=\frac{2\pi}{\lambda}\Delta n_{\text{eff}} \text{ with } \Delta n_{\text{eff}}=\iint \Delta n_z|\Upsilon|^2 dS,$$

where Υ is the normalized electric field of the optical mode. The integration is carried out over the entire optical mode. Substituting the value of Δn_z derived in Equation 14.40, we obtain

$$\Delta n_{\text{eff}}=\frac{1}{2}r_{33}n_e^3\iint E_z|\Upsilon|^2 dS. \qquad (14.41)$$

Multiplying and dividing this equation by v/g, where v is the applied voltage and g is the electrode gap, we obtain

$$\Delta n_{\text{eff}}=\frac{1}{2}r_{33}n_e^3\frac{v}{g}\left[\frac{g}{v}\iint E_z|\Upsilon|^2 dS\right]=\frac{1}{2}r_{33}n_e^3\frac{v}{g}\Gamma. \qquad (14.42)$$

Γ is known as the overlap integral and is expressed as

$$\Gamma=g\iint \frac{E_z}{v}|\Upsilon|^2 dS. \qquad (14.43)$$

Γ is proportional to the overlap of the magnitude squared normalized optical mode electric field and the normalized z component of the applied modulating field. With this definition, we can express the K coefficients defined in Equations 14.11 and 14.13 as

$$K=\frac{1}{2}r_{33}n_e^3\frac{\Gamma}{g}. \qquad (14.44)$$

Therefore, the V_π of a push–pull driven $LiNbO_3$ directional coupler modulator is given as

$$V_\pi=\frac{\sqrt{3}}{2}\frac{\lambda_0}{L}\frac{g}{r_{33}n_e^3\Gamma}. \qquad (14.45)$$

Similarly, that of a push–pull driven $LiNbO_3$ Mach–Zehnder modulator is

$$V_\pi=\frac{1}{2}\frac{\lambda_0}{L}\frac{g}{r_{33}n_e^3\Gamma}. \qquad (14.46)$$

Obviously in order to have a low drive voltage, a large electro-optic coefficient, a large index of refraction, a large overlap integral, a long electrode length, and a small electrode gap are desired. A simple back of the envelope calculation indicates that unless L is very long, it is not possible to achieve low drive voltage modulators. However, long electrode length necessitates a traveling wave configuration as described earlier. For this reason, all practical LiNbO₃ electro-optic modulators are traveling wave modulators. Such modulators are described in the next section.

14.7.2.2 TRAVELING WAVE LINBO₃ ELECTRO-OPTIC MODULATORS

A schematic of a typical LiNbO₃ traveling wave modulator is shown in Figure 14.20 [44]. The optical structure is a Mach–Zehnder interferometer. The electrode length is in the order of centimeters; hence, velocity matching is essential in such a modulator.

The dielectric constant of LiNbO₃ shows a large amount of dispersion going from microwave to optical frequencies due to large ionic contribution to its dielectric constant. For z-cut crystals, the microwave relative dielectric constants of LiNbO₃ parallel and perpendicular to the crystal are $\varepsilon_{ry}=43$ and $\varepsilon_{rz}=28$, respectively. For a transmission line running along the y direction, one can define an effective relative dielectric constant under quasi-static approximation as [45]

$$\varepsilon_{reff} = \sqrt{\varepsilon_{ry}\varepsilon_{rz}}, \qquad (14.47)$$

which is around 35. The most commonly used electrode for such modulators is the coplanar transmission line (CPW). The effective microwave index of a CPW on LiNbO₃ with zero conductor thickness and air as the top dielectric is [46]

$$n_\mu = \sqrt{\frac{\varepsilon_{reff}+1}{2}} = \sqrt{\frac{\sqrt{\varepsilon_{ry}\varepsilon_{rz}}+1}{2}}. \qquad (14.48)$$

This value is larger than 4. On the other hand, the commonly assumed effective index of an optical mode in a Ti indiffused LiNbO₃ optical waveguide is 2.15. As a result, an electrical signal applied to the electrode will travel significantly slower than the optical wave. Therefore, velocity matching in LiNbO₃ modulators requires increasing the velocity of propagation of the microwave electrode.

The most common way of achieving this is to use a SiO₂ buffer layer under the electrode and to increase the thickness of the conductors. The characteristic impedance, Z_0, and the velocity of propagation, v_μ, of a transmission line are given as

$$Z_0 = \sqrt{\frac{l}{C}}; \quad v_\mu = \frac{1}{\sqrt{lC}} \qquad (14.49)$$

where l and C are the inductance and capacitance per unit length of the transmission line, respectively. Under quasi-static approximation, l is found using the exact same line geometry in which the dielectrics are replaced by air. Since electro-optic materials used are nonmagnetic, l of this air line is the same as the l of the original line. But the capacitance per unit length of the air line, C_a, is different. Such a line with uniform air dielectric supports a TEM mode with the velocity of propagation the same as the speed of light in air, c. Hence

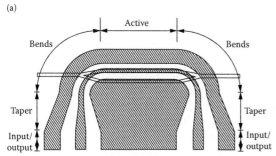

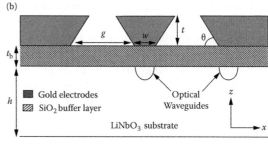

Figure 14.20 (a) Top view and (b) cross-sectional schematics of a z-cut y propagating LiNbO₃ traveling wave modulator. (Gopalakrishnan, G.K. et al., J. Lightwave Technol., Figure 1. ©1994 IEEE.)

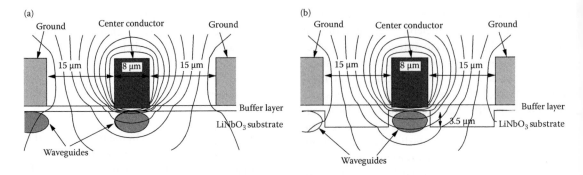

Figure 14.21 Cross-sectional schematic and potential distribution around the center conductor of (a) a conventional modulator and (b) a ridge structure. (Noguchi, K. et al., *IEEE Photon. Technol. Lett.*, Figure 3. ©1993 IEEE.)

$$c = \frac{1}{\sqrt{lC_a}} \text{ or } l = \frac{1}{c^2 C_a}. \qquad (14.50)$$

Combining these equations together, we find that

$$v_\mu = c\sqrt{\frac{C_a}{C}} \text{ and } Z_0 = \frac{1}{c\sqrt{CC_a}}. \qquad (14.51)$$

The relative dielectric constant of SiO_2 is 3.9, which allows filling part of the transmission line with a low index medium. This in turn reduces C; hence, v_μ and Z_0 are increased. Placing a dielectric like SiO_2 under the metal electrode also helps to reduce the overlap of the optical mode with the metal electrode as described earlier. This helps to reduce the optical propagation loss significantly especially for z-cut devices. However, thick SiO_2 layers are not desirable since part of the electrode voltage that drops across this layer reduces the electric field intensity in $LiNbO_3$ and hence, the modulation efficiency of the device. Typical SiO_2 layer thickness is about 1 μm. The additional velocity increase is typically achieved by increasing the conductor thickness. Increasing the conductor thickness increases the electric field strength in the slots between the conductors. This in turn increases C_a; hence, v_μ increases and Z_0 decreases. Typical conductor thicknesses range from 10 to 20 μm. For such thick electrodes, the slope of the sidewalls starts to affect the electrode characteristics [44]. Although it is possible to achieve velocity matching this way, the undesirable side effect is the reduction in the characteristic impedance of the electrode. It is possible to adjust the gap and the width of the center conductor of the CPW to match v_μ and Z_0

simultaneously, but this requires rather narrow gap and width values [47]. As a result, microwave electrode loss increases and becomes the limiting factor for the bandwidth. This difficulty was solved by introducing the ridge structure [48].

Figure 14.21 illustrates the basic idea behind this approach [48]. Compared to the conventional modulator, the ridge removes the high microwave dielectric constant $LiNbO_3$ between the conductors. As a result, C reduces and both v_μ and Z_0 are increased simultaneously. This allows velocity matching without sacrificing impedance matching. Furthermore, a high dielectric constant in the ridge helps to confine the electric field lines under the electrode so that the field becomes almost vertical under the electrode. As a result, the desired vertical component of the electric field overlaps better with the optical mode improving the efficiency of the modulator. This is especially true for z-cut devices. Detailed analysis of this modulator geometry is reported in Reference [49].

Traveling wave $LiNbO_3$ modulators with very wide bandwidths have been demonstrated. Frequency dependence of the modulation response of a modulator fabricated using the ridge structure is shown in Figure 14.22.

For this modulator, $t_b = 1.0 \, \mu m$, ridge height, $t_r = 4.0 \, \mu m$, $t_m = 20 \, \mu m$, $L = 2 \, cm$, $W = 8 \, \mu m$, and $G = 25 \, \mu m$. The measured electrical and optical bandwidths are 75 and 110 GHz, respectively and V_π is 5.1 V [50]. If the length of the same device is increased to 3 cm, V_π decreases to 3.5 V but the electrical and optical bandwidths also decrease to 30 and 45 GHz, respectively. Recently, careful loss measurements up to 110 GHz revealed that up to 20 GHz loss is dominated by conductor

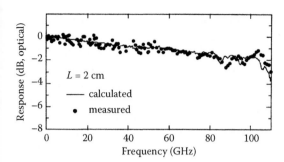

Figure 14.22 Modulation response as a function of frequency for a broadband LiNbO₃ traveling wave modulator employing ridge structure. (Noguchi, K. et al., *Electron. Lett.*, Figure 9. ©1998 IEEE.)

losses [51]. Above 20 GHz, dielectric and radiation losses become important. The SiO_2 buffer layer is found to have a loss tangent four times higher than that of the LiNbO₃ substrate [51]. Therefore, the quality of this buffer layer needs to be carefully controlled. Another loss component is the coupling-to-substrate modes in a CPW structure [52]. The usual technique to eliminate this loss is to thin the LiNbO₃ substrate. Typical substrate thickness used is less than 0.5 mm. The thinner the substrate, the higher the frequency at which this coupling occurs. Moreover, keeping a large part of the substrate covered with metal helps to eliminate this coupling. CPW electrodes provide an advantage in this regard. In the measurements in Reference [51], several narrowband electrode loss increases at 85,

95, and 105 GHz were reported. This observation was attributed to coupling-to-substrate modes in the electrical connectors.

The recent results in the bandwidth of LiNbO₃ modulators are impressive. However, drive voltages required to drive such modulators at high frequencies are still high. Present day electronics is expected to generate about 3.5 V drive voltages at 40 GHz and even less at higher frequencies. This means the drive voltages of existing modulators need to be reduced even further. One obvious way to reduce drive voltages is to increase the electrode length. In a recent work, a reflection type traveling wave modulator was reported [53]. This approach reflects the electrical and optical signals from the cleaved and polished edge of the substrate. The reflection doubles the interaction length with the electrode. For an electrode and interaction length of 5.3 and 10.6 cm, respectively, V_π was 0.89 V at 1.3 μm.

Another approach is to use a thin film of LiNbO₃ on a low dielectric constant material. Figure 14.23a shows the cross-sectional profile of such a modulator [54]. A thin film LiNbO₃ of thickness T_{LN} is attached to a LiNbO₃ substrate using a low dielectric constant adhesive of thickness T_{ad}. Both the LiNbO₃ substrate and thin film are x-cut. Figure 14.23b shows the scanning electron microscope (SEM) image of the fabricated structure.

The presence of a low dielectric constant layer reduces the capacitance of the CPW electrode, which allows the reduction of the electrode gap, G. This in turn reduces V_π. T_{ad} was set at 50 μm.

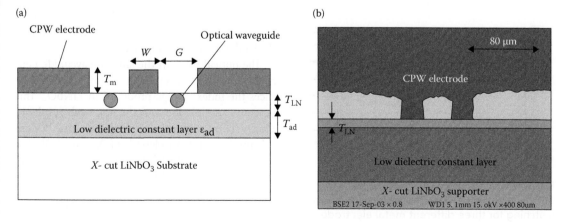

Figure 14.23 (a) Cross-sectional profile of an x-cut LiNbO₃ thin film modulator, (b) SEM image of the fabricated structure. (Kondo, J. et al., *IEEE Photon. Technol. Lett.* ©2005 IEEE.)

The variation of the $V_\pi L$ product, characteristic impedance, and T_{LN} required for velocity matching for three different metal electrode thickness values are shown in Figure 14.24.

As expected, as G decreases, $V_\pi L$ decreases. However, small gaps increase electrode capacitance and thin film thickness should be reduced to remove the high dielectric constant $LiNbO_3$ and reduce electrode capacitance to the value required for velocity matching. Hence, G was chosen as 25 μm, which requires about 7 μm thick thin film for 20 μm thick metal. This yields a $V_\pi L$ product of 7.7 V cm and V_π becomes 1.7 V for a 4.4 cm long electrode. Z_0 is 40 Ω, which is acceptable in a 50 Ω system. The fabricated device with $W = 35$ μm had V_π of 1.8 V. The reflection coefficient of the electrode was less

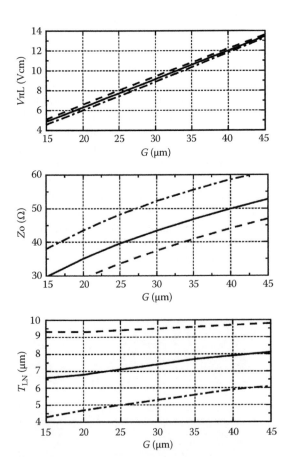

Figure 14.24 Variation of $V_\pi L$ product, characteristic impedance, and T_{LN} required for velocity matching for three different metal electrode thickness values of 10, 20, and 30 μm. (From Kondo, J. et al., *IEEE Photon. Technol. Lett.*, 17, 10, 2077–2079, 2005. With permission.)

than −10 dB up to 50 GHz and the 3 dB electrical-to-optical bandwidth is given as 25 GHz. The device was successfully used for 40 Gb/s transmission.

In this case, a thin film was obtained using precise polishing and lapping; its thickness was 7.1 ± 0.25 μm over a 3″ wafer. Thin $LiNbO_3$ films can also be manufactured based on ion slicing. This approach uses two $LiNbO_3$ wafers. One of them is ion implanted with either hydrogen or helium. The depth of penetration depends on ion energy. These are the lightest ions possible and can penetrate into $LiNbO_3$ an appreciable distance at high ion energies. Significant damage results at a depth where ions finally stop due to nuclear interactions with $LiNbO_3$ ions. A SiO_2 layer is deposited on the other $LiNbO_3$ wafer. Then, the implanted and SiO_2 covered wafers are bonded to each other. This is followed by thermal treatment. During this thermal treatment, a thin single crystal $LiNbO_3$ layer above the damaged region in the implanted wafer splits off. Therefore, a thin single crystal $LiNbO_3$ layer on SiO_2 on a $LiNbO_3$ wafer like the one shown in Figure 14.23a is formed. At this point, further annealing is used to strengthen the bonding. Finally, a chemical mechanical polish is used to obtain a smooth surface. However, in this approach, thickness of the single crystal $LiNbO_3$ layer is limited to about 700 nm due to limited ion penetration depth.

Yet another approach to integrate $LiNbO_3$ with other materials is to bond other materials directly on to a $LiNbO_3$ substrate [55]. In one example, a silicon layer on silicon on insulator (SOI) wafer is bonded to $LiNbO_3$. This is followed by removing the silicon substrate and SiO_2 layers of the SOI wafer as shown in Figure 14.25. Then, an optical waveguide is formed in the silicon and modulator electrodes are formed on $LiNbO_3$.

The optical wave in the silicon waveguide penetrates into the $LiNbO_3$ lower cladding and experiences the index change produced in $LiNbO_3$. This generates modulation and the approach results in Mach–Zehnder intensity modulators with a 26 V cm $V_\pi L$ product at a wavelength of 3.39 μm [55]. The propagation loss of such silicon on $LiNbO_3$ waveguides were given as 2.5 dB/cm. Using this approach, other materials can be bonded on $LiNbO_3$ possibly providing optical gain or sources. The drawback is reduced optical mode overlap with $LiNbO_3$ due to high index contrast between silicon and $LiNbO_3$ and limited wafer size of $LiNbO_3$ substrates.

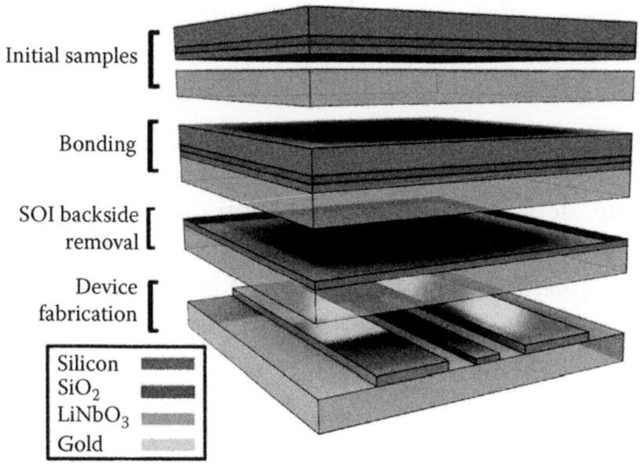

Initial samples [

Bonding [

SOI backside removal [

Device fabrication [

| Silicon |
| SiO$_2$ |
| LiNbO$_3$ |
| Gold |

Figure 14.25 Process for preparation of silicon on LiNbO$_3$.

LiNbO$_3$ is susceptible to what is known as photorefractive effect [56]. Under optical illumination, electrons from ionized impurity centers (typically Fe^{2+}) are excited to the conduction band. Then, they move around until they are recaptured by traps. This creates charge separation, which creates large electric fields inside the material. These electric fields change the index through the electro-optic effect. Propagation loss also changes. This is a time-dependent process depending on the charge dynamics and changes in the illumination levels. The effect is most significant at wavelengths shorter than about 870 nm [57] and becomes less of an issue at infrared wavelengths like 1.55 μm for typical power levels used in communication applications. It is shown that photorefractive sensitivity of Ti indiffused LiNbO$_3$ modulators can be reduced significantly using high temperature O$_2$ anneals [58]. Annealed modulators were operated at 1.3 μm at input optical powers as high as 400 mW up to 168 h with only a 3.5° drift in the modulator bias point. There was no measurable change in the bias point at 100 mW. It is also shown that annealing at high temperatures in N$_2$ and Ar ambience increases the sensitivity to the photorefractive effect significantly [58].

Bias stability of modern day LiNbO$_3$ devices is very good. They can operate at constant bias for thousands of hours under the control of automatic biasing circuitry. However, when bias voltages change suddenly, bias drifts with time constants from hours to many days may be observed. The major source of bias point drift is attributed to flow and redistribution of electrical charge under the application of applied voltages. The conductivity

of SiO$_2$ buffer and LiNbO$_3$ are very high but finite. Usually, a certain amount of low mobility charge exists in these materials at their interface or on their surface. These charges could be process related or can be generated with optical or thermal excitation during the operation of the device. When voltages are applied to the electrodes, the charge starts to move and modify the electric field applied to the optical guides. As a result, modulation characteristics change. This behavior can be modeled using an electrical network approach [59]. Each layer in the device can modeled by an equivalent resistance representing the flow of charge through it and each interface can modeled by a capacitance representing the potential for charge storage. Such models are able to describe experimental observations successfully [59]. The usual method of preventing such drift involves controlling the resistivity of the buffer, matching it to the resistivity of waveguide layers, and modifying the LiNbO$_3$ surface before depositing the buffer layer.

The electro-optic effect in LiNbO$_3$ depends on the polarization of light; high-speed modulators are polarization dependent. However, polarization-independent LiNbO$_3$ have been demonstrated. They usually take advantage of the anisotropic electro-optic tensor of LiNbO$_3$. One approach relies on TE/TM mode coupling via off diagonal elements of the electro-optic tensor [60]. In another approach, a Δβ coupler is used [61]. There have been other novel approaches using crossing waveguides and specially designed directional couplers [62,63]. Digital optical switches can also be used as polarization-independent modulators [64]. Polarization-independent modulators usually require special crystal cuts and electrode geometries. As a result, the electro-optic coefficient used is not the largest, and the electrode is not suitable for high-speed operation. This makes the efficiency and speed of such devices low.

14.7.3 III–V compound semiconductor electro-optic modulators

14.7.3.1 ELECTRO-OPTIC EFFECT IN III–V COMPOUND SEMICONDUCTORS

III–V compound semiconductors such as GaAs, InP, and their alloys have excellent optical properties due to their direct and tunable bandgap and are materials of choice in many opto-electronic components such as lasers and detectors. They also lack inversion symmetry and possess an electro-optic coefficient. It is also possible to utilize the other electro-optic effects described earlier if the wavelength of operation is close to the material bandgap. As a result, they are also used in modulator applications. The most commonly used optical structure is a Mach–Zehnder interferometer. There are many different ways of making optical waveguides in III–V compound semiconductors. It is possible to adjust the index of refraction of these materials by controlling the composition of their alloys. For example, increasing Al composition x in $Al_xGa_{1-x}As$ compound semiconductor decreases its index of refraction. Furthermore, $Al_xGa_{1-x}As$ is lattice matched to GaAs for all x values. By growing such layers epitaxially using techniques such as molecular beam epitaxy (MBE) or metal organic chemical vapor deposition (MOCVD), it is possible to sandwich a higher index material between two lower index materials. This forms a slab waveguide and provides waveguiding in the vertical direction. The most common approach to provide lateral waveguiding is to etch a rib and form what is known as a rib waveguide. The effective index under the rib is higher than the effective index outside the rib in the etched regions. This provides a lateral index step and a two-dimensional waveguide is formed. There are many other ways to provide waveguiding that include proton implantation, buried heterostructures, pn junctions, and disordering. However, rib waveguides are almost the universal choice since vertical and lateral index profiles and dimensions can be independently and precisely controlled. Typical rib widths are in the 2–4 μm range and rib heights are typically less than 1 μm.

III–V compound semiconductors have a or zinc blende crystal structure. When an external modulating field is applied, the index ellipsoid is perturbed and using Equations 14.34 and 14.35 and Figure 14.15, it can be expressed as

$$\begin{pmatrix} x & y & z \end{pmatrix} \begin{bmatrix} \dfrac{1}{n^2} & r_{41}E_z & r_{41}E_y \\ r_{41}E_z & \dfrac{1}{n^2} & r_{41}E_x \\ r_{41}E_y & r_{41}E_x & \dfrac{1}{n^2} \end{bmatrix} \begin{pmatrix} x \\ y \\ z \end{pmatrix} = 1.$$

(14.52)

The refractive index value for compound semiconductors is around 3.5 at 1.55 μm. This value can be adjusted over a wide range by changing the composition of the material. The electro-optic coefficient is given as $r_{41} = 1.4 \times 10^{-12}$ m/V [65]. In this case, depending on the direction of the modulating electric field, different possibilities exist. For example, if the modulating field is z directed and $E_x = E_y = 0$, Equation 14.52 reduces to

$$\begin{pmatrix} x & y & z \end{pmatrix} \begin{bmatrix} \dfrac{1}{n^2} & r_{41}E_z & 0 \\ r_{41}E_z & \dfrac{1}{n^2} & 0 \\ 0 & 0 & \dfrac{1}{n^2} \end{bmatrix} \begin{pmatrix} x \\ y \\ z \end{pmatrix} = 1.$$

(14.53)

In this case, the index ellipsoid is no longer diagonal. It can be diagonalized using a simple coordinate transformation in the xy plane as shown in Figure 14.26. The new $x'y'$ axes are simply rotated 45° with respect to the xy axes. In the new x', y', z coordinate system, the index ellipsoid can be written as

$$\begin{pmatrix} x' & y' & z \end{pmatrix} \begin{bmatrix} \dfrac{1}{n^2} - r_{41}E_z & 0 & 0 \\ 0 & \dfrac{1}{n^2} + r_{41}E_z & 0 \\ 0 & 0 & \dfrac{1}{n^2} \end{bmatrix} \begin{pmatrix} x' \\ y' \\ z \end{pmatrix} =$$

$$\begin{pmatrix} x' & y' & z \end{pmatrix} \begin{bmatrix} \dfrac{1}{(n+\Delta n_{x'})^2} & 0 & 0 \\ 0 & \dfrac{1}{(n+\Delta n_{y'})^2} & 0 \\ 0 & 0 & \dfrac{1}{n^2} \end{bmatrix} \begin{pmatrix} x' \\ y' \\ z \end{pmatrix} = 1.$$

(14.54)

In this case, there is no index change for an optical wave polarized in the z direction. The index of refraction for an optical wave polarized in the x' direction and propagating in the y' direction can be found using Equation 14.54 as

$$\frac{1}{(n+\Delta n_{x'})^2} = \frac{1}{n^2} - r_{41}E_z \text{ or } (n+\Delta n_{x'}) = \frac{n}{\sqrt{1 - r_{41}n^2E_z}}.$$

Since $r_{41}n^2E_z \ll 1$, $(n+\Delta n_{x'}) \approx n(1 + 1/2\, r_{41}n^2E_z)$; hence

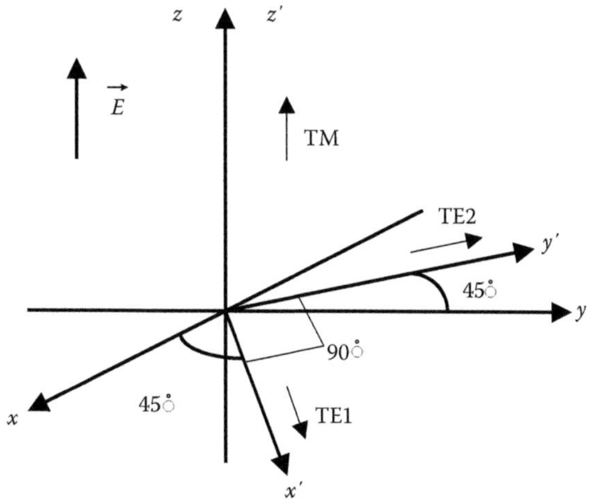

Figure 14.26 The axes of the index ellipsoid for a III–V compound semiconductor in the case of a z-directed external electric field. The dielectric tensor is diagonal in x, y, z and x′,y′,z′ coordinate systems in the absence and the presence of the electric field, respectively. TE1, TE2, and TM show the direction of the main optical field of the two orthogonal TE and TM modes, respectively.

$$\Delta n_{x'} = \frac{1}{2} r_{41} n^3 E_z. \tag{14.55}$$

Similarly, the index of refraction for an optical wave polarized in the y' direction and propagating in the x' direction can be found as

$$\Delta n_{y'} = -\frac{1}{2} r_{41} n^3 E_z. \tag{14.56}$$

In practice, the z-axis is along the 001 crystal plane. Then, the x′- and y′- axes are along the 110 and 1 $\overline{1}$0 crystal axis. Therefore, in such a material, a vertically applied electric field in the 001 direction increases the index of refraction by $\Delta n_{x'}$ in the 110 direction and decreases by $\Delta n_{y'}$ in the 1 $\overline{1}$0 direction. In other words, the index increases along one of the two mutually orthogonal directions parallel to the surface of the crystal. It decreases by the same amount in the other orthogonal direction. These directions correspond to the cleavage planes of the 001 oriented material. No index change is observed in the 001 direction, which implies that a vertically applied electric field to an 001 oriented crystal will only modulate the TE mode of the optical waveguide in which the main electric field component of the optical mode is either in the 110 or

1 $\overline{1}$0 directions; in other words, it is tangential to the surface. No modulation will result for the TM mode, which has its main electric field component in the 001 direction, i.e., normal to the surface of the crystal. The electro-optic coefficient in compound semiconductors is about 20 times less than that of LiNbO$_3$; however, the net index change for a given electric field is only about five times less due to higher index of refraction of the semiconductor. It is possible to enhance this index modulation using other physical effects described earlier.

14.7.3.2 LUMPED III–V COMPOUND SEMICONDUCTOR MODULATORS

Electric fields that exist in the depletion regions of either *pn* junctions or Schottky barriers are used to modify the index of compound semiconductors. Figure 14.27 shows a phase modulator using the electric field in the reverse biased *pn* junction of a GaAs/AlGaAs bulk heterostructure.

The high index GaAs layer of thickness d in between two lower index AlGaAs layers provides vertical waveguiding. For lateral confinement, etching the layers outside the rib of width w forms a rib waveguide. Ohmic contacts are formed at the top p$^+$-GaAs contact layer and at the backside of the n$^+$-substrate. A dielectric isolation layer, such as SiO$_2$, covers the sidewalls and the etched regions. By reverse biasing the *pn* junction, a large electric field is generated in the depletion region. Typically, the core of the waveguide—in this case, the GaAs layer—is undoped. Hence, a fairly constant electric field is generated across the core of the waveguide. This electric field generates an index change through the electro-optic effects described earlier. These are the linear electro-optic effect, electro-refractive effect, plasma effect, and band-filling

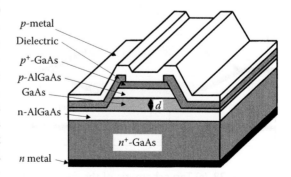

Figure 14.27 Schematic of a III–V compound semiconductor waveguide phase modulator.

effects. QWs instead of bulk material are also used to enhance the electro-refractive effect. We can express the resulting effective index change using Equations 14.29, 14.42, and 14.55 as

$$\Delta n_{eff} = \left(\frac{1}{2} r_{41} n^3 \frac{V}{g} \Gamma + G(\lambda) \left(\frac{V}{g} \Gamma \right)^2 \right). \quad (14.57)$$

In Equation 14.57, it is assumed that the linear electro-optic and electro-refractive effects are the dominant effects. The basic phase modulator shown in Figure 14.27 can easily be extended into an amplitude modulator using either a Mach–Zehnder or directional coupler design. The presence of doped layers may increase the optical insertion loss. One of the doped layers can be eliminated by replacing one side of the junction with a Schottky barrier. As a matter of fact, totally undoped material can be used if we replace both doped layers with Schottky barriers. One such modulator was fabricated using substrate removal [66]. The top and cross-sectional schematic of a substrate-removed modulator is shown in Figure 14.28.

In this case, an undoped GaAs/AlGaAs epilayer grown on a GaAs substrate is etched to form rib waveguides. An AlAs etch stop layer is also grown between the substrate and the epilayer. Then, a Schottky barrier is formed on top of the rib. Next, the entire wafer is glued onto a transfer substrate using a polymer bonding layer. In this case, the polymer is benzocyclobutene (BCB). After this bonding, the growth substrate is etched away. The substrate etch stops on the AlAs etch stop layer. Next, the etch stop layer is also etched away. This exposes the backside of the epilayer. Another Schottky electrode is formed on the backside. Hence, in between two electrodes on either side of the epilayer, two back-to-back Schottky diodes are formed. A voltage applied to the electrodes reverse biases one of the Schottky diodes and the electric field formed in its depletion region generates index changes. The thin undoped GaAs/AlGaAs layers self-deplete due to Fermi level pinning at the surfaces and behave effectively as a dielectric material with electro-optic properties. This makes the electrode gap the same as the thickness of the epilayer, which is only 1.95 μm. Such a small gap results in low V_π and can be kept very uniform since it is obtained by epitaxial growth rather than by lithography. Another big advantage of such a design is the ability to bias both arms of the interferometer independently. As a result, true push–pull operation is possible. This creates a zero-chirp modulator. Using appropriate bias on both arms, any amount of chirp can be obtained. The modulator reported in Reference [66] had a V_π of 3.7 for a 1 cm long electrode when driven push–pull as shown in Figure 14.29a. Phase modulators fabricated using the same technology also validated this result. Phase modulators combined with facet reflections form Fabry–Perot resonators. Transmission through such a resonator is shown in Figure 14.29b. Fringe visibility of this measurement also gives the on chip propagation

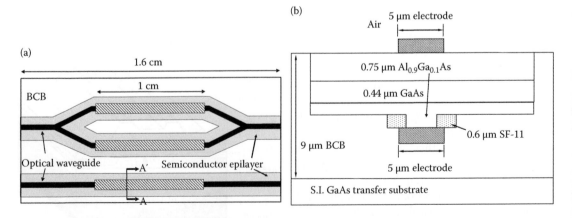

Figure 14.28 (a) Top schematic of the device reported in [66] and (b) the cross-sectional schematic of the optical waveguide along A–A′. (Shin, J.H. et al., *IEEE Photonics Technol. Lett.*, Figure 1. ©2006 IEEE.)

(a)

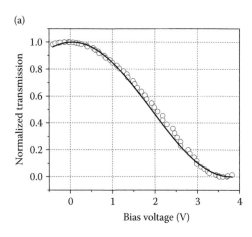

(b)

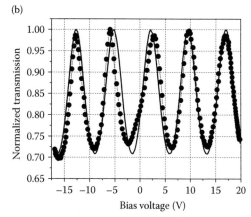

Figure 14.29 (a) Measured transfer function of the modulator shown in Figure 14.28. The curve fit to the well-known Mach–Zehnder transfer function is also shown. (b) Measurement and fit of the Fabry–Perot transmission as a function of the applied voltage for a phase modulator. For both cases, the electrode length and the total device length were 1 and 1.6 cm, respectively. (Shin, J.H. et al., *IEEE Photonics Technol. Lett.*, Figure 2. ©2006 IEEE.)

loss. In this case, the experimentally determined value is 2.9 dB/cm.

This approach can be further refined to reduce V_π significantly. One approach is to increase vertical confinement and decrease the electrode gap. Figure 14.30 shows a phase modulator in a substrate removed waveguide [67,68]. This waveguide is formed by removing a GaAs epilayer from its growth substrate and gluing it on top of a transfer substrate using BCB, very much like the example

described earlier. In this case, GaAs epilayer is only 0.3 μm thick. This figure also shows the intensity contours of the optical mode. As seen, the mode can be confined very tightly in the vertical direction due to very high index steps. However, electro-optical modulation requires external electrodes to apply the modulating electric field. Usually, this is achieved using metal electrodes, but in the phase modulator shown in Figure 14.30, this approach is not acceptable. Even though the vertical optical

(a)

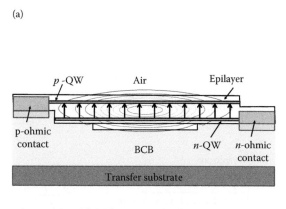

(b)

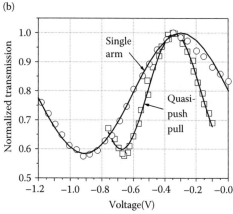

Figure 14.30 (a) Cross-sectional schematic of a phase modulator made out of a substrate removed waveguide with buried doped semiconductor electrodes. (b) Normalized transmission of a Mach–Zehnder intensity modulator with 7 mm long electrodes made out of the phase modulator shown in (a). (Shin, J.H. and Dagli, N., *IEEE J. Sel. Top. Quantum Electron.*, Figure 1. © 2013 IEEE; Reprinted with permission from Shin, J.H. et al., *Appl. Phys. Lett.*, 92, 201103, 2008, Figure 2. Copyright 2008 by the American Physical Society.)

confinement is very strong, the mode still extends a submicron distance outside the epilayer. If metal electrodes are placed on both sides of the epilayer, the metal overlaps strongly with the optical mode and optical propagation loss becomes excessively large. This makes the waveguide unsuitable for any practical application. The difficulty is solved by using buried electrodes made out of n- and p-doped QWs as shown in Figure 14.30. These electrodes form a p-i-n diode in which the p and n regions are contacted on the sides of the optical waveguide. Under reverse bias, a very large electric field overlapping very well with the optical mode can be applied to the submicron i region. Since buried electrodes are contacted on the sides of the optical waveguide, the ohmic contacts and the contact metals do not overlap with the optical mode. There is some FC absorption loss due to doped buried electrodes. But having FCs confined within the QWs reduces the overlap of the FCs with the optical mode; hence, it keeps the FC absorption loss minimal. In this case, the electrode gap is the separation between the n- and p-doped QWs. This separation, which is the gap of the modulator electrode, can be made very small and uniform since it is determined by the material growth. In the device reported in Reference [68], this gap was reduced to 0.15 μm with low propagation loss resulting in a practical device. The transfer function of a Mach–Zehnder modulator with arms made out of the phase modulators shown in Figure 14.30a is reproduced in Figure 14.30b. Reported V_π was 0.3 V for a 7-mm long electrode at 1.55 μm. This corresponds to a modulation efficiency of 0.21 V cm, which is much less than LiNbO$_3$ can offer. However, coupling in and out of such compact waveguides is challenging.

Low loss coupling requires mode transformers. Since substrate is removed, mode transformation from the semiconductor waveguide into a much bigger polymer waveguide is possible [69]. The extinction ratio of the transfer function shown in Figure 14.30b is low mainly due to unmodulated light trapped in BCB between the epilayer and the transfer substrate. Although this part is not a waveguide and cannot guide radiation, the index difference between BCB and the substrate/epilayer is so large that leakage is not complete. As a result, some portion of the light that couples to this part emerges at the output and increases the off state power since it is not modulated. Improving coupling using mode transformers should improve the extinction ratio. In Reference [67], a thin metal layer was deposited on the surface of the transfer substrate. This layer absorbed most of this stray light and the extinction ratio improved to 15 dB.

These results were obtained using bulk material. As discussed before, electro-optic efficiency can be improved further by using QWs. Therefore, efficiency was expected to improve on using an MQW of appropriate composition in the core of the waveguide. One such result was reported in Reference [70].

Figure 14.31 shows the top schematic and cross-sectional schematic of one of the arms of an intensity modulator made this way. This is very similar to the bulk GaAs/AlGaAs device described earlier. In this case, an InP-based material system is used and a p-i-n diode exists in the waveguide. The core of the waveguide contains n- and p-doped layers that act as buried electrodes and an InGaAlAs/InAlAs MQW. Ohmic contacts are formed at these layers on the sides away from the optical mode. The epilayer exists only in areas where optical waveguiding is needed; it is removed elsewhere. There is also an isolation implant on the Y-branches. This isolates the arms electrically; the arms can be

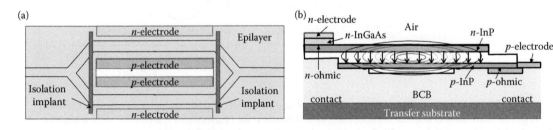

Figure 14.31 (a) Top schematic of the fabricated Mach–Zehnder intensity modulator. (b) Cross-sectional profile of one of the arms, which is a phase modulator. (Dogru, S. and Dagli, N., *IEEE/OSA J. Lightwave Technol.*, Figure 1. ©2014 IEEE.)

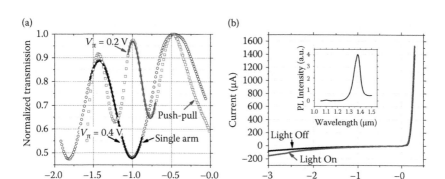

Figure 14.32 (a) Normalized optical transmission of a Mach–Zehnder modulator with a 3 mm long electrode under single arm and push–pull drive at 1.55 μm. (b) Current–voltage characteristics of one arm of the modulator with a 3 mm long electrode with and without 1.55 μm radiation. Insert shows PL spectra of the MQW. (Dogru, S. and Dagli, N., *IEEE/OSA J. Lightwave Technol.*, Figures 4 and 5. ©2014 IEEE.)

biased independently. By applying a reverse bias to the *p-i-n* diode, a very large electric field overlapping very well with the optical mode is generated. The electrode gap is the same as the thickness of the *i* region of the *p-i-n* diode, which is only 160 nm thick and mostly consists of the MQW of photo luminescence (PL) peak 1.37 μm. This corresponds to a detuning of 180 nm between the MQW PL peak and the operating wavelength. Hence, MWQ absorption and its contribution to propagation loss are minimized.

Figure 14.32a shows the transfer function of a modulator with a 3-mm long electrode for single arm and push–pull drives. V_π is 0.4 V under single arm drive and around 1.2 V reverse bias. It goes down to 0.2 V around 0.8 V reverse bias as expected. This corresponds to a modulation efficiency of 0.06 V cm. Figure 14.32b shows the current–voltage (*IV*) characteristics of one of the modulator arms of a modulator with a 3-mm long electrode. The insert shows the PL spectra of the MQW. The PL peak is at 1.37 μm. The *IV* with and without 1.55 μm radiation in the modulator is shown and the expected diode behavior is observed. Reverse leakage current up to 2 V reverse bias is less than 50 μA. This indicates that current-related index changes do not contribute to modulation and voltages up to 1.5 V can be comfortably applied. The change in the current with 1.55 μm radiation is negligible up to −1.5 V and increases for higher reverse bias. This shows the presence of photo-detected current; hence, absorption is very low at moderate reverse

biases but increases at higher reverse biases due to QCSE.

14.7.3.3 TRAVELING WAVE III–V COMPOUND SEMICONDUCTOR MODULATORS

Traveling wave designs are used for high-speed operation. The index of bulk compound semiconductors is isotropic. Furthermore, the refractive index variation between microwave and optical frequencies is rather small. The optical index of refraction is around 3.4 and the relative dielectric constant at microwave frequencies is about 13, which corresponds to an index of 3.6. Since it is easier to form electrodes on the surface of a crystal, the most commonly used electrode is either a coplanar waveguide (CPW) or a coplanar strip (CPS) structure. The optical signal is entirely confined in the semiconductor, which has a refractive index of about 3.4. On the other hand, the microwave and millimeter wave electric field fringes into the air and experiences an effective index between that of the air and the semiconductor. For example, for a coplanar line of zero conductor thickness, the effective dielectric constant is the arithmetic mean of the dielectric constants of the air and the semiconductor as indicated earlier in Equation 14.48. Therefore, the effective dielectric constant is around 7, which corresponds to a microwave index of about 2.65. Therefore, there is in about 38% of index mismatch between the optical and microwave signals. This requires about 23% velocity reduction. Therefore, in III–V compound semiconductors, velocity matching requires slowing down of the microwave signal. The most commonly used technique to slow down the microwave signal is to use a slow wave transmission line [10,71–73]. Such lines are obtained by periodically loading a uniform transmission line. The loading element is typically a capacitor. This can be achieved using either doped or undoped epitaxial layers.

Figure 14.33 shows the schematic of a traveling wave Mach–Zehnder modulator using doped layers [71]. The optical structure is a Mach–Zehnder interferometer utilizing multimode interference sections at the input and output for power splitting and combining. GaAs/AlGaAs epitaxial layer is grown on a semi insulating (SI) GaAs substrate. Underneath the GaAs core, there is a buried n^+-layer, which acts as a ground plane. The main electrode is a coplanar strip line. This electrode is periodically loaded by narrow and small capacitive elements. In Reference [71], the capacitive elements used are the reverse biased capacitance of a Schottky-i-n junction as shown in Figure 14.33. The advantage of this approach is that the large vertical electric field existing in the reverse biased Schottky-i-n junction can be utilized. This field overlaps very well with the optical mode enabling low drive voltages. Furthermore, these capacitive elements do not carry any of the axial currents in the transmission line. Hence, increase in the microwave loss of the electrode due to loading is marginal.

Such devices with a total of 1-cm electrode length and loading segment lengths of 0.5, 0.6, and 0.7 mm were fabricated and characterized [71]. DC V_π values at 1.15 µm for 0.5, 0.6, and 0.7 mm segment-length electrodes were 5.7, 4.25, and 4.24 V, respectively. Corresponding bandwidths were >26.5, 25.0, and 22.5 GHz, respectively. Recently, such a device demonstrated an electrical 3 dB bandwidth of 50 GHz and V_π of 13 V at 1530 nm [72].

Another approach to design a GaAs traveling wave modulator is to use unintentionally doped epitaxial layers [10,74,75]. Such GaAs/AlGaAs layers self-deplete due to Fermi level pinning at the surface and the depletion originating at the SI substrate interface and behave very similar to low loss dielectric materials. As a result, optical and

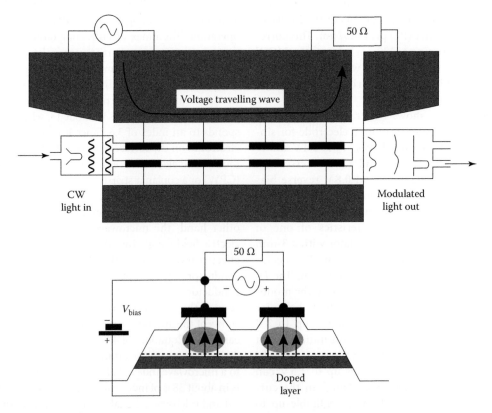

Figure 14.33 Schematic of traveling wave Mach-Zehnder modulator using capacitively loaded coplanar strip line. (Walker, R.G., Proceedings of *IEEE/LEOS'95 8th Annual Meeting*, Figure 1. ©1995 IEEE.)

microwave losses become very low. The required velocity slowing can be achieved using a properly designed electrode. A schematic of such a device is shown in Figure 14.34. The waveguide used here is the same as the waveguide shown in Figure 14.28b except that the metal electrode widths are reduced to 3 μm. 2 μm wide ribs are etched 0.65 μm into the top clad of a 0.75 μm $Al_{0.9}Ga_{0.1}As$/0.44 μm GaAs/0.75 μm $Al_{0.9}Ga_{0.1}As$ epilayer. 2 μm waveguide width was chosen for ease of fabrication. The etch depth is the maximum allowable for single mode operation. The epilayer is removed from the substrate it was grown on and glued onto a transfer substrate using the polymer BCB. It is undoped and etched away in areas where optical guiding is not needed. The remaining epilayer self-depletes and behaves as a dielectric electro-optic material with slight optical loss. Removing the growth substrate allows the placement of metal electrodes on both sides of the epilayer. The overlap of the optical mode and the electrodes is negligible; the electrodes do not increase the optical loss. The worst case of chip propagation loss was 2.9 dB/cm.

These devices had simple electrodes suitable for low-speed designs. Realizing the high-speed version of this modulator requires a traveling wave design due to the length of the electrode. In such designs, the modulator electrode is designed as a microwave transmission line. Such an approach is used in this work and schematic of the resulting capacitively loaded traveling wave electrode modulator is shown in Figure 14.34a. This electrode consists of a regular ground–signal–ground (GSG) coplanar transmission waveguide (CPW) loaded periodically with small capacitances. The loading capacitances are formed between the electrodes on the top and the bottom of the optical waveguide. The top and the bottom electrodes are connected to ground and signal lines of the CPW using short stems. These electrodes are called T-rails due to their shape. The connection is done such that modulating electric field direction reverses between the arms of the modulator resulting in a true push–pull operation. If the period of loading d is such that the Bragg frequency of the loaded line is much larger than the frequency of interest, the loading is mainly capacitive and the inductance per unit length remains unchanged. The microwave phase velocity and characteristic impedance of the capacitively loaded line are given by [76]

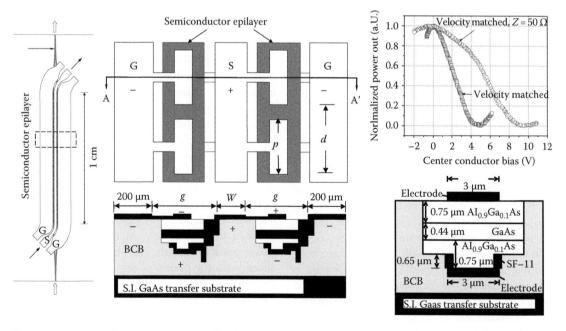

Figure 14.34 (a) Schematic top view of a GaAs/AlGaAs modulator using unintentionally doped layers along with the cross-sectional schematic in AA′. (b) Details of the optical waveguide along with optical transfer functions of both types of modulators. (Shin, J.H. et al., *IEEE Photonics Technol. Lett.*, Figures 1 and 2. ©2007 IEEE.)

$$v_{ph} = \frac{1}{\sqrt{L_0(C_0 + 2\Delta C)}} \text{ and } Z = \sqrt{\frac{L_0}{C_0 + 2\Delta C}}, \quad (14.58)$$

where L_0 and C_0 are the inductance and capacitance of the unloaded CPW and ΔC is the additional capacitance per arm due to loading. ΔC is a fraction of the capacitance between the electrodes sandwiching the optical waveguide C_T such that $\Delta C = FC_T = (p/d)C_T$. $F = p/d$ is the fill factor of the electrodes, i.e., the fraction of the interferometer arm that is electro-optically active. As seen in Figure 14.34, the appropriate fill factor is achieved by segmenting the electrode on the waveguides. This segmentation also helps to keep the microwave loss low since no axial currents can flow along the segmented T-rails. Axial currents only flow along the conductors of the CPW where the electric field and current density is low. The operating voltage of a velocity matched modulator can be expressed as

$$V_\pi = \frac{\lambda_0}{2l} \frac{t}{n_e^3 r_{41}} \frac{1}{\Gamma \cdot F} = \frac{V_{\pi 0}}{F}, \quad (14.59)$$

where l is the electrode length, Γ is the overlap of the vertical electric field with the optical mode, n_e is the effective index, r_{41} is the electro-optic coefficient of the material, and t is the thickness of the epilayer. As mentioned earlier, $V_{\pi 0}$ values as low as 3.7 V cm have been obtained earlier using this design [66].

For the waveguide shown in Figure 14.28, Γ improves very little for rail widths larger than 3 μm. So, electrode width is chosen as 3 μm. The group velocity of the optical waveguide used is 8.43 cm/ns at 1.55 μm. Therefore, using Equation 14.58, a velocity matched 50 Ω electrode requires $L_0 = 5.9$ nH/cm and $C_0 + 2\Delta C = 2.4$ pF/cm. Hence, the unloaded line dimensions should be chosen such that $L_0 = 5.9$ nH/cm with C_0 and the propagation loss as low as possible. Another constraint is the arm separation of the interferometer. If this is large, input and output Y-branches tend to be long and the modulator becomes long. This typically limits the width of the CPW signal line to less than 50 μm. The ground conductor widths of the CPW were chosen as 200 μm since line properties do not change for larger widths. The loss coefficient of the

unloaded CPW saturates if the gap and W product is larger than a few thousands of μm². Using these constraints, W and the gap were chosen as 40 and 80 μm, respectively, resulting in $L_0 = 5.9$ nH/cm and $C_0 = 0.73$ pF/cm. The required segmentation F, and the resultant V_π were 0.47 and 8.4 V cm, respectively. Another possible design is to have only a velocity matched electrode. If the electrode is terminated by its characteristic impedance, there will be no standing wave on the electrode and bandwidth will be maximized. Based on Equation 14.58, it can be seen that velocity matching is possible with larger ΔC values if L_0 is reduced. This comes at the expense of reduced electrode impedance. Again, while keeping the ground signal width at 200 μm, a range of signal line widths and gaps were explored and W and gap were chosen as 50 and 20 μm, respectively, resulting in $L_0 = 3.62$ nH/cm and $C_0 = 1.05$ pF/cm. For the velocity matched configuration, the segmentation F, and drive voltage was 0.85 and 4.6 V cm, respectively. The characteristic impedance was 30 Ω.

The optical transfer functions of the two different Mach–Zehnder modulators are shown in Figure 14.34. The measured drive voltage for the velocity matched and the velocity matched 50 Ω designs were 5 and 10 V cm, respectively. This is larger than the design values of 4.6 and 8.4 V cm.

The S-parameters of the electrodes along with extracted electrode parameters are shown in Figure 14.35 up to 40 GHz. In both cases, S-parameters show periodic oscillations due to interference arising from the reflections at both ends of the electrode. In the microwave measurements, TLM calibration method is used and an unloaded CPW fabricated on a chip is used as a line. This makes the reference impedance level for velocity matched 50 Ω and velocity matched only designs about 97 and 60 Ω, respectively. That is why at frequencies when s_{11} becomes very small due to destructive interference, the line appears to be impedance matched to these values. At the same time, loss also appears to be high. The phase velocity mainly depends on the period of the oscillations, is very smooth and does not show any dispersion. For both designs, the phase velocity is 9.5 cm/ns. 50 Ω and velocity matched designs had about 57 and 46 Ω impedance. The higher than expected velocity and impedance values are due to the misalignment of the top and bottom electrodes as explained before. Reduced field strength reduces the loading

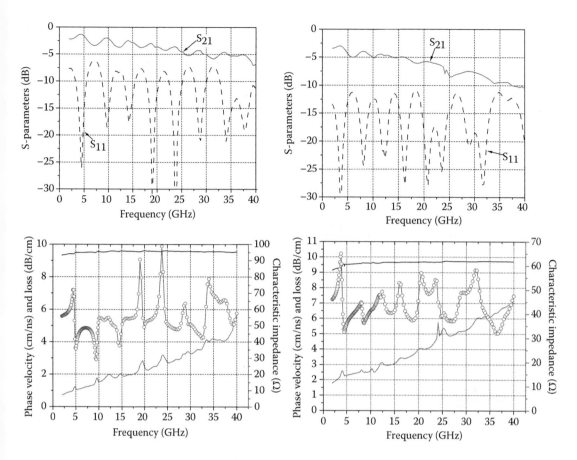

Figure 14.35 S-parameter measurements, microwave phase velocity, characteristic impedance (circle symbols), and microwave loss as a function of microwave frequency for the (a) velocity matched and (b) 50 Ω impedance design. (Shin, J.H. et al., *IEEE Photonics Technol. Lett.*, Figure 3. ©2007 IEEE.)

capacitance, increasing the velocity and the impedance. The microwave loss of the velocity matched 50 Ω and velocity matched only designs at 40 GHz were 6 and 7 dB/cm, respectively. Loss difference is due to different unloaded CPW dimensions and impedance levels of the two different designs. The loss coefficient of the electrode can be approximated as

$$\alpha \approx \frac{R}{2Z} = \frac{R}{2\sqrt{\dfrac{L_0}{C_0 + 2\Delta C}}} = \frac{R}{2Z_0}\sqrt{1 + \frac{2\Delta C}{C_0}} = \alpha_0\sqrt{1 + \frac{2\Delta C}{C_0}}.$$

In the loaded line design, the electrode resistance per unit length changes very little between loaded and unloaded lines. Velocity matched only designs had higher α_0 due to unloaded CPW dimensions and higher $2\Delta C / C_0$ ratio due to increased

loading. As a result, it has higher attenuation. Such low electrode loss indicates that both designs have electrical-to-electrical bandwidths of about 40 GHz.

It is possible to improve on these results significantly using doping and InP material system. This material system allows for the incorporation of MQWs of appropriate composition to improve electro-optic efficiency for 1.5 μm operation [77].

Figure 14.36 shows the modulator described here. It is a traveling wave electro-optic intensity modulator. The optical structure is a Mach–Zehnder interferometer, which consists of two Y-branches and arms made out of single-mode optical waveguides. Waveguides are contained in the epilayer, which exists only as two narrow stripes located within the gaps of the electrode. Figure 14.36 also shows the details of the optical waveguide used in the modulator as well as the

(a)

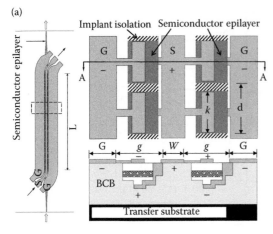

(b)

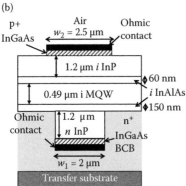

(c)

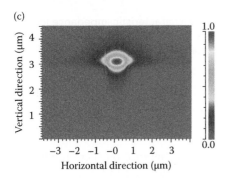

Figure 14.36 (a) The top and cross-sectional schematics of a traveling wave electro-optic intensity modulator. (b) Cross-sectional profile of the optical waveguide used in the modulator. (c) Optical mode pattern of the fundamental mode of the waveguide. (After Dogru, S. and Dagli, N., *Opt. Lett.*, 39, 20, 6074–6077, 2014. With permission of OSA.)

fundamental mode pattern. This is also one of the phase modulators used in the Mach–Zehnder interferometer arms. Its core is a 0.49 μm thick undoped MQW of 33×10 nm $In_{0.53}Al_{0.08}Ga_{0.39}As$

QWs and 32×5 nm $In_{0.52}Al_{0.48}As$ barriers. Room temperature photoluminescence peak of this MQW stack is at 1.37 μm. This is 180 nm away from the operating wavelength of 1.55 μm; hence, absorption due to MQW is minimal and not a significant contributor to propagation loss. The MQW core is clad above and below with 1.2 μm thick InP layers. One of the InP layers is n-doped to $5 \times 10^{17}\,cm^{-3}$. The other is undoped. Thin InAlAs layers are the barriers to the first and last QWs. These are thicker than the barriers used in the MQW stack and act as etch stop layers during fabrication. The design of the waveguide is such that the optical mode is entirely contained within the InP cladding and MQW stack. The epilayer also has p^+ and n^+ $In_{0.53}Ga_{0.47}As$ layers on the undoped and n-doped InP claddings. Therefore, it contains a p-i-n diode formed by p^+ $In_{0.53}Ga_{0.47}As$, the undoped InP/MQW stack, and n InP/n^+ $In_{0.53}Ga_{0.47}As$. n^+ and p^+ $In_{0.53}Ga_{0.47}As$ layers are also used for low resistance ohmic contact formation. $In_{0.53}Ga_{0.47}As$ layers and the ohmic contacts do not overlap with the optical mode. Furthermore, there is no p doping within the waveguide. Hence, the propagation loss is mainly due to scattering from etched sidewalls and FC absorption due to n InP. Both of these loss components are low and experimentally measured optical propagation loss is about 2.5–4 dB/cm depending on the waveguide width. Waveguide width mainly affects the loss due to sidewall roughness scattering by changing the mode overlap with etched sidewalls. Width of the waveguides used in the modulator is 2 μm. Furthermore, the big size of the optical mode results in efficient coupling with lensed fibers. Measured coupling loss on waveguides without antireflection (AR) coated facets is about 3–5 dB per facet depending on the waveguide width. Coupling loss can be reduced by about 3 dB per facet by using AR coating and simple mode transformers.

The electrode is a modified coplanar transmission line in ground–signal–ground configuration. The epilayer containing the waveguide is periodically implanted with boron. This implant makes the epilayer highly resistive and breaks the waveguide into sections electrically isolated from one another. The period of this segmentation is d and the length of the ohmic contacts on the electrically active epilayer is k. Therefore, only the k/d fraction of the optical waveguide contributes to modulation. This ratio is defined as the fill factor.

In this case, the index changes are due to FC depletion, linear and quadratic electro-optic (LEO) effects. V_π of the interferometer of length L can be expressed as

$$V_\pi = \frac{\dfrac{\lambda}{2L}t - 2K_N\Delta N^x\Gamma_N t}{n_e^3\left(r_{41}\Gamma_{LEO} + 2RE_B\Gamma_{QEO}\right)}.$$

Using $\lambda = 1.55$ μm, $L = 1$ cm, $t = 2.0$ μm, $K_N\Delta N^x = 10^{-3}$ for $N = 10^{17}$ cm^{-3} [78], $\Gamma_N = 1.5\times10^{-2}$, $n_e = 3.3$, $r_{41} = 1.4\times10^{-12}$ m/V, $R = 4.1\times10^{-19}$ m^2/V^2 [79], $E_B = 1\times10^7$ V/m, $\Gamma_{LEO} = 0.6$ and $\Gamma_{QEO} = 0.54$, we obtain $V_\pi = 0.5$ V.

The measured transfer function of a 1-cm long electrode device is shown in Figure 14.37 both under single arm and push–pull drive conditions. Fits to data using the well-known Mach–Zehnder transfer function are also shown. There is very little modulation up to about 4 V reverse bias. This is due to depletion of charge in the unintentionally doped i MQW and InP layers. Once this charge depletes, a high electric field is set up on the i layers and efficient modulation starts. Based on curve fitting, a V_π value of 1.6 V is obtained around 7 V bias and under single arm drive. This value reduces to 0.77 V around the same bias under the push–pull drive. Modulation becomes more efficient under increased reverse bias due to increased bias field. Increase bias field increases the effective LEO coefficient due to MQW and modulation becomes more efficient. However, peak transmission also drops. This is due to the quantum confined Stark effect (QCSE), which moves the tail of the MQW absorption to longer wavelengths.

The effect of increased absorption is also evident in the IV characteristics of the device shown in Figure 14.38. Increase in reverse current under higher reverse bias when there is 1.55 μm radiation in the device is due to photo-detected current caused by MQW absorption. This increases with increasing reverse bias due to QCSE. Around 7 V reverse bias, this current is less than 25 μA and the total reverse current is less than 50 μA. The measured small signal modulation response of this modulator under zero bias is shown in Figure 14.38. The 3 dB bandwidth corresponding to 3 dB reduction in electrical-to-electrical response is 40 GHz. The 3 dB reduction in electrical-to-optical response, which corresponds to 6 dB reduction in this figure is more than 67 GHz. This is the most commonly quoted bandwidth. Both measured V_π and bandwidth values are very close to the design targets. These values make the reported device the lowest drive voltage and widest bandwidth modulator ever fabricated.

It is also possible to realize this device on a growth substrate without substrate removal [80,81]. This is possible since the mode is entirely confined within the epilayer. Wide bandwidth operation requires growth on an SI substrate. One drawback of this approach is the inability to segment the doped contact layer at the bottom of the epilayer. Hence, microwave loss of the electrode increases since microwave currents are allowed to flow through this doped layer. The result is a reduction in the electrical bandwidth. Nevertheless, electrical bandwidths approaching 20 GHz is

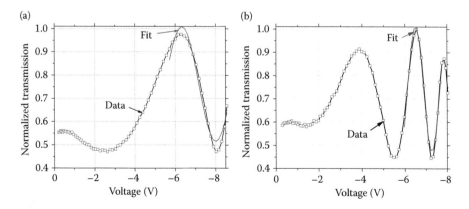

Figure 14.37 (a) Transfer function of a 1 cm long electrode device under single arm and (b) push–pull drive at 1.55 μm. (Dogru, S. and Dagli, N., *Opt. Lett.*, 39, 20, 6074–6077, 2014. With permission of Optical Society of America.)

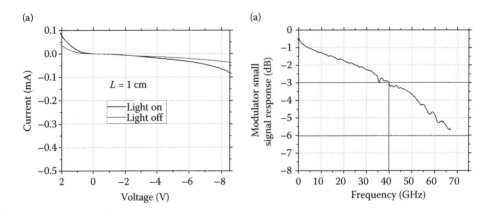

Figure 14.38 (a) IV characteristics of a 1 cm long electrode device with and without 1.55 μm radiation. (b) Small signal modulation response of this device up to 67 GHz. (Dogru, S. and Dagli, N., *Opt. Lett.* ©2014 IEEE.)

possible [82], making this device a candidate for a range of applications.

Polarization-independent operation of III–V electro-optic modulators can be achieved either by designing the material using appropriate amount of strain, or using novel device ideas. Strain may provide equal electro-optic coefficients for TE and TM modes, making the operation the same for two orthogonal polarizations. In bulk material, it is possible to get TE/TM coupling for certain applied field orientations. In one approach, polarization independence was obtained using near degenerate TE and TM modes and controlling the coupling between them [83]. In this case, electrodes were suitable for traveling wave operation [84]. Such a device has potential for broadband polarization-independent operation.

14.7.4 Polymer modulators

Organic polymers have many attractive features for integrated optical applications [85]. It is possible to form multilayer polymer stacks by spin deposition and curing on a large variety of substrates or even on fabricated circuits [86]. They can be patterned using several different techniques such as photobleaching and reactive ion etching. Organic polymers present good optical properties, such as low propagation loss and low index of refraction very close to that of the single mode fiber. They also have a low dielectric constant, which is important for high-speed devices. These properties resulted in passive low loss polymer optical waveguides that can couple to single-mode optical fibers very efficiently [87,88].

Polymers can also be made electro-optic and used in active devices. Certain molecules known as chromophores possess large ground state electric dipole moments and exhibit large optical nonlinearities. These microscopic nonlinearities can be converted into a macroscopic nonlinearity by mixing chromophores with polymers and aligning their dipole moments. The most common method of doing this is to take a chromophoric polymer film and applying a strong electric field across it while keeping the film heated at a high temperature near its glass transition temperature. At such high temperatures, randomly aligned individual chromophores are able to move in the polymer and align themselves with the externally applied electric field due to their dipole moments. After this alignment, if the film is cooled to room temperature while the applied electric field is still present, ordering of chromophores is achieved. This creates a macroscopic nonlinearity that can be used for electro-optic modulation. The process is known as high-temperature poling. Poling temperatures and fields depend on the particular film, but are typically about 100–200°C and 100–200 V/μm.

The most common techniques of poling a polymer film is either corona or electrode poling. In corona poling, corona discharge is used to create large poling fields over large areas. In electrode poling, one approach is to spin coat the lower cladding and core on a ground electrode. Then poling electrodes are formed on top of the cladding. After poling, these electrodes are removed and the upper cladding is spin coated. Alternatively, the whole stack can be spin coated, followed by poling electrode formation, poling and removal of poling electrodes. The conductivity of the upper and lower cladding regions of the polymer waveguide should be larger than the core of the waveguide for most of the poling voltage to drop across the electro-optic core and result in effective poling. Usually, the electro-optic coefficient increases linearly with poling field but so does the optical loss and birefringence.

After poling, electro-optic activity is observed. For modulating fields applied in the poling field direction, two different electro-optic coefficients are observed depending on the polarization of the optical mode. The coefficient for TM polarization, i.e., when the optical field is polarized in the modulating field direction, is r_{33}. The coefficient for TE polarization, i.e., when the optical field is polarized perpendicular to the modulating field direction is r_{13}. Typically, for polymers, r_{33} is about three times larger than r_{13}. At the present time, for most of the reported modulators, r_{33} values range between 1 and 20 pm/V, although values as high as 67 pm/V have been reported [89]. The index change for an applied electric field is given by Equation 14.42. Polymer indices of refraction are around 1.6, which makes a polymer with $r_{33} \cong 12$ pm/V equivalent to bulk GaAs as far as index change is concerned. To take advantage of the high r_{33}, the modulating field should be applied in the poling field direction and the optical mode should be polarized in the same direction. For these considerations, the most commonly used electrode is a microstrip and the optical polarization is TM.

Figure 14.39 shows a traveling wave polymer modulator [92]. The optical structure is the Mach–Zehnder interferometer. The optical waveguides were fabricated by spin coating three layers of polymers that act as claddings and the core of the optical waveguide onto a high resistivity Si wafer, which was coated with a patterned gold plated film. Exposing the polymer film to a high intensity light at the appropriate wavelength reduces the refractive index in the exposed areas, which is known as photobleaching. After poling, photobleaching was used to form channel optical waveguides. The thickness of the polymer stack was 6.5 μm. Tapered coplanar lines were used to couple in and out of the 12 mm long microstrip electrode. V_π of this modulator was 10 V. Low dielectric constant and low dielectric constant dispersion of polymers from microwave to optical frequencies offer an advantage in the design of traveling wave modulators. It is possible to get very good velocity matching using a microstrip electrode. For example, using a commonly quoted value of 1.6 for the optical index, 2.9 for the microwave relative dielectric constant, and a 10 μm thick polymer stack, a 50 Ω microstrip line would have about 25 μm strip width and less than 5% velocity mismatch. In the case of the modulator shown in Figure 14.39, the index mismatch was estimated to be 0.03. The bandwidth is wider than 40 GHz. Other polymer modulators with bandwidths wider than 40 GHz have also been reported [90,91]. Furthermore, polymer modulators operating at W band (75–110 GHz) have been reported demonstrating the intrinsic high frequency response of electro-optic polymers [90,91].

If corona poling or a single poling electrode is used, chromophores in both arms of the Mach–Zehnder are aligned in the same direction. This makes it difficult to make a push–pull modulator

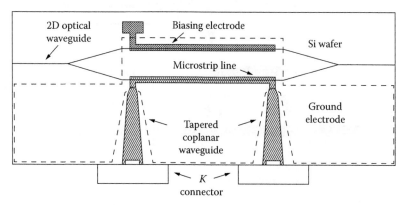

Figure 14.39 Schematic view of a polymeric traveling wave modulator. (Reprinted with permission from Teng, C.C., *Appl. Phys. Lett.*, 60, 1538–1540, 1992, Figure 1. Copyright 1992, American Institute of Physics.)

using a single microstrip electrode and usually only one arm of the interferometer is phase modulated [92,93]. It is possible to use two electrodes driven 180° out of phase on both arms of the interferometer to get push–pull operation and halve the drive voltage. This requires a broadband 3 dB microwave hybrid and makes only halve the power available to each arm without considering the loss of the hybrid. Another way of obtaining push–pull operation is to pole the arms of the Mach–Zehnder in opposite directions as shown in Figure 14.40 [94]. In this case, poling electrodes had to be closely spaced in order to form a 50 Ω microstrip line. The polymer stack is about 10 μm thick, which makes the required microstrip width about 30 μm, and implies that the separation between the arms of the interferometer should be less than that. Therefore, the poling electrodes should be less than 30 μm apart. Since voltages approaching 1000 V must be applied to create poling fields in the range of 100 V/μm across the 10 μm film, the air breakdown between closely spaced poling electrodes becomes an issue. In this case, the difficulty was addressed by using a fused silica piece on top of the poling electrodes to modify the field distribution in the gap and the sample was placed in an inert gas (SF_6) ambience to increase the breakdown field. Modulators with 2 cm long microstrip electrodes had V_π values of 40 and 20 V for one arm and push–pull operation, respectively, at 1.3 μm [94]. This corresponds to an r_{33} value of about 7 pm/V. The typical loss of a pigtailed modulator was about 12 dB. 6 dB of the loss was attributed to the on-chip

optical loss of the 3 cm long interferometer and the rest was due to mode mismatch between the fiber and the optical waveguide. The additional processing associated with poling electrode formation can be eliminated if device electrodes can be used as poling electrodes. By applying 180° out of phase pulsed voltages to the electrodes, it is possible to pole both arms of the interferometer sequentially [95]. In this way, the maximum voltage between the electrodes is reduced from $2V_p$ in the case of DC poling to V_p in the case of pulse poling, where V_p is the poling voltage applied across one arm of the interferometer. The field strength between the device electrodes during poling can be further reduced by increasing the separation between electrodes. One approach uses parallel microstrip lines of 100 Ω characteristic impedance each [96]. In this way, it was possible to get a separation of 200 μm between the arms of the interferometer.

Chromophores used in electro-optic polymers have certain optical transitions in the wavelength range used in optical applications. For example, it is reported that disperse red 19 (DR19) chromophores have a peak absorption wavelength around 470 nm corresponding to $\pi-\pi^*$ transition [97]. Therefore, any optical energy present around this wavelength excites the chromophore, which loses its orientation when it relaxes. This results in photoinduced relaxation, which reduces the electro-optic coefficient, absorption, and the optical index. This effect is shown to be polarization dependent due to the large optical dipole moment associated with the chromophore. At wavelengths away from

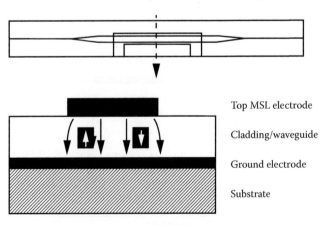

Figure 14.40 Top and cross-sectional view of the device showing the optical push–pull operation in a polymer electro-optic Mach–Zehnder modulator and a single microstrip line driving circuit. The white and gray arrows represent poling and modulation field directions, respectively. (Wang, W. et al., *IEEE Photonics Technol. Lett.*, Figure 1. ©1999 IEEE.)

the absorption peak, this effect decreases significantly. However, optical nonlinearity of the material results in second harmonic generation, which in turn generates optical power within the absorption band even for near infrared excitations [98]. Typically, second harmonic generation efficiency in commonly used optical waveguides is very low due to poor phase matching. Nevertheless, this places an upper limit on the maximum power handling capacity of the optical waveguide. For example, at 1320 nm for 10 mW in a 10 μm² cross-section waveguide, the device lifetime is estimated to be larger than 10⁴ h for a 30% decay in the electro-optic efficiency [97]. On the other hand, lifetime at 1064 nm was estimated to be ~1 h due to 532 nm second harmonic [97]. Recently, double-end cross-linked polymers were found to be stable when exposed to 250 mW at 1.32 μm for 1 week [99]. They were also found thermally stable when baked at 100°C in air for more than 2000 h. r_{33} of these materials are quoted to be 6 pm/V.

Long-term bias stability is another very important concern. Polymer layers have finite conductivity that effects electrical response under DC and AC applied biases. This effect was studied using an equivalent network approach [59,100,101]. Every layer of the optical waveguide can be modeled as a parallel RC circuit. The resistance and capacitance of each layer depend on its conductivity, σ, and dielectric constant, ε, respectively. At low frequencies and low DC biases, the fraction of the applied voltage across the electro-optically active core layer depends on the relative conductivity of the layers. For low bias operation, conductivity of the electro-optic layer is desired to be much smaller than the other layers. This condition is also required for effective poling. However, it is shown that low conductivity would generate a bias drift that depends on the amplitude of the AC voltage swing. This could be a problem in practical operations requiring continuous bias point monitoring and adjustment. The effect has a ω⁻¹ dependence and can be made zero by matching the relaxation constants (σ/ε) for all layers [100].

Currently, the field of electro-optic polymers is a very intense research area. One way forward is to improve material stability and the electro-optic properties of polymers. Another possibility seems to be elimination of high-temperature electric field poling. This would enable room temperature processing, but requires electro-optically active, or ordered, material to start with. Such a material can be achieved by depositing the polymer film monolayer by monolayer using Langmuir–Blodgett films and relying on self-ordering of the chromophores in each monolayer.

14.7.5 Silicon modulators

Recently, a wide range of guided wave components have been demonstrated in crystalline silicon [102]. A significant advantage of these components is the possibility of integration with CMOS-based electronics and very large-scale photonic/electronic integration. This is the silicon photonics field that offers significant cost reduction and performance enhancement. However, it is not sufficient to have passive guided wave components, such as waveguides, couplers, arrayed waveguide gratings, and so on. Active components such as modulators and switches that are controlled by external voltages are also needed. Bulk crystalline silicon has inversion symmetry and lacks electro-optic coefficient. However, it is possible to use other electro-optic effects to generate loss and index modulation in silicon. In particular, the Franz–Keldysh effect and carrier injection or depletion-related effects can be used [103]. The Frantz–Keldysh effect-related index change is about 10⁻⁵ at 1.07 μm but is too small at regular communication wavelengths. On the other hand, carrier-related index changes could be strong but comes with an associated absorption change. For crystalline silicon, the changes in the real and imaginary parts of the index $n = n' - jn''$ at 1.55 μm are given as

$$\Delta n'(1.55 \, \mu m) = -2.37 \times 10^{-23} \Delta N^{1.08} - 3.93 \times 10^{-18} \Delta P^{0.815},$$

$$\Delta n''(1.55 \, \mu m) = 4.92 \times 10^{-26} \Delta N^{1.2} + 1.96 \times 10^{-24} \Delta P^{1.1}.$$

Figure 14.41 shows the real and imaginary parts of the silicon index due to changes in electron and hole concentrations at 1.55 μm. For most device applications, a phase change of π is necessary. Carrier injection or depletion to get a π phase shift can be calculated using these formulas as $L_\pi = \lambda / 2\Delta n'$. The loss change over this length in dB units is $\Delta \alpha = 4.34 (2\pi L_\pi / \lambda) \Delta n''$.

Hole concentration changes are more effective for index changes compared to electron

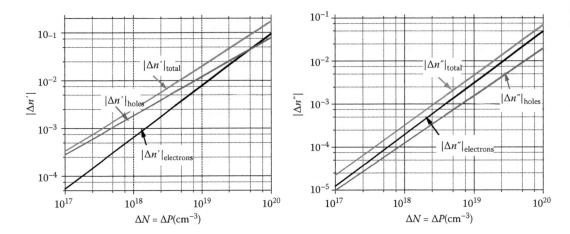

Figure 14.41 Real and imaginary parts of the silicon index due to changes in the electron and hole concentrations at 1.55 µm.

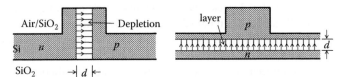

Figure 14.42 Horizontal and vertical pn junctions for carrier depletion type of modulators.

concentration changes. There is also an accompanying significant loss change. As electron and hole concentration increases, index decreases and loss increases. Carrier concentration changes can be realized using a *pn* junction. Forward biasing the *pn* junction increases carrier injection and hence decrease in index is possible. Under large forward biases, significant carrier injection and index change is possible. But in this case, carrier reduction is due to electron–hole recombination and is limited by minority carrier lifetime to nanosecond levels. It is also possible to start with heavily doped material and decrease the carriers through depletion under reverse bias. In this case, intrinsic response times can be of the order of sub picosecond, but ultimate speed of response is limited by the RC time constant of the *pn* junction. Modulators of both types have been realized. For depletion-type modulators, *pn* junction can be fabricated vertically or horizontally as shown in Figure 14.42.

This figure also shows the most commonly used silicon waveguide. In this case, silicon is on 1–2 µm thick SiO_2. Typical silicon thickness is in the 220–240 nm range to keep the waveguide single mode in the vertical direction. The rib waveguide etch is typically kept shallow (70–90 nm) to

reduce the scattering loss due to etched sidewalls and the waveguide width is around 0.5 µm. Hence, the width-to-height ratio is about 2. For either junction type in this waveguide to get the largest effective index change, overlap of the optical mode with the depletion region should be maximized. This overlap can be increased by increasing the mode confinement. For the typical SOI waveguide, the index difference is much higher in the vertical direction. Therefore, mode confinement is stronger in this direction, which makes the overlap of the optical mode with the depletion layer of a vertical *pn* junction higher. The ratio of the overlap factors for the vertical and horizontal *pn* junctions is about the width to height ratio or a factor of 2. Therefore, it is possible to get a factor of two higher effective index change in the vertical junction. Lateral confinement can also be increased by etching the rib deeper but this increases scattering loss as well as the resistance of the *p* and *n* layers. Ultimately, careful optimization of doping profiles and levels is needed [104]. Both types of junctions were used in Mach–Zehnder type of intensity modulators. Horizontal *pn* junctions are more suitable for guided wave types of modulators where ohmic contacts to *p* and *n* regions can be made on both sides of the waveguide away from the optical mode. Vertical *pn* junctions are usually used in microdisk modulators that allow an ohmic contact to the top doped layer in the middle of the disk where there is negligible optical field. In both cases, modulator performance optimization requires the optimization of the junction geometry.

For the horizontal type *pn* junctions used in guide wave modulators, this is done in Reference [105] considering several different junction geometries. Table 14.1 shows the different phase shifter configurations considered.

In configurations B and C, the waveguide is *p*-doped and *p* concentration is significantly lower than *n* concentration. Hence, under reverse bias, the depletion region extends mainly into the waveguide. The junction lies inside the rib waveguide in type A. Phase shifters of type D consists of a higher doped 60 nm wide stripe inside the low *p*-doped waveguide to provide enhancement in efficiency. Type E refers to a similar configuration where a 100 nm *pn*-doped stripe is implemented in the center of the rib waveguide for higher modal overlap. In this case, the waveguide is broader and shallow-etched for minimization of absorption losses.

Figure 14.43 illustrates the important characteristic curves of the phase shifters. In Figure 14.43b, the absorption losses are shown including linear waveguide loss as a function of applied reverse bias. For the phase shifters type A, B, C, and D, the linear waveguide loss was set to the typical value of 2.0 dB/cm. On the other hand, the broader, shallow etched waveguides used for type E typically exhibit less linear loss of 0.5 dB/cm. Since the modulator structures are usually several mm long, it is very critical for the phase shifter to introduce as low absorption losses as possible. In Figure 14.43a, the $V_\pi L$ product calculated from the voltage required for π phase change is shown. For *pn*-type phase shifters, the trade-off between FC absorption and $V_\pi L$ product is obvious. Loss and $V_\pi L$ product show opposite trends as a function of applied voltage. Type A is commonly used and has low $V_\pi L$ product but high loss. For these types of modulators, a typical drive voltage length product is in the 1.5 –3 V cm range [104,106]. This may suggest a 1.5 V_π device with 1-cm long electrode but the optical loss due to undepleted doped layers is in the 10–15 dB/cm range. This makes a low V_π device unpractical. Reducing the electrode length reduces loss at the expense of V_π. Such devices have been demonstrated and can be practical in application when a low on/off ratio is sufficient. A 1–3 dB extinction ratio can be achieved by applying only a fraction of V_π, which can be supplied from a CMOS circuitry that can be integrated with such a modulator. Type E analyzed in Table 14.1 looks promising. This design has a heavily doped *pn* junction stripe in the middle of the rib for higher confinement. The background doping is kept low to minimize absorption losses. The waveguide is also shallow etched to minimize the linear loss. This design can provide a 1.5 V_π device having a 1 cm long electrode

Table 14.1 Different phase shifter configurations considered for phase shifter optimization in horizontal pn junctions

Ph. Sh. type	Scheme	dopants in p	dopants in n	dop. n-stripe	dop. n-stripe	Rib width (nm)	Rib height (nm)	Etch depth (nm)
A		5e17 cm^{-3}	1e18 cm^{-3}	–	–	500	250	200
B		3e17 cm^{-3}	1e18 cm^{-3}	–	–	400	220	100
C		1e17 cm^{-3}	5e17 cm^{-3}	–	–	400	220	100
D		7e16 cm^{-3}	5e17 cm^{-3}	2e17 cm^{-3}	–	400	220	100
E		3e16 cm^{-3}	3e16 cm^{-3}	2e18 cm^{-3}	2e18 cm^{-3}	700	220	70

Source: Petousi, D. et al., *J. Lightwave Technol.* ©2013 IEEE.

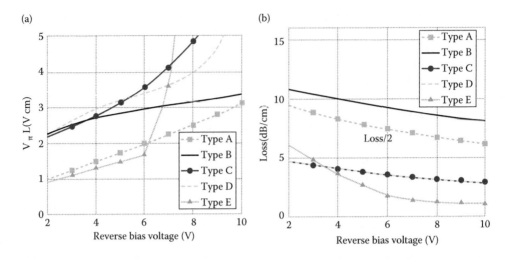

Figure 14.43 (a) $V_\pi L$ product and (b) propagation loss of the phase shifters given in Table 14.1 as a function of applied reverse bias. (Petousi, D. et al., *J. Lightwave Technol.* ©2013 IEEE.)

with about 5 dB loss. However, the fabrication of such a doping profile is challenging.

A modulator containing type B phase shifter was reported in Reference [107] using a self-aligned *pn* junction approach. Accurately positioning the junction using a self-aligned fabrication process has significant advantages. It reduces performance variations caused by process tolerances. This in turn increases device yield, which reduces the chip cost. Furthermore, older generation fabrication tools with lower alignment accuracy can be used providing further cost reduction. Basic fabrication steps outlining this approach are shown in Figure 14.44. The active waveguide region is first doped *p*-type by implanting boron ions through a photoresist mask. Next, a silicon dioxide layer is then deposited and patterned. This patterned layer is used first as a hard mask to etch the optical waveguides. After etching the waveguides, another photoresist pattern is formed. This combined with the silicon dioxide layer acts as a mask for the subsequent phosphorus implant. It is followed by a boron implantation through another photoresist mask and the waveguide shown in Figure 14.45 results.

The waveguide height and width are 220 and 400 nm, respectively. The slab height is 100 nm. The separations between the highly doped regions and the waveguide edge are 500 and 450 nm for n^+ and p^+, respectively. Holes are etched through a 1 μm thick silicon dioxide top cladding layer down to the highly doped regions to form ohmic contacts. Coplanar waveguide electrodes are used to drive the device

at high speed. An electrode thickness of 1.3 μm is selected to reduce electrode loss and allow for an increased electrode bandwidth. In this design, the *pn* junction will always be at the waveguide edge. Hence, there is no need for very high alignment accuracy. The position of the *pn* junction requires that the doping concentrations should be arranged such that the depletion extends mainly under the rib under reverse bias conditions. This can be achieved by making the *n* concentration larger than the *p* concentration. In the device reported in Reference [107], *p* and *n* concentrations were 3×10^{17} and 1.5×10^{18} cm^{-3}, respectively, making it equivalent to the type B device shown in Table 14.1.

The modulation efficiency for an Mach Zehnder modulator (MZM) fabricated this way with 3.5 mm phase shifters is measured to be approximately 2.7 V cm. Propagation loss is given as 4 dB/mm. $V_\pi L$ product matches well with the result of the analysis given in Figure 14.43a, but loss value is significantly higher than what is shown in Figure 14.43b. This again shows the difficulty of getting a low V_π with low loss. High-speed performance has been analyzed by applying a 40 Gbit/s electrical pseudo random bit sequence (PRBS) data stream amplified to 6.5 V peak-to-peak to the device. This resulted in open eyes with a 10 dB extinction ratio. Electrical bandwidth of such Mach–Zehnder modulators can be improved using a traveling wave electrode [106].

Compact and lower drive voltage modulators can be fabricated using ring resonators. It is possible to tune the bandstop resonance and quality factor of a

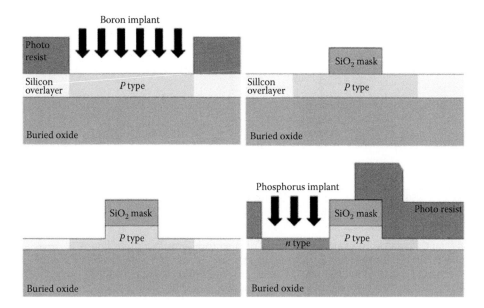

Figure 14.44 Basic steps of the self-aligned fabrication process to form the type B phase shifter given in Table 14.1. (Thomson, D. J. et al., *IEEE J. Sel. Top. Quantum Electron.* ©2013 IEEE.)

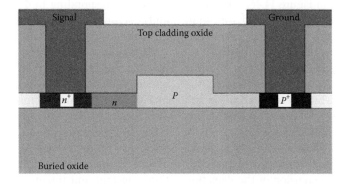

Figure 14.45 Resulting phase shifter following the process shown in Figure 14.44. This is the type B phase shifter given in Table 14.1. (Thomson, D. J. et al., IEEE J. Sel. Top. Quantum Electron. ©2013 IEEE.)

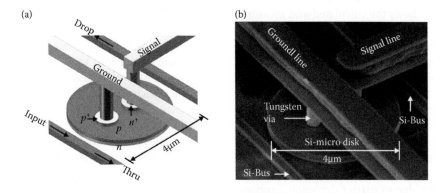

Figure 14.46 (a) Schematic of a silicon disk resonator modulator containing a vertical pn junction. (b) Scanning electron microscope of the fabricated device. (Watts, M.R. et al., *Opt. Express*, 19, 22, 21989–22003, 2011. With permission of Optical Society of America.)

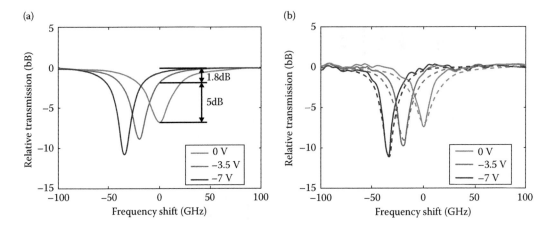

Figure 14.47 (a) Calculated transmission spectra of the modulator shown in Figure 14.46 as a function of applied voltage. (b) Measured transmission spectra of the same device as a function of applied voltage. Calculation result shown in (a) is reproduced as dashed lines. (Watts, M.R. et al., *Opt. Express*, 19, 22, 21989–22003, 2011. With permission of Optical Society of America.)

single disk or ring resonator resonances by changing the index and loss of the ring or disk. An example of such a device is shown in Figure 14.46. The microdisk contains vertical *pn* junction formed using multiple implants [108]. In this case, the optical mode is contained along the outer perimeter of the disk.

This allows making ohmic contacts within the disk as shown in Figure 14.46. Furthermore, the disk is deeply etched, which permits very low disk diameter without significant radiation loss. In this case, disk diameter is only 4 µm, which results in a very compact device. Under a reverse bias, depletion of the *pn* junction in the ring creates index and loss change due to extension of the depletion region. These changes change the resonant frequency and quality factor of the disk resonances. This effect is shown in Figure 14.47.

This figure shows the calculated and measured transmission spectra of the device shown in Figure 14.46 and described in Reference [108]. As the applied voltage increases, the depletion layer of the vertical *pn* junction widens and more carriers are swept out in the disk. The associated index change increases the effective index of the disk mode and shifts the resonance frequency to lower frequencies. Higher voltages mean more depletion, carriers sweep out, and there is more index change resulting in more resonant frequency shift. At the same time, due to carrier depletion, propagation loss decreases and quality factor of the resonances improves. This is observed as sharpening of resonances. In this case, a resonant frequency

shift of 21 GHz is observed under a 3.5 V reverse bias. Figure 14.47a shows that operating at the frequency corresponding to 0 frequency shift at 0 V and applying a 3.5 V reverse bias, it is possible to get an on/off or extinction ratio of 5 dB with a 1.8 dB additional insertion loss. Such performance could be sufficient for short reach applications such as optical interconnects. Measurement results shown in Figure 14.47b match closely with calculations. This modulator was used in a 10 Gb/s NRZ transmission experiment with a 1.8 V swing and low bit error rate was demonstrated [108]. The bandwidth of such resonators is limited extrinsically by the RC time constant and intrinsically by the resonance built-up times. A combination of these two effects usually limits the bandwidth to around 25 GHz. The biggest drawback of this approach is the limited optical bandwidth of the modulator and sensitivity of the resonance frequency to external effects such as temperature and frequency fluctuations as well as fabrication tolerances. These arise due to the resonant nature of these modulators. As a matter of fact, such microdisks or rings are efficient sensors. In order to utilize these devices in real-life applications, some tuning mechanism is needed. This is typically achieved using thermal tuning [109]. In a recent demonstration, compensation for thermal drift of 7.5°C with low additional energy consumption was reported [109]. This is one area that requires further work to make these devices viable in applications that use large numbers of resonators operating at large number of wavelengths.

14.7.6 Graphene modulators

Graphene, a recently discovered new material, is an example of new two-dimensional materials suitable for modulator applications. It is a 0.34 nm thick monoatomic sheet of carbon atoms arranged in a hexagonal lattice [110]. It can be deposited or transferred on a wide range of substrates making its integration possible on different material platforms. One of the very important attributes of graphene is its very high and tunable conductivity. This tuning can be achieved either through doping or moving the Fermi level, E_f, using external gate voltages. Tunable conductivity also makes the dielectric constant and index of refraction of graphene tunable. Therefore, it is possible to generate both phase shifts and absorption changes in graphene electrically, which is the basis of optical modulation. Figure 14.48 shows the energy band diagram of graphene.

Graphene does not have a bandgap and the dispersion diagram is linear over the typical energy ranges used in practice. In pristine graphene, the Fermi level E_f is at Dirac point. This point is usually taken as the reference point for E_f. It is possible to move E_f up or down by applying external electric fields. This changes the carrier concentration and conductivity. A theoretical analysis describes the graphene conductivity due to intraband and interband transitions as [111]

$$\sigma_{\text{intra}}(\omega, E_f) = \frac{-jq^2}{\pi\hbar^2(\omega - j2\tau)}\left[\int_0^\infty E\left(\frac{\partial f_d(E)}{\partial E} - \frac{\partial f_d(-E)}{\partial E}\right)dE\right],$$

$$\sigma_{\text{inter}}(\omega, E_f) = \frac{-jq^2(\omega - j2\tau)}{\pi\hbar^2}\left[\int_0^\infty \frac{f_d(-E) - f_d(E)}{(\omega - j2\tau)^2 - 4(E/\hbar)^2}dE\right]$$

where $f_d(E) = 1/\left(e^{(E-E_f)/kT} + 1\right)$ is the Fermi function and τ is a broadening term. In the following calculations, $\hbar/\tau = 5$ meV. Using these equations, conductivity of the graphene can be determined. This allows us to calculate the dielectric constant of graphene as $\varepsilon(\omega, E_f) = 1 - j\sigma(\omega, E_f)/\omega\varepsilon_0\delta$, where $\delta = 3.4\times10^{-10}$ m is the thickness of a monolayer of graphene. The results of a numerical calculation showing the variation of the conductivity and dielectric constant of graphene as a function of E_f is shown in Figure 14.49. This calculation is done at $\lambda = 1.55$ μm corresponding to a photon energy of 0.81 eV. Since $\varepsilon(\omega, E_f) = \varepsilon(\omega, -E_f)$ and $\sigma(\omega, E_f) = \sigma(\omega, -E_f)$, the exact same behavior is obtained for negative E_f values. In other words, it does not matter if E_f moves up or down in the band diagram. It is observed that around 0.4 eV where $2E_f \approx hf$, there are very strong conductivity and dielectric constant changes. Magnitude of the imaginary part of the dielectric constant that accounts for material loss decreases very rapidly. This is due to blocking of the interband absorption as shown in Figure 14.48. For $2E_f > hf$, there is low residual absorption due to intraband processes; hence, the imaginary part of the dielectric constant remains low and changes very little. But the real part that accounts for the index of the material keeps changing significantly. It should be noted that both the absorption and index changes are very high; indeed, much higher than the commonly used materials such as semiconductors, polymers, and ferroelectrics. This presents a unique opportunity for device applications that are described in the next section. Changing E_f also changes the charge, $n_s(E_f)$, and electric field, ε, on

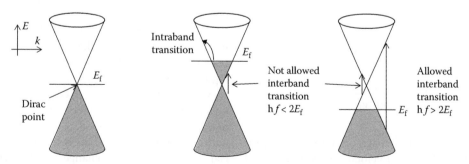

Figure 14.48 Energy band diagram of graphene. Shaded areas indicate filled states at 0 K. (a) Pristine graphene, (b) n-type graphene, and (c) p-type graphene. Intraband transitions are transitions within the same band. Interband transitions are between the bands. The length of the vertical arrows connecting the bands is the photon energy. If the energy of photon is less than, interband transitions are not possible.

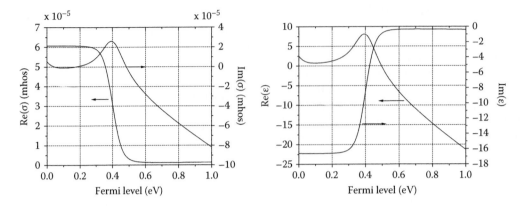

Figure 14.49 Variation of conductivity and dielectric constant of graphene as a function of E_f at $\lambda = 1.55\,\mu m$. Only positive E_f values are shown. The results are the same for negative E_f values.

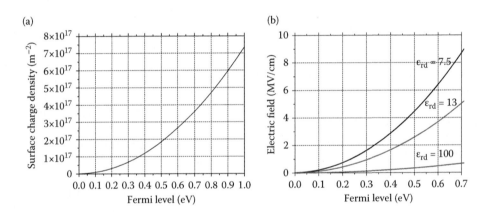

Figure 14.50 (a) Surface charge density and (b) electric field on graphene as a function of Fermi level for three different dielectrics.

the graphene. These can be calculated using the following formulas:

$$n_s(E_f) = \int_0^\infty D(E)f(E)\,dE = \frac{2}{\pi\hbar^2 v_F^2} \int_0^\infty E \frac{1}{e^{\frac{E-E_f}{kT}}+1}\,dE$$

$$= \frac{2}{\pi\hbar^2 v_F^2}\left(\frac{E_f^2}{2}\right) = \frac{E_f^2}{\pi\hbar^2 v_F^2} \quad \text{and} \quad \varepsilon = \frac{qn_s}{\varepsilon_d}.$$

Here, $D(E)$ is the density of states of the graphene, v_F is the Fermi velocity in graphene, and ε_d is the dielectric constant of the dielectric on graphene. In the calculations, v_F is taken as 10^6 m/s. Variations of surface charge density and electric field on graphene for different dielectrics are shown in Figure 14.50.

Electric field decreases as the dielectric constant of the dielectric increases. At the desired E_f, the electric field should be less than the breakdown field, ε_{bd}, of the dielectric used. It is well known that high dielectric constant materials have lower breakdown fields. This in turn translates to lower operating voltage. For example, in Figure 14.50, $\varepsilon_{rd} = 7.5$ corresponds to Si_3N_4 with a $\varepsilon_{bd} = 11$ MV/cm. Therefore, E_f can be moved comfortably by 0.7 eV or more for Si_3N_4. On the other hand, $\varepsilon_{rd} = 100$ corresponds to sputtered $BaTi_3O_4$ with $\varepsilon_{bd} = 0.8$ MV/cm. Therefore, in this material, somewhat less E_f change can be achieved. However, based on Figure 14.49, changes up to $E_f = 0.6$ eV seems to be sufficient to observe very large loss and index changes in graphene. Hence, it is possible to use a wide range of dielectrics in combination with graphene.

14.7.6.1 UTILIZATION OF GRAPHENE IN DEVICE APPLICATIONS

Although there is a large publication body on theoretical graphene modulators, so far very limited experimental demonstration of graphene modulators of any kind has been reported [112,113].

Figure 14.51 shows one example [112]. A 50 nm thick Si layer was used to connect a 250 nm thick Si waveguide and one gold electrode. Both the silicon layer and the waveguide were doped with boron to reduce the sheet resistance. A spacer of 7 nm thick Al_2O_3 was then uniformly deposited on the surface of the waveguide by atomic layer deposition. A graphene sheet grown by chemical vapor deposition was then transferred onto the Si waveguide. To further reduce the access resistance of the device, the counter electrode was extended toward the bus waveguide by depositing a platinum (10 nm) film on top of the graphene layer. The distance between the platinum electrode and the waveguide was kept at 500 nm, so that the optical modes of the waveguide do not overlap with the platinum contact. Excess graphene was removed by oxygen plasma, leaving graphene only on top of the waveguide and between the waveguide and the platinum electrode.

The cross-sectional view of the device structure and the optical field distribution of the guided mode are also shown in Figure 14.51. Graphene only interacts with the tangential (in-plane) electric field of electromagnetic waves; the graphene modulator is polarization sensitive, as are conventional semiconductor-based electro-optical modulators.

Figure 14.52 displays the waveguide transmission of 1.53 µm at different drive voltages, V_D. At low drive voltage (-1 V $\leq V_D \leq 3.8$ V), the Fermi level $E_F(V_D)$ of graphene is close to the Dirac point ($E_F(V_D) < hf_0/2$), and interband transitions occur when electrons are excited by the incoming photons (hf_0). The optical absorption of graphene is determined by the position of the Fermi level. By applying a drive voltage between the graphene and the waveguide, the Fermi level of the graphene is tuned and the total transmission is modulated. With the current waveguide design, the modulation depth is as high as 0.1 dB per µm, resulting in a graphene EA modulator with a footprint of merely 25 µm². At large negative V_D (< -1 V), the Fermi level is lowered below the transition threshold ($E_F(V_D) = hf_0/2$) due to positive charge accumulation. As a result, there are no electrons available for interband transitions, and hence, the graphene appears transparent. On the other hand, at large positive V_D (> 3.8 V), all electron states are filled up, and no interband transitions are allowed. Ideally, there should be a sharp change in transmission at $E_F(V_D) = hf_0/2$. In reality, this transition was broadened due to defects in the graphene, and shifted to higher voltage owing to natural doping from the substrate.

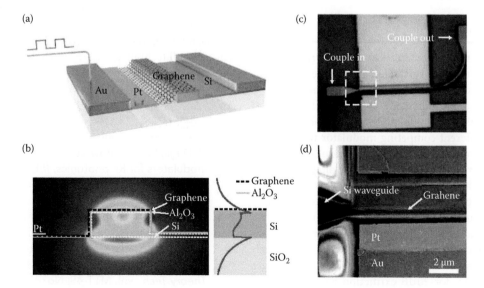

Figure 14.51 Three-dimensional schematic, cross-sections and pictures of the modulator. (Reprinted by permission from Macmillan Publishers Ltd. *Nature*, Liu, M. et al., *Nature*, vol. 474, pp. 64–67, copyright 2011.)

(a)

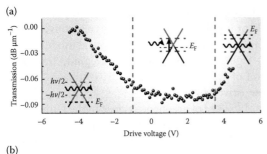

(b)

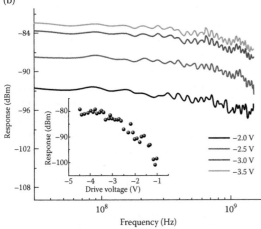

Figure 14.52 (a) Transmission through the device shown in Figure 14.51 at different voltages. (b) Frequency response of the device. (Reprinted by permission from Macmillan Publishers Ltd. *Nature*, Liu, M. et al., *Nature*, vol. 474, pp. 64–67, Figure 3. copyright 2011.)

14.8 SUMMARY

Optical modulators are key components for fiber optic networks. This chapter described the basic modulator specifications, phase and different types of amplitude modulators, as well as physical effects used in modulators. In addition, examples from different competing modulator technologies were given. Of these, electro-absorption modulators offer compact size, low drive voltage, and high bandwidth. However, modulation characteristics depend critically on wavelength and temperature. Furthermore, material preparation requires critical layer thickness and composition control. Lumped EA modulator modules with electrical bandwidths approaching 60 GHz and drive voltages around 2 V for 20 dB extinction have been demonstrated at 1.55 μm. Their fiber-to-fiber insertion loss is around 10 dB and polarization-independent operation has been achieved. Drive voltage and

bandwidth can be improved further using TW-EA modulator approach. LiNbO$_3$ has the most mature technology. LiNbO$_3$ traveling wave modulators are commonly used in industrial applications. They have higher drive voltages, which requires a modulator driver. Such modulators with electrical bandwidths as high as 75 GHz and V_π around 5 V have been demonstrated. Their fiber-to-fiber insertion loss is around 5 dB. Further V_π reduction is being pursued using thin LiNbO$_3$ films on different substrates. III–V compound semiconductor electro-optic modulators offer excellent bandwidth and drive voltages at the expense of higher insertion loss Electrical bandwidths exceeding 67 GHz have been realized. V_π values as low as 0.2 V were demonstrated. Wide bandwidth modulators with 0.77 V V_π and bandwidths exceeding 67 GHz were demonstrated. However, these devices have fiber-to-fiber insertion loss in the 10–15 dB range. Efforts to reduce the fiber-to-fiber insertion using novel processing techniques are under way. Polymers offer the promise of low cost technology and presently, are an active research area. High-speed capability of electro-optic polymers has been demonstrated. Polymer modulators with electrical bandwidths exceeding 40 GHz have been reported. V_π values are around 10 V and fiber-to-fiber insertion loss is about 10 dB. Silicon modulators provide a significant advantage in integration especially with CMOS and could be useful in short range applications such as optical interconnects. Newly emerging material systems such as graphene offer significant potential along with significant challenges. In all technologies, drive voltage reduction while improving the speed of operation remains a challenge.

REFERENCES

1. N. Dagli, Wide bandwidth lasers and modulators for RF photonics, *IEEE Trans. Microwave Theory Tech.*, vol. MTT-47, No. 7, pp. 1151–1171, July 1999.
2. C. Cox III, E. Ackerman, R. Helkey and G. Betts, Techniques and performance of intensity modulation direct detection analog optical links, *IEEE Trans. Microwave Theory Tech.*, vol. MTT-45, No. 8, pp. 1375–1383, August 1997.
3. D. J. Blumenthal, A. Carena, V. Curri and S. Humphries, All-optical label swapping with

wavelength conversion for WDM-IP networks with subcarrier multiplexed addressing, *IEEE Photonics Technol. Lett.*, vol.11, No.11, pp.1497–1499, November 1999.

4. H. Zmuda and E. N. Toughlian, eds., *Photonic Aspects of Modern Radar*, Artech House, Boston, 1994.

5. G. P. Agrawal, *Fiber Optic Communication Systems*, Wiley-Interscience, Hoboken, NJ, 1992.

6. H. A. Haus, *Waves and Fields in Optoelectronics*, Prentice-Hall, Chapter 7, 1984.

7. F. Koyama and K. Iga, Frequency chirping in external modulators, *J. Lightwave Technol.*, vol. 6, No. 1, pp. 87–93, January 1988.

8. K. Kubota, J. Noda and O. Mikami, Traveling wave optical modulator using a directional coupler LiNbO$_3$ waveguide, *IEEE J. Quantum Electron.*, vol. QE-16, No. 7, pp. 754–760, July 1980.

9. W. W. Rigrod and I. P. Kaminow, Wide-band microwave light modulation, *Proc. IEEE*, vol. 51, pp. 137–140, 1963.

10. R. Spickermann, S. R. Sakamoto and N. Dagli, GaAs/AlGaAs traveling wave electrooptic modulators, *Optoelectronic Integrated Circuits Conference, SPIE International Symposium on Optoelectronics'97*, Paper 33, San Jose, CA, February 8–14, 1997.

11. R. Spickermann, S. R. Sakamoto and N. Dagli, In traveling wave modulators which velocity to Match?, *IEEE/LEOS 1996 Annual Meeting*, Paper WM3, Boston, MA, November 18–21, 1996.

12. L. A. Coldren and S. W. Corzine, *Diode Lasers and Photonic Integrated Circuits*, Wiley Interscience, Hoboken, NJ, 1995.

13. W. Franz, Influence of an electric field on an optical absorption edge, *Z. Naturforsch.*, vol. 13a, p. 484, 1958.

14. L. V. Keldysh, The effect of a strong electric field on the optical properties of insulating crystals, *Sov. Phys. JETP*, vol. 7, 1958.

15. A. Anselm, *Introduction to the Semiconductor Theory*, Prentice Hall, p. 447, 1981.

16. S. Wang, *Fundamentals of Semiconductor Theory and Device Physics*, Prentice Hall, p. 619, 1989.

17. M. Shinado and S. Sugano, *J. Phys. Soc. Jpn.*, vol. 21, p. 1936, 1966.

18. T. Ido, H. Sano, S. Tanaka, D. J. Moss and H. Inoue, Performance of strained InGaAs/InAlAs multiple quantum well electroabsorption modulators, *J. Lightwave Technol.*, vol. 14, No. 10, pp. 2324–2331, October 1996.

19. G. Bastard, E. E. Mendez, L. L. Chang and L. Esaki, Variational calculations on a quantum well in an electric field, *Phys. Rev. B*, vol. 28, pp. 3241–3245, 1983.

20. J. I. Pankove, *Optical Processes in Semiconductors*, Dover, pp. 74–75, 1971.

21. R. G. Hunsperger, *Integrated Optics: Theory and Technology*, Springer-Verlag, pp. 75–79, 1982.

22. A. Alping and L. A. Coldren, Electrorefraction in GaAs and InGaAsP and its application to phase modulators, *J. Appl. Phys.*, vol. 61, pp. 2430–2433, April 1987.

23. J. G. Mendoza-Alvarez, L. A. Coldren, A. Alping, R. H. Yan, T. Hausken, K. Lee and K. Pedrotti, Analysis of depletion edge translation lightwave modulators, *J. Lightwave Technol.*, vol. 6, No. 6, pp. 793–808, June 1988.

24. J. E. Zucker, I. Bar-Joseph, B. I. Miller, U. Koren and D. S. Chemla, Quaternary wells for electro-optic intensity and phase modulation at 1.3 and 1.55 µm, *Appl. Phys. Lett.* vol. 54, No. 1, pp. 10–12, January 1989.

25. J. G. Mendoza-Alvarez, R. H. Yan and L. A. Coldren contribution of the band filling effect to the effective refractive index change in DH GaAs/AlGaAs phase Modulators, *J. Appl. Phys.*, vol. 61, 1987.

26. J. G. Mendoza-Alvarez, F. D. Nunes and N. B. Pate, Refractive index dependence on free carriers for GaAs, *J. Appl. Phys.*, vol. 51, pp. 4365–4367, 1980.

27. M. Born and E. Wolf, *Principles of Optics*, Pergamon Press, 1980.

28. I. P. Kaminow, *An Introduction to Electro Optic Devices*, Academic Press, 1974.

29. K. Wakita, I. Kotaka, H. Asai, S. Nojima and O. Mikami, High efficiency electroabsorption in quaternary AlGaInAs quantum well optical modulators, *Electron. Lett.*, vol. 24, pp. 1324–1326, 1988.

30. T. Ido, H. Sano, D. J. Moss, S. Tanaka and A. Takai, Strained InGaAs/InAlAs MQW electroabsorption modulators with large bandwidth and low driving voltage, *IEEE*

Photon. Technol. Lett., vol. 6, pp. 1207–1209, October 1994.

31. F. Devaux, F. Dorgeuille, A. Ougazzaden, F. Huet, M. Carenco, M. Henry, Y.Sorel, J. F. Kerdiles and E. Jeanney, 20 GBit/s operation of a high efficiency InGaAsP/InGaAsP MQW electroabsorption modulator with 1.2V drive voltage, *IEEE Photon. Technol. Lett.*, vol. 5, No. 11, pp. 1288–1290, 1993.

32. K. Wakita, I. Kotaka, O. Motomi, H. Asai, Y. Kawamura and M. Naganuma, High speed InGaAlAs/InAlAs multiple quantum well optical modulators, *J. Lightwave Technol.*, vol. 8, No. 7, pp. 1027–1032, July 1990.

33. K. Wakita, K. Yoshino, I. Kotaka, S. Kondo and Y. Noguchi, Blue chirp electroabsorption modulators with very thick quantum wells, *IEEE Photon. Technol. Lett.*, vol. 8, No. 9, pp. 1169–1171, September 1996.

34. T. Ido, S. Tanaka, M. Suzuki, M. Koizumi, H. Sano and H. Inoue, Ultra high speed multiple quantum well electro absorption optical modulators with integrated waveguides, *J. Lightwave Technol.*, vol. 14, No. 9, pp. 2026–2034, September 1996.

35. N. Mineo, K. Yamada, K. Nakamura, S. Sakai and T. Ushikobo, 60 GHz band electroabsorption modulator module, *Optical Fiber Conference*, Paper ThH4, San Jose, CA, 1998.

36. O. Motomi, I. Kotaka, K. Wakita, S. Nojima, K. Kawano, Y. Kawamura and H. Asai, 40 GHz bandwidth InGaAs/InAlAs multiple quantum well optical intensity modulator, *Appl. Opt.*, vol. 31, pp. 2030–2035, 1992.

37. K. Satzke, D. Baums, U. Cebulla, H. Haisch, D. Kaiser, E. Lach, E. Kuhn, J. Weber, R. Weinmann, P. Widemann and E. Zielinski, Ultrahigh bandwidth (42 GHz) polarization independent ridge waveguide electroabsorption modulator based on tensile strained InGaAsP MQW, *Electron. Lett.*, vol. 31, pp. 2030–2032, 1995.

38. K. K. Loi, X. B. Mei, J. H. Hodiak, C. W. Tu and W. S. C. Chang, 38 GHz bandwidth 1.3 μm MQW electroabsorption modulators for RF photonic links, *Electron. Lett.*, vol. 34, pp. 1018–1019, May 1998.

39. K. Wakita, I. Kotaka, K. Yoshino, S. Kondo and Y. Noguchi, Polarization independent electroabsorption modulators using strain compensated InGaAs/InAlAs MQW

structures, *IEEE Photon. Technol. Lett.*, vol. 7, No. 12, pp. 1418–1420, December 1995.

40. F. Devaux, S. Chelles, A. Ougazzaden, A. Mircea, M. Carre, F. Huet, A. Carenco, Y. Sorel, J. F. Kerdiles and M. Henry, Full polarization insensitivity of a 20 Gb/s strained MQW electroabsorption modulator, *IEEE Photon. Technol. Lett.*, vol. 6, No. 10, pp. 1203–1205, October 1994.

41. H. H. Liao, X. B. Mei, K. K. Loi, C. W. Tu, P. M. Asbeck and W. S. C. Chang, Microwave structures for traveling-wave MQW electroabsorption modulators for wide band 1.3 μm photonic links, *Proceedings of the SPIE Optoelectronic Integrated Circuits*, vol. 3006, pp. 291–300, San Jose, CA, February 12–14, 1997.

42. S. Z. Zhang, Y. J. Chiu, P. Abraham and J. E. Bowers, 25 GHz polarization insensitive electroabsorption modulators with traveling wave electrodes, *IEEE Photonics Technol. Lett.*, vol. 11, No. 2, pp. 191–193, Februray 1999.

43. K. Kawano, M. Kohtoku, M. Ueki, T. Ito, S. Kondoh, Y. Noguchi and Y. Hasumi, Polarization insensitive traveling wave electrode electroabsorption (TW-EA) modulator with bandwidth over 50 GHz and driving voltage less than 2V, *Electron. Lett.*, vol. 33, No. 18, pp. 1580–1581, August 1997.

44. G. K. Gopalakrishnan, W. K. Burns, R. W. McElhanon, C. H. Bulmer and A. S. Greenblatt, Performance and modeling of broadband LiNbO$_3$ traveling wave optical intensity modulators, *J. Lightwave Technol.*, vol. 12, No. 10, pp. 1807–1819, October 1994.

45. B. T. Szentkuti, Simple analysis of anisotropic microstrip lines by a transform method, *Electron. Lett.*, vol. 12, pp. 672–673, 1976.

46. K. C. Gupta, R. Garg, I. Bahl, and P. Bhartia, *Microstrip Lines and Slotlines*, Artech House, Boston, 1996.

47. X. Zhang and T. Miyoshi, Optimum design of coplanar waveguide for LiNbO$_3$ optical modulator, *IEEE Trans. Microwave Theory Tech.*, vol. MTT-43, No. 3, pp. 523–528, March 1995.

48. K. Noguchi, O. Mitomi, K. Kawano and Y. Yanagibashi, Highly efficient 40-GHz bandwidth Ti:LiNbO$_3$ optical modulator employing ridge structure, *IEEE Photon. Technol. Lett.*, vol. 5, pp. 52–54, 1993.

49. O. Mitomi, K. Noguchi and H. Miyazawa, Design of ultra broad band LiNbO$_3$ optical modulators with ridge structure, *IEEE Trans. Microwave Theory Tech.*, vol. MTT-43, No. 9, pp. 2203–2207, September 1995.

50. K. Noguchi, O. Mitomi and H. Miyazawa, Millimeter-wave Ti:LiNbO$_3$ optical modulators, *J. Lightwave Technol.*, vol. 16, No. 4, pp. 615–619, April 1998.

51. K. Noguchi, H. Miyazawa and O. Mitomi, Frequency dependent propagation characteristics of coplanar waveguide electrode on 100GHz TiLiNbO$_3$ optical modulator, *Electron. Lett.*, vol. 34, No. 7, pp. 661–663, January 1998.

52. G. K. Gopalakrishnan, W. K. Burns and C. H. Bulmer, Electrical loss mechanism in traveling wave switch/modulators, *Electron. Lett.*, vol. 28, No. 2, pp. 207–209, 1992.

53. W. K. Burns, M. M. Howerton, R. P. Moeller, A. S. Greenblatt and R. W. McElhanon, Broad band reflection traveling-wave LiNbO$_3$ modulator, *IEEE Photon. Technol. Lett.*, vol. 10, pp. 805–806, June 1998.

54. J. Kondo, K. Aoki, A. Kondo, T. Ejiri, Y. Iwata, A. Hamajima, T. Mori, Y. Mizuno, M. Imaeda, Y. Kozuka, O. Mitomi and M. Minakata, High-speed and low-driving-voltage thin-sheet X-Cut LiNbO$_3$ modulator with laminated low-dielectric-constant adhesive, *IEEE Photon. Technol. Lett.*, vol. 17, No. 10, pp. 2077–2079, October 2005.

55. J. Chiles and S. Fathpour, Mid-infrared integrated waveguide modulators based on silicon-on-lithium-niobate photonics, *Optica*, vol. 1, No. 5, pp. 350–355, November 2014.

56. A. M. Glass, The photorefractive effect, *Opt. Eng.*, vol. 17, pp. 470–479, 1978.

57. T. Fujiwara, S. Sato and H. Mori, Wavelength dependence of photorefractive effect in Ti indiffused LiNbO$_3$ waveguides, *Appl. Phys. Lett.*, vol. 54, pp. 975–977, 1989.

58. G. E. Betts, F. J. O'Donnell and K. G. Ray, Effect of annealing photorefractive damage in Titanium indiffused LiNbO$_3$ modulators, *IEEE Photon. Technol. Lett.*, vol. 6, pp. 211–213, 1994.

59. S. K. Korotky and J. J. Veselka, An RC network analysis of long term Ti:LiNbO$_3$ bias stability, *J. Lightwave Technol.*, vol. 14, No. 12, pp. 2687–2697, December 1996.

60. H. F. Taylor, Polarization independent guided wave optical modulators and switches, *J. Lightwave Technol.*, vol. 3, pp. 1277–1280, December 1985.

61. P. Granestrand, L. Thylen and B. Stoltz, Polarization independent switch and polarization splitter employing $\Delta\beta$ and $\Delta\kappa$ modulation, *Electron. Lett.*, vol. 24, No. 18, pp. 1142–1143, September 1988.

62. L. McCaughan, Low loss polarization independent electrooptic switches, *J. Lightwave Technol.*, vol. 2, pp. 51–55, February 1984.

63. R. C. Alferness, Polarization independent optical directional coupler switching using weighted coupling, *Appl. Phys. Lett.*, vol. 35, pp. 748–750, 1979.

64. Y. Silberberg, P. Perlmutter and J. E. Baran, Digital optical switch, *Appl. Phys. Lett.*, vol. 51, No. 16, pp. 1230–1232, 19 October 1987.

65. H. G. Bach, J. Kauser, H. P. Nolting, R. A. Logan and F. K. Reinhart, Electro-optical light modulation in InGaAsP/InP double heterostructure diodes, *Appl. Phys. Lett.*, vol. 42, pp. 692–694, 1983.

66. J. H. Shin, S. Wu and N. Dagli, Bulk undoped GaAs-AlGaAs substrate-removed electrooptic modulators with 3.7-V-cm drive voltage at 1.55 μm, *IEEE Photonics Technol. Lett.*, vol. 18, No. 21, pp. 2251–2253, November 2006.

67. J. H. Shin and N. Dagli, Ultra-low drive voltage substrate removed GaAs/AlGaAs electro-optic modulators at 1550 nm, *IEEE J. Sel. Top. Quantum Electron.*, vol. 19, No. 6, p. 3400408, November–December 2013.

68. J. H. Shin, Y.-C. Chang and N. Dagli, 0.3V drive voltage GaAs/AlGaAs substrate removed Mach-Zehnder intensity modulators, *Appl. Phys. Lett.*, vol. 92, p. 201103, 2008.

69. S. Dogru and N. Dagli. Mode transformers for substrate removed waveguides, *Proceedings of IEEE Photonics Society Annual Meeting (IEEE Photonics 2013)*, Paper MF1.3, pp. 102–103, Bellevue, Washington,s 8–12 September 2013.

70. S. Dogru and N. Dagli, 0.2V drive voltage substrate removed electro-optic Mach-Zehnder modulators with MQW cores at 1.55 μm, *IEEE/OSA J. Lightwave Technol.*, vol. LT-32, No. 3, pp. 435–439, February 2014.

71. R. G. Walker, High speed III-V electrooptic waveguide modulators, *IEEE J. Quantum Electron.*, vol. 27, pp. 654–667, March 1991.

72. R. G. Walker, Electro-optic modulation at mm-wave frequencies in GaAs/AlGaAs guided wave devices, *Proceedings of IEEE/LEOS'95 8th Annual Meeting*, Paper IO4.2, pp. 118–119, San Francisco, CA, October 30–November 2 1995.

73. R. Spickermann, S. R. Sakamoto, M. G. Peters and N. Dagli, GaAs/AlGaAs traveling wave electrooptic modulator with electrical bandwidth greater than 40 GHz, *Electron. Lett.*, vol. 32, No. 12, pp. 1095–1096, June 1996.

74. R. Spickermann, S. R. Sakamoto, M. G. Peters and N. Dagli, GaAs/AlGaAs traveling wave electrooptic modulator with electrical bandwidth greater than 40 GHz, *Electron. Lett.*, vol. 32, No. 12, pp. 1095–1096, June 1996.

75. J. H. Shin, S. Wu, and N. Dagli, 35 GHz bandwidth, 5 V-cm drive voltage, bulk GaAs substrate removed electro optic modulators, *IEEE Photonics Technol. Lett.*, vol. 19, No. 18, pp. 1362–1364, September 15 2007.

76. J. H. Shin, C. Ozturk, S. R. Sakamoto, Y. J. Chiu and N. Dagli, Novel T-rail electrodes for substrate removed low-voltage, high-speed GaAs/AlGaAs electro-optic modulators, *IEEE Trans. Microwave Theory Tech.*, vol. MTT-53, No. 2, pp. 636–643, February 2005.

77. S. Dogru and N. Dagli, 0.77-V drive voltage electro-optic modulator with bandwidth exceeding 67 GHz, *Opt. Lett.*, vol. 39, No. 20, pp. 6074–6077, 15 October 2014.

78. B. R. Bennett, R. A. Soref and J. A. Del Alamo, Carrier-induced change in refractive index of InP, GaAs and InGaAsP, *IEEE J. Quantum Electron.*, vol. 26, No. 1, pp. 113–122, 1 January 1990.

79. S. Dogru, J. H. Shin and N. Dagli, InGaAlAs/InAlAs multi quantum well substrate removed electro-optic modulators, *Proceedings of IEEE Photonics Society Annual Meeting (IEEE Photonics 2011)*, Paper ThJ2, pp. 739–740, Arlington, VA, 9–13 October 2011.

80. N. K. Kim, P. Bhasker, S. Dogru and N. Dagli, InP based very-low voltage electro-optic intensity modulators in conventional waveguides, *Integrated Photonics Research, Silicon and Nano Photonics Conference Proceedings*, Paper IM2B.2, Boston, MA, June 27–July 1, 2015.

81. N. K. Kim, P. Bhasker, S. Dogru and N. Dagli, Very low voltage InGaAlAs/InAlAs MQW core electro-optic modulators fabricated with conventional processing, *Proceedings of IEEE Photonics Society Annual Meeting (IEEE Photonics 2015)*, Paper MC3.2, pp. 92–93, Reston, VA, 4–8 October 2015.

82. N. K. Kim and N. Dagli, Series connected low voltage III-V electro-optic modulators, *Proceedings of IEEE Photonics Society Annual Meeting (IEEE Photonics 2016)*, Paper WA2.1, Waikoloa, Hawaii, 2–6 October 2016.

83. R. Spickermann, M. Peters and N. Dagli, A polarization independent GaAs/AlGaAs electrooptic modulator, *IEEE J. Quantum Electron.*, vol. QE-32, No. 5, pp. 764–769, May 1996.

84. R. Spickermann and N. Dagli, Experimental analysis of millimeter wave coplanar waveguide slow wave structures on GaAs, *IEEE Trans. Microwave Theory Tech.*, vol. MTT-42, No. 10, pp. 1918–1924, October 1994.

85. L. A. Hornak, Ed., *Polymers for Lightwave and Integrated Optics*, Marcel Dekker, New York, 1992.

86. S. Kalluri, M. Ziari, A. Chen, V. Chuyanov, W. H. Steir, D. Chen, B. Jalali, H.R. Fetterman and L. R. Dalton, Monolithic integration of waveguide polymer electrooptic modulators on VLSI circuitry, *IEEE Photonics Technol. Lett.*, vol. 8, pp. 644–646, May 1996.

87. C. F. Kane and R. R. Krchnavek, Benzocyclobutene optical waveguides, *IEEE Photonics Technol. Lett.*, vol. 7, No. 5, pp. 535–537, May 1995.

88. G. Fishbeck, R. Moosburger, C. Kostrzewa, A. Achen, and K. Petermann, Single mode optical waveguides using a high temperature stable polymer with low losses in the 1.55 μm range, *Electron. Lett.*, vol. 33, No. 6, pp. 518–19, March 1997.

89. G. F. Lipscomb, A. F. Garito and R. S. Narang, A large linear electro-optic effect in a polar organic crystal 2-methyl-4-nitroaniline, *Appl. Phys. Lett.*, vol. 38, No. 9, pp. 663–665, 1981.

90. D. Chen, H. Fetterman, A. Chen, W. H. Steier, L. R. Dalton, W. Wang and Y. Shi,

Demonstration of 110 GHz electrooptic polymer modulators, *Appl. Phys. Lett.*, vol. 70, No. 25, pp. 3335–3337, 1997.

91. D. Chen, D. Bhattacharta, A. Udupa, B. Tsap, H. Fetterman, A. Chen, S. S. Lee, J. Chen, W. H. Steier and L. R. Dalton, High frequency polymer modulators with integrated finline transitions and low V_π, *IEEE Photonics Technol. Lett.*, vol. 11, No. 1, pp. 54–56, January 1999.

92. C. C. Teng, Traveling wave polymeric intensity modulator with more than 40 GHz 3 dB electrical bandwidth, *Appl. Phys. Lett.*, vol. 60, pp. 1538–1540, 1992.

93. G. D. Girton, S. L. Kwiatkowski, G. F. Lipscomb and R. S. Lytel, 20 GHz electrooptic polymer Mach-Zehnder modulator, *Appl. Phys. Lett.*, vol. 58, pp. 1730–1732, 1991.

94. W. Wang, Y. Shi, D. J. Olson, W. Lin and J. Bechtel, Push pull poled polymer Mach Zehnder modulators with a single microstrip line electrode, *IEEE Photonics Technol. Lett.*, vol. 11, No. 1, pp. 51–53, January 1999.

95. T. T. Tumolillo and P. R. Ashley, A novel pulse poling technique for EO polymer waveguide devices using device electrode poling, *IEEE Photon. Technol. Lett.*, vol. 4, No. 2, pp. 142–145, February 1992.

96. K. H. Hahn, D. W. Dolfi, R. S. Moshrefzadeh, P. A. Pedersen and C. V. Francis, Novel two arm microwave transmission line for high speed electrooptic polymer modulators, *Electron. Lett.*, vol. 30, No. 15, pp. 1220–1222, July 1994.

97. Y. Shi, D. J. Olson and J. Bechtel, Photoinduced molecular alignment relaxation in poled electrooptic polymer thin films, *Appl. Phys. Lett.*, vol. 68, No. 8, pp. 1040–1042, 1996.

98. M. Mortazavi, H. Yoon and C. Teng, Optical power handling properties of polymeric nonlinear optical waveguides, *J. Appl. Phys.*, vol. 74, pp. 4871–4876, 1993.

99. Y. Shi, W. Wang, W. Lin, D. J. Olson and J. H. Bechtel, Double end cross linked electrooptic polymer modulators with high optical power handling capability, *Appl. Phys. Lett.*, vol. 70, No. 11, pp. 1342–1344, 1997.

100. Y. Shi, W. Wang, W. Lin, D. J. Olson and J. H. Bechtel, Long term stable direct current bias operation in electrooptic polymer modulators with an electrically compatible multilayer structure, *Appl. Phys. Lett.*, vol. 71, No. 16, pp. 2236–2238, 1997.

101. H. Park, W. Y. Hwang and J. J. Kim, Origin of direct current drift in electrooptic polymer modulator, *Appl. Phys. Lett.*, vol. 70, No. 21, pp. 2796–2798, 1997.

102. S. Rumley, D. Nikolova, R. Hendry, Q. Li, D. Calhoun and K. Bergman, silicon photonics for exascale systems, *J. Lightwave Technol.*, vol. 33, No. 3, pp. 547–562, October 2015, DOI: 10.1109/JLT.2014.2363947.

103. R. A. Soref and B. R. Bennett, Electrooptical effects in silicon, *IEEE J. Quantum Elect.*, vol. QE-23, pp. 123–129, January 1987.

104. X. Xiao, H. Xu, X. Li, Z. Li, T. Chu, Y. Yu and J. Yu, High-speed, low-loss silicon Mach–Zehnder modulators with doping optimization, *Opt. Express*, vol. 21, No. 4, pp. 4116–4125, February 2013.

105. D. Petousi, L. Zimmermann, K. Voigt and K. Petermann, Performance limits of depletion-type silicon Mach–Zehnder modulators for telecom applications, *J. Lightwave Technol.*, vol. 31, No. 22, pp. 3556–3562, November 2013.

106. M. Streshinsky, R. Ding, Y. Liu, A. Novack, Y. Yang, Y. Ma, X. Tu, E. K. S. Chee, A. E. J. Lim, P. G. Q. Lo, T. B. Jones and M. Hochberg, Low power 50 Gb/s silicon traveling wave Mach-Zehnder modulator near 1300 nm, *Opt. Express*, vol. 21, No. 25, pp. 30350–30357, December 2013.

107. D. J. Thomson, F. Y. Gardes, S. Liu, H. Porte, L. Zimmermann, J.-M. Fedeli, Y. Hu, M. Nedeljkovic, X. Yang, P. Petropoulos and G. Mashanovich, High performance Mach–Zehnder-based silicon optical modulators, *IEEE J. Sel. Top. Quantum Electron.*, vol. 19, No. 6, November/December 2013.

108. M. R. Watts, W. A. Zortman, D. C. Trotter, R. W. Young and A. L. Lentine, Vertical junction silicon microdisk modulators and switches, *Opt. Express*, vol. 19, No. 22, pp. 21989–22003, October 2011.

109. E. Timurdogan, C. M. Sorace-Agaskar, J. Sun, E. S. Hosseini, A. Biberman and M. R. Watts, An ultralow power athermal silicon modulator, *Nat. Commun.*, vol. 5, p. 4008, 2014. DOI: 10.1038/ncomms5008.

110. F. Xia, H. Yan, and P. Avouris, The interaction of light and graphene: Basics, devices, and applications, *Proc. IEEE*, vol. 101, No. 7, pp. 1717–1731, 2013.

111. V. P, Gusynin, S. G. Sharapov, and J. P. Carbotte. Magneto-optical conductivity in graphene. *J. Phys.: Condens. Matter*, vol. 19, No. 2, p. 026222, 2007.

112. M. Liu, X. Yin, E. Ulin-Avila, B. Geng, T. Zentgraf, L. Ju, F. Wang and X. Zhang, A graphene-based broadband optical modulator, *Nature*, vol. 474, pp. 64–67, 2011.

113. Y. T. Hu, M. Pantouvaki, S. Brems, I. Asselberghs, C. Huyghebaert, M. Geisler, C. Alessandri, R. Baets, P. Absil, D. Van Thourhout and J. Van Campenhout, Broadband 10Gb/s graphene electro-absorption modulator on silicon for chip-level optical interconnects, *Proceedings of International 2014 Electron Devices Meeting (IEDM 2014)*, pp. 5.6.1–5.6.4, 2014. DOI: 10.1109/IEDM.2014.7046991.

15

Optical amplifiers

JOHAN NILSSON
University of Southampton

JESPER LÆGSGAARD
Department of Photonics Engineering

ANDERS BJARKLEV
Technical University of Denmark

15.1 BACKGROUND

From the first demonstrations of optical communication systems, a primary drive in the research activity has been directed towards constant increase in system capacity. Over the past 25 years, it has been an alternating activity to overcome the fundamental fiber limitations of either attenuation or dispersion, and on the basis of the limiting term of the transmission link, systems have been denoted as either *loss limited* or *dispersion limited*. In the mid-1980s the international development had reached a state at which not only dispersion-shifted fibers were available but also spectrally pure signal sources emerged. The long-haul optical communication systems were, therefore, clearly loss limited and their problems had to be overcome by periodic regeneration of the optical signals at repeaters applying conversion to an intermediate electrical signal, which was a complex technology with lack of flexibility.

The technological challenge was to develop a practical way of obtaining the needed gain, that is, to develop relatively simple and flexible optical amplifiers, which would be superior to the electrical regenerators. Several means of doing this had been suggested in the 1960s and 1970s, including direct use of the transmission fiber as amplifying medium through nonlinear effects [1], semiconductor amplifiers with common technical basis in the components used for signal sources [2], or doping optical waveguides with an active material (rare-earth (RE) ions) that could provide the gain [3]. First, however, with the pronounced technological need for optical amplifiers and the spectacular results on erbium-doped fiber amplifiers (EDFAs) [4], an intense worldwide research activity on optical amplifiers was initiated. This resulted in the appearance of commercially available packaged EDFA modules in 1990.

EDFAs operate at signal wavelengths around 1550 nm, i.e., in the so-called third optical transmission window of silica fibers used in telecommunication systems. With the availability of efficient optical amplifiers in this wavelength region, which allowed for cost-effective transmission of optical signals (almost independent of modulation format and bit-rate), the development of multi-channel—wavelength-division-multiplexed (WDM)—systems became possible and a very rapid development of improved amplifier technologies has followed over the past 10 years. EDFAs have truly revolutionized optical telecommunications. By contrast, other optical amplification technologies such as Raman amplification, semiconductor amplifiers, planar amplifiers, and various RE-doped amplifiers for different wavelength regimes have yet to make a significant impact on deployed optical communication systems.

In this chapter, we will provide a short review of the general properties and limitations of optical amplifiers. Thereafter, the specific technologies available for the realization of optical amplifiers are reviewed, and their key parameters are compared. However, since the area of optical amplifiers is extremely broad, we do not attempt to cover all details of the latest development; rather, a selection of recent research and development results will be presented. This selection includes an overall discussion of amplification bands, cladding-pumped amplifiers for high-power amplification, recent results on high-concentration erbium doping in silica glass, and planar optical amplifiers in silica-on-silicon technology.

By and large, we restrict our discussion to amplifiers used in optical transmission systems. There are many other uses for optical amplifiers, in other types of system for other types of application. While all optical amplifiers used in optical transmission systems are waveguiding ones (fiber amplifiers in particular), there are also nonguiding ("bulk") amplifiers. These are primarily used for high-power amplification and amplification of high-energy pulses. Bulk amplifiers have many disadvantages compared to waveguiding ones, including a low gain efficiency and generally a low gain. Furthermore, in the case of a crystalline host,

the bandwidth is narrow. Bulk amplifiers are not used in optical transmission systems and will not be discussed here. See, for example, Reference [5] for further details.

15.2 GENERAL AMPLIFIER CONCEPTS

In this section, we discuss some basic concepts common to all optical amplifiers. Due to the general nature of this first part of the chapter, this first part will follow the description given in Reference [6]

Ideally, an optical amplifier would amplify the signal by adding, in phase, a well-defined number of photons to each incident photon. This means that a bit sequence (or analogue optical signal) simply would increase its electromagnetic field strength, but not change its shape by passage through the optical amplifier. Figure 15.1 shows the basics of optical amplification. In a perfect amplifier, this process would take place independent of the wavelength, state of polarization, intensity, (bit) sequence, and optical bandwidth of the incident light signal, and no interaction would take place between signals, if more than one signal were amplified simultaneously. In practice, however, the optical gain depends not only on the wavelength of the incident signal, but also on the electromagnetic field intensity, and thus the power, at any point inside the amplifier. Details of wavelength and intensity dependence of the optical gain depend on the amplifying medium.

An amplifying medium needs to be pumped. For example, a sufficient number of erbium ions must be excited so that a population inversion is reached, otherwise the erbium ions will attenuate

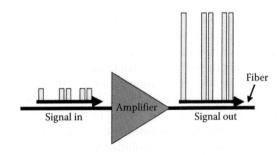

Figure 15.1 Optical amplification. In this case, the amplifier is fiber coupled, so that the input and output signals propagate in an optical fiber.

rather than amplify the beam. All amplifiers described here need to be optically pumped, with the exception of semiconductor optical amplifiers. Laser diodes are the only acceptable pump sources for telecom applications, because of their efficiency, reliability, and small size. The development of suitable pump diodes has therefore been an integral part of the development of optical amplifiers. However, this will not be discussed here.

Next, some general concepts of practical optical amplifiers will be outlined. First, we consider a case in which the amplifying medium is modelled as a homogeneously broadened two-level system. In such a medium, the gain coefficient (i.e., the gain per unit length) can be written as [7]:

$$g(\omega) = \frac{g_0}{1 + (\omega - \omega_0)^2 T_2^2 + P_s/P_{sat}}. \quad (15.1)$$

Here g_0 is the peak value of the gain coefficient determined by the pumping level of the amplifier, and ω is the optical angular frequency of the incident signal, related to its vacuum wavelength λ by $\omega = 2\pi c/\lambda$. Furthermore, ω_0 is the atomic transition angular frequency. P_s is the optical power of the signal and P_{sat} is the saturation power, which depends on amplifying medium parameters such as fluorescence lifetime and the transition cross section at the signal frequency. The parameter T_2 in Equation 15.1 is normally denoted as the dipole relaxation time [7].

Two important amplifier characteristics are described in Equation 15.1: First, if the signal power ratio obeys $P_s/P_{sat} \ll 1$ throughout the amplifier, the amplifier is said to be operated in the unsaturated region. The gain coefficient is then maximal when the incident angular frequency ω coincides with the atomic transition angular frequency ω_0 The gain reduction for angular frequencies different from ω_0 is generally given by a more complex function than the Lorentzian profile, but this simple example allows us to define the general property of the gain bandwidth. This is normally defined as the full width at half maximum (FWHM) value of the gain coefficient spectrum $g(\omega)$. For the Lorentzian spectrum, the gain bandwidth is given by $\Delta\omega_g = 2/T_2$ (Figure 15.2).

From a communication system point of view, it is more natural to use the related concept of amplifier bandwidth (often determined by the 3 dB

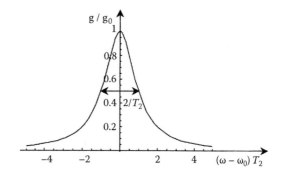

Figure 15.2 Lorentzian gain profile with bandwidth $\Delta\omega_g = 2/T_2$ (FWHM, full width at half maximum). The unsaturated gain is shown, with $P_s \ll P_{sat}$.

points) instead of the gain bandwidth, which is related to a point within the amplifying medium. The difference becomes clear, when we consider the linear gain of the amplifier G defined as

$$G = \frac{P_s^{out}}{P_s^{in}} \quad (15.2)$$

where P_s^{in} is the input power and P_s^{out} is the output power of a continuous wave (cw) signal being amplified. The amplifier gain G may be found by using the relation

$$\frac{dP}{dz} = gP \quad (15.3)$$

where $P(z)$ is the optical power at a distance z from the amplifier input end. If the gain coefficient $g(\omega)$, for simplicity, is considered constant along the amplifier length, the solution of Equation 15.3 is an exponentially growing signal power $P(z) = P_s^{in} \exp(gz)$ For an amplifier length L, we then find that

$$G(\omega) = \exp[g(\omega)L]. \quad (15.4)$$

Equation 15.4 illustrates the frequency dependence of the amplifier gain G, and shows that $G(\omega)$ decreases much faster than $g(\omega)$ with the signal detuning $(\omega - \omega_0)$, because of the exponential dependence of G on g. Having said that, in many cases the gain spectrum is tophat-like (e.g., supergaussian) rather than Lorentzian, and then, the difference between gain bandwidth and amplifier bandwidth, which is smaller, is reduced. The limited amplifier bandwidth results in signal distortion in the cases where a broadband optical signal is transmitted through the amplifier. In practical optical transmission systems, a very important consequence of a restricted bandwidth relates to WDM signal amplification: different signals at different wavelengths are amplified by different amounts if the amplifier bandwidth is too narrow. The deviation from a flat gain is referred to as gain ripple. The problem of gain ripple is exacerbated in a long-distance transmission system employing many cascaded amplifiers. Any systematic differences in amplification of the different channels then add up, leading to high requirements on gain flatness. To meet this requirement many amplifiers incorporate gain-flattening filters that equalize the gain over a range of wavelengths [8]. While most gain-flattening filters are static, dynamic ones have been developed for demanding applications. In any case, the bandwidth of optical amplifiers is extremely large and one of their principal attractions, maybe around 1 THz even in the simplest amplifier and exceeding 10 THz in state-of-the art, gain-flattened, ones.

Another limitation of the nonideal amplifier is expressed in the power dependence of the gain coefficient. This property, which is known as gain saturation, is included in the example of Equation 15.1, and it appears in all cases where the term P_s/P_{sat} is non-negligible. Since the gain coefficient is reduced when the signal power P_s becomes comparable to the saturation power P_{sat}, the amplification factor (or amplifier gain) G will also decrease. This limits an amplifier's maximum signal output power, known as the saturated output power. Gain saturation might be seen as a serious limitation for multi-wavelength optical communication systems. However, gain-saturated amplifiers have practical use because of their self-regulating effect on the signal output power in links of many concatenated amplifiers. Furthermore, the saturated output power of modern EDFAs is high enough not to be a practical limitation.

Besides the bandwidth and gain saturation limitations of practical optical amplifiers, there is another important limitation in practical amplified systems. Optical amplifiers, in general, will add spontaneously emitted or scattered photons to the signal during the amplification process,

and this will consequently lead to a degradation of the signal-to-noise ratio (SNR). The SNR degradation is quantified through a parameter F, normally denoted as the amplifier noise figure, which is defined as the SNR ratio between input and output [9]:

$$F = \frac{SNR_{in}}{SNR_{out}}. \qquad (15.5)$$

It should be noted that it is common practice to refer the SNR to the electrical power generated when the optical signal is converted to electrical current by using a photodetector. The noise figure as defined in Equation 15.5, therefore, in general, would depend on several detector parameters, which determine the shot noise and thermal noise associated with the practical detector. However, the influence of detector parameters will not help us to clarify the amplifier noise properties, and it is, consequently, advantageous to consider an ideal detector, whose performance is limited by shot noise only [9].

In practice, the spontaneous emission is reduced by optical filtering of the received signal, by rejecting light at frequencies away from the signal. Therefore, the SNR generally also will be dependent on the bandwidth of the optical filters and the spectral power distribution of the spontaneous emission from the amplifier. However, since this filtering is a process independent of the amplifier properties, it is also common practice to eliminate this ambiguity by considering an ideal filter [7]. Such an ideal filter is introduced only to allow the signal and the spontaneous emission within the signal bandwidth to pass to the detector. Therefore, it will only be the spontaneous emission spectral power density at the signal wavelength that enters the ideal detector, and the noise figure becomes independent of the spectral shape of the spontaneous emission. Note that since the amplifier deteriorates the SNR, that is, $SNR_{in} > SNR_{out}$, the noise figure will always obey the relation $F > 1$. In fact, conventional, so-called phase-insensitive, amplifiers like the EDFA always have a noise figure of at least 2 (i.e., 3 dB) in the limit of high gain. This is known as the quantum limit. We emphasize that the common interpretation that the output SNR is always at least 3 dB worse than the input SNR is only correct for high-gain amplifiers when the input SNR is shot-noise limited. Though the coherent state emitted from an ideal laser is shot-noise limited, far from all input signals fulfil this. In particular, it is not true for an input signal that has already been amplified.

The noise contribution of an amplifier can also be understood as arising from a fundamentally stochastic nature of the amplification process (stimulated emission). Absorption can also occur in an amplifier, again in a fundamentally stochastic manner that adds to the noise. In order to minimize the noise contribution of an amplifier, one should maximize the stimulated emission to absorption ratio—preferably there should be no absorption at all.

Though the SNR at the output of the amplifier is degraded, the high power of the amplified signal means that additional noise added further down the transmission line will have less impact. Thus, the amplifier improves the SNR of the transmission line as a whole.

A phase-insensitive amplifier operates independently of the optical phase of the signal. In contrast, its counterpart, the phase-sensitive amplifier, provides amplification that does depend on the phase of the signal. A phase-sensitive amplifier does not necessarily degrade the SNR (i.e., it can have a noise figure of unity) [10]. Phase-sensitive amplifiers are used for squeezed-state amplification in research applications, but are not in common use because of practical difficulties.

For a practical communication system, the amplifier spontaneous emission may have another influence besides that described through the noise figure (i.e., besides the effect that it adds fluctuations to the amplified optical signal power, which are converted to current fluctuations during the photo-detection process). This additional influence is that the spontaneous emission, which is emitted from the amplifier input end, may enter the signal source (a semiconductor laser), where it can result in performance disturbances. Therefore, it is often necessary to include isolation between amplifier and light source to avoid additional noise in the system. Therefore, it must be considered that the optical communication system has to be protected against undesired emission from the amplifier. These may also transmit residual pump power onto the transmitter and/or detector. The actual effect of this will be strongly dependent on the spectral properties of the pump light and whether the signal source or detector is sensitive to such radiation.

Another property that has to be evaluated for the optical amplifier is the polarization sensitivity. A high polarization sensitivity means that the amplifier gain G differs for different polarization states of the input signal. Since optical communication systems normally do not include polarization control, and because the polarization state is likely to vary due to external factors such as mechanical pressure, acoustic waves, and temperature variations, the amplifier polarization sensitivity is an undesired property in most cases. Therefore, for amplifiers that are inherently polarization sensitive, it has been a primary goal to reduce or even eliminate the amplifier output power variation due to changes in the signal input polarization state. Fortunately, the dominating EDFA has a negligible intrinsic polarization dependence of the gain.

All of the limiting properties which we have previously discussed are not surprisingly distinctly different for different types of amplifier. This is also the case for the crosstalk limitation that relates to multi-channel applications of optical amplifiers. In contrast to the ideal case, where all signal channels (or wavelengths) are amplified by the same amount, undesired nonlinear phenomena may introduce inter-channel crosstalk (i.e., the modulation of one channel is affected or modified by the presence of another signal channel). We will return to the more specific nature of these nonlinear phenomena in connection with the discussion of the different optical amplifiers.

Other factors such as polarization mode dispersion, multi-path interference, and nonlinearities may also degrade an amplifier.

The final limiting factors that should be mentioned here are closely related to the physical environment in which the amplifier is placed. These may include sensitivity toward vibrations, radioactive radiation, chemicals, pressure, and temperature. However, since at least amplifiers for optical communication systems generally are placed in controlled environments, we will not go further into a detailed discussion of such properties.

The relative importance of the different limiting factors as just discussed depends on the actual amplifier application. One application is the replacement of electronic regenerators; in such cases the amplifiers are placed at a considerable distance from the transmitter and receiver and they are denoted as in-line amplifiers (Figure 15.3a). The optical amplifier may also be used to increase the transmitter power by placing an amplifier just after the transmitter (Figure 15.3b). Such amplifiers are called power amplifiers, or boosters, because their main purpose is to boost the transmitted power. Long-distance systems may also be improved by the inclusion of the so-called preamplifiers, which are placed just before the receiver (Figure 15.3b). Effectively, they improve the sensitivity of the receiver, so that lower-power signals can be detected, thus increasing the distance over which a signal can be transmitted. Furthermore, optical amplifiers can be used in local area networks in which they can compensate for distribution losses (Figure 15.3c). Thereby, the number of nodes in the networks may be increased. Amplifiers can also compensate for the loss of various components, thus enabling the use of lossy optical components such as optical add–drop multiplexers and switches (Figure 15.3d).

It is interesting to note that the most important parameter of an amplifier, its gain, is not a limiting factor, at least not for EDFAs in optical communication systems. They can easily reach a gain of 10,000 (40 dB). However, because of noise limits and nonlinear limits in the transmission system,

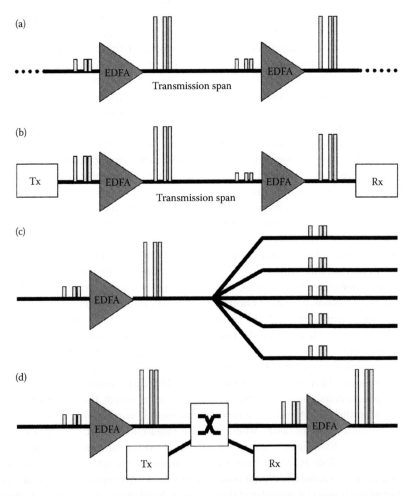

Figure 15.3 Amplification of optical signals with erbium-doped fiber amplifiers (EDFAs) in various optical transmission fibers. (a) In-line amplifiers are used periodically in a link comprising several fiber spans. (b) A booster amplifier is used after a transmitter (Tx) and a pre-amplifier is used before a receiver (Rx) in a simple repeaterless link. (c) A signal is amplified before being split up in a distribution network. (d) Amplifiers enable the use of optical add–drop multiplexers and other lossy optical components. Similarly, signals can be amplified in high-power EDFAs to the point where nonlinear optical effects can be utilized, e.g., for all-optical regeneration.

there is no need for such high gain in a well-designed system. Thus, EDFAs can be designed to reach the gain required in an optically amplified system relatively easily.

15.3 DIFFERENT CLASSES OF OPTICAL AMPLIFIERS

The most important factors that limit the performance of practical optical amplifiers have now been identified. The next task is to investigate the technologies available for obtaining real-world amplifiers. Basically, six different ways of obtaining optical amplification are considered: fiber-Raman and fiber-Brillouin amplifiers, semiconductor optical amplifiers (SOAs), RE-doped fiber amplifiers, RE-doped waveguide amplifiers, and optical parametric amplifiers (OPAs). The aim of the following description is to present the basic physical properties of these amplifiers. The properties of the different amplifiers are summarized in Table 15.1.

15.3.1 Fiber-Raman amplifiers

Raman amplification is currently a very active field of research. The fiber-Raman amplifier works through the nonlinear process of stimulated Raman scattering (SRS) occurring in the fiber itself. Like most optical amplifiers, a fiber Raman amplifier must be optically pumped, preferably by a semiconductor laser diode (LD) or a fiber laser. As early as 1962, it was observed that for very intense pump waves a new phenomenon of SRS can occur in which the Stokes wave grows rapidly inside the medium such that most of the pump energy appears in it. SRS is an interaction between light and the vibrational modes of silica molecules that converts a fraction of the incident power from one optical beam (the pump) to another optical beam (the signal, or Stokes wave) at a frequency down-shifted by an amount determined by the vibrational modes (phonon energy) of the medium. A phonon is a quantized vibration of the surrounding medium. SRS involves the so-called optical phonons. Optical phonons of a given energy exist with a broad range of momenta, so that momentum conservation (phase matching) is guaranteed. As a consequence, Stokes waves are generated both co-propagating and counter-propagating with the incident beam. See, e.g., Reference [1] for a discussion of SRS.

In an amplifier context, if the signal angular frequency ω_s and pump angular frequency ω_p *are* chosen with a difference $\Omega_R = \omega_p - \omega_s$ (the Stokes shift) corresponding to the vibrational energy of the molecules, the signal may experience Raman gain via SRS. Thus, with a proper choice of pump wavelength, Raman amplification is possible at any signal wavelength (e.g., 1300 nm, 1550 nm, …), limited only by the transparency of the material. Since phase-matching is automatically fulfilled, Raman gain is independent of the pump direction. Another significant feature of the Raman gain in silica fibers is that it extends over a large frequency range (up to 40 THz) with a broad dominant peak near 13 THz with a typical FWHM width of 6 THz (Figure 15.4). This is due to the amorphous nature of fused silica in which the molecular vibrational frequencies spread out into bands overlapping each other and creating a continuum (depending on the fiber core composition). The bandwidth of a Raman amplifier would normally be narrower than the intrinsic Raman gain, but it can be extended by using multiple pumps at different wavelengths. Each pump creates Raman gain with a peak down-shifted by ~13 THz. With an appropriate choice of pump wavelengths, the Raman gain spectra of the individual pumps will overlap. Thus, the combined amplification spectrum can be significantly broader, with a low ripple. This is shown in Figure 15.5. Thus, Raman amplification bandwidths exceeding 10 THz can be obtained in silica fibers, and with careful optimization, the gain ripple can be below 0.5 dB [12]. In this kind of system there is significant stimulated Raman scattering between the pumps, that redistributes the pump power and therefore needs to be taken into account [13]. However, the pump beams can be time multiplexed in order to suppress the interaction [14]. Also SRS between signals at different wavelengths is a concern in high-power, broadband systems. For bandwidths beyond ~10 THz, the signal and pump beams would have to overlap spectrally with silica fibers. This leads to severe signal degradation, and is why the bandwidth of silica fiber Raman amplifiers is limited to ~10 THz, even though amplification can be obtained at any wavelength with the right pump.

The large bandwidth of fiber Raman amplifiers and a very large saturation power (typically 0.1–1 W) makes them attractive for optical communication systems. Also their noise properties are good,

Table 15.1 Characteristics of optical amplifiers for telecom applications, with an emphasis on commercially available amplifiers

	EDFA	EDWA	Other RE-doted amplifiers	Fiber Raman	Semiconductor	Fiber OPA	Fiber Brillouin	Comment
Amplifying medium	Er-doped fiber	Er-doped planar waveguide	Rare-earth doped fiber or waveguide	Any fiber; Lumped Raman amplifier should use optimized fiber; distributed Raman amplification uses transmission fiber	Semiconductor, e.g., InGaAsP for 1550 nm operation	Dispersion-shifted-fiber (zero-dispersion wavelength must coincide with pump wavelength)	Any fiber	30dB sufficient for telecom applications; even 10–20dB is sufficient in many cases
Status, deployment	Widely deployed in commercial systems	Commercially available but not deployed	Some amplifiers commercially available (e.g., Tm and Pr-doped fluoride fiber amplifiers); not deployed	Lumped Raman amplifiers not deployed; small-scale deployment of distributed Raman amplification	Commercially available but not deployed	Not commercially available or deployed; not active R&D area	Not commercially available or deployed; not active R&D area	For nonfiber amplifiers, fiber coupling loss degrades the noise figure
Pumping	Optical, at 980 or 1480 nm; typical pump power a few hundred milliwatts	Optical, at 980 or 1480 nm; typical pump power a few hundred milliwatts	Optical; typical pump power a few hundred milliwatts	Optical, e.g. at ~1450 nm for amplification at 1550 nm; typical pump power 1 W	Electrical; typical pump current a few hundred milliamperes	Optical, at Wavelength near signal wavelength; typical pump power >1 W	Optical, ~10 GHz above signal frequency	Generally maximum output power depends on pump power, which is high for Raman and parametric amplifiers
Gain	20–30 Db typical, 40dB readilyobtainable	20dB (typical)	20dB (typical)	15–20dB optimum for distributed Raman amplification	10–30dB (typical)	40dB obtainable; uni-directional if pump is uni-directional	Sufficient; uni-directional if pump is uni-directio	
Noise performance	Good (e.g., 4 dB noise figure)	Reasonable (noise figure typically 5–7 dB)	Varying (good in many cases)	Reasonable for lumped RFAs; excellent for distributed Raman amplification	Reasonable (6–10dB noise figure typical)	Good	Poor	
Maximum output power	From 10 dBm up to 23–27 dBm in high-power EDFAs, limited by pump power (50 dBm has been demonstrated [11])	Typically 10–15 dBm	Varying	25 dBm	8–17 dBm (typical)	Typically up to 30 dBm	Depends on pump power	

(*Continued*)

Table 15.1 (Continued) Characteristics of optical amplifiers for telecom applications, with an emphasis on commercially available amplifiers

	EDFA	EDWA	Other RE-doted amplifiers	Fiber Raman	Semiconductor	Fiber OPA	Fiber Brillouin	Comment
Operating wavelengths	1480–1620 nm (C-band 1528–1562 nm, L-band 1570–1620 nm, S-band 1480–1520 nm	1530–1560 nm	Tm, 1450–1520 nm (S-band); Pr, 1300 nm (2nd telecommunications window); Nd, 1300 nm	Any wavelength, determined by pump wavelength and fiber composition	Any wavelength (determined by bandgap of given material and structure); 1200–1650 nm demonstrated	Any wavelength (near zero dispersion wavelength of fiber)	Any wavelength, determined by pump wavelength and fiber composition	
Bandwidth	1–10 THz(single band) 8–18 THz (split band)	1–4 THz	1–2 THz	~3 THz with single-wavelength pumping, up to, 10 THz with broadband pumping; 20 THz possible with tellurite fibers	Up to, ~10 THz	Typically 1–2 THz (25 THz with pulsed pumping)	Intrinsically below 0.1 GHz	
Polarization dependence of gain	Negligible	Small with appropriate waveguide design	Negligible in fibers (small in appropriately designed planar waveguides)	Small with polarization multiplexed or scrambled pump, otherwise high	Small with appropriate waveguide design (e.g., 0.5 dB)	Intrinsically high, but smaller with a polarization multiplexed or scrambled pump	Small with polarization multiplexed or scrambled ump, otherwise high (	
Gain response time	0.1–1 ms	~1 ms	Varying, 0.1 ms typical	Instantaneous	Sub-nanosecond	Instantaneous	Nanosecond	Depends on gain medium, waveguide design, and operating conditions
Extractable energy stored in gain medium	Large	Large	Large	None	Small	None	None	The small energy stored in SOAs and their short response time leads to signal distortion in saturation
Transient behavior	Transients potentially severe because saturated gain, often suppressed with electronic pump power control	Gain often not saturated, in which case transients are small	Varying	Gain normally not saturated, transients therefore small	OK in special designs (gain-clamped, "linear", amplifiers)	Gain normally not saturated, transients therefore small		
Gain efficiency	High (a few dB Mw⁻¹ in C-band)	Reasonable	Varying	Low(up to ~50 dB·W⁻¹ for distributed Raman amplification in standard single-mode fiber	High	Low	High	Gain efficiency less important for high-power devices
Power conversion efficiency	Good	Often poor	Varying	Good	Good	Good	Good	Important for high-power devices

(Continued)

Table 15.1 (*Continued*) Characteristics of optical amplifiers for telecom applications, with an emphasis on commercially available amplifiers

	EDFA	EDWA	Other RE-doted amplifiers	Fiber Raman	Semiconductor	Fiber OPA	Fiber Brillouin	Comment
WDM capability?	OK	OK	OK	OK	OK in special designs (gain-clamped, "linear", amplifiers)	OK	No	Besides broadband gain, unsaturated operation or slow gain dynamics (preferably with transient control) required for WDM amplification
Fiber coupling loss	Low (fusion splicing)	Moderate	Low (splicing)	Low (fusion splicing)	High	Low(fusion splicing)	Low (fusion splicing)	
Typical length of waveguide (or fiber)	1–50m	0.1–1 m (long waveguides in coiled geometry to reduce device size)	1 m	Kilometer	Millimeter	0.1–1 km (typical)	<1 km	Depends on concentration for RE-doped devices; nonlinear processes occur faster at higher powers, allowing for shorter devices
Cost	Low, and low relative to performance ($1000 EDFAs advertised)	Potentially very low	High	Relativel	Potentially very low (currently advertised below $1000)	High		Low-cost simpler amplifiers geared towards metro; highend EDFAs and Raman for long haul
Example of suppliers	Numerous	Inplane Photonics NKT Integration Teem Photonics	NEL	IPG Photonics Licomm MPB Communications Xtera (distributed Raman amplification)	Alcatel Genoa Kamelian Opto Speed	None	None	

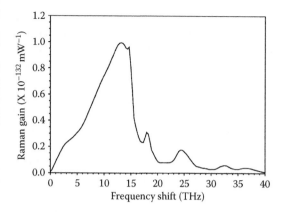

Figure 15.4 Raman gain in fused silica for parallel polarizations, with a pump wavelength of 1064 nm. The Raman gain coefficient scales approximately inversely with the pump wavelength. (After Stolen, R. H., *Proc. IEEE.*, 69(10), 1232–1236, 1980.)

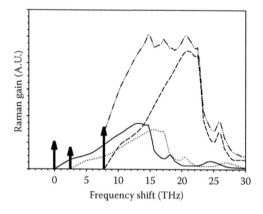

Figure 15.5 Schematic illustration of how several pumps (three in this case) at different wavelengths combine to create a broadened Raman gain.

with a noise figure close to 3 dB being obtainable. While stimulated Raman scattering leads to gain, the pump beam will also cause spontaneous Raman scattering. The scattered light is Stokes shifted by the Raman frequency shift, so it will overlap spectrally with the signal. Since spontaneous Raman scattering is a stochastic process, it effectively adds noise to the signal, though the noise addition can be close to the 3 dB quantum limit.

SRS requires long fibers and high pump powers because it is a weak process. Starting with the Raman gain coefficient shown in Figure 15.4, the actual Raman gain can be obtained by dividing the gain coefficient by the effective area (e.g., 50 μm²) and multiplying it by the effective fiber length (up to ~20 km). Despite the weakness of SRS, with the present development of powerful solid-state pump lasers, sufficient power levels are obtainable from fiber lasers as well as directly from laser diodes. A pump power of 1 W can be enough to generate a gain of up to 50 dB in standard single-mode fiber. For amplification at 1550 nm, a pump wavelength of ~1450 nm is preferable, and suitable laser diodes have been developed at that wavelength. In addition, since SRS will be more efficient if the pump and signal beams are more tightly focused, efficient Raman amplification is possible in fibers shorter than 10 km with optimized designs. In highly nonlinear fibers such as small-core holey fibers, device lengths can be of the order of 100 m.

The weak nature of SRS is actually an advantage for distributed Raman amplification: since Raman gain takes place directly in a silica fiber, it can also occur in the transmission fiber itself. Thus, it is possible to turn the transmission fiber itself into an amplifier, and with an appropriate pump source, it is possible to distribute the gain over several tens of kilometers of transmission fiber. That way amplification can be realized many kilometers away from the pump injection point (often at the receiver end of a transmission link). This leads to a better SNR, since the signal starts to be amplified before it has experienced the full loss of the transmission link (in the absence of Raman gain). Most of the recent record-breaking transmission experiments have indeed utilized distributed Raman amplification in the transmission line [16].

Raman amplification is limited by Rayleigh scattering. At too high gain multiple Rayleigh scattering leads to multi-path interference, which degrades the signal. Optimum distributed Raman on–off gain is in the range 15–20 dB.

Polarization is another concern for Raman amplification. The gain in a polarization orthogonal to the pump polarization is only a few percent of the gain for parallel polarization [17–19]. Thus, the polarization properties must also be taken into account when evaluating the Raman gain in a fiber [19]. Because polarization control is not normally employed in optical communication systems, this property introduces an additional complexity. In practice, the polarization dependence can be mitigated with pump polarization scrambling or multiplexing.

The numbers quoted so far refer to silica and germanosilicate fibers. Recently, however, Raman amplifiers have been demonstrated in tellurite fibers. Tellurite is a nonsilica glass with a significantly broader bandwidth. An amplification bandwidth of 20 THz, from 1490 to 1650 nm, has so far been demonstrated, which is a record for a single-band cw amplifier [20].

15.3.2 Brillouin amplifiers

Because of severe limitations such as narrow linewidth, Brillouin fiber amplifiers are not practical and will not be treated henceforth. See, e.g., Reference [1] for details on Brillouin amplifiers.

15.3.3 Semiconductor optical amplifiers

The SOA is fabricated in a material and structure very different from the silica transmission fiber. It incorporates a waveguide within a semiconductor gain medium. For 1550 nm amplification, InGaAsP SOAs are often used. The major attraction of SOAs is the potential for low cost and the direct electrical pumping.

Since an SOA is not a fiber device, the coupling of light from fiber to SOA and vice versa is difficult. There are inevitably significant coupling losses which degrade the noise figure. Furthermore, intrinsically, the SOA itself also will experience relatively large feedback due to reflections occurring at the cleaved end facets (32% reflectivity due to refractive index differences between semiconductor and air), resulting in sharp and highly disadvantageous gain reduction between the cavity resonances of the Fabry–Perot resonator. Therefore, it is generally necessary to design travelling wave (TW) type SOAs by suppressing the reflections from the end facets; a common solution is the inclusion of antireflection coating of the end facets. It turns out that to avoid the amplifier bandwidth being determined by the cavity resonances rather than the gain spectrum itself, it becomes necessary to require that the facet reflectivities satisfy the condition [7]:

$$G\sqrt{R_1 R_2} < 0.17. \qquad (15.6)$$

Here, G is the single-pass amplification factor, and R_1 and R_2 are the power reflection coefficients at the input and output facets, respectively. For an SOA designed to provide a 30 dB gain, this condition will mean that $\sqrt{R_1 R_2} < 0.17 \times 10^{-3}$, which is difficult to obtain in a predictable and reproducible manner by antireflection coating alone. Additional methods to reduce the reflection feedback include designs in which the active-region stripe is tilted from the facet normal and introduction of a transparent window region (nonguiding) between the active-layer ends and the antireflection coated facets. Thereby, reflectivities as small as 10^{-4} may be provided, so that the SOA bandwidth can be determined by the amplifying medium rather than by the narrower resonance peaks of the cavity. Typically, 3 dB amplifier bandwidths of 60–80 nm may be obtained. Another attractive spectroscopic property of SOAs is that, with appropriate composition and design, gain can be realized over a very wide range of wavelengths from 1200 to 1650 nm (though limited to, say, 80 nm in a given device with a specific bandgap).

The gain in an SOA depends on the carrier population, which changes with the signal power and the injection current. To reiterate, the SOA is pumped electrically. By contrast, alternative optical amplifiers are optically pumped. Typical pump currents are a few hundred milliamperes. An important property is the very short effective lifetime of the injected carriers (e.g., of the order of 100 ps). This property becomes specifically relevant for multi-channel applications of SOAs, where crosstalk limitations arise. This crosstalk originates from two nonlinear phenomena: cross-saturation and four-wave mixing. The former appears because the signal in one channel through stimulated recombinations affects the carrier population and, thereby, the gain of other channels. This significant problem may only be reduced by operating the amplifier well below saturation, but this is not an easy task due to the relatively limited saturation output power of the order of 10 mW. Four-wave mixing also appears because stimulated recombinations affect the carrier number. More specifically, the carrier population may be found to oscillate at the beat frequencies between the different channels, whereby both gain and refractive index are modulated. The multi-channel signal, therefore, creates gain and index gratings, which will introduce inter-channel crosstalk by scattering a part of the signal from one channel to another.

Despite these difficulties, the so-called gain-clamped ("linear") SOAs for multi-wavelength amplification have been realized [21,22]. Because of their linear (power-independent) characteristics, channel cross talk and four-wave mixing can be avoided. For instance, a linear SOA with 10 dBm output power, up to 25 dB gain, and 8 dB noise figure is offered commercially by Genoa [23].

As with other types of optical amplifier, the stimulated emission is accompanied by spontaneous emission, which leads to noise. The noise properties of the SOA are determined by two factors. One is the emission due to spontaneous decays and the other is the result of nonresonant internal losses α_{int} (e.g. free carrier absorption or scattering loss), which reduce the available gain from g to $(g-\alpha_{int})$. The nonzero absorption then increases the noise figure. Also residual facet reflectivities increase the noise figure, via loss of signal input power and via multi-path interference. Typical values of the noise figure for SOAs are 6–10 dB.

Another undesirable characteristic of early SOAs is the polarization sensitivity, which appears because the amplifier gain differs for the transverse-electric (TE) and transverse-magnetic (TM) modes in the semiconductor waveguide structure. However, intense research in the early 1990s has reduced the problem, and SOAs with polarization sensitivity reduced to less than 0.5 dB have been reported [24]. Other methods of using serial or parallel coupled amplifiers or two passes through the same amplifier are also suggested. Although such schemes increase complexity, cost, and stability requirements, they may add attractive properties applicable in optical signal processing (e.g., in optical wavelength conversion Reference [24]).

To conclude the discussion of the SOA, it is important to note that the major drawbacks are polarization sensitivity, interchannel crosstalk, and large coupling losses. These drawbacks have been overcome to some extent, but the resulting performance is still significantly worse than that of EDFAs. Therefore, the impact of SOAs on real systems has been small so far. In favor of the SOA are the large amplifier bandwidth, the possibility of amplification at any wavelength in the range 1200–1650 nm (at present), and the possibility of monolithic optoelectronic integration, especially within the receiver, where the input signal powers are weak enough to avoid undesired nonlinearities. It should also be mentioned that the SOA itself can be used as an amplifier and a detector at the same time, because the voltage across the pn-junction depends on the carrier density, which again interacts with the optical input signal. Promising results for transparent channel drops, channel monitoring, and automatic gain control have been demonstrated [24]. Note that possible future applications of SOAs are not limited to amplification alone; they can also be used as wavelength conversion elements and optical gates [24].

15.3.4 Rare-earth-doped fiber amplifiers

Parallel to the maturation of SOAs, another development has taken place, with a revolutionary impact on optical communication systems. With a point of reference in work on RE-doped glass lasers initiated as early as 1963 [3], the first low-loss, single-mode, RE-doped fiber amplifiers (as possible useful devices for telecommunication applications) were demonstrated in 1987 [4]. Progress since then has multiplied to the extent that amplifiers today offer far-reaching new opportunities in telecommunication networks. We will in the following sections discuss some of the recent results on RE-doped fiber amplifiers.

The possible operational wavelengths of RE-doped fiber amplifiers are determined by the emission spectra of the RE ions, moderately dependent on the host material in which they are embedded. Only a few RE materials become relevant for optical communication purposes, primarily erbium, which may provide amplification in the 1550 nm wavelength band (the third telecom window), and praseodymium, which may be operated around the 1300 nm band (the second telecom window). Note that also ytterbium and thulium have shown important features in amplifier development over the past few years.

The absorption spectrum holds accurate information about the location of possible pump wavelengths that can be used to excite the RE ions to higher energy levels. From this higher energy level the ion can relax to the ground state, transferring its packet of energy either radiatively or nonradiatively. The most prominent

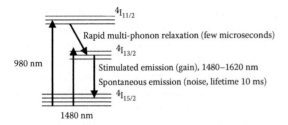

Figure 15.6 Energy levels of Er^{3+}. Each level is further split by the so-called Stark effect, as schematically illustrated. Within each level, the Stark sub-levels are in thermal equilibrium.

nonradiative de-excitation mechanism, known as multi-phonon relaxation, involves the creation of phonons. The maximum energy of a phonon is limited in any given host, for example, in silica to approximately 1100 cm^{-1} (corresponding to a frequency of 33 THz). Thus, the larger the energy gap that is to be bridged via multi-phonon relaxation, the more phonons need to be created. However, this becomes an increasingly improbable process. At energy gaps exceeding around six times the maximum phonon energy, multi-phonon relaxation becomes insignificant. Then, other processes, notably radiative decay, will dominate the relaxation process. Figure 15.6 shows the energy levels, possible pumping

wavelengths (980 and 1480 nm), and emission processes for erbium. Quite fortuitously, the phonon energy in silica is such that the 980 nm pump level ($^{4}I_{11/2}$) relaxes via rapid multi-phonon relaxation to the metastable level ($^{4}I_{13/2}$), while multi-phonon relaxation from $^{4}I_{13/2}$ is virtually absent. Instead, the decay from the meta-stable level to the lower energy level (the ground state $^{4}I_{15/2}$) is radiative. Thus, it leads to the emission of a photon, either via spontaneous or stimulated emission. Spontaneous emission always takes place, when an amplifying medium such as a collection of ions is in an excited state; therefore, spontaneously emitted light cannot be avoided in a fiber amplifier. Stimulated emission is the process that allows signal amplification to take place and, therefore, is the desired property of the fiber amplifier. The process may be explained as follows: a photon incident on the medium, with an energy equal to the difference in energy of the ground state and an excited metastable state, promotes de-excitation, with the creation of a photon that is in phase with the incident photon.

Radiative transitions within triply ionized RE ions are actually relatively weak because they are "forbidden" for reasons of symmetry (the normally dominant electrical dipole transition is not allowed for the relevant intra-4f transitions). Because of the weak nature, the involved lifetimes in the upper

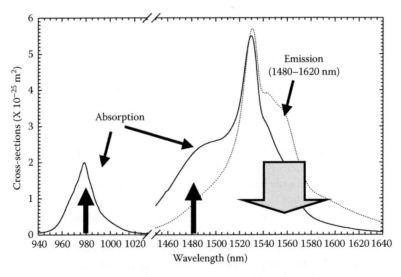

Figure 15.7 Absorption cross-sections (from the ground state) and emission cross-sections (from the metastable state) for Er^{3+}. Possible pump wavelengths and emission bands are indicated.

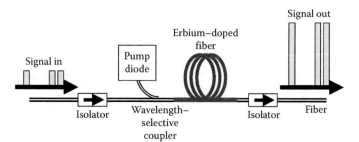

Figure 15.8 Schematic of an erbium-doped fiber amplifiers.

laser level (a measure of the spontaneous decay) are many orders of magnitude larger than the non-radiative lifetimes of the higher energy levels, and many orders of magnitude larger than the radiative lifetime of states that can decay via electric dipole transitions (typical, e.g., for transition metals). In Er^{3+}, the upper-level lifetime is typically 10–14 ms.

The absorption and emission cross-sections for Er^{3+} are shown in Figure 15.7. Their size is the measure of the probability for a photon to interact with an Er^{3+} ion in the ground state or metastable state. They are wavelength dependent, and allow absorption and stimulated emission in Er^{3+} to be described quantitatively.

EDFAs are commercially available and are widely deployed. For this reason, we will describe the EDFA in some detail.

EDFAs have revolutionized optical communication by removing the loss limit from optical transmission systems, enabling simultaneous amplification of a large number of optical channels, and facilitating the use of lossy components, as well as nonlinear components that only work at high powers. Hence, more complex links and networks with vastly higher capacity are made possible. As shown in Figure 15.7, EDFAs operate in the third transmission window around 1550 nm. Figure 15.8 illustrates a simple EDFA. An input signal is launched into the erbium-doped fiber via an isolator and a wavelength-selective coupler. A pump beam from a pump diode is combined with the signal beam in the coupler, and also launched into the erbium-doped fiber. There, it is absorbed by the erbium ions, which thereby are excited to the metastable state. Thus, the erbium-doped fiber can amplify the signal via stimulated emission. The amplified signal exits the amplifier

through a second isolator. The isolators only transmit light in one direction, and protect the amplifier from external feedback from reflections. Though the isolators and pump diode are discrete components, they are all fiber pigtailed so that the amplifier can be fusion-spliced together.

Real EDFAs are more complicated than the simple one shown in Figure 15.8. They often consist of two amplifier stages, and may incorporate a gain-equalizing filter (Figure 15.9). Furthermore, the input and output signal powers are monitored, and the pump power is controlled to maintain a constant gain. The pump power control must be fast enough to suppress transients that can arise in amplified systems and destroy them. Telemetry systems need to be implemented, too. For demanding applications, the erbium-doped fiber may be temperature controlled.

An EDFA may be fabricated using a silica glass host, and semiconductor pump sources have been successfully used to pump amplifiers in the 800, 980, and 1480 nm absorption bands. In addition, pumping at shorter wavelengths is also possible, but less interesting due to the lack of practical pump sources [25]. Basic differences exist between the application of the three mentioned pump choices. First, it should be noted that amplification occurs according to a three-level scheme when 800 or 980 nm pumping is applied, but the erbium ion works as a two-level system when 1480 nm pumping is used. Furthermore, 800 nm pumping is much less efficient than 980 nm pumping due to a pronounced excited-state absorption (ESA) of pump photons [25]. Because of these differences, today only 980 and 1480 nm pumping are considered in practical system applications. In cases, where very low noise figures are required, 980 nm pumping is the preferred choice, since the three-level nature of the

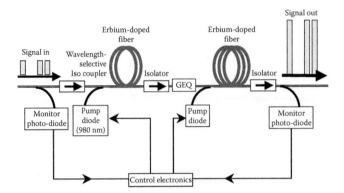

Figure 15.9 Two-stage erbium-doped fiber amplifier with gain-flattening filter (GEQ) and control electronics.

system makes excitation of essentially all erbium ions (full inversion) possible, resulting in a noise figure very close to 3 dB. For 1480 nm pumping only around 70% of the erbium ions may be excited to the upper laser level. Ions in the ground state absorb the signal and thus degrade the noise figure. Noise figures down to 5 dB are generally obtainable with 1480 nm pumping. However, 1480 nm pumping still provides highly efficient amplifiers.

EDFAs readily provide small-signal gain in the order of 30–40 dB, and values as high as 54 dB have been achieved [26]. The EDFA also provides high saturation powers, and signal output powers of more than 500 mW have been demonstrated; multi-watt output power is even possible in cladding-pumped, ytterbium-sensitized, EDFAs [11]. In such power amplifier applications, 80% or more of the pump photons may be converted to signal photons, indicating the high efficiency of the EDFA. A 30 nm bandwidth of the EDFA may easily be obtained, and is available in commercial amplifiers.

The long lifetime in the upper laser level makes the EDFA an excellent energy reservoir, and it is important to note that, with the power levels used in modern optical communication systems, one signal pulse will only interact with a very small fraction of the erbium ions within the EDFA. The following pulse is, therefore, very unlikely to interact with the same erbium ions, and it will remain unaffected by other pulses. For this reason, in practice no crosstalk will be seen in fiber amplifiers. The EDFA is also essentially insensitive to polarization variations of the signal, and only systems with cascaded amplifiers have lately demonstrated degradation due to polarization effects. A typical

gain variation of 0.1 dB was found [27], when the polarization state of the signal is switched between orthogonal linearly polarized states. Finally, it should be mentioned that since the EDFA is an optical fiber itself, very low coupling losses (comparable to ordinary splice losses) are involved in the application of EDFAs.

It should be stressed that EDFAs work very well, and well-engineered EDFAs are available commercially. Performance-wise, EDFAs can provide more power, and more gain, than what is required in a telecom system. The noise figure can be very near the quantum limit of 3 dB of traditional (phase-insensitive) amplifiers. There is essentially no polarization dependence. They are very efficient. In these aspects, there is no need, or it is fundamentally impossible, to improve the EDFA.

EDFAs with larger bandwidth would arguably be desirable. Most EDFAs, and all of those used in real systems, are made with silica fibers, often Er-doped aluminosilicate. Therefore, this article is focused on silica EDFAs. Other types of EDFA, made, e.g., with fluoride, tellurite, or bismuth glass may offer some performance advantages, especially in terms of bandwidth. Unfortunately fibers from such glasses are vastly inferior to silica fibers in properties such as durability and strength, and are much more difficult to fabricate and splice. Furthermore fluoride glasses are hygroscopic. In fact, silica EDFAs have been engineered to the point where they outperform other types of EDFA in all important respects, and the over 30 nm bandwidth available in typical broadband commercial silica EDFAs is sufficient for the vast majority of today's transmission systems. Furthermore bandwidths of up to 48 nm have been demonstrated in an antimony silicate EDFA operating at around 1550 nm [28]. This bandwidth is still not quite as large as that of the SOA, but the bandwidth can be extended if one considers other amplification bands: during the first years of the EDFA research and commercial employment, the primary interest was focused on the spectral band from ~1530 to ~1560 nm, because limited lengths of erbium-doped fibers allow for very efficient amplification in this frequency range, and because this range coincides with the loss minimum of silica fibers. Hence, this band has been denoted as the "conventional" band or just C-band. This use of names naturally indicates that other bands have emerged. Indeed, differently designed EDFAs allow for amplification in the 1570–1620 nm wavelength range, where the loss of silica fibers is still quite low. Such EDFAs are generally said to operate in the long-wavelength or L-band. For example, Codemard et al. [29] report an unflattened amplification band from 1555 to 1620 nm in a cladding-pumped ytterbium-sensitized phosphosilicate EDFA. L-band amplification can be realized by operating the EDFA at a low inversion level (with only a small fraction of the Er ions excited). This leads to a ground-state absorption in the C-band, which suppresses gain there. At longer wavelengths, the transition becomes quasi-four level (with insignificant ground-state absorption), so a low but finite gain can be realized even with a low inversion level. Long fibers must then be used in order to reach a sufficient gain.

Amplifiers operating on the short-wavelength side of the C-band (from around 1450 to 1520 nm) are defined by the International Telecommunications Union (ITU) to belong to the class of S-band amplifiers. Sometimes the range 1450–1480 nm is called the S^+-band. Transmission in the S-band, at wavelengths between the second and third telecom windows, is possible due to the low loss at those wavelengths in newly developed low-OH silica fibers. High-gain EDFA have also been demonstrated in the S^+-band, down to 1480 nm [30]. In order to realize an S^+-band EDFA one must operate at a high inversion level in order to bleach the large erbium absorption in the S^+-band. Furthermore, a distributed filter is required to suppress the C-band gain, which otherwise saturates the amplifier. Note that S^+-band EDFAs are generally considerably more complicated than C- and L-band EDFAs. All these types of amplifier have been realized with silica fibers.

Thulium-doped fiber amplifiers (TDFAs) are in fact dominating in the S-band while S-band EDFAs are a later development. Typically, TDFAs provide amplification in the 1450–1480 nm window [31], though TDFAs gain-shifted to longer wavelengths within the S-band have been demonstrated. Thus, TDFAs can access the whole S-band. Another option for extending the gain to longer wavelengths is to cascade a TDFA with a discrete Raman amplifier, e.g., to a total bandwidth of 40 nm [32]. Still, compared to (C-band) EDFAs, TDFAs have several disadvantages. The energy-level diagram of triply ionized thulium ions (Tm^{3+}) is quite complicated, and many levels play important roles in S-band TDFAs. One problem is that the upper laser level is quenched in silica fibers by multi-phonon relaxation. Therefore, nonsilica glasses with lower maximum phonon energy such as fluoride and, recently, tellurite glasses are used instead. Another problem is that the lower laser level needs to be depopulated in order to maintain population inversion and avoid the so-called self-termination. Because of this, complicated pumping schemes, including two-wavelength pumping schemes, are employed. Nevertheless, TDFAs are commercially available, and high-gain (20–30 dB), high-output-power (e.g., 20 dBm), and low-noise operation (~5 dB noise figure) has been demonstrated.

Though transmission at around 1550 nm is currently the dominating technology, historically, large-scale deployment of systems in the second transmission window at 1300 nm predates this development. As a result, there has been a large demand for amplifiers at 1300 nm. Unfortunately, RE-doped fiber amplifiers for the second transmission window at 1300 nm have to deal with a number of difficulties, and their applicability is quite limited. The neodymium-doped fiber amplifier is limited by excited-state absorption to wavelengths that are too long (around 1340 nm and beyond) for real systems (at ~1310 nm), and so the best RE candidate in the 1300 nm wavelength band is the praseodymium-doped fiber amplifier (PDFA). The primary difficulty is that also praseodymium has to be placed in a host material other than silica, with a lower phonon energy, since otherwise, the upper energy level of the transition in question will be quenched by multi-phonon relaxation, and no net gain will appear [25]. Until now the most promising host materials have been different compositions of fluoride glasses, with problems such as low strength and hygroscopy.

Another problem is that the 1300 nm amplifiers normally have lower pump efficiency than EDFAs. However, pump efficiency as high as 4 dBm·W^{-1} for praseodymium-doped amplifiers is shown together with noise figures in the 5–8 dB range.

15.3.5 Rare-earth-doped waveguide amplifiers

RE-doped waveguide amplifiers are only fundamentally different from fiber amplifiers in the respect that the RE material is embedded in a planar optical waveguide and not a fiber. This means at first glance that only another waveguide geometry has to be considered. However, several important factors have to be considered in the design of RE-doped waveguide amplifiers. First, the background loss is several orders of magnitude larger in a planar waveguide (around 0.01–0.05 dB·cm^{-1}) than in a fiber, and second, a much higher RE-dopant concentration is necessary to achieve high gain in the relatively short planar waveguides. This raises the problem of energy transfer between the RE ions and results in a lower amplifier efficiency. Not surprisingly, the focus has been on the erbium-doped planar waveguide amplifiers in silica, and integration with a 980/1530 nm wavelength division multiplexer was presented early in the development [33]. In this component a maximum gain of 13.5 dB for a 600 mW pump power was shown, but gain values up to 20 dB have been demonstrated in other amplifiers. Recent results by Hübner et al. [34] also demonstrate that boundaries between active and passive regions may be fabricated with transition losses below 0.03 dB/transition and back reflections suppressed more than 70 dB. However, it is not realistic to expect that the RE-doped waveguide amplifier will completely replace EDFAs as high-performance amplifiers in optical communication systems, but their advantages have to be found in the ability for integration of several functionalities, e.g., splitting and amplification on the same component. In addition, planar waveguide amplifiers can be realized in a wider range of materials than fibers. We may consider host materials such as lithium niobate, which is electro-optic, and new applications besides optical communication. Furthermore, Nd-doped planar waveguides in fluoro-aluminate glass remain a possibility for 1310 nm amplification [35]. We

will return to more detailed descriptions of properties of Er-doped planar waveguide amplifiers and high-concentration doping in the following sections.

15.3.6 Optical parametric amplifiers

An OPA utilizes a so-called wavemixing process, in which pump photons at one wavelength are converted directly to signal photons at another wavelength through a lossless nonlinear interaction in a medium. At the same time, idler photons are created in an idler beam at yet another wavelength to satisfy requirements on energy and momentum conservation. It is referred to as a parametric process as it originates from light-induced modulation of a medium parameter such as the refractive index. Its origin lies in the nonlinear response of bound electrons of the material to an applied optical field. Optical parametric amplification is traditionally realized via three-wave mixing (or three-photon mixing) in electro-optic materials such as lithium niobate with a second-order nonlinearity. However, it is also possible in media without a second-order nonlinearity, such as silica [36,37]. In this case, a third-order nonlinearity (the nonlinear refractive index) leads to amplification. This is now a four-photon process (i.e., a four-wave-mixing process), involving two pump photons, a signal photon, and an idler photon. Two pump photons, from a single or from two different pump beams, are annihilated, and a signal photon and an idler photon are created. Thus, the idler becomes an image of the signal. This allows OPAs to be used for signal wavelength conversion. Because of energy conservation, the signal and idler photons are created at optical frequencies located symmetrically on opposite sides of the frequency (or frequencies) of the pump beam(s). Though four-wave mixing in silica is intrinsically a much weaker process than three-wave mixing in electro-optic materials, the low loss (allowing for long interaction lengths) and tight modal confinement of silica fibers enhances nonlinear interactions and thus compensates for the intrinsic weakness of the four-wave mixing process.

Interestingly, OPAs can be configured both as conventional phase-insensitive amplifiers as well as phase-sensitive ones [38]. In the case of a phase-sensitive fiber OPA, powerful pump as well as idler beams need to be injected into the amplifier. The pump and idler beams interact to create gain for a signal beam of a particular optical phase. Any signal beam with opposite phase would be attenuated, in a process in which a signal and an idler photon combine to create two pump photons. The difficulties in tracking and controlling optical phases are major practical drawbacks of phase-sensitive OPAs.

There is no idler beam injected into a phase-insensitive OPA. Instead, the idler is created in the four-wave mixing process, with an optical phase that depends on the phase of the signal. Because of this, the signal can be amplified independently of its phase, and there is no need to control the optical phase of any beam. Nevertheless, there are many challenges with a phase-insensitive OPA (which also apply to phase-sensitive ones). The main difficulty is that phase matching of the interacting pump, signal, and idler waves has to be maintained along the fiber. This means that fibers with tailored dispersion properties must be used, and the pump wavelength must be near the zero-dispersion wavelength of the fiber. Four-wave mixing depends on the polarization of the interacting waves, so in a typical system without control of signal polarization, the pump beam should be unpolarized, polarization scrambled, or polarization multiplexed. Furthermore, a high-pump-power requirement is a serious limitation. Still, for example, the pump power needed to reach a certain gain with OPAs may be less than half of what is needed in a Raman amplifier. Thus, an OPA has a higher gain efficiency, though its power conversion efficiency may be lower. In addition, an OPA does not require long fibers as a Raman amplifier does. Phase-insensitive fiber OPAs were first demonstrated with 49 dB peak fiber gain for 32.2 dBm of launched pump power [36]. The amplification bandwidth was approximately 10 nm on both sides of the pump wavelength of 1562 nm (parametric gain occurs symmetrically, in wavelength ranges on both sides of the pump wavelength in a phase-insensitive parametric fiber amplifier). The noise figure was inferred to be similar to that of an EDFA. The bandwidth of OPAs increases with pump power, and Ho et al. recently demonstrated a pulsed fiber OPA with more than 200 nm of total bandwidth using only 20 m of highly nonlinear fiber [39]. High gain, high output power, and low noise

figure (e.g., 4 dB) have been confirmed in several recent experiments.

Parametric amplifiers have several attractions, but they are at present not developed to the level of erbium-doped, semiconductor, or Raman optical amplifiers. See, e.g., Reference [40] for a recent summary on fiber OPAs.

15.4 CHALLENGES IN FIBER AMPLIFIER APPLICATION AND DEVELOPMENT

To reiterate, the performance of EDFAs is sufficient for transmission applications, and there is not much reason to improve performance. A major challenge is instead to cut cost, e.g., by reducing the cost of pump diodes and other EDFA components. A lower cost would increase the proliferation of EDFAs in the network, and include uses for which the specifications are not so demanding. This allows for further cost reductions. Other cost-reduction strategies include a higher degree of parallelization and integration (e.g., by employing planar technology), as well as cladding pumping, both to be discussed further below.

The only performance limitation of EDFAs is its limited wavelength range. Still, the EDFA bandwidth is sufficient for today's systems. Nevertheless, the limited bandwidth may well become a bottleneck some day. Since the available amplification bandwidth—as indicated—may not be extended significantly beyond the C-or L-band using established EDFA technology, and since the RE-doped S-band amplifiers still are complex (including S-band EDFAs), researchers have begun seeking alternative enabling techniques wider than those that can be managed by erbium-(or even thulium-) doped amplifiers. In this context, broadband, wavelength-agile technologies such as SOAs and Raman amplifiers, discussed above, are attractive options. Also from a cost perspective, SOAs are potentially very attractive, but they still require further development before they can be widely deployed.

Distributed Raman amplification directly in the transmission line can provide the best overall system performance, and offer the longest reach of optical networks. Still, the cost, dominated by the pumping system, must be reduced before Raman-amplified systems will be widely deployed.

15.5 CLADDING-PUMPING

Early fiber amplifiers were pumped by large laser systems requiring many kilowatts of electrical power, with an overall EDFA power conversion efficiency ("wallplug efficiency") of the order of one in a million. EDFAs only became practical with the realization of suitable semiconductor laser diodes for pumping. These pump-LDs improved the wall-plug efficiency to the per cent level and above. In most EDFAs, the optical pump beam is injected into the core, which implies that a (transversally) single-mode pump source must be used. Though single-mode pump diodes are being developed rapidly, they are still limited in power to less than a watt. While an EDFA is typically pumped by more than one LD, this power limitation still means that, in practice, traditional core-pumped EDFAs are limited in power to less than 1 W (i.e., 30 dBm). This power is more than enough for most applications in optical communications, but higher powers can be used in free-space communication and in distribution networks in which the transmitted power (e.g., 1 W from an EDFA) is divided and distributed in a large number of different fiber branches.

Cladding-pumping is a method of overcoming the power limitation of core-pumped EDFAs. Cladding-pumped fiber devices utilize multi-mode pump-LDs rather than single-mode ones. Multi-mode diodes are more powerful and often cheaper than single-mode ones. They come in many shapes for a wide range of power ranges, up to several kilowatts, but the simplest, lowest-cost multimode diode is the broad-stripe diode, typically generating up to a few watts of output power.

Standard RE-doped fibers cannot be cladding-pumped. Instead, special fibers known as double-clad fibers are used. These comprise a secondary, multi-mode waveguide into which the multi-mode pump beam can be launched. Typically, this structure is realized with a secondary cladding ("outer cladding") surrounding the original cladding ("inner cladding"). The outer cladding has a lower refractive index than the inner cladding, so that the pump beam can be guided within the inner cladding via total internal reflection. There is still a core for guiding signal light. It resides in the inner cladding and is similar to the core of a core-pumped EDFA, i.e., it is doped with erbium (or another RE) and has a higher refractive index than

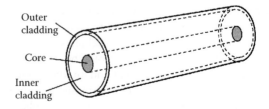

Figure 15.10 Schematic drawing of a double-clad fiber, comprising an RE-doped core, and an inner and an outer cladding.

the surrounding inner cladding (Figure 15.10). A single-mode core is preferred for amplifiers in optical transmission systems. The pump light still reaches the RE-doped core and excites the RE ions, so that gain is created and the signal can be amplified.

Thus, a cladding-pumped fiber amplifier has a potential not only for higher power, but also lower cost, than a core-pumped one. However, there are challenges with this technology. While the pump has a high power, its intensity (power density) is much lower than it is with core pumping. This is a problem since erbium ions require a minimum pump intensity to reach gain. A related problem is that because the overlap between the pump beam and the RE-doped region is small, the pump absorption is low. Since the intrinsic absorption of Er^{3+} is already quite low, and since the maximum Er^{3+} concentration is relatively limited, this can lead to excessively long Er-doped fibers, e.g., 100 m or even more if special care is not taken. For this reason a small inner cladding is preferred for EDFAs, significantly smaller than 100 µm diameter [41], but this makes it more difficult to launch the pump light into the fiber. Another proposed [42] and recently demonstrated [43] design modification is to dope the erbium in a ring around the core, since this reduces reabsorption and signal-induced gain-compression. The most common approach, however, is to co-dope (sensitize) the erbium with ytterbium [44]. In an ytterbium-sensitized EDFA, pump photons are initially absorbed by Yb^{3+} ions, which are thus excited. Then, the energy is transferred nonradiatively from Yb^{3+} ions to Er^{3+} ions, resulting in de-excitation of the Yb^{3+} ions and excitation of the Er^{3+} ions. Ytterbium has a larger absorption cross-section than erbium, as well as a much broader absorption band, from 910 to 980 nm. Significantly, the broad absorption can alleviate the requirement of pump wavelength stabilization via temperature control of the pump-LD. Furthermore, Yb^{3+} is comparatively resistant to a reduction of efficiency at high Yb concentrations (concentration quenching), and can therefore be incorporated in much higher concentrations than Er^{3+}. In total, this means that a high gain can be obtained with much lower pump intensities than without sensitization, and even in fibers with a large inner cladding, a fully adequate pump absorption of several decibels/meter can be reached. However, so far, efficient Yb-sensitized EDFAs require a host glass with a high phosphorus content. Unfortunately, this leads to a relatively narrow gain spectrum, so Yb-sensitized EDFAs have not been able to cover the whole C-band. This is a very important aspect. Ytterbium-free EDFAs do not suffer from this problem—their gain spectrum even defined the extent of the C-band. But, because of the poor pump absorption, Yb-free cladding-pumped EDFAs are relatively inefficient and of low output power. With cladding-pumping, the L-band actually appears more attractive than the C-band, for Yb-sensitized EDFAs because the gain spectrum is wider there, and for Yb-free EDFAs because the pump absorption is higher.

The pump launch is another challenge with cladding-pumped EDFAs. Wavelength-selective fused fiber couplers used to combine the signal and pump beams in core-pumped EDFAs do not work with multi-mode beams. Micro-optic couplers do work and have been used for cladding-pumping. However, these rely on lenses and wavelength-selective reflectors with free-space (unguided) beam propagation to combine the beams, and require precise assembly and alignment. They are much more complex and therefore more costly than fused fiber couplers. Their power-handling capability is also worse, which is a distinct disadvantage for cladding-pumped EDFAs designed for high powers.

Instead of launching the pump through the end of the erbium-doped fiber, together with the signal, one can launch it through the side of the fiber. We will describe three such side-pumping schemes.

One scheme utilizes several (typically three) parallel fibers combined into a common fiber assembly. The fibers are embedded in a polymer coating with a low refractive index so that the fibers themselves can guide pump light. Furthermore, the fibers are in optical contact with each other, meaning that pump light can couple from one fiber into another. One (or several) of the fibers contain a core doped with erbium, typically sensitized with ytterbium. Figure 15.11 shows an example of such a multi-fiber assembly, known as a GTwave fiber. In the GTwave fiber, the fiber with a doped core can be spliced to any fiber carrying a signal. The signal is then launched into the erbium-doped core and amplified in the GTwave fiber. The fibers without a core constitute pump fibers, to which pigtailed pump diodes can be spliced, as many as one to each end of each pump fiber. Once in the pump fiber, the pump light will couple over to the Er-doped fiber and excite the ions in the core.

GTwave fiber devices take simplicity to the limit in that no other components, bar pump-LDs, are required to realize an EDFA (though practical amplifiers do include isolators and, optionally, gain-flattening filters). Another attractive feature is the possibility to include several parallel Er-doped fibers in a multi-port amplifier for parallel amplification. Amplification in eight parallel fibers has been demonstrated [45]. Single-channel GTwave amplifiers can reach well over 1 W of output power with high gain and noise characteristics typical for EDFAs. Broadband versions have been realized both in the C-and L-band [46,47].

V-groove side-pumping (VSP) is another approach to cladding-pumping, that offers many of the advantages that GTwave fibers provide, with comparable performance [47,48]. An additional advantage is that VSP amplifiers can be made with any double-clad fiber. On the other hand, while

Figure 15.11 A GTwave fiber assembly consisting of a single pump fiber and a single signal fiber, surrounded by a polymer coating. Pump light (red) is launched into the pump fiber, and couples over to the signal fiber. Here, it excites the erbium in the core, leading to amplification of the signal light.

conceptually simple, the fabrication of the v-groove as well as the pump launch are relatively complex issues. A v-groove is fabricated in the side of a double-clad fiber, penetrating into the inner cladding but leaving the (erbium-doped) core intact. Then, pump light from a laser diode is launched through the opposite side of the fiber, and hits the v-groove from within the fiber. The light is then deflected off the v-groove facet by ~90°, along the fiber axis and inside the inner cladding, via total internal reflection. Thus, the pump light is launched into the inner cladding of the double-clad fiber, while signal-carrying fibers can be spliced to its ends.

Finally, multimode couplers offer a convenient and widely used way of launching pump light into a double-clad fiber through its side. For example, the tapered fiber bundle is a multi-port version of such a coupler [49].

Despite the success of GTwave, VSP, and multi-mode couplers, the difficulty of combining high output power, efficient operation, and broadband gain (over the whole C-band) in cladding-pumped EDFAs remains. High-power cladding-pumped ytterbium-doped fiber lasers emitting at 980 nm have therefore been developed as a pump source for EDFAs [50–52]. This is an alternative way of combining cladding-pumping with EDFAs that is compatible with conventional core-pumped EDFAs, including fused fiber couplers. The EDFA is powered by laser diodes, albeit in an indirect way. Thus, the high-power advantage of cladding-pumping can be realized without compromising amplification bandwidth or reliability. Cladding-pumped ytterbium-doped fiber lasers can generate several watts of output power at 980 nm in a single-mode beam.

15.6 PLANAR ERBIUM-DOPED OPTICAL AMPLIFIERS

The development of integrated optical amplifiers has been intensified due to the demand for more compact and lower-cost solutions for the DWDM systems, in which the vast majority of erbium-doped fiber-based amplifiers have found commercial use. The commercialization of other planar lightwave components (PLCs) such as AWGs, power splitters and optical switch blocks has resulted in a maturing of the PLC manufacturing with silica-on-silicon being the most widely used technology. The

next logical step in the evolution of optical components and subsystems is to combine functions realized in passive PLCs with amplifiers, realizing loss-less or amplifying components of high complexity and functionality that at the same time can be mass produced and offer the overall component size to be reduced significantly compared to the fiber-or bulk optic based counterparts.

In recent years, the development of the planar amplifiers has thus moved from the university and research labs into industry, where it is presently being commercialized in a number of different technologies such as erbium-doped silica-on-silicon, ion-exchange in erbium-doped glass, and as SOAs. At the same time, the fiber-based optical amplifiers are decreasing rapidly in size, the smallest now being less than the size of a credit card. Comparing the technologies, the silica-on-silicon technology potentially has the advantages of the erbium-doped fiber-based amplifier (low noise figure, low polarization dependence, and low gain cross-modulation). In addition, it allows monolithic integration of components known from the passive PLC technology. The main obstacles for the planar erbium-doped amplifier remain the propagation loss and the limited length of the waveguide that requires the erbium concentration to be increased with a factor of 10–100 compared to that in normal erbium-doped fiber. As we describe in Section 15.7, the efficiency of the amplifier strongly depends on the erbium concentration and at high concentration it is reduced due to the ion–ion energy transfer that is increasing rapidly with concentration and even more as the erbium ions tend to form clusters at high concentrations. This is known as concentration quenching.

The propagation loss of high-quality waveguides has been reduced to a minimum of 1–3 dB m^{-1}, which is still orders of magnitude larger than for fibers, but sufficiently low to allow the use of a long waveguide and hence a relatively low erbium concentration. The fact that the propagation loss is non-negligible, however, means that a planar erbium-doped waveguide amplifier will never be as efficient as the fiber-based amplifiers. On the other hand, the pump laser technology has led to high-power pump lasers at significantly reduced prices. This enables the less efficient planar amplifier technology to be used instead of the fiber-based technology, if it offers other advantages in terms of functionality, performance, manufacturability, cost, or size.

The focused development of the erbium-doped silica-on-silicon technology has led to planar amplifiers with sufficiently high gain and low noise figures to be applicable in commercial long-haul or metro systems. On the picture on Figure 15.12, a wafer with four PLC chips, each containing ten individual waveguide amplifiers, is shown. Each of the amplifiers consists of an erbium-doped waveguide section that is curled up in a spiral to achieve sufficient length. 980/1550 nm combiners on all amplifier inputs and outputs are included for pump multiplexing and removal of excess pump power.

These waveguide amplifiers were produced by a modified plasma-enhanced chemical vapor deposition (PECVD) process, in which the erbium precursors and other co-dopants are introduced directly into the core deposition process. This technique allows detailed control of the glass composition, and enables highly uniform glass layers to be deposited. In contrast to fabrication processes not involving etching of the waveguide core, such as the ion exchange process, the PECVD-based silica-on-silicon manufacturing process allows the erbium doping to be confined to the waveguide core, where the pump intensity is high. Hence, the erbium ions are easily inverted, leading to a low threshold pump

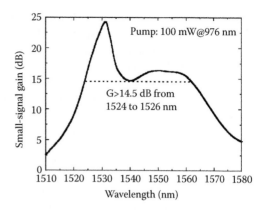

Figure 15.13 A typical gain spectrum for a waveguide chip obtained for a pump power of 100 mW at 976 nm. The results are kindly provided by NKT Integration A/S. Similar gain spectra can be obtained with EDFAs, albeit typically with a higher gain. For WDM applications, the strong gain variations would have to be equalized with a gain-flattening filter.

power for amplification. A typical gain spectrum for such a waveguide chip is shown in Figure 15.13 for a pump power of 100 mW at 976 nm. For this device, a net gain exceeding 14.5 dB has been achieved over the entire C-band. The performance of such amplifiers typically results in noise figures between 4 and 4.5 dB at low input signal powers.

As for all erbium-doped amplifiers, the performance is strongly dependent on the selected amplifier length, erbium concentration and pump power. For optimized waveguide parameters, gain in excess of 16 dB over the C-band is reached for 100 mW of pump power, and for 200 mW pump power more than 25 dB of gain can be achieved over the C-band, while maintaining the low noise figures. Such gain values will suffice for most applications, and thus represent a viable alternative to the fiber-based amplifiers.

15.7 RARE-EARTH INCORPORATION IN SILICA GLASS

Practical use of RE-doped silica for amplification purposes typically requires RE concentration levels above 10^{18}–10^{19} cm^{-3}, and elimination of quenching processes. The main sources of quenching are multi-phonon processes associated with O–H bonds and the so-called "concentration-quenching" arising from electrostatic interactions between RE ions. These effects lead to the

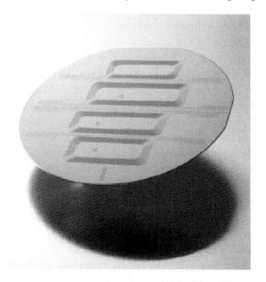

Figure 15.12 Picture of a wafer with four PLC chips, each containing ten individual waveguide amplifiers. Each amplifier consists of a curled-up erbium-doped waveguide section, and 980/1550 nm combiners on all inputs and outputs for pump multiplexing and removal of excess pump power. The picture is kindly provided by NKT Integration A/S.

requirement of a "dry" silica (i.e., with a low hydrogen content) with well-dispersed RE ions. The content of hydrogen in as-fabricated samples can be substantial depending on the fabrication process, but it may be straightforwardly eliminated by subsequent heat treatment and chlorine drying if necessary [53]. Concentration quenching is a more difficult problem to overcome. The root of the problem is that excited RE ions can de-excite each other through electrostatic dipole-dipole interactions, whose magnitudes decay with distance as r^{-6} so that the interaction between neighboring RE ions in the glass network may be substantial. This problem is particularly severe in the case of erbium, where the energy of the upper laser level ($I_{13/2}$) is close to half the energy of the $I_{9/2}$ level so that a near-resonant upconversion process can take place between close-lying Er ions. Because of its fundamental nature this problem sets an ultimate limit to the useful RE concentration in many systems such as Er-and Nd-doped silica glasses. The problem is magnified by the low solubility of RE ions in the pure SiO_2 matrix, which leads to strong RE clustering during post-production annealing, and in practice it is this effect which limits the usable RE concentrations [54,55]. This is particularly critical in the case of integrated optics, where the limitations on device dimensions put a lower limit (around 10^{19} cm^{-3}) on the Er concentrations needed to obtain useful signal amplification.

Co-doping of silica glass with aluminium or phosphorus increases the RE solubility and is therefore often used to counter RE cluster formation. Such glasses are known as aluminosilicate and phosphosilicate. Aluminium doping has proven very effective, ensuring essentially complete cluster dissolution at RE concentration levels in the range 10^{19}–10^{20} cm^{-3} (approximately 0.1–1% by weight) when the Al/RE concentration ratio is around 10 or higher. Furthermore, Al doping broadens the emission spectrum of erbium, which is highly desirable. Using P as a co-dopant a slightly higher ratio of about 15 is needed [56]. Nevertheless, with sufficient P content, phosphosilicate glass has a very high RE solubility. This is also true for phosphate glasses (free from silica), in which RE contents of 35% by weight have been realized (however with severe quenching). However, the gain spectrum of phosphosilicate EDFAs is narrower than that of aluminosilicate EDFAs.

At RE concentration levels encountered in Er-doped amplifiers (say, up to 1% by weight), the macroscopic properties of the silica glass are not strongly affected by the co-doping and integration with other glass-based components is a straightforward matter. This suggests that the mechanism underlying the cluster dissolution is a local interaction between RE and co-dopant ions rather than an overall modification of the SiO_2 network structure. Whereas the effect of P codoping is not well understood on a microscopic level there is some experimental [57] and theoretical [58] evidence for this hypothesis in the case of Al codoping. General thermodynamic arguments suggest that cluster dissolution should be favored by a high fictive temperature of the glass [58], but no systematic experimental data exist to test this hypothesis. In addition, thermodynamic arguments may not apply to glasses fabricated by low-temperature processes such as sol–gel or PECVD. In fact, for Er-doped PECVD glass it is an advantageous strategy to keep the processing temperatures as low as possible in order to limit Er diffusion and thereby cluster formation [55].

An important effect of the local environment around RE dopants is to modify the level spacings and optical matrix elements within the 4f manifold due to ligand-field effects. The site-dependent Stark shifts are an advantage for device fabrication, since they facilitate the manufacturing of amplifiers with wide and flat gain curves suitable for operation at multiple wavelengths. It has been found that the addition of Al leads to an increase of the inhomogeneous broadening by more than a factor of two compared to pure or P-doped silica [59]. The ligand fields also lead to a slight mixing of the f-orbitals with states of even parity, in effect making f–f transitions dipole allowed. As it turns out, this mechanism controls the oscillator strengths of the f-shell transitions. The optical intensities of the f-shell transitions are usually rationalized through the so-called Judd–Ofelt theory [60,61], which expresses all oscillator strengths in terms of three phenomenological parameters which are related to the matrix elements of the local electrostatic field, but which are most often used as fitting parameters. The Judd–Ofelt theory is approximate, and the underlying assumptions were only partially justified in a recent theoretical analysis of the local fields and electronic structure around an Er impurity [58]. However, in most practical cases the theory is able to describe all oscillator strengths of RE impurities to an accuracy of 10–30% with suitable parameter choices [62–64]. Most systematic investigations of the dependence of Judd–Ofelt parameters on chemical composition of silicate glasses have focused on glass containing significant amounts (> 10 mol. %) of dopants such as Na, Al, B, Mg or Ca [62,63,65,66]. Even for such mixed glasses the Judd–Ofelt parameters show only slow and moderate (within a factor of 2–3) variation with composition, suggesting that the possibilities for engineering these quantities by varying impurity concentrations is limited in the lightly doped, so-called high-silica, glasses commonly used for fiber amplifiers. Interestingly, the Judd–Ofelt parameters found in Al-codoped fiber preforms [64] differ substantially from those found in heavily doped silicates, possibly due to a more covalent character of the Er–O bonding in the weakly Al-doped silica glass.

15.8 SUMMARY

EDFAs have revolutionized optical communication systems and are widely deployed. They have removed the loss limit and thereby extended the optical reach to over 10,000 km. They have also enabled mass deployment of WDM systems, and transmission of several terabits per second has been demonstrated in laboratories. The use of lossy components, including optical switches for all-optical networks, is made possible by EDFAs. The only limitation of EDFAs is their limited bandwidth, but they have the potential to cover over 140 nm (~18 THz) in split band configurations and even the more standard 30 nm bandwidth is fully sufficient for systems considered today.

Distributed Raman amplification can provide the best system performance, but deployment is so far very limited. Several amplifier technologies (semiconductor, Raman, OPAs) can operate at any wavelength that is considered for telecoms. Bandwidths of 20 THz or more have been realized in a continuous band. The rapid progress in this field continues towards better optical amplifiers that will find even more uses in the telecom systems of tomorrow. In particular metro applications benefit from the ongoing cost reduction and the development of simple amplifiers [67].

ACKNOWLEDGMENTS

We would like to acknowledge the significant help provided by Dr Thomas Feuchter from NKT Integration A/S for his assistance in writing Section 15.6, and providing the most recent results on planar erbium-doped optical amplifiers.

REFERENCES

1. Agrawal, G. P. 2000. *Nonlinear Fiber Optics.* 3rd edn. (San Diego, CA: Academic).
2. Simon, J. C. 1983. Semiconductor laser amplifier for single mode optical fiber communications. *J. Opt. Commun.* 4, 51–62.
3. Koester, C. J., and Snitzer, E. 1964. Amplification in a fiber laser. *Appl. Opt.* 3, 1182.
4. Mears, R. J., Reekie, L., Jauncey, I. M., and Payne, D. N., 1987. Low-noise erbium-doped fibre amplifier operating at 1.54 µm. *IEEE Electron. Lett.* 23, 1026.
5. Siegman, A. E. 1986. *Lasers* (Mill Valley, CA: University Science Books).
6. Bjarklev, A. 1997. Optical amplifiers. *The Communications Handbook*, ed., J. D. Gibson (Boca Raton, FL: CRC Press) Chapter 62.
7. Agrawal, G. P. 1992. *Fiber-Optic Communication Systems* (New York: Wiley).
8. Tachibana, M., Laming, R. I., Morkel, P. R., and Payne, D. N. 1991. Erbium-doped fiber amplifier with flattened gain spectrum. *IEEE Photon. Technol. Lett.* 3, 118–120.
9. Yariv, A. 1990. Signal-to-noise considerations in fiber links with periodic or distributed optical amplification. *Opt. Lett.* 15, 01064–1066.
10. Kumar, P., Kath, W. L., and Li, R.-D. 1994. Phase-sensitive optical amplifiers. *Integrated Photonics Research IPR'94* (San Francisco, CA) 1:SaB1.
11. Nilsson, J., Jeong, Y., Alegria, C., Selvas, R., Sahu, J., Williams, R., Furusawa, K., Clarkson, W., Hanna, D., Richardson, D., and Monro, T. 2003. Beyond 1 kW with fiber lasers and amplifiers. *OSA Trends in*

Optics and Photonics (TOPS), Optical Fiber Communication Conference, Technical Digest (Washington, DC: Optical Society of America), vol. 86, pp. 686–686.

12. Emori, Y., Kado, S., and Namiki, S. 2002. Broadband flat-gain and low-noise Raman amplifiers pumped by wavelength-multiplexed high-power laser diodes. Opt. Fiber Technol. 8, 107–122.

13. Kidorf, H., Rottwitt, K., Nissov, M., Ma, M., and Rabarijaona, E. 1999. Pump interactions in a 100-nm bandwidth Raman amplifier. IEEE Photon. Technol. Lett. 11, 530–532.

14. Mollenauer, L. F., Grant, A. R., and Mamyshev, P. V. 2002. Time-division multiplexing of pump wavelengths to achieve ultrabroadband, flat, backward-pumped Raman gain. Opt. Lett. 27, 592–594.

15. Stolen, R. H. 1980. Nonlinearity in fiber transmission. Proc. IEEE. 69(10), 1232–1236.

16. Session, W. E. 2003. Raman transmission. OSA Trends in Optics and Photonics (TOPS) Optical Fiber Communication Conference Technical Digest (Washington, DC: Optical Society of America), vol. 86. pp. 326–336.

17. Stolen, R. H. 1979. Polarization effects in fiber Raman and Brillouin laser. IEEE J. Quantum Electron. 15, 1157.

18. Dougherty, D. J., Kärtner, F. X., Haus, H. A., and Ippen, E. P. 1995. Measurement of the Raman gain spectrum of optical fibers. Opt. Lett. 20, 31–33.

19. Lin, Q., and Agrawal, G. P. 2003. Vector theory of stimulated Raman scattering and its application to fiber-based Raman amplifiers. J. Opt. Soc. Am. B20, 1616–1631.

20. Mori, A., and Shimizu, M. 2003. Ultra-wideband tellurite-based fiber Raman amplifiers. OSA Trends in Optics and Photonics (TOPS), Optical Fiber Communication Conference, Technical Digest (Washington, DC: Optical Society of America), vol. 86, pp. 427–429.

21. Spiekman, L. H., van den Hoven, G. N., van Dongen, T., Sander-Jochem, M. J. H., Binsma, J. J. M., Wiesenfeld, J. M., Gnauck, A. H., and Garrett, L. D. 2000. Recent advances in SOA's in WDM applications.

Proceedings of European Conference on Optical Communication (ECOC 2000) (München), vol. 1 (VDE), pp. 35–38.

22. Tangdiongga, E., Crijns, J. J. J., Spiekman, L. H., van den Hoven, G. N., and de Waardt, H. 2002. Performance analysis of linear optical amplifiers in dynamic WDM systems. IEEE Photon. Technol. Lett. 14, 1196–1198.

23. Genoa corporation. Product information. www.genoa.com.

24. Stubkjaer, K. E., Mikkelsen, B., Durhuus, T., Joergensen, C. G., Joergensen, C., Nielsen, T. N., Fernier, B., Doussiere, P., Leclerc, D., and Benoit, J. 1993. Semiconductor optical amplifiers as linear amplifiers, gates and wavelength converters. ECOC'93: Proceedings 19th European Conference on Optical Communication, 1:TuC5.

25. Bjarklev, A. 1993. Optical Fiber Amplifiers: Design and System Applications (Boston, MA: Artech).

26. Hansen, S. L., Dybdal, K., and Larsen, C. C. 1992. Upper gain limit in Er-doped fiber amplifiers due to internal Rayleigh back-scattering. Optical Fiber Communications, OFC'92 (San Jose, CA) I:TuL4.

27. Mazurczyk, V. J., and Zyskind, J. L. 1993. Polarization hole burning in erbium doped fiber amplifiers. CLEO'93: Conference on Lasers and Electro-Optics (Baltimore, MD, May 2–7, 1993) CPD26.

28. Goforth, D. E., Minelly, J. D., Ellison, A. J. E., Wang, D. S., Trentelman, J. P., and Nolan, D. A. 2000. Proceedings of NFOEC, 2000, Paper B1.

29. Codemard, C., Soh, D. B. S., Ylä-Jarkko, K., Sahu, J. K., Laroche, M., and Nilsson, J. 2003. Cladding-pumped L-band phosphosilicate erbium–ytterbium co-doped fiber amplifier. Topical Meeting on Optical Amplifiers and their Applications (Otaru, Japan, July 6–9, 2003).

30. Arbore, M. A., Zhou, Y., Keaton, G., and Kane, T. 2002. 36 dB gain in S-band EDFA with distributed ASE suppression. Proceedings of Topical Meeting on Optical Amplifiers and Their Applications Post-deadline (Vancouver, July 14–17, 2002), Paper PDP4.

31. Kani, J. and Jinno, M. 1999. Wideband and flat-gain optical amplification from 1460 to 1510 nm by serial combination of a thulium-doped fluoride fibre amplifier and fibre Raman amplifier. *IEEE Electron. Lett.* 35, 1004–1006.

32. Masum-Thomas, J., Crippa, D., and Maroney, A. 2001. A 70 nm wide S-band amplifier by cascading TDFA and Raman fibre amplifier. *OFC'2001*. Paper WDD9.

33. Hattori, K., Kitagawa, T., Oguma, M., Ohmori, Y., and Horiguchi, M. 1994. Erbium-doped silica-based planar-waveguide amplifier integrated with a 980/1530-nm WDM coupler. *Optical Fiber Communications, OFC'94* (San Jose, CA) 1:FB2.

34. Hübner, J., and Guldberg-Kjsr, S. 2001. Active and passive silica waveguide integration. *Proc. ECOC'2001* (Amsterdam, September 2001).

35. Harwood, D. W. J., Fu, A., Taylor, E. R., Moore, R. C., West, Y. D., and Payne, D. N. 2000. A 1317 nm neodymium doped fluoride glass waveguide laser. *European Conference on Optical Communication 2000* (München), Paper 6.4.8.

36. Hansryd, J., and Andrekson, P. A. 2001. Broad-band continuous-wave-pumped fiber optical parametric amplifier with 49-dB gain and wavelength-conversion efficiency. *IEEE Photon. Technol. Lett.* 13, 194–196.

37. Southampton Photonics, Inc. *Product information.* www.southamptonphotonics.com.

38. Hansryd, J., Andrekson, P. A., Westlund, M., Li, J., and Hedekvist, P. O. 2002. Fiber-based optical parametric amplifiers and their applications. *IEEE J. Sel. Top. Quantum Electron.* 8, 506–520.

39. Ho, M.-C., Uesaka, K., Marhic, M. E., Akasaka, Y., and Kazovsky, L. G. 2001. 200-nm-bandwidth fiber optical amplifier combining parametric and Raman gain. *J. Lightwave Technol.* 19, 977–981.

40. Marhic, M. 2003. Toward practical fiber optical parametric amplifiers. *OSA Trends in Optics and Photonics (TOPS) Optical Fiber Communication Conference, Technical Digest* (Washington, DC: Optical Society of America), vol. 86, pp. 564–565.

41. Minelly, J. D., Chen, Z. J., Laming, R. I., and Caplen, J. E. 1995. Efficient cladding pumping of an Er^{3+} fibre. *Proceedings European Conference on Optical Communication (ECOC 1995)* (Brussels, 17–21 September 1995), Paper Th.L.1.2, pp. 917–290.

42. Nilsson, J. 1998. Cladding-pumped erbium-doped fiber amplifiers for low-noise high-power WDM and analogue CATV boosters—new design using ring-doping. *Optical Fiber Communication Conference, OSA Technical Digest Series* (Washington, DC: Optical Society of America), vol. 2, pp. 38–39.

43. Bousselet, P., Leplingard, F., Simonneau, C., Moreau, C., Gasca, L., Provost, L., Bettiati, M., and Bayart, D. 2001. 30% power conversion efficiency from a ring-doping all silica octagonal Yb-free double-clad fiber for *WDM applications in the C* band. *Proceedings of Topical Meeting on Optical Amplifiers and Their Applications (OAA 2001) Post-deadline*, Paper PD1.

44. Minelly, J. D., Barnes, W. L., Laming, R. I., Morkel, P. R., Townsend, J. E., Grubb, S. G., and Payne, D. N. 1993. Diode-array pumping of Er^{3+}/Yb^{3+} co-doped fiber lasers and amplifiers. *IEEE Photon. Technol. Lett.* 5, 301–303.

45. Alam, S. U., Nilsson, J., Turner, P. W., Ibsen, M., Grudinin, A. B., and Chin, A. 2000. Low cost multi-port reconfigurable erbium doped cladding pumped fiber amplifier. *European Conference on Optical Communication 2000* (München), Paper 5.4.3.

46. Codemard, C., Yla-Jarkko, K., Singleton, J., Turner, P. W., Godfrey, I., Alam, S.-U., Nilsson, J., Sahu, J. K., and Grudinin, A. B. 2002. Low noise, intelligent cladding pumped L-band EDFA. *Proceedings of European Conference on Optical Communication* (Copenhagen, September 8–12, 2002) Post-deadline, vol. 3, Paper 1.6.

47. Goldberg, L., and Koplow, J. 1998. Compact, side-pumped 25 dBm Er/Yb co-doped double cladding fibre amplifier. *Electron. Lett.* 34, 2027–2028.

48. Keopsys, S. A. *Product information.* www.keopsys.com.

49. DiGiovanni, D. J., and Stentz, A. J. 1999. Tapered fiber bundles for coupling light into and out of cladding-pumped fiber devices. US Patent Specification 5864644.

50. Zenteno, L. A., Minelly, J. D., Liu, A., Ellison, J. G., Crigler, S. G., Walton, D. T., Kuksenkov, D. V., and Deineka, M. J. 2001. 1W single-transverse-mode Yb-doped double-clad fibre laser at 978 nm. *Electron. Lett.* 37, 819–820.

51. Fu, L. B., Selvas, R., Ibsen, M., Sahu, J. K., Alam, S.-U., Nilsson, J., Richardson, D. J., Payne, D. N., Codemard, C., Goncharev, S., Zalevsky, I., and Grudinin, A. B. 2002. An 8-channel fibre-DFB laser WDM-transmitter pumped with a single 1.2 W Yb-fiber laser operated at 977 nm. *Proceedings of European Conference on Optical Communication* (Copenhagen, September 8–12, 2002), vol. 3, Paper 8.3.5.

52. Ylä-Jarkko, K. H., Selvas, R., Soh, D. B. S., Sahu, J. K., Codemard, C., Nilsson, J., Alam, S.-U., and Grudinin, A. B., 2003. A 3.5 W 977 nm cladding-pumped jacketed-air clad ytterbium-doped fiber laser. *ASSP 2003* Post-deadline, Paper PDP2.

53. Stone, B. T., and Bray, K. L. 1996. Fluorescence properties of Er^{3+}-doped sol–gel glasses. *J. Non-Cryst. Solids* 197, 136–144.

54. Sen, S., and Stebbins, J. F. 1995. Structural role of Nd^{3+} and Al^{3+} cations in SiO_2 glass: A ^{29}Si MAS-NMR spin-lattice relaxation, ^{27}Al NMR and EPR study. *J. Non-Cryst. Solids.* 188, 54–62.

55. Sckerl, M. W., Guldberg-Kjsr, S., Rysholt Poulsen, M., Shi, P., and Chevallier, J. 1999. Precipitate coarsening and self-organization in erbium-doped silica. *Phys. Rev. B.* 59, 13494.

56. Arai, K., Namikawa, H., Kumata, K., Honda, T., Ishii, Y., and Handa, T. 1986. Aluminium or phosphorus co-doping effects on the fluorescence and structural properties of a neodymium-doped silica glass. *J. Appl. Phys.* 59, 3430–3436.

57. Arai, K., Yamasaki, S., Isoya, J., and Namikawa, H. 1996. Electron-spin-echo envelope-modulation study of the distance between Nd^{3+} ions and Al^{3+} ions in the codoped SiO_2 glasses. *J. Non-Cryst. Solids.* 196, 216–220.

58. Lægsgaard, J. 2002. Dissolution of rare-earth clusters in SiO_2 by Al codoping: A microscopic model *Phys. Rev. B.* 65, 174114.

59. Zemon, S., Lambert, G., Andrews, L. J., Miniscalco, W. J., Hall, B. T., Wei, T., and Folweiler, R. C. 1999. Characterization of Er^{3+}-doped glasses by fluorescence line narrowing. *J. Appl. Phys.* 69, 6799–6811.

60. Judd, B. R. 1962. Optical absorption intensities of rare-earth ions. *Phys. Rev.* 127, 750.

61. Ofelt, G. S. 1962. *J. Chem. Phys.* 37, 511.

62. Takebe, H., Morinaga, K., and Izumitani, T. 1994. Correlation between radiative transition probabilities of rare-earth ions and composition in oxide glasses. *J. Non-Cryst. Solids.* 178, 58–63.

63. Li, H., Li, L., Vienna, J. D., Qian, M., Wang, Z., Darb, J. G., and Peeler, D. K. 2000. Neodymium(III) in alumino-borosilicate glasses. *J. Non-Cryst. Solids.* 278, 35–57.

64. Zemon, S., Pedersen, B., Lambert, G., Miniscalco, W. J., Andrews, L. J., Davies, R. W., and Wei, T. 1991. Excited-state absorption cross sections in the 800-nm band for Er-doped Al/P silica fibers: Measurements and amplifier modeling. *IEEE Photon. Technol. Lett.* 3, 621–624.

65. Tanabe, S., and Hanada, T. 1996. Local structure and 1.5 μm quantum efficiency of erbium doped glasses for optical amplifiers. *J. Non-Cryst. Solids.* 196, 101–105.

66. Hehlen, M. P., Cockroft, N., Gosnell, T. R., and Bruce, A. J. 1997. Spectroscopic properties of Er^{3+} and Yb^{3+} doped soda-lime silicate and aluminosilicate glasses. *Phys. Rev. B.* 56, 9302–9318.

67. Clesca, B. 2002. Optical amplification techniques tussle for metro dominance. *Lightwave Europe*, pp. 22–23.

FURTHER READING

Agrawal, G. P. 2000. *Nonlinear Fiber Optics.* 3rd edn. (San Diego, CA: Academic).

Anders, B. 1993. *Optical Fiber Amplifiers: Design and System Applications* (Boston, MA: Artech).

Becker, P. C., Olsson, N. A., and Simpson, J. R. 1999. *Erbium-Doped Fiber Amplifiers: Fundamentals and Technology* (New York: Academic).

Desurvire, E. 1994. *Erbium-Doped Fiber Amplifiers: Principles and Applications* (New York: Wiley).

Desurvire, E., Bayart, D., Desthieux, B., and Bigo, S. 2002. *Erbium-Doped Fiber Amplifiers Device and System Developments* (New York: Wiley).

Digonnet, M. J. F., ed. 1993. *Rare Earth Doped Fiber Lasers and Amplifiers* (New York: Dekker).

Ellis, A., and Minelly, J. D. 2002. New materials for optical amplifiers. *Optical Fiber Telecommunications IV A—Components*, eds., I. P. Kaminow and T. Li (San Diego, CA: Academic).

France, P. W., ed. 1991. *Optical Fiber Lasers and Amplifiers* (Glasgow: Blackie–Boca Raton, FL: CRC Press).

Siegman, A. E. 1986. *Lasers* (Mill Valley, CA: University Science Books).

Srivastava, A. K., and Sun, Y. 2002. Advances in erbium-doped fibre amplifiers. *Optical Fiber Telecommunications IVA—Components*, eds., I. P. Kaminow and T. Li (San Diego, CA: Academic).

Sudo, S., ed. 1997. *Optical Fiber Amplifiers: Materials, Devices, and Applications* (Norwood, MA: Artech).

Zyskind, J. L., Nagel, J. A., and Kidorf, H. D. 1997. Erbium-doped fiber amplifiers for optical communications. *Optical Fiber Telecommunications IIIB*, eds., I. P. Kaminow and T. L. Koch (San Diego, CA: Academic).

16

Ultrafast optoelectronics

GÜNTER STEINMEYER
Max Born Institute

The science and technology of ultrafast pulses has, in recent years, come of age, becoming ever more important in the real world. In the early days of lasers, efforts concentrated on getting lasers to operate as continuous light sources. Compared to conventional light sources, continuous-wave (cw) lasers perfectly concentrate light in the plane transverse to the propagation direction, which gives rise to the extreme high brightness of laser beams.

Nevertheless, a continuous laser ignores the third dimension along the propagation direction for further possible concentration of energy. Using mode-locking, the light field circulating through a mode-locked laser can be temporally focused within one millionth of the total cavity length, which translates to a pulse duration of a few femtoseconds compared to a total cavity round-trip time of several nanoseconds. The peak power, even in an oscillator pulse, readily reaches hundreds of kilowatts. Employing the method of chirped-pulse amplification (CPA), one can easily generate pulses with multiple gigawatt peak power, which exceeds the continuous capabilities of several nuclear power plants for the duration of a few femtoseconds. The advantages of having lasers operating with very short pulses are currently being realized more and more in both industry and science. An ultrashort pulse can process material surfaces by action of its extreme electric field strength without causing internal heating or melting. This nonthermal interaction may be applied to ablate material for analysis or to create gaseous phases as desired. In spectroscopy, very fast pulses can be used to investigate very fast chemical reactions or trigger fast photochemical reactions and observe features that would be impossible to observe using slower processes, leading to far greater understanding of the science and technology involved. This enables the interim states of such reactions to be observed, whereas previously only the starting materials and end products would be apparent. To make use of this, it is necessary to be able to generate ultrashort pulses and also to be able to measure the pulses that have been created. This chapter describes the science and technology in this rapidly developing field.

16.1 INTRODUCTION

The technology of ultrafast optoelectronics involves devices for emitting, modulating, transmitting, or sensing (and even harvesting) light. At a fundamental stage, these devices require interaction of light with electrical current, effectively converting photons into electrons or vice versa. The temporal response of, e.g., an optoelectronic emitter is, therefore, generally limited by the fastest available rise time of a current-pulse generator. Similarly, the direct detection of light in a photodetector can only be accomplished directly with a few picosecond temporal resolution [74].

It is not only the device limitations as, on the electronics side, further constraints can be imposed by parasitic inductances or capacitances in the electronic circuitry and high-frequency attenuation mechanisms in cables.

Some devices, such as streak cameras [70], can merge the generation of photoelectrons and their temporal resolution into one device and overcome some limitations. Nevertheless, even these fastest

direct optoelectronic detection devices are typically limited to a response time of the order of 1 ps.

The examples discussed so far rely on a direct interaction of light with an electronic current. In the following, we will describe ways to circumvent the electronic bandwidth problem. The basic technologies used for ultrafast optoelectronics are also now being used in the latest high-speed optical communication systems. In early fiber-optic data links, light was converted directly back into an electronic current prior to any processing. In ultrafast systems, this direct detection approach has been replaced by all-optical means of processing photonic data streams. In parallel with enhancing the speed of a single optical channel, further major improvement in terms of data capacity have been achieved by the method of wavelength-division multiplexing (WDM). This permits terabit/s rates by simultaneously transmitting many channels at different optical wavelengths through a single fiber. Unfortunately, of course, increasing the data rate of a "single wavelength" channel inevitably involves broadening the bandwidth of this channel, requiring greater spectral separation between each carrier in a WDM link.

In this chapter, we will introduce methods of providing the fundamental optoelectronic functions, viz. emitting, modulating, transmitting, and sensing light, all with a temporal response or resolution of a few femtoseconds. These methods are ultimately limited only by the duration of the optical cycle itself and the inevitably huge spectral broadening when only a few cycles are present in a single pulse. We will refer to these schemes as *ultrafast optoelectronic devices*, even though the response of the basic optoelectronic components, where photons and electrons interact, can be inherently slow. As indicated above, the methods described split the optoelectronic process into two steps, an ultrafast all-optical step, which ensures sufficient bandwidth and a second slower direct step where efficient conversion between photons and electrons occurs at a strongly reduced bandwidth.

Before continuing our technical discussion in detail, it is appropriate here to indicate how this chapter is organized.

First, we will review methods for the generation of femtosecond pulses. Short light pulses can experience reshaping effects, which may strongly modify their pulse shape and these effects become increasingly severe the shorter the pulses are. Compared to electronic pulses, however, these effects only start to become significant at terahertz rather than gigahertz frequency and also mainly affect the spectral phase rather than the amplitude. Therefore, particular attention will be paid to methods that allow compensation of dispersive pulse broadening. Second, this discussion will be followed by an overview of femtosecond pulse characterization methods. Finally, in Section 16.5, we will address the methods to modulate phase and amplitude of femtosecond pulses. Together with the characterization methods, ultrafast optoelectronics nowadays allows for synthesis of desired pulse shapes and control of optical waveforms, very similar to the generation of arbitrary electronic waveforms, so the technology behind this will be presented.

16.2 ULTRAFAST LASER PULSE GENERATION

Compared to that practically possible with electronics, the available bandwidth in optical systems is enormous. This is a clear driving force behind all-optical telecommunications. Fiber-optic systems can provide terabit/s of transmission capacitance over transatlantic distances [2,75]. From elementary Fourier theorems, it is also clear that a wide optical bandwidth must be used to support extremely short pulses. The laser material with the widest known gain bandwidth is titanium-doped sapphire [22]. The 650–1100 nm gain capability allows us to generate directly pulses of about 5 fs pulse duration, corresponding to less than two optical cycles of the electric field. As an alternative to direct generation in lasers, the only method allowing an even wider bandwidth is optical parametric chirped pulse amplification (OPCPA) [120,121], which we discuss further below.

Laser operation is generally sustained by an optical cavity to provide optical feedback into the gain material. The photons circulating in the cavity experience laser gain but losses from the cavity can occur due imperfect reflections but primarily from output coupling [83]. Laser gain saturation favors equal filling of the cavity with photons, i.e., cw operation of the laser [56]. In order to generate a short pulse, the energy content of the optical cavity has to be temporally confined into an interval as short as possible. This requires us to introduce a condition where lasing action occurs preferentially in bursts of short pulses. One means of achieving

this is to insert an intracavity amplitude modulator, which opens and closes in synchronism with the light traveling through the cavity [84] (Figure 16.1).

This process generates a gain advantage for those photons that travel through during the fully open state of the modulator, so this produces a regularly spaced stream of very narrow optical pulses at the output. The process is known as mode-locking because it produces a relatively broad optical spectrum consisting of a finely spaced manifold of very many narrow-linewidth laser cavity modes, each related in phase and corresponding to a resonant optical mode of the laser cavity.

If conventional electro-optic or acousto-optic modulators are used, this whole scheme is always limited to some extent by the electronic pulse width of the modulator driver, but even so optical pulses shorter than the electronic driving pulses can still be generated at the minimum attenuation point of the modulator cycle. Following the philosophy mentioned in our introduction, it is, however, preferentially desirable to eliminate electronic bandwidth limitations by switching to an all-optical modulator.

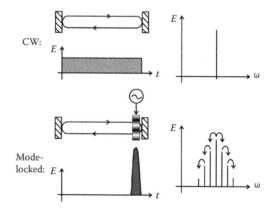

Figure 16.1 Schematic of cw (top) and mode-locked (bottom) operation of a laser. In the cw case, the cavity is equally filled with photons, i.e., the energy density E is constant. Insertion of a modulator synchronously driven at the cavity round-trip frequency focuses light into a small time slot centered at the fully open position of the modulator (bottom). The two situations are also depicted in the frequency domain on the right side. Continuous operation results in single-longitudinal-mode operation of the laser. The modulator creates sidebands at the neighboring modes, effectively transferring energy to the spectral wings. This is called mode-locking.

16.2.1 Saturable absorber mode-locking

The simplest mode-locking schemes using an all-optical modulator rely directly on saturable absorption of an optical material [38,48,67]. Typically, either organic dyes or semiconductor materials can be used. At high intensities, the absorption of these materials bleaches out because the majority of electrons or holes has been excited from the ground state to higher energy states. The absorption recovers quickly after the pulse passes, so such *an all-optical modulator* is automatically self-synchronous with the light in the cavity. To start the process, any small initial power spike inside the cavity will experience less loss than the rest of the cavity's energy content, so this spike becomes bigger on every cycle, until all the energy is concentrated in a small time slot. Unfortunately, the relaxation of the bleaching process is not arbitrarily fast, but again, pulses shorter than the modulation time constants can be produced. As with an active electro-optic modulator, it is not even necessary for the relaxation time of the absorber to be faster than the cavity round trip, but nevertheless the all-optical method still experiences a limitation from response-bandwidth effects. Typically, this prevents generation of pulses much shorter than a picosecond. There are, however, some ingenious schemes that overcome some of the limitations of a slow absorber and extend the operation of saturable absorber mode-locking well into the femtosecond range [27,37,51].

In the last two decades, a number of different approaches for implementing the saturable absorber in solid-state lasers and fiber lasers have been explored. The most widespread method is the semiconductor absorber mirror (SESAM) [122–125]. The SESAM relies on the saturable absorption of a single (or even several) quantum well(s), embedded in a resonant [126] or antiresonant [127,128] Gires–Tournois [129] interferometer structure, viz. an interferometer consisting of a high reflector and a partial reflector. Although the resonant variant allows the high modulation depths required for use with fiber lasers, the antiresonant versions of SESAM are typically employed in solid-state lasers, usually having much higher intracavity pulse energies.

In a SESAM, the quantum well structure needs to be suitably designed to provide saturable absorption over a reasonable, but still fairly narrow wavelength range (typically 50–100 nm). More recently, carbon nanotubes (CNTs) have been found to provide very similar properties [130] as SESAMs over an optical bandwidth of up to a whole octave [131]. Moreover, CNTs have found to be applicable to nearly all-solid-state [131] and fiber lasers [132]. Even more recently, the remarkable new material graphene has emerged as yet another alternative to implement a saturable absorber for mode-locking a laser [133,134].

Many of these approaches are fairly specialized and rely on the interplay of several optical mechanisms inside the cavity. An alternative approach came from the use of the so-called *reactive nonlinearities* (e.g., the Kerr nonlinearity, see Figure 16.2). In contrast to the amplitude changes caused by saturable absorption devices, these alternatives influence the phase of the light only. This self-phase modulation causes larger phase delay for high light intensities compared to low light intensities [1]. As no electron carrier dynamics are involved in the process, the mechanism can be very fast with a typical response time of less than 1 fs. The "reactive" character of the

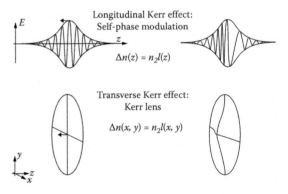

Figure 16.2 The nonlinear optical Kerr effect is caused by a dependence of the index of refraction on intensity. Along the axis of propagation z, this causes a phase retardation of the most intense part of the temporal pulse profile due to self-phase modulation. In the plane perpendicular to z, the retardation causes a deformation of the phase fronts. In the central part of the spatial beam profile, the phase fronts experience an additional curvature, i.e., the Kerr effect causes an effect similar to a lens. Therefore, the transverse Kerr effect is also referred to as a Kerr lens.

nonlinearity, however, now requires a conversion mechanism to transfer the nonlinear phase modulation into an effective amplitude modulation to form an *effective saturable absorber.*

16.2.2 Passive mode-locking based on reactive nonlinearities

We shall now describe how such an *effective saturable absorber* can be constructed, using a combination of a phase-to-amplitude converter and a reactive nonlinearity. Several types of effective saturable absorbers have been proposed. One method is to place the self-phase modulator in a second cavity, which is coupled to the gain cavity. The coupled cavity acts as a nonlinear mirror, providing high reflectivity for the more intense center part of the pulse and low reflectivity everywhere else. This method is called additive-pulse mode-locking [46]. Another method is *Kerr-lens mode-locking* (*KLM*) [54,78,89]. Here, the transverse effect of the Kerr nonlinearity is used instead (see Figure 16.2). When the most intense axial region of an optical beam is phase retarded compared to the outer region, the effect corresponds to the focusing action of a lens. A suitable arrangement of intracavity apertures (field stops) can then be used as spatial filters to translate the Kerr nonlinearity into an effective absorber. The focused high-intensity light experiences fewer losses at these apertures due to its smaller diameter. Alternatively, one can also arrange the cavity mirror design to provide a better spatial overlap with the pump light when the additional nonlinear lens is in place. Pulses as short as approximately 5 fs have been produced with this method [23,95].

Other variants of the effective saturable absorber are now particularly popular for mode-locking fiber lasers where the transverse effect of a Kerr lens cannot be exploited. These methods include *nonlinear polarization rotation* (*NPR*, [135,136]) and the *nonlinear amplifying loop mirror* (*NALM*, [137]). The former method exploits the fact that short pulses may experience an additional polarization rotation while propagating through an optical fiber. This NPR effect is caused by the all-optical Kerr effect, which causes different phase shifts for light with parallel and orthogonal polarization. Suitably combined with a linear rotation effect of opposite sign and a polarizer, the same effect as in *KLM* can be obtained.

For the *NALM* method, one of the laser end mirrors is effectively replaced by a nonlinear Sagnac interferometer [137,138].

16.2.3 Amplification

Typical KLM lasers can deliver pulses with time durations in the range from 5 to 100 fs at energies of a few nanojoules and at a repetition rate of 100 MHz. Pulse energies can be increased either by using a longer cavity [15], by additional extracavity amplification [63], or by cavity dumping [8]. All these methods allow for a few 10 nJ pulse energy at megahertz repetition rates. Further increase of the pulse energy into the microjoule range, whereas still maintaining femtosecond time duration would increase the peak power of the pulse into a regime where most optical materials are likely to encounter damage. For achieving greater amplification, the pulse, therefore, needs to be "stretched" before amplification and then be recompressed to its original duration after the amplifier. The design of the stretcher and compressor will be treated in the following section. The method is called CPA [6,93] and has been demonstrated with 15 fs pulses of millijoule energy at a 10 kHz repetition rate [5,114]. Even stronger amplification and reduction of repetition rate can lead to hundreds of joules pulse energy at 450 fs pulse duration [73]. The latter system reaches a peak power of 1.5 PW (1.5×10^{15} W). The focused intensity reaches 6×10^{20} W cm^{-2}. The electric field strength in the focus of such a laser pulse exceeds typical interatomic binding forces by about three orders of magnitude [11].

Although the peak powers demonstrated in [73] remained unsurpassed for more than a decade, there are currently several facilities offering petawatt powers. In 2015, e.g., the 2 PW laser for fast ignition experiment laser has been commissioned, redefining the world record in terms of laser peak powers [139]. Moreover, initiatives are underway to increase the obtainable peak powers to 10 PW and beyond. More importantly, however, the more recent PW laser facilities use significantly shorter pulses to obtain the high peak power. Although the first demonstration in [73] generated 450 fs pulses, plans for a 10 PW system to be installed at the European Union's Extreme Light Infrastructure Beamlines facility in Dolní Břežany near Prague, foresee generation of 150 fs pulses. Then, pulse energies only need to be scaled by a factor of 3.5 to obtain 10 times higher peak powers [140]. In a second laser to be installed at the same facility, an even more radical 30 fs pulse duration is planned to obtain the same peak power as in Reference [73] with a pulse energy of only 8% of that of the former design [141].

16.2.4 Wavelength conversion—from terahertz to x-rays

So far, we have concentrated on direct oscillator and amplifier schemes. Laser materials having the wide optical bandwidth as needed for short-pulse generation are only available in the near-infrared (IR) and part of the visible spectral range [16,26,27,117]. Another approach for the generation of short pulses is the conversion of femtosecond radiation by nonlinear optical mechanisms. One mechanism is *frequency doubling* in nonlinear optical crystals. For frequency doubling of extremely short pulses, very thin crystals with a typical thickness of a few micrometers have to be employed [30]. Another method is *optical parametric amplification*, which can be used both in the near-IR and visible spectral range. The parametric process splits an input photon into two photons of lower energy, with wavelengths depending on the phase-matching conditions in the nonlinear optical material. Beta-barium borate has been shown to provide extremely wideband phase matching, which has been used to generate pulses well below 5 fs [81,115]. There is a wealth of other methods leading deeper into the ultraviolet (UV), including Raman sideband generation [110], high-harmonic generation [11,57,60], and Thomson scattering [79]. Wavelength conversion is also a very important mechanism to generate radiation of longer wavelengths than directly available from oscillators. Again, parametric processes can be used here [55]. Another important case is *terahertz radiation* [88]. All these mechanisms allow access to wavelength regions that are not readily covered by wideband laser materials.

16.2.5 Optical parametric chirped-pulse amplification

The general idea of CPA mentioned above can also be transferred to nonlaser gain processes, such as parametric amplification, combining the methods

from Sections 16.2.4 and 16.2.5. This method, called OPCPA was first discussed by Dubietis et al. in the 1990s [120,121,142,143]. Compared to direct parametric amplification, the seed pulses are now prestretched into the range of 100 ps–1 ns prior to amplification, and similar to CPA, this effect is then reverted after amplification. In OPCPA, one can, therefore, employ pump lasers with rather long-pulse durations. Highly energetic pulses for this application can often be generated by relatively simple Q-switching. Compared to their laser equivalents, OPCPAs feature a number of advantages, including a very high amplification in a single amplification stage, adaptability over a wider range of wavelengths and much lower thermal effect problems as negligible energy loss occurs in the parametric process. Moreover, extremely wide bandwidths can be hosted, in particular, in noncollinear schemes where several 100 nm wide near-ideal phase-matching conditions can be achieved, when arranged with a particular "magic angle" between pump and signal wave [144]. The specific noncollinear arrangement in parametric amplifiers is often also referred to as NOPA. The OPCPA concept has been scaled toward PW peak powers [145–149] and typically enables to reach shorter pulse durations together with high peak powers compared to traditional CPA. Another advantage is a better suppression of pre-pulses and postpulses [145].

16.3 FEMTOSECOND PULSE PROPAGATION EFFECTS AND DISPERSION COMPENSATION

In microwave electronic systems, a severe limitation is imposed by high-frequency damping mechanisms. In optics, limitations due to spectral absorption over short optical paths are often not a concern or can be easily avoided. Many dielectric media, like glasses and crystals, are transparent in the range of 150–1000 THz [102]. Limitations typically only arise in optical amplification or nonlinear optical conversion. In a 10 THz window in the near-IR (1.55 µm wavelength), exceptionally low losses of 0.3 dB km^{-1} have been demonstrated in silica fibers [69]. In this exact wavelength region, Er-doped glass amplifiers [10,17] can easily be embedded into optical telecommunication systems. Compared to electronic systems, optical bandwidth is, therefore, abundant and a much lesser concern.

16.3.1 Group delay dispersion as leading-order propagation effect

If one tries to launch a 100 fs pulse train into an optical fiber, one finds that due to dispersion, the extremely short pulses will broaden to picoseconds after only a few meters of propagation. Dispersion causes different spectral components of the pulse to propagate at different group velocities, which induces broadening of short pulses during propagation. Compensation of dispersive effects, therefore, poses a ubiquitous problem not only in telecommunication systems [2] but also with ultrashort pulse generation systems [106]. A pulse with angular carrier frequency $\omega = 2\pi c/\lambda$ experiences a *group delay* (GD):

$$\mathrm{GD}(\omega) = l\frac{\mathrm{d}}{\mathrm{d}\omega}\frac{\omega}{c}n(\omega) \qquad (16.1)$$

when propagating through a dispersive medium with index n and length l. The GD determines the propagation time of a pulse and must not be confused with the phase delay ln/c. To first order, pulse broadening is governed by the *group delay dispersion* (GDD):

$$\mathrm{GDD}(\omega) = l\frac{\mathrm{d}^2}{\mathrm{d}\omega^2}\frac{\omega}{c}n(\omega). \qquad (16.2)$$

Gaussian pulses of duration τ_0 are stretched to a duration τ, where

$$\tau = \tau_0\sqrt{1 + \left[\frac{|\mathrm{GDD}|^2}{\tau_0}\right]^2} \qquad (16.3)$$

when propagating through a material with dispersion GDD. Equation 16.3 is a useful relation to estimate the severity of pulse broadening in optical systems [1]. Optical materials can exhibit either negative or positive dispersion and this can actually change sign as the wavelength varies (see Figure 16.3). Fused silica, e.g., shows zero dispersion at 1.3 μm with positive dispersion below and negative dispersion above this wavelength. At the *zero-dispersion wavelength* of a material, broadening effects due to GDD are eliminated; but similar effects are caused by

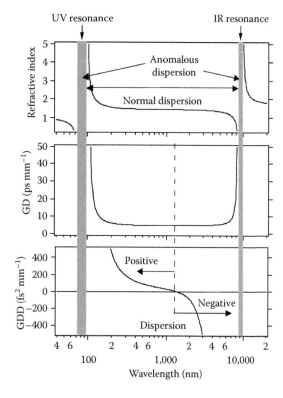

Figure 16.3 Material dispersion. The dispersion of a dielectric material (simplified model of fused silica) with a UV resonance at 80 nm and a vibronic IR resonance at 10 μm is schematically shown. The top figure shows the refractive index itself. For the entire range between the resonances, dn/dλ ≤ 0 holds, which is referred to as normal dispersion. Only on resonance, we find anomalous dispersion dn/dλ > 0. The resulting GD and GDD are also shown. The GDD is the leading term responsible for pulse reshaping during propagation. Note that regions of positive and negative GDD do not coincide with those of normal and anomalous dispersion.

higher order derivatives of the refractive index, e.g., *third-order dispersion*:

$$\mathrm{TOD}(\omega) = l\frac{\mathrm{d}^3}{\mathrm{d}\omega^3}\frac{\omega}{c}n(\omega). \qquad (16.4)$$

Positive and negative GDD must not be confused with normal or anomalous first-order dispersion, which refers to the sign of dn/dω. In a system with positive dispersion, blue spectral components will be retarded relative to red components. This causes a variation of the pulse carrier frequency with time, which is usually referred to as a chirp. Depending on the sign of the dispersion, one talks about positive or negative chirp. If no other mechanism is present, both signs of dispersion are totally equivalent in terms of pulse broadening. However, self-phase modulation will typically generate a positive chirp. As self-phase modulation also introduces spectral broadening, its combination with negative dispersion can be used for pulse compression schemes. Balancing of a positive chirp, generated by the nonlinear optical process of self-phase modulation and negative material dispersion can lead to self-stabilizing optical pulses called *solitons*. These solitons can propagate over great distances through dispersive systems without changing their pulse shape [1,36].

The discussion earlier makes it clear that control and engineering of dispersion is of paramount importance for ultrafast optical systems and for telecommunications. Dispersion management is particularly important for long-distance fiber links [86,87]. With ever wider bandwidth becoming accessible, compensation of higher order dispersion becomes a consideration [72]. Pulse compression is a particularly useful mechanism for ultrashort pulse generation [1,34]. Passively mode-locked lasers (described above in its basic function) make extensive use of recompression of pulses, employing self-phase modulation in the laser crystal together with negative dispersion in the cavity [12,39]. Only by fully exploiting this mechanism can the shortest pulses be generated. In general, an ultrafast pulse compressor can only be built with negative dispersion, which unfortunately is not a characteristic available for optical materials below 1 μm wavelength. This calls for alternative concepts to compensate for material dispersion and chirps caused by nonlinear optical mechanisms.

One can classify sources of dispersion into *bulk* or *material dispersion* (i.e., from homogeneous materials like glasses and crystals), *geometrical dispersion* (prism and grating arrangements), *dispersion from interferometric effects*, and *microstructured dispersion* [fiber Bragg gratings, chirped mirrors, chirped quasi-phase matched crystals, arrayed waveguide gratings (AWGs)]. Bulk dispersion has already been treated earlier and reference data for many materials can be found in Reference [102].

16.3.2 Geometrical dispersion—prism and grating compressors

In the following, we will first address geometrical dispersion as can be produced by prism [28,80] and grating sequences [99]. When a short pulse is sent into a prism or on to a grating, its spectral components are angularly dispersed and sent into different directions (see Figure 16.4). A second prism of opposite alignment can then be used to make the spectrally dispersed beams parallel again. On their propagation between both prisms, the outer rays have experienced a delay relative to the center ones. It is important to note that this "parabolic" spectral delay is equivalent to negative GDD. It can, therefore, be used to compensate positive material dispersion. Pairs of Brewster-cut prisms can compensate dispersion without introducing losses and have been very successfully used inside laser cavities [28].

The major shortcoming of the geometrical approach, however, is that it introduces higher order dispersion terms. For prism compressors, a careful choice of the prism material can give vanishing third-order dispersion in the wavelength range above 800 nm [58]. In particular, fused silica prism pairs introduce vanishing third-order aberrations in the Ti:sapphire laser wavelength range, as was used for the first demonstration of sub-10 fs pulse generation with this laser [118].

Diffraction grating sequences may be used instead for the same purpose (see Figure 16.4), i.e., an arrangement that is also known as the Treacy compressor. These sequences are of extreme importance for CPA [64,93], which allows for amplification of pulses from the oscillator up to the millijoule level, or, at a reduced repetition rate, even to the joule level. To prevent optical damage in the amplifier chains, the oscillator pulse is stretched into the picosecond range before amplification. This reduces its peak power by the stretching ratio and also prevents significant nonlinear optical effects. After amplification, the pulse can then be recompressed into the femtosecond range, using a grating sequence with exactly the opposite dispersion

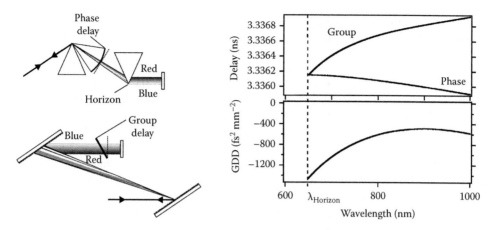

Figure 16.4 Geometric dispersion as caused by prism and grating sequences. Both prisms and gratings exhibit angular dispersion. A beam with broad input spectrum is dispersed into different direction of propagations, with all beams originating at one and the same location at the tip of the first prism. Positions with equal phase delay therefore describe a circle centered at the prism tip. To leading order, the phase delay relative to the center beam is parabolic. The second prism only serves to render all beams parallel again. Typically, such an arrangement is used in double pass with a retro-reflecting mirror as shown. The beam propagating exactly from tip to tip marks the short-wavelength horizon of the prism compressor. A parabolic phase delay front corresponds to a linear GD (as shown in the grating compressor). An exact calculation for the group and phase delay of a prism compressor (fused silica Brewster prisms with 1 m apex distance) based on [80] is also shown. The resulting GDD is depicted below. Note the strong higher order dispersion of this approach.

of the stretcher plus any amplifier material dispersion. This restores a short pulse duration and allows for the generation of extremely high peak powers. The *stretcher* [119] employed in CPA is typically a grating sequence, which incorporates a telescope with −1 magnification. The telescope exactly inverts the dispersion of a compressor in all orders. The trick is now to slightly unbalance a stretcher and a matched compressor and to accommodate for material dispersion using the difference between stretcher and compressor dispersion. Second-order dispersion (Equation 16.2) is adjusted by a difference in grating distances; third-order dispersion (Equation 16.3) can be zeroed out by adjusting grating angles [59]. Finally, fourth-order aberrations can be compensated by the use of gratings having different line spacings [90]. Typically, aberrations of the telescope have to be compensated by using suitably corrected optics [13]. This approach has been used for the demonstration of 15 fs amplified pulse duration [5,114]. Other approaches exist that introduce a controlled amount of imaging aberration in the stretcher's telescope to achieve compensation up to fourth order [94].

16.3.3 Microstructured dispersion— chirped mirrors and similar devices

One of the major shortcomings of the geometrical dispersion compensation approach is that, with the few exceptions already noted, they typically allow only for the compensation of second-order dispersion. Geometrical dispersion compensation schemes are, therefore, limited to approximately 100 THz bandwidth. However, the idea of the prism compressor can be readily extended to compensation of arbitrary dispersion: rather than using free-space propagation of laser beams, one could imagine coupling each and every spectral component into an individual fiber of precisely engineered length. A discrete approach would certainly be cumbersome but integrated optical devices similar in function have been demonstrated and are referred to as AWGs [21,85,97]. The main use of these devices is channel multiplexing in telecom systems; nevertheless, their use for dispersion compensation has recently been pointed out [72]. This type of device is shown in Figure 16.5

and it is probably the most pictorial example for microstructured dispersion compensation, even though its application is not very widespread yet.

Rather than directly introducing a wavelength dependent propagation length, several other methods for arbitrary dispersion compensation are possible. These are also shown in Figure 16.5. One of these approaches is *chirped mirrors* [52,66,91,96]. These dielectric mirrors consist of alternating pairs of transparent high-index and low-index layers. The same effect can be achieved with chirped gratings in optical fibers by modifying the refractive index with exposure to UV light through a variable-pitch periodic mask [42,53]. The portions of the fiber that have been exposed to the short-wavelength radiation show a modified index of refraction. Even though the index differences are much smaller in the fiber grating approach, they provide the same functionality as a distributed Bragg reflector if the period of the index modulation is chirped along the fiber [24,71]. A Bragg mirror reflects light when all Fresnel reflections at the high-/low-index interfaces constructively interfere. This is the case when the optical thickness of all layers is chosen to be equal to a quarter of the light wavelength. Varying the optical layer

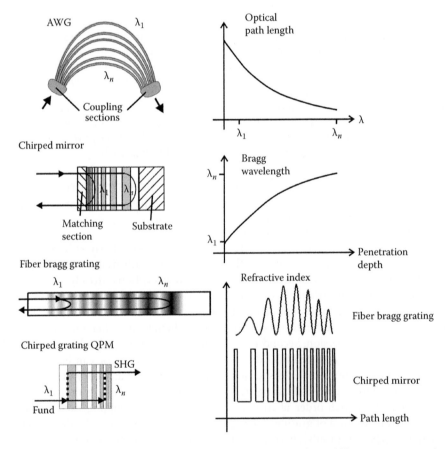

Figure 16.5 Microstructured dispersion compensation. Shown are four different concepts that can compensate for arbitrarily shaped dispersion. Top: AWGs. The coupling sections provide wavelength-dependent coupling into wave guides of different lengths. This makes the path length a designable function of wavelength. Second: chirped mirrors. Here, the Bragg wavelength is varied over the mirror stack, making the penetration depth a function of wavelength. Impedance matching sections are required to reduce detrimental interferometric effects. Third: fiber Bragg gratings. They work similarly to chirped mirrors, but achieve impedance matching by a different apodization method, as shown on the right. Bottom: chirped grating quasi-phase matching. Here, the input light is converted into the second harmonic in a quasi-phase-matched crystal. The poling period of the crystal determines the conversion wavelength, which again allows control of the total GD as a function of wavelength.

thickness along the mirror structure then results in a dependence of the wavelength of peak reflectivity, the so-called Bragg wavelength λ_B, on the penetration depth z into the mirror structure. Chirping the mirror structure, therefore, allows the generation of a desired group delay GD(ω). It is obvious that the Bragg wavelength does not have to be varied linearly with penetration depth; in fact, any single-valued function can be tailored into the Bragg-wavelength chirp function.

It needs to be mentioned that this simple picture is often distorted by other contributions to the dispersion. In chirped mirrors, the top reflection at the interface to air gives rise to undesired interferences, which spoil the dispersion characteristics of the mirror and give rise to strong spectral fluctuations of the dispersion. A solution to this problem is impedance matching from the ambient medium to the mirror stack. In deposited mirror structures, a partial solution can be provided by *double chirping* [52,66]. Other methods have been proposed to overcome these dispersion fluctuations [65,98]. In fiber optics, the UV exposure level can be slowly reduced to give a slow increase of the index modulation in the initial part of the fiber Bragg grating. This is referred to as *apodization* [3].

A novel approach to compensation of arbitrary dispersion is also offered by the method of *quasi-phase matching* (see Figure 16.5). Here, the wavelength of conversion of a broadband nonlinear optical conversion process is varied with propagation distance. This means that short wavelengths are converted, e.g., into the second harmonic, at a different propagation distance compared to longer wavelengths. This offers a means to tailor processes like second-harmonic generation (SHG) in order to support extremely broad bandwidths [31,45].

16.3.4 Interferometric effects—Gires–Tournois interferometers

The dispersion of chirped mirrors arises from change in Bragg wavelength of alternating high- and low-index layers. An alternative approach to achieve dispersion compensation is the use of a *Gires–Tournois interferometer* (GTI) [34]. Such an interferometer consists of a partial reflector and a high reflector. This combination reflects all incoming light and has a spectrally flat reflection amplitude response. Its phase response, however, exhibits resonances spaced by $\Delta\nu = c/2L$, similar to a Fabry–Perot interferometer. The spectral phase is a periodic function with regions of negative and positive dispersions. The GTI can be implemented from two air-spaced discrete components but also using a monolithic mirror structure with a relatively thick spacer layer between quarter-wave sections for the partial and high reflector. These structures have been successfully used in femtosecond oscillators [35,41]. Compared to chirped mirrors, they typically exhibit a lower bandwidth, but they can provide larger values of negative GDD. Their manufacture is not quite so demanding as for chirped mirrors and they can also exhibit very high values of reflectivity [35]. This concept is, therefore, interesting for lasers having much greater intracavity dispersion and working at longer pulse durations than cavities with chirped mirrors.

16.3.5 The carrier-envelope phase

The carrier-envelope phase has proven to be a decisive parameter for ultrahigh speed (attosecond) spectroscopy [149,150] and precision frequency metrology [151,152]. For a pulse propagating inside a laser oscillator cavity, the round-trip time of the envelope is ruled by the group velocity, whereas the underlying carrier is influenced by the phase velocity. The difference between these two velocities gives rise to a phase difference $\Delta\varphi_{GPO}$

$$\Delta\varphi_{GPO} = -\omega \int_0^L \left(\frac{1}{v_g} - \frac{1}{v_p} \right) dx$$
$$= \int_0^L \frac{\omega^2}{c} \frac{dn(\omega, x)}{d\omega} dx. \qquad (16.5)$$

per cavity round-trip [153,154]. Here, v_g defines the group velocity, v_p is the phase velocity, ω is the angular frequency, and $n(x)$ is the intracavity index of refraction along the coordinate x. The phase difference is usually referred to as the group-phase offset and may take large values $\gg 2\pi$. It is often more convenient to simply take the fractional part of Equation 16.5, which is called the carrier-envelope offset phase. The difference between the two velocities in Equation 16.5 relates to the linear dispersion $dn/d\omega$ inside the intracavity media. The optical carrier and the envelope pattern inevitably propagate through the dispersive gain medium, and the position of the peak electric field shifts, relative to the position of maximum intensity, is determined by the envelope. The relative phase shift between carrier and envelope is highly susceptible to environmental influences, e.g., temperature of the gain medium or even the external air pressure [155]. Quantum noise also has another important influence on this quantity [156]. The carrier-envelope frequency of laser oscillators is typically measured in an $f-2f$ interferometer [153], but other variants like the $0-f$ interferometer [157] provide similar performance. These interferometers provide an alternating radio-frequency signal at a detector, which needs to be stabilized to a reference signal, typically derived from the repetition rate of the laser. For the stabilization, one can either employ a feedback scheme [158], which acts back on the pump power of the laser or a feed-forward scheme [159]. The latter scheme has a number of advantages and has been proven to deliver the smallest remaining jitter between carrier and envelope [160].

For amplified sources, slightly different variants of the $f-2f$ interferometer are typically employed, which rely on spectral interferometry [161,162] rather than radio frequency heterodyning. Another variant is the stereo above-threshold ionization technique developed by Paulus et al. [163], which provides remarkable signal-to-noise ratio at kHz repetition rates.

16.4 MEASUREMENT OF OPTICAL WAVEFORMS WITH FEMTOSECOND RESOLUTION

16.4.1 Autocorrelation

The major problem in monitoring or characterizing optical waveforms lies in the fact that ultrashort optical pulses are among the shortest of manmade events and there is no shorter controllable event that could be used to sample the waveform. Because of this fundamental limitation, all early characterization methods employed *autocorrelation* [18], a method that characterizes a laser pulse using the same pulse as both the sample pulse to be measured and a reference pulse, after splitting it into two. Using a portion of the input pulse as the reference sample, another portion of the same pulse is delayed relative to the first portion and then optically mixed (i.e., multiplied) with this reference pulse (Figure 16.6). Technically, the multiplication of the two optical signals is done using a nonlinear optical effect such as SHG or *two-photon absorption* [77]. As a result, the autocorrelation trace

$$\text{AC}(\delta t) = \int_{-\infty}^{\infty} I(t)I(t-\delta t)\mathrm{d}t \qquad (16.6)$$

is measured as a function of time, where $I(t) = E(t)$ $E^*(t)$ is the optical intensity [40,47]. This type of autocorrelation is deemed to be *background free*, as it will measure zero signal for large delays δt $\to \pm\infty$. A background-free autocorrelator uses a *noncollinear beam geometry*, in such a way that SHG requires one photon from each of the two beams while SHG from each individual beam is not phase matched. This background-free set-up allows for large dynamic ranges, but is typically not the preferred set-up for use in the sub-10 fs regime. In a *noncollinear set-up*, an additional problem occurs due to "*beam smearing*" caused by the finite crossing angle of the two beams [7]. Therefore, a *collinear set-up* is preferred. This then yields the *interferometric autocorrelation (IAC)* trace [19]:

$$\text{IAC}(\delta t) = \int_{-\infty}^{\infty} |[E(t)+E(t-\delta t)]^2|^2 \, \mathrm{d}t. \qquad (16.7)$$

Unfortunately, no way exists to retrieve the original pulse profile from any measured autocorrelation traces without additional knowledge. Being inspired by an expected theoretical description of the mode-locking process, one can sometimes assume an expected pulse shape and then retrieval is simple. This is, unfortunately, not a valid assumption in the sub-10 fs regime with its complex pulse shapes. In this regime, simple analytical functions can no longer be assumed for *decorrelation* of the measured autocorrelation function. Additionally, the sub-10 fs regime is very demanding, and pulse shaping by spectral filtering or dispersion in the beam splitters and nonlinear crystal has to be kept to a minimum. Wherever possible, this regime calls for the use of metal-coated reflective optics.

Methods have been discussed to solve the problem of decorrelation [8,68]. Decorrelation methods require additional experimental information,

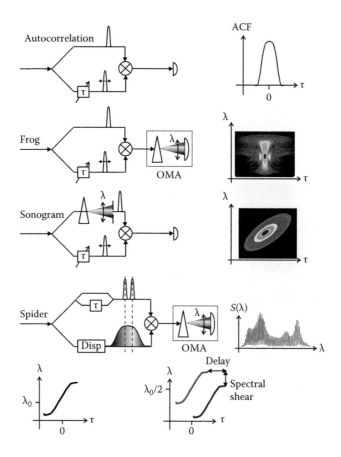

Figure 16.6 Pulse characterization methods. From top to bottom: autocorrelation. The input pulse is split into two identical replicas. One of them serves as the reference sample and is temporarily delayed relative to the other. Multiplication of both replicas in a process such as SHG or two-photon absorption delivers the autocorrelation function versus delay time shown on the right. This allows a coarse estimation of the pulse duration. FROG additionally spectrally disperses the autocorrelation function and delivers a wavelength-resolved autocorrelation function called an FROG trace. Other than simple autocorrelation, FROG allows complete reconstruction of the input pulse profile. Related is the sonogram technique, which spectrally resolves one of the input replicas instead. SPIDER also creates two replicas of the input pulse at a fixed delay and mixes these two with a chirped copy of the input pulse. The resulting two upconverted replicas are spectrally sheared with respect to each other. The resulting spectral interference pattern $S(\lambda)$ allows reconstruction of the spectral phase of the input pulse.

which, in the simplest form, can be provided by a simultaneous measurement of the power spectrum of the laser. Decorrelation methods employ a computer optimization strategy to find a simultaneous fit to the measured spectrum and autocorrelation. This removes the arbitrariness of assuming a particular pulse shape for pulse retrieval, but, for reliable operation, requires data with excellent signal-to-noise ratio. A practical example based on decorrelation of the IAC trace (Equation 16.7) and the spectrum is shown in Figure 16.7.

16.4.2 Frequency-resolved optical gating and sonogram

We shall now describe advanced methods called *frequency-resolved optical gating* (FROG) and *sonogram* to determine the pulse shape. One can conceptually understand the FROG method [100,101] and the *sonogram technique* [14,76] as a further extension of the decorrelation methods. For FROG, the autocorrelation of Equation 16.6 is spectrally resolved for each and every delay step

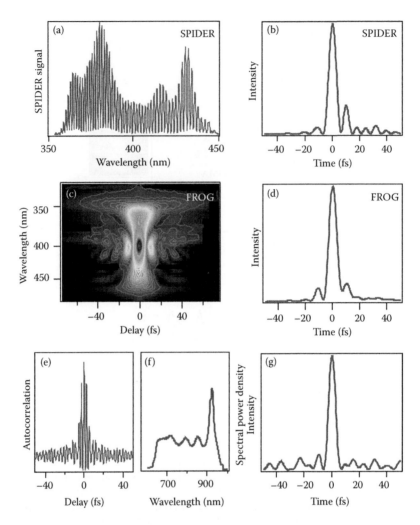

Figure 16.7 Examples of measurements of pulses from a Ti:sapphire laser. Top left: SPIDER measurement and reconstructed pulse [33]. Below: FROG trace and reconstructed pulse measured under nearly identical conditions at the same laser [32]. Bottom: iteratively reconstructed pulse shape from interferometric autocorrelation and power spectrum of the laser [95]. Note that all three measurements gave compatible pulse durations of about 6 fs or slightly below.

(see Figure 16.6). The autocorrelation is then sampled on a $\{\delta t, \omega\}$ grid. The autocorrelation spectrogram of the electric field E of the pulse

$$I_{FROG}^{SHG}(\delta t, \omega) = \left| \int_{-\infty}^{+\infty} E(t)E(t-\delta t)\exp(-i\omega t)\,dt \right|^2 \quad (16.8)$$

is called an *FROG trace*. The sonogram technique is very similar in principle, but cuts out a spectral slice of one of the two replica pulses using a narrowband filter function $F(\Omega-\omega)$ centered at a frequency offset Ω from the pulse center frequency. This filtered replica is then cross-correlated with the other unfiltered replica yielding the *sonogram trace*:

$$`I_{sonogram}(\delta t, \omega)$$
$$= \left| \int_{-\infty}^{+\infty} E(\Omega)F(\Omega-\omega)\exp(-i\Omega t)\,d\Omega \right|^2. \quad (16.9)$$

Both these techniques, FROG and sonogram, record data on a two-dimensional array, rather than recording two one-dimensional data traces as in the decorrelation. Examples of such measurements are shown in Figure 16.7. The excess data give rise to an increased robustness of the two-dimensional methods, which result in an improved immunity toward noise. Moreover, FROG and sonogram provide built-in consistency checks (marginals), which allow one to detect experimental flaws, e.g., due to limited phase-matching bandwidth or spectral filtering in the set-up. Beyond SHG-FROG, which can be simply understood as an extension of autocorrelation methods, a wide variety of FROG methods have been described. Most notable is *self-diffraction FROG (SD FROG)*, which is very important for measurements on amplifier systems [50]. Here, the FROG trace is given by

$$I_{\text{FROG}}^{\text{SD}}(\delta t, \omega)$$
$$= \left| \int_{-\infty}^{+\infty} E(t)^2 E(t-\delta t)^* \exp(-i\omega t)\,\mathrm{d}t \right|^2 . \quad (16.10)$$

The main advantage of SD FROG is that, unlike the as the SHG variant, it does not suffer from limited phase-matching bandwidths and it can be used in the deep UV. SD FROG generates a sinusoidal intensity pattern in a dielectric medium by crossing two beams under a small angle. This pattern creates an index grating via the Kerr effect and causes self-diffraction of both beams. One of the self-diffracted beams is then spectrally resolved and detected, similar to the SHG variant. This FROG variant, however, requires significantly higher pulse energies and cannot be used for the oscillator-level pulse energies. FROG can also be used in a cross-correlation variant to characterize one unknown pulse with the aid of another known one. This method is called XFROG and is of particular interest for the characterization of pulses in the UV and IR spectral range [61,62]. This variant of FROG has been demonstrated with the extremely complex pulse shapes of white-light continua generated in microstructure fibers [111].

Similar to the one-dimensional methods, FROG and the sonogram technique use an optimization strategy for pulse retrieval. Mathematically, it can be strictly shown that knowledge of the FROG trace of a pulse unambiguously defines the pulse's amplitude and phase except for *time reversal*, i.e., one does not know what is the front and what is the back of the pulse. This ambiguity can be removed with an additional measurement. In contrast to the one-dimensional case, a solution always exists, even though it can be a time-consuming computation to find this solution. Recent improvements of the FROG technique have led to very sophisticated retrieval procedures, which can rapidly retrieve the pulse from the FROG trace and allow update rates up to several hertz [49].

16.4.3 Spectral phase interferometry for direct electric field reconstruction

All techniques described so far involve auto- or cross-correlation, together with spectral resolution to remove any ambiguity in the pulse reconstruction. Recently, a completely different technique based on *spectral interferometry* [29] emerged for the characterization of femtosecond pulses. This technique is called *spectral phase interferometry for direct electric-field reconstruction* (SPIDER) [44]. The spectrum $S(\omega)$ of two identical pulses $I(t)$ with respective temporal delay ΔT is the spectrum of the single pulse $\tilde{I}(\omega)$, multiplied by a spectrally oscillating term. Measuring the spectral fringe spacing $\Delta\omega = 2\pi/\Delta T$ of $S(\omega)$ allows determination of the temporal spacing ΔT of the two pulses. If these two pulses are identical, the fringe spacing of $S(\omega)$ is also strictly constant over the entire spectrum. A spectral interferogram between a chirped and an unchirped pulse, however, allows not only determination of the delay between the pulses, but also the difference in chirp between the two pulses (Figure 16.6). This is the fundamental idea of spectral interferometry. SPIDER generates two delayed replicas of the pulse to be measured. It also generates a third pulse from the input pulse. The third is strongly chirped by sending it through a grating sequence, or through a highly dispersive glass block. The dispersion creates a GD between the red and blue Fourier components of the third pulse (Equation 16.1). This chirped pulse is then used to frequency shift the two replicas of the input pulse using sum-frequency generation. Because of their temporal delay and the strong chirp on the upconverter pulse, both replicas are shifted in frequency by different amounts. Measuring the spectral interferogram of the upconverted replicas allows sampling their relative phase delay as a function of frequency. This is now exactly the information needed to reconstruct the spectral phase. Together with an independent measurement of the amplitude spectrum, this yields a complete description of the pulse. Again, this technique has been demonstrated using pulses from a Ti:sapphire laser (see Figure 16.7).

The SPIDER method has been demonstrated with sub-6 fs pulses from Ti:sapphire lasers, with compressed pulses from an amplifier system [20] and with optical parametric amplifiers [115]. One of the major advantages of SPIDER is that it offers a direct reconstruction of the pulse profile, rather than requiring computationally intensive optimization strategies. Typically, the acquisition speed is only limited by the read-out speed of the charge-coupled device (CCD) array used in the spectrograph. Acquisition and reconstruction rates of up to 20 Hz have been demonstrated [82], which makes SPIDER an ideal online tool for aligning complex femtosecond laser systems. SPIDER can also be used in combination with pulse shapers [9]. Other than evolutionary strategies, which try to compensate phase distortions of a pulse by optimizing, e.g., its second-harmonic efficiency, SPIDER provides enough information to set directly the spectral phase to generate a bandwidth-limited pulse. Given the much more concise data of the SPIDER method, this should allow for a direct and much more rapid phase adaption than lengthy evolutionary optimization strategies. The rapid data acquisition capabilities of the SPIDER method can also be exploited in another way to measure spatially resolved temporal pulse profiles. SPIDER can be adapted to spatially resolve measurements using an imaging spectrograph together with a two-dimensional array. This set-up spatially resolves temporal pulse profiles along the axis defined by the entrance slit of the spectrograph. Rather than the methods discussed so far that integrate over the spatial beam profile, the spatial resolution enables building of an ultrafast camera. Such an ultrafast camera can detect differences in pulse width between beam center and off-center regions.

Some more recent developments include spatial shearing variants, such as the spatially encoded arrangement for SPIDER (SEA-SPIDER) [164]. In these variants of SPIDER, the pulse fronts of the interfering pulses are tilted orthogonal to the spectral plane in the analyzing spectrograph, which essentially yields a two-dimensional SPIDER trace. The information on the fringe spacing $\Delta\omega$ is now redundantly contained in every pixel line of a two-dimensional monitoring camera chip, placed in the Fourier plane of an imaging spectrometer. This redundancy greatly increases the reliability of the SPIDER method. A very similar idea, called two-dimensional shearing interferometry, has been developed in parallel [165]. Both these methods are particularly suitable for characterizing extremely short pulses. In fact, *SEA-SPIDER* has been used to characterize some of the shortest pulses achieved to date, which reached a remarkably short pulse duration corresponding to only a single cycle of the carrier field [166].

16.4.4 Dispersion scan

FROG qualifies as a tomographic technique, which does not immediately measure the pulse shape, yet allows its algorithmic reconstruction from measured data. In FROG, one spectrally resolves the signal from a nonlinear process, typically SHG, while varying the delay between two replicas of the pulse. In principle, the same information on the pulse can be scanned by changing another parameter, e.g., the GDD of the pulse, cf. Equation 16.2. In fact, varying the dispersion by a set of wedge prisms and overcompensating for their average GD by chirped mirrors, one can obtain the so-called *d-scan* traces [167], which formally contain identical information as FROG traces. An advantage over FROG is that the beam path is completely collinear, so no geometrical smearing may appear, a problem that was discussed above for noncollinear autocorrelators. *D-scan* therefore combines some of the virtues of FROG, in particular the redundancy of the two-dimensional measurements, with the advantages of interferometric autocorrelation. This makes the *d-scan* method attractive for characterization of few-cycle pulses.

16.4.5 Multiphoton intrapulse interference phase scan (MIIPS)

Multiphoton intrapulse interference phase scan (MIIPS) [168–170] is a method for simultaneously measuring and compensating femtosecond laser pulses using an adaptive pulse shaper, which is an integral part of the set-up. *MIIPS* pursues an iterative approach to maximize the conversion efficiency of the shaped output pulse in a nonlinear optical process, which is SHG in the simplest case. Once the optimum has been reached, one can safely assume that the phase of the output pulse is flat and the pulse is close to the Fourier limit. The beauty of *MIIPS* lies in the fact that one can easily generate arbitrarily shaped and rather complicated pulses while still maintaining full control of the exact generated pulse shape. Such pulses or pulse sequences are often important for coherent control experiments or high speed spectroscopy for chemical and/or biological applications. In contrast, however, the method seems to be less suitable for extremely short pulses.

16.4.6 Interferometric FROG

In autocorrelation measurements, it has always been customary to switch to a collinear arrangement when pulse durations are smaller than 100 fs. Using the same polarization in both correlator arms, one observes interference oscillations when scanning the arm length. Although it is no problem to record these fringes with a photodiode or multiplier in an autocorrelator, it requires synchronization of a camera with the exact movement of the translation stage to measure clean *interferometric FROG* traces. Details on how this can be accomplished can be found in Reference [171]. *Interferometric FROG* provides more information than noncollinear FROG, as it contains the regular FROG trace as well as a second one stemming from the oscillatory part. Both traces enable independent retrieval of the pulse shape, and in combination of the information in both traces, very complex pulses can be reconstructed [172]. Other than regular FROG, its interferometric variant does not suffer from beam smearing, i.e., it can be applied to few-cycle pulses. Another advantage of the collinear geometry is the fact that tight focusing on the nonlinear crystal can be employed, enabling superior signal-to-noise ratios even for pulses with relatively low peak powers. Moreover, even third-harmonic generation (THG) can be used with oscillator pulses for *interferometric THG-FROG* [173].

16.4.7 The coherent artifact

Regardless of the interrogation method used, problems in the interpretation of the results may arise when pulse instabilities occur. In an unstable pulse train with megahertz repetition rates, two successive pulses may already look completely different. As all techniques (FROG, SPIDER, or d-scan trace) discussed above can at best resolve dynamics at the millisecond range, they will inevitably only measure an *average* signal. Because all these techniques involve a nonlinear optical process, the average

FROG trace, e.g., will not be identical to the FROG trace of the average pulse [174,175]. This problem is also well known from autocorrelation, where one sometimes observes bimodal autocorrelation traces consisting of a broad pedestal and a cusp-like temporally narrow spike near zero delay [175]. The relevant pulse duration is that relating to the pedestal, but often the much shorter so-called coherence spike is erroneously interpreted as a short pulse. Unfortunately, similar misinterpretations can appear for all the methods discussed in this chapter, so great experimental care is needed. Fortunately, according to References [174,175], it appears that tomographic methods, like FROG or d-scan, do at least contain some redundancy which can help to indicate warning of a possible misinterpretation as a short pulse. SPIDER, in contrast, does not have such a built-in warning and would only indicate the width of the coherence spike. Fortunately, mode-locked solid-state lasers rarely ever exhibit a coherent artifact, but the behavior seems to be much more widespread with mode-locked semiconductor lasers [125]. Similar instabilities may also appear in compression experiments. If the suspicion of a coherence spike arises, one should therefore always measure with two different techniques, ideally including at least one tomographic one.

16.4.8 Conclusion: Which pulse measurement method should I use?

Given the plethora of methods that have been developed for characterizing short pulses, it appears difficult to decide which method to use. For many less critical applications, simple autocorrelation measurements may often still provide sufficient information, e.g., if the routine performance of a commercial laser needs to be checked. Measuring the pulse duration after initial installation and comparing with later measurements will fairly reliably indicate a degradation of the laser.

Unfortunately, autocorrelation does not provide detailed information about the pulse shape. If such information is necessary to understand the physical mechanisms occurring, then methods like FROG, SPIDER, d-scan, etc., are required. Then, one needs to consider the complexity and expected duration of the pulses under test. When pulses are complex, then FROG is probably the best suited method, preferably its X-FROG and interferometric FROG variants. These methods can measure very complicated pulse structures, but they are comparatively slow, both in the measurement process and in the iterative retrieval procedure.

If a quick update of the pulse shape is required then SPIDER (or one of its variants) is definitely the way to go. SPIDER traces can be measured at kHz acquisition rates, and retrieval is analytic and only requires millisecond times on a modern computer. SPIDER is also much more suitable for short pulses in the few-cycle range, but will fail if the pulses are very complex and probably even more when there are pulse train instabilities.

If both complications, complexity and few-cycle character, come together, it seems that there is no suitable method anymore. However, for this situation, interferometric FROG may offer a viable compromise. Although a decision on the right characterization techniques may be difficult, one can immediately discard certain techniques once the measurement problem is clearly defined. Moreover, one should only resort to complicated set-ups if there is no other way.

As said before, if the only problem is detection of a slow degradation of the pulse shape of a commercial laser then autocorrelation may suffice.

16.5 PHASE AND AMPLITUDE MODULATION OF SHORT OPTICAL PULSES

We shall now describe how short pulses can be modulated in phase and amplitude.

Grating or prism sequences, such as those introduced in Section 16.3.2, can also be employed to adjust dispersion in an adaptive way. A very common setup is the so-called *4f zero dispersion delay line* shown in Figure 16.8 [107,108]. This setup is very similar to the previously described stretcher, but operates at exactly −1 magnification equivalent to an effective grating distance of zero. A first grating disperses the input pulse, then a lens at distance *f* from the grating creates a spectrally dispersed picture of the input. A second identical lens-grating system reimages the spectrally dispersed picture back onto one point at a distance *f* from the second grating. Provided proper alignment, this setup is totally neutral in terms of dispersion, i.e., the shape of a pulse propagating through a *4f* assembly is not modified. However, as the pulse is spectrally dispersed in the Fourier plane in the center of the setup, a phase or amplitude modulator array at this location allows manipulation of the waveform.

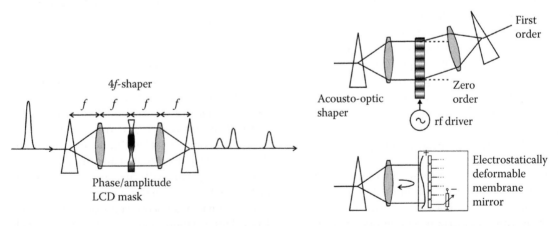

Figure 16.8 Ultrafast amplitude and phase modulators. Liquid crystal phase and amplitude shaper. The input pulse is spectrally dispersed and imaged onto the array with an adjustable retardation/absorption profile. This allows the generation of controllable pulse sequences from an input pulse. The same function can also be achieved by an acousto-optic device. An ultrasonic wave is written into an acousto-optic deflector, where control of the acoustic waveform allows control of the deflection of the optical Fourier components into the first order of the device. Phase control can also be achieved with micromachined mirror membranes, which can be electrostatically controlled.

16.5.1 Liquid crystal arrays

Several approaches exist for the technical implementation of the modulator array. These can be categorized into phase, amplitude, and combined amplitude/phase modulators. Another important aspect is pixelation and the number of pixels or degrees of freedom of such a device. Historically, *liquid crystal arrays* were first used in a 4f shaper [108]. Liquid crystals can be used for *phase modulators* and in combination with polarizers they may also serve as *amplitude modulators*. Combined devices consist of two liquid crystal arrays and polarizers and allow the control of both amplitude and phase. Most devices are pixelated with pixel numbers between 128 [109] and 640 [92]. The pixels typically consist of stripes of a few 10 μm width and can be individually electronically addressed. Liquid crystal phase masks have been used for a variety of applications. Used in a phase shaper, they can compensate for arbitrary dispersion. In this regard, the shaper can be understood as a programmable, microstructured dispersive device. It is particularly useful when dispersion is not very well known or where it may change over time. Recently, as an example, an adaptive pulse compressor was used to compress pulses to sub-6 fs pulse duration [112,113]. Despite the capability to impose an arbitrary spectral modulation on the pulse, one has to emphasize that a phase modulator can never compress pulses beyond the limit imposed by a Fourier transform of the spectral amplitude profile.

16.5.2 Acousto-optic pulse shaping

The same functionality can be achieved with an *acousto-optic approach*. The acousto-optic device replaces the liquid crystal array and serves both as amplitude and phase modulator. The acoustic wave transmitted into the acousto-optic modulator is generated by an arbitrary waveform generator, synchronized with the kilohertz repetition rate of the laser source. This acoustic wave generates an index pattern in the acousto-optic crystal. This pattern determines the amplitude and phase of spectral components that are diffracted into the first order of the acousto-optic deflector [25,43,103]; see Figure 16.8. The main application of this technique is with amplified *kilohertz-repetition-rate* pulses for amplitude shaping. A very similar device is the "dazzler" [105]. Here, the interaction of acoustic wave and optical wave is collinear rather than perpendicular. The acoustic wave transfers the input pulse entering in one axis of an acousto-optic crystal into the perpendicular axis. Again, an arbitrary waveform allows one to control the location and efficiency for each and every spectral component of the pulse. This allows compensation for the dispersion and spectral narrowing effects in amplifiers [105]. Note that both acousto-optic approaches are suited only for kilohertz repetition rates.

16.5.3 Flexible membrane mirrors

A further approach to phase shaping is to place a mirror with adjustable shape in the Fourier plane of the 4f set-up. Such *flexible membrane mirrors* [104] consist of a thin metal-coated silicon nitride membrane. The membrane is suspended over a number of planar electrodes, which act as *electrostatic actuators* (Figure 16.8). The membrane is electrically grounded, and a voltage is applied to the actuators, which locally deforms the membrane shape. Typically, these devices have around 10–20 electrodes, and deformations of a few microns can be achieved. For applications in dispersion compensation this complexity is normally more than sufficient. Pulses as short as 7 fs have been generated with this approach [4,9,116].

16.6 SUMMARY

In this chapter, we have highlighted the methods to generate and measure ultrafast optical pulses, thereby circumventing the fundamental electronic bandwidth limitations in conventional optoelectronics. The methods described allow for the generation and the characterization of optical pulses much shorter than a picosecond. We have examined propagation effects in optical media and their compensation. Finally, we also addressed setups that allow for a manipulation of amplitude and phase of an optical pulse with femtosecond temporal resolution. It is important to note that all these methods are based on rather conventional and relatively slow optoelectronic building blocks, such as cw laser pump diodes for light generation, photodiodes, photomultipliers, or CCD cameras for photodetection, and liquid crystal arrays for the modulation of optical short pulses. The trick lies in exploiting physical optics either by using an ultrafast nonlinear optical effect or by WDM.

This allows for ultrafast control, without having to control or measure electronic currents on a femtosecond time scale. Telecommunications is a paradigm for the possibilities enabled by optical multiplexing. Transferring more and more of the functionality of a communication network into photonic rather than electronic components has the potential for greatly increased capacities in the near future. Elementary ultrafast optoelectronic building blocks have been introduced in this chapter. These subsystems open up an avenue for many more applications of photonics in the ultrafast realm.

REFERENCES

1. Agrawal, G.P. 1989. *Nonlinear Fiber Optics* (New York: Academic).
2. Agrawal, G.P. 1997. *Fiber-Optic Communication Systems* (New York: Wiley-Interscience).
3. Albert, J. et al. 1995. Apodization of the spectral response of fiber Bragg gratings using a phase mask with variable diffraction efficiency. *Electron. Lett.*, 31, 222–223.
4. Armstrong, M.R., Plachta, P., Ponomarev, E.A., and Miller, R.J.D. 2001. Versatile 7-fs optical parametric pulse generation and compression by use of adaptive optics. *Opt. Lett.*, 26, 1152–1154.
5. Backus, S., Bartels, R., Thompson, S., Dollinger, R., Kapteyn, H.C., and Murnane, M.M. 2001. High-efficiency, single-stage 7-kHz high-average-power ultrafast laser system. *Opt. Lett.*, 26, 465–467.
6. Backus, S., Durfee, C.G. III, Murnane, M.M., and Kapteyn, H.C. 1998. High power ultrafast lasers. *Rev. Sci. Instrum.*, 69, 1207–1223.
7. Baltuska, A., Pshenichnikov, M.S., and Wiersma, D.A. 1999. Second-harmonic generation frequency-resolved optical gating in the single-cycle regime. *IEEE J. Quant. Electron.*, 35, 459–478.
8. Baltuska, A., Wei, Z., Pshenichnikov, M.S., Wiersma, D.A., and Szipocs, R. 1997. All-solid-state cavity dumped sub-5-fs laser. *Appl. Phys. B*, 65, 175–188.
9. Baum, P., Lochbrunner, S., Gallmann, L., Steinmeyer, G., Keller, U., and Riedle, E. 2002. Real-time characterization and optimal phase control of tunable visible pulses with a flexible compressor. *Appl. Phys. B*, 74, S219–S224.
10. Becker, P.C., Olsson, N.A., and Simpson, J.R. 1999. *Erbium-Doped Fiber Amplifiers: Fundamentals and Technology* (New York: Academic).
11. Brabec, T., and Krausz, F. 2000. Intense few-cycle laser fields: Frontiers of nonlinear optics. *Rev. Mod. Phys.*, 72, 545–591.
12. Brabec, T., Spielmann, C., and Krausz, F. 1991. Mode locking in solitary lasers *Opt. Lett.*, 16, 1961–1963.
13. Cheriaux, G., Rousseau, P., Salin, F., Chambaret, J.P., Walker, B., and Dimauro, L.F. 1996. Aberration-free stretcher design for ultrashort-pulse amplification. *Opt. Lett.*, 21, 414–416.
14. Chilla, J.L.A., and Martinez, O.E. 1991. Direct determination of the amplitude and the phase of femtosecond light pulses. *Opt. Lett.*, 16, 39–41.
15. Cho, S.H. et al. 2001. Generation of 90-nJ pulses with a 4-MHz repetition-rate Kerr-lens mode-locked Ti:Al_2O_3 laser operating with net positive and negative intracavity dispersion. *Opt. Lett.*, 26, 560–562.
16. Chudoba, C. et al. 2001. All-solid-state Cr: Forsterite laser generating 14-fs pulses at 1.3 μm. *Opt. Lett.*, 26, 292–294.
17. Desurvire, E. 1994. *Erbium-Doped Fiber Amplifiers: Principles and Applications* (New York: Wiley-Interscience).
18. Diels, J.-C., and Rudolph, W. 1996. *Ultrashort Laser Pulse Phenomena* (San Diego, CA: Academic).
19. Diels, J.-C., Fontaine, J.J., McMichael, I.C., and Simoni, F. 1985. Control and measurement of ultrashort pulse shapes (in amplitude and phase) with femtosecond accuracy. *Appl. Opt.*, 24, 1270–1282.
20. Dorrer, C. et al. 1999. Single-shot real-time characterization of chirped pulse amplification systems using spectral phase interferometry for direct electric-field reconstruction. *Opt. Lett.*, 24, 1644–1646.

21. Dragone, C. 1991. An N×N optical multiplexer using a planar arrangement of 2 star couplers. *IEEE Photon. Technol. Lett.*, 3, 812–815.

22. Eggleston, J.M., DeShazer, L.G., and Kangas, K.W. 1988. Characteristics and kinetics of laser-pumped Ti:sapphire oscillators. *IEEE J. Quant. Electron.*, 24, 1009–1015.

23. Ell, R. et al. 2001. Generation of 5-fs pulses and octave-spanning spectra directly from a Ti:sapphire laser. *Opt. Lett.*, 26, 373–375.

24. Farries, M.C., Sugden, K., Reid, D.C.J., Bennion, I., Molony, A., and Goodwin, M.J. 1994. Very broad reflection bandwidth (44 nm) chirped fibre gratings and narrow bandpass-filters produced by the use of an amplitude mask. *Electron. Lett.*, 30, 891–892.

25. Fetterman, M.R., Goswami, D., Keusters, D., Yang, W., Rhee, J.-K., and Warren, W.S. 1998. Ultrafast pulse shaping: amplification and characterization. *Opt. Express*, 3, 366–375.

26. Fork, R.L., Cruz, C.H.B., Becker, P.C., and Shank, C.V. 1987. Compression of optical pulses to six femtoseconds by using cubic phase compensation. *Opt. Lett.*, 12, 483–485.

27. Fork, R.L., Greene, B.I., and Shank, C.V. 1981. Generation of optical pulses shorter than 0.1 ps by colliding pulse modelocking. *Appl. Phys. Lett.*, 38, 617–619.

28. Fork, R.L., Martinez, O.E., and Gordon, J.P. 1984. Negative dispersion using pairs of prisms. *Opt. Lett.*, 9, 150–152.

29. Froehly, C., Lacourt, A., and Vienot, J.C. 1973. Notions de réponse impulsionelle et de fonction de transfert temporelles des pupilles optiques, justifications expérimentales et applications. *Nouv. Rev. Optique*, 4, 183–196.

30. Fürbach, A., Le, T., Spielmann, C., and Krausz, F. 2000. Generation of 8-fs pulses at 390 nm. *Appl. Phys. B*, 70, S37–S40.

31. Gallmann, L., Steinmeyer, G., Keller, U., Imeshev, G., Fejer, M.M., and Meyn, J.-P. 2001. Generation of sub-6-fs blue pulses by frequency doubling with quasi-phase-matching gratings. *Opt. Lett.*, 26, 614–616.

32. Gallmann, L., Sutter, D.H., Matuschek, N., Steinmeyer, G., and Keller, U. 2000. Techniques for the characterization of sub-10-fs optical pulses: A comparison. *Appl. Phys. B*, 70, S67–S75.

33. Gallmann, L. et al. 1999. Characterization of sub-6-fs optical pulses with spectral phase interferometry for direct electric-field reconstruction. *Opt. Lett.*, 24, 1314–1316.

34. Gires, F., and Tournois, P. 1964. Interférom-sitrè utilisable pour la compression d'impulsions lumineuses modulées en fréquence. *C. R. Acad. Sci. Paris*, 258, 6112–6115.

35. Golubovic, B. et al. 2000. Double Gires–Tournois interferometer negative dispersion mirror for use in tunable mode-locked lasers. *Opt. Lett.*, 25, 275–277.

36. Hasegawa, A. 1989. *Optical Solitons in Fibers* (Berlin: Springer).

37. Haus, H.A. 1975. Theory of mode locking with a slow saturable absorber. *IEEE J. Quant. Electron.*, 11, 736–746.

38. Haus, H.A. 1975. Theory of modelocking with a fast saturable absorber. *J. Appl. Phys.*, 46, 3049–3058.

39. Haus, H.A., Fujimoto, J.G., and Ippen, E.P. 1991. Structures for additive pulse modelocking. *J. Opt. Soc. Am. B*, 8, 2068–2076.

40. Haus, H.A., Shank, C.V., and Ippen, E.P. 1975. Shape of passively mode-locked laser pulses. *Opt. Commun.*, 15, 29.

41. Heppner, J., and Kuhl, J. 1985. Intracavity chirp compensation in a colliding pulse mode-locked laser using thin-film interferometers. *Appl. Phys. Lett.*, 47, 453.

42. Hill, K.O., and Meltz, G. 1997. Fiber Bragg grating technology fundamentals and overview. *IEEE J. Lightwave Technol.*, 15, 1263–1276.

43. Hillegas, C.W., Tull, J.X., Goswami, D., Strickland, D., and Warren, W.S. 1994. Femtosecond laser pulse shaping by use of microsecond radio-frequency pulses. *Opt. Lett.*, 737–739.

44. Iaconis, C., and Walmsley, I.A. 1998. Spectral phase interferometry for direct electric field reconstruction of ultrashort optical pulses. *Opt. Lett.*, 23, 792–794.

45. Imeshev, G., Arbore, M.A.,, Fejer, M.M., Galvanauskas, A., Fermann, M., and Harter, D. 2000. Ultrashort-pulse second-harmonic generation with longitudinally nonuniform quasi-phase-matching gratings: Pulse compression and shaping. *J. Opt. Soc. Am. B*, 17, 304–318.

46. Ippen, E.P., Haus, H.A., and Liu, L.Y. 1989. Additive pulse modelocking. *J. Opt. Soc. Am. B*, 6, 1736–1745.

47. Ippen, E.P., and Shank, C.V. 1975. Dynamic spectrometry and subpicosecond pulse compression. *Appl. Phys. Lett.*, 27, 488.

48. Ippen, E.P., Shank, C.V., and Dienes, A. 1972. Passive modelocking of the cw dye laser. *Appl. Phys. Lett.*, 21, 348–350.

49. Kane, D.J. 1999. Recent progress toward real-time measurement of ultrashort laser pulses. *IEEE J. Quant. Electron.*, 35, 421–431.

50. Kane, D.J., and Trebino, R. 1993. Characterization of arbitrary femtosecond pulses using frequency-resolved optical gating. *IEEE J. Quant. Electron.*, 29, 571–578.

51. Kärtner, F.X., Jung, I.D., and Keller, U. 1996. Soliton modelocking with saturable absorbers. *IEEE J. Sel. Top. Quant. Electron.*, 2, 540–556.

52. Kärtner, F.X. et al. 1997. Design and fabrication of double-chirped mirrors. *Opt. Lett.*, 22, 831–833.

53. Kashyap, R. 1999. *Fiber Bragg Gratings* (New York: Academic).

54. Keller, U., 'tHooft, G.W., Knox, W.H., and Cunningham, J.E. 1991. Femtosecond pulses from a continuously self-starting passively mode-locked Ti:sapphire laser. *Opt. Lett.*, 16, 1022–1024.

55. Kobayashi, T., and Shirakawa, A. 2000. Tunable visible and near-infrared pulse generator in a 5 fs regime. *Appl. Phys. B*, 70, S239–S246.

56. Krausz, F., Brabec, T., and Spielmann, C. 1991. Self-starting passive modelocking. *Opt. Lett.*, 16, 235–237.

57. L'Huillier, A., Lompre, L.-A., Mainfray, G., and Manus, C. 1992. *High-Order Harmonic Generation in Rare Gases*, eds. A. L'Huillier, L.-A. Lompre, G. Mainfray, and C. Manus, 139–206 (New York: Academic).

58. Lemoff, B.E., and Barty, C.P.J. 1993. Cubic-phase-free dispersion compensation in solid-state ultrashort-pulse lasers. *Opt. Lett.*, 18, 57–59.

59. Lemoff, B.E., and Barty, C.P.J. 1993. Quintic-phase-limited, spatially uniform expansion and recompression of ultrashort optical pulses. *Opt. Lett.*, 18, 1651–1653.

60. Lewenstein, M., Balcou, P., Ivanov, M.Y., L'Huillier, A., and Corkum, P.B. 1994. Theory of high-harmonic generation by low-frequency laser fields. *Phys. Rev. A*, 49, 2117–2132.

61. Linden, S., Giessen, H., and Kuhl, J. 1998. XFROG—A new method for amplitude and phase characterization of weak ultrashort pulses. *Phys. Status Solidi B*, 206, 119–124.

62. Linden, S., Kuhl, J., and Giessen, H. 1999. Amplitude and phase characterization of weak blue ultrashort pulses by downconversion. *Opt. Lett.*, 24, 569–571.

63. Liu, Z., Izumida, S., Ono, S., Ohtake, H., and Sarukura, N. 1999. High-repetition-rate, high-average-power, mode-locked Ti:sapphire laser with an intracavity continuous-wave amplification scheme. *Appl. Phys. Lett.*, 74, 3622–3623.

64. Maine, P., Strickland, D., Bado, P., Pessot, M., and Mourou, G. 1988. Generation of ultrahigh peak power pulses by chirped pulse amplification. *IEEE J. Quant. Electron.*, 24, 398–403.

65. Matuschek, N., Gallmann, L., Sutter, D.H., Steinmeyer, G., and Keller, U. 2000. Back-side coated chirped mirror with ultra-smooth broadband dispersion characteristics. *Appl. Phys. B*, 71, 509–522.

66. Matuschek, N., Kärtner, F.X., and Keller, U. 1999. Analytical design of double-chirped mirrors with custom-tailored dispersion characteristics. *IEEE J. Quant. Electron.*, 35, 129–137.

67. Mocker, H.W., and Collins, R.J. 1965. Mode competition and self-locking effects in a Q-switched ruby laser. *Appl. Phys. Lett.*, 7, 270–273.

68. Naganuma, K., Mogi, K., and Yamada, H. 1989. General method for ultrashort light pulse chirp measurement. *IEEE J. Quant. Electron.*, 25, 1225–1233.

69. Nagel, S.R., MacChesney, J.B., and Walker, K.L. 1985. *Optical Fiber Communications*, vol. 1, ed. T. Li (Orlando, FL: Academic). Chapter 1.

70. Nordlund, T.M. 1991. *Streak Cameras for Time-Domain Fluorescence*, ed. T.M. Nordlund (New York: Plenum).

71. Ouellette, F. 1987. Dispersion cancellation using linearly chirped Bragg grating filters in optical wave-guides. *Opt. Lett.*, 12, 847–849.

72. Parker, M.C., and Walker, S.D. 2001. Multiple-order adaptive dispersion compensation using polynomially-chirped grating devices. *Appl. Phys. B*, 73, 635.

73. Pennington, D.M. et al. 2000. Petawatt laser system and experiments. *IEEE J. Sel. Top. Quant. Electron.*, 6, 676–688.

74. Prein, S., Diddams, S., and Diels, J.C. 1996. Complete characterization of femtosecond pulses using an all-electronic detector. *Opt. Commun.*, 123, 567–573.

75. Ramaswami, R., and Sivarajan, K. 1998. *Optical Networks: A Practical Perspective* (San Mateo, CA: Morgan Kaufmann).

76. Reid, D.T. 2000. Algorithm for complete and rapid retrieval of ultrashort pulse amplitude and phase from a sonogram. *IEEE J. Quant. Electron.*, 35, 1584–1589.

77. Reid, D.T., Padgett, M., McGowan, C., Sleat, W.E., and Sibbett, W. 1997. Light-emitting diodes as measurement devices for femtosecond laser pulses. *Opt. Lett.*, 22, 233–235.

78. Salin, F., Squier, J., and Piche, M. 1991. Mode of Ti:Al$_2$O$_3$ lasers and self-focusing: A Gaussian approximation. *Opt. Lett.*, 16, 1674–1676.

79. Schoenlein, R.W. et al. 1996. Femtosecond x-ray pulses at 0.4 Å generated by 90° Thomson scattering: A tool for probing the structural dynamics of materials. *Science*, 274, 236–238.

80. Sherriff, R.E. 1998. Analytic expressions for group-delay dispersion and cubic dispersion in arbitrary prism sequences. *J. Opt. Soc. Am. B*, 15, 1224–1230.

81. Shirakawa, A., Sakane, I., Takasaka, M., and Kobayashi, T. 1999. Sub-5-fs visible pulse generation by pulse-front-matched noncollinear optical parametric amplification *Appl. Phys. Lett.*, 74, 2268–2270.

82. Shuman, T.M., Anderson, M.E., Bromage, J., Iaconis, C., Waxer, L., and Walmsley, I.A. 1999. Real-time SPIDER: Ultrashort pulse characterization at 20 Hz. *Opt. Express*, 5, 134–143.

83. Siegman, A.E. 1986. *Lasers* (Mill Valley, CA: University Science Books).

84. Siegman, A.E., and Kuizenga, D.J. 1974. Active mode-coupling phenomena in pulsed and continuous lasers. *Optoelectronics*, 6, 43–66.

85. Smit, M.K. 1988. New focusing and dispersive planar component based on optical phased array. *Electron. Lett.*, 24, 385–386.

86. Smith, N.J., Forysiak, W., and Doran, N.J. 1996. Reduced Gordon-Haus jitter due to enhanced power solitons in strongly dispersion managed systems. *Electron. Lett.*, 32, 2085–2086.

87. Smith, N.J., Knox, F.M., Doran, N.J., Blow, K.J., and Bennion, I. 1996. Enhanced power solitons in optical fibres with periodic dispersion management. *Electron. Lett.*, 32, 54–55.

88. Smith, P.R., Auston, D.H., and Nuss, M.C. 1988. Subpicosecond photoconducting dipole antennas. *IEEE J. Quant. Electron.* 24, 255.

89. Spence, D.E., Kean, P.N., and Sibbett, W. 1991. 60-fsec pulse generation from a self-mode-locked Ti:sapphire laser. *Opt. Lett.*, 16, 42–44.

90. Squier, J., Barty, C.P.J., Salin, F., LeBlanc, C., and Kane, S. 1998. Using mismatched grating pairs in chirped pulse amplification systems. *Appl. Opt.*, 37, 1638–1641.

91. Stingl, A., Spielmann, C., and Krausz, F. 1994. Generation of 11-fs pulses from a Ti:sapphire laser without the use of prisms. *Opt. Lett.*, 19, 204–206.

92. Stobrawa, G., Hacker, M., Feurer, T., Zeidler, D., Motzkus, M., and Reichel, F. 2001. A new high-resolution femtosecond puls shaper. *Appl. Phys. B*, 72, 627–630.

93. Strickland, D., and Mourou, G. 1985. Compression of amplified chirped optical pulses. *Opt. Commun.*, 56, 219–221.

94. Sullivan, A., and White, W.E. 1995. Phase control for production of high-fidelity optical pulses for chirped-pulse amplification. *Opt. Lett.*, 20, 192–194.

95. Sutter, D.H. et al. 1999. Semiconductor saturable-absorber mirror-assisted Kerr-lens mode-locked Ti:sapphire laser producing pulses in the two-cycle regime. *Opt. Lett.*, 24, 631–633.

96. Szipöcs, R., Ferencz, K., Spielmann, C., and Krausz, F. 1994. Chirped multilayer coatings for broadband dispersion control in femtosecond lasers. *Opt. Lett.*, 19, 201–203.

97. Takahashi, H., Nishi, I., and Hibino, Y. 1992. 10 GHz spacing optical frequency-division multiplexer based on arrayed waveguide grating. *Electron. Lett.*, 28, 380–382.

98. Tempea, G. 2001. Tilted-front-interface chirped mirrors. *J. Opt. Soc. Am. B*, 18, 1747–1750.

99. Treacy, E.B. 1969. Optical pulse compression with diffraction gratings. *IEEE J. Quant. Electron.*, 5, 454–458.

100. Trebino, R., DeLong, K.W., Fittinghoff, D.N., Sweetser, J., Krumbügel, M.A., and Richman, B. 1997. Measuring ultrashort laser pulses in the time-frequency domain using frequency-resolved optical gating. *Rev. Sci. Instrum.*, 68, 1–19.

101. Trebino, R., and Kane, D.J. 1993. Using phase retrieval to measure the intensity and phase of ultrashort pulses: Frequency-resolved optical gating. *J. Opt. Soc. Am. A*, 10, 1101–1111.

102. Tropf, W.J., Thomas, M.E., and Harris, T.J. 1995. *Properties of Crystals and Glasses*, eds. W.J. Tropf, M.E. Thomas, and T.J. Harris, 33.1–33.101 (New York: McGraw-Hill).

103. Tull, J.X., Dugan, M.A., and Warren, W.S. 1997. High resolution acousto-optic shaping of unamplified and amplified femtosecond laser pulses. *J. Opt. Soc. Am. B*, 14, 2348.

104. Vdovin, G.V. 1995. Spatial light modulator based on the control of the wavefront curvature. *Opt. Commun.*, 115, 170–178.

105. Verluise, F., Laude, V., Cheng, Z., Spielmann, C., and Tournois, P. 2000. Amplitude and phase control of ultrashort pulses by use of an acousto-optic programmable dispersive filter: Pulse compression and shaping. *Opt. Lett.*, 25, 575–577.

106. Walmsley, I.A., Waxer, L., and Dorrer, C. 2001. The role of dispersion in optics *Rev. Sci. Instrum.*, 72, 1–29.

107. Weiner, A.M. 2000. Femtosecond pulse shaping using spatial light modulators. *Rev. Sci. Instrum.*, 71, 1929–1960.

108. Weiner, A.M., Heritage, J.P., and Kirschner, E.M. 1988. High-resolution femtosecond pulse shaping. *J. Opt. Soc. Am. B*, 5, 1563–1572.

109. Weiner, A.M., Leaird, D.E., Patel, J.S., and Wullert, J.R. 1992. Programmable shaping of femtosecond optical pulses by use of 128-element liquid crystal phase modulator. *IEEE J. Quant. Electron.*, 28, 908–920.

110. Wittmann, M., Nazarkin, A., and Korn, G. 2001. Synthesis of periodic femtosecond pulse trains in the ultraviolet by phase-locked Raman sideband generation. *Opt. Lett.*, 26, 298–300.

111. Xu, L., Kimmel, M.W., O'Shea, P., Trebino, R., Ranka, J.K., Windeler, R.S., and Stentz, A.J. 2000. Measuring the intensity and phase of an ultrabroadband continuum. In *Proceedings of the Ultrafast Phenomena XII*, Charleston, SC, July 9–13, 129–131.

112. Xu, L., Li, L.M., Nakagawa, N., Morita, R., and Yamashita, M. 2000. Application of a spatial light modulator for programmable optical pulse compression to the sub-6-fs regime. *IEEE Photon. Technol. Lett.*, 12, 1540–1542.

113. Xu, L., Nakagawa, N., Morita, R., Shigekawa, H., and Yamashita, M. 2000. Programmable chirp compensation for 6-fs pulse generation with a prism-pair-formed pulse shaper. *IEEE J. Quant. Electron.*, 36, 893–899.

114. Yamakawa, K., Aoyama, M., Matsuoka, S., Takuma, H., Barty, C.P.J., and Fittinghoff, D. 1998. Generation of 16-fs, 10-TW pulses

at a 10-Hz repetition rate with efficient Ti:sapphire amplifiers. *Opt. Lett.*, 23, 525–527.

115. Zavelani-Rossi, M. et al. 2001. Pulse compression over a 170-THz bandwidth in the visible by use of only chirped mirrors. *Opt. Lett.*, 26, 1155–1157.

116. Zeek, E., Bartels, R., Murnane, M.M., Kapteyn, H.C., Backus, S., and Vdovin, G. 2000. Adaptive pulse compression for transform-limited 15-fs high-energy pulse generation. *Opt. Lett.*, 25, 587–589.

117. Zhang, Z., Nakagawa, T., Torizuka, K., Sugaya, T., and Kobayashi, K. 1999. Self-starting modelocked Cr^{4+}:YAG laser with a low-loss broadband semiconductor saturable-absorber mirror. *Opt. Lett.*, 24, 1768–1770.

118. Zhou, J., Taft, G., Huang, C.-P., Murnane, M.M., Kapteyn, H.C., and Christov, I.P. 1994. Pulse evolution in a broad-band-width Ti:sapphire laser. *Opt. Lett.*, 19, 1149–1151.

119. Martinez, O.E. 1987. 3000 times grating compressor with positive group velocity dispersion: Application to fiber compensation in 1.3–1.6 μm region. *IEEE J. Quant. Electron.*, 23, 59–65.

120. Dubietis, A. et al., 1992. Powerful femtosecond pulse generation by chirped and stretched pulse parametric amplification in BBO crystal. *Opt. Commun.*, 88, 433.

121. Cerullo, G., Baltuška, A., Mücke, O.D., and Vozzi, C. 2011. Few-optical-cycle light pulses with passive carrier-envelope phase stabilization. *Laser Photonics Rev.*, 5(3), 323–351.

122. Keller, U. et al. 1996. Semiconductor saturable absorber mirrors (SESAM's) for femtosecond to nanosecond pulse generation in solid-state lasers. *IEEE J. Sel. Top. Quant. Electron.*, 2, 435.

123. Keller, U. 2004. Ultrafast solid-state lasers. *Prog. Opt.*, 46, 1.

124. Keller, U. 2003. Recent developments in compact ultrafast lasers. *Nature*, 424, 831.

125. Keller, U., and Tropper, A.C. 2006. Passively modelocked surface-emitting semiconductor lasers. *Phys. Rep.*, 429, 67.

126. Moenster, M., Griebner, U. Richter, W., and Steinmeyer, G. Resonant saturable absorber mirrors for dispersion control in ultrafast lasers. *IEEE J. Sel. Top. Quant. Electron.*, 43, 174.

127. Keller, U., Miller, D.A.B., Boyd, G.D., Chiu, T.H., Ferguson, J.F., and Asom, M.T. 1992. Solid-state low-loss intracavity saturable absorber for Nd:YLF lasers: an antiresonant semiconductor Fabry–Perot saturable absorber. *Opt. Lett.*, 17, 505.

128. Spühler, G.J. et al. 2005. Semiconductor saturable absorber mirror structures with low saturation fluence. *Appl. Phys. B*, 81, 27–32.

129. Gires, F., and Tournois, P. 1964. Interféromètre utilisable pour la compression d'impulsions lumineuses modulées en fréquence. *C. R. Acad. Sci. Paris*, 258, 6112.

130. Schibli, T.R. et al. 2005. Ultrashort pulse-generation by saturable absorber mirrors based on polymer-embedded carbon nanotubes. *Opt. Express*, 13, 8025.

131. Cho, W.B. et al. 2010. Boosting the non linear optical response of carbon nanotube saturable absorbers for broadband mode-locking of bulk lasers. *Adv. Funct. Mater.*, 20, 1937.

132. Kieu, K., and Mansuripur, M. 2007. Femtosecond laser pulse generation with a fiber taper embedded in carbon nanotube/polymer composite. *Opt. Lett.*, 32, 2242-2244.

133. Bao, Q. et al. 2009. Atomic-layer graphene as a saturable absorber for ultrafast pulsed lasers. *Adv. Funct. Mater.*, 19, 3077.

134. Sun, Z. et al. 2010. Graphene mode-locked ultrafast laser. *ACS Nano*, 4, 803–810.

135. Tamura, K. et al. 1992. Self-starting additive pulse mode-locked erbium fibre ring laser. *Electron. Lett.*, 28, 2226.

136. Haus, H.A., Tamura, K., Nelson, L.E., and Ippen, E.P. 1995. Stretched-pulse additive pulse mode-locking in fiber ring lasers: Theory and experiment. *IEEE J. Quantum Electron.*, 31, 591.

137. Fermann, M.E., Haberl, F., Hofer, M., and Hochreiter, H. 1990. Nonlinear amplifying loop mirror. *Opt. Letter*, 15, 752.

138. Doran, N.J., and Wood, D. 1988. Nonlinear-optical loop mirror. *Opt. Lett.*, 13, 56.

139. Osaka University, Press Release. 2015. World-largest petawatt laser completed, delivering 2,000 trillion watts output. http://www.ile.osaka-u.ac.jp/en/information/information/2015/150727.pdf.

140. Photonics.com. 2014. National energetics to build 10 PW research laser. http://www.photonics.com/Article.aspx?AID=56706.

141. Optics.org. 2015. ELI pump laser installed at Lawrence Livermore. http://optics.org/news/6/3/3.

142. Vaupel, A., Bodnar, N., Webb, B., Shah, L., and Richardson, M. 2014. Concepts, performance review, and prospects of table-top, few-cycle optical parametric chirped-pulse amplification. *Opt. Eng.*, 53(5), 051507.

143. Dubietis, A., Butkus, R., and Piskarskas, A.P., 2006. Trends in chirped pulse optical parametric amplification. *IEEE J. Sel. Top. Quant. Electron.*, 12, 163.

144. Gale, G.M., Cavallari, M., and Hache, F. 1998. Femtosecond visible optical parametric oscillator. *J. Opt. Soc. Am. B*, 15(2), 702–714.

145. Kitagawa, Y. et al. 2004. Prepulse-free petawatt laser for a fast ignitor. *IEEE J. Quant. Electron.*, 40(3), 281.

146. Danson, C.N. et al. 2005. Vulcan petawatt: design, operation, and interactions at 5×10^{20} W cm^{-2}. *Laser Part. Beams*, 23, 87.

147. Waxer, L.J. et al. 2005. High-energy petawatt capability for the Omega laser. *Opt. Photon. News*, 16, 30.

148. Lozhkarev, V.V. et al. 2006. 200 TW 45 fs laser based on optical parametric chirped pulse amplification. *Opt. Express*, 14, 446.

149. Krausz, F., and Ivanov, M. 2009. Attosecond physics. *Rev. Mod. Phys.*, 81, 163.

150. Hentschel, M. et al. 2001. Attosecond metrology. *Nature*, 414, 509–513.

151. Udem, T., Holzwarth, R., and Hänsch, T.W. 2002. Optical frequency metrology. *Nature*, 416, 233–237.

152. Hänsch, T.W. 2006. Nobel lecture: Passion for precision. *Rev. Mod. Phys.*, 78, 1297.

153. Telle, H.R., Steinmeyer, G., Dunlop, A.E., Stenger, J., Sutter, D.H., Keller, U. 1999. Carrier-envelope offset phase control: A novel concept for absolute optical frequency measurement and ultrashort pulse generation. *Appl. Phys. B*, 69, 327.

154. Helbing, F.W., Steinmeyer, G., and Keller, U. 2003. Carrier-envelope offset phase-locking with attosecond timing jitter. *IEEE J. Sel. Top. Quant. Electron.*, 9, 1030.

155. Telle, H.R., Lipphardt, B., and Stenger, J. 2002. Kerr-lens, mode-locked lasers as transfer oscillators for optical frequency measurements. *Appl. Phys. B*, 74, 1–6.

156. Borchers, B., Anderson, A., and Steinmeyer, G. 2014. On the role of shot noise in carrier-envelope phase stabilization. *Laser Photon. Rev.*, 8, 303.

157. Fuji, T. et al. 2005. Attosecond control of optical waveforms. *New J. Phys.*, 7, 116.

158. Jones, D.J. et al. 2000. Carrier-envelope phase control of femtosecond mode-locked lasers and direct optical frequency synthesis. *Science*, 288, 635.

159. Koke, S., Grebing, C., Frei, H., Anderson, A., Assion, A., and Steinmeyer, G. 2010. Direct frequency comb synthesis with arbitrary offset and shot-noise-limited phase noise. *Nat. Photonics*, 4, 462–465.

160. Borchers, B., Koke, S., Husakou, A., Herrmann, J., and Steinmeyer, G. 2011. Carrier-envelope phase stabilization with sub-10 as residual timing jitter. *Opt. Lett.*, 36, 4146–4148.

161. Mehendale, M., Mitchell, S.A., Likforman, J.-P., Villeneuve, D.M., and Corkum, P.B. 2000. *Opt. Lett.*, 25, 1672.

162. Kakehata, M. et al. 2001. Single-shot measurement of carrier-envelope phase changes by spectral interferometry. *Opt. Lett.*, 26, 1436–1438.

163. Paulus, G.G. et al. 2001. Absolute-phase phenomena in photoionization with few-cycle laser pulses. *Nature*, 414, 182.

164. Wyatt, A.S., Walmsley, I.A., Stibenz, G., Steinmeyer, G. 2006. Sub-10 fs pulse characterization using spatially encoded arrangement for spectral phase interferometry for direct electric field reconstruction. *Opt. Lett.*, 31, 1914–1916.

165. Birge, J.R., and Kärtner, F.X. 2008. Analysis and mitigation of systematic errors in spectral shearing interferometry

of pulses approaching the single-cycle limit [Invited]. *J. Opt. Soc. Am. B*, 25, A111–A119.

166. Balciunas, T. et al. 2015. A strong-field driver in the single-cycle regime based on self-compression in a kagome fibre. *Nat. Commun.*, 6, 6117.

167. Miranda, M. et al. 2012. Characterization of broadband few-cycle laser pulses with the d-scan technique. *Opt. Express*, 20, 18732–18743.

168. Dantus, M., Lozovoy, V.V., and Pastirk, I. 2003. Measurement and repair: The femtosecond wheatstone bridge. *OE Mag.*, 9, 15–17.

169. Lozovoy, V.V., Pastirk, I., and Dantus, M. 2004. Multiphoton intrapulse interference 4: Characterization and compensation of the spectral phase of ultrashort laser pulses. *Opt. Lett.*, 29, 775–777.

170. Xu, B., Gunn, J.M., Dela Cruz, J.M., Lozovoy, V.V., and Dantus, M. 2006. Quantitative investigation of the MIIPS method for phase measurement and compensation of femtosecond laser pulses. *J. Opt. Soc. Am. B*, 23, 750–759.

171. Stibenz, G., and Steinmeyer, G. 2005. Interferometric frequency-resolved optical gating. *Opt. Express*, 13, 2617.

172. Stibenz, G., and Steinmeyer, G. 2006. Structures of interferometric frequency-resolved optical gating. *IEEE J. Sel. Top. Quant. Electron.*, 12, 286.

173. Das, S.K. et al. 2011. Highly efficient THG in TiO_2 nanolayers for third-order pulse characterization. *Opt. Express*, 19, 16985–16995.

174. Ratner, J., Steinmeyer, G., Wong, T.C., Bartels, R., and Trebino, R. 2012. Coherent artifact in modern pulse measurements. *Opt. Lett.*, 37, 2874–2876.

175. Rhodes, M., Steinmeyer, G., Ratner, J., and Trebino, R. 2013. Pulse-shape instabilities and their measurement. *Laser Photon. Rev.*, 7, 557.

Integrated optics

NIKOLAUS BOOS
EADS Eurocopter SAS

CHRISTIAN LERMINIAUX
Corning SA–CERF

17.1 INTRODUCTION

Integrated optics, which has been a research topic for about 20 years, deals with compact single function devices and the integration of multiple optical functionalities on a single chip or into a single package. Ultimately, this is seen as a way to reduce footprint and cost with respect to conventional bulk and micro-optic components, and also to obtain components with increased performance or even new functionalities.

The fabrication technologies have benefited from the experience in the semiconductor industry. However, there are a few main differences between integrated electronics and integrated optics: routing of photons on a chip needs structures with sizes of the order of millimeters, which results in a lower circuit density with respect to electronics. Also, for a given optical function there may be one preferred material, which currently puts a limitation on the development and deployment of monolithic integrated optical circuits. Hybrid integration combines the best materials on a common platform and represents an intermediate solution to this problem. Both monolithic and hybrid integration approaches will probably coexist depending on application, performance and cost.

At present, integrated optics is the technology of choice for a few functions in the optical network. With the increasing maturity of the fabrication technology and the growth in bandwidth and ongoing standardization of optical networks, integrated optic devices will become more cost effective than discrete components.

This chapter is organized as follows: we will start with a description of waveguides and basic elements for phase and polarization control, waveguide couplers and interferometers.

Fabrication processes and properties of common opto-electronic materials will be sketched in Section 17.3, followed by device packaging and the techniques for function and material integration.

Section 17.5 gives a brief overview of optical networks and the functions therein, followed by a selection of recent publications on integrated devices to realize these network elements. We will conclude with all-optical components, which will slowly replace electronics in future generation networks.

17.2 THE INTEGRATED OPTICS TOOLBOX—WAVEGUIDES AND BASIC DEVICES

17.2.1 Waveguides and requirements

Planar optical waveguides are the keys for the construction of integrated optical circuits. In Chapters A1.5 and A2.4 a detailed formalism for the propagation of an electromagnetic wave in dielectric media based on Maxwell's equations and the boundary conditions for the electrical and magnetic fields has been developed for

- Slab waveguides, in which light is confined in only one direction.
- Rectangular waveguides, in which light is confined in two directions. As an exact analytical solution for the electric and magnetic fields cannot be given in this case, a variety of approximation and numerical methods [1,2] exist. For example, tightly confined modes are needed for low threshold III–V lasers or merely guiding light efficiently from one point to another point, and they represent the foundation of planar integrated optics. Cross-sections for commonly used rectangular waveguides are shown in Figure 17.1.

One basic requirement for large-scale integration is that the overall device losses should be low, which in turn translates into an upper limit for the propagation losses through straight and curved

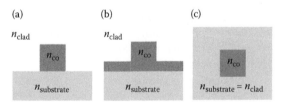

Figure 17.1 Cross-sections through common waveguides. (a) The raised strip guide is formed by an etching process removing the higher index material n_{co} on both sides of the guide. (b) The rib or ridge waveguide is similar to (a), except that the higher index material n_{co} is not completely removed. (c) The channel waveguide can be formed by ion implantation, ion exchange/diffusion processes or over-cladding a waveguide as depicted in (a).

waveguides. Exact values depend on the functionality and final device size, but typically <0.1 dB cm^{-1} is needed for silica waveguides. There are essentially four loss contributions:

- Absorption in the material through atomic/molecular transitions in amorphous materials, and through excitation of electrons from the valence to the conduction band (interband absorption), or electrons within the conduction band (free carrier absorption) in semiconductor materials.
- Scattering occurs at defects or material inhomogeneities within the guide, or through scattering at the surfaces through roughness induced in the waveguide fabrication process.
- Radiation losses arise in waveguide bends with constant bend radius and in waveguides with discontinuities in the bend radius (transition losses). Bending losses can be minimized by increasing the bend radius R and confining the mode well by increasing the relative index contrast Δ

$$\Delta = (n_{core} - n_{clad}) / n_{core}$$

Transition losses can be minimized through a lateral offset between waveguide sections or by varying the bend radius adiabatically.

- Substrate leakage—as will be mentioned in Section 17.3.1, silica waveguides are commonly fabricated on a silicon substrate, from which they are optically separated via a buffer layer. If this buffer layer is not thick enough, the high index of silicon gives rise to leakage of light from the core into the substrate. As a rule of thumb, the buffer layer should be at least twice as thick as the core layer.

Other losses that play a role in interconnecting optical building blocks arise from a mismatch of the respective mode fields as well as angular or lateral misalignments (see Section 17.4.1).

In the remainder of this section we will give an overview on

- Dynamic index control in waveguides
- Waveguide couplers and power splitters
- Interferometric devices—*Mach–Zehnder interferometers*, the *arrayed waveguide grating* (AWG) and *ring resonators*

- Modelling of integrated functions

17.2.2 Dynamic phase and polarization control in optical waveguides

Functional devices often need a dynamic control of the properties of the guided mode (effective index, polarization state), and depending on the application different physical phenomena are used.

The *thermo-optic effect* uses the temperature dependence of the waveguide index dn/dT to obtain a phase shift

$$\Delta\phi = \frac{2}{\lambda}\frac{dn}{dT}\Delta T \times L$$

where L is the length over which the waveguide temperature is changed by ΔT.

In practice, this effect is used in silica ($dn/dT = 1 \times 10^{-5}$ K^{-1} [3]), polymer ($dn/dT = -2 \dots -3 \times 10^{-4}$ K^{-1} [4]), Si (1.8×10^{-4} K^{-1} [5]) or lithium niobate (LiNbO$_3$, $dn/dT = 5 \times 10^{-5}$ K^{-1}) waveguides for switching or modulating devices with tuning speeds of the order of milliseconds (polymer, silica) to microseconds (silicon).

17.2.2.1 ELECTRO-OPTIC EFFECTS

- Current injection into a semiconductor junction or application of a reverse voltage changes the electron/hole density in the valence and conduction bands of a semiconductor. As a consequence, the absorption for photons with a wavelength close to the band gap energy will change, which in turn results in an index variation, since both parameters are linked through the Kramers–Kronig relations [6,7].
- Application of an electric field to a dielectric medium leads to an instantaneous induced polarization P [8]

$$P = \varepsilon_0 \times \left(\chi(1)E + \chi(2)EE + \chi(3)EEE + \cdots\right)$$

$$\equiv P_{Linear} + P_{Nonlinear}$$

where $\chi(i)$ are the tensors of the linear ($i=1$), quadratic ($i=2$) and higher order susceptibilities. $\chi(1)$ accounts for linear phenomena like absorption and reflection.

The second order susceptibility $\chi(2)$ in materials without inversion symmetry leads to the linear electro-optic (Pockels) effect, and the associated index change is usually expressed as

$$\Delta\left(\frac{1}{n^2}\right)_i = \sum_{j=1...3} r_{ij} E_j$$

with $j=1,\ldots,3$ denoting x-, y- and z-axes and r_{ij} being the electro-optic tensor.

Lithium niobate (LiNbO$_3$, $r_{33}=30.9$ pm V^{-1} [9]) is widely used for high speed modulation and switching. Induced $\chi(2)$ is possible through poling, which has been reported in silica glasses ($r=1$ pm V^{-1} [10]) and polymer ($r_{33}=13$ pm V^{-1} [11]).

Third order susceptibility $\chi(3)$ allows the control of the refractive index through the electrical field of a second optical signal (all-optical processing). The index change can be parametrized as

$$\Delta n = n_2 I$$

with I being the power density of the optical control signal. In silica, n_2 is of the order of 10^{-16} cm^2 W^{-1} [8,10], and therefore is too small for integrated devices. On the other hand, semiconductors exhibit a very large nonlinearity (n_2 of the order of 10^5–10^6 cm^2 W^{-1}) used for switching (Section 17.5.4) and all-optical signal processing (Section 17.5.6).

We also refer to Chapters B1.2, B4 and B6 for a more detailed treatment of electro-optic control.

The *acousto-optic* effect: mechanical strain from a surface or volume acoustic wave induces index changes, which in turn alter the phase of light. Depending on the interaction length between the acoustic wave and the optical mode, one distinguishes between Raman–Nath (short interaction length) and Bragg type modulators.

Magneto-optic control uses the Faraday effect, i.e., linearly polarized light is rotated in the presence of a magnetic field. The Faraday effect can be used for polarization-splitting and nonreciprocal devices such as optical isolators and circulators.

17.2.3 Waveguide couplers and power splitters

Waveguide couplers and power splitters are needed for the construction of interferometric devices, which will be described in Section 17.2.4.

Directional couplers: when waveguides are brought into proximity in such a way that their mode fields partially overlap, power is transferred from one waveguide to another (Figure 17.2). For

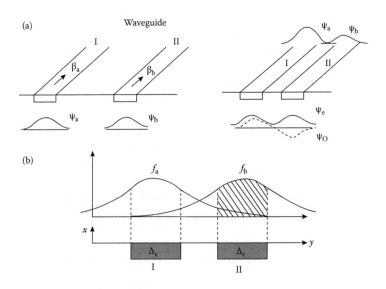

Figure 17.2 Directional coupler [13]. (a) Propagating modes in two uncoupled (left) and coupled (right) waveguides. In analogy to a coupled pendulum, the coupled system has both symmetric ψ_e and asymmetric ψ_o fundamental modes (having different propagation constants) which are excited and lead to a periodic energy exchange between modes ψ_a and ψ_b. (b) Cross section through a directional coupler. The coupling coefficient κ is proportional to the overlap integral of the shaded areas. (Reproduced by permission of The McGraw-Hill Companies.)

identical lossless waveguides and co-propagating modes, the power coupling ratio is given by (I_0 is the incoming power, I_1 and I_2 are the powers at the end of the coupler)

$$I_1 / I_0 = \sin^2(\kappa z) \equiv \sin\left(\frac{\pi}{2L_c} z\right)$$

$$I_2 / I_0 = \cos^2(\kappa z) \equiv \cos^2\left(\frac{\pi}{2L_c} z\right)$$

The coupling coefficient κ (Figure 17.2b) increases with the decreasing gap between the coupled waveguides, and by adjusting the ratio of propagation length z to coupling length L_c (defined as the length needed for complete power transfer), any splitting ratio can be designed. However, the wavelength dependent mode field diameter makes the coupling ratio also wavelength dependent; this can be avoided in more sophisticated designs with tapered asymmetric coupling regions [12]. For a treatment of counter-propagating modes, see Reference [13].

On the other hand, the operation of Y-splitters (Figure 17.3) depends only on their symmetry and thus the splitting ratio is wavelength independent. Power splitters (1×2^n) can be built by concatenation of Y-splitters.

A simpler way to obtain a large number of input or output ports is the star coupler [14], in which a multimode slab waveguide is placed between a fan of input and output waveguides (Figure 17.4). A uniform splitting ratio is achieved by adequate mutual coupling between input waveguides, and couplers with up to 144×144 ports have been published [15].

Multimode interference couplers [16] are based on self-imaging, i.e., the fact that single or multiple images are produced when higher order modes are excited within a wide multimode waveguide, and these modes interfere (constructively or destructively) along the propagation direction of the light. An example of a 1×4 splitter is depicted in Figure 17.5.

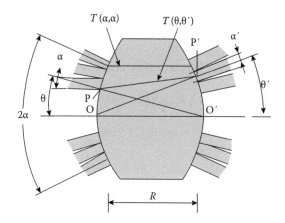

Figure 17.4 Star coupler [14], consisting of an array of input and output waveguides interconnected by a multimode slab guide. Light entering the multimode slab waveguide will diffract and excite fundamental and higher order modes, resulting in mixing and excitation of the guided modes of the output waveguide array. (Reproduced by permission of IEEE.)

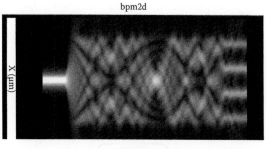

Figure 17.5 Multimode interference MMI coupler [16]. The diagram shows the amplitude distribution in a 1×4 splitter as obtained from beam propagation modelling (BPM) [196]. Light entering from the left will excite multiple modes in the middle section, which interfere and create local minima and maxima. Finally, constructive interference is obtained in multiple focal points, and the fundamental mode in the output waveguides is excited.

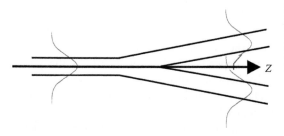

Figure 17.3 Y-splitter—the fundamental mode propagating in the left input waveguide will be equally split between the two output guides. For a symmetric device, the splitting ratio is 50%/50% (3 dB). 2^n splitter trees can be obtained through cascading of Y-splitters.

17.2.4 Interferometric devices—Mach–Zehnder interferometers, the arrayed waveguide grating, ring resonators

The Mach–Zehnder interferometer is a key element for filters, switches and modulators, which will be discussed later in Section 17.5. A schematic is shown in Figure 17.6—light is split in a first directional coupler, passes through the delay lines and then recombines. The transmitted optical power is a function of the phase difference, and is given by

$$I_1 / I_0 = \sin^2(\phi / 2) \text{ "bar" path}$$

$$I_2 / I_0 = \cos^2(\phi / 2) \text{ "cross" path}$$

in the case of perfect 50%/50% (3 dB) directional couplers.

For the "cross" path, there will be a constructive interference for $\phi = m,\ldots, 2\pi$ and destructive interference for $\phi = (2m+1)\pi$. Tunable devices can be made by dynamic control of ϕ via the effects as described in Section 17.2.2.

For broadband devices (i.e., devices having a flat spectral response over a wide passband), such as switches, the phase difference will be within $[0, \pi]$ ("symmetric MZI"). If spectral features within the

passband are desired (as in the case of the equalizing filters described in Section 17.5.2), the phase difference will rather be chosen to be within $[m \cdot 2\pi, (m+1)2\pi]$. The fixed part $\phi = m \cdot 2\pi$ will be generated by a fixed delay of geometrical length in one arm ("asymmetric MZI")

$$\Delta L = \frac{m\lambda}{n}$$

and the additional 2π will be kept tunable. The device has a periodic response, which will be repeated for every free spectral range (FSR)

$$\text{FSR} \equiv \Delta\lambda = \frac{n\Delta L}{m(m+1)}$$

The arrayed waveguide grating (AWG) is the analogue to a bulk diffraction grating and can be used in optical networks as a multiplexer/demultiplexer (Figure 17.7a) in wavelength division multiplexed (WDM) transmission systems, or as a wavelength selective building block for add–drop devices or optical crossconnects (OXCs) (Sections 17.5 and C1).

The AWG [17,18] (Figure 17.7b) consists of input/output waveguides, two star couplers and an array of waveguides with a constant path length difference ΔL between adjacent guides. Light coming into the first "free propagating region" (FPR) radiates and excites all waveguides in the grating. After propagation through the array, the light will constructively interfere in one focal point of the second star coupler if the grating condition

$$m\lambda = n\Delta L$$

is fulfilled. n is the effective index of the mode guided in the waveguide array, λ is the wavelength, m is the grating order. The location of the focal point depends on the wavelength as the phase delay between adjacent guides is given by $\Delta L/\lambda$, and the passbands have a Gaussian shape as shown in Figure 17.7c.

A ring resonator consists of couplers and a feedback loop of geometrical length L and is the analogue to the free-space Fabry–Perot interferometer.

The two-port resonator shown in Figure 17.8a is the simplest case: it has a unity power transmission and a phase response, making it useful in the synthesis of phase compensating devices [100].

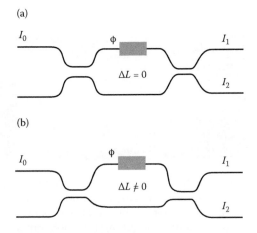

Figure 17.6 Symmetric (a) and asymmetric (b) Mach–Zehnder interferometer consisting of two directional couplers, two interferometer arms and a dynamic phase shifter. The path from input 0 to output 1 is called "bar", the path 0 → 2 "cross".

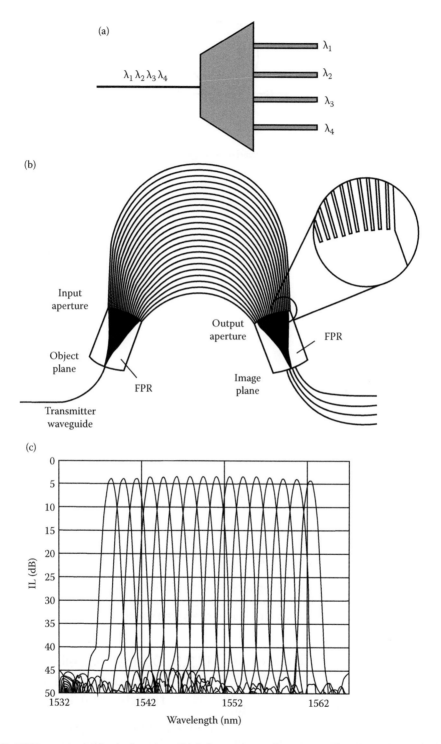

Figure 17.7 AWG router. (a) Basic function. (b) Schematic configuration of an AWG [17]: light enters the input star coupler (free propagating region FPR), will then diffract and then excites guided modes in each of the waveguides in the array. In the arrayed waveguide section there is a constant delay ΔL between adjacent arms, and finally constructive interference is obtained in the image plane (second star coupler). (c) Spectrum of a 1×16 demultiplexer with Gaussian passband and <5 dB fiber–fiber losses [197]. (Reproduced by permission of IEEE.)

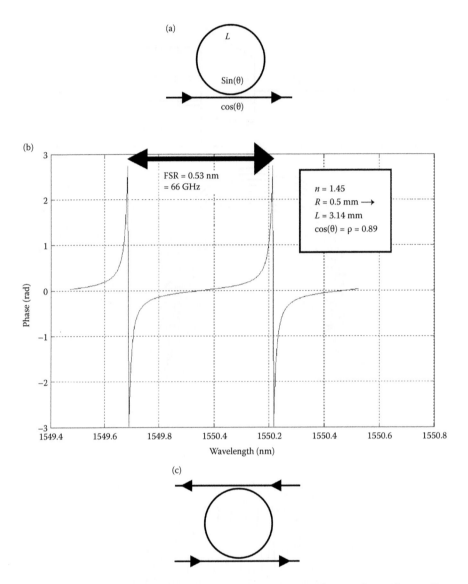

Figure 17.8 Ring resonator. (a) Two-port ring resonator consisting of a coupler with coupling angle θ and a feedback loop of length L. (b) The phase response calculated from the transfer function given in Table 17.1 has a resonance peak which is repeated every free spectral range FSR. (c) Open ring resonator.

Open ring resonators (Figure 17.8c) have resonances in both amplitude and phase response and may be used in $N \times N$ switch matrices [19] or for bistable devices having a nonlinear medium in the feedback loop [2].

In analogy to the Mach–Zehnder interferometer, the FSR of a ring resonator is given by

$$\text{FSR} = \frac{\lambda^2}{nL}$$

17.2.5 Optical modelling—beam propagation methods and the transfer matrix formalism

For the modelling of simple components or optical integrated circuit building blocks, the beam propagation method (BPM) and mode solvers are used:

- In BPM, the field is propagated stepwise through slices of the known waveguide structure, and in each step a phase correction is applied if the propagation constant of the guided mode changes through variations of the refractive index or the waveguide geometry. BPM is extensively treated in [2,7,20].

- Mode solvers calculate the field distributions and propagation constants for the electrical and magnetical eigenmodes, and are often used as starting conditions for a subsequent beam propagation modelling. Methods solving the full Maxwell vector equations or semivectorial approximations are employed. The accuracy of different methods is compared in Reference [7].

The applicability of the BPM is limited by the computation time needed for the simulation, so that alternative tools are needed for circuits with a higher degree of complexity. One way to achieve this is to use a transfer matrix formalism relating the amplitudes at the inputs and outputs of a device (Figure 17.9), and the unitarity of the matrix assumes that filters are lossless. $H_{11}(\omega)$ is the complex transfer function giving access to the filter's power transmission $|H_{11}(\omega)|^2$ and phase)$=\phi(\omega)=\arctan\left[\mathrm{Im}(H_{11}(\omega))/\mathrm{Re}(H_{11}(\omega))\right]$.

Transfer matrices and transfer functions of some devices described in this section are listed in Table 17.1. The amplitude and phase characteristics of a more complex filter are then obtained from the multiplication of the elementary matrices, and examples are the serial and parallel lattice filters to be described in Section 17.5.

Finally, the transfer function can be expressed as

$$H_{11}(\omega) = \frac{A(\omega)}{B(\omega)}$$

where the polynomials $A(\omega)$ and $B(\omega)$ have only zeros. Depending on the numbers of zeros or poles of $H_{11}(\omega)$, filters are classified as [21,22]:

- Moving average (MA), also called finite impulse response (FIR), filters have only zeros and consist only of forward paths. Examples are the AWG and the Mach–Zehnder interferometer.
- Autoregressive filters (AR) have only poles, and the two-port ring resonator (Figure 17.8a) is one example.
- Autoregressive moving average filters (ARMA) have both zeros and poles. Infinite impulse response filters (IIR) represent a sub-category with at least one pole.

For a detailed treatment of the transfer matrix formalism, see References [20] (Chapter 6.5) and [22] (Chapter 3.3), with application to the synthesis of optical filters in [23–25].

17.3 INTEGRATED OPTICS MATERIALS AND FABRICATION TECHNOLOGY

A major challenge facing integrated optical circuit developers is the fact that devices can be fabricated in different materials, and the ultimate technology choice will be based on the performance, manufacturability and cost.

Hybrid integration (Section 17.4) allows combining the best suitable materials on one platform, but for monolithic circuits a compromise may be needed.

This section will give an overview of the commonest materials, their properties and the fabrication processes currently deployed in integrated optics; we refer to Chapter 2 for further reading.

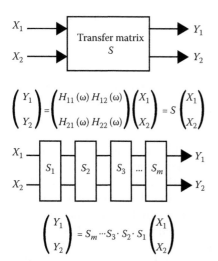

Figure 17.9 Device with two input and output ports and unitary transfer matrix S (top); serial filter through concatenated two port devices (bottom).

Table 17.1 Transfer matrices and transfer functions for waveguide devices described in sections 17.2.3 and 17.24

Sketch	Matrix/transfer function	Interpretation
Directional coupler	$\begin{pmatrix} \cos(\theta) & -j\sin(\theta) \\ -j\sin(\theta) & \cos(\theta) \end{pmatrix}$ $\equiv \begin{pmatrix} \cos(\kappa z) & -j\sin(\kappa z) \\ -j\sin(\kappa z) & \cos(\kappa z) \end{pmatrix}$	θ = coupling angle ($\pi/4$ for 3 dB coupler)
Phase shifter/delay line	$\begin{pmatrix} e^{-j\phi} & 0 \\ 0 & 1 \end{pmatrix} \begin{pmatrix} e^{-j2\,nL/\lambda} & 0 \\ 0 & 1 \end{pmatrix}$	Propagation of a mode with effective index n, wavelength λ through a section of length L
Mach–Zehnder	$1/\sqrt{2}\begin{pmatrix} 1 & -j \\ -j & 1 \end{pmatrix}$ $\times \begin{pmatrix} e^{-j\phi} & 0 \\ 0 & 1 \end{pmatrix} 1/\sqrt{2}\begin{pmatrix} 1 & -j \\ -j & 1 \end{pmatrix}$	For 3 dB couplers, $\theta = \pi/4$
Phasar/AWG	$H(\omega) = \sum_{l=1}^{N} A(l) \exp\left(-j\omega \dfrac{n_{\text{eff}} l \Delta L}{c}\right)$	N = number of waveguides in the arrayed waveguide grating. $A(l)$ is the amplitude in guide l, following a Gaussian distribution
Open ring resonator	$H(\omega) = \dfrac{\rho - \exp[-j(\omega T + \phi)]}{1 - \rho \exp[-j(\omega T + \phi)]}$ $= \dfrac{\cos(\theta) - \exp[-j(\omega T + \phi)]}{1 - \cos(\theta) - \exp[-j(\omega T + \phi)]}$	ρ is the through path amplitude transmission of the coupler. T is the roundtrip time through the feedback loop
Parallel lattice filter/ tapped delay line filter (MA)	$H(\omega) = \sum_{k=0}^{N-1} h_k e^{-jkT_k\omega}$	N = number of delay lines. T_k is the time delay through arm k.

17.3.1 Silica

Most of the commercially available passive opto-electronic circuits use doped silica waveguides, as propagation losses below 0.1 dB cm^{-1} are easily achieved in the 1.3 and 1.5 μm telecom windows. Pioneering work in this field has been done in the early 1970s–1980s by NTT, and today devices are commercialized by many vendors (NEL, SDL/PIRI, IONAS/NKT Integration, Agere, Hitachi, etc.).

A typical fabrication process based on deposited thin films is shown in Figure 17.10: a thin layer of silica is deposited on a planar silica or silicon substrate via chemical vapor deposition (CVD) processes such as flame hydrolysis (FHD) [3,26] or plasma enhanced deposition (PECVD) [27] of gaseous precursors. In the case of the silicon substrate, a pure SiO$_2$ buffer layer with 10–20 μm thickness optically isolates the waveguide layer. The index of the waveguide layer is raised from the initial silica index ($n = 1.444$ at 1550 nm) above the index of the cladding by adding dopants such as germanium, phosphorus or titanium during thin film deposition, and index contrasts between 0.34% Δ

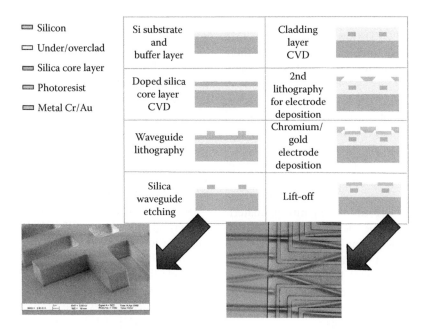

Silicon	Si substrate and buffer layer		Cladding layer CVD
Under/overclad			
Silica core layer	Doped silica core layer CVD		2nd lithography for electrode deposition
Photoresist			
Metal Cr/Au	Waveguide lithography		Chromium/ gold electrode deposition
	Silica waveguide etching		Lift-off

Figure 17.10 Fabrication process for silica waveguides [198]. The buffer layer separates the core layer from the substrate. Waveguides are formed via etching, covered by a silica cladding layer, and electrodes for tunable components are usually deposited via physical vapor deposition.

(matched to standard single mode fiber) and 2% Δ are obtained with losses as low as 0.017 dB cm^{-1} for 0.75% Δ [26]. Mechanical stresses built up during waveguide deposition can be released by annealing at high temperature [28], and stress due to the mismatch of the thermal expansion coefficients of the substrate and waveguide materials can be avoided by a proper choice of dopant levels and waveguide geometries [29,30]. Waveguide patterning is done by photolithography and etching [31,32, Chapter V-2], and finally a cladding covers the waveguides. Additional metallization layers can be deposited for the powering of thermo-optic phase shifters ($dn/dT = 1 \times 10^{-5}$ K^{-1} [3]).

An alternative to thin film deposition is to fabricate the waveguides through ion exchange [33]: alkali ions, such as Na$^+$ or K$^+$, present in the glass are replaced by another cation such as Ag$^+$ or Tl$^+$, which locally increase the index proportional to their concentration. The process for fabricating buried channel waveguides usually consists of two steps: first, a glass substrate with an appropriate lithographic mask is immersed in a molten salt containing the silver or thallium ions, but no sodium or potassium ions. There will be an inter-diffusion of the species, resulting in a channel waveguide at the surface of the substrate. In the second step, an electric field is applied leading to the migration of the exchanged ions deeper into the substrate. The resulting waveguides have an index profile, which is determined by the diffusion characteristics of the Ag$^+$ or Tl$^+$ ions. Passive directional couplers with <0.15 dB cm^{-1} propagation loss have been reported in Reference [34].

At present, ion exchange is primarily used for fabricating waveguides in lithium niobate (Section 17.3.4) and rare-earth doped glasses (Section 17.3.6).

The availability of strong laser sources has allowed us to study the photo-refractivity in glasses: in GeO$_2$ doped films, permanent index changes can be induced by irradiating with 240 nm [35] or 157 nm laser light [36], and the sensitivity is usually enhanced by deuterium loading [37]. Index changes Δn between -6×10^{-3} [38,39] and $+3 \times 10^{-3}$ [37] have been reported, allowing direct laser writing of waveguides [35], Bragg gratings [40] and waveguide couplers [41].

17.3.2 Silicon oxynitride

Higher index contrasts and thus integration densities are available from silicon oxynitride (SiO$_x$N$_y$ or SiON) films, which cover the index range between

SiO_2 ($n=1.45$) and Si_3N_4 ($n=2.0$) and are deposited by PECVD or LPCVD (low pressure CVD) [27]. Although not yet commercialized in volume, there is an industry pull mainly from IBM and Kymata/Alcatel.

Reduction of the stress-related birefringence after waveguide fabrication has been reported by having waveguides with a rectangular cross-section [42], or by an extra Si_3N_4 birefringence compensating layer underneath the channel waveguide [43] as shown in Figure 17.11.

Propagation losses of <0.1 dB cm^{-1} are achievable for lower index oxy-nitride materials, and thermo-optic devices (the coefficient $dn/dT=1.2\times10^{-5}$ K^{-1} being close to that of silica) have been made out of SiON waveguides with 3.3% Δ index contrast [44]. In this particular case, coupling losses due to the mismatch of the mode field diameter have been reduced by attaching a fiber with smaller core diameter.

Silicon-rich nitride (SRN) [45] deposited by LPCVD allows us to obtain index contrasts of $\Delta n=0.6$, and thus a bending radii of 40 μm. Although the reported propagation losses are still of the order of 0.6 dB cm^{-1} (measured in a one-dimensional slab waveguide), SRN will be a promising candidate for high density integrated optics.

17.3.3 Silicon-on-insulator

Silicon-on-insulator (SOI) integrated photonic circuits may have a potential for true monolithic integration with electronics due to their compatibility with the micro-electronics CMOS process, which can provide gigabits/second electronic circuitry with a low noise. As we will see in Section 17.4, silicon is also the preferred substrate material for hybrid integrated circuits. The main commercial source for SOI devices is Bookham (transceivers, multi-channel monitors).

SOI layers can be fabricated by the bonding of thermal oxide layers (BESOI [46]), oxygen ion implantation [47,48] and sputtering, CVD or evaporation [49].

Single mode rib waveguides [50] with dimensions comparable to single mode fiber are fabricated via etching techniques, and propagation losses of <0.2 dB cm^{-1} [51] at 1550 nm and <0.1 dB cm^{-1} at 1300 nm [52] are obtained. The cross-section of a SOI rib waveguide is shown in Figure 17.12.

The high index of $n=3.45$ allows high confinement of the optical mode. This makes Si a suitable material for photonic bandgap structures [53], for which an index contrast of >2 is required for guiding [54].

Modulators have used the thermo-optic coefficient ($dn/dT=1.8\times10^{-4}$ K^{-1} [5]). Carrier injection is the most important electro-optic effect in Si [55,56], and index changes of the order of $\Delta n=1.5\times10^{-3}$ have been demonstrated [57,48].

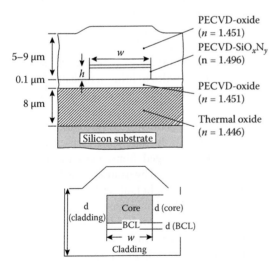

PECVD-oxide ($n=1.451$)

PECVD-SiO_xN_y ($n=1.496$)

PECVD-oxide ($n=1.451$)

Thermal oxide ($n=1.446$)

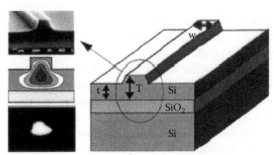

Figure 17.11 PECVD waveguides with compensated birefringence: rectangular waveguide cross section [42] (top, indices are at 1300 nm); guide with Si_3N_4 birefringence compensating layer BCL underneath the channel waveguide [43] (bottom). (Reproduced by permission of IEEE.)

Figure 17.12 Silicon-on-insulator (SOI) single mode rib waveguide [46] fabricated by wafer bonding and etching. Schematic (right), scanning electron micrograph (left top), modeled (left middle) and experimental guided mode (left bottom). (Reproduced by permission of IEEE.)

17.3.4 Lithium niobate

Lithium niobate (LiNbO$_3$, often abbreviated as LN) is a uniaxial crystal with large birefringence ($n_{extraordinary}=2.2$, $n_{ordinary}=2.286$) and a large linear electro-optic (Pockels) effect of $r_{33}=30.9$ pm V^{-1} [9]. Wafers with 100 mm diameter are commercially available [58]; common techniques for fabricating waveguides are thermal in-diffusion of titanium or proton exchange [59] and both processes do not seriously disrupt the lattice structure of the host material.

Ti-diffused channel waveguides typically have index contrasts of $\Delta n=0.5\%$–1% [60] and <0.1 dB cm^{-1} propagation losses [61], but index contrasts of up to 7% are possible with proton exchange [59].

The large EO effect of LN is used for external modulation of laser sources, and bandwidths of up to 100 GHz are reported on probed devices [62]. For achieving these high modulation bandwidths, special designs for the electrodes are required to match the velocities of the electrical and optical mode [58]. Modulators are treated in detail in Chapter B4. Main industry players in the modulator segment are JDS Uniphase and Corning OTI.

Polarization converters and tunable wavelength filters via acousto-optic effects are reported in Reference [63, Chapter 17.8].

17.3.5 Polymer

While not traditionally a high performance optical material, polymer has a number of attributes that make it an interesting material for optical circuits [64] and structural devices [65].

For optical waveguides, photosensitive optical polymers based on various monomers (including acrylates, polyimides, cyclobutenes) have been developed and are commercially available. Upon exposure, these monomers form crosslinked networks and materials with indices between 1.3 and 1.6 are available [4]. Mixing of two polymers which can be co-polymerized allows tailoring of the refractive index with an accuracy of 10^{-4}.

Adding photoinitiators to the material allows fabrication of waveguides by conventional mask photolithography (Figure 17.13) or direct waveguide writing, although etching or moulding processes are also possible [4].

Standard materials offer waveguide propagation losses of 0.2 dB cm^{-1}, and the large temperature dependence of the refractive index $dn/dT= -2, \cdots, -3 \times 10^{-4}$ K^{-1} [4] makes polymers particularly attractive for thermally tunable devices with low power consumption such as photoinscribed Bragg gratings and Mach–Zehnder switches, but also passive AWGs [66]. Main commercial products are thermo-optic switches (Akzo Nobel/JDS Uniphase).

The wide index range allows us to use polymers in hybrid silica/polymer devices [67], and the negative dn/dT can compensate the positive dn/dT of silica for athermalizing temperature sensitive silica waveguide components [68–70].

Electro-optic coefficients of $r_{33}=10$–15 pm V^{-1} have been achieved in polymer waveguides containing chromophores aligned in an electric field. Although photochemical stability and loss (typically 1 dB cm^{-1}) still remain to be addressed, modulating devices with >110 GHz bandwidth have been reported [71].

Waveguides with integrated structures to grip optical fibers with <1 µm alignment tolerance can be fabricated by etching [64], but also LIGA moulding processes [65], and optical backplanes have been reported in Reference [72].

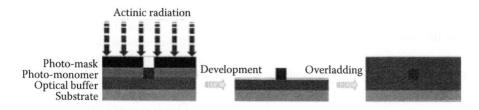

Figure 17.13 Schematic of the photolithography process for patterning of polymer waveguides [64]. (Reproduced by permission of SPIE.)

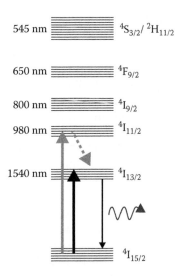

545 nm — $^4S_{3/2}/\,^2H_{11/2}$

650 nm — $^4F_{9/2}$

800 nm — $^4I_{9/2}$

980 nm — $^4I_{11/2}$

1540 nm — $^4I_{13/2}$

— $^4I_{15/2}$

Figure 17.14 Simplified representation of the Er³+ energy levels. Optical pumping into the excited $^4I_{11/2}$ and $^4I_{13/2}$ states requires semiconductor lasers operating at 980 or 1480 nm [73]. (Reproduced by permission of the Materials Research Society.)

17.3.6 Active waveguides

Active waveguides have been obtained from the implantation of rare-earth ions into a host material [73], ion exchange in Er/Yb co-doped glasses [74] and through PECVD thin film deposition [75].

Er³+ has a transition in the 1.54 µm telecom window, and optical pumping from the ground to the excited states is possible by 980 or 1480 nm pump lasers (Figure 17.14) [73]. Net gains of 20 dB can be obtained with 120 mW pump power [63, Chapter 6] (see this reference for a state-of-the-art on this subject). TEEM Photonics and NKT Integration commercialize Er³+-doped planar amplifiers.

In a similar way, Nd³+ doping in silica waveguides provides a gain at 1.05 nm wavelength [76].

Active Er-doped devices have also been fabricated in LiNbO₃ [7,77, Chapter 6], and other oxide materials, ceramics and Si [78].

17.3.7 Indium phosphide gallium arsenide

Among the III-V semiconductors, InP plays a major role, as it is the current material of choice for the fabrication of lasers and detectors, which in turn can be monolithically integrated with passive InP waveguides. The possibility of matching the lattice constant of the quarternary alloy In$_x$Ga$_{1-x}$As$_{1-y}$P$_y$ to InP over a large window of compositions results in a wide range of bandgaps, thus operating wavelengths between 1000 and 1700 nm are obtained. However, the fabrication process is quite expensive and wafer sizes are limited to 2–3 in. Although we will illustrate a few monolithic InP circuits integrating passive and active device functionalities in Section 17.5, most of the InP-based components today are combined with other passive components through hybrid integration (Section 17.4.).

GaAs has a major role in the development of high speed electronics, thus up to 6 in. wafers are available. The bandgap of GaAlAs/GaAs and InGaAs/GaAs can be engineered for obtaining laser operation in the 780 and 980 nm wavelength regions, respectively, the latter being used for optical pumping of Er³+-doped glasses.

We refer to Chapters B1.1, B1.2 and B2 for further reading.

17.3.8 Conclusion

Table 17.2 summarizes the key properties and primary phase tuning mechanism of the materials presented in this section. Each of the materials has its advantages and disadvantages, so that there is no clear winner. The choice of the material depends primarily on the specific needs for a given functionality.

17.4 DEVICE PACKAGING AND FUNCTION INTEGRATION

The first part of this section is about packaging of opto-electronic circuits, which fulfil a single or multiple functions in the network. The second part will discuss briefly the different technologies for the integration of multiple functions on one chip, and we will focus more specifically on hybrid integration, as this is the main integration technique today.

17.4.1 Device packaging

There are different levels of opto-electronic integration in a communication system [79] (Figure 17.15):

Table 17.2 Key properties of the materials described in section 17.3

	Propagation loss	Coupling loss	Refractive index n	Primary phase tuning mechanism	Primary application
Silica	<0.05 dB cm^{-1}	Low-medium	1.444	Thermo-optic	Passive components, thermo-optic tunable devices
Silicon oxynitride	<0.1–0.2 dB cm^{-1} for n around 1.45 0.6 dB cm^{-1} for n around 2.0	Low-medium	1.444–2.0	Thermo-optic	Passive components, thermo-optic tunable devices
Silicon-on-insulator SOI	<0.2 dB cm at 1550 nm	High	3.45	Thermo-optic, electro-optic	Passive components, photonic bandgap devices
Lithium niobate LN	<0.1 dB cm^{-1}	Medium	2.0	Primarily electro-optic, but also thermo-optic and acousto-optic	High speed gigabit/second modulators and switches
Polymer	<0.2 dB cm^{-1}	Low-medium	1.3–1.6	Thermo-optic, electro-optic (chromophores)	Passive components, Thermo-optic tunable devices/switches with low power consumption
Er^{3+} doped glasses	<0.1 dB cm^{-1}	Low	1.49–1.50		Pre- and post- amplifiers
InP	ca. 3 dB cm^{-1}	High	3.1	Electro-optic, $\chi(3)$ nonlinearity	Lasers, detectors, semiconductor optical amplifiers, gigabit/second optical gates and switches
GaAs	ca. 0.5 dB cm^{-1}	Medium	3.3737		Pump lasers (980 nm)

For the coupling loss to a standard single mode fiber SMF28 [82] we indicate only a range (high >3 dB/interface, medium 1–3 dB/interface, low <1 dB/interface), as exact coupling loss values (see Section 17.4.1 and Table 17.3) depend on mode field diameters which in turn can be adapted.

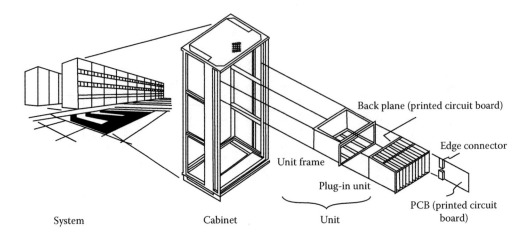

Figure 17.15 Packaging levels in a generic communication system [79]. (Reproduced by permission of John Wiley & Sons, Inc.)

- A system consists of several racks or cabinets of equipment
- A rack or cabinet contains frames, which in turn have units combining a multitude of optical and electronic functions
- A single or a few optical functions together with control electronics are packaged on a board

The purpose of packaging is (as an example, see Figure 17.16).

1. To connect a fiber to the monolithic or hybrid integrated optic *chip* without adding loss.
2. The monolithic or hybrid integrated optical circuit can in turn contain dynamic elements (phase shifters, active opto-electronic devices)

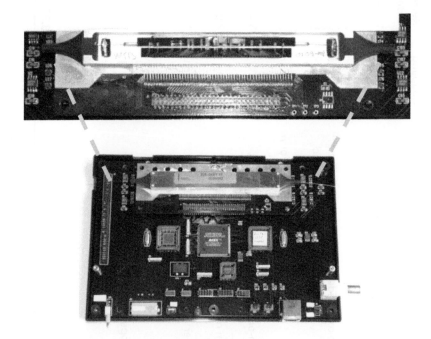

Figure 17.16 Example design for a opto-electronic component consisting of a silica waveguide chip in its housing (top) and electronics control board [135]. (Reproduced by permission of IEEE.)

which need to have an *electrical interconnection to control electronics*.

3. The fiber-optical circuit ensemble needs to be fixed into a *housing* or *package* without degradation of the mechanical stability or optical performance.
4. The housing or package ensures *reliability* under different operating or environmental conditions.

Coupling from a fiber to an optical integrated circuit can be accomplished [80].

- From the waveguide surface via prisms or surface gratings.
- Through edge coupling into the waveguide cross-section via focusing lenses, or through direct attachment of the fiber to the chip. For the latter, fibers can be placed into V-grooves fabricated into Si substrates [81], or actively aligned to the chip which is the standard method used today.

Edge coupling losses arise from the mismatch of the fiber and waveguide mode fields, as well as from angular or lateral misalignments and are given by the overlap integral

$$\text{Loss} = -10\,\text{dB} \times \log_{10}\left[\frac{\left(\int_{\text{area}} \phi_1\phi_2\,dA\right)^2}{\left(\int_{\text{area}} \phi_1^2\,dA\right)\left(\int_{\text{area}} \phi_2^2\,dA\right)}\right]$$

with $\phi_1 \cdot \phi_2$ being the mode fields and A the waveguide cross-section. Approximations for the various contributions are given in Table 17.3.

Table 17.3 Coupling loss mechanisms and formulas

Loss mechanism	Loss (dB)	Schematic
Mode mismatch	$-20 \log_{10}\left[\dfrac{2w_1 \cdot w_2}{w_1^2 + w_2^2}\right]$ w = mean field radius (1/e amplitude)	
Transverse offset	$-10 \log_{10}\left[\dfrac{x^2}{w^2}\right] \approx 4.343\left[\dfrac{x}{w}\right]$ where $w = \sqrt{(w_1^2 + w_2^2)/2}$ and x is the offset	
Angular misalignment	$-10 \log_{10}\left\{\exp\left[-\dfrac{1}{2}\left(\dfrac{2\,n_{\text{eff}}w\theta}{\lambda}\right)^2\right]\right\}$ $\approx 8.69\left(\dfrac{n_{\text{eff}}w\theta}{\lambda}\right)^2$ where $w = \sqrt{(w_1^2 + w_2^2)/2}$ and θ is the angular misalignment (rad)	

As an example, the mode-field mismatch of a standard single mode fiber (mean field radius $w_0=4.6\,\mu m$ [82]) to a planar waveguide ($w_0=3.3\,\mu m$) results in losses of the order of 0.5 dB. To keep the excess loss from lateral misalignment to below 0.3 dB, the fiber needs to be positioned with <1.0 μm accuracy. The fibers are usually attached by optical adhesives or directly fused to the chip using a CO_2 laser [83,84].

Edge coupling of III–V components such as lasers, detectors or semiconductor optical amplifiers (SOAs) to fibers is more difficult, because their small modefield diameter of 1–2 μm would immediately lead to increased mode-field mismatch losses and also slightly reduced alignment tolerances.

In this case, loss reduction is possible by using fibers with anamorphic lenses [85] or enlarging the mode field of the active component with a taper [86,87] (Figure 17.17). These relaxed alignment tolerances are the keys to high yield low cost passive alignment techniques as used in hybrid integration.

For electrical interconnection of active semiconductor and dynamic elements mainly two technologies are used in opto-electronics [88–90]:

- Shorter interconnection (typically 100 μm), and thus higher electrical bandwidth, is possible with the flip chip process, which is moreover a parallel process. The electrical interconnection can be made with higher

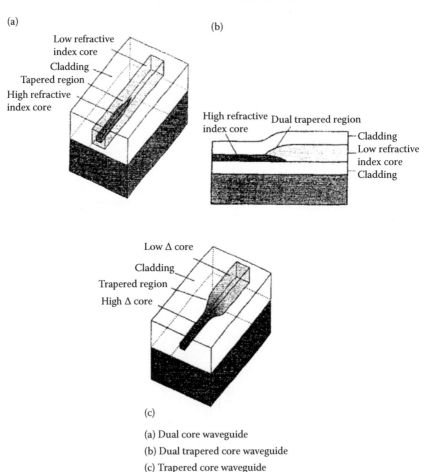

(a)

Low refractive index core
Cladding
Tapered region
High refractive index core

(b)

High refractive index core Dual trapered region
Cladding
Low refractive index core
Cladding

Low Δ core
Cladding
Trapered region
High Δ core

(c)

(a) Dual core waveguide
(b) Dual trapered core waveguide
(c) Trapered core waveguide

Figure 17.17 Mode-field converting waveguides [86]. The principle is to have the high index waveguide (with small mode-field diameter) connected through an intermediate taper to a low index waveguide (with larger mode-field diameter), which is then coupled to a single mode fiber. (Reproduced by permission of John Wiley & Sons, Inc.)

density anywhere on the device surface, and the flip chip (or solder bump self-alignment technique) is also one of the passive assembly techniques used in hybrid integration. *The package or housing of the device* needs to enclose and protect the optical circuit without performance degradation through

- Thermal effects—heat is generated by the active components, so that a good dissipation is essential for maintaining the circuit at stable temperature. This is achieved by
  - Having laser submounts with high conductivities (diamond, silicon carbide (SiC), alumina (Al_2O_3), aluminium nitrite (AlN), silicon (Si)) or Si substrates for thermo-optic and hybrid components.
  - Thermo-electric coolers or passive heat dissipation.
- Stress related effects, as they induce birefringence and change the response of an optical circuit based on phase control. Potential issues are addressed by using substrates and packaging materials with adapted thermal expansion coefficients and filled adhesives with low Young's modulus.

At present, commonly used housing materials are aluminium, Kovar, Invar and organic/inorganic composites with near zero expansion coefficient [79]. Bending losses or breakage of the fiber attached to the optical circuit can be avoided by keeping the fiber bend radii above 2–3 cm.

Opto-electronic components are required to have failure-free performance over a range of operating temperatures (typically −5 to +70°C) over their lifetime (miminum 25 years). For assuring a reliable operation, testing procedures are defined [91], the purpose of which is to

- Verify reliability against thermal shocks, thermal cycles (−40 to 70°C), damp heat (85°C/85% relative humidity) and moisture.
- Screen devices through thermal cycling and short term storage at high temperature.
- Estimate life expectancy through accelerated ageing (70 or 85°C, 2000–5000 h). The principle is to test several identical components at different temperatures and to estimate the mean time to failure (MTTF) from an Arrhenius relation.

Depending on the application, operating environment and chip materials, the OE components can be packaged hermetically or nonhermetically.

17.4.2 Function integration and hybrid technology

The integration of optical functions can be accomplished mainly in three different ways:

1. Precision placement of the various elements into one package and interconnection via free-space optical links. This method has limitations for the complexity of the OE circuit, and the cost reduction potential through mass production is questionable.
2. In monolithic integration, the optical and electrical functions are fabricated—in analogy to electronic LSI—on the same semiconductor substrate such as InP or GaAs. The main drawback is that currently best performances for different optical functions may be obtained with different materials (see Section 17.3), for example modulators with $LiNbO_3$, passive wavelength multiplexers with silica waveguides and so on. Also, the achievable integration density is limited by optical constraints (waveguide bend radii and the size of the fiber/chip interconnection) rather than the size of the electronic circuitry.
3. In hybrid integration, active devices and electronics are assembled together with passive lightwave optical circuits on a common platform, which allows combining the best available circuits. Furthermore, the yield for this technology may be higher than for monolithic integration, since the components can be selected for performance prior to the assembly. The remainder of the section reviews the critical technologies needed for surface-hybrid integration (for the various types of hybrid integration, see Reference [86]).

A generic hybrid integrated circuit is shown in Figure 17.18, and the main technical ingredients for functional integration are

1. A platform onto which the active and passive optical circuits are integrated. Among the materials used in packaging (diamond, SiC,

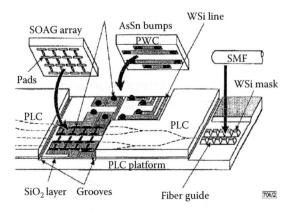

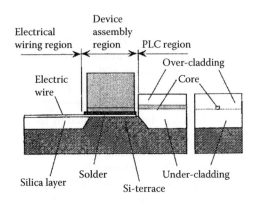

Figure 17.18 Hybrid integrated circuit [168] showing passive planar circuits (PLC), electronic transmission WSi lines, single mode fiber (SMF) with passive fiber guides and active circuits (semiconductor optical amplifier gates SOAG) with AuSn solder bumps for flip chip self-alignment. (Reproduced by permission of IEEE.)

Figure 17.19 Cross-section of the NTT PLC platform: the Si terrace and the deposited silica buffer layer are planarized, thus defining the vertical plane for subsequent passive assembly of the active components [63].

Al_2O_3, AlN, Si), Si is considered to be the best material for a number of reasons: low cost, good thermal conductivity (1.57 W cm^{-1} K^{-1}) for evacuation of thermal load from active components, a thermal expansion coefficient (CTE $= 2.33 \times 10^{-6}$ K^{-1}) almost matched to that of InP components (CTE $= 4.5 \times 10^{-6}$ K^{-1}); availability of processes for fabrication of electrical circuits and Si alignment features. Moreover, passive silica waveguide components are usually fabricated on Si substrates.

There are mainly two different types of platform:

a. The Si optical bench has fiber V-grooves fabricated by anisotropic etching as well as other alignment features for active components, but has no passive waveguides.

b. NTT has developed a planar lightwave circuit (PLC) platform [92,93], which also integrates passive silica waveguide components. A cross-section of the structure is depicted in Figure 17.19.

2. In the integration step, the optical building blocks need to be interconnected with low optical loss due to physical misalignment, i.e., within the roughly 1 μm accuracies mentioned in Section 17.4.1. Both active and passive alignment methods [63, Chapter 10] are used today:

In an active alignment process—for example coupling of a fiber to a laser diode—the optimum position is found through a measurement of the transmitted laser power, and sub-micrometer accuracy can be achieved with commercial precision alignment stages.

In a passive alignment process, the accuracy is defined through alignment features fabricated on both the OE building blocks and the integration platform:

1. The index method uses microscopes and precision manipulators for accurate alignment in the horizontal (platform) plane; an example is shown in Reference [94].

2. Mechanical contact alignment uses topographic features such as stops, standoffs, notches, pedestals (Figure 17.20) for alignment in both horizontal and vertical directions. The major difficulty is the precise definition of the vertical plane during the waveguide fabrication process. Accuracies of 0.8 μm are reported for lasers coupled to fibers [95], passive alignment of single mode fibers to both facets of a semiconductor optical amplifier is shown in [96], and integration of both diode laser and photodiode with <3 μm precision has been demonstrated in Reference [97].

3. The flip chip or solder bump technique uses precisely known volumes of solder confined between wettable metal pads on either side of the bond. Raising the

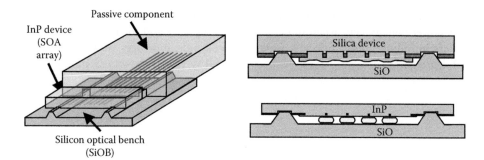

Figure 17.20 Mechanical contact alignment: notches fabricated into the active InP and passive silica waveguide components allow sub-micron alignment on a common silicon optical bench [199].

Figure 17.21 Flip chip solder bond—solder bumps with precise volumes (typically AuSn or PbSn) are fabricated on wettable pads, then brought into contact and heated above the melting point. The surface tension acts as a restoring force, resulting in alignment.

temperature above the solder melting point will simultaneously align the structures through surface tension and provide an electrical interconnection [90,98] (Figure 17.21). Solder bumps can be fabricated by vapor or liquid phase deposition [99], and a variety of active/passive circuits have been published with alignment accuracies of up to $\pm 1\,\mu m$ [98].

4. A combination of mechanical and solder bump alignment has been reported for transceiver modules [100,101].

3. A reliable mechanical and electrical interconnection between the opto-electronic circuits and the platform. Materials and processes commonly used are [98,102,103]

1. In the flip chip process, eutectic 63% Sn/37% Pb (melting point 183°C) or 80% Au/20% Sn (melting point 281°C) solder bumps are fabricated on Ti or Cr adhesion layers.

2. Bonds can also be formed through a thermo-compression process at lower temperatures using Au/Sn or Au/In alloys.

3. Metal-filled epoxies with good thermal and electrical conductivity are suitable for curing at 150°C–200°C.

17.5 OPTICAL NETWORKS AND INTEGRATED OPTICAL FUNCTIONS

With the increasing demand for communications bandwidth, digital optical transmission has become more and more important over the last few years. Currently, systems with capacities of hundreds of gigabits/second are commercially available, and transmission of 10.2 Tbits^{-1} over 100 km of single fiber has been demonstrated in laboratory experiments [104]. In practice, this is done by simultaneously carrying optical signals with slightly different wavelengths over a single fiber (wavelength division multiplexing, WDM). In parallel, the bit rates for single channels have steadily increased to 40 Gbit s^{-1}, and in future systems of higher capacities will be achieved through time domain multiplexing (TDM) of several signals with identical wavelength into one high speed data stream.

The ultimate system figure of merit in digital transmission is the bit error rate (BER), which is defined as the probability of detecting a "1" although a "0" was received and vice versa. Typical goals are BER $< 10^{-9}$ for voice traffic and BER $< 10^{-12}$ for data transmission. For a fixed bit rate, the BER mainly depends on the received power and one speaks about power penalty (PP) when imperfect components are introduced into a network resulting in a BER degradation.

For optical networks, we refer to Section C1, but for simplicity and for illustrating the use of integrated optical devices we will consider a generic WDM system as shown in Figure 17.22:

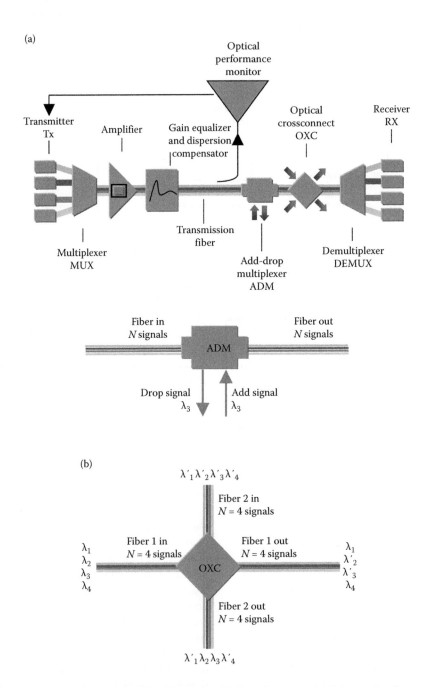

Figure 17.22 Schematic of a wavelength division multiplexed system. (a) Schematic of an optical add–drop multiplexer. As an example, there are four signals $\lambda_1,\ldots,\lambda_4$ on the fiber. The signal λ_3 is dropped and replaced by λ'_3. (b) Optical cross connect with two input and two output fibers, each carrying four signals. In this example, signals 2 and 3 are exchanged between fibers 1 and 2.

- Light from a semiconductor *transmitter laser Tx is modulated*. In a WDM system, several signals are combined into the transmission fiber by an optical multiplexer MUX.
- The propagating signals are amplified every 70–100 km; mainly Er^{3+} *doped fiber amplifiers* are deployed [105], and distributed *Raman amplification* [106, Chapter 8] in the transmission fiber is used to increase the system reach. Er^{3+} doped amplifiers need semiconductor pump lasers operating at 980 and 1480 nm; Raman amplified systems are pumped at 14xx nm wavelength.
- The gain of an EDFA is wavelength dependent, which leads to significant variations in signal-to-noise ratio (OSNR) of the received channels. The OSNR in turn depends on the power per channel and the number of amplifiers in the system. Nowadays, the channel power is periodically flattened with (static) thin film filters, but upcoming reconfigurable systems will require *dynamic gain equalization*.
- The finite spectral width of the source laser together with the positive fiber chromatic dispersion (i.e., the variation of the group velocity with wavelength) and nonlinear effects (self-phase modulation (SPM), cross-phase modulation (XPM), four-wave mixing) in the transmission fiber lead to a spreading of the optical pulse, which can be reshaped in *chromatic dispersion compensating devices*.

In addition to that, the transmission fiber may not have perfect circular symmetry, giving rise to different propagation velocities as a function of the incoming polarization state (*polarization mode dispersion (PMD)*).

- More complex optical networks contain *add–drop multiplexers* (ADMs) and/or *optical cross-connect* (OXC) switches for interconnecting fiber links and for provisioning a path through the network:
  - ADMs (Figure 17.22a) have $M=2$ fiber ports. Their function is to locally drop and add a few wavelengths out of a stream of N signals carried on a fiber while directly passing on the other data streams. ADMs can be static or reconfigurable.
  - OXCs (Figure 17.22b) have $M > 2$ aggregate ports and allow routing of any of the N

signals from any input fiber to any output fiber.
  - The hearts of reconfigurable ADMs and OXCs are *space switches*: for the ADM, an array of N 1×2 or 2×2 switches is required, and the OXC needs a matrix of dimension $N\times M$.
- At the end of the optical link, the signals are demultiplexed, and the *receiver Rx* converts the signals back into the electrical domain. The signal can be either terminated or regenerated; in the latter case, a fresh copy of the signal is produced through optical–electrical–optical (O–E–O) conversion and fed into the next link.

As a single fiber carries several gigabits/second of information, the system operators have to guarantee the reliability of the transmission system against failures such as a fiber break. *Network protection* techniques ensure reliability by providing redundant capacity in the network, and by using protection switches for re-routing the traffic in case of failures (Figure 17.23).

Transmitter laser wavelength drifts and power variations are likely to cause an increase in BER, and it is therefore important for network operators to *monitor the performance* of the communication channels in order to guarantee quality of service (QoS) to their users.

Wavelength conversion of an optical signal is needed, for example, in OXCs if a signal gets switched onto a fiber in which a certain wavelength is already occupied.

Imperfect filters, dispersion, nonlinear effects and accumulated noise from Er^{3+} amplifiers limit the transparent length, over which a signal can be transmitted with a reasonable OSNR of about 1500–2000 km, and regeneration of the signal is required. As mentioned before, nowadays this is usually done at the receiver end in the electrical domain, but *regeneration in the optical domain* [107] has been demonstrated [108], resulting in extended reach [109]. Optical regeneration may eliminate a part of the cost of the current O–E–O conversion.

In current systems, a light path is set up for the duration of the communication (circuit switching). Future systems may evolve—in analogy to ATM or SONET packet switching in the electrical domain—to *optical packet switching* in which data packets are optically routed through the network

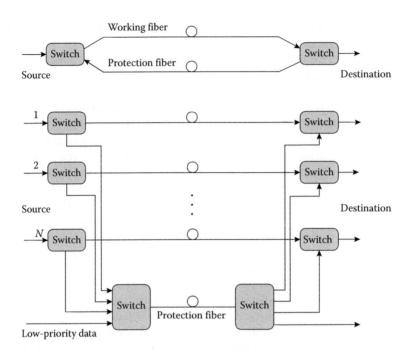

Figure 17.23 Protection in optical point–point networks by switching traffic from a working fiber on a protection fiber: 1:1 (top) and 1:*N* (bottom) protection [142]. The 1:*N* scheme is more bandwidth efficient as *N* working fibers share only one protection fiber.

based on information (such as the destination address) carried within the packet.

After the basic toolbox and technology overview in Sections 17.2–17.4, this last part will allude to examples of integrated devices. We will concentrate on some of the optical network functions mentioned above, and for avoiding duplication, we will not treat lasers, modulators, receivers and amplifiers for which we refer to Chapters B1.2, B2, B4 and B6.

For most of the examples, several technologies are available and integrated waveguide devices represent one option, and the user will make his final choice based on performance, cost and maturity of the technology. Some of the functions are commercially available today, but most are still in research and will emerge over the next few years with the deployment of future generation systems.

17.5.1 Optical multiplexers

Optical filters for WDM systems should have low insertion loss, low polarization and temperature dependence of both transmission wavelength and transmission loss, and they should not distort the incoming signals due to their average chromatic dispersion (nonlinear group delay of the filter) or dispersion variation within the filter bandwidth (dispersion slope) [110]. Technologies available today are mainly dielectric thin film filters, fiber Bragg gratings and planar devices.

Early wideband filters for separation of signals in the 1.3 μm/1.5 μm telecommunication windows have used single stage asymmetric Mach–Zehnder interferometers as described in Chapter 17.2.4 [3], and wider passbands have been obtained by concatenation of three Mach-Zehnder filters [111]. Similar architectures are used for separating odd from even channels (interleaver) [112,113].

Planar narrowband filters use the AWG (Section 17.2.4) with typical channel spacings of 100 or 200 GHz (see also as an example, the 16 channel device in Figure 17.7c). Commercial AWGs with Gaussian passband have fiber–fiber losses of the order of 4–5 dB, and in research devices with <1.2 dB insertion loss [114] and up to 1010 channels on a 10 GHz grid [115] (concatenated devices) have been demonstrated.

A wide transmission passband rather than a Gaussian passband is preferred for accommodating fluctuations in the transmitter laser Tx frequency and minimizing distortions due to filter concatenation [116]. Usual techniques modify the mode profile at the first AWG star coupler by MMI couplers [117] or Y-splitters -[118], or interleave two gratings resulting in two separate focal points [119,120]. Dynamic shaping of the passband is also possible [121].

The silica dn/dT leads to a temperature dependence of the transmission spectrum, and for silica waveguide devices compensation methods [70,122,123] have been published. Polarization dependence has been eliminated by converting a TE mode into a TM mode in the middle of the device [124] and minimizing stress birefringence in the fabrication process [29,30].

Both chromatic dispersion and dispersion slopes across filters also lead to signal degradation; however, the dispersion of the AWG filter is negligible and strictly zero if the component is symmetric and the loss can be neglected [125] (note that the transfer function in Table 17.1 corresponds to that of an AR/FIR fitter without poles).

Other applications use very high resolution AWGs for converting a single data pulse from the frequency domain into the spatial domain. Examples are

- Shaping of ps pulses, which was reported first using dispersive bulk gratings and phase masks in the grating's focal plane [126] and later on using fully integrated AWGs with phase shifter arrays [2].
- Dispersion compensation [127] (Section 17.5.3) and optical code division multiplexing [128].

17.5.2 Dynamic gain equalization

Dynamic gain equalization is starting to be implemented in current systems, and a variety of technologies are being investigated today:

- Diffractive micro-electro-mechanical systems (MEMS) [129,130].
- Bulk optic diffraction gratings in combination with liquid crystal spatial light modulators (LC-SLMs) [131,132].
- Bulk and fiber acousto-optic gratings [133].
- Waveguides with surface electro-optic switchable Bragg gratings [134].

- We will describe integrated optic solutions in more detail, and there are basically two types.
- Fourier filters—multiple sinusoidal transmission functions are generated by a multitude of asymmetric Mach–Zehnder interferometers (Chapter 17.2.4) having path length differences ΔL. The ΔL results in different free spectral ranges, and superposition of the terms leads to the desired spectrum.

The Mach–Zehnder devices can be arranged in parallel ("tapped delay line filter") or by concatenation ("serial lattice filter") [22, Chapters 4.3 and 4.5].

As an example, a schematic tapped delay line gain flattening filter [135] with eight arms (resulting in seven interference terms) is shown in Figure 17.24 (top), and its transfer function is listed in Table 17.1. Tailoring the delay T_k between adjacent arms and the amplitudes h_k allows synthesis of targets as in Figure 17.24 (bottom). The typical insertion loss of such a device is 4 dB.

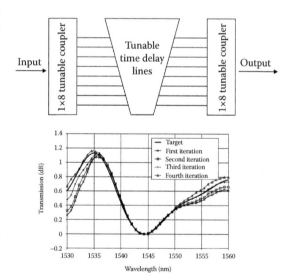

Figure 17.24 Tapped delay line filter for dynamic gain flattening. The device consists of a tunable 1×8 splitter made from cascaded MZIs, the tunable delay lines and an 8×1 combiner (top) architecture. Feedback is given to an electronic control circuit, and the measured transmission after four iterations (triangles) is closer than 0.1 dB to the target (thick solid line) [135]. The transmission is normalized to 0 dB. (Reproduced by permission of IEEE.)

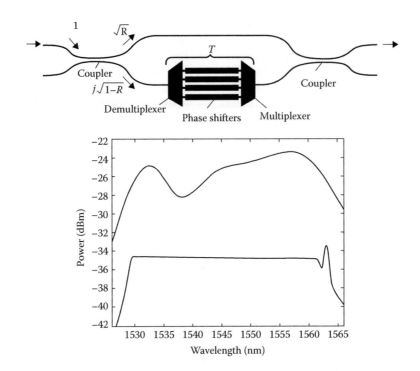

Figure 17.25 Channelized dynamic gain flattening filter [140,141]—device architecture (top) and result of automatic flattening of the emission from an erbium doped amplifier (bottom), the upper and lower curves corresponding to before and after flattening. (Reproduced by permission of IEEE.)

Other applications of tapped delay lines are narrowband filters with small free spectral range [136] and multi-channel selectors [137]. Serial lattice gain flattening filters have been published using SiON waveguide technology [138], and silica waveguides [139].

- Channelized devices, which equalize individual wavelengths or groups of channels. An example is shown in Figure 17.25 (top) [140]: a portion $\sqrt{R}$ of the light propagates through the upper unfiltered arm, and $\sqrt{(1-R)}$ will go through the lower arm. This lower arm contains a demultiplexer for selecting a single wavelength or a band of wavelengths and an array of phase shifters. Depending on the setting of the phase shifters, constructive or destructive interference between upper and lower arms results for each spectral band. Devices have been fabricated in both InP [140] and silica waveguides [141], and for the latter device a typical spectrum is shown in Figure 17.25 (bottom). The insertion loss through the device is around 6.5 dB.

17.5.3 Chromatic and polarization mode dispersion compensation

Spreading of the light pulses propagating through a fiber arises from chromatic dispersion of the fiber itself, from nonlinear effects leading to a modulation of the phase, and from the chromatic dispersion of filters present in the link. As a consequence, the pulses in adjacent bit periods will start to overlap (inter-symbol interference (ISI)) leading to an increased BER.

For a signal with bit rate B and a fiber of length L (km) with dispersion D (ps nm^{-1} km^{-1}), the condition

$$B^2 LD < 11,000\,\mathrm{ps\,nm^{-1}(Gbit/s)^{-2}}$$

has to be fulfilled for a PP of <1 dB [142]. In other words, a 10 Gbit s^{-1} signal tolerates an accumulated dispersion LD of 1100 ps nm^{-1}, but for a 40 Gbit s^{-1} signal, the limit is only 69 ps nm^{-1}.

In practice, the chromatic dispersion of standard single mode fiber [82] $D=+17$ ps (nm km)$^{-1}$

is compensated by periodically adding a section of dispersion compensating fiber (DCF) having negative dispersion, thus bringing the cumulated dispersion back to tolerable values. However, the tight limits for 40 Gbit s^{-1} signals and variations in the fiber dispersion (through fabrication tolerances and daily temperature changes) may necessitate tunable dispersion compensation devices operating on single channels or groups of channels.

In addition to the dispersion D, the dispersion slope $dD/d\lambda$ across a single WDM channel needs to be compensated for ultra-high speed systems (>100 Gbit s^{-1}) as well.

Technologies for dispersion compensation are chirped fiber Bragg gratings [143], micro-optic Gires–Tournois interferometers [144] and virtual phased arrays [145], and in the following we will describe the integrated optic devices, which offer an opportunity for integration with the multiplexers at the end of the transmission link.

Planar dispersion compensators have been studied based on both Mach–Zehnder filters and ring resonators, corresponding to the FIR and IIR categories (Section 17.2.5).

An FIR dispersion slope compensator is shown in Figure 17.26 (top) [146], and in this case integration with an AWG multiplexer allows us to compensate 16 WDM channels simultaneously.

For each channel, the dispersion equalizer consists of five asymmetric Mach–Zehnder filters and six tunable couplers. The asymmetric MZIs provide a wavelength sensitive splitting and a delay between the signals propagating through the upper and lower arms, resulting in a wavelength dependent delay. The device itself is fabricated in 1.5% Δ silica waveguides, and the thermo-optic phase shifters allow tuning of the dispersion characteristics. The measured versus required delay for transmission over 640 km dispersion shifted fiber (DSF) is shown in Figure 17.26 (bottom).

An alternative architecture for compensation of both dispersion and dispersion slope is based on a pair of AWGs and a filter (Figure 17.27) [127]: the AWG separates the incoming signal into its frequency components, and the spatial phase filter will apply the appropriate delay. Designs for compensation of up to 260 ps nm^{-1} are proposed, but losses are expected to be of the order of 15 dB due to the specific design of the AWG and coupling losses from the AWG to the spatial phase filter.

IIR filters are particularly interesting, since the poles in their transfer function lead to a non-linear phase response and thus naturally provide dispersion.

In particular, the example shown in Figure 17.28 (top, see also the transfer function in Table 17.1) [147] represents a two stage all-pass filter which is theoretically lossless, and the dispersion can be tailored by adjusting the power coupling ratios κ_1, κ_2 and the phase shifts ϕ_1, ϕ_2. A dispersion of ±4000 ps nm^{-1} has been demonstrated over a bandwidth of 4.5 GHz (Figure 17.28, bottom), and more recently [148] the bandwidth has been increased to 13.8 GHz (dispersion ±2000 ps nm^{-1}) by using a silica waveguide fabrication process with a 2% Δ index contrast and thus shorter feedback path.

PMD arises from ellipticity of the transmission fiber as well as from polarization dependent performance of the components within the transmission link, and as a result TE and TM modes experience a differential delay. PMD is a time dependent statistical phenomenon and needs to be compensated adaptively.

Early PMD compensators [149] were based on a series of three squeezed polarization maintaining fibers, and a planar solution is described in [150] (Figure 17.29, top). The compensator splits the incoming signal into TE and TM components at the first polarization beam splitter (using stress birefringence) [151], then TM is converted into TE allowing interference between TE and TM modes in the subsequent serial MZI. In this particular case, the group delay difference for the two polarization states is compensated in a fixed 7.5 ps delay line, and then the polarization modes are recombined. The systems benefit is demonstrated in Figure 17.29 (bottom), showing the BER for a transmission link with (□) and without compensator (Δ).

17.5.4 Space switches and all-optical switching

In the beginning of this section we mentioned multiple applications for switches in an optical network:

- Space switches in OXCs and add–drop multiplexers enable dynamic provisioning or reconfiguration of an optical WDM lightpath (circuit switching), thus replacing manual fiber

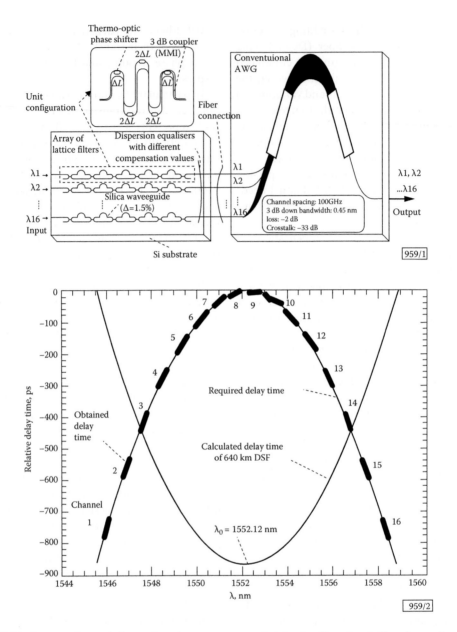

Figure 17.26 Sixteen channel dispersion slope compensating lattice filter (top). The dispersion slope is compensated separately for each wavelength through a five stage lattice filter. The bottom figure shows the delay of the transmission fiber, the required delay and the experimental delay (full symbols) [146]. (Reproduced by permission of IEEE.)

patch panels. For OXCs, the number of input/output ports needed is difficult to estimate, but the many wavelengths carried on one fiber (at present usually $N=40$–80) and the large number of fibers M will drive the demand for a high port count (256×256 and larger) devices. The switching speed required for this application is of the order of 1–10 ms.

- Fiber optic networks can encounter two types of failure: problems with the transmission equipment or path interruptions due to fiber break. Both issues can be addressed by switching the traffic onto a second unused fiber which is dedicated (1:1 protection) or shared with other traffic streams (Figure 17.23). Protection switches are required to have a commutation time of <10 ms.

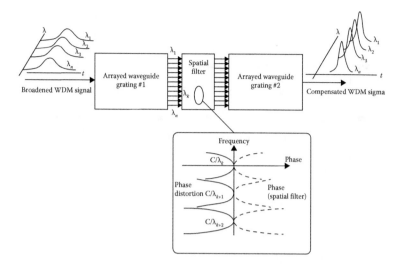

Figure 17.27 Dispersion compensator using arrayed waveguide gratings [127]. AWG 1 will spread the spectrum, and for each WDM channel there is a spatial filter (shown as an inset) which will correct for the phase distortion/chromatic dispersion. AWG 2 then recombines the compensated signals. (Reproduced by permission of IEEE.)

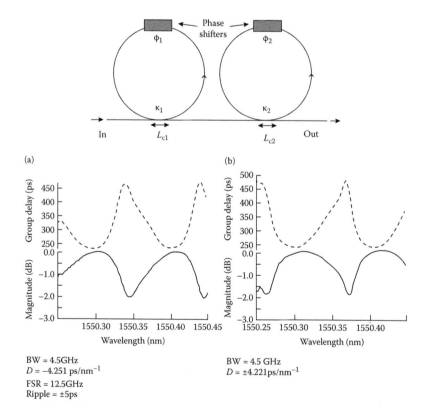

Figure 17.28 Two stage optical AR filter based on ring resonators (top) [147] for compensation of both positive and negative chromatic dispersion. Each stage has a feedback path with radius $R=2.2\,\text{mm}$, resulting in a free spectral range of about 0.12 nm (15 GHz). Measured group delay and normalized insertion loss are shown for negative (bottom, left) and positive (bottom, right) dispersion. (Reproduced by permission of the Optical Society of America.)

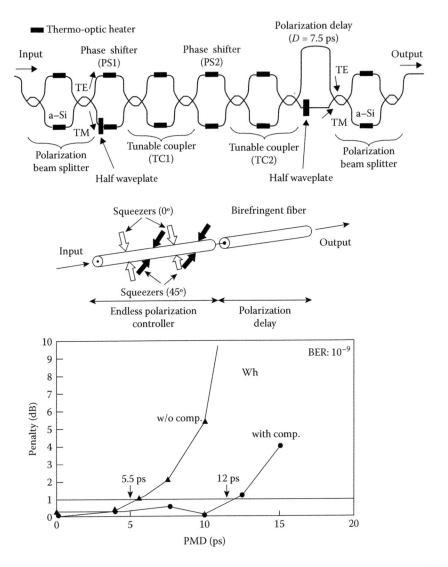

Figure 17.29 Integrated planar lightwave circuit PMD compensator and equivalent circuit [150] (top). Measured power penalty as a function of the polarization mode dispersion for a fixed bit error rate (BER) of 10^{-9}. (Reproduced by permission of IEEE.)

- Future systems will use high speed switches for switching and routing of data packets rather than provisioning a dedicated lightpath. The required switching speed is inversely proportional to the bit rate, i.e., it is of the order of a few nanoseconds.

Electronic data processing seems to be currently limited to around 40 Gbit s^{-1}, and there is an interest performing switching (and more generally signal processing such as multiplexing, routing, wavelength conversion, optical logic) in the all-optical domain [152]. As an example, we will briefly mention all-optical switches at the end of this section, and we refer to Chapter C1.3 and Reference [63, Chapter 9] for all-optical time >1 Tbits^{-1} systems.

In addition to the commutation time, other important switch characteristics are extinction (or ON–OFF) ratio, insertion loss, crosstalk (i.e., leaking of light from one path into another), polarization dependent loss and power consumption.

The basic elements are 1×2 or 2×2 switches, and larger size devices can be realized by appropriate

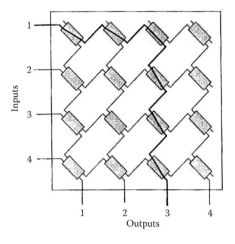

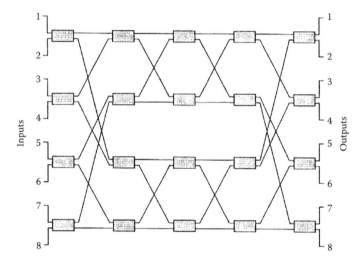

Figure 17.30 4×4 switch: crossbar (top) and Beneš architectures (bottom) [142].

cascading of these devices. As an example out of many possible arrangements, the crossbar and the Beneš architectures [142] are shown in Figure 17.30, and the choice between architectures will depend on criteria such as

- The number of 1×2s or 2×2s: An $n \times n$ crossbar switch consists of n^2 elementary switches, and Beneš switches need only $n/2 \times (2 \times \log_2(n) - 1)$.
- Loss and loss uniformity—in the Beneš switches, each signal crosses the same number of 2×2s, but in the crossbar switches there is a shortest path (going through one switch only) and a longest path (through $2n - 1$ elements).
- The number of waveguide crossovers—this will lead to crosstalk between the lightpaths.

Crossbar switches have no crossovers, in contrast to Beneš switches.

- Switches can be blocking or nonblocking. In the latter, a connection can be made between any unused input port and any unused output port. If this new connection can be made without interrupting the traffic on the other channels, the switch is called "wide sense non-blocking" (the crossbar switch belongs to that category); otherwise it is called "rearrangeably nonblocking" (e.g., Beneš switches).

Other architectures are described in [142,153, Chapter 3.7].

The elementary 1×2 or 2×2 optical switches themselves use a wide range of operating principles, which can be roughly classified into:

		Operating principle	
		Interferometric	**Noninterferometric**
Mechanical	Switching mechanism		Mechanical Mechanical micro-optic systems, MEMS
Nonmechanical		Mach–Zehnder interferometer: thermo-optic effect in doped silica and polymer waveguides, electrooptic LN switches	Y-branch devices with polymer thermo-optic, LN and InP electro-optic effects
		Hybrid silica/polymer tunable vertical coupler switch	Total internal reflection: silica—air interfaces, InP electro-optic
		All-optical $\chi(3)$ switches in GaAs, InP	Polarization switching in liquid crystals
			Semiconductor optical amplifier (SOA) ON–OFF gate

Mechanical switches route light through a physical displacement of the light path: the fundamental switching elements are movable mirrors or prisms, which are mechanically actuated. Another approach is to steer beams from N input fibers on N output fibers, the difficulty being that an analogue control of the beam position is required.

Main advantages of mechanical switches are low insertion losses and high crosstalk between channels. However, their commutation speed is of the order of tens of milliseconds to seconds, and their size is limited to about 32×32.

An emerging technology is MEMS, which combine the optical performance of mechanical switches with the benefits of integrated optics (compactness, low cost volume production, optical prealignment) [154–156]. Schematic drawings and SEM pictures of the two-dimensional (2D) matrix and the movable mirrors are shown in Figure 17.31. The mirrors are fabricated via Si surface micro-machining; the translation movement of a scratch drive is translated into a rotation, and as a result the mirror flips between the two states. For 8×8 2D devices insertion losses of 3.1–3.5 dB with <1 ms switching speed have been reported, and large scale cross connects with < 6 dB losses and 512×512 ports or higher are expected to be constructed out of 32×32 modules [155].

Higher port count monolithic devices are accessible through MEMS mirrors with two rotational degrees of freedom. The continuous rotation is used for steering the beam in all three dimensions, and the MEMS panel is placed between an array

of input/output fibers and a fixed reflecting mirror. Switches with 112×112 ports have been demonstrated [157].

In general, mechanical switches use expanded beam optics to minimize coupling losses.

Nonmechanical noninterferometric switches depend on the control of waveguide index or polarization state.

In a fixed symmetric Y-branch (Section 17.2.3), light is equally divided between the output waveguides, but dynamic control of the refractive indices allows us to "push" or "pull" light into one or the other output.

Layout and switching characteristics for a 1×2 digital optical polymer waveguide switch are shown in Figure 17.32 [7]: increasing the temperature of the upper arm will decrease the waveguide index, thus coupling more light into the lower arm. Typical power consumption is around 50–80 mW with about 1 ms commutation time.

Electro-optic switches can use the same principle, but provide higher speeds. Hundreds of megahertz have been reached using Ti:LiNbO$_3$ waveguides [158], and crosstalk has been enhanced to 30–50 dB by concatening 1×2 switches ("dilated switch"). Although power consumption is low, the main inconvenience is the applied voltage of about 25 V. Monolithic 4×8 and 6×6 LN switches are reported in [159].

Crossconnects based on EO-InP switches combined with SOAs are particularly promising for nanosecond packet switching applications, as they provide high extinction ratio and can be lossless.

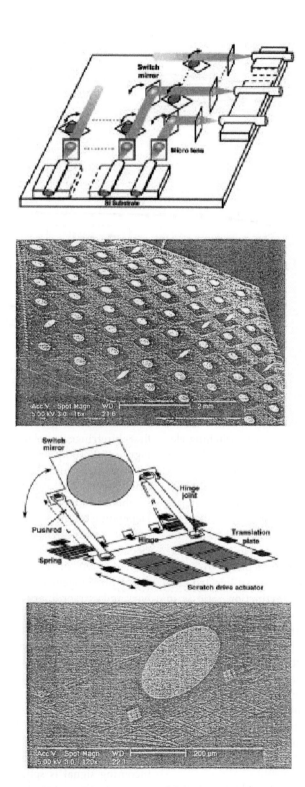

Figure 17.31 Free space MEMS switch: schematic [200] and scanning electron microscope picture [155] of a switch matrix (above). Schematic drawing of a single mirror and SEM close-up (next page). (Reproduced by permission of IEEE and the Optical Society of America.)

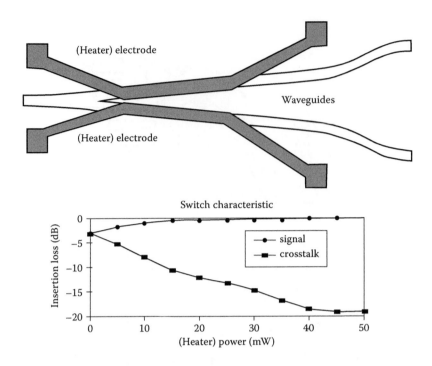

Figure 17.32 1×2 digital optical Y-branch switch: layout with waveguides and heating electrodes for TO effect, extinction as a function of the applied power [7]. (Reproduced by permission of Marcel Dekker Inc.)

Figure 17.33a shows an elementary switching element [160]: in the Y-junction the switching is done by current injection and the associated index change (Section 17.2.2). The travelling wave amplifier TWA compensates for loss when the switch is in the "ON" state and suppresses crosstalk from other channels when the switch is "OFF". Fiber-to-fiber performance of a 4×4 crossbar switch operating at 1.31 μm wavelength is depicted in Figure 17.33b—the device is lossless for an injected current of >200 mA, and the extinction ratio is in excess of 40 dB. Polarization sensitivity is an issue but can be reduced to <1 dB by using square shaped waveguide cross-sections [161].

Switches using total internal reflection (TIR) at silica waveguide–air interfaces are reported in [162–164]. The principle is shown in Figure 17.34 (top): a drop of index matching fluid can be moved within a narrow trench, leading to transmission or TIR.

An example of asymmetric Y-couplers and electro-optic induced TIR is shown in Figure 17.34 (bottom) [165].

Polarization rotation in liquid crystal (LC) spatial light modulators in conjunction with birefringent crystals results in space switching. The first birefringent polarization beam splitter in Figure 17.35 (bottom) [166] will separate the beam into ordinary and extraordinary beams having orthogonal polarizations, and the arrayed λ/2 plate symbolizes a matrix of liquid crystal modulators. Depending on the orientation of the LC molecules (which in turn can be changed by application of an electric field) the polarization of incoming light is rotated by 90° or remains unchanged. The subsequent birefringent crystals will recombine ordinary and extraordinary rays.

SOAs can be used as a gate (ON/OFF switch) by the variation of the applied bias voltage: a low voltage results in low population inversion, and an incoming signal gets absorbed. Application of the bias voltage leads to an increase in population inversion, resulting in the transmission and gain of the signal. The commutation speed is of the order of 1 ns [142]. 1×4 [167], 4×4 [168] and 8×8 space switches [169] operating at 10 Gbit s⁻¹ have been demonstrated based on a broadcast-and-select (the incoming signal is split, then the SOA ON/OFF gates pass through or block the signals) architecture. Devices have been fabricated through hybrid integration of the SOA gates on SiO_2 motherboards.

(a)

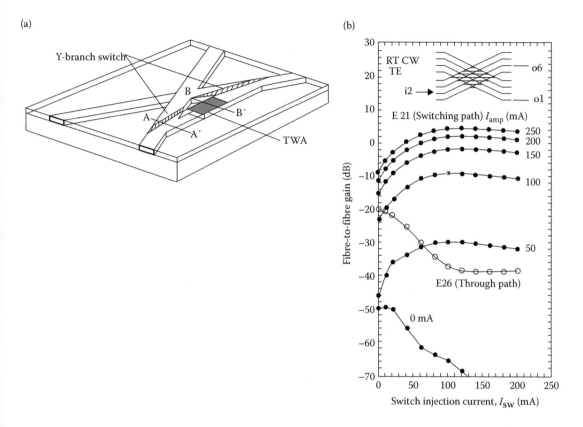

(b)

Figure 17.33 (a) Elementary semiconductor switch element consisting of Y-branches and a semiconductor travelling wave amplifier for loss compensation. (b) Performance of a 4×4 switch: the device is lossless for >200 mA injection current, and the extinction ratio is >40 dB [160]. (Reproduced by permission of World Scientific Publishing Co Ltd.)

Another application for SOA gates is to select a set of wavelengths from a WDM signal, and both hybrid [170] and monolithic integrated InP [171] devices have been demonstrated (Figure 17.36): the incoming WDM signal is wavelength demultiplexed in a first arrayed waveguide grating, passed through or blocked in the SOAs and then recombined in a second arrayed waveguide grating.

Nonmechanical interferometric switches are based on tunable directional couplers or Mach–Zehnder interferometers (Sections 17.2.3 and 17.2.4) and used in thermo-optic, linear (Pockels) or nonlinear (Kerr) electro-optic effects for achieving the required phase shift.

Monolithic 16×16 crossbar switch matrices using silica waveguide MZIs and the thermo-optic effect fabricated on a 6 in. substrate are described in [172] (Figure 17.37). The high extinction ratio of 55 dB is achieved by concatenation of two MZIs

("dilated switch"), and the average insertion loss is 6.6 dB. The electrical power consumption per switch point is 450 mW resulting in a total of 17 W. However, in recent 2×2 and 8×8 switches a reduction to 45 mW per stage has been demonstrated by micro-machining grooves on either side of the waveguide and insertion of a thick silica buffer layer for enhanced thermal isolation between the waveguide and the Si substrate [173,174].

Polymer waveguides and their large dn/dT further reduce the power consumption, and for a 2×2 device, 5 mW [175] have been reported in the 1.3 μm telecom window.

The vertical coupler 1×2 switch shown in Figure 17.38 [67] combines the low propagation loss in silica with the low switching power of polymer devices: the structure consists of a vertically stacked lower doped silica waveguide layer, a thin silica cladding and the polymer waveguides. When the electrodes are not powered (OFF), input light

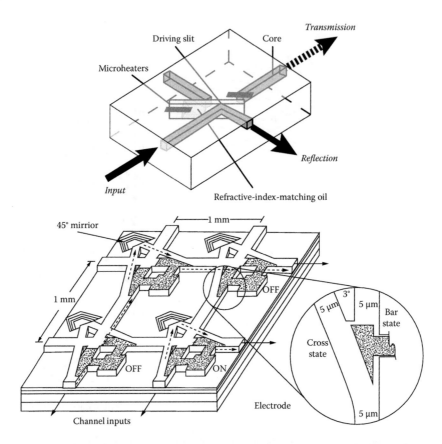

Figure 17.34 Total internal reflection switch with silica waveguides (top, [163]). The left part of the driving slit is filled with index matching oil. Application of heat and capillarity will drive the oil into the intersection of waveguides and switches from reflection to transmission. TIR EO switch (bottom) [165]. (Reproduced by permission of the Optical Society of America and the American Institute of Physics.)

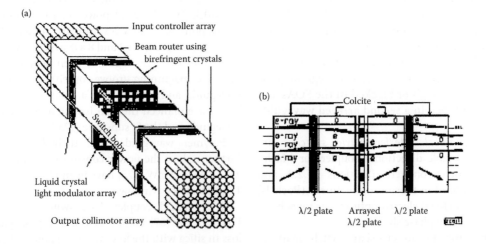

Figure 17.35 Liquid crystal free space optical switch (top) with input/output collimator arrays and polarization based switching (bottom) [166]: the incoming beam is split into o and eo polarizations. The liquid crystals represent a switchable array of halfwave plate, i.e., they turn (or not) the polarization by 90°. The right part is symmetric to the left part. (Reproduced with permission from IEEE.)

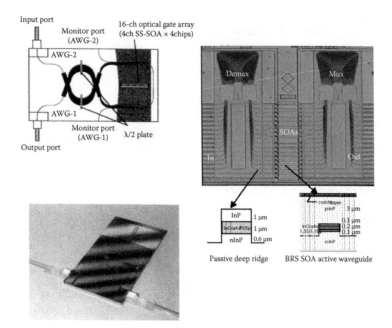

Figure 17.36 Hybrid (left, [173]) and monolithic InP (right, [171]) channel selector. In both cases, a DEMUX will separate the wavelengths, which are gated through InP SOAs. The second MUX then recombines the channels. (Reproduced by permission of IEEE.)

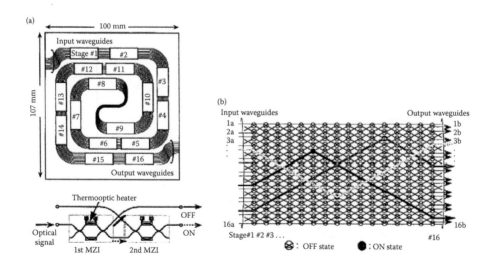

Figure 17.37 Silica waveguide thermo-optic switch (16×16) [172]: layout on the 6 in. wafer (top) and architecture (bottom). Each switch point consists of a two stage Mach–Zehnder for enhanced ON–OFF extinction ratio. (Reproduced by permission of IEEE.)

coupled to the lower silica waveguide will propagate through this silica waveguide (bar state). When the electrodes are sufficiently powered (ON), the input light is coupled to the polymer waveguide in the first element and coupled back to another parallel silica waveguide in the second element (cross state). The reported switching power was <80 mW and crosstalk <20 dB.

For electro-optic MZI LiNbO$_3$ waveguide switches see Reference [176].

Forward current injection into InGaAsP/In structures leads to a negative index change

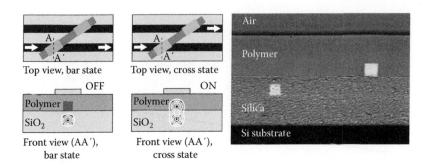

Figure 17.38 Vertical coupler switch (VCS) [67]: schematic layout of the 1×2 hybrid polymer/silica VCS and calculated field distribution (left); cross-section of a fabricated vertical coupled structure (right). (Reproduced by permission of VDE Verlag.)

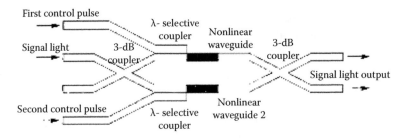

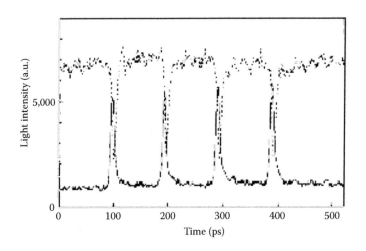

Figure 17.39 Monolithic all-optical symmetric Mach–Zehnder switch using nonlinearity in GaAs [179]. Device layout with inputs for the signal and the two control pulses defining a gate (top); switching characteristics (bottom). (Reproduced by permission of the American Institute of Physics.)

of −0.4 rad mA^{-1} [177], and application of a reverse voltage [178] increases the refractive index, leading to a positive phase change of 0.25 rad V^{-1}.

All-optical switching is a very attractive concept for ultra-high speed systems (>100 Gbit s^{-1}) as bandwidth-limited electronic processing is replaced by the manipulation of an optical data signal via a second optical control pulse [7, Chapter 11.3] (see Chapters A2.4 and B6).

Third order nonlinearity $\chi(3)$ (Section 17.2.2) in GaAs has been used in the monolithic device shown in Figure 17.39 (top) [179]: when the first control pulse is absorbed in the upper nonlinear waveguide the Mach–Zehnder interferometer becomes asymmetric and the incoming signal is switched from one output to the other. The second control pulse then cancels the index change from the first pulse, thus switching the signal back. The control pulses with an energy of <7 pJ and a length of 1.3 ps came from a Ti:sapphire laser, and switching speeds were 8 ps (Figure 17.39, bottom).

17.5.5 Integrated add–drop multiplexers and optical crossconnects

Add–drop multiplexers are wavelength selective switches allowing dropping of one or more signals from an incoming data stream and adding signals to the outgoing data stream. Different monolithic arrangements based on AWGs (Chapter 17.2.4) and switches are shown in Figure 17.40 (top) [17], and configurations (a) and (b) use the same AWG for demultiplexing and multiplexing the incoming traffic into individual wavelength channels. The loop back configuration (a) has been chosen for the monolithic InP component in Figure 17.40 (bottom) [180], which is designed for four channels on a 200 GHz grid. The component is extremely compact (3×6 mm^2) and uses electro-optic MZI 2×2 switches for passing through or adding/dropping traffic on each of the four channels.

Another device architecture for ADM is the serial lattice filter (Chapter 17.5.3) made of SiON waveguides (Figure 17.41, top) [144]: there are 12 asymmetric Mach–Zehnder stages resulting in a chip size of 6×65 mm^2. The device operates on a 200 GHz grid and can dynamically add/drop one out of the incoming eight channels. The 12 MZI stages give sufficient finesse for suppressing crosstalk from adjacent channels as shown in the theoretical filter response (Figure 17.41, bottom). Experimental on-chip losses at 1550 nm were about 2 dB, isolation of the through channels from the dropped wavelengths was around 20 dB and tuning was accomplished with thermo-optic phase shifters.

Polymer waveguides with their $10\times$ higher thermo-optic coefficient dn/dT and the possibility of writing Bragg gratings (see Chapter 17.3.5) provide another possible architecture when combined with four-port circulators (Figure 17.42) [64]. The function of the Bragg grating is to separate the drop wavelength by reflecting it back to the left circulator and to pass all other wavelengths through. The main challenge is the fabrication of the Bragg grating with uniform and strong reflection within a narrow wavelength band, no out-of-band spectral features, low insertion loss and polarization dependence.

Other material systems for AD filters are Si waveguide ring resonators [181] and acousto-optic LiNbO$_3$ Mach–Zehnder switches [63,182, Chapter 7.8].

The integrated 2×2 optical cross connect shown in Figure 17.43 [183] consists of AWGs giving simultaneous access to all 16 wavelengths and an array of thermo-optic 2×2 switches: light entering AWG1 is demultiplexed and enters the left hand side of the switches. When the switches are "OFF", the signals will be recombined in AWG3; otherwise they are routed to AWG4. The second data stream coming into the OXC/AWG2 will arrive at the right hand side of the TO switches and go through to AWG4 ("OFF") or is exchanged to AWG3 ("ON"). The switches consist of two stage MZIs for reaching an extinction ratio of <−28 dB. A transmission spectrum is shown in Figure 17.43 (bottom). The size of the device was 87×74 mm^2 and insertion losses between 7.8 and 10.3 dB.

A simpler arrangement (Figure 17.44, top) [184] uses only two AWGs. The AWGs have an interleaved chirped grating, thus producing two separate images for each wavelength in different grating orders (denoted by ω_0, ω_{-1}, ω_{+1}). Per channel, there are three connecting waveguides with InP phase shifters between the two AWGs, allowing the dynamic re-routing of wavelength channels to either line 1 or line 2 of the second AWG. The footprint of the device is 4.2×9.8 mm^2 [177], and experimental on chip spectra for a 2×2 crossconnect with six channels are shown in Figure 17.44 (bottom). Reported on chip losses were about 10 and 19 dB including fiber/chip coupling.

17.5.6 Integrated monitoring devices

Among the different network management functions [142, Chapter 10.1], performance monitoring deals with providing a guaranteed QoS to the network users. As an example, a drift of the transmitter laser Tx wavelength due to temperature variations or ageing can lead to a decrease in

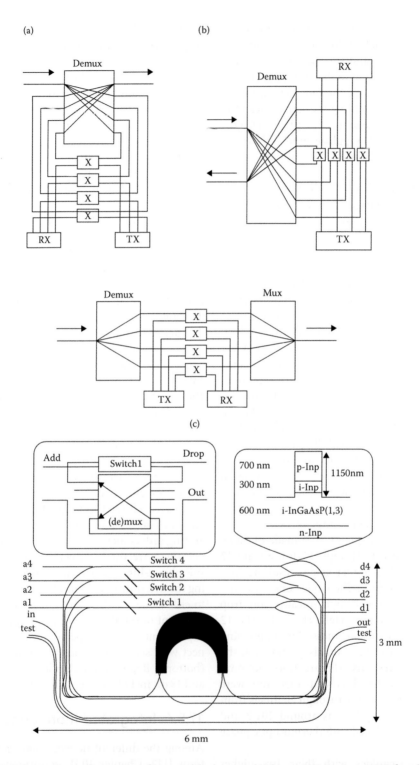

Figure 17.40 Schematic configurations for an add–drop multiplexer ADM using AWG for demultiplexing/ multiplexing and space switches [17] (top): the conventional architecture (c) consists of two AWGs, and the fold back (a)/loop back (b) architectures use the same AWG for multiplexing and demultiplexing (bottom). Monolithic InP components for four channels with 400 GHz spacing [180]. (Reproduced by permission of IEEE.)

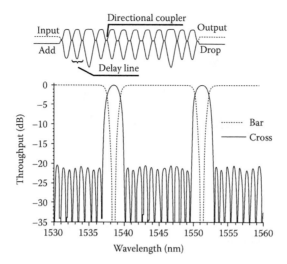

Figure 17.41 Serial lattice add–drop filter with 12 asymmetric Mach–Zehnder stages (top), theoretical device response for a drop (solid) and through channels (dashed) [64]. (Reproduced by permission of IEEE.)

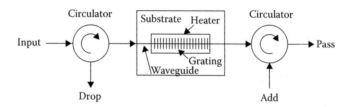

Figure 17.42 Optical add–drop filter using a thermo-optic tunable polymer Bragg grating and a pair of three port circulators [64]. The incoming signals will propagate through the left circulator into the Bragg grating, which reflects the drop wavelength back and passes the other channels through. In the same way, the add signals are reflected back and recombine with the through channels. (Reproduced by permission of SPIE.)

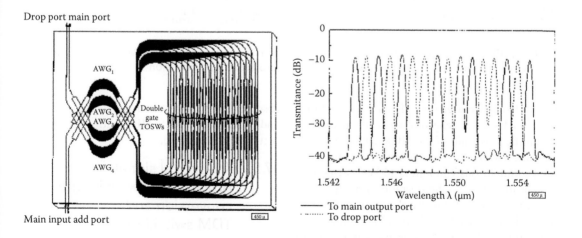

Figure 17.43 Integrated silica waveguide 2×2 optical crossconnect for 16 channels for each input/fiber [183]: layout with four AWGs and double stage TO switches, transmission spectrum for channels passing through (solid line) and dropped channels (dotted). (Reproduced by permission of IEEE.)

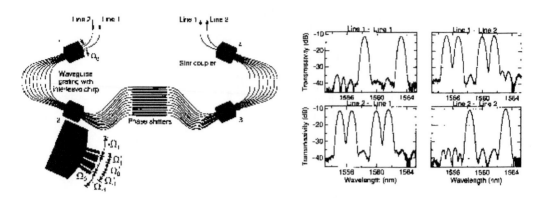

Figure 17.44 Monolithic 2×2 crossconnect with two AWGs: design [184] and on-chip spectra of an InP device [177]. (Reproduced by permission of IEEE.)

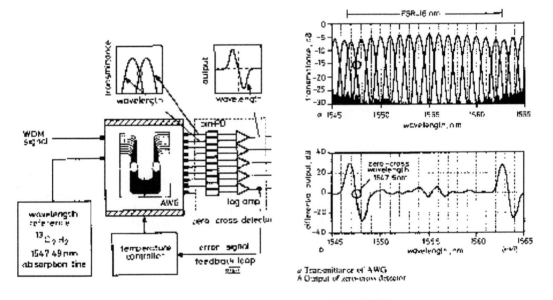

Figure 17.45 Wavelength monitoring circuit [185]: the large passbands of the AWG overlap produce an electrical differential signal for zero detection. (Reproduced by permission of IEEE.)

received signal power (and thus increased BER), when the laser wavelength is not centered any longer on the passbands of the MUX/DEMUX filters. The silica waveguide component in Figure 17.45 [185] simultaneously monitors 16 WDM channels present in a system. The operation principle is to feed a WDM signal (the wavelength is marked by the circle in Figure 17.45, bottom) into an arrayed waveguide grating with large passbands and to compare the received power on adjacent outputs. Misalignment of the laser frequency will result in a positive or negative differential signal. A molecular

absorption line serves as a wavelength reference, and a resolution of 10 MHz with sub-gigahertz accuracy was reported.

17.5.7 All-optical signal processing—wavelength conversion and TDM switching

Like the all-optical switches described in Section 17.5.4, all-optical signal processing [152] devices rely on third order nonlinearity χ(3). For practical all-optical devices, the nonlinearity has to be large enough to achieve a π phase shift with control pulses having picosecond length and picojoule energy, but at the same time the losses need to be sufficiently low. The semiconductor optical amplifier can overcome this trade-off, thus playing an important role in wavelength conversion, all-optical time division multiplexing and regeneration [171,186].

For an SOA in the gain regime (i.e., with inverted population densities between valence and conduction bands), the following nonlinear phenomena occur.

- Cross gain modulation (XGM)—an incoming photon will depopulate the conduction band, thus changing inversion andgain. One application of XGM is wavelength conversion, the conceptof which is shown in Figure 17.46: for an incoming "1" at the signal wavelength λ_{signal} the gain for the probe signal at wavelength λ_{probe} will drop, producing a "0".
- Associated with the change in inversion is avariation in index and thus phase (cross phase modulation XPM), and this phase change is used in waveguide interferometers for signal processing:

- Time division demultiplexing: the monolithic InP MZI with SOAs in both arms depicted in Figure 17.47 [187] (the architecture being the same as the GaAs all-optical passive switch in Section 17.5.4/Figure 17.39) represents a device for demultiplexing a 40 Gbit s^{-1} stream into 10 Gbit s^{-1} streams. Initially, the Mach–Zehnder is symmetric, but control pulse 1 will introduce a phase change in the upper SOA and will switch the signals from one output port to the other. As a result, the input data stream is demultiplexed. Control pulse 2 is introduced with an appropriate timing delay and resets the MZI to the initial state. The subsequent control pulses are needed for achieving switching speeds beyond the limit of the SOA carrier lifetimes (30–300 ps) [152].
- Wavelength conversion via XPM is explained schematically in Figure 17.48a: the switching Signal λ_{switch} will introduce a de-phasing in the interferometer, thus impressing the data pattern on the continuous-wave probe wavelength λ_{signal} The result is one inverted and one non-inverted signal as shown on the right hand side. The corresponding monolithic

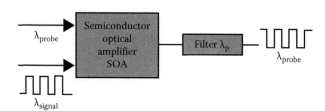

Figure 17.46 Wavelength conversion by cross-gain modulation (XGM) in a nonlinear semiconductor optical amplifier SOA. An incoming "1" on the signal wavelength will reduce the gain at the wavelength of the probe laser, producing an inverted signal.

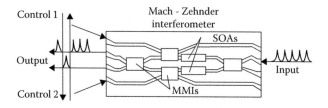

Figure 17.47 OTDM device [187]. The architecture is similar to the one in Figure 17.39, but this time nonlinear InGaAsP SOAs in the 1.5 μm telecom window are used. (Reproduced by permission of IEEE.)

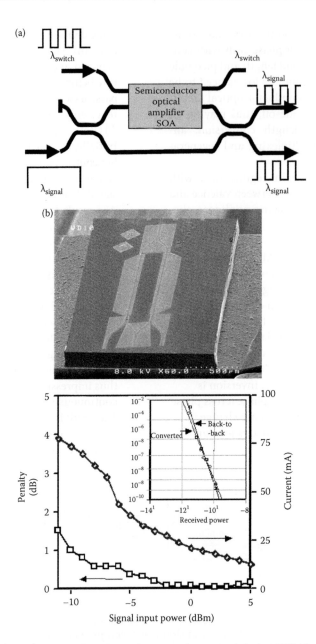

Figure 17.48 (a) Wavelength conversion through cross phase modulation XPM in nonlinear SOAs: the switching signal λ_{switch} will introduce a de-phasing in the interferometer, and the result is one inverted and one noninverted signal at wavelength λ_{signal}. (b) Monolithic InP cross phase modulation wavelength converter and performance in a 10 Gbit s^{-1} system; the inset compares the BER for the initial ("back-to-back") and wavelength converted signal [186]. (Reproduced by permission of VDE Verlag.)

integrated InP device is shown in Figure 17.48b together with the performance at 10 Gbit s^{-1} modulation [186], comparing the BER for the initial ("back-to-back") and wavelength converted signal. Besides monolithic solutions, hybrid integrated circuits had been published earlier [188].

XPM and XGM wavelength conversion are both suitable for single wavelengths, but not in WDM systems as—besides the signal λ_{switch}—all the other data channels also contribute to XGM and XPM. Despite the fact that XPM devices have a more complicated layout and are difficult to fabricate, they have significant advantages: higher extinction

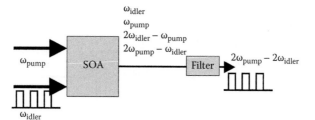

Figure 17.49 Four-wave mixing in an SOA. The nonlinearity of the SOA will cause generation of sum and difference frequencies, and the signal is converted from frequency ω_i to $2\omega_{pump} - \omega_{idler}$.

ratio for the wavelength converted signal, availability of both inverted and noninverted data outputs and regeneration of the signal (see Section 17.5.8). Also, the direct amplitude modulation in XGM introduces chirp (i.e., the instantaneous phase of the signal is a function of the amplitude), but chirp can be controlled in XPM devices.

- Four wave mixing (FWM), which also arises in passive nonlinear transmission media such as fiber, causes three waves at frequencies ω_1, ω_2, and ω_3 to generate a fourth wave at frequency $\omega_1 + \omega_2 - \omega_3$. Pumping at wavelength $\omega_{pump} = \omega_1 = \omega_2$ and providing an idler at $\omega_{idler} = \omega_3$ results in a wavelength converted signal at $\omega_s = 2\omega_{pump} - \omega_{idler}$, as schematically shown in Figure 17.49.

In contrast to XPM and XGM, four wave mixing can convert several signals at once.

For an exhaustive treatment of all-optical time division multiplex system techniques, in particular using semiconductor and also fiber-based devices, see Reference [63, Chapter 9] and Chapters A2.4, B.6 and C3.5.

For nonlinear optics, see Chapters A2.4 and Reference [8], and for SOAs in particular, see Reference [7, Chapter 11].

17.5.8 All-optical regeneration

Degradation of the OSNR in a system arises from the accumulation of amplified spontaneous emission (ASE) from the line amplifiers, chromatic dispersion of the transmission fiber and components, PMD and third order nonlinearity of the fiber (self phase and cross phase modulation [142, 189]). This leads to a transparent reach of about 1500–2000 km, and the signal needs to be regenerated before further

transmission. In [109], a re-circulating loop network with a reach of 69,000 km had been demonstrated using electronically driven modulators for reshaping the 20 Gbit s^{-1} signals for every 100 km.

All-optical regeneration could be an interesting alternative to electronic regeneration for different reasons: it is expected to be more cost effective through elimination of expensive O–E–O converters, moreover all-optical regeneration allows the processing of low as well as high bit rate signals rather than being specifically designed for 10 or 40 Gbit s^{-1} Also, 40 Gbit s^{-1} is believed to be the limitation for electronic regeneration but optical regeneration may perform at bit rates above this.

For regenerating a signal, three levels of manipulation are considered [107]:

- Optical re-amplification of the optical signal (1R regeneration).
- Amplification and re-shaping (2R).
- Amplification, re-shaping and retiming fully recovers the signal (3R). Retiming is necessary when accumulation of jitter, i.e., random signal delay through phase modulation in the fiber or in nonlinear devices (such as wavelength converters) is an issue.

1R regeneration can be done all-optically in a linear amplifier (see Chapters A1.6 and B5) or through OE-conversion followed by an electrical amplification and EO-conversion.

2R signal reshaping and amplification can be implemented with an amplifier and a nonlinear gate as sketched in Figure 17.50a: the nonlinear gate modulates a "clean" CW signal having the same wavelength as the input data. Besides fiber based nonlinear devices, the SOA Mach–Zehnder interferometer previously used for XPM

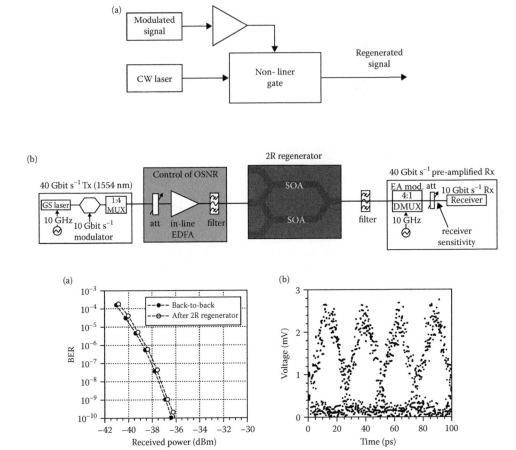

Figure 17.50 (a) 2R regenerator. The nonlinear all-optical gate modulates a "clean" CW signal having the same wavelength as the modulated input signal. (b) Set-up for 2R regeneration at 40 Gbit s⁻¹ (top) [190] and system performance (bottom): bit error rate versus received power in a back-to-back configuration and after the 2R regenerator. The penalty of <0.2 dB indicates that the signals are regenerated almost with their initial signal-to-noise ratio. The bottom right figure is the oscilloscope trace of the received signal. (Reproduced by permission of the Optical Society of America.)

(Section 17.5.7) provides a nonlinear amplitude response and can thus be used as a gate, as demonstrated in [190] for regeneration of 40 Gbit s⁻¹ signals. In the SOA, the presence of the CW laser reduces the carrier lifetime through stimulated emission [152], thus enabling operation at high bit rates.

Figure 17.50b, [190] shows the set-up for a 2R experiment at 40 Gbit s⁻¹ as well as the BER results for a system with a 2R regenerator compared to the back-to-back operation of the Tx/Rx. The penalty is <0.2 dB, indicating that the 2R regenerator restores the signal almost to its initial quality.

3R regeneration (Figure 17.51a) requires—besides 2R amplification and reshaping—the

extraction of a jitter-free clock signal from the incoming data stream. The clock signals are then sent to the nonlinear optical gate, which is modulated by the initial data stream, and as a result a jitter-free signal is generated at the 3R output.

The main difficulty is the extraction of the clock signal, which can be done opto-electronically or all-optically via, for example, self-pulsating laser diodes (SP-LDs) [191,192]. The SP-LD initially has a free-running repetition rate, and the basic principle (Figure 17.51b) of clock extraction is that the SP-LD will change and lock its repetition rate to the data bit rate when optical data signals are injected into the laser. The oscilloscope traces in Figure 17.51c, [193] show the incoming 5 Gbit s⁻¹

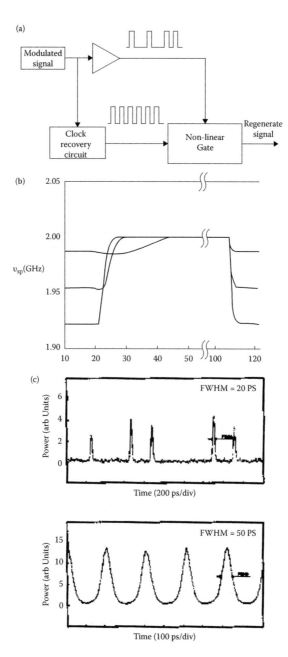

Figure 17.51 (a) Schematic 3R regenerator, consisting of a (noisy) input signal, from which the clock recovery circuit will extract a regular train of signals. The clock signals in turn will gate the all-optical nonlinear circuit, so that the data are not only re-amplified and re-shaped but also re-timed. (b) Self-pulsating laser diode [192] for clock recovery: the figure shows how laser diodes with different free-running frequencies will lock their repetition rate when synchronization pulses of precisely 2 GHz are injected. (c) Self-pulsating laser diode as a 5 Gbit s^{-1} clock recovery circuit: the clock signal (a train of "1" pulses) is extracted from an incoming signal consisting of "1"s and "0"s (top) [193]. (d) 3R regeneration (40 Gbit s^{-1}) [195]. In the experimental set-up (top), the nonlinear gate consists of a fiber-based Mach–Zehnder interferometer with a nonlinear SOA in one arm. BER (bottom) as a function of the received power, comparing the regenerated signals (open symbols) to a back-to-back experiment (full symbols). For a BER of 10^{-9}, the power penalty is about 2.2 dB. The insets are oscilloscope traces of the initial and regenerated signals. (Reproduced by permission of IEEE.)

(Continued)

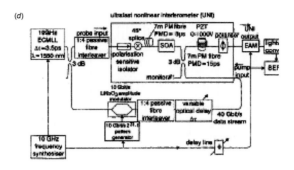

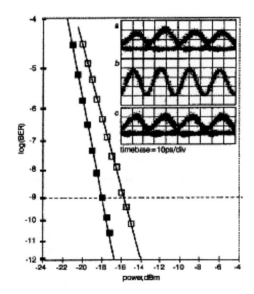

Figure 17.51 (Continued) (a) Schematic 3R regenerator, consisting of a (noisy) input signal, from which the clock recovery circuit will extract a regular train of signals. The clock signals in turn will gate the all-optical nonlinear circuit, so that the data are not only re-amplified and re-shaped but also re-timed. (b) Self-pulsating laser diode [192] for clock recovery: the figure shows how laser diodes with different free-running frequencies will lock their repetition rate when synchronization pulses of precisely 2 GHz are injected. (c) Self-pulsating laser diode as a 5 Gbit s^{-1} clock recovery circuit: the clock signal (a train of "1" pulses) is extracted from an incoming signal consisting of "1"s and "0"s (top) [193]. (d) 3R regeneration (40 Gbit s^{-1}) [195]. In the experimental set-up (top), the nonlinear gate consists of a fiber-based Mach–Zehnder interferometer with a nonlinear SOA in one arm. BER (bottom) as a function of the received power, comparing the regenerated signals (open symbols) to a back-to-back experiment (full symbols). For a BER of 10^{-9}, the power penalty is about 2.2 dB. The insets are oscilloscope traces of the initial and regenerated signals. (Reproduced by permission of IEEE.)

data consisting of 1s and 0s (top) and the extracted clock signal consists of a train of "1" pulses.

In system experiments, 3R regeneration has been demonstrated for up to 40 Gbit s^{-1} bit rate using polarization rotation in a nonlinear fiber [194] and later on using a fiber based interferometer including an SOA as the nonlinear decision gate [195]. The latter set-up and experimental results are shown in Figure 17.51d: the inserts

represent oscilloscope traces of the incoming and regenerated signals, and the BER diagrams show a PP of about 2.2 dB.

17.6 CONCLUSION

Integrated optics deals with compact single function devices, devices with integrated multiple functions (either on a single chip or in a single package) or devices with a functionality which cannot be achieved with common bulk or thin film optic devices. Besides these advantages in functionality or performance, the main driver for integrated optic devices is the desire for device footprint and reduced cost. The latter will be achieved through ongoing investments and standardization.

Tremendous work has been done over the last 20 years, but there are still a few words of caution:

- The technology is much less mature than electronics.
- As pointed out in Section 17.3, there are multiple materials and technologies: that unlike in electronics where Si has an outstanding role, at present there is no material that is capable of addressing all needs in terms of device performance and cost simultaneously. InP may take the lead as it is suitable for the fabrication of both active and passive functions, but fundamental issues such as wafer size and materials compatibility will need to be addressed.
- Another difference with respect to electronics relates to the density of the optical circuits: the typical size of an optical chip is a few mm^2, which considerably limits the number of functions on a single substrate. The connection of the optical circuit to the outside world is done by aligning single mode fiber—the typical fiber diameter of 125 μm limits the number of I/O connections.
- Many current high speed devices (such as the switches in Section 17.5.4, or the optical cross-connect in Section 17.5.5) use the electro-optic effect in SOAs, but all-optical devices (in analogy to the "all-electronic" transistor) are still a subject of fundamental research, and thus are far from deployment.

However, there are key advantages of optical devices: the capability of achieving modulation speeds beyond 40 Gbit s^{-1}, the absence of electromagnetic interference, and finally optical devices may be used for sensing applications in environments in which electrical charges are prohibited.

Today, integrated optics is the technology of choice for a few functions such as filters, switches and gain equalizers. In future telecommunication applications, we will see an increasing deployment due to the advantages in device functionality, footprint, device cost and finally modulation speed resulting in higher transmission capacities.

REFERENCES

1. Ladouceur, F. and Love, J. D. 1996. *Silica-Based Buried Channel Waveguides and Devices* (London: Chapman and Hall).
2. Okamoto, K. 2000. *Fundamentals of Optical Waveguides*. 2nd ed., English Translation (San Diego, CA: Academic).
3. Kawachi, M. 1990. Silica waveguides on silicon and their application to integrated-optic components. *Opt. Quantum Electron.*, 22, 391.
4. Eldada, L. and Shacklette, L. W. 2000. Advances in polymer integrated optics. *IEEE J. Sel. Top. Quantum Electron.*, 6, 54.
5. Cocorullo, G. and Rendina, I. 1992. Thermo-optical modulation at 1.5 μm in silicon. *Electron. Lett.*, 28, 83.
6. Jackson, J. D. 1998. *Classical Electrodynamics*. 3rd ed. (New York: Wiley).
7. Murphy, E. J. (ed.). 1999. *Integrated Optical Circuits and Components Design and Applications* (New York: Dekker).
8. Sutherland, R. L. 1996. *Handbook of Nonlinear Optics* (New York: Dekker).
9. Alferness, R. 1988. *Titanium-Diffused Lithium Niobate Waveguide Devices Guided Wave Optoelectron*, ed. T. Tamir (New York: Springer).
10. Ikushima, A. J., Fujiwara, T. and Saito, K. 2000. Silica glass: A material for photonics. *J. Appl. Phys.*, 88, 1201.
11. Wang, W., Shi, Y., Olson, D. J., Lin, W. and Bechtel, J. H. 1997. Polymer integrated modulators for photonic data link applications. *Proc. SPIE—The International Soc. for Opt. Eng.*, 2997, 114.
12. Takagi, A., Jinguji, K. and Kawachi, M. 1992. Silica-based waveguide-type wavelength-insensitive couplers (WINCs) with series-tapered coupling structure. *J. Lightwave Technol.*, 10, 1814.
13. Nishihara, H., Haruna, M. and Suhara, T. 1989. *Optical Integrated Circuits*. 2nd ed. (New York: McGraw-Hill).
14. Dragone, C. 1989. Efficient $N \times N$ star couplers using Fourier optics. *IEEE J. Lightwave Technol.*, 7, 479.

15. Okamoto, K., Okazaki, H., Ohmori, Y. and Kato, K. 1992. Fabrication of large scale integrated-optic $N \times N$ star couplers. *IEEE Photon. Technol. Lett.*, 4, 1032.

16. Soldano, L. B. and Pennings, E. C. M. 1995. Optical multi-mode interference devices based on self-imaging: Principles and applications. *J. Lightwave Technol.*, 13, 615.

17. Smit, M. K. and Van Dam, C. 1996. PHASAR-based WDM-devices: Principles design and applications. *IEEE J. Sel. Top.Quantum Electron.*, 2, 236.

18. Takahashi, H., Oda, K., Toba, H. and Inoue, Y. 1995. Transmission characteristics of arrayed waveguide $N \times N$ wavelength multiplexer. *J. Lightwave Technol.*, 13, 447.

19. Soref, R. A. and Little, B. E. 1998. Proposed *N*-wavelength *M*-fibre WDM crossconnect switch using active microring resonators. *IEEE Photon. Technol. Lett.*, 10, 1121.

20. März, R. 1995. *Integrated Optics Design and Modeling* (Boston, MA: Artech).

21. Lenz, G., Eggleton, B. J. and Madsen, C. K. 1999. Optical filter dispersion in WDM systems: A review. *OSA Trends in Optics and Photonics WDM Components*, vol. 29, p. 246.

22. Madsen, C. K. and Zhao, J. H. 1999. *Optical Filter Design and Analysis: A Signal Processing Approach* (New York: Wiley).

23. Jinguji, K. 1998. Broadband programmable optical frequency filter. *Electron. Commun. Jpn Part 2 (Electronics)*, 81, 1.

24. Madsen, C. K. and Zhao, J. H. 1996. A general planar waveguide autoregressive optical filter. *J. Lightwave Technol.*, 14, 437.

25. Madsen, C. K. 2000. General IIR optical filter design for WDM applications using all-pass filters. *J. Lightwave Technol.*, 18, 860.

26. Himeno, A., Kato, K. and Miya, T. 1998. Silica-based planar lightwave circuits. *IEEE J. Sel. Top. Quantum Electron.*, 4, 913.

27. Martinu, L. and Poitras, D. 2000. Plasma deposition of optical films and coatings: A review. *J. Vacuum Sci. Technol. A*, 18, 2619.

28. Bushan, B., Murarka, S. P. and Gerlach, J. 1990. Stress in silicon dioxide films deposited using chemical vapour deposition techniques and the effect of annealing on these stresses. *J. Vacuum Sci. Technol. B*, 8, 1068.

29. Kilian, A., Kirchof, J., Przyrembel, G. and Wischmann, W. 2000. Birefringence free planar optical waveguide made by flame hydrolysis deposition (FHD) through tailoring of the overcladding. *J. Lightwave Technol.*, 18, 193.

30. Ojha, S. M., Cureton, C., Bricheno, T., Day, S., Moule, D., Bell, A. J. and Taylor, J. 1998. Simple method of fabricating polarisation-insensitive and very low crosstalk AWG grating devices. *Electron. Lett.*, 34, 78.

31. Coburn, J. W., and Winters, H. F. 1979. Plasma etching—A discussion of mechanisms. *J. Vacuum Sci. Technol.*, 16, 391.

32. Vossen, J.-L. and Kern, W. 1978. *Thin Film Processes* (Orlando, FL: Academic).

33. Ramaswamy, R. U. and Srivastava, R. 1988. Ion-exchanged glass waveguides: A review. *J. Lightwave Technol.*, 6, 984.

34. Cheng, H. C. and Ramaswamy, R. U. 1990. A dual wavelength directional coupler demultiplexer by ion exchange in glass. *IEEE Photon. Technol. Lett.*, 2, 637.

35. Mizrahi, V., Lemaire, P. J., Erdogan, T., Reed, W. A., DiGiovanni, D. J. and Atkins, R. M. 1993. Ultraviolet laser fabrication of ultrastrong optical fibre gratings and of germania-doped channel waveguides. *Appl. Phys. Lett.*, 63, 1727.

36. Herman, P. R., Chen, K., Li, J., Wie, M., Ihlemann, J. and Marowsky, G. 2001. F2 lasers: Precise shaping and trimming of photonic components. In *2001 Digest of LEOS Summer Topical Meetings: Advanced Semiconductor Lasers and Applications/ Ultraviolet and Blue Lasers and Their Applications/Ultralong Haul DWDM Transmission and Networking/WDM Components*.

37. Strasser, T. A., Erdogan, T., White, A. E., Mizrahi, V. and Lemaire, P. J. 1994. Ultraviolet laser fabrication of strong nearly polarization-independent Bragg reflectors in germanium-doped silica waveguides on silica substrates. *Appl. Phys. Lett.*, 65, 3308.

38. Bazylenko, M. V., Gross, M., Chu, P. L. and Moss, D. 1996. Photosensitivity of Ge-doped silica deposited by hollow cathode PECVD. *Electron. Lett.*, 32, 1198.

39. Bazylenko, M. V., Gross, M. and Moss, D. 1997. Mechanisms of photosensitivity in germanosilica films. *J. Appl. Phys.*, 81, 7497.

40. Maxwell, G. D., Ainslie, B. J., Williams, D. L. and Kashyap, R. 1993. UV written 13 dB reflection filters in hydrogenated low loss planar silica waveguides. *Electron. Lett.*, 29, 425.

41. Maxwell, G. D. and Ainslie, B. J. 1995. Demonstration of a directly written directional coupler using UV-induced photosensitivity in a planar silica waveguide. *Electron. Lett.*, 31, 95.

42. Hoffmann, M., Kopka, P. and Voges, E. 1997. Low-loss fibre-matched low-temperature PECVD waveguides with small-core dimensions for optical communication systems. *IEEE Photon. Technol. Lett.*, 9, 1238.

43. de Ridder, R. M., Warhoff, K., Driessen, A., Lambeck, P. V. and Albers, H. 1998. Silicon oxynitride planar waveguiding structures for application in optical communication. *IEEE J. Sel. Top. Quantum Electron.*, 4, 930.

44. Offrein, B. J., Horst, F., Bona, G. L., Salemink, H. W. M., Germann, R. and Beyeler, R. 1999. Wavelength tunable 1-from-16 and flat passband 1-from-8 add–drop filters. *IEEE Photon. Technol. Lett.*, 11, 1440.

45. Mertens, H., Andersen, K. N. and Svendsen, W. E. 2002. Optical loss analysis of silicon rich nitride waveguides. *ECOC'02: 28th European Conference on Optical Communication poster*, vol. 1, p. 38.

46. Jalali, B., Yegnanarayanan, S., Yoon, T., Yoshimoto, T., Rendina, I. and Coppinger, F. 1998. Advances in silicon-on-insulator optoelectronics. *IEEE J. Sel. Top. Quantum Electron.*, 4, 93.

47. Rickman, A. G. and Reed, G. T. 1994. Silicon-on-insulator optical rib waveguides: Loss mode characteristics bends and y-junctions. *IEE Proc. Optoelectron.*, 141, 391.

48. Tang, C. K., Kewell, A. K., Reed, G. T., Rickman, A. G and Namavar, F. 1996. Development of a library of low-loss silicon-on-insulator optoelectronic devices. *IEE Proc. Optoelectron.*, 143, 312.

49. Cocorullo, G., Della Corte, F. G., de Rosa, R., Rendina, I., Rubino, A. and Terzini, E. 1998. Amorphous silicon-based guided-wave passive and active devices for silicon integrated optoelectronics. *IEEE J. Sel. Top. Quantum Electron.*, 4, 997.

50. Soref, R. A., Schmidtchen, J. and Petermann, K. 1991. Large single-mode rib waveguides in GeSi–Si and Si-on-SiO$_2$. *IEEE J. Quantum Electron.*, 27, 1971.

51. Bestwick, T. 1998. ASOC™-a silicon-based integrated optical manufacturing technology. *Proceedings of 48th Electronic Components and Technology Conference (Cat No 98CH36206)*, p. 566.

52. Fischer, U., Zinke, T., Kropp, J.-R., Arndt, F. and Petermann, K. 1996. 0.1 dB/cm waveguide losses in single-mode SOI rib waveguides. *IEEE Photon. Technol. Lett.*, 8, 647.

53. Joannopoulos, J. D., Villeneuve, P. R. and Fan, S. 1997. Photonic crystals: Putting a new twist on light. *Nature*, 386, 143.

54. Berger, V. 1999. From photonic band gaps to refractive index engineering. *Opt. Mater.*, 11, 131.

55. Pirnat, T. and Friedman, L. 1991. Electro-optic mode-displacement silicon light modulator. *J. Appl. Phys.*, 70, 3355.

56. Soref, R. A. and Bennett, B. R. 1987. Electrooptical effects in silicon. *IEEE J. Quantum Electron.*, QE-23, 123.

57. Jackson, S. M., Hewitt, P. D., Reed, G. T., Tang, C. K., Evans, A. G. R., Clark, J., Aveyard, C. and Namavar, F. 1998. A novel optical phase modulator design suitable for phased arrays. *J. Lightwave Technol.*, 16, 2016.

58. Wooten, E. L. et al. 2000. A review of lithium niobate modulators for fibre-optic communications systems. *IEEE J. Sel. Top. Quantum Electron.*, 6, 69.

59. Kip, D. 1998. Photorefractive waveguides in oxide crystals: Fabrication properties and applications. *Appl. Phys. B (Lasers and Opt.)*, B67, 131.

60. Korkishko, Y. N. and Federov, V. A. 1999. *Ion Exchange in Single Crystals for Integrated Optics and Optoelectronics* (Cambridge: Cambridge International Science Publishing).

61. Hofmann, D., Schreiber, G., Haase, C., Herrmann, H., Grundkotter, W., Ricken, R. and Sohler, W. 1999. Quasi-phase-matched

difference-frequency generation in periodically poled Ti:LiNbO$_3$ channel waveguides. *Opt. Lett.*, 24, 896.

62. Noguchi, K., Mitomi, O., and Miyazawa, H. 1996. Low-voltage and broadband Ti:LiNbO3 modulators operating in the millimeter wavelength region. *OFC'96: Optical Fibre Communication 1996 Technical Digest Series Conference Edition (IEEE Cat No96CH35901)*, 2, 205.

63. Grote, N. and Venghaus, H. (ed.). 2001. *Fibre Optic Communication Devices* (Heidelberg: Springer).

64. Eldada, L., Blomquist, R., Shacklette, L. W. and McFarland, M. J. 2000. High-performance polymeric componentry for telecom and datacom applications. *Opt. Eng.*, 39, 596.

65. Bauer, H.-D., Ehrfeld, W., Harder, M., Paatzsch, T., Popp, M. and Smaglinski, I. 2000. Polymer waveguide devices with passive pigtailing: An application of LIGA technology. *Synth. Met.*, 115, 13.

66. Viens, J.-F., Callender, C. L., Noad, J. P. and Eldada, L. 2000. Compact wide-band polymer wavelength-division multiplexers. *IEEE Photon. Technol. Lett.*, 12, 1010.

67. Keil, N. et al. 2000. Thermo-optic vertical coupler switches using hybrid polymer/silica integration technology. *Electron. Lett.*, 36, 430; Keil, N. et al., 2000. Thermo-optic switches using vertically coupled polymer/silica waveguides. *Proceedings of 26th European Conference on Optical Communications*, p. 101.

68. Inoue, Y., Kaneko, A., Hanawa, F., Takahashi, H., Hattori, K. and Sumida, S. 1997. Athermal silica-based arrayed-waveguide grating multiplexer. *Electron. Lett.*, 33, 1945.

69. Kokubun, Y., Takizawa, M. and Taga, S. 1994. Three-dimensional athermal waveguides for temperature independent lightwave devices. *Electron. Lett.*, 30, 1223.

70. Moroni, M. and Vallon, S. 1999. Athermalized polymer overclad integrated planar optical waveguide device and its manufacturing method. EP 1026526 A1.

71. Dalton, L., Harper, A., Ren, A., Wang, F., Todorova, G., Chen, J., Zhang, C. and Lee, M. 1999. Polymeric electro-optic modulators: from chromophore design to integration with semiconductor very large scale integration electronics and silica fibre optics. *Ind. Eng. Chem. Res.*, 38, 8.

72. Wiesmann, R., Kalveram, S., Rudolph, S., Johnck, M. and Neyer, A. 1996. Singlemode polymer waveguides for optical backplanes. *Electron. Lett.*, 32, 2329.

73. Kik, P. G. and Polman, A. 1998. Erbium-doped optical-waveguide amplifiers on silicon. *MRS Bull.*, 23, 48.

74. Shooshtari, A., Meshkinfam, P., Touam, T., Andrews, M. and Najafi, S. 1998. Ion-exchanged Er/Yb phosphate glass waveguide amplifiers and lasers. *Opt. Eng.*, 37, 1188.

75. Poulsen, M. 2002. Private communication.

76. Bonar, J., Bebbington, J. A., Aitchison, J. S., Maxwell, G. D. and Ainslie, B. J. 1995. Aerosol doped Nd planar silica waveguide laser. *Electron. Lett.*, 31, 99.

77. Baumann, I., Bosso, S., Brinkmann, R., Corsini, R., Dinand, M., Greiner, A., Schafer, K., Sochtig, J., Sohler, W., Suche, H. and Wessel, R. 1996. Er-doped integrated optical devices in LiNbO$_3$. *IEEE J. Sel. Top. Quantum Electron.*, 2, 355.

78. Polman, A. 1997. Erbium implanted thin film photonic materials. *J. Appl. Phys.*, 82, 1.

79. Mickelson, A. R., Basavanhally, N. R. and Yung-Cheng, L. (ed.). 1997. *Optoelectronic Packaging* (New York: Wiley).

80. Hunsperger, R. G. 1995. *Integrated Optics Theory and Technology*. 4th ed. (Heidelberg: Springer).

81. Kaufmann, H., Buchmann, P., Hirter, R., Melchior, H. and Guekos, G. 1986. Self-adjusted permanent attachment of fibres to GaAs waveguide components. *Electron. Lett.*, 22, 642.

82. GR-20-CORE. *Generic Requirements for Optical Fibre and Optical Fibre Cable.* http://www.telcordia.com.

83. Paris, B. 2000. Multiple planar complex optical devices and the process of manufacturing the same. EP 1170606 A1.

84. Shimizu, N., Imoto, N. and Ikeda, M. 1983. Fusion splicing between optical circuits and optical fibres. *Electron. Lett.*, 19, 96.

85. Modavis, R. and Webb, T. 1994. Anamorphic microlens for coupling optical fibres to elliptical light beams. US 5455879 A1.

86. Kobayashi, M. and Kato, K. 1994. Hybrid optical integration technology. *Electron. Commun. Jn Part 2 (Electron.)*, 77, 67.

87. Lealman, I. F., Kelly, A. E., Rivers, L. J., Perrin, S. D. and Moore, R. 1998. Improved gain block for long wavelength (1,55mum) hybrid integrated devices. *Electron. Lett.*, 34, 2247.

88. Büttgenbach, S. 1993. *Mikromechanik* (Stuttgart: Teubner Studienbücher).

89. Krishnamoorthy, A. V. and Goossen, K. W. 1998. Optoelectronic-VLSI: Photonics integrated with VLSI circuits. *IEEE J. Sel. Top. Quantum Electron.*, 4, 899.

90. Wale, M. and Goodwin, M. 1992. Flip-chip bonding optimizes opto-ICs. *IEEE Circuits Devices Mag.*, 8, 25.

91. GR-468-CORE. *Generic Reliability Assurance Requirements for Optoelectronic and Electro-Opto-Mechanical Devices Used in Telecommunications.* http://www.telcordia.com.

92. Mino, S., Yoshino, K., Yamada, Y., Terui, T., Yasu. M. and Moriwaki, K. 1995. Planar lightwave circuit platform with coplanar waveguide for opto-electronic hybrid integration. *J. Lightwave Technol.*, 13, 2320.

93. Yoshida, J. 1999. Hybridization of active and passive optical devices toward multifunctional optical modules. *ECOC'99 Proceedings 25th European Conference on Optical Communication*, p. 170.

94. Terui, H., Shimokozono, M., Yanagisawa, M., Hashimoto, T., Yamada, Y. and Horiguchi, M. 1996. Hybrid integration of eight channel PD-array on silica-based PLC using micromirror fabrication technique. *Electron. Lett.*, 32, 1662.

95. Lai, Q., Hunziker, W. and Melchior, H. 1996. Silica on Si waveguides for self-aligned fibre array coupling using flip-chip Si V-groove technique. *Electron. Lett.*, 32, 1916.

96. Collins, J. V., Lealman, I. F., Kelly, A. and Ford, C. W. 1997. Passive alignment of second generation optoelectronic devices. *IEEE J. Sel. Top. Quantum Electron.*, 3, 1441.

97. Choi, M. H., Koh, H. J., Yoon, E. S., Shin, K. C. and Song, K. C. 1999. Self-aligning silicon groove technology platform for the low cost optical module. *1999 Proceedings of Conference on 49th Electronic Components and Technology*, p. 1140.

98. Qing Tan and Lee, Y. C. 1996. Soldering technology for optoelectronic packaging. *1996 Proceedings of 46th Electronic Components and Technology Conference*, p. 26.

99. Rinne, G. A. 1997. Solder bumping methods for flip chip packaging. *1997 Proceedings of 47th Electronic Components and Technology Conference*, p. 240.

100. Jackson, K. P., Flint, E. B., Cina, M. F., Lacey, D., Trewhella, J. M., Caulfield, T. and Sibley, S. 1992. A compact multichannel transceiver module using planar-processed optical waveguides and flip-chip optoelectronic components. *1992 Proceedings of 42nd Electronic Components and Technology Conference*, p. 93.

101. Lee, S.H., Joe, G.C., Park, K.S., Kim, H.M., Kim, D.G. and Park, H.M. 1995. Optical device module packages for subscriber incorporating passive alignment techniques. *1995 Proceedings of 45th Electronic Components and Technology Conference*, p. 841.

102. Basavanhally, N. 1993. Application of soldering technologies for opto-electronic component assembly. *Advances in Electronic Packaging 1993 Proceedings of 1993 ASME International Electronics Packaging Conference*, p. 1149.

103. Lee, C. C., Wang, C. Y. and Matijasevic, G. 1993. Advances in bonding technology for electronic packaging. *Trans. ASME J. Electron. Packaging*, 115, 201.

104. Bigo, S. et al. 2001. 10.2Tbit/s (256×427 Gbit/s PDM/WDM) transmission over 100km TeraLight™ fibre with 128bit/s Hz spectral efficiency. *OFC 2001. Optical Fibre Communication Conference and Exhibit Technical Digest Postconference Edition Postdeadline Papers*, p. PD 25.

105. Desurvire, E. 1994. *Erbium-Doped Fibre Amplifiers: Principles and Applications.* 1st ed. (New York: Wiley-Interscience).

106. Agrawal, G. 1995. *Nonlinear Fibre Optics.* 2nd ed. (San Diego, CA: Academic).

107. Simon, J. C., Billess, L., Dupas, A. and Bramerie, L. 1999. All optical regeneration techniques. *ECOC'99 25th European Conference on Optical Communication*, p. 256.

108. Pender, W. A., Watkinson, P. J., Greer, E. J. and Ellis, A. D. 1995. 10 Gbit/s all-optical regenerator. *Electron. Lett.*, 31, 1587.

109. Ellis, A. D. and Widdowson, T. 1995. 690 node global OTDM network demonstration. *Electron. Lett.*, 31, 1171.

110. Kuznetsov, M., Froberg, N. M., Henion, S. R. and Rauschenbach, K. A. 1999. Power penalty for optical signals due to dispersion slope in WDM filter cascades. *IEEE Photon. Technol. Lett.*, 11, 1411.

111. Hida, Y., Jinguji, K. and Takato, N. 1998. Wavelength demulti/multiplexers with non-sinusoidal filtering characteristics composed of point-symmetrically connected Mach-Zehnder interferometers. *Electron. Commun. Jpn Part 2 (Electronics)*, 81, 19.

112. Chiba, T., Arai, H., Ohira, K., Nonen, H., Okano, H. and Uetsuka, H. 2001. Novel architecture of wavelength interleaving filter with Fourier transform-based MZIs. *OFC 2001: Optical Fibre Communication Conference and Exhibit Technical Digest Postconference Edition*, p. WB5.

113. Oguma, M., Jinguji, K., Kitoh, T., Shibata, T. and Himeno, A. 2000. Flat-passband interleave filter with 200 GHz channel spacing based on planar lightwave circuit-type lattice structure. *Electron. Lett.*, 36, 1299.

114. Sugita, A., Kaneko, A., Okamoto, K., Itoh, M., Himeno, A. and Ohmori, Y. 1999. Fabrication of very low insertion loss (~0.8 dB) arrayed-waveguide grating with vertically tapered waveguides. *ECOC'99: Proceedings of 25th European Conference on Optical Communication*, p. 4.

115. Takiguchi, K. 2001. Recent advances in PLC functional devices 2001. *Digest of LEOS Summer Topical Meetings: Advanced Semiconductor Lasers and Applications/Ultraviolet and Blue Lasers and Their Applications/Ultralong Haul DWDM Transmission and Networking/WDM Components*.

116. Khrais, N. N. and Wagner, R. E. 1998. General (de)multiplexer cascade model for transparent digital transmission. *J. Opt. Commun.*, 19, 75.

117. Amersfoort, M. and Soole, J. 1995. Passband flattening of integrated optical filters. US 5629992 A1.

118. Dragone, C. 1994. Frequency routing device having a wide and substantially flat passband. US 5412744 A1.

119. Rigny, A., Bruno, A. and Sik, H. 1997. Multigrating method for flattened spectral response wavelength multi/demultiplexer. *Electron. Lett.*, 33, 1701.

120. Trouchet, D. 1996. Multiplexer/demultiplexer with flattened spectral response. EP0816877 A1.

121. Doerr, C. R., Stulz, L. W., Cappuzzo, M., Laskowski, E., Paunescu, A., Gomez, L., Gates, J. V., Shunk, S., Chandrasekhar, S. and Kim, H. 1999. 40-channel programmable integrated add–drop with flat through-spectrum. *Proceedings of ECOC'99 25th European Conference on Optical Communication*, p. 46.

122. Heise, G., Schneider, H. W. and Clemens, P. C. 1998. Optical phased array filter module with passively compensated temperature dependence. *24th European Conference on Optical Communication ECOC'98 (IEEE Cat No 98TH8398) Proceedings of ECOC'98— 24th European Conference on Optical Communication*, p. 319.

123. Kaneko, A., Kamei, S., Inoue, Y., Takahashi, H. and Sugita, A. 2000. Athermal silica-based arrayed-waveguide grating (AWG) multi/demultiplexers with new low loss groove design. *Electron. Lett.*, 36, 318.

124. Inoue, Y., Ohmori, Y., Kawachi, M., Ando, S., Sawada, T. and Takahashi, H. 1994. Polarization mode converter with polyimide half waveplate in silica-based planar lightwave circuits. *IEEE Photon. Technol. Lett.*, 6, 626.

125. Vieira Segatto, M. E., Maxwell, G. D., Kashyap, R. and Taylor, J. R. 2001. High-speed transmission and dispersion characteristics of an arrayed-waveguide grating. *Opt. Commun.*, 195, 151.

126. Weiner, A. M. and Kan'an, A. M. 1998. Femtosecond pulse shaping for synthesis processing and time-to-space conversion of ultrafast optical waveforms. *IEEE J. Sel. Top. Quantum Electron.*, 4, 317.

127. Tsuda, H., Takenouchi, H., Hirano, A., Kurokawa, T. and Okamoto, K. 2000. Performance analysis of a dispersion compensator using arrayed-waveguide gratings. *J. Lightwave Technol.*, 18, 1139.

128. Tsuda, H., Takenouchi, H., Ishii, T., Okamoto, K., Goh, T., Sato, K., Hirano, A., Kurokawa, T. and Amano, C. 1999. Photonic spectral encoder/decoder using an arrayed-waveguide grating for coherent optical code division multiplexing. *OSA Trends in Optics and Photonics WDM Components*, vol. 29, p. 206.

129. Godil, A. A. 2002. Diffractive MEMS for optical networks. *Electron. Eng. Design*, 74, 43.

130. Gorecki, C. 2001. Recent advances in silicon guided-wave MOEMS: From technology to application. *Opto-Electron. Rev.*, 9, 248.

131. Huang, T., Huang, J., Liu, Y., Xu, M., Yang, Y., Li, M., Mao. C. and Chiao, J.C. 2001. Performance of a liquid-crystal optical harmonic equalizer. *OFC 2001: Optical Fibre Communication Conference and Exhibit Technical Digest Postconference Edition Postdeadline Papers*, p. PD29.

132. Ranalli, A. R., Scott, B. A. and Kondis, J. P. 1999. Liquid crystal-based wavelength selectable cross-connect. *Proceedings of ECOC'99 25th European Conference on Optical Communication*, p. 68.

133. Dimmick, T. E., Kakarantzas, G., Birks, T. A., Diez, A. and Russell, P. S. J. 2000. Compact all-fibre acoustooptic tunable filters with small bandwidth-length product. *IEEE Photon. Technol. Lett.*, 12. 1210.

134. Yeralan, S., Gunther, J., Ritums, D. L., Cid, R., Storey, J., Ashmead, A. C. and Popovich, M. M. 2001. Switchable Bragg grating devices for telecommunications applications. *Proc. SPIE — The Int. Soc. for Opt. Eng.*, 4291, 79.

135. Vallon, S., Cayrefourcq, I., Chevallier, P., Landru, N., Alibert, G., Laborde, P., Little, J., Ranalli, A. and Boos, N. 2001. Tapped delay line dynamic gain flattening filter. *2001 Digest of LEOS Summer Topical Meetings: Advanced Semiconductor Lasers and Applications/Ultraviolet and Blue Lasers and Their Applications/Ultralong Haul DWDM Transmission and Networking/WDM Components).*

136. Sasayama, K., Okuno, M. and Habara, K. 1991. Coherent optical transversal filter using silica-based waveguides for high-speed signal processing. *J. Lightwave Technol.*, 9, 1225.

137. Sasayama, K., Okuno, M. and Habara, K. 1994. Photonic FDM multichannel selector using coherent optical transversal filter. *J. Lightwave Technol.*, 12, 664.

138. Offrein, B. J., Horst, F., Bona, G. L., Germann, R., Salemink, H. W. M. and Beyeler, R. 2000. Adaptive gain equalizer in high-index-contrast SiON technology. *IEEE Photon. Technol. Lett.*, 12, 504.

139. Li, Y. P. and Henry, C. H. 1996. Silica-based optical integrated circuits. *IEE Proc. Optoelectron.*, 143, 263.

140. Doerr, C. R., Joyner, C. H. and Stulz, L. W. 1998. Integrated WDM dynamic power equalizer with potentially low insertion loss. *IEEE Photon. Technol. Lett.*, 10, 1443.

141. Doerr, C. R., Stulz, L. W., Pafchek, R., Gomez, L., Cappuzzo, M., Paunescu, A., Laskowski, E., Buhl, L., Kim, H. K. and Chandrasekhar, S. 2000. An automatic 40-wavelength channelized equalizer. *IEEE Photon. Technol. Lett.*, 12, 1195.

142. Ramaswami, R. and Sivarajan, K. N. 1997. *Optical Networks* (San Mateo, CA: Morgan Kaufman).

143. Ouellette, F., Cliche, J.-F. and Gagnon. S. 1994. All-fibre devices for chromatic dispersion compensation based on chirped distributed resonant coupling. *J. Lightwave Technol.*, 12, 1728.

144. Madsen, C. K. and Lenz, G. 2000. A multi-channel dispersion slope compensating optical allpass filter. *Optical Fibre Communication Conference Technical Digest Postconference Edition Trends in Optics and Photonics (IEEE Cat No 00CH37079)*, p. 94.

145. Shirasaki, M. 1997. Chromatic-dispersion compensator using virtually imaged phased array. *IEEE Photon. Technol. Lett.*, 9, 1598.

146. Takiguchi, K., Okamoto, K. and Goh, T. 2001. Integrated optic dispersion slope equaliser for N*20 Gbit/s WDM transmission. *Electron. Lett.*, 37, 701.

147. Madsen, C. K., Lenz, G., Nielsen, T. N., Bruce, A. J., Cappuzzo, M. A. and Gomez, L. T. 1999. Integrated optical allpass filters for dispersion compensation. *OSA Trends in Optics and Photonics WDM Components*, vol. 29, p. 142.

148. Madsen, C. K. 2001. Tunable dispersion compensators based on optical allpass filters. *Digest of LEOS Summer Topical Meetings: Advanced Semiconductor Lasers and Applications/Ultraviolet and Blue Lasers and Their Applications/Ultralong Haul DWDM Transmission and Networking/WDM Components (IEEE Cat No 01TH8572)*.

149. Noe, R., Heidrich, H. and Hoffmann, D. 1988. Endless polarization control systems for coherent optics. *J. Lightwave Technol.*, 6, 1199.

150. Saida, T., Takiguchi, K., Kuwahara, S., Kisaka, Y., Miyamoto, Y., Hashizume, Y., Shibata, T. and Okamoto, K. 2002. Planar lightwave circuit polarization-mode dispersion compensator. *IEEE Photon. Technol. Lett.*, 14, 507.

151. Okuno, M., Sugita, A., Jinguji, K. and Kawachi, M. 1994. Birefringence control of silica waveguides on Si and its application to a polarization-beam splitter/switch. *J. Lightwave Technol.*, 12, 625.

152. Cotter, D., Manning, R. J., Blow, K. J., Ellis, A. D., Kelly, A. E., Nesset, D., Phillips, I. D., Poustie, A. J. and Rogers, D. C. 1999. Nonlinear optics for high-speed digital information processing. *Science*, 286, 1523.

153. Kaminow, I. P. and Koch, T. L. (ed.). 1997. *Optical Fibre Telecommunications IIIA* (San Diego, CA: Academic).

154. Bishop, D. 2000. Silicon micromachines for lightwave networks. *Photonics in Switching Topical Meeting OSA Trends in Optics and Photonics Series*, vol. 32, p. 11.

155. Lin, Y., Goldstein, E. L., Lunardi, L. M. and Tkach, R. W. 1999. Optical crossconnects for high-capacity lightwave networks. *J. High Speed Networks*, 8, 17.

156. Lin, L. Y. and Goldstein, E. L. 2000. MEMS for optical switching. *Photonics in Switching Topical Meeting OSA Trends in Optics and Photonics Series*, vol. 32, p. 23.

157. Neilson, D. T. et al. 2000. Fully provisioned 112*112 micro-mechanical optical crossconnect with 358 Tb/s demonstrated capacity. *Optical Fibre Commun. Conference Technical Digest Postconference Edition Trends in Optical and Photonics (IEEE Cat No 00CH37079)*, vol. 37, p. 202.

158. Krähenbühl, R. and Burns, W. K. 2000. Enhanced crosstalk suppression for Ti:LiNbO$_3$ digital optical switches. *Photonics in Switching Topical Meeting OSA Trends in Optics and Photonics Series*, vol. 32, p. 160.

159. Chen, A., Irvin, R. W., Murphy, E. J., Grencavich, R., Murphy, T. O. and Richards, G. W. 2000. High performance LiNbO$_3$ switches for multiwavelength optical networks. *Photonics in Switching Topical Meeting OSA Trends in Optics and Photonics Series*, vol. 32, p. 163.

160. Kirihara, T. and Inoue, H. 1996. InP-based optical switch arrays using semiconductor optical amplifiers. *Int. J. High Speed Electron. Systems*, 7, 85.

161. van Berlo, W., Janson, M., Lundgren, L., Morner, A.-C., Terlecki, J., Gustavsson, M., Granestrand, P. and Svensson, P. 1995. Polarization-insensitive monolithic 4*4 InGaAsP–InP laser amplifier gate switch matrix. *IEEE Photon. Technol. Lett.*, 7, 1291.

162. Fouquet, J. E. 2000. Progress in optical cross-connects for circuit-switched applications. *Photonics in Switching Topical Meeting OSA Trends in Optics and Photonics Series*, vol. 32, p. 14.

163. Sakata, T., Togo, H. and Shimokawa, F. 2000. Reflection-type 2*2 optical waveguide switch using the Goos-Hanchen shift effect. *Appl. Phys. Lett.*, 76, 2841.

164. Venkatesh, S., Haven, R., Chen, D., Reynolds, H. L., Harkins, G., Close, S., Troll, M., Fouquet, J. E., Schroeder, D. and McGuire, P. 2001. Recent advances in bubble-actuated cross-connect switches. *Technical Digest CLEO/Pacific Rim 2001 4th Pacific Rim Conference on Lasers and Electro-Optics (Cat No 01TH8557)*, p. 1.

165. Betty, I., Rousina-Webb, R. and Wu, C. 2000. A robust, low-crosstalk, InGaAs–InP total-internal-reflection switch for optical cross-connect. *Photonics in Switching Topical Meeting OSA Trends in Optics and Photonics Series*, vol. 32, p. 5.

166. Noguchi, K. 1997. Optical multichannel switch composed of liquid-crystal light-modulator arrays and birefringent crystals. *Electron. Lett.*, 33, 1627.

167. Dorgeuille, F., Ambrosy, A., Grieshaber, W., Pommereau, F., Boubal, F., Rabaron, S., Gaborit, F., Guillemot, I., Poucheron, C., Le Bris, J., Blume, O., Lauckner, J., Luz, G., Matthles, K., Ruess, K., Schilling, M., Schneider, S., Noire, L., Tregoat, D. and Artigue, C. 1999. Loss-free 1*4 opto-hybrid space switch based on an array of 4 gain-clamped SOA gates. *Proceedings of ECOC'99. 25th European Conference on Optical Communication*, p. 176.

168. Sasaki, J., Hatakeyama, H., Tamanuki, T., Kitamura, S., Yamaguchi, M., Kitamura, N., Shimoda, T., Kitamura, M., Kato, T. and Itoh, M. 1998. Hybrid integrated 4*4 optical matrix switch using self-aligned semiconductor optical amplifier gate arrays and silica planar lightwave circuit. *Electron. Lett.*, 34, 986.

169. Dorgeuille, F., Noirie, L., Faure, J. P., Ambrosy, A., Rabaron, S., Boubal, F., Schilling, M. and Artigue, C. 2000. 1.28Tbit/s throughput 8*8 optical switch based on arrays of gain-clamped semiconductor optical amplifier gates. *Optical Fibre Communication Conference Technical Digest Postconference Edition. Trends in Optics and Photonics (IEEE Cat. No. 00CH37079)*, vol. 37, p. 221.

170. Kasahara, R., Yanagisawa, M., Sugita, A., Ogawa, I., Hashimoto, T., Suzaki, Y. and Magari, K. 1999. Fabrication of compact optical wavelength selector by integrating arrayed-waveguide-gratings and optical gate array on a single PLC platform. *Proceedings of ECOC'99: 25th European Conference on Optical Communication*, p. 122.

171. Renaud, M., Keller, D., Sahri, N., Silvestre, S., Prieto, D., Dorgeuille, F., Pommereau, F., Emery, J. Y., Grard, E. and Mayer, H. P. 2001. SOA-based optical network components. *Proceedings of 51st Electronic Components and Technology Conference (Cat. No. 01CH37220) 2001*, p. 433.

172. Goh, T., Yasu, M., Hattori, K., Himeno, A., Okuno, M. and Ohmori, Y. 1998. Low-loss and high-extinction-ratio silica-based strictly nonblocking 16*16 thermooptic matrix switch. *IEEE Photon. Technol. Lett.*, 10, 810.

173. Kasahara, R., Yanagisawa, M., Sugita, A., Goh, T., Yasu, M., Himeno, A. and Matsui, S. 1999. Low-power consumption silica-based 2*2 thermooptic switch using trenched silicon substrate. *IEEE Photon. Technol. Lett.*, 11, 1132.

174. Sohma, S., Goh, T., Okazaki, H., Okuno, M. and Sugita, A. 2002. Low switching power silica-based super high delta thermo-optic switch with heat insulating grooves. *Electron. Lett.*, 38, 127.

175. Hida, Y., Onose, H. and Imamura, S. 1993. Polymer waveguide thermooptic switch with low electric power consumption at 1.3 μm. *IEEE Photon. Technol. Lett.*, 5, 782.

176. Murphy, E. J. 1997. Photonics switching. In *Optical Fibre Telecommunications III*, ed. B. I. P. Kaminow and T. L. Koch (New York: Academic), p. 463.

177. Doerr, C. R., Joyner, C. H., Stulz, L. W. and Monnard, R. 1998. Wavelength-division multiplexing cross connect in InP. *IEEE Photon. Technol. Lett.*, 10, 117.

178. Vinchant, J.-F., Cavailles, J. A., Erman, M., Jarry, P. and Renaud, M. 1992. InP/GaInAsP guided-wave phase modulators based on carrier-induced effects: Theory and experiment. *J. Lightwave Technol.*, 10, 63.

179. Nakamura, S., Tajima, K. and Sugimoto, Y. 1995. High-repetition operation of a symmetric Mach-Zehnder all-optical switch. *Appl. Phys. Lett.*, 66, 2457.

180. Vreeburg, C. G. M., Uitterdijk, T., Oei, Y. S., Smit, M. K., Groen, F. H., Metaal, E. G., Demeester, P. and Frankena, H. J. 1997. First InP-based reconfigurable integrated add-drop multiplexer. *IEEE Photon. Technol. Lett.*, 9, 188.

181. Little, B. E., Foresi, J. S., Steinmeyer, G., Thoen, E. R., Chu, S. T., Hans, H. A., Ippen, E. P., Kimerling, L. C. and Greene, W. 1998.

Ultra-compact Si–SiO$_2$ microring resonator optical channel dropping filters. *IEEE Photon. Technol. Lett.*, 10, 549.

182. Wehrmann, F., Harizi, C., Herrmann, H., Rust, U., Sohler, W. and Westenhofer, S. 1996. Integrated optical wavelength selective acoustically tunable 2∗2 switches (add–drop multiplexers) in LiNbO$_3$. *IEEE J. Sel. Top. Quantum Electron.*, 2, 263.

183. Okamoto, K., Okuno, M., Himeno, A. and Ohmori, Y. 1996. 16-channel optical add/drop multiplexer consisting of arrayed-waveguide gratings and double-gate switches. *Electron. Lett.*, 32, 1471.

184. Doerr, C. R. 1998. Proposed WDM cross connect using a planar arrangement of waveguide grating routers and phase shifters. *IEEE Photon. Technol. Lett.*, 10, 528.

185. Teshima, M., Koga, M. and Sato, K. 1995. Multiwavelength simultaneous monitoring circuit employing wavelength crossover properties of arrayed-waveguide grating. *Electron. Lett.*, 31, 1595.

186. Janz, C., Dagens, B., Emery, J.-Y., Renaud, M. and Lavigne, B. 2000. Integrated SOA-based interferometers for all-optical signal processing. *Proceedings of 26th European Conference on Optical Communication*, p. 115.

187. Fischer, S., Duelk, M., Puleo, M., Girardi, R., Gamper, E., Vogt, W., Hunziker, W., Gini, E. and Melchior, H. 1999. 40-Gb/s OTDM to 4∗10 Gb/s WDM conversion in monolithic InP Mach–Zehnder interferometer module. *IEEE Photon. Technol. Lett.*, 11, 1262.

188. Ueno, Y., Nakamura, S., Hatakeyama, H., Tamanuki, T., Sasaki, T. and Tajima, K. 2000. 168-Gb/s OTDM wavelength conversion using an SMZ-type all-optical switch. *Procedings of 26th European Conference Optical Communication*, p. 13.

189. Kazovski, L., Benedetto, S. and Willner, A. E. 1996. *Optical Fibre Communication Systems* (Boston, MA: Artech).

190. Wolfson, D., Hansen, P. B., Kloch, A., Fjelde, T., Janz, C., Coquelin, A., Guillemot, I., Garorit, F., Poingt, F. and Renaud, M. 1999. All-optical 2R regeneration at 40 Gbit/s in an SOA-based Mach–Zehnder interferometer. *OFC/IOOC'99: Optical Fibre Communication Conference and International Conference on Integrated Optical and Optical Fibre Communications (Cat. No 99CH36322)*, p. PD 36.

191. Mirasso, C. R. et al. 1999. Self-pulsating semiconductor lasers: Theory and experiment. *IEEE J. Quantum Electron.*, 35, 764.

192. Rees, P., McEvoy, P., Valle, A., O'Gorman, J., Lynch, S., Landais, P., Pesquera, L. and Hegarty, J. 1999. A theoretical analysis of optical clock extraction using a self-pulsating laser diode. *IEEE J. Quantum Electron.*, 35, 221.

193. Barnsley, P. E., Wickes, H. J., Wickens, G. E. and Spirit, D. M. 1991. All-optical clock recovery from 5 Gb/s RZ data using a self-pulsating 1.56 μm laser diode. *IEEE Photon. Technol. Lett.*, 3, 942.

194. Pender, W. A., Widdowson, T. and Ellis, A. D. 1996. Error free operation of a 40 Gbit/s all-optical regenerator. *Electron. Lett.*, 32, 567.

195. Phillips, I. D., Ellis, D., Thiele, J., Manning, R. J. and Kelly, A. E. 1998. 40 Gbit/s all-optical data regeneration and demultiplexing with long pattern lengths using a semiconductor nonlinear interferometer. *Electron. Lett.*, 34, 2340.

196. Vallon, S. 2002. Private communiation.

197. Bourdon, G. 2002. Private communication.

198. Alibert, G. 2002. Private communication.

199. Delprat, D. 2002. Private communication.

200. Lin, L. Y., Goldstein, E. L. and Tkach, R. W. 1998. Free-space micromachined optical switches with submillisecond switching time for large-scale optical crossconnects. *IEEE Photon. Technol. Lett.*, 10, 525.

FURTHER READING

The textbooks in [2,13,20,22,80] develop the theory of waveguides, couplers and splitters as well as electro-optic and magneto-optic control. For approximation methods in channel waveguides and numerical techniques see in particular References [1,2].

Material properties of nonlinear materials are found in Reference [8], with details on nonlinear semiconductor devices in [7, Chapter 11] and applications of nonlinear devices for TDM systems in [63, Chapter 9].

Okamoto [2] and Madsen [22] treat authoritatively single mode planar waveguide building blocks and integrated devices, and [22] is the key reference for the design of optical filters. [20] and [2] have chapters on BPM, and [7, Chapter 12] focuses on numeric design tools and their accuracy.

The recent books in [7, 63] have specific chapters on hybrid and monolithic integration technique as well as lithium niobate components, rare-earth doped glass waveguides and integrated InP devices.

Packaging of opto-electronic devices is discussed in Reference [79] in specific chapters on laser packaging, optical interconnection techniques and interconnection loss budgets.

Optical network components, propagation in fibre and network architectures are described from a practical perspective in Reference [142], and detailed theoretical treatment of (single and multi-channel) signal propagation in fibre and noise in systems can be found in Reference [189]. Finally, Agrawal [106] treats nonlinear effects in fibre and applications such as Raman amplification and nonlinear optic fiber devices.

18

Infrared devices and techniques

ANTONI ROGALSKI AND KRZYSZTOF CHRZANOWSKI
Military University of Technology

18.1 INTRODUCTION

Looking back over the past 1000 years, we notice that infrared (IR) radiation itself was unknown until 200 years ago when Herschel's experiment with a thermometer was first reported. Herschel built a crude monochromator that used a thermometer as a detector so that he could measure the distribution of energy in sunlight [1]. Following the works of Kirchhoff, Stefan, Boltzmann, Wien, and Rayleigh, Max Planck culminated the effort with the well-known Planck's law.

Traditionally, IR technologies are connected with controlling functions and night-vision problems, with earlier applications connected simply with detection of IR radiation, and later by forming IR images from temperature and emissivity differences (systems for recognition and surveillance, tank sight systems, anti-tank missiles, air–air missiles). The years during World War II saw the origins of modern IR techniques. Recent success in applying IR technology to remote sensing problems has been made possible by the successful development of high-performance IR detectors over the past five decades. Most of the funding has been provided to fulfil military needs, but peaceful applications have increased continuously, especially in the last decade of the 20th century. These include medical, industry, earth resources, and energy conservation applications. Medical applications include thermography in which IR scans of the body detect cancers or other traumas that raise the body surface temperature. Earth resource determinations are done by using IR images from satellites in conjunction with field observation for calibration (in this manner, e.g., the area and content of fields and forests can be determined). In some cases, even the state of health of a crop can be determined from space. Energy conservation in homes and industry has been aided by

Table 18.1 Division of the IR radiation region

Region (abbreviation)	Wavelength range (μm)
Near-infrared (NIR)	0.78–1
Short wavelength IR (SWIR)	1–3
Medium wavelength IR (MWIR)	3–6
Long wavelength IR (LWIR)	6–15
Very long wavelength IR (VLWIR)	15–30
Far-infrared (FIR)	30–100
Submillimeter (Sub-mm)	100–1000

the use of IR scans that can determine the points of maximum heat loss. Demands to use these technologies are quickly growing due to their effective applications, for example, in global monitoring of environmental pollution and climate changes, long time prognoses of agriculture crop yield, chemical process monitoring, Fourier transform IR spectrometry, IR astronomy, car driving, IR imaging in medical diagnostics, and others.

The IR range covers all electromagnetic radiation longer than the visible, but shorter than millimeter waves. Many proposals for the division of the IR range have been published. The division shown in Table 18.1 is used by the military community and is based on the limits of spectral bands of commonly used IR detectors. A wavelength of 1 μm is the sensitivity limit for popular Si detectors. Similarly, a wavelength of 3 μm is the long wavelength sensitivity of PbS and InGaAs detectors; a wavelength of 6 μm is the sensitivity limit of InSb, PbSe, PtSi detectors and HgCdTe detectors are optimized for the 3–5 μm atmospheric window; and finally, a wavelength of 15 μm is the long wavelength sensitivity limit of HgCdTe detectors optimized for the 8–14 μm atmospheric window.

18.2 IR SYSTEM FUNDAMENTALS

18.2.1 Thermal emission

All objects are composed of continually vibrating atoms, with higher energy atoms vibrating more frequently. The vibration of all charged particles, including these atoms, generates electromagnetic waves. The higher the temperature of an object, the faster the vibration, and thus, higher the spectral radiant energy. As a result, all objects are continually emitting radiation with a wavelength

distribution that depends upon the temperature of the object and its spectral emissivity, $\varepsilon(\lambda)$.

Radiant emission is usually treated in terms of the concept of a black body [2]. A black body is an object that absorbs all incident radiation and conversely, according to the Kirchhoff law, is a perfect radiator. The energy emitted by a black body is the maximum theoretically possible for a given temperature. The radiative power (or number of photons emitted) and its wavelength distribution are given by the Planck radiation law:

$$W(\lambda,T) = \frac{2\pi hc^2}{\lambda^5}\left[\exp\left(\frac{hc}{\lambda kT}\right)-1\right]^{-1} \text{ W cm}^{-2}\mu\text{m}^{-1},$$

(18.1)

$$P(\lambda,T) = \frac{2\pi c}{\lambda^4}\left[\exp\left(\frac{hc}{\lambda kT}\right)-1\right]^{-1} \text{ photons s}^{-1}\text{cm}^{-2}\mu\text{m}^{-1},$$

(18.2)

where λ is the wavelength, T is the temperature, h is Planck's constant, c is the velocity of light, and k is Boltzmann's constant.

Figure 18.1 shows a plot of these curves for a number of black body temperatures. As the temperature increases, the amount of energy emitted at any wavelength increases too, and the wavelength of peak emission decreases. The latter is given by the Wien displacement law:

$$\lambda_{\text{mw}}T = 2898 \ \mu\text{m K for maximum watts.}$$

$$\lambda_{\text{mp}}T = 3670 \mu\text{m K for maximum photons.}$$

The loci of these maxima are shown in Figure 18.1. Note that for an object at an ambient temperature of 290 K, λ_{mw} and λ_{mp} occur at 10.0 and 12.7 μm, respectively. We need detectors operating near 10 μm if we expect to "see" room temperature objects such as people, trees, and trucks without the aid of reflected light. For hotter objects such as engines, maximum emission occurs at shorter wavelengths. Thus, the waveband 2–15 μm in the IR or thermal region of the electromagnetic spectrum contains the maximum radiative emission for thermal imaging purposes.

18.2.2 Atmospheric transmission

Most of the infrared system applications require transmission through air, but the radiation is attenuated by the processes of scattering and

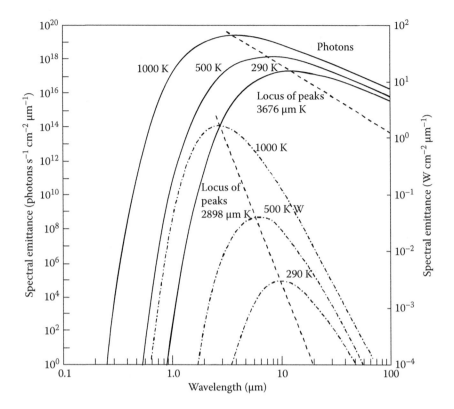

Figure 18.1 Planck's law for spectral emittance.

absorption. Scattering causes a change in the direction of a radiation beam; it is caused by absorption and subsequent reradiation of energy by suspended particles.

For larger particles, scattering is independent of wavelength. However, for small particles, with size comparable to the wavelength of the radiation, the process is known as Rayleigh scattering and it exhibits a λ^{-4} dependence. Therefore, scattering by gas molecules is negligibly small for wavelengths longer than 2 µm. Smoke and light mist particles are usually small with respect to IR wavelengths, therefore, IR radiation can penetrate further through smoke and mists than visible radiation. However, rain, fog particles, and aerosols are larger and consequently scatter IR and visible radiation to a similar degree.

Figure 18.2 shows a plot of the transmission through 6000 ft of air as a function of wavelength. Specific absorption bands of water, carbon dioxide, and oxygen molecules are indicated that restrict atmospheric transmission to two windows at 3–5 and 8–14 µm. Ozone, nitrous oxide, carbon

monoxide, and methane are less important IR absorbing constituents of the atmosphere.

18.2.3 Scene radiation and contrast

The total radiation received from any object is the sum of the emitted, reflected, and transmitted radiation. Objects that are not black bodies emit only the fraction $\varepsilon(\lambda)$ of blackbody radiation; the remaining fraction, $1-\varepsilon(\lambda)$, is either transmitted or, for opaque objects, reflected. When the scene is composed of objects and backgrounds of similar temperatures, reflected radiation tends to reduce the available contrast. However, reflections of hotter or colder objects have a significant effect on the appearance of a thermal scene. The powers of 290 K black body emission and ground-level solar radiation in MWIR and LWIR bands are given in Table 18.2. We can see that while reflected sunlight has a negligible effect on 8–14 µm imaging, it is important in the 3–5 µm band.

A thermal image arises from temperature variations or differences in emissivity within a scene. Thermal contrast, C, is one of the important

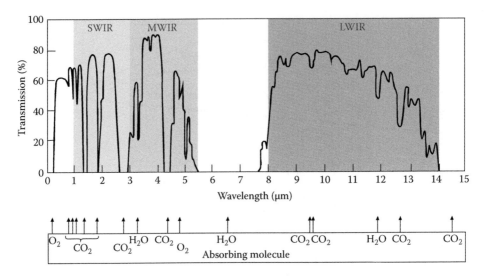

Figure 18.2 Transmission of the atmosphere for a 6000 ft horizontal path at sea level containing 17 mm of precipitate water. (Reproduced from Hudson, R. D., *Infrared System Engineering*, Wiley, New York, 1969.)

Table 18.2 Power available in each MWIR and LWIR imaging band

IR region (μm)	Ground-level solar radiation (W m^{-2})	Emission from 290 K black body (W m^{-2})
3–5	24	4.1
8–13	1.5	127

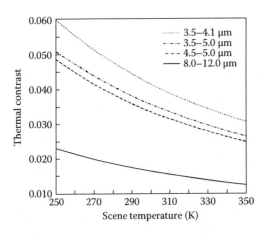

Figure 18.3 Spectral photon contrast in the MWIR and LWIR bands.

parameters for IR imaging devices. It is the ratio of the derivative of spectral radiant exitance, W, to the spectral radiant exitance

$$C = \frac{\partial W/\partial T}{W}.$$

The contrast in a thermal image is small when compared with the visible image contrast due to differences in reflectivity. Figure 18.3 shows a plot of C for several MWIR subbands and the 8–12 μm LWIR spectral band. The contrast in a thermal image is small when compared with visible image contrast due to differences in reflectivity. On the other hand, in the MWIR bands at 300 K, the contrast is 3.5%–4% compared to 1.6% for the LWIR band. Thus, while the LWIR band may have higher sensitivity for ambient temperature objects, the MWIR band has the greater contrast.

18.2.4 Choice of IR band

The SWIR wavelength band offers unique imaging advantages over visible and thermal bands. Like visible cameras, the images are primarily created by reflected broadband light sources, so SWIR images are easier for viewers to understand. Most materials used to make windows, lenses, and coatings for visible cameras are readily useable for SWIR cameras, keeping costs down. Ordinary glass transmits radiation to about 2.5 μm. SWIR cameras can image many of the same light sources, such as Yag laser wavelengths. Thus, with safety concerns shifting laser operations to the "eyesafe" wavelengths where beams will not focus on the retina (beyond 1.4 μm), SWIR cameras are in a unique position to replace visible cameras for many tasks. Due to the reduced Rayleigh scatter of light at longer wavelengths by particulates in the air, such as dust or fog, SWIR cameras can see through haze better than visible cameras.

In general, the 8–14-μm band is preferred for high performance thermal imaging because of its higher sensitivity to ambient temperature objects and its better transmission through mist and smoke. However, the 3–5 μm band may be more appropriate for hotter objects, or if sensitivity is less important than contrast. There may be additional differences, for example, the advantage of the MWIR band is the smaller diameter of the optics required to obtain a certain resolution; some detectors may operate at higher temperatures (thermoelectric cooling) than is usual in the LWIR band where cryogenic cooling is required (about 77 K).

Summarizing, MWIR and LWIR μm spectral bands differ substantially with respect to background flux, scene characteristics, temperature contrast, and atmospheric transmission under diverse weather conditions. Factors that favor MWIR applications are higher contrast, superior clear weather performance (favorable weather conditions, e.g., in most countries of Asia and Africa), higher transmittivity in high humidity, and higher resolution due to ~3× smaller optical diffraction. Factors that favor LWIR applications are better performance in fog and dust conditions, winter haze (typical weather conditions, e.g., in West Europe, North United States, Canada), higher immunity to atmospheric turbulence, and reduced sensitivity to solar glints and fire flares. The possibility of achieving higher signal-to-noise ratio (SNR) due to greater radiance levels in LWIR spectral range is not persuasive because the background photon fluxes are higher to the same extent, and also because of readout limitation possibilities. Theoretically, in staring arrays, charge can be integrated for the full frame time, but because of restrictions in the charge-handling capacity of the readout cells, it is much less compared to the frame time, especially for LWIR detectors for which background photon flux exceeds the useful signals by orders of magnitude.

18.2.5 Detectors

The figure of merit used for detectors is detectivity. It has been found in many instances that this parameter varies inversely with the square root of both the detector's sensitive area, A, and the electrical bandwidth, Δf. In order to simplify the comparison of different detectors, the following definition has been introduced [4]

$$D^* = \frac{(A\Delta f)^{1/2}}{\Phi_e}(\text{SNR}), \qquad (18.3)$$

where Φ_e is the spectral radiant incident power. $D*$ is defined as the rms (root-mean-square) SNR in a 1 Hz bandwidth per unit rms incident radiation power per square root of detector area. $D*$ is expressed in cm $\text{Hz}^{1/2}$ W^{-1}, which is recently called "Jones." Spectral detectivity curves for a number of commercially available IR detectors are shown in Figure 18.4. Interest has centered mainly on the wavelengths of the two atmospheric windows 3–5 and 8–14 μm, though in recent years, there has been increasing interest in longer wavelengths stimulated by space applications.

Progress in IR detector technology is connected mainly to semiconductor IR detectors, which are included in the class of photon detectors. In the class of photon detectors, the radiation is absorbed within the material by interaction with electrons. The observed electrical output signal results from the changed electronic energy distribution. Photon detectors show a selective wavelength dependence of the response per unit incident radiation power. They exhibit both perfect signal-to-noise performance and a very fast response. But to achieve this, the photon detectors require cryogenic cooling. Cooling requirements are the main obstacle to the more widespread use of IR systems based on semiconductor photodetectors making them bulky, heavy, expensive, and inconvenient to use. Depending on the nature of interaction, the class of photon detectors is further sub-divided into different types. The most important are intrinsic detectors (HgCdTe, InGaAs, InSb, PbS, PbSe), extrinsic detectors (Si:As, Si:Ga), photoemissive (metal silicide Schottky barriers) detectors, quantum well detectors (GaAs/AlGaAs QWIPs), type-II

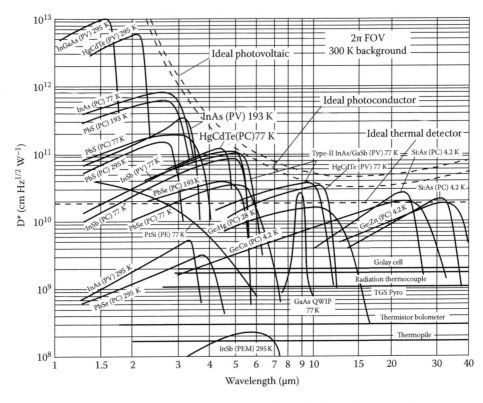

Figure 18.4 Comparison of the $D*$ of various commercially available IR detectors when operated at the indicated temperature. The chopping frequency is 1000 Hz for all detectors except the thermopile (10 Hz), thermocouple (10 Hz), thermistor bolometer (10 Hz), Golay cell (10 Hz), and pyroelectric detector (10 Hz). Each detector is assumed to view a hemispherical surround at a temperature of 300 K. Theoretical curves for the background-limited $D*$ for ideal photovoltaic and photoconductive detectors and thermal detectors are also shown.

superlattice III–V photodiodes and nBn barrier detectors.

The second class of IR detectors is composed of thermal detectors. In a thermal detector, the incident radiation is absorbed to change the temperature of the material, and the resultant change in some physical properties is used to generate an electrical output. The detector element is suspended on lags, which are connected to the heat sink. Thermal effects are generally wavelength independent; the signal depends upon the radiant power (or its rate of change) but not upon its spectral content. In pyroelectric detectors, a change in the internal spontaneous polarization is measured, whereas in the case of bolometers, a change in the electrical resistance is measured. In contrast to photon detectors, thermal detectors typically operate at room temperature. They are usually characterized by modest sensitivity and slow response but they are cheap and easy to use. Bolometers, pyroelectric detectors, and thermopiles are the most useful devices with IR technology. Typical values of detectivities of thermal detectors at 10 Hz change in the range between 10^8 and 10^9 cm Hz$^{1/2}$ W^{-1}.

Up until the nineties of the 20th century, thermal detectors have been considerably less exploited in commercial and military systems in comparison with photon detectors. The reason for this disparity is that thermal detectors are popularly believed to be rather slow and insensitive in comparison with photon detectors. As a result, the worldwide effort to develop thermal detectors was extremely small relative to that of photon detectors. In the last two decades, however, it has been shown that extremely good imagery can be obtained from large thermal detector arrays operating uncooled at TV frame rates. The speed of thermal detectors is quite adequate for nonscanned imagers with two-dimensional (2D) detectors. The moderate sensitivity of thermal detectors can be compensated by a large number of elements in 2D electronically scanned arrays. With large arrays of thermal detectors, the best values of NEDT (noise equivalent differential temperature, explained later in Section 18.2.9), below 0.1 K, could be reached because effective noise bandwidths less than 100 Hz can be achieved.

Initially developed for the military market by U.S. defense companies, IR uncooled cameras are now widely used in many commercial applications. Currently, microbolometer detectors are produced in larger volumes than all other IR array technologies together. In large volume production for automobile drivers, the cost of uncooled imaging systems is below $1000. The global IR market value was estimated to be nearly $321.4 million in 2013. It is expected to reach $704.8 million by 2020, at a CAGR (compound annual growth rate) of 11.9% from 2014 to 2020 [5].

18.2.6 Cooling

The signal output of a photon detector is so small that at ordinary temperatures, it is swamped by the thermal noise due to random generation and recombination of carriers in the semiconductor. In order to reduce the thermal generation of carriers and minimize noise, photon detectors must be cooled and therefore be encapsulated. The method of cooling varies according to the operating temperature and the system's logistical requirements.

The two technologies currently available for addressing the cooling requirements of IR and visible detectors are closed cycle refrigerators and thermoelectric coolers. Closed cycle refrigerators can achieve the cryogenic temperatures required for cooled IR sensors, while thermoelectric coolers are generally the preferred approach to temperature control uncooled visible and IR sensors. The major difference between thermoelectric and mechanical cryocoolers is the nature of the working fluid. A thermoelectric cooler is a solid-state device that uses charge carriers (electrons or holes) as the working fluid, whereas mechanical cryocoolers use a gas such as helium as the working fluid. The selection of a cooler for a specific application depends on the cooling capacity, operating temperature, procurement, cost and maintenance, and servicing requirements. A survey of currently operating cryogenic systems for commercial, military, and space applications are summarized in Figure 18.5. Table 18.3 presents advantages and disadvantages of different cryocoolers for space applications.

Most 8–14-μm detectors operate at approximately 77 K and can be cooled by liquid nitrogen. Cryogenic liquid pour-filled Dewars are frequently used for cooling detectors in laboratories. They are rather bulky and need to be refilled with liquid nitrogen every few hours. For many applications, especially in the field, LN_2 (liquid nitrogen) pour-filled Dewars are impractical, so many manufacturers are turning to alternative coolers that do not require cryogenic liquids or solids. It is more

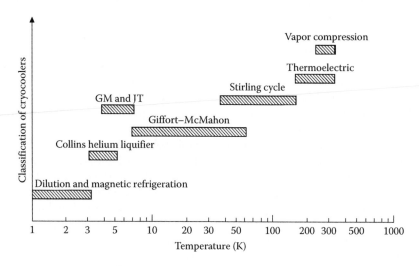

Figure 18.5 Temperature ranges for commercial refrigerators.

Table 18.3 Cryocoolers for space applications

Cooler	Typical temperature (K)	Typical heat lift	Advantages	Disadvantages
Radiator	80	0.5 W	Reliable, low vibration, long lifetime	Complicates orbit
Stirling-1 stage	80	0.8 W	Efficient, heritage	Vibrations
Stirling-2 stage	20	0.06 W	Intermediate temp	Under development
Pulse tube	80	0.8 W	Lower vibrations, efficiency comparable to Stirling	Difficulties in scaling down to small sizes while maintaining high efficiency, larger diameter cold finger
Joule–Thompson	4	0.01 W	Low vibrations	Requires hybrid design
Sorption	10	0.1 W	Low vibrations	Low efficiency, under development
Brayton	65	8 W	High capacity	Complex
ADR	0.05	0.01 mW	Only way to reach these temps	Large magnetic field
Peltier	170	1 W	Light weight	High temp, low efficiency

convenient to use Stirling and Joule–Thompson minicoolers (see Figure 18.6).

Stirling coolers operate by making a working gas undergo a Stirling cycle, which consists of two constant volume processes and two isothermal processes. Devices consist of a compressor pump and a displacer unit with a regenerative heat exchanger, known as a regenerator. Stirling engines require several minutes of cool-down time; the working fluid is helium. The development of two-stage devices has extended the lower temperature range from 60–80 to 15–30 K.

Both Joule–Thompson and engine-cooled detectors are housed in precision-bore Dewars into which the cooling device is inserted (see Figure 18.6). The detector, mounted in the vacuum space at the end of the inner wall of the Dewar and surrounded by a cooled radiation shield compatible with the convergence angle of the optical system, looks out through an IR window. In some Dewars, the electrical leads to detector elements are embedded in the inner wall of the Dewar to protect them from damage due to vibration. After 3000–5000 h of operation, Stirling coolers require factory service to maintain performance. Because the Dewar

and cooler is one integrated detector assembly unit, the entire unit must be serviced together.

Split Stirling coolers are also fabricated. The detector is mounted on the Dewar bore, and the cold finger of the cooler is thermally connected to the Dewar by a bellows. A fan is necessary to dissipate the heat. The cooler can be easily removed from the detector/Dewar for replacement.

The moving displacer in the Stirling cryocoolers has several disadvantages. It is a source of vibration, has a limited lifetime, and contributes to axial heat conduction as well as to shuttle heat loss. Pulse tube coolers are similar to Stirling coolers. However, their thermodynamic processes are quite different. The proper gas motion in phase with the pressure is achieved by the use of an orifice, along with a reservoir volume to store the gas during a half cycle. The reservoir volume is large enough that negligible pressure oscillation occurs in it during the oscillating flow. The oscillating flow through the orifice separates the heating and cooling effects just as the displacer does for the Stirling refrigerator.

Since there are no moving parts at the cold-end, reliability is theoretically higher than that for Stirling cycle machines. Efficiencies approaching

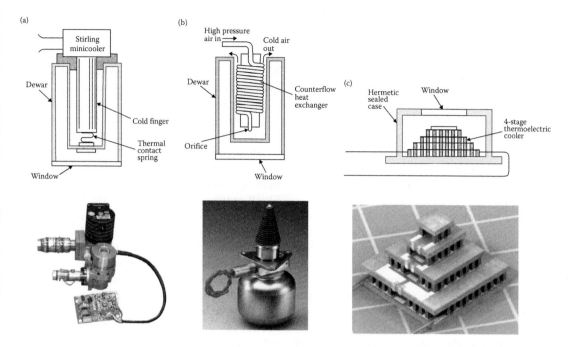

Figure 18.6 Three ways of cooling IR detectors: (a) Stirling cycle engine, (b) Joule–Thompson cooler, and (c) four-stage thermoelectric cooler.

Stirling cycle coolers can be achieved and several recent missions have demonstrated their usefulness in space.

The design of Joule–Thompson (J–T) coolers is based on the fact that as a high-pressure gas expands upon leaving a throttle valve, it cools and liquefies (leading to isenthalpic cooling). The coolers require a high-pressure gas supply from bottles and compressors. Although this is an irreversible process, with correspondingly low efficiency, J–T coolers are simple, reliable, and have low electrical and mechanical noise levels.

Using compressed air, temperatures of the order of 80 K can be achieved in 1 or 2 min. The gas used must be purified to remove water vapor and carbon dioxide that could freeze and block the throttle valve. Specially designed J–T coolers using argon are suitable for ultrafast cool-down (a few seconds cooling time). Recent advances in J–T cryocoolers have been associated with the use of mixed gases as the working fluid rather than pure gases.

Thermoelectric (TE) cooling of detectors is simpler and less costly than closed-cycle cooling. Thermoelectric coolers work by exploiting the Peltier effect that refers to the creation of a heat flux at the junction of two dissimilar conductors in the presence of current flow.

Commercially available coolers do not go beyond six stages. They are based on alloys of bismuth telluride and antimony telluride materials. Detectors are usually mounted in a hermetic encapsulation with a base designed to make good contact with a heat sink. TE coolers can achieve temperatures of up to ≈200 K, have about a 20-year operating life, are small and rugged, and have low input power (<1 W for a two-stage device and <3 W for a three-stage device). Their main disadvantage is low efficiency (see Table 18.3).

The TE coolers used for IR FPA (focal plane array, explained later in Section 18.3.2) operation include one-stage (TE1, down to−20°C or 253 K), two-stage (TE2, down to−40°C or 233 K), three-stage (TE3, down to −65°C or 208 K), and four-stage (TE4, down to −80°C or 193 K) processes. Peltier coolers are also the preferred approach to temperature control at the required level, for example, for uncooled visible and IR sensors.

18.2.7 IR optics

The optical block in an IR system creates an image of the observed objects in the plane of the detector (detectors). In the case of a scanning imager, the optical scanning system creates an image with the number of pixels much greater than the number of elements of the detector. In addition, optical elements like windows, domes, and filters can be used to protect the system from the environment or to modify the detector spectral response.

There is no essential difference in design rules of optical objectives for visible and IR ranges. The designer of IR optics is only more limited because there are significantly fewer materials suitable for IR optical elements, in comparison with those for the visible range, particularly for wavelengths over 2.5 μm.

There are two types of IR optical elements: reflective elements and refractive elements. As the names suggest, the role of reflective elements is to reflect incident radiation and the role of refractive elements is to refract and transmit incident radiation.

Mirrors used extensively inside IR systems (especially in scanners) are most often reflective elements that serve manifold functions in the IR systems. Elsewhere they need a protective coating to prevent them from tarnishing. Spherical or aspherical mirrors are employed as imaging elements. Flat mirrors are widely used to fold optical paths, and reflective prisms are often used in scanning systems.

Four materials are most often used for mirror fabrication: optical crown glass, low-expansion borosilicate glass (LEBG), synthetic fused silica, and Zerodur. Less popular in use are metallic substrates (beryllium, copper) and silicon carbide. Optical crown glass is typically applied in nonimaging systems. It has a relatively high thermal expansion coefficient and is employed when thermal stability is not a critical factor. LEBG, known by the Corning brand name Pyrex, is well suited for high quality front surface mirrors designed for low optical deformation under thermal shock. Synthetic fused silica has a very low thermal expansion coefficient. Zerodur is a registered trademark of Schott AG for lithium-aluminosilicate glass-ceramic characterized by near zero thermal expansion and very good other optical properties. However, Zerodur is very expensive comparing to common borosilicate glass.

Metallic coatings are typically used as reflective coatings for IR mirrors. There are three types of metallic coatings used most often: protected aluminum, protected silver, and protected gold. They offer high reflectivity (over about 97%) in the 3–15 μm spectral range. Bare aluminum has a very high reflectance value but oxidizes over time. Protected aluminum is a bare aluminum coating with a dielectric overcoat that arrests the oxidation process. Silver offers better reflectance in the near-IR region than aluminum and high reflectance across a broad spectrum. Gold is a widely used material and offers consistently very high reflectance (about 99%) in the 0.8–50 μm range. However, gold is soft (it cannot be touched to remove dust) and is most often used in the laboratory.

Most glasses used to manufacture optical elements for the visible and near-IR range transmit light up to about 2.2 μm well and can be used for SWIR optics. Thermal imagers use almost exclusively two spectral bands: 3–5 μm or 8–14 μm. Therefore, typically, for IR optics, suitable materials are considered that transmit IR radiation in the spectral range from 2 to 14 μm.

The list of potential materials that could be used to manufacture IR refractive optics is quite long: AMTIR-1 (amorphous material transmitting infrared radiation), barium fluoride (BaF_2), cadmium telluride (CdTe), calcium fluoride (CaF_2), cesium bromide (CsBr), cesium iodide (CsI), fused silica-IR grade, gallium arsenide (GaAs), germanium (Ge), lithium fluoride (LiF), magnesium fluoride (MgF_2), potassium bromide (KBr), potassium chloride (KCl), silicon (Si), sodium chloride (NaCl), thallium bromoiodide (KRS-5), zinc selenide (ZnSe), and zinc sulfide (ZnS). However, in this chapter, only the most popular materials used to manufacture refractive optical objectives for thermal imagers will be discussed. The basic parameters of these materials are presented in Table 18.4 and their IR transmission is shown in Figure 18.7.

Germanium is a silvery metallic-appearing solid of veryhigh refractive index, n, (>4) that enables design of high-resolution optical systems using a minimal number of germanium lenses. Its useful transmission range is from 2 to about 15 μm. It is quite brittle and difficult to cut but accepts a very good polish. Germanium is nonhygroscopic and nontoxic, has good thermal conductivity, excellent surface hardness, and good strength. Additionally, due to its very high refractive index, antireflection coatings are essential for any germanium transmitting optical system. Germanium has a low dispersion and is unlikely to need color correcting except in the highest resolution systems. A significant disadvantage of germanium is the serious dependence of its refractive index on temperature, so germanium lenses may need to be athermalized. In spite of the high material price and cost of antireflection

Table 18.4 Principal characteristics of some IR materials

Material	Waveband (μm)	$n_{4\mu m}$, $n_{10\mu m}$	dn/dT (10^{-6} K^{-1})	Density (g/cm³)	Other characteristics
Ge	2–12	4.0245, 4.0031	424 (4 μm), 404 (10 μm)	5.32	Brittle, semiconductor, can be diamond-turned, visibly opaque, hard
Chalcogenide glasses	3–12	2.5100, 2.4944	55 (10 μm)	4.63	Amorphous IR glass, can be slumped to near-net shape
Si	1.2–7.0	3.4289 (4 μm)	159 (5 μm)	2.329	Brittle, semiconductor, diamond-turned with difficulty, visibly opaque, hard
GaAs	3–12	3.304, 3.274	150	5.32	Brittle, semiconductor, visibly opaque, hard
ZnS	3–13	2.251, 2.200	43 (4 μm), 41 (10 μm)	4.08	Yellowish, moderate hard and moderate strong, can be diamond-turned, scatters short wavelengths
ZnSe	0.55–20	2.4324, 2.4053	63 (4 μm), 60 (10 μm)	5.27	Yellow-orange, relatively soft and weak, can be diamond-turned, very low internal absorption and scatter
CaF$_2$	3–5	1.410	−8.1 (3.39 μm)	3.18	Visibly clear, can be diamond-turned, mildly hygroscopic
Sapphire	3–5	1.677(n_o), 1.667(n_e)	6 (o), 12 (e)	3.99	Very hard, difficult to polish due to crystal boundaries
BF7 (Glass)	0.35–2.3		3.4	2.51	Typical optical glass

Source: Couture, M. E., *Proc. SPIE*, 4369, 649–661, 2001.

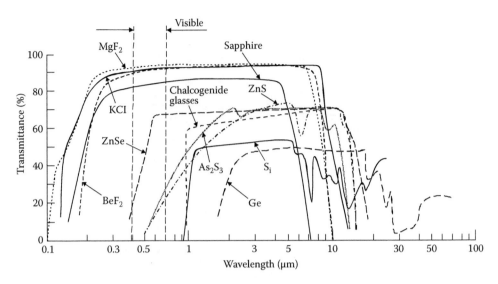

Figure 18.7 Transmission range of IR materials. (Reproduced from Couture, M. E., *Proc. SPIE*, 4369, 649–661, 2007.)

coatings, germanium is a favorite choice of optical designers of high-performance IR objectives for thermal imagers.

IR chalcogenide classes offer good transmission from about 1 μm to about 13 μm (from SWIR to LWIR range). Physical properties such as low dn/dT and low dispersion enable optical designers to engineer color-correcting optical systems without thermal defocusing. A moldable feature of these glasses allows the cost-effective manufacture of complex lens geometries in medium to large volumes. Further on, these glasses can also be processed using conventional grinding and polishing techniques and single point diamond turning if higher performance is to be achieved. Due to these features, IR chalcogenide glasses caused a revolution in the manufacturing of optics for thermal imagers in the past decades by enabling the mass manufacture of low cost, good optical performance optical objectives. Now these glasses compete with germanium as the most popular IR optical material. Most popular brands of IR chalcogenide glasses are: AMTIR from Amorphous Materials Inc., GASIR from Umicore Inc., and IRG glasses from Schott. It should however be noted that it is more difficult to fabricate high accuracy lenses using chalcogenide glasses than germanium.

Physical and chemical properties of silicon are very similar to the properties of germanium. It has a high refractive index (≈3.45), is brittle, does not cleave, takes an excellent polish and has large

dn/dT. Similar to germanium, silicon optics must have antireflection coatings. Silicon offers two transmission ranges: 1–7 and 25–300 μm. Only the first range is used in typical IR systems. The material is significantly cheaper than germanium, ZnSe, and ZnS. It is used mostly for IR systems operating in the 3–5 μm band. Due to its low density, silicon is ideal for MWIR objectives with weight constraints.

ZnSe is an optical material with optical properties mostly similar to germanium but of wider transmission range (from about 0.55 μm to about 20 μm) and a refractive index of about 2.4. It is partially translucent in the visible range and reddish in color. Due to the relatively high refractive index, antireflection coatings are necessary. The chemical resistance of the material is excellent. It is a popular material for lenses for both LWIR and MWIR objectives and for broadband IR windows.

ZnS offers relatively good transmission in the range from about 3 to 13 μm. It exhibits exceptional high fracture strength, high hardness, and high chemical resistance. Due to high resistance to rain erosion and high-speed dust abrasion, ZnS is popular for windows or external lenses in thermal imagers used in high-speed airborne applications.

Ordinary glass does not transmit radiation beyond 2.5 μm in the IR region. Fused silica is characterized by a very low thermal expansion coefficient that makes optical systems particularly useful in changing environmental conditions. It offers a transmission range from about 0.3 to 3 μm. Because of low reflection losses due to the low refractive index (≈1.45), antireflection coatings are not needed. However, an antireflection coating is recommended to avoid ghost images. Fused silica is more expensive than BK-7, but still significantly cheaper than Ge, ZnS, and ZnSe, and is a popular material for lenses of IR systems with bands located below 3 μm.

The alkali halides have excellent IR transmission; however, they are either soft or brittle and many of them are attacked by moisture, making them generally unsuitable for industrial applications. For a more detailed discussion on IR materials, see References. [7,8].

18.2.8 Night-vision system concepts

Night-vision systems can be divided into two categories: those depending upon the reception and processing of radiation reflected by an object and those that operate with radiation internally generated by an object. The latter systems are described in Section 18.2.9. These devices gather existing ambient light (starlight, moonlight, or IR light) through the front lens. This light, which is made up of photons, goes into a photocathode tube. The photocathode tube changes the photons to electrons.

The human visual perception system is optimized to operate in daytime illumination conditions. The visual spectrum extends from about 420 to 700 nm and the region of greatest sensitivity is near the peak wavelength of sunlight at around 550 nm. However, at night, fewer visible light photons are available and only large, high-contrast objects are visible. It appears that the photon rate in the region from 800 to 900 nm is five to seven times greater than in the visible region around 500 nm. Moreover, the reflectivity of various materials (e.g., green vegetation, because of its chlorophyll content) is higher between 800 and 900 nm than at 500 nm. It means that at night, more light is available in the NIR than in the visual region and that against certain backgrounds, more contrast is available.

The early concepts of image intensification were not basically different from those of today. However, the early devices suffered from two major deficiencies: poor photocathodes and poor coupling. Later development of both cathode and coupling technologies changed the image intensifier into a much more useful device. The concept of image intensification by cascading stages was suggested independently by a number of workers in the early 1930s.

A considerable improvement in night-vision capability can be achieved with night viewing equipment that consists of an objective lens, image intensifier, and eyepiece (see Figure 18.8). Improved

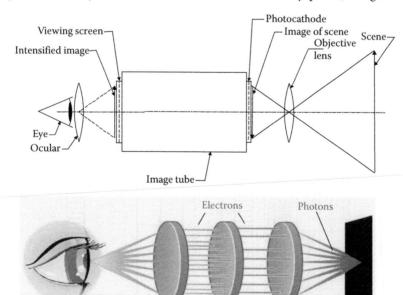

Figure 18.8 Image intensifier.

visibility is obtained by gathering more light from the scene with an objective lens than the unaided eye; by using a photocathode that has higher photosensitivity and broader spectral response than the eye; and by amplification of photo events for visual sensation.

18.2.9 Thermal imaging system concepts

Thermal imaging is a technique for converting a scene's thermal radiation pattern (invisible to the human eye) into a visible image. Its usefulness is due to the following aspects:

- It is a totally passive technique and allows day and night operation.
- It is ideal for detection of hot or cold spots, or areas of different emissivities, within a scene.
- Thermal radiation can penetrate smoke and mist more readily than visible radiation.
- It is a real-time, remote sensing technique.

The thermal image is a pictorial representation of temperature difference. Displayed on a scanned raster, the image resembles a television picture of the scene and can be computer processed to color-coded temperature ranges. Originally developed (in the 1960s) to extend the scope of night-vision systems, thermal imagers at first provided an alternative to image intensifiers. As the technology matured, its range of application has expanded and now extends into fields that have little or nothing to do with night vision (e.g., stress analysis, medical diagnostics). In most present-day thermal imagers, an optically focused image is scanned electronically across detectors (many elements or 2D array) the output of which is converted into a visual image. The optics, mode of scanning, and signal processing electronics are closely interrelated. The number of picture points in the scene is governed by the nature of the detector (its performance) or the size of the detector array. The effective number of picture points or resolution elements in the scene is steadily increased.

Detectors are only a part of usable sensor systems. Military sensor systems include optics, coolers, pointing and tracking systems, electronics, communication, processing together with information-extraction subsystems, and displays (see Figure 18.9). Hence, the process of developing a sensor system is significantly more challenging than fabricating a detector array.

Noise equivalent difference temperature (NEDT) is a commonly reported figure of merit for thermal imagers. In spite of its widespread use in IR literature, it is applied to different systems, in different conditions, and with different meanings.

NEDT of a detector represents the temperature change, for incident radiation, that gives an output signal equal to the rms noise level. While normally thought of as a system parameter, detector NEDT and system NEDT are the same except for system losses. NEDT is defined as

$$\mathrm{NEDT} = \frac{V_\mathrm{n}(\partial T/\partial Q)}{(\partial V_\mathrm{s}/\partial Q)} = V_\mathrm{n}\frac{\Delta T}{\Delta V_\mathrm{s}}, \qquad (18.4)$$

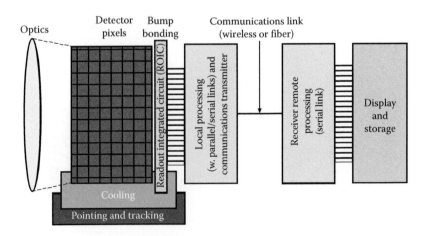

Figure 18.9 Schematic representation of an imaging system showing important subsystems.

where V_n is the rms noise, Q is the spectral photon flux density (photons/cm²s) incident on a focal plane, and ΔV_s is the signal measured for the temperature difference ΔT.

It can be shown that NEDT is given by [9]

$$\text{NEDT} = \frac{4f_\#^2(\Delta f)^{1/2}}{A^{1/2}t_{op}M^*}, \qquad (18.5)$$

where $f_\#$ is the f-number of the detector optics ($f_\# = f/D$; f is the focal length and D is the diameter of the lens), t_{op} is the transmission of the optics, and $M*$ is the figure of merit that includes not only the detector performance $D*$ but also the spectral dependence of the emitted radiation (L), $(\partial L/\partial T)\lambda$, and the atmospheric transmission t_{at}. It is given by the following equation:

$$M^* = \int_0^\infty \left(\frac{\partial L}{\partial T}\right)_\lambda t_{at\lambda} D_\lambda^* \, d\lambda. \qquad (18.6)$$

Usually, the performance of MW and LWIR FPAs is limited by the readout circuits [by storage capacity of the readout integrated circuit (ROIC)]. In this case [10]

$$\text{NEDT} = \left(t_{op} C \eta_{BLIP} \sqrt{N_w}\right)^{-1}, \qquad (18.7)$$

where N_w is the number of photogenerated carriers integrated for one integration time, t_{int}

$$N_w = \eta A_d t_{int} Q_B, \qquad (18.8)$$

where A_d is the detector area, η_{BLIP} is ratio of photon noise to composite FPA noise, and Q_B is the photon flux density.

It results from the aforementioned formulas that the charge handling capacity of the readout, the integration time linked to the frame time, and dark current of the sensitive material become the major issues of IR FPAs. The NEDT is inversely proportional to the square root of the integrated charge and therefore, the greater the charge, the higher the performance. The well charge capacity is the maximum amount of charge that can be stored in the storage capacitor of each cell. The size of the unit cell is limited to the dimensions of the detector element in the array.

To receive best sensitivity (lowest NEDT), the spectral integral in Equation 18.6 should be maximized. This can be obtained when the peak of the spectral responsivity and the peak of the exitance contrast coincide. However, the thermal imaging system may not satisfy these conditions because of other constraints, such as atmospheric/obscurant transmittance effects or available detector characteristics. Dependence on the square root of bandwidth is intuitive, since the rms noise is proportional to $(\Delta f)^{1/2}$. In addition, a better NEDT results from lower $f/\#$. A lower $f/\#$ number results in more flux captured by the detector, which increases SNR for a given level.

The dependence of NEDT on the detector area is critical. The inverse-square-root dependence of NEDT on detector area results as an effect of two terms: increasing of rms noise as the square root of the detector area and proportional increasing of the signal voltage to the area of the detector. The net result is that NEDT$\propto 1/(A_d)^{1/2}$. While the thermal sensitivity of the imager is better for larger detectors, the spatial resolution is poorer for larger detectors (pixels). Another parameter, the minimum resolvable difference temperature (MRDT), considers both thermal sensitivity and spatial resolution, and is more appropriate for design (for more information, see Section 18.3.2).

The previous considerations are valid if we assume that the temporal noise of the detector is the main source of noise. However, this assertion is not true for staring arrays, where the nonuniformity of the detector's response is a significant source of noise. This nonuniformity appears as a fixed pattern noise (spatial noise). It is defined in various ways in the literature; however, the most common definition is that it is the dark signal nonuniformity arising from electronic source (i.e., other than thermal generation of the dark current); for example, clock breakthrough or from offset variations in row, column, or pixel amplifiers/switches. Hence, estimation of IR sensor performance must include a treatment of spatial noise that occurs when FPA nonuniformities cannot be compensated correctly.

Mooney et al. [11] have given a comprehensive description of the origin of spatial noise. The total noise of a staring array is the composite of the temporal noise and the spatial noise. The spatial noise is the residual nonuniformity u after application of nonuniformity compensation, multiplied by the signal electrons N. Photon noise, equal to $N^{1/2}$, is the dominant temporal noise for the high

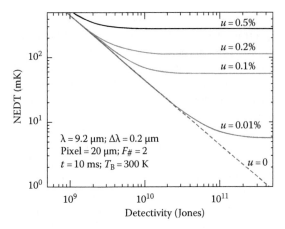

Figure 18.10 NEDT as a function of detectivity. The effects of nonuniformity are included for u=0.01%, 0.1%, 0.2%, and 0.5%. Note that for $D*$ > 10^{10} cmHz$^{1/2}$ W^{-1}, detectivity is not the relevant figure of merit.

IR background signals for which spatial noise is significant. Then, the total NEDT is

$$\text{NEDT}_{\text{total}} = \frac{(N+u^2N^2)^{1/2}}{\partial N/\partial T} = \frac{\left(1/N+u^2\right)^{1/2}}{(1/N)(\partial N/\partial T)},$$

(18.9)

where $\partial N/\partial T$ is the signal change for a 1 K source temperature change. The vdenominator, $(\partial N/\partial T)/N$, is the fractional signal change for a 1 K source temperature change. This is the relative scene contrast.

The dependence of the total NEDT on detectivity for different residual nonuniformity is plotted in Figure 18.10 for 300 K scene temperature and set of parameters shown as insert in the figure. When the detectivity is approaching a value above 10^{10} cmHz$^{1/2}$ W^{-1}, the FPA performance is uniformity limited prior to correction and thus, essentially independent of the detectivity. An improvement in nonuniformity from 0.1% to 0.01% after correction could lower the NEDT from 63 to 6.3 mK.

18.3 IR SYSTEMS

This section briefly concentrates on selected IR systems and is arranged in order of increasing complexity: smart weapon seekers, FLIRs (forward looking infrareds), and space-based systems. For more information on IR systems, please refer to *The Infrared and Electro-Optical Systems Handbook* (executive editors: Joseph S. Accetta and David L. Shumaker).

18.3.1 Image intensifier systems

As is marked in Section 18.2.8, the image intensifier tube is a vacuum tube device for increasing the intensity of available light in an optical system to allow use under low-light conditions, such as at night (see Figure 18.8) [12]. Typical spectral sensitivity curves of various photocathodes charge with typical transmittance of window materials together with a list of important photocathodes are shown in Figure 18.11.

As is shown in Figure 18.7, the image intensifier system is built from three main blocks: optical objective, multichannel plate (MCP), and optical ocular. An MCP is a secondary electron multiplier consisting of an array of millions of very thin glass channels (of internal diameter ≈10 μm, each capillary works as an independent electron multiplier) bundled in parallel and sliced in the form of a disk (see Figure 18.12). Secondary electrons are accelerated by the voltage applied across both ends of the MCP. This process is repeated many times along the channel wall and as a result, a great number of electrons are output from the MCP. Furthermore, the electron flux can be reconverted into an optical image by using a phosphor coating as the rear electrode to provide electroluminescence; this combination provides an image intensifier.

Image intensifiers are classed by generation (Gen) numbers. Gen0 refers to the technology of World War II, employing fragile, vacuum-enveloped photon detectors with poor sensitivity and little gain. Further evolution of image intensifier tubes is presented in Table 18.5. Gen1 represents the technology of the early Vietnam era, the 1960s. In this era, the first passive systems, able to amplify ambient starlight, were introduced. Though sensitive, these devices were large and heavy. Gen1 devices used tri-alkali photocathodes to achieve a gain of about 1000. By the early 1970s, the MCP amplifier was developed comprising more than two million microscopic conducting channels of hollow glass, each of

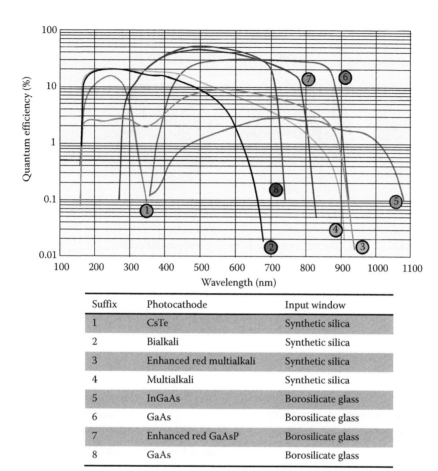

Suffix	Photocathode	Input window
1	CsTe	Synthetic silica
2	Bialkali	Synthetic silica
3	Enhanced red multialkali	Synthetic silica
4	Multialkali	Synthetic silica
5	InGaAs	Borosilicate glass
6	GaAs	Borosilicate glass
7	Enhanced red GaAsP	Borosilicate glass
8	GaAs	Borosilicate glass

Figure 18.11 Spectral sensitivity curves of various photocathodes. (Reproduced from http://www.hamamatsu.com/resources/pdf/etd/II_TII0004E02.pdf.)

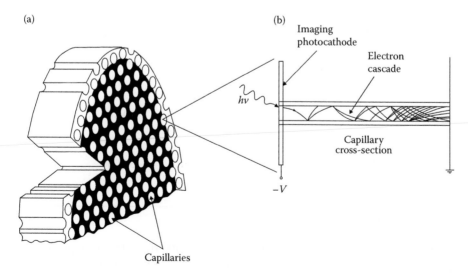

Figure 18.12 Schematic presentation of a microchannel plate: (a) cutaway view and (b) a single capillary.

Table 18.5 Image intensifier tubes—history and basic parameters

GenI	GenII	GenIII	GenIII filmless

• Vietnam War • SbCsi SbNa$_2$KCs photocathodes (S10,S20) • Electrostatic inversion • Photosensivity up to 200 µA/lm	• 1970s • Multialkali photocathodes (S25) • Microchannel plate (MCP) • Photosensivity up to 700 µA/lm • Operational live to 4000 h	• Early 1980s • GaAs photocathodes • Ion barrier to microchannel plate • Photosensivity up to 1800 µA/lm • Operational live to 10,000 h	• Late 1990s • Multialkali photocathodes • "Filmless" tube • Photosensivity up to 2200 µA/lm • Operational time to 10,000 h

Gen. No.	Photocathode material	Photocathode sensitivity (µA/lm)	Design type	Luminance gain (lm/ lm)	Resolution (lp/mm)	SNR
0	S1	<60	Inverter tube	<200	20–60	—
I	S20	<160	Inverter tube	<800	20–60	—
I+	S20	<160	Cascade inverter tube	<20,000	20–30	5–8
II	S25	<350	Inverter MCP tube	<50,000	24–43	12–17
II$^+$	Improved S25	<700	Proximity focus MCP tube	<70,000	43–81	16–24
III	GaAs/GaAsP	<1600	Proximity focus MCP tube with protecting film	<70,000	36–64	18–25
III$^+$ Thin film	GaAs/GaAsP	<1800	Proximity focus MCP thin film tubes	<70,000	57–71	24–28
III Filmless	GaAs/GaAsP	<2200	Proximity focus MCP filmless tubes	<80,000	57–71	24–31

Source: Chrzanowski, K., Opto-Electron. Rev., 21, 153–182, 2013.

about 10 µm in diameter, fused into a disk-shaped array. Coupling the MCP with multialkali photocathodes, capable of emitting more electrons per incident photon, produced GenII. GenII devices boasted amplifications of 20,000 and operational lives to 4000 h. Interim improvements in bias voltage and construction methods produced the GenII+ version. Substantial improvements in gain and bandwidth in the 1980s heralded the advent of GenIII. Gallium arsenide photocathodes and internal changes in the MCP design resulted in gains ranging from 30,000 to 50,000 and operating lives of 10,000 h.

Figure 18.13 shows the response of a typical GenIII image intensifier superimposed on the night sky radiation spectrum [14]. This figure also shows the Commission Internationale de l'Eclairage (CIE) photopic curve illustrating the spectral response of the human visual perception system, and the GenII response.

Many candidate technologies could form the basis of a GenIII, ranging from enhanced current designs to completely different approaches. Among these are devices with a new photocathode that extend spectral response to 1.6 μm and the use of an amplifying mechanism other than MCPs. Other potential candidates include lightweight systems that fuse the outputs from image intensifiers and thermal imagers, and those that couple electron-bombarded CCD (charge coupled device) arrays—providing sensitivity in the NIR and MWIR regions—with miniature flat-panel displays.

Various implementations of image intensifier tubes have been realized. Phosphor output image intensifiers were reviewed in depth by Csorba [15]. The image is focused onto a semitransparent photocathode and photoelectrons are emitted with a spatial intensity distribution that matches the focused image. In image intensifiers, the electrons are then accelerated towards a phosphor screen where they reproduce the original image with enhanced intensity. Three common forms of image tubes are shown in Figure 18.14.

In a "proximity-focused" tube, a high electric field (typically 5 kV), and a short distance between the photocathode and the screen, limit spreading of electrons to preserve an image. This form of tube is compact, the image is free from distortion, and only a simple power supply is required. However, the resolution of such a tube is limited by the field strength at the photocathode and the resolution is highest when the distance between the cathode and the screen is small.

An electrostatically focused tube is based upon a system of concentric spheres (cathode and anode,

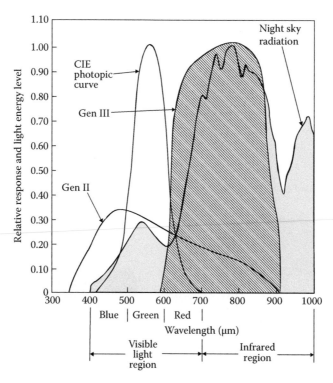

Figure 18.13 Image intensifier tube spectral response curves. (Reproduced from Cameron, A. A., *Proc. SPIE*, 1290, 16–19, 1990.)

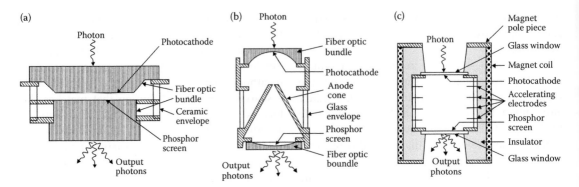

Figure 18.14 Cross-sectional diagrams of a variety of image intensifier types: (a) proximity focused, (b) electronically focused, and (c) magnetically focused. (Reproduced from Csorba, I. P., *Image Tubes*, Sams, Indianapolis, IN, 1985.)

typical bias voltage of 15 kV). In practice, the electrodes depart radically from the simple spherical concept. Additional electrodes can be introduced to provide focusing control and reduce the image distortion, while fiber optic windows at the input and output can be used to improve image quality and provide a better matching to objective and coupling optics. Power suppliers are very simple and lightweight, so this type of tube is widely used in portable applications.

A magnetically focused system gives very high-resolution imagers with little or no distortion. The focusing coil, however, is usually heavy and power consuming. For the best picture quality, the power suppliers for both tube and coil must be stable. This type of tube is used in applications where resolution and low distortion are vital and weight and power consumption do not create unacceptable problems.

Image intensifiers were primarily developed for night time viewing and surveillance under moonlight or starlight. At present, image intensifier applications have spread from night time viewing to various fields including industrial product inspection and scientific research, especially when used with CCD cameras—the so-called intensified CCD or ICCD (see Figure 18.15a). Gate operation models are also useful for observation and motion analysis of high-speed phenomena (high-speed moving objects, fluorescence lifetime, bioluminescence, and chemiluminescence images). Figure 18.15b shows an example of a GenIII night vision goggle.

Image intensifiers are widespread in many military applications. The advent of night vision devices and helmet-mounted displays places additional constraints on the helmet, which is

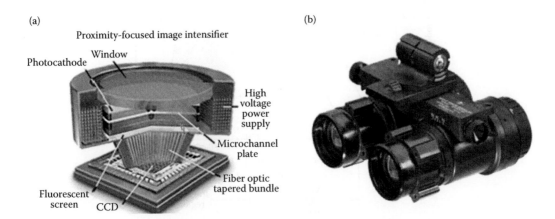

Figure 18.15 Night vision device: (a) proximity-focused image intensifier and (b) GenIII night vision goggle AN/AVS-9 (ITT Night Vision).

Figure 18.16 Tales TopOwl helmet incorporates an optical combiner assembly for each eye, allowing the pilot to view the cockpit and the outside world directly with the night imagery superimposed on it. TopOwl has a 40° FOV and a total headborne weight of 2.2 kg in full configuration. (Reproduced from http://optronique.net/ defense/wp-content/uploads/2012/07/TopOwl-Datasheet.pdf.)

now an important element of the cockpit display system, providing weapon aiming, and other information—such as aircraft attitude and status—to the pilot. For example, Figure 18.16 illustrates TopOwl imaging system developed for airborne applications by Tales [16]. The helmet-mounted sight and display system incorporates a night vision system with a 100% overlapped projection of a binocular image on the visor. TopOwl projects the night scene and associated symbology onto two circular reflective surfaces with a fully overlapped, 40°, binocular FOV (field of view). Standard symbology is used to display flight and weapon management data, helping to reduce crew workload.

18.3.2 Thermal imaging systems

The basic concept of a modern thermal imager system is to form a real image of the IR scene, detect the variation in the imaged radiation, and, by suitable electronic processing, create a visible representation of this variation analogous to conventional television cameras.

Due to existing terminology confusion in the literature, we can find at least 11 different terms used as synonyms of the earlier defined thermal imaging systems: thermal imager, thermal camera, thermal imaging camera, FLIR, IR imaging system, thermograph, thermovision, thermal viewer, IR viewer, IR imaging radiometer, thermal viewer, thermal data viewer, and thermal video system. The only real difference between the aforementioned terms is that the designations "thermograph," "IR imaging radiometer," and "thermovision" usually refer to thermal cameras used for measurement applications, while the other terms refer to thermal cameras used in observation applications. For example, thermographic imagers supply quantitative temperature, while radiometers provide quantitative radiometric data on the scene (such as radiance or irradiance) or process these data to yield information about temperatures.

Thermal imagers have various applications, depending on the platform and user [17]. Most of them are used in military applications. They often have multiple fields of view which are user switchable during operation, which gives both a wide, general surveillance mode as well as a high magnification and narrow field for targeting, designating, or detailed intelligence gathering. Many military thermal imagers are integrated with a TV camera and a laser range finder. A TV color camera is used during day time conditions due to its superior image quality. Nonmilitary uses include generic search and track, snow rescue, mountain rescue, illegal border crossing detection and pilot assistance at night or in bad weather, forest fire detection, firefighting, inspection and discreet surveillance, and evidence gathering. A small but increasing group of thermal imagers enables noncontact temperature measurement and these cameras are used in areas of industry, science, and medicine.

The term "focal plane array" (FPA) refers to an assemblage of individual detector picture elements ("pixels") located at the focal plane of an imaging system. Although the definition could include one-dimensional ("linear") arrays as well as two-dimensional (2D) arrays, it is frequently applied to the latter. Usually, the optics part of an optoelectronic images device is limited only to focusing of the image onto the detectors array. These so-called staring arrays are scanned electronically usually using circuits integrated with the arrays. The architecture of detector-readout assemblies has assumed a number of forms. The types of readout integrated circuits (ROICs) include the function of pixel deselecting, antiblooming on each pixel, subframe imaging, output preamplifying, and may include yet other functions. IR imaging systems, which use 2D arrays, belong to the so-called second generation systems.

The simplest scanning linear array used in thermal imaging systems, the so-called focal plane array (FPA), consists of a row of detectors (Figure 18.17a). An image is generated by scanning the scene across the strip using, as a rule, a mechanical scanner. At standard video frame rates, at each pixel (detector), a short integration time has been applied and the total charge is accommodated. A staring array is a 2D array of detector pixels (Figure 18.17b) scanned electronically.

The scanning system, which does not include multiplexing functions in the focal plane, belongs to the class of first-generation systems. A typical example of this kind of detector is a linear photoconductive array (PbS, PbSe, HgCdTe) in which an electrical contact for each element of a multielement array is brought off the cryogenically cooled focal plane to the outside, where one electronic channel is used at ambient temperature for each detector element. The U.S. common module HgCdTe arrays employ 60, 120, or 180 photoconductive elements depending on the application.

The second-generation systems (full-framing systems), being developed at present, have at least three orders of magnitude more elements ($<10^6$) on the focal plane than first-generation systems and the detector elements are configured in a 2D array. These staring arrays are scanned electronically by circuits integrated with the arrays. The ROICs perform, for example, pixel deselecting, antiblooming on each pixel, subframe imaging, output preamplifying and some other functions. The optics merely focuses the IR image onto the matrix of sensitive elements.

Intermediary systems are also fabricated with multiplexed scanned photodetector linear arrays in use and with, as a rule, time delay and integration (TDI) functions. Typical examples of these systems are HgCdTe multilinear 288×4 arrays fabricated by Sofradir, both for 3–5 and 8–10.5 μm bands with signal processing in the focal plane (photocurrent integration, skimming, partitioning, TDI function, output preamplification, and some others).

A number of architectures are used in the development of IR FPAs. In general, they may be classified as hybrid and monolithic ones, but these distinctions are often not as important as proponents and critics state them to be. The central design questions involve performance advantages versus ultimate producibility. Each application may favor a different approach depending on the technical requirements, projected costs, and schedule.

In the monolithic approach (see Figure 18.18a), some of the multiplexing is done in the detector

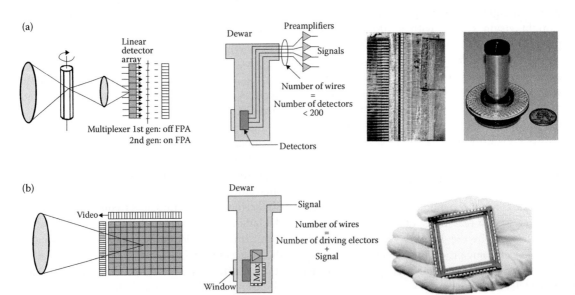

Figure 18.17 (a) Scanning and (b) staring focal plane arrays.

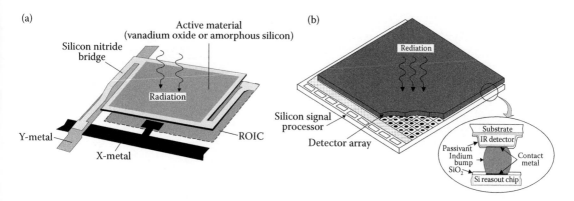

Figure 18.18 IR focal plane arrays: (a) monolithic (microbolometer) and (b) hybrid.

material itself rather than in an external read-out circuit. The basic element of monolithic CCD devices is a metal-insulator-semiconductor (MIS) structure. An MIS capacitor detects and integrates the IR-generated photocurrent. Although efforts have been made to develop monolithic FPAs using narrow-gap semiconductors, silicon-based FPA technology with Schottky-barrier detectors is the only technology matured to a level of practical use.

Hybrid FPA detectors and multiplexers are fabricated on different substrates and are typically mated with each other by flip-chip bonding (Figure 18.18b). In this case, we can optimize the detector material and multiplexer independently. Other advantages of hybrid FPAs are near 100% fill factor and increased signal-processing area on the multiplexer chip. In the flip-chip bonding, the detector array is typically connected by pressure contacts via indium bumps to the silicon multiplet pads. The detector array can be illuminated from either the frontside or backside (with photons passing through the transparent detector array substrate). In general, the latter approach is most advantageous. When using opaque materials, substrates must be thinned to 10 μm in order to obtain sufficient quantum efficiencies and reduce the crosstalk. In some cases, the substrates are completely removed.

Two types of silicon addressing circuit have been developed: CCDs and complementary metal-oxide-semiconductor (CMOS) switches. In CCD addressing circuits, the photogenerated carriers are first integrated in the well formed by a photogate and subsequently transferred to slow (vertical) and fast (horizontal) CCD shift registers [19].

An attractive alternative to the CCD readout is coordinative addressing with CMOS switches. The configuration of CCD devices requires specialized processing, unlike CMOS imagers that can be built on fabrication lines designed for commercial microprocessors. CMOS have the advantage that existing foundries, intended for application specific integrated circuits (ASICs), can be readily used by adapting their design rules. Design rules of 14 nm are currently in production, with preproduction runs of 10 nm design rules. As a result of such fine design rules, more functionality has been designed into the unit cells of multiplexers with smaller unit cells, leading to large array sizes.

A typical CMOS multiplexer architecture consists of fast (column) and slow (row) shift registers at the edges of the active area, and pixels are addressed one by one through the selection of a slow register, while the fast register scans through a column, and so on. Each image sensor is connected in parallel to a storage capacitor located in the unit cell. A column of diodes and storage capacitors is selected one at a time by a digital horizontal scan register and a row bus is selected by the vertical scan register. Therefore, each pixel can be individually addressed.

The minimum resolvable difference temperature (MRDT) is often the preferred figure of merit for imaging sensors. This figure of merit comprises both resolution and sensitivity of the thermal imager. MRDT enables us to estimate probability of detection, recognition, and identification of targets knowing MRDT of the evaluated thermal imager. The MRDT is a subjective parameter that describes the ability of the imager–human system for detection of low contrast details of the tested object. It is measured as a minimum temperature difference between the bars of the standard 4-bar

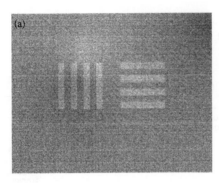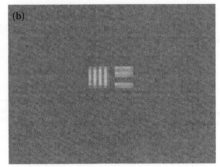

Figure 18.19 Image of a standard four-bar target during MRTD measurement: (a) target of low spatial frequency for low temperature difference and (b) target of high spatial frequency for high temperature difference between target and background (image magnified four times).

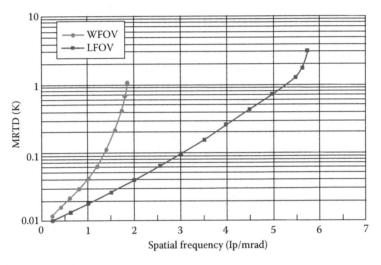

Figure 18.20 MRTD of an exemplary cooled staring thermal imager with two field of view (WFOV and LFOV).

target and the background required to resolve the thermal image of the bars by an observer versus spatial frequency of the target (see Figure 18.19). Military standards determining testing of thermal imaging systems usually specify that MRTD values for a set of spatial frequencies of the tested imager must be lower than certain values if the imager is to pass the test. The measurement results of typical military thermal imagers for airborne surveillance are shown in Figure 18.20.

Nowadays, classical MRTD is considered as the most important parameter of thermal imagers and is typically used for range predictions of real targets. However, it was reported many times that the MRTD concept, when applied to undersampled imagers, generates incorrect range predictions, particularly the detection range. The biggest

problem is low accuracy of performance modeling over Nyquist frequency.

There are at least three competing solutions to eliminate the aforementioned limitation of the MRTD concept and to improve accuracy of range prediction: triangle orientation discrimination (TOD), minimum temperature difference perceived (MTDP), and dynamic MRTD. Therefore, it can be expected that in the future, MRTD will be replaced by a new parameter as a main criterion for evaluation of thermal imagers.

18.3.3 IR cameras versus FLIR systems

Historically, a "camera" includes neither the storage medium nor the display, while the "camera system' includes the complete package. At present, manufacturers offer an optional recording medium (usually CD-ROM), display, and electronics for the display. For example, Figure 18.21 is a photograph of the FLIR P660 IR. This camera has a high resolution 640×480 thermal imagery with 30 mK sensitivity (NEDT), interchangeable lenses, and the flexibility of a tiltable, high-fidelity color LCD. The camera can be used by anyone who does thermal inspections, or who needs to accurately measure small objects from far away.

Figure 18.22 shows a representative camera architecture with three distinct hardware pieces: a camera head (which contains the optics, including collecting, imaging, zoom, focusing, and spectral filtering assembles), electronics/control processing box, and the display (see Figure 18.9). Electronics and motors to control and drive moving parts must be included. The control electronics usually consist of communication circuits, bias generators, and clocks. Usually, the camera's sensor (FPA) needs cooling and therefore, some form of cooler is included, along with its closed-loop cooling control electronics. Signals from the FPA is of low voltage and amperage and requires analog preprocessing (including amplification, control, and correction), which is located physically near the FPA and included in the camera head. Often, the A/D is also included here. For user convenience, the camera head often contains the minimum hardware needed to keep volume, weight, and power to a minimum.

Typical costs of cryogenically cooled imagers around $50,000 restrict their installation to critical military applications allowing conducting of operations in complete darkness. Moving from cooled to uncooled operation (e.g., using a silicon microbolometer) reduces the cost of an imager to below $10,000. Less expensive IR cameras present a major departure from the camera architecture presented in Figure 18.22.

Cameras usually produce high-quality images with NEDTs of 20–50 mK. Details and resolution vary by optics and focal planes. A good camera produces an image akin to that of a black and white television.

"FLIR" is an archaic 1960s jargon for forward-looking IR to distinguish these systems from IR line scanners, which look down rather than forward. Conversely, most sensors that do look forward are not considered to be FLIRs (e.g., cameras and astronomical instruments). The term "FLIR" should be eliminated from IR technospeak, but is still used and is likely to remain in the jargon for a while.

It is difficult to explain the difference between a camera and an FLIR system. In general, FLIRs are designed for specific applications and specific platforms, their optics is integrated into the package, and they are used mostly by people. Cameras usually rely on "imaging" of a "target" and they are designed for generic purposes, without much consideration for form and fit; they can be used with many different fore optics and are often used by computers and machines (not just people).

The term "FLIR" usually implies military or paramilitary use, air-based units, and scanners. The FLIR provides automatic search, acquisition, tracking, precision navigation, and weapon delivery functions. A typical FLIR comprises four line replaceable units, such as an FLIR optical assembly mounted on a gyro-stabilized platform, an electronics module containing all necessary electronics circuits and a cryogenically cooled detector array, a power supply unit, and a control and processing assembly.

In the 1960s the earliest FLIRs were linear scanners. In the 1970s, first-generation common modules (including a Dewar containing 60, 120, or 180 discrete elements of photoconductive HgCdTe) were introduced. The next generation of FLIRs employed a dense linear array of photovoltaic HgCdTe, usually 2(4)×480 or 2(4)×960 elements in TDI for each element. At present, these systems are replaced by full-framing FLIRs that employ staring arrays (HgCdTe, InSb, and

Figure 18.21 FLIR P660 IR camera.

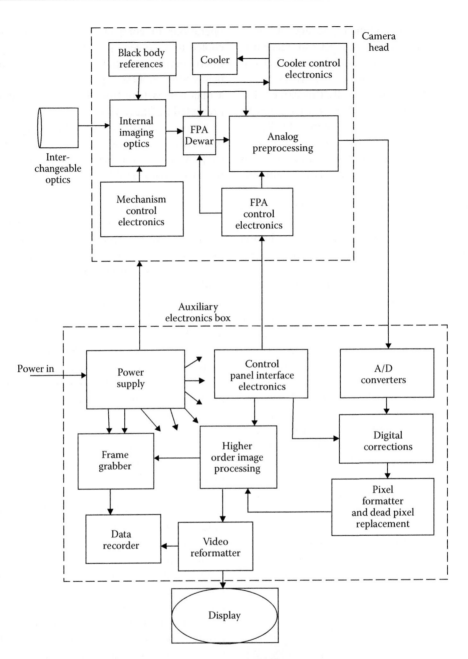

Figure 18.22 Representative IR camera system architecture. (Reproduced from Miller, J. L., *Principles of Infrared Technology*, Van Nostrand Reinhold, New York, 1994.)

QWIP). For example, Figure 18.23 shows the eLRAS3 (Long-Range Scout Surveillance System) fabricated by Raytheon. eLRAS3 provides the real-time ability to detect, recognize, identify, and geo-locate distant targets outside the threat acquisition. In addition, Raytheon's high-definition resolution FLIR (also called 3rd Gen FLIR) combines HgCdTe LWIR and MWIR IR arrays.

In several discrete packages, FLRs are usually referred to as line replaceable units (LRUs) such as scanner head, power supply, image processor, recorder, display, and controls. They have the form of boxes spread around the host platform. The controls and display must be mounted in the cockpit with the humans. A representative FLIR architecture with the video signal output (to support LRU

Figure 18.23 3rd Gen eLRAS3 FLIR system produced by Raytheon.

for image and higher-order processing) is shown in Figure 18.24. Many systems depart significantly from this architecture [18].

FLIRs usually use telescopes in the sense that the lens system is focused at a distance very much larger than the focal length. Characteristics such as FOV, resolution, element size, and spatial frequency are expressed in angular units. By convention, FOV is expressed in degrees, resolution in milliradians, spatial frequency in cycles per milliradian, and noise in units of temperatures.

Worldwide, there are several hundreds of different FLIR systems in operation. The most important of them are described in the literature [7,19]. Several FLIRs integrate a laser ranger or target designator.

Recent outgrowths from military FLIRs are the infrared search and track (IRST) systems. They are a subset or class of passive systems whose objective is to reliably detect, locate, and continuously track IR-emitting objects and targets in the presence of background radiation and other disturbances. They are used in a radar-like manner (usually with a radar-like display) to detect and track objects. Most of the current research in IRST systems is concentrated in signal processing to extract target tracks from severe clutter.

Another group of outgrowths from military thermal imagers are airborne line scanners. These are one-dimensional scanning systems that enable the creation of a two-dimensional thermal image of the observed scenery only when the system is moving. In contrast to typical thermal imagers with FOV not higher than about 40°, the airborne line scanners can provide an FOV up to about 180°.

Due to wide FOV, airborne thermal scanners are widely used in military aerial reconnaissance.

18.3.4 Space-based systems

The formation of NASA (National Aeronautics and Space Administration) in 1958, and development of the early planetary exploration program, was primarily responsible for the development of the modern optical remote sensing systems, as we know them today. During the 1960s, optical mechanical scanner systems became available that made possible acquisition of image data outside the limited spectral region of the visible and NIR available with film. "Eye in The Sky" was the first successfully flown long-wavelength sensor launched in 1967. A major milestone was the development of the Landsat Multispectral Scanner, because it provided the first multispectral synoptic in digital form. The period following the launch of Landsat-1 in 1972 stimulated the development of a new series of airborne and spaceborne sensors. Since that time, hundreds of space-based sensors have been put into orbit.

The main advantages of space IR sensors are as follows [7]:

- The ability to tune the orbit to cover a ground swath in an optimal spatial or temporal way
- A lack of atmospheric effects on observation
- Global coverage
- The ability to engage in legal clandestine operations

Hitherto, anti-satellite weapons do not exist, so satellites are relatively safe from attack. The disadvantages of satellite systems are protracted and excessive costs of fabrication, launch and maintenance of satellites. Moreover, such operations as repair and upgrade are difficult, expensive, and usually not possible.

The space-based systems installed on space platfrms usually perform one of the following functions: military/intelligence gathering, astronomy, earth environmental/resources sensing or weather monitoring. Hence, these functions can be classified as forms of earth remote sensing and astronomy.

Figure 18.25 shows a representative space sensor architecture. It should be stressed, however, that many individual space sensors do not have this exact architecture.

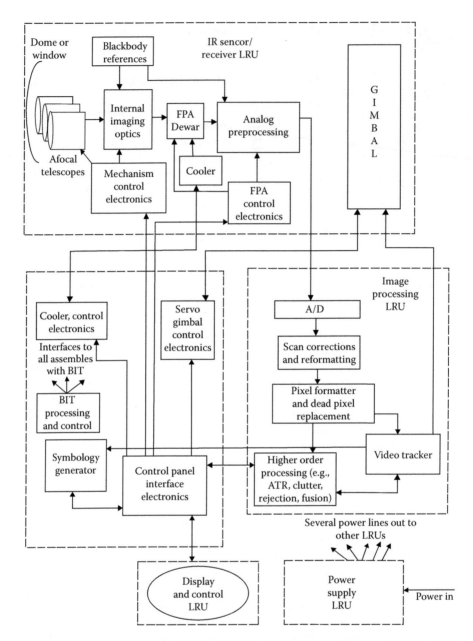

Figure 18.24 Representative FLIR system architecture. (Reproduced from Miller, J. L., *Principles of Infrared Technology*, Van Nostrand Reinhold, New York, 1994.)

Intelligence and military services from wealthy nations have long employed space-based sensors to acquire information. A satellite-borne IR warning receiver, designed to detect intercontinental ballistic missiles, is a strategic system that protects a large area, or nation. The US spends about $10 billion per year on space reconnaissance. Although the Cold War is over, the long-term strategic monitoring to access military and economic might is still important. Intelligence gathering of crop data and weather trends from space has also been used by hunger relief organizations to more effectively forecast droughts and famines. The military also has space-based surveillance for missile launches and additionally, a space base that provides excellent viewing geometries for global events such as nuclear explosions and environmental changes that the military is concerned about.

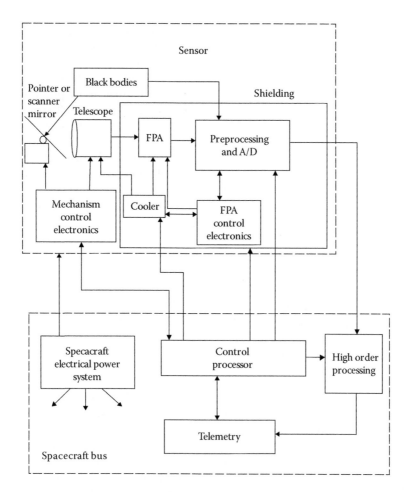

Figure 18.25 Representative space sensor architecture. (Reproduced from Miller, J. L., *Principles of Infrared Technology*, Van Nostrand Reinhold, New York, 1994.)

Imaging with IR FPAs provides increasingly detailed and quantitative information about relatively cool objects in the space of our galaxy and beyond. Dwarf stars, for example, or giant Jupiter-like planets in other distant solar systems, do not emit much visible and ultraviolet (UV) light, so they are extremely faint at these wavelengths. Moreover, the longer IR wavelengths can penetrate dusty and optically opaque nebulous molecular clouds in interstellar space where new stars and planetary systems are forming.

There are several unique reasons for conducting astronomy in space [17]:

- To eliminate the influence of absorption, emission, and scattering of IR radiation
- To answer basic cosmological and astronomical questions (e.g., formation of stars, protoplanetary disks, extra-solar planets, brown dwarfs,

dust and interstellar media, protogalaxies, the cosmic distance scale and ultra-luminous galaxies)
- To observe the earth's environment (detecting subtle changes indicating environmental stresses and trends)

NASA has historically been the leading US agency for promoting the development of long wavelength detector technologies. Figure 18.26 shows the sensitivities of active or currently planned far-IR/sub-mm spectroscopic facilities in the near future. Table 18.6 describes briefly several airborne and spaceborne platform missions.

As Figure 18.26 shows, the James Webb Space Telescope (JWST) operates at wavelengths below ~27 μm. The Atacama Large Millimeter/submillimeter Array (ALMA) operating through a number of submillimeter atmospheric windows as well

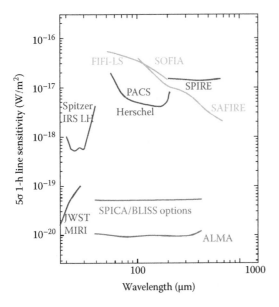

Figure 18.26 Sensitivity of far-IR spectroscopy platforms. (Reproduced from http://www.ipac.caltech.edu/pdf/FIR-SMM_Crosscutting_Whitepaper.pdf.)

as ~650 μm, has sensitivities at least 100 times higher than Herschel, spanning the intervening 60–650 μm wavelength range. JWST is currently scheduled to start operations within the next year. The Space Infrared Telescope for Cosmology and Astrophysics (SPICA), with launch envisioned in 2017, will provide two-to-three orders of magnitude increase in sensitivity in comparison with Herschel that will bring far-IR/sub-mm sensitivity into line with those of JWST and ALMA. The ambitious requirements of future space missions are summarized in Table 18.6.

18.3.5 Smart weapon seekers

The seeker is the primary homing instrument for a smart weapons that include missiles, bombs, artillery projectiles, and standoff cruise missiles. They are developed to increase the accuracy of munitions.

Taking into account which part of the electromagnetic spectrum is used, smart weapon seekers can be divided into:

- Electro-optical seekers
- Microwave seekers
- Radio controlled seekers

In this section, electro-optical seekers that dominate in the group of short-range smart munitions will be discussed.

Electro-optical seekers are basically homing instruments that use optical radiation in the range from far-IR to UV. The most popular group of electro-optical seekers employs IR radiation. However, there are seekers that use visible or UV band too. More advanced seekers can use several spectral bands of electromagnetic radiation: IR/UV, IR/microwave, etc.

In general, seekers can be divided into two groups:

- Passive seekers
- Active seekers

Passive seekers do homing using optical radiation emitted by the target or radiation emitted by natural sources and reflected by the target. Passive seekers can be later divided into two basic groups: nonimaging seekers and imaging seekers. Active seekers use optical radiation emitted by external platforms or by themselves to irradiate the target and carry out homing.

Passive nonimaging seekers use a circular optical plate, with adjacent transparent and nontransparent parts, called a reticle that is fixed at the image plane of the imaging optics of the head of the missile (Figure 18.27). A single IR detector of the size a bit larger than the reticle is placed just behind it. Location of the point image of the target on the reticle plate changes, even when the target does not change its position, due to rotation of the reticle or rotation of the imaging optics. Therefore, radiation emitted by the target generates electrical pulses at the detector output. Pulse duration and phase of these pulses give information about the angular position of the target (Figure 18.28).

The grandfather of all passive IR seekers—the Sidewinder seeker developed in the 1950s—was a passive nonimaging seeker; it employed vacuum tubes and a lead salt single-element detector. In the next decades it was found that, despite their simplicity, passive nonimaging seekers are very effective for guiding missiles when the target is on a uniform background. Therefore, at present, the majority of currently used short-range smart missiles use this type of seekers. However, the effectiveness of passive nonimaging seekers decreases significantly for targets on nonuniform

Table 18.6 Far-IR spectroscopy platforms

Spitzer Space Telescope 2003		The *Spitzer Space Telescope* was launched in August 2003. It is the last of NASA's "great observatories" in space. Spitzer is much more sensitive than earlier IR missions and studies the universe at a wide range of IR wavelengths. Spitzer concentrates on the study of brown dwarfs, super planets, protoplanetary and planetary debris disks, ultraluminous galaxies, active galaxies, and deep surveys of the early universe
SOFIA 2005		SOFIA (*Stratospheric Observatory For Infrared Astronomy*) was finally completed in 2005. SOFIA, a joint project between NASA and the German Space Agency, incorporates a 2.5 m optical/infrared/sub-millimeter telescope mounted in a Boeing 747. Designed as a replacement for the successful Kuiper Airborne Observatory, SOFIA is the largest airborne telescope in the world
Herschel Space Observatory 2009		The *Herschel Space Observatory*, carried into orbit in May 2009, is a European Space Agency IR-submillimeter mission. Herschel performs spectroscopy and photometry over a wide range of IR wavelengths and is used to study galaxy formation, interstellar matter, star formation and the atmospheres of comets and planets. The Herschel Observatory is capable of seeing the coldest and dustiest objects in space. It is the largest space telescope ever launched carrying a single mirror 3.5 m in diameter
ALMA 2011		The *Atacama Large Millimeter/submillimeter Array* (*ALMA*) is the result of an international partnership between Europe, North America, East Asia, and the Republic of Chile to build the largest astronomical project in existence. It is an astronomical interferometer, comprising an array of 66 12 and 7 m diameter radiotelescopes observing at millimeter and submillimeter wavelengths. It is being built on the Chajnantor plateau at 5000 m altitude in the Atacama desert of northern Chile. ALMA is expected to provide insight on star birth during the early universe and detailed imaging of local star and planet formation. Costing more than a billion dollars, it is the most ambitious ground-based telescope currently under construction. ALMA began scientific observations in the second half of 2011 and became fully operational in March 2013

(Continued)

Table 18.6 (*Continued*) Far-IR spectroscopy platforms

James Webb Space Telescope 2018		The *James Webb Space Telescope* (*JWST*) is a large, IR-optimized space telescope, scheduled for launch in 2018. It is a visible/IR space mission that will have extremely good sensitivity and resolution, giving us the best views yet of the sky in the near-mid IR. JWST will be used to study the early universe and the formation of galaxies, stars, and planets. Webb will have a large mirror, 6.5 m in diameter, and a sunshield the size of a tennis court. Both the mirror and sunshade will not fit onto the rocket fully open, so both will fold up and open once Webb is in outer space. Webb will reside in an orbit about 1.5 million km from the earth
SPICA/BLISS 2017		The *Background-Limited Infrared-Submillimeter Spectrograph* (*BLISS*) is a far-IR spectrograph concept for *Space Infrared Telescope for Cosmology and Astrophysics* (*SPICA*). The SPICA mission is a future Japanese infrared astronomical satellite, with launch envisioned in 2017, to explore the universe with a cooled, large telescope The philosophy of BLISS is to provide a rapid survey spectroscopy capability over the full far-IR range. The baseline approach is a suite of broadband grating spectrometer modules with transition-edge superconducting (TES) bolometers SPICA will use a cooled telescope (3.5 m diameter primary, ~5 K) to achieve sensitivities currently inaccessible to existing facilities operating over this wavelength range (SOFIA, Herschel)

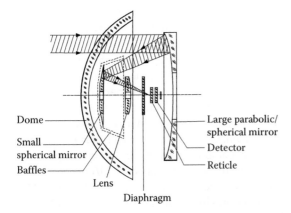

Figure 18.27 Optical diagram of a typical passive nonimaging seeker.

background like typical ground military targets or in the presence of countermeasures. Therefore, the trend of future systems is toward passive imaging seekers.

Passive imaging seekers typically have a thermal imager (SWIR camera or VIS/NIR camera) in their optoelectronic head. The location of a target is determined from analysis of the image generated by the imaging module. Thermal imagers or SWIR imagers, due to their ability to generate high contrast images in difficult atmospheric conditions, are preferred solutions. VIS/NIR cameras are rarely used due to their vulnerability to atmospheric conditions and problems in recognizing low contrast targets. The most advanced imaging seekers in fact use two imaging modules: the

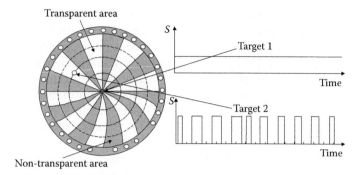

Figure 18.28 Exemplary reticle and the signal generated at the detector output by a few targets at different location.

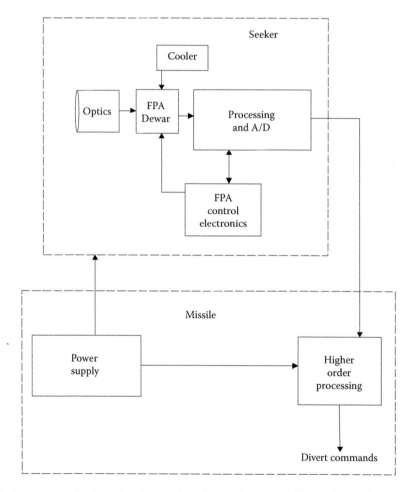

Figure 18.29 Representative imaging (staring) seeker architecture. (Reproduced from Miller, J. L., *Principles of Infrared Technology*, Van Nostrand Reinhold, New York, 1994.)

MWIR imager and the SWIR imager or the MWIR imager and the UV imager.

A representative architecture of a staring seeker is shown in Figure 18.29. To keep seeker volume, weight and power requirements low, only the minimum hardware needed to sense the scene is included. From the figure, it can be seen that the seeker's output goes into a missile-based processor, behind the FPA in the seeker, to perform tracking and aimpoint selection. The concept of the seeker's operation includes a standby turn-on, followed by a commit, which cools the FPA. At the beginning, the seeker is locked onto its target by an external sensor or a human. Next, the missile is launched and flies out locked onto its target, matching any target movement. Finally, when the target is close and imaged, the missile chooses an aimpoint and conducts final maneuvers to get to the target or selects a point and time to fuse and explode.

Due to the use of advanced image analysis, imaging seekers can attack targets located in nonuniform background and in the presence of advanced countermeasures. Thus, some of the modern air-to-ground missiles use this method to attack and destroy ground targets; particularly, large nonmovable targets like bridges, bunkers, and buildings. However, there are some techno-commercial challenges. First, it is difficult to design small size, reliable MWIR/SWIR imagers capable of generating high quality images under harsh environmental conditions that are met in most smart munitions (ultrahigh acceleration, extreme temperatures). Second, there is also a commercial challenge in designing passive imaging seekers as such seekers cannot be too expensive because they are disposable devices. Therefore, imaging modules used in passive imaging seekers are often strongly simplified versions of typical thermal imagers/SWIR imagers or UV imagers used for surveillance applications. Imaging modules that generate low resolution image (below 320×240 pixels—see Figure 18.30) are frequently

Figure 18.30 Image of an aircraft generated by a low resolution MWIR imager (320 × 240 pixels).

met. Due to the aforementioned technological problems, passive imaging-based MWIR/SWIR/UV modules in their optical heads are rarely used and are still at a development stage.

Active electro-optical seekers typically use lasers that operate in the SWIR range (1064 nm or in 1550 nm band). These lasers emit radiation that irradiate the target of interest and enable homing on this target.

Active laser guided seekers can be divided into two subclasses based on their method of homing. The first group are seekers homing in on the target irradiated by a target designator using radiation reflected by the target (Figure 18.31a). The second group are seekers in homing on the target indicated by a target designator using the radiation of a laser beam that irradiates back of the seeker head (Figure 18.31b).

Seekers homing in on the target irradiated by pulsed laser designators are the most popular group of active electro-optical seekers. These seekers enable very accurate location of small targets in a highly nonuniform background and are particularly well suited for air-to-ground missiles or bombs. However, warning systems or other countermeasures can significantly reduce the effectiveness of these seekers because it is relatively easy to detect being illuminated by pulsed laser radiation.

Seekers irradiated with a laser beam are kept on their flight to the target within the beam emitted by the laser designator that irradiates the target. Laser radiation, which gives information on target location, comes directly from the illuminator to the sensors at the back of the missile, not after the reflection by the target as in the previous method. Therefore, low-power designators can be used here and the effectiveness of the warning systems is reduced. However, a big drawback of this group of seekers is a fact that the seeker must be guided by aircraft that shoot the missile until it hits the target.

Munitions, and particularly missiles, that use smart seekers, are very expensive. The typical price is in the range of over one hundred thousand US$, sometimes even in the millions of US$. It is a fair price to use such disposable weapons to destroy enemy aircraft, helicopters, and tanks but there are doubts about the economic sense of using such expensive weapons in asymmetric wars against lightly armored guerillas. Therefore, low-cost smart munitions are a noticeable trend in smart weapon technology. Another big trend is seekers of improved intelligence capable of finding the

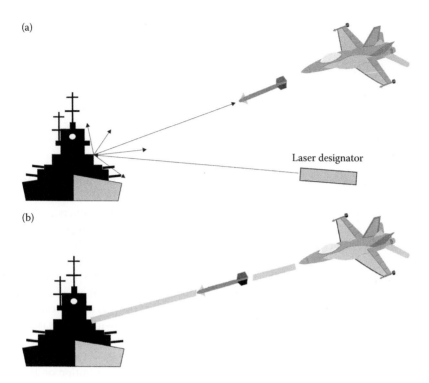

Figure 18.31 Principle of work of active laser guided seekers: (a) seeker homing in on the irradiated target and (b) seeker irradiated with a laser beam.

true target against modern countermeasures. Dual passive imaging seekers that employ two spectral bands (MWIR/LWIR or MWIR/UV) to eliminate countermeasures are pioneer examples of new more advanced seekers. Active laser seekers that combine classical laser homing methods with an additional imaging module are another example of the new generation of advanced smart seekers.

The new generation of standoff weapons relies on real-time target recognition, discrimination, tracking, navigation, and night vision. It is predicted that smart weapons will tend to replace the radar emphasis as stealth platforms are increasingly used for low-intensity conflicts. It is more difficult to perform IR missile warning than radar-guided missile warning.

18.4 NONCONTACT THERMOMETERS

IR thermometers measure temperature indirectly in two stages [21]:

- Measurements of radiation power in one or more spectral bands

- Determination of an object temperature on the basis of the measured radiometric signals

Even simple IR thermometers usually consist of five or more blocks (see Figure 18.32). An optical objective is usually used to increase the amount of radiation emitted by the tested object and limit thermometer FOV. The signal at the output of the detector is typically amplified, converted into a more convenient electronic form, and finally digitized. A separate visualization block is typically used for presentation of the measurement results.

IR thermometers can be divided into a few groups according to different criteria: presence of an additional co-operating source, number of system spectral bands, number of measurement points, width of system spectral bands, and transmission media.

In a passive system, the object temperature is measured knowing the radiation power emitted by the object in one or more spectral bands. With an active system, we can get some information about the emissive properties of the tested object by using an additional co-operating source that emits radiation directed at the tested object and measuring the

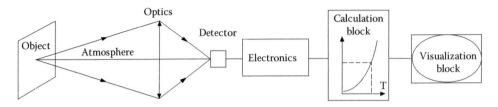

Figure 18.32 General diagram of a simple noncontact thermometer.

reflected radiation. Active thermometers are more sophisticated and more expensive. So far, only in a few applications can they really offer better accuracy than passive systems. Therefore, nowadays, almost all practical noncontact thermometers are passive ones.

In the passive single-band systems, the object temperature is determined using a system calibration chart derived from radiometric calculation of the output signal as a function of black body temperature. The temperature of nonblack body objects can be corrected only if their emissivity over the spectral band is known.

The ratio of the power emitted by a graybody at two different wavelengths does not depend on the object emissivity but only on the object temperature. In passive dual-band systems, the object temperature is usually determined using a calibration chart that represents a ratio of the emitted power in two bands as a function of the object temperature.

At present, at least 99.9% of systems available commercially on the market are passive single-band systems; passive dual-band systems are rather rarely used; passive multiband systems are still at a laboratory stage of development.

According to the number and location of measurement points, IR thermometers can be divided into pyrometers, line scanners, and thermal cameras.

Pyrometers enable temperature measurement of only a single point or rather of a single sector (usually a circle or a square) of the surface of the tested object (see Figure 18.33a). Line scanners enable temperature measurement of many points located along a line. Scanners are typically used to measure temperature of moving objects (see Figure 18.33b). Due to the movement of the target of interest, the scanners can create a two-dimensional image of the temperature distribution on the surface of the tested object. Thermal cameras enable temperature measurement of thousands of points located within typically a rectangle or square and create a 2D image of the temperature distribution on this area.

Most commercially available noncontact thermometers are pyrometers. They are small, light, and low-cost systems that have found numerous applications in industry, science, etc. Majority of them are handheld, quasi-universal devices integrated with a laser sight as shown in Figure 18.34a [22]. On the market, there are also compact, rugged, industrial

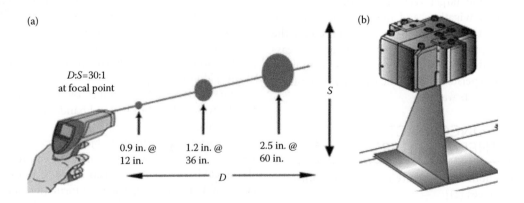

Figure 18.33 Principle operation of (a) pyrometer and (b) line scanners.

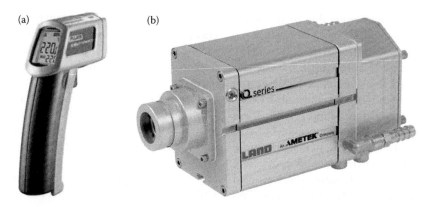

Figure 18.34 Photos of two pyrometers: (a) Fluke 62 handheld pyrometer (http://www.fluke.com) and (b) Land IQ industrial pyrometer.

pyrometers optimized for harsh industrial conditions (see Figure 18.34b).

Pyrometers offer noncontact temperature measurement in the range from about −40°C to over 2000°C using different spectral bands from the LWIR band to ultra narrow bands in the visible or near-IR ranges. Most pyrometers are devices optimized for measurement of low temperatures (from about −20°C to 250°C) using thermopile detectors.

Line scanners are especially suitable for temperature measurement of moving objects and have found applications in the automotive industry, welding, robotics, etc. Unlike pyrometers—that measure a single point—line scanners measure very quickly multiple temperature points across a scan line, with high resolution and in ultra-wide FOV. On the market, there are line scanners

capable of measuring 150 lines per second, with 1024 pixels in a single line, and in 90° FOV (see Figure 18.35) [23]. High measurement speed, high image resolution, and ultra-wide FOV are reasons why line scanners are able to withstand competition from thermal cameras.

Thermal cameras offer the greatest capabilities of all the discussed types of noncontact thermometers. For several decades, high price was a factor that limited significantly mass applications of thermal cameras. In the past, thermal cameras were divided into two basic groups: MWIR thermal cameras based on thermoelectrically/Stirling/liquid nitrogen cooled FPAs sensitive in the 3–5 μm spectral band and LWIR thermal cameras based on Stirling/liquid nitrogen cooled FPAs sensitive in the 8–12 μm band. MWIR

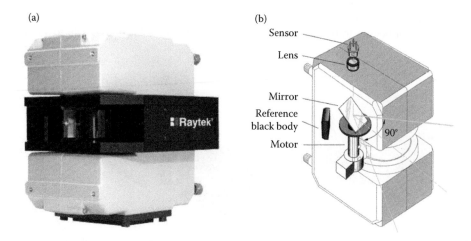

Figure 18.35 MP150 line scanner from Raytek: (a) photo and (b) block diagram.

cameras were more numerous due to the lower cost of these sensors.

Emergence at the beginning of the 2000s of low-cost LWIR noncooled arrays (microbolometers) has totally changed the situation. Nowadays, the market is totally dominated by noncooled LWIR thermal cameras (see Figure 18.36) [24,25]. These cameras have found numerous applications such as control of electrical supply lines, heat supply lines, in civil engineering, environmental protection, nondestructive testing, and many others. Cooled thermal cameras built using MWIR or LWIR FPAs are mostly met in high end expensive applications like spectral selective measurement of metals, polymers, glasses, through flames, etc.

LWIR/MWIR thermal cameras can potentially enable measurement of temperature up to 3000°C but cameras that operate at the SWIR/NIR spectral range can offer much lower internal measurement errors at higher temperatures over about 400°C. Costs of SWIR cameras is much higher than cost of VIS/NIR cameras built using low-cost CMOS/CCD sensors. Therefore, thermal cameras using a narrow spectral band in the NIR range dominate among thermal cameras optimized for testing targets at high temperatures over about 600°C (see Figure 18.36b). The latter thermal cameras should be treated as special versions of typical CCD/CMOS cameras optimized for radiometric measurements.

It should be noted that LWIR/MWIR thermal cameras dedicated for noncontact temperature measurements differ significantly from LWIR/MWIR thermal cameras for surveillance applications. Different FPA sensors, different electronics and optics are typically used to design thermal cameras that belong to these two groups.

Moreover, speed of measurement for thermal cameras is typically kept below about nine frames per second to fulfil requirements on thermal cameras for nonmilitary applications kept by export control agencies.

Noncontact thermometers such as pyrometers, line scanners, and thermal cameras represent a technology that has found mass application in almost every form of human activity. The number of these devices sold on the market is increasing every year. The price varies from several dozen US$ for some simple pyrometers to tens of thousands of US$ for high-end thermal cameras.

However, in spite of the high number of available thermometers, accurate noncontact measurement of temperature is a real challenge in many applications. The reason is that though internal errors of typical noncontact temperature meters are relatively low (below 1%–2%), external errors due to improperly estimated emissivity of the tested object, radiation reflected by this object, or due to noncorrected atmospheric attenuation and emission, can be ten times (or more) higher than the internal errors. Therefore, at least a basic knowledge about the methods of noncontact temperature measurement is crucial to keep these external errors at a minimal level.

18.5 RADIOMETERS

In general, the IR thermometers discussed in Section 18.4 can be treated as a class of radiometers because they determine temperature on the basis of the signal generated by the radiant flux coming to the detector. However, IR thermometers are designed to measure only temperature and it is usually not possible to use them to measure radiant

(a)　　　　　(b)

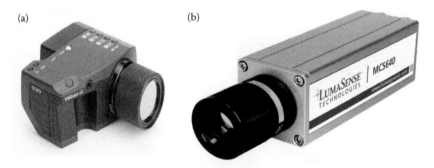

Figure 18.36 Photos of two thermal cameras: (a) VigoCAM5 thermal camera built using a 640×480 LWIR array of bolometers and (b) MSC640 thermal camera built using an NIR CMOS imaging sensor.

flux. In our definition, a radiometer is an instrument designed to measure quantities of optical radiation or radiant properties of materials in the optical range from UV to far IR. It should be noted that photometric meters can be considered as specific radiometers optimized to measure quantities of visible radiation (light) when the receiving sensor simulates spectral sensitivity of the human eye. However, in this chapter, photometers are excluded from further analysis.

Optical radiometers can be divided into several groups according to three main criteria:

• Number of measured quantities
• Number of spectral bands
• Number of measurement points

A universal radiometer should enable measurement of a long list of quantities of optical radiation (radiant power, radiant energy, radiant intensity, radiance, irradiance, radiant exposure) and measurement of radiant properties of materials (like emissivity, reflectance, and transmittance) in a set of spectral bands of optical radiation (UV,VIS, NIR, SWIR, MWIR, LWIR, VLWIR).

It is technically possible to develop a universal radiometric system. However, such a test system would be bulky, modular, composed from many blocks, and extremely expensive. Therefore, there are no truly universal radiometers on a market. However, semi-universal radiometric systems of more limited performance are commonly found on the market. These radiometers are based on a concept of a quasi-universal electronic meter with several exchangeable optical heads (see Figure 18.37).

Optical meters capable of measuring power, pulse energy, and irradiance of VIS/NIR/SWIR radiation are the prime examples of radiometers from the latter group (see Figure 18.38). Their extreme versatility is achieved by a modular approach coupled with an extensive selection of accessories and powerful application of software packages that enables users to tailor a turnkey system to their exact requirements as well as insure expandability in the future. Another even more popular type of radiometers are specialized, compact optical meters optimized to measure a single optical quantity (power meters, pulse energy meter, wavelength meters, etc.).

According to the classification by number of spectral bands, optical radiometers can be divided into four groups: single-band radiometers, dual-band radiometers, multiband radiometers, and spectroradiometers.

Optical meters equipped with an optical head based on a single detector can be considered as single-band radiometers. If an optical meter is equipped with several optical heads of different

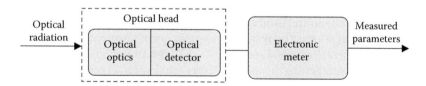

Figure 8.37 Block diagram of a typical semi-universal optical radiometer.

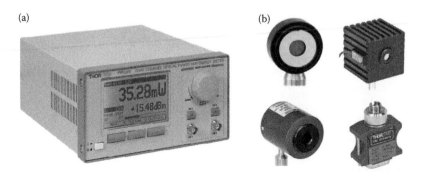

Figure 18.38 PM320E optical meter produced by Thorlabs Inc.: (a) photo of a universal electronic meter and (b) photo of several optical heads.

spectral bands, then such a device can be considered as a dual-band or multiband radiometer. When the spectral bands are narrow and their numbers is high enough, then the multiband radiometers are termed spectroradiometers. In contrast to the situation in noncontact thermometry, where multiband thermometers are very rarely met, spectroradiometers are a popular group of optical radiometers and have found wide areas of applications.

The key component of any spectroradiometer is a module that can be termed the spectral band selector. Its task is to select the desired spectral band from the incoming radiation. This task is achieved by using several methods: variable wavelength filters, monochromators, and Fourier transform (FT) interferometers.

The transmission wavelength of circular (linear) variable filters (VFs) changes continuously (discretely) with the position of the fraction of the filter. Simplicity of design is a great advantage of this type of spectroradiometer—it enables design of small size, reliable, high speed, and mobile systems. However, using the VFs, it is not possible to achieve good spectral resolution; resolution of spectroradiometers built using VFs is not better than about 2% of the wavelength.

A monochromator is an optical instrument that uses a dispersing component (a grating or a prism) and transmits to the exit slit (optionally directly to the detector) only a selected spectral fraction of the radiation incoming to the entrance slit. The center wavelength of the transmitted spectral band can be changed within the instrument's spectral region by rotating the dispersing element. Very good spectral resolution can be achieved using grating monochromators (as high as 0.1% of the wavelength). However, the monochromator transmits the optical radiation

only in a very narrow spectral band. Moerover, grating monochromators are designed using low speed optics. As a result, the optical detectors at the output of the monochromators receive only a very small fraction of the radiation incoming to the monochromator input. Therefore, spectroradiometers built using grating monochromators suffer poor sensitivity. Cooled optical detectors of ultrahigh sensitivity are often used in such spectroradiometers to reduce the aforementioned drawback, however with a limited effect.

The Michelson interferometer is the spectral band selector in FT spectroradiometers. The interferometer is usually built as an optical instrument consisting of a beam splitter and two flat mirrors arranged so as to recombine the two separated beams back on the same spot at the beam splitter. One of the mirrors moves linearly in order to produce variable optical interference (see Figure 18.39b) [26].

The Michelson interferometer can also be seen as a modulator. From a constant spectral radiation input, a temporal modulation occurs at the detector having a unique modulation frequency for each wavelength of radiation. The modulation frequency can be scaled via the velocity of the mirror movement. This modulated signal registered by the detector is called the interferogram. It is digitized at the rate of at least twice the maximum modulation frequency and a mathematical operation, the Fourier transform, is applied to retrieve the spectral distribution of the input radiation. A calibration with a known source is required in order to obtain quantitative radiometric results.

FT spectroradiometers are characterized by very good spectral resolution and very good sensitivity, better than offered by spectroradiometers built using grating monochromators or variable

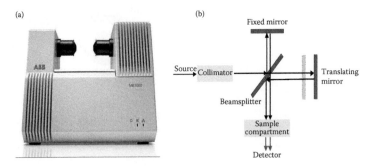

Figure 18.39 The ABB MB3000 Series FT-IR Laboratory Analyzer: (a) photo and (b) schematic block diagram (http://www.abb.com).

wavelength filters. These features enable design of high-speed, high spectral resolution FT spectroradiometers using noncooled or thermoelectrically cooled detectors (typically HgCdTe detectors) instead of the bulky liquid nitrogen cooled detectors needed in the VF or dispersive spectroradiometers. Therefore, FT spectroradiometers have found mass applications in many areas, but are particularly popular as spectral analyzers in industry.

Performance of the FT spectroradiometers can be severely reduced even by a very small nonalignment of the optical system, which makes them inherently sensitive to shocks and vibrations. Therefore, in the last few decades, FT spectroradiometers have been considered rather as laboratory type equipment that cannot be used in field applications. However, at present, this opinion is outdated as there are on the market fully mobile FT spectroradiometers that can be used in military, meteorological, and environmental applications.

From the point of view of measuring points, all radiometers can be divided into two basic groups: spot radiometers and imaging radiometers.

The great majority of commercially available radiometers are systems enabling measurement of the spectral distribution of radiation emitted/reflected or transmitted by a single spot and these systems can be termed spot radiometers. The radiometers that have been discussed so far belong to this group. The second group are imaging radiometers that generate images at one or more spectral bands and can be further divided into: single-band, dual-band, multiband, multispectral, and hyperspectral imaging radiometers.

Basically all electro-optical imagers (thermal imagers, SWIR imagers, VIS/NIR cameras, UV cameras) designed to measure absolute distribution of radiance on the surface of the analyzed target can be considered as single-band imaging radiometers.

Surveillance satellites are often equipped with multiband imaging radiometers that offers two-dimensional images of the earth at several spectral bands, for example, visible, near-IR, and long wavelength IR region. These imaging radiometers are often built as several single-band imaging radiometers that share common broadband optics.

Multispectral imaging radiometers are a special subgroup of multiband imaging devices that offer images at a significant number of relatively narrow spectral bands, but the spectral bands do not form a continuous spectrum. Landsat satellites are good examples of multispectral multiband imaging radiometers. The Landsat 8 satellite can deliver images in eleven spectral bands from visible to far IR (see Table 18.7) [27]. Please note that there are some empty spectral regions that are not covered by Landsat 8 satellite.

Hyperspectral imaging radiometers are the most advanced subgroup of the multiband imaging radiometers that deliver two-dimensional images at so high a number of narrow spectral bands that they form a continuous spectrum. In other words, these imaging radiometers deliver a spectrum for each pixel of generated images. The output data is typically shown in the form of so-called hyperspectral cubes (x, y, λ), where x, y are spatial coordinates and λ is wavelength (see Figure 18.40) [28].

Table 18.7 Spectral bands of Landsat 8 satellite

No.	Name of spectral band	Wavelength (µm)
1	Coastal aerosol	0.43–0.45
2	Blue	0.45–0.51
3	Green	0.53–0.59
4	Red	0.64–0.67
5	Near-IR	0.85–0.88
6	SWIR 1	1.57–1.65
7	SWIR 2	2.11–2.29
8	Panchromatic	0.50–0.68
9	Cirrus	1.36–1.38
10	Thermal infrared (TIRS) 1	10.60–11.19
11	Thermal infrared (TIRS) 2	11.50–12.51

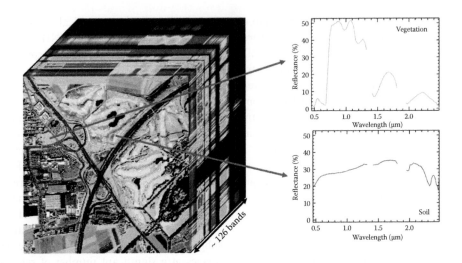

Figure 18.40 Hyperspectral data cube acquired with an imaging hyperspectral radiometer HyMap.

Hyperspectral imaging radiometers can be designed using four distinct methods of acquiring hyperspectral cubes (x, y, λ):

- Spatio-spectral scanning
- Spatial scanning
- Spectral scanning
- Nonscanning

The spectral scanning system and spatial scanning system are needed in hyperspectral imaging radiometers built using spatio-spectral scanning method. An imaging radiometer built as a scanning grating monochromator with a linear imaging detector at its exit slit moving in space can be treated as an example of this method. The radiometer originally generates only a linear spatial image at a single wavelength but a two-dimensional spatial image is achieved by moving the radiometer; data about the complete spectrum is achieved due to scanning of the grating monochromator. A very long time period is needed to acquire a two-dimensional spatial image at complete spectrum (complete hyperspectral cube (x, y, λ) for all combinations of the coordinates x, y, λ is shown in Figure 18.41) because spectral scanning along the λ-axis must be repeated for many discrete spatial values of the y-axis assuming that the linear detector is oriented along the x-axis. The acquisition time can be shortened by many times if both spectral scanning and spatial scanning work at the same time but then this method generates an incomplete hyperspectral cube (x, y, λ) because information about all combinations of x, y, λ is not delivered.

Hyperspectral imaging radiometers built using the spatial scanning method need only the spatial scanning system that creates a two-dimensional spatial image from the original one-dimensional spatial image. A nonscanning grating monochromator with a 2D imaging detector at its exit plane moving in space is an example of this method. This imaging detector captures a spectrally dispersed image of a linear target. Two-dimensional image of the scenery of interest is achieved by moving the radiometer. This method is particularly suited for space radiometers due to their natural movement relative to the earth.

Hyperspectral imaging radiometers built by using a spectrally tunable optical filter placed before the FPA are example of radiometers using spectral scanning. At one discrete moment of time, this imaging radiometer generates a complete 2D image of the observed scenery. Spectral scanning due to filter tuning generates a series of such 2D spatial images for different wavelengths.

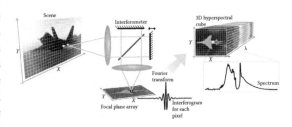

Figure 18.41 Spectral scanning method used by hyperspectral imaging FT spectroradiometers.

Different types of optical tunable filters are used: liquid crystal, acousto-optic, or thin-film filters tunable by adjusting the angle of incidence. Light weight, small dimensions, and low power requirements are a big advantage of imaging radiometers built using tunable optical filters. However, these imaging radiometers suffer from rather poor sensitivity, poor spectral resolution, and too narrow spectral band. These disadvantages are not present in imaging FT spectroradiometers that can also be considered as a special case of imaging radiometers built using spectral scanning method. These imaging spectroradiometers generate a 2D spatial image of the scene of interest and an interferogram for each pixel of this image. Mathematical transformations of the acquired image data generate a hyperspectral 3D cube as shown in Figure 18.41 [26].

Spectral scanning in FT imaging spectroradiometers is indirectly done by moving a mirror in the interferometer. The captured interferogram can be converted into a spectrum using a Fourier transform. FPA receives the full spectrum emitted by the scene of interest and this feature makes this method particularly sensitive. Ultra high spectral resolution and wide spectral ranges can be achieved, too.

The spectral scanning method is optimal for typical applications at earth conditions where both the radiometer and the target are static or are moving but not along the desired axis. Therefore, these imaging radiometers dominate on the market.

All three methods discussed so far need a relatively long time (hundreds times more than the time needed to create a single image frame by electronic imaging sensors) to generate the hyperspectral cube and this long acquisition time is a significant disadvantage of these methods. The nonscanning method can potentially deliver a complete 3D hyperspectral cube at a time period comparable to the integration time of electronic imaging sensors (below 50 ms). Next, nonscanning hyperspectral imaging radiometers can potentially measure a spectrum of the scenery that emits low amount of optical radiation due to higher light efficiency. This method called "snapshot hyperspectral imaging" can potentially bring a revolution in hyperspectral imaging comparable to the replacement of scanning thermal imagers based on discrete/linear IR detectors with staring thermal imagers built using two-dimensional IR FPAs [29]. However, there are many technological problems that have to be faced while practically implementing this method in commercial products. So far, there is a significant interest in nonscanning imaging radiometers but there are no such imaging radiometers offered on the international market.

As it was stated earlier, successful design of a nonscanning hyperspectral imaging radiometer is a big technological challenge. However, design of scanning hyperspectral imaging radiometers is difficult, too. Hyperspectral imaging radiometers differ significantly from nonimaging spectroradiometers. The first systems must use a modified optical system with a corrected curved output field and astigmatism to generate sharp images of the input slit corresponding to the wavelength focused on different parts of the array detector, which enables simultaneous measurement of the spectrum. There are also big differences in complexity of electronics and software needed for data processing. Therefore, hyperspectral imaging can be considered as the most technologically challenging part of optical radiometry.

18.6 LIGHT DETECTION AND RANGING (LIDAR)

LIDAR is an acronym for light detection and ranging, an active optical remote technique in which a beam of light is used to make range-resolved measurements. The LIDAR emits a light beam that interacts with the medium or object under study in much the same way that sonar uses sound pulses, or radar uses radio waves. In radar, radio waves are transmitted into the atmosphere, which scatters some of the power back to the radar's receiver. Similar to radar technology, the range to an object is determined by measuring the time delay between transmission of a pulse and detection of the reflected signal. The LIDAR also transmits and receives electromagnetic radiation, but at higher frequency. Analysis of the backscattered light allows some property of the medium or object to be determined. LIDARs operate in the UV, visible, and IR region of the electromagnetic spectrum. Typically, wavelengths vary to suit the target from about 10 μm to approximately 250 nm.

LIDAR is popularly used as a technology to make high-resolution maps, with applications in agriculture, geomatics, archeology, geography, geology, seismology, forestry, physics, astronomy (remote sensing and metrology), etc. Few military applications are known to be in place and are classified, but a considerable amount of research is underway in their use for imaging. Higher resolution systems collect enough detail to identify targets, such as tanks. Examples of military applications of LIDAR include the airborne laser mine detection system for counter-mine warfare. At present, LIDAR systems are also used to do standoff detection for the discrimination of biological warfare agents and to provide the earliest possible standoff warning of a biological attack.

LIDAR was developed in the early 1960s, shortly after the invention of the laser, and it was initially used for mapping particles in the atmosphere. During the 1980s, the development of GPS (Global Positioning System) opened up the applications to moving sensors (airborne LIDAR). By the early 1990s, decimeter accuracies were achievable.

In the LIDAR approach, a laser radiation is transmitted into the atmosphere and backscattered radiation is detected as a function of time by the optical receiver. The return time of the reflected or scattered pulses provides range information. In a LIDAR arrangement, the backscattered light is collected by a telescope, usually placed coaxially with the laser emitter. The signal is then focused on a photodetector through a spectral filter, adapted to the laser wavelength.

A LIDAR instrument can be conveniently divided into three subsystems [30]: the transmitter, the receiver, and the detector (see Figure 18.42). The transmitter is the subsystem that generates light pulses and directs them into the atmosphere. The receiver collects and processes the scattered laser light and then directs it to a photodetector.

The signal detection and recording section takes the light from the receiver system and produces a permanent record of the backscattering intensity, and possibly wavelength and/or polarization, as a function of altitude. In modern LIDARs, detection and recording are achieved electronically.

Different kinds of lasers are used depending on the power and wavelength required. The lasers may be either continuous wave or pulsed. Inexpensive lasers emitting light in the spectral range between 0.6 and 1.0 μm are the most common for nonscientific LIDAR applications. Their maximum power is limited by the need to make them eyesafe (visible light can be focused and is easily absorbed by the eye). Eye safety is often a requirement for most applications. Common alternatives are 1.55 μm lasers, which are eyesafe at much higher power levels since this wavelength is not focused by the eye. Airborne topographic mapping LIDARs generally use 1.064 μm diode pumped YAG lasers, while bathymetric systems generally use 0.532 μm frequency doubled diode pumped YAG lasers (this light penetrates water with much less attenuation than does 1.064 μm). Laser settings include the laser repetition rate (which controls the data collection speed). Better target resolution is achieved with shorter pulses, provided the LIDAR receiver detectors and electronics have sufficient bandwidth.

In general, there are two kinds of LIDAR detection schemes: "incoherent" or direct energy detection (which is principally an amplitude measurement) and coherent detection (which is best for Doppler, or phase sensitive measurements). Coherent systems being more sensitive than direct detection, allows

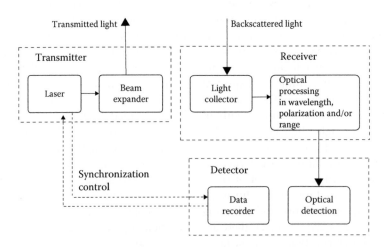

Figure 18.42 Schematic of a simple LIDAR.

them to operate at a much lower power but at the expense of more complex transceiver requirements.

Two main photodetector technologies are used in LIDARs: solid-state photodetectors, such as silicon avalanche photodiodes or photomultipliers (visible and UV). 3D imaging can be achieved using both scanning and nonscanning systems. "3D gated viewing laser radar" is a nonscanning laser ranging system that applies a pulsed laser and a fast gated camera. Imaging LIDAR can also be performed using arrays of high-speed detectors and modulation sensitive detector arrays typically built on single chips using CMOS and hybrid CMOS/CCD fabrication techniques. Using this technique, many thousands of pixels/channels may be acquired simultaneously.

Airborne topographic LIDAR systems are the most common LIDAR systems used for generating digital elevation models for large areas. The combination of an airborne platform and a scanning LIDAR sensor is an effective and efficient technique for collecting elevation data across tens to thousands of square miles. A basic LIDAR system involves a laser range finder reflected by a rotating mirror. The laser is scanned around the scene being digitized, in one or two dimensions, gathering distance measurements at specified angle intervals. The schematic idea shown in Figure 18.43 is fairly straightforward [31]: measure the time that it takes a laser pulse to strike an object and return to the sensor (which itself has a known location due to direct georeferencing systems), determine the distance using the travel time, record the laser angle, and then, from this information, compute where the reflecting object (e.g., ground, tree, car, etc.) is located in three dimensions.

LIDAR systems allow scientists and mapping professionals to examine both natural and artificial environments with accuracy, precision, and flexibility. For example, Figure 18.44 shows LIDAR collected by the National Oceanic and Atmospheric Administration (NOAA) survey aircraft (top) over Bixby Bridge in Big Sur, California [32]. Here, LIDAR data reveals a top-down (bottom left) and profile view of Bixby Bridge.

Typically, light is reflected via backscattering. Different types of physical processes in the atmosphere are related to different types of light scattering. Choosing different types of scattering processes allows atmospheric composition, temperature, and wind to be measured. The scattering is essentially caused by Rayleigh scattering on nitrogen and oxygen molecules, Mie scattering on aerosols (dusts, water droplets, etc.), Raman scattering, and fluorescence. At low attitudes, Mie scattering is predominant because of the higher cross section and the high aerosol concentration. Based on different kinds of backscattering, the LIDAR can be accordingly called Rayleigh LIDAR, Mie LIDAR, Raman

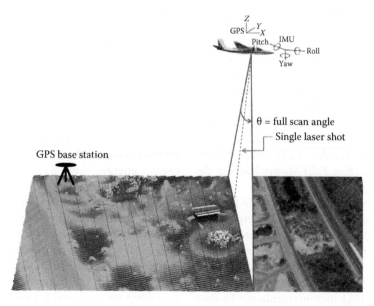

Figure 18.43 Schematic diagram of an airborne LIDAR performing line scanning resulting in parallel lines of measured points.

Figure 18.44 LIDAR data is often collected by NOAA survey aircraft (top) over Bixby Bridge in Big Sur, California.

LIDAR, Na/Fe/K fluorescence LIDAR, and so on. Suitable combinations of wavelengths can allow for remote mapping of atmospheric contents by identifying wavelength-dependent changes in the intensity of the returned signal.

In general, a signal is produced by direct absorption, fluorescence, or Raman scattering. Absorption techniques are most straightforward and widely applied. In the atmosphere, for example, long-path absorption spectrometry is used in two wavelength bands—the IR, where many molecules have characteristic fingerprints, and the UV to visible range.

In a typical case, the laser is alternatively tuned to a wavelength within the absorption band of interest (at λ_{on}) and then to a wavelength with negligible absorption (at λ_{off}), so that difference in the signal returned either from a surface or from airborne or waterborne particles is recorded. By dividing the two LIDAR signals by each other, most troublesome and unknown parameters are eliminated and the gas concentration as a function of the range along the beam can be evaluated. Such applications require tunable lasers, either tunable diode lasers in the IR or Nd:YAG dye lasers in the UV to visible range.

Differential absorption LIDAR (DIAL) measurements utilize two or more closely spaced (<1 nm) wavelengths to factor out surface reflectivity as well as other transmission losses, since these factors are relatively insensitive to wavelength. When tuned to the appropriate absorption lines of a particular gas, DIAL measurements can be used to determine the concentration (mixing ratio) of that particular gas in the atmosphere.

The principle for different absorption LIDAR (DIAL) is schematically represented in Figure 18.45. Let us now assume that a wavelength couple (λ_{on}, λ_{off}) is sent simultaneously into the atmosphere. As λ_{on} and λ_{off} have been chosen close enough to exhibit the same scattering properties, the first chimney plume will cause an increase in the backscattering signal, because the concentration of aerosols is larger, but the same increase for both pulses. Conversely, the second chimney plume, which contains a certain quantity of the pollutant, will absorb the backscattered signal at the λ_{on}-wavelength much more strongly than at the λ_{off}-one. From this difference, and using the Beer–Lambert law, one can deduce the specific concentration of the pollutant under investigation versus range. For typical pollutants, such as sulfur dioxide, nitrous oxide, ozone, and mercury, detection ranges for the part-per-billion detection level are between 0.5 and 5 km.

The main alternative to direct-absorption spectrometry is Raman scattering. This occurs when photons are inelastically scattered from molecules, exciting them in the process and releasing some photon energy. Thus, Raman return signals are at a different, longer wavelength than the exciting wavelength. Raman cross sections are much smaller than absorption cross sections, so the Raman technique works well using high-power lasers only for higher concentrations (hundreds of parts per million) and distances of less than 1 km. Water vapor profiles can be obtained in vertical soundings up to several kilometers in height, and pressure profiles up to tens of kilometers are measurable using Raman signals from atmospheric N_2.

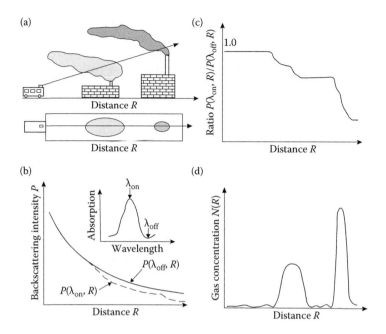

Figure 18.45 Illustration of the principle of different absorption LIDAR (DIAL): (a) pollution measurement situation, (b) back-scattered laser intensity for the on- and off-resonance wavelengths, (c) ratio (DIAL) curve and (d) evaluated gas concentration. (Reproduced from Svanberg, S., *Applied Laser Spectroscopy*, Plenum, New York, 1990.)

The other major technique, fluorescence spectrometry, has limited use in atmospheric measurement because the return signal is too weak. The technique is, however, an excellent way to monitor solid targets in the biosphere, such as oil, spills, algae bloom patches, and forest area. The fluorescent signal from plants originates from the excitation of chlorophyll and other leaf pigments. Fluorescence LIDAR is also a powerful technique for measurements at mesospheric heights where the pressure is low and the fluorescence is not quenched by collisions. This technique has been used extensively to monitor layers of various alkali and alkaline earth atoms (Li, Na, K, Ca, and Ca$^+$) at a height of about 100 km [33,34].

In addition to monitoring pollutants, LIDAR is widely used to measure wind velocities via Doppler shifts. The improved laser stability has resulted in LIDAR being applied to more ambitious projects, including the study of winds in the stratosphere. Doppler systems are also now beginning to be successfully applied in the renewable energy sector to acquire wind speed, turbulence, wind veer, and wind shear data.

The main advantage of LIDAR is that it can map the location of chemicals over a wide region.

Due to the rapid nature of laser pulses, the time resolution is critical (a few nanoseconds) to get good spatial resolution. Overall, DIAL systems can provide 2D or 3D information of air pollutants. However, most of the existing LIDAR systems have not met the pragmatic deployment requirements of users in industry or government. LIDAR systems are usually complex, large, expensive, and require highly skilled personnel for their operation.

18.7 IR GAS SENSORS

The late 19th and early 20th century saw the observation spectra of various gases, even the resolving of the rotational fine structure associated with certain simple molecules. Measurements were, however, very difficult due to the lack of suitable devices. A major advance occurred with the development of nondispersive IR (NDIR) technique in 1943 [35], when useful IR detectors became available (in the 1950s), and after the development of the multilayer thin film interface filter for wavelength selection in the IR region (in the 1950s). At present, IR gas detection is a well-developed measurement technology. In general, these instruments are among the most user friendly and require the least amount of maintenance.

Optical gas detection using absorption spectroscopy is based on the application of the Beer–Lambert law:

$$I = I_0 \exp(-\alpha l), \qquad (18.10)$$

where I is the light transmitted through the gas cell, I_0 is the light incident on the gas cell, α is the absorption coefficient of the sample (typically with units of cm^{-1}), and l is the cell's optical path length (typically with units of cm). The absorption coefficient α is the product of the gas concentration (e.g., in atm—the partial pressure in atmospheres) and the specific absorptivity of the gas (e.g., in cm^{-1} atm^{-1}). Typical absorption spectra for five gases in the IR spectral region are shown in Figure 18.46.

The Beer–Lambert law applies for monochromatic radiation; when using light sources that are broader than absorption lines, the width of the source must be accounted for. The law also assumes that there are no chemical changes in the sample—at high concentrations, dimer formation can alter spectra, but this is a minor effect for most gases at standard temperature and pressure.

The most commonly used techniques in gas sensing based on measurement of optical absorption at specific wavelengths are: NDIR technique, spectrophotometry, tunable diode laser spectroscopy (TDLS), and photoacoustic spectroscopy (PAS) [36].

The NDIR sensor, commonly referred simply as the IR sensor, can detect gases in inert atmospheres, are not susceptible to poisons, and can be made very specific to a particular target gas. The limitation of NDIR technology for gas detection is dependent on the uniqueness of the absorption spectrum of a particular gas. NDIR sensors are also extremely stable, quick to respond to gas, and can tolerate long calibration intervals. IR sensors are commonly used to detect methane, carbon dioxide, and nitric oxides in both portable and fixed gas detection instrumentation.

There are a number of ways by which various IR components can be arranged to produce a gas analyzer. IR gas sensors utilize only part of the IR spectrum, corresponding to wavelengths that are absorbed by the gas to be detected. The optical bandwidth of a laser source is sufficiently narrow for it to be used directly, but with wideband sources, as thermal sources or even light emitting diodes (LEDs), some additional wavelength selection in the optical path is required. The optical bandwidth of a sensor is typically in the tens or low hundreds of nanometers to match the absorption band of the gas. Wavelength selection is achieved using dispersive elements such as prisms or diffraction gratings, or nondispersive element such as a multilayer thin film filter.

The design may be relatively simple, or very complicated depending on the type of analyzer used for the applications. Figure 18.47 illustrates some of the basic features of IR analyzers.

The basic design is shown in Figure 18.47a, which consists of an IR source, band-pass filter, and the interaction with the gas sample and the detector. The detector is selectively sensitized to the absorption wavelength of the gas whose presence is to be detected by the use of a narrow-band-pass

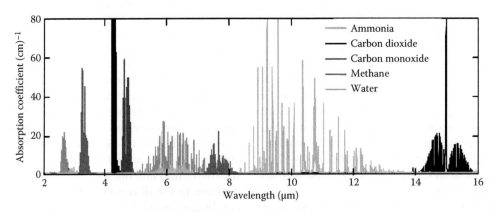

Figure 18.46 Absorption spectra for five gases in the mid-IR region of the spectrum (all at 100% vol).

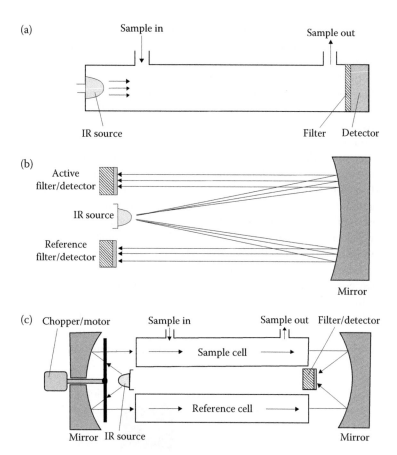

Figure 18.47 Configurations of gas analyzers: (a) a basic gas detector layout, (b) a two-detector layout, and (c) double beams with chopper layout. (Reproduced from Chou, J., *Hazardous Gas Monitors*, McGraw-Hill, New York, 2000.)

optical filter. Clearly then, increased gas concentration in the optical path between source and filtered detector leads to a depression in signal level. The band-pass filter could be placed in front of the light source, instead of placing it in front of the detector.

In practice, in order to reduce false alarms and introduce a level of quantification, it is necessary to provide some calibration. Dependent on the application and instrument manufacturer, this may take the form of a reference chamber containing a known concentration of the gas, or measurement of a reference wavelength just slightly outside the absorption band and/or dual matched detectors.

Figure 18.47b shows another popular design with two detectors. Modulated flashing IR sources are reflected back to the detectors. In this arrangement, the active detector has a filter for the target gas, while the reference detector has a filter with different wavelength. In such a way, the active

detector is used to detect the target gas and the reference detector is used to ignore the target gas. In actual operation, the reference detector provides a base point value (or zero point) while the active detector is used to provide the signal. An advantage of this design is compensation of changes that occur in the detector's sensitivity with time (e.g., change in the intensity of the light source).

The design illustrated in Figure 18.47c uses two tubes or cells. One is a reference cell that is filled with a pure target or reference gas, while the other is a sampling cell in which the sample gas passes through. Additionally, a chopper in the form of a disk with a number of slots in it is used. As the chopper rotates, it alternately allows the light beam to pass through the sample and reference cells. The detector gets its base reading from the reference cell.

There are many light sources available, ranging from a regular incandescent light bulb to specially designated heating filaments and electronically generated sources. The last sources are used to generate enough radiation at the wavelength of interest for the purpose of detecting the specific target gas. A heated wire filament, similar to that in a pen flashlight, is used in the 1–5 μm spectral range for the detection of most hydrocarbons, carbon dioxide, and carbon monoxide. Alternatives include glowbars (rods of silicon carbide) or coils, typically of nichrome alloy resistance wire with high emissivity in the MWIR region. Much recent research has been concentrated on the development of sources that are both more spectrally efficient and capable of more rapid modulation frequencies. The new sources can be categorized as thin incandescent membranes based on MEMS technology, some of which have engineered high emissivity surfaces and LEDs [36].

The selection of the optimum operating conditions for a gas analyzer is always a trade-off between the required sensitivity and an operational system.

In conventional absorption spectroscopy, using broadband incoherent radiation sources, the wavelength resolution is determined by the resolving power of the spectral analyzer or spectrometer. Laser absorption spectroscopy, on the other hand, uses coherent light sources, whose line widths can be ultra-narrow and whose spectral densities can be made many orders of magnitude larger (~10^9 W cm^{-2}MHz) than those of incoherent light sources. As can be seen from Figure 18.48, the whole spectral range from the visible to the IR can be covered by semiconductor lasers of various compounds, which mainly are gallium arsenide, indium phosphide, antimonides, and lead salts.

Up to the end of the twentieth century, a drawback for industrial applications of gas analyzers based on semiconductor lasers was the lack of high-quality, high-power laser diodes for many spectral regions of interest. These lasers relied on direct band-to-band transitions in the bulk material or analogous transition in type I quantum wells.

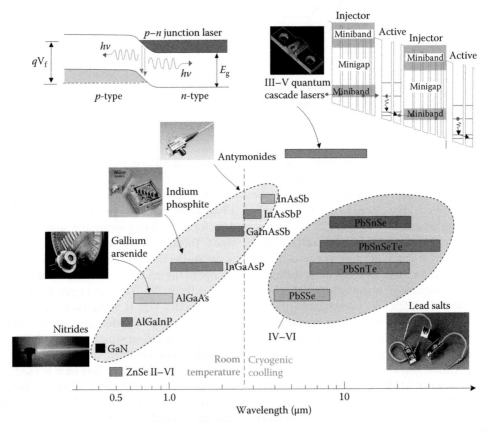

Figure 18.48 Coverage of the spectral range from the visible to the IR by tunable diode lasers of different material systems.

Since the mid-1990s, lead salt and antimonide lasers have been facing competition from quantum cascade lasers (QCLs) [38] and type II quantum well lasers (interband cascade lasers—ICLs) [39]. QCLs operate on transitions within the conduction subbands of multiple quantum wells and have been pioneered by Bell Laboratories. Operation in the IR region above room temperature has been reported. These new types of laser are available commercially/on research basis and they appear to offer the prospect of more robust construction and higher temperature operation than is possible with the materials that have been used up to this point. Trace-gas optical spectroscopic sensors using QCLs and ICLs are ultrasensitive and can detect molecular species at concentration levels from the percent level down to parts per trillion (ppt).

Today, TDLS is a widely used technique for environmental monitoring, remote sensing, and process gas analytics. It has rapidly become the most commonly used laser-based technique for quantitative measurements of species in the gas phase. It is a highly selective, sensitive, and versatile fast operation technique (up to MHz) for measuring trace species. The diode laser source is ideal for optical spectroscopy because of its narrow linewidth, tunability, stability, compactness, and ability to operate at room temperature.

The basic TDLS measurement setup is fairly simple, as shown in Figure 18.49—it contains a laser diode at the right wavelength that is tunable, a gas cell and a detector. The exact wavelength of the laser can be tuned slightly over the absorption line from ν_1 to ν_2 by changing the laser temperature and/or current. The laser light passes through the gas sample and the laser power transmitted through the sample is detected as a function of the laser wavelength. When the laser emission wavelength coincides with a resonant absorption in the molecule, we see a sharp absorption signal. Direct absorption measurements have to resolve small changes ΔI in a large signal offset I_0. The highest sensitivity is achieved at low gas pressure, when absorption lines are not substantially pressure broadened.

To increase the SNR, an additional noise suppression can be achieved by the application of modulation techniques. In modulation spectroscopy, the laser injection current is modulated at ω_m, while the laser wavelength is tuned repeatedly over the selected absorption line to accumulate the signal from the lock-in amplifier with a digital signal averager (Figure 18.50). This produces a derivative line profile with an amplitude proportional to the species concentration.

The benefits of modulation spectroscopy are twofold: (i) offsets are eliminated (zero baseline technique) as it produces a derivative signal, directly proportional to the species concentration and (ii) it allows narrow-band detection of the signal at a frequency at which the laser noise is reduced.

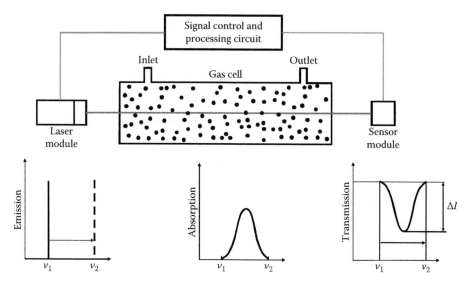

Figure 18.49 Basic setup for laser spectroscopy.

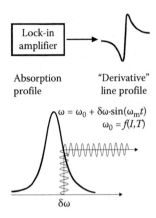

Figure 18.50 Wavelength modulation spectroscopy.

To provide high sensitivity, long-path absorption cells, L, are required. This method is often used in open path measurement techniques such as remote sensing, where L are usually from tens of centimeters to several meters (see Figure 18.51a).

However, most conventional trace-gas sensors have gas cells inside them. To increase their sensitivity and reduce the size of the apparatus, multipath cells are used (see Figure 18.51b). The laser light enters the gas cell and is reflected many times by mirrors that are located at either end of the cell. In this way, we can obtain a long effective light path of tens to hundreds of meters with a gas cell whose actual length is less than 1 m.

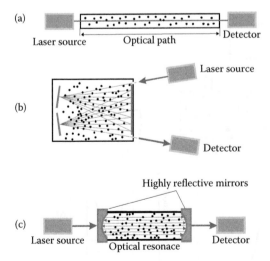

Figure 18.51 Different approaches of gas cells. (a) Long path, (b) multipath, and (c) optical cavity.

Another method that uses an optical cavity is shown in Figure 18.51c. This method utilizes optical resonance with highly reflective mirrors and traps laser light in a cell. We can obtain a very long effective light path that exceeds several kilometers.

18.8 CONCLUSIONS

This chapter is about general IR engineering, technology, practices, and principles as they apply to modern IR techniques. The information contained here is critical in the day-to-day life of engineering practitioners and those on the periphery of IR techniques. It serves rather as a guide for those wishing to "catch up" on the general information in the field—managers, students, and technicians. We abstain from the traditional development of equations and conclusions from first principles—they are presented in many other texts. Therefore, the chapter functions as a compilation of the state of the art, and provides background information for the reader devoted to IR devices and techniques. It is concentrated on the physical fundamentals of IR radiation and practical instruments for military applications, metrology, astronomy, environmental monitoring, surveillance, commercial, and industrial uses.

REFERENCES

1. Herschel, W. 1800. Experiments on the refrangibility of the invisible rays of the sun. *Philos. Trans. R. Soc. London* 90, 284–292.
2. Ross, W. 1994. *Introduction to Radiometry and Photometry* (Boston, MA: Artech).
3. Hudson, R. D. 1969. *Infrared System Engineering* (New York: Wiley).
4. Rogalski, A. 2010. *Infrared Detectors* (Boca Raton, FL: CRC Press).
5. IR Detector Market by Technology (MCT, InGaAs, Pyroelectric, Thermopile, Microbolometer), Application (People & Motion Sensing, Temperature Measurement, Fire & Gas Detection, Spectroscopy), Spectral Range (S/M/LWIR) & by Geography - Global Forecast to 2013–2020, http://www.electronics-ca.com/infrareddetectors-market-report.html..
6. Couture, M. E. 2001. Challenges in IR optics. *Proc. SPIE* 4369, 649–661.

7. Harris, D. C. 1999. *Materials for Infrared Windows and Domes* (Bellingham, WA: SPIE Optical Engineering Press).

8. Smith, W. J. 2000. *Modern Optical Engineering* (New York: McGraw-Hill).

9. Lloyd, J. M. 1975. *Thermal Imaging Systems* (New York: Plenum).

10. Kozlowski, L. J. and Kosonocky, W. F. 1995. Infrared detector arrays, in *Handbook of Optics*. Chapter 23, ed. M. Bass, E. W. Van Stryland, D. R. Williams and W. L. Wolfe, pp. 23.1–23.37 (New York: McGraw-Hill).

11. Mooney, J. M., Shepherd, F. D., Ewing, W. S., and J. Silverman. 1989. Responsivity non-uniformity limited performance of infrared staring cameras. *Opt. Eng.* 28, 1151–1161.

12. Chrzanowski, K. 2013. Review of night vision technology. *Opto-Electron. Rev.* 21, 153–182.

13. Hamamatsu. Photon is our Business, http://www.hamamatsu.com/resources/pdf/etd/II_TII0004E02.pdf.

14. Cameron, A. A. 1990. The development of the combiner eyepiece night vision goggle. *Proc. SPIE* 1290, 16–19.

15. Csorba, I. P. 1985. *Image Tubes* (Indianapolis, IN: Sams).

16. TopOwl®. Designed by pilots, for pilots, http://optronique.net/defense/wp-content/uploads/2012/07/TopOwl-Datasheet.pdf.

17. Miller, J. L. 1994. *Principles of Infrared Technology* (New York: Van Nostrand Reinhold).

18. STANAG No. 4349. *NATO Standardization Agreement: Measurement of the Minimum Resolvable Temperature Difference (MRTD) of Thermal Cameras.*

19. Campana, S. B. (ed.) 1993. *The Infrared and Electro-Optical Systems Handbook*, vol. 5, Passive Electro-Optical Systems (Bellingham, WA: SPIE Optical Engineering Press).

20. Far-Infrared/Submillimeter Astronomy from Space Tracking an Evolving Universe and the Emergence of Life. A White Paper and Set of Recommendations for The Astronomy & Astrophysics Decadal Survey of 2010, https://asd.gsfc.nasa.gov/cosmology/spirit/FIR-SIM_Crosscutting_White_Paper.pdf.

21. Chrzanowski, K. 2001. *Non-Contact Thermometry—Measurement Errors, Research and Development Treaties*, vol. 7 (Warsaw: SPIE Polish Chapter).

22. Infrared Temperature Measurement. Combustion & Environmental Monitoring, http://www.landinst.com.

23. Noncontact Infrared Temperature Measurement, http://www.raytek.com/Raytek/enr0/ProductsAndAccessories/InfraredLineScanners/InfraredLineScannersSeries/.

24. VigoCam5 Thermal Camera, http://www.vigo.com.pl.

25. MIKRON Cameras, http://www.lumasense.com.

26. ABB MB3000 PH High Performance and Versatile FT IR Spectrometer, http://www.abb.com.

27. Landsat 8 Overview, https://landsat.gsfc.nasa.gov/landsat-8/landsat-8-overview/.

28. The German Remote Sensing Data Center, http://www.dlr.de/eoc/en/desktopdefault.aspx/tabid-5278/8856_read-15911/.

29. Schlerf, M., Rock, G., Lagueux, P., Ronellenfitsch, F., Gerhards, M., Hoffmann, L., and Udelhoven, T. 2012. A Hyperspectral Thermal Infrared Imaging Instrument for Natural Resources Applications, *Remote Sens.* 4, 3995–4009.

30. Argall, P. S. and Sica, R. J. 2003. Lidar (Laser Radar), in *The Optics Encyclopedia. Vol. 2*, ed. Th. G. Brown, K. Creath, H. Kogelnik, M. A. Kriss, J. Schmit, and M. J. Weber, pp. 1305–1322 (Berlin: Wiley-VCH).

31. An Introduction to Lidar Technology, Data, and Applications, https://coast.noaa.gov/data/digitalcoast/pdf/lidar-101.pdf.

32. What is LIDAR? http://oceanservice.noaa.gov/facts/lidar.html.

33. Svanberg, S. 1990. Environmental monitoring using optical techniques, in *Applied Laser Spectroscopy*, ed. W. Demtröder and M. Inguscio, pp. 417–434 (New York: Plenum).

34. Wolf, J. P., Kölsch, H. J., Rairoux, P., and Wöste, L. 1990. Remote detection of atmospheric pollutants using differential absorption lidar techniques, in *Applied Laser Spectroscopy*, ed. W. Demtröder and M. Inguscio, pp. 435–467 (New York: Plenum).

35. Luft, K. V. 1943. Über eine neue Methode der Registrierenden Gasanalyse mit Hilfe der Absorption Ultraroter Strahlen ohne Spektrale Zerlegu. *Z. Tech. Phys.* 24, 97–104.

36. Hodgkinson, J. and Tatam, R. P. 2013. Optical gas sensing: A review. *Meas. Sci. Technol.* 24, 012004 (pp. 59).

37. Chou, J. 2000. *Hazardous Gas Monitors* (New York: McGraw-Hill).

38. Capasso, F., Gmachl, C., Paiella, R., Tredicucci, A., Hutchinson, A. L., Sivco, D. L., Baillargeon, J. N., and Cho, A. Y. 2000. New frontiers in quantum cascade lasers and applications. *IEEEJ. Sel. Top. Quantum Electron.* 6, 931–947.

39. Yang, R. Q., Bradshaw, J. L., Bruno, J. D., Pham, J. T., and Wortman, D. E. 2002. Mid-infrared type II interband cascade lasers. *IEEE J. Quantum Electron.* 38, 547–558.

Organic light emitting devices

MARTIN GRELL
University of Sheffield

19.1 INTRODUCTION AND HISTORIC DEVELOPMENT

In many organic molecules, the absorption of a photon of given wavelength creates an excited state of the molecule (an "exciton") which in turn is capable of re-emitting light of longer wavelength. This is known as fluorescence. The common feature of fluorescent dyes is the presence of alternating single- and double-bonds between carbon atoms ("conjugated" units), resulting in delocalized π-electron clouds. A wide range of fluorescent organic dyes, spanning the entire visible and near-IR spectrum, is now available, e.g., for dye laser applications.

Also, the semiconducting properties of a number of organic materials have long been known, and were studied first in crystalline phthalocyanine [1]. Since the 1970s, the amorphous polymeric organic photoconductor poly(vinyl carbazole) (PVK) has been widely studied and is now commonly used for electrophotography (generally, in the form of a charge transfer complex with the electron acceptor 2,5,7-trinitrofluorenone (TNF)). PVK is conceptually a "hybrid" between low-molecular weight and polymeric organic semiconductors. The major difference between a mainchain conjugated polymer such as a polyene or a poly(*para*-phenylene vinylene) (PPV), and a sidechain conjugated polymer such as PVK is that in the mainchain polymer, the π-conjugation extends over more than one repeat unit, leading to a conjugated unit called the "effectively conjugated segment" (ECS) longer than one repeat unit, and with rather different properties.

The capability of semiconducting organic materials to sustain and transport charge carriers (known as electron/hole polarons, or radical anions/cations) opens the possibility to generate excitons by the combination of an electron and a hole polaron rather than the absorption of a photon. The subsequent light emission from an electrically generated exciton is known as electroluminescence (EL). For semiconducting organic crystals, this was first reported by Pope, and Helfrich and Schneider, in the 1960s [2]. In 1983, Partridge reported for the first time on EL from a semiconducting polymer [3]; namely, PVK. Partridge used the common thin film device architecture as shown in Figure 19.1. However, OLEDs based on PVK displayed poor brightness and efficiency. This was mainly due to the difficulty to inject electrons into PVK. Ideally, equal number of holes and electrons need to be injected for efficient EL. In 1987, a major breakthrough was reported by Tang and van Slyke of the Kodak group [4]. They introduced OLEDs with a multilayer organic semiconductor architecture. This allowed for better electron/hole balance and lead to devices with much improved brightness and efficiency.

Tang and van Slyke worked with highly fluorescent low-molecular weight organic molecules, and Partridge with a sidechain conjugated polymer. The mainchain conjugated semiconducting polymers that were studied in the 1970s and 1980s such as polyacetylene and polydiacetylene are not fluorescent, because in these polyenes, strong electron–electron interactions break parity alternation between subsequent excited states.

Thus, in polyenes, the first excited state has no dipole-allowed (fluorescent) transition to the groundstate. In the late 1980s, Friend, Burroughes, Bradley et al. at Cambridge, experimented with poly(para-phenylene vinylene) (PPV), a conjugated polymer containing a phenyl ring in the mainchain. When they tried to establish the dielectric breakdown characteristics of PPV with a view towards its use as gate insulator in organic transistors, they were surprised to find electroluminescence (EL). They published their findings in 1990 [5]. With hindsight, it is surprising that EL from PPV was not discovered earlier, since PPV

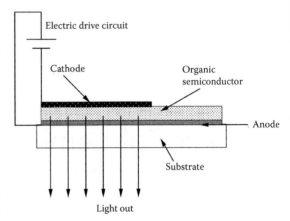

Figure 19.1 The basic OLED architecture. The thickness of the active organic semiconductor film is typically in the order of 100 nm. One of the electrodes (typically, the anode) needs to be transparent to allow for the coupling-out of light.

chemistry and its photoconducting and fluorescent properties had been established previously, much of it due to the work of Hörhold et al. in Jena [6]. Similarly, another celebrated breakthrough in organic electronics, the discovery of "synthetic metals" by Heeger, MacDiarmid, Shirakawa et al. [7], was accidental due to the excessive addition of a catalyst.

The discovery of conjugated mainchain polymer EL has triggered a massive academic and industrial research effort, with the aim of developing and establishing a new flat panel display technology that in principle can replace both cathode ray tube and liquid crystal displays. In recent years, both the development of organic logic circuits based on organic field effect transistors (OFETs), and of organic photovoltaic devices has attracted increasing attention.

19.2 COMMON OLED MATERIALS AND THEIR PROPERTIES

19.2.1 Common organic semiconductors

Before discussing OLEDs in detail, it is instructive to have a list of common organic semiconductors that have played an important role in the development of the field. Figure 19.2 presents a compilation of a few examples that were selected from the wide variety reported in the literature. Commonly used acronyms are given and some basic properties of the materials are discussed.

1. *Low-molecular weight materials*: TPD (*N,N'*-bis-(*m*-tolyl)-*N,N'*diphenyl-1,1-biphenyl-4,4'-diamine) is a hole transporting and weakly luminescent organic semiconductor. 6T (hexi-thiophene) is one representative of the thio-phene family of organic semiconductors which are known for their fast hole mobilities. PBD (2-(biphenyl-4-yl)-5-(4-*tert*-butylphenyl)-1,3,4-oxadiazole) is an electron conductor. Both TPD and PBD have been used as carrier injection layers in multilayer device architectures.

Alq₃ (tris(8-quinolinolato)aluminum(III)) is an organometallic complex with efficient green electroluminescence and remarkable stability. btp₂Ir(acac) (bis(2-(2′-benzothienyl)-pyridinato-*N,C*-3′)iridiumacetylacetonate) is the representative of a family of phospho-rescent dyes that have been used successfully as triplet-harvesting emitters in efficient electrophosphorescent devices. ADS053 RE is the trade name for the red-emitting organolanthanide tris(dinapthoylmethane) mono(phenanthroline)-europium(III). Organolanthanides transfer both singlet- and triplet-excitons to an excited atomic state of

Figure 19.2 Selected examples of organic semiconductors and dyes that have played an important role in the development of organic light emitting devices

1. Low-molecular weight organic semiconductors

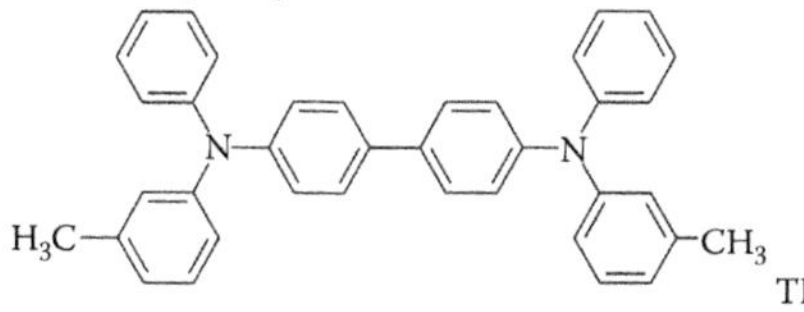

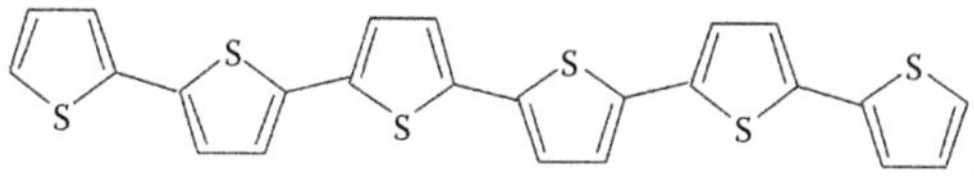

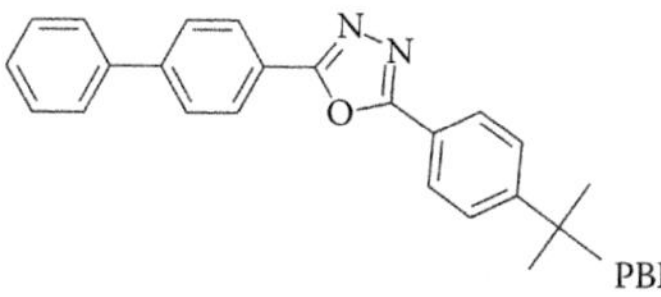

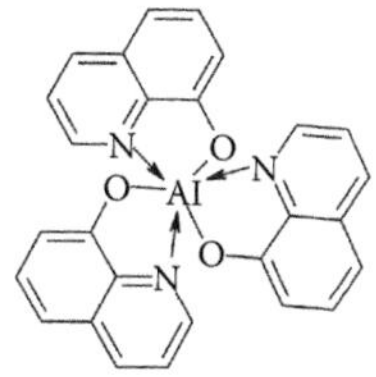

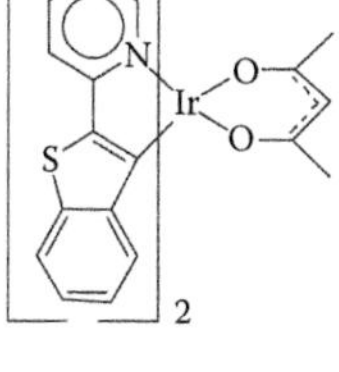

(Continued)

Figure 19.2 (Continued) Selected examples of organic semiconductors and dyes that have played an important role in the development of organic light emitting devices

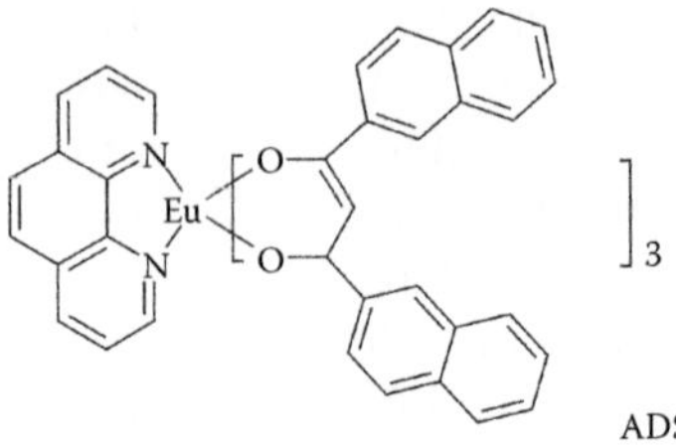

ADS053 RE

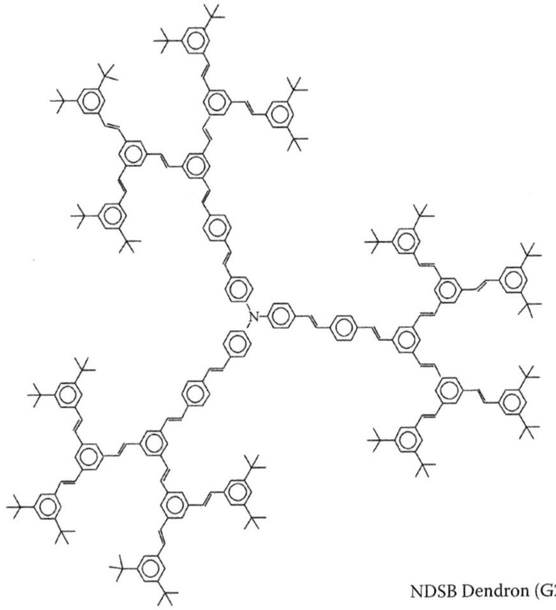

NDSB Dendron (G2)

2. Polymeric organic semiconductors

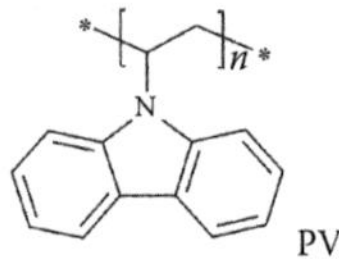

PVK

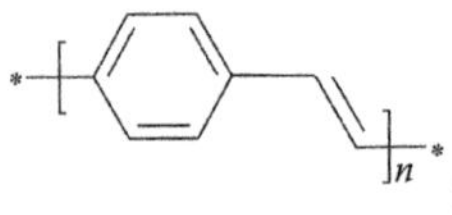

PPV

(Continued)

Figure 19.2 (Continued) Selected examples of organic semiconductors and dyes that have played an important role in the development of organic light emitting devices

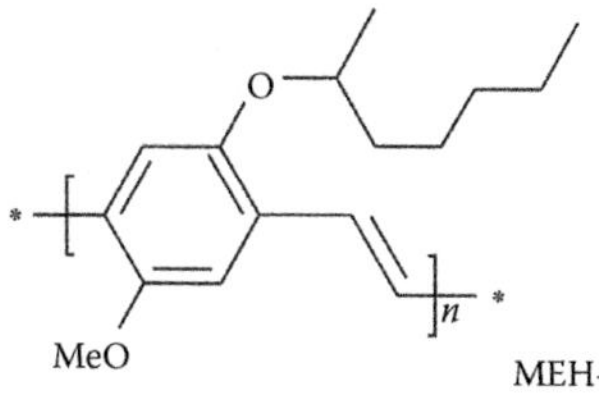

MEH–PPV

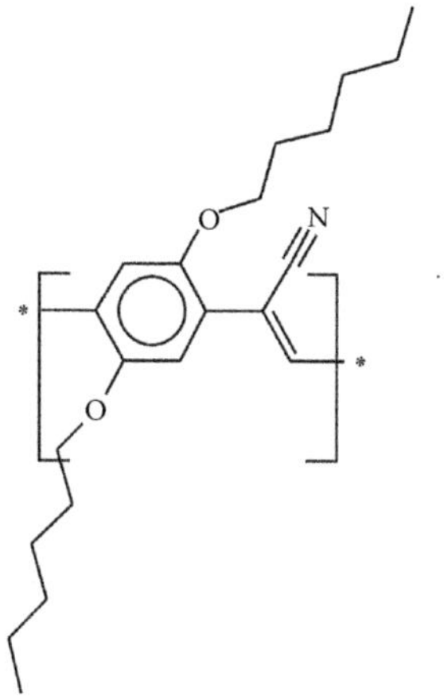

CN–PPV

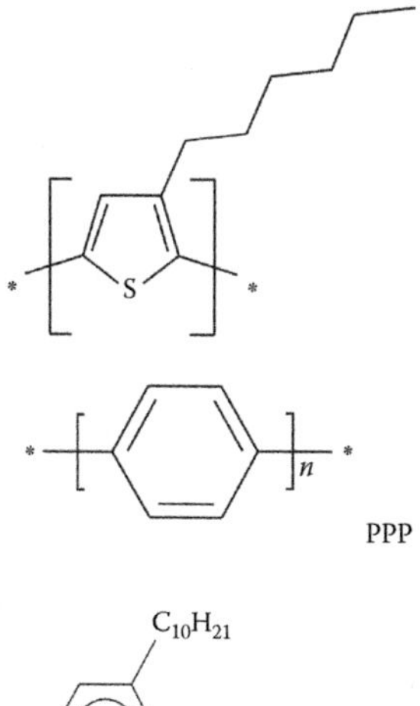

PHT

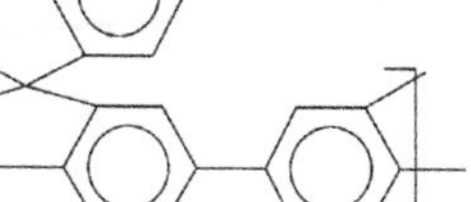

PPP

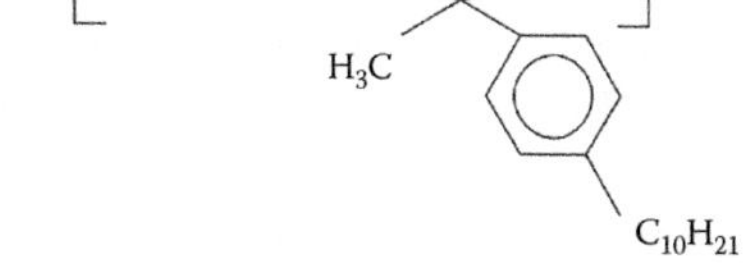

MeLPPP

(Continued)

Figure 19.2 (Continued) Selected examples of organic semiconductors and dyes that have played an important role in the development of organic light emitting devices

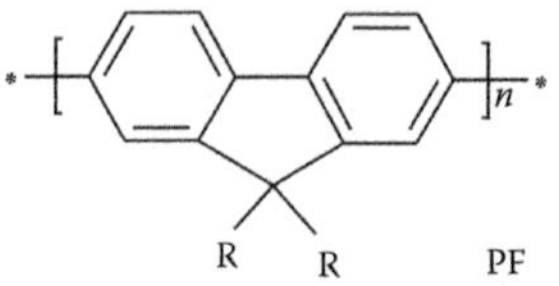

the central lanthanide, resulting in very narrow emission lines, i.e., spectrally pure colors. NDSB Dendron (G2) is a nitrogen-cored distyryl benzene second-generation dendrimer. The core displays visible absorption and emission, the *meta*-linked dendronic sidegroups have a bandgap in the UV and for the purposes of charge injection, transport, and light emission can be considered as inert.

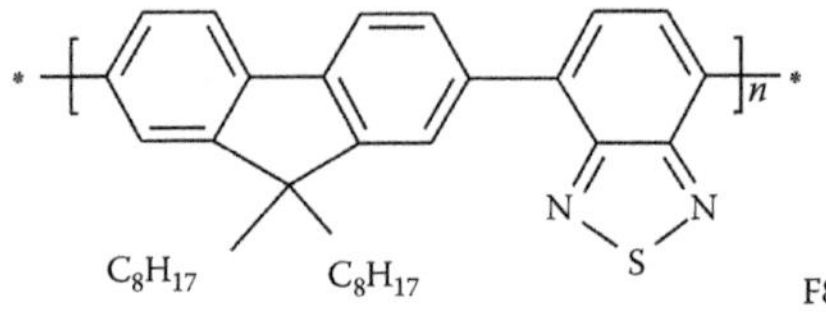

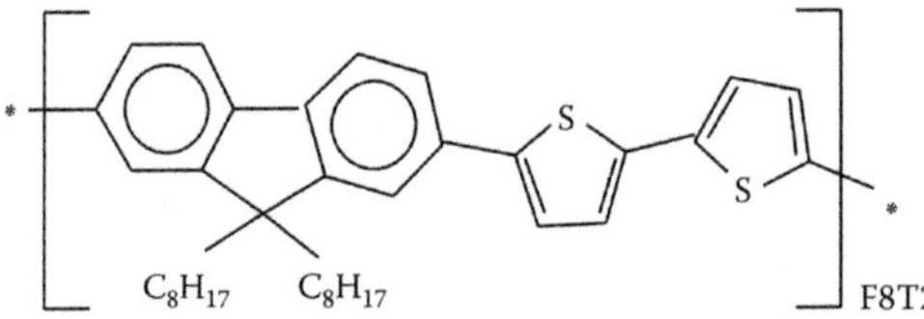

2. *Polymeric materials*: PVK (poly(vinyl carbazole)) is a hole transporting and weakly emissive polymeric semiconductor that has been extensively used as photoconductor in photocopiers (in the form of a charge transfer complex with trinitrofluorenone). Conceptually, it is a hybrid between low-molecular weight organic semiconductors and conjugated polymers: The non-conjugated backbone gives it the film forming properties of a polymer, but the semiconducting units are isolated and retain the properties of low-molecular weight carbazole. PPV (poly(*para*-phenylene vinylene)) is a semiconducting, mainchain conjugated and highly emissive polymer. In the form shown in Figure 19.2, it is insoluble, and it is derived in situ via thermal conversion of a precursor. MEH-PPV (poly[2-methoxy-5-(2′-ethyl-hexyloxy)]-1,4-phenylene vinylene) is a soluble derivative of PPV with somewhat

smaller bandgap due to the alkoxy substitution of the benzene ring. Both PPV and MEH-PPV are hole transporting polymers. CN-PPV (poly(2,5-hexyloxy *para*-phenylene) cyanovinylene) is an electron-transporting, low bandgap polymer due to the high electron affinity cyano substitution at the vinylene bond. PHT (poly(hexyl thiophene)) is weakly luminescent, but displays fast hole mobility, in particular, in its regioregular form. PPP (poly(*para*-phenylene)) is a blue emitter, but insoluble in the form shown in Figure 19.2. Substituting PPP with alkyl sidechains affects solubility, but also twists the rings away from coplanar arrangement. Methylated ladder-type PPP (MeLPPP) enforces coplanar ring arrangement by chemical bonds, and is a blue emitter showing homogeneous broadening only. Poly((9,9-dialkyl)2,7-fluorene)s (PFs) are efficient blue emitters with high carrier mobility. The related copolymer F8BT (poly(9,9-dioctylfluorene-*alt*-benzothiadiazole)) displays a lower bandgap (in the green) and electron transporting properties, as well as partial compatibility with PFs. Consequently, PF/F8BT blends have emerged as alternative to multilayer architectures. F8T2 (poly(9,9-dioctylfluorene-*alt*-dithiophene) is a hole-transporting fluorene copolymer with lower ionization potential than homo-PF.

19.2.2 Comparing organic with inorganic semiconductors

The molecular nature of organic semiconductors leads to a number of significant physical differences between organic and inorganic semiconductors. The most important are summarized below.

1. *Excitations are localized*: Wavefunction coherence in a conjugated polymer extends over a few (order 5) repeat units (the effectively conjugated segment (ECS) [6]), but not further. Consequently, excitations are localized on the rather short scale of the ECS. The wave vector k is not a good quantum number for a localized excitation. Charged excitations are generally more akin to the concept of the "radical ion," as familiar from solution-based chemistry, than the concept of a "polaron" in solid state physics (nevertheless, the term "hole" is commonly used for radical cations). Neutral excitations or "excitons" are best described as Frenkel, not Mott-Wannier, excitons.

   Localization leads to a strong coupling between excitations and local molecular geometry. In the excited ECS, electron clouds and bond lengths are redistributed. In the case of charged excitations, these geometric relaxations can break the local symmetry, and thus activate vibronic bands in the infrared that are symmetry-forbidden (i.e., Raman-but not IR-active) in the ground state ("IRAVs"). In the case of neutral excitations (excitons), the fluorescence emitted on radiative decay to the ground state may display a relatively large Stokes shift and pronounced vibronic structure.

   Typical exciton binding energies, E_b, are in the order 0.2–0.5 eV, and typical exchange energies, i.e., the energetic difference between singlet- and triplet-excitons, are 0.5–1 eV.

2. *Excitations are one-dimensional*: Electronic transition moments are strong and highly directional, parallel to the chain or molecular axis. Some conjugated polymers display liquid crystalline self-organization at elevated temperatures, i.e., the spontaneous parallel alignment of an ensemble of chains [8]. Light polarized parallel to chain alignment then interacts very strongly with the polymer. Also, in aligned polymers, the mobility of charge carriers (radical ions) is enhanced when compared to the non-aligned polymer. Mobility is fastest parallel to the alignment direction, but even perpendicular to the alignment direction, the mobility is faster for an aligned than a non-aligned material [9,10].

3. *There are no "dangling bonds"*: Even a very thin organic film always consists of complete molecules, with the chemical coordination at the film surface equal to that in the bulk. This is very different in vapor-deposited inorganic semiconductor films, where atoms at the surface usually are not chemically coordinated in the same way as in the bulk. These "dangling bonds" distort the band structure at the film surface from its bulk properties. In contrast, surface- and bulk-ionization potentials, and electron affinities of organic semiconductors are generally equal.

19.2.3 Controlling the bandgap

The relation between optical emission spectra and their perception as colors by human vision is essential for display technology. This shall not be discussed in detail here, a good review is in [11]. The basic facts are that color perception changes in the order blue–green–yellow–orange–red as wavelength increases from 400 to >700 nm, and that colors are perceived pure when the emission band is narrow. It is therefore essential to control the bandgap of the emissive semiconductors in a display to reproduce the full color gamut. It is one of the key assets of organic semiconductors that their bandgap can be controlled systematically via modification of their chemical structure. Chemical bandgap tuning can be classified roughly into three types: steric, electronic via sidechain variation and electronic via copolymerization. These are discussed in the following.

19.2.3.1 THE STERIC CONTROL OF BANDGAP

The degree of π-electron delocalization along a polymer backbone is governed by the conformation and configuration of the respective molecule. Conformation can be quantified, e.g., by a dihedral angle φ that describes the relative rotation of a stiff moiety with respect to a neighboring stiff moiety around a flexible chemical bond connecting the two. The so-called rotational potential $E_{rot}(\varphi)$ describes the relative energy of

Table 19.1 Absorption maxima for *para*-phenylene based oligomers versus number of benzene rings

Number of benzene rings	Oligo-PP (eV)	Oligofluorene (eV)	Oligo-MeLPPP (eV)
3	4.44		3.70
4	4.25		
5	4.15		3.18
6	4.03	3.56	
7			3.00
8		3.43	
9			
10		3.35	

Source: Grimme, J., Kreyenschmidt, M., Uckert, F., Müllen, K., and Scherf, U. *Adv. Mater.* 7, 292, 1995; Klärner, G. and Miller, R.D., *Macromolecules* 31, 2007, 1998.

the conformation as a function of dihedral angle. Generally, the π-overlap is optimized if successive units are coplanar ($\varphi=0°$), but disappears if they are orthogonal ($\varphi=90°$). Hence, in conjugated molecules, there is a contribution to the rotational potential that favors coplanar conformation. However, e.g., in coplanar biphenyl, there would be a clash between the H-atoms in the neighboring rings for the coplanar arrangement that would be extremely costly in energy. The relative minimum for $E_{rot}(\varphi)$ is therefore not at $\varphi=0°$; in the case of biphenyl, $\varphi_{min}=23°$ [12]. With the attachment of sidechains, these steric interactions increase, and the twist angle increases from 23° to 45° for *ortho*-dihexyl-substituted polyphenylene. This leads to a "blueshift" (larger bandgap) due to reduced π-conjugation. A prominent example for steric bandgap tuning are the poly(alkyl thiophene)s (PATs), where bandgap tuning throughout the entire visible spectrum by steric effects alone was demonstrated. This is reviewed in [13]; however, PATs are rarely used in OLEDs as their PL quantum efficiencies are very low. Another prominent example for the control of bandgap by more or less bulky sidechains is the case of (non-emissive) poly(diacetylenes) (PDAs), as reviewed in [14].

A method to improve π-overlap, rather than disrupt it by sidechains, is to force rings into a coplanar arrangement by chemical bonds. Let us look at oligomers of *para*-phenylene (PPP), dioctyl fluorene (PFO), and ladder-type *para*-phenylene (MeLPPP), all of which have in common a backbone of *para*-linked benzene rings. In PPP, every ring can twist out-of-plane with respect to its neighbor. In PFO, rings are fused pairwise into a coplanar moiety, there is only one possible rotation per pair, i.e., per two rings. In MeLPPP, all rings are forced into a coplanar arrangement by chemical bonds, there is no degree of conformational freedom in the backbone. Table 19.1 shows the location of absorption maximum (as a measure of bandgap) in dependence of the number of benzene rings for different oligomers (unsubstituted oligo-*para*-phenylene/fluorene-endcapped dihexylfluorene/oligo-ladder-*para*-phenylene).

It is evident that for a given number of rings, the more the backbone is planarized, the smaller is the bandgap.

19.2.3.2 THE ELECTRONIC CONTROL OF THE BANDGAP

By introducing either electron-withdrawing or electron-donating chemical groups into a conjugated molecule, the electron affinity and ionization potential are affected, and hence the bondgap changes. Such groups can be introduced in two ways, namely as sidechains or in the mainchain. We discuss the following examples: Alkoxy-chains attached to PPV rings (MEH–PPV, sidechain modification), CN-groups attached to alkoxy-PPV vinylene bonds (CN–PPV, a case somewhat intermediate between sidechain and mainchain modification), and fluorene copolymers (mainchain modification).

MEH–PPV is an example for alkoxy-substituted PPVs. Its sidechains make MEH–PPV soluble in organic solvents such as THF or chloroform. Sidechains also somewhat isolate chain backbones from each other in the solid film which increases quantum yield over unsubstituted PPV. It was therefore a welcome step forward in the development of conjugated polymers. The alkoxy-linkage of its sidechains to the phenyl backbone ring also changes the electronic structure of the backbone. Alkoxy links have a tendency to donate electrons to the backbone which changes the shape and location of the HOMO. As a result, emission is red-shifted compared to PPV, from green to orange.

The case of CN–PPV is somewhat intermediate between sidechain and mainchain modification. In addition to the alkoxy-sidechains in MEH–PPV, highly electron-withdrawing cyano groups are attached to the vinylene bonds. This leaves the conjugated backbone highly electron deficient, thus considerably increasing the electron affinity (by about 0.6 eV [13]). CN–PPVs emit in the red.

Another approach to bandgap control is copolymerization of different conjugated units into the polymer backbone. Copolymers between alkane- and alkoxy-substituted PPV-type polymers are discussed in [13]. Here, we focus on copolymers of fluorene. Polyfluorenes display a "blue" bandgap that is almost indifferent to the type of sidechain attached. Note that sidechains are attached pairwise at the 9-position of the fluorene ring which itself is not part of the conjugated backbone. Consequently, there is little electronic impact of the sidechains on the backbone properties. However, strictly alternating copolymers of fluorenes with comonomers having different electronic properties were prepared, such as F8BT and F8T2. For both F8BT and F8T2, the resulting bandgap is reduced, and they both emit in the green–yellow region. The reduction in the size of bandgap has different reasons: In the case of F8BT, the benzothiadiazole comonomer has a higher electron affinity E_a than fluorene, thus leading to a polymer with higher E_a. In the case of F8T2, the two thiophene groups have a lower ionization potential I_p than fluorene, thus leading to a polymer with lower I_p. Thus, copolymerization not only allows control of bandgap, but also I_p and E_a in a predictable manner, and a large number of fluorene copolymers have been synthesized and studied. For a review, see [17]. Both the polymers have found interesting applications: some of the most efficient organic EL devices have been built from blends of a minority amount of F8BT as electron injecting/transporting material in PFO as hole injecting/transporting material. The miscibility (or at least, slow separation on the spincoating timescale) of PFO and F8BT is highly exceptional, and allows for the preparation of single layer devices with the properties of double layer devices. F8T2, on the other hand, is an excellent material for p-type OFET channels [9].

19.3 DEVICE PREPARATION AND CHARACTERIZATION

Practical OLEDs conventionally use a film of indium tin oxide (ITO) on a glass substrate as anode. ITO is a highly doped semiconductor that exhibits almost metallic conductivity (~20 Ω/), and is transparent to allow for the out-coupling of light. Onto the anode, one or several organic layers are prepared. Nowadays, ITO is usually coated with a PEDOT/PSS synthetic metal film, for reasons to be discussed in Section 19.4.2. For a review of the remarkable properties of PEDOT/PSS, see [18]. Then, one or several organic semiconducting layers will be deposited; their functions are discussed in Section 19.4.

Obviously, it is tempting to replace ITO completely by flexible sheets of PEDOT/PSS prepared on a plastic substrate which are now commercially available. For some small, low-resolution displays this is already possible. However, the most conductive PEDOT/PSS sheets to date still display a sheet resistance about one order of magnitude larger than ITO. This leads to a drop in the applied voltage across a large display and compromises the addressing of pixels far away from the voltage source.

The preparation of polymeric and low-molecular weight organic semiconductor layers is typically very different, and gives rise to two separate "cultures" of organic semiconductor research. Polymers are typically processed from solution. Typical processes are spincasting, which is the conventional laboratory technique, or ink jet printing, which is of increasing interest for industrial production. Low-molecular weight molecules are typically processed by evaporation.

19.3.1 Solution processing

A key asset of organic semiconducting polymers is that they can be molecularly engineered to be soluble. This allows for the preparation of good quality, uniform thin films over large areas by the spincoating (or spincasting) technique. Much of the interest and momentum in semiconducting polymer research results from this cheap and easy technique to make quality films of arbitrary size.

A typical spincasting solution has a concentration of 5–20 mg of polymer per milliliter of an organic solvent, such as toluene, xylene, tetrahydrofurane (THF), or chloroform. Before use, all solutions should be filtered through a microporous filter, with pore sizes in the range 0.2–0.45 μm.

The principle of a spincoater is shown in Figure 19.3.

A drop of polymer solution is placed onto a substrate that is held down on a turntable by vacuum suction. As the turntable starts rotating, the solution spreads out into a thin film that wets the whole substrate with uniform thickness. Typical "spin speeds" are in the range 1000–4000 rpm, with spin times in the order of 1 min. Under rotation, the solvent evaporates and leaves behind a thin, uniform polymer film. In principle, there is no limit to the size of the film. In industry, films in the size of TV screens are spincoated routinely by automated systems. The thickness d of the resulting film is controlled by solution concentration c, viscosity η and spin speed ω according to Equation 19.1

$$d \propto \frac{c\eta(c)}{\sqrt{\omega}} \qquad (19.1)$$

The proportionality constant depends strongly on solvent, substrate, and other factors. Since η strongly depends on concentration, concentration has a much stronger impact on the thickness of the resulting film than spin speed. d typically will be in the order of 100 nm, with thickness control within a few nanometer via spin speed. Note that the thickness and quality of the resulting films depend on the solvent used. Sometimes, in particular when solvents of low volatility were used, it is advisable to dry films after spinning at moderately elevated temperature (40°C–60°C) under vacuum.

Often, multilayer architectures of several layers with different organic semiconductors are required. To make these by spincasting, subsequent layers have to be spun from mutually exclusive solvents (so called "orthogonal" solvents), otherwise the first layer dissolves on spinning the second. This can be a challenge to material chemistry. The conjugated polymer community has recently started to use polymer blends deposited from a common solvent in one spin cycle, instead of multilayer architectures.

Drawbacks of the spincoating technique are the need for soluble materials, the sometimes limited options for deposition of multilayer devices, the waste of material that flies off the edge of the substrate on spinning, poor quality of films if the solvent evaporates very fast or does not wet the substrate well, and its limited use for low-molecular weight materials.

19.3.2 Vapor deposition

For low-molecular weight materials, the method of choice for device fabrication is often vapor deposition instead of spincoating. The apparatus is shown in Figure 19.4.

In a vapor deposition chamber, high vacuum ($<10^{-6}$ bar) is required. This is usually established with a two-stage pumping system, comprising a rotary pump for a rough pre-vacuum, and an oil diffusion or turbopump for high vacuum. Material is placed into a metal "boat" and is heated to evaporate and to condense on the substrate. This technique is applicable only to small molecules; polymers do not evaporate. A quartz microbalance is used to monitor the thickness and the deposition rate; this can be fed back via a controller to the sample heater to tune to a programmable deposition rate. Co-evaporation of different materials at the same time is an option, but to arrive at defined blend compositions, careful calibration of the heaters is required. Multilayer architectures, however,

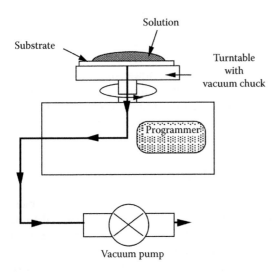

Figure 19.3 A spincoater.

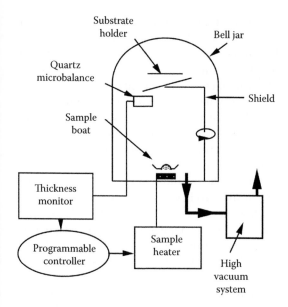

Figure 19.4 An evaporator system for the vapor deposition of small-molecule organic semiconductors, or metals.

are easily made. Tang and van Slyke's first double-layer organic light emitting device was made by vapor deposition [4].

In some cases, exceptional control over the morphology of resulting films can be exercised by evaporation rate, and most importantly, type and temperature of the substrate. Higher temperature allows molecules to reorganize into more ordered structures immediately after deposition, but before a dense film is formed. The resulting films may have different crystal structures, and different orientation of the crystal axis with respect to the substrate; but these may be specific to certain substrates, which are not always useful for device fabrication. Control of morphology via evaporation conditions has been explored in much detail, e.g., for hexithiophenes [19].

Evaporators are also useful for the deposition of metal cathodes, which act as electrical contacts for devices, and can be applied to vapor deposited organic devices without breaking the vacuum. It is a drawback of the generally simpler and cheaper spincasting procedure that the subsequent evaporation of metal cathodes requires an additional high vacuum processing step. Recent polymer OLED research has therefore explored alternative metal deposition procedures that require no vacuum steps [20].

19.3.2.1 DEVICE CHARACTERIZATION

The electrical and optical properties of OLEDs are essential for their applications, and are characterized by a wide variety of methods. The most basic characterization of a conjugated material is by recording its absorption and photoluminescence (PL) spectra, and determining photoluminescence quantum yield (photons out per photons in). The relevant instrumentation and interpretation of such spectra has been discussed widely in the literature [21]. Instead, we discuss here the basic techniques specific to the use of such materials in OLEDs.

19.3.2.2 ELECTRICAL CHARACTERIZATION

The current–voltage (C/V) characteristic of an OLED is determined simply by connecting the device electrodes to an electrical source-measure unit. Typically, voltage is ramped up continuously up to the order of $30\,V$ or more, although for an OLED with low onset voltage, up to $10\,V$ or less may be sufficient and higher voltages will lead to a damaging level of current. The typical range of current densities in an OLED under above voltage cycle will be 0.1 (or less) mA cm^{-2} at the light emission "onset," up to the order of 100 mA cm^{-2} under the highest reasonable (i.e., non-destructive) drive voltages. Current sensitivity for a typical sample device, with typically a few square millimeter active area, should be at least 1 µA.

Often, while undergoing C/V characterization, the luminescent intensity emitted from an OLED is detected simultaneously by a photodiode. Luminescence/voltage characteristics is typically plotted into the same graph as current/voltage, resulting in current/voltage/luminescence ($C/V/L$) characteristics. However, the response of the photodiode is usually not calibrated, and luminescence results are presented in "arbitrary units" (a.u.). Note that there is no physical current or luminescence "onset" voltage for an OLED; practically, however, an "onset" voltage is often reported. This only makes sense if a definition of "onset" is given.

19.3.2.3 QUANTIFYING "BRIGHTNESS"

One of the most interesting properties of an OLED is how "bright" it is, and how efficiently it converts electrical energy into light. The most straightforward quantity to characterize efficiency is the internal quantum efficiency η_{int}. η_{int} is the ratio of photons generated per two charge carriers injected (two because an exciton is formed from two carriers). To optimize η_{int}, we have to

- *Balance carrier injection*: An excess of one type of carriers means not all injected carriers can find a partner of opposite sign to form an exciton, and will exit the device at the opposite electrode as a "blind" current. Unpaired carriers contribute to the number of carriers injected, but not the luminescence. Methods to optimize carrier balance are discussed in Section 19.4.2.
- *Emit as many as possible photons from each generated exciton*: Methods to maximize photon yield are discussed in Section 19.5.

The "external quantum efficiency" η_{ext} is defined as the number of photons coupled out from a device per two carriers injected. Remarkably, η_{ext} is often considerably smaller than η_{int}, often in the order $\eta_{ext} \approx (1/8)\,\eta_{int}$. Approaches to optimize η_{ext} are discussed in Section 19.5.5.

Both internal and external quantum efficiencies are defined in terms of the fundamental electrical and photophysical properties of OLED operation. Their measurement, however, can be rather intricate. There is no direct approach to η_{int}. To measure η_{ext}, we have to count photons, e.g., in an "integrating sphere" or with calibrated photometers.

Instead, device performance is often reported in terms of photometric quantities. Photometric quantities are "physiological" units, i.e., they consider the response of the human eye as well as purely physical quantities such as the power of radiant flux. The most important photometric quantities are the luminous flux F with unit lumen (lm), and the luminous intensity I with unit candela (cd). The luminous intensity I is related to the luminous flux F by $I = dF/d\Omega$, with the solid angle Ω in sterad [11]. To give a characteristic that is independent of the arbitrary size of the OLED, the luminous intensity per unit area or "Luminance," $L = I/A$ in units cd m^{-2}, is used.

The relation between the physiological quantity luminous flux F (in lm) and the physical quantity radiant flux per unit wavelength P (in W/nm) is given by Equation 19.2 [11]

$$F = K \int_{\lambda_1}^{\lambda_2} V(\lambda)P(\lambda)\,d\lambda \qquad (19.2)$$

where $V(\lambda)$ is the dimensionless "photopic efficacy" that describes the spectral sensitivity of the human eye, and $P(\lambda)$ is the radiant flux per unit wavelength (in W nm^{-1}). Note that radiant flux is a physical unit and is measured in watts, while luminous flux in lm is the corresponding physiological unit. The interval λ_1 to λ_2 is the wavelength interval wherein $P(\lambda)$ is different from zero, and K is the "absolute luminous efficiency," $K = 678.8$ lm W^{-1}. $V(\lambda)$ is non-zero within the range of visible wavelengths ($\approx$380–750 nm) only, and is normalized to 1 at the wavelength of maximum sensitivity of the human eye, $\lambda_{max} = 555$ nm; thus $V(\lambda) \leq 1$. The above discussion applies to the bright-adapted eye ("photopic vision") which is relevant for display technology. In the dark-adapted state ("scotopic vision") a different K and $V(\lambda)$ apply.

Conveniently, calibrated cameras that give luminous intensity in cd are commercially available. Therefore, device efficiency is often expressed in terms of luminous intensity/electric current through the device (unit cd A^{-1}), rather than in terms of η_{ext}.

Another common quantity to characterize device efficiency is the power efficiency expressed in lumen/watt (lm W^{-1}). The "watt" in lm W^{-1} here refers to a watt of electric energy driven through the device, i.e., current $\times$ drive voltage, not a watt of radiant flux. To optimize power efficiency, we have to optimize not only η_{ext} but also have to minimize the drive voltage required. Section 19.4.1 discusses the method of achieving it.

19.4 PHYSICS OF OLED OPERATION

In fluorescence, excitons are created by the absorption of light, while in EL, excitons are created by electron and hole polaron "capture." Polarons first have to be injected from the electrodes, and migrate towards each other. They then form an exciton that sometimes can decay under the emission of light. The variety of electrical and photophysical processes involved are summarized in the Figure 19.5. We discuss the most important of these processes in detail.

19.4.1 Charge carrier injection

The first step towards exciton formation in an OLED is the injection of holes from the anode/ electrons from the cathode under an applied

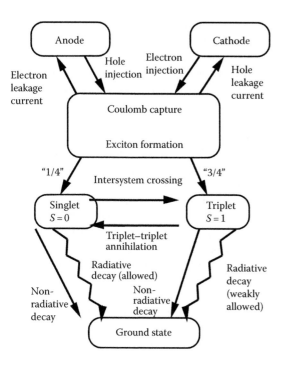

Figure 19.5 A chart describing the formation and decay of excitons in organic EL devices. (Adapted from Bradley, D.D.C., *Curr. Opin. Solid State Mater. Sci.*, 1, 789, 1996.)

voltage. Good injection is a considerable challenge, in particular, since we need to inject carriers of both signs into the device. Carrier injection is controlled by the workfunction Φ of a metal electrode relative to the electron affinity E_a of the semiconductor for electron injection, and relative to the ionization potential I_p of the semiconductor for hole injection.

A level diagram shown in Figure 19.6 is often used to illustrate carrier injection. Note that due to the molecular nature of organic semiconductors, there are no surface "dangling bonds" that can distort bulk energy levels. In Figure 19.6, left, there is an "injection barrier" of 0.5 eV for holes from ITO into PPV, and 0.3 or 1.7 eV for electron injection from Ca or Al into PPV, respectively. For electron injection from ITO, there would be a large barrier of 2.2 eV. Thus, the use of electrodes made of unlike metals defines a forward and reverse bias for the OLED.

The right-hand side part of Figure 19.6 shows the same device (assuming a Ca cathode) under a forward bias. Carriers can now overcome injection barriers by tunneling, with tunneling distances $t_{h/e}$ for holes/electrons, respectively, given by Equation 19.3

$$t_{h/e} = d\,\frac{\Delta V_{h/e}}{V_{bias}} \qquad (19.3)$$

with d the semiconductor film thickness, ΔV the respective injection barriers, and V_{bias} the applied forward bias. The resulting tunneling current density/voltage characteristic $j(V_{bias})$ is described by the equation of Fowler and Nordheim ("F–N tunneling"). For a detailed discussion, see, e.g., [23]; however, it is obvious that injection barriers should be minimized or entirely absent for good

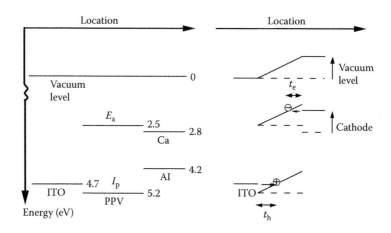

Figure 19.6 Energy levels in a single layer organic EL device with PPV emissive layer. Left: no bias voltage applied. Right: a voltage is applied in forward bias.

carrier injection. A metal–semiconductor junction is termed "ohmic" if current density across the device is limited by the transport of carriers across the semiconducting film, rather than by the injection barriers at the interface. In that case, carrier transport is termed "space charge limited conduction" (SCLC). The problem of carrier injection can be considered solved in the case of SCLC. Practically, injection barriers of 0.3 eV mostly lead to ohmic behavior [24].

Generally, ohmic injection requires high workfunction anodes and low workfunction cathodes. Considerable effort has been devoted to increase the workfunction of the transparent ITO anode by a variety of physicochemical treatment cycles [25]. Recently, it has become common to coat ITO with a thin film of the high workfunction synthetic metal PEDOT/PSS [18] ($\Phi = 5.2eV$). As cathodes, low workfunction materials such as Ca are commonly used. These require protection from ambient atmosphere, otherwise they would rapidly degrade. This can be provided by encapsulation, or by capping with a more stable metal such as Al.

To discriminate experimentally between barrier-type and ohmic injection, it is common to compare $j(V_{bias})$ characteristics of devices of different thickness d. F–N tunneling depends only on the applied field $E = V_{bias}/d$, thus in a plot of j versus E, all characteristics will coincide regardless of d. In the case of SCLC, j will follow Child's law, $j \sim V_{bias}^2 / d^3$. Thus in the plot j versus E, the characteristics will not coincide for different d, but will in the plot jd versus E. The situation is more complex when injection of both electrons and holes may occur. Therefore, the above experimental procedure is to be carried out in symmetric devices, e.g., Au/semiconductor/Au for holes and Ca/semiconductor/Ca for electrons. This will ensure single carrier currents.

19.4.2 Charge carrier transport

After injection, carriers drift across a device under the pull of the local field, which depends on the applied bias voltage, device thickness, and distribution of space charges in the device. The carrier drift velocity v depends on field E as $v = \mu E$ with the carrier mobility μ. Experimentally, mobilities can be determined by the time-of-flight (TOF) technique [10], analysis of the $j(V)$ characteristics of space charge limited currents [26], or the

analysis of output characteristics of organic field effect transistors [9,27].

In single crystals of low-molecular weight organic semiconductors, at low temperatures sometimes a band-like (coherent) charge carrier transport is observed with very high charge carrier mobilities in the order of 100 cm² V⁻¹ s⁻¹ [28]. However, this requires elaborate single crystal preparation that is inconsistent with conventional layer deposition onto device substrates such as ITO. In practical situations, one generally finds (incoherent) hopping type carrier transport, which can be described as a directed random walk through a medium characterized by both energetic and positional disorder. This is quantitatively described by a model developed by Bässler in Marburg [23]. Bässler arrives at a rather complex equation to describe both field- and temperature-dependence of carrier mobility μ, which is further complicated in the presence of traps. At constant temperature, the logarithm of mobility is predicted to be proportional to $\sqrt{E}$. As long as $[\mu(E) - \mu(E \rightarrow 0)] \ll \mu(E \rightarrow 0)$, this can be approximated by Equation 19.4

$$\mu(E) = \mu(E \rightarrow 0) + k\sqrt{E} \qquad (19.4)$$

where k can be both positive or negative, depending on the relative size of disorder parameters describing positional and energetic disorder, respectively. Hence, experimentalists often plot mobility data versus $\sqrt{E}$ [10].

Mobilities in the incoherent transport regime are typically much smaller than the coherent transport, ranging roughly from 1 to 10⁻⁸ cm² V⁻¹ s⁻¹ for common materials. However, due to the short distance charge carriers have to travel in a typical OLED (film thickness ~100 nm), OLEDs can operate well with rather low charge carrier mobility. For example, the hole mobility in both PPV and dialkoxy-PPV is in the order 10⁻⁶ cm² V⁻¹ s⁻¹ [26,29]. Nevertheless, PPV and its derivatives are highly successful OLED polymers. The situation is quite different for organic field effect transistors (OFETs), because carriers have to travel from source to drain across a channel of at least several micrometers.

Practical materials are often characterized by the presence of charge carrier traps. Traps are localized sites with either electron affinity higher than the electron affinity of the bulk material (electron traps), or the ionization potential lower than that of the bulk semiconductor (hole traps). Common trap sites are impurities (e.g., catalyst residues), chemical defects (e.g., sites that have oxidized during OLED operation), and grain boundaries in partially crystalline materials. In the presence of traps, it is not appropriate to describe carrier transport in terms of a single mobility. Instead, we find dispersive charge carrier transport. In a time-of-flight mobility experiment, no clear transient can be assigned in the case of dispersive transport. Instead, transit times vary widely (over several orders of magnitude) due to "late arrival" of carriers that had been trapped for various lengths of time. Typically, in the case of dispersive transport, the apparent mobility deduced from the time-of-flight experiment is longer when a thick sample is used. This ambiguity signals the breakdown of the assumption of a defined carrier mobility.

Traps are roughly classified as either shallow or deep. Shallow traps are those with trap depths of only a few kT, so that trapped carriers can be thermally reactivated into the conduction band, while deep traps are considerably deeper than kT and will release carriers only when the external bias becomes sufficiently high (in effect, the carrier has to be re-injected back into the transport band against an injection barrier defined by the trap depth). Deep traps are a serious problem, as trapped carriers effectively have zero mobility and thus contribute to space charge but not to the current. The general observation that organic semiconductors do intrinsically carry only one type of carriers (i.e., an organic semiconductor is either hole or electron transporting, but only very few are "ambipolar") is thought to be the result of carrier-specific deep traps. To avoid deep traps, materials have to be as pure as possible, the formation of defects during operation has to be minimized by the exclusion of oxygen and water (i.e., device encapsulation), and some materials have to be prepared as amorphous ("glassy") films rather than semicrystalline, because crystallite grain boundaries represent deep trap sites. These requirements define some of the extraordinary challenges on organic semiconductor chemistry.

Colloquially, the terms hole (electron) transporting layer (HTL/ETL) are used to denote a material that at the same time facilitates low-barrier or barrier free hole (electron) injection and hole (electron) transport without deep traps.

Strictly, injection and trap-free transport are different phenomena; however, they are both closely related to the location of I_p or E_a, respectively. For example, in the case of an ETL, electron injection will be facilitated by high E_a. At the same time, an impurity will act as electron trap only if the E_a of impurities is higher than that of the host material. Thus, high host E_a makes it less likely that an impurity can act as electron trap.

19.4.3 Electrical device optimization

Assuming that carrier traps can be avoided, it is fair to say that the minimization of injection barriers is much more important for the optimization of OLED efficiency than charge carrier mobility. However, it is generally difficult to achieve ohmic (i.e., barrier free) injection at both electrodes for a given organic semiconductor. The breakthrough towards efficient organic EL devices comes from a simple but ingenious idea. Tang and van Slyke from the Kodak group have manufactured a bilayer device consisting of a low-ionization potential, hole transporting diamine layer and a high electron affinity, electron transporting Alq$_3$ layer, which is also an efficient green emitter [4]. They also used extremely thin layers of organic semiconductors (order of 100 nm) which lead to higher field at a given drive voltage, and thus lower onset voltage and higher efficiency.

Figure 19.7 shows the level diagram for a (fictitious) double-layer device consisting of an HTL and an ETL. Both layers are assumed to have a bandgap $(I_p - E_a) = 2.5$ eV, however, the HTL has lower I_p than the ETL, and the ETL has higher E_a than the HTL.

It is immediately obvious that a single-layer device using either HTL or ETL alone would necessarily have one large (1 eV) injection barrier. In the double-layer architecture both barriers are moderate (0.5 eV). Since tunneling depends exponentially on barrier height, the double layer architecture leads to much improved and more balanced injection.

In addition to the injection barrier, both holes and electrons encounter another internal barrier at the HTL/ETL interface. This additional barrier is not detrimental to device performance though. Instead, it can help to improve the balance between electron and hole currents. Assuming a slightly smaller injection barrier for holes than for

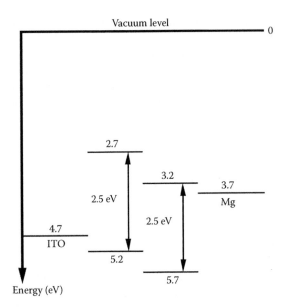

Figure 19.7 Level diagram of a fictitious double-layer device using ITO anode and Mg cathode.

electrons, or higher hole mobility than electron mobility, even in a bilayer device we would expect a carrier imbalance with a larger hole current than the electron current. However, since holes encounter an internal barrier, they do not simply cross the device and leave at the cathode as a "blind" leakage current. Instead, they accumulate at the interface, where they represent a positive space charge.

The effect of the field resulting from that space charge is to improve charge carrier balance. Firstly, it impedes the further injection of majority carriers (holes) from the anode, and secondly, it enhances the injection of minority carriers (electrons) from the cathode. Also, excitons will form at the internal interface, far away from the electrodes. Cathodes in particular have been associated with exciton "quenching" (i.e., radiationless exciton decay); this is avoided by placing exciton formation at the center of the device rather than close to the cathode.

As Tang and van Slyke used small molecules, bilayers could readily be manufactured by subsequent evaporation. This approach has been extended to sophisticated multilayer architectures, e.g., by the group of Kido in Yamagata. They have demonstrated some of the brightest and most efficient OLEDs to date (140,000 cd m^{-2} and 7.1% external quantum efficiency) [30].

With polymeric organic semiconductors, vapor deposition is not an option, devices have to be prepared by spincasting instead. Multilayer architectures are harder to realize with spincasting than with vapor deposition, because of the need for "orthogonal" solubilities. To sidestep solubility problems, in principle, a precursor route may be employed, where the first layer is prepared from a soluble precursor polymer which then is converted in situ into a conjugated and completely insoluble polymer. This has been successfully employed for hole-transporting PPV/electron transporting CN–PPV double-layer polymer OLEDs [31]. However, the precursor route requires lengthy in situ thermal conversion under high vacuum and has generally fallen out of favor with the advent of soluble conjugated polymers.

Recently, a very favorable approach has emerged that combines the ease of injection into a multi-component device with the simplicity of solution processing. In that approach, a single layer of a blend of a hole-transporting and an electron-transporting conjugated polymer, namely poly(dioctyl fluorene) (PFO) and F8BT, is spincast in one single preparation step. As spincasting implies the very rapid formation of a solid film from solution, the two polymers have little time to phase separate and a solid film may result wherein both polymers remain intimately mixed. Such a mixture has been termed "bulk heterojunction." Holes are injected and transported into the (majority component) PFO, but can be transferred easily to F8BT, as it has similar ionization potential. However, F8BT has poor hole mobility due to hole-specific traps. Instead, it has rather high electron affinity and displays comparatively good (albeit dispersive) electron transport [33]. Thus electrons are mobile on the F8BT chain until they encounter a trapped hole. With some further device improvements, highly efficient (4.1 cd A^{-1}) and low onset voltage (≈ 3 V) OLEDs have been prepared from such blends [34]. The preparation and morphology control of hole/electron transporting blends is the focus of much current research, mainly with a view to photovoltaic applications of organic semiconductors [32].

19.4.4 Exciton formation

When both hole and electron polarons have been injected into a device, and these drift towards each other under the applied voltage, one expects the formation of hole polaron/electron polaron pairs that are bound to each other. Such bound pairs are termed excitons, and can in some cases be identical to excitons that are formed in an organic semiconductor after the absorption of a photon of light, and subsequent relaxation into the lowest vibrational state of the first excited electronic state. Just as some organic materials display fluorescence (i.e., radiative decay of the excited state), we may find electroluminescence (EL) in such materials.

At first sight, it appears that exciton formation in multilayer architectures is hindered by the internal barrier that carriers of either type encounter at the HTL/ETL interface. However, this is generally not the case. Excitons in organic semiconductors generally display exciton binding energies E_b of a few tenths of an eV [35]. When a carrier has to overcome an internal barrier to form an exciton, it requires a certain amount of energy; however, on exciton formation, E_b is instantly "refunded"— effectively, the internal barrier is reduced by E_b. Thus, majority carriers remain stuck at an internal barrier and redistribute the internal field in the favorable way discussed earlier, until a minority carrier arrives at the interface. As soon as a minority carrier is available, exciton formation is then helped by the effective barrier reduction E_b. High E_b also stabilizes excitons against dissociation and non-radiative decay.

In "bulk heterojunction" blends, one carrier has to transfer from one chain to another to form an exciton. This will be the type of carrier for which the energy level offset of either the ionization potentials ($|\Delta I_p|$) or the electron affinities ($|\Delta E_a|$) is smaller. The smaller of the two offsets (min($|\Delta E_p|$, $|\Delta E_a|$)) defines the energetic cost of carrier transfer. Two very different scenarios emerge for the case min ($|\Delta E_p|$, $|\Delta E_a|$) < E_b as opposed to min ($|\Delta I_p|$, $|\Delta E_a|$) > E_b, In the former case, formation of excitons from polarons is favored, while in the latter case, the dissociation of existing excitons into polarons is preferred. In the case of F8/F8BT blends that had been introduced previously, exciton formation is clearly favored, and such blends are useful for OLED applications. In other hole/electron transport material blends, such as poly(alkyl thiophene)/perylene tetracarboxyl diimide blends, exciton dissociation is favored, which makes such blends attractive for use in photovoltaic devices [32]. While measurements of $|\Delta I_p|$, $|\Delta E_a|$, and E_b with sufficient precision to predict exciton formation or dissociation are usually not

available, there is a simple experimental approach to decide which is the case: if in a blend fluorescence intensity is much reduced compared to the pure components, excitons are separated efficiently due to the presence of the blend partner.

19.5 OPTIMIZING EFFICIENCY

The discussion in Section 19.4 outlines the strategy towards OLED devices with balanced carrier injection and quantitative exciton formation that can be driven at low voltage. The (formidable) challenge that then remains is to maximize the amount of light generated from the excitons. It is obvious that we require a material with a high luminescence quantum yield. The otherwise excellent thiophene-based organic semiconductors fail this criterion and are not suitable for efficient OLED devices. But even given a high luminescence yield, an extraordinary challenge remains that is rooted in the basic properties of excitons.

There is a fundamental difference between the formation of excitons by absorption of a photon, and exciton formation by the combination of electron and hole polarons. The presence of a non-vanishing optical dipole transition moment between a ground state and an excited state that allows for the absorption of a photon implies a difference in the orbital angular momenta of ground and excited states, thus a difference in the orbital quantum number l ("selection rule" $\Delta l = 1$). The overall spin of the resulting exciton, however, will be $S = 0$, just as for the ground state. Such excitons are termed "singlet" excitons, and correspond to the electron/hole spin combination $(1/\sqrt{2})(|\uparrow\downarrow\rangle - |\downarrow\uparrow\rangle)$. Fluorescence is the re-emission of a photon under return of the singlet excited to the molecular ground state. Unit angular momentum is supplied to the photon from the difference in orbital angular momenta between the singlet excited and the ground state.

When electrically injected electrons and holes combine into excitons, their spins can combine in one of four possible ways. One of those is the singlet combination as discussed as earlier, but there is three more possible polaron spin combinations, namely $(|\uparrow\uparrow\rangle, |\downarrow\downarrow\rangle, (1/\sqrt{2})(|\uparrow\downarrow\rangle + |\downarrow\uparrow\rangle))$. These three correspond to so-called "triplet" excitons with $S = 1$. Triplet excitons have no dipole-allowed (fluorescent) relaxation to the ground state, because there is no orbital angular momentum difference between a triplet excited and the molecular ground state.

Consequently, electroluminescence- and photoluminescence-quantum yields $\eta_{EL/PL}$ are related via Equation 19.5

$$\eta_{EL} = \frac{\sigma_S/\sigma_T}{\sigma_S/\sigma_T+3}\eta_{PL} \qquad (19.5)$$

with $\sigma_{S/T}$ the polaron capture cross-section for singlet and triplet exciton formation, respectively. The assumption that the exciton formation cross-section is independent of the relative orientation of polaron spins ($\sigma_S = \sigma_T$) leads to the prediction that only one-fourth of all formed excitons will be singlet excitons, thus $\eta_{EL} = (1/4)\eta_{PL}$. This limit would apply even in the case of an electrically "ideal" device with ohmic and perfectly balanced injection of electrons and holes, trap free transport, and "quantitative" exciton formation without leakage currents. Clearly, this limit is undesireable, and the OLED community has devised several approaches to overcome it. Some of these are discussed in the following.

19.5.1 Enhanced singlet exciton formation

While some experimental studies based on comparisons of EL and PL quantum efficiencies appeared to confirm $\eta_{EL}=(1/4)\eta_{PL}$ [36], other studies found high EL quantum efficiencies consistent with a singlet/triplet formation ratio $\approx$1:1 [37], implying $\sigma_S \approx 3\sigma_T$. To determine singlet/triplet formation ratio directly rather than inferring them from EL/PL efficiencies, Vardeny et al. [38] carried out a systematic magnetic resonance study on a number of organic semiconductors with bandgaps in the visible. They found that σ_S/σ_T was indeed generally greater than 1, namely between 2 and 5 for different materials. $\sigma_S/\sigma_T \approx (2-5)$ corresponds to $\eta_{EL} \approx (0.4-0.6)\,\eta_{PL}$, instead of $0.25\eta_{PL}$.

Following a later study on an oligomer series [39], Vardeny et al. now interpret their results in terms of the effective conjugation length of the organic semiconductor, with σ_S/σ_T increasing with conjugation length. This implies larger σ_S/σ_T for polymeric than for low-molecular weight organic semiconductors—a finding that potentially can influence the future direction of an entire industry.

Indeed, Friend et al. found a marked violation of the naïve 1-singlet/3-triplet rule for a polymeric organic semiconductor, but not for a low-molecular weight analogue [40]. They interpret this as the result of the relatively long quantum coherence in a polymeric organic semiconductor. Oppositely charged polarons "feel" their Coulomb attraction with or without quantum coherence. However, in polymers we have quantum coherence over one conjugation length; consequently, polarons can "sense" each others' spin from a relatively long distance. Thus, if singlet formation is preferred over triplet formation, polymers have a chance of avoiding triplet formation while electron and hole are still widely separated, i.e., only weakly bound electrostatically. In small molecules, quantum coherence ends at the end of a molecule. Once a hole and an electron are on the same molecule, it is not possible to avoid exciton formation, regardless of spin statistics.

19.5.2 Electrophosphorescence

As an alternative to the enhanced singlet formation cross-section in polymers, in particular, the low-molecular weight OLED community has developed the concept of "harvesting" triplets for light emission by using phosphorescence. Phosphorescence bypasses the $\Delta l = 1$ selection rule that normally restricts emissive transitions to singlet excitons. In phosphorescence, the angular momentum necessary for the emission of a photon is supplied from the triplet spin ($S = 1$) rather than from the orbital angular momentum difference of excited/ground state wave functions. To transfer angular momentum from the triplet spin angular momentum to a photon, a spin–orbit coupling term $L \cdot S$ is required in the molecular Hamiltonian. Spin–orbit coupling can be of substantial magnitude only if orbitals with higher angular momentum L are present in the molecule. Phosphorescence is therefore typically linked to the presence of atoms with "high" (i.e., higher than carbon) order number in a molecule ("heavy atom effect"). The phosphorescence transition moment is considerably weaker than for fluorescence, leading to excited state lifetimes typically in the range of microseconds or more ("weakly allowed transition"), as compared to lifetimes in the nanosecond range for fluorescence.

In a typical electrophosphorescent device, a wide bandgap host semiconductor is "doped" with a small percentage of a phosphorescent emitter. The excitation is transferred from the "host" to the "guest" via excitonic energy transfer. Forrest et al. at Princeton have developed a range of green, yellow, orange and red organoiridium complexes [42], which are exemplified by the particularly efficient red phosphor $btp_2Ir(acac)$. When doped into a wide bandgap host, electrophosphorescence with >80% internal quantum efficiency and $60\,lm\,W^{-1}$ is observed [43]. Using pure $btp_2Ir(acac)$ without host matrix has resulted in less efficient devices [44].

Given the fact that polymeric EL devices are considerably more efficient in the formation of singlets than the naïve 1:3 expectation, one may question the need for electrophosphorescence towards enhanced efficiency. However, in particular, in the red, electrophosphorescence is an attractive approach. Due to the response characteristics of the human eye, red dyes must show very narrow emission peaks, otherwise color purity will be compromised. Iridium-based phosphors display considerably narrower emission bands than typical fluorescent dyes, and are thus particularly useful as red emitters. These phosphors also display relatively short triplet lifetimes ($4\,\mu s$), thus avoiding problems associated with triplet–triplet annihilation at high brightness, which had been encountered with longer lifetime phosphors [45]. However, electrophosphorescence is ambitious for blue emission due to the need for a high bandgap host semiconductor.

19.5.3 Organolanthanides

Another approach to "triplet harvesting" is represented by the organolanthanide dyes. Organolanthanides are somewhat similar to organometallic phosphors, however, the central metal atom is a lanthanide such as europium (Eu) or terbium (Tb). The red dye "ADS053RE" is a typical example. Organolanthanides owe their properties to the unique electronic structure of the lanthanides (or "rare earth" metals), which is reflected in their positioning in the periodic table. Up to lanthanium (atomic number 57), the 4f electronic shell remains empty, instead the 5s, 5p, and even the 6s shell are filled first. Only for the elements between cerium (Ce, atomic number 58) and luthetium (Lu, atomic number 71), it becomes energetically more favorable to fill the 4f shell rather than adding more electrons in the

sixth shell. The outer (fifth and sixth) shells remain unchanged throughout the rare earths. Hence, all rare earths are chemically very similar. Electronic transitions in the isolated, but incomplete 4f shells are therefore not affected by the chemical bonding.

In a dye such as ADS053RE, the organic ligand can absorb light (typically in the blue or near UV), or can be excited electrically. The exciton is then passed to the central lanthanide and excites an electron of the lanthanide 4f shell; notably this works for both singlet and triplet excitons. The intramolecular excitation transfer is schematically shown in Figure 19.8.

Note that the observed emission comes from the radiative decay of the excited 4f state: it is an atomic and not a molecular transition. This is the marked difference between organolanthanides and conventional organometallic phosphors. The excited state has a long lifetime of about 1 ms, which makes organolanthanides particularly well suited for passive matrix displays (see Chapter 7 in Volume 2). Nevertheless, due to the localized and isolated nature of the excited state, organo-lanthanides do not suffer from triplet–triplet annihilation. η_{int} well in excess of 25% can be achieved, and since emission comes from an atomic transition, bands are extremely narrow (FWHM $\approx 10\,nm$). This results in very pure colors (green from terbium (Tb), red from europium (Eu)). Organolanthanide-based OLEDs were developed mainly by Christou et al. at OPSYS in Oxford [46].

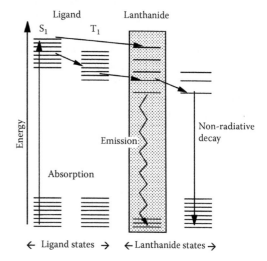

Figure 19.8 An excitation migrates through the molecular (ligand) and atomic (lanthanide) levels in an organolanthanide.

Two major drawbacks for organolanthanide applications are the following. Firstly, for electrical excitation, carriers have to be injected into a rather large bandgap material (deep blue or near UV) even if the emitted light is red. Larger bandgaps generally make good, well-balanced carrier injection harder and lead to reduced power efficiencies and higher onset voltages. Also, as in all transfer-based concepts, the generation of blue light is somewhat problematic. Secondly, the bonding between lanthanide and organic shell has considerable ionic (as opposed to covalent) character. The organic shell acquires a partial negative charge, the lanthanide a partial positive charge. When an electron is injected to the ligand in addition to its partial negative charge, it becomes rather instable against degradation. Consequently, until now, even for encapsulated organolanthanide devices, device lifetimes are poor.

19.5.4 Conjugated dendrimers

Another recent approach to improved efficiency light emitting devices is the use of dendrons with a conjugated core surrounded by non-conjugated dendrimers. A schematic representation of the dendron concept is given in Figure 19.9.

Therein, the conjugated core can be either fluorescent or phosphorescent. The dendron concept seeks to combine the advantages of conjugated polymers and low-molecular weight materials. Dendrimers can be processed from solution and form films in a manner similar to polymers. However, due to the dendronic sidegroups, individual chromophores are shielded from each other. This avoids some of the problems encountered when conjugated polymers are being used. Firstly, conjugated polymers often (but not always) display inhomogeneous broadening of emission spectra.

This is the result of a statistical distribution of effectively conjugated segment lengths, and is detrimental to color purity. Secondly, due to interchain interactions like aggregation and excimer formation, quantum efficiency may be reduced, and excimer emission may again compromise color purity. The major drawback of the dendrimer approach is the much reduced charge carrier mobility due to the increasing separation between conjugated units. Mobility μ scales with the separation between conjugated groups D according to $\mu \sim D^2 \exp(-D/R_0)$.

Samuel et al. studied dendrimers with a core consisting of three distyrylbenzene groups grouped around a central nitrogen, and dendrimeric sidegroups consisting of *meta*-linked vinylene phenylene groups, up to third generation [47]. OLEDs made from higher generation dendrimers displayed narrow EL spectra that approached solution PL spectra of the conjugated core, and external quantum efficiencies rose steeply with dendrimer generation. This is the result of a successful isolation of the emissive core groups from each other. Carrier injection was not affected by dendrimer generation, however, charge carrier mobility decreased dramatically. For second- and third-generation dendrimers, that did display narrow spectra and improved efficiency, mobility was in the order of only 10^{-8} cm^2 V^{-1} s^{-1} [48]. Consequently, in transient EL studies under pulsed drive schemes, increasing EL rise times were found for higher generation. Thus, dendrimer-based OLEDs will not be suitable for fast devices suitable for data communication purposes, or for applications that require high current densities (e.g., organic injection lasers). However, the development of dendrimers with phosphorescent cores [49] may lead to OLEDs with very high efficiencies.

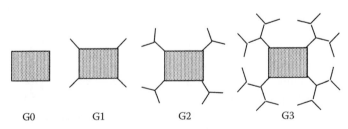

G0   G1   G2   G3

Figure 19.9 Schematic representation of zeroeth-to third-generation conjugated dendrimers. Shaded: conjugated core. Thin lines: dendronic sidegroups (non-conjugated).

19.5.5 Light outcoupling

After the efficient electrical generation of a photon, this photon must still leave the device to be observed by the viewer of the display. Practically, this outcoupling is often very inefficient, as mentioned in Section 19.3.3.

The main loss mechanism here is in-plane waveguiding within the device. To resolve this problem, devices have been designed that suppress waveguiding. With a photoresist-based technique, Samuel et al. [50] have manufactured a corrugated anode that Bragg-scatters light out of in-plane modes and thus, out of the device and towards the observer. In this way, η_{ext} was improved twofold. A more sophisticated approach for improved outcoupling is to establish a resonant vertical cavity mode in a device. Such cavities have spectrally narrow modes and a strong directional (non-Lambertian) emission characteristic. Thus, spectrally pure colors can be generated even from broadband emitters. The principle of EL from resonant cavities has been established [51], however, device manufacture is difficult. This is due to the need to incorporate a dielectric mirror into the device architecture. Resonant cavities have proven a powerful tool for the investigation of fundamental phenomena in quantum optics of organic semiconductors ("strong coupling" [52]), but for practical devices probably will not be cost efficient. They may, however, play an important role in future developments of organic injection lasers.

19.5.6 Examples of high performance OLEDs

After discussing several strategies to optimize device performance, here follows a summary description of some of the most efficient OLEDs demonstrated so far and their performance parameters. This list does not claim to be complete and will probably be outdated quickly.

Approach a: Conjugated polymer, solution processed.
  Device architecture: Synthetic metal coated anode, electron blocking layer, emissive blend layer.
  Emissive layer: Polyfluorene–poly(fluorine-*alt*-benzothiadiazol) blend.

Luminescence data: Low onset (3 V and 0.04 mA cm^{-2} for 0.1 cd m^{-2}), efficiency 4.1 cd A^{-1} at 16.1 mA cm^{-2} and approximately 1000 cd. Brightness 3500 cd m^{-2} at 12.5 V. Reference [34].

Approach b: Small molecule electrophosphorescence, vapor deposited.
  Device architecture: Multilayer (hole injection layer, emissive layer, electron injection layer).
  Emissive layer: Iridium-containing organometallic phosphor doped into electron transporting host layer.
  Luminescence data: 19% external quantum efficiency, 60 ± 5 lm W^{-1} at 1.3 cd m^{-2} and 0.0015 mA cm^{-2}. 13.7% external quantum efficiency and ≈ 20 lm W^{-1} at ≈ 1000 cd m^{-2} and 2.1 mA cm^{-2}. Reference [43].

Approach c: Phosphorescent dendrimer, solution processed.
  Device architecture: Single-layer blend.
  Emissive layer: First generation Iridium-containing dendrimer doped into hole conducting host.
  Luminance data: Maximum external quantum efficiency 8.1% at 13.4 V, 13.1 mA cm^{-2}, giving 28 cd A^{-1} and 3450 cd m^{-2}. Maximum power efficiency 6.9 lm W^{-1} at 12 V, 5 mA cm^{-2}, and 1475 cd m^{-2}. Reference [53].

Approach d: Fluorescent low-molecular weight dye, vapor deposited.
  Device architecture: Multilayer (hole injection layer, hole transport layer, dye-doped emitting layer, electron transport layer).
  Emissive layer: Aluminium(III)quinolinolato organometallic complex doped with Coumarin 6 laser dye. Luminance data: 24 cd A^{-1}, 7.1% external quantum efficiency, 140,000 cd m^{-2} at 12 V. Reference [30].

Approach e: Organolanthanide complex, hybrid solution processed/vapor deposited.
  Device architecture: Double layer, solution-processed blend plus evaporated hole-blocking layer.
  Emissive layer: Blend of PVK polymeric hole transporter, low-molecular weight PBD electron transporter, and emissive organoeuropium complex.

Luminance data: Efficiency 0.73 cd A^{-1} at 23 cd m^{-2} at 16 V drive; brightness 417 cd m^{-2} at 175 mA cm^{-2} and 0.24 cd A^{-1} efficiency at 25 V drive. Reference [46]. Organolanthanide EL with higher efficiency (2.2 lm W^{-1}, 7 cd A^{-1} at 70 cd m^{-2} under 10 V drive) used to be claimed on the webpage of an industrial developer of organolanthanide EL (OPSYS of Oxford, UK), but has now been withdrawn from the webpage.

19.6 CONCLUSION

From the performance data listed in Section 19.4.6, it is obvious that OLEDs are now a competitive option for display applications, and even room lighting.

However, these performance data need to be put into perspective. A single organic LED is not a display that communicates information. Instead, we require pixellated arrays of OLEDs, which for high end applications should be able to display large, full-color pictures at video rates (we are talking about a television screen). Also, one point that has not been discussed here is the all-important issue of device lifetime.

The engineering of efficient, high resolution, full-color pixellated displays at a competitive price, and approaches to improved lifetimes and their rapid testing via accelerated ageing protocols, are no less formidable challenges than the demonstration of new, highly efficient device concepts. Resolving these challenges requires resources that are generally not available in the academic research environment, and thus were addressed mainly by the industrial players in the field. Chapter 7 in Volume 2 covers device engineering issues from an industrial perspective. From that contribution, it will become apparent that these practical challenges have by and large been addressed and resolved now.

In conclusion, OLEDs will outcompete established display technologies—such as cathode ray tube technology and liquid crystal displays—in the near future.

REFERENCES

1. Eley D D 1948 *Nature* **162** 819; Vartanyan A T 1948 *Zh. Fiz. Khim.* **22** 769.
2. Pope M, Kallmann H and Magnante P 1963 *J. Chem. Phys.* **38** 2042; Helfrich W and Schneider W G 1965 *Phys. Rev. Lett.* **14** 229.
3. Partridge R H 1983 *Polymer* **24** 733, 739, 748, 755.
4. Tang C W and van Slyke S A 1987 *Appl. Phys. Lett.* **51** 913.
5. Burroughes J H, Bradley D D C, Brown A R, Marks R N, Mackay K, Friend R H, Burns P L and Holmes A B 1990 *Nature* **347** 539.
6. Hörhold H H, Helbig M, Raabe D, Opfermann J, Scherf U, Stockmann R and Weiß D 1987 *Z. Chem.* **27** 126.
7. Chiang C K, Fincher C R, Park Y W, Heeger A J, Shirakawa H, Louis E J, Gau S C and MacDiarmid A G 1977 *Phys. Rev. Lett.* **39** 1098.
8. Grell M, Bradley D D C, Inbasekaran M and Woo E P 1997 *Adv. Mater.* **9** 798.
9. Sirringhaus H, Wilson R J, Friend R H, Inbasekaran M, Wu W, Woo E P, Grell M and Bradley D D C 2000 *Appl. Phys. Lett.* **77** 406.
10. Redecker M, Bradley D D C, Inbasekaran M and Woo E P 1999 *Appl. Phys. Lett.* **74** 1400.
11. Sheppard J J Jr 1968 *Human Color Perception* (New York: Elsevier) (ISBN 67025430).
12. Baker G L and Pasco S T 1997 *Synth. Met.* **84** 275.
13. Krft A, Grimsdale A C and Holmes A B 1998 *Angew. Chem. Int. Ed. Engl.* **37** 402.
14. Batchelder D N 1988 *Contemporary Physics* **29** 3.
15. Grimme J, Kreyenschmidt M, Uckert F, Müllen K and Scherf U 1995 *Adv. Mater.* **7** 292.
16. Klärner G and Miller R D 1998 *Macromolecules* **31** 2007.
17. Bernius M, Inbasekaran M, Woo E, Wu W and Wujkowski L 2000 *J. Mater. Sci.: Mater. El.* **11** 111.
18. Groenendaal B L, Jonas F, Freitag D, Pielartzik H and Reynolds J R 2000 *Adv. Mater* **12** 481.
19. Andreev A, Matt G, Brabec C J, Sitter H, Badt D, Seyringer H and Sariciftci N S 2000 *Adv. Mater.* **12** 629; Yanagi H and Okamoto S 1997 *Appl. Phys. Lett.* **71** 2563.
20. Frey G L, Reynolds K J and Friend R H 2002 *Adv. Mater.* **14** 265.

21. Barashkov N N and Gunder O A 1994 *Fluorescent Polymers* (New York: Ellis Horwood) (ISBN 0133235106).

22. Bradley D D C 1996 *Curr. Opin. Solid State Mater. Sci.* **1** 789.

23. Bässler H 1993 *Phys. Status Solidi B* **175** 15; Bässler H 2000 *Semiconducting Polymers*, ed Hadziioannou G and van Hutten P F (Weinheim: Wiley-VCH) (ISBN 3-52729507-0).

24. Malliaras G G and Scott J C 1999 *J. Appl. Phys.* **85** 7426.

25. Kim J S, Granstrom M, Friend R H, Johansson N, Salaneck W R, Daik R, Feast W J and Cacialli F 1998 *J. Appl. Phys.* **84** 6859.

26. de Blom P W M, de Jong M J M and Vleggar J J M 1996 *Appl. Phys. Lett.* **68** 3308; de Blom P W M, de Jong M J M and Liedenbaum C T H F 1998 *Polym. Adv. Technol.* **9** 390.

27. Horrowitz G 1999 *J. Mater. Chem.* **9** 2021.

28. Warta W and Karl N 1985 *Phys. Rev. B* **32** 1172; Schön J H, Kloc C and Batlogg B 2001 *Phys. Rev. Lett.* **86** 3843.

29. Campbell A J, Bradley D D C and Lidzey D G 1997 *Appl. Phys. Lett.* **82** 6326.

30. Kido J and Matsumoto T 1998 *Appl. Phys. Lett.* **73** 2866; Kido J and Iizumi Y 1998 *Appl. Phys. Lett.* **73** 2721.

31. Becker H, Burns S E and Friend R H 1997 *Phys. Rev. B* **55** 1.

32. Dittmer J J, Marseglia E A and Friend R H 2000 *Adv. Mater.* **12** 1270.

33. Campbell A J and Bradley D D C 2001 *Appl. Phys. Lett.* **79** 2133.

34. Morgado J, Friend R H and Cacialli F 2002 *Appl. Phys. Lett.* **80** 2436.

35. Bredas J L, Cornil J and Heeger A J 1996 *Adv. Mater.* **8** 447.

36. Baldo M A, O'Brien D F, Thompson M E and Forrest S R 1999 *Phys. Rev. B* **60** 14422.

37. Cao Y, Parker I D, Yu G, Zhang C and Heeger A J 1999 *Nature* 6718 **397** 414.

38. Wohlgenannt M, Tandon K, Mazumdar S, Ramasesha S and Vardeny Z V 2001 *Nature* 6819 **409** 494.

39. Wohlgenannt M, Jiang X M, Vardeny Z V and Janssen R A J 2002 *Phys. Rev. Lett.* **88** art. no. 197401.

40. Wilson J S, Dhoot A S, Seeley A J A B, Khan M S, Köhler A and Friend R H 2001 *Nature* 6858 **413** 828.

41. Sirringhaus H, Brown P J, Friend R H, Nielsen M M, Bechgaard K, Langeveld-Voss B M W, Spiering A J H, Janssen R A J, Meijer E W, Herwig P and de Leeuw D M 1999 *Nature* **401** 685.

42. Lamansky S, Djurovich P, Murphy D, Abdel-Razzaq F, Lee H E, Adachi C, Burrows P E, Forrest S R and Thompson M E 2001 *JACS* **123** 4304.

43. Adachi C, Baldo M A, Thompson M E and Forrest S R 2001 *J. Appl. Phys.* **90** 5048.

44. Adachi C, Baldo M A, Forrest S R, Lamansky S, Thompson M E and Kwong R C 2001 *Appl. Phys. Lett.* **78** 1622.

45. Baldo M A, Thompson M E and Forrest S R 2000 *Nature* **403** 750.

46. Male N A H, Salata O V and Christou V 2002 *Synth. Met.* **126** 7; Moon D G, Salata O V, Etchells M, Dobson P J and Christou V 2001 *Synth. Met* **123** 355; Christou V, Salata O V and Bailey N J 2000 *Abstr. Pap. Am. Chem. Soc.* **219**, 788-ORGN (Pt 2); Christou V 2000 *Abstr. Pap. Am. Chem. Soc.* **219**, 99-INOR (Pt 1).

47. Lupton J M, Samuel I D W, Beavington R, Burn P L and Bassler H 2001 *Adv. Mater.* **13** 258.

48. Lupton J M, Samuel I D W, Beavington R, Frampton M J, Burn P L and Bassler H 2001 *Phys. Rev. B* **63** 155206.

49. Lupton J M, Samuel I D W, Frampton M J, Beavington R and Burn P L 2001 *Adv. Func. Mater.* **11** 287.

50. Lupton J M, Matterson B J, Samuel I D W, Jory M J and Barnes W L 2000 *Appl. Phys. Lett.* **77** 3340.

51. Fisher T A, Lidzey D G, Pate M A, Weaver M S, Whittaker D M, Skolnick M S and Bradley D D C 1995 *Appl. Phys. Lett.* **67** 1355.

52. Lidzey D G, Bradley D D C, Skolnick M S, Virgili T, Walker S and Whittaker D M 1998 *Nature* 6697 **395** 53–55.

53. Markham J P J, Lo S C, Magennis S W, Burn P L and Samuel I D W 2002 *Appl. Phys. Lett.* **80** 2645.

20

Microstructured optical fibers

JESPER LÆGSGAARD AND ANDERS BJARKLEV
Technical University of Denmark

TANYA MONRO
University of Adelaide
and
University of South Australia

20.1 INTRODUCTION

In the mid-1990s, a new class of optical fibers emerged: the microstructured optical fiber (MOF) [1,2]. In these fibers, light is guided by a complex microstructure that often, but not always, includes air holes running along the fiber length. Due to their enormous scientific and technological potential, these fibers have been a major field of study in academic and industrial photonic research during the past two decades. This chapter provides a brief overview of the major types and applications of MOFs, along with a short discussion on their modeling and fabrication.

MOFs, which are often alternatively called *photonic crystal fibers* (PCFs), are classified according to the guiding principle, which can be *index-guiding*, *photonic bandgap guiding* (PBG), or *inhibited coupling* (IC) *guiding*. Index-guiding MOFs guide light using the principle of modified total internal reflection. A microstructure typically consisting of air holes acts to lower the effective refractive index in the cladding region, and thus, light is confined to the solid core, which has a relatively higher index. Some examples are shown in Figure 20.1a and b. In such fibers, the hole-to-hole spacing is typically labeled Λ and d is the hole diameter. The fibers can be made entirely from a single material, typically pure undoped silica, although index-guiding MOFs have also been fabricated in chalcogenide glass [3] and in polymers [4]. The basic operation of index-guiding fibers does not depend on having a periodic array of holes; in fact, the holes can even be arranged randomly [5].

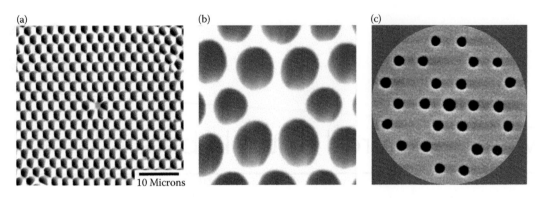

Figure 20.1 Some typical microstructured silica optical fibers. (a) A small-core index-guiding MOF (picture supplied by the ORC, Southampton). (b) A polarization-maintaining index-guiding MOF (picture provided by Crystal Fiber A/S). (c) A bandgap-guiding fiber (picture provided by Crystal Fiber A/S).

The effective refractive index of the cladding can vary strongly as a function of the wavelength of light guided by the fiber. For this reason, it is possible to design fibers with spectrally unique properties that are not possible in conventional solid optical fibers. For instance, index-guiding MOFs with mode areas ranging over 3 orders of magnitude can be designed simply by scaling the dimensions of the structure [6]. Small-mode-area fibers can be used for devices based on nonlinear effects [7], whereas large-mode fibers allow high power delivery [8]. In addition, these fibers can exhibit optical properties not readily attainable in conventional fibers, including endlessly single-mode guidance [9] and an anomalous dispersion well below 1.3 μm [10]. Dispersion and birefringence are two properties that can depend strongly on the cladding configuration, particularly when the hole-to-hole separation is small. By exploiting the innate flexibility provided by the choice of hole arrangement, it is thus possible to design fibers with a wide range of characteristics. Note that the modes of all single-material index-guiding MOFs are leaky modes because the core index is the same as that beyond the finite microstructured cladding, and for some designs, this can lead to significant confinement loss [11].

The second guidance mechanism in microstructured fibers, PBG, can occur if the air holes that define the cladding region are arranged on a strictly periodic lattice. For such structures, photonic bandgaps may appear [1,12]. These are effective index regions, below the effective cladding index, in which no periodic cladding modes are allowed. By breaking the periodicity of the cladding (e.g., by adding an extra air hole to form a low-index core region), it is possible to introduce a mode that is only allowed in the low-index core region, while being forbidden in the cladding region because of the photonic bandgap. This core mode will, therefore, be guided along the fiber, because of the photonic bandgap of the cladding region. If the core mode has an effective index that is either below or above the effective index range covered by the photonic bandgap at the particular wavelength, the core mode will not be guided. Examples of PBG fibers include all-solid fibers with a triangular array of up-doped regions [13], hollow-core silica–air microstructures [12], and fibers with honeycomb air hole claddings, and a smaller air hole in the core center [14]. The latter fiber type, depicted in Figure 20.1c, was historically the first kind of PBG fiber to be fabricated. Here, the necessary photonic bandgaps are formed for comparatively small air hole diameters [15]. However, the aforementioned PBG fiber types have found wider practical applications.

The waveguiding mechanism is illustrated in Figure 20.2, for a structure consisting of circular high-index inclusions in a silica background. Such structures can be manufactured by doping the silica or by infiltrating an index-guiding MOF with a high-index liquid [16–18]. In Figure 20.2, the dark lines illustrate the boundaries of an effective index region, the photonic bandgap, where no modes are found in the periodic cladding structure. The cladding modes at higher effective indices are seen to be localized inside the high-index regions, whereas the

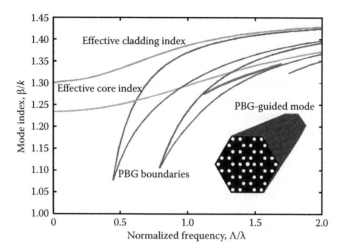

Figure 20.2 Illustration of the two lowest-frequency PBGs of a honeycomb microstructured fiber with a cladding air filling fraction of 30% and a defect hole with same size as the cladding holes. Within the primary PBG, a single degenerate mode is found. This defect mode may not propagate in the cladding structure (due to the PBG effect). The mode is strongly localized to the region containing the extra air hole that forms the core. The inset shows schematically a honeycomb fiber with the core region formed through the use of an extra air hole.

low-index cladding modes extend over the whole cladding structure. Due to the presence of a core defect, in this case, a missing high-index inclusion, a localized core mode can be trapped in the bandgap region as indicated by the thick line traversing the rightmost bandgap. Note that in contrast to planar or fully three-dimensional PBG structures, there are no forbidden frequency regions in a fiber geometry; instead, the bandgaps appear in the distribution of cladding-mode effective indices for a given frequency. A defining feature of PBG fibers is the restriction of mode guidance to finite frequency intervals, as is also seen in Figure 20.2. Depending on the PBG fiber design, one or several of these transmission windows may be present.

The final guidance mechanism to be discussed in this chapter is the so-called inhibited coupling (IC) guiding. IC fibers are similar to PBG fibers in that they guide in a (hollow) core having, on average, a lower refractive index than the cladding. However, in the IC fibers, no true photonic bandgap exists in the cladding, that is, there are bands of cladding states overlapping the effective indices of the guided core modes. The fibers guide light by a reduced photonic density of states at these effective index values, and especially due to a strongly reduced mode overlap between core and cladding modes having similar effective indices [19]. The IC guidance principle has

in recent years been utilized to strongly improve the bandwidth and short-wavelength loss properties of hollow-core MOFs [20].

20.2 MODELING MICROSTRUCTURED FIBERS

The presence of wavelength-scale holes in microstructured optical fibers leads to challenges in the accurate modeling of their optical properties. A wide variety of techniques can be used, ranging from effective step-index fiber models to approaches that incorporate the full complexity of the fiber cross section. Here these approaches will be reviewed and assessed in terms of their suitability for modeling optical properties of microstructured optical fibers such as their mode area, chromatic dispersion, form birefringence, and confinement loss. Some of the issues associated with designing and modeling practical microstructured fibers are discussed.

20.2.1 Effective index methods

The complex nature of the cladding structure of the microstructured optical fiber does not generally allow for the direct use of analysis methods from traditional fiber theory. However, for index-guiding MOFs, a simpler scalar model, based on an effective index of the cladding, has proven to give a good qualitative description of their operation. Birks et al. first proposed this effective index approach in 1997 [9]. The fundamental idea is to set the effective index of the cladding equal to the highest effective index found among the modes of a periodic cladding structure, without the core defect. This index may be straightforwardly calculated by a variety of full-vectorial numerical methods. Figure 20.3 shows the effective cladding index calculated by a plane-wave method for a d/Λ value of 0.5, assuming a constant silica refractive index of 1.45. At long wavelengths, the field penetrates the air holes, and the effective index resembles a geometric average of the microstructure refractive indices. The resulting index contrast to the core (with an index of 1.45) is much larger than what could be obtained with conventional doping techniques, even for this rather moderate air hole size. In the short-wavelength limit, however, the effective cladding index asymptotically approaches the silica index, because the field at shorter wavelengths may partly, and eventually fully, avoid the air holes, as is illustrated in the inset of Figure 20.3.

This procedure allows the effective cladding index of the fundamental cladding mode to be determined as a function of the wavelength. The next step of the method is then to model the fiber as a standard step-index fiber, employing the

strongly wavelength-dependent effective cladding index. The core of the equivalent step-index fiber is assumed to have the refractive index of pure silica, while the core radius is typically taken to be 0.62 times the typical center-to-center cladding hole distance Λ (This assumes that the core is created by the omission of one cladding air hole.).

Despite ignoring the spatial distribution of the refractive index profiles within MOFs, the effective index method can provide some insight into MOF operation. For example, it correctly predicts the endlessly single-mode guidance regime in small-hole MOFs. This method has also been used as a basis for the approximate dispersion and bending analysis presented in [21]. However, this reduced model cannot accurately predict modal properties such as dispersion, birefringence, or other polarization properties that depend critically on the hole configuration within the cladding. Note that when dispersion predictions are required, the effective index approach allows for the inclusion of material dispersion properties through the usual Sellmeier formula. It is also noteworthy that for structures with relatively large air holes, it may be relevant to approximate the fiber to an isolated strand of silica surrounded by air [10].

However, one difficulty that arises when using this approach is the question of how to define the properties of the equivalent step-index fiber. One method for making this choice was described

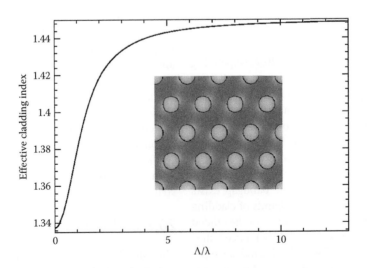

Figure 20.3 Effective cladding index of a triangular air hole structure with $d/\Lambda=0.5$ as a function of normalized frequency, Λ/λ. Insert shows the field energy distribution across the structure for $\Lambda/\lambda=0.73$.

earlier. In this work, the core radius was taken to be 0.62 Λ; the results obtained using the effective index method were made to agree well with full simulations via appropriate choice of this constant of proportionality. Koshiba and Saitoh found that the well-known single-mode criterion for step-index fibers gave accurate results within the effective index model if a core radius of Λ√3 was used [22]. However, for different structures or wavelengths, different choices can become necessary. This restricts the usefulness of this approach, because it is typically necessary to determine the best choice of equivalent structure by referring to results from a more complete numerical model. Reference [23] explores the possibility of choosing the step-index fiber parameters in a more general fashion by allowing a wavelength-dependent core or cladding index. However, to date, no entirely satisfactory method for ascribing parameters to the equivalent fiber has been found.

20.2.2 Numerical methods

Due to the complex spatial structure of MOFs, full numerical simulations are generally required to obtain quantitatively accurate predictions of their optical properties. Whereas the scalar approximation is highly useful for the modeling of standard optical fibers, the high index contrast between silica and air means that it is not *a priori* justified when dealing with MOFs. However, if the guided-mode optical fields are small at the silica–air boundaries, the scalar approximation may give good results [24]. This is for instance found to be the case in MOFs with large core areas. Due to the existence of efficient full-vectorial mode solvers, there is not much incentive in terms of computational speed to adopt the scalar approximation, although it does allow for a somewhat easier implementation of, for example, finite-difference-based mode solvers [24]. However, the scaling properties inherent in the scalar wave equation can sometimes be useful for obtaining design guidelines, even for fibers that are clearly outside its range of validity [25]. Finally, it is possible to design all-silica PBG fibers using a microstructure of doped silica regions with weak index contrasts. In such cases, the scalar approximation is well justified, in spite of the structural complexity [13].

A large variety of numerical methods have been applied to MOF design over the years. Early work

utilized plane-wave techniques borrowed from the concurrent studies of two- and three-dimensional PBG structures, to account for the periodic cladding structures [26–28]. However, MOF geometries are not truly periodic, as the cladding microstructure is typically surrounded by an outer cladding of solid silica, and the plane-wave methods are not easily adapted to model properties such as the leakage loss that result from the finite cladding size. Real-space methods such as the finite-difference [29] and finite-element methods [30], in combination with absorbing boundaries using the technique of perfectly matched layers have been shown to give good results for leakage losses as well as other modal properties. Finite-element methods in particular have proven highly accurate and computationally efficient, even for very complex microstructures, and have become the preferred modeling tool due to their great flexibility and generality. The various types of hollow-core MOFs are especially quite challenging for other methods due to the combination of very thin glass bridges and nodes with very large open spaces in the transverse structure.

An alternative technique that is also able to combine high accuracy with good computational efficiency is the multipole approach [11]. This method is suitable for studying effects caused by the finite cladding region, because it does not make use of periodic boundary conditions. It expresses the modal fields in terms of Bessel function decompositions that are based in each of the cladding air holes, which gives an accurate representation of the fiber symmetries, and therefore the properties such as birefringence, which are strongly symmetry dependent. A variant of this method was also used for the challenging task of proving that the fundamental mode in a step-index MOF with infinite cladding is, in principle, guided at all wavelengths, just as it is the case for a standard step-index fiber [31]. The main drawback of the method is that it is difficult to adapt to microstructures that do not consist of an array of circular holes.

20.3 HIGHLY NONLINEAR INDEX-GUIDING MOFs

20.3.1 Nonlinear silica MOFs

One of the most interesting properties of index-guiding MOFs is that very large effective index

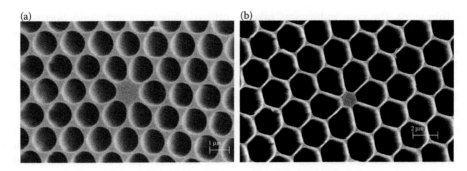

Figure 20.4 Two typical highly nonlinear silica MOFs made at the ORC, Southampton. In each case, a small core diameter combines with a large air-filling fraction to result in a fiber that confines light tightly within the solid central core region. (a) $d/\Lambda \approx 0.85$. (b) $d/\Lambda \approx 0.95$.

contrasts between core and cladding can be obtained by reducing the pitch and increasing the air-filling fraction of the microstructure. As a consequence, the core size may be scaled down, which has two important implications: First, the nonlinear coefficients of the fibers can be strongly increased compared to what is possible with standard fibers. Second, the waveguide dispersion may become very large, opening up unique possibilities for dispersion engineering. Two examples of manufactured structures are shown in Figure 20.4, which illustrates the very high air-filling fractions obtainable in the cladding without compromising the regularity of the periodic structure.

20.3.1.1 DESIGN CONSIDERATIONS

Even though silica is not intrinsically a highly nonlinear material, its nonlinear properties can be utilized in silica optical fibers, if high light intensities are guided within the core. This is because the extremely low loss of silica allows very long propagation lengths. One commonly used measure of fiber nonlinearity is the effective nonlinearity γ [32], which is given by

$$\gamma = \frac{2\pi n_2}{\lambda A_{\text{eff}}}$$

where n_2 is the nonlinear coefficient of the material ($n_2 \approx 2.2 \times 10^{-20}\,\text{m}^2\,\text{W}^{-1}$ for pure silica), A_{eff} the effective mode area, and λ the optical wavelength. For example, a standard Corning SMF28 fiber has $\gamma \approx 1\ \text{W}^{-1}\,\text{km}^{-1}$. By modifying conventional fiber designs, values of γ as large as $20\,\text{W}^{-1}\,\text{km}^{-1}$ have been achieved [33]. By contrast, in index-guiding MOFs,

effective nonlinearities as high as $\gamma \sim 60\ \text{W}^{-1}\,\text{km}^{-1}$ was demonstrated at a wavelength of 1550 nm [34]. This result is near to the limit in mode area that can be achieved in pure silica MOFs (A_{eff} can be as small as $1.7\,\mu\text{m}^2$ at 1550 nm; see Figure 20.5) [35]. Hence, nonlinearities more than 50 times higher than in standard telecommunications fibers and two times higher than the large numerical aperture conventional designs are possible.

To ascertain the range of effective mode areas that can be achieved using silica glass at 1550 nm, consider the extreme case of a rod of diameter Λ suspended in air. As the diameter of the rod is reduced, the mode becomes more confined, and the effective mode area decreases as shown by the dashed line in Figure 20.5. Once the core size becomes significantly smaller than the optical wavelength, the rod becomes too small to confine

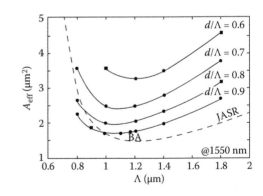

Figure 20.5 Effective mode area (A_{eff}) as a function of the hole-to-hole Λ spacing for a range of fiber designs. JASR corresponds to the case of an air-suspended core of diameter Λ (ORC, Southampton).

the light well and the mode broadens again. Hence, there is a minimum effective mode area that, for a given wavelength, depends only on the refractive index of the rod. For silica, this minimum effective mode area is ~1.5 μm². Figure 20.5 also shows the effective mode area as a function of the hole-to-hole spacing (Λ) for a range of MOFs. These MOF structures also exhibit a minimum effective mode area due to the same mechanism described here. The fiber with the largest air-filling fraction (d/Λ=0.9) has a minimum effective mode area of 1.7 μm², only slightly larger than the air-suspended rod.

As shown in Figure 20.5, the hole-to-hole spacing (Λ) can be chosen to minimize the value of the effective mode area, and this is true regardless of the air-filling fraction. However, it is not always desirable to use structures with the smallest effective mode area, because they typically exhibit higher confinement losses [35]. A relatively modest increase in the structure scale in this small-core regime can lead to dramatic improvements in the confinement of the mode without compromising the achievable effective nonlinearity significantly.

As mentioned earlier, the low loss is a critically important feature in enabling the use of silica fibers for nonlinear optics applications. Fiber losses as low as 0.28 dB km⁻¹ has been demonstrated in pure silica MOFs with a mode area comparable to that of standard fibers [36]. However, when the core diameter is reduced to scales comparable to (or less than) the wavelength of light guided within the fiber, confinement loss arising from the leaky nature of the modes can contribute significantly to the overall fiber loss [35]. Indeed, the small-core MOFs fabricated to date have been typically more lossy than their larger-core counterparts. In this small-core regime, unless many (more than six) rings of holes are used, the mode can *see over* the finite cladding region. Thus, it is important to be able to reliably calculate the confinement loss characteristics of MOFs. Here, we briefly outline some general design rules for designing low-loss high-nonlinearity MOFs described in [35].

Lower confinement loss and tighter mode confinement can always be achieved by using fiber designs with larger air-filling fractions. Finally, for all fiber designs, it is always possible to reduce the confinement loss by adding more rings of holes to the fiber cladding. In the limit of core dimensions that are much smaller than the wavelength guided by the fiber, many rings (>6) of air holes

are required to ensure low-loss operation, which increases the complexity of the fabrication process. Note that although fiber loss limits the effective length of any nonlinear device, for highly nonlinear fibers, short lengths (<10 m) are typically required, and thus, loss values on the order of 1–10 dB km⁻¹ can be readily tolerated. In addition, note that reducing the core diameter to dimensions comparable to the wavelength of light generally increases the fiber loss for another reason: in relatively small-core fibers, light interacts much more with the air/glass boundaries near the core, and thus, the effect of surface roughness can become significant [37].

20.3.1.2 APPLICATIONS OF SMALL-CORE MOFs

The interest in high nonlinearities stems mainly from applications in telecommunication where fast signal processing at low optical power levels is becoming increasingly important. Although device applications such as 2R regeneration [38] and all-optical time-domain demultiplexing [39] were demonstrated quite early on, MOFs have not made a big impact on telecommunication and are today being challenged by integrated optics materials platforms such as silicon and chalcogenide soft glasses, which promise even higher nonlinear coefficients, along with reduced device footprints.

A commercially much more interesting application for small-core MOFs is within nonlinear frequency conversion. This application stems from the need to convert the powerful near-infrared fiber lasers based on rare-earth laser ions such as Yb and Er into other wavelength bands, especially those below 1 μm, which are of great interest for one- and two-photon biomedical imaging. For frequency conversion, the interesting aspect of small-core MOFs is their unusual dispersion properties. The large nonlinear coefficients can actually be seen as a drawback, because plenty of power is available from today's picosecond and femtosecond fiber lasers, so small MOF cores become a limiting factor for power scaling.

The breakthrough for nonlinear applications of microstructured fibers came with the experimental demonstration of supercontinuum generation in microstructured silica fibers reported by Ranka et al. [40] in 2000. The possibility of shifting a zero-dispersion wavelength (ZDW) to ~800 nm opened up the possibility of launching femtosecond pulses from Ti:sapphire lasers into a fiber with anomalous dispersion, with pulse breakup and continuum formation into the visible range as a result. Ranka's results spurred a huge research activity in subsequent years [41], which eventually has resulted in the appearance of commercial supercontinuum sources pumped by Yb-doped picosecond fiber lasers. These sources provide extremely broad and flat spectra spanning from the blue end of the visible region, between 400 and 500 nm, and toward the mid-infrared region, between 2200 and 2400 nm, where the infrared loss edge of silica eventually limits the spectrum. Such spectra are obtained in MOFs with ZDWs slightly below 1000 nm, to obtain low anomalous dispersion at the pump wavelength, which is typically 1064 nm. Pulse breakup occurs through modulational instability, and is followed by the formation of a large number of very short solitons, along with dispersive waves below the pump wavelength [41]. It has been shown that these dispersive waves can be trapped behind the solitons, and

continuously blue-shifted by cross-phase modulation effects in such a way that their group velocities are always matched to the solitons, which redshift due to the Raman effect [42]. Therefore, it is really the group-velocity curve that determines the shape of the final supercontinuum spectrum, and proper design of this curve is of paramount importance. Typically, one aims to match the group velocity of the solitons at the loss edge (2200–2400 nm) to that of waves at the shortest wavelength desired, usually below 500 nm. In Figure 20.6, spectra from three different commercial supercontinuum sources are shown. The difference in the extent into the visible region is mainly determined by this group-velocity phase matching, which shifts with the ZDW, so that fibers with a shorter ZDW also have group-velocity matching to shorter visible wavelengths.

Supercontinuum sources have seen a wide range of applications, including optical coherence tomography [43] and frequency metrology [44]. However, the most important application is perhaps within the rapidly growing field of biomedical imaging, especially confocal microscopy, where

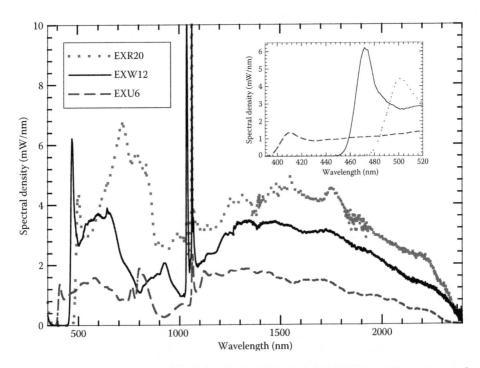

Figure 20.6 Supercontinuum spectra from three different commercial sources. The main panel shows the full spectra, whereas the inset shows a close-up of the short-wavelength edge. The main difference is the extent of the visible edge of the spectrum, which is determined by different group-velocity matching properties between solitons at the long-wavelength transmission edge (2200–2400 nm) and dispersive waves in the visible region. Data provided by NKT Photonics A/S.

fluorescent markers are introduced in biological samples, and excited by focused laser light [45,46]. Thus, a wavelength-flexible light source with good spatial coherence can give a very large degree of flexibility in the number and range of fluorophores, which can be used for imaging the movement of, for example, different proteins in a biological specimen. Commercial use of supercontinuum sources in this area took off in 2008 with their incorporation in the Leica TCS SP5 X confocal microscope. Coupled with a tunable acousto-optic filter, the supercontinuum replaces a large battery of lasers at different excitation wavelengths, which was previously a necessity in such instruments.

The main drawback of present-day supercontinuum sources is their inherently noisy spectrum, which stems partly from the initial modulational instability breakup of the pump pulse, which is seeded by noise, and partly from the complex and "turbulent" nature of the subsequent nonlinear propagation [41,47]. It is expected that the utilization of fiber-based femtosecond pumps will allow reduction of this noise in the coming years. Some authors have also discussed lifetime effects, caused by chemical modifications of the silica glass due to the intense pump power [48]. In this regard, it is noteworthy that the supercontinuum sources are typically operated at repetition rates in the tens of megahertz, while delivering more than 1 W of visible power, and 5–10 W in total over the whole spectrum. Even if lifetime effects are not a serious problem for most laboratory uses of supercontinuum sources, they may be problematic for industrial applications. Therefore, research into improving the resilience of the base material (pure silica) toward prolonged exposure to intense near-infrared and visible radiation is ongoing.

A promising alternative to supercontinuum sources for generating low-noise visible radiation with clean temporal profiles is the so-called Cherenkov sources [49,50]. In these sources, femtosecond near-infrared pump pulses are injected into a MOF featuring anomalous dispersion at the pump wavelength. The femtosecond pulse undergoes soliton compression to few-cycle duration, and this is accompanied by the emission of a short-wavelength pulse, subject to the criterion that the phase velocity of the emitted dispersive wave should be matched to that of the soliton in a co-moving reference frame. This radiation is conventionally termed "Cherenkov" radiation, due

to certain formal similarities to the more widely known Cherenkov radiation produced by superluminal charged particles [51]. Thus, the wavelength of the Cherenkov radiation is determined by the fiber dispersion properties, and once again the highly flexible design space for the dispersion curve is the crucial advantage of MOFs in this context. It has been experimentally shown that high-power Cherenkov wave generation at wavelengths below 400 nm is possible with Er-based pump lasers [52]. Fully fiber-integrated low-noise systems have also been demonstrated [53,54]. The main challenge for this technology is currently the limited wavelength flexibility imposed by the phase-matching criterion which is fixed once and for all by the chosen fiber design. Figure 20.7 illustrates the range of visible and near-infrared wavelengths accessible when pumping with an Er-based laser in different fiber types, including both standard fibers and large- and small-core MOFs. It is seen that the use of a small-core MOF is instrumental in reaching the short-wavelength edge of the visible spectrum.

Yet another option for frequency conversion is the technique of four-wave mixing, which allows for the generation of "signal" and "idler" sidebands to a strong pump signal, such as Yb or Er fiber laser pulses. MOF technology allows phase matching of Yb-based pumps to a wide range of signal/idler wavelengths, especially when pumping is just below the ZDW. An example is shown in Figure 20.8 for a standard design with a triangular array of air holes and a core area around $10\,\mu m^2$. It is notable that a tuning of the pump wavelength over only a few tens of nanometers allows one to span an idler wavelength range from below 750 nm to above 1000 nm, with the signal wavelength reaching into the common telecom bands. This wide range of phase-matched wavelengths make small-core MOFs useful building blocks for partly or fully fiberized parametric oscillators and amplifiers. Of particular interest is the generation of short-pulsed radiation in the biological transparency window between 700 and 1000 nm, which may be applied in a number of biomedical imaging modalities such as two-photon fluorescence spectroscopy and coherent anti-Stokes Raman spectroscopy (CARS). As early as 2005, Deng et al. showed a fiber-optic parametric oscillator (FOPO) featuring idler wavelengths tunable between 930 and 990 nm, with a corresponding signal wavelength between 1060 and 1200 nm [55]. In this work, the tunability was

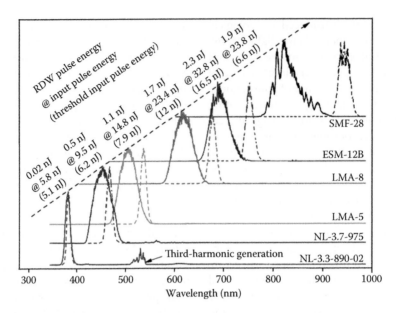

Figure 20.7 Spectra of visible Cherenkov radiation obtained from pump pulses at 1550 nm in different fibers, from standard step-index fiber (SMF28), down to a PCF with 3 μm core and a large air-filling fraction of the cladding. (Reproduced from Tu, H. et al., *Opt. Express*, 20, 23188–23196, 2013.)

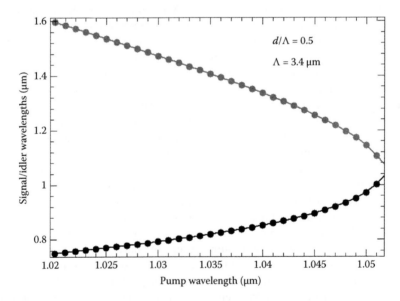

Figure 20.8 Phase-matched signal and idler wavelength, as a function of pump wavelength, for four-wave mixing in a PCF with $d/\Lambda = 0.5$, $\Lambda = 3.4$ μm.

based on tuning the pump laser between 1020 and 1038 nm. For CARS imaging, Zhai et al. demonstrated a FOPO tunable between 840 and 930 nm by employing two different MOFs in combination with varying lengths of the dispersive cavity [56]. A similar tuning technique was recently used to demonstrate a fully fiber-integrated FOPO tunable

between 988 and 1046 nm in the idler, and 1085–1151 nm in the signal [57]. Although parametric sources are presently less commercially developed than the supercontinuum sources discussed earlier, they show greater promise with respect to the generation of high spectral power densities over a wide range of input/output pulse durations. One limiting factor for power scaling is the relatively low core area of the necessary MOFs; however as described in Section 20.5, PBG fiber technology may have the potential to lift some of these limitations.

20.3.2 Nonsilica MOFs

As described earlier, using single-material silica MOFs, effective nonlinearities as high as $\gamma \approx 60$ W^{-1} km^{-1} are possible. Moving to glasses with a higher refractive index than silica, it is possible to access material nonlinearities that are orders of magnitude larger than that of silica. For example, the chalcogenide glass As$_2$S$_3$ has a refractive index of ~2.4 at 1550 nm and is 100 times more nonlinear than silica glass (n_2(As$_2$S$_3$) ~2×10^{-19} W^{-1} km^{-1} [58]). The Schott lead-glass SF57 has a refractive index of 1.8 at 1550 nm and is 20 times more nonlinear than silica (n_2(SF57) $\approx 4 \times 10^{-19}$ W^{-1}km^{-1} [59]). Recall that for silica, the theoretical lower bound for the effective mode area is ~1.5 μm^2. For the higher index SF57 glass, the minimum effective mode area is reduced to ~0.75 μm^2. Hence, in addition to providing high intrinsic material nonlinearity, such glasses also offer improvements in terms of mode confinement relative to silica.

Conventional fibers made using As$_2$S$_3$ have been used to reduce the power levels and fiber lengths required for all-optical switching [58]. Further improvements are to be expected, when such a highly nonlinear glass is combined with the tight mode confinement offered by an MOF structure. Although compound glasses are clearly attractive for nonlinear devices, the application of compound glass fibers has been limited because it is difficult to fabricate low-loss single-mode fibers using conventional techniques. Single-material fiber designs avoid core/cladding interface problems, and thus should potentially allow low-loss fibers to be drawn from a wide range of novel glasses. As we demonstrate here, MOF technology provides a powerful new technique for producing single-mode compound glass fibers. We briefly review some recent results using SF57 glass from [60].

Like many compound glasses, SF57 has a low softening temperature (~520°C) and so it is possible to extrude the MOF preform directly from bulk glass, and the cross section of a fiber fabricated from an extruded preform is shown in Figure 20.9.

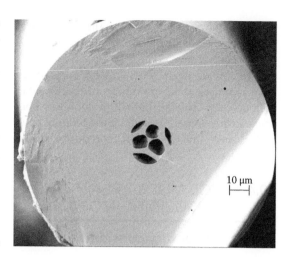

Figure 20.9 Scanning electron microscope image (SEM) of the extruded SF57 microstructured fiber from [60]. Picture supplied by the ORC, Southampton.

The core diameter is ≈2 μm and the core is suspended by three ≈2 μm long supports that are less than 400 nm thick. These supporting struts allow the solid core to guide light by helping to isolate the core from the outer solid regions of the fiber cross section. Although single-material fibers support only leaky modes, it is possible to design low-loss fibers of this type (see Reference [61]). This can be done by ensuring that the supporting struts are long and fine enough that they act purely as structural members that isolate the core from the external environment.

Extrusion offers a controlled and reproducible method for fabricating complex structured preforms with good surface quality. In addition, extrusion can be used to produce structures that could not be created with capillary stacking approaches, and so a significantly broader range of properties should be accessible in extruded MOFs.

The measured effective nonlinearity for this fiber is $\gamma \approx 550$ W^{-1} km^{-1}, and assuming the value of n_2 for this material given earlier, this implies that $A_{\mathrm{eff}} \approx 3.0 \pm 0.3$ μm^2. This mode area is ≈30 times smaller than the conventional SMF28 fiber, and the intrinsic material nonlinearity of the SF57 glass provides an n_2 that is ≈20 times larger than silica, and so the effective fiber nonlinearity is more than 500 times larger than that of standard single-mode optical fibers. These results represent the first demonstration of single-mode guidance and nonlinearity enhancement in a nonsilica MOF.

Possibilities exist to increase the nonlinearity still further through the use of glasses with still higher nonlinearity than SF57 (such as As_2S_3), and to design fibers with smaller cores. Such fibers promise a route toward record effective fiber nonlinearities, paving the way to nonlinear fiber devices with unprecedentedly low operating powers (1–10 mW) and remarkably short device lengths (0.1–1 m).

20.4 LARGE-MODE-AREA INDEX-GUIDING MOFs

20.4.1 Elementary considerations

Index-guiding MOFs offer an alternative route toward large mode areas [8]. The development of large-mode area microstructured fibers (LMA-MOFs) is important for a wide range of practical applications most notably those requiring the generation, amplification, and delivery of high-power optical beams. For many of these applications, spatial mode quality is a critical issue and such fibers should preferably support just a single transverse mode. Relatively large-moded single-mode fibers can be made using conventional fiber doping techniques such as modified chemical vapor deposition simply by reducing the NA of the fiber and increasing the fiber core size. However, the minimum NA that can be reliably achieved is restricted by the accuracy of the control of the refractive index difference between the core and the cladding. By contrast, in an endlessly-single-mode MOF, the fiber pitch can in principle be scaled to any value, without leaving the single-mode regime. This can be understood within the framework of the effective index model, as a consequence of the effective cladding index tending toward the index of the base material in the limit of large pitch. The approach to the base material index happens in such a way as to make the effective V-parameter converge toward a constant value, which may be below the single-mode cutoff value of 2.4 if the fiber has the proper design—specifically, the d/Λ ratio should be below 0.42 [62]. In practice, the manufacture of large-pitch MOFs with such values of d/Λ is not a problem, and the core size of MOFs can therefore be scaled up until one encounters the limits imposed by bend loss, as described in the next section.

For amplifier fibers with active cores, an additional problem is that the core has to be doped with at least the active material (typically, Yb or Er), and therefore, the index difference between the doped core and undoped cladding material must be controlled. It would therefore seem that one is back to the limitations imposed by conventional preform-manufacture technology. However, the stack-and-draw approach commonly used for MOF fabrication offers a solution. Rods of silica with different refractive indices (e.g., rare-earth doped and fluorine-doped silica) may be stacked together and drawn down to a single cane, which then becomes the core rod in the final MOF preform. When finally drawn into a fiber, the different core materials will form a nanoscale structure, with an effective refractive index, which can be controlled through the stacking pattern, and doping radii of the original stacked rods. In this way, the control of the core refractive index relative to the surrounding pure silica can be improved by about an order of magnitude that is sufficient to leverage the advantages of the special MOF properties. A further advantage is that the air hole size can be easily adjusted during the final drawing process to compensate for small deviations of the target core index [63]. This is typically achieved by slightly pressurizing the air holes during the draw, which allows the inflation of their diameter in a well-controlled manner.

20.4.1.1 BEND LOSS LIMITATIONS

Because LMA-MOFs rely on a very small effective index contrast between core and cladding, the fibers can be sensitive to macro-bending (this is discussed further later in this section). The fiber properties can therefore be also highly sensitive to the precise details of the fiber structure, and so accurate fabrication techniques are required. Figure 20.10 shows cross-sectional pictures of two typical LMA-MOFs with regularly arranged air holes. The models described in Section 20.2 can be applied to model the optical properties of these fibers. Polarization effects are typically less important in this class of fibers, and it is often sufficient to use a scalar model.

Macroscopic bend loss ultimately limits the practicality of such large-mode fibers, and so understanding bend loss is important in the design of this class of fiber. Two distinct bend loss mechanisms have been identified in conventional fibers: transition loss and pure bend loss [64]. As light travels into a curved fiber, the mode distorts,

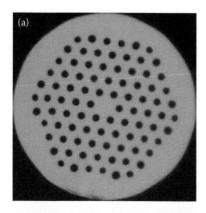
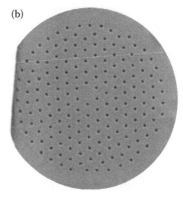

Figure 20.10 Full cross section of two typical large-mode-area MOFs with different air-filling fractions (both have core diameters of 15 μm). (a) Photograph provided by Crystal Fiber A/S. (b) Photograph provided by the ORC, Southampton.

causing a transition loss (analogous to a splice loss). Pure bend loss occurs continually along any curved section of fiber: at some radial distance (r_c in Figure 20.12d), the tails of the mode need to travel faster than the speed of light to negotiate the bend, and are thus lost.

Using the simplified effective index method described in Section 20.2, it is possible to derive useful formulas for the pure bend loss in these fibers by applying standard results for the power loss coefficient of standard step-index fibers (see References [63,65]). Utilizing this approach, the pure macro-bending loss can be derived from the coefficients of the Bessel function of the equivalent step-index fiber, as done in Reference [66].

Figure 20.11 illustrates the bending radius dependence of the operational windows for a specific LMA-MOF with air hole diameter $d=2.4$ μm and hole-to-hole spacing $\Lambda=7.8$ μm. As can be seen in Figure 20.11, a short-wavelength loss edge is evident for LMA-MOFs. This is in contrast to the standard fiber case, where only a long-wavelength loss edge is found. Therefore, noting that the effective cladding index only depends on the ratio between pitch and wavelength, when scaling up the pitch, macro-bending loss will ultimately limit the mode area of a practically useful fiber.

Figure 20.11 further includes the predictions of the effective index model, and as can be seen, the model is capable of predicting accurately the

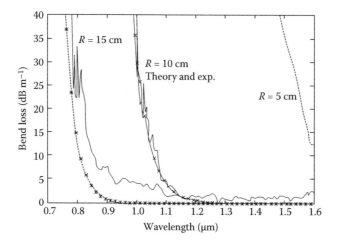

Figure 20.11 Calculated and measured values of spectral bending loss of LMHF. For operation around 1.5 μm, the critical bending radius is ~6 cm. (From Sørensen, T. et al., *Proceedings of ECOC'2001*, Amsterdam, The Netherlands, 2001.)

spectral location of the short-wavelength bend loss edge [8]. The long-wavelength bend loss edge of this specific fiber is positioned at mid-infrared wavelengths for all bending radii. Although this figure presents results for one particular LMA-MOF, a number of generalizations can be made. Unsurprisingly, the critical bending radius and the spectral width and location of the operational window depend strongly on hole size (d) and hole-to-hole spacing (Λ). Generally, larger holes result in broader operational windows, whereas the hole-to-hole distance roughly determines the center position of the window (as a first approximation, the minimum bend loss occurs at a wavelength around $\Lambda/2$) [66].

The results described earlier focused on pure bend losses, which will be the dominant form of bend loss in long fiber lengths. For shorter lengths of fiber, it can be important to investigate the impact of transition loss too. Transition and pure bend losses can be distinguished experimentally by progressively wrapping a fiber around a drum of radius R_0 [64]. The fiber experiences a sharp change of curvature as it enters and leaves the drum surface, which results in transition losses at these points. As the angle is increased, the length of the curved section (and the pure bend loss) increases linearly. Figure 20.12 shows the measured loss as a function of angle for $R_0 = 14.5\,\text{mm}$ (taken from Reference [67]). Each data set shows two regions: the curved section of each plot is the transition region, while the pure bend loss dominates as the length of the bent fiber increases. Results for two different angular orientations of the same fiber illustrate that the geometry of the cladding structure has a noticeable effect on the bend loss characteristics. Hence, in order to understand and predict such effects, it is necessary to use a numerical method that accounts for the complex structure, because effective index methods cannot account for orientationally dependent behavior.

One way of modeling the propagation of light in a fiber with a radius of curvature R_0 is to scale the refractive index using the transformation: $(1 + (2r \cos \alpha)/R_0)^{1/2}$, where the coordinates are defined in Reference [68]. Using this transformation, the modes of the bent microstructured fiber can then be calculated [67]. Note that the slant introduced by this transformation cannot be described in all the models. To calculate the modes shown in Figure 20.12, the hybrid orthogonal function model [6] was used: modal intensities are shown for (a) a straight fiber, and (b) and (c) for fiber bent in horizontal and vertical directions, respectively.

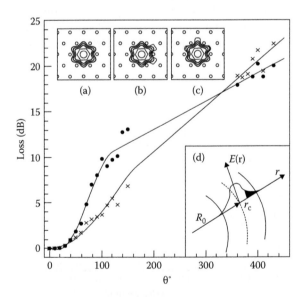

Figure 20.12 Loss as a function of angle for two fiber orientations (see text and Reference [67]) for the MOF in Figure 10.10 (right). (a) Calculated mode profile for this MOF; (b) and (c) calculated mode for a bend in horizontal and vertical directions, respectively (contours every 2 dB); (d) slice through mode propagating around bend of radius R_0.

Using these modes, the transition loss can be calculated as the overlap between the straight and bent modes, and for pure bend loss, it is assumed that the fraction of energy in the guided mode at $r > r_c$ is lost over some distance scale. This model gives good quantitative agreement with experimental data, and allows trends relating to the angular orientation of the fiber to be identified.

20.4.2 Active large-mode-area fibers

The capillary-stacking techniques that are generally used to make LM-MOFs can be readily adapted to allow the incorporation of high-NA air-clad inner claddings within jacketed all-glass structures [69]. This technique was described by DiGiovanni et al. [70], who outlined how an outer cladding having a very high air-filling fraction may result in an effective refractive index below 1.35. The optical characteristics of the light guided within the core of the optical fiber are essentially independent of the second outer cladding, and the fiber can become insensitive to the external environment. The advantage in this connection is the large index step between air and silica, which makes microstructured fibers with extremely high NA values (compared to standard fibers) possible. Fibers with NAs as high as 0.9 have been demonstrated, and

NA values of 0.5–0.6 are routinely achieved in production of fibers. High-NA fibers (typically multimode) collect light very efficiently from a very broad space angle and distribute light in a broad angle at the output end. An example of a high-NA multimode fiber is shown in Figure 20.13. Such fibers can in themselves find applications where weak signals are to be collected efficiently, or high optical powers need to be delivered without particular requirements on beam quality. However, the main potential of the high NA values obtainable with MOF technology lies within the field of high-power, rare-earth-doped (e.g., Yb^{3+}, Nd^{3+}), LMA devices. In the so-called double-clad (DC) amplifier fibers, an outer "pump core" with a large NA is used to efficiently collect pump light from a large array of laser diodes. The pump light is absorbed in a smaller rare-earth doped low-NA "signal core," where a single- or few-moded signal may be amplified. These amplifiers are therefore able to massively improve the limited brightness of the highly energy-efficient laser diodes that are in the market today. MOF technology here brings the combined advantages of allowing a very large NA for the pump core, and simultaneously, a very small NA for the signal core [71]. A further advantage is that the air holes defining the large-NA pump core are much more power tolerant than the low-index polymer claddings, which are used in conventional

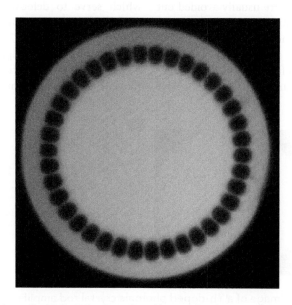

Figure 20.13 Cross-section of a high-NA pure-silica fiber with a large multimoded core. The NA is higher than 0.55. Photograph provided by Crystal Fiber A/S.

DC fibers. For this reason, high-power fiber amplifiers have become a major application for MOF technology.

The capabilities of DC-MOFs are particularly suited to minimize nonlinear optical effects within the amplifier. For amplification of short pulses (nanoseconds or shorter), both spectral broadening from self-phase modulation, and Raman scattering effects may limit the output power. In the case of narrow-band continuous-wave sources, Brillouin scattering is a major limitation. All of these effects can be mitigated by increasing the core area, and/or shortening the length of the amplifier fiber. MOF technology facilitates both strategies—single-mode signal cores with very large areas can be obtained as described earlier; at the same time, the opportunity of obtaining a high NA for the outer cladding helps to reduce the area of the pump core, for a given brightness of the pump diodes, and thus shorten the amplifier by increasing the pump absorption in the signal core. DC-MOFs are often fabricated with pump/ signal core area ratios of only 10–30, whereas standard DC fibers often have ratios well above 100. The area of the signal core is limited to around 1–$2000\,\mu m^2$ by bending effects. For areas of this size, bend-induced modal deformation as depicted in Figure 20.12 significantly reduces the effective mode area for coil diameters as large as 40 cm [72], negating the advantages of further core area scaling. Larger coil diameters are usually avoided out of concern for the footprint of the packaged laser/ amplifier system. If further scaling of the effective mode area is desired, a better alternative is to abandon flexible fibers in favor of stiff microstructured silica rods, with outer diameters of about

1.5 mm, and typical lengths of 1–1.5 m. In this way, one obviously avoids any bending effects, and the very large outer diameter also serves to isolate the guided mode from roughness scattering at the outer surface. An example of such a structure is shown in Figure 20.14 [73]. Microstructured rod amplifiers have been scaled to core sizes of up to $135\,\mu m$ [74–76], which implies effective mode areas approaching $10,000\,\mu m^2$. Commercially available off-the-shelf products currently feature core sizes of 55–$85\,\mu m$, and corresponding mode field diameters of 45–$65\,\mu m$. For these extreme microstructures, it has proven advantageous to utilize the photonic bandgap effect to discriminate against higher-order modes [63]. An example of such a rod structure is shown in Figure 20.15. The light rings surrounding some of the cladding air holes are regions of doped glass with increased refractive index. They form a band of high-index cladding states, which can be designed to cut out the higher-order modes of the core, while leaving the fundamental mode well-guided. Because the high-index areas surround the air holes, their size and therefore the position of the high-index cladding states can be adjusted very accurately in the production process by controlling the air hole size [63]. An alternative approach is represented by the so-called large-pitch fibers (LPF), an example of which is shown in Figure 20.16. They feature a very open inner cladding structure with few air holes, which serve to delocalize higher-order modes, while leaving the fundamental mode well confined to the active core area [77,78]. This is sufficient to discriminate against higher-order modes, since amplification is proportional to the overlap with the rare-earth doped core region. The advantage

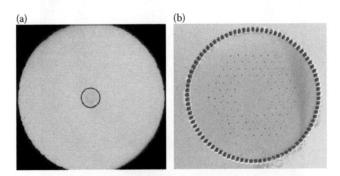

Figure 20.14 Microscope image of a Yb-doped photonic crystal rod amplifier. (a) Shows the full extent of the rod, while (b) shows a close-up of the microstructured region, with inner and outer cladding. (Reproduced from Limpert, J. *Opt. Express*, 14, 2715–2720, 2006.)

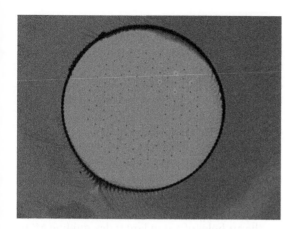

Figure 20.15 Microscope image of a photonic crystal rod amplifier with a photonic-bandgap cladding structure. The light rings surrounding the air holes are Ge-doped sections with raised refractive index, which serve as resonant filters for unwanted higher-order modes in the core. Picture provided by NKT Photonics A/S.

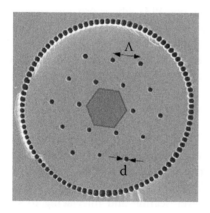

Figure 20.16 SEM micrograph of a large-pitch double clad photonic crystal rod amplifier with a pitch of 30 μm. The green hexagon indicates the extent of the Yb-doped active core. (Reproduced from Jansen, F. et al., *Opt. Express*, 20, 3997–4008, 2012.)

of the LPF design is its structural simplicity; however it does rely on a very precise control of the core refractive index. Recent research has shown that control to an order of a few times 10^{-5} is feasible, which has enabled the fabrication and single-mode operation of rod structures with mode field diameters up to 130 μm [77]. Both bandgap and LPF amplifiers have proven useful for scaling up peak and average powers of ultrafast laser systems. An example is the recent report of a chirped-pulse

amplification 200 fs laser system featuring 7.8 GW peak power after compression from a single LPF rod amplifier with a mode field diameter of 80 μm, and 22 GW peak power with 230 W average power from a beam combination of four such amplifiers [79]. Laser systems of this kind feature many bulk-optics components such as grating stretchers and compressors, and phase shaping tools. Therefore, the advantage of MOF technology in this case does not lie in the prospect of monolithic fiber integration, but rather in the superior mode stability of the rod amplifiers compared to bulk solid-state amplifiers [79].

Regarding the mitigation of Brillouin scattering effects in narrow-band amplifiers, a third strategy is to structure the acoustic properties of the core so as to suppress guidance of acoustic modes, or broaden the acoustic gain spectrum, with a resulting lowering of the gain peaks. This strategy in combination with the inner/outer core area scalings described earlier has been used to obtain 1.5 MW peak powers for 1.55 ns pulses [80], and kW power levels of continuous-wave lasers with line widths in the order of 1 GHz [81], all in flexible fibers. Alternatively, using 100 μm-core rod amplifiers was also shown to allow 1.35 MW peak power of subnanoseconds pulses with perfect single-mode beam quality at the output [74].

A fundamental issue with large-core amplifiers is that operation at high average power leads to a heating of the core, which will induce a refractive-index gradient through the thermo-optic effect. Therefore, even a fiber/rod, which is single-mode by design will eventually become multimoded at elevated power levels, at least in that part of the amplifier where the pump-signal power transfer is strongest. Furthermore, thermo-optic effects can induce instabilities in the spatial mode structure in such multimoded fibers, which typically limits the average power to a few hundred W [82,83]. Thermo-optic effects may also adversely affect the fundamental guided mode through mode shrinking and resonant couplings to delocalized cladding modes at certain power levels. These effects have most clearly been demonstrated in LPF fibers [76], however the fibers may be designed to have a smooth modal shape at maximum operating power. In the case of photonic bandgap fibers, similar effects are seen, with the narrow wavelength region of good single-mode guidance shifting with operating power. Also in this case, fibers may easily be designed so as to provide single-mode operation with a well-guided fundamental mode in the "hot" fiber, that is, during high-power operation. In one experiment, it was shown that a correct positioning of the bandgap increased the modal instability power threshold by about 40% [84].

20.5 PBG FIBERS

While the index-guiding MOFs discussed in the previous sections can to a first approximation be regarded as a variant of the traditional step-index fiber with a larger and wavelength-dependent index contrast between core and cladding, the PBG fibers constitute a fundamentally different class of waveguides. In PBG fibers, light is not confined to a region of higher refractive index, but is rather localized at a defect placed in a PBG material, which suppresses transverse propagation at the frequency of the guided mode. Due to this fundamental difference, PBG fibers attracted considerable academic interest in the early stage and several kinds of PBG fibers were realized experimentally within 2–3 years after the first successful fabrication of an index-guiding microstructured optical fiber [12,14]. In this section, we discuss the two main categories of PBG fibers: *solid-core PBG fibers,* which feature spectral filtering and peculiar dispersion properties compared with index-guiding fibers, and *hollow-core PBG fibers,* which allow low-loss single-mode guidance in an air core, and thus, represent one of the greatest breakthroughs of MOF technology.

For the PBG-guiding mechanism to operate, the bandgap must extend over the whole plane perpendicular to the direction of propagation. In silica–air structures, such a gap can only be achieved for a finite longitudinal propagation constant, the minimum value required being dependent on the fiber structure. The simple triangular arrangement of cladding holes (commonly used for making index-guiding microstructured fibers) only provides a complete bandgap for air-filling fractions of 30% or higher. In contrast, arranging the holes on a honeycomb lattice makes it possible to open up complete gaps at much lower fill fractions (<1%) [15]. For this reason, the first PBG fiber experimentally fabricated was based on the honeycomb lattice [14]. However, practical applications for these fibers did not materialize, although improved fabrication methods made it possible to realize hollow-core PBG fibers based on triangular cladding structures with extremely large air holes shortly afterward [12]. An alternative approach to bandgap formation allows for relatively easy fabrication of solid-core PBG fibers. In this approach, bandgaps are formed in a cladding containing an array (typically triangular) of high-index inclusions. The bandgaps appear between cladding modes localized in the high-index inclusions, and states extending over the whole cladding structure, but with some structure around the high-index inclusions to orthogonalize to the localized states. Such fibers may be fabricated either by infiltrating the air holes of a standard index-guiding MOF with a high-index fluid [16,85], by manufacturing an all-silica fiber with doped (typically by Ge) high-index inclusion [13,86], or even by making a microstructure of different glass types [87]. The all-silica approach has the advantage that a silica structure free of air holes makes splicing to other (standard) fibers considerably easier.

Solid-core PBG fibers provide large anomalous waveguide dispersion even for medium-sized cores, which is in contrast to index-guiding MOFs, where small cores are a requirement to achieve this property. As a result, dispersion compensation of standard fibers at short wavelengths is an obvious application for these fibers. This has most clearly been demonstrated by their use as intracavity dispersion compensation elements in femtosecond fiber laser cavities operating in the Yb band around 1030 nm [88,89]. The all-silica nature and relatively large core size of the fibers facilitates power scaling and interfacing to standard fibers. An example of such a fiber structure, and its transmission and dispersion properties, is shown in Figure 20.17 The spectral filtering properties of this fiber were found to be instrumental in stabilizing a passively mode-locked cavity [89]. The filtering effect is also highly useful for suppressing unwanted wavelengths [90]. This has been used to great effect in connection with Yb-doped amplifiers for narrow-band continuous-wave signals at a wavelength of 1178 nm [91,92]. Such sources are of interest for creating artificial guide stars at astronomical observatories. After frequency doubling to 579 nm, the light excites sodium atoms in the uppermost atmosphere, thus creating a localized light source in the sky. The challenge is that the Yb gain is very small at 1178 nm, and lasing or even amplified spontaneous emission at other

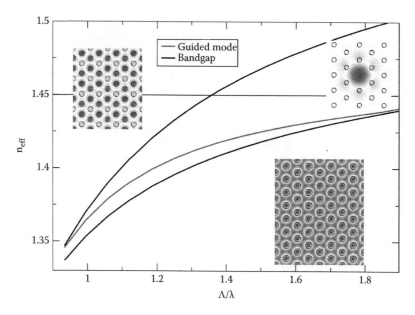

Figure 20.17 Boundaries of the photonic bandgap in a solid-core PBG fiber (black lines), and effective index curve of the fundamental guided mode (red line). The fiber has circular inclusions with a refractive index of 1.6 in the cladding. Upper inset shows the field distribution of a cladding mode above the bandgap, whereas the lower inset illustrates the field distribution of a cladding mode below the bandgap.

wavelengths with stronger gain must be suppressed, which is conveniently done by proper design of the photonic bandgaps.

PBG effects may also be utilized to suppress mode guidance at the wavelengths where a strong core signal would generate Raman components. This is of interest for power scaling of short-pulsed fiber amplifiers, and in the longer term, also for realizing cladding-pumped Raman amplifiers [93]. Raman-suppressing large-core fibers have been realized by a hybrid approach in which rows of Ge-doped high-index inclusions inserted into a conventional cladding structure with air holes provide a PBG filtering effect, so that the fiber guides through both index-guiding and the PBG mechanism [94,95]. The fibers are therefore termed *hybrid MOFs*. The bandgap effects have been shown to lead to novel phase-matching mechanisms for parametric processes; such processes have also been observed experimentally [96]. A further application especially for hybrid fibers could therefore be wavelength conversion of high-power pulses, for example, from the Yb band into the biological transparency window below 1000 nm.

Perhaps the most important aspect of PBG fiber technology is the possibility of creating structures that predominantly guide the light in air. This is possible if a cladding structure can be fabricated in which modes with an effective index of 1 fall within the bandgap, or, in other words, where the effective index of the cladding mode at the lower boundary of the gap falls below the light line. The triangular lattice structure has this property for sufficiently large air-filling fractions. In Figure 20.18, the bandgap diagram for a structure with a filling fraction of 70% is shown [97]. It can be seen that several bandgaps cross the air line. In order to confine an air-guided mode within these bandgaps, a core defect consisting of a rather large air hole must be introduced. In the inset of Figure 20.18, a structure is shown in which the core defect has been obtained by replacing the central cladding hole and its six nearest neighbors by a single air hole. This structure was found to support both a fundamental and a second-order guided mode, though not at the same frequency. The traces of the two modes are shown in the upper inset of Figure 20.18: inside the bandgap, the modes are bound, while outside, they appear as leaky resonances within the bands of cladding modes. For both modes, guidance only takes place in a rather narrow frequency interval. Furthermore, the guidance

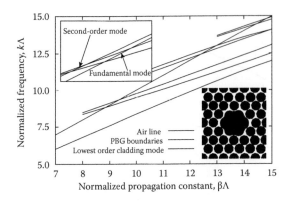

Figure 20.18 PBG boundaries and defect mode traces (upper inset) for the air-guiding PBG fiber with a filling fraction of 70%. The defect modes are confined when they fall inside the PBG, and leaky elsewhere. The fiber microstructure is shown in the lower inset. (Reproduced from Broeng, J. et al., *Opt. Lett.*, 25, 96–98, 2000.)

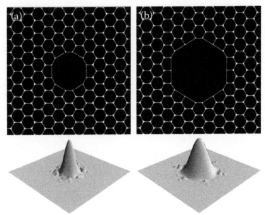

Figure 20.19 Illustration of a 7-cell (a) and a 19-cell (b) hollow-core PBG fiber structure, and examples of calculated mode profiles. (Reproduced from Humbert, G. et al., *Opt. Express*, 12, 1477–1484, 2004.)

properties of the structure were found to depend strongly on the radius of the core defect. For these reasons, the fabrication of air-guiding fibers for practical applications at specific wavelengths requires a high degree of control over the manufacturing process.

Modern hollow-core PBG fibers feature air-filling fractions above 95%, that is, much larger than the example in Figure 20.18. As a result, wider bandgaps may be obtained. However, typically only one useful bandgap exists, which is in contrast to solid-core PBG fibers, where several bandgaps may be utilized. The design in Figure 20.18 is commonly known as a "7-cell" core, since the core air hole replaces seven cladding unit cells. This terminology carries over to other core designs, such as the 19-cell design [98], where an extra ring of cladding air holes is removed to form a larger core. Examples of high-quality hollow-core PBG structures with 7-cell and 19-cell cores are shown in Figure 20.19.

Experimentally, air-guiding fibers were first realized by Cregan et al. [12]. In fibers with pitches around 5 μm and air-filling fraction of 30–50%, several air-guiding transmission bands were observed over a fiber length of a few centimeters. An important development was the demonstration by Venkataraman et al., of a hollow-core fiber with a loss as low as 13 dB km⁻¹ in 2002 [99]. Subsequent research has brought the loss in 7-cell cores below 10 dB km⁻¹ in few-moded fibers [100,101] and less

than 20 dB km⁻¹ in single-mode polarization-maintaining fibers [102]. In 19-cell fibers, losses as low as 1.2 dB km⁻¹ have been demonstrated [103], however these fibers feature a large number of well-guided core modes. The loss appears to be limited by scattering from surface roughness at the core wall, and scales with λ^{-3}, where λ is the wavelength.

The initial motivation for developing hollow-core PBG fibers was to facilitate the transmission of telecommunication signals by reduction of the fiber nonlinear coefficient, and perhaps also improve loss and dispersion properties. Indeed, the hollow-core fibers allow a reduction of the nonlinear coefficient by about three orders of magnitude compared to solid-core fibers with a similar core area, and one order of magnitude compared to large-core fibers [104]. However, the impossibility of obtaining very low loss in a single-mode fiber has so far precluded transmission applications of these fibers. Recent research has shown that higher-order modes in 19-cell cores may be suppressed by the presence of resonant satellite cores in the cladding, in combination with bending [105,106]. However, development of a fiber useful for telecom still seems to be some way off in the future. Interestingly, however, spatial multiplexing in multimode fibers has recently seen substantial research interest in the telecommunications community. If this research proves fruitful, it could ultimately facilitate the use of hollow-core transmission fibers, as the loss in 19-cell or even larger hollow cores could be scaled to values competitive with standard fibers [103], and the strongly reduced nonlinear coefficient would provide a decisive advantage.

There exist several other applications for hollow-core PBG fibers. Perhaps the most important is their use in fiber-optical gyroscopes, where the suppression of light–matter interactions in a polarization-maintaining single-mode fiber is very important[107,108]. Another application is compression of high-power laser pulses [109,110], where the large anomalous dispersion of the guided mode, the very low nonlinear coefficient, and the possibility of monolithic integration into fiber-based laser/amplifier systems [111] provide the motivation. Further applications include low-loss/high-power transmission at wavelengths where the absorption loss of silica is high [112], gas-sensing applications and spectroscopy, especially for metrology [113], and spectral filtering exploiting the narrowness of the guided-mode transmission bands [12]. Applications for laser-assisted atom transport or even as particle accelerators have also been proposed [114].

20.6 IC FIBERS

The hollow-core PBG fibers have provided a major technological breakthrough by allowing low-loss guidance at visible or near-infrared wavelengths in a single- or few-moded waveguide, which is outside the reach of conventional fiber technology. Their main limitation is the fact that they only support a single transmission band of limited spectral width. In addition, transmission at visible and UV wavelength is adversely affected by the wavelength dependence of PBG fiber loss discussed in the previous section. The development of hollow-core fibers guiding by IC has helped to overcome these limitations.

The earliest example of a fiber guiding by the IC mechanism was published as early as 2002 [115]. However, at that time, the nature of the guidance mechanism was not fully understood. In 2007, the guidance mechanism was resolved as being due to a strongly reduced overlap between cladding modes and core-confined modes, which allowed low-loss guidance of the latter even in the absence of a photonic bandgap, hence the term *inhibited coupling* [19]. The characteristic features of IC fibers that allow this coupling mechanism is

a large core, typically a few tens of microns wide, and a cladding structure that resembles a network of intersecting silica planes with sub-micrometer widths. Such structures may be obtained by drawing at lower temperatures and higher tensions than for other MOF types, to yield structures with little deformation at the intersections of the silica bridges. Two examples of contemporary IC fiber structures are shown in Figure 20.20 [116].

The earliest examples of IC fibers had propagation losses of the order of 0.5–1 dB/m at near-infrared or visible wavelengths. The spectral dependence of the loss somewhat resembles PBG fibers in that spectral regions of low-loss guidance are separated by "forbidden" bands with high loss. In contrast to hollow-core PBG fibers, several useful transmission bands may be found, which in total can span several octaves. The positions of the transmission bands may be predicted by a simple criterion of antiresonance with modes in the surrounding silica bridges, and IC fibers can, to a first approximation, be modeled by a single silica capillary ring, for which analytical solutions may be established [117,118].

In recent years, it has been realized that the loss of IC fibers can be significantly reduced by the introduction of *hypocycloid* core structures, which feature a core wall having an inward curvature, as shown in Figure 20.20a. This structure minimizes the overlap between the guided core modes and the silica bridges, and in particular, the nodes in the silica network, on which one often finds localized modes that can couple strongly to the core modes. Using hypocycloid cores, losses as low as 17 dB km^{-1} have been obtained at a wavelength of 1 µm [117], and very recently 70 dB km^{-1} has been reached at 532 nm [119]. These results actually make IC fibers superior to hollow-core PBG fibers at short wavelengths in terms of loss. In addition, hypocycloid cores have been found to improve suppression of higher-order modes by increasing their propagation loss [117,120].

The core areas of these fibers are significantly larger than in hollow-core PBG fibers, often reaching several thousand µm^2, which implies extremely low values of dispersion and nonlinear coefficients. Remarkably, the mode guidance is highly robust against bending, with critical (3 dB) bending radii at the sub-cm level being reported [121].

Due to the combination of extremely low nonlinear coefficients, and conversely extremely high

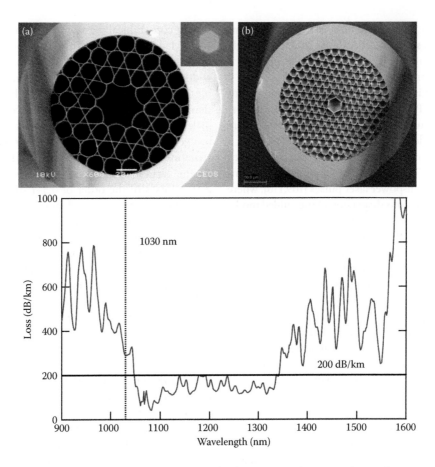

Figure 20.20 SEM images of IC fibers with hypocycloid core (a) and a more classical structure with straight core walls (b). (c) The loss curve of the hypocycloid-core fiber, featuring a minimal loss well below 100 dB/km. (Reproduced from Heckl, O. et al., *Opt. Express*, 19, 19142–19149, 2011.)

power tolerance with practically useful single-mode guidance, bend resistance, and low losses, IC fibers represent a real technological breakthrough, whose eventual significance cannot yet be fully assessed. Some of the most promising applications are within high-power pulse delivery [122], and gas-based nonlinear optics [123], in particular nonlinear frequency conversion, which can be taken into power regimes previously unheard of within fiber optics. As an example, Raman cascades extending from 2300 nm to 325 nm have been obtained in a H_2-filled IC fiber, by pumping at 1064 nm, which implies that both Stokes and anti-Stokes lines are excited efficiently [19]. In another line of research, Cherenkov radiation has been generated from 176 to 550 nm using a Ti:Sapphire laser at 800 nm as a pump source, reaching μJ energies in the UV spectral region [124]. In both of these approaches, it is an important advantage of the IC fibers that the very low mode overlap with silica yields extraordinary resistance to photochemically induced damage from short-wavelength radiation.

20.7 CONCLUSIONS

After 20 years of microstructured fiber research and development, it is tempting to try to sum up their scientific and commercial impact. Such an assessment cannot be complete, because several MOF types are still making major strides in terms of design and performance developments, IC fibers and large-core amplifiers being the most prominent examples. Nevertheless, many of the original expectations for MOF applications can now be judged in the light of an extended body of concluded research.

The first striking point to notice is that MOFs have so far not made any significant impact on telecommunications. This is in sharp contrast to early expectations, which envisioned hollow-core photonic bandgap fibers, or at least large-core endlessly-single-mode MOFs being used for transmission, and small-core MOFs being used for nonlinear optical signal processing and dispersion compensation. The fact that none of these visions have been fulfilled may be put down to a variety of factors, both relating to the fiber properties and to the general nature and development of the telecommunications industry. Hollow-core PBG fibers did not reach the low losses for single-mode fibers needed for the long-haul transmission lines that might have greatly benefited from their low nonlinearity. Regarding solid-core designs, the possible improvements over standard fiber technology appear too marginal to justify a more advanced technology platform, splicing complications, and so on. The same seems to be true for the dispersion-compensating fibers that were investigated at an early stage. Regarding all-optical signal processing, none of the technologies investigated have as yet been able to compete commercially with ultrafast electronics components.

The current push in telecommunications research for improved capacity through various kinds of space-division multiplexing may imply important future roles for MOF technology. First, design and manufacture of multicore fibers with sufficiently low intercore coupling will benefit from MOF design and fabrication methods. Second, techniques for mode division multiplexing, if successful, could make hollow-core PBG fibers attractive for long-haul communication, because the loss in a multimode hollow-core fiber could be reduced to very low levels, while at the same time leveraging the benefits of a strongly reduced nonlinear coefficient.

In other areas, however, the promises of MOF technology have come to fruition. This is especially the case for nonlinear frequency conversion, where supercontinuum generation extending into the visible region was demonstrated in 2000 and is now commercialized by a number of companies. Devices based on other frequency conversion principles, such as Cherenkov radiation or four-wave mixing are also on the horizon, fuelled by the impressive developments of fiber-based ps and fs lasers. In this way, MOFs are, and will increasingly be, a crucial ingredient in expanding the scope of fiber-based laser systems. At the same time, large-core MOFs are helping to push the limits of fiber lasers by enabling very large active cores with single-mode behavior, in combination with efficient collection of pump light. It is by now conceivable that short-pulse amplification may ultimately be performed in single-mode cores having a diameter exceeding the cladding diameter of a standard telecommunications transmission fiber, and with amplifier lengths at the level of one meter or shorter. Further power scaling may be enabled by the possibility of fabricating multiple amplifier cores within one microstructured rod element, which will greatly facilitate beam combination approaches. These developments are poised to greatly impact the manufacturing industry, where precision cutting, welding, and scribing are examples of industrial processes where fiber-based laser systems are currently replacing traditional gas or solid-state lasers or other methods. On the low-power side, the use of hollow-core PBG fibers for dispersive pulse compression may pave the way for fully fiber-integrated fs laser devices, which in combination with nonlinear frequency conversion techniques will be of crucial importance for taking various biomedical imaging applications out of the laboratory and into a clinical setting.

The impressive progress within hollow-core fibers by the development of IC fibers in recent years, will undoubtedly also impact the aforementioned fields, due to the perspectives of single-mode delivery of tremendous powers, as well as fiber-integrated nonlinear frequency conversion at unprecedented power levels, without the usual limitations in frequency range imposed by the transmission loss and damage thresholds of silica glass. As the designs and performance of these fibers are rapidly evolving at present, the full scientific and commercial perspectives are yet to be determined.

In summary, one may state that the general MOF fabrication technology has already been shown to enable a range of different designs and applications, which could not have been foreseen from the outset. Although many concepts have already been brought to maturity, there is no reason to expect that we have seen the last radical invention arising from the current or future design and manufacturing capabilities. The field of MOF research should thus be exciting to follow for many years to come.

ACKNOWLEDGMENTS

Tanya Monro would like to thank a number of her colleagues at the ORC, University of Southampton, who have made many important contributions to the research described within this paper. In particular, warm thanks to Kentaro Furusawa, Vittoria Finazzi, Joanne Baggett, Periklis Petropoulous, Walter Belardi, Ju Han Lee, Jonathan Price and David Richardson. Tanya Monro also acknowledges the support of a Royal Society University Research Fellowship.

Jesper Lægsgaard and Anders Bjarklev would like to thank all our colleagues at DTU Fotonik and at NKT Photonics A/S for the collaboration and results which enter this work. Special thanks should be given to Dr. Jens Kristian Lyngsø, Dr. Thomas Tanggaard Alkeskjold and Mr René; Engel Kristiansen at NKT Photonics A/S, as well as Prof. Jes Broeng at DTU Fotonik for particularly great help during the writing of this manuscript.

REFERENCES

1. Birks, T. A., Roberts, P. J., Russell, P. St J. et al. 1995. Full 2-D photonic band gaps in silica/air structures. *Electron. Lett.* 31: 1941–1943.
2. Knight, J. C., Birks, T. A., Atkin, D. M. et al. 1996. Pure silica single-mode fibre with hexagonal photonic crystal cladding. *Optical Fiber Communication Conference* (San Jose, CA, February 25, vol. 2), p. CH35901.
3. Monro, T. M., West, Y. D., Hewak, D. W. et al. 2000. Chalcogenide holey fibres. *Electron. Lett.* 36: 1998–2000.
4. van Eijkelenborg, M., Large, M., Argyros, A. et al. 2001. Microstructured polymer optical fibre. *Opt. Express* 9: 319–327.
5. Monro, T. M., Bennett, P. J., Broderick, N. G. R. et al. 2000. Holey fibers with random cladding distributions. *Opt. Lett.* 25: 206–208.
6. Monro, T. M., Richardson, D. J., Broderick, N. G. R. et al. 1999. Holey optical fibers: An efficient modal model. *J. Lightwave Technol.* 17: 1093–1101.
7. Broderick, N. G. R., Monro, T. M., Bennett, P. J. et al. 1999. Nonlinearity in holey optical fibers: Measurement and future opportunities. *Opt. Lett.* 24: 1395–1397.
8. Knight, J. C., Birks, T. A., Cregan, R. F. et al. 1998. Large mode area photonic crystal fibre. *Electron. Lett.* 34: 1347–1348.
9. Birks, T. A., Knight, J. C., and Russell, P. St J. 1997. Endlessly single-mode photonic crystal fiber. *Opt. Lett.* 22: 961–963.
10. Knight, J. C., Arriaga, J., Birks, T. A. et al. 2000. Anomalous dispersion in photonic crystal fiber. *IEEE Photonics Technol. Lett.* 12: 807–809.
11. White, T. P., McPhedran, R. C., deSterke, C. M. et al. 2001. Confinement losses in microstructured optical fibers. *Opt. Lett.* 26: 1660–1662.
12. Cregan, R. F., Mangan, B. J., Knight, J. C. et al. 1999. Single-mode photonic band gap guidance of light in air. *Science* 285: 1537–1539.
13. Riishede, J., Lægsgaard, J., Broeng, J. et al. 2004. All-silica photonic bandgap fibre with zero dispersion and a large mode area at 730 nm. *J. Opt. A: Pure Appl. Opt.* 6: 667–670.
14. Knight, J. C., Broeng, J., Birks, T. A. et al. 1998. Photonic band gap guidance in optical fibers. *Science* 282: 1476–1478.
15. Barkou, S. E., Broeng, J., and Bjarklev, A. 1999. Silica-air photonic crystal fiber design that permits waveguiding by a true photonic bandgap effect. *Opt. Lett.* 24: 46–48.
16. Larsen, T. T., Bjarklev, A., and Hermann, D. S. 2003. Optical devices based on liquid crystal photonic bandgap fibres. *Opt. Express* 20: 2589–2596.
17. Lægsgaard, J. 2004. Gap formation and guided modes in photonic-bandgap fibres with high-index rods. *J. Opt. A: Pure Appl. Opt.* 6: 798–804.
18. Steinvurzel, P., Kuhlmey, B. T., White, T. P. et al. 2004. Long wavelength anti-resonant guidance in high index inclusion microstructured fibers. *Opt. Express* 12:5424–5433.
19. Couny, F., Benabid, F., Roberts, P. J. et al. 2007. Generation and photonic guidance of multi-octave optical-frequency combs. *Science* 318: 1118–1121.
20. Wang, Y. Y., Wheeler, N. V., Couny, F. et al. 2011. Low-loss broadband transmission in hypocycloid-core Kagome hollow-core photonic crystal fiber. *Opt. Lett.* 36: 669–671.

21. Bjarklev, A., Broeng, J., Barkou, S. E. et al. 1998. Dispersion properties of photonic crystal fibres. *24th European Conference on Optical Communication* (ECOC98, Madrid, September 20–24, vol. 1), pp. 135–136.

22. Koshiba, M., and Saitoh, K. 2004. Applicability of classical optical fiber theories to holey fibers. *Opt. Lett.* 29:1739–1741.

23. Riishede, J., Libori, S. B., Bjarklev, A. et al. 2001. Photonic crystal fibers and effective index approaches. *Proceedings 27th European Conference on Optical Communication (ECOC 2001)* Th.A1.5.

24. Riishede, J., Mortensen, N. A., and Lægsgaard, J. 2003. A "poor man's approach" to modelling micro-structured optical fibres. *J. Opt. A: Pure Appl. Opt.* 5: 534–538.

25. Birks, T. A., Bird, D. M., Hedley, T. D. et al. 2004. Scaling laws and vector effects in bandgap guiding fibres. *Opt. Express* 12: 69–74.

26. Meade, R. D., Rappe, A. M., Brommer, K. D. et al. 1993. Accurate theoretical analysis of photonic band-gap materials. *Phys. Rev. B* 48: 8434–8437.

27. Silvestre, E., Andrés, M. V., and Andrés, P. 1998. Biorthonormal-basis method for the vector description of optical-fiber modes. *J. Lightwave Technol.* 16:923–928.

28. Ferrando, A., Silvestre, E., Andrés, P. et al. 2001. Designing the properties of dispersion-flattened photonic crystal fibers. *Opt. Express* 9: 687–697.

29. Zhu, Z. M., and Brown, T. G. 2002. Full-vectorial finite-difference analysis of micro-structured optical fibers. *Opt. Express* 10: 853–864.

30. Brechet, F., Marcou, J., Pagnoux, D. et al. 2000. Complete analysis of the characteristics of propagation into photonic crystal fibers, by the finite element method. *Opt. Fiber Technol.* 6: 181–191.

31. Wilcox, S., Botten, L. C., de Sterke, C. M. et al. 2005. Long wavelength behavior of the fundamental mode in microstructured optical fibers. *Opt. Express* 13: 1978–1984.

32. Agrawal, G. P. 1989. *Nonlinear Fiber Optics* (New York: Academic).

33. Okuno, T., Onishi, M., Kashiwada, T. et al. 1999. Silica-based functional fibers with enhanced nonlinearity and their applications. *IEEE J. Sel. Top. Quantum Electron.* 5: 1385–1391.

34. Belardi, W., Lee, J. H., Furusawa, K. et al. 2002. A 10 Gbit/s tuneable wavelength converter based on four-wave mixing in highly nonlinear holey fibre. *Proceedings of ECOC*, Copenhagen, Denmark, September 8–12, 2002, Postdeadline Paper PD1.2.

35. Finazzi, V., Monro, T. M., and Richardson, D. J. 2002. Confinement loss in highly nonlinear holey optical fibers. *Proceedings of OFC*, OSA *Technical Digest* (Anaheim, California, March 17, 2002), pp. 524–525.

36. Tajima, K., Zhou, J., Kurokawa, K. et al. 2003. Low water peak photonic crystal fibres. *Proceedings of ECOC* (Rimini, Italy, September 21–25, 2003), pp. 42–43. Paper Th4.1.6.

37. Roberts, P. J., Couny, F., Sabert, H. et al. 2005. Loss in solid-core photonic crystal fibers due to interface roughness scattering. *Opt. Express* 20: 7779–7793.

38. Petropoulos, P., Monro, T. M., Belardi, W. et al. 2001. 2R-regenerative all-optical switch based on a highly nonlinear holey fiber. *Opt. Lett.* 26: 1233–1235.

39. Hansen, K. P., Jensen, J. R., Jacobsen, C. et al. 2002. Highly nonlinear photonic crystal fiber with zero-dispersion at 1.55 μm. *OFC 02 Post Deadline*, Anaheim, CA, March 17, Paper FA9.

40. Ranka, J. K., Windeler, R. S., and Stentz, A. J. 2000. Visible continuum generation in air silica microstructure optical fibers with anomalous dispersion at 800 nm. *Opt. Lett.* 25: 25–27.

41. Dudley, J. M., Genty, G., and Coen, S. 2006. Supercontinuum generation in photonic crystal fiber. *Rev. Mod. Phys.* 78: 1135–1184.

42. Gorbach, A. V., and Skryabin, D. V. 2007. Light trapping in gravity-like potentials and expansion of supercontinuum spectra in photonic-crystal fibres. *Nat. Photonics* 1: 653–657.

43. Povazay, B., Bizheva, K., Unterhuber, A. et al. 2002. Submicrometer axial resolution optical coherence tomography. *Opt. Lett.* 27: 1800–1802.

44. Diddams, S. A., Jones, D. J., Ye, J. et al. 2000. Direct link between microwave and optical frequencies with a 300 THz femtosecond laser comb. *Phys. Rev. Lett.* 84: 5102–5105.

45. Shi, K. B., Li, P., Yin, S. Z. et al. 2004. Chromatic confocal microscopy using supercontinuum light. *Opt. Express* 12: 2096–2101.

46. Selchi, S., Bertani, F. R., and Ferrari, L. 2011. Supercontinuum ultra wide range confocal microscope for reflectance spectroscopy of living matter and material science surfaces. *AIP Adv.* 1:032143.

47. Møller, U., Sørensen, S. T., Jakobsen, C. et al. 2012. Power dependence of supercontinuum noise in uniform and tapered PCFs. *Opt. Express* 20: 2851–2857.

48. Stone, J. M., Wadsworth, W. J., and Knight, J. C. 2013. 1064 nm laser-induced defects in pure SiO_2 fibers. *Opt. Lett.* 38: 2717–2719.

49. Lu, F., Deng, Y., and Knox, W. H. 2005. Generation of broadband femtosecond visible pulses in dispersion-micro-managed holey fibers. *Opt. Lett.* 30: 1566–1568.

50. Tu, H., and Boppart, S. A. 2009. Optical frequency up-conversion by supercontinuum-free widely-tunable fiber-optic Cherenkov radiation. *Opt. Express* 17:9858–9872.

51. Akhmediev, N., and Karlsson, M. 1995. Cherenkov radiation emitted by solitons in optical fibers. *Phys. Rev. A* 51: 2602–2607.

52. Tu, H., Lægsgaard, J., Zhang, R. et al. 2013. Bright broadband coherent fiber sources emitting strongly blue-shifted resonant dispersive wave pulses. *Opt. Express* 20: 23188–23196.

53. Liu, X., Lægsgaard, J., Møller, U. et al. 2012. All-fiber femtosecond Cherenkov radiation source. *Opt. Lett.* 37: 2769–2771.

54. Liu, X., Lægsgaard, J., Villanueva, G. E. et al. 2013. Low-noise operation of all-fiber femtosecond Cherenkov laser. *IEEE Phot. Technol. Lett.* 25: 892–895.

55. Deng, Y., Lin, Q., Lu, F. et al. 2005. Broadly tunable femtosecond parametric oscillator using a photonic crystal fiber. *Opt. Lett.* 30: 1234–1236.

56. Zhai, Y. H., Goulart, C., Sharping, J. E. et al. 2011. Multimodal coherent anti-Stokes Raman spectroscopic imaging with a fiber optical parametric oscillator. *Appl. Phys. Lett.* 98: 191106.

57. Zhang, L., Yang, S., Li, P. et al. 2013. An all-fiber continuously time-dispersion-tuned picosecond optical parametric oscillator at 1 µm region. *Opt. Express* 21: 25167–25173.

58. Asobe, M. 1997. Nonlinear optical properties of chalcogenide glass fibers and their application to all-optical switching. *Opt. Fiber Technol.* 3: 142–148.

59. Friberg, S. R., and Smith, P. W. 1987. Nonlinear optical-glasses for ultrafast optical switches. *IEEE J. Quantum Electron.* 23: 2089–2094.

60. Monro, T. M., Kiang, K. M., Lee, J. H. et al. 2002. Highly nonlinear extruded single-mode holey optical fibers. *Proceedings of OFC, OSA Technical Digest* (Anaheim, CA, March 17), pp. 315–317.

61. Poladian, L., Issa, N. A., and Monro, T. M. 2002. Fourier decomposition algorithm for leaky modes of fibres with arbitrary geometry. *Opt. Express* 10: 449–454.

62. Renversez, G., Bordas, F., and Kuhlmey, B. T. 2005. Second mode transition in microstructured optical fibers: Determination of the critical geometrical parameter and study of the matrix refractive index and effects of cladding size. *Opt. Lett.* 30: 1264–1266.

63. Alkeskjold, T. T., Laurila, M., Scolari, L. et al. 2011. Single-mode ytterbium-doped large-mode-area photonic bandgap rod fiber amplifier. *Opt. Express* 19: 7398–7409.

64. Gambling, A., Matsumura, H., Ragdale, C. M. et al. 1978. Measurement of radiation loss in curved single-mode fibres. *Microwaves Opt. Acoust.* 2: 134–140.

65. Sakai, J. I., and Kimura, T. 1978. Bending loss of propagation modes in arbitrary-index profile fibers. *Appl. Opt.* 17: 1499–1506.

66. Sørensen, T., Broeng, J., Bjarklev, A. et al. 2001. Macrobending loss properties of photonic crystal fibres with different air filling fractions. *Proceedings of ECOC'2001*, Amsterdam, The Netherlands, September 30–October 4.

67. Baggett, J. C., Monro, T. M., Furusawa, K. et al. 2002. Distinguishing transition and pure bend losses in holey fibers. *Paper CMJ6 CLEO* 2002 (Long Beach, California, May 19–22).

68. Marcuse, D. 1982. Influence of curvature on the losses of doubly clad fibres. *Appl. Opt.* 21: 4208–4213.

69. Sahu, J. K., Renaud, C. C., Furusawa, K. et al. 2001. Jacketed air-clad cladding pumped ytterbium-doped fibre laser with wide tuning range. *Electron. Lett.* 37: 1116–1117.

70. DiGiovanni, D. J., and Windeler, R. S. 1999. Article comprising an air-clad optical fiber. US Patent 5 907 652, May 25, 1999.

71. Furusawa, K., Malinowski, A. N., Price, J. H. V. et al. 2001. A cladding pumped ytterbium-doped fiber laser with holey inner and outer cladding. *Opt. Express* 9: 714–720.

72. Fini, J. M. 2007. Intuitive modeling of bend distortion in large-mode-area fibers. *Opt. Lett.* 32: 1632–1634.

73. Limpert, J., Schmidt, O., Rothhardt, J. et al. 2006. Extended single-mode photonic crystal fiber lasers. *Opt. Express* 14: 2715–2720.

74. Di Teodoro, F., Hemmat, M. K., Morais, J. et al. 2010. High peak power operation of a 100 μm-core, Yb-doped rod-type photonic crystal fiber amplifier *Proceedings of SPIE*, 7580. Paper 758006.

75. Limpert, J., Stutzki, F., Jansen, F. et al. 2012. Yb-doped large-pitch fibres: Effective single-mode operation based on higher-order mode delocalization. *Light: Sci. Appl.* 1: e8.

76. Jansen, F., Stutzki, F., Otto, H. J. et al. 2012. Thermally induced waveguide changes in active fibers. *Opt. Express* 20: 3997–4008.

77. Baumgartl, M., Jansen, F., Stutzki, F. et al. 2011. High average and peak power femtosecond large-pitch photonic-crystal-fiber laser. *Opt. Lett.* 36: 244–246.

78. Stutzki, F., Jansen, F., Eidam, T. et al. 2011. High average power large-pitch fiber amplifier with robust single-mode operation. *Opt. Lett.* 36: 689–691.

79. Klenke, A., Hädrich, S., Eidam, T. et al. 2014. 22 GW peak-power fiber chirped-pulse-amplification system. *Opt. Lett.* 39: 6875–6878.

80. Di Teodoro, F., Morais, J., McComb, T. S. et al. 2013. SBS-managed high-peak-power nano-second-pulse fiber-based master oscillator power amplifier. *Opt. Lett.* 38: 2162–2164.

81. Sipes, D. L., Tafoya, J. D., Schulz, D. S. et al. 2014. *Proceedings of SPIE* 8961. Paper 896114.

82. Eidam, T., Wirth, C., Jauregui, C. et al. 2011. Experimental observations of the threshold-like onset of mode instabilities in high power fiber amplifiers. *Opt. Express* 19: 13218–13224.

83. Smith, A. V., and Smith, J. J. 2011. Mode instability in high power fiber amplifiers. *Opt. Express* 19: 10180–10192.

84. Laurila, M., Jørgensen, M. M., Hansen, K. R. et al. 2012. Distributed mode filtering rod fiber amplifier delivering 292W with improved mode stability. *Opt. Express* 20: 5742–5743.

85. Jasapara, J., Her, T. H., Bise, R. et al. 2003. Group-velocity dispersion measurements in a photonic bandgap fiber. *J. Opt. Soc. Am. B* 20: 1611–1615.

86. Argyros, A., Birks, T. A., Leon-Saval, S. G. et al. 2005. Photonic bandgap with an index step of one percent. *Opt. Express* 13: 309–314.

87. Luan, F., George, A. K., Hedley, T. D. et al. 2004. All-solid photonic bandgap fiber. *Opt. Lett.* 29: 2369–2371.

88. Nielsen, C. K., Jespersen, K. G., and Keiding, S. R. 2006. A 158 fs 5.3 nJ fiber-laser system at 1 μm usin bg photonic bandgap fibers for dispersion control and pulse compression. *Opt. Express* 14: 6063–6068.

89. Liu, X., Lægsgaard, J., and Turchinovich, D. 2010. Self-stabilization of a mode-locked femtosecond fiber laser using a photonic bandgap fiber. *Opt. Lett.* 35, 913–915.

90. Olausson, C. B., Falk, C. I., Lyngsø, J. K. et al. 2008. Amplification and ASE suppression in a polarization-maintaining ytterbium-doped all-solid photonic bandgap fibre. *Opt. Express* 16: 13657–13662.

91. Olausson, C. B., Shirakawa, A., Chen, M. et al. 2010. 167 W power scalable ytterbium-doped photonic bandgap fiber amplifier at 1178 nm. *Opt. Express* 18: 16345–16352.

92. Chen, M., Shirakawa, A., Olausson, C. B. et al. 2015. 87 W, narrow-linewidth, lineary-polarized 1178 nm photonic bandgap fiber amplifier. *Opt. Express* 23: 3134–3141.

93. Ji, J. H., Codemard, C. A., Sahu, J. K. et al. 2010. Design, performance, and limitations of fibers for cladding-pumped Raman lasers. *Opt. Fiber Technol.* 16: 428–441.

94. Cerqueira, S. A., Luan, F., Cordeiro, C. M. B. et al. 2006. Hybrid photonic crystal fiber. *Opt. Express* 14: 926–931.

95. Alkeskjold, T. T. 2009. Large-mode-area ytterbium-doped fiber amplifier with distributed narrow spectral filtering and reduced bend sensitivity. *Opt. Express* 17: 16394–16405.

96. Petersen, S. R., Alkeskjold, T. T., and Lægsgaard, J. 2013. Degenerate four-wave mixing in large mode area hybrid photonic crystal fibers. *Opt. Express* 21: 18111–18124.

97. Broeng, J., Barkou, S. E., Søndergaard, T. et al. 2000. Analysis of air-guiding photonic bandgap fibers. *Opt. Lett.* 25: 96–98.

98. Humbert, G., Knight, J. C., Bouwmans, G. et al. 2004. Hollow core photonic crystal fibersfor beam delivery. *Opt. Express* 12: 1477–1484.

99. Smith, C. M., Venkataraman, N., Gallagher, M. T. et al. 2003. Low-loss hollow-core silica/air photonic bandgap fibre. *Nature* 424: 657–659.

100. Amezcua-Correa, R., Gérôme, F., Leon-Saval, S. G. et al. 2008. Control of surface modes in low loss hollow-core photonic bandgap fibers. *Opt. Express* 16: 1142–1149.

101. Mangan, B. J., Lyngsø, J. K., and Roberts, P. J. 2008. Realization of low loss and polarization maintaining hollow core photonic crystal fibers. *CLEO/QELS Proceedings* 1–9: 2016–2017.

102. Lyngsø, J. K., Jakobsen, C., Simonsen, H. R. et al. 2012. Truly single-mode polarization maintaining hollow core PCF. *Proc. SPIE* 8421: 84210C.

103. Roberts, P. J., Couny, F., Sabert, H. et al. 2005. Ultimate low loss of hollow-core photonic crystal fibres. *Opt. Express* 13: 236–244.

104. Lægsgaard, J., Mortensen, N. A., Riishede, J. et al. 2003. Material effects in air-guiding photonic bandgap fibers. *J. Opt. Soc. Am. B* 20: 2046–2051.

105. Fini, J. M., Nicholson, J. W., Windeler, R. S. et al. 2013 Low-loss hollow-core fibers with improved single-modedness. *Opt. Express* 21: 6233–6242.

106. Fini, J. M., Nicholson, J. W., Mangan, B. et al. 2014. Polarization maintaining single-mode low-loss hollow-core fibres. *Nat. Commun.* 5: 5085.

107. Kim, H. K., Digonnet, M. J. F., and Kino, G. S. 2006. Air-core photonic-bandgap fiber-optic gyroscope. *J. Lightwave Technol.* 24: 3169–3174.

108. Terrel, M. A., Digonnet, M. J. F., and Fan, S. 2012. Resonant fiber optic gyroscope using an air-core fiber. *J. Lightwave Technol.* 30: 931–937.

109. De Matos, C. J. S., Taylor, J. R., Hansen, T. P. et al. 2003. All-fiber chirped pulse amplification using highly-dispersive air-core photonic bandgap fiber. *Opt. Express* 11: 2832–2837.

110. Limpert, J., Schreiber, T., Nolte, S. et al. 2003. All fiber chirped pulse amplification system based on compression in air-guiding photonic bandgap fiber. *Opt. Express* 11: 3332–3337.

111. Turchinovich, D., Liu, X., and Lægsgaard, J. 2008. Monolithic all-PM femtosecond Yb-fiber laser stabilized with a narrow-band fiber Bragg grating and pulse-compressed in a hollow-core photonic crystal fiber. *Opt. Express* 16: 14004–14014.

112. Shephard, J. D., MacPherson, W. N., Maier, R. R. J. et al. 2005. Single-mode mid-IR guidance in a hollow-core photonic crystal fiber. *Opt. Express* 13: 7139–7144.

113. Henningsen, J., Hald, J., and Peterson, J. C. 2005. Saturated absorption in acetylene and hydrogen cyanide in hollow-core photonic bandgap fibers. *Opt. Express* 13: 10475–10482.

114. Xintian, E. L. 2001. Photonic band gap fiber accelerator. *Phys. Rev. Spec. Top.—Accel. Beams* 4: 051301.

115. Benabid, F., Knight, J. C., Antonopoulos, G. et al. 2002. Stimulated Raman scattering in hydrogen-filled hollow-core photonic crystal fiber. *Science* 298: 399–402.

116. Heckl, O., Saraceno, C. J., Baer, C. R. E. et al. 2011. Temporal pulse compression in a xenon-filled Kagome-type hollow-core photonic crystal fiber at high average power. *Opt. Express* 19: 19142–19149.

117. Debord, B., Alharbi, M., Bradley, T. et al. 2013. Hypocycloid-shaped hollow-core photonic crystal fiber Part I: Arc curvature effect on confinement loss. *Opt. Express* 21: 28597–28608.

118. Marcatili, E. A. J., and Schmeltzer, R. A. 1964. Hollow metallic and dielectric waveguides for long distance optical transmission and lasers. *Bell Syst. Tech. J.* 43: 1783–1809.

119. Debord, B., Alharbi, M., Benoît, A. et al. 2014. Ultra low-loss hypocycloid-core Kagome hollow-core photonic crystal fiber for green spectral-range applications. *Opt. Lett.* 39: 6245–6248.

120. Wang, Y. Y., Peng, X., Alharbi, M. et al. 2012. Design and fabrication of hollow-core photonic crystal fibers for high-power ultrashort pulse transportation and pulse compression. *Opt. Lett.* 37: 3111–3113.

121. Bradley, T. D., Wang, Y., Alharbi, M. et al. 2013. Optical properties of low-loss (70 dB/km) hypocycloid core Kagome hollow-core photonic crystal fiber for Rb and Cs based optical applications. *J. Lightwave Technol.* 31: 2752–2755.

122. Debord, B., Alharbi, M., Vincetti, L. et al. 2014. Multi-meter fiber-delivery and pulse self-compression of milli-Joule femtosecond laser and fiber-aided laser-micromachining. *Opt. Express* 22: 10735–10746.

123. Russell, P. S. J., Hölzer, P., Chang, W. et al. 2014. Hollow-core photonic crystal fibres for gas-based nonlinear optics. *Nat. Photonics* 8:278–286.

124. Mak, K. F., Travers, J. C., Hölzer, P. et al. 2013. Tunable vacuum-UV to visible ultrafast pulse source based on gas-filled Kagome-PCF. *Opt. Express* 21: 10942–10953.

21

Engineered optical materials

PETER G. R. SMITH AND CORIN B. E. GAWITH
University of Southampton

21.1 INTRODUCTION

The purpose of this chapter is to review recent developments in engineered optical materials, and in particular nonlinear crystals. This definition can be taken to mean conventional materials that are structurally altered to give new and enhanced optical properties. To distinguish such materials from other structured materials such as doped semiconductors, or countless other examples, we will also limit the definition to mean only those that are noncentrosymmetric. Thus, it includes periodically poled ferroelectrics and polar crystals. The artificially engineered nonlinear opticals materials covered in this chapter were mostly developed for their nonlinear optical properties, but are now seeing applications in other areas of optics such as electro-optics. This review is written for the general laser scientist who wishes to learn more about what can be achieved with these materials and not for those specialists already working in quasi-phase-matched materials.

21.1.1 Overview

To date, the most widely used engineered nonlinear optical material is periodically poled lithium niobate, commonly known by the acronym PPLN. Figure 21.1 shows the number of journal publications per annum over the past 12 years on PPLN and closely related materials such as periodically poled lithium tantalate (PPLT) and doped variants of PPLN. It is clear that there has been a dramatic development of this technology in the last decade. The original concept of quasi-phase-matching (QPM) dates back to the paper by Armstrong [1],

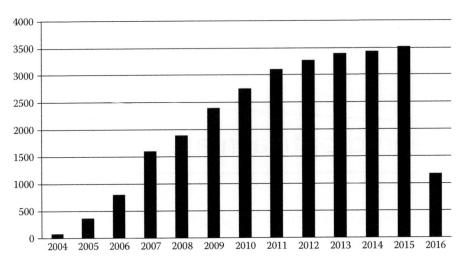

Figure 21.1 The number of publications per annum on periodically poled materials.

after which various techniques such as periodic poling during crystal growth and ion diffusion to induce domain inversion were used. However, the great upsurge in interest came following the successful demonstration of pulsed electrical field poling by Yamada [2]. The majority of these publications have made use of the QPM property of domain reversal gratings to allow efficient nonlinear optical interactions. This QPM technique overcomes the inherent dispersion of materials by periodic reversals of the nonlinear coefficient of the material. A detailed derivation of QPM will presented in Section 21.5.

Periodically poled lithium niobate and its close relatives do not hold a monopoly on QPM, and indeed QPM can be carried out in materials by periodically modulating the strength of the nonlinearity rather than by inverting the sign of the nonlinearity. This technique is particularly attractive in the area of poled glasses where it is often simpler to periodically degrade the nonlinearity by ultraviolet (UV) or electron beam exposure [3,4]. The poled glass materials are very promising from a technological perspective because of their compatibility with optical transmission fiber. However, to date, the nonlinearity is significantly lower than that achievable in ferroelectrics, significantly less than 1 pm/V compared with >10 pm/V, and thus, applications are limited. Another research area for QPM is in GaAs where various techniques have been used

to make structurally reversed materials for use in long wavelength generation. Techniques include direct bonding of stacked materials [5] and templated growth [6].

Many other ferroelectrics can be used to produce QPM materials, in particular the potassium titanyl phosphate (KTP) family materials have a number of advantages, particularly for high-energy applications where their higher nanosecond-pulse optical-damage thresholds and smaller temperature coefficients provide advantages over PPLN despite lower nonlinearity. More exotic ferroelectric materials have also seen applications where larger transmission ranges are desirable, and in particular $BaMgF_2$ for UV and Strontium Barium Niobate (SBN) for longer wavelengths. Interested readers are directed to recent reviews [7,8].

Quasi-phase-matched materials find application as bulk crystals for many applications, but equally there are many applications where waveguide devices are more desirable, as they are able to prevent the deleterious effects of diffraction. By tightly confining the optical power within the waveguide core it is possible to achieve both higher absolute conversion efficiencies and lower input powers.

Table 21.1 shows the most common QPM materials and their principle advantages. Note that it has become a commonplace to use additional dopants, for example, MgO in $LiNbO_3$ to improve optical properties.

Table 21.1 The most common QPM materials and their principle advantages

Material	Transmission range (μm)	Advantages
LiNbO$_3$	0.38–5.2	Cheap, high nonlinearity, available in large sizes (up to 5 cm)
LiTaO$_3$	0.28–5.5	Better UV transmission, slight smaller nonlinearity than LiNbO$_3$
KTP	0.35–3.3	Better power handling in the Visible region and for nanosecond pulses. Smaller temperature coefficients than LiNbO$_3$ or LiTaO$_3$, relatively more expensive

21.1.2 Overview of applications of QPM materials

Having introduced the most common QPM materials, it is now appropriate to discuss the applications of these materials in optics and elsewhere. The vast majority of applications involve the use of the materials for optical frequency conversion, however, more recently, the properties of domain engineering have started to receive attention for other optical applications. In particular, researchers have begun to exploit other properties of domain engineered materials such as their differential etching, electro-optic or pyroelectric properties. These novel applications will be reviewed later in Section 21.8.

The desirable nonlinear optical properties of QPM materials have provided most of the impetus behind their development in recent years. They have the enormous advantage of freeing nonlinear device performance from the limitations resulting from the naturally available birefringence of conventional crystals. Subject only to the proviso that the structure must permit fabrication, QPM materials allow efficient nonlinear optical interactions throughout the whole transmission range of the material. Furthermore, by choosing appropriate grating design, it is possible to tailor the nonlinear response to match the particular geometry, spectrum, temporal and chirp properties, temperature bandwidth, and configuration of the nonlinear interaction. This flexibility provides a tremendous opportunity for new and exciting device performance, particularly for short-pulse applications.

The commercial development of QPM materials has led to a number of companies offering devices and crystals, and a Google search (2016) reveals a list including HC Photonics, Raicol, Covesion, Gooch & Housego, C2C Link, SRICO, and NTT Electronics.

21.2 SECOND-ORDER NONLINEAR PROCESSES

The majority of nonlinear interactions in QPM materials make use of the second-order nonlinearity $\chi(2)$, which in general mediate the interactions of three interacting waves in a material. The second-order nonlinearity may be used in a number of processes depending on the frequencies of the three waves and on the boundary conditions in the interaction (such as feedback mirrors) and input waves. The most common interaction is second harmonic generation (SHG) in which a new wave is created with half of the input wavelength or equivalently twice the frequency, the most familiar example of which is probably in frequency doubling of the 1064 nm Nd:YAG laser to give visible green at 532 nm. SHG can be seen as a special case of a three-wave interaction in which the two input fields are degenerate. In general, these $\chi(2)$ processes are called optical parametric interactions, and they provide a rich set of possibilities for applications in optics. Closely related to SHG are the processes of sum and difference frequency generation (SFG

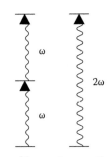

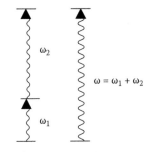

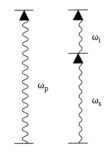

Figure 21.2 Illustration of energy conservation in $\chi(2)$ nonlinear materials.

conversion of energy. This distinction is important; in the SHG, SFG, and DFG processes, the output frequencies created are set by the input frequencies in contrast to parametric processes in which any energy-conserving process is possible (i.e., any pair of photons for which the sum of their energies adds up to the input energy). In a parametric process, the phase-matching requirements decide which wave sees coherent growth. Figure 21.2 shows schematically the various nonlinear mixing processes of SHG, sum frequency generation, and generalized optical parametric generation.

The optical parametric approach can be used to make devices, which in many ways mirror the operation of lasers with the parametric process providing a gain medium through parametric amplification rather than the stimulated emission process that occurs in a conventional laser. These devices are called optical parametric oscillators (OPOs) and can provide a widely tunable light output throughout the visible and near infrared (IR), and present a valuable enabling tool in spectroscopy and remote sensing. OPOs may be characterized by the feedback schemes used into singly resonant, in which a single output wave is fed back into the cavity, and doubly resonant, in which feedback occurs at both output wavelengths. Useful nomenclature is the use of self-explanatory term "pump" for the input laser pump. In addition, the terms "signal" and "idler" are widely used for the two output waves, with signal usually referring to the more energetic of the two output waves, although it is occasionally used to refer to the resonated wave, even if it is of longer wavelength. Continuing the similarity with conventional lasers, it is possible also to use parametric devices as amplifiers known as optical parametric amplifiers or at very high gains in which amplification of parametric fluorescence is used as an optical parametric generator. Figure 21.3 shows a schematic of an OPO.

OPOs provide versatile sources with a key feature being that they are able to produce light at

and DFG, respectively) in which the output occurs at the sum or difference frequency of two input waves. In addition to SHG, SFG, and DFG, there also exist optical parametric processes in which an input photon is split to create two output photons at longer wavelength, subject to the constraint of

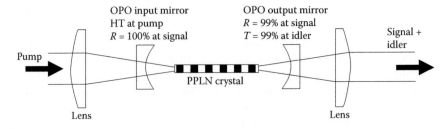

Figure 21.3 Schematic of an optical parametric oscillator.

almost any wavelength within the transmission of the material. To produce light efficiently, the process must also be phase matched, and thus, it is possible to change the wavelength of an OPO by changing angle, QPM period, temperature, or pump wavelength. Thus, OPOs are valuable for converting light from available laser transitions to longer wavelengths at which no suitable laser transitions occur.

21.3 MATERIALS

Nonlinear optics is concerned with materials in which the response of the material is nonlinear function of the incident electric fields. It is useful to think of the response of material in terms of the polarization set up due to the applied electric fields. In general, the induced polarization will depend on the state of the material (temperature, thermal history, pressure, strain, etc.) and also on the frequencies of the applied field(s). The polarization response will normally show a strong dependence on frequency: at low frequencies, it will have contributions from orientational, ionic, and electronic motion, whereas at higher frequencies (e.g., in the visible optical), only the electrons in the material are capable responding quickly enough to applied optical fields.

It is conventional to describe the polarization in terms of a power series in applied electric field as in Equation 21.1. The polarization can be seen to comprise a number of terms, the first of which is the linear polarization and is responsible for the linear refractive index. The variation of refractive index with wavelength, or equivalently frequency, is known as dispersion, and plays an important role in wavelength conversion.

$$P = \epsilon_0 [X^{(1)} + X^{(2)} E^2 + X^{(2)} E^3 + \cdots] E \tag{21.1}$$

The values of the various nonlinear terms in the susceptibility will depend on the material and will vary with electronic configuration, density, and particularly crystal structure. In general, the coefficients χ will be tensor quantities as they can relate any the generated polarization to any of the fields present. However, it is possible to make an important general observation regarding the symmetry of the crystal. If the material is centrosymmetric, then it will not have any even order nonlinearities, so that $\chi(2)$, $\chi(4)$, etc. will be zero. The reason for this is that the nonlinear susceptibility must display the symmetry properties of the crystal medium itself, and in a centrosymmetric crystal, the inversion operation ($x \to -x$, $y \to -y$, $z \to -z$) leaves the crystal unaltered. The same operation on vector quantities such as the polarization P or E reverses their sign, and thus, it follows from applying inversion to either side of Equation 21.1 that all even order nonlinear coefficients must be zero.

This requirement that the material be non-centrosymmetric provides the unifying feature of these materials, and it is the possibility of structurally modifying them, for example, through periodic reversal that enables their unique potential. All polar materials are non-centrosymmetric, common examples are quartz and gallium arsenide, and are thus able to exhibit second-order nonlinearity. However, in these polar materials it is relatively difficult to create spatially structured or reversed patterns, and so the majority of activities has concentrated on the use of ferroelectric materials such as lithium niobate. Patterned domain inversion in materials such as lithium niobate is essential for the fabrication of engineered optical materials and can be accomplished in a large number of different ways. The most comprehensive reference on ferroelectrics is probably the book by Lines and Glass [9], which provides a complete discussion of these materials.

21.4 FERROELECTRIC MATERIALS

It is useful to review some of the most important properties of ferroelectric materials to gain an understanding of their uses as engineered optical materials. The ferroelectric materials are a subset of polar materials, in simple terms being those that possess a spontaneous electric dipole moment within the unit cell that is capable of being inverted by the application of a sufficiently large external electric field. In reality, the definition is more complex, and for a full discussion, one should consult standard texts such as Lines and Glass [9]. As each unit cell in a ferroelectric has a permanent electric dipole moment, the whole crystal will have what is known as a spontaneous polarization. In the unit cell of a ferroelectric, the mean position of negative charge is displaced from the mean position of positive charge leading to the permanent dipole moment.

Within a ferroelectric crystal a region within which all the unit cells have the same polarity is called a domain. Commercially purchased ferroelectric crystals are normally purchased in a single domain state. In general, ferroelectric crystals will exhibit one or more structural phase transitions between the ferroelectric state and other ferroelectric or paraelectric states. These phase transitions can have a dramatic effect on a crystal, and great care must be exercised when taking a crystal through a phase transition. The temperature at which the material becomes a ferroelectric is known as the Curie temperature. For lithium niobate, for example, the Curie temperature is 1145°C compared to a melting point of 1240°C. A comprehensive treatment of lithium niobate can be found in the book by Prokhorov and Kuz'minov [10].

Another important property of ferroelectrics is pyroelectricity, which is the appearance of charge on certain surfaces of the material as it is heated or cooled. This pyroelectric charge is caused by variation of the spontaneous polarization with temperature. The relative position of different ions will change with temperature, causing a change in dipole moment. In a free crystal, the spontaneous polarization is neutralized on the crystal's surface by free charge, and as the spontaneous polarization varies, so charges appear on the surfaces.

The application of a sufficiently large electric field to a ferroelectric can cause the domain structure to reverse, and thus the mean positions of the positive and negative charges are swapped. This movement of electrical charge constitutes a displacement current, and thus, an electrical current must be supplied by the external poling circuit. The amount of charge supplied controls the poled area (A) so that $Q = 2P\,sA$. For lithium niobate, $Ps = 72$ $\mu C/cm^2$. The subject of domain inversion in ferroelectrics is well researched and can be found in a number of books [9,11]. However, despite all this research, practical methods of controlling domain inversion remain something of a black art.

In ferroelectric materials, the domain formation is strongly influenced by the underlying

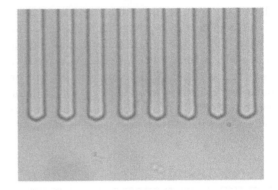

Figure 21.4 Photomicrograph of +z face of PPLN crystal showing domains terminating at ±120o to the x-direction (vertical in image).

crystal symmetry. A widely adopted model for the electric field poling process involves growth of domains from one polar face of the crystal to the other followed by sideways growth of the domain wall. It is widely noticed that the domain formation habit is strongly influenced by the symmetry of the crystal. A striking example comes in $LiNbO_3$ with its 3 m symmetry. It can be seen from Figure 21.4, which shows an enlargement of a section of PPLN crystal, that the domain walls are predominantly found to lie along the x-direction (vertical in the image) and at ±120° to that axis. Therefore, in designing a PPLN grating, the sample is usually oriented so that light travels along the y-direction, with the domain walls running along the x-direction. The underlying symmetry properties of the crystal dominate the poling process, and from the image, it is clear that at the ends of gratings bars, the domains terminate in facets at ±120° despite the fact that the electrodes that produced the PPLN had squared-off ends! Consequently, it is important that the desired pattern be compatible with the preferred domain habit of the material.

21.5 QUASI-PHASE-MATCHED MATERIALS

The QPM technique has a long history. It was first proposed in 1962 by Armstrong [1] but was largely eclipsed by the development of birefringent phase matching 1962 [12,13]. QPM continued to attract interest throughout the 1970s, concentrating largely on materials with domain formed during crystal growth. An excellent article covering the history and development of QPM materials by one of the major figures can be found in the work of Byer [14].

The realization that in-diffusion of dopants could be used to reverse the domain direction close to the crystal surface and thus make periodic structures led to a rapid growth in research into QPM waveguides in materials such as LiNbO$_3$ and KTP. However, the critical step came in the work by Yamada [2], in which bulk domain inversion by using a periodic electrode was used for the first time to form PPLN. It is interesting that although this process was revolutionary in optics, the idea of forming a domain pattern in a ferroelectric by the use of patterned electrodes and a high-field pulse was well established for ferroelectric memories in the 1970s [11].

The basic idea of QPM is to periodically reverse the direction of the nonlinearity in the material so that the phase of the nonlinearly generated light adds in phase with the light generated earlier in the crystal. The phase mismatch occurs because the different wavelengths in the nonlinear interaction have different refractive indices due to dispersion and thus different phase velocities. As we shall see later, the phase of the nonlinearly generated light depends upon the phase of driving fields and the direction of the nonlinearity, and thus, periodic reversal of the nonlinearity allows constructive growth of the generated light.

In many ways, it is easier to understand these effects via a mathematical derivation of the equations governing the second-order interactions. From a personal perspective, the clearest exposition of this topic can be found in the chapter written by Byer in the book *Nonlinear Optics* edited by Harper and Wherrett [15]. The derivation is standard and the starting point is the wave equation, which can be manipulated to become

$$\nabla^2(E) - \mu_0 \sigma \frac{\partial E}{\partial t} - \mu_0 \epsilon_0 \frac{\partial^2 E}{\partial t^2}$$

$$= \mu_0 \sigma \frac{\partial^2 P}{\partial t^2} \qquad (21.2)$$

The right-hand side of this equation contains the polarization term, P, and it is this term that causes the generation of new frequencies through the susceptibility expansion given by Equation 21.1. By substituting traveling wave fields in the z-direction,

and using slowly varying envelopes for the electric field and polarization, and further by making the slowly varying envelope approximation, it is possible to find three coupled equations, which describe the second-order optical nonlinearity.

$$\frac{\partial E_p}{\partial z} = \frac{i\omega_p}{n_p c} dE_s E_i \exp^{-i\Delta kz} \qquad (21.3)$$

$$\frac{\partial E_s}{\partial z} = \frac{i\omega_s}{n_s c} dE_p E_i^* \exp^{i\Delta kz} \qquad (21.4)$$

$$\frac{\partial E_i}{\partial z} = \frac{i\omega_i}{n_i c} dE_p E_s^* \exp^{i\Delta kz} \qquad (21.5)$$

These equations relate the three waves in the material (generally pump, signal, and idler E_p, E_s, E_i, respectively). In the equations, $n\alpha$ is the refractive index of wavelength $\lambda\alpha$, c is the speed of light, $\omega\alpha$ is the angular frequency, and d, or more accurately d_{eff}, is the nonlinear coefficient. The wavevector mismatch Δk is given by $\Delta k = kp - ks - ki$. As usual, $k\alpha = 2\pi n_\alpha/\lambda_\alpha$.

These three equations are the governing equations for nonlinear optics; they allow for a tremendous richness of solution, especially if mirrors are used around the nonlinear crystal as in the case of an OPO. These equations can be solved in a variety of situations, and the interested reader is referred to the extensive literature, for example, in the works of Shen [16], Yariv [17], Byer [15], Risk [38], etc.

There are some general observations that can be made about these equations and their derivation. The first concerns conservation of energy; in their derivation, it is necessary to constrain the frequencies included to satisfy the time dependence of the equations. This constraint $\omega_p = \omega_s + \omega_i$ is equivalent to the conservation of energy in the process.

The second observation is that each equation has a spatial dependence of $\exp(i\Delta kz)$, where Δk is the phase mismatch. Consideration of any of Equations 21.3 through 21.5 shows that the rate of change of that field component depends upon the product of the nonlinearity, the strengths of fields other two fields, and, because of the $\exp(i\Delta kz)$ term, a periodic phase factor.

To illustrate how QPM works consider a simple second harmonic process. In this case, we have two fundamental fields interacting to produce a new field at 2ω. If we relabel the general fields as $E(\omega)$ for the fundamental and $E(2\omega)$ as the second harmonic, and simplify to bring out the essential concepts, we end up with the following pair of equations:

$$\frac{\partial E_{(\omega)}}{\partial z} = ikE_{(2\omega)}E^*_{(\omega)}\exp^{i\Delta kz} \qquad (21.6)$$

$$\frac{\partial E_{(2\omega)}}{\partial z} = ikE_{(\omega)}E^*_{(\omega)}\exp^{-i\Delta kz} \qquad (21.7)$$

If the field $E\omega$ remains constant (as would be the case for low conversion efficiency and thus low depletion of the fundamental), then it is possible to simply integrate Equation 21.7 to get the field strength $E(2_\omega)(L)$ at the end of the crystal (L).

$$E(2\omega)_L = \int_0^F dE(2\omega)$$

$$= ikE^2(\omega)\int_0^L \exp^{-i\Delta kz}dz \qquad (21.8)$$

Considering the real and imaginary parts of the integral separately means that the output is given by the integral of sine and cosine over many periods, which has no cumulative contribution. If significant conversion efficiency is desired, it is necessary to find a way to deal with the phase mismatch. In birefringent phase matching, different polarizations are used to access appropriate refractive indices in a birefringent material to give zero phase mismatch $\Delta k = 0$.

In the QPM approach, the material is periodically altered to modify Equation 21.7 so that the nonlinearity is now a function of z. This seemingly simple modification provides the power of the QPM technique.

$$E(2\omega)_L = \int_0^E dE(2\omega)$$

$$= iE^2(\omega)\int_0^L k(z)\exp^{-i\Delta kz}dz \qquad (21.9)$$

The nonlinearity is made into a periodic function of z with a period that matches $\exp(i\Delta k)$. The most common technique is to make $d(z)$ take positive and negative values by inverting the nonlinearity through periodic poling. This means that the

integral over z becomes the integral of a rectified sinusoidal function, which has a cumulative contribution with z.

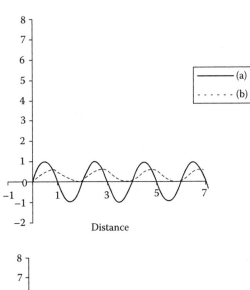

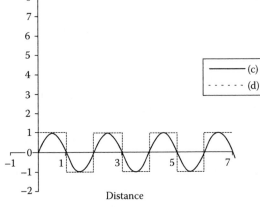

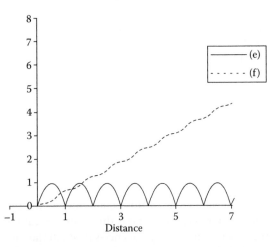

Figure 21.5 Electric field envelope distributions in quasi-phase matching.

Figure 21.5 illustrates this effect. Curve a shows the real part of the left-hand side of the integrand of Equation 21.8, and curve b shows the integral with respect to z. It is seen that there is no constructive growth with distance. Curve c again shows the integrand of Equation 21.8, and curve d shows an appropriate $\kappa(z)$, which matches the polarity of the generated wave. Curve e shows the product of curves c and d which becomes a rectified sine wave, and finally, curve f shows the integral of curve e. Curve f shows a constructive growth in the second harmonic field strength along the crystal. Thus, we can clearly see that by reversing the sign of κ with an appropriate period, we can get a useful quasi-phase-matched output.

The characteristic length over which constructive addition of the fields occurs is known as the coherence length $l_c = \pi/\Delta k$. As is clear from Figure 21.5, the QPM period is $\Lambda = 2lc = 2\pi/\Delta k$. It is worth mentioning that some authors call Λ the coherence length. To calculate the coherence length for a given material, it is necessary to know the variation of refractive index with wavelength. This is most commonly described by a Sellmeier equation, which provides a power series type expression for $n(\lambda, T)$ as a function of wavelength (and often temperature). An example is the one for LiNbO$_3$ developed from QPM experimental data by Jundt [18], which was used to generate Figure 21.6, which shows coherence length versus wavelength for SHG at a temperature of 150°C.

The QPM technique does result in a reduced nonlinear coefficient because there is a degree of cancellation within each coherence length,

resulting in a $d_{\text{eff}} = (2/\pi)d$. For LiNbO$_3$, this means that $d_{\text{eff}} \approx 16$ pm/V compared to $d33 \approx 30$ pm/V.

Any periodic modulation of the nonlinearity can be used for QPM, the optimal is periodic inversion matching the phase mismatch factor. However, it is also possible to use higher order phase matching (in which the inversion has a period that is an integral multiple of the phase mismatch period) and even a modulation scheme in which the nonlinearity is periodically erased. This latter scheme is commonly employed in poled glasses where inversion is harder to achieve.

The order of phase matching for a given period can be expressed as $\Lambda = nl_c$, where n is the order. To be efficient, n must clearly be integer. For a simple 50:50 mark-space ratio grating, only odd order QPM will be obtained, and in fact, this provides a sensitive probe of the quality of a QPM structure. If higher order QPM is used, the d_{eff} becomes $(2/n)\pi$.

For each nonlinear process, there will be a bandwidth associated over which an efficient interaction will occur. In general, there will be bandwidths associated with wavelength and temperature (and indeed any other parameter that affects the refractive index). An example of a phase-matching curve is shown in Figure 21.7, together with a theoretical fit. The PPLN in this example had a period of 6.4 μm and a length of 3.2 mm. The laser was a Nd:YLF operating at 1047 nm. The theoretical shape is a sinc^2 function, which can simply be derived from Equation 21.7. The first zero in the efficiency occurs when the light generated in the second half of the crystal exactly cancels that generated in the first half. A useful approximate condition for phase

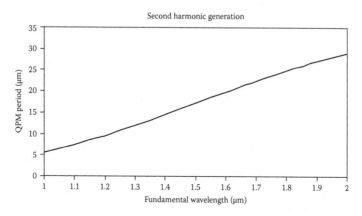

Figure 21.6 The QPM period in LiNbO$_3$ for SHG as a function of fundamental wavelength, calculated at 150°C.

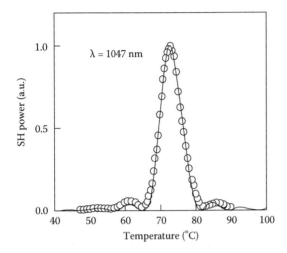

Figure 21.7 Temperature tuning curve showing sinc2 response for second harmonic generation in PPLN at low conversion efficiency.

matching is that $\Delta k \cdot L < \pi/4$. Thus, it is clear that the bandwidth is inversely related to the crystal length. General expressions for QPM bandwidths can be found in the paper by Fejer [19].

21.6 NONLINEAR PROCESSES

The three coupled equations provide a way of characterizing and understanding the various nonlinear processes possible in a nonlinear material, but generally there is little difference between a correctly designed QPM material and a birefringently phase-matched material. Consequently, as there are many good discussions of nonlinear optics, there is little point in discussing them further here. On a purely personal basis, the author would suggest the books on nonlinear optics already mentioned as well as the one by Koechner [20].

All of the standard nonlinear conversion processes have been demonstrated in QPM materials, where their higher nonlinearity and availability in longer lengths make them a material of choice. In addition, they also provide attractive materials for the observation of other nonlinear processes, such as cascaded nonlinearity and spatial soliton behavior.

In real-world situations, it is almost always the case that the maximum conversion efficiency is desired, and this leads to the use of focused beams or the use of waveguides to maximize the efficiency. In this regime, the plane wave treatments do not strictly hold; however, they can still often be applied provided that care is used. Three cases are worth looking at the following:

1. The simplest case is a plane wave interaction. In this case, the conversion efficiency scales with length squared. If loosely focused Gaussian laser beams are used, then the conversion efficiency becomes

$$\eta = \frac{2\omega^2 d_{\text{eff}}^2 L^2 I(\omega)}{n^3 c^3 \varepsilon_0} \sin c^2\left(\frac{\delta k L}{2}\right) \quad (21.10)$$

This can be applied to an unfocused Gaussian beam such that $I = P/A$ where P is the power and the area $A = \pi\omega 2/2$. In this equation, $\delta k = \Delta k - 2\pi/\Lambda_{\text{grating}}$.

2. If optimal conversion is desired, then it is necessary to move to tighter focusing. This case received extensive attention by Boyd and Kleinman [21]. In this case, there is an issue that focusing to a tighter spot size will result in a higher power at the beam waist, but that this tight focusing will cause the beam to diverge more strongly, and thus become larger at the front and rear of the crystal. The analysis by Boyd and Kleinman shows that there is an optimal focused spot size for a given length of crystal. In this case, it is found that the conversion efficiency scales as L rather than L^2. Thus, in this case, the conversion efficiency takes on units of %/W cm, and a useful (very approximate) number to remember is that for PPLN this number becomes of order 1%/W cm (obviously depending on wavelength). Thus, with a 1 cm crystal, and say 200 mW of input power, the efficiency will be of order 0.2% = 0.4 mW. If the crystal is 4 cm long and the input power is 400 mW, then this increases to 1.6% = 6.4 mW. A full calculation for PPLN will lead to a slightly higher number and depends quite critically on the details of the laser source. It is important to stress that these example numbers are "ball-park" and act only as a guide.

Figure 21.8 shows a conversion efficiency curve for frequency doubling of a 946 nm Nd:YAG laser. The experiment used optimum focusing and generated up to 450 mW of blue light at 473 nm. To achieve this result, the fundamental laser was used in a relaxation oscillation mode to increase the peak power and hence conversion efficiency.

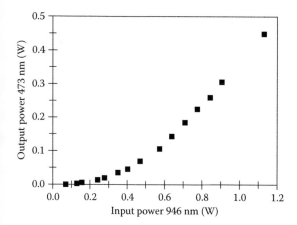

Figure 21.8 Blue light second harmonic generation.

3. The third case is waveguide in which the optical confinement provided by the waveguiding from the core compensates the diffraction of the beam. Consequently, a higher intensity results, but at the expense of greater complication in fabrication. The expression for conversion efficiency now becomes more complex, and a calculation must take into account the spatial overlap integral between the nonlinearity and the guided modes. In this case, the efficiency becomes once more proportional to L^2 and will be seen quoted as %/W cm². In PPLN waveguides, this efficiency can be of order 100%/W cm², although in this case launched powers are likely to be far smaller. However, the higher efficiency more than compensates, and in a proton-exchange PPLN waveguide 99% pump depletion has been demonstrated by Parameswaran [22].

In conclusion, unfocused beams result in a simple expression for the conversion efficiency, and this regime is often used to characterize material. The optimal focusing regime is ideal for maximizing conversion efficiency of high-power lasers, but is subject to not exceeding the damage threshold of the material! Waveguides are ideal for low-power conversion but are considerably more complicated to fabricate and their analysis is more complex.

21.7 PERIODICALLY POLED LITHIUM NIOBATE

The choice of periodically poled material for a given application depends on a number of factors: the most common are availability of starting material, cost of starting material, ease of periodic poling, optical transmission, obtainable nonlinearity, damage threshold, short pulse walk-off, etc. The order here is deliberate: the first two factors tend to dominate which materials are routinely used, and indeed one of the great advantages of periodic poling is their potential for engineering to fit an application. The great success of lithium niobate as a QPM material comes about to a large extent because of its low cost. As an approximate example, a 76-mm-diameter lithium niobate wafer costs around $100–200, whereas KTP will cost around $100 for a 10–15 mm square piece. Thus, the cost per area for KTP is around 50 times higher than for lithium niobate. Although the exact multiplier can be debated, there is no doubt that the fact that lithium niobate is produced in quantities of several tons per annum for the surface acoustic wave industry has been one of the major factors in its dominance.

Periodically poled lithium niobate has proved extremely versatile. It has been successfully operated from the blue [2,23,24] through to the mid-IR [25]. It can be fabricated with into gratings with periods from a few microns upward and in lengths of up to 5 cm. PPLN samples are usually fabricated by the electrical poling technique, where the coercive field is around 22 kV/mm. Most workers place a large ballast resistor in series with the crystal to limit the current flowing in the sample, and then apply a high voltage in excess of the coercive field. By limiting the time for which the field is applied, it is possible to control the total charge, and thus poled area. An alternative approach is to control the current in the sample by varying the voltage; this can lead to higher yield and allows excellent control of the poled area.

PPLN has been fabricated in thicknesses of up 5 mm but is more commonly available in 1 mm and in 500 μm thickness. This limitation on aperture is not normally too restrictive for optical access but for nanosecond pulses in which the damage threshold becomes energy dependent (around 2 J/cm²) it can be a problem and workers have adopted the approach of bonding samples together to increase the working aperture. In longer pulse regimes and for continuous wave (cw) lasers, the damage threshold depends more on the intensity, a reasonable figure for which is 150 MW/cm². In very short-pulse applications, the damage threshold is often found to depend most strongly on the average power through the crystal, and intensities as high as 2 GW/cm² have been used without damage.

The development of processes for periodically poling of magnesium oxide lithium niobate (commonly termed MgO:PPLN) has led to a significant improvement in performance compared to regular PPLN, allowing larger apertures and high-power handling, for example, an OPO operating with 52 mJ pulse energy in a 5 mm by 5 mm crystal [26]. The addition of MgO doping reduces the photorefractive effect in the material, allowing operation at higher optical powers, and importantly, with better performance at visible and near-IR wavelengths.

Lithium niobate has a high refractive index ($n_o = 2.2$, $n_e = 2.1$ at 1064 nm), leading to strong Fresnel reflections (approx 14% per surface), and thus, it is often antireflection coated. In OPO applications, this process can actually be more expensive than the crystal itself as complex coatings with multiple transparency bands are needed. In an actual experiment, the PPLN sample is usually mounted in an oven with feedback temperature control. PPLN has relatively large temperature coefficients, which can be either viewed as conveniently useful for tuning of OPO systems or inconveniently expensive for frequency doubling applications.

It is normal to run a PPLN sample at a temperature in excess of 100°C and often around 150°C to prevent the buildup of photorefractive damage. This effect is caused by charge migration within the crystal, which causes index modulation via the electro-optic effect. The effect is reduced in PPLN relative to unpoled lithium niobate [27], but still remains a major barrier to its application in visible light generation. Another limitation to high-power visible operation is the so-called green induced infrared absorption (GRIIRA) effect [28] that is seen in high-power harmonic generation experiments into the visible. In this effect, the green light generated in doubling causes an increase in the IR absorption at 1 micron, resulting in heating and eventually in catastrophic damage to the crystal. This limits the visible power to <1 W in the green in PPLN.

With the development of MgO:PPLN, it is possible to use the crystals at much lower temperatures (down to room temperature), but in general, the crystal still needs to be temperature controlled to achieve phase matching, and thus, it is convenient to design crystals to operate at around 50°C,

so temperature can be controlled simply by using a heater and feedback.

The procedure for periodic poling is generally as follows:

1. Selection of appropriate wafers
2. Cleaning
3. Definition of electrodes
4. Electrical poling
5. Visual characterization
6. Cutting and polishing
7. Coating

The definition of the electrodes is usually done using photolithographic patterning, a mask is used to fabricate a pattern in photoresist, which is then used to form the electrode structure. A number of different electrode types can be used, ranging from simple metal electrodes deposited over photoresist, through bar electrodes underneath resist, to contact pressed metal electrodes. It appears possible to get good quality poling with any of these processes, and it is often simpler to use a liquid or gel electrode, although intriguingly the patterned electrode must be placed on the −z face with liquid electrodes, whereas the +z face is normally

(a) Photo-resist

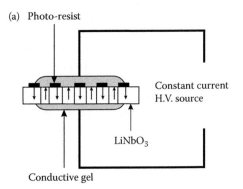

Conductive gel

(b)

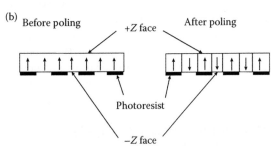

Figure 21.9 Schematic of poling method using conductive gel electrodes patterned on −z face of sample.

patterned for metal electrodes. The poling process itself remains something of an art, with different groups adopting varying approaches to PPLN fabrication. Figure 21.9 shows schematics of the various steps in the poling process.

PPLN samples are often designed to have multiple gratings with different periods running in parallel tracks along the crystal. This allows tuning of the phase matching by lateral translation of the crystal within the optical beam. An extension of this idea is the use of fanned gratings, in which continuous variation of the period can be achieved. PPLN crystals are almost always housed in small ovens to control the temperature to within a fraction of a degree. The temperature is then optimized to give maximum conversion efficiency for harmonic generation or is deliberately altered to tune the output wavelengths in an OPO.

21.7.1 Bandwidth and short-pulse operation

The bandwidth of a given grating depends upon the wavelengths used in the interaction and the length of the device [19]. It is possible to artificially increase the bandwidth of a grating by deliberately chirping the grating. Such chirped gratings can be advantageously used in recompressing chirped pulses [29]. When dealing with short pulses, it is important to consider the variation in group velocity within the crystal at the different wavelengths and also the spectral acceptance bandwidth of the crystal for the bandwidth of the pulses. These effects can prove advantageous and can, for example, provide a method for compressing pulses within an OPO [30]. In short-pulse applications, the temporal walk-off and bandwidth effects can provide the limit on crystal length, and thus, when considering different materials, it is important to use an appropriate figure of merit to compare the different materials. The figure of merit must be related carefully to the problem in question, as it may often be the case that a material with a lower nonlinear coefficient but a lower temporal walk-off of pulses will give better performance.

For many applications, variants of lithium niobate are attractive, and there has been considerable recent activity in producing periodically poled materials with superior properties. One of the major drawbacks to PPLN is the photorefractive damage that occurs in visible operation and to counter this, a number of groups have worked on producing magnesium-doped materials. Another important recent development has been the use of stoichiometric lithium niobate for periodic poling. Unlike conventional congruent lithium niobate, which has a deficiency of lithium ions from its growth, this material has close to ideal stoichiometry, which results in enhanced nonlinear optical properties and a much lower coercive field. With stoichiometric material, it has been possible to pole samples with much larger apertures allowing high-power operation.

Another QPM material, which has similar properties to lithium niobate, is lithium tantalate. It is also produced on a large scale for the surface acoustic wave industry, and thus provides an attractive alternative to PPLN. PPLT, as it is known, benefits from better transparency in the near UV (down to 280 nm compared to 380 nm for PPLN). It does, however, have a smaller d_{33} coefficient and is somewhat more expensive than lithium niobate, so it tends to receive less use.

The other major family of QPM materials is that of KTP. In this material family, it is possible to substitute rubidium for potassium and an arsenate group for the phosphate, to produce RTA, and also RTP and KTA. KTP has many advantages, such as higher damage thresholds and lower temperature coefficients, but against this must be offset the higher price and smaller set of growers for the material. The other major problem seen in the KTP family is the higher ionic conductivity, which can make electrical poling problematic. This can be addressed by several routes, including the use of rubidium ion exchange into the top surface of the crystal [31] and low-temperature poling [32] to counter the higher conductivity, or by the use of the hydrothermal growth process, which naturally results in lower conductivity crystals. Recent efforts to overcome these problems in flux-grown KTP have made use of dopants such as rubidium to control the conductivity [33]. Recently, PPKTP has been fabricated with ultrafine periods (800 nm) allowing for the first demonstration of a backward wave OPO, a unique mirrorless device [34].

It is hard to compare PPKTP and PPLN as QPM materials; each has advantages and disadvantages, and the correct choice depends on the application, and is often quite subtle.

21.8 APPLICATIONS OF PPLN AND RELATED MATERIALS

The number of papers published on PPLN in recent years provides a clear demonstration of the excitement and potential of PPLN and related materials. These applications span a number of industries, wavelength ranges, and temporal configurations. These materials can be seen to provide a way of spanning the gaps in the visible and near-IR spectrum at which regular lasers cannot operate, and to provide for applications that need wide tunability. In an even more general sense, these nonlinear materials provide an optical equivalent for the frequency mixing functions so widely used in Radio Frequency (RF) and microwave electronics.

This review is organized into broad industrial areas and will not pretend to be exhaustive, but is intended to whet the appetite of potential end users of QPM materials.

21.8.1 Aerospace

This area is probably the most heavily researched to date, with military applications leading the way. Current military systems tend to operate using 1.064 μm Nd:YAG lasers, and there is a general desire to provide laser systems for longer wavelengths—first, to make filtering or observation more difficult, then to exploit the better transmission characteristics of the atmosphere in longer wavelength bands, and also simply to benefit from the better eye safety obtained at wavelengths such as 1.5 μm.

21.8.1.1 REMOTE SENSING

This area provides one of the great opportunities for QPM materials, by providing sources tunable to specific wavelength for the detection of trace gases, or atmospheric scattering. OPOs provide flexible and versatile sources for these applications. It is possible to use even simpler systems, such as difference frequency generation, for making low-cost and low-power consumption sources.

21.8.1.2 TELECOMMS

This has become a very hot topic for QPM materials in the past few years. The nonlinear conversion of light allows users to provide a wide range of optical processing functions. Applications that are currently receiving attention include wavelength conversion for wavelength division multiplexed (WDM) systems, mid-span spectral inversion, non-linear dispersion compensation, and temporal pulse manipulation for all-optical switching and TDM applications. As these applications need to work with low optical powers, they tend to use a waveguide format, which utilizes either proton exchange or titanium diffusion to make the waveguides [35–37]. Waveguide QPM devices are now sold commercially (e.g., by HC Photonics, SRICO, and NTT Electronics) and are typically based on proton exchange; however, excellent performance has been demonstrated using other approaches, including zinc diffusion and physical machining.

21.8.1.3 DISPLAYS

Although to date the success in this area has been limited, it remains one with a great potential. Laser-based light shows and displays are already one of the major "public" applications of the laser. QPM materials can play an important role in helping to provide red, green, and blue sources for laser projector systems. A number of schemes are possible, ranging from high-power OPOs with suitable powers through to fiber laser-based systems. The requirements on the materials are stringent, needing several watts of each of color at optimal wavelengths. To date, problems such as GRIIRA in PPLN and gray tracking in KTP provide barriers to the commercial use of existing QPM materials, but this can be expected to change in the future. An example of harmonic generation to create green light is shown in Figure 21.10.

21.8.1.4 STORAGE AND BLUE LASERS

Historically, the need for blue sources for optical disc technology provided a strong impetus for QPM material development; but, with the recent development of blue semiconductor lasers, it now seem less likely that QPM materials will find use for readout lasers. However, it may well be the case that for higher power applications, such as DVD writing and for laser marking, where 10–100 mW is attractive that QPM blue and near-UV sources may become commercial. Further interest in this area comes from the recent increases in fiber laser power such that doubling, tripling, and even frequency quadrupling in QPM materials could provide a viable alternative to gas laser in the blue and near UV. The book by Risk [38] provides an excellent coverage of the applications of blue lasers and a comparison of QPM sources to other types of laser.

Figure 21.10 An example of frequency doubling of a 1064 nm laser to 532 nm. (Image courtesy of Covesion Limited.)

21.8.1.5 SPECTROSCOPY AND SCIENTIFIC APPLICATIONS

This provides a tremendous opportunity for QPM materials as a a platform both for investigating nonlinear optics and for creating light sources of unparalleled versatility. A good example of such applications of nonlinear crystals is in sources based on frequency doubling of fiber lasers, where mode-locked femtosecond sources based on Erbium-doped fiber can be frequency doubled to around 760 nm where they provide the same wavelengths as mode-locked Ti:sapphire lasers but at a fraction of the cost. Similarly, although many conventional nonlinear materials can be used for making OPOs, it is likely that higher nonlinearities of periodically poled materials will see them becoming increasingly important in commercial systems. In addition to their use as IR sources for spectroscopy, there are a host of applications in quantum optics experiments where the ability to tailor the nonlinearity is likely to lead to new devices and experiments. This trend particularly applies in periodically

poled waveguides and poled glass fibers where the high-degree confinement leads to high conversion processes and efficient capture of generated light [39].

21.8.2 Quantum science

A full description of nonlinear optics is necessarily quantum mechanical, and indeed, the unique properties of light generated via nonlinear optics is attracting great interest. A good example is spontaneous parametric downconversion, where a more energetic photon is split into two less energetic (longer wavelength) photons. In such a process, the two generated photons are inherently correlated, as they are derived from the same starting photon. Such pair photons are inherently entangled, and thus can be used for a variety of applications, including heralded single photon production, and in experiments in quantum key distribution.

Optical wavelength conversion does not provide the only application area for domain engineered materials. Particularly interesting are applications in modulators and beam deflectors where domain engineering can allow new functionality and confer advantages. One example is the waveguide lithium niobate modulators, in which it has been shown that domain reversal can be used to provide phase matching between the RF drive and the optical wave without having to use segmented electrodes [40]. Domain engineering also provides ways of making new devices for bulk beam modulation through the creation of electro-optic Bragg gratings [41]. Closely related, although not yet demonstrated, would be the development of very fine period domain gratings within waveguides to make switchable retroreflective Bragg gratings, as these would require a period of a fraction of a micron, so they are perhaps not likely to be realized at any time soon.

Another class of domain engineered devices that do not require gratings are the beam-deflector devices based on refraction [42]. These devices are based on prisms formed from poled electro-optic material and rely on the fact that the change in refractive index in the up and down regions is of opposite sign. Thus, by passing a beam through one or more interfaces, the beam will be deflected by Snell's law allowing the creation of very fast all solid-state deflectors and scanners.

21.8.3 Future of QPM materials

It should be clear from this discussion that QPM materials provide one of the most exciting and vibrant areas of optical research, and it is likely that they will gradually come to replace many of the uses of birefringently phase-matched materials. The field is maturing, as the advantages of QPM are compelling, but clearly many challenges remain. In particular, it is clear that all of the existing QPM materials have drawbacks when used in certain ways, and it is likely that new materials will be developed, either as variants of existing materials (such as the recent works on stoichiometric lithium niobate) or in new ways of making composite materials (QPM GaAs), or even in wholly new nonlinear materials. Another important area of QPM research will be the development of more sophisticated grating designs optimized for high efficiency and specific spectral and temporal response.

REFERENCES

1. J.A. Armstrong, N. Bloembergen, J. Ducuing, P.S. Pershan, Interactions between light waves in a nonlinear dielectric, *Phys. Rev.*, 127(6): 1918–1939, 1962.
2. M. Yamada, N. Nada, M. Saitoh, K. Watanabe, First-order quasiphase-matched LiNbO$_3$ waveguide periodically poled by applying an external field for efficient blue second-harmonic generation, *Appl. Phys. Lett.*, 62: 435–436, 1993.
3. P.G. Kazansky, V. Pruneri. Electric-field poling of quasi-phase-matched optical fibers, *J. Opt. Soc. Am. B*, 14(11): 3170–3179, 1997.
4. A. Canagasabey, C. Corbari, A.V. Gladyshev, F. Liegeois, S. Guillemet, Y. Hernandez, M.V. Yashkov, A. Kosolapov, E.M. Dianov, M. Ibsen, P.G. Kazansky, High-average-power second-harmonic generation from periodically poled silica fibers, *Opt. Lett.*, 34: 2483–2485, 2009.
5. L. Gordon, G.L. Woods, R.C. Eckardt, R.R. Route, R.S. Feigelson, M.M. Fejer, R.L. Byer, Diffusion-bonded stacked GaAs for quasi-phase-matched 2nd-harmonic generation of a carbon-dioxide laser, *Elec. Lett.*, 29(22): 1942–1944, 1993.
6. C.B. Ebert, L.A. Eyres, M.M. Fejer, J.S. Harris, MBE growth of an-tiphase GaAs films using GaAs/Ge/GaAs heteroepitaxy, *J. Cryst. Growth*, 201: 187–193, 1999.
7. D.S. Hum, M.M. Fejer, Quasi-phasematching, *C. R. Phys.* 8, 2007.
8. V. Pasiskevicius, G. Strömqvist, F. Laurell, C. Canalias, Quasi-phase matched nonlinear media: Progress towards nonlinear optical engineering, *Opt. Mat.*, 34: 513–523, 2012.
9. M.E. Lines, A.M. Glass, *Principles and Applications of Ferroelectrics and Related Materials*, Oxford University Press, Oxford, 1977.
10. A.M. Prokhorov, Yu. S. Kuz'minov, *Physics and Chemistry of Crystalline Lithium Niobate*, Institute of Physics, Bristol, 1990.
11. J.C. Burfoot, G.W. Taylor, *Polar Dielectrics and their Applications*, Macmillan, London, 1979.
12. J.A. Giordmaine, Mixing of light beams in crystals, *Phys. Rev. Lett.*, 8(1): 19–20, 1962.
13. P.D. Maker, R.W. Terhune, M. Nisenoff, C. M. Savage, Effects of dispersion and focusing on the production of optical harmonics, *Phys. Rev. Lett.*, 8(1): 21–22, 1962.
14. R.L. Byer, Quasi–phasematched nonlinear interactions and devices, *J. Nonlinear Opt. Phys. Mater.*, 6(4): 549–592, 1997.
15. R.L. Byer, Chapter 2—Parametric oscillators and nonlinear materials. In *Nonlinear Optics*, P. G. Harper and B. S. Wherrett, eds., 47–160, Academic Press, London, 1977.
16. Y.R. Shen, *The Principles of Nonlinear Optics*, Reprint edition, John Wiley & Sons, New York, 2002.
17. A. Yariv, *Quantum Electronics*, 3rd ed., John Wiley & Sons, New York, 1989.
18. D.H. Jundt, Temperature-dependent Sellmeier equation for the index of refraction, n_e, in congruent lithium niobate, *Opt. Lett.*, 22(20): 1553–1555, 1997.
19. M.M. Fejer, G.A. Magel, D.H. Jundt, R.L. Byer, Quasi-phasematched second harmonic generation: Tuning and tolerances, *IEEE J. Quant. Electron.*, 28: 2631–2654, 1992.
20. W. Koechner, *Solid State Laser Engineering*, Springer–Verlag, New York, 1999.
21. G.D. Boyd, D.A. Kleinman, Parametric interaction of focused Gaussian light beams, *J. Appl. Phys.*, 19(8): 3597–3639, 1968.

22. K.R. Parameswaran, J.R. Kurz, R.V. Roussev, M.M. Fejer, Observation of 99% pump depletion in single-pass second-harmonic generation in a periodically poled lithium niobate waveguide, *Opt. Lett.*, 27(1): 43–45, 2002.

23. G.W. Ross, M. Pollnau, P.G.R. Smith, W.A. Clarkson, P.E. Britton, D.C. Hanna, Generation of high-power blue light in periodically poled LiNbO$_3$, *Opt. Lett.*, 23(3): 171–173, 1998.

24. K. Yamamoto, K. Mizuuchi, Y. Kitaoka, M. Kato, Highly efficient quasi-phase-matched 2nd-harmonic generation by frequency-doubling of a high-frequency superimposed laser-diode, *Opt. Lett.*, 20(3): 273–275, 1995.

25. M.A. Watson, M.V. O'Connor, P.S. Lloyd, D.P. Shepherd, D.C. Hanna, C.B.E. Gawith, L. Ming, P.G.R. Smith, O. Balachninaite, Extended operation of synchronously pumped optical parametric oscillators to longer idler wavelengths, *Opt. Lett.*, 27(23): 2106–2108, 2002.

26. J. Saikawa, M. Fujii, H. Ishizuki, T. Taira, 52 mJ narrow-bandwidth degenerated optical parametric system with a large-aperture periodically poled MgO:LiNbO$_3$ device, *Opt. Lett.*, 31: 3149–3151, 2006.

27. B. Sturman, M. Aguilar, F. AgulloLopez, V. Pruneri, P.G. Kazansky, Photorefractive nonlinearity of periodically poled ferroelectrics, *J. Opt. Soc. Am. B*, 14(10): 2641–2649, 1997.

28. Y. Furukawa, K. Kitamura, A. Alexandrovski, et al., Green-induced infrared absorption in MgO doped LiNbO$_3$, *Appl. Phys. Lett.*, 78(14): 1970–1972, 2001.

29. M.A. Arbore, O. Marco, M.M. Fejer, Pulse compression during second-harmonic generation in aperiodic quasi-phase-matching gratings, *Opt. Lett.*, 22(12): 865–867, 1997.

30. L. Lefort, K. Puech, S.D. Butterworth, Y.P. Svirko, D.C. Hanna, Generation of femtosecond pulses from order-of-magnitude pulse compression in a synchronously pumped optical parametric oscillator based on periodically poled lithium niobate, *Opt. Lett.*, 24(1): 28–30, 1999.

31. F. Laurell, H. Karlsson, Electric field poling of flux grown KTiOPO$_4$, *Appl. Phys. Lett.*, 71(24): 3474–3476, 1997.

32. G. Rosenman, A. Skliar, D. Eger, M. Oron, M. Katz, Low temperature periodic electrical poling of flux-grown KTiOPO$_4$ and isomorphic crystals, *Appl. Phys. Lett.*, 73(25): 3650–3652, 1998.

33. Q. Jiang, P.A. Thomas, K.B. Hutton, R.C.C. Ward, Rb-doped potassium titanyl phosphate for periodic ferroelectric domain inversion, *J. Appl. Phys.*, 92(5): 2717–2723, 2002.

34. C. Canalias, V. Pasiskevicius, Mirrorless optical parametric oscillator, *Nat. Photonics*, 1: 459–462, 2007.

35. M.H. Chou, K.R. Parameswaran, M.M. Fejer, I. Brener, Multiple-channel wavelength conversion by use of engineered quasi phase-matching structures in LiNbO$_3$ waveguides, *Opt. Lett.*, 24(16): 1157–1159, 1999.

36. J. Amin, V. Pruneri, J. Webjorn, P.S. Russell, D.C. Hanna, J.S. Wilkin-son, Blue light generation in a periodically poled Ti:LiNbO$_3$ channel waveguide, *Opt. Commun.*, 135(1–3): 41–44, 1997.

37. D. Hofmann, G. Schreiber, C. Haase, H. Herrmann, W. Grundkotter, R. Ricken, W. Sohler, Quasi-phase-matched difference-frequency generation in periodically poled Ti: LiNbO$_3$ channel waveguides, *Opt. Lett.*, 24(13): 896–898, 1999.

38. W.P. Risk, T.R. Gosnell, A.V. Nurmikko, *Compact blue-green laser*, Cambridge University Press, Cambridge, 2003.

39. G. Bonfrate, V. Pruneri, P.G. Kazansky, P. Tapster, J.G. Rarity, Para-metric fluorescence in periodically poled silica fibers, *Appl. Phys. Lett.*, 75(16): 2356–2358, 1999.

40. W. Wang, R. Tavlykaev, R.V. Ramasawamy, Bandpass traveling-wave Mach-Zehnder modulator in LiNbO$_3$ with domain reversal, *IEEE Photon. Technol. Lett.*, 9(5): 610–612, 1997.

41. M. Yamada, M. Saitoh, H. Ooki, Electric-field induced cylindrical lens, switching and deflection devices composed of the inverted domains in LiNbO$_3$ crystals, *Appl. Phys. Lett.*, 69(24): 3659–3661, 1996.

42. J. Li, H.C. Cheng, M.J. Kawas, D.N. Lambeth, T.E. Schlesinger, D.D. Stancil, Electrooptic wafer beam deflector in LiTaO$_3$, *IEEE Photon. Technol. Lett.*, 8(11): 1486–1488, 1996.

22

Silicon photonics

SASAN FATHPOUR
University of Central Florida

22.1 INTRODUCTION

Silicon (Si) photonics is the science and technology of using the group-IV element of the periodic table as the principle material for optoelectronic integrated circuits (OEICs). Silicon has long had photonic applications. Notably, it has been the dominant material of choice for photovoltaic cells, visible and short near-infrared (IR) (<~1 μm wavelength) photodetectors, as well as for X-ray detectors in conjunction with scintillators. However, the younger field of silicon photonics is less related to these traditional applications. Rather, it refers to performing a wide range of passive and active optical functionalities on an integrated circuit at wavelengths within the material's transparency range, i.e., above the bandgap cutoff of ~1.1 μm. Particularly, the telecommunication wavelengths of 1.3–1.6 μm and more recently the longer wavelengths of the mid-IR—up to ~6.7 μm beyond which the material becomes lossy—are the realm of silicon photonics. Based on some interesting waveguide platforms, silicon is recognized as an ideal material for low loss and ultracompact passive integrated optics at these wavelength ranges. There are also means to achieve optical modulators on the material. Heterogeneous integration of Si with devices on germanium (Ge), silicon–germanium (Si–Ge) alloys, and III–V compound semiconductor materials has allowed achieving detection and generation and amplification of optical waves in the near IR range.

From a larger perspective, Si is arguably one of the most thoroughly studied materials in the human history. Originally fueled by the material's very amazingly unique properties for making transistors and electronic integrated circuits, the mature processing techniques developed for microelectronic were readily available for photonic device and circuit processing. An attractive feature of silicon photonics has always been the possibility of such seamless integration with electronic circuitry on the same inexpensive substrate. This opportunity has rendered silicon photonics as an established technology for electronic–photonic integrated circuits at telecommunication wavelength for optical interconnect applications. Such optical transceivers are perhaps the most important silicon-photonic products commercially available for some years for short-haul data communication applications.

In addition, silicon has a strong third-order optical nonlinearity and these properties are further enhanced at above ~2.2 μm wavelength (the material's two-photon bandgap). Nonlinear integrated optics at both near- and mid-IR has, therefore, been one of the potential applications of silicon photonics. Beyond these, silicon photonics may find applications in biomedical sensing and environmental monitoring. This chapter aims at reviewing this field with emphasis on some educational and historic background of the field, state-of-the-art waveguide platforms, heterogeneous integration and optical transceivers, nonlinear optics, and mid-IR platforms.

There are a handful of books published on silicon photonics, and the reader is encouraged to consult them for a more detailed study of the field (e.g., Pavesi and Lockwood 2004; Reed and Knights 2004; Reed 2008; Khriachtchev 2009; Fathpour and Jalali 2012). Several review articles have also been published throughout the years (e.g., Soref 1993; Jalali et al. 1996, 1998; Lipson 2005; Jalali and Fathpour 2006; Soref 2006a; Jalali 2008).

22.2 HISTORICAL PERSPECTIVE

Historically, silicon used to be dismissed for photonic functionalities, simply because the material is a very poor light emitter. For example, some of the most important classes of photonic devices, i.e., light emitting diodes and diode lasers, are not easily attainable in the material. Instead, all group III–V OEICs were being aggressively pursued in the 1980s and the early 1990s, with an eye on their clear advantages in terms of light generation, detection, and modulation. With the much higher inherent speed of III–V transistors, compared to Si-based transistors, the hope was to monolithically integrate III–V electronics and photonics on the same chips. However, this vision of III–V "homogenization" of electronics and photonics did not materialize for a variety of reasons. Most notably, the silicon-based complementary metal oxide semiconductor (CMOS) technology demonstrated unbeatable performance

for digital—and later analog—electronics, especially when large-scale integration is concerned. Si CMOS advantages include negligible static power, high input impedance, self-isolation, ease of device layout, very high yield and scalability of processing, and abundance and low cost of the material. It is noteworthy that silicon is the second-most abundant element in the Earth crust after oxygen, and Si wafers, typically made by the Czochralski process, are fairly inexpensive. Group III–V CMOS, on the other hand, did not succeed even at the research stage, primarily because of the lack of native oxides for the compound semiconductors. Also, III–V optoelectronics was not capable of providing tightly confined and low-loss passive optical waveguides with the high yield and low cost demanded for medium- and larger-scale integration, i.e., when the number of photonic devices exceeded a few.

In parallel, integrated optics on dielectrics, most notably silica, has been pursued for decades. The so-called planar lightwave circuits (PLC) are fairly inexpensive, easy to process, and low loss. However, the low-contrast passive devices on dielectrics are very bulky (not compatible with the size of semiconductor lasers and CMOS electronics), cannot be bent at short radii without significant loss, and cannot be manipulated electrically for active devices like modulators and detectors.

The prevalence of silicon electronics imposed a paradigm shift onto photonics too and led to the seed ideas of silicon photonics. It was argued first in the mid-1980s that even though light emission was not yet possible in silicon, several other photonic functionalities are feasible (Soref and Lorenzo 1985, 1986; Lorenzo and Soref 1987; Soref and Bennett 1987; Soref 1993). Early works used the carrier plasma effect for waveguiding, but in the 1990s the silicon-on-insulator (SOI) wafers became available and emerged as the versatile platform for passive integrated optics (Jalali et al. 1998). For instance, directional couplers, arrayed-waveguide gratings (AWGs), multimode interferometers, optical delay lines, gratings, and optical mode converters were demonstrated on SOI. Si optical modulators, microring resonators, and the first works on Ge waveguide photodetectors appeared in the early 2000s (Jalali and Fathpour 2006). In the same time period, the first experiments on nonlinear silicon photonics were conducted. Since the mid-2000s, the first commercial optical transceivers appeared and since then the silicon photonic industry has been maturing. Also, foundry services have been initiated to promote fabless research and development (R&D). The first experiments on futuristic biomedical and mid-IR applications of silicon photonics emerged at the same time (Jalali et al. 2006). All of these efforts have made silicon photonics one of the most booming fields of optics and photonics in the last decade both in academic research and industrial R&D and commercialization.

22.3 WAVEGUIDE PLATFORMS

In one of the very first Si waveguides, the carrier plasma effect was exploited to create a bottom cladding layer for optical waveguides. Essentially, the index contrast between n-doped Si layers, epitaxially grown on n$^+$-doped Si substrates, forms slab waveguides, on which ridge (or rib) waveguides can be formed by lithography and etching techniques (Figure 22.1a) (Lorenzo and Soref 1987). Expectedly, however, the low doping-induced contrast of ~1.5×10^{-2} does not allow compact waveguides, and the dimensions are larger than those of III–V heterostructure waveguides. Also, the highly doped substrate, with a donor concentration of 2×10^{19} cm^{-3}, causes high carrier-induced optical loss for the guided wave.

Much higher index contrast and hence more compact waveguides can be realized if a thin film of Si can be somehow placed on top of a buried insulting layer with lower refractive index. Silicon dioxide (SiO$_2$)—with an index, n, of ~1.44 versus ~3.48 of Si at ~1.55 μm wavelength—is an excellent choice (other choices are sapphire, silicon nitride (SiN), lithium niobate (LiNbO$_3$), and even air, as discussed later). Si-on-SiO$_2$, called SOI, has been

the most developed and most common waveguide platform, particularly for telecommunication wavelengths to the extent that silicon photonics has almost become synonymous with SOI photonics. The interest for developing SOI wafers was originally for electronic applications but coincidentally they happened to be useful for photonics too. Separation by IMplantation of OXygen, Bond and Etch-back SOI, ion implantation followed by thermal splitting (the SmartCut process) are some of the methods developed for realizing SOI wafers (Celler and Cristoloveanu 2003). However, the SmartCut process of Soitec Incorporation has been more popular because of the uniform thin films it can provide.

Earlier SOI-based photonic works were on above-micron-size ridge waveguides, with typical dimensions shown in Figure 22.1b, and with linear propagation loss of ~0.5 dB/cm or lower (Fathpour et al. 2006). Some workers still prefer such larger cross sections because the waveguides possess lower propagation losses and size compatibility with III–V lasers and yet are single mode (see Section 22.4.1). However, most of other researchers have shifted to submicron waveguides with typical heights of 200–250 nm and widths of 400–500 nm. As shown in Figure 22.1c, both ridge and channel waveguides are feasible on nowadays standard 220-nm-thick Si films on about 2-μm-thick SiO$_2$ insulator, also known as the buried oxide (BOX) layer. There is typically a tradeoff between the waveguides' width and etch depth, on the one hand, and the propagation loss, on the other hand, depending on the optical intensity at the scattering nonsmooth sidewalls. Accordingly, shallow ridge waveguides may have lower propagation loss than

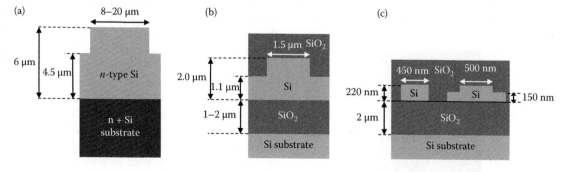

Figure 22.1 Schematic of different silicon photonic waveguide platforms: (a) historic dopant-based waveguides, (b) typical micron-size SOI ridge waveguides, and (c) *de facto* standard submicron SOI channel and ridge (rib) waveguides.

channel counterparts, but the waveguide mode penetrates laterally into the slab region and the bending loss is higher than channel waveguides. However, for active devices, such as optical modulators, attaining p–i–n diodes in parallel to the waveguides becomes necessary (see Section 22.4.3). Then, having a thin slab region for doping Si for making metallic contacts to them is required and using ridge SOI waveguides is inevitable.

In the last decade, a handful of companies and organizations have been attempting to standardize SOI waveguide processing steps and design libraries. In Europe, for instance, the ePIXfab foundry has established a standard process for both channel and ridge waveguides for passive and active silicon photonics, respectively. The dimensions of single-mode waveguides shown in Figure 22.1c are actually borrowed from their standard dimensions (ePIXfab 2014). The propagation loss of ePIXfab's channel and ridge waveguides is 2.5–3.0 and 1.8–2.5 dB/cm, respectively. Multiproject wafer (MPW) or "shuttle" runs in ePIXfab foundry services have been accessible to fabless researchers around the globe in the last few years. In the United States, Luxtera Corporation has been working on standard computer-aided design libraries, and the OpSIS organization has made an attempt in shuttle runs (OpSIS 2014).

The described SOI waveguide technology has undoubtedly established itself as a versatile benchmark platform for silicon photonics in the telecommunication wavelengths. More recently, there has been an interest for mid-IR wavelengths, defined loosely as the atmospheric window of 3–5 μm, and sometimes the wider 2.2–6.5 μm range. Unfortunately, the BOX layer of SOI is optically lossy in the 2.5–2.9 μm and above 3.5 μm wavelengths. Alternative bottom cladding materials include sapphire ($n \approx 1.76$), SiN ($n \approx 2$), air ($n=1$)

(Soref et al. 2006b), as well as lithium niobate, LiNbO$_3$ ($n \approx 2.2$) (Chiles and Fathpour 2014). These platforms are discussed in detail in Section 22.10.

22.4 PHOTONIC DEVICES BASED ON THE SOI PLATFORM

Several passive and active photonic devices have been studied in the last 20 years, mostly on the SOI platform. It is impossible to review all of these work here. Rather, modeling of SOI optical waveguides is succinctly provided in Section 22.4.1 and is followed by a short review of other passive and active photonic devices based on SOI in Sections 22.4.2 and 22.4.3, respectively.

22.4.1 SOI optical waveguides

Optical waveguides are the most fundamental devices for any integrated photonic technology. The SOI waveguide platform was reviewed in Section 22.3 from a structural viewpoint. Here, the optical properties of the waveguides are introduced. Special attention is paid to the single-mode behavior and polarization dependency.

Arbitrary planar, three-dimensional (3D) waveguides are generically introduced in Figure 22.3 with refractive indices n_0, n_1, and n_2, for top cladding, and core and bottom cladding regions, respectively, and for ridge width $W=2a\lambda$, ridge height $H=2b\lambda$ and slab height $h=rH=2br\lambda$, where λ is the vacuum wavelength of the guided wave. In principle, these general index profiles and wavelength-normalized dimensions should suffice to cover all types of silicon-based waveguide platforms shown in Figures 22.1 and 22.2. For instance, $r=0$ corresponds to channel waveguides. Developing a comprehensive theory (preferably analytical or semianalytical) that can formulate the guided mode properties of such structures is definitely very appealing and has been extensively investigated for at least 45 years (Marcatili 1969). Unfortunately, solutions to the wave equation in the Cartesian coordinates with the appropriate boundary conditions of Maxwell's equations become so complicated for 3D structures that universal exact solutions are formidable and even approximate solutions, such as Kumar's and the effective index methods (Okamoto 2006), are cumbersome and one has to pay a lot of attention to the simplifying assumptions. This is a more pressing

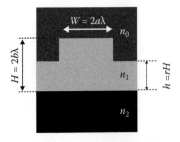

Figure 22.2 Generic 3D waveguide structure and notations for modeling purposes.

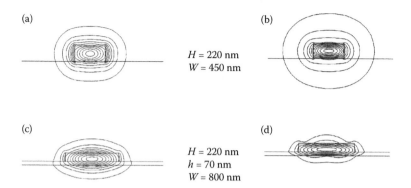

(a)

(b)

$H = 220$ nm
$W = 450$ nm

(c)

(d)

$H = 220$ nm
$h = 70$ nm
$W = 800$ nm

Figure 22.3 Field profiles of fundamental TE and TM modes in typical 220-nm-thick SOI channel and rib submicron waveguides. (a) Channel TE, (b) channel TM, (c) rib TE and (d) rib TM. (Courtesy of M. Malinowski of CREOL.)

issue in high-contrast waveguides such as those used in silicon photonics.

Nonetheless, a plethora of classic works on the fundamentals of optical waveguides can be employed, to a certain extent, as guidelines for evaluating the single-mode condition. For example, assuming that $r \geq 0.5$ (i.e., shallow-etched ridge waveguides) and b is large (i.e., dimensions larger than the wavelength), an approximate condition for single-mode operation exists. That is, the only modes will be the fundamental transverse-electric (TE) and transverse-magnetic (TM) modes if (Soref et al. 1991)

$$\frac{W}{H} = \frac{a}{b} \leq 0.3 + \frac{r}{\sqrt{1-r^2}}$$

For instance, the waveguide of Figure 22.1b has $r = 0.55$ and $H = 2$ μm. Therefore, for $W \leq 1.9$ μm the waveguide is estimated to remain single mode. The above single-mode condition appears to be independent of the index profile. However, the mentioned condition for large b is quantitatively related to the index profile, i.e., $4\pi b$ should be much larger than

$$q_{TE} = \frac{1}{\sqrt{n_1^2 - n_0^2}} + \frac{1}{\sqrt{n_1^2 - n_2^2}}$$

for the TE modes and be much larger than

$$q_{TM} = \frac{(n_0/n_1)^2}{\sqrt{n_1^2 - n_0^2}} + \frac{(n_0/n_2)^2}{\sqrt{n_1^2 - n_2^2}}$$

for the TM modes. These conditions are easily satisfied in our example, where with $n_1 = 3.48$ and

$n_0 = n_2 = 1.44$, leading to $q_{TE} \approx 0.63$, $q_{TM} \approx 0.11$, and $b \approx 0.65$ at $\lambda = 1.550$ μm. It is notable that the rather large waveguide of Figure 22.1b is actually single mode. This nonintuitive feature emphasizes the much more forgiving nature of 3D waveguides for single-mode behavior in terms of waveguide height, as opposed to 2D (slab) waveguides. To elaborate this statement, it is instructive to examine the corresponding single-mode condition for symmetric 2D waveguides with slab height, d, is $V = 2\pi \cdot d(n_1^2 - n_2^2)^{1/2}/\lambda < \pi$ (Okamoto 2006). Accordingly, for the $SiO_2/Si/SiO_2$ slab structure and at $\lambda = 1.55$ μm, $d < \sim 250$ nm is required for the cutoff of higher-order modes in the direction perpendicular to the slab. This d range is much thinner than $H = 2$ μm in our example and is indeed one reason 220-nm SOI has become popular, so not only 3D but also 2D (slab) waveguides will be single mode on them.

For the submicron waveguides of Figure 22.1c, the above formulation becomes inapplicable. In practice, fully numerical methods ought to be employed. The versatile finite element method (FEM) can be adopted to solve the wave equation in the frequency domain (i.e., the Helmholtz equation). Under slowly varying envelope approximation (SVEA), the Helmholtz equation can be solved more conveniently. For axially varying waveguides and devices, such as bent and tapered waveguides, the FEM stationary analysis under SVEA can be further simplified by the beam-propagation method (BPM) (Okamoto 2006), at the cost of losing some accuracy. Certain approximations can further make BPM converge faster. They include scalar field (i.e., neglecting polarization effects), paraxiality (i.e., propagation restricted to a narrow

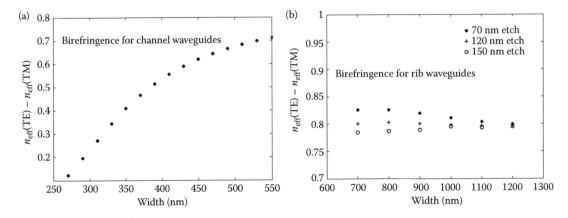

Figure 22.4 Birefringence of (a) channel and (b) rib 220-nm-thick SOI waveguides. (Courtesy of M. Malinowski of CREOL.)

range of angles), semivectorial (i.e., each polarization simulated separately), and full vectorial (i.e., both polarizations and the coupling between them simulated) approximations. The finite-difference time domain method can also be used to directly solve Maxwell's equations in the time domain, a computational intensive method but useful for studying ultrafast pulse propagations. Several commercial software packages are available for all the described methods.

Figures 22.3 and 22.4 present some basic properties of 220-nm SOI ($H=220$ nm) channel and ridge waveguides passivated with SiO_2 ($n_0=n_2=1.44$). The radio frequency (RF) module of COMSOL software package was used to obtain these plots. The FEM simulator directly solves Helmholtz equation under the SVEA with the appropriate boundary conditions. Figure 22.3 shows the E-and H-field contours for TE and TM modes of channel ($W=450$ nm) and rib ($W=800$ and $h=70$ nm) single-mode waveguides, respectively. A larger W was chosen for the rib waveguides, as the modes tend to become leaky into slab modes for smaller values. For instance, the TM modes start to leak for $W<{\sim}600$ nm. The channel waveguide starts supporting the second TE and TM modes at W of ~500 and ~800 nm, respectively.

One important property of any optical waveguide is its effective index, n_{eff}, and the birefringence (difference) of the indices between the TE and TM modes, $n_{eff}(TE)-n_{eff}(TM)$ (Okamoto 2006). This polarization sensitivity in integrated waveguides in general stems from mechanical stress in the waveguiding layer and/ or cross-sectional geometry of the waveguide.

Mechanical stress is generally negligible in standard SOI waveguides. The COMSOL-simulated geometry dependence of birefringence for channel and rib waveguides discussed here is plotted versus W in Figures 22.4a and b, respectively. Expectedly, the more rectangular the channel waveguides, the higher the birefringence. However, by symmetry zero birefringence would occur in channel waveguides when $W=H=220$ nm, a geometrical condition that does not support a mode. Because the rib waveguides are far from square ($W/H >{\sim}600$ nm/220 nm), the evident birefringence in the ~0.77 to ~0.83 range is a weak function of W and h. Zero birefringence can be achieved in waveguides with higher H (Chan et al. 2005; Lim et al. 2007).

22.4.2 Other passive photonic sevices on SOI

Based on the SOI platform, several types of passive devices, other than optical waveguides, have been demonstrated. They include, but not limited to, optical mode converters and grating couplers for coupling light in and out of the photonic chips, directional couplers, Y-junction splitters and combiners, multimode interference (MMI) splitters, star couplers, AWG, grating waveguides, Fabry–Perot (FP) waveguide resonators, microring resonators, etc. Scanning-electron micrographs (SEMs) of some of these devices are presented in Figure 22.5. It is beyond the scope of this monograph to review all of these devices. The reader is referred to several review papers and book chapters on the concept, design, and performance of

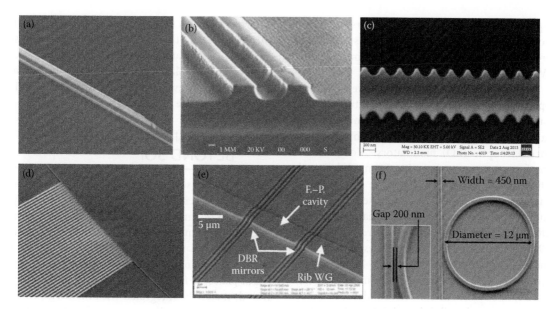

Figure 22.5 SEM images of various passive photonic devices on SOI: (a) 3D mode converters. (Courtesy of Mellanox Corp., former Kotura), (b) directional couplers (Trinh et al. 1995), (c) grating waveguides (Courtesy of S. Khan of CREOL), (d) input section of an AWG (Courtesy of B. Jalali of UCLA), (e) FP waveguide resonators (Pruessnera et al. 2006), and (f) microring resonators and bus waveguide (Xu et al. 2005).

these devices (Jalali et al., 1996, 1998; Reed and Knights 2004; Lipson 2005; Jalali and Fathpour 2006; Reed 2008). The key advantage of SOI for all these passive devices, compared to PLC and optical fiber-based counterparts, is the compactness that becomes feasible by the high index contrast of the SOI platform, allowing submicron waveguide cross-sections and sharp bends with negligible loss. Bending loss of waveguides is another challenging analysis and design problem studied since the very early days of integrated photonics (Marcatili and Miller 1969; Heiblum and Harris 1975). Modeling of bending loss in modern submicron waveguides typically requires detailed full-vectorial FEM simulations (Jedidi and Pierre 2007). In practice, losses per 90° bend of 0.086±0.005 dB for a bending radius of 1 μm and 0.013±0.005 dB for a bend radius of 2 μm have been reported in 445 nm×220 nm channel waveguides (Vlasov et al. 2004).

22.4.3 Active photonic devices on SOI

A key difference between silicon photonics, on the one hand, and PLCs and optical fibers, on the other hand, is the existence of charge carriers in silicon. This feature has its own pros and cons.

The advantage is that using p–n junction diodes or metal-oxide-semiconductor capacitor to manipulate optical properties of Si for a range of devices by injection, depletion, or accumulation of carriers (it will be discussed later, however, that the existence of free carriers created by two-photon absorption can become problematic for nonlinear-optic applications). Typically, a p–n junction straddles the waveguide region. Two major configurations are shown in Figure 22.6. In the state of the art, more elaborate design of the doping profiles across the waveguide structure may be devised for the optimal performance of the targeted device applications. One example device is variable-optical attenuators (VOAs) used for automatic channel

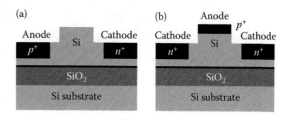

Figure 22.6 Two configurations of SOI waveguides with p–n junction diode for carrier injection and/or depletion.

equalization in add/drop multiplexers for metropolitan area networks. Indeed, VOAs are considered the first commercial silicon photonic products (Whiteman et al. 2003).

Perhaps the most important active SOI device operating on similar device architectures is optical modulators. Silicon is a centrosymmetric crystal and lacks the Pockels effect commonly used in electrooptic (EO) modulators. Modulation of the free-carrier concentration for modulating the phase and/or intensity of light, the so-called free-carrier plasma effect, was proposed in some of the earliest works as a viable alternative to the EO effect (Lorenzo and Soref 1987; Soref and Bennett 1987; Soref 1993). In typical devices, $p-n$ junction diodes are incorporated into Mach–Zehnder interferometers (MZI) and sometimes microrings and used to modulate the refractive index. Numerous works on such all-Si modulators have been reported, a review of which can be found elsewhere (Reed et al. 2010). More recently, modulators with adequately high bit rates (~50 Gb/s) and low half-voltage length-product $V\pi \cdot L$ (~2.8 V cm) have been reported in MZIs (Thomson et al. 2012; Tu et al. 2013). However, the devices only offer modest (3–5.5 dB) modulation depth or extinction ratio. Heterogeneous platforms based on EO-active materials to address this shortcoming of SOI optical modulators are discussed in Section 22.9.

22.5 SILICON PHOTONICS BEYOND SOI

A fully SOI-based solution for photonics would certainly be very rewarding and revolutionary. Numerous researchers around the globe have been intensely working on advancing the field on various fronts with the ultimate hope of achieving a reliable technology that can be utilized as a universal platform for various optical functionalities and ultimately be integrated with microelectronic circuits monolithically on a single silicon chip. The endeavor has certainly been successful to the extent that a handful of companies have been marketing a variety of OEIC products for some years, most notably optical transceivers for data

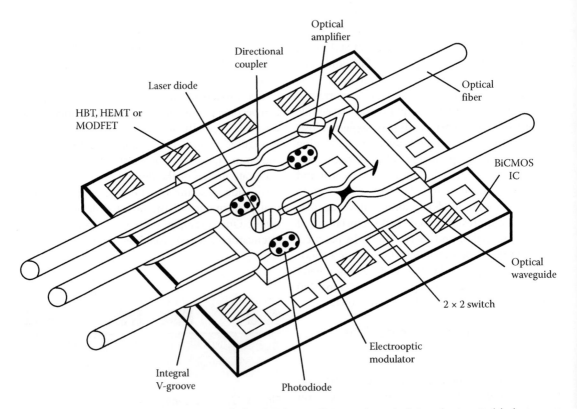

Figure 22.7 One of the original proposals for OEICs on silicon, acknowledging the inevitable heterogeneous destiny of the technology (Soref 1993).

communications (datacom), Ethernet, and other short-haul applications.

However, it was clear from the beginning that there is no "pure silicon" photonics, i.e., other materials are inevitably required. First, silicon does not absorb light at wavelengths above the material's bandgap (~1.1 µm). Instead, other materials, e.g., Ge and SiGe, photodetectors have been employed, as reviewed in Section 22.6. Accordingly, it has been argued that group-IV photonics is a more accurate name than silicon photonics.

Second, the bandgap of bulk crystalline Si is indirect, i.e., the light emission process is phonon mediated and the associated spontaneous recombination lifetime is very long, leading to a very low internal quantum efficiency (Jalali and Fathpour 2006; Jalali et al. 2008). Consequently, achieving lasing in bulk Si has not been possible. Several approaches have been alternatively pursued. Unfortunately, for a variety of reasons, room temperature electrically injected silicon-based lasers remain elusive to date (Jalali and Fathpour 2006). A pragmatic solution is hybridization of silicon photonic chips with III–V lasers, as reviewed in Section 22.7.

In reality and with the lack of any other practical solution, the silicon-photonic companies typically heterogeneously integrate III–V lasers and germanium photodetectors, one way or the other, and silicon is mostly used for passive devices and modulators. Figure 22.7 shows one of the original concepts of a silicon-based "superchip" (Soref 1993). Clearly, incorporating laser diodes, HBTs, HEMTs, and BiCMOS devices on other materials are implicitly envisioned, i.e., the need for heterogeneous integration was acknowledged from the early

days of silicon photonics. This historic vision is to a large extent in agreement with the state of the art. For instance, Figure 22.8 shows a schematic of one of Luxtera Corporation's early optical transceiver chips and how it incorporates nonsilicon devices (Gunn 2006). As depicted in Figure 22.9, it can be argued that "groups III-to-V" or "groups III–IV–V" photonics are semantically more accurate than silicon or group IV photonics for the established commercial technology. It is noted that the schematics of Figure 22.9 are very crude and generic and many variations exist, as discussed later in Section 22.7.

Third, despite a lot of effort for using crystalline silicon as an active region for nonlinear optical processes due to its high third-order optical susceptibility, the omnipresent optical loss induced by two-photon absorption (TPA) and free-carrier absorption (FCA) has been very problematic (Claps 2004; Liang and Tsang 2004). Using thin films of other nonlinear materials "hybridized" on silicon substrates has been chipping away another core prospect of the SOI platform. These heterogeneous platforms for third-order nonlinear applications are reviewed in Section 22.8.

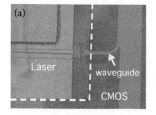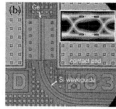

Figure 22.8 Images of optoelectronic chips produced by Luxtera Corporation and how (a) III–V lasers and (b) Ge photodetectros are integrated with Si waveguides (Gunn 2006).

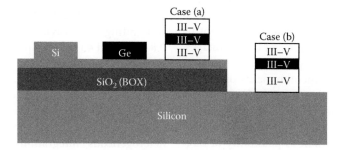

Figure 22.9 Simple schematic of the SOI waveguide platform, and how germanium and III–V compound semiconductor layers ought to typically be integrated in order to achieve hybrid photodetectors and lasers, respectively, suggesting that in reality there exists no pure silicon or even group IV photonics, rather "groups III-to-V photonics" or "groups III–IV–V photonics" are semantically more accurate.

Fourth, although all-Si optical modulators operating based on the free-carrier plasma effect have come a long way and their performance has been steadily improving (see Section 22.4.3), the inherent tradeoff between modulation bandwidth and depth remains a challenge and the devices cannot compete with telecom EO modulators in this regard. Hybridization of thin films of other materials known for their strong EO effect, e.g., LiNbO$_3$, on silicon substrates could be a possible route forward. Such materials also possess strong second-order optical nonlinearity and could enable compact chips for entangled photon generation and other quantum-optic-related applications. Efforts on such material heterogenization are reviewed in Section 22.9.

Fifth, silicon is an ideal material for nonlinear integrated optics at wavelengths above ~2.2 μm. That is, the wavelength range at which TPA vanishes and the material becomes transparent even at high optical intensities, until multiphonon absorption kicks in and the loss increases above ~6.5 μm. The 2.2–6.5 μm window covers the longer portion of the traditionally defined short-wavelength IR (SWIR of 1.4–3.0 μm) range as well as the shorter portion of the mid-wavelength IR (MWIR of 3.0–8.0 μm). Using silicon for integrable devices in the atmospheric window of 3–5 μm, and sometimes the wider 2.2–6.5 μm range, has been loosely defined as mid-IR silicon photonics. However, the BOX layer of the standard SOI technology is lossy across most of this range and hence other heterogeneous platforms such as silicon-on-sapphire, silicon-on-nitride, and air-clad waveguides have been developed, as reviewed in Section 22.10.

All in all, with the lack of any integrated optical material system for a whole range of applications, researchers have been reluctantly accepting that, despite their challenges and disadvantages, heterogeneous platforms appear to be the practical way forward. Still, silicon remains one of the most attractive substrate materials due to its low cost, availability of large wafers, ease of handling and thermal cycling, and well-established processing techniques, as well as possibility of integrating with CMOS electronics. The philosophy is to use silicon as much as possible but hybridize it with other materials when better performance can justify the challenges of hybridization.

22.6 GERMANIUM PHOTODETECTORS ON SILICON

The least disputed shortcoming of the benchmark SOI platform is its incapability of absorbing photons with $\lambda > \sim 1.1$ μm. For the O- to U-bands of telecom wavelengths (1.26–1.67 μm), germanium's cutoff wavelength of ~1.8 μm appeals as a perfect choice. Ge-on-Si photodetectors have been in the making for three decades (Luryi et al. 1984; Temkin et al. 1986; Jalali et al. 1992, 1994). A review of the works prior to 2010 can be found elsewhere (Michel et al. 2010). More recent progress has been made in the performance of the devices (Chen et al. 2011; DeRose et al. 2011; Liao et al. 2011; Tsuchizawa et al. 2011).

The historic challenge for Ge photodetecros has been the material's large lattice mismatch with silicon (~4%). Thick Ge layers would lead to surface roughness, threading dislocations and increased dark photocurrent. Also, Ge does not have a stable oxide and passivating it to avoid surface carrier recombination—which further increases the dark current—is another challenge. Dislocations can be controlled by graded SiGe buffered layers or in thin strained layers, both of which can reduce the dark current. Waveguide photodiodes with thin strained Ge layers on Si are more appealing for integrated chips. The small waveguide dimensions also result in shorter carrier transit time and increased detection bandwidth. The tradeoff is the length of the devices that can be tens of micrometers long in order to increase the responsivity. One of the lowest reported dark current densities is ~40 mA/cm^2, achieved in a 1.3-μm-wide and 4-μm-long waveguide at −1 V of bias (DeRose et al. 2011). The corresponding responsivity and 3-dB bandwidth are 0.8 A/W and 45 GHz, respectively. Evidently and despite all the efforts, relatively high dark current remains a disadvantage for Ge-on-Si waveguide photodetectors as compared with normal-incidence counterparts as well as III–V detectors.

To highlight the fabrication challenges of heterogeneous integration of Ge on Si substrates, the processing steps of one of the earlier high-speed photodetectors (40 Gb/s) by Intel Corporation is presented here (Yin et al. 2007). Figure 22.10 shows schematics of the device,

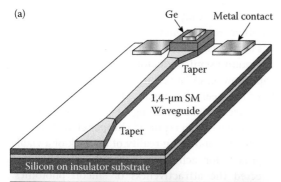

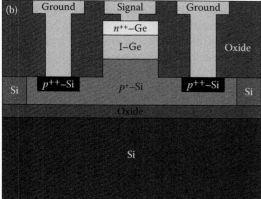

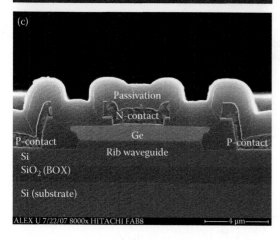

Figure 22.10 (a) Side-view, (b) cross-section, and (c) micrograph of one of the early high-speed Ge waveguide photodiodes developed at Intel Corporation (Yin et al. 2007).

how it is integrated with a SOI waveguide and its lateral cross-section. The following processing steps were conducted in the selected region where the photodetector was intended to be fabricated: (a) implanting the substrate with boron to reduce parasitic capacitance, (b) oxide deposition and patterning for defining the Ge region,

(c) selective epitaxial growth of 1.3-μm-thick Ge, (d) chemical–mechanical polishing (CMP) of ~0.5 μm of the grown Ge, (e) thermal annealing to reduce the threading dislocation density, (f) n-doping of the Ge layer, (g) selective excess p^{++}-doping of silicon for improved ohmic contacts, (h) dopant activation by rapid thermal annealing, (i) metallization and patterning to form the diode contacts.

The extra steps for Ge hybridization are the selective epitaxial growth, CMP, and thermal annealing. Such extra, sometimes complicated, fabrication steps are typically the price tag for any heterogeneous technology, more examples of which are discussed subsequently. Meanwhile, one advantage of Ge over other materials is that, despite its large lattice mismatch, thin films of it can be epitaxially grown on Si with little complications. Not all other materials can be as easily grown on Si, and hence other heterogeneous integration techniques, such as wafer-and flip-chip-bonding, ought to be devised as seen in the following sections.

22.7 III–V LASERS ON SILICON

Attaining lasing is perhaps the holy grail of silicon photonics. Unlike compound semiconductors, bulk crystalline Si is an indirect bandgap material and light emission is mediated by phonon emission or absorption. This second-order emission process is very weak (with an internal quantum efficiency of ~10^{-6}) and the spontaneous recombination lifetime is in the millisecond range (Jalali and Fathpour 2006). In principle, silicon can provide high material gain at high carrier injection levels for lasing. However, at high injection levels, FCA and Auger recombination prevent achieving any net optical gain (Jalali and Fathpour 2006; Jalali et al. 2008).

Since the early 1990s, several approaches have been pursued to increase the quantum efficiency of luminescence in bulk Si. They include, but not limited to, Erbium (Er)-doped Si, Si-rich oxide or Si nanocrystals (NCs), Er-doped Si NCs, deposition of boron dopant with SiO_2 nanoparticles mix on Si, surface texturing, dislocation loops, and stimulated Raman scattering. However, for a range of specific reasons beyond the scope of this paper and reviewed elsewhere (Jalali and Fathpour 2006), unfortunately none of these approaches have been able to offer room temperature electrically injected lasers. Since then, an interesting development has been room temperature lasing in strained and n-type-doped Ge on Si (Camacho-Aguilera et al. 2012; Kasper et al. 2013). Strain makes Ge an "almost" direct bandgap material and n-doping populates the engineered direct conduction band. Evidence of room temperature electrically injected lasers at a high doping level of 4×10^{19} cm^{-3} and tensile strain of about 0.2% in the 1520–1700 nm range was reported in 2012 (Camacho-Aguilera et al. 2012). However, a high optical loss of 90 cm^{-1}, induced by high doping levels right in the middle of the active region of the waveguides, results in high threshold currents and the approach does not

appear to be ready for prime-time applications yet. Another recent photonic material development has been germanium tin (GeSn), a group IV binary compound semiconductor alloy with direct bandgap below 0.8 eV at certain compositions. No GeSn laser has been demonstrated yet but electroluminescence devices spanning ~1530–1900 nm wavelengths have been reported (Kasper et al. 2013).

The disappointing reality of the lack of a viable approach for achieving Si-based lasers has not affected the attractiveness of silicon photonics. Instead, hybrid and heterogeneous approaches have been aggressively pursued. Figure 22.11 summarizes results from some of the key such hybrid laser approaches by the various groups discussed below.

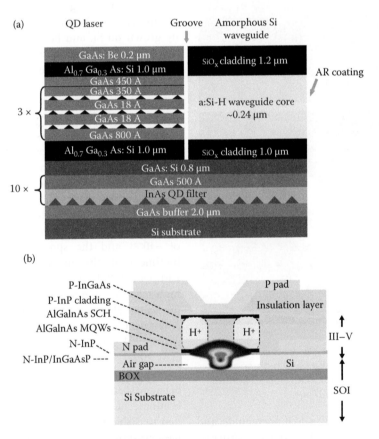

Figure 22.11 Schematics of four heterogeneous approaches to integrate III–V laser on Si substrates: (a) monolithic epitaxial growth of InGaAs/GaAs quantum-dot regions buffered by 10 layers of inactive dots to filter propagation of threading dislocations into the active region (Yang and Bhattacharya 2008), (b) evanescently coupled InP lasers bonded on SOI waveguides (Heck et al. 2013), (c) tape-waveguide vertical coupling form III–V gain regions into SOI (Keyvaninia et al. 2013a), and (d) bonding III–V gain regions into trenches made in silicon for planar coupling between the two (Creazzo et al. 2013).

(Continued)

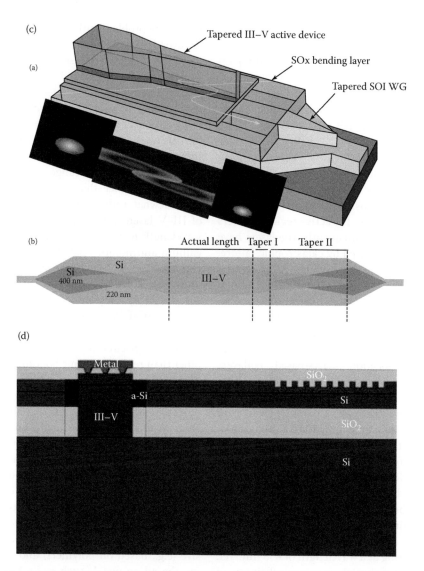

Figure 22.11 (Continued) Schematics of four heterogeneous approaches to integrate III–V laser on Si substrates: (a) monolithic epitaxial growth of InGaAs/GaAs quantum-dot regions buffered by 10 layers of inactive dots to filter propagation of threading dislocations into the active region (Yang and Bhattacharya 2008), (b) evanescently coupled InP lasers bonded on SOI waveguides (Heck et al. 2013), (c) tape-waveguide vertical coupling form III–V gain regions into SOI (Keyvaninia et al. 2013a), and (d) bonding III–V gain regions into trenches made in silicon for planar coupling between the two (Creazzo et al. 2013).

The most straightforward hybrid approach that comes to mind is to flip-chip bond or solder-based assemble a completely grown and processed functional III–V laser onto the silicon photonic circuit and couple the laser output into silicon waveguides via some coupling mechanism. These approaches are indeed as old as silicon photonics itself (Terui et al. 1985; Joppe et al. 1991; Friedrich et al. 1992; Sasaki et al. 2001). Variations of this approach are fairly established and adopted by the silicon photonic industry, e.g., Luxtera Corporation (Gunn 2006), NEC Corporation (Chu et al. 2009), Toshiba Corporation (Ohira et al. 2010), and Fujitsu Laboratories (Tanaka et al. 2012). Perhaps such more straightforward per-device flip-chip bonding approaches suffice the present low-volume, low-throughput needs of the industry but for future high-volume demanding markets, heterogeneous monolithic integration of III–V wafers and Si wafers, or already processed circuits, is more desirable.

Epitaxial growth of III–V heterostructures on Si substrates has been pursued for over twenty five years (Chen et al. 1988). Similar to Ge-on-Si epitaxial growth, however, the challenge is the large lattice mismatch (4.1%) between GaAs and Si, as well as the very large (250%) difference of their thermal expansion coefficients. Direct growth of GaAs-based heterostructures on Si creates very high densities of surface misfit dislocations (~10^9 cm^{-2}), which transform into threading dislocations propagating into the active region of lasers during the growth. Consequently, the lasers have very short lifetime caused by catastrophic failure. It turns out that the lattice mismatch of Ge and GaAs is very small (0.077%) and their thermal expansion coefficients match over large temperature ranges. Hence, the same graded SiGe epitaxial growth technology discussed for SiGe photodetectors can be utilized as buffer layers in order to achieve low-dislocation GaAs-based epilayers on Si. A challenge has been the growth of a polar (GaAs) over a nonpolar (Si) surface, a phenomenon long known as antiphase disorder (Anderson 1962). Certain substrate preparation and initial growth precautions can significantly alleviate the issue (Aspnes and Stunda 1981; Fischer et al. 1985; Petroff 1986; Reddy et al. 1987). AlGaAs/GaAs quantum well lasers on relaxed SiGe buffers layers on Si have been reported (Groenert et al. 2003). Surface threading dislocation densities were as low as 2×10^6 cm^{-2} allowing continuous-wave, room-temperature lasing at a wavelength of 858 nm. Yet, relatively low differential quantum efficiency of 0.24 and high threshold current density of 577 A/cm^2 impaired the devices.

Similarly, other approaches for growing GaAs-based laser on Si, including strained-layer superlattices (SLS) (Fischer et al. 1986; Sakai et al. 1986) and more recently quantum dot dislocation filters (see Figure 22.11a) (Yang et al. 2007; Yang and Bhattacharya 2008), typically suffer from high threshold current and short lifetimes. At any rate, GaAs-based lasers, even their longer-wavelength quantum-dot types, become increasingly less efficient light emitters at wavelengths above silicon's bandgap, and InP-based lasers are more suitable for integrating with Si and for telecom wavelengths. In addition to the thermal expansion coefficient difference and polar/nonpolar antiphase disorder

issues, the problem with direct InP growth on Si is the even larger (8.1%) lattice mismatch between the materials, prohibiting using graded SiGe buffers. Even employing InGaAs SLS buffer layers is not as effective as in GaAs. Initial works resulted in high dislocations densities in the order of 10^8–10^9 cm^{-2} (Crumbaker et al. 1989). Progress was later made for this method (Sugo et al. 1990; Wuu et al. 1990; Itakura et al. 1991), yet the dislocation densities remain in the 10^7 cm^{-2} range, which is an order of magnitude higher than lattice-matched growth. A somewhat radical solution for direct growth of III–V lasers emitting in the 1550 nm telecom wavelength is using another material system, e.g., gallium antimonide (GaSb)-based lasers. Strained GaInSb/AlGaAsSb quantum-well lasers have been grown on Si, but at room temperature the devices could operate only under pulsed bias and with a high threshold current density of 5 kA cm^{-2} (Cerutti et al. 2010).

The tough circumstances may leave no option other than relying on some bonding techniques. As discussed, a very practical, yet low yield and high cost, approach is flip-chip bonding of completely processed individual lasers on silicon photonic chips. A more monolithic approach for large-scale integration is to first conduct wafer (full or partial) or die bonding of III–V epilayers onto silicon photonic chips, followed by processing the III–V devices. An advantage is that aligning III–V devices with Si waveguides during lithography is typically easier this way than during chip bonding or assembly.

Indirect bonding methods, including adhesive (e.g., based on polymers) and eutectic bonding (based on metals such as In-Au and Pb) may be convenient ways to effectively "glue" the two wafers. However, bonding by polymer materials, such as the benzocyclobutene (BCB) family (Roelkens et al. 2006), may be fine for lab demonstration but is not reliable for practical applications due to weak formed bonds. Also, metallic layers of eutectic bonding cause optical loss.

A more reliable approach is direct wafer (or fusion) bonding of InP and Si layers. The idea can be traced back to Lord Rayleigh (Rayleigh 1936), who bonded optically polished glass plates at room temperature. In modern semiconductor processing, two atomically smooth (<1 nm roughness) and clean surfaces are brought to physical contact at low temperatures (to form atomic bonds with no worry of the lattice-mismatch hurdle) and subsequently the bonds are strengthened by thermal annealing. The research in this area was motivated by bonding two Si wafers in the 1980s for electronic applications, pioneered by researchers at IBM (Lasky et al. 1985) and Toshiba (Shimbo et al. 1986). Interestingly, the former group (Lasky et al. 1985) reports one of the very first SOI wafers achieved by the aforementioned bonding and etchback technique.

Taking advantage of hydrophilic surfaces (e.g., SiO_2) or assisting the process by oxygen plasma can enhance the initial bonding, as follows. One of the earliest III–V lasers bonded to Si wafers was achieved at Ecole Centrale de Lyon and LETI-CEA in France (Seassal et al. 2001), where optically pumped InP microdisk lasers were bonded to Si wafers. Two SiO_2 layer, formed by plasma-enhanced chemical vapor deposition (PECVD) on both InP and Si wafers, subsequently polished, cleaned, bonded, and annealed at 200°C for 60 minutes to increase the bonding strength. The InP substrate was subsequently removed by selective wet etching using an $In_{0.53}Ga_{0.47}$. As sacrificial (etch-stop) layer, exposing the heterolayers for microdisk laser processing, i.e., 6-μm-diameter disks were patterned by standard dry etching techniques. An electrically injected InAsP/InGaAsP microdisk laser was later demonstrated (Rojo et al. 2006). One issue with this approach is the quality of the PECVD oxide and the strength of the created bonds and their stability at elevated temperatures (Pasquariello and Hjort 2002). Low-heat dissipation of the underlying dielectric SiO_2 layer (see case (a) in Figure 22.9), leading to degraded laser performance, is another disadvantage of using hydrophilic SiO_2 bonding layers.

One challenge for wafer-bonding approaches is how to couple light into Si waveguides. Based on the concept of vertical coupling via evanescent coupling [originally proposed and demonstrated in the context of III–V microdisk lasers stacked on another III–V bus waveguide (Seung et al. 2003)], the mentioned French research groups managed to couple light from optically pumped microdisk lasers into silicon strip waveguides by decreasing the total SiO_2–SiO_2 hydrophilic surfaces from 800 nm down to 300 nm (Hattori et al. 2006).

Pioneered by a group at Uppsala University, Sweden, plasma-assisted direct bonding of semiconductors has been shown to be an interesting technique to avoid SiO_2 hydrophilic layers and

their complications (Pasquariello and Hjort 2002). Based on this bonding technique, researchers at the University of California, Santa Barbara (UCSB), demonstrated optically pumped (Park et al. 2005) and later the first electrically injected (Fang et al. 2006) evanescently coupled edge-emitting FP AlGaInAs/InP lasers coupled into Si waveguides (see Figure 22.11b). A hybrid optical mode, mostly residing in the SOI waveguide, has partial overlap with the laser separate confinement structure, allowing coupling between the two regions. Since then and in collaboration with Intel Corporation, the UCSB group has impressively matured this platform and managed to demonstrate more complicated photonic circuitry, i.e., demonstrate and integrate lasers (including FP, microdisk, microring, and distributed Bragg reflector types), semiconductor optical amplifiers, electroabsorption and phase modulators, as well as photodetectors. Detailed reviews of these works can be found elsewhere (Liang and Bowers 2010; Heck et al. 2013). Also, other researchers have shown very similar lasers with InGaAsP, rather than AlGaInAs, multiquantum well active regions (Sun et al. 2009).

Meanwhile, arguing that the evanescently coupled lasers should better have more confinement in the InP-based gain region than the Si waveguides, research groups at Ghent University and IMEC, Belgium, in collaboration with the mentioned French groups, demonstrated somewhat different electrically pumped evanescently coupled lasers (Van Campenhout et al. 2007). Another approach pursued by the same groups is using III–V gain regions and distributed reflectors on SOI waveguides linked by tapered adiabatic mode converters for realizing tunable lasers (see Figure 22.11c) (Keyvaninia et al. 2013a,b). Finally, Skorpios Technologies has envisioned and developed a planar platform in which unprocessed III–V epilayers are metal bonded into grooves made in silicon substrate, such that the processed laser is eventually butt coupled into SOI waveguides (Creazzo et al. 2013) (see case (b) in Figures 22.9 and 22.11d). This approach avoids the inherent shortcoming of vertically coupled lasers in terms of guided optical mode distribution between Si and InP waveguides and offers efficient direct optical coupling between the two planar (horizontally aligned) regions. Other advantages include better heat dissipation due to the omission of the BOX underneath the laser, allowing laser operation up to 80°C, as well as hermetically sealing the lasers in SiO_2 encapsulation, eliminating the need for their packaging.

To summarize this section, it is apparent that any existing and perhaps upcoming laser technology on Si has its own pros and cons. Thus, it is hard to judge which of so many competing approaches will prevail. But it is likely that some heterogeneous integration methods, based on III–V lasers pre- or postprocess bonded to Si, will be the most relevant solutions in the foreseeable future.

22.8 THIRD-ORDER NONLINEAR PHOTONICS ON SILICON

Silicon has a centrosymmetric lattice structure, thus it inherently lacks second-order optical nonlinearity. It has high third-order optical nonlinearity (both Kerr and Raman types), though. Specifically, silicon waveguides have a much high nonlinear parameter $\gamma = (\omega \times n_2)/(c \times A_{eff})$ compared to traditional integrated-optics platforms, e.g., silica fibers or planar waveguides. Here, n_2 is the nonlinear refractive index, λ and ω are the wavelength and angular frequency of light, respectively, and c is the speed of light in vacuum. From the early 2000s, pioneered by B. Jalali's group at UCLA, both Raman- and Kerr effect-based nonlinear effects have been explored in silicon photonics (Jalali and Fathpour 2006). Numerous studies have been published in the area of nonlinear silicon photonics by several groups, a review of which can be found elsewhere (Leuthold et al. 2010).

An omnipresent problem in the telecommunication wavelengths, however, is the TPA at required high optical intensities and the even higher FCA induced by it (Claps 2004; Liang and Tsang 2004). Certain approaches can reduce the carrier lifetime, alleviate the problem, and modestly improve the nonlinear device performances. Active carrier sweep-out in reverse-biased waveguide diodes has been proposed (Claps 2004) and demonstrated (Rong et al. 2005). Energy harvesting based on two-photon photovoltaic effect in order to achieve energy-efficient nonlinear devices has also been demonstrated (Fathpour et al. 2006, 2007). Nonetheless, in all of these sweep-out-based solutions, the carrier lifetime remains relatively high and FCA cannot be significantly subdued due to free-carrier screening of the junction electric field (Dimitropoulos et al. 2005). Difficulties of coupling light into higher order modes of specially designed waveguides notwithstanding, cladding pumping of nonlinear silicon photonic devices is another potential solution (Krause et al. 2007). Also, introducing midgap states through high-energy irradiation of ions, e.g., helium, is another means of modestly reducing the carrier lifetime, although it comes at the expense of higher linear propagation loss. Therefore, only small values of net gain in Raman amplifiers have been observed (Liu and Tsang 2006).

Several researchers have totally moved to other nonlinear materials such as hydrogenated amorphous silicon (a-Si:H) (Narayanan and Preble 2010; Kuyken et al. 2011; Grillet et al. 2012), silicon nitride (SiN) and Hydex (Moss et al. 2013), chalcogenide glass (Eggleton et al. 2011), and tantalum pentoxide (Ta_2O_5) (Rabiei et al. 2013a, 2014) waveguides on silicon substrates to totally circumvent the nonlinear loss problem of SOI waveguides. The argument is that the improved nonlinear figure of merit, FOM = $n_2/(\beta_{TPA} \times \lambda)$ in these materials— with negligible TPA coefficient, β_{TPA}—is more important than the previously argued high γ of Si. The common idea for all of these heterogeneous platforms is to use deposition, spinning or other methods of thin film formation to create waveguides on a dielectric bottom cladding layer (usually SiO_2) on Si substrates and exploit the materials' nonlinear optical properties for functions such as supercontinuum generation, four-wave mixing, self-phase modulation, cross-phase modulation, third-harmonic generation, and stimulated Raman scattering, etc.

Table 22.1 summarizes n_2 and β_{TPA} in all the materials discussed in the previous paragraph in comparison to SiO_2, as a benchmark (Fathpour 2015). The literature of a-Si:H is inconsistent and seemingly is very dependent on the deposition condition, but very high values of n_2 have been claimed (Narayanan and Preble 2010, Grillet et al. 2012). However, a tradeoff between n_2 and β_{TPA} is observable and overall the considerable value of β_{TPA} [which can be much higher than crystalline Si (Narayanan and Preble 2010) limits the FOM (e.g., ~5 in Grillet et al. 2012)]. The other materials of interest in Table 22.1 have negligible β_{TPA} and hence measuredly or potentially possess very high FOM. Among them, Ta_2O_5 has an n_2 three times higher than SiN and comparable to chalcogenide glasses. Meanwhile, its refractive index of ~2.1 is higher than Hydex's (~1.7) and its damage

Table 22.1 Nonlinear refractive indices and two-photon absorption coefficients of various materials

Material	n_2 ($\times 10^{-20}$ m²/W)	β_{TPA} (cm/GW)
SiO$_2$	3.2	Negligible
Hydex	11	Negligible
SiN	26	Negligible
Ta$_2$O$_5$	72.3	Negligible
Chalcogenide	~100	6.2×10^{-4}
Crystalline Si	250	0.7
a-Si:H	2, 100–4, 200	0.25–4.1

Source: Fathpour, S., Nanophotonics 4, 143–164, 2015.

threshold is much higher than that of chalcogenide glasses. Reduction of the waveguide dimensions in low-loss Ta$_2$O$_5$ submicron waveguides, fabricated by selective oxidation of Ta (Rabiei et al. 2013a, 2014), reduces A_{eff}. Consequently, the γ of Ta$_2$O$_5$ waveguides can be much higher than the competing materials of Table 22.1.

22.9 SECOND-ORDER NONLINEAR PHOTONICS ON SILICON

There have been some reports on breaking the lattice symmetry of bulk Si by applying stress on Si waveguides to take advantage of the induced second-order optical nonlinearity as well as linear EO (or Pockels) effect (Jacobsen et al. 2006; Chmielak et al. 2011; Cazzanelli et al. 2012). In this respect, inducing electronically tunable birefringence in SOI waveguides has been demonstrated by blanketing them with a piezoelectric transducer made of lead zirconium titanate (PZT) (Tsia et al. 2008a,b). However, the amount of induced birefringence is in the range of 3×10^{-5}, which is large enough for electrically tuning the phase-mismatch of TE and TM modes in parametric nonlinear optical effects, but not enough to take advantage of the effect for second-order nonlinear effect itself. It is true that the induced strain can be higher using stressed cladding layers, such as SiN, particularly in smaller waveguide cross-sections (Jacobsen et al. 2006; Chmielak et al. 2011; Cazzanelli et al. 2012). However, overall the extent to which second-order optical nonlinearity and/or Pockels effect can be induced by such methods is questionable considering that single-element Si is a nonpolar crystal.

SOI-based modulators were discussed in Section 22.4.3. As mentioned therein, although high-speed and low-voltage all-Si MZI modulators have been demonstrated, the devices only offer modest modulation depths of a few dB. Such low modulation depth may be enough for near-term short-haul links but this certainly remains a shortcoming for the long-term requirements of the industry, where at least 7 dB of modulation depth at high speeds is demanded (Reed et al. 2010). Indeed, the modulation depth times the bit rate may be a better figure of merit, which is ~150 to ~280 dB Gb/s in the aforementioned state of the art (Thomson et al. 2012; Tu et al. 2013). In clear contrast, the benchmark LiNbO$_3$ modulators boast 20 dB or better modulation depth (the industry standard for digital communications) (Wooten et al. 2000) and 40 Gb/s or higher bit rates are commonplace in commercial long-haul telecom systems, giving a figure of merit of at least 800 dB Gb/s. Up to 100 GHz modulation bandwidth has been long measured in LiNbO$_3$ with careful RF design of traveling-wave electrodes (Noguchi et al. 1998), hence even higher modulation depth times bit rates are feasible in principle, the complications of detection and electronic circuitry for modulation and detection notwithstanding.

It again appears that heterogeneous integration of Si with another material with strong second-order optical nonlinearity, as well as linear EO or electro-absorption (EA) effect, may be the way forward. Nitride semiconductors have high second-order nonlinearities. GaN has been bonded to Si by PECVD SiO$_2$ with measured second-order nonlinear susceptibility, $\chi^{(2)}$, of ~16 pm/V (Xiong et al. 2011). However, as mentioned in Section 22.7, the quality of PECVD oxide is low for bonding. Also, the uniformity of the thin GaN films obtained by CMP, after removing the (111) Si substrate, on which they were originally grown, is an issue. Sputtering of AlN on Si wafers coated with SiO_2 has also been reported (Pernice et al. 2012). However, the crystalline quality of sputtered AlN is expectedly low, leading to a weak $\chi^{(2)}$ of ~4 pm/V. Nonlinear polymer materials have shown strong EO effect and, their long-term reliability notwithstanding, have potential for high-performance modulators in Si waveguides coated with them (Leuthold et al. 2013).

Telecom wavelengths are far from silicon's bandgap. Hence, EA modulators based on the Franz–Keldysh effect (FKE) are not possible on Si. But both the quantum-confined Stark effect (QCSE) and FKE in SiGe can be exploited at longer wavelengths. $Si_{0.6}Ge_{0.4}$–Si quantum well waveguide EA modulators based on QCSE operating at 1.15 μm wavelength were demonstrated at the University of Michigan (Qasaimeh et al. 1998). Later, another team employed Ge–$Si_{0.1}Ge_{0.9}$ quantum wells in surface-illuminated modulators to red shift the operating wavelength to ~1.45 μm (Kuo et al. 2005). More recently, GeSi (Ge epitaxial layer comprising <1% of silicon) has been used to exhibit strong FKE around 1.55 μm and demonstrate EA modulators with 10 dB extinction ratio (Liu et al. 2008). Mellanox Technologies has been recently maturing this technology to integrate it into their OEICs (Feng et al. 2013). Another approach, pursued at UCSB, is to take advantage of the quantum-well intermixing technique in order to shift the bandgap of InP-based heterostructures and achieve EA modulators on the same epilayers used for lasers in a hybrid OEIC on Si substrates (Heck et al. 2013).

An alternative approach, pursued at the author's research group at CREOL, is heterogeneous integration of $LiNbO_3$ on Si (Rabiei et al. 2013b). With high nonlinear optical coefficients (e.g., $d_{33} \approx -25.2$ pm/V), $LiNbO_3$ is an ideal material for nonlinear integrated photonics. Also, single-crystalline $LiNbO_3$ has large EO coefficients (r_{33}=31 pm/V and r_{13}=8 pm/V) and a wide transparency range (0.4–5 μm wavelength) (Wong 2002). Indeed, standard $LiNbO_3$ waveguides are widely regarded as the best vehicle for EO modulation in the photonic industry (Wooten et al. 2000) with impressively high-modulation bandwidths (over 100 GHz; Noguchi et al. 1998). As mentioned before, $LiNbO_3$ modulators undoubtedly offer higher performance (in terms of modulation bandwidth and modulation depth) compared to all-Si-based optical modulators. Reliable $LiNbO_3$-on-Si thin films as well as low-loss submicron waveguides, microring resonators, and record-low $V\pi \cdot L$ modulators have been demonstrated (Rabiei et al. 2013b). $V\pi \cdot L$ as low as 4 V-cm is obtained in push–pull configured MZI modulators, which is three times lower than the best commercial devices. Calculations show that $V\pi \cdot L$ as low as 1.2 V-cm is possible. Modulation depth is expectedly very high (~20 dB), which is a clear advantage over free-carrier-modulated all-Si modulators.

22.10 MID-IR INTEGRATED PHOTONIC PLATFORMS

Optics in the MWIR or mid-IR wavelengths, particularly the atmospheric transmission window of 3–5 μm, has been researched on for over 60 years (Gebbie et al. 1951). Mid-IR silicon photonics has several potential biochemical, biomedical, and communication potential applications (Jalali et al. 2006; Soref 2010; Fathpour and Jalali 2012). The first demonstration of mid-IR Raman amplifiers (Raghunathan et al. 2007; Borlaug et al. 2009) suggested that nonlinear silicon photonic devices can be achieved. Indeed, TPA vanishes in the mid-IR regime (above ~2.2 μm to be more precise) and three-photon absorption is negligible at very high intensities (Raghunathan et al. 2006). Thus, the discussed omnipresent problem of nonlinear silicon devices in the near-IR, i.e., TPA-induced FCA, is inherently nonexistent in the mid-IR. Analytic and numerical works suggest that high-performance 3–5 μm nonlinear silicon photonic devices can be attained provided that low-loss waveguides and sources with high beam-quality are available (Ma and Fathpour 2012, 2013).

However, the standard SOI platform is generally not ideal for mid-IR photonics because the bottom cladding material, SiO_2, is very lossy over 2.5–2.9 μm and above 3.5 μm wavelengths (Soref et al. 2006b). As shown in Figure 22.12a and b, using SiN (transparent up to 6.6 μm) or air (suspended silicon membrane, transparent up to 6.9 μm) as the bottom waveguide cladding layer are alternatives. Silicon-on-sapphire (transparent up to 4.3 μm) and Si-on-$LiNbO_3$ (transparent up to 5.2 μm) are other

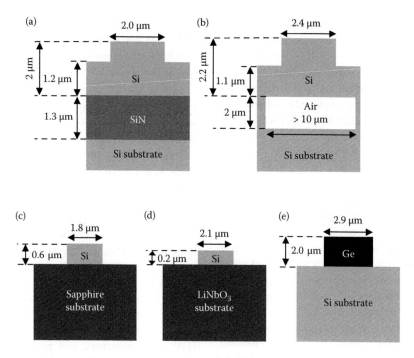

Figure 22.12 Five different mid-IR waveguide technologies on silicon, their schematics and their typical waveguide dimensions.

possibilities (Figure 22.12c and d). Also as shown in Figure 22.12e, Ge waveguides directly formed on Si (transparent up to 6.9 μm), have been proposed (Soref et al. 2006b). All of these waveguide platforms have been demonstrated and briefly reviewed below.

Silicon-on-sapphire (SOS) wafers were demonstrated long before SOI wafers (Manasevit and Simpson 1964). Both SOS and SOI wafers were interested for CMOS electronics because the insulating substrates contribute to lower parasitic capacitance, higher transistor speed, lower power, better linearity, and more isolation for electronic circuitry (Culurciello 2010). Si can be grown on sapphire substrates by chemical vapor deposition and solid-phase epitaxy methods, although the quality of the attained films, their defect and optical loss is very much influenced by the growth method and conditions. Available SOS wafers, originally developed for microelectronic applications, led to the demonstration of first Si-based mid-IR waveguides (Baehr-Jones et al. 2010; Mashanovich et al. 2011; Wong et al. 2012). Low-loss waveguides and grating couplers on SOS were later demonstrated (Shankar et al. 2013). A typical waveguide structure is shown in Figure 22.12c.

The author's group has also fabricated SOS waveguides. However, we have not been able to achieve low propagation loss (~12 dB/cm at 1.55 μm). The reason for this is not clear, but private communication with other researchers, who have been using available SOS wafers in the market, suggests that quality variation of the Si thin layer grown epitaxially on sapphire substrates can cause this. Indeed, there has been some report on inherent twining defects in SOS (Imthurn 2014). At any rate and as mentioned, sapphire is transparent only up to 4.3 μm and the platforms of Figure 22.12 offer wider transparency ranges.

Reduced-pressure chemical vapor deposition was employed to grow a 2-μm-thick monocrystalline Ge relaxed layer on a Si and ~3-μm-wide stripes were fabricated by photolithography (Figure 22.12e) (Chang et al. 2012). These waveguides are inherently large due to the not very high index contrast between Ge and Si (~4 vs. 3.4) but for the same reason have relatively low propagation loss of ~3.6 dB/cm at 5.8 μm wavelength. More recently, Ge-on-Si thermo-optic phase shifters have been reported (Malik et al. 2014).

Silicon-on-nitride (SON) waveguides have been recently demonstrated at the author's lab (Khan et al.

2013). The waveguide structures shown in Figure 22.12b resemble those of SOI waveguides except that the BOX layer is replaced by a SiN layer formed by deposition and bonding techniques. In principle, the discussed ion implantation and wafer slicing steps used for SOI wafers can be employed to achieve SON wafers. But with the availability of SOI wafers, perhaps there is no need to repeat those steps in order to attain Si thin films. Rather, back-side etching of the Si substrate and the BOX layer can expose the thin Si layer. Standard techniques were then used to fabricate ridge waveguides. At 3.39 μm wavelength, the propagation loss of the characterized waveguides is 5.1 dB/cm for the TM mode.

Silicon's inherent transparency range of up to 6.9 μm can be fully exploited with air-clad waveguides. Suspended membrane waveguides for mid-IR were first demonstrated by local removal of BOX in SOI waveguides via forming an array of periodical holes in the top silicon layer adjacent to the rib waveguides (Chang et al. 2012). However, due to the via forming and isotropic wet etching steps, the attained membranes become too wide to be mechanically and thermally stable, i.e., they are prone to vibrations, bowing, and low thermal dissipation. A more reliable and robust process, developed at CREOL, is based on fusion bonding an inverted SOI wafer to a bulk silicon substrate prepatterned with trenches (Chiles et al. 2013). The SOI substrate and BOX are subsequently removed and a thin Si film on the buried trench is exposed. The air-filled trench acts as the bottom waveguide cladding when waveguide ridges are formed by standard lithographic techniques. The waveguides have a TM-mode low propagation loss of 2.8 dB/ cm at 3.39 μm. This is the first demonstration of a homogeneous all-silicon optical platform. The depth of the trench can be freely controlled and the formed fusion-sealed internal channel can be potentially used for microfluidic or gas-sensing applications.

The last mid-IR platform discussed here is based on bonding a Si thin film on a $LiNbO_3$ substrate (Chiles and Fathpour 2014). The motivation for developing this silicon-on-lithium-niobate or SiLN platform at CREOL is primarily to achieve EO modulator in the mid-IR range. According to mode simulations and at 3.4 μm wavelength, ~24% of the optical mode resides in $LiNbO_3$. Hence, relatively strong EO modulators can be envisaged. These first demonstrated mid-IR EO modulators in attaining a modulation depth of 8 dB, a $V\pi \cdot L$ of 26 V.cm at low frequencies, a propagation loss of 2.5 dB/cm and an insertion loss of ~3 dB, all characterized at a wavelength of 3.4 μm.

22.11 CONCLUDING REMARKS

Herbert Spencer, the famous 19th century English polymath once wrote: "From the remotest past which Science can fathom, up to the novelties of yesterday, that in which Progress essentially consists, is the transformation of the homogeneous to the heterogeneous" (Spencer 1891). Such a transformation appears to be the destiny of progress in the science and technology of electronic–photonic integrated circuits. It may be unwise to make predictions for any field of science and engineering, but the following remarks are cautiously made. It appears that in the foreseeable future, heterogeneous technologies, that combine group IV and III–V semiconductors, and perhaps $LiNbO_3$ and other materials such as nitride semiconductors and high-index dielectrics, will most likely dominate integrated optoelectronics. Electronics will most likely be based on Si CMOS and lasers will most likely be on III–Vs. Prevailing materials and device platforms for other functionalities are much more debatable and predicting them is more difficult, hence less assertive adjectives are chosen in the following. Photodetectors will likely be on Ge, but perhaps on III–Vs. Modulators could be based either on free-carrier-plasma effect in Si, EO or EA effect in III–Vs or EA in GeSi, or EO in $LiNbO_3$. Passive devices (waveguides, filters, etc.) will likely be on Si due to its inexpensiveness and low loss.

ACKNOWLEDGMENTS

Some of the sections in this chapter are partially based on material used in the review article: S. Fathpour, "Emerging Heterogeneous Integrated Photonic Platforms on Silicon," *Nanophotonics*, vol. 4, pp. 143–164, May 2015.

REFERENCES

Anderson, R. L., Experiments on Ge-GaAs heterojunctions, *Solid-State Electron.* 5 (1962): 341–344.

Aspnes, D. E., Stunda, A. A., Chemical etching and cleaning procedures for Si, Ge, and some III-V compound semiconductors, *Appl. Phys. Lett.* 39 (1981): 316–318.

Baehr-Jones, T., Spott, A., Ilic, R. et al., Silicon-on-sapphire integrated waveguides for the mid-infrared, *Opt. Express* 18 (2010): 12127–12135.

Borlaug, D., Fathpour, S., Jalali, S., Extreme value statistics in silicon photonics, *IEEE Photonics J.* 1 (2009): 33–39.

Camacho-Aguilera, R. E., Cai, Y., Patel, N. et al., An electrically pumped germanium laser, *Opt. Express* 20 (2012): 11316–11320.

Cazzanelli, M., Bianco, F., Borga, E., Second-harmonic generation in silicon waveguides strained by silicon nitride, *Nat. Mater.* 11 (2012): 148–154.

Celler, G. K., Cristoloveanu, S., Frontiers of silicon-on-insulator, *J. Appl. Phys.* 93 (2003): 4955–4978.

Cerutti, L., Rodriguez, J. B., Tournie, E., GaSb-based laser, monolithically grown on silicon substrate, emitting at 1.55 μm at room temperature, *IEEE Photon. Tech. Lett.* 22 (2010): 553–555.

Chan, S. P., Png, C. E., Lim, S. T. et al., Single-mode and polarization-independent silicon-on-insulator waveguides with small cross section, *IEEE J. Lightwave Technol.* 23 (2005): 2103–2111.

Chang, Y. C., Paeder, V., Hvozdara, L. et al., Low-loss germanium strip waveguides on silicon for the mid-infrared, *Opt. Lett.* 37 (2012): 2883–2885.

Chen, H. Z., Palaski, J., Yariv, A. et al., High-frequency modulation of AlGaAs/GaAs lasers grown on Si substrate by molecular beam epitaxy, *Appl. Phys. Lett.* 52 (1988): 605–607.

Chen, L., Doerr, C. R., Buhl, L. et al., Monolithically integrated 40-wavelength demultiplexer and photodetector array on silicon, *IEEE Photonics Technol. Lett.* 23 (2011): 869–871.

Cheng, Z., Chen, X., Wong, C. Y. et al., Mid-infrared suspended membrane waveguide and ring resonator on silicon-on-insulator, *IEEE Photonics J.* 4 (2012): 1510–1519.

Chiles, J., Khan. S., Ma. J., Fathpour. S., High-contrast, all-silicon waveguiding platform for ultra-broadband mid-Infrared photonics, *Appl. Phys. Lett.* 103 (2013): 151106.

Chiles, J., Fathpour, S., Mid-infrared integrated waveguide modulators based on silicon-on-lithium-niobate photonics, *Optica* 1 (2014): 350–355.

Chmielak, B., Waldow, M., Matheisen, C. et al., Pockels effect based fully integrated, strained silicon electro-optic modulator, *Opt. Express* 19 (2011): 17212–17219.

Chu, T., Fujioka, N., Ishizaka, M., Compact, lower-power-consumption wavelength tunable laser fabricated with silicon photonic-wire waveguide micro-ring resonators, *Opt. Express* 17 (2009): 14063–14068.

Claps, R., Raghunathan, V., Dimitropoulos, D. et al., Influence of nonlinear absorption on Raman amplification in silicon waveguides, *Opt. Express* 12 (2004): 2774–2780.

Creazzo, T., Marchena, E., Krasulick, S. B. et al., Integrated tunable CMOS laser, *Opt. Express* 21 (2013): 28048–28053.

Crumbaker, T. E., Lee, H. Y., Hafich, M. J. et al., Growth on InP on Si substrates by molecular beam epitaxy, *Appl. Phys. Lett.* 54 (1989): 140–142.

Culurciello, E., *Silicon-on-sapphire circuits and systems: Sensor and biosensor interfaces* (New York: McGraw-Hill, 2010).

DeRose, C. T., Douglas, C., Trotter, D. C. et al., Ultra compact 45 GHz CMOS compatible Germanium waveguide photodiode with low dark current, *Opt. Express* 19 (2011): 24897–24904.

Dimitropoulos, D., Fathpour, S., Jalali, B., Intensity dependence of the carrier lifetime in silicon Raman lasers and amplifiers, *Appl. Phys. Lett.* 87 (2005): 261108.

Eggleton, B. J., Luther-Davies, B., Richardson, K., Chalcogenide photonics, *Nat. Photonics* 5 (2011): 141–148.

ePIXfab website (Accessed September 30, 2014). http://www.epixfab.eu/technologies.

Fang, A. W., Park, H., Cohen, O. et al., Electrically pumped hybrid AlGaInAs-silicon evanescent laser, *Opt. Express* 14 (2006): 9203–9210.

Fathpour, S., Tsia, K. K., Jalali, B., Energy harvesting in silicon Raman amplifiers, *Appl. Phys. Lett.* 89 (2006): 061109.

Fathpour, S., Tsia, K. K., Jalali, B., Two-photon photovoltaic effect in silicon, *IEEE J. Quantum Electron.* 43 (2007): 1211–1217.

Fathpour, S., Jalali, B. eds., *Silicon photonics for telecommunications and biomedicine* (Boca Raton, FL: Taylor and Francis, 2012).

Fathpour, S., Emerging heterogeneous integrated photonic platforms on silicon, *Nanophotonics* 4 (2015): 143–164.

Feng, D., Qian, W., Liang, H. et al., High-speed GeSi electro-absorption modulator on the SOI waveguide platform, *IEEE J. Sel. Top. Quantum Electron.* 19 (2013): 3401710.

Fischer, R., Masselink, W. T., Klem, J. et al., Growth and properties of GaAs/AlGaAs on nonpolar substrates using molecular beam epitaxy, *J. Appl. Phys.* 58 (1985): 374–381.

Fischer, R., Kopp, W., Morkoc, H. et al., Low threshold laser operation at room-temperature in GaAs/(Al, Ga)As structures grown directly on (100) Si, *Appl. Phys. Lett.* 48 (1986): 1360–1361.

Friedrich, E. L., Oberg, M. G., Broberg, B. et al., Hybrid integration of semiconductor lasers with Si-based single-mode ridge waveguides, *J. Lightwave Technol.* 10 (1992): 336–340.

Gebbie, H. A., Harding, W. R., Hilsum, C. et al., Atmospheric transmission in the 1 to 14 µm region, *Proc. R. Soc. London, Ser. A*, 206 (1951): 87–107.

Grillet, C., Carletti, L., Monat, C., Amorphous silicon nanowires combining high nonlinearity, FOM and optical stability, *Opt. Express* 20 (2012): 22609–22615.

Groenert, M. E., Leitz, C. W., Pitera, A. J. et al., Monolithic integration of room-temperature cw GaAs/AlGaAs lasers on Si substrates via relaxed graded GeSi buffer layers, *J. Appl. Phys.* 93 (2003): 362–367.

Gunn, C., CMOS photonics for high-speed interconnects, *IEEE Micro* 26 (2006): 58–66.

Hattori, H. T., Seassal, C., Touraille, E. et al., Heterogeneous integration of Microdisk lasers on silicon strip Waveguides for Optical Interconnects, *IEEE Photon. Technol. Lett.* 18 (2006): 223–225.

Heck, M. J. R., Bauters, J., Davenport, M. et al., Hybrid silicon photonic integrated circuit technology, *IEEE J. Sel. Top. Quantum Electron.* 19 (2013): 6100117.

Heiblum, M., Harris, J., Analysis of curved optical waveguides by conformal transformation, *IEEE J. Quantum Electron.* 11 (1975): 75–83.

Imthurn, G., The history of silicon-on-sapphire. White paper for Peregrine Semiconductor Corporation, 2007 (Accessed September 15, 2014). http://www.psemi.com/articles/History_SOS_73-0020-02.pdf.

Itakura, H., Suzuki, T., Jiang, Z. K. et al., Effect of InGaAs/InP strained layer superlattice in InP-on-Si, *J. Cryst. Growth* 115 (1991): 154–157.

Jacobsen, R. S., Andersen, K. N., Borel, P. I. et al., Strained silicon as a new electro-optic material, *Nature* 441 (2006): 199–202.

Jalali, B., Levi, A. F. J., Ross, F. et al., SiGe waveguide photodetectors grown by rapid thermal chemical vapour deposition, *Electron. Lett.* 28 (1992): 269–271.

Jalali, B., Naval, L., Levi, A. F. J., Si-based receivers for optical data links, *J. Lightwave Technol.* 12 (1994): 930–935.

Jalali, B., Trinh, P. D., Yegnanarayanan, S. et al., Guided-wave optics in silicon-on-insulator technology, *IEE. Proc. Optoelectron.* 143 (1996): 307–311.

Jalali, B., Yegnanarayanan, S., Yoon, T. et al., Advances in silicon-on-insulator optoelectronics, *IEEE J. Sel. Top. Quantum Electron.* 4 (1998): 938–947.

Jalali, B., Fathpour, S., Silicon photonics, *J. Lightwave Technol.* 24 (2006): 4600–4615.

Jalali, B., Raghunathan, V., Shori, R. et al., Prospects for silicon mid-IR Raman lasers, *IEEE J. Sel. Top. Quantum Electron.* 12 (2006): 1618–1627.

Jalali, B., Can silicon change photonics? *Physica Status Solidi A.* 205 (2008): 213–224.

Jalali, B., Dimitropoulos, D., Raghunathan, V. et al., Silicon lasers, in *Silicon Photonics: State of the Art*, ed. G. T. Reed, pp. 147–189 (West Sussex: John Wiley and Sons, 2008).

Jedidi, R., Pierre, R., High-order finite-element method for the computation of bending loss in optical waveguides, *J. Lightwave Technol.* 25 (2007): 2618–2630.

Joppe, J. L., de Krijger, A. J. T., Noordman, O. F. J., Hybrid integration of laser diode and monomode high contrast slab waveguide on silicon, *Electron. Lett.* 27 (1991): 162–163.

Kasper, E., Kittler, M., Oehme, M. et al., Germanium tin: Silicon photonics toward the mid-infrared, *Photonics Res.* 1 (2013): 69–76.

Keyvaninia, S., Roelkens, G., Van Thourhout, D. et al., Demonstration of a heterogeneously integrated III-V/SOI single wavelength tunable laser, *Opt. Express* 21 (2013a): 3784–3792.

Keyvaninia, S., Verstuyft, S., Van Landschoot, L. et al., Heterogeneously integrated III-V/silicon distributed feedback lasers, *Opt. Lett.* 38 (2013b): 5434–5437.

Khan, S., Chiles, J., Ma, J. et al., Silicon-on-nitride waveguides for mid-and near-infrared integrated photonics, *Appl. Phys. Lett.* 102 (2013): 121104.

Khriachtchev, L. ed., *Silicon Nanophotonics* (Singapore: World Scientific Publishing, 2009).

Krause, M., Renner, H., Fathpour, S. et al., Raman-gain enhancement in cladding-pumped silicon waveguides, *IEEE J. Quantum Electron.* 44 (2008): 692–704.

Kuo, Y., Lee, Y. K., Ge, Y. et al., Strong quantum-confined Stark effect in germanium quantum-well structures on silicon, *Nature* 437 (2005): 1334–1336.

Kuyken, B., Ji, H., Clemmen, S., Nonlinear properties of and nonlinear processing in hydrogenated amorphous silicon waveguides, *Opt. Express* 19 (2011): 146–153.

Lasky, J. B., Stiffler, S. R., White, F. R. et al., Silicon-on-insulator (SOI) by bonding and etch-back, *Proceedings of IEEE International Electron Device Meeting*, 684–687, 1985.

Liang, T. K., Tsang, H. K., Role of free carriers from two-photon absorption in Raman amplification in silicon-on-insulator waveguides, *Appl. Phys. Lett.* 84 (2004): 2745–2747.

Liang, D., Bowers, J. E., Recent progress in lasers on silicon, *Nat. Photon.* 4 (2010): 511–517.

Liao, S., Feng, N. N., Feng, D. et al., 36 GHz submicron silicon waveguide germanium photodetector, *Opt. Express* 19 (2011): 10967–10972.

Lim, S. T., Png, C. E., Ong, E. A. et al., Single mode, polarization-independent submicron silicon waveguides based on geometrical adjustments, *Opt. Express* 15 (2007): 11061–11072.

Liu, Y., Tsang, H. K., Nonlinear absorption and Raman gain in helium-ion-implanted silicon waveguides, *Opt. Lett.* 31 (2006): 1714–1716.

Liu, J. F., Beals, M., Pomerene, A., Waveguide-integrated, ultralow-energy GeSi electro-absorption modulators, *Nat. Photonics* 2 (2008): 433–437.

Leuthold, J., Koos, C., Freude, W., Nonlinear silicon photonics, *Nat. Photonics* 4 (2010): 535–544.

Leuthold, J., Koos, C., Freude, W. et al., Silicon-organic hybrid electro-optical devices, *IEEE J. Sel. Top. Quantum Electron.* 19 (2013): 3401413.

Lipson, M., Guiding, modulating, and emitting light on silicon—Challenges and opportunities, *J. Lightwave Technol.* 23 (2005): 4222–4238.

Lorenzo, J. P., Soref, R. A., 1.3 μm electro-optic silicon switch, *Appl. Phys. Lett.* 51 (1987): 6–8.

Luryi, S., Kastalsky, A., Bean, J. C., New infrared detector on a silicon chip, *IEEE Trans. Electron. Dev.* 31 (1984): 1135–1139.

Ma, J., Fathpour, S., Pump-to-stokes relative intensity noise transfer and analytical modeling of mid-infrared silicon Raman lasers, *Opt. Express* 20 (2012): 17962–17972.

Ma, J., Fathpour, S. Noise characteristics of mid-and near-infrared nonlinear silicon photonic devices, *IEEE J. Lightwave Technol.* 31 (2013): 3181–3187.

Malik, A., Dwivedi, S., Van Landschoot, L., Ge-on-Si and Ge-on-SOI thermo-optic phase shifters for the mid-infrared, *Opt. Express* 22 (2014): 28479–28488.

Manasevit, H. M., Simpson, W. I., Single-crystal silicon on a sapphire substrate, *J. Appl. Phys.* 35 (1964): 1349–1351.

Marcatili, E. A. J., Dielectric rectangular waveguide and directional coupler for integrated optics, *Bell Syst. Tech. J.* 48 (1969): 2071–2102.

Marcatili, E. A. J., Miller, S. E., Improved relations describing directional control in lectromagnetic wave guidance, *Bell Syst. Tech. J.* 48 (1969): 2161–2188.

Mashanovich, G. Z., Miloševic, M. M., Nedeljkovic, M. et al., Low loss silicon waveguides for the mid-infrared, *Opt. Express* 19 (2011): 7112–7119.

Michel, J., Liu, J., Kimerling, L. C., High-performance Ge-on-Si photodetectors, *Nat. Photonics* 4 (2010): 527–534.

Moss, D. J., Morandotti, R., Gaeta, A. L. et al., New CMOS-compatible platforms based on silicon nitride and Hydex for nonlinear optics, *Nat. Photonics* 7 (2013): 597–607.

Narayanan, K., Preble, S. F., Optical nonlinearities in hydrogenated-amorphous silicon waveguides, *Opt. Express* 18 (2010): 8998–9005.

Noguchi, K., Mitomi, O., Miyazawa, H., Millimeter-wave Ti: LiNbO$_3$ optical modulators, *IEEE J. Lightwave Technol.* 16 (1998): 615–619.

Ohira, K., Kobayashi, K., Iizuka, N. et al., On-chip optical interconnection by using integrated III–V laser diode and photodetector with silicon waveguide, *Opt. Express* 18 (2010): 15440–15447.

Okamoto, K., *Fundamentals of Optical Waveguides* (Burlington, MA: Academic Press, 2006).

OpSIS website (Accessed September 30, 2014). http://opsisfoundry.org/.

Park, H., Fang, A. W., Kodama, S. et al., Hybrid silicon evanescent laser fabricated with a silicon waveguide and III–V offset quantum wells, *Opt. Express* 13 (2005): 9460–9464.

Pasquariello, D., Hjort, K., Plasma-assisted InP-to-Si low temperature wafer bonding, *IEEE J. Sel. Top. Quantum Electron.* 8 (2002): 118–131.

Pavesi, L., Lockwood, D. J. eds., *Silicon Photonics* (Berlin, Germany: Springer-Verlag, 2004).

Pernice, W. H. P., Xiong, C., Schuck, C. et al., Second harmonic generation in phase matched aluminum nitride waveguides and micro-ring resonators, *Appl. Phys. Lett.* 100 (2012): 223501.

Petroff, P. M., Nucleation and growth of GaAs on Ge and the structure of antiphase boundaries, *J. Vac. Sci. Tech. B.* 4 (1986): 874–877.

Pruessnera, M. W., Stievater, T. H., Rabinovich, W. S., High-finesse micromachined Fabry–Perot cavities with silicon/air DBR mirrors, *LEOS 2006. 19th Annual Meeting of the IEEE,* 46–47, 2006.

Qasaimeh, O., Bhattacharya, P., Croke, E. T., SiGe–Si quantum-well electroabsorption modulators, *IEEE Photon. Technol. Lett.* 10 (1998): 807–809.

Rabiei, P., Ma, J., Khan, S. et al., Submicron tantalum pentoxide optical waveguide and microring resonators fabricated by selective oxidation of refractory metal, *Opt. Express* 21 (2013a): 6967–6972.

Rabiei, P., Ma, J., Khan, S. et al., Heterogeneous lithium niobate photonics on silicon substrates, *Opt. Express* 23 (2013b): 25573–25581.

Rabiei, P., Rao, A., Ma, J. et al., Low-loss and high index-contrast tantalum pentoxide microring resonators and grating couplers on silicon substrates, *Opt. Lett.* 39 (2014): 5379–5383.

Raghunathan, V., Shori, R., Stafsudd, O. M. et al., Nonlinear absorption in silicon and the prospects of mid-infrared silicon Raman laser, *Phys. Status Solidi A* 203 (2006): R38–R40.

Raghunathan, V., Borlaug, D., Rice, R. et al., Demonstration of a mid-infrared silicon Raman amplifier, *Opt. Express,* 15 (2007): 14355–14362.

Rayleigh, L., A study of glass surfaces in optical contact, *Proc. Phys. Soc. A* 156 (1936): 326–349.

Reddy, U. K., Houdre, R., Munns, G. et al., Investigation of GaAs/(Al, Ga)As multiple quantum wells grown on Ge and Si substrates by molecular-beam epitaxy. *J. Appl. Phys.* 62 (1987): 4858–4862.

Reed, G. T., Mashanovich, G., Gardes, F. Y. et al., Silicon optical modulators, *Nat. Photonics* 4 (2010): 518–526.

Reed, G. T., Knights, A. P., *Silicon Photonics: An Introduction* (West Sussex: John Wiley and Sons, 2004).

Reed, G. T. ed., *Silicon Photonics: The State of the Art* (West Sussex: John Wiley and Sons, 2008).

Roelkens, G., Van Thourhout, D., Baets, R. et al., Laser emission and photodetection in an InP/InGaAsP layer integrated on and coupled to a silicon-on-insulator waveguide circuit, *Opt. Express* 14 (2006): 8154–8159.

Rojo, R. P., Van Campenhout, J., Regreny, P. et al., Heterogeneous integration of electrically driven microdisk based laser sources for optical interconnects and photonic ICs, *Opt. Express* 14 (2006): 3864–3871.

Rong, H., Jones, R., Liu, A. et al., A continuous-wave Raman silicon laser, *Nature* 433 (2005): 725–728.

Sakai, S., Soga, T., Takeyasu, M. et al., Room-temperature laser operation of AlGaAs-GaAs double heterostructures fabricated on Si substrates by metalorganic chemical vapor-deposition, *Appl. Phys. Lett.* 48 (1986): 413–414.

Sasaki, J., Itoh, M., Tamanuki, T. et al., Multiple-chip precise self-aligned assembly for hybrid integrated optical modules using Au–Sn solder bumps, *IEEE Trans. Adv. Packag.* 24 (2001): 569–575.

Seassal, C., Rojo-Romeo, P., Letartre, X. et al., InP microdisk lasers on silicon wafers: CW room temperature operation at 1.6 µm, *Electron. Lett.* 37 (2001): 222–223.

Seung, C. J., Djordjev, K., Choi, S. J. et al., Microdisk lasers vertically coupled to output waveguides, *IEEE Photon. Technol. Lett.* 15 (2003): 1330–1332.

Shankar, R., Bulu, I., Loncar, M., Integrated high-quality factor silicon-on-sapphire ring resonators for the mid-infrared, *Appl. Phys. Lett.* 102 (2013): 051108.

Shimbo, M., Furukawa, K., Fukuda, K. et al., Silicon-to-silicon direct bonding method, *J. Appl. Phys.* 60 (1986): 2987–2989.

Soref, R. A., Lorenzo, J. P., Single-crystal silicon—A new material for 1.3 and 1.6 μm integrated-optical components, *Electron. Lett.* 21 (1985): 953–954.

Soref, R. A., Lorenzo, J. P., All-silicon active and passive guided-wave components for λ = 1.3 and 1.6 μm, *IEEE J. Quantum Electron.* 22 (1986): 873–879.

Soref, R. A., Bennett, B. R., Electrooptical effects in silicon, *IEEE J. Quantum Electron.* 22 (1987): 23, 123–129.

Soref, R. A., Schmidtchen, J., Petermann, K., Large single-mode rib waveguides in GeSi and Si-on-SiO$_2$, *IEEE J. Quantum Electron.* 27 (1991): 1971–1974.

Soref, R. A., Silicon-based optoelectronics, *Proc. IEEE* 81 (1993): 1687–1706.

Soref, R. A., The past, present, and future of silicon photonics, *IEEE J. Sel. Top. Quantum Electron.* 12 (2006a): 1678–1687.

Soref, R. A., Emelett, S. J., Buchwald, W. R., Silicon waveguided components for the long-wave infrared region, *J. Optics A* 8 (2006b): 840–848.

Soref, R., Mid-infrared photonics in silicon and germanium, *Nat. Photonics* 4 (2010): 495–497.

Spencer, H., Progress: Its law and cause, in *Essays: Scientific, Political, and Speculative*, vol. I, p.35 (New York: D. Appleton and Company, 1891).

Sugo, M., Takanashi, Y., Al-Jassim, M. M. et al., Heteroepitaxial growth and characterization of InP on Si substrates, *J. Appl. Phys.* 68 (1990): 540–547.

Sun, X., Zadok, A., Shearn, M. J. et al., Electrically pumped hybrid evanescent Si/InGaAsP lasers, *Opt. Lett.* 34 (2009): 1345–1347.

Tanaka, S., Jeong, S. H., Sekiguchi, S. et al., High-output-power, single-wavelength silicon hybrid laser using precise flip-chip bonding technology, *Opt. Express* 20 (2012): 28057–28069.

Temkin, H., Bean, J. C., Pearsall, T. P. et al., High photoconductive gain in Ge$_x$Si$_{1-x}$/Si strained-layer superlattice detectors operating at 1.3 μm, *Appl. Phys. Lett.* 49 (1986): 155–157.

Terui, H., Yamada, Y., Kawachi, M. et al., Hybrid integration of a laser diode and high-silica multimode optical channel waveguide on silicon, *Electron. Lett.* 21 (1985): 646–648.

Thomson, D. J., Gardes, F. Y., Fedeli, J. M. et al., 50-Gb/s silicon optical modulator, *IEEE Photon. Technol. Lett.* 24 (2012): 234–236.

Trinh, P.D., Yegnanarayanan, S., Jalali, B., Integrated optical directional couplers in silicon-on-insulator, *Electron. Lett.* 31 (1995): 2097.

Tsia, K. K., Fathpour, S., Jalali, B., Electrical control of parametric processes in silicon waveguides, *Opt. Express* 16 (2008a): 9838–9843.

Tsia, K. K., Fathpour, S., Jalali, B., Electrical tuning of birefringence in silicon waveguides, *Appl. Phys. Lett.* 92 (2008b), 061109.

Tsuchizawa, T., Yamada, K., Watanabe, T. et al., Monolithic integration of silicon-, germanium-, and silica-based optical devices for telecommunications applications, *IEEE J. Sel. Top. Quantum Electron.* 17 (2011): 516–525.

Tu, X., Liow, T. Y., Song, J. et al., 50-Gb/s silicon optical modulator with traveling-wave electrodes, *Opt. Express* 21 (2013): 12776–12782.

Van Campenhout, J., Romero, P. R., Regreny, P. et al., Electrically pumped InP-based microdisk lasers integrated with a nanophotonic silicon-on-insulator waveguide circuit, *Opt. Express* 15 (2007): 6744–6749.

Vlasov, Y. A., McNab, S. J., Losses in single-mode silicon-on-insulator strip waveguides and bends, *Opt. Express* 12 (2004): 1622–1631.

Whiteman, R. R., Knights, A. P., George, D. et al., Recent progress in the design, simulation, and fabrication of small cross-section silicon-on-insulator VOAs, *Proc. SPIE 4997, Photonics Packaging and Integration III*, 146, June 2003.

Wong, K., *Properties of Lithium Niobate* (London: INSPEC, 2002).

Wong, C. Y., Cheng, Z., Chen, X. et al., Characterization of mid-infrared silicon-on-sapphire microring resonators with thermal tuning, *IEEE Photonics J.* 4 (2012): 1095–1102.

Wooten, E. L., Kissa, K. M., Yi-Yan, A., A review of lithium niobate modulators for fiber-optic communications systems, *IEEE J. Sel. Top. Quantum Electron.* 6 (2000): 69–72.

Wuu, D. S., Horng, R. H., Lee, M. K., Indium phosphide on silicon heteroepitaxy: Lattice deformation and strain relaxation, *J. Appl. Phys.* 68 (1990): 3338–3342.

Xiong, C., Pernice, W., Ryu, K. K., Integrated GaN photonic circuits on silicon (100) for second harmonic generation, *Opt. Express* 19 (2011): 10462–10470.

Xu, Q., Schmidt, B., Pradhan, S. et al., Micrometre-scale silicon electro-optic modulator, *Nature* 435 (2005): 325–327.

Yang, J., Mi, Z., Bhattacharya, P., Grooved-coupled InGaAs/GaAs quantum dot laser/waveguide on silicon, *J. Lightwave Technol.* 25 (2007): 1826–1831.

Yang, J., Bhattacharya, P., Integration of epitaxially-grown InGaAs/GaAs quantum dot lasers with hydrogenated amorphous silicon waveguides on silicon, *Opt. Express* 16 (2008): 5136–5140.

Yin, T., Cohen, R., Morse, M. M. et al., 31 GHz Ge n-i-p waveguide photodetectors on silicon-on-insulator substrate, *Opt. Express* 15 (2007): 13965–13971.

23

Nanoplasmonic optoelectronics

ROBERT G. W. BROWN
University of California, Irvine

23.1 INTRODUCTION AND MOTIVATION

Nanoplasmonic optoelectronics, a subject too immature for the first edition of the book, but now well developed in the research laboratory and poised to play an important role in the future development of novel optoelectronic devices and systems, is the focus of this forward-looking chapter. We do not cover here nano-optoelectronics, with its focus on quantum dots, wires, and well, especially when used in laser diodes, light emitting diodes (LEDs), and photodetectors of various

kinds, as these aspects at the nanolevel have been intensively researched and reviewed (Saleh and Teich, 2007) these past 20+ years and described in Chapters 10–19 in this handbook.

Much of the motivation driving the development of nanoscale optoelectronics and plasmonic optoelectronics is that of SWAP-C, meaning Size, Weight, Power, and Cost to the Military and Aerospace communities, and, Cooling (rather than Cost) for others. Many applications and potential applications of optoelectronics, particularly in mobile devices that are hand held or flown continuously, demand very small size devices and systems, the lowest possible weight of such devices, together with the smallest possible power consumption, and, if possible, no cooling systems being required. For most applications, cost is also a key driver, although the fabrication and testing costs of nanoscale devices are relatively very large indeed. This high cost will have to change for many of the devices described here to become accepted as components in novel ultrasmall-scale optosystems in the future.

The structure of this chapter is outlined earlier, starting with the multifaceted tool kit of physics and mathematics that under pins all nanoplasmonics, followed by a brief look at appropriate plasmonic materials and some of the challenges involved with them. We look briefly at key fabrication and test methods, before focusing on the content of this chapter, nanoplasmonic light sources, modulators, detectors, optics (metamaterials), and potentially advantageous applications that we can expect in the future. Extreme opposites are found in the potential applications, e.g., from use at the single biological-unit scale (DNA, viruses, cells, etc.), to uses in unmanned aerial vehicles (UAVs) and space-borne sensing probes.

23.2 THE TOOL KIT

The essential physics of plasmonics is beautifully reviewed by Maier (2007), starting from Maxwell's equations and the criteria for the existence of a plasmon (a cloud of electrons) propagating at the interface between a dielectric and a conductor, evanescently confined in the perpendicular direction. The dielectric has a positive real dielectric constant, whereas the metal must have the real part of its dielectric constant being negative at the wavelength or frequency of interest. For gold, this implies infrared wavelengths. For aluminum and silver, visible and ultraviolet (UV) wavelengths are appropriate; materials and their properties, including graphene are discussed in the next section.

Immediately, here at the outset, we distinguish carefully between decades of research on surface plasmons operating on a scale of microns to millimeters, and the focus of this review, which is *localized* surface plasmons, operating on a scale of nanometers to a micron or so, because of the involvement of nanoparticles and arrays of nanoparticles (Zayats et al., 2005). "Plasmons" in this review means localized plasmons. We will also encounter "plasmon-polaritons" the propagating entanglement

of plasmons and photons, also assumed here at the nanoscale.

Maier continues by describing in detail the polarization properties of plasmons, excitation, and lifetime of plasmons, propagation dispersion relations, waveguides, etc., and all the essential physics before outlining basic device architectures and early applications interests such as surface-enhanced Raman scattering (SERS) and metamaterials. An up-to-date and comprehensive basic treatment of this subject is given by Novotny and Hecht (2012), in their excellent book.

Two other plasmonics-focused books are well worth looking at closely: *Nanophotonics* by Prasad (2004) and *Surface Plasmon Nanophotonics*, edited by Brongersma and Kik (2010). Additionally, throughout the development of nanoplasmonic optoelectronics, some fine review papers have also been published (e.g., Maier and Atwater, 2005; Berini, 2009; Flory et al., 2011; Halas et al., 2011; Stockman, 2011; Zhang and Zhang, 2012).

Beyond the basic underpinning physics set out by Maier (2007) and Novotny and Hecht (2012), there are more advanced physics concepts that appear repeatedly as we attempt to use nanoplasmonics/polaritonics in new optoelectronic devices. This is our new tool kit as we develop novel optoelectronic devices, system, and applications in the future:

1. At the quantum level, we become aware of Fano factors and their role in shaping plasmon resonances (Luk'yanchuk et al., 2010) and the ability to use classical physics to describe plasmons all the way down to the nanometer scale, but not in the subnanometer region (Duan et al., 2012). Even wave particle duality of plasmons has been explored (Kolesov et al., 2009). Quantum plexitonics and quantum plasmonics have emerged in recent years (Manjavacas et al., 2011; Marinica et al., 2012), involving strongly interacting plasmons and excitons, and strong nonlinear effects in a nanoparticle dimer. Control of spontaneous emission is possible plasmonically (Belacel et al., 2013).

2. Quantum optics—single photons upon demand can be realized through nanoplasmonic architectures (Chang et al., 2006).

3. Nonlinear optics—both $\chi^{(3)}$ (Suh et al., 2011) and $\chi^{(2)}$ (Zhang et al., 2011a), extreme nonlinear-optical effects have been reported using nanoplasmonics, offering considerable potential

for novel low-power optoelectronic devices to be described later. The subject was excellently reviewed by Kauranen and Zayats (2012).

4. The roles of hot electrons and Schottky barriers become crucial when we enter the realm of detector design and performance. The physics is well described by Knight et al. (2013) and Manjavacas et al. (2014); enhanced photodetector device potential is demonstrated by Brown and Stanley (2011) and Knight et al. (2011). Responsivities in excess of 25× over planar diodes have been demonstrated, and we expect to exceed this further as nanofabrication techniques are improved in the future. Chalabi and Brongersma (2013) consider artificial photosynthesis based on this new capability. Atwater and Polman (2010) suggested improved photovoltaic devices using plasmonic light trapping, enhanced solar cell performance.

5. Precision arrays of nanoparticles (and nanopinholes, Ebbesen et al., 1998; Garcia-Vidal et al., 2005) and their cooperative resonances (Niklasson and Craighead, 1985) become central to the design of metamaterials and transformation optics (Kildishev and Shalaev, 2011), which is beyond the scope of this optofocused review chapter, but also metamaterial optics (Brown, 2012a), and plasmon-enhanced photodetectors (see Brown, 2011; Knight et al., 2011), which utilize plasmonic perfect absorbers (e.g., Liu et al., 2010a; Aydin et al., 2011). Here, we encounter the potential for total light absorption, creation of massive refractive indexes, extraordinary photodetection capabilities, etc., all of which will be described further later.

6. Interference of plasmons. Plasmons can be directed to interfere with each other (Choi et al., 2014), indeed, plasmonic interferometers are possible for ultrasensitive applications at the nanoscale (Gao et al., 2011).

7. Electronic modulation of plasmon and plasmon-resonance properties. It is all well and good to be able to create specific plasmon properties, but often in optoelectronic devices and systems we wish to vary or modulate properties. Techniques have emerged allowing active spatial control of plasmonic fields (Gjonaj et al., 2011), also plasmon resonance, through varying electric potential (Lioubimov et al., 2004; Noginova, 2011).

8. Spasers. One of the most interesting areas of recent plasmonics, surface plasmon lasers. A spaser is the nanoplasmonic counterpart of a laser. Stockman and his collaborators have led the charge in this area, from the first paper (Bergman and Stockman, 2003), to a layman's explanation (Stockman, 2008) and practical demonstrations of a spaser-based nanolaser (Noginov et al., 2009; Lu et al., 2014). We explain considerably more about this work later in this chapter.

From this new optoelectronic tool kit you can immediately see that at the nanoscale, we have all the essential ingredients to create the light sources, modulators, detectors, and other sensors upon which the whole of optoelectronics is traditionally based. In the future, the scale of optoelectronic devices and systems might be truly micro/nanoscopic because of these recent discoveries and developments.

We now explore these new tool kit items through the remainder of this chapter, some novel applications and possibilities also, as we look to future decades of optoelectronics R&D.

23.3 LOCALIZED SURFACE PLASMONS—ESSENTIAL MATH

Localized surface plasmons, the core essence of this chapter, are nonpropagating excitations of metal nanostructures' conduction electrons when coupled to an incident electromagnetic field. Full details have been discussed at length and with rigor by Maier (2007) and Novotny and Hecht (2012). Here, we summarize the basic, essential, and frequently used calculations.

Being small metallic structures of varying shape, less in size than the wavelength of light, nanostructures respond to incident electromagnetic radiation (EMR) in the manner known in great detail to the light scattering field for over 100 years through Mie scattering and absorption, etc., described in detail by Bohren and Huffman (1983).

The essential optical properties start from a knowledge of the real and imaginary parts of the complex refractive index, $RI = n + ik$, and the real and imaginary parts of the complex dielectric function, or relative permittivity, $\varepsilon = \varepsilon' + i\varepsilon''$.

For nonmagnetic materials ($\mu = \mu_0$), RI and ε are related through the equations:

$$\varepsilon' = \frac{\varepsilon'}{\varepsilon_0} = n^2 - k^2,$$

$$\varepsilon'' = \frac{\varepsilon''}{\varepsilon_0} = 2nk,$$

$$n = \sqrt{\frac{\sqrt{\varepsilon'^2 + \varepsilon''^2} + \varepsilon'}{2}}$$

$$k = \sqrt{\frac{\sqrt{\varepsilon'^2 + \varepsilon''^2} - \varepsilon'}{2}}$$

Much of the RI data we need is listed by Palik (1985), but as discussed later, thin film data can vary significantly from bulk materials data, and for precision designs and analysis we need to measure thin film data in the spectral region of interest. From RI we can determine ε parameters that we need.

Next of interest is the polarizability, α, of the nanoparticles:

$$\alpha = 4\pi a^3 \frac{\varepsilon - \varepsilon_m}{\varepsilon + 2\varepsilon_m}$$

where "a" represents the particle's radius. ε_m is the relative permittivity of the medium surrounding the particle.

Knowing the nanoparticle's polarizability permits us to calculate the scattering and absorption cross-sections, C_{sca} and C_{abs}, respectively:

$$C_{sca} = \frac{k^4}{6\pi}|\alpha|^2 = \frac{8\pi}{3}k^4 a^6 \left|\frac{\varepsilon - \varepsilon_m}{\varepsilon + 2\varepsilon_m}\right|^2$$

$$C_{abs} = k\,\mathrm{Im}[\alpha] = 4\pi k a^3 \,\mathrm{Im}\left|\frac{\varepsilon - \varepsilon_m}{\varepsilon + 2\varepsilon_m}\right|^2$$

where $k = 2\pi/\lambda$, and λ is the wavelength of the incident electromagnetic radiation.

When dealing with nanoparticles of nonspherical shape, such as rods, ellipsoids, and the like, the shape of the particle is important in the calculations of the polarizability, α, and therefore of the absorption and scattering cross sections C_{sca} and C_{abs}. The polarizability is modified as shown in the following equation:

$$\alpha_i = 4\pi a_1 a_2 a_3 \frac{\varepsilon(\omega) - \varepsilon_m}{3\varepsilon_m + 3L_i(\varepsilon(\omega) - \varepsilon_m)}$$

where L_i is a geometrical factor given by

$$L_i = \frac{a_1 a_2 a_3}{2} \int_0^\infty \frac{dq}{(a_i^2 + q)f(q)}$$

and

$$f(q) = \sqrt{(q + a_1^2)(q + a_2^2)(q + a_3^2)}$$

The geometrical factors L_i sum to unity; for a sphere each factor equals 1/3.

More advanced calculations and effects such as the lifetime of plasmons, propagation of plasmons along chains of nanoparticles, and localized plasmons in gain media have been surveyed by Maier (2007). Use of nanoparticles in plasmonic antennas has been surveyed at length by Novotny and Hecht (2012). Often in this application we need to know the plasma frequency, ω_p, of the plasmonic material, which can be estimated by the following equation:

$$\omega_p = \left[(n_e \cdot e^2)/(m \cdot \varepsilon_0)\right]^{1/2} \quad \{\text{SI Units}\}$$

where n_e is the number density of electrons, e is the electronic charge, m is the effective mass of the electrons, and ε_0 is the permittivity of free space.

Finally, we note that for arrays of coupled plasmon-resonant nanoparticles, the transmission and reflection coefficients can be estimated by the following equations (Niklasson and Craighead, 1985; Brown, 2012a); thus, absorption coefficient can also be estimated as $T + R + A = 1$:

$$T = \left[1 + 16\pi^2 \frac{\mathrm{Im}\{\alpha_{eff}\}}{lc^2 \cdot \lambda \cdot (N_S + N_M)} + 64 \cdot \frac{\pi^4}{\left[lc^2 \cdot \lambda \cdot (N_S + N_M)\right]^2} \cdot |\alpha_{eff}|^2\right]^{-1}$$

$$R = \frac{64 \cdot \pi^4}{\left[lc^2 \cdot (N_S + N_M)\right]^2} \cdot |\alpha_{eff}|^2 \cdot T$$

where lc is the lattice dimension, λ = EMR wavelength,

$\propto_{eff}$ = effective polarizability, N_S = refractive index of substract, and

N_M = refractive index of medium surrounding the nanoparticles (nanorods).

For simple plasmonic calculations, the above equations suffice, but more often than not we encounter complicated 3D arrangements of nanoparticles and then full-3D electromagnetic modeling software is used to estimate plasmonic/optoelectronic properties, e.g., FDTD (finite difference time domain) techniques (as sold by Lumerical, Inc., and other companies).

23.4 MATERIALS SELECTION

The traditional materials for plasmonics are metals, with dielectrics. The importance of the real part of the metal's dielectric constant being negative at the wavelength or frequency of interest for plasmonic functionality was emphasized earlier. Gold is frequently used in the infrared region, aluminum and silver are often used in the visible and UV regions. Aluminum is especially of interest at visible wavelengths (Knight et al., 2012, 2014). Metal nanostructures are fairly straightforward to fabricate (see Section 23.5), so there has been good experimental progress from many laboratories around the world in recent years. An introductory book covers most of the key concepts of physics and engineering (Pelton and Bryant, 2013).

The key drawback with metals for nanoplasmonics is their Ohmic loss, which can be substantial. Nevertheless, metals have been used for the majority of nanoplasmonics studies published to date. Khurgin (2015) discussed this in detail.

There are, however, alternative materials for nanoplasmonics. Doped oxides and doped nitrides have been explored (West et al., 2010; summarized by Boltasseva and Atwater, 2011). Transparent conducting oxides such as Al:ZnO, Ga:ZnO, and indium tin oxide (ITO) enable plasmonic metamaterials operating in the near-IR; transition metal nitrides such as TiN or ZrN operate well in the visible region.

Graphene is increasingly popular in such studies (Grigorenko et al., 2012) and will be discussed in much greater detail later in this section. Even DNA has been used as the basis for plasmonic nanostructures (Tan et al., 2011). Eutectic liquid alloys are also found to have plasmonic properties (Blaber et al., 2012).

When using metals for nanoplasmonic studies, we usually start with the dielectric properties at the intended wavelength or frequency of operation. The standard sources of this information are Palik (1985), and at least two papers in Applied Optics (Ordal et al., 1983; Rakić et al., 1998). Online, it is useful to search for information at refractiveindex.info, a refractive index database.

However, most importantly for precision structural designs and performance expectations, nanoscopic materials properties can *differ substantially* from the bulk properties reported in these standard sources for data. Figure 23.1 shows differences supplied to the author by Dr. David Peters of Sandia National Labs.

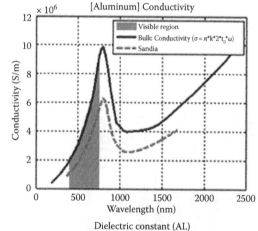

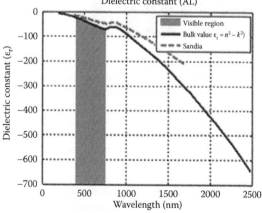

Figure 23.1 Palik's bulk values of conductivity and dielectric constant for aluminum across the visible region (blue traces) together with few nanometer-scale thin-film measurements by Sandia National Labs (dashed red traces). (Courtesy of Dr. David Peters, Sandia National Laboratory, with permission.)

Nanoplasmonics is the basis for the construction of entirely new materials that do not appear to exist naturally in nature: metamaterials and negative refractive index materials (Kildishev and Shalaev, 2011). Although this optics topic is beyond the scope of this optoelectronics-focused chapter, nevertheless we note the interface of these two subjects, such as described by Hess et al. (2012). We also note that plasmonic metamaterials may become active and tunable, as reviewed by Boardman et al. (2010), and form low-loss negative index materials in the visible spectrum (Aslam and Güney, 2011). Metamaterials have been made with *zero* refractive index (Vesseur et al., 2013) and can have tailored nonlinear optical response (Husu et al., 2012). Very high refractive indexes may also be created in metamaterials: Choi et al. (2011) in the THz region and Brown (2012a) in the visible and IR regions.

Graphene plasmonics (thoroughly reviewed by Bonaccorso et al., 2010; Koppens et al., 2011; Vakil and Engheta, 2011; Bao and Loh, 2012; Grigorenko et al., 2012; Tassin et al., 2012) can now be contrasted to the metal/oxide/nitride plasmonic approaches just reviewed here. Graphene offers a substantially different platform for the creation of optoelectronic devices and systems (see Figure 23.2). Perhaps the most important difference to the above approaches is that the bandgap structure of graphene can be varied through doping, and this opens up a variety of different plasmonic schemes. Plasmons in graphene can be tunable by electrostatic doping, and have long lifetimes compared to noble metal plasmons. Low loss at IR frequencies has been investigated (Jablan et al., 2009).

At the most fundamental level in graphene, plasmon localization has been explored by Zhou et al. (2012), quantum finite size effects by Thongrattanasiri et al. (2012b), and hot carrier transport by Song et al.

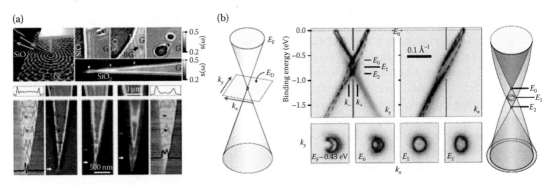

Figure 23.2 Launching and imaging graphene plasmons. (a) Top left: schematic of an infrared nano-imaging experiment. The blue and green arrows label the incoming and back-scattered light, respectively. Top right: images of various interference patterns close to graphene edges (blue dashed lines), defects (green dashed lines), or boundaries between monolayer (G) and bilayer (BG) graphene (white dashed line). Scale bar represents normalized near-field amplitude of excited plasmons. Arrows show features connected to plasmon interference. Bottom: controlling the plasmon wavelength in scattering experiments. The control is based on strong dependence of graphene plasmons on dielectric properties of substrate. From left to right the excitation wavelength changed from 9200 to 10,152 nm, the permittivity of substrate from 2.9 to 0.7, and the plasmon wavelength from 200 to 700 nm. (b) Plasmaron satellite bands in the Angle-resolved photoemission spectroscopy spectrum of graphene on SiC. Left: The Dirac energy spectrum of graphene in a noninteracting picture as a function of the two components of momentum, k_x and k_y. E_D indicates the energy of the "Dirac point," where the upper and lower Dirac cones touch. E_0, E_1, and E_2 are the characteristic energies of the reconstructed Dirac crossing. Middle-top panel: Experimental spectrum of doped graphene perpendicular and parallel to the Γ–K direction. The dashed lines are guides to the dispersion of the observed hole and plasmaron bands. The red lines are at $k = 0$. Middle-bottom panels: Constant energy cuts of the spectral function at different binding energies. Right panel: Schematic Dirac spectrum in the presence of electron–electron interactions. (From Grigorenko, A. N. et al., *Nat. Phot.*, 6, 749, 2012, with permission.)

(2011), all being foundation stones for future opto-electronic device developments. Waveguiding has been explored by Christensen et al. (2011). Plasmons in graphene can be controlled electrically (Emani et al., 2012; Kim et al., 2012) and by spatial patterning (Yan et al., 2012).

Graphene photodetectors have been widely explored, from the total absorption of light (Thongrattanasiri et al., 2011) and observations of strong plasmonic enhancement of photovoltage (Echtermeyer et al., 2011), to exploitation of these effects in experimental graphene-antenna sandwich photodetectors (Fang et al., 2012a), and multi-color photodetection (Liu et al., 2011). Applications of graphene plasmonics as diverse as biosensors (Salihoglu et al., 2012) and THz metamaterials (Ju et al., 2011) have been reported.

As a material, graphene offers considerable promise for plasmonic optoelectronics, the details of which we explore in some detail in future sections of this chapter.

23.5 FABRICATION TECHNIQUES

Such is the fundamental importance of fabrication processes and excellence to nanoplasmonic device performance that we present here a brief mention of the main fabrication and test techniques for nanoplasmonic devices.

Despite considerable understanding and impressive demonstrations of nanoplasmonics over a wide variety of wavelengths, there remain some challenging aspects of this business; notably control of the surfaces and interfaces, bonds, discontinuities, and defects at the 1-nm scale that can turn fine theory into poor performance unless handled and controlled with extreme care in the sequence of fabrication processes in the clean room. Outside of a nanoparticle, it is often the first nanometer (~5 atoms!) that counts, for that controls nearly everything about the performance of the plasmonic optodevice architecture.

Most nanofabrication work is done in clean room fabrication/test facilities such as commonly found in optoelectronic device development laboratories: e-beam lithography for patterning the nanostructures, photoresist processes, and etching for the creation of the basic structures, metallization for applying electrical contacts to such structures, other coating techniques for dielectric, semiconductor, and/or filter layers, etc. These are

standard techniques, beyond the scope of this chapter to repeat here, and may be comprehensively understood from Madou's excellent multivolume nanofabrication descriptions (Madou, 2011).

In addition to these standard techniques, various other approaches have been developed recently, notably self-templating and self-organization of nanoparticles, to avoid the need for precision lithography, which is challenging over large areas, e.g., beyond making a single device. Scale-up to large-area nanodevice fabrication for volume manufacturing remains a serious challenge today, but approaches employing nanoimprinting are starting to bear fruit.

Excellent precision and larger-scale device-fab are central to future nanoplasmonic optoelectronics, so we now briefly outline a variety of approaches currently in the research lab that need to be developed to their full potential as soon as possible. There are perhaps four main themes: (1) nanolithography, (2) self-assembly, (3) polymers and biology, and (4) nanoimprinting.

23.5.1 Nanolithography

Photoresist technology limits the ultimate resolution in nano-lithography, so much effort is given to making improvements where possible, such as described by Fourkas (2011). In the future, we will need not only nanoscale features, but also tightly packed arrays of such features, hence interest in the most fundamental of photoresist materials and processes.

Electron beam lithography and focused ion beam (FIB) milling regularly achieve 10 nm structures, but they suffer from inherent limitations when a precision of ~1 nm is required. Under these circumstances, the combined approach of FIB and He-ion lithography shows promise (Kollman et al., 2014).

The smallest possible metallic nanorods are extremely important, so establishing the limit near 10 nm for physical vapor deposition was most important, as described by Niu et al. (2013). Creating the smallest possible inorganic crystals with predetermined shapes is also important, as explored by Gonzalez et al. (2011).

In graphene plasmonics, a different control limit is vitally important, namely that of precision local doping, thus we note the achievements of

Goncher et al. (2013), showing a possible route to nanoscopic control of electron and hole doping in graphene via specific substrate architecture.

There is still very much a need for developments in individual nanostructure creation over relatively large areas, so the low cost, scalable tip-based nanopatterning developed by Shim et al. (2011) for sub-50-nm resolution over centimeter-scale areas is noteworthy. Additionally, we may wish electronically to move and reposition nanostructures, such as suggested by Nickel et al. (2012). At larger scales of say 10 cm², high-throughput soft-interference nanolithography may offer a way forward (Henzie et al., 2007), who demonstrated 100-nm hole-arrays over such areas.

23.5.2 Self-assembly

Instead of lithographically patterning and etching overgrowth and other similar approaches to nanofab, the self-assembly of preordained nanoparticles offers fascinating alternatives. Spontaneous organization of the building blocks ordered by thermodynamic and other constrains is a start (Grzelczak et al., 2010). Ten-nanometer dense hole arrays have been achieved in self-assembled monolayers of nanoparticles as etch-masks (Wen et al., 2012). Plasmonic modes have been engineered by the templated self-assembly of nanoclusters (Fan et al., 2012a). Control of the configuration of nanoparticle clusters is becoming possible through self-organization (Galván-Moya et al., 2014), as is larger-scale tunable plasmon-resonance properties (Zhang et al., 2011b). Free-standing plasmonic nanorod superlattices have been constructed (Ng et al., 2011), and even 3D self-assembled nanorod arrays with smectic-B ordering has been achieved (Hamon et al., 2012).

23.5.3 Polymers and biology

Polymer and biological techniques represent an entirely different class of approaches to the nanoparticle precision-array challenge. Bottom-up metamaterial fabrication involving self-assembly of structures with nanoscale features seems attractive using block copolymer self-assembly (Hur et al., 2011). After a structure has been made, one of the polymers is dissolved away, leaving a 3D structure that can be filled by metal. Another copolymer

approach has been used to create 3D plasmonic nanoclusters (Urban et al., 2013).

Biologically, there are some fascinating possibilities ahead. Peptide sequencing effects can be used to control the size, structure, and function of nanoparticles on a sub-10-nm scale as demonstrated at the 2–3 nm scale by Coppage et al. (2012). Artificial DNA structures have been created for novel nanoelectronic and nano-optical devices (Kershner et al., 2009), whose work enables complex arrangements of nanotubes, nanowires, and/or quantum dots, but can equally well be extended to the nanoplasmonic challenges we have encountered so far, such as demonstrated by Tan et al. (2011). Plasmonic nanoarrays have even been genetically engineered, as shown by Forestiere et al. (2012).

23.5.4 Nanoimprinting

A key step to volume production of practical nanoplasmonic optoelectronic structures will be to develop the above and other fabrication techniques to cover large areas, perhaps on many different kinds of substrate materials, e.g., Si, GaAs, GaP, etc. One promising approach, currently with some fundamental reproduction precision challenges, is that of nanoimprinting. Soft nanoimprint lithography has demonstrated defect-free arrays of nanowires over 2-in. wafer substrates (Pierret et al., 2010). Single-layer and multilayer plasmonic metamaterials have been fabricated by production-capable nanoimprint lithography (Bergmair et al., 2011). Sub-10-nm patterns have been demonstrated with step-and-repeat nanoimprinting (Peroz et al., 2012), and sub-100-nm metal nanodot arrays have been created using nanostamping (Lee et al., 2011a). Optical force stamping has also been employed (Nedev et al., 2011), for creating arbitrary patterns of colloidal nanoparticles. Large-area "nanocoining" has been used to create millions of nanostructures per second over hundreds of square millimeters (Zdanowicz et al., 2012). 3D large-area negative-index metamaterials have also been created by nanotransfer printing (Chanda et al., 2011).

The nanoimprinting business is now becoming truly commercial in scale and capability, evidenced by the involvement of leading companies such as Suss, Papenheim et al. (2016).

23.6 NANO-OPTO SOURCES

The trend in recent years has been towards the subwavelength miniaturization of light sources such as semiconductor LEDs and laser diodes. Nanocavities, noninversion lasing, Purcell factors (control of spontaneous emission in resonant cavities), and many other new concepts to consider have all emerged in this change toward the smallest possible lasers. Among hundreds of research papers moving ever smaller, the following address important issues and provide some impressive demonstrations: Park et al. (2008a), Noh et al. (2011), Ding and Ning (2012), and Andersen et al. (2011).

The underlying technology for future nano-optoelectronic devices and system architectures has also been advanced through creation of nanomembranes (Park et al., 2014) upon which to build novel structures.

En-route to nanoplasmonic lasers we have seen confined surface plasmon-polariton amplifiers (Kéna-Cohen et al., 2013), with high gain of 93 dB/mm, which will be important for compensating plasmonic-materials losses in plasmonic devices and systems, as we noted previously. We have also seen electrically pumped polariton lasers (Schneider et al., 2013) and electrically injected exciton–polariton lasers (Bhattacharya et al., 2013); ultralow laser thresholds are a primary aim of this field of research.

23.6.1 Plasmonic lasers

We focus on spasers in the next section, but it is worth noting other plasmonic-based light emission and lasing structures ahead of that discussion.

Ultra-short extreme-ultraviolet light pulses is a key need for time-resolved spectroscopy used in many applications involving atoms, molecules, and solids; plasmonic generation of XUV in metallic waveguides via field-enhancement using surface-plasmon-polaritons may, therefore, become important (Park et al., 2011).

Plasmonic green nanolasers based on metal oxide semiconductor structures have been demonstrated (Wu et al., 2011), using a bundle of green-emitting InGaN/GaN nanorods strongly coupled to a gold plate through a SiO_2 dielectric nanogap layer. Confinement of the plasmonic field was $<8.0 \times 10^{-4} \mu m^3$. Lasing in metal–insulator–metal

subwavelength plasmonic waveguides has been used to create plasmon mode lasers at wavelengths near 1.5 μm (Hill et al., 2009).

Plasmon lasers using cadmium sulfide nanowires separated from a silver surface by a 5-nm-thick insulating layer have been shown to exhibit high gain (Oulton et al., 2009). Room temperature operation of plasmon-based CdS/silver structures employing total-internal reflection was later demonstrated by much the same team (Ma et al., 2011).

Plasmonic bowtie lasers, employing an EM hotspot supported by discrete metallic (gold) nanoparticles, permit the creation of ultrasmall mode volumes (Suh et al., 2012). But the authors' suggestion that such structures might readily be integrated with Si-based photonic structures is open to debate, as gold so readily diffuses into silicon at temperature above 100°C, and is not permitted at all in many Si-fabs. Aluminum might be a better material choice for near infrared operation and Si compatibility.

In addition to these plasmonic lasers, hybrid plasmonic-photonic crystal-coupled nanolasers have also been demonstrated (Zhang et al., 2014).

23.6.2 Spasers

Surface plasmon amplification by stimulated emission of radiation—the spaser—has excited considerable global interest in recent years since the inaugural research paper (Bergman and Stockman, 2003). Numerous papers then discussed issues such as the stimulated emission of surface plasmon polaritons (Noginov et al., 2008)—and the spaser as a nanoscale quantum generator and ultrafast amplifier (Stockman, 2010). By combining metamaterial and spaser ideas, Zheludev et al. (2008) were able to propose the lasing spaser as a coherent source of electromagnetic radiation that is fueled by plasmonic oscillations. The field was nicely reviewed at an introductory level at that time (Stockman, 2008).

Spasers are in some respects distinctly different from previous lasers such as vertical cavity surface emitting laser diodes (VCSELs) and such differences, advantages, and disadvantages have been carefully analyzed and presented by Khurgin and Sun (2014).

Of the greatest importance was the practical demonstration of an operating spaser.

Noginov et al. (2009) reported spasing action at 531 nm wavelength using 44-nm-diameter nanoparticles comprising a gold core and dye-doped silica shell to completely overcome the loss of localized surface plasmons by gain. The oscillating surface plasmon mode provides the feedback needed for stimulated emission of localized surface plasmons.

A room temperature semiconductor spaser operating near 1.5 μm wavelength was later demonstrated by Flynn et al. (2011) by sandwiching a gold-film plasmonic waveguide between optically pumped InGaAs quantum well gain media. Such a device, already capable of generating pulsed >25 mW peak power when fully developed may find applications in the telecommunications industry.

Room temperature plasmonic lasing in a continuous wave operating mode from an InGaN/GaN single nanorod was demonstrated with low threshold (Hou et al. 2014). The device was optically pumped by a laser diode and employed the nanorod on a thin SiO_2 spacer-layer on a silver film. Such a geometry allowed accurate control of the surface plasmon coupling and opportunity for its optimization.

Recently, all-color plasmonic nanolasers with ultralow thresholds have been demonstrated (Lu et al., 2014). Broadband tunable performance emitting in the full visible spectrum was achieved. These nanolasers were based on a single metal oxide semiconductor nanostructure platform comprising InGaN/GaN nanorods supported on an Al_2O_3-capped epitaxial Ag film. The possibility of "thresholdless" lasing for the blue plasmonic nanolaser was evident.

In the future we expect to see many new developments in the spaser field, and this important class of device is becoming a foundation stone of nano-optoelectronic devices and systems, alongside and in competition with VCSELs for different applications.

23.6.3 Graphene electromagnetic radiation nanosources

The role of graphene in various optoelectronic devices has been frequently reviewed in recent years by Grigorenko et al. (2012), Bonaccorso et al. (2010), Vakil and Engheta (2011), Koppens et al. (2011), Tassin et al. (2012), Bao and Loh (2012) and Long et al. (2011) and for that reason we do not embark here on yet another review.

Instead, in this section we focus on graphene as an emitter of EM radiation, noting immediately graphene's lack of a natural bandgap, the need for doping, which can be plasmon induced (Fang et al., 2012b), and the achievement of flashes of Terahertz emission (Gierz and Cavalleri, 2013), flashes of infrared (Lagatsky et al., 2013; Mary et al., 2013; Perakis, 2012), and perhaps flashes of visible radiation in the future.

The success story to date for graphene in lasers is that of use of the material (carbon nanotubes also) as saturable-absorber material in mode-locked lasers (Hasan et al., 2009; Lagatsky et al., 2013; Mary et al., 2013).

From this overview you will understand that graphene lasers, intrinsically emitting continuous wave or controlled pulses of coherent infrared or visible light, have not yet been created; perhaps something for the future, depending on bandgap creation and many other challenges being overcome, although we might surmise from Li et al. (2012b), who reported population inversion and stimulated emission at femto-second time scales that the advent of such lasers is not a long time away.

23.7 NANO-OPTO MODULATORS

Compared to nanoplasmonic optosources and nanoplasmonic optodetectors, nanoplasmonic optomodulators are perhaps underresearched to date. But this state will change as nanoplasmonic optoelectronics heads more firmly towards applications, especially in telecommunications, which depends heavily on modulator performances.

In a sense, the graphene layers used in mode-locked laser, just described, are "modulators" but in this brief section, we look at how plasmonic modulators are used to frequency-modulate light.

Plasmon–plasmon interactions can be used to control light at the nanoscale (Akimov and Chu, 2012), perhaps at THz frequencies, and compatible with CMOS, by using highly doped silicon as the active material (instead of a metal).

Graphene has a clear potential role in nanoplasmonic optomodulator devices. Electrical modulation of Fano resonance in plasmonic nanostructures using graphene has been explored with this and other applications in mind (Emani et al., 2014).

Graphene modulators have been achieved using graphene laid over a silicon waveguide (Liu et al., 2011). Modulation frequencies up to 1.2 GHz were reported, using an on-chip area of only 25 µm²; some orders of magnitude smaller than conventional modulators. Furthermore, the modulator-fabrication was compatible with CMOS chip manufacture.

Broadband modulators involving excitation of plasmonic waves by guided-mode resonances have been described by Gao et al. (2012). The resonant wavelength can be tuned over a wide wavelength range by a small change in the Fermi energy level of the graphene.

Despite the "youth" of plasmonic modulators and their development, nevertheless, the fundamental limits and near-optimal design of graphene modulators and nonreciprocal devices have already been explored (Tamagnone et al., 2014). Figures of merit have been proposed, together with absolute upper performance bounds. We confidently expect to see an array of new modulators, rotators, and isolators using graphene emerging in future years based on this work.

23.8 NANO-OPTO DETECTORS

The essence of nanoplasmonic enhanced photodetection lies in the properties of the nanoantenna (see Figure 23.3), the nano-Schottky barrier, and properties of hot electrons in the nanostructure, also plasmon resonances due to coupled nanoparticles in regular geometric arrays. The techniques we review here have been applied across much of the electromagnetic spectrum using different plasmonic materials, from the UV and visible, through all infrared regions, and out to terahertz and millimeter waves. Solar cells have been of major interest, nanoscale avalanche photodiodes (APDs) also (Hayden et al., 2006).

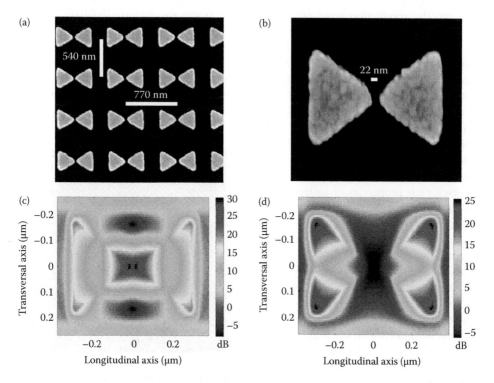

Figure 23.3 (a, b) Scanning electron microscope (SEM) image of particle antennas at two different magnifications. (c) Electric field intensity at resonance at a plane located 10 nm below the antenna calculated using finite difference time domain (FDTD) simulations for longitudinal mode. (d) Electric field intensity at resonance at a plane located 10 nm below the antenna calculated using FDTD simulations for transversal mode. The image is presented in logarithmic scale relative to the excitation field. (From Nevet, A. et al., *Nano Lett.*, 10, 1848, 2010, with permission.)

23.8.1 Metal nanoparticle/hole plasmonic photodetectors

We discussed nanoantennas above (Novotny and Hecht, 2012). Perfect absorption of light into a material via plasmon resonances becomes possible using a variety of nanoplasmonic array structures, for this is the first step toward high efficiency photodetection (Liu et al., 2010a; Aydin et al., 2011; Hedayati et al., 2011; Zhang et al., 2011c; Chen et al., 2012; Buscema et al., 2014), the business of nanostructured materials for photon detection having been reviewed by Konstantatos and Sargent (2010). Plasmons have been used to enhance absorption directly into nanopillar detectors via nanoholes (Senanayake et al., 2011).

The efficient harvesting of hot electrons is the next step in the process, and to that end a substantial amount of investigation has occurred. The excitation of plasmons in nanostructures generates short-lived highly energetic carriers that can be injected into the conduction band of a semiconductor and used to drive photodetection. The extraction of these carriers in the most efficient way is of major concern, as discussed by Chalabi and Brongersma (2013), Wang and Melosh (2011), Manjavacas et al. (2014), Knight et al. (2013), de Arquer et al. (2013), Mubeen et al. (2011), Li and Valentine (2014), and Lee et al. (2011c).

Another aspect, not always appreciated even by those designing and demonstrating nanoplasmonic photodetectors, is that of using the resonant-array excitation properties of collections of nanoparticles to enhance the detector's responsivity. The theory for such collective excitation was set out by Niklasson and Craighead (1985) and references therein, then apparently rediscovered later by Giannini et al. (2010), Brown and Stanley (2011), and Diedenhofen et al. (2011). Plasmon resonant absorption offers very considerable responsivity enhancements in such photodetectors compared to standard planar designs (Brown and Stanley, 2011).

We have already seen such developments moved toward extreme speed photodetection (Berrier et al., 2010), and resonance effects being used in early on-chip devices (Cao et al., 2010c; Pernice et al., 2012; Sun et al., 2013).

In the early days of nanoplasmonics, integrated color pixels integrated in 0.18 μm CMOS were investigated, albeit with almost no reference to plasmon operation being employed (although plasmon operation was clearly suspected; Catrysse and Wandell, 2003). Ten years later, in a different department at Stanford, a similar approach was taken, now explicitly with plasmons, and this time now for photodetection (Chalabi et al., 2014).

Photodetection with active optical antennas was described by Knight et al. (2011) using regular geometric gold nanoparticle arrays, and by Brown and Stanley (2011), who additionally employed array resonant plasmonic absorption to create high responsivity in infrared detection (notably in APD structures and infra-red detectors). Here we saw antennas, hot electrons, Schottky barriers, and resonant photodetection all coming together for new advantages in signal strength and polarization selection, etc., see Figure 23.4. Although initially focused in the infrared regions, later work focused across the visible region using aluminum as the nanoplasmonic material (Knight et al., 2012). Narrowband photodetection in the near-infrared using plasmon-based devices was also explored (Sobhani et al., 2013).

Ultrahigh responsivities—up to some two to three orders-of-magnitude increase over the usual performance—have been demonstrated by close attention to fabrication details (Tsai et al., 2011). Interest in exploitation of this style of photodetector in the telecoms regime, around 1.5 and 1.33 μm,

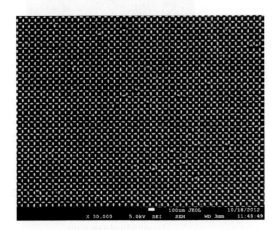

Figure 23.4 A square array of gold nanobars, ~21 nm by 47 nm in dimensions, for nanophotodetection in the infrared region. SEM picture by Dr. Yuwei Fan at California Nanosystems Institute, University of California, Los Angeles, developed from collaborative research with the chapter author, who designed and specified this array.

has been developed (Goykhman et al., 2011; Li et al., 2011), as has the application to camera technology (Lee et al., 2011c). We expect to see considerable further effort in this style of photodetector in years to come. Even "invisible" photodetectors can be fabricated in this manner! (Fan et al., 2012b).

Further application of this plasmonic detector capability has extended the frequency range considerably lower, down to the terahertz region (Shaner et al., 2007; Kim et al., 2008; Park et al., 2012), and even further down to the millimeter-wave and sub-millimeter-wave regions (Dyer et al., 2011).

23.8.2 Graphene plasmonic photodetection

As you might expect, graphene is an interesting material for photodetection applications; see, for example, Tielrooij et al. (2013).

As with the metal plasmonic structures just discussed, there are key roles for hot carriers and their efficient extraction is of paramount importance (Liu et al., 2012). Once again, periodically patterned plasmonic material, now doped-graphene, is capable of complete optical absorption; arrays of such nanodisks can exhibit 100% absorption (Thongrattanasiri et al., 2012a). Furthermore, strong plasmonic enhancement of photovoltage is observed in graphene (Echtermeyer et al., 2011), as is sought-after increased responsivity (Freitag et al., 2013).

The plasmon resonance in graphene, as above, can be electrically controlled (Kim et al., 2012), both in graphene–gold nanostructures, and in graphene–silicon heterojunctions (An et al., 2013). Graphene-enabled silver nanoantenna sensors have achieved 26× responsivity over bare silver (Reed et al., 2012), whereas hybrid graphene–quantum-dot phototransistors have exhibited gains of 10^8 electrons per photon and responsivity of 10^7 A/W (Konstantatos et al., 2012). These are extraordinary achievements, exceeding even the best expectations of metal-nanoplasmonic detectors (Brown and Stanley, 2011).

A wide-bandgap metal–semiconductor–metal nanostructure has been made entirely from graphene on SiC (Hicks et al., 2012), while all-graphene photodetectors have also been produced (Withers et al., 2013). Graphene does not behave like conventional semiconductors when exposed to light, but the mechanisms for response are only just

being clarified (Sun and Chang, 2014): photoelectric, photo-thermoelectric, and photo-bolometric. Nevertheless, graphene photodetectors with ultra-broadband and high responsivity properties at room temperature have been demonstrated (Liu et al., 2014).

Plasmon-resonant-enhanced multicolor photodetection (Liu et al., 2011) and broadband high response from pure monolayer graphene photodetection have been observed (Zhang et al., 2013b). Mid-infrared graphene detectors using antenna-enhancement have also been demonstrated (Yao, 2014). Infrared detection using graphene is an important area (Jablan et al., 2009; Mousavi et al., 2013) but perhaps even more important to the THz community is its use there because traditional detectors are relatively poor and physically large. To this end, THz graphene detectors employing plasmonics have been intensively pursued (Ju et al., 2011; Crassee et al., 2012; Gan et al., 2012) and nicely reviewed by Low and Avouris (2014).

It seems that graphene and its alternatives (Jamieson,2014) probably have a lot to offer the optoelectronics community in the years to come.

23.8.3 Solar cells using plasmonic photodetection

One of the biggest potential application areas for ultrahigh efficiency plasmonic photodetection might be in the solar cell industry if the scale-up to production quantities and at acceptable costs can be achieved. For this reason, there has been intensive effort in plasmonic solar cell design, fab, and test. Here we subdivide that progress into four brief sections for clarity: (1) general plasmonic photovoltaic issues, (2) plasmon-sensitized solar cells, (3) plasmon-enhanced silicon solar cells, and (4) plasmon-graphene solar cells.

23.8.3.1 GENERAL PLASMONIC PHOTOVOLTAIC ISSUES

We are fortunate that plasmonic solar cells have been the subject of at least three fine reviews over the years (Catchpole and Polman, 2008; Atwater and Polman, 2010; Pillai and Green, 2010). Key issues include light scattering and absorption techniques and structures, questions about how much guided modes can enhance absorption in thin solar cells (Saeta et al., 2009), light-trapping in ultrathin plasmonic solar cells (Ferry et al.,

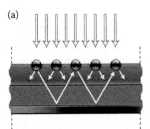 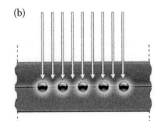 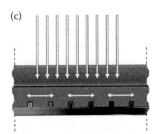

Figure 23.5 Plasmonic light-trapping geometries for thin-film solar cells. (a) Light trapping by scattering from metal nanoparticles at the surface of the solar cell. Light is preferentially scattered and trapped into the semiconductor thin film by multiple and high-angle scattering, causing an increase in the effective optical path length in the cell. (b) Light trapping by the excitation of localized surface plasmons in metal nanoparticles embedded in the semiconductor. The excited particles' near-field causes the creation of electron–hole pairs in the semiconductor. (c) Light trapping by the excitation of surface plasmon polaritons at the metal/semiconductor interface. A corrugated metal back surface couples light to surface plasmon polariton or photonic modes that propagate in the plane of the semiconductor layer. (From Atwater, H. A.; Polman, A., *Nat. Mater.*, 9, 205, 2010, with permission.)

2010), light trapping beyond the ray optic limit (Callahan et al., 2011), whether random can beat regular structure for trapping light (Battaglia et al., 2012), and the best design principles for ultrahigh efficiency (Polman and Atwater, 2012), see Figure 23.5. There are important issues about all charge carriers being derived from plasmons (Mubeen et al., 2013) and about efficient extraction of photogenerated carriers (Mubeen et al., 2014).

23.8.3.2 PLASMON-SENSITIZED SOLAR CELLS

Specific issues and approaches regarding plasmon-sensitized solar cells include plasmonic nano-structure design for efficient light coupling into solar cells (Ferry et al., 2008), how front side plasmonic nanostructures enhance solar cell efficiency (Diukman and Orenstein, 2011), and semiconductor nanowire optical antenna solar absorbers (Cao et al., 2010b). Dye sensitization of solar cells started at least as early as 1991 (O'Regan and Grätzel, 1991) and more recently has morphed into study of enhanced charge carrier generation in dye-sensitized solar cells by nanoparticle plasmons (Hägglund et al., 2008a).

Plasmons can also assist in charge separation-plasmon-induced photoelectro-chemistry, which may find application in photovoltaic cells to optimize incident photon to current conversion efficiency (Tian et al., 2005). Plasmon sensitization may also find application in "environmentally friendly" solar cells (Su et al., 2012).

23.8.3.3 PLASMON-ENHANCED SILICON SOLAR CELLS

Plasmonic crystal back reflectors within a-Si:H solar cells can create enhanced absorption of near infrared photons and so can plasmonic nanoparticles on ITO (Biswas et al., 2009). Indeed, the principle use of plasmons in silicon solar cells is to increase the absorption of solar radiation. To this end many schemes have been developed and published (Hägglund et al., 2008b; Wang et al., 2011a; Wang et al., 2012; Lee et al., 2014). The principles and practice of light super-absorption in plasmonic structures were described and discussed in some detail above.

23.8.3.4 PLASMON-GRAPHENE SOLAR CELLS

Last and by no means the least, given the intense interest in graphene in the present era, you might expect it to feature in the solar cell development story. It does. A fine example is given by Miao et al. (2012), but as plasmonic optoelectronics is not part of that development, we discuss it no further here, despite close parallels in Schottky barrier involvement to the above descriptions.

23.9 NANO-OPTICS AND NLO

Although our focus in this chapter is in nano-optoelectronics, the intimate relationship of optics with opto requires at least a brief consideration of nanoplasmonic optics. We will not cover optical metamaterials and negative index materials, etc., but we look briefly at more conventional optics and nonlinear optics (NLO) with respect to nanoplasmonics.

At the most fundamental physics level, nanoplasmonic optics is well-reviewed in the standard textbooks (Maier, 2007; Novotny and Hecht, 2012). Virtually everything we do in nano-optics relies on the plasmonic-optical properties of nanoparticles (Kelly et al., 2003), the particles' shape, size, and dielectric surround material. Beyond that, we note that it is possible to bend the trajectories of propagating plasmons in plasmon lenses (Zentgraf et al., 2011) and in chain-arrays of nanoparticles (Brongersma et al., 2000). Stockman (2004) conceived tapered plasmonic waveguides to control direction and for the nanofocussing of optical energy into giant local fields at the waveguide-tip.

In optics, we are interested in the controlled absorption, reflection, and transmission of light. Enhanced absorption using plasmonics was discussed at length above with respect to nanophotodetectors and has further been developed for ultra-broad band use (Cui et al., 2012) and as angularly insensitive protective ultrathin films against the UV region (Hedayati et al., 2014). Brown (2012a) made use of controllable plasmonic absorption and transmission to design high refractive index metamaterial structures in the visible and infrared regions; also bandpass and polarization filters. Concerning reflection, metamaterial mirrors for optoelectronic devices were explored by Esfandyarpour et al. (2014) for solar cell performance enhancement.

Holography has benefited from nanoplasmon structures. Surface-plasmon holographic beam-shaping permits control of amplitude and phase of free space beams. Surface plasmon polaritons may be coupled into free space beams, airy beams, and vortex beams (Dolev et al., 2012). Wide angle holography becomes possible using dipole nanoantenna arrays (Yifat et al., 2014). Ultrathin meta-surface holograms operating in the visible, yet being only ~30 nm thick offer many new opto-device development opportunities (Ni et al., 2013).

Color filtering and tuning is a natural domain for nanoplasmonics; experiments with Si and Ag nanoparticles have demonstrated both effects and potential applications (Cao et al., 2010a; Xu et al., 2010; Tanabe and Tasuma, 2012). The bar-array used by Xu et al. (2010) is reminiscent of the approach by Catrysse and Wandell (2003) whose interest was CMOS-compatible ultrathin color filters to cover pixels in an imager. Further related development of this aim at CMOS image sensor applications is to be found in Yokogawa et al. (2012), but now using nano-pinhole arrays to perform the color discriminations. Transmissive ultrathin nanostructured metals for plasmonic subtractive color filters were described by Zeng et al. (2013). Subtractive color processing is generally required for these approaches and because of nonlinearities, it is not the easiest precision color-processing approach. An alternative plasmonic color-filter approach enabling standard red, green, blue color processing using LED sources in high-efficiency liquid crystal display and liquid crystal on silicon displays was proposed by Brown et al. (2014).

23.9.1 NLO and nonlinear plasmonics

Kauranen and Zayats (2012) recently reviewed the basic physics of nonlinear plasmonics in a comprehensive and valuable paper. At the most fundamental level we are concerned with quantum plasmonics (Marinica et al., 2012) and nonlinear effects in the field enhancement of plasmonic particle dimers as this underpins the nonlinear plasmonics subject. Subnanometer-spaced nanoparticles are particularly useful for nonlinear plasmonics, using power levels in excess of ~10^9 W/cm^2.

The basic nonlinear optical properties of nano-metal particles have been studied for many years (Heilweil and Hochstrasser, 1985; Hache et al., 1986; Antoine et al., 1997; Ganeev et al., 2004; Hamanaka et al., 2004) and in the reported results we see increasing understanding and quantification-particularly of third-order nonlinear susceptibilities, $\chi^{(3)}$, mostly concerning gold nanoparticles, but sometimes for Ag nanoparticles also. Gehr and Boyd (1996) reviewed the optical and NLO properties of such nanoparticle materials, whereas Shalaev et al. (1996) focused more on their nonlinear optical properties in greater theoretical depth.

Local electric field enhancements and large third-order optical nonlinearity in nanocomposite materials are now well understood (Prot et al., 2002). Third-harmonic generation from single gold nanoparticles has been explored thoroughly (Lippitz et al., 2005) and third-harmonic-upconversion enhancement from a single semiconductor nanoparticle coupled to a plasmonic antenna has been achieved (Aouani et al., 2014).

Not only can third-order effects be used to create light, but they can be used to absorb light in extremely high nonlinear absorption {Im($\chi^{(3)}$)} rectangular grid geometries using 3D bowtie nanoantennas (Suh et al., 2011), with reported two-orders-of-magnitude-enhanced absorption over other metal nanoparticle–dielectric composites.

Broadband near-infrared plasmonic nanoantennas for higher-harmonic generation have been developed, using trapezoidal nanoantenna shapes derived directly from standard microwave antenna designs (Navarro-Cia and Maier, 2012). Nanoclusters Fano-resonance manipulation has been used in designing high-performance third-order nonlinear media and the enhancement of four-wave mixing (Zhang et al., 2013a). Even the long-standing challenge of effective optical limiting, to protect against high-power pulsed laser attack (Hagan, 2001), has been attacked with a nonlinear nanoplasmonic optoelectronics approach (Brown and Foote, 2014).

Although the bulk of nanoplasmonic nonlinear optics activity has been concerned with $\chi^{(3)}$, Zhang et al. (2011a) showed that 3D nanostructures can be highly efficient generators of second harmonic light, i.e., based on $\chi^{(2)}$, through using nanocups dielectric nanoparticles on which a hemispherical layer of metal is deposited.

23.10 NANO-OPTOELECTRONICS APPLICATIONS—THE FUTURE

Throughout the course of this chapter, potential applications for nanoplasmonic optoelectronics have been mentioned frequently. The range of applications is as diverse as for today's standard optoelectronics devices and systems, from data storage to medical diagnostics, from displays to telecommunications, from computing to imaging systems—i.e., in nearly every aspect of human activity, as optoelectronics surrounds and interacts with you all the time, every day—in TVs and computer displays, in DVD and blue-ray players, in fiber telecommunications and cell-phone displays, in biomedical diagnostics and research, in room and street lighting using LEDs, in solar power stations, etc. How did we ever manage life without optoelectronics?

And so it may be also with nanoplasmonic optoelectronics in the future, as all the optoelectronic functionalities, components, and systems we use today are open to miniaturization employing nano-optoelectronic components.

Here, we can look only very briefly at some of these potential applications: data storage, biological and chemical sensing, protective absorbing coatings, efficient color filters and displays, imaging systems, and ultracomputing. There will be many other applications, yet to be invented and conceived in detail.

Data storage utilizing plasmonic optoelectronics has been discussed in terms of nanoantennas for boosting optical data storage from 28.6 GB per disk to maybe 75 GB per disk at DVD/Blu-ray disc sizes (Roxworthy, 2014), but even more ambitious plans to create five-dimensional optical recording mediated by surface plasmons in gold nanorods are also underway (Zijlstra et al., 2009), hoping one day to extend optical data recording beyond 10^{12} bits (Tbit) per cm^3.

Surface plasmons have long been used for biological and chemical reaction sensing in Otto and Kretschmann configurations (e.g., Cullen et al., 1987/88), but now surface plasmon polaritons offer the potential to interface electrical and optical/optoelectronic devices. Already on-chip single plasmon detection has been demonstrated, offering exquisite sensitivity for future detection of diabetes, cholesterol, antigen-antibody binding reactions, and the like (Heeres et al., 2010).

Protective absorbing coatings for UV protection and plasmon-based absorptive coatings for photodetector enhancements have been a major focus of this chapter and are extensively mentioned in Section 23.8. Solar cells, infrared and THz detectors, and APDs all benefit significantly from nanoplasmonics. In a thicker form, such protective coatings can be the basis for optical metamaterials exhibiting broadband, very high refractive index properties for lenses, and other optical components in the visible and infrared regions (Brown, 2012a).

Efficient color filters and displays have also been much discussed throughout this chapter, especially in the section titled Nano-optics and NLO. Ultrathin high-performance visible region bandpass filters for detector arrays and displays are expected; plasmonic-based edge filters have been demonstrated already. R, G, B liquid-crystal/LED displays using such bandpass filters seem to offer significantly increased efficiency (Brown et al., 2014).

Imaging systems have frequently been mentioned in this chapter—from plasmon-enhanced detectors that might be used in focal plane arrays (Section 23.8) especially in the infrared and THz regions where they offer significant performance improvements over today's standard technologies to employment of plasmonic filters as laser-line optical limiters (Brown and Foote, 2014). It can only be a matter of a short time before plasmonic optofilters are used often in detector and imaging applications.

Telecommunications and information encryption are of great concern at present and forms the basis of modern communications and information protection. The potential subpicosecond switching/modulation speeds associated with plasmons will benefit optical telecommunications in the future. As an aside, it is interesting to note interest at present in temporal cloaking at telecommunications data rates (>12 GB/s) (Lukens et al., 2013), which can hide time events from unwanted eavesdroppers. Plasmonic optoelectronic keys for ultrasecure information encryption have already been demonstrated (Gu et al., 2012), as has three-dimensional orientation-unlimited polarization encryption by a single optically configured vectorial beam employing gold nanorod arrays (Li et al., 2012a).

Ultracomputing: The most radical vision of the nanoplasmonic future is that of ultrafast and ultrasmall computers because we already have most of the key functional components in plasmonic form; it is purely a matter of time and considerable investment to realize this vision of whole computers based on plasmonics (if competitive with as yet unknown alternative approaches), well beyond today's challenges of Moore's law with silicon-based computing. See Figure 23.6 for the positioning of ultrasmall and ultrafast nanoplasmonics with respect to previous optoelectronic and electronic platforms.

Nanoplasmonic computing draws together much, perhaps even all of the optoplasmonic discussion in this chapter; an entirely novel kind of computing system and set of principles and architectures based on optoplasmonics. Novel laser/spaser sources will find a role, especially with their femtosecond switching potential (MacDonald et al., 2009). Efficient coupling of these sources to

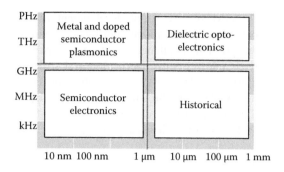

Figure 23.6 The different regions of device switching-speed and device-size depend on the materials properties of semiconductors (electronics), dielectric insulators (optoelectronics), and metals or doped semiconductors (plasmonics). The crossed vertical and horizontal solid lines indicate the operational extent of different-technologies. Plasmonics provides a link between optoelectronics and nanoelectronics. The vertical axis of the figure denotes device switching speed; the horizontal axis denotes characteristic device spatial scale. (Adapted from Brongersma, M. L.; Shalaev, V. M., *Science*, 328, 440, 2010.)

chain nanoparticle waveguides (Brongersma et al., 2000; Chang et al., 2007) will be used extensively to route bits (Maier et al., 2003) and results of logic operations. The substantial Ohmic losses (in metals) will need to be compensated by local gain media (Citrin, 2006). Plasmon interference gates will form the basis of plasmonic circuits (Heeres et al., 2013) and asynchronous propagation and gate operations will likely replace the clocked-logic operations we use in today's silicon-based logic circuitry. Much of this is set out in more detail by Brown (2012b).

We have already seen demonstration of plasmon-based logic gates (Hyde et al., 2007; Fu et al., 2012) and the early stages in development of plasmonic optical nanocircuits (Fedyanin et al., 2012; Saari et al., 2013; Dai et al., 2014). Even controlled-NOT gates operating with single photons have been demonstrated (Pooley et al., 2012) as the fundamental unit for building quantum computers.

Electrically driven optical nanocircuits have been attempted (Huang, 2014) and cascaded logic gates in nanophotonic plasmon networks have been achieved (Wei et al., 2011). Silicon-based plasmonics for on-chip photonics has been explored (Dionne et al., 2010).

The next 10–20 years of development in this optoelectronic plasmonic computing field seems to be exceptionally challenging in terms of increasingly complex circuitry fabrication and scale-up, but the potential benefits and competitiveness with CMOS (Miller, 2010; Brown, 2012b) as we look past the limitations of Moore's law look to be tantalizing if the investment levels are good enough.

REFERENCES

Akimov, Y. A and H. S. Chu, Plasmon-plasmon interaction: Controlling light at nanoscale, *Nanotechnology*, 23, 444004 (2012).

An, X., F. Liu, Y. J. Jung et al., Tunable graphene-silicon heterojunctions for ultrasensitive photodetection, *Nano Lett.*, 13, 909 (2013).

Andersen, M. L., S. Stobbe, A. S. Sørensen et al., Strongly modified plasmon-matter interaction with mesoscopic quantum emitters, *Nat. Phys.*, 7, 215 (2011).

Antoine, R., P. F. Brevet, H. H. Girault et al., Surface plasmon enhanced non-linear optical response of gold nanoparticles at the air/toluene interface, *Chem. Comm.*, 19, 1901 (1997). doi:10.1039/A704846G.

Aouani, H., M. Rahmani, M. Navarro-Cia et al., Third-harmonic-upconversion enhancement from a single semiconductor nanoparticle coupled to a plasmonic antenna, *Nat. Nanotech.*, 9, 290 (2014). doi:10.1038/nnano.2014.27.

Aslam, M. I. and D. Ö Güney, Surface plasmon driven scalable low-loss negative-index metamaterial in the visible spectrum, *Phys. Rev. B*, 84, 195465 (2011).

Atwater, H. A. and A. Polman, Plasmonics for improved photo-voltaic devices, *Nat. Mater.*, 9, 205 (2010).

Aydin, K., V. E. Ferry, R. M. Briggs et al., Broadband polarization independent resonant light absorption using ultrathin plasmonic superabsorbers, *Nat. Comm.*, 2, 517 (2011). doi:10.1038/ncomms1528.

Bao, Q. and K. P. Loh, Graphene photonics, plasmonics, and broadband optoelectronic devices, *ACS Nano*, 6, 3677 (2012).

Battaglia, C., C.-M. Hsu, K. Söderström et al., Light trapping in solar cells: Can periodic beat random?, *ACS Nano*, 6, 2790 (2012). doi:10.1021/nn300287j.

Belacel, C., B. Habert, F. Bigourdan et al., Controlling spontaneous emission with plasmonic optical patch antennas, *Nano Lett.*, 13, 1516 (2013). doi:10.1021/nl3046602.

Bergmair, I., B. Dastmalchi, M. Bergmair et al., Single and multilayer metamaterials fabricated by nanoimprint lithography, *Nanotechnology*, 22, 325301 (2011).

Bergman, D. J. and M. I. Stockman, Surface plasmon amplification by stimulated emission of radiation: Quantum generation of coherent surface plasmons in nanosystems, *Phys. Rev. Lett.*, 90, 027402 (2003).

Berini, P., Long-range surface plasmon polaritons, *Adv. Opt. Photon.* 1, 484 (2009).

Berrier, A., R. Ulbricht, M. Bonn et al., Ultrafast active control of localized surface plasmon resonances in silicon bowtie antennas, *Opt. Express*, 18, 23226 (2010).

Bhattacharya, P., B. Xiao, A. Das et al., Solid state electrically injected exciton-polariton laser, *Phys. Rev. Lett.*, 110, 206403 (2013).

Biswas, R., D. Zhou, B. Curtin et al., Surface plasmon enhancement of optical absorption of thin film a-Si:H solar cells, IEEE Photovoltaics Specialists Conference (PVSC), 7–12 June 2009, Philadelphia, PA, p. 000557 (2009).

Blaber, M. G., C. J. Engel, S. R. C. Vivekchand et al., Eutectic liquid alloys for plasmonics, *Nano Lett.*, 12, 5275 (2012).

Boardman, A. D., V. V. Grimalsky, Y. S. Kivshar et al., Active and tunable metamaterials, *Laser Photon. Rev.*, 5, 287 (2010). doi:10.1002/lpor.201000012.

Bohren, C. F. and D. R. Huffman, *Absorption and Scattering of Light by Small Particles*, John Wiley & Sons, New York (1983).

Boltasseva, A. and H. A. Atwater, Low-loss plasmonic metamaterials, *Science*, 331, 290 (2011).

Bonaccorso, F., Z. Sun, T. Hassan et al., Graphene photonics and optoelectronics, *Nat. Photon.*, 4, 611 (2010).

Brongersma, M. L., J. W. Hartman, and H. A. Atwater, Electromagnetic energy transfer and switching in nanoparticle chain arrays below the diffraction limit, *Phys. Rev. B*, 62, R16356 (2000).

Brongersma, M. L. and P. G. Kik (Eds), *Surface Plasmon Nanophotonics*, Springer, Dordrecht (2010).

Brongersma, M. L. and V. M. Shalaev, The case for plasmonics, *Science*, 328, 440 (2010).

Brown, R. G. W., High refractive index, polarization insensitive nano-rod-based plasmonic metamaterials for lenses, US Patent 8,681,428 (2012a).

Brown, R. G. W., Localized plasmon-polariton (LPP) logic gates and computing, US Patent 8,633,729 (2012b).

Brown, R. G. W. and B. D. Foote, Plasmonic optical limiters: Ultra-fast, ultra-non-linear plasmonic film EO-system protection against giant laser pulses, Rockwell Collins Inc., IP Disclosure 14-TU-00519 (2014).

Brown, R. G. W., J. E. Melzer and B. D. Foote, Nanoplasmonic reflective filter and display employing same, Rockwell Collins Inc., IP Disclosure 14-TU-00404; US Patent 9,65,817 (2014).

Brown, R. G. W. and J. H. Stanley, Nano-structure arrays for EMR imaging, US Patent 8,492,727 (2011).

Buscema, M., D. J. Groenendijk, S. I. Blanter et al., Fast and broadband photoresponse of few-layer black phosphorous field-effect transistors, *Nano Lett.*, 14, 3347 (2014). doi:10.1021/nl5008085.

Callahan, D. M., J. N. Munday and H. A. Atwater, Solar cell light trapping beyond the ray optic limit, *Nano Lett.*, 12, 214 (2011).

Cao, L., P. Fan, E. S. Barnard et al., Tuning the color of silicon nanostructures, *Nano Lett.*, 10, 2649 (2010a).

Cao, L., P. Fan, A. P. Vasudev et al., Semiconductor nanowire optical antenna solar absorbers, *Nano Lett.*, 10, 439 (2010b).

Cao, L., J.-S. Park, P. Fan et al., Resonant germanium nanoantenna photodetectors, *Nano Lett.*, 10, 1229 (2010c).

Catchpole, K. R. and A. Polman, Plasmonic solar cells, *Opt. Express*, 16, 21793 (2008).

Catrysse, P. B. and B. A. Wandell, Integrated color pixels in 0.18-μm complementary metal oxide semiconductor technology, *J. Opt. Soc. Am. A* 20, 2293 (2003).

Chalabi, H. and M. L. Brongersma, Harvest season for hot electrons, *Nat. Nanotech.*, 8, 229 (2013).

Chalabi, H., D. Schoen and M. L. Brongersma, Hot-electron photodetection with a plasmonic nanostripe antenna, *Nano Lett.*, 14, 1374 (2014). doi:10.1021/nl4044373.

Chanda, D., K. Shigeta, S. Gupta et al., Large-area flexible 3D optical negative index metamaterial formed by nanotransfer printing, *Nat. Nanotech.*, 6, 402 (2011).

Chang, D. E., A. S. Sørensen, P. R. Hemmer et al., Quantum optics with surface plasmons, *Phys. Rev. Lett.*, 97, 053002 (2006).

Chang, D. E., A. S. Sørensen, P. R. Hemmer et al., Strong coupling of single emitters to surface plasmons, *Phys. Rev. B*, 76, 035420 (2007).

Chen, K., R. Adato and H. Altug, Dual-band perfect absorber for multispectral plasmon-enhanced infrared spectroscopy, *ACS Nano*, 6, 7998 (2012).

Choi, M., S. H. Lee, Y. Kim et al., A terahertz material with unnaturally high refractive index, *Nature*, 470, 369 (2011).

Choi, D., C. K Shin, D. Yoon et al., Plasmonic optical interference, *Nano Lett.*, 14, 3374 (2014).

Christensen, J., A. Manjavacas, S. Thongrattanasiri et al., Graphene plasmon waveguiding and hybridization in individual and paired nanoribbons, *ACS Nano*, 6, 431 (2011). doi:10.1021/nn2037626.

Citrin, D. S., Plasmon-polariton transport in metal-nanoparticle chains embedded in a gain medium, *Opt. Lett.*, 31, 98 (2006).

Coppage, R., J. M. Slocik, B. D. Briggs et al., Determining peptide sequence effects that control the size, structure, and function of nanoparticles, *ACS Nano*, 6, 1625 (2012). doi:10.1021/nn204600d.

Crassee, I., M. Orlita, M. Potemski et al., Intrinsic terahertz plasmons and magnetoplasmons in large scale monolayer graphene, *Nano Lett.*, 12, 2470 (2012).

Cui, Y., K. H. Fung, J., Xu et al., Ultrabroadband light absorption by a sawtooth anisotropic metamaterial slab, *Nano Lett.*, 12, 1443 (2012).

Cullen, D. C., R. G. W. Brown and C. R. Lowe, Detection of immuno-complex formation via surface plasmon resonance on gold-coated diffraction gratings, *Biosensors*, 3, 211 (1987/88).

Dai, W.-H., F.-C. Lin, C.-B. Huang et al., Mode-conversion in high definition plasmonic optical nanocircuits, *Nano Lett.*, 14, 3881 (2014). doi:10.1021/nl501102n.

de Arquer, P. F. G., A. Mihi, D. Kufer et al., Photoelectric energy conversion of plasmon-generated hot carriers in metal-insulator-semiconductor structures, *ACS Nano*, 7, 3581 (2013). doi:10.1021/nn400517w.

Diedenhofen, S. L., O. T. A. Janssen, G. Grzela et al., Strong geometrical dependence of the absorption of light in arrays of semiconductor nanowires, *ACS Nano*, 5, 2316 (2011). doi:10.1021/nn103596n.

Ding, K. and C. Z. Ning, Metallic subwavelength-cavity semiconductor nanolasers, *Light: Sci. Appl.*, 1, e20 (2012). doi:10.1038/lsa.2012.20. See also K. Ding et al., *Opt. Express*, 21, 4728 (2013).

Dionne, J. A., L. A. Sweatlock, M. T. Sheldon et al., Silicon-based plasmonics for on-chip photonics, *IEEE Sel. Top. Quantum Electron.*, 16, 295 (2010).

Diukman, I., and M. Orenstein, How front side plasmonic nanostructures enhance solar cell efficiency, *Solar Energy Mater. Solar Cells*, 95, 2628 (2011).

Dolev, I., I. Epstein and A. Arie, Surface-plasmon holographic beam shaping, *Phys. Rev. Lett.*, 109, 203903 (2012).

Duan, H., A. I. Fernández-Dominguez, M. Bosman et al., Nanoplasmonics: Classical down to the nanometer scale, *Nano Lett.*, 12, 1683 (2012). doi:10.1021/nl3001309.

Dyer, G. C., G. R. Aizin, J. L. Reno et al., Novel tunable millimeter-wave grating-gated plasmonic detectors, *IEEE J. Sel. Top. Quant. Electron.*, 17, 85 (2011).

Ebbesen, T. W., H. J. Lezec, H. F. Ghaemi et al., Extraordinary optical transmission through sub-wavelength hole arrays, *Nature*, 391, 667 (1998).

Echtermeyer, T. J., L. Britnell, P. K. Jasnos et al., Strong plasmonic enhancement of photovoltage in graphene, *Nature Comm.*, (2011). doi:10.1038/ncomms1464.

Emani, N. K., T.-F. Chung, A. V. Kildishev et al., Electrical modulation of Fano resonance in plasmonic nanostructures using graphene, *Nano Lett.*, 14, 78 (2014).

Emani, N. K., T.-F. Chung, X. Ni et al., Electrically tunable damping of plasmonic resonances with graphene, *Nano Lett.*, 12, 5202 (2012).

Esfandyarpour, M., E. C. Garnett, Y. Cui et al., Metamaterial mirrors in optoelectronic devices, *Nat. Nanotech.*, 9, 542 (2014).

Fan, J. A., K. Bao, L. Sun et al., Plasmonic mode engineering with templated self-assembled nanoclusters, *Nano Lett.*, 12, 5318 (2012a).

Fan, P., U. K. Chettiar, L. Cao et al., An invisible metal-semiconductor photodetector, *Nat. Photon.*, 6, 380 (2012b).

Fang, Z., Z. Liu, Y. Wang et al., Graphene-antenna sandwich photodetector, *Nano Lett.*, 12, 3808 (2012a).

Fang, Z., Y. Wang, Z. Liu et al., Plasmon-induced doping of graphene, *ACS Nano*, 6, 10222 (2012b). doi:10.1021/nn304028b.

Fedyanin, D. Y., A. V. Krasavin, A. V. Arsenin et al., Surface plasmon polariton amplification upon electrical injection in highly integrated plasmonic circuits, *Nano Lett.*, 12, 2459 (2012).

Ferry, V. E., L. A. Sweatlock, D. Pacifici et al., Plasmonic nanostructure design for efficient light coupling into solar cells, *Nano Lett.*, 8, 4391 (2008).

Ferry, V. E., M. A Verschuuren, H. B. T. Li et al., Light trapping in ultrathin plasmonic solar cells, *Opt. Express*, 18, A237 (2010).

Flory, F., L. Escoubas and G. Berginc, Optical properties of nanostructured materials: A review, *J. Nanophoton.*, 5, 052502 (2011).

Flynn, R. A., C. S. Kim, I. Vurgaftman et al., A room-temperature semiconductor spaser operating near 1.5μm, *Opt. Express*, 19, 8954 (2011).

Forestiere, C., A. J. Pasquale, A. Capretti et al., Genetically engineered plasmonic nanoarrays, *Nano Lett.*, 12, 2037 (2012). doi:10.1021/nl300140g.

Fourkas, J. T., Rapid lithography: New photoresists achieve nanoscale resolution, *Opt. Photon. News*, 22, June issue, 24 (2011).

Freitag, M., T. Low and P. Avouris, Increased responsivity of suspended graphene photodetectors, *Nano Lett.*, 13, 1644 (2013). doi:10.1021/nl4001037.

Fu, Y., X. Hu, C. Lu et al., All-optical logic gates based on nanoscale plasmonic slot waveguides, *Nano Lett.*, 12, 5784 (2012).

Galván-Moya, J. E., T. Atlantzis, K. Nelissen et al., Self-organization of highly symmetric nanoassemblies: A matter of competition, *ACS Nano*, 8, 3869 (2014). doi:10.1021/nn500715d.

Gan, C. H., H. S. Chu and E. P. Li, Synthesis of highly confined surface plasmon modes with doped graphene sheets in the midinfrared and terahertz frequencies, *Phys. Rev. B*, 85, 125431 (2012).

Ganeev, R. A., M. Baba, A. I. Ryasnyansky et al., Characterization of optical and nonlinear optical properties of silver nanoparticles prepared by laser ablation in various liquids, *Opt. Comm.*, 240, 437 (2004).

Gao, Y., Q. Gan, Z. Xin et al., Plasmonic Mach-Zender interferometer for ultrasensitive on-chip biosensing, *ACS Nano*, 5, 9836 (2011). doi:10.1021/nn2034204.

Gao, W., J. Shu, C. Qiu et al., Excitation of plasmonic waves in graphene by guided-mode resonances, *ACS Nano*, 6, 7806 (2012). doi:10.1021/nn301888e.

Garcia-Vidal, F. J., L. Martin-Moreno and J. B. Pendry, Surfaces with holes in them: New plasmonic metamaterials, *J. Opt. A: Pure Appl. Opt.*, 7, S97 (2005).

Gehr, R. J. and R. W. Boyd, Optical properties of nanostructured optical materials, *Chem. Mater.*, 8, 1807 (1996).

Giannini, V., G. Vecchi and J. Gómez Rivas, Lighting up multipolar surface plasmon polaritons by collective resonances in arrays of nanoantennas, *Phys. Rev. Lett.*, 105, 266801 (2010).

Gierz, I. and A. Cavalleri, Graphene can emit laser flashes, www.mpg.de/7583916/graphene-terahertz-laser (2013).

Gjonaj, B., J. Aulbach, P. M. Johnson et al., Active spatial control of plasmonic fields, *Nat. Photon.*, 5, 360 (2011).

Goncher, S. J., L. Zhao, A. N. Pasupathy et al., Substrate level control of the local doping in graphene, *Nano Lett.*, 13, 1386 (2013).

Gonzalez, E., J. Arbiol and V. F. Puntes, Carving at the nanoscale: Sequential galvanic exchange and Kirkendall growth at room temperature, *Science*, 334, 1377 (2011).

Goykhman, I., B. Desiatov, J. Khurgin et al., Locally oxidized silicon surface-plasmon Schottky detector for telecom regime, *Nano Lett.*, 11, 2219 (2011). doi:10.1021/nl200187v.

Grigorenko, A. N., M. Polini and K. S. Novoselov, Graphene plasmonics, *Nat. Phot.*, 6, 749 (2012).

Grzelczak, M., J. Vermant, E. M. Furst et al., Directed self-assembly of nanoparticles, *ACS Nano*, 7, 3591 (2010).

Gu, M., X. Li, T.-H. Lan et al., Plasmonic keys for ultra-secure information encryption, *SPIE* (2012). doi:10.1117/2.1201211.004538.

Hache, F., D. Ricard and C. Flytzanis, Optical non-linearities of small metal particles: Surface-mediated resonance and quantum size effects, *J. Opt. Soc. B*, 3, 1647 (1986).

Hagan, D. J., Optical Limiting, in the *Handbook of Optics*, Vol. IV, M. Bass (Ed.), McGraw-Hill (2001).

Hägglund, C., M. Zäch and B. Kasemo, Enhanced charge carrier generation in dye sensitized solar cells by nanoparticle plasmons, *Appl. Phys. Lett.*, 92, 013113 (2008).

Hägglund, C., M. Zäch, G. Petersson et al., Electromagnetic coupling of light into a silicon solar cell by nanodisk plasmons, *Appl. Phys. Lett.*, 92, 053110 (2008).

Halas, N. J., S. Lal, W.-S. Chang et al., Plasmons in strongly coupled metallic nanostructures, *Chem. Rev.*, 111, 3913 (2011).

Hamanaka, Y., K. Fukuta, A. Nakamura et al., Enhancement of third-order nonlinear optical susceptibilities in silica-capped Au nanoparticle films with very high concentrations, *Appl. Phys. Lett.*, 84, 4938 (2004).

Hamon, C., M. Postic, E. Mazari et al., Three-dimensional self-assembling of gold nanorods with controlled macroscopic shape and local smectic B order, *ACS Nano*, 6, 4137 (2012).

Hasan, T., Z. Sun, F. Wang et al., Nanotube-polymer composites for ultrafast photonics, *Adv. Mater.*, 21, 3874 (2009).

Hayden, O., R. Agarwal and C. M Lieber, Nanoscale avalanche photodiodes for highly sensitive and spatially resolved photodetection, *Nat. Mater.*, 5, 352 (2006).

Hedayati, M. K., M. Javaherirahim, B. Mozooni et al., Design of a perfect black absorber at visible frequencies using plasmonic metamaterials, *Adv. Mater.*, 23, 5410 (2011).

Hedayati, M. K., A. U. Zillohu, T. Strunskus et al., Plasmonic tunable metamaterial absorber as ultraviolet protection film, *Appl. Phys. Lett.*, 104, 041103 (2014).

Heeres, R. W., S. N. Dorenbos, B. Koene et al., On-chip single plasmon detection, *Nano Lett.*, 10, 661 (2010).

Heeres, R. W., L. P. Kouwenhoven and V. Zwiller, Quantum interference in plasmon circuits, *Nat. Nanotech.*, 8, 719 (2013).

Heilweil, E. J. and R. M. Hochstrasser, Nonlinear spectroscopy and picosecond transient grating study of colloidal gold, *J. Chem. Phys.*, 82, 4762 (1985).

Henzie, J., M. H. Lee and T. W. Odom, Multiscale patterning of plasmonic metamaterials, *Nat. Nanotech.*, 2, 549 (2007).

Hess, O., J. B. Pendry, S. A. Maier et al., Active nanoplasmonic metamaterials, *Nat. Mater.*, 11, 573 (2012).

Hicks, J., A. Tejeda, A. Taleb-Ibrahimi et al., A wide-bandgap metal-semiconductor-metal nanostructure made entirely from graphene, *Nat. Phys.*, 9, 49 (2012). doi:10.1039/nphys2487.

Hill, M. T., M. Marell, E. S. P. Leong et al., Lasing in metal-insulator-metal sub-wavelength plasmonic waveguides, *Opt. Express*, 17, 11107 (2009).

Hou, Y., P. Renwick, B. Liu et al., Room temperature plasmonic lasing in a continuous wave operation mode from an InGaN/GaN single nanorod with a low threshold, *Sci. Rep.*, 4, 5014 (2014). doi:10.1038/srep05014.

Huang, K. C. Y, M.-K. Seo, T. Sarmiento et al., Electrically driven subwavelength optical nanocircuits, *Nat. Photon.*, 8, 244 (2014). doi:10.1038/nphoton.2014.2.

Hur, K., Y. Francescato, V. Giannini et al., Three-dimensionally isotropic negative refractive index materials from block copolymer self-assembled chiral gyroid networks, *Angew. Chem.*, 123, 12191 (2011).

Husu, H., R. Siikanen, J. Mäkitalo et al., Metamaterials with tailored nonlinear optical response, *Nano Lett.*, 12, 673 (2012). doi:10.1021/nl203524k.

Hyde, R. A., E. K. Y. Jung, N. P. Myhrvold et al., *Plasmon gate*, US Patent Application, US 2007/0292076 (2007).

Jablan, M., H. Buljan and M. Soljačic, Plasmonics in graphene at infrared frequencies, *Phys. Rev. B*, 80, 245435 (2009).

Jamieson, F., Graphene: Overcoming the hype, *Trans. Mater. Res.*, 1, 2, 020204 (2014).

Ju, L., B. Geng, J. Horng et al., Graphene plasmonics for tunable terahertz metamaterials, *Nat. Nanotech.*, 6, 630 (2011).

Kauranen, M. and A. V. Zayats, Nonlinear plasmonics, *Nat. Photon.*, 6, 737 (2012).

Kelly, K. L., E. Coronado, L. L. Zhao et al., The optical properties of metal nanoparticles: The influence of size, shape, and dielectric environment, *J. Phys. Chem. B*, 107, 668 (2003).

Kéna-Cohen, S., P. N. Stavrinou, D. D. C. Bradley et al., Confined surface plasmon-polariton amplifiers, *Nano Lett.*, 13, 1323 (2013).

Kershner, R. J., L. D. Bozano, C. M. Micheel et al., Placement and orientation of individual DNA shapes on lithographically patterned surfaces, *Nat. Nanotech.*, 4, 557 (2009).

Khurgin, J. B., How to deal with the loss in plasmonics and metamaterials, *Nat. Nanotech.*, 10, 2 (2015).

Khurgin, J. B. and G. Sun, Comparative analysis of spasers, vertical-cavity surface-emitting lasers and surface-plasmon emitting diodes, *Nat. Photon.*, 8, 468 (2014). doi:10.1038/nphoton.2014.94.

Kildishev, A. V. and V. M. Shalaev, Transformation optics and metamaterials, *Phys.-Uspekhi*, 54, 53 (2011).

Kim, J., H. Son, D. J. Cho et al., Electrical control of optical plasmon resonance with graphene, *Nano Lett.*, 12, 5598 (2012). doi:10.1021/nl302656d.

Kim, S., J. D. Zimmerman, P. Focardi et al., Room temperature terahertz detection based on bulk plasmons in antenna-coupled GaAs field effect transistors, *Appl. Phys. Lett.*, 92, 253508 (2008).

Knight, M. W., N. S. King, L. Liu et al., Aluminum for plasmonics, *ACS Nano*, 8, 834 (2014).

Knight, M. W., L. Liu, Y. Wang et al., Aluminum plasmonic nanoantennas, *Nano Lett.*, 12, 6000 (2012). doi:10.1021/nl303517v.

Knight, M. W., H. Sobhani, P. Nordlander et al., Photodetection with active optical antennas, *Science*, 332, 702 (2011).

Knight, M. W., Y. Wang, A. Urban et al., Embedding plasmonic nanostructure diodes enhances hot electron emission, *Nano Lett.*, 13, 1687 (2013). doi:10.1021/nl400196z.

Kolesov, R., B. Grotz, G. Balasubramanian et al., Wave-particle duality of single surface plasmon polaritons, *Nat. Phys.*, 5, 470 (2009).

Kollman, H., X. Piao, M. Esmann et al., Towards plasmonics with nanometer precision: Nonlinear optics of Helium-ion milled gold nanoantennas, *Nano Lett.*, 14, 4778 (2014). doi:10.1021/nl5019589.

Konstantatos, G., M. Badioli, L. Gaudreau et al., Hybrid graphene-quantum dot phototransistors with ultrahigh gain, *Nat. Nanotech.*, 7, 363 (2012). doi:10.1038/nnano.2012.60.

Konstantatos, G. and E. H. Sargent, Nanostructured materials for photon detection, *Nat. Nanotech.*, 5, 391 (2010).

Koppens, F. H., D. E. Chang, F. J. Garcia de Abajo, Graphene plasmonics: A platform for strong light-matter interactions, *Nano Lett.*, 11, 3370 (2011).

Lagatsky, A. A., Z. Sun, T. S. Kulmala et al., 2 μm solid-state laser mode-locked by single-layer graphene, *Appl. Phys. Lett.*, 102, 013113 (2013).

Lee, S. H., B. Cho, S. Yoon et al., Printing of sub-100-nm metal nanodot arrays by carbon nanopost stamps, *ACS Nano*, 5, 5543 (2011a).

Lee, Y. K., C. H. Jung, J. Park et al., Surface plasmon-driven hot electron flow probed with metal-semiconductor nanodiodes, *Nano Lett.*, 11, 4251 (2011b).

Lee, S. J., Z. Ku, A. Barve et al., A monolithically integrated plasmonic infrared quantum dot camera, *Nat. Comm.*, 2, 286 (2011c). doi:10.1038/ncomms1283.

Lee, D. H., J. Y. Kwon, S. Maldonado et al., Extreme light absorption by multiple plasmonic layers on upgraded metallurgical grade silicon solar cells, *Nano Lett.*, 14, 1961 (2014). doi:10.1021/nl4048064.

Li. S. Q., P. Guo, L. Zhang et al., Infrared plasmonics with indium-tin-oxide nanorod arrays, *ACS Nano*, 5, 9161 (2011).

Li, X., T.-H. Lan, C.-H. Tien et al., Three-dimensional orientation-unlimited polarization encryption by a single optically configured vectorial beam, *Nat. Comm.*, 3, 998 (2012a). doi: 10.1038/ncomms2006.

Li, T., L. Luo, M. Hupalo et al., Femtosecond population inversion and stimulated emission of dense Dirac fermions in graphene, *Phys. Rev. Lett.*, 108, 167401 (2012b). See also, Perakis, I. E., *Physics*, 5, 43 (2012) for explanatory remarks.

Li, W. and J. G. Valentine, Metamaterial perfect absorber based hot electron photodetection, *Nano Lett.*, 14, 3510 (2014).

Lioubimov, V., A. Kolomenskii, A. Mershin et al., Effect of varying electric potential on surface-plasmon resonance sensing, *Appl. Opt.*, 43, 3426 (2004).

Lippitz, M., M. A. van Dijk, and M. Orrit, Third-harmonic generation from single gold nanoparticles, *Nano Lett.*, 5, 799 (2005).

Liu, C.-H., Y.-C. Chang, T. B. Norris et al., Graphene photodetectors with ultra-broadband and high responsivity at room-temperature, *Nat. Nanotech.*, 9, 273 (2014). doi:10.1038/nnano.2014.31.

Liu, Y., R. Cheng, H. Zhou et al., Plasmon resonance enhanced multicolor photodetection by graphene, *Nat. Comm.*, 2, 579 (2011). doi:10.1038/ncomms1589.

Liu, C.-H., N. M. Dissanayake, S. Lee et al., Evidence for extraction of photoexcited hot carriers from graphene, *ACS Nano*, 6, 7172 (2012). doi:10.1021/nn302227r.

Liu, N., M. Mesch, T. Weiss et al., Infrared perfect absorber and its application as plasmonic sensor, *Nano Lett.*, 10, 2342 (2010a).

Liu. M., X. Yin, E. Ulin-Avila et al., A graphene-based broadband optical modulator, *Nature*, 474, 64 (2010b).

Long, J., B. Jeng, J. Horng et al., Graphene plasmonics for tunable terahertz metamaterials, *Nat. Nanotech.*, 6, 630–634 (2011). doi:10.1038/nnano.2011.146.

Low, T. and P. Avouris, Graphene plasmonics for terahertz to mid-infrared applications, *ACS Nano*, 8, 1086 (2014).

Lu, Y.-J., C.-Y. Wang, J. Kim et al., All-color plasmonic nanolasers with ultralow thresholds: Auto-tuning mechanism for single mode lasing, *Nano Lett.*, 14, 4381 (2014). doi:10.1021/nl501273u.

Lukens, J. M., D. E. Leaird and A. M. Weiner, A temporal cloak at telecommunication data rate, *Nature*, 498, 205 (2013). doi:10.1038/nature12224.

Luk'yanchuk, B., N. I. Zheludev, S. A Maier et al., The Fano resonance in plasmonic nanostructures and metamaterials, *Nat. Mater.*, 9, 707 (2010).

Ma, R.-M., R. F. Oulton, V. J. Sorger et al., Room-temperature sub-diffraction-limited plasmon laser by total internal reflection, *Nat. Mater.*, 10, 110 (2011).

MacDonald, K. F., Z. L. Sámson, M. I. Stockman et al., Ultrafast active plasmonics, *Nat. Photon.*, 3, 55 (2009).

Madou, M. J., *Fundamentals of Microfabrication and Nanotechnology*, Third Edition, Three-Volume set, CRC Press, Boca Raton, FL (2011).

Maier, S. A., *Plasmonics: Fundamentals and Applications*, Springer, New York (2007).

Maier, S. A. and H. A. Atwater, Plasmonics: Localization and guiding of electromagnetic energy in metal/dielectric structures, *J. Appl. Phys.*, 98, 011101 (2005).

Maier, S. A., P. G. Kik and H. A. Atwater, Optical pulse propagation in metal nanoparticle chain waveguides, *Phys. Rev. B*, 67, 205402 (2003).

Manjavacas, A., F. J. Garcia de Abajo and P. Nordlander, Quantum plexitonics: Strongly interacting plasmons and excitons, *Nano Lett.*, 11, 2318 (2011).

Manjavacas, A., J. G. Liu, V. Kulkarni et al., Plasmon-induced hot carriers in metallic nanoparticles, *ACS Nano*, 8, 7630 (2014). doi:10.1021/nn502445f.

Marinica, D. C., A. K. Kazansky, P. Nordlander et al., Quantum plasmonics: Nonlinear effects in the field enhancement of a plasmonic nanoparticle dimer, *Nano Lett.*, 12, 1333 (2012). doi:10.1021/nl300269c.

Mary, R., G. Brown, S. J. Beecher et al., 1.5 GHz picosecond pulse generation from a monolithic waveguide laser with a graphene-film saturable absorber, *Opt. Express*, 21, 7943 (2013).

Miao, X., S. Tongay, M. K. Petterson et al., High efficiency graphene solar cells by chemical doping, *Nano Lett.*, 12, 2745 (2012).

Miller, D. A. B., Are optical transistors the logical next step?, *Nat. Photon.*, 4, 3 (2010).

Mousavi, S. H., I. N. Kholmanov, K. B. Alici et al., Inductive tuning of Fano-resonant meta-surfaces using plasmonic resonance of graphene in mid-infrared, *Nano Lett.*, 13, 1111 (2013). doi:10.1021/nl304476b.

Mubeen, S., G. Hernandez-Sosa, D. Moses et al., Plasmonic photosensitization of a wide band-gap semiconductor: Converting plasmons to charge carriers, *Nano Lett.*, 11, 5548 (2011).

Mubeen, S., J. Lee, W.-r. Lee et al., On the plasmonic photovoltaic, *ACS Nano*, 8, 6066 (2014).

Mubeen, S., J. Lee, N. Singh et al., An autonomous photosynthetic device in which all charge carriers derive from surface plasmons, *Nat. Nanotech.*, 8, 247 (2013).

Navarro-Cia, M. and S. A. Maier, Broad-band near-infrared plasmonic nanoantennas for higher harmonic generation, *ACS Nano*, 6, 3537 (2012). doi:10.1021/nn300565x.

Nedev, S., A. S. Urban, A. A. Lutich et al., Optical force stamping lithography, *Nano Lett.*, 11, 5066 (2011). doi:10.1021/nl203214n.

Nevet, A., N. Berkovitch, A. Hayat et al., Plasmonic nanoantennas for broad-band enhancement of two-photon emission from semiconductors, *Nano Lett.*, 10, 1848 (2010).

Ng, K. C., I. B. Udagedara, I. D. Rukhlenko et al., Free-standing plasmonic-nanorod super-lattice sheets, *ACS Nano*, 6, 925 (2011). doi:10.1021/nn204498j.

Ni, X., A. V Kildishev and V. M. Shalaev, Metasurface holograms for visible light, *Nat. Comm.* 4 (2013). doi:10.1038/ncomms3807.

Nickel, A., R. Ohmann, J. Meyer et al., Moving nanostructures: Pulse induced positioning of supramolecular assemblies, *ACS Nano*, 7, 191 (2012). doi:10.1021/nn303708h.

Niklasson, G. A. and H. G. Craighead, Optical response and fabrication of regular arrays of ultrasmall gold particles, *Thin Solid Films*, 125, 165 (1985).

Niu, X., S. P. Stagon, H. Huang et al., Smallest metallic nanorods using physical vapor deposition, *Phys. Rev. Lett.*, 110, 136102 (2013).

Noginova, N., A. V. Yakim, J. Soimo et al., Light-to-current and current-to-light coupling in plasmonic systems, *Phys. Rev. B*, 84, 035447 (2011).

Noginov, M. A., G. Zhu, A. M. Belgrave et al., Demonstration of spaser-baser nanolaser, *Nature*, 460, 1110 (2009).

Noginov, M. A., G. Zhu, M. Mayy et al., Stimulated emission of surface plasmon polaritons, *Phys. Rev. Lett.*, 101, 226806 (2008).

Noh, H., J.-K. Yang, S. F. Liew et al., Control of lasing in biomimetic structures with short-range order, *Phys. Rev. Lett.*, 106, 183901 (2011).

Novotny, L. and B. Hecht, *Principles of Nano-Optics*, Second Edition, Cambridge University Press, Cambridge (2012).

Ordal, M. A. L. L. Long, R. J. Bell et al., Optical properties of the metals Al, Co, Cu, Au, Fe, Pb, Ni, Pd, Pt, Ag, Ti, and W in the infrared and far infrared, *Appl. Opt.*, 22, 1099 (1983).

O'Regan B. and M. Grätzel, A low cost, high efficiency solar cell based on dye-sensitized colloidal TiO_2 films, *Nature*, 353, 737 (1991).

Oulton, R. F., V. J. Sorger, T. Zentgraf et al., Plasmon lasers at deep subwavelength scale, *Nature*, 461, 629 (2009).

Palik, E. D., *Handbook of Optical Constants of Solids*, Academic Press, Orlando, FL (1985).

Papenheim, M., A. Mayer, S. Wang et al., Flat and highly flexible composite stamps for nanoimprint, their preparation and their limits. *J. Vac. Sci. Technol. B*, 34, 06K406 (2016).

Park, I.-Y., S. Kim, J. Choi et al., Plasmonic generation of ultrashort extreme-ultra-violet light pulses, *Nat. Photon.*, 5, 677 (2011). doi:10.1038/nphoton.2011.258.

Park, K. Y., C. S. Meierbachtol, N. Wiwatcharagoses et al., Surface plasmon-assisted terahertz imaging array, *IEEE Electronic Components and Technology Conference (ECTC)*, 29 May–1 June 2012, San Diego, CA, p. 1846 (2012).

Park, T.-H., N. Mirin, B. Lassiter et al., Optical properties of a nanosized hole in a thin metallic film, *ACS Nano*, 2, 25 (2008a).

Park, H.-G., M.-K. Seo, S.-H. Kim et al., Electrically pumped photonic crystal nanolasers, *Opt. Photon. News*, 19, May issue, 40 (2008b).

Park, S. H., G. Yuan, D. Chen et al., Wide band-gap III-Nitride nanomembranes for opto-electronc applications, *Nano Lett.*, 14, 4293 (2014). doi:10.1021/nl5009629.

Pelton, M. and G. Bryant, *Introduction to Metal-Nanoparticle Plasmonics*, John Wiley & Sons, Hoboken, NJ (2013).

Perakis, I. E., Stimulated near-infrared light emission in graphene, *Physics*, 5, 43 (2012).

Pernice, W. H. P., C. Schuck, O. Minaeva et al., High-speed and high-efficiency travelling wave single-photon detectors embedded in nanophotonic circuits, *Nat. Comm.*, (2012). doi:10.1038/ncomms2307.

Peroz, C., S. Dhuey, M. Cornet et al., Single digit nanofabrication by step-and-repeat nano-imprint lithography, Nanotechnology, 23, 015305 (2012).

Pierret, A., M. Hocevar, S. L. Diedenhofen et al., Generic nano-imprint process for fabrication of nanowire arrays, *Nanotechnology*, 21, 065305 (2010).

Pillai, S. and M. A. Green, Plasmonics for photovoltaic applications, *Solar Energy Mater. Solar Cells*, 94, 1481 (2010).

Polman, A. and H. A. Atwater, Photonic design principles for ultra-high efficiency photovoltaics, *Nat. Mater.*, 11, 174 (2012).

Pooley, M. A., D. J. P. Ellis, R. B. Patel et al., Controlled-NOT gate operating with single photons, *Appl. Phys. Lett.*, 100, 211103 (2012).

Prasad, P. N., *Nanophotonics*, Wiley-Interscience, New York (2004).

Prot, D., D. B. Stout, J. Lafait et al., Local electric filed enhancements and large third-order optical nonlinearity in nanocomposite materials, *J. Opt. A: Pure Appl. Opt.*, 4, S99 (2002).

Rakić, A. D., A. B. Djurišić, J. M. Elazar et al., *Optical properties of metallic films for vertical-cavity optoelectronic devices*, Appl. Opt., 37, 5271 (1998).

Reed, J. C., Z. Hai, A. Y. Zhu et al., Graphene-enabled silver nanoantenna sensors, *Nano Lett.*, 12, 4090 (2012). doi:10.1021/nl301555t.

Roxworthy, B. J., A. M. Bhuiya, V. V. G. K. Inavalli et al., Multifunctional plasmonic film for recording near-field optical intensity, *Nano Lett.*, 14, 4687 (2014). doi:10.1021/nl501788a.

Saari, J. I., M. K. Krause, B. R. Walsh et al., Terahertz bandwidth all-optical modulation and logic using multiexcitons in semiconductor nanocrystals, *Nano Lett.*, 13, 722 (2013). doi:10.1021/nl3044053.

Saeta, P. N., V. E Ferry, D. Pacifici et al., How much can guided modes enhance absorption in thin solar cells? *Opt. Express*, 17, 20975 (2009).

Saleh, B. E. A. and M. C. Teich, *Fundamentals of Photonics*, Second Edition, Wiley-Interscience, New York (2007).

Salihoglu, O., S. Balci and C. Kocabas, Plasmon-polaritons on graphene-metal surface and their use in biosensors, *Appl. Phys. Lett.*, 100, 213110 (2012).

Schneider, C. A. Rahimi-Iman, N. Y. Kim et al., An electrically pumped polariton laser, *Nature*, 497, 348 (2013).

Senanayake, P., C.-H. Hung, J. Shapiro et al., Surface plasmon-enhanced nanopillar photodetectors, *Nano Lett.*, 11, 5279 (2011). doi:10.1021/nl202732r.

Shalaev, V. M., E. Y. Poliakov and V. A. Markel, Small-particle composites: II. Nonlinear optical properties, *Phys. Rev. B*, 53, 2437 (1996).

Shaner, E. A., M. C. Wanke, A. D. Grine et al., Enhanced responsivity in membrane isolated split-grating-gate plasmonic terahertz detectors, *Appl. Phys. Lett.*, 90, 181127 (2007).

Shim, W., A. B. Braunschweig, X. Liao et al., Hard-tip, soft-spring lithography, *Nature*, 469, 516 (2011).

Sobhani, A., M. W. Knight, Y. Wang et al., Narrowband photodetection in the near-infrared with a plasmon-induced hot electron device, *Nat. Comm.*, 4, 1643 (2013). doi:10.1038/ncomms2642.

Song, J. C. W., M. S. Rudner, C. M. Marcus et al., Hot carrier transport and photocurrent response in graphene, *Nano Lett.*, 11, 4688 (2011).

Stockman, M. I., Nanofocussing of optical energy in tapered plasmonic waveguides, *Phys. Rev. Lett.*, 93, 137404 (2004).

Stockman, M. I., Spasers explained, *Nat. Photon.*, 2, 327 (2008).

Stockman, M. I., The spaser as a nanoscale quantum generator and ultrafast amplifier, *J. Opt.*, 12, 024004 (2010).

Stockman, M. I., Nanoplasmonics: Past, present and glimpse into future, *Opt. Express*, 19, 22029 (2011).

Su, Y.-H., Y.-F. Ke, S.-L. Cai et al., Surface Plasmon resonance of layer-by-layer gold nanoparticles induced photoelectric current in environmentally-friendly plasmon-sensitized solar cell, *Light: Sci. Appl.*, 1, e14 (2012). doi:10.1038/lsa.2012.14.

Suh, J. Y., M. D. Huntington, C. H. Kim et al., Extraordinary nonlinear absorption in 3D bowtie nanoantennas, *Nano Lett.*, 12, 269 (2011). doi:10.1021/nl2034915.

Suh, J. Y., C. H. Kim, W. Zhou et al., Plasmonic bowtie nanolaser arrays, *Nano Lett.*, 12, 5769 (2012). doi:10.1021/nl303086r.

Sun, Z. and H. Chang, Graphene and graphene-like two-dimensional materials in photodetection: Mechanisms and methodology, *ACS Nano*, 8, 4133 (2014).

Sun, J., E. Timurdogan, A. Yaacobi et al., Large-scale nanophotonic phased arrays, *Nature*, 493, 195 (2013).

Tamagnone, M., A. Fallahi, J. R. Mosig et al., Fundamental limits and near-optimal design of graphene modulators and non-reciprocal devices, *Nat. Photon.*, 8, 556 (2014). doi:10.1038/nphoton.2014.109.

Tan, S. J., M. J. Campolongo, D. Luo et al., Building plasmonic nanostructures with DNA, *Nat. Nanotech.*, 6, 268 (2011).

Tanabe, I. and T. Tasuma, Plasmonic manipulation of color and morphology of single silver nanospheres, *Nano Lett.*, 12, 5418 (2012).

Tassin, P., T. Koschny, M. Kafesaki et al., A comparison of graphene, superconductors and metals as conductors for metamaterials and plasmonics, *Nat. Photon.*, 6, 259 (2012).

Thongrattanasiri, S., F. H. L. Koppens and F. J. Garcia de Abajo, Total light absorption in graphene, arXiv:1106.4460v1 [physics.optics] (2011).

Thongrattanasiri, S., F. H. L. Koppens and F. J. Garcia de Abajo, Complete optical absorption in periodically patterned graphene, *Phys. Rev. Lett.*, 108, 047401 (2012a).

Thongrattanasiri, S., A. Manjavacas and F. J. Garcia de Abajo, Quantum finite-size effects in graphene plasmons, *ACS Nano*, 6, 1766 (2012b). doi:10.1021/nn204780e.

Tian, Y. and T. Tatsuma, Mechanisms and applications of plasmon-induced charge separation at TiO$_2$ films loaded with gold nanoparticles, *J. Am. Chem. Soc.*, 127, 7632 (2005).

Tielrooij, K. J., J. C. W. Song, S. A. Jensen et al., Photoexcitation cascade and multiple hot-carrier generation in graphene, *Nat. Phys.*, 9, 248 (2013). doi:10.1038/nphys.2564.

Tsai, D.-S., C.-A. Lin, W.-C. Lien et al., Ultra-high-responsivity broadband detection of Si metal-semiconductor-metal Schottky photodetectors improved by ZnO nanorod arrays, *ACS Nano*, 5, 7748 (2011). doi:10.1021/nn203357e.

Urban, A. S., X. Shen, Y. Wang et al., Three-dimensional plasmonic nanoclusters, *Nano Lett.*, 13, 4399 (2013). doi:10.1021/nl4Q2231z.

Vakil, A., and N. Engheta, Transformation optics using graphene, *Science*, 332, 1291 (2011).

Vesseur, E. J. R., T. Coonen, H. Caglayan et al., Experimental verification of n=0 structures for visible light, *Phys. Rev. Lett.*, 11, 013902 (2013).

Wang, F. and N. A. Melosh, Plasmonic energy collection through hot carrier extraction, *Nano Lett.*, 11, 5426 (2011).

Wang, Y., T. Sun, T. Paudel et al., Metamaterial plasmonic absorber structure for high efficiency amorphous silicon solar cells, *Nano Lett.*, 12, 440 (2011a).

Wang, K. X., Z. Yu, V. Liu et al., Absorption enhancement in ultra-thin crystalline silicon solar cells with anti-reflection and light-trapping nanocone gratings, *Nano Lett.*, 12, 1616 (2012). doi:10.1021/nl204550q.

Wei, H., Z. Wang, X. Tian et al., Cascaded logic gates in nanophotonic plasmon networks, *Nat. Comm.*, 2, 387 (2011b). doi:10.1038/ncomms1388.

Wen, T., R. A. Booth and S. A. Majetich, Ten-nanometer dense hole arrays generated by nanoparticle lithography, *Nano Lett.*, 12, 5873 (2012). doi:10.1021/nl3032372.

West, P. R., S. Ishii, G. V. Naik et al., Searching for better plasmonic materials, *Laser Photon. Rev.*, 4, 795 (2010).

Withers, F., T. H. Bointon, M. F. Craciun et al., All-graphene photodetectors, *ACS Nano*, 7, 5052 (2013).

Wu, C.-Y., C.-T. Kuo, C.-Y. Wang et al., Plasmonic green nanolaser based on metal-oxide-semiconductor structure, *Nano Lett.*, 11, 4256 (2011). doi:10.1021/nl2022477.

Xu, T., Y.-K. Wu, X. Luo et al., Plasmonic nanoresonators for high-resolution colour filtering and spectral imaging, *Nat. Comm.*, 1, 59 (2010). doi:10.1038/ncomms1058.

Yan, H., X. Li, B. Chandra et al., Tunable infrared plasmonic devices using graphene/insulator stacks, *Nat. Nanotech.*, 7, 330 (2012). doi:10.1038/nnano.2012.59.

Yao, Y., R. Shankar, P. Rauter et al., High-responsivity mid-infrared graphene detectors with antenna-enhanced photocarrier generation and collection, *Nano Lett.*, 14, 3749 (2014). doi:10.1021/nl500602n.

Yifat, Y., M. Eitan, Z. I. Iluz et al., Highly efficient and broadband wide-angle holography using patch-dipole nano-antenna reflectarrays, *Nano Lett.*, 14, 2485 (2014). doi:10.1021/nl5001696.

Yokogawa, S., S. P. Burgos and H. A. Atawater, Plasmonic color filters for CMOS image sensor applications, *Nano Lett.*, 12, 4349 (2012). doi:10.1021/nl302110z.

Zayats, A. V., I. I. Smolyaninov and A. A. Maradudin, Nano-optics of surface plasmon polaritons, *Phys. Rep.*, 408, 131 (2005).

Zdanowicz, E., T. A. Dow and R. O. Scattergood, Rapid fabrication of nanostructured surfaces using nanocoining, *Nanotechnology*, 23, 415303 (2012).

Zeng, B., Y. Gao and F. J. Bartoli, Ultrathin nanostructured metals for highly transmissive plasmonic subtractive color filters, *Sci. Rep.*, (2013) doi:10.1038/srep02840.

Zentgraf, T., Y. Liu, M. H. Mikkelson et al., Plasmonic Luneburg and Eaton lenses, *Nat. Nanotech.*, 6, 151 (2011).

Zhang, Y., N. K. Grady, C. Ayala-Orozco et al., Three dimensional nanostructures as highly efficient generators of second harmonic light, *Nano Lett.*, 11, 5519 (2011a).

Zhang, X.-Y., A. Hu, T. Zhang et al., Self-assembly of large-scale and ultrathin silver nanoplate films with tunable plasmon resonance properties, *ACS Nano* (2011b). doi:10.1021/nn203336m.

Zhang, B., Y. Zhao, Q. Hao et al., Polarization-independent dual-band infrared perfect absorber based on a metal-dielectric-metal elliptical nanodisk array, *Opt. Express*, 19, 15221 (2011c).

Zhang, Y., T. Liu, B. Meng et al., Broadband high photoresponse from pure monolayer graphene photodetector, *Nat. Comm.*, 4, 1811 (2013a). doi:10.1038/ncomms2830.

Zhang, Y., F. Wen, Y.-R. Zhen et al., Coherent Fano resonances in a plasmonic nanocluster enhance optical four-wave mixing, *Proc. Nat. Acad. Sci.*, 110, 9215 (2013b).

Zhang, J. and L. Zhang, Nanostructures for surface plasmons, *Adv. Opt. Photon.*, 4, 157 (2012).

Zhang, T., S. Callard, C. Jamois et al., Plasmonic-photonic crystal coupled nanolaser, *Nanotechnology*, 25, 315201 (2014).

Zheludev, N. I., S. L. Prosvirnin, N. Papasimakis et al., Lasing spaser, *Nat. Photon.*, 2, 351 (2008).

Zhou, W., J. Lee, J. Nanda et al., Atomically localized plasmon enhancement in monolayer graphene, *Nat. Nanotech.*, 7, 161 (2012).

Zijlstra, P., J. W. M. Chon and M. Gu, Five-dimensional optical recording mediated by surface plasmons in gold nanorods, *Nature*, 459, 410 (2009).

Index

Printed and bound by CPI Group (UK) Ltd, Croydon, CR0 4YY

24/10/2024

01778286-0018